Student Manual to Accompany Atkins' Physical Chemistry

ELEVENTH EDITION

Peter Bolgar

Haydn Lloyd

Aimee North

Vladimiras Oleinikovas

Stephanie Smith

and

James Keeler

Department of Chemistry
University of Cambridge
UK

OXFORD
UNIVERSITY PRESS

Great Clarendon Street, Oxford, OX2 6DP,
United Kingdom

Oxford University Press is a department of the University of Oxford.
It furthers the University's objective of excellence in research, scholarship,
and education by publishing worldwide. Oxford is a registered trade mark of
Oxford University Press in the UK and in certain other countries

Eighth edition 2006
Ninth edition 2010
Tenth edition 2014

Impression: 2

Published in the United States of America by Oxford University Press
198 Madison Avenue, New York, NY 10016, United States of America

British Library Cataloguing in Publication Data
Data available

ISBN 978-0-19-880777-3

Printed in Great Britain by
Bell & Bain Ltd., Glasgow

Table of contents

Preface

This manual provides detailed solutions to the (a) *Exercises* and the odd-numbered *Discussion questions* and *Problems* from the 11th edition of *Atkins' Physical Chemistry*.

Conventions used is presenting the solutions

We have included page-specific references to equations, sections, figures and other features of the main text. Equation references are denoted [14B.3b–*595*], meaning eqn 14B.3b located on page 595 (the page number is given in italics). Other features are referred to by name, with a page number also given.

Generally speaking, the values of physical constants (from the first page of the main text) are used to 5 significant figures except in a few cases where higher precision is required. In line with the practice in the main text, intermediate results are simply truncated (not rounded) to three figures, with such truncation indicated by an ellipsis, as in 0.123...; the value is used in subsequent calculations to its full precision.

The final results of calculations, generally to be found in $\boxed{\text{a box}}$, are given to the precision warranted by the data provided. We have been rigorous in including units for all quantities so that the units of the final result can be tracked carefully. The relationships given on the back of the front cover are useful in resolving the units of more complex expressions, especially where electrical quantities are involved.

Some of the problems either require the use of mathematical software or are much easier with the aid of such a tool. In such cases we have used *Mathematica* (Wolfram Research, Inc.) in preparing these solutions, but there are no doubt other options available. Some of the *Discussion questions* relate directly to specific section of the main text in which case we have simply given a reference rather than repeating the material from the text.

Acknowledgements

In preparing this manual we have drawn on the equivalent volume prepared for the 10th edition of *Atkins' Physical Chemistry* by Charles Trapp, Marshall Cady, and Carmen Giunta. In particular, the solutions which use quantum chemical calculations or molecular modelling software, and some of the solutions to the *Discussion questions*, have been quoted directly from the solutions manual for the 10th edition, without significant modification. More generally, we have benefited from the ability to refer to the earlier volume and acknowledge, with thanks, the influence that its authors have had on the present work.

This manual has been prepared by the authors using the LaTeX typesetting system, in the implementation provided by MiKTeX (**`miktex.org`**); the vast majority of the figures and graphs have been generated using PGFPlots. We are grateful to the community who maintain and develop these outstanding resources.

Finally, we are grateful to the editorial team at OUP, Jonathan Crowe and Roseanna Levermore, for their invaluable support in bringing this project to a conclusion.

Errors and omissions

In such a complex undertaking some errors will no doubt have crept in, despite the authors' best efforts. Readers who identify any errors or omissions are invited to pass them on to us by email to **pchem@ch.cam.ac.uk**.

1 The properties of gases

1A The perfect gas

Answers to discussion questions

D1A.1 An equation of state is an equation that relates the variables that define the state of a system to each other. Boyle, Charles, and Avogadro established these relations for gases at low pressures (perfect gases) by appropriate experiments. Boyle determined how volume varies with pressure ($V \propto 1/p$), Charles how volume varies with temperature ($V \propto T$), and Avogadro how volume varies with amount of gas ($V \propto n$). Combining all of these proportionalities into one gives

$$V \propto \frac{nT}{p}$$

Inserting the constant of proportionality, R, yields the perfect gas equation

$$V = R\frac{nT}{p} \quad \text{or} \quad pV = nRT$$

Solutions to exercises

E1A.1(a) From the inside the front cover the conversion between pressure units is: 1 atm $\equiv$ 101.325 kPa $\equiv$ 760 Torr; 1 bar is 10^5 Pa exactly.

(i) A pressure of 108 kPa is converted to Torr as follows

$$108\ \text{kPa} \times \frac{1\ \text{atm}}{101.325\ \text{kPa}} \times \frac{760\ \text{Torr}}{1\ \text{atm}} = \boxed{810\ \text{Torr}}$$

(ii) A pressure of 0.975 bar is 0.975×10^5 Pa, which is converted to atm as follows

$$0.975 \times 10^5\ \text{Pa} \times \frac{1\ \text{atm}}{101.325\ \text{kPa}} = \boxed{0.962\ \text{atm}}$$

E1A.2(a) The perfect gas law [1A.4–8], $pV = nRT$, is rearranged to give the pressure, $p = nRT/V$. The amount n is found by dividing the mass by the molar mass of Xe, 131.29 g mol^{-1}.

$$p = \overbrace{\frac{(131\ \text{g})}{(131.29\ \text{g mol}^{-1})}}^{n} \frac{(8.2057 \times 10^{-2}\ \text{dm}^3\ \text{atm K}^{-1}\ \text{mol}^{-1}) \times (298.15\ \text{K})}{1.0\ \text{dm}^3}$$

$$= 24.4\ \text{atm}$$

So $\boxed{\text{no}}$, the sample would not exert a pressure of 20 atm, but $\boxed{24.4\ \text{atm}}$ if it were a perfect gas.

E1A.3(a) Because the temperature is constant (isothermal) Boyle's law applies, pV = const. Therefore the product pV is the same for the initial and final states

$$p_f V_f = p_i V_i \quad \text{hence} \quad p_i = p_f V_f / V_i$$

The initial volume is 2.20 dm^3 greater than the final volume so $V_i = 4.65 + 2.20 = 6.85\ \text{dm}^3$.

$$p_i = \frac{V_f}{V_i} \times p_f = \frac{4.65\ \text{dm}^3}{6.85\ \text{dm}^3} \times (5.04\ \text{bar}) = 3.42\ \text{bar}$$

(i) The initial pressure is $\boxed{3.42\ \text{bar}}$

(ii) Because a pressure of 1 atm is equivalent to 1.01325 bar, the initial pressure expressed in atm is

$$\frac{1\ \text{atm}}{1.01325\ \text{bar}} \times 3.40\ \text{bar} = \boxed{3.38\ \text{atm}}$$

E1A.4(a) If the gas is assumed to be perfect, the equation of state is [1A.4–8], $pV = nRT$. In this case the volume and amount (in moles) of the gas are constant, so it follows that the pressure is proportional to the temperature: $p \propto T$. The ratio of the final and initial pressures is therefore equal to the ratio of the temperatures: $p_f/p_i = T_f/T_i$. The pressure indicated on the gauge is that in excess of atmospheric pressure, thus the initial pressure is $24 + 14.7 = 38.7\ \text{lb in}^{-2}$. Solving for the final pressure p_f (remember to use absolute temperatures) gives

$$\begin{aligned} p_f &= \frac{T_f}{T_i} \times p_i \\ &= \frac{(35 + 273.15)\ \text{K}}{(-5 + 273.15)\ \text{K}} \times (38.7\ \text{lb in}^{-2}) = 44.4...\ \text{lb in}^{-2} \end{aligned}$$

The pressure indicated on the gauge is this final pressure, minus atmospheric pressure: $44.4... - 14.7 = \boxed{30\ \text{lb in}^{-2}}$. This assumes that (i) the gas is behaving perfectly and (ii) that the tyre is rigid.

E1A.5(a) The perfect gas law $pV = nRT$ is rearranged to give the pressure

$$\begin{aligned} p &= \frac{nRT}{V} \\ &= \overbrace{\frac{255 \times 10^{-3}\ \text{g}}{20.18\ \text{g mol}^{-1}}}^{n} \times \frac{(8.3145 \times 10^{-2}\ \text{dm}^3\ \text{bar K}^{-1}\ \text{mol}^{-1}) \times (122\ \text{K})}{3.00\ \text{dm}^3} \\ &= \boxed{0.0427\ \text{bar}} \end{aligned}$$

Note the choice of R to match the units of the problem. An alternative is to use $R = 8.3154\ \mathrm{J\,K^{-1}\,mol^{-1}}$ and adjust the other units accordingly, to give a pressure in Pa.

$$p = \frac{[(255 \times 10^{-3}\ \mathrm{g})/(20.18\ \mathrm{g\,mol^{-1}})] \times (8.3145\ \mathrm{J\,K^{-1}\,mol^{-1}}) \times (122\ \mathrm{K})}{3.00 \times 10^{-3}\ \mathrm{m^3}}$$
$$= \boxed{4.27 \times 10^5\ \mathrm{Pa}}$$

where $1\ \mathrm{dm^3} = 10^{-3}\ \mathrm{m^3}$ has been used along with $1\ \mathrm{J} = 1\ \mathrm{kg\,m^2\,s^{-2}}$ and $1\ \mathrm{Pa} = 1\ \mathrm{kg\,m^{-1}\,s^{-2}}$.

E1A.6(a) The vapour is assumed to be a perfect gas, so the gas law $pV = nRT$ applies. The task is to use this expression to relate the measured mass density to the molar mass.

First, the amount n is expressed as the mass m divided by the molar mass M to give $pV = (m/M)RT$; division of both sides by V gives $p = (m/V)(RT/M)$. The quantity (m/V) is the mass density ρ, so $p = \rho RT/M$, which rearranges to $M = \rho RT/p$; this is the required relationship between M and the density.

$$M = \frac{\rho RT}{p} = \frac{(3.710\ \mathrm{kg\,m^{-3}}) \times (8.3145\ \mathrm{J\,K^{-1}\,mol^{-1}}) \times ([500 + 273.15]\ \mathrm{K})}{93.2 \times 10^3\ \mathrm{Pa}}$$
$$= 0.255...\ \mathrm{kg\,mol^{-1}}$$

where $1\ \mathrm{J} = 1\ \mathrm{kg\,m^2\,s^{-2}}$ and $1\ \mathrm{Pa} = 1\ \mathrm{kg\,m^{-1}\,s^{-2}}$ have been used. The molar mass of S is $32.06\ \mathrm{g\,mol^{-1}}$, so the number of S atoms in the molecules comprising the vapour is $(0.255... \times 10^3\ \mathrm{g\,mol^{-1}})/(32.06\ \mathrm{g\,mol^{-1}}) = 7.98$. The result is expected to be an integer, so the formula is likely to be $\boxed{S_8}$.

E1A.7(a) The vapour is assumed to be a perfect gas, so the gas law $pV = nRT$ applies; the task is to use this expression to relate the measured data to the mass m. This is done by expressing the amount n as m/M, where M is the the molar mass. With this substitution it follows that $m = MPV/RT$.

The partial pressure of water vapour is 0.60 times the saturated vapour pressure

$$m = \frac{MpV}{RT}$$
$$= \frac{(18.0158\ \mathrm{g\,mol^{-1}}) \times (0.60 \times 0.0356 \times 10^5\ \mathrm{Pa}) \times (400\ \mathrm{m^3})}{(8.3145\ \mathrm{J\,K^{-1}\,mol^{-1}}) \times ([27 + 273.15]\ \mathrm{K})}$$
$$= 6.2 \times 10^3\ \mathrm{g} = \boxed{6.2\ \mathrm{kg}}$$

E1A.8(a) Consider $1\ \mathrm{m^3}$ of air: the mass of gas is therefore 1.146 kg. If perfect gas behaviour is assumed, the amount in moles is given by $n = pV/RT$

$$n = \frac{pV}{RT} = \frac{(0.987 \times 10^5\ \mathrm{Pa}) \times (1\ \mathrm{m^3})}{(8.3145\ \mathrm{J\,K^{-1}\,mol^{-1}}) \times ([27 + 273.15]\ \mathrm{K})} = 39.5...\ \mathrm{mol}$$

(i) The total amount in moles is $n = n_{O_2} + n_{N_2}$. The total mass m is computed from the amounts in moles and the molar masses M as

$$m = n_{O_2} \times M_{O_2} + n_{N_2} \times M_{N_2}$$

These two equations are solved simultaneously for n_{O_2} to give the following expression, which is then evaluated using the data given

$$n_{O_2} = \frac{m - M_{N_2}n}{M_{O_2} - M_{N_2}} = \frac{(1146\ \text{g}) - (28.02\ \text{g mol}^{-1}) \times (39.5...\ \text{mol})}{(32.00\ \text{g mol}^{-1}) - (28.02\ \text{g mol}^{-1})} = 9.50...\ \text{mol}$$

The mole fractions are therefore

$$x_{O_2} = \frac{n_{O_2}}{n} = \frac{9.50...\ \text{mol}}{39.5...\ \text{mol}} = \boxed{0.240} \qquad x_{N_2} = 1 - x_{O_2} = \boxed{0.760}$$

The partial pressures are given by $p_i = x_i p_{\text{tot}}$

$$p_{O_2} = x_{O_2} p_{\text{tot}} = 0.240(0.987\ \text{bar}) = \boxed{0.237\ \text{bar}}$$

$$p_{N_2} = x_{N_2} p_{\text{tot}} = 0.760(0.987\ \text{bar}) = \boxed{0.750\ \text{bar}}$$

(ii) The simultaneous equations to be solved are now

$$n = n_{O_2} + n_{N_2} + n_{Ar} \qquad m = n_{O_2}M_{O_2} + n_{N_2}M_{N_2} + n_{Ar}M_{Ar}$$

Because it is given that $x_{Ar} = 0.01$, it follows that $n_{Ar} = n/100$. The two unknowns, n_{O_2} and n_{N_2}, are found by solving these equations simultaneously to give

$$n_{N_2} = \frac{100m - n(M_{Ar} + 99M_{O_2})}{100(M_{N_2} - M_{O_2})}$$
$$= \frac{100 \times (1146\ \text{g}) - (39.5...\ \text{mol}) \times [(39.95\ \text{g mol}^{-1}) + 99 \times (32.00\ \text{g mol}^{-1})]}{100 \times [(28.02\ \text{g mol}^{-1}) - (32.00\ \text{g mol}^{-1})]}$$
$$= 30.8...\ \text{mol}$$

From $n = n_{O_2} + n_{N_2} + n_{Ar}$ it follows that

$$n_{O_2} = n - n_{Ar} - n_{N_2}$$
$$= (39.5...\ \text{mol}) - 0.01 \times (39.5...\ \text{mol}) - (30.8...\ \text{mol}) = 8.31...\ \text{mol}$$

The mole fractions are

$$x_{N_2} = \frac{n_{N_2}}{n} = \frac{30.8...\ \text{mol}}{39.5...\ \text{mol}} = \boxed{0.780} \qquad x_{O_2} = \frac{n_{O_2}}{n} = \frac{8.31...\ \text{mol}}{39.5...\ \text{mol}} = \boxed{0.210}$$

The partial pressures are

$$p_{N_2} = x_{N_2} p_{\text{tot}} = 0.780 \times (0.987\ \text{bar}) = \boxed{0.770\ \text{bar}}$$

$$p_{O_2} = x_{O_2} p_{\text{tot}} = 0.210 \times (0.987\ \text{bar}) = \boxed{0.207\ \text{bar}}$$

Note: the final values are quite sensitive to the precision with which the intermediate results are carried forward.

E1A.9(a) The vapour is assumed to be a perfect gas, so the gas law $pV = nRT$ applies. The task is to use this expression to relate the measured mass density to the molar mass.

First, the amount n is expressed as the mass m divided by the molar mass M to give $pV = (m/M)RT$; division of both sides by V gives $p = (m/V)(RT/M)$. The quantity (m/V) is the mass density ρ, so $p = \rho RT/M$, which rearranges to $M = \rho RT/p$; this is the required relationship between M and the density.

$$M = \frac{\rho RT}{p} = \frac{(1.23\,\text{kg}\,\text{m}^{-3}) \times (8.3145\,\text{J}\,\text{K}^{-1}\,\text{mol}^{-1}) \times (330\,\text{K})}{20.0 \times 10^3\,\text{Pa}} = \boxed{0.169\,\text{kg}\,\text{mol}^{-1}}$$

The relationships $1\,\text{J} = 1\,\text{kg}\,\text{m}^2\,\text{s}^{-2}$ and $1\,\text{Pa} = 1\,\text{kg}\,\text{m}^{-1}\,\text{s}^{-2}$ have been used.

E1A.10(a) Charles' law [1A.3b–7] states that $V \propto T$ at constant n and p, and $p \propto T$ at constant n and V. For a fixed amount the density ρ is proportional to $1/V$, so it follows that $1/\rho \propto T$. At absolute zero the volume goes to zero, so the density goes to infinity and hence $1/\rho$ goes to zero. The approach is therefore to plot $1/\rho$ against the temperature (in °C) and then by extrapolating the straight line find the temperature at which $1/\rho = 0$. The plot is shown in Fig 1.1.

$\theta/°\text{C}$	$\rho/(\text{g}\,\text{dm}^{-3})$	$(1/\rho)/(\text{g}^{-1}\,\text{dm}^3)$
−85	1.877	0.532 8
0	1.294	0.772 8
100	0.946	1.057 1

The data are a good fit to a straight line, the equation of which is

$$(1/\rho)/(\text{g}^{-1}\,\text{dm}^3) = 2.835 \times 10^{-3} \times (\theta/°\text{C}) + 0.7734$$

The intercept with $1/\rho = 0$ is found by solving

$$0 = 2.835 \times 10^{-3} \times (\theta/°\text{C}) + 0.7734$$

This gives $\boxed{\theta = -273\,°\text{C}}$ as the estimate of absolute zero.

E1A.11(a) (i) The mole fractions are

$$x_{H_2} = \frac{n_{H_2}}{n_{H_2} + n_{N_2}} = \frac{2.0\,\text{mol}}{2.0\,\text{mol} + 1.0\,\text{mol}} = \boxed{\tfrac{2}{3}} \qquad x_{N_2} = 1 - x_{H_2} = \boxed{\tfrac{1}{3}}$$

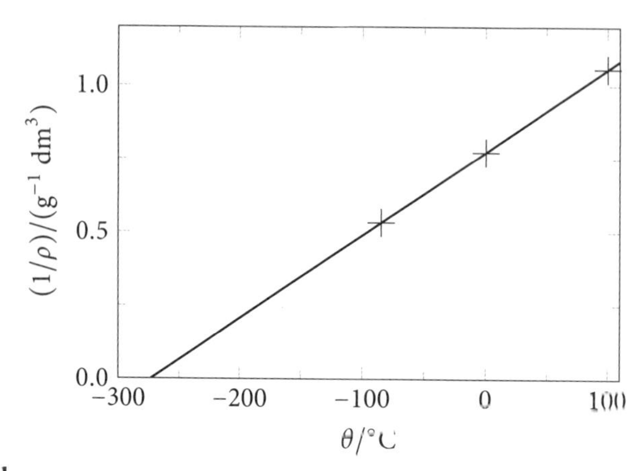

Figure 1.1

(ii) The partial pressures are given by $p_i = x_i p_{\text{tot}}$. The total pressure is given by the perfect gas law: $p_{\text{tot}} = n_{\text{tot}}RT/V$

$$p_{\text{H}_2} = x_{\text{H}_2} p_{\text{tot}} = \frac{2}{3} \times \frac{(3.0\,\text{mol}) \times (8.3145\,\text{J}\,\text{K}^{-1}\,\text{mol}^{-1}) \times (273.15\,\text{K})}{22.4 \times 10^{-3}\,\text{m}^3}$$
$$= \boxed{2.0 \times 10^5\,\text{Pa}}$$

$$p_{\text{N}_2} = x_{\text{N}_2} p_{\text{tot}} = \frac{1}{3} \times \frac{(3.0\,\text{mol}) \times (8.3145\,\text{J}\,\text{K}^{-1}\,\text{mol}^{-1}) \times (273.15\,\text{K})}{22.4 \times 10^{-3}\,\text{m}^3}$$
$$= \boxed{1.0 \times 10^5\,\text{Pa}}$$

Expressed in atmospheres these are 2.0 atm and 1.0 atm, respectively.

(iii) The total pressure is

$$\frac{(3.0\,\text{mol}) \times (8.3145\,\text{J}\,\text{K}^{-1}\,\text{mol}^{-1}) \times (273.15\,\text{K})}{22.4 \times 10^{-3}\,\text{m}^3} = \boxed{3.0 \times 10^5\,\text{Pa}}$$

or 3.00 atm.

Alternatively, note that 1 mol at STP occupies a volume of 22.4 dm^3, which is the stated volume. As there are a total of 3.0 mol present the (total) pressure must therefore be 3.0 atm.

Solutions to problems

P1A.1 (a) The expression ρgh gives the pressure in Pa if all the quantities are in SI units, so it is helpful to work in Pa throughout. From the front cover, 760 Torr is exactly 1 atm, which is 1.01325×10^5 Pa. The density of 13.55 $\text{g}\,\text{cm}^{-3}$ is equivalent to $13.55 \times 10^3\,\text{kg}\,\text{m}^{-3}$.

$$p = p_{\text{ex}} + \rho gh$$
$$= 1.01325 \times 10^5\,\text{Pa} + (13.55 \times 10^3\,\text{kg}\,\text{m}^{-3}) \times (9.806\,\text{m}\,\text{s}^{-2})$$
$$\times (10.0 \times 10^{-2}\,\text{m}) = \boxed{1.15 \times 10^5\,\text{Pa}}$$

(b) The calculation of the pressure inside the apparatus proceeds as in (a)

$$p = 1.01325 \times 10^5\ \text{Pa} + (0.9971 \times 10^3\ \text{kg m}^{-3}) \times (9.806\ \text{m s}^{-2}) \times (183.2 \times 10^{-2}\ \text{m}) = 1.192... \times 10^5\ \text{Pa}$$

The value of R is found by rearranging the perfect gas law to $R = pV/nT$

$$R = \frac{pV}{nT} = \frac{(1.192... \times 10^5\ \text{Pa}) \times (20.000 \times 10^{-3}\ \text{m}^3)}{[(1.485\ \text{g})/(4.003\ \text{g mol}^{-1})] \times ([500 + 273.15]\ \text{K})} = \boxed{8.315\ \text{J K}^{-1}\ \text{mol}^{-1}}$$

P1A.3 The perfect gas law $pV = nRT$ implies that $pV_m = RT$, where V_m is the molar volume (the volume when $n = 1$). It follows that $p = RT/V_m$, so a plot of p against T/V_m should be a straight line with slope R.

However, real gases only become ideal in the limit of zero pressure, so what is needed is a method of extrapolating the data to zero pressure. One approach is to rearrange the perfect gas law into the form $pV_m/T = R$ and then to realise that this implies that for a real gas the quantity pV_m/T will tend to R in the limit of zero pressure. Therefore, the intercept at $p = 0$ of a plot of pV_m/T against p is an estimate of R. For the extrapolation of the line back to $p = 0$ to be reliable, the data points must fall on a reasonable straight line. The plot is shown in Fig 1.2.

p/atm	$V_m/(\text{dm}^3\ \text{mol}^{-1})$	$(pV_m/T)/(\text{atm dm}^3\ \text{mol}^{-1}\ \text{K}^{-1})$
0.750 000	29.8649	0.082 001 4
0.500 000	44.8090	0.082 022 7
0.250 000	89.6384	0.082 041 4

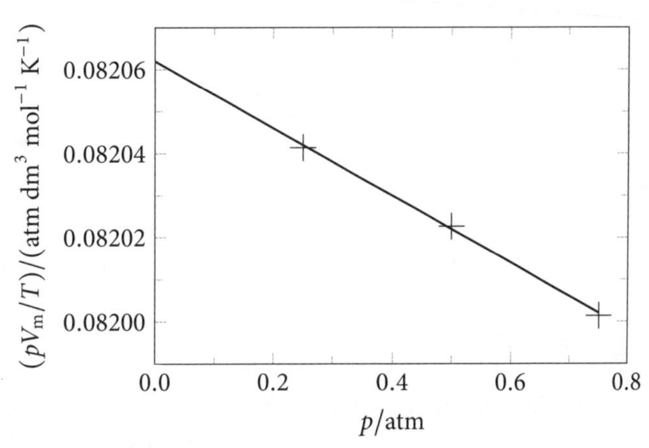

Figure 1.2

The data fall on a reasonable straight line, the equation of which is

$$(pV_m/T)/(\text{atm dm}^3\text{ mol}^{-1}\text{ K}^{-1}) = -7.995\times10^{-5}\times(p/\text{atm}) + 0.082062$$

The estimate for R is therefore the intercept, $\boxed{0.082062\text{ atm dm}^3\text{ mol}^{-1}\text{ K}^{-1}}$. The data are given to 6 figures, but they do not fall on a very good straight line so the value for R has been quoted to one fewer significant figure.

P1A.5 For a perfect gas $pV = nRT$ which can be rearranged to give $p = nRT/V$. The amount in moles is $n = m/M$, where M is the molar mass and m is the mass of the gas. Therefore $p = (m/M)(RT/V)$. The quantity m/V is the mass density ρ, and hence

$$\boxed{p = \rho RT/M}$$

It follows that for a perfect gas p/ρ should be a constant at a given temperature. Real gases are expected to approach this as the pressure goes to zero, so a suitable plot is of p/ρ against p; the intercept when $p = 0$ gives the best estimate of RT/M. The plot is shown in Fig. 1.3.

p/kPa	$\rho/(\text{kg m}^{-3})$	$(p/\rho)/(\text{kPa kg}^{-1}\text{ m}^3)$
12.22	0.225	54.32
25.20	0.456	55.26
36.97	0.664	55.68
60.37	1.062	56.85
85.23	1.468	58.06
101.30	1.734	58.42

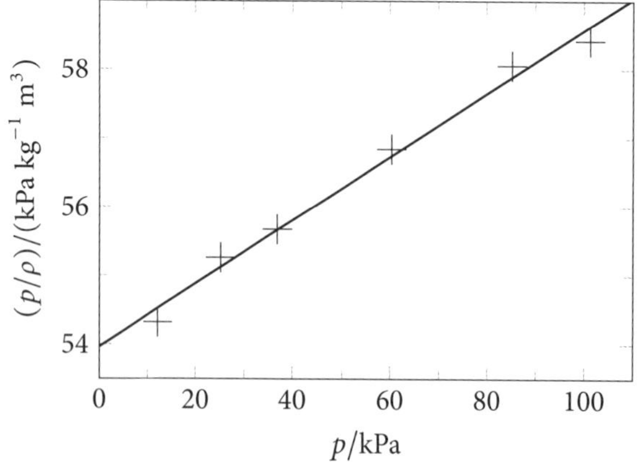

Figure 1.3

The data fall on a reasonable straight line, the equation of which is

$$(p/\rho)/(\text{kPa kg}^{-1}\text{ m}^3) = 0.04610\times(p/\text{kPa}) + 53.96$$

The intercept is $(p/\rho)_{\lim p\to 0}$, which is equal to RT/M.

$$M = \frac{RT}{(p/\rho)_{\lim p\to 0}} = \frac{(8.3145\,\text{J K}^{-1}\,\text{mol}^{-1}) \times (298.15\,\text{K})}{53.96 \times 10^3\ \text{Pa kg}^{-1}\ \text{m}^3} = 4.594\times10^{-2}\,\text{kg mol}^{-1}$$

The estimate of the molar mass is therefore $\boxed{45.94\ \text{g mol}^{-1}}$.

P1A.7 (a) For a perfect gas $pV = nRT$ so it follows that for a sample at constant volume and temperature, $p_1/T_1 = p_2/T_2$. If the pressure increases by Δp for an increase in temperature of ΔT, then with $p_2 = p_1 + \Delta p$ and $T_2 = T_1 + \Delta T$ is follows that

$$\frac{p_1}{T_1} = \frac{p_1 + \Delta p}{T_1 + \Delta T} \qquad \text{hence} \qquad \Delta p = \frac{p_1 \Delta T}{T_1}$$

For an increase by 1.00 K, $\Delta T = 1.00$ K and hence

$$\Delta p = \frac{p_1 \Delta T}{T_1} = \frac{(6.69 \times 10^3\ \text{Pa}) \times (1.00\ \text{K})}{273.16\ \text{K}} = \boxed{24.5\ \text{Pa}}$$

Another way of looking at this is to write the rate of change of pressure with temperature as

$$\frac{\Delta p}{\Delta T} = \frac{p_1}{T_1} = \frac{6.69 \times 10^3\ \text{Pa}}{273.16\ \text{K}} = 24.5...\ \text{Pa K}^{-1}$$

(b) A temperature of 100.00 °C is equivalent to an increase in temperature from the triple point by 100.00 + 273.15 − 273.16 = 99.99 K

$$\Delta p' = \Delta T' \times \left(\frac{\Delta p}{\Delta T}\right) = (99.99\ \text{K}) \times \frac{6.69 \times 10^3\ \text{Pa}}{273.16\ \text{K}} = 2.44... \times 10^3\ \text{Pa}$$

The final pressure is therefore 6.69 + 2.44... = $\boxed{9.14\ \text{kPa}}$.

(c) For a perfect gas $\Delta p/\Delta T$ is independent of the temperature so at 100.0 °C a 1.00 K rise in temperature gives a pressure rise of $\boxed{24.5\ \text{Pa}}$, just as in (a).

P1A.9 The molar mass of SO_2 is $32.06+2\times16.00 = 64.06\ \text{g mol}^{-1}$. If the gas is assumed to be perfect the volume is calculated from $pV = nRT$

$$V = \frac{nRT}{p} = \overbrace{\left(\frac{200 \times 10^6\ \text{g}}{64.06\ \text{g mol}^{-1}}\right)}^{n} \frac{(8.3145\,\text{J K}^{-1}\,\text{mol}^{-1}) \times ([800 + 273.15]\ \text{K})}{1.01325 \times 10^5\ \text{Pa}}$$
$$= \boxed{2.7 \times 10^5\ \text{m}^3}$$

Note the conversion of the mass in t to mass in g; repeating the calculation for 300 t gives a volume of $\boxed{4.1 \times 10^5\ \text{m}^3}$.

The volume of gas is therefore between $\boxed{0.27\ \text{km}^3 \text{ and } 0.41\ \text{km}^3}$.

P1A.11 Imagine a column of the atmosphere with cross sectional area A. The pressure at any height is equal to the force acting down on that area; this force arises from the gravitational attraction on the gas in the column *above* this height – that is, the 'weight' of the gas.

Suppose that the height h is increased by dh. The force on the area A is reduced because less of the atmosphere is now bearing down on this area. Specifically, the force is reduced by that due to the gravitational attraction on the gas contained in a cylinder of cross-sectional area A and height dh. If the density of the gas is ρ, the mass of the gas in the cylinder is $\rho \times A\,dh$ and the force due to gravity on this mass is $\rho g A\,dh$, where g is the acceleration due to free fall. The change in pressure dp on increasing the height by dh is this force divided by the area, so it follows that

$$dp = -\rho g dh$$

The minus sign is needed because the pressure decreases as the height increases.

The density is related to the pressure by starting from the perfect gas equation, $pV = nRT$. If the mass of gas is m and the molar mass is M, it follows that $n = m/M$ and hence $pV = (m/M)RT$. Taking the volume to the right gives $p = (m/MV)RT$. The quantity m/V is the mass density ρ, so $p = (\rho/M)RT$; this is rearranged to give an expression for the density: $\rho = Mp/RT$.

This expression for ρ is substituted into $dp = -\rho g dh$ to give $dp = -(Mp/RT)g dh$. Division by p results in separation of the variables $(1/p)\,dp = -(M/RT)g dh$. The left-hand side is integrated between p_0, the pressure at $h = 0$ and p, the pressure at h. The right-hand side is integrated between $h = 0$ and h

$$\int_{p_0}^{p} \frac{1}{p} dp = \int_0^h -\frac{Mg}{RT} dh$$

$$[\ln p]_{p_0}^{p} = -\frac{Mg}{RT}[h]_0^h$$

$$\ln \frac{p}{p_0} = -\frac{Mgh}{RT}$$

The exponential of each side is taken to give

$$p = p_0 e^{-h/H} \qquad \text{with} \qquad H = \frac{RT}{Mg}$$

It is assumed that g and T do not vary with h.

(a) The pressure decrease across such a small distance will be very small because $h/H \ll 1$. It is therefore admissible to expand the exponential and retain just the first two terms: $e^x \approx 1 + x$

$$p = p_0(1 - h/H)$$

This is rearranged to give an expression for the pressure decrease, $p - p_0$

$$p - p_0 = -p_0 h/H$$

If it is assumed that p_0 is one atmosphere and that $H = 8$ km,

$$p - p_0 = -p_0 h/H = -\frac{(1.01325 \times 10^5 \text{ Pa}) \times (15 \times 10^{-2} \text{ m})}{8 \times 10^3 \text{ m}} = \boxed{-2 \text{ Pa}}$$

(b) The pressure at 11 km is calculated using the full expression

$$p = p_0 e^{-h/H} = (1 \text{ atm}) \times e^{-(11 \text{ km})/(8 \text{ km})} = \boxed{0.25 \text{ atm}}$$

P1A.13 Imagine a volume V of the atmosphere, at temperature T and pressure p_{tot}. If the concentration of a trace gas is expressed as X parts per trillion (ppt), it means that if that gas were confined to a volume $X \times 10^{-12} \times V$ at temperature T is would exert a pressure p_{tot}. From the perfect gas law it follows that $n = pV/RT$, which in this case gives

$$n_{\text{trace}} = \frac{p_{\text{tot}}(X \times 10^{-12} \times V)}{RT}$$

Taking the volume V to the left gives the molar concentration, c_{trace}

$$c_{\text{trace}} = \frac{n_{\text{trace}}}{V} = \frac{X \times 10^{-12} \times p_{\text{tot}}}{RT}$$

An alternative way of looking at this is to note that, at a given temperature and pressure, the volume occupied by a gas is proportional to the amount in moles. Saying that a gas is present at X ppt implies that the volume occupied by the gas is $X \times 10^{-12}$ of the whole, and therefore that the amount in moles of the gas is $X \times 10^{-12}$ of the total amount in moles

$$n_{\text{trace}} = (X \times 10^{-12}) \times n_{\text{tot}}$$

This is rearranged to give an expression for the mole fraction x_{trace}

$$x_{\text{trace}} = \frac{n_{\text{trace}}}{n_{\text{tot}}} = X \times 10^{-12}$$

The partial pressure of the trace gas is therefore

$$p_{\text{trace}} = x_{\text{trace}} p_{\text{tot}} = (X \times 10^{-12}) \times p_{\text{tot}}$$

The concentration is $n_{\text{trace}}/V = p_{\text{trace}}/RT$, so

$$c_{\text{trace}} = \frac{n_{\text{trace}}}{V} = \frac{X \times 10^{-12} \times p_{\text{tot}}}{RT}$$

(a) At 10 °C and 1.0 atm

$$c_{CCl_3F} = \frac{x_{CCl_3F} \times 10^{-12} \times p_{tot}}{RT}$$
$$= \frac{261 \times 10^{-12} \times (1.0\ atm)}{(8.2057 \times 10^{-2}\ dm^3\ atm\ K^{-1}\ mol^{-1}) \times ([10 + 273.15]\ K)}$$
$$= \boxed{1.1 \times 10^{-11}\ mol\ dm^{-3}}$$

$$c_{CCl_2F_2} = \frac{x_{CCl_2F_2} \times 10^{-12} \times p_{tot}}{RT}$$
$$= \frac{509 \times 10^{-12} \times (1.0\ atm)}{(8.2057 \times 10^{-2}\ dm^3\ atm\ K^{-1}\ mol^{-1}) \times ([10 + 273.15]\ K)}$$
$$= \boxed{2.2 \times 10^{-11}\ mol\ dm^{-3}}$$

(b) At 200 K and 0.050 atm

$$c_{CCl_3F} = \frac{x_{CCl_3F} \times 10^{-12} \times p_{tot}}{RT}$$
$$= \frac{261 \times 10^{-12} \times (0.050\ atm)}{(8.2057 \times 10^{-2}\ dm^3\ atm\ K^{-1}\ mol^{-1}) \times (200\ K)}$$
$$= \boxed{8.0 \times 10^{-13}\ mol\ dm^{-3}}$$

$$c_{CCl_2F_2} = \frac{x_{CCl_2F_2} \times 10^{-12} \times p_{tot}}{RT}$$
$$= \frac{509 \times 10^{-12} \times (0.050\ atm)}{(8.2057 \times 10^{-2}\ dm^3\ atm\ K^{-1}\ mol^{-1}) \times (200\ K)}$$
$$= \boxed{1.6 \times 10^{-12}\ mol\ dm^{-3}}$$

1B The kinetic model

Answer to discussion questions

D1B.1 The three assumptions on which the kinetic model is based are given in Section 1B.1 on page 11.

1. The gas consists of molecules in ceaseless random motion obeying the laws of classical mechanics.
2. The size of the molecules is negligible, in the sense that their diameters are much smaller than the average distance travelled between collisions; they are 'point-like'.
3. The molecules interact only through brief elastic collisions.

An elastic collision is a collision in which the total translational kinetic energy of the molecules is conserved.

None of these assumptions is strictly true; however, many of them are good approximations under a wide range of conditions including conditions of ambient temperature and pressure. In particular,

(a) Molecules are subject to laws of quantum mechanics; however, for all but the lightest gases at low temperatures, non-classical effects are not important.

(b) With increasing pressure, the average distance between molecules will decrease, eventually becoming comparable to the dimensions of the molecules themselves.

(c) Intermolecular interactions, such as hydrogen bonding, and the interactions of dipole moments, operate when molecules are separated by small distances. Therefore, as assumption (2) breaks down, so does assumption (3), because the molecules are often close enough together to interact even when not colliding.

D1B.3 For an object (be it a space craft or a molecule) to escape the gravitational field of the Earth it must acquire kinetic energy equal in magnitude to the gravitational potential energy the object experiences at the surface of the Earth. The gravitational potential between two objects with masses m_1 and m_2 when separated by a distance r is

$$V = -\frac{Gm_1m_2}{r}$$

where G is the (universal) gravitational constant. In the case of an object of mass m at the surface of the Earth, it turns out that the gravitational potential is given by

$$V = -\frac{GmM}{R}$$

where M is the mass of the Earth and R its radius. This expression implies that the potential at the surface is the same as if the mass of the Earth were localized at a distance equal to its radius.

As a mass moves away from the surface of the Earth the potential energy increases (becomes less negative) and tends to zero at large distances. This change in potential energy must all be converted into kinetic energy if the mass is to escape. A mass m moving at speed v has kinetic energy $\frac{1}{2}mv^2$; this speed will be the *escape velocity* v_e when

$$\tfrac{1}{2}mv_e^2 = \frac{GmM}{R} \qquad \text{hence} \qquad v_e = \sqrt{\frac{2GM}{R}}$$

The quantity in the square root is related to the acceleration due to free fall, g, in the following way. A mass m at the surface of the Earth experiences a gravitational *force* given GMm/R^2 (note that the force goes as R^{-2}). This force accelerates the mass towards the Earth, and can be written mg. The two expressions for the force are equated to give

$$\frac{GMm}{R^2} = mg \qquad \text{hence} \qquad \frac{GM}{R} = gR$$

This expression for GM/R is substituted into the above expression for v_e to give

$$v_e = \sqrt{\frac{2GM}{R}} = \sqrt{2Rg}$$

The escape velocity is therefore a function of the radius of the Earth and the acceleration due to free fall.

The radius of the Earth is 6.37×10^6 m and $g = 9.81$ m s^{-2} so the escape velocity is 1.11×10^4 m s^{-1}. For comparison, the mean speed of He at 298 K is 1300 m s^{-1} and for N_2 the mean speed is 475 m s^{-1}. For He, only atoms with a speed in excess of eight times the mean speed will be able to escape, whereas for N_2 the speed will need to be more than twenty times the mean speed. The fraction of molecules with speeds many times the mean speed is small, and because this fraction goes as e^{-v^2} it falls off rapidly as the multiple increases. A tiny fraction of He atoms will be able to escape, but the fraction of heavier molecules with sufficient speed to escape will be utterly negligible.

Solutions to exercises

E1B.1(a) (i) The mean speed is given by [1B.9–*16*], $v_{\text{mean}} = (8RT/\pi M)^{1/2}$, so $v_{\text{mean}} \propto \sqrt{1/M}$. The ratio of the mean speeds therefore depends on the ratio of the molar masses

$$\frac{v_{\text{mean,H}_2}}{v_{\text{mean,Hg}}} = \left(\frac{M_{\text{Hg}}}{M_{\text{H}_2}}\right)^{1/2} = \left(\frac{200.59 \text{ g mol}^{-1}}{2 \times 1.0079 \text{ g mol}^{-1}}\right)^{1/2} = \boxed{9.975}$$

(ii) The mean translational kinetic energy $\langle E_k \rangle$ is given by $\frac{1}{2}m\langle v^2 \rangle$, where $\langle v^2 \rangle$ is the mean square speed, which is given by [1B.7–*15*], $\langle v^2 \rangle = 3RT/M$. The mean translational kinetic energy is therefore

$$\langle E_k \rangle = \frac{1}{2}m\langle v^2 \rangle = \frac{1}{2}m\left(\frac{3RT}{M}\right)$$

The molar mass M is related to the mass m of one molecule by $M = mN_A$, where N_A is Avogadro's constant, and the gas constant can be written $R = kN_A$, hence

$$\langle E_k \rangle = \frac{1}{2}m\left(\frac{3RT}{M}\right) = \frac{1}{2}m\left(\frac{3kN_A T}{mN_A}\right) = \frac{3}{2}kT$$

The mean translational kinetic energy is therefore independent of the identity of the gas, and only depends on the temperature: it is the same for H_2 and Hg.

This result is related to the principle of equipartition of energy: a molecule has three translational degrees of freedom (x, y, and z) each of which contributes $\frac{1}{2}kT$ to the average energy.

E1B.2(a) The rms speed is given by [1B.8–15], $v_{rms} = (3RT/M)^{1/2}$.

$$v_{rms,H_2} = \left(\frac{3RT}{M_{H_2}}\right)^{1/2} = \left(\frac{3 \times (8.3145\,\mathrm{J\,K^{-1}\,mol^{-1}}) \times (293.15\,\mathrm{K})}{2 \times 1.0079 \times 10^{-3}\,\mathrm{kg\,mol^{-1}}}\right)^{1/2}$$
$$= \boxed{1.90\,\mathrm{km\,s^{-1}}}$$

where $1\,\mathrm{J} = 1\,\mathrm{kg\,m^2\,s^{-2}}$ has been used. Note that the molar mass is in $\mathrm{kg\,mol^{-1}}$.

$$v_{rms,O_2} = \left(\frac{3 \times (8.3145\,\mathrm{J\,K^{-1}\,mol^{-1}}) \times (293.15\,\mathrm{K})}{2 \times 16.00 \times 10^{-3}\,\mathrm{kg\,mol^{-1}}}\right)^{1/2} = \boxed{478\,\mathrm{m\,s^{-1}}}$$

E1B.3(a) The Maxwell–Boltzmann distribution of speeds, $f(v)$, is given by [1B.4–14]. The fraction of molecules with speeds between v_1 and v_2 is given by the integral

$$\int_{v_1}^{v_2} f(v)\,\mathrm{d}v$$

If the range $v_2 - v_1 = \delta v$ is small, the integral is well-approximated by

$$f(v_{mid})\,\delta v$$

where v_{mid} is the mid-point of the velocity range: $v_{mid} = \frac{1}{2}(v_2 + v_1)$. In this exercise $v_{mid} = 205\,\mathrm{m\,s^{-1}}$ and $\delta v = 10\,\mathrm{m\,s^{-1}}$.

$$\text{fraction} = f(v_{mid})\,\delta v = 4\pi \times \left(\frac{M}{2\pi RT}\right)^{3/2} v_{mid}^2 \exp\left(\frac{-Mv_{mid}^2}{2RT}\right)\delta v$$
$$= 4\pi \times \left(\frac{2 \times 14.01 \times 10^{-3}\,\mathrm{kg\,mol^{-1}}}{2\pi \times (8.3145\,\mathrm{J\,K^{-1}\,mol^{-1}}) \times (400\,\mathrm{K})}\right)^{3/2} \times (205\,\mathrm{m\,s^{-1}})^2$$
$$\times \exp\left(\frac{-(2 \times 14.01 \times 10^{-3}\,\mathrm{kg\,mol^{-1}}) \times (205\,\mathrm{m\,s^{-1}})^2}{2 \times (8.3145\,\mathrm{J\,K^{-1}\,mol^{-1}}) \times (400\,\mathrm{K})}\right) \times (10\,\mathrm{m\,s^{-1}})$$
$$= \boxed{6.87 \times 10^{-3}}$$

where $1\,\mathrm{J} = 1\,\mathrm{kg\,m^2\,s^{-2}}$ has been used. Thus, 0.687% of molecules have velocities in this range.

E1B.4(a) The mean relative speed is given by [1B.11b–16], $v_{rel} = (8kT/\pi\mu)^{1/2}$, where $\mu = m_A m_B/(m_A + m_A)$ is the effective mass. Multiplying top and bottom of the expression for v_{rel} by N_A and using $N_A k = R$ gives $v_{rel} = (8RT/\pi N_A \mu)^{1/2}$ in which $N_A\mu$ is the molar effective mass. For the relative motion of N_2 and H_2 this effective mass is

$$N_A\mu = \frac{M_{N_2}M_{H_2}}{M_{N_2} + M_{H_2}} = \frac{(2 \times 14.01\,\mathrm{g\,mol^{-1}}) \times (2 \times 1.0079\,\mathrm{g\,mol^{-1}})}{(2 \times 14.01\,\mathrm{g\,mol^{-1}}) + (2 \times 1.0079\,\mathrm{g\,mol^{-1}})} = 1.88...\,\mathrm{g\,mol^{-1}}$$

$$v_{rel} = \left(\frac{8RT}{\pi N_A \mu}\right)^{1/2} = \left(\frac{8 \times (8.3145\,\mathrm{J\,K^{-1}\,mol^{-1}}) \times (298.15\,\mathrm{K})}{\pi \times (1.88... \times 10^{-3}\,\mathrm{kg\,mol^{-1}})}\right)^{1/2} = \boxed{1832\,\mathrm{m\,s^{-1}}}$$

The value of the effective mass μ is dominated by the mass of the lighter molecule, in this case H_2.

E1B.5(a) The most probable speed is given by [1B.10–*16*], $v_{\rm mp} = (2RT/M)^{1/2}$, the mean speed is given by [1B.9–*16*], $v_{\rm mean} = (8RT/\pi M)^{1/2}$, and the mean relative speed between two molecules of the same mass is given by [1B.11a–*16*], $v_{\rm rel} = \sqrt{2}v_{\rm mean}$.

$M_{\rm CO_2} = 12.01 + 2 \times 16.00 = 44.01\ {\rm g\,mol^{-1}}$.

$$v_{\rm mp} = \left(\frac{2RT}{M}\right)^{1/2} = \left(\frac{2 \times (8.3145\,{\rm J\,K^{-1}\,mol^{-1}}) \times (293.15\,{\rm K})}{44.01 \times 10^{-3}\,{\rm kg\,mol^{-1}}}\right)^{1/2} = \boxed{333\,{\rm m\,s^{-1}}}$$

$$v_{\rm mean} = \left(\frac{8RT}{\pi M}\right)^{1/2} = \left(\frac{8 \times (8.3145\,{\rm J\,K^{-1}\,mol^{-1}}) \times (293.15\,{\rm K})}{\pi \times (44.01 \times 10^{-3}\,{\rm kg\,mol^{-1}})}\right)^{1/2} = \boxed{376\,{\rm m\,s^{-1}}}$$

$$v_{\rm rel} = \sqrt{2}v_{\rm mean} = \sqrt{2} \times (376\,{\rm m\,s^{-1}}) = \boxed{531\,{\rm m\,s^{-1}}}$$

E1B.6(a) The collision frequency is given by [1B.12b–*17*], $z = \sigma v_{\rm rel} p/kT$, with the relative speed for two molecules of the same type given by [1B.11a–*16*], $v_{\rm rel} = \sqrt{2}v_{\rm mean}$. The mean speed is given by [1B.9–*16*], $v_{\rm mean} = (8RT/\pi M)^{1/2}$. From the *Resource section* the collision cross-section σ is $0.27\ {\rm nm^2}$.

$$z = \frac{\sigma v_{\rm rel} p}{kT} = \frac{\sigma p}{kT} \times \sqrt{2} \times \left(\frac{8RT}{\pi M}\right)^{1/2}$$

$$= \frac{(0.27 \times 10^{-18}\,{\rm m^2}) \times (1.01325 \times 10^5\,{\rm Pa})}{(1.3806 \times 10^{-23}\,{\rm J\,K^{-1}}) \times (298.15\,{\rm K})} \times \sqrt{2}$$

$$\times \left(\frac{8 \times (8.3145\,{\rm J\,K^{-1}\,mol^{-1}}) \times (298.15\,{\rm K})}{\pi \times (2 \times 1.0079 \times 10^{-3}\,{\rm kg\,mol^{-1}})}\right)^{1/2}$$

$$= \boxed{1.7 \times 10^{10}\,{\rm s^{-1}}}$$

where $1\ {\rm J} = 1\ {\rm kg\,m^2\,s^{-2}}$ and $1\ {\rm Pa} = 1\ {\rm kg\,m^{-1}\,s^{-2}}$ have been used. Note the conversion of the collision cross-section σ to ${\rm m^2}$: $1\ {\rm nm^2} = (1 \times 10^{-9})^2\ {\rm m^2} = 1 \times 10^{-18}\ {\rm m^2}$.

E1B.7(a) The mean speed is given by [1B.9–*16*], $v_{\rm mean} = (8RT/\pi M)^{1/2}$. The collision frequency is given by [1B.12b–*17*], $z = \sigma v_{\rm rel} p/kT$, with the relative speed for two molecules of the same type given by [1B.11a–*16*], $v_{\rm rel} = \sqrt{2}v_{\rm mean}$. The mean free path is given by [1B.14–*18*], $\lambda = kT/\sigma p$

(i) The mean speed is calculated as

$$v_{\rm mean} = \left(\frac{8RT}{\pi M}\right)^{1/2} = \left(\frac{8 \times (8.3145\,{\rm J\,K^{-1}\,mol^{-1}}) \times (298.15\,{\rm K})}{\pi \times (2 \times 14.01 \times 10^{-3}\,{\rm kg\,mol^{-1}})}\right)^{1/2} = \boxed{475\,{\rm m\,s^{-1}}}$$

(ii) The collision cross-section σ is calculated from the collision diameter d as $\sigma = \pi d^2 = \pi \times (395 \times 10^{-9}\ {\rm m})^2 = 4.90... \times 10^{-19}\ {\rm m^2}$. With this value the mean free path is calculated as

$$\lambda = \frac{kT}{\sigma p} = \frac{(1.3806 \times 10^{-23}\,{\rm J\,K^{-1}}) \times (298.15\,{\rm K})}{(4.90... \times 10^{-19}\,{\rm m^2}) \times (1.01325 \times 10^5\,{\rm Pa})} = 82.9 \times 10^{-9}\,{\rm m} = \boxed{82.9\,{\rm nm}}$$

where $1\ {\rm J} = 1\ {\rm kg\,m^2\,s^{-2}}$ and $1\ {\rm Pa} = 1\ {\rm kg\,m^{-1}\,s^{-2}}$ have been used.

(iii) The collision rate is calculated as

$$z = \frac{\sigma v_{\text{rel}} p}{kT} = \frac{\sigma p}{kT} \times \sqrt{2} \times \left(\frac{8RT}{\pi M}\right)^{1/2}$$
$$= \frac{(4.90... \times 10^{-19}\,\text{m}^2) \times (1.01325 \times 10^5\,\text{Pa})}{(1.3806 \times 10^{-23}\,\text{J K}^{-1}) \times (298.15\,\text{K})} \times \sqrt{2}$$
$$\times \left(\frac{8 \times (8.3145\,\text{J K}^{-1}\,\text{mol}^{-1}) \times (298.15\,\text{K})}{\pi \times (2 \times 14.01 \times 10^{-3}\,\text{kg mol}^{-1})}\right)^{1/2}$$
$$= \boxed{8.10 \times 10^9\,\text{s}^{-1}}$$

An alternative for the calculation of z is to use [1B.13–*18*], $\lambda = v_{\text{rel}}/z$, rearranged to $z = v_{\text{rel}}/\lambda$

$$z = \frac{v_{\text{rel}}}{\lambda} = \frac{\sqrt{2}v_{\text{mean}}}{\lambda} = \frac{\sqrt{2} \times (475\,\text{m s}^{-1})}{82.9 \times 10^{-9}\,\text{m}} = \boxed{8.10 \times 10^9\,\text{s}^{-1}}$$

E1B.8(a) The container is assumed to be spherical with radius r and hence volume $V = \frac{4}{3}\pi r^3$. This volume is expressed in terms the the required diameter $d = 2r$ as $V = \frac{1}{6}\pi d^3$. Rearrangement of this expression gives d

$$d = \left(\frac{6V}{\pi}\right)^{1/3} = \left(\frac{6 \times 100\,\text{cm}^3}{\pi}\right)^{1/3} = 5.75...\,\text{cm}$$

The mean free path is given by [1B.14–*18*], $\lambda = kT/\sigma p$. This is rearranged to give the pressure p with λ equal to the diameter of the vessel

$$p = \frac{kT}{\sigma d} = \frac{(1.3806 \times 10^{-23}\,\text{J K}^{-1}) \times (298.15\,\text{K})}{(0.36 \times 10^{-18}\,\text{m}^2) \times (5.75... \times 10^{-2}\,\text{m})} = \boxed{0.20\,\text{Pa}}$$

Note the conversion of the diameter from cm to m.

E1B.9(a) The mean free path is given by [1B.14–*18*], $\lambda = kT/\sigma p$.

$$\lambda = \frac{kT}{\sigma p} = \frac{(1.3806 \times 10^{-23}\,\text{J K}^{-1}) \times (217\,\text{K})}{(0.43 \times 10^{-18}\,\text{m}^2) \times (0.05 \times 1.01325 \times 10^5\,\text{Pa})}$$
$$= \boxed{1.4 \times 10^{-6}\,\text{m} = 1.4\,\mu\text{m}}$$

Solutions to problems

P1B.1 A rotating slotted-disc apparatus consists of a series of disks all mounted on a common axle (shaft). Each disc has a narrow radial slot cut into it, and the slots on successive discs are displaced from one another by a certain angle. The discs are then spun at a constant angular speed.

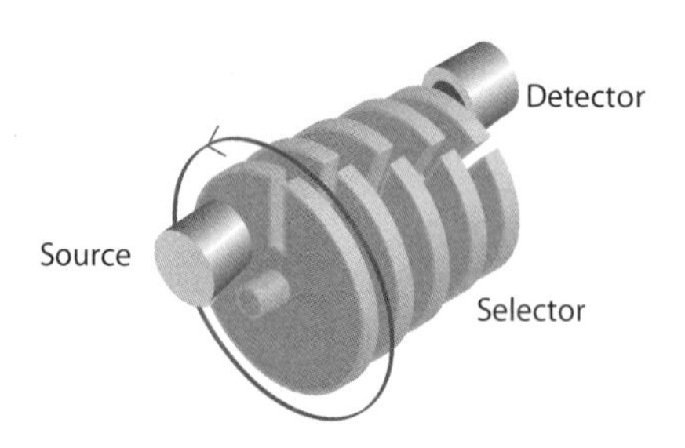

Imagine a molecule moving along the direction of the axle with a certain velocity such that it passes through the slot in the first disc. By the time the molecule reaches the second disc the slot in that disc will have moved around, and the molecule will only pass through the slot if the speed of the molecule is such that it arrives at the second disc at just the time at which the slot appears in the path of the molecule. In this way, only molecules with a specific velocity (or, because the slot has a finite width, a small range of velocities) will pass through the second slpt. The velocity of the molecules which will pass through the second disc is set by the angular speed at which the discs are rotated and the angular displacement of the slots on successive discs.

The angular velocity of the discs is $2\pi\nu$ rad s^{-1} so in time t the discs move through an angle $\theta = 2\pi\nu t$. If the spacing of the discs is d, a molecule with velocity v_x will take time $t = d/v_x$ to pass from one disc to the next. If the second slit is set at an angle α relative to the first, such a molecule will only pass through the second slit if

$$2\pi\nu\left(\frac{d}{v_x}\right) = \alpha \qquad \text{hence} \qquad v_x = \frac{2\pi\nu d}{\alpha}$$

If the angle α is expressed in degrees, $\alpha = \pi(\alpha^\circ/180^\circ)$, this rearranges to

$$v_x = \frac{2\pi\nu d}{\pi(\alpha^\circ/180^\circ)} = \frac{360^\circ \nu d}{\alpha^\circ}$$

With the values given the velocity of the molecules is computed as

$$v_x = \frac{360^\circ \nu d}{\alpha^\circ} = \frac{360^\circ \nu(0.01\ \text{m})}{2^\circ} = 180\nu(0.01\ \text{m})$$

The Maxwell–Boltzmann distribution of speeds in one dimension is given by [1B.3–*13*]

$$f(v_x) = \left(\frac{m}{2\pi kT}\right)^{1/2} \mathrm{e}^{-mv_x^2/2kT}$$

The given data on the intensity of the beam is assumed to be proportional to $f(v_x)$: $I \propto f(v_x) = Af(v_x)$. Because the constant of proportionality is not known and the variation with v_x is to be explored, it is convenient to take logarithms to give

$$\ln I = \ln[Af(v_x)] = \ln A + \ln\left(\frac{m}{2\pi kT}\right)^{1/2} - \frac{mv_x^2}{2kT}$$

A plot of $\ln I$ against v_x^2 is expected to be a straight line with slope $-m/2kT$; such a plot is shown in Fig. 1.4.

ν/Hz	v_x/m s^{-1}	$v_x^2/(10^4\,\mathrm{m^2\,s^{-2}})$	$I(40\,\mathrm{K})$	$\ln I(40\,\mathrm{K})$	$I(100\,\mathrm{K})$	$\ln I(100\,\mathrm{K})$
20	36	0.13	0.846	−0.167	0.592	−0.524
40	72	0.52	0.513	−0.667	0.485	−0.724
80	144	2.07	0.069	−2.674	0.217	−1.528
100	180	3.24	0.015	−4.200	0.119	−2.129
120	216	4.67	0.002	−6.215	0.057	−2.865

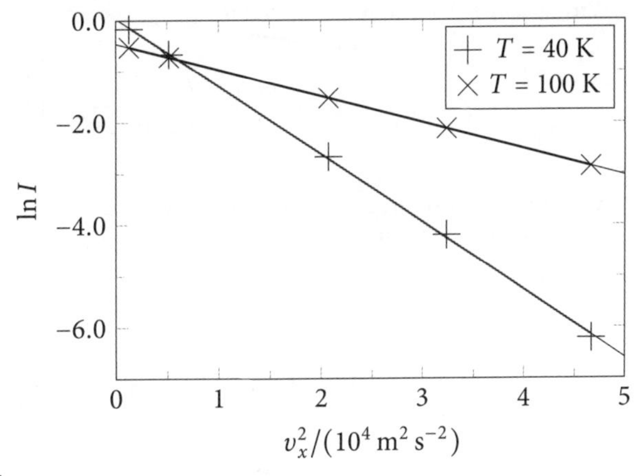

Figure 1.4

At both temperatures the data fall on reasonable straight lines, with slope −1.33 at 40 K and −0.516 at 100 K.

If the Maxwell–Boltzmann distribution applies the expected slope at 40 K is computed as

$$-\frac{m}{2kT} = -\frac{M}{2RT} = -\frac{83.80\times10^{-3}\,\mathrm{kg\,mol^{-1}}}{2\times(8.3145\,\mathrm{J\,K^{-1}\,mol^{-1}})\times(40\,\mathrm{K})} = -1.26\times10^{-4}\,\mathrm{m^{-2}\,s^2}$$

where $R = N_\mathrm{A}k$ has been used. The expected slope of the above graph is therefore −1.26, which compares reasonably well with that found experimentally.

At 100 K the expected slope is

$$-\frac{83.80\times10^{-3}\,\mathrm{kg\,mol^{-1}}}{2\times(8.3145\,\mathrm{J\,K^{-1}\,mol^{-1}})\times(100\,\mathrm{K})} = -5.04\times10^{-5}\,\mathrm{m^{-2}\,s^2}$$

Again, the expected slope −0.504 compares reasonably well with that found experimentally.

P1B.3 The Maxwell–Boltzmann distribution of speeds in one dimension (here x) is given by [1B.3–*13*]

$$f(v_x) = \left(\frac{m}{2\pi kT}\right)^{1/2} \mathrm{e}^{-mv_x^2/2kT}$$

The first task is to find an expression for the mean speed, which is found using [1B.6–15], $\langle v^n \rangle = \int_0^\infty v^n f(v)\,dv$. In this case

$$\langle v_x \rangle = \int_0^\infty v_x \left(\frac{m}{kT}\right)^{1/2} e^{-mv_x^2/2kT}\,dv$$

The required integral is of the form of G.2 from the *Resource section*

$$\int_0^\infty x e^{-ax^2}\,dx = \frac{1}{2a}$$

With $a = m/2kT$ the mean speed is

$$v_{\text{mean}} = \langle v_x \rangle = \left(\frac{m}{kT}\right)^{1/2} \left(\frac{1}{2(m/2kT)}\right) = \left(\frac{kT}{2\pi m}\right)^{1/2}$$

After the beam emerges from the velocity selector, $f(v_x)$ is zero for $v_x > v_{\text{mean}}$. The probability distribution is therefore changed and so needs to be re-normalized such that

$$K_x \int_0^{v_{\text{mean}}} e^{-mv_x^2/2kT}\,dv_x = 1$$

This integral is best evaluated using mathematical software which gives

$$\int_0^{v_{\text{mean}}} e^{-mv_x^2/2kT}\,dv = \left(\frac{\pi kT}{2m}\right)^{1/2} \operatorname{erf}\!\left(\frac{1}{2\sqrt{\pi}}\right)$$

where $\operatorname{erf}(x)$ is the *error function*. The normalized distribution is therefore

$$f_{\text{new}}(v_x) = \left(\frac{2m}{\pi kT}\right)^{1/2} \frac{1}{\operatorname{erf}(\frac{1}{2\sqrt{\pi}})} e^{-mv_x^2/2kT}$$

The new mean speed is computed using this distribution; again this intergral is best evaluated using mathematical software. Note that the integral extends up to v_{mean}

$$v_{\text{mean, new}} = \left(\frac{2m}{\pi kT}\right)^{1/2} \frac{1}{\operatorname{erf}(\frac{1}{2\sqrt{\pi}})} \int_0^{v_{\text{mean}}} v_x e^{-mv_x^2/2kT}\,dv_x$$

$$= (1 - e^{1/4\pi}) \left(\frac{2kT}{\pi m}\right)^{1/2} \frac{1}{\operatorname{erf}(\frac{1}{2\sqrt{\pi}})} = (1 - e^{1/4\pi})2\overbrace{\left(\frac{kT}{2\pi m}\right)^{1/2}}^{v_{\text{mean}}} \frac{1}{\operatorname{erf}(\frac{1}{2\sqrt{\pi}})}$$

$$= (1 - e^{1/4\pi})2v_{\text{mean}} \frac{1}{\operatorname{erf}(\frac{1}{2\sqrt{\pi}})}$$

The error function is evaluated numerically to give $\boxed{v_{\text{mean, new}} \approx 0.493\ v_{\text{mean}}}$.

P1B.5 The Maxwell–Boltzmann distribution of speeds in three dimensions is given by [1B.4–*14*]

$$f(v) = 4\pi\left(\frac{M}{2\pi RT}\right)^{3/2} v^2 e^{-Mv^2/2RT}$$

with M the molar mass. The most probable speed is given by [1B.10–*16*], $v_{mp} = (2RT/M)^{1/2}$. If the interval of speeds, Δv is small, the fraction of molecules with speeds in this range, centred at speed v_{mp} is well-approximated by $f(v_{mp})\Delta v$.

The required fraction of molecules with speeds in the range Δv around $n \times v_{mp}$ compared to that centred around v_{mp} is given by

$$\frac{f(n\times v_{mp})\Delta v}{f(v_{mp})\Delta v} = \frac{(n\times v_{mp})^2}{v_{mp}^2}\frac{e^{-M(nv_{mp})^2/2RT}}{e^{-Mv_{mp}^2/2RT}} = n^2 e^{-Mv_{mp}^2(n^2-1)/2RT}$$

In taking the ratio, with the exception of the term v^2, all of the terms in $f(v)$ which multiply the exponential cancel. In this expression the term v_{mp} is replaced by $(2RT/M)^{1/2}$ to give

$$\frac{f(n\times v_{mp})\Delta v}{f(v_{mp})\Delta v} = n^2 e^{-Mv_{mp}^2(n^2-1)/2RT} = n^2 e^{-M(2RT/M)(n^2-1)/2RT} = n^2 e^{(1-n^2)}$$

For $n = 3$ this expression evaluates to $\boxed{3.02\times10^{-3}}$ and for $n = 4$ it evaluates to $\boxed{4.89\times10^{-6}}$. These numbers indicate that very few molecules have speeds several times greater than the most probable speed.

P1B.7 The key idea here is that for an object to escape the gravitational field of the Earth it must acquire kinetic energy equal in magnitude to the gravitational potential energy the object experiences at the surface of the Earth. The gravitational potential energy between two objects with masses m_1 and m_2 when separated by a distance r is

$$V = -\frac{Gm_1m_2}{r}$$

where G is the (universal) gravitational constant. In the case of an object of mass m at the surface of the Earth, it turns out that the gravitational potential energy is given by

$$V = -\frac{GmM}{R}$$

where M is the mass of the Earth and R its radius. This expression implies that the potential at the surface is the same as if the mass of the Earth were localized at a distance equal to its radius.

As a mass moves away from the surface of the Earth the potential energy increases (becomes less negative) and tends to zero at large distances. If the mass is to escape its kinetic energy must be greater than or equal to this change in potential energy. A mass m moving at speed v has kinetic energy $\frac{1}{2}mv^2$; this speed will be the *escape velocity* v_e when

$$\tfrac{1}{2}mv_e^2 = \frac{GmM}{R} \qquad \text{hence} \qquad v_e = \left(\frac{2GM}{R}\right)^{1/2}$$

The quantity in the square root is related to the acceleration due to free fall, g, in the following way. A mass m at the surface of the Earth experiences a gravitational *force* given GMm/R^2 (note that the force goes as R^{-2}). This force accelerates the mass towards the Earth, and can be written mg. The two expressions for the force are equated to give

$$\frac{GMm}{R^2} = mg \qquad \text{hence} \qquad \frac{GM}{R} = gR \tag{1.1}$$

This expression for GM/R is substituted into the above expression for v_e to give

$$v_e = \left(\frac{2GM}{R}\right)^{1/2} = (2Rg)^{1/2}$$

The escape velocity is therefore a function of the radius of the Earth and the acceleration due to free fall.

The quoted values for the Earth give

$$v_e = \sqrt{2Rg} = \sqrt{2 \times (6.37 \times 10^6 \text{ m}) \times (9.81 \text{ m s}^{-2})} = \boxed{1.12 \times 10^4 \text{ m s}^{-1}}$$

For Mars, data is not given on the acceleration due to free fall. However, it follows from eqn 1.1 that $g = GM/R^2$, and hence

$$\frac{g_{\text{Mars}}}{g_{\text{Earth}}} = \frac{M_{\text{Mars}}}{M_{\text{Earth}}} \left(\frac{R_{\text{Earth}}}{R_{\text{Mars}}}\right)^2$$

The acceleration due to freefall on Mars is therefore computed as

$$\begin{aligned} g_{\text{Mars}} &= g_{\text{Earth}} \frac{M_{\text{Mars}}}{M_{\text{Earth}}} \left(\frac{R_{\text{Earth}}}{R_{\text{Mars}}}\right)^2 \\ &= (9.81 \text{ m s}^{-2}) \times (0.108) \times \left(\frac{6.37 \times 10^6 \text{ m}}{3.38 \times 10^6 \text{ m}}\right)^2 = 3.76... \text{ m s}^{-2} \end{aligned}$$

The escape velocity on Mars is therefore

$$v_e = \sqrt{2Rg} = \sqrt{2 \times (3.38 \times 10^6 \text{ m}) \times (3.76... \text{ m s}^{-2})} = \boxed{5.04 \times 10^3 \text{ m s}^{-1}}$$

The mean speed is given by [1B.9–*16*], $v_{\text{mean}} = (8RT/\pi M)^{1/2}$. This expression is rearranged to give the temperature T at which the mean speed is equal to the escape velocity

$$T = \frac{v_e^2 \pi M}{8R}$$

For H_2 on the Earth the calculation is

$$T = \frac{(1.12 \times 10^4 \text{ m s}^{-1})^2 \times \pi \times (2 \times 1.0079 \times 10^{-3} \text{ kg mol}^{-1})}{8 \times (8.3145 \text{ J K}^{-1} \text{ mol}^{-1})} = 1.19 \times 10^4 \text{ K}$$

The following table gives the results for all three gases on both planets

planet	v_e/m s^{-1}	$T/10^4$ K (H_2)	$T/10^4$ K (He)	$T/10^4$ K (O_2)
Earth	1.12×10^4	1.19	2.36	18.9
Mars	5.04×10^3	0.242	0.481	3.84

The fraction of molecules with speed greater than v_e is found by integrating the Maxwell–Boltzmann distribution from this speed up to infinity:

$$\text{fraction with speed} \geq v_e = F = \int_{v_e}^{\infty} 4\pi \left(\frac{M}{2\pi RT}\right)^{3/2} v^2 e^{-Mv^2/2RT}\, dv$$

This integral is best computed using mathematical software, to give the following results for the fraction F; an entry of zero indicates that the calculated fraction is zero to within the machine precision.

planet	T/K	$F(H_2)$	F(He)	$F(O_2)$
Earth	240	0	0	0
	1500	1.49×10^{-4}	9.52×10^{-9}	0
Mars	240	1.12×10^{-5}	5.09×10^{-11}	0
	1500	0.025	4.31×10^{-2}	4.61×10^{-14}

These results indicate that the lighter molecules have the greater chance of escaping (because they are moving faster on average) and that increasing the temperature increases the probability of escaping (again becuase this increases the mean speed). Escape from Mars is easier than from the Earth because of the lower escape velocity, and heavier molecules are seemingly very unlikely to escape from the Earth.

P1B.9 The Maxwell–Boltzmann distribution of speeds in three dimensions is given by [1B.4–*14*]

$$f(v) = 4\pi \left(\frac{M}{2\pi RT}\right)^{3/2} v^2 e^{-Mv^2/2RT}$$

The fraction with speed between v_1 and v_2 is found by integrating the distribution between these speeds; this is best done using mathematical software

$$\text{fraction with speed between } v_1 \text{ and } v_2 = \int_{v_1}^{v_2} 4\pi \left(\frac{M}{2\pi RT}\right)^{3/2} v^2 e^{-Mv^2/2RT}\, dv$$

At 300 K and with $M = 2 \times 16.00$ g mol^{-1} the fraction is $\boxed{0.0722}$ and at 1000 K the fraction is $\boxed{0.0134}$.

P1B.11 Two hard spheres will collide if their line of centres approach within $2r$ of one another, where r is the radius of the sphere. This distance defines the collision diameter, $d = 2r$, and the collision cross-section is the area of a circle with this radius, $\sigma = \pi d^2 = \pi(2r)^2$. The pressure is computed from the other parameters using the perfect gas law: $p = nRT/V$.

The collision frequency is given by [1B.12b–17], $z = \sigma v_{\rm rel} p/kT$, with the relative speed for two molecules of the same type given by [1B.11a–16], $v_{\rm rel} = \sqrt{2}v_{\rm mean}$. The mean speed is given by [1B.9–16], $v_{\rm mean} = (8RT/\pi M)^{1/2}$.

Putting this all together gives

$$z = \frac{\sigma v_{\rm rel} p}{kT} = \frac{\pi(2r)^2}{kT} \times \sqrt{2} \times \left(\frac{8RT}{\pi M}\right)^{1/2} \times \frac{nRT}{V}$$
$$= \pi(2r)^2 \times \sqrt{2} \times \left(\frac{8RT}{\pi M}\right)^{1/2} \times \frac{nN_{\rm A}}{V}$$

where to go to the second line $R = N_{\rm A}k$ has been used. The expression is evaluated to give

$$z = \pi(2\times(0.38\times10^{-9}\ {\rm m}))^2 \times \sqrt{2} \times \left(\frac{8\times(8.3145\ {\rm J\,K^{-1}\,mol^{-1}})\times(298.15\ {\rm K})}{\pi\times(16.0416\times10^{-3}\ {\rm kg\,mol^{-1}})}\right)^{1/2}$$
$$\times \frac{(0.1\ {\rm mol})\times(6.0221\times10^{23}\ {\rm mol^{-1}})}{1\times10^{-3}\ {\rm m^3}} = \boxed{9.7\times10^{10}\ {\rm s^{-1}}}$$

1C Real gases

Answer to discussion questions

D1C.1 Consider three temperature regions:

(1) $T < T_{\rm B}$. At very low pressures, all gases show a compression factor, $Z \approx 1$. At high pressures, all gases have $Z > 1$, signifying that they have a molar volume greater than a perfect gas, which implies that repulsive forces are dominant. At intermediate pressures, most gases show $Z < 1$, indicating that attractive forces reducing the molar volume below the perfect value are dominant.

(2) $T \approx T_{\rm B}$. $Z \approx 1$ at low pressures, slightly greater than 1 at intermediate pressures, and significantly greater than 1 only at high pressures. There is a balance between the attractive and repulsive forces at low to intermediate pressures, but the repulsive forces predominate at high pressures where the molecules are very close to each other.

(3) $T > T_{\rm B}$. $Z > 1$ at all pressures because the frequency of collisions between molecules increases with temperature.

D1C.3 The van der Waals equation ‘corrects’ the perfect gas equation for both attractive and repulsive interactions between the molecules in a real gas; see Section 1C.2 on page 23 for a fuller explanation.

The Berthelot equation accounts for the volume of the molecules in a manner similar to the van der Waals equation but the term representing molecular attractions is modified to account for the effect of temperature. Experimentally it is found that the van der Waals parameter a decreases with increasing temperature. Theory (see Focus 14) also suggests that intermolecular attractions can decrease with temperature. This variation of the attractive interaction with temperature can be accounted for in the equation of state by replacing the van der Waals a with a/T.

Solutions to exercises

E1C.1(a) The van der Waals equation of state in terms of the volume is given by [1C.5a–23], $p = nRT/(V - b) - an^2/V^2$. The parameters a and b for ethane are given in the *Resource section* as $a = 5.507\ \text{atm dm}^6\ \text{mol}^{-2}$ and $b = 6.51 \times 10^{-2}\ \text{dm}^3\ \text{mol}^{-1}$.

With these units it is convenient to use $R = 8.2057 \times 10^{-2}\ \text{dm}^3\ \text{atm K}^{-1}\ \text{mol}^{-1}$.

(i) $T = 273.15\ \text{K}$, $V = 22.414\ \text{dm}^3$, $n = 1.0\ \text{mol}$

$$p = \frac{nRT}{V - nb} - \frac{an^2}{V^2}$$
$$= \frac{(1.0\ \text{mol}) \times (8.2057 \times 10^{-2}\ \text{dm}^3\ \text{atm K}^{-1}\ \text{mol}^{-1}) \times (273.15\ \text{K})}{(22.414\ \text{dm}^3) - (1.0\ \text{mol}) \times (6.51 \times 10^{-2}\ \text{dm}^3\ \text{mol}^{-1})}$$
$$- \frac{(5.507\ \text{atm dm}^6\ \text{mol}^{-2}) \times (1.0\ \text{mol})^2}{(22.414\ \text{dm}^3)^2} = \boxed{0.99\ \text{atm}}$$

(ii) $T = 1000\ \text{K}$, $V = 100\ \text{cm}^3 = 0.100\ \text{dm}^3$, $n = 1.0\ \text{mol}$

$$p = \frac{nRT}{V - nb} - \frac{an^2}{V^2}$$
$$= \frac{(1.0\ \text{mol}) \times (8.2057 \times 10^{-2}\ \text{dm}^3\ \text{atm K}^{-1}\ \text{mol}^{-1}) \times (1000\ \text{K})}{(0.100\ \text{dm}^3) - (1.0\ \text{mol}) \times (6.51 \times 10^{-2}\ \text{dm}^3\ \text{mol}^{-1})}$$
$$- \frac{(5.507\ \text{atm dm}^6\ \text{mol}^{-2}) \times (1.0\ \text{mol})^2}{(0.100\ \text{dm}^3)^2} = \boxed{1.8 \times 10^3\ \text{atm}}$$

E1C.2(a) Recall that $1\ \text{atm} = 1.01325 \times 10^5\ \text{Pa}$, $1\ \text{dm}^6 = 10^{-6}\ \text{m}^6$, and $1\ \text{Pa} = 1\ \text{kg m}^{-1}\ \text{s}^{-2}$

$$a = (0.751\ \text{atm dm}^6\ \text{mol}^{-2}) \times \frac{1.01325 \times 10^5\ \text{Pa}}{1\ \text{atm}} \times \frac{10^{-6}\ \text{m}^6}{1\ \text{dm}^6} = 0.0761\ \text{Pa m}^6\ \text{mol}^{-2}$$
$$= 0.0760\ \text{kg m}^{-1}\ \text{s}^{-2}\ \text{m}^6\ \text{mol}^{-2} = \boxed{0.0761\ \text{kg m}^5\ \text{s}^{-2}\ \text{mol}^{-2}}$$

$$b = (0.0226\ \text{dm}^3\ \text{mol}^{-1}) \times \frac{10^{-3}\ \text{m}^3}{1\ \text{dm}^3} = \boxed{2.26 \times 10^{-5}\ \text{m}^3\ \text{mol}^{-1}}$$

E1C.3(a) The compression factor Z is defined in [1C.1–*20*] as $Z = V_\text{m}/V_\text{m}^\circ$, where V_m° is the molar volume of a perfect gas under the same conditions. This volume is computed from the equation of state for a perfect gas, [1A.4–*8*], as $V_\text{m}^\circ = RT/p$, hence $Z = pV_\text{m}/RT$ [1C.2–*20*].

(i) If V_m is 12% smaller than the molar volume of a perfect gas, it follows that $V_m = V_m^\circ(1-0.12) = 0.88V_m^\circ$. The compression factor is then computed directly as

$$Z = \frac{V_m}{V_m^\circ} = \frac{0.88 \times V_m^\circ}{V_m^\circ} = \boxed{0.88}$$

(ii) From [1C.2–*20*] it follows that $V_m = ZRT/p$

$$V_m = \frac{ZRT}{p} = \frac{0.88 \times (8.2057 \times 10^{-2}\,\mathrm{dm^3\,atm\,K^{-1}\,mol^{-1}}) \times (250\,\mathrm{K})}{15\,\mathrm{atm}}$$
$$= \boxed{1.2\,\mathrm{dm^3\,mol^{-1}}}$$

Because $Z < 1$, implying that $V_m < V_m^\circ$, attractive forces are dominant.

E1C.4(a) The van der Waals equation of state in terms of the volume is given by [1C.5a–*23*], $p = nRT/(V-b) - an^2/V^2$. The molar mass of N_2 is $M = 2\times 14.01\,\mathrm{g\,mol^{-1}} = 28.02\,\mathrm{g\,mol^{-1}}$, so it follows that the amount in moles is

$$n = m/M = (92.4\,\mathrm{kg})/(0.02802\,\mathrm{kg\,mol^{-1}}) = 3.29... \times 10^3\,\mathrm{mol}$$

The pressure is found by substituting the given parameters into [1C.5a–*23*], noting that the volume needs to be expressed in $\mathrm{dm^3}$

$$p = \frac{nRT}{V-nb} - \frac{an^2}{V^2}$$
$$= \frac{(3.29... \times 10^3\,\mathrm{mol}) \times (8.2057 \times 10^{-2}\,\mathrm{dm^3\,atm\,K^{-1}\,mol^{-1}}) \times (500\,\mathrm{K})}{(1000\,\mathrm{dm^3}) - (3.29... \times 10^3\,\mathrm{mol}) \times (0.0387\,\mathrm{dm^3\,mol^{-1}})}$$
$$- \frac{(1.352\,\mathrm{atm\,dm^6\,mol^{-2}}) \times (3.29... \times 10^3\,\mathrm{mol})^2}{(1000\,\mathrm{dm^3})^2} = \boxed{140\,\mathrm{atm}}$$

E1C.5(a) (i) The pressure is computed from the equation of state for a perfect gas, [1A.4–*8*], as $p = nRT/V$

$$p = \frac{nRT}{V} = \frac{(10.0) \times (8.2057 \times 10^{-2}\,\mathrm{dm^3\,atm\,K^{-1}\,mol^{-1}}) \times ([27+273.15]\,\mathrm{K})}{4.860\,\mathrm{dm^3}}$$
$$= \boxed{50.7\,\mathrm{atm}}$$

(ii) The van der Waals equation of state in terms of the volume is given by [1C.5a–*23*], $p = nRT/(V-b) - an^2/V^2$. This is used to calculate the pressure

$$p = \frac{nRT}{V-nb} - \frac{an^2}{V^2}$$
$$= \frac{(10.0\,\mathrm{mol}) \times (8.2057 \times 10^{-2}\,\mathrm{dm^3\,atm\,K^{-1}\,mol^{-1}}) \times ([27+273.15]\,\mathrm{K})}{(4.860\,\mathrm{dm^3}) - (10.0\,\mathrm{mol}) \times (0.0651\,\mathrm{dm^3\,mol^{-1}})}$$
$$- \frac{(5.507\,\mathrm{atm\,dm^6\,mol^{-2}}) \times (10.0\,\mathrm{mol})^2}{(4.860\,\mathrm{dm^3})^2} = 35.2... = \boxed{35.2\,\mathrm{atm}}$$

The compression factor Z is given in terms of the molar volume and pressure by [1C.2–20], $Z = pV_m/RT$. The molar volume is V/n

$$Z = \frac{pV_m}{RT} = \frac{pV}{nRT}$$

$$= \frac{(35.2...\ \text{atm}) \times (4.860\ \text{dm}^3)}{(10.0\ \text{mol}) \times (8.2057 \times 10^{-2}\ \text{dm}^3\ \text{atm}\ \text{K}^{-1}\ \text{mol}^{-1}) \times (300.15\ \text{K})} = \boxed{0.695}$$

E1C.6(a) The relation between the critical constants and the van der Waals parameters is given by [1C.6–26]

$$V_c = 3b \quad p_c = \frac{a}{27b^2} \quad T_c = \frac{8a}{27Rb}$$

All three critical constants are given, so the problem is over-determined: any pair of the these expressions is sufficient to find values of a and b. It is convenient to use $R = 8.2057 \times 10^{-2}\ \text{dm}^3\ \text{atm}\ \text{K}^{-1}\ \text{mol}^{-1}$ and volumes in units of dm^3.

If the expressions for V_c and p_c are used, a and b are found in the following way

$$V_c = 3b \quad \text{hence} \quad b = V_c/3 = (0.0987\ \text{dm}^3\ \text{mol}^{-1})/3 = 0.0329\ \text{dm}^3\ \text{mol}^{-1}$$

$$p_c = \frac{a}{27b^2} = \frac{a}{27(V_c/3)^2} \quad \text{hence} \quad a = 27(V_c/3)^2 p_c$$

$$a = 27(V_c/3)^2 p_c = 27([0.0987\ \text{dm}^3\ \text{mol}^{-1}]/3)^2 \times (45.6\ \text{atm})$$

$$= 1.33\ \text{atm}\ \text{dm}^6\ \text{mol}^{-2}$$

There are three possible ways of choosing two of the expressions with which to find a and b, and each choice gives a different value. For a the values are 1.33, 1.74, and 2.26, giving an average of $\boxed{1.78\ \text{atm}\ \text{dm}^6\ \text{mol}^{-2}}$. For b the values are 0.0329, 0.0329, and 0.0429, giving an average of $\boxed{0.0362\ \text{dm}^3\ \text{mol}^{-1}}$.

In Section 1C.2(a) on page 23 it is argued that $b = 4V_{\text{molec}}N_A$, where V_{molec} is the volume occupied by one molecule. This volume is written in terms of the radius r as $4\pi r^3/3$ so it follows that $r = (3b/16\pi N_A)^{1/3}$.

$$r = \left(\frac{3b}{16\pi N_A}\right)^{1/3} = \left(\frac{3 \times (0.0362\ \text{dm}^3\ \text{mol}^{-1})}{16\pi \times (6.0221 \times 10^{23}\ \text{mol}^{-1})}\right)^{1/3} = 1.53 \times 10^{-9}\ \text{dm} = \boxed{153\ \text{pm}}$$

E1C.7(a) (i) In Section 1C.1(b) on page 20 it is explained that at the Boyle temperature $Z = 1$ and $dZ/dp = 0$; this latter condition corresponds to the second virial coefficient, B or B', being zero. The task is to find the relationship between the van der Waals parameters and the virial coefficients, and the starting point for this is the expressions for the product pV_m is each case ([1C.5b–24] and [1C.3b–21])

$$\text{van der Waals:} \quad p = \frac{RT}{(V_m - b)} - \frac{a}{V_m^2} \quad \text{hence} \quad pV_m = \frac{RTV_m}{(V_m - b)} - \frac{a}{V_m}$$

$$\text{virial: } pV_\text{m} = RT\left(1 + \frac{B}{V_\text{m}}\right)$$

The van der Waals expression for pV_m is rewritten by dividing the denominator and numerator of the first fraction by V_m to give

$$pV_\text{m} = \frac{RT}{(1 - b/V_\text{m})} - \frac{a}{V_\text{m}}$$

The dimensionless parameter b/V_m is likely to be $\ll 1$, so the approximation $(1-x)^{-1} \approx 1 + x$ is used to give

$$pV_\text{m} = RT(1 + b/V_\text{m}) - \frac{a}{V_\text{m}} = RT\left[1 + \frac{1}{V_\text{m}}\left(b - \frac{a}{RT}\right)\right]$$

Comparison of this expression with the virial expansion shows that

$$B = b - \frac{a}{RT}$$

It therefore follows that the Boyle temperature, when $B = 0$, is $T_\text{b} = a/Rb$. For the van der Waals parameters from the *Resource section*

$$T_\text{b} = \frac{a}{Rb} = \frac{6.260 \text{ atm dm}^6 \text{ mol}^{-2}}{(8.2057 \times 10^{-2} \text{ dm}^3 \text{ atm K}^{-1} \text{ mol}^{-1}) \times (5.42 \times 10^{-2} \text{ dm}^3 \text{ mol}^{-1})}$$
$$= \boxed{1.41 \times 10^3 \text{ K}}$$

(ii) In Section 1C.2(a) on page 23 it is argued that $b = 4V_\text{molec}N_\text{A}$, where V_molec is the volume occupied by one molecule. This volume is written in terms of the radius r as $4\pi r^3/3$ so it follows that $r = (3b/16\pi N_\text{A})^{1/3}$.

$$r = \left(\frac{3b}{16\pi N_\text{A}}\right)^{1/3} = \left(\frac{3 \times (5.42 \times 10^{-2} \text{ dm}^3 \text{ mol}^{-1})}{16\pi \times (6.0221 \times 10^{23} \text{ mol}^{-1})}\right)^{1/3}$$
$$= 1.75 \times 10^{-9} \text{ dm} = \boxed{175 \text{ pm}}$$

E1C.8(a) The reduced variables are defined in terms of the critical constants,[1C.8–*26*]

$$V_\text{r} = V_\text{m}/V_\text{c} \quad p_\text{r} = p/p_\text{c} \quad T_\text{r} = T/T_\text{c}$$

If the reduced pressure is the same for two gases (1) and (2) it follows that

$$\frac{p^{(1)}}{p_\text{c}^{(1)}} = \frac{p^{(2)}}{p_\text{c}^{(2)}} \quad \text{hence} \quad p^{(2)} = \frac{p^{(1)}}{p_\text{c}^{(1)}} \times p_\text{c}^{(2)}$$

and similarly

$$T^{(2)} = \frac{T^{(1)}}{T_\text{c}^{(1)}} \times T_\text{c}^{(2)}$$

These relationships are used to find the pressure and temperature of gas (2) corresponding to a particular state of gas (1); it is necessary to know the critical constants of both gases.

(i) From the tables in the *Resource section*, for H_2 $p_c = 12.8$ atm, $T_c = 33.23$ K, and for NH_3 $p_c = 111.3$ atm, $T_c = 405.5$ K. Taking gas (1) as H_2 and gas (2) as NH_3, the pressure and temperature of NH_3 corresponding to $p^{(H_2)} = 1.0$ atm and $T^{(H_2)} = 298.15$ K is calculated as

$$p^{(NH_3)} = \frac{p^{(H_2)}}{p_c^{(H_2)}} \times p_c^{(NH_3)} = \frac{1.0\text{ atm}}{12.8\text{ atm}} \times (111.3\text{ atm}) = \boxed{8.7\text{ atm}}$$

$$T^{(NH_3)} = \frac{T^{(H_2)}}{T_c^{(H_2)}} \times T_c^{(NH_3)} = \frac{298.15\text{ K}}{33.23\text{ K}} \times (405.5\text{ K}) = \boxed{3.6 \times 10^3\text{ K}}$$

(ii) For Xe $p_c = 58.0$ atm, $T_c = 289.75$ K.

$$p^{(Xe)} = \frac{p^{(H_2)}}{p_c^{(H_2)}} \times p_c^{(Xe)} = \frac{1.0\text{ atm}}{12.8\text{ atm}} \times (58.0\text{ atm}) = \boxed{4.5\text{ atm}}$$

$$T^{(Xe)} = \frac{T^{(H_2)}}{T_c^{(H_2)}} \times T_c^{(Xe)} = \frac{298.15\text{ K}}{33.23\text{ K}} \times (289.75\text{ K}) = \boxed{2.6 \times 10^3\text{ K}}$$

(iii) For He $p_c = 2.26$ atm, $T_c = 5.2$ K.

$$p^{(He)} = \frac{p^{(H_2)}}{p_c^{(H_2)}} \times p_c^{(He)} = \frac{1.0\text{ atm}}{12.8\text{ atm}} \times (2.26\text{ atm}) = \boxed{0.18\text{ atm}}$$

$$T^{(He)} = \frac{T^{(H_2)}}{T_c^{(H_2)}} \times T_c^{(He)} = \frac{298.15\text{ K}}{33.23\text{ K}} \times (5.2\text{ K}) = \boxed{47\text{ K}}$$

E1C.9(a) The van der Waals equation of state in terms of the molar volume is given by [1C.5b–24], $p = RT/(V_m - b) - a/V_m^2$. This relationship is rearranged to find b

$$p = \frac{RT}{V_m - b} - \frac{a}{V_m^2} \quad \text{hence} \quad p + \frac{a}{V_m^2} = \frac{RT}{V_m - b}$$

$$\text{hence} \quad \frac{pV_m^2 + a}{V_m^2} = \frac{RT}{V_m - b} \quad \text{hence} \quad \frac{V_m^2}{pV_m^2 + a} = \frac{V_m - b}{RT}$$

$$\text{hence} \quad b = V_m - \frac{RTV_m^2}{pV_m^2 + a}$$

With the data given

$$b = V_m - \frac{RTV_m^2}{pV_m^2 + a} = (5.00 \times 10^{-4}\text{ m}^3\text{ mol}^{-1}) - \frac{(8.3145\text{ J K}^{-1}\text{ mol}^{-1}) \times (273\text{ K}) \times (5.00 \times 10^{-4}\text{ m}^3\text{ mol}^{-1})^2}{(3.0 \times 10^6\text{ Pa}) \times (5.00 \times 10^{-4}\text{ m}^3\text{ mol}^{-1})^2 + (0.50\text{ m}^6\text{ Pa mol}^{-2})}$$

$$= \boxed{4.6 \times 10^{-5}\text{ m}^3\text{ mol}^{-1}}$$

where 1 Pa = 1 kg m^{-1} s^{-2} and 1 J = 1 kg m^2 s^{-2} have been used.

The compression factor Z is defined in [1C.1–20] as $Z = V_m/V_m^\circ$, where V_m° is the molar volume of a perfect gas under the same conditions. This volume is computed from the equation of state for a perfect gas, [1A.4–8], as $V_m^\circ = RT/p$, hence $Z = pV_m/RT$, [1C.2–20]. With the data given

$$Z = \frac{pV_m}{RT} = \frac{(3.0\times10^6\ \text{Pa})\times(5.00\times10^{-4}\ \text{m}^3\ \text{mol}^{-1})}{(8.3145\ \text{J K}^{-1}\ \text{mol}^{-1})\times(273\ \text{K})} = \boxed{0.66}$$

Solutions to problems

P1C.1 The virial equation is given by [1C.3b–21], $pV_m = RT(1 + B/V_m + \ldots)$, and from the *Resource section* the second virial coefficient B for N_2 at 273 K is $-10.5\ \text{cm}^3\ \text{mol}^{-1}$. The molar mass of N_2 is $2\times14.01 = 28.02\ \text{g mol}^{-1}$, hence the molar volume is

$$V_m = \frac{V}{n} = \frac{V}{m/M} = \frac{2.25\ \text{dm}^3}{(4.56\ \text{g})/(28.02\ \text{g mol}^{-1})} = 13.8...\ \text{dm}^3\ \text{mol}^{-1}$$

This is used to calculate the pressure using the virial equation. It is convenient to use $R = 8.2057\times10^{-2}\ \text{dm}^3\ \text{atm K}^{-1}\ \text{mol}^{-1}$ and express all the volumes in dm^3

$$p = \frac{RT}{V_m}\left(1 + \frac{B}{V_m}\right)$$

$$= \frac{(8.2057\times10^{-2}\ \text{dm}^3\ \text{atm K}^{-1}\ \text{mol}^{-1})\times(273\ \text{K})}{13.8...\ \text{dm}^3\ \text{mol}^{-1}}\left(1 + \frac{-1.05\times10^{-2}\ \text{dm}^3\ \text{mol}^{-1}}{13.8...\ \text{dm}^3\ \text{mol}^{-1}}\right)$$

$$= \boxed{1.62\ \text{atm}}$$

P1C.3 The virial equation is [1C.3b–21], $pV_m = RT(1 + B/V_m + C/V_m^2 + \ldots)$. The compression factor is defined in [1C.1–20] as $Z = V_m/V_m^\circ$, and the molar volume of a perfect gas, V_m° is given by $V_m^\circ = RT/p$.
It follows that

$$V_m = (RT/p)(1 + B/V_m + C/V_m^2) = V_m^\circ(1 + B/V_m + C/V_m^2)$$

$$\text{hence } Z = \frac{V_m}{V_m^\circ} = 1 + \frac{B}{V_m} + \frac{C}{V_m^2}$$

To evaluate this expression, the molar volume is approximated by the molar volume of a perfect gas under the prevailing conditions

$$V_m^\circ = \frac{RT}{p} = \frac{(8.2057\times10^{-2}\ \text{dm}^3\ \text{atm K}^{-1}\ \text{mol}^{-1})\times(273\ \text{K})}{100\ \text{atm}} = 0.224...\ \text{dm}^3\ \text{mol}^{-1}$$

This value of the molar volume is then used to compute Z; note the conversion of all the volume terms to dm^3

$$Z = 1 + \frac{B}{V_m} + \frac{C}{V_m^2}$$

$$= 1 + \frac{-21.3\times10^{-3}\ \text{dm}^3\ \text{mol}^{-1}}{0.224...\ \text{dm}^3\ \text{mol}^{-1}} + \frac{1200\times10^{-6}\ \text{dm}^6\ \text{mol}^{-2}}{(0.224...\ \text{dm}^3\ \text{mol}^{-1})^2} = 0.928... = \boxed{0.929}$$

The molar volume is computed from the compression factor

$$Z = \frac{V_m}{V_m^{\circ}} = \frac{V_m}{RT/p}$$

$$\text{hence } V_m = \frac{ZRT}{p} = \frac{0.928... \times (8.2057 \times 10^{-2}\,\text{dm}^3\,\text{atm}\,\text{K}^{-1}\,\text{mol}^{-1}) \times (273\,\text{K})}{100\,\text{atm}}$$

$$= \boxed{0.208\,\text{dm}^3\,\text{mol}^{-1}}$$

P1C.5 In Section 1C.1(b) on page 20 it is explained that at the Boyle temperature $Z = 1$ and $\mathrm{d}Z/\mathrm{d}p = 0$; this latter condition corresponds to the second virial coefficient, B or B', being zero. The Boyle temperature is found by setting the given expression for $B(T)$ to zero and solving for T

$$0 = a + be^{-c/T^2} \quad \text{hence} \quad -a/b = e^{-c/T^2}$$

Taking logarithms gives $\ln(-a/b) = -c/T^2$ hence

$$T = \left(\frac{-c}{\ln(-a/b)}\right)^{1/2} = \left(\frac{-1131\,\text{K}^2}{\ln[-(-0.1993\,\text{bar}^{-1})/(0.2002\,\text{bar}^{-1})]}\right)^{1/2}$$

$$= \boxed{501.0\,\text{K}}$$

P1C.7 (a) The molar mass M of H_2O is $18.02\,\text{g}\,\text{mol}^{-1}$. The mass density ρ is related to the molar density ρ_m by $\rho_m = \rho/M$, and the molar volume is simply the reciprocal of the molar density $V_m = 1/\rho_m = M/\rho$

$$V_m = \frac{M}{\rho} = \frac{18.02 \times 10^{-3}\,\text{kg}\,\text{mol}^{-1}}{133.2\,\text{kg}\,\text{m}^{-3}} = 1.352... \times 10^{-4}\,\text{m}^3\,\text{mol}^{-1}$$

The molar volume is therefore $\boxed{0.1353\,\text{dm}^3\,\text{mol}^{-1}}$

(b) The compression factor Z is given by [1C.2–*20*], $Z = pV_m/RT$

$$Z = \frac{pV_m}{RT} = \frac{(327.6\,\text{atm}) \times (0.1352...\,\text{dm}^3\,\text{mol}^{-1})}{(8.2057 \times 10^{-2}\,\text{dm}^3\,\text{atm}\,\text{K}^{-1}\,\text{mol}^{-1}) \times (776.4\,\text{K})} = \boxed{0.6957}$$

(c) The virial equation (up to the second term) in terms of the molar volume is given by [1C.3b–*21*]

$$pV_m = RT\left(1 + \frac{B}{V_m}\right)$$

Division of each side by p gives

$$V_m = \frac{RT}{p}\left(1 + \frac{B}{V_m}\right)$$

The quantity RT/p is recognised as the molar volume of a perfect gas, V_m°, so it follows that

$$V_m = V_m^\circ\left(1+\frac{B}{V_m}\right) \text{ hence } \frac{V_m}{V_m^\circ} = Z = \left(1+\frac{B}{V_m}\right)$$

In *Problem* P1C.4 it is shown that B is related to the van der Waals constants by $B = b - a/RT$; using this, it is then possible to compute the compression factor

$$B = b - \frac{a}{RT} = (0.03049\ \text{dm}^3\ \text{mol}^{-1}) - \frac{(5.464\ \text{atm dm}^6\ \text{mol}^{-2})}{(8.2057\times10^{-2}\ \text{dm}^3\ \text{atm K}^{-1}\ \text{mol}^{-1})\times(776.4\ \text{K})}$$
$$= -0.552...\ \text{dm}^3\ \text{mol}^{-1}$$

$$Z = 1 + \frac{B}{V_m} = 1 + \frac{-0.552...\ \text{dm}^3\ \text{mol}^{-1}}{0.1352...\ \text{dm}^3\ \text{mol}^{-1}} = \boxed{0.5914}$$

P1C.9 According to Table 1C.4 on page 25, for the Dieterici equation of state the critical constants are given by

$$p_c = \frac{a}{4e^2b^2} \quad V_c = 2b \quad T_c = \frac{a}{4bR}$$

From the *Resource section* the values for Xe are $T_c = 289.75$ K, $p_c = 58.0$ atm, $V_c = 118.8\ \text{cm}^3\ \text{mol}^{-1}$. The coefficient b is computed directly from V_c

$$b = V_c/2 = (118.8\times10^{-3}\ \text{dm}^3\ \text{mol}^{-1})/2 = \boxed{0.0594\ \text{dm}^3\ \text{mol}^{-1}}$$

The expressions for p_c and V_c are combined to eliminate b

$$p_c = \frac{a}{4e^2b^2} = \frac{a}{4e^2V_c^2/4}$$

This is then rearranged to find a

$$a = p_c e^2 V_c^2 = (58.0\ \text{atm})\times e^2\times(118.8\times10^{-3}\ \text{dm}^3\ \text{mol}^{-1})^2$$
$$= \boxed{6.049\ \text{atm dm}^6\ \text{mol}^{-2}}$$

Alternatively, the expressions for T_c and V_c are combined to eliminate b

$$T_c = \frac{a}{4bR} = \frac{a}{4RV_c/2}$$

This is then rearranged to find a

$$a = 2T_cV_cR$$
$$= 2\times(289.75\ \text{K})\times(118.8\times10^{-3}\ \text{dm}^3\ \text{mol}^{-1})$$
$$\times(8.2057\times10^{-2}\ \text{dm}^3\ \text{atm K}^{-1}\ \text{mol}^{-1}) = \boxed{5.649\ \text{atm dm}^6\ \text{mol}^{-2}}$$

The two values of a are not the same; their average is 5.849 atm dm^6 mol^{-2}.

From Table 1C.4 on page 25 the expression for the pressure exerted by a Dieterici gas is

$$p = \frac{nRT\exp(-a/[RTV/n])}{V - nb}$$

With the parameters given the exponential term evaluates to

$$\exp\left(\frac{-(5.849 \text{ atm dm}^6 \text{ mol}^{-2})}{(8.2057\times10^{-2} \text{ dm}^3 \text{ atm K}^{-1} \text{ mol}^{-1})\times(298.15 \text{ K})\times(1.0 \text{ dm}^3)/(1.0 \text{ mol})}\right)$$
$$= 0.787...$$

and hence the pressure evaluates to

$$p = \frac{(1.0 \text{ mol})\times(8.2057\times10^{-2} \text{ dm}^3 \text{ atm K}^{-1} \text{ mol}^{-1})\times(298.15 \text{ K})\times(0.787...)}{(1.0 \text{ dm}^3) - (1.0 \text{ mol})\times(0.0594 \text{ dm}^3 \text{ mol}^{-1})}$$
$$= \boxed{20.48 \text{ atm}}$$

P1C.11 The van der Waals equation in terms of the molar volume is given by [1C.5b–*24*], $p = RT/(V_m - b) - a/V_m^2$. Multiplication of both sides by V_m gives

$$pV_m = \frac{RTV_m}{(V_m - b)} - \frac{a}{V_m}$$

and then division of the numerator and denominator of the first fraction by V_m gives

$$pV_m = \frac{RT}{(1 - b/V_m)} - \frac{a}{V_m}$$

The approximation $(1-x)^{-1} \approx 1+x+x^2$ is the used to approximate $1/(1-b/V_m)$ to give

$$pV_m = RT\left(1 + \frac{b}{V_m} + \frac{b^2}{V_m^2}\right) - \frac{a}{V_m}$$

The terms in $1/V_m$ and $1/V_m^2$ are gathered together to give

$$pV_m = RT\left(1 + \frac{1}{V_m}\left[b - \frac{a}{RT}\right] + \frac{b^2}{V_m^2}\right)$$

This result is then compared with the virial equation in terms of the molar volume, [1C.3b–*21*]

$$pV_m = RT\left(1 + \frac{B}{V_m} + \frac{C}{V_m^2}\right)$$

This comparison identifies the virial coefficients as

$$B = b - \frac{a}{RT} \qquad C = b^2$$

From the given value $C = 1200\,\text{cm}^6\,\text{mol}^{-2}$ it follows that $b = \sqrt{C} = 34.64\,\text{cm}^3\,\text{mol}^{-1}$. Expressed in the usual units this is $b = \boxed{0.03464\,\text{dm}^3\,\text{mol}^{-1}}$. The value of a is found by rearranging $B = b - a/RT$ to

$$a = RT(b - B) = (8.2057 \times 10^{-2}\,\text{dm}^3\,\text{atm}\,\text{K}^{-1}\,\text{mol}^{-1}) \times (273\,\text{K}) \times [(0.03464\,\text{dm}^3\,\text{mol}^{-1}) - (-21.7 \times 10^{-3}\,\text{dm}^3\,\text{mol}^{-1})]$$
$$= \boxed{1.262\,\text{atm}\,\text{dm}^6\,\text{mol}^{-2}}$$

P1C.13 In Section 1C.2(b) on page 24 it is explained that critical behaviour is associated with oscillations in the isotherms predicted by a particular equation of state, and that at the critical point there is a point of inflexion in the isotherm. At this point it follows that

$$\frac{dp}{dV_m} = 0 \qquad \frac{d^2p}{dV_m^2} = 0$$

The procedure is first to find expressions for the first and second derivatives. Then these are both set to zero give two simultaneous equations which can be solved for the critical pressure and volume.

$$\frac{dp}{dV_m} = -\frac{RT}{V_m^2} + \frac{2B}{V_m^3} - \frac{3C}{V_m^4} = 0 \qquad \frac{d^2p}{dV_m^2} = \frac{2RT}{V_m^3} - \frac{6B}{V_m^4} + \frac{12C}{V_m^5} = 0$$

The first of these equations is multiplied through by V_m^4 and the second by V_m^5 to give

$$-RTV_m^2 + 2BV_m - 3C = 0 \qquad 2RTV_m^2 - 6BV_m + 12C = 0$$

The first equation is multiplied by 2 and added to the second, thus eliminating the terms in V_m^2 and giving

$$4BV_m - 6C - 6BV_m + 12C = 0 \qquad \text{hence} \qquad V_m = 3C/B$$

This expression for V_m is then substituted into $-RTV_m^2 + 2BV_m - 3C = 0$ to give

$$-RT\frac{(3C)^2}{B^2} + 2B\frac{3C}{B} - 3C = 0$$

A term $3C$ is cancelled and the equation is multiplied through by B^2 to give

$$-RT(3C) + 2B^2 - B^2 = 0 \qquad \text{hence} \qquad T = B^2/3RC$$

Finally the pressure is found by substituting $V_m = 3C/B$ and $T = B^2/3RC$ into the equation of state

$$p = \frac{RT}{V_m} - \frac{B}{V_m^2} + \frac{C}{V_m^3}$$
$$= \frac{B^2R}{3RC}\frac{B}{3C} - \frac{B^3}{9C^2} + \frac{CB^3}{27C^3} = \frac{B^3}{9C^2} - \frac{B^3}{9C^2} + \frac{B^3}{27C^2} = \frac{B^3}{27C^2}$$

In summary, the critical constants are

$$\boxed{V_m = 3C/B \qquad T = B^2/3CR \qquad p = B^3/27C^2}$$

P1C.15 The virial equation in terms of the pressure, [1C.3a–*21*], is (up to the second term)

$$pV_m = RT(1 + B'p)$$

The mass density ρ is given by m/V, and the mass m can be written as nM, where n is the amount in moles and M is the molar mass. It follows that $\rho = nM/V = M/V_m$, where V_m is the molar volume. Rearranging gives $V_m = M/\rho$: measurements of the mass density therefore lead to values for the molar volume.

With this substitution for the molar volume the virial equation becomes

$$\frac{pM}{\rho} = RT(1 + B'p) \qquad \text{hence} \qquad \frac{p}{\rho} = \frac{RT}{M}(1 + B'p)$$

Therefore a plot of p/ρ against p is expected to be a straight line whose slope is related to B'; such a plot is shown in Fig. 1.5.

p/kPa	$\rho/(\mathrm{kg\,m^{-3}})$	$(p/\rho)/(\mathrm{kPa\,kg^{-1}\,m^3})$
12.22	0.225	54.32
25.20	0.456	55.26
36.97	0.664	55.68
60.37	1.062	56.85
85.23	1.468	58.06
101.30	1.734	58.42

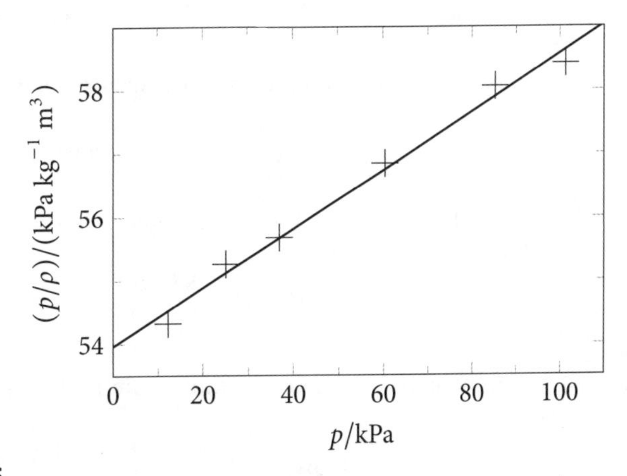

Figure 1.5

The data fall on a reasonable straight line, the equation of which is

$$(p/\rho)/(\mathrm{kPa\,kg^{-1}\,m^3}) = 0.04610 \times (p/\mathrm{kPa}) + 53.96$$

The slope is $B'RT/M$

$$\frac{B'RT}{M} = 0.04610\ \mathrm{kg^{-1}\,m^3}$$

For methoxymethane, CH_3OCH_3, $M = 2 \times 12.01 + 6 \times 1.0079 + 16.00 = 46.0674\ \mathrm{g\,mol^{-1}}$.

$$B' = \frac{(0.04610\ \mathrm{kg^{-1}\,m^3}) \times (46.0674 \times 10^{-3}\ \mathrm{kg\,mol^{-1}})}{(8.3145\ \mathrm{J\,K^{-1}\,mol^{-1}}) \times (298.15\ \mathrm{K})} = 8.57 \times 10^{-7}\ \mathrm{m^3\,J^{-1}}$$

The units of the result can be simplified by noting that $1\ \mathrm{J} = 1\ \mathrm{kg\,m^2\,s^{-2}}$, so $1\ \mathrm{m^3\,J^{-1}} = 1\ \mathrm{m\,kg^{-1}\,s^2}$. Recall that $1\ \mathrm{Pa} = 1\ \mathrm{kg\,m^{-1}\,s^{-2}}$, so the units of the B' are $\mathrm{Pa^{-1}}$, an inverse pressure, as expected: $B' = 8.57 \times 10^{-7}\ \mathrm{Pa^{-1}}$ or $\boxed{B' = 0.0868\ \mathrm{atm^{-1}}}$.

The virial coefficient B is found using the result from *Problem* P1C.14, $B = B'RT$

$$\begin{aligned} B &= B'RT \\ &= (0.0868\ \mathrm{atm^{-1}}) \times (8.2057 \times 10^{-2}\ \mathrm{dm^3\,atm\,K^{-1}\,mol^{-1}}) \times (298.15\ \mathrm{K}) \\ &= \boxed{2.12\ \mathrm{dm^3\,mol^{-1}}} \end{aligned}$$

P1C.17 A gas can only be liquefied by the application of pressure if the temperature is below the critical temperature, which for N_2 is 126.3 K.

P1C.19 The compression factor is given by [1C.1–*20*], $Z = V_m/V_m^\circ = V_m p/RT$. The given equation of state is rearranged to give an expression for V_m after putting $n = 1$

$$p(V - nb) = nRT \quad \text{becomes} \quad p(V_m - b) = RT \quad \text{hence} \quad V_m = \frac{RT}{p} + b$$

It follows that the compression factor is given by

$$Z = \frac{V_m p}{RT} = \frac{(RT/p + b)p}{RT} = \boxed{1 + \frac{bp}{RT}}$$

If $V_m = 10b$ it follows from the previous equation that

$$\frac{V_m p}{RT} = \frac{10bp}{RT} = 1 + \frac{bp}{RT} \quad \text{hence} \quad b = \frac{RT}{9p}$$

With this expression for b the compression factor is computed from $Z = 1 + bp/RT$ as

$$Z = 1 + \frac{bp}{RT} = 1 + \frac{RT}{9p}\frac{p}{RT} = 1 + \frac{1}{9} = \boxed{1.11}$$

P1C.21 The virial equation in terms of the molar volume, [1C.3b–*21*], is (up to the third term)

$$pV_m = RT\left(1 + \frac{B}{V_m} + \frac{C}{V_m^2}\right)$$

For part (a) only the first two terms are considered, and it then follows that a plot of pV_m against $1/V_m$ is expected to be a straight line with slope BRT; such a plot is shown in Fig. 1.6.

p/MPa	$V_m/(\mathrm{dm^3\,mol^{-1}})$	$(pV_m)/(\mathrm{MPa\,dm^3\,mol^{-1}})$	$(1/V_m)/(\mathrm{dm^{-3}\,mol})$
0.4000	6.2208	2.4883	0.16075
0.5000	4.9736	2.4868	0.20106
0.6000	4.1423	2.4854	0.24141
0.8000	3.1031	2.4825	0.32226
1.000	2.4795	2.4795	0.40331
1.500	1.6483	2.4725	0.60669
2.000	1.2328	2.4656	0.81116
2.500	0.98357	2.4589	1.0167
3.000	0.81746	2.4524	1.2233
4.000	0.60998	2.4399	1.6394

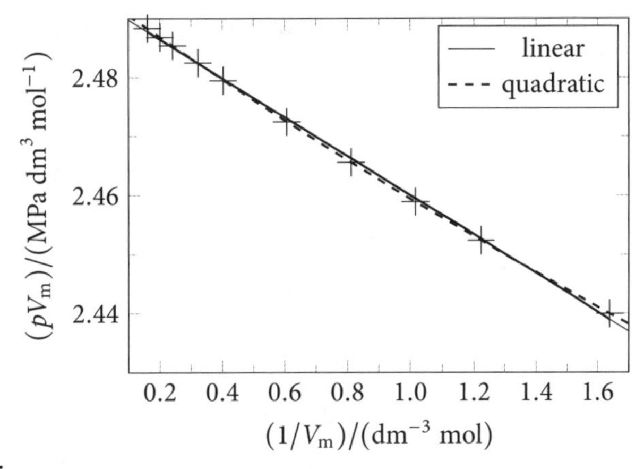

Figure 1.6

The data fall on a reasonable straight line, the equation of which is

$$(pV_m)/(\mathrm{MPa\,dm^3\,mol^{-1}}) = -0.03302 \times (1/V_m)/(\mathrm{dm^{-3}\,mol}) + 2.4931$$

The slope is BRT

$$BRT = (-0.03302\ \mathrm{MPa\,dm^6\,mol^{-2}})$$

It is convenient to convert to atm giving $BRT = (-0.3259\ \mathrm{atm\,dm^6\,mol^{-2}})$ hence

$$\begin{aligned} B &= \frac{(-0.3259\ \mathrm{atm\,dm^6\,mol^{-2}})}{RT} \\ &= \frac{(-0.3259\ \mathrm{atm\,dm^6\,mol^{-2}})}{(8.2057\times10^{-2}\ \mathrm{dm^3\,atm\,K^{-1}\,mol^{-1}})\times(300\ \mathrm{K})} \\ &= \boxed{-0.01324\ \mathrm{dm^3\,mol^{-1}}} \end{aligned}$$

For part (b) the data points are fitted to polynomial of order 2 in $1/V_m$ using mathematical software; the data are a slightly better fit to such a function (see the dashed line in the graph above) which is

$$(pV_m)/(\text{MPa dm}^3\,\text{mol}^{-1}) =$$
$$0.002652 \times (1/V_m)^2/(\text{dm}^{-6}\,\text{mol}^2) - 0.03748 \times (1/V_m)/(\text{dm}^{-3}\,\text{mol}) + 2.494$$

The coefficient of the term in $(1/V_m)^2$ is CRT

$$CRT = (0.002652\ \text{MPa dm}^9\,\text{mol}^{-3})$$

It is convenient to convert to atm giving $CRT = (0.02617\ \text{atm dm}^9\,\text{mol}^{-3})$ hence

$$C = \frac{(0.02617\ \text{atm dm}^9\,\text{mol}^{-3})}{RT}$$
$$= \frac{(0.02617\ \text{atm dm}^9\,\text{mol}^{-3})}{(8.2057 \times 10^{-2}\ \text{dm}^3\,\text{atm K}^{-1}\,\text{mol}^{-1}) \times (300\ \text{K})}$$
$$= \boxed{1.063 \times 10^{-3}\ \text{dm}^6\,\text{mol}^{-2}}$$

P1C.23 The van der Waals equation of state in terms of the molar volume is given by [1C.5b–24], $p = RT/(V_m - b) - a/V_m^2$. This equation is a cubic in V_m, as is seen by multiplying both sides by $(V_m - b)V_m^2$ and then gathering the terms together

$$pV_m^3 - V_m^2(pb + RT) + aV_m - ab = 0$$

From the *Resource section* the van der Waals parameters for Cl_2 are

$$a = 6.260\ \text{atm dm}^6\,\text{mol}^{-2} \qquad b = 5.42 \times 10^{-2}\ \text{dm}^3\,\text{mol}^{-1}$$

It is convenient to convert the pressure to atm

$$p = (150 \times 10^3\ \text{Pa}) \times (1\ \text{atm})/(1.01325 \times 10^5\ \text{Pa}) = 1.4804\ \text{atm}$$

and to use $R = 8.2057 \times 10^{-2}\ \text{dm}^3\,\text{atm K}^{-1}\,\text{mol}^{-1}$; inserting all of these values and the temperature gives the polynomial

$$1.4804V_m^3 - 20.5946V_m^2 + 6.260V_m - 0.3393 = 0$$

The roots of this polynomial are found numerically using mathematical software and of these roots only $\boxed{V_m = 13.6\ \text{dm}^3\,\text{mol}^{-1}}$ is a physically plausible value for the molar volume.

The molar volume of a perfect gas under corresponding conditions is

$$V_m = \frac{RT}{p} = \frac{(8.2057 \times 10^{-2}\ \text{dm}^3\,\text{atm K}^{-1}\,\text{mol}^{-1}) \times (250\ \text{K})}{1.48\ \text{atm}} = 13.9\ \text{dm}^3\,\text{mol}^{-1}$$

The molar volume of the van der Waals gas is about 2% smaller than that of the perfect gas.

Answers to integrated activities

I1.1 The Maxwell–Boltzmann distribution of speeds in three dimensions is given by [1B.4–*14*]

$$f(v) = 4\pi\left(\frac{M}{2\pi RT}\right)^{3/2} v^2 e^{-Mv^2/2RT}$$

with M the molar mass. The most probable speed is found by taking the derivative of $f(v)$ with respect to v, and setting this to zero; calculating the derivative requires the use of the chain rule

$$\frac{df(v)}{dv} = 4\pi\left(\frac{M}{2\pi RT}\right)^{3/2}\left[2ve^{-Mv^2/2RT} + v^2\left(\frac{-2Mv}{2RT}\right)e^{-Mv^2/2RT}\right] = 0$$

The multiplying constant and factors of v and $e^{-Mv^2/2RT}$ are cancelled (these do not correspond to maxima) leaving

$$2 - \frac{Mv^2}{RT} = 0 \qquad \text{hence} \qquad \boxed{v = \left(\frac{2RT}{M}\right)^{1/2}}$$

Inspection of the form of the distribution shows that this is a maximum.

The average kinetic energy is calculated from the average of the square of the speed: $\langle E_k\rangle = \frac{1}{2}m\langle v^2\rangle$. The task is therefore to calculate this average using the Maxwell–Boltzmann distribution: the required integral is

$$\langle v^2\rangle = \int_0^\infty v^2 f(v)\,dv = 4\pi\left(\frac{M}{2\pi RT}\right)^{3/2}\int_0^\infty v^4 e^{-Mv^2/2RT}\,dv$$

This integral is of the form of G.8 from the *Resource section*

$$\int_0^\infty x^{2m}e^{-ax^2}\,dx = \frac{(2m-1)!!}{2^{m+1}a^m}\left(\frac{\pi}{a}\right)^{1/2}$$

with $m = 2$, $(2m-1)!! = 3 \times 1 = 3$, $2^{m+1} = 8$, $a^m = a^2$ and $a = M/2RT$.

$$\langle v^2\rangle = 4\pi\left(\frac{M}{2\pi RT}\right)^{3/2} \times \frac{3}{8} \times \left(\frac{4R^2T^2}{M^2}\right) \times \left(\frac{2RT\pi}{M}\right)^{1/2}$$
$$= \frac{3RT}{M} = \frac{3kT}{m}$$

To go to the last line from the previous one involves a deal of careful algebra, and for the final step $R = N_A k$ and $M = mN_A$ have been used, with m the mass of the molecule.

With this result

$$\langle E_k\rangle = \frac{1}{2}m\langle v^2\rangle = \frac{1}{2}m\left(\frac{3kT}{m}\right) = \frac{3}{2}kT$$

which is in accord with the equipartition principle.

I1.3 In Section 1C.2(a) on page 23 it is argued that $b = 4V_{\text{molec}}N_A$, where V_{molec} is the volume occupied by one molecule. The collision cross-section σ is defined in terms of a collision diameter d as $\sigma = \pi d^2$, and in turn the diameter is interpreted as twice the radius of the colliding spheres: $d = 2r$. It follows that $r = (\sigma/4\pi)^{1/2}$

$$b = 4V_{\text{molec}}N_A$$

$$= 4\left(\frac{4}{3}\pi r^3\right)N_A = \frac{16\pi N_A}{3}\left(\frac{\sigma}{4\pi}\right)^{3/2}$$

$$= \frac{16\pi(6.0221\times10^{23}\ \text{mol}^{-1})}{3}\left(\frac{0.46\times10^{-18}\ \text{m}^2}{4\pi}\right)^{3/2}$$

$$= 7.1\times10^{-5}\ \text{m}^3\ \text{mol}^{-1} = \boxed{0.071\ \text{dm}^3\ \text{mol}^{-1}}$$

Internal energy

2A Internal energy

Answers to discussion questions

D2A.1 In physical chemistry, the universe is considered to be divided into two parts: the system and its surroundings. In thermodynamics, the system is the object of interest which is separated from its surroundings, the rest of the universe, by a boundary. The characteristics of the boundary determine whether the system is open, closed, or isolated.

An open system has a boundary that permits the passage of both matter and energy. A closed system has a boundary that allows the passage of energy but not of matter. Closed systems can be either adiabatic or diathermic. The former do not allow the transfer of energy as a result of a temperature difference, but the latter do. An isolated system is one with a boundary that allows neither the transfer of matter nor energy between the system and the surroundings.

D2A.3 Table 2A.1 on page 39 lists four varieties of work: expansion, surface expansion, extension, and electrical. There is also work associated with processes in magnetic and gravitational fields which we will not describe in detail.

D2A.5 An isothermal expansion of a gas may be achieved by making sure that the gas and its container are in thermal contact with a large 'bath' which is held at a constant temperature – that is, a thermostat.

Solutions to exercises

E2A.1(a) *The chemist's toolkit* 7 in Topic 2A gives an explanation of the equipartition theorem. The molar internal energy is given by

$$U_\mathrm{m} = \tfrac{1}{2} \times (\nu_\mathrm{t} + \nu_\mathrm{r} + 2\nu_\mathrm{v}) \times RT$$

where ν_t is the number of translational degrees of freedom, ν_r is the number of rotational degrees of freedom and ν_v is the number of vibrational degrees of freedom. As each gas molecule can move independently along the x, y and z axis, the number of translational degrees of freedom is three.

(i) Molecular iodine is a diatomic molecule, therefore it has two degrees of rotational freedom. On account of its heavy atoms, molecular iodine is likely to have one degree of vibrational freedom at room temperature. Therefore, the molar internal energy of molecular iodine gas at room temperature is

$$U_m = \tfrac{1}{2} \times (3+2+2) \times RT = \tfrac{7}{2} \times (8.3145\,\mathrm{J\,K^{-1}\,mol^{-1}}) \times (298.15\,\mathrm{K})$$
$$= \boxed{8.7\,\mathrm{kJ\,mol^{-1}}}$$

(ii) and (iii) Both methane (tetrahedral) and benzene (planar) have three degrees of rotational freedom. At room temperature it is unlikely that any of their vibrational modes would be excited, therefore both are expected to have approximately the same internal energy at room temperature:

$$U_m = \tfrac{1}{2} \times (3+3+0) \times RT = 3 \times (8.3145\,\mathrm{J\,K^{-1}\,mol^{-1}}) \times (298.15\,\mathrm{K})$$
$$= \boxed{7.4\,\mathrm{kJ\,mol^{-1}}}$$

E2A.2(a) A state function is a property with a value that depends only on the current state of the system and is independent of how the state has been prepared. Pressure, temperature and enthalpy are all state functions.

E2A.3(a) The system is expanding against a constant external pressure, hence the expansion work is given by [2A.6–*40*], $w = -p_{ex}\Delta V$. The change in volume is the cross-sectional area times the linear displacement

$$\Delta V = (50\,\mathrm{cm^2}) \times (15\,\mathrm{cm}) = 750\,\mathrm{cm^3} = 7.5 \times 10^{-4}\,\mathrm{m^3}$$

The external pressure is 1.0 atm = 1.01325×10^5 Pa, therefore the expansion work is

$$w = -(1.01325 \times 10^5\,\mathrm{Pa}) \times (7.5 \times 10^{-4}\,\mathrm{m^3}) = \boxed{-76\,\mathrm{J}}$$

Note that the volume is expressed in $\mathrm{m^3}$. The relationships 1 Pa = 1 $\mathrm{kg\,m^{-1}\,s^{-2}}$ and 1 J = 1 $\mathrm{kg\,m^2\,s^{-2}}$ are used to verify the units of the result.

E2A.4(a) For all cases $\Delta U = 0$, because the internal energy of a perfect gas depends on the temperature alone.

(i) The work of reversible isothermal expansion of a perfect gas is given by [2A.9–*41*]

$$w = -nRT \ln\left(\frac{V_f}{V_i}\right)$$
$$= -(1.00\,\mathrm{mol}) \times (8.3145\,\mathrm{J\,K^{-1}\,mol^{-1}}) \times (293.15\,\mathrm{K}) \times \ln\left(\frac{30.0\,\mathrm{dm^3}}{10.0\,\mathrm{dm^3}}\right)$$
$$= -2.68 \times 10^3\,\mathrm{J} = \boxed{-2.68\,\mathrm{kJ}}$$

Note that the temperature is expressed in K in the above equation. Using the First Law of thermodynamics, [2A.2–*38*], gives

$$q = \Delta U - w = 0 - (-2.68\,\mathrm{kJ}) = \boxed{+2.68\,\mathrm{kJ}}$$

(ii) The final pressure of the expanding gas is found using the perfect gas law, [1A.4–8]

$$p_f = \frac{nRT}{V_f} = \frac{(1.00 \text{ mol}) \times (8.3145 \text{ J K}^{-1} \text{ mol}^{-1}) \times (293.15 \text{ K})}{(30.0 \times 10^{-3} \text{ m}^3)}$$
$$= 8.12... \times 10^4 \text{ Pa}$$

This pressure equals the constant external pressure against which the gas is expanding, therefore the work of expansion is

$$w = -p_{ex} \times \Delta V = (8.12... \times 10^4 \text{ Pa}) \times (30.0 \times 10^{-3} \text{ m}^3 - 10.0 \times 10^{-3} \text{ m}^3)$$
$$= -1.62 \times 10^3 \text{ J} = \boxed{-1.62 \text{ kJ}}$$

and hence $q = \boxed{+1.62 \text{ kJ}}$

(iii) Free expansion is expansion against zero force, so $\boxed{w = 0}$ and therefore $\boxed{q = 0}$ as well.

E2A.5(a) For a perfect gas at constant volume $p_i/T_i = p_f/T_f$ therefore,

$$p_f = p_i \times \frac{T_f}{T_i} = (1.00 \text{ atm}) \times \left(\frac{400 \text{ K}}{300 \text{ K}}\right) = \boxed{1.33 \text{ atm}}$$

The change in internal energy at constant volume is given by [2A.15b–45]

$$\Delta U = nC_{V,m}\Delta T = (1.00 \text{ mol}) \times \left(\frac{3}{2} \times 8.3145 \text{ J K}^{-1} \text{ mol}^{-1}\right) \times (400 \text{ K} - 300 \text{ K})$$
$$= +1.25 \times 10^3 \text{ J} = \boxed{+1.25 \text{ kJ}}$$

The volume of the gas is constant, so the work of expansion is zero, $\boxed{w = 0}$. The First Law of thermodynamics gives $q = \Delta U - w = +1.25 \text{ kJ} - 0 = \boxed{+1.25 \text{ kJ}}$.

E2A.6(a) (i) The work of expansion against constant external pressure is given by [2A.6–40]

$$w = -p_{ex}\Delta V$$
$$= -(200 \text{ Torr}) \times \left(\frac{133.3 \text{ Pa}}{1 \text{ Torr}}\right) \times (3.3 \times 10^{-3} \text{ m}^3) = \boxed{-88 \text{ J}}$$

Note that the pressure is expressed in Pa and the change in volume in m^3, to give the work in J.

(ii) The work done in a reversible isothermal expansion is given by [2A.9–41], $w = -nRT \ln(V_f/V_i)$. The amount in moles of methane is

$$n = \frac{m}{M} = \frac{(4.50 \text{ g})}{(16.0416 \text{ g mol}^{-1})} = 0.280... \text{ mol}$$
$$w = -(0.280... \text{ mol}) \times (8.3145 \text{ J K}^{-1} \text{ mol}^{-1}) \times (310 \text{ K})$$
$$\times \ln\left(\frac{[12.7 + 3.3] \text{ dm}^3}{12.7 \text{ dm}^3}\right) = \boxed{-1.7 \times 10^2 \text{ J}}$$

Note that the modulus of the work done in a reversible expansion is greater than the work for expansion against constant external pressure because the latter is an irreversible process.

Solutions to problems

P2A.1 From the equipartition theorem (*The chemist's toolkit* 7 in Topic 2A), the molar internal energy is given by

$$U_{\rm m} = \tfrac{1}{2} \times (\nu_{\rm t} + \nu_{\rm r} + 2\nu_{\rm v}) \times RT$$

where $\nu_{\rm t}$ is the number of translational degrees of freedom, $\nu_{\rm r}$ is the number of rotational degrees of freedom and $\nu_{\rm v}$ is the number of vibrational degrees of freedom. Each gas molecule can move independently along the x, y and z axis giving rise to three translational degrees of freedom. Carbon dioxide is a linear molecule therefore it has two rotational degrees of freedom. None of the vibrational modes of carbon dioxide are likely to be significantly excited at room temperature.

$$U = \tfrac{1}{2} \times (3 + 2 + 0) \times RT = \tfrac{5}{2} \times (8.3145\,{\rm J\,K^{-1}\,mol^{-1}}) \times (298.15\,{\rm K})$$
$$= \boxed{6.2\,{\rm kJ\,mol^{-1}}}$$

P2A.3 The definition of work is given by [2A.4–39], $dw = -|F|dz$. Integrating both sides gives:

$$\int {\rm d}w = \int_0^l F(x)\,{\rm d}x$$
$$w = \int_0^l k_{\rm f}(x)x\,{\rm d}x = \int_0^l \left(a - bx^{\frac{1}{2}}\right) x\,{\rm d}x$$
$$= \left[\frac{1}{2}ax^2 - \frac{2}{5}bx^{\frac{5}{2}}\right]_0^l = \boxed{\frac{1}{2}al^2 - \frac{2}{5}bl^{\frac{5}{2}}}$$

Note that the second term arises from the non Hooke's Law behaviour of the elastomer, reducing the overall work done.

P2A.5 (a) The natural logarithm can be expanded using the Taylor series as $\ln(1 + \nu) \approx \nu + \nu^2/2! + \nu^3/3! + \ldots$, which, for $\nu << 1$, can be approximated as $\ln(1 + \nu) \approx \nu$, and similarly, $\ln(1 - \nu) \approx -\nu$. Therefore,

$$F = \frac{kT}{2l}[\ln(1+\nu) - \ln(1-\nu)] \approx \frac{kT}{2l}[\nu - (-\nu)] = \frac{\nu kT}{l}$$

Because $\nu = n/N$, it follows that

$$F = \frac{\nu kT}{l} = \frac{nkT}{Nl}$$

(b) Hooke's law predicts $F = {\rm const} \times x$, that is the restoring force is directly proportional to the displacement. Using $n = x/l$, the expression for the force obtained in part (a) is rewritten as

$$\frac{nkT}{Nl} = \frac{kTx}{Nl^2} \equiv {\rm const} \times x$$

Therefore Hooke's law applies and kT/Nl^2 is the force constant.

P2A.7 The van der Waals equation of state is given by [1C.5b–24],

$$p = \frac{nRT}{V - nb} - \frac{n^2 a}{V^2}$$

The expansion work is given by [2A.6–40], $\mathrm{d}w = -p_{\mathrm{ex}}\mathrm{d}V$. For a reversible expansion p_{ex} is always equal to the pressure of the gas so

$$\mathrm{d}w = -p_{\mathrm{gas}}\,\mathrm{d}V = -\left(\frac{nRT}{V} - \frac{n^2 a}{V^2}\right)\mathrm{d}V$$

Integrating both sides give

$$\begin{aligned} w &= -\int_{V_\mathrm{i}}^{V_\mathrm{f}} \frac{nRT}{V - nb} - \frac{n^2 a}{V^2}\,\mathrm{d}V \\ &= -nRT\int_{V_\mathrm{i}}^{V_\mathrm{f}} \frac{\mathrm{d}V}{V - nb} + n^2 a \int_{V_\mathrm{i}}^{V_\mathrm{f}} \frac{\mathrm{d}V}{V^2} \\ &= -nRT\ln\left(\frac{V_\mathrm{f} - nb}{V_\mathrm{i} - nb}\right) - n^2 a\left(\frac{1}{V_\mathrm{f}} - \frac{1}{V_\mathrm{i}}\right) \end{aligned}$$

The work done during the isothermal reversible expansion in the various cases is calculated below and the results portrayed in Fig. 2.1.

(a) A perfect gas.

$$w = -nRT\ln\left(\frac{V_\mathrm{f}}{V_\mathrm{i}}\right) = -(1.0\ \mathrm{mol}) \times (8.3145\ \mathrm{J\,K^{-1}\,mol^{-1}}) \times (298\ \mathrm{K})$$
$$\times \ln\left(\frac{2.0\ \mathrm{dm^3}}{1.0\ \mathrm{dm^3}}\right) = \boxed{-1.7\ \mathrm{kJ}}$$

(b) A van der Waals gas in which repulsive forces dominate.
Using the general expression derived above with $a = 0$ and $b = 5.11 \times 10^{-2}\ \mathrm{dm^3\,mol^{-1}}$

$$w = -(1.0\ \mathrm{mol}) \times (8.3145\ \mathrm{J\,K^{-1}\,mol^{-1}}) \times (298\ \mathrm{K})$$
$$\times \ln\left(\frac{(2.0\ \mathrm{dm^3}) - (1.0\ \mathrm{mol}) \times (5.11 \times 10^{-2}\ \mathrm{dm^3\,mol^{-1}})}{(1.0\ \mathrm{dm^3}) - (1.0\ \mathrm{mol}) \times (5.11 \times 10^{-2}\ \mathrm{dm^3\,mol^{-1}})}\right) = \boxed{-1.8\ \mathrm{kJ}}$$

(c) A van der Waals gas in which attractive forces dominate.
Using the general expression derived above with $a = 4.2\ \mathrm{dm^6\,atm\,mol^{-2}}$ and $b = 0$. Constant a in SI units is

$$a = (4.2\ \mathrm{dm^6\,atm\,mol^{-2}}) \times \left(\frac{1\ \mathrm{m^6}}{10^6\ \mathrm{dm^6}}\right) \times \left(\frac{1.01325 \times 10^5\ \mathrm{Pa}}{1\ \mathrm{atm}}\right)$$
$$= 0.425...\ \mathrm{m^6\,Pa\,mol^{-2}}$$

$$w = -(1.0\ \mathrm{mol}) \times (8.3145\ \mathrm{J\,K^{-1}\,mol^{-1}}) \times (298\ \mathrm{K}) \times \ln\left(\frac{2.0\ \mathrm{dm^3}}{1.0\ \mathrm{dm^3}}\right)$$
$$-\ (1.0\ \mathrm{mol})^2 \times (4.25... \times 10^5\ \mathrm{dm^6\,Pa\,mol^{-2}}) \times \left(\frac{1}{2.0\ \mathrm{dm^3}} - \frac{1}{1.0\ \mathrm{dm^3}}\right)$$
$$= -1.71...\mathrm{kJ} + 0.212...\mathrm{kJ} = \boxed{-1.5\ \mathrm{kJ}}$$

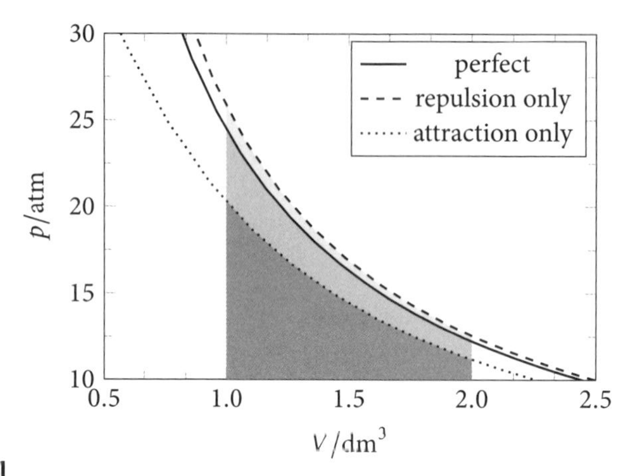

Figure 2.1

P2A.9 (a) The virial equation of state is given by [1C.3b–*21*]. The first three terms are

$$p = RT\left(\frac{1}{V_m} + \frac{B}{V_m^2} + \frac{C}{V_m^3}\right)$$

Using $V_m = V/n$ gives

$$p = nRT\left(\frac{1}{V} + \frac{nB}{V^2} + \frac{n^2C}{V^3}\right)$$

The work of expansion is calculated as

$$\begin{aligned} w &= -\int_{V_i}^{V_f} p\mathrm{d}V = -\int_{V_i}^{V_f} nRT\left(\frac{1}{V} + \frac{nB}{V^2} + \frac{n^2C}{V^3}\right)\mathrm{d}V \\ &= -nRT\ln\frac{V_i}{V_f} + n^2RTB\left(\frac{1}{V_f} - \frac{1}{V_i}\right) + \frac{1}{2}n^3RTC\left(\frac{1}{V_f^2} - \frac{1}{V_i^2}\right) \end{aligned}$$

From Table 1C.1, at 273 K, $B = -21.7\ \mathrm{cm^3\,mol^{-1}}$ and from the problem $C = 1.2 \times 10^3\ \mathrm{cm^6\,mol^{-2}}$, therefore

$$\begin{aligned} nRT &= (1.0\ \mathrm{mol}) \times (8.3145\ \mathrm{J\,K^{-1}\,mol^{-1}}) \times (273\ \mathrm{K}) = 2.26... \times 10^3\ \mathrm{J} \\ n^2RTB &= n \times (nRT) \times B = (1.0\ \mathrm{mol}) \times (2.26... \times 10^3\ \mathrm{J}) \\ &\quad \times (-21.7\ \mathrm{cm^3\,mol^{-1}}) \\ &= -4.92... \times 10^4\ \mathrm{J\,cm^3} \end{aligned}$$

$$\begin{aligned} \tfrac{1}{2}n^3RTC &= \tfrac{1}{2} \times n^2 \times (nRT) \times C \\ &= \tfrac{1}{2} \times (1.0\ \mathrm{mol})^2 \times (2.26... \times 10^3\ \mathrm{J}) \times (1.2 \times 10^3\ \mathrm{cm^6\,mol^{-2}}) \\ &= 1.36... \times 10^6\ \mathrm{J\,cm^6} \end{aligned}$$

$$w = -(2.26... \times 10^3 \text{ J}) \times \ln\left(\frac{1000 \text{ cm}^3}{500 \text{ cm}^3}\right)$$
$$+ (-4.92... \times 10^4 \text{ J cm}^3) \times \left(\frac{1}{1000 \text{ cm}^3} - \frac{1}{500 \text{ cm}^3}\right)$$
$$+ (1.36... \times 10^6 \text{ J cm}^6) \times \left(\frac{1}{(1000 \text{ cm}^3)^2} - \frac{1}{(500 \text{ cm}^3)^2}\right)$$
$$= \boxed{-1.5 \text{ kJ}}$$

(b) For a perfect gas, the work of reversible isothermal expansion is given by [2A.9–*41*]

$$w = -nRT\ln\left(\frac{V_f}{V_i}\right)$$
$$= -(1.0 \text{ mol}) \times (8.3145 \text{ J K}^{-1} \text{ mol}^{-1}) \times (273 \text{ K}) \times \ln\left(\frac{1000 \text{ cm}^3}{500 \text{ cm}^3}\right)$$
$$= -1.6 \times 10^3 \text{ J} = \boxed{-1.6 \text{ kJ}}$$

2B Enthalpy

Answers to discussion questions

D2B.1 The enthalpy is defined in terms of the internal energy as $H = U + pV$. It follows that the enthalpy change is related to the internal energy change by $\Delta H = \Delta U + \Delta(pV)$. Apart from in some special cases, $\Delta(pV)$ is generally non-zero, and so ΔH and ΔU have different values.

As is discussed in Topic 2B, ΔH is the heat associated with a process at constant pressure, whereas ΔU is the heat at constant volume. They differ because work of expansion may be involved in constant-pressure processes, but not in constant-volume processes.

Solutions to exercises

E2B.1(a) The heat transferred under constant pressure equals the change in enthalpy of the system, [2B.2b–*47*], $q_p = \Delta H$. The relationship between the change in enthalpy, change in temperature and the heat capacity is given by [2B.6b–*49*]

$$C_{p,\text{m}} = \frac{\Delta H}{n\Delta T} = \frac{(229 \text{ J})}{(3.0 \text{ mol}) \times (2.55 \text{ K})} = 29.9... \text{ J K mol}^{-1} = \boxed{30 \text{ J K}^{-1} \text{ mol}^{-1}}$$

For a perfect gas $C_{p,\text{m}} - C_{V,\text{m}} = R$, [2B.9–*49*], therefore

$$C_{V,\text{m}} = C_{p,\text{m}} - R = (29.9... \text{ J K}^{-1} \text{ mol}^{-1}) - (8.3145 \text{ J K}^{-1} \text{ mol}^{-1}) = \boxed{22 \text{ J K}^{-1} \text{ mol}^{-1}}$$

E2B.2(a) In the reaction 2 moles of gas are formed from 4 moles of gas. The relationship between ΔH and ΔU is given by [2B.3–48]

$$\Delta H_m - \Delta U_m = \Delta n_g RT = (-2.0)\times(8.3145\,\mathrm{J\,K^{-1}\,mol^{-1}})\times(298\,\mathrm{K}) = \boxed{-5.0\,\mathrm{kJ\,mol^{-1}}}$$

E2B.3(a) (i) The heat capacity can be expressed as $C_p = a + bT$ where $a = 20.17\,\mathrm{J\,K^{-1}}$ and $b = 0.3665\,\mathrm{J\,K^{-2}}$. Integrating the relationship $\mathrm{d}H = C_p\mathrm{d}T$ on both sides gives

$$\int_{T_1}^{T_2} \mathrm{d}H = \int_{T_1}^{T_2} C_p \mathrm{d}T = \int_{T_1}^{T_2} (a + bT)\mathrm{d}T = \left[aT + \tfrac{1}{2}bT^2\right]_{T_1}^{T_2}$$

$$\begin{aligned} H(T_2) - H(T_1) &= na(T_2 - T_1) + \tfrac{1}{2}nb(T_2^2 - T_1^2) \\ &= (20.17\,\mathrm{J\,K^{-1}}) \times (373.15\,\mathrm{K} - 298.15\,\mathrm{K}) \\ &\quad + \tfrac{1}{2} \times (0.3665\,\mathrm{J\,K^{-2}}) \times [(373.15\,\mathrm{K})^2 - (298.15\,\mathrm{K})^2] \\ &= +10.7...\,\mathrm{kJ} = \boxed{+10.7\,\mathrm{kJ}} \end{aligned}$$

Under constant pressure conditions $\Delta H = \boxed{q_p = +10.7\,\mathrm{kJ}}$.
The work of expansion against constant pressure p_{ex} is given by [2A.6–40], $w = -p_{ex}\Delta V = -p_{ex}(V_f - V_i)$. Assume that the gas is in mechanical equilibrium with its surroundings, therefore p_{ex} is the same as the pressure of the gas, p. The initial and final volumes are calculated from T_f and T_i by $V_f = nRT_f/p$ and $V_i = nRT_i/p$, therefore $V_f - V_i = (T_f - T_i)nR/p$. Hence

$$\begin{aligned} w &= -p \times \frac{nR}{p}(T_f - T_i) = -nR\Delta T \\ &= -(1.00\,\mathrm{mol}) \times (8.3145\,\mathrm{J\,K^{-1}\,mol^{-1}}) \times (373.15\,\mathrm{K} - 298.15\,\mathrm{K}) \\ &= -6.23... \times 10^2\,\mathrm{J} = \boxed{-624\,\mathrm{J}} \end{aligned}$$

$$\Delta U = q + w = (+10.7...\,\mathrm{kJ}) + (-0.623...\,\mathrm{kJ}) = \boxed{+10.1\,\mathrm{kJ}}$$

(ii) The energy and enthalpy of a perfect gas depends on the temperature alone, hence ΔH and ΔU is the same as above, $\boxed{\Delta H = \Delta U = +10.7\,\mathrm{kJ}}$. Under constant volume conditions there is no expansion work, $\boxed{w = 0}$, therefore the heat is equal to the change in internal energy, $q_V = \Delta U = \boxed{+10.1\,\mathrm{kJ}}$.

E2B.4(a) Under constant pressure $q_p = \Delta H$, therefore

$$\begin{aligned} q_p = \Delta H = nC_{p,m}\Delta T &= (3.0\,\mathrm{mol}) \times (29.4\,\mathrm{J\,K\,mol^{-1}}) \times (285\,\mathrm{K} - 260\,\mathrm{K}) \\ &= 2.20...\,\mathrm{kJ} = \boxed{+2.2\,\mathrm{kJ}} \end{aligned}$$

The definition of enthalpy is given by [2B.1–46], $H = U + pV$. For a change at constant pressure, it follows that $\Delta H = \Delta U + p\Delta V$, where $\Delta V = V_f - V_i$.

If the gas is assumed to be perfect, then $V_f = nRT_f/p$ and $V_i = nRT_i/p$, so $\Delta V = (T_f - T_i)nR/p$. Hence $p\Delta V = nR(T_f - T_i) = nR\Delta T$.

$$\begin{aligned}\Delta U &= \Delta H - nR\Delta T\\ &= (2.20...\ \text{kJ}) - (3.0\ \text{mol}) \times (8.3145\ \text{J K}^{-1}\ \text{mol}^{-1}) \times (285\ \text{K} - 260\ \text{K})\\ &= \boxed{+1.6\ \text{kJ}}\end{aligned}$$

Solutions to problems

P2B.1 10 g of benzene is $n = m/M = (10\ \text{g})/(78.1074\ \text{g mol}^{-1}) = 0.128...\ \text{mol}$. The heat needed to vaporize 10 g of benzene under constant pressure is $q_p = n \times \Delta_{vap}H_m = (0.128...\ \text{mol}) \times (30.8\ \text{kJ mol}^{-1}) = 3.94...\ \text{kJ}$. A current I flowing through a potential $\Delta\phi$ corresponds to a power of $I\Delta\phi$. If the current flows for a time interval Δt the energy is power × time , that is $q = I\Delta\phi\Delta t$. Therefore $\Delta t = q/(I\Delta\phi)$.

$$\Delta t = \frac{q_p}{I\Delta\phi} = \frac{(3.94... \times 10^3\ \text{J})}{(0.50\ \text{A}) \times (12\ \text{V})} = 657\ \text{s} = \boxed{11\ \text{min}}$$

The relationships 1 A = 1 C s^{-1} and 1 V = 1 J C^{-1} are used in the last step.

P2B.3 Fitting the data set using a computer program to an expression in the form of $C^{\ominus}_{p,m}(T) = a+bT+cT^{-2}$ yields $a = 48.0\ \text{J K}^{-1}\ \text{mol}^{-1}$, $b = 6.49\times10^{-3}\ \text{J K}^{-2}\ \text{mol}^{-1}$ and $c = -9.33 \times 10^5\ \text{J K mol}^{-1}$. The change in enthalpy in response to a change in temperature at constant pressure is given by [2B.6a–49], $dH_m = C_{p,m}dT$. Substituting in the expression for $C_{p,m}$ and integrating both sides gives

$$dH_m = (a + bT + cT^{-2})\,dT$$

$$\int_{(T_1)}^{(T_2)} dH_m = \int_{T_1}^{T_2} (a + bT + cT^{-2})\,dT$$

By using Integral A.1 in the *Resource section* for each term, it follows that

$$\begin{aligned}H_m(T_2) - H_m(T_1) &= a(T_2 - T_1) + \tfrac{1}{2}b(T_2^2 - T_1^2) - c\left(\frac{1}{T_2} - \frac{1}{T_1}\right)\\ &= (48.0\ \text{J K}^{-1}\ \text{mol}^{-1}) \times (1500\ \text{K} - 298.15\ \text{K})\\ &\quad + \tfrac{1}{2} \times (6.49 \times 10^{-3}\ \text{J K}^{-2}\ \text{mol}^{-1})\\ &\quad \times [(1500\ \text{K})^2 - (298.15\ \text{K})^2]\\ &\quad - (-9.33 \times 10^5\ \text{J K mol}^{-1}) \times \left(\frac{1}{1500\ \text{K}} - \frac{1}{298.15\ \text{K}}\right)\\ &= \boxed{62.2\ \text{kJ}}\end{aligned}$$

P2B.5 Since the volume is fixed no expansion work is done, hence $\boxed{w = 0}$ and $\boxed{\Delta U = q_V = +2.35\ \text{kJ}}$. From $H = U + pV$ it follows that $\Delta H = \Delta U + V\Delta p$ at constant volume. Using the van der Waals equation of state, [1C.5b–*24*],

$$p = \frac{RT}{V_m - b} - \frac{a}{V_m^2} \quad \text{it follows that} \quad \Delta p = \frac{R\Delta T}{V_m - b}$$

Note that the term in a cancels because the volume is constant. Therefore

$$\Delta H = \Delta U + v\Delta p = \Delta U + \frac{RV\Delta T}{V_m - b}$$

From the given data $V_m = V/n = (15.0\ \text{dm}^3)/(2.0\ \text{mol}) = 7.5\ \text{dm}^3\ \text{mol}^{-1}$, $\Delta T = 341\ \text{K} - 300\ \text{K} = 41\ \text{K}$ and from Table 1C.3 $b = 4.29 \times 10^{-2}\ \text{dm}^3\ \text{mol}^{-1}$. Therefore

$$\Delta H = (+2.35 \times 10^3\ \text{J}) + \frac{(8.3145\ \text{J K}^{-1}\ \text{mol}^{-1}) \times (15.0\ \text{dm}^3) \times (41\ \text{K})}{(7.5\ \text{dm}^3\ \text{mol}^{-1}) - (4.29 \times 10^{-2}\ \text{dm}^3\ \text{mol}^{-1})}$$
$$= 3.03... \times 10^3\ \text{J} = \boxed{3.0\ \text{kJ}}$$

2C Thermochemistry

Answers to discussion questions

D2C.1 The evaporation of liquid water is an endothermic process. As the water evaporates from the strips of cloth there will be a cooling of the surrounding air as energy as heat is transferred from the air to the evaporating water.

D2C.3 The standard state of a substance is the pure substance at a pressure of 1 bar and a specified temperature. The term reference state generally refers to elements and is the thermodynamically most stable state of the element at the temperature of interest. The distinction between standard state and reference state for elements may seem slight but becomes clear for those elements that can exist in more than one form at a specified temperature. So an element can have more than one standard state, one for each form that exists at the specified temperature.

Standard states are used in thermodynamics to provide a fixed point from which the thermodynamic property can be computed under non-standard conditions – for example at a different pressure. From tabulated data at the standard state other data can be therefore be inferred, thus limiting the amount of information which needs to be tabulated.

The identity of a reference state is needed so that consistent tables of data referred to these agreed reference states can be drawn up.

Solutions to exercises

E2C.1(a) Tetrachloromethane is vaporized at constant pressure, therefore $q = \Delta H$.

$$q = \Delta H = n\Delta_{\text{vap}}H^{\ominus} = (0.75\ \text{mol}) \times (30.0\ \text{kJ mol}^{-1}) = \boxed{+22.5\ \text{kJ}}$$

The work of expansion under constant pressure is given by [2A.6–40], $w = -p_{ex}\Delta V$. Note that $\Delta V = V_f$ because the final state (gas) has a much larger volume than the initial state (liquid). The perfect gas law is used to calculate V_f.

$$w = -p_{ex}\Delta V = -p_{ex}(V_f - V_i) \approx -p_{ex}V_f = -p_{ex} \times \frac{nRT}{p_{ex}} = -nRT$$
$$= -(0.75\ \text{mol}) \times (8.3145\ \text{J K}^{-1}\ \text{mol}^{-1}) \times (250\ \text{K}) = -1.55...\ \text{kJ} = \boxed{-1.6\ \text{kJ}}$$

The First Law of thermodynamics [2A.2–38] gives

$$\Delta U = q + w = (+22.5\ \text{kJ}) + (-1.55...\ \text{kJ}) = \boxed{+21\ \text{kJ}}$$

E2C.2(a) The equation for combustion of ethylbenzene is $C_8H_{10}(l) + 10.5\,O_2(g) \longrightarrow 8\,CO_2(g) + 5\,H_2O(l)$. The standard enthalpy of combustion is calculated using [2C.5a–55] and using values for the standard enthalpies of formation from Table 2C.7.

$$\begin{aligned}\Delta_c H^\ominus &= \sum_{\text{products}} \nu\Delta_f H^\ominus - \sum_{\text{reactants}} \nu\Delta_f H^\ominus \\ &= 8\Delta_f H^\ominus(CO_2,g) + 5\Delta_f H^\ominus(H_2O,l) - \tfrac{21}{2}\Delta_f H^\ominus(O_2,g) - \Delta_f H^\ominus(C_8H_{10},l) \\ &= 8 \times (-393.51\ \text{kJ mol}^{-1}) + 5 \times (-285.83\ \text{kJ mol}^{-1}) - 0 - (-12.5\ \text{kJ mol}^{-1}) \\ &= \boxed{-4.57 \times 10^3\ \text{kJ mol}^{-1}}\end{aligned}$$

E2C.3(a) The standard enthalpy of formation of HCl(aq) is $\Delta_r H^\ominus$ for the reaction

$$\tfrac{1}{2}H_2(g) + \tfrac{1}{2}Cl_2(g) \longrightarrow HCl(aq)$$

Because HCl is a strong acid, the reaction is effectively

$$\tfrac{1}{2}H_2(g) + \tfrac{1}{2}Cl_2(g) \longrightarrow H^+(aq) + Cl^-(aq)$$

By definition, $\Delta_f H^\ominus(H^+, aq) = 0$, so $\Delta_r H^\ominus$ for this reaction is $\Delta_f H^\ominus(Cl^-, aq)$.

$$\Delta_r H^\ominus = \Delta_f H^\ominus(Cl^-, aq) = \boxed{-167\ \text{kJ mol}^{-1}}$$

E2C.4(a) The equation for the combustion of naphthalene is $C_{10}H_8(s) + 12\,O_2(g) \longrightarrow 10\,CO_2(g) + 4\,H_2O(l)$. $\Delta_c H^\ominus$ is computed using the thermochemical data from Tables 2C.6 and 2C.7 and equation [2C.5a–55]

$$\begin{aligned}\Delta_c H^\ominus &= \sum_{\text{products}} \nu\Delta_f H^\ominus - \sum_{\text{reactants}} \nu\Delta_f H^\ominus \\ &= 10\Delta_f H^\ominus(CO_2,g) + 4\Delta_f H^\ominus(H_2O,l) \\ &\quad - 12\Delta_f H^\ominus(O_2,g) - \Delta_f H^\ominus(C_{10}H_8,s) \\ &= 10 \times (-393.51\ \text{kJ mol}^{-1}) + 4 \times (-285.83\ \text{kJ mol}^{-1}) \\ &\quad - 0 - (+78.53\ \text{kJ mol}^{-1}) = -5.16 \times 10^3\ \text{kJ mol}^{-1}\end{aligned}$$

In a bomb calorimeter the heat is at constant volume and is given by $q_V = n\Delta_c U^\ominus$. $\Delta_c U^\ominus$ is related to $\Delta_c H^\ominus$ by [2B.3–48], $\Delta_r H^\ominus = \Delta_r U^\ominus + \Delta n_g RT$, where Δn_g is the change in the amount of gas molecules in the reaction. In this case $\Delta n_g = 10\ \text{mol} - 12\ \text{mol} = -2\ \text{mol}$, hence

$$\begin{aligned}\Delta_c U^\ominus &= \Delta_c H^\ominus - \Delta n_g RT\\ &= (-5.15... \times 10^6\ \text{J}) - (-2.0\ \text{mol}) \times (8.3145\ \text{J K}^{-1}\ \text{mol}^{-1}) \times (298\ \text{K})\\ &= -5.15... \times 10^3\ \text{kJ mol}^{-1}\end{aligned}$$

The heat released in the bomb calorimeter on combustion of 120 mg of naphthalene ($M = 128.1632\ \text{g mol}^{-1}$) is

$$q_V = n\Delta_c U^\ominus = \left(\frac{0.120\ \text{g}}{128.1632\ \text{g mol}^{-1}}\right) \times (-5.15... \times 10^3\ \text{kJ mol}^{-1}) = -4.82...\text{kJ}$$

Therefore the calorimeter constant is

$$C = \frac{|q_V|}{\Delta T} = \frac{(4.82...\ \text{kJ})}{(3.05\ \text{K})} = 1.58...\ \text{kJ K}^{-1} = \boxed{1.58\ \text{kJ K}^{-1}}$$

The chemical equation for combustion of phenol is $C_6H_6O(s) + 7\,O_2(g) \longrightarrow 6\,CO_2(g) + 3\,H_2O(l)$. Following the same logic as above, $\Delta_c H^\ominus = -3054\ \text{kJ mol}^{-1}$, $\Delta n_g = -1$ mol and $\Delta_c U^\ominus = -3.05... \times 10^3\ \text{kJ mol}^{-1}$. The heat released on combustion of 150 mg of phenol ($M = 94.10\ \text{g mol}^{-1}$) is

$$q_V = n\Delta_c U^\ominus = \left(\frac{0.150\ \text{g}}{94.1074\ \text{g mol}^{-1}}\right) \times (-3.05... \times 10^3\ \text{kJ mol}^{-1}) = -4.86...\ \text{kJ}$$

Therefore $\Delta T = |q_V|/C = (4.86...\ \text{kJ})/(1.58...\text{kJ K}^{-1}) = \boxed{+3.07\ \text{K}}$

E2C.5(a) (i) Reaction(3) is reaction(2) – 2 × reaction(1), therefore

$$\begin{aligned}\Delta_r H^\ominus(3) &= \Delta_r H^\ominus(2) - 2\Delta_r H^\ominus(1)\\ &= (-483.64\ \text{kJ mol}^{-1}) - 2 \times (-184.62\ \text{kJ mol}^{-1})\\ &= \boxed{-114.40\ \text{kJ mol}^{-1}}\end{aligned}$$

The relationship between $\Delta_r H$ and $\Delta_r U$ is given by [2B.3–48], $\Delta_r H = \Delta_r U + \Delta n_g RT$ where Δn_g is the change in the amount of gas molecules in the reaction. For this reaction $\Delta n_g = 2\ \text{mol} + 2\ \text{mol} - 4\ \text{mol} - 1\ \text{mol} = -1\ \text{mol}$

$$\begin{aligned}\Delta_r U^\ominus &= \Delta_r H^\ominus - \Delta n_g RT\\ &= (-114.40 \times 10^3\ \text{J mol}^{-1})\\ &\quad - (-1.0\ \text{mol}) \times (8.3145\ \text{J K}^{-1}\ \text{mol}^{-1}) \times (298\ \text{K}) = \boxed{-112\ \text{kJ mol}^{-1}}\end{aligned}$$

(ii) Reaction(1) represents the formation of 2 moles of HCl(g) from its elements in their reference states, therefore the standard enthalpy of formation of HCl(g) is

$$\Delta_f H^\ominus(\text{HCl, g}) = \tfrac{1}{2}\Delta_r H^\ominus(1) = \tfrac{1}{2}\times(-184.62\ \text{kJ mol}^{-1}) = \boxed{-92.31\ \text{kJ mol}^{-1}}$$

Reaction(2) represents the formation of 2 moles of $H_2O(g)$ from its elements in their reference states, therefore

$$\Delta_f H^\ominus(H_2O,g) = \tfrac{1}{2}\Delta_r H^\ominus(2) = \tfrac{1}{2}\times(-483.64\,\text{kJ mol}^{-1}) = \boxed{-241.82\,\text{kJ mol}^{-1}}$$

E2C.6(a) The relationship between $\Delta_r H$ and $\Delta_r U$ is given by [2B.3–*48*], $\Delta_r H = \Delta_r U + \Delta n_g RT$ where Δn_g is the change in the amount of gas molecules in the reaction. For this reaction $\Delta n_g = 2\text{ mol} + 3\text{ mol} - 3\text{ mol} = +2\text{ mol}$

$$\begin{aligned}\Delta_r H^\ominus &= \Delta_r U^\ominus + \Delta n_g RT\\ &= (-1373\times10^3\,\text{J mol}^{-1}) + (+2.0\,\text{mol})\times(8.3145\,\text{J K}^{-1}\,\text{mol}^{-1})\times(298\,\text{K})\\ &= \boxed{-1368\,\text{kJ mol}^{-1}}\end{aligned}$$

E2C.7(a) (i) The standard reaction enthalpy at 298 K is calculated from the standard enthalpies of formation at 298 K using [2C.5a–*55*]

$$\begin{aligned}\Delta_r H^\ominus(298\,\text{K}) &= \sum_{\text{products}} \nu\Delta_f H^\ominus - \sum_{\text{reactants}} \nu\Delta_f H^\ominus\\ &= \Delta_f H^\ominus(\text{CO},g) + \Delta_f H^\ominus(\text{H}_2,g) - \Delta_f H^\ominus(\text{H}_2\text{O},g) - \Delta_f H^\ominus(\text{graphite},s)\\ &= (-110.53\,\text{kJ mol}^{-1}) + 0 - (-241.82\,\text{kJ mol}^{-1}) - 0\\ &= \boxed{+131.29\,\text{kJ mol}^{-1}}\end{aligned}$$

The relationship between $\Delta_r H$ and $\Delta_r U$ is given by [2B.3–*48*], $\Delta_r H = \Delta_r U + \Delta n_g RT$ where Δn_g is the change in the amount of gas molecules in the reaction. For this reaction $\Delta n_g = 1\text{ mol} + 1\text{ mol} - 1\text{ mol} = +1\text{ mol}$

$$\begin{aligned}\Delta_r U^\ominus(298\,\text{K}) &= \Delta_r H^\ominus(298\,\text{K}) - \Delta n_g RT\\ &= (+131.29\times10^3\,\text{J mol}^{-1})\\ &\quad - (+1.0\,\text{mol})\times(8.3145\,\text{J K}^{-1}\,\text{mol}^{-1})\times(298\,\text{K})\\ &= \boxed{+128.81\,\text{kJ mol}^{-1}}\end{aligned}$$

(ii) The difference of the molar heat capacities of products and reactants is calculated using [2C.7b–*55*] and the data from Table 2C.4

$$\begin{aligned}\Delta_r C_p^\ominus &= \sum_{\text{products}} \nu C_{p,\text{m}}^\ominus - \sum_{\text{reactants}} \nu C_{p,\text{m}}^\ominus\\ &= C_{p,\text{m}}^\ominus(\text{CO},g) + C_{p,\text{m}}^\ominus(\text{H}_2,g) - C_{p,\text{m}}^\ominus(\text{H}_2\text{O},g) - C_{p,\text{m}}^\ominus(\text{graphite},s)\\ &= (29.14\,\text{J K}^{-1}\,\text{mol}^{-1}) + (28.824\,\text{J K}^{-1}\,\text{mol}^{-1}) - (33.58\,\text{J K}^{-1}\,\text{mol}^{-1})\\ &\quad - (8.527\,\text{J K}^{-1}\,\text{mol}^{-1}) = +15.86\,\text{J K}^{-1}\,\text{mol}^{-1}\end{aligned}$$

It is assumed that all heat capacities are constant over the temperature range of interest, therefore the integrated form of Kirchhoff's Law is applicable, [2C.7d–56]

$$\begin{aligned}\Delta_r H^\ominus(478\text{ K}) &= \Delta_r H^\ominus(298\text{ K}) + \Delta T \Delta_r C_p^\ominus \\ &= (+131.29\times10^3\text{ J mol}^{-1}) \\ &\quad + (478\text{ K} - 298\text{ K}) \times (+15.86\text{ J K}^{-1}\text{ mol}^{-1}) \\ &= \boxed{+134.1\text{ kJ mol}^{-1}} \\ \Delta_r U^\ominus(478\text{ K}) &= \Delta_r H^\ominus(478\text{ K}) - \Delta n_g RT \\ &= (+134.1\times10^3\text{ J mol}^{-1}) \\ &\quad - (+1.0\text{ mol}) \times (8.3145\text{ J K}^{-1}\text{ mol}^{-1}) \times (478\text{ K}) \\ &= \boxed{+130\text{ kJ mol}^{-1}}\end{aligned}$$

E2C.8(a) The reaction equation $C(\text{graphite, s}) + O_2(g) \longrightarrow CO_2(g)$ represents the formation of $CO_2(g)$ from its elements in their reference states, therefore $\Delta_r H^\ominus(298\text{ K}) = \Delta_f H^\ominus(CO_2, g) = -393.51\text{ kJ mol}^{-1}$. The variation of standard reaction enthalpy with temperature is given by Kirchhoff's Law, [2C.7a–55]

$$\Delta_r H^\ominus(T_2) = \Delta_r H^\ominus(T_1) + \int_{T_1}^{T_2} \Delta_r C_p^\ominus\,dT$$

The difference of the molar heat capacities of products and reactants is given by [2C.7b–55]

$$\begin{aligned}\Delta_r C_p^\ominus &= \sum_{\text{products}} \nu C_{p,m}^\ominus - \sum_{\text{reactants}} \nu C_{p,m}^\ominus \\ &= C_{p,m}^\ominus(CO_2, g) - C_{p,m}^\ominus(O_2, g) - C_{p,m}^\ominus(\text{graphite, s})\end{aligned}$$

The heat capacities in Table 2B.1 are expressed in the form of $C_{p,m}^\ominus = a + bT + c/T^2$ therefore $\Delta_r C_p^\ominus = \Delta a + \Delta bT + \Delta c/T^2$ where $\Delta a = a(CO_2, g) - a(O_2, g) - a(\text{graphite, s})$ and likewise for Δb and Δc.

$$\begin{aligned}\Delta a &= (44.22\text{ J K}^{-1}\text{ mol}^{-1}) - (29.96\text{ J K}^{-1}\text{ mol}^{-1}) - (16.86\text{ J K}^{-1}\text{ mol}^{-1}) \\ &= -2.60\text{ J K}^{-1}\text{ mol}^{-1} \\ \Delta b &= (8.79\times10^{-3}\text{ J K}^{-2}\text{ mol}^{-1}) - (4.18\times10^{-3}\text{ J K}^{-2}\text{ mol}^{-1}) \\ &\quad - (4.77\times10^{-3}\text{ J K}^{-2}\text{ mol}^{-1}) = -0.16\times10^{-3}\text{ J K}^{-2}\text{ mol}^{-1} \\ \Delta c &= (-8.62\times10^5\text{ J K mol}^{-1}) - (-1.67\times10^5\text{ J K mol}^{-1}) \\ &\quad - (-8.54\times10^5\text{ J K mol}^{-1}) = +1.59\times10^5\text{ J K mol}^{-1}\end{aligned}$$

Integrating Kirchhoff's Law gives

$$\begin{aligned}
\Delta_r H^\ominus(T_2) &= \Delta_r H^\ominus(T_1) + \int_{T_1}^{T_2} \left(\Delta a + \Delta b T + \frac{\Delta c}{T^2}\right) dT \\
&= \Delta_r H^\ominus(T_1) + \Delta a(T_2 - T_1) + \tfrac{1}{2}\Delta b(T_2^2 - T_1^2) - \Delta c\left(\frac{1}{T_2} - \frac{1}{T_1}\right) \\
&= (-393.51 \times 10^3\ \mathrm{J\,mol^{-1}}) + (-2.60\ \mathrm{J\,K^{-1}\,mol^{-1}}) \times (500\ \mathrm{K} - 298\ \mathrm{K}) \\
&\quad + \tfrac{1}{2} \times (-0.16 \times 10^{-3}\ \mathrm{J\,K^{-2}\,mol^{-1}}) \times [(500\ \mathrm{K})^2 - (298\ \mathrm{K})^2] \\
&\quad - (1.59 \times 10^5\ \mathrm{J\,K\,mol^{-1}}) \times \left(\frac{1}{500\ \mathrm{K}} - \frac{1}{298\ \mathrm{K}}\right) \\
&= \boxed{-394\ \mathrm{kJ\,mol^{-1}}}
\end{aligned}$$

Solutions to problems

P2C.1 At constant pressure the temperature rise is given by [2B.7–49], $q_p = C_p\Delta T$. The heat capacity is approximated as $nC_{p,m}^\ominus(\mathrm{H_2O, l})$ where n is the amount in moles. Therefore

$$\Delta T = \frac{q_p}{nC_{p,m}^\ominus} = \frac{(1.0 \times 10^7\ \mathrm{J})}{\left(\frac{65000\ \mathrm{g}}{18.0158\ \mathrm{g\,mol^{-1}}}\right) \times (75.29\ \mathrm{J\,K^{-1}\,mol^{-1}})} = 36.8...\ \mathrm{K} = \boxed{37\ \mathrm{K}}$$

Assuming that $\mathrm{H_2O}(298\ \mathrm{K, l}) \longrightarrow \mathrm{H_2O}(298\ \mathrm{K, g})$ is the main process responsible for heat loss, and using data from Table 2C.1, the amount of water to evaporate is

$$n = \frac{q_p}{\Delta_{vap}H^\ominus} = \frac{(1.0 \times 10^7\ \mathrm{J})}{(44.016 \times 10^3\ \mathrm{J\,mol^{-1}})} = 2.27... \times 10^2\ \mathrm{mol}$$

The corresponding mass is

$$m = nM = (2.27... \times 10^2\ \mathrm{mol}) \times (18.0158\ \mathrm{g\,mol^{-1}}) = \boxed{4.1\ \mathrm{kg}}$$

P2C.3 (a) The combustion equation for cyclopropane is $\mathrm{C_3H_6(g)} + 4.5\,\mathrm{O_2(g)} \longrightarrow 3\,\mathrm{CO_2(g)} + 3\,\mathrm{H_2O(l)}$. Using data from Table 2C.7 and applying [2C.5a–55] for this case

$$\Delta_c H^\ominus = 3\Delta_f H^\ominus(\mathrm{CO_2, g}) + 3\Delta_f H^\ominus(\mathrm{H_2O, l}) - \Delta_f H^\ominus(\mathrm{C_3H_6, g})$$

Therefore

$$\begin{aligned}
\Delta_f H^\ominus(\mathrm{C_3H_6, g}) &= 3\Delta_f H^\ominus(\mathrm{CO_2, g}) + 3\Delta_f H^\ominus(\mathrm{H_2O, l}) - \Delta_c H^\ominus \\
&= 3 \times (-393.51\ \mathrm{kJ\,mol^{-1}}) + 3 \times (-285.83\ \mathrm{kJ\,mol^{-1}}) - (2091\ \mathrm{kJ\,mol^{-1}}) \\
&= +52.9...\ \mathrm{kJ\,mol^{-1}} = \boxed{+52.98\ \mathrm{kJ\,mol^{-1}}}
\end{aligned}$$

(b) The reaction is cyclopropane $\longrightarrow$ propene

$$\begin{aligned}
\Delta_r H^\ominus &= \Delta_f H^\ominus(\text{propene, g}) - \Delta_f H^\ominus(\text{cyclopropane, g}) \\
&= (+20.42\ \mathrm{kJ\,mol^{-1}}) - (+52.9...\ \mathrm{kJ\,mol^{-1}}) = \boxed{-32.56\ \mathrm{kJ\,mol^{-1}}}
\end{aligned}$$

P2C.5 The combustion of 0.825 g of benzoic acid (M = 122.1174 g mol^{-1}) is used to determine the calorimeter constant. The heat released in the calorimeter is

$$q_V = n\Delta_c U^{\ominus} = \left(\frac{0.825\text{ g}}{122.1174\text{ g mol}^{-1}}\right) \times (-3251\text{ kJ mol}^{-1}) = -21.9\ldots\text{ kJ}$$

Therefore the calorimeter constant is

$$C = \frac{|q_V|}{\Delta T} = \frac{(21.9\ldots\text{ kJ})}{(1.940\text{ K})} = 11.3\ldots\text{ kJ K}^{-1}$$

It follows that the heat released during the combustion of ribose is $|q_V| = C\Delta T$ = $(11.3\ldots\text{ kJ K}^{-1}) \times (0.910\text{ K}) = 10.3\ldots\text{ kJ}$, and so the standard internal energy of combustion is

$$\Delta_c U^{\ominus} = \frac{|q_V|}{n} = \frac{-(10.3\ldots\text{ kJ})}{(0.727\text{ g})/(150.129\text{ g mol}^{-1})} = -2.12\ldots \times 10^3\text{ kJ mol}^{-1}$$

The combustion reaction is $C_5H_{10}O_5(s) + 5\,O_2(g) \longrightarrow 5\,CO_2(g) + 5\,H_2O(l)$. There is no change in the number of gaseous species when going from reactants to products, therefore $\Delta_c U^{\ominus} = \Delta_c H^{\ominus} = -2.12\ldots \times 10^3\text{ kJ mol}^{-1}$. Applying [2C.5a–55] for this reaction and using the values for the standard enthalpies of formation from Table 2C.7 gives

$$\begin{aligned}
\Delta_c H^{\ominus} &= 5\Delta_f H^{\ominus}(CO_2, g) + 5\Delta_f H^{\ominus}(H_2O, l) - 5\Delta_f H^{\ominus}(O_2, g) \\
&\quad - \Delta_f H^{\ominus}(C_5H_{10}O_5, s) \\
\Delta_f H^{\ominus}(C_5H_{10}O_5, s) &= 5\Delta_f H^{\ominus}(CO_2, g) + 5\Delta_f H^{\ominus}(H_2O, l) \\
&\quad - 5\Delta_f H^{\ominus}(O_2, g) - \Delta_c H^{\ominus} \\
&= 5 \times (-393.51\text{ kJ mol}^{-1}) + 5 \times (-285.83\text{ kJ mol}^{-1}) \\
&\quad - 0 - (-2.12\ldots \times 10^3\text{ kJ mol}^{-1}) \\
&= \boxed{-1.27 \times 10^3\text{ kJ mol}^{-1}}
\end{aligned}$$

P2C.7 The chemical reaction for the combustion is $C_{60}(s) + 60\,O_2(g) \longrightarrow 60\,CO_2(g)$. The standard internal energy of combustion is

$$\Delta_c U^{\ominus} = (-36.0334\text{ kJ g}^{-1}) \times (60 \times 12.01\text{ g mol}^{-1}) = -2.59\ldots \times 10^4\text{ kJ mol}^{-1}$$

There is no change in the number of gaseous species when going from reactants to products, therefore $\Delta_c U^{\ominus} = \Delta_c H^{\ominus} = -2.59\ldots \times 10^4\text{ kJ mol}^{-1}$ = $\boxed{-25966\text{ kJ mol}^{-1}}$. Applying [2C.5a–55] for this case gives

$$\Delta_c H^{\ominus} = 60\Delta_f H^{\ominus}(CO_2, g) - 60\Delta_f H^{\ominus}(O_2, g) - \Delta_f H^{\ominus}(C_{60}, s)$$

Therefore

$$\begin{aligned}
\Delta_f H^{\ominus}(C_{60}, s) &= 60\Delta_f H^{\ominus}(CO_2, g) - 60\Delta_f H^{\ominus}(O_2, g) - \Delta_c H^{\ominus} \\
&= 60 \times (-393.51\text{ kJ mol}^{-1}) - 0 - (-2.59\ldots \times 10^4\text{ kJ mol}^{-1}) \\
&= \boxed{+2355.1\text{ kJ mol}^{-1}}
\end{aligned}$$

P2C.9 The reaction equation for combustion of methane is $CH_4(g) + 2O_2(g) \longrightarrow CO_2(g) + 2H_2O(g)$. The standard enthalpy of combustion can be calculated using [2C.5a–55] and using the values for the standard enthalpies of formation from Tables 2C.6 and 2C.7

$$\begin{aligned}\Delta_c H^\ominus &= \sum_{\text{products}} \nu\Delta_f H^\ominus - \sum_{\text{reactants}} \nu\Delta_f H^\ominus \\ &= \Delta_f H^\ominus(CO_2,g) + 2\Delta_f H^\ominus(H_2O,g) - 2\Delta_f H^\ominus(O_2,g) - \Delta_f H^\ominus(CH_4,g) \\ &= (-393.51\ \text{kJ mol}^{-1}) + 2\times(-241.82\ \text{kJ mol}^{-1}) - 0 - (-74.81\ \text{kJ mol}^{-1}) \\ &= -802.3\ \text{kJ mol}^{-1}\end{aligned}$$

The variation of standard reaction enthalpy with temperature is given by Kirchhoff's Law, [2C.7a–55]

$$\Delta_r H^\ominus(T_2) = \Delta_r H^\ominus(T_1) + \int_{T_1}^{T_2} \Delta_r C_p^\ominus\, dT$$

The difference of the molar heat capacities of products and reactants is given by [2C.7b–55]

$$\begin{aligned}\Delta_r C_p^\ominus &= \sum_{\text{products}} \nu C_{p,m}^\ominus - \sum_{\text{reactants}} \nu C_{p,m}^\ominus \\ &= C_{p,m}^\ominus(CO_2,g) + 2C_{p,m}^\ominus(H_2O,g) - 2C_{p,m}^\ominus(O_2,g) - C_{p,m}^\ominus(CH_4,g)\end{aligned}$$

The heat capacities are expressed in the form of $C_{p,m}^\ominus = \alpha + \beta T + \gamma T^2$ therefore $\Delta_r C_p^\ominus = \Delta\alpha + \Delta\beta T + \Delta\gamma T^2$ where $\Delta\alpha = \alpha(CO_2,g) + 2\alpha(H_2O,g) - 2\alpha(O_2,g) - \alpha(CH_4,g)$ and likewise for $\Delta\beta$ and $\Delta\gamma$.

$$\begin{aligned}\Delta\alpha &= [(26.86) + 2\times(30.36) - 2\times(25.72) - (14.16)]\ \text{J K}^{-1}\,\text{mol}^{-1} \\ &= +21.98\ \text{J K}^{-1}\,\text{mol}^{-1} \\ \Delta\beta &= [(6.97) + 2\times(9.61) - 2\times(12.98) - (75.5)]\times 10^{-3}\ \text{J K}^{-2}\,\text{mol}^{-1} \\ &= -75.27\times 10^{-3}\ \text{J K}^{-2}\,\text{mol}^{-1} \\ \Delta\gamma &= [(-0.82) + 2\times(1.184) - 2\times(-3.862) - (-17.99)]\times 10^{-6}\ \text{J K}^{-3}\,\text{mol}^{-1} \\ &= +27.262\times 10^{-6}\ \text{J K}^{-3}\,\text{mol}^{-1}\end{aligned}$$

Integrating Kirchhoff's Law gives

$$\begin{aligned}\Delta_c H^\ominus(T_2) &= \Delta_c H^\ominus(T_1) + \int_{T_1}^{T_2} (\Delta\alpha + \Delta\beta T + \Delta\gamma T^2)\, dT \\ &= \Delta_c H^\ominus(T_1) + \Delta\alpha(T_2 - T_1) + \tfrac{1}{2}\Delta\beta(T_2^2 - T_1^2) + \tfrac{1}{3}\Delta\gamma(T_2^3 - T_1^3) \\ &= (-802.3\ \text{kJ mol}^{-1}) + (+21.98\ \text{J K}^{-1}\,\text{mol}^{-1})\times(500\ \text{K} - 298\ \text{K}) \\ &\quad + \tfrac{1}{2}\times(-75.27\times 10^{-3}\ \text{J K}^{-2}\,\text{mol}^{-1})\times[(500\ \text{K})^2 - (298\ \text{K})^2] \\ &\quad + \tfrac{1}{3}\times(27.262\times 10^{-6}\ \text{J K}^{-3}\,\text{mol}^{-1})\times[(500\ \text{K})^3 - (298\ \text{K})^3] \\ &= \boxed{-803\ \text{kJ mol}^{-1}}\end{aligned}$$

P2C.11 (a) The heat released in the calorimeter by the combustion of glucose is

$$|q_\text{v}| = c\Delta T = (641\ \text{J K}^{-1}) \times (7.793\ \text{K}) = 4.99...\ \text{kJ}$$

Therefore the standard internal energy of combustion is

$$\Delta_\text{c}U^\ominus = \frac{-|q_\text{v}|}{n} = \frac{-(4.99...\ \text{kJ})}{(0.3212\ \text{g})/(180.1548\ \text{g mol}^{-1})} = -2.80... \times 10^3\ \text{kJ mol}^{-1}$$
$$= \boxed{-2.80 \times 10^3\ \text{kJ mol}^{-1}}$$

The chemical equation for the combustion of glucose is $C_6H_{12}O_6(s) + 6\,O_2(g)$
$\longrightarrow 6\,CO_2(g) + 6\,H_2O(l)$. There is no change in the number of gaseous species when going from reactants to products, therefore $\Delta_\text{c}U^\ominus = \Delta_\text{c}H^\ominus = -2.80... \times 10^3\ \text{kJ mol}^{-1} = \boxed{-2.80 \times 10^3\ \text{kJ mol}^{-1}}$. Applying [2C.5a–55] for this reaction and using the values for the standard enthalpies of formation from Table 2C.7 gives

$$\begin{aligned}
\Delta_\text{c}H^\ominus &= 6\Delta_\text{f}H^\ominus(CO_2,g) + 6\Delta_\text{f}H^\ominus(H_2O,l) - 6\Delta_\text{f}H^\ominus(O_2,g) \\
&\quad - \Delta_\text{f}H^\ominus(C_6H_{12}O_6,s) \\
\Delta_\text{f}H^\ominus(C_6H_{12}O_6,s) &= 6\Delta_\text{f}H^\ominus(CO_2,g) + 6\Delta_\text{f}H^\ominus(H_2O,l) \\
&\quad - 6\Delta_\text{f}H^\ominus(O_2,g) - \Delta_\text{c}H^\ominus \\
&= 6 \times (-393.51\ \text{kJ mol}^{-1}) + 6 \times (-285.83\ \text{kJ mol}^{-1}) \\
&\quad - 0 - (-2.80... \times 10^3\ \text{kJ mol}^{-1}) \\
&= \boxed{-1.27 \times 10^3\ \text{kJ mol}^{-1}}
\end{aligned}$$

(b) The reaction equation corresponding to anaerobic oxidation is $C_6H_{12}O_6(s) \longrightarrow 2\,C_3H_6O_3(s)$. The standard enthalpy of reaction is

$$\begin{aligned}
\Delta_\text{r}H^\ominus &= 2\Delta_\text{f}H^\ominus(C_3H_6O_3(s)) - \Delta_\text{f}H^\ominus(C_6H_{12}O_6(s)) \\
&= 2 \times (-694.0\ \text{kJ mol}^{-1}) - (-1.27 \times 10^3\ \text{kJ mol}^{-1}) \\
&= -1.13... \times 10^2\ \text{kJ mol}^{-1}
\end{aligned}$$

The amount of energy released in aerobic oxidation exceeds that of anaerobic glycolysis by

$$2.80... \times 10^3\ \text{kJ mol}^{-1} - 1.13... \times 10^2\ \text{kJ mol}^{-1} = \boxed{2.69 \times 10^3\ \text{kJ mol}^{-1}}$$

2D State functions and exact differentials

Answers to discussion questions

D2D.1 In Section 3E.1(b) on page 106 it is shown that for a van der Waals gas the internal pressure is given by $\pi_T = a/V_\text{m}^2$. This internal pressure is defined as $\pi_T = (\partial U/\partial V)_T$, and so indicates the required change of internal energy with volume. The van der Waals constant a is invariably positive, so the implication is that the internal energy increases as the gas expands. This is interpreted as being a result of the favourable (attractive) interactions between the molecules becoming less significant as they move further apart.

The internal pressure goes inversely with the molar volume: the larger the molar volume the further apart the molecules are and therefore their interactions become relatively less important.

Solutions to exercises

E2D.1(a) The molar volume of a perfect gas at 400 K is calculated as

$$V_m = \frac{RT}{p} = \frac{(8.3145\,\mathrm{J\,K^{-1}\,mol^{-1}}) \times (400\,\mathrm{K})}{(1.00\,\mathrm{bar}) \times [(10^5\,\mathrm{Pa})/(1.00\,\mathrm{bar})]} = 3.32... \times 10^{-2}\,\mathrm{m^3\,mol^{-1}}$$

The van der Waals parameter a of water vapour is found in Table 1C.3, and needs to be converted to SI units

$$a = (5.464\,\mathrm{dm^6\,atm\,mol^{-2}}) \times \left(\frac{10^{-6}\,\mathrm{m^6}}{1\,\mathrm{dm^6}}\right) \times \left(\frac{1.01325 \times 10^5\,\mathrm{Pa}}{1\,\mathrm{atm}}\right)$$
$$= 0.553...\,\mathrm{m^6\,Pa\,mol^{-2}}$$

Therefore the internal pressure is

$$\pi_T = \frac{a}{V_m^2} = \frac{(0.553...\,\mathrm{m^6\,Pa\,mol^{-2}})}{(3.32... \times 10^{-2}\,\mathrm{m^3\,mol^{-1}})^2} = \boxed{501\,\mathrm{Pa}}$$

E2D.2(a) The internal energy of a closed system of constant composition is a function of temperature and volume. For a change in V and T, $\mathrm{d}U$ is given by [2D.5–*61*], $\mathrm{d}U = \pi_T\mathrm{d}V + C_V\mathrm{d}T$. At constant temperature, this reduces to $\mathrm{d}U = \pi_T\mathrm{d}V$. Substituting in the given expression for π_T for a van der Waals gas and using molar quantities

$$\mathrm{d}U_m = \frac{a}{V_m^2}\mathrm{d}V_m$$

This expression is integrated between $V_{m,i}$ and $V_{m,f}$ to give

$$\int_{V_{m,i}}^{V_{m,f}} \mathrm{d}U_m = \int_{V_{m,i}}^{V_{m,f}} \frac{a}{V_m^2}\mathrm{d}V_m$$

hence

$$\Delta U_m = -\left.\frac{a}{V_m^2}\right|_{V_{m,i}}^{V_{m,f}} = -a\left(\frac{1}{V_{m,f}} - \frac{1}{V_{m,i}}\right)$$

The van der Waals parameter a for nitrogen gas is found in Table 1C.3, and needs to be converted to SI units

$$a = (1.352\ \text{dm}^6\ \text{atm mol}^{-2}) \times \left(\frac{10^{-6}\ \text{m}^6}{1\ \text{dm}^6}\right) \times \left(\frac{1.01325 \times 10^5\ \text{Pa}}{1\ \text{atm}}\right)$$
$$= 0.136...\ \text{m}^6\ \text{Pa mol}^{-2}$$

$$\Delta U_\text{m} = -(0.136...\ \text{m}^6\ \text{Pa mol}^{-2})\left(\frac{1}{20.00 \times 10^{-3}\ \text{m}^3\ \text{mol}^{-1}} - \frac{1}{1.00 \times 10^{-3}\ \text{m}^3\ \text{mol}^{-1}}\right)$$
$$= +1.30... \times 10^2\ \text{J mol}^{-1} = \boxed{+130\ \text{J mol}^{-1}}$$

The work done by an expanding gas is given by [2A.5a–39], $\text{d}w = -p_\text{ex}\text{d}V$. For a reversible expansion p_ex is the pressure of the gas, hence

$$w = -\int p\,\text{d}V_\text{m}$$

Substituting in the expression for the pressure of a van der Waals gas, [1C.5b–24]

$$w = -\int \frac{RT}{V_\text{m} - b} - \frac{a}{V_\text{m}^2}\,\text{d}V_\text{m} = -\int \frac{RT}{V_\text{m} - b}\,\text{d}V_\text{m} + \int \frac{a}{V_\text{m}^2}\,\text{d}V_\text{m}$$
$$= -\int \frac{RT}{V_\text{m} - b}\,\text{d}V_\text{m} + \Delta U_\text{m}$$

The second term is identified as ΔU_m from the above. According to the First Law, $\Delta U = q + w$, the first term in the expression above must be $-q$, therefore

$$q = \int_{V_\text{m,i}}^{V_\text{m,f}} \frac{RT}{V_\text{m} - b}\,\text{d}V_\text{m} = RT\ln(V_\text{m} - b)\big|_{V_\text{m,i}}^{V_\text{m,f}} = RT\ln\left(\frac{V_\text{m,f} - b}{V_\text{m,i} - b}\right)$$
$$= (8.3145\ \text{J K}^{-1}\ \text{mol}^{-1}) \times (298\ \text{K}) \times \ln\left(\frac{20.00\ \text{dm}^3 - 3.87 \times 10^{-2}\ \text{dm}^3}{1.00\ \text{dm}^3 - 3.87 \times 10^{-2}\ \text{dm}^3}\right)$$
$$= +7.51... \times 10^3\ \text{J mol}^{-1} = \boxed{+7.52\ \text{kJ mol}^{-1}}$$

where the value for b is taken from the Resource section. From the First Law the work done is $w = -q + \Delta U_\text{m}$, hence

$$w = -q + \Delta U_\text{m} = -7.51... \times 10^3\ \text{J mol}^{-1} + 1.30... \times 10^2\ \text{J mol}^{-1} = \boxed{-7.39\ \text{kJ mol}^{-1}}$$

E2D.3(a) The volume of the liquid can be written as

$$V = V'(a + bT + cT^2)$$

where $a = 0.75$, $b = 3.9 \times 10^{-4}\ \text{K}^{-1}$ and $c = 1.48 \times 10^{-6}\ \text{K}^{-2}$. The expansion coefficient is defined in [2D.6–62] as $\alpha = (1/V)(\partial V/\partial T)_p$. The derivative with respect to T is

$$\left(\frac{\partial V}{\partial T}\right)_p = V'(b + 2cT)$$

Therefore

$$\alpha = \frac{1}{V'(a+bT+cT^2)} \times [V'(b+2cT)] = \frac{b+2cT}{a+bT+cT^2}$$

Evaluating this expression at 320 K gives

$$\alpha_{320} = \frac{(3.9\times10^{-4}\,\mathrm{K}^{-1}) + 2\times(1.48\times10^{-6}\,\mathrm{K}^{-2})\times(320\,\mathrm{K})}{(0.75)+(3.9\times10^{-4}\,\mathrm{K}^{-1})\times(320\,\mathrm{K})+(1.48\times10^{-6}\,\mathrm{K}^{-2})\times(320\,\mathrm{K})^2}$$

$$= \boxed{+1.3\times10^{-3}\,\mathrm{K}^{-1}}$$

E2D.4(a) The isothermal compressibility is defined in [2D.7–62], $\kappa_T = -(1/V)(\partial V/\partial p)_T$, therefore at constant temperature $\mathrm{d}V/V = -\kappa_T \mathrm{d}p$. This question is concerned with changes in density, so the next step is to rewrite the volume in terms of the density, ρ. If the mass is m, $V = m/\rho$, and therefore $\mathrm{d}V = (-m/\rho^2)\mathrm{d}\rho$. Therefore

$$\frac{\mathrm{d}V}{V} = \frac{1}{V}\left(-\frac{m}{\rho^2}\mathrm{d}\rho\right) = -\left(\frac{\rho}{m}\right)\left(\frac{m}{\rho^2}\right)\mathrm{d}\rho = -\frac{1}{\rho}\mathrm{d}\rho$$

It therefore follows that

$$\frac{\mathrm{d}\rho}{\rho} = \kappa_T \mathrm{d}p \quad \text{and hence} \quad \mathrm{d}p = \frac{1}{\kappa_T \rho}\mathrm{d}\rho$$

This expression gives the relationship between the change in pressure and the change in density. Approximating $\mathrm{d}\rho$ by $\delta\rho$ and $\mathrm{d}p$ by δp for sufficiently small changes gives

$$\delta p = \frac{1}{\kappa_T} \times \frac{\delta\rho}{\rho} = \left(\frac{1}{4.96\times10^{-5}\,\mathrm{atm}^{-1}}\right) \times (1.0\times10^{-3}) = \boxed{+20\,\mathrm{atm}}$$

E2D.5(a) The difference $C_{p,\mathrm{m}} - C_{V,\mathrm{m}}$ is given by [2D.11–63], $C_{p,\mathrm{m}} - C_{V,\mathrm{m}} = \alpha^2 T V_\mathrm{m}/\kappa_T$. In this expression the molar volume is found from the mass density ρ and the molar mass M by $V_\mathrm{m} = M/\rho$. The values of α and κ are available in the *Resource section*, as is the mass density.

$$C_{p,\mathrm{m}} - C_{V,\mathrm{m}} = \frac{\alpha^2 T V_\mathrm{m}}{\kappa_T} = \frac{\alpha^2 T M}{\kappa_T \rho}$$

$$= \frac{(12.4\times10^{-4}\,\mathrm{K}^{-1})^2 \times (298\,\mathrm{K}) \times (78.1074\,\mathrm{g\,mol}^{-1})}{[(92.1\times10^{-6}\,\mathrm{bar}^{-1})\times(1\,\mathrm{bar}/10^5\,\mathrm{Pa})]\times(0.879\times10^6\,\mathrm{g\,m}^{-3})}$$

$$= \boxed{+44.2\,\mathrm{J\,K}^{-1}\,\mathrm{mol}^{-1}}$$

The units are $\mathrm{K}^{-1}\,\mathrm{Pa\,m}^3\,\mathrm{mol}^{-1} = \mathrm{K}^{-1}\,(\mathrm{N\,m}^{-2})\,\mathrm{m}^3\,\mathrm{mol}^{-1} = \mathrm{K}^{-1}\,\mathrm{N\,m\,mol}^{-1}$ $= \mathrm{J\,K}^{-1}\,\mathrm{mol}^{-1}$

Solutions to problems

P2D.1 The expansion coefficient is defined in [2D.6–62] as $\alpha = (1/V)(\partial V/\partial T)_p$. For sufficiently small changes the derivatives are approximated by finite changes to give

$$\alpha = \frac{1}{V}\left(\frac{\delta V}{\delta T}\right)$$

Using the relationship given in the hint, $\delta V = A\delta r$

$$\alpha = \frac{1}{V}\left(\frac{A\delta r}{\delta T}\right) \quad \text{therefore} \quad \delta r = \frac{\alpha V}{A}\delta T$$

From Table 2D.1 $\alpha = 2.1 \times 10^{-4}\ \text{K}^{-1}$. For a temperature rise of 1.0 °C

$$\Delta r = \frac{(2.1 \times 10^{-4}\ \text{K}^{-1}) \times (1.37 \times 10^{9}\ \text{km}^3)}{(361 \times 10^{6}\ \text{km}^2)} \times (1.0\ \text{K}) = 7.96... \times 10^{-4}\ \text{km}$$
$$= \boxed{0.80\ \text{m}}$$

As $\Delta r \propto \Delta T$, for a temperature rise of 2.0 °C, $\Delta r = \boxed{1.6\ \text{m}}$, and for a temperature rise of 3.5 °C, $\Delta r = \boxed{2.8\ \text{m}}$ is expected. This calculation assumes that the total surface area of the oceans remains constant and that $\alpha(H_2O) = \alpha(\text{ocean})$.

P2D.3 (a) If $V = V(p, T)$ then

$$\mathrm{d}V = \left(\frac{\partial V}{\partial p}\right)_T \mathrm{d}p + \left(\frac{\partial V}{\partial T}\right)_p \mathrm{d}T$$

If $p = p(V, T)$ then

$$\mathrm{d}p = \left(\frac{\partial p}{\partial V}\right)_T \mathrm{d}V + \left(\frac{\partial p}{\partial T}\right)_V \mathrm{d}T$$

(b) Dividing the expression for $\mathrm{d}V$ by V gives

$$\frac{1}{V}\mathrm{d}V = \frac{1}{V}\left(\frac{\partial V}{\partial p}\right)_T \mathrm{d}p + \frac{1}{V}\left(\frac{\partial V}{\partial T}\right)_p \mathrm{d}T$$

Noting the definition of κ_T, [2D.7–62], $\kappa_T = -(1/V)(\partial V/\partial p)_T$ and that of α, [2D.6–62], $\alpha = (1/V)(\partial V/\partial T)_p$, and rewriting $(1/V)\mathrm{d}V$ as $\mathrm{d}\ln V$ gives

$$\mathrm{d}\ln V = -\kappa_T \mathrm{d}p + \alpha \mathrm{d}T$$

Dividing the expression for $\mathrm{d}p$ by p gives

$$\begin{aligned}
\frac{1}{p}\mathrm{d}p &= \frac{1}{p}\left(\frac{\partial p}{\partial V}\right)_T \mathrm{d}V + \frac{1}{p}\left(\frac{\partial p}{\partial T}\right)_V \mathrm{d}T \\
\mathrm{d}\ln p &= \frac{1}{p}\left(\frac{\partial p}{\partial V}\right)_T \mathrm{d}V - \frac{1}{p}\left(\frac{\partial p}{\partial V}\right)_T\left(\frac{\partial V}{\partial T}\right)_p \mathrm{d}T \\
&= \frac{1}{p}\left(\frac{\partial p}{\partial V}\right)_T\left[\mathrm{d}V - \left(\frac{\partial V}{\partial T}\right)_p \mathrm{d}T\right] \\
&= \left(-\frac{1}{V}\right)\frac{1}{p}\left(\frac{1}{-\frac{1}{V}\left(\frac{\partial V}{\partial p}\right)_T}\right)\left[\mathrm{d}V - \left(\frac{\partial V}{\partial T}\right)_p \mathrm{d}T\right] \\
&= \frac{1}{p\kappa_T}\left[-\frac{\mathrm{d}V}{V} + \frac{1}{V}\left(\frac{\partial V}{\partial T}\right)_p \mathrm{d}T\right] \\
&= \frac{1}{p\kappa_T}\left(-\frac{\mathrm{d}V}{V} + \alpha \mathrm{d}T\right)
\end{aligned}$$

Going from line 1 to 2, the identity $(1/x)\mathrm{d}x = \mathrm{d}\ln x$ and Euler's chain relationship are used. Going from line 3 to 4 the reciprocal relationship is used. The definition of κ_T is used to go from line 4 to 5 and to go from line 5 to 6 the definition of α is used.

P2D.5 The isothermal compressibility is defined in [2D.11–63], $\kappa_T = -(1/V)(\partial V/\partial p)_T$. Using the reciprocal identity this is equivalent to

$$\kappa_T = -\frac{1}{V}\frac{1}{(\partial p/\partial V)_T}$$

The pressure of a van der Waals gas is $p = nRT/(V-nb) - n^2a/V^2$, therefore

$$\left(\frac{\partial p}{\partial V}\right)_T = -\frac{nRT}{(V-nb)^2} + \frac{2n^2a}{V^3} = \frac{-nRTV^3 + 2n^2a(V-nb)^2}{(V-nb)^2V^3}$$

It follows that

$$\left(\frac{\partial V}{\partial p}\right)_T = \frac{(V-nb)^2V^3}{-nRTV^3 + 2n^2a(V-nb)^2}$$

and hence

$$\kappa_T = -\frac{1}{V}\left(\frac{\partial V}{\partial p}\right)_T = \frac{V^2(V-nb)^2}{nRTV^3 - 2n^2a(V-nb)^2}$$

The expansion coefficient is defined in [2D.6–62] as $\alpha_T = (1/V)(\partial V/\partial T)_p$, which is rewritten using the reciprocal identity as

$$\alpha = \frac{1}{V}\frac{1}{(\partial T/\partial V)_p}$$

Rearranging the van der Waals equation of state gives

$$T = \frac{p}{nR}(V-nb) + \frac{na}{RV^2}(V-nb)$$

hence

$$\left(\frac{\partial T}{\partial V}\right)_p = \overbrace{\frac{p}{nR} + \frac{na}{RV^2}}^{A} - \frac{2na}{RV^3}(V - nb) = \frac{T}{V - nb} - \frac{2na}{RV^3}(V - nb)$$
$$= \frac{RTV^3 - 2na(V - nb)^2}{(V - nb)RV^3}$$

where the term A is identified from the previous equation as being equal to $T/(V - nb)$.Using the reciprocal identity

$$\left(\frac{\partial V}{\partial T}\right)_p = \frac{(V - nb)RV^3}{RTV^3 - 2na(V - nb)^2}$$

therefore

$$\alpha = \frac{1}{V}\left(\frac{\partial V}{\partial T}\right)_p = \frac{(V - nb)RV^2}{RTV^3 - 2na(V - nb)^2}$$

It follows that

$$\frac{\kappa_T}{\alpha} = \frac{V^2(V - nb)^2}{nRTV^3 - 2n^2a(V - nb)^2} \times \frac{(V - nb)RV^2}{RTV^3 - 2na(V - nb)^2}$$
$$= \frac{(V - nb)}{nR}$$

For a molar quantity $V \to V_m$ and $n \to 1$ to give $\kappa_T/\alpha = (V_m - b)/R$ or $\boxed{\kappa_T R = \alpha(V_m - b)}$.

P2D.7 The equation of state of this gas is rearranged to give

$$V = \frac{nRT}{p} + nb$$

From this it follows that

$$\left(\frac{\partial V}{\partial T}\right)_p = \frac{nR}{p}$$

Substituting this result in the expression for μ gives

$$\mu = \frac{1}{C_p}\left[T\left(\frac{\partial V}{\partial T}\right)_p - V\right]$$
$$= \frac{1}{C_p}\left[T\left(\frac{nR}{p}\right) - \left(\frac{nRT}{p} + nb\right)\right] = -\frac{nb}{C_p}$$

The heat capacity and the van der Waals parameter b are always positive, therefore the Joule–Thomson coefficient μ is negative for this gas. This means that the temperature of the gas will increase when expanding in an isenthalpic process.

P2D.9 From [2D.13–63], $dH = -\mu C_p dp + C_p dT$, it follows that at constant temperature

$$\mu = -\frac{1}{C_p}\left(\frac{\partial H}{\partial p}\right)_T$$

The partial derivative $(\partial H/\partial p)_T$ is the slope of the plot of H against p measured at constant temperature.

(a) The plot for 300 K is shown in Fig. 2.2.

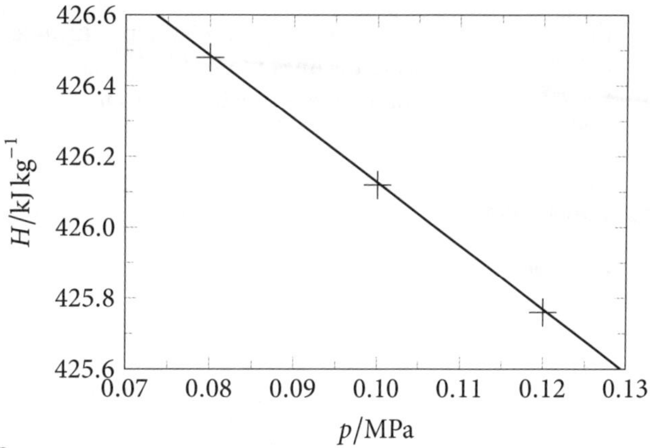

Figure 2.2

The slope of the plot is −17.93. The Joule–Thomson coefficient is determined from the slope,

$$\mu = -\frac{(-17.93\ \text{kJ kg}^{-1}\ \text{MPa}^{-1})}{(0.7649\ \text{kJ K}^{-1}\ \text{kg}^{-1})} = \boxed{23\ \text{K MPa}^{-1}}$$

(b) The plot for 350 K is shown in Fig. 2.3.

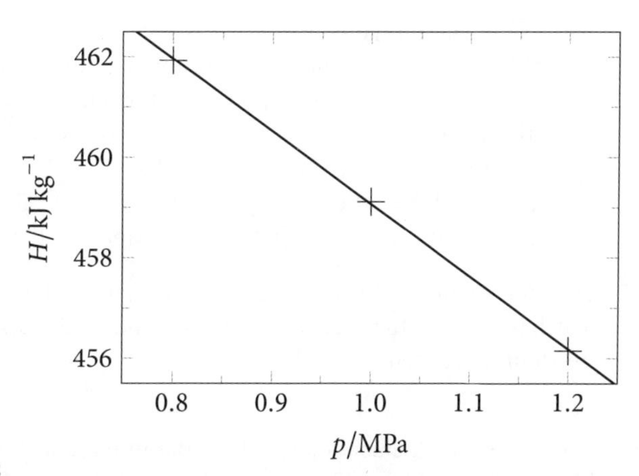

Figure 2.3

The slope of the plot is −14.46. The Joule–Thomson coefficient is determined from the slope,

$$\mu = -\frac{(-14.46\ \mathrm{kJ\,kg^{-1}\,MPa^{-1}})}{(1.0392\ \mathrm{kJ\,K^{-1}\,kg^{-1}})} = \boxed{14\ \mathrm{K\,MPa^{-1}}}$$

2E Adiabatic changes

Answers to discussion questions

D2E.1 Energy as heat is not supplied to the system during an adiabatic expansion; therefore, the pressure declines more steeply than in an isothermal expansion because the temperature decreases in the former.

Solutions to exercises

E2E.1(a) Ammonia and methane are nonlinear polyatomic molecules which have three degrees of translational and three degrees of rotational freedom. From the equipartition theorem (*The chemist's toolkit* 7 in Topic 2A)

$$C_{V,\mathrm{m}} = \tfrac{1}{2} \times (\nu_\mathrm{t} + \nu_\mathrm{r} + 2\nu_\mathrm{v}) \times R$$

where ν_t is the number of translational degrees of freedom, ν_r is the number of rotational degrees of freedom and ν_v is the number of vibrational degrees of freedom.

(i) Without any vibrational contribution the calculation is the same for both molecules. $C_{V,\mathrm{m}} = \tfrac{1}{2} \times (3+3+0) \times R = 3R$. For a perfect gas, [2B.9–49], $C_{p,\mathrm{m}} = C_{V,\mathrm{m}} + R$, therefore

$$\gamma = \frac{C_{V,\mathrm{m}} + R}{C_{V,\mathrm{m}}} = \frac{3R + R}{3R} = \frac{4}{3}$$

(ii) If the vibrational contribution is included, the molecules have different values of γ. The number of vibrational modes for a nonlinear polyatomic molecule is $\nu_\mathrm{v} = 3N - 6$, where N is the number of atoms in the molecule. Therefore, for ammonia $\nu_\mathrm{v} = 3N - 6 = 6$ and for methane $\nu_\mathrm{v} = 3N - 6 = 9$. This gives $C_{V,\mathrm{m}}(\mathrm{NH_3}) = \tfrac{1}{2} \times (3 + 3 + 2 \times 6) \times R = 9R$, therefore $C_{p,\mathrm{m}}(\mathrm{NH_3}) = 10R$ and $\gamma = 10/9$. For methane $C_{V,\mathrm{m}}(\mathrm{CH_4}) = \tfrac{1}{2} \times (3+3+2\times 9) \times R = 12R$, $C_{p,\mathrm{m}}(\mathrm{CH_4}) = 13R$ and $\gamma = 13/12$.

The experimental values of γ for ammonia and methane are the same, $\gamma = 1.31$, which is closer to the value calculated without taking the vibrational degrees of freedom into account.

E2E.2(a) For a reversible adiabatic expansion the initial and final states are related by [2E.2a–68], $(T_\mathrm{f}/T_\mathrm{i}) = (V_\mathrm{i}/V_\mathrm{f})^{1/c}$, where $c = C_{V,\mathrm{m}}/R$. For a monoatomic ideal gas $C_{V,\mathrm{m}} = \frac{3}{2}R$, so $c = \frac{3}{2}$. Therefore

$$T_\mathrm{f} = T_\mathrm{i}\left(\frac{V_\mathrm{i}}{V_\mathrm{f}}\right)^{\frac{1}{c}} = (273.15\ \mathrm{K}) \times \left(\frac{1.0\ \mathrm{dm^3}}{3.0\ \mathrm{dm^3}}\right)^{\frac{2}{3}} = \boxed{1.3 \times 10^2\ \mathrm{K}}$$

E2E.3(a) For a reversible adiabatic expansion the initial and final states are related by [2E.3–*68*], $p_\mathrm{i}V_\mathrm{i}^\gamma = p_\mathrm{f}V_\mathrm{f}^\gamma$, where γ is the ratio of heat capacities, $\gamma = C_{p,\mathrm{m}}/C_{V,\mathrm{m}}$. The initial volume of the sample is

$$V_\mathrm{i} = \frac{nRT}{p} = \frac{(1.0\ \mathrm{mol}) \times (8.3145\ \mathrm{J\,K^{-1}\,mol^{-1}}) \times (300\ \mathrm{K})}{(4.25\ \mathrm{atm}) \times [(1.01325 \times 10^5\ \mathrm{Pa})/(1\ \mathrm{atm})]} = 5.79... \times 10^{-3}\ \mathrm{m^3}$$
$$= 5.79...\ \mathrm{dm^3}$$

For a perfect gas $C_{p,\mathrm{m}} - C_{V,\mathrm{m}} = R$, hence

$$\gamma = \frac{C_{p,\mathrm{m}}}{C_{V,\mathrm{m}}} = \frac{C_{V,\mathrm{m}} + R}{C_{V,\mathrm{m}}} = \frac{(20.8\ \mathrm{J\,K^{-1}\,mol^{-1}}) + (8.3145\ \mathrm{J\,K^{-1}\,mol^{-1}})}{(20.8\ \mathrm{J\,K^{-1}\,mol^{-1}})} = 1.39...$$

The final volume is given by $V_\mathrm{f} = (5.79...\ \mathrm{dm^3}) \times (4.25/2.50)^{1/1.39...} = 8.46...\ \mathrm{dm^3}$. Thus $\boxed{V_\mathrm{f} = 8.46\ \mathrm{dm^3}}$.

The initial and final states are also related by $(T_\mathrm{f}/T_\mathrm{i}) = (V_\mathrm{i}/V_\mathrm{f})^{1/c}$ where $c = C_{V,\mathrm{m}}/R$. For this gas $c = (20.8\ \mathrm{J\,K^{-1}\,mol^{-1}})/(8.3145\ \mathrm{J\,K^{-1}\,mol^{-1}}) = 2.50...$ and hence

$$T_\mathrm{f} = T_\mathrm{i}\left(\frac{V_\mathrm{i}}{V_\mathrm{f}}\right)^{1/c} = (300\ \mathrm{K}) \times \left(\frac{5.79...\ \mathrm{dm^3}}{8.46\ \mathrm{dm^3}}\right)^{1/2.50...} = 2.57... \times 10^2\ \mathrm{K} = \boxed{258\ \mathrm{K}}$$

The work done by a perfect gas during adiabatic expansion is given by [2E.1–*67*]

$$w_\mathrm{ad} = C_V\Delta T = C_V(T_\mathrm{f} - T_\mathrm{i}) = (20.8\ \mathrm{J\,K^{-1}}) \times (2.57... \times 10^2\ \mathrm{K} - 300\ \mathrm{K}) = \boxed{-877\ \mathrm{J}}$$

E2E.4(a) The work done in a reversible adiabatic expansion is given by [2E.1–*67*], $w_\mathrm{ad} = C_V\Delta T$. The task is to find ΔT. The initial and final states in a reversible adiabatic expansion are related by [2E.2b–*68*], $V_\mathrm{i}T_\mathrm{i}^c = V_\mathrm{f}T_\mathrm{f}^c$, where $c = C_{V,\mathrm{m}}/R$. If the gas is assumed to be perfect, $C_{p,\mathrm{m}} - C_{V,\mathrm{m}} = R$ [2B.9–*49*], and so

$$c = \frac{C_{V,\mathrm{m}}}{R} = \frac{C_{p,\mathrm{m}} - R}{R} = \frac{(37.11\ \mathrm{J\,K^{-1}\,mol^{-1}}) - (8.3145\ \mathrm{J\,K^{-1}\,mol^{-1}})}{(8.3145\ \mathrm{J\,K^{-1}\,mol^{-1}})} = 3.46...$$

The temperature change is given by

$$\Delta T = T_\mathrm{f} - T_\mathrm{i} = T_\mathrm{i}\left(\frac{V_\mathrm{i}}{V_\mathrm{f}}\right)^{1/c} - T_\mathrm{i} = T_\mathrm{i}\left[\left(\frac{V_\mathrm{i}}{V_\mathrm{f}}\right)^{1/c} - 1\right]$$

Therefore the work done is

$$w_{\text{ad}} = C_V\Delta T = nC_{V,\text{m}}\Delta T = nC_{V,\text{m}}T_{\text{i}}\left[\left(\frac{V_{\text{i}}}{V_{\text{f}}}\right)^{1/c} - 1\right]$$
$$= n(C_{p,\text{m}} - R)T_{\text{i}}\left[\left(\frac{V_{\text{i}}}{V_{\text{f}}}\right)^{1/c} - 1\right]$$
$$= \left(\frac{2.45\ \text{g}}{44.01\ \text{g mol}^{-1}}\right) \times [(37.11\ \text{J K}^{-1}\ \text{mol}^{-1})$$
$$- (8.3145\ \text{J K}^{-1}\ \text{mol}^{-1})] \times (300.15\ \text{K})$$
$$\times \left[\left(\frac{500\ \text{cm}^3}{3.00 \times 10^3\ \text{cm}^3}\right)^{1/3.46...} - 1\right] = \boxed{-194\ \text{J}}$$

E2E.5(a) The initial and final states in a reversible adiabatic expansion are related by [2E.3–68], $p_{\text{i}}V_{\text{i}}^{\gamma} = p_{\text{f}}V_{\text{f}}^{\gamma}$, therefore

$$p_{\text{f}} = p_{\text{i}}\left(\frac{V_{\text{i}}}{V_{\text{f}}}\right)^{\gamma} = (67.4\ \text{kPa}) \times \left(\frac{0.50\ \text{dm}^3}{2.00\ \text{dm}^3}\right)^{1.4} = \boxed{9.7\ \text{kPa}}$$

Solutions to problems

P2E.1 The work done in a reversible adiabatic expansion is given by [2E.1–67], $w_{\text{ad}} = C_V\Delta T$. The initial and final states in a reversible adiabatic expansion are related by [2E.2b–68], $V_{\text{i}}T_{\text{i}}^{c} = V_{\text{f}}T_{\text{f}}^{c}$, where $c = C_{V,\text{m}}/R$. If the gas is assumed to be perfect, $C_{p,\text{m}} - C_{V,\text{m}} = R$ [2B.9–49], and so

$$c = \frac{C_{V,\text{m}}}{R} = \frac{C_{p,\text{m}} - R}{R} = \frac{(35.06\ \text{J K}^{-1}\ \text{mol}^{-1}) - (8.3145\ \text{J K}^{-1}\ \text{mol}^{-1})}{(8.3145\ \text{J K}^{-1}\ \text{mol}^{-1})} = 3.21...$$

The final temperature is

$$T_{\text{f}} = T_{\text{i}}\left(\frac{V_{\text{i}}}{V_{\text{f}}}\right)^{1/c} = (298\ \text{K}) \times \left(\frac{0.50\ \text{dm}^3}{2.00\ \text{dm}^3}\right)^{1/3.21...} = 1.93... \times 10^2\ \text{K} = \boxed{194\ \text{K}}$$

Therefore the work done is

$$w_{\text{ad}} = C_V\Delta T = nC_{V,\text{m}}\Delta T = n(C_{p,\text{m}} - R)(T_{\text{f}} - T_{\text{i}})^{1/c}$$
$$= (1.00\ \text{mol}) \times [(35.06\ \text{J K}^{-1}\ \text{mol}^{-1}) - (8.3145\ \text{J K}^{-1}\ \text{mol}^{-1})]$$
$$\times [(1.93... \times 10^2\ \text{K}) - (298\ \text{K})]^{1/3.21...} = \boxed{-2.79\ \text{kJ}}$$
$$\Delta U = w_{\text{ad}} = \boxed{-2.79\ \text{kJ}}$$

Integrated activities

I2.1 A state function is a thermodynamic property, the value of which is independent of the history of the system. Examples of state functions are the properties of pressure, temperature, internal energy, enthalpy as well as the properties of entropy, Gibbs energy, and Helmholtz energy to be discussed fully in Focus 3. The differentials of state functions are exact differentials. Hence, the mathematical properties of exact differentials can be used to draw far-reaching conclusions about the relations between physical properties and establish connections that were unexpected but turn out to be very significant.

One practical importance of these results is that the value of a property of interest can be obtained from the combination of measurements of other properties without actually having to measure the required property itself, the measurement of which might be very difficult.

I2.3 The change in reaction enthalpy with respect to temperature is described by Kirchhoff's Law, [2C.7a–55]. Assuming that the heat capacities are independent of temperature, the integrated form of Kirchoff's Law is applicable and is given by [2C.7d–56]

$$\Delta_r H^\ominus(T_2) = \Delta_r H^\ominus(T_1) + (T_2 - T_1)\Delta_r C_p^\ominus$$

If $\Delta_r C_p^\ominus$ is negative, then the reaction enthalpy will decrease with increasing temperature, whereas if $\Delta_r C_p^\ominus$ is positive, then the reaction enthalpy will increase with increasing temperature.

(a)

$$\begin{aligned}\Delta_r C_p^\ominus &= \sum_{\text{products}} \nu C_{p,\text{m}}^\ominus - \sum_{\text{reactants}} \nu C_{p,\text{m}}^\ominus \\ &= 2C_{p,\text{m}}^\ominus(\text{H}_2\text{O},\text{g}) - C_{p,\text{m}}^\ominus(\text{O}_2,\text{g}) - 2C_{p,\text{m}}^\ominus(\text{H}_2,\text{g}) \\ &= 2\times(4R) - \left(\tfrac{7}{2}R\right) - 2\times\left(\tfrac{7}{2}R\right) = -2\tfrac{1}{2}R\end{aligned}$$

$\Delta_r C_p^\ominus$ is negative, therefore the standard reaction enthalpy of reaction will decrease with increasing temperature.

(b)

$$\begin{aligned}\Delta_r C_p^\ominus &= C_{p,\text{m}}^\ominus(\text{CO}_2,\text{g}) + 2C_{p,\text{m}}^\ominus(\text{H}_2\text{O},\text{g}) - 2C_{p,\text{m}}^\ominus(\text{O}_2,\text{g}) - C_{p,\text{m}}^\ominus(\text{CH}_4,\text{g}) \\ &= \left(\tfrac{7}{2}R\right) + 2\times(4R) - 2\times\left(\tfrac{7}{2}R\right) - (4R) = +\tfrac{1}{2}R\end{aligned}$$

$\Delta_r C_p^\ominus$ is positive, therefore the standard reaction enthalpy of reaction will increase with increasing temperature.

(c)

$$\begin{aligned}\Delta_r C_p^\ominus &= 2C_{p,\text{m}}^\ominus(\text{NH}_3,\text{g}) - C_{p,\text{m}}^\ominus(\text{N}_2,\text{g}) - 3C_{p,\text{m}}^\ominus(\text{H}_2,\text{g}) \\ &= 2\times(4R) - \left(\tfrac{7}{2}R\right) - 3\times\left(\tfrac{7}{2}R\right) = -6R\end{aligned}$$

$\Delta_r C_p^\ominus$ is negative, therefore the standard reaction enthalpy of reaction will decrease with increasing temperature.

I2.5 The definition of C_V is given by [2A.14–43], $C_V = (\partial U/\partial T)_V$. It follows that

$$\left(\frac{\partial C_V}{\partial V}\right)_T = \left(\frac{\partial}{\partial V}\left(\frac{\partial U}{\partial T}\right)_V\right)_T$$

Using the property of partial differentials given in the problem gives

$$\left(\frac{\partial}{\partial V}\left(\frac{\partial U}{\partial T}\right)_V\right)_T = \left(\frac{\partial}{\partial T}\left(\frac{\partial U}{\partial V}\right)_T\right)_V$$

In Section 2D.2 on page 60 it is explained that, for a perfect gas, $(\partial U/\partial V)_T = 0$, therefore it follows from the above that $(\partial C_V/\partial V)_T = 0$.

I2.7 (a) The work of isothermal reversible expansion is given by the integral $w = \int -p\,\mathrm{d}V$, where p is the pressure of the gas [2A.8b–41]. The equation of state of a van der Waals gas is $p = nRT/(V - nb) - an^2/V^2$. The task is to evaluate the integral, between V_i and V_f, using this expression for the pressure

$$\begin{aligned} w &= -\int_{V_\mathrm{i}}^{V_\mathrm{f}} \frac{nRT}{V - nb} - \frac{an^2}{V^2}\,\mathrm{d}V \\ &= -nRT\ln(V - nb) - \frac{an^2}{V}\Bigg|_{V_\mathrm{i}}^{V_\mathrm{f}} \\ &= nRT\ln\frac{V_\mathrm{i} - nb}{V_\mathrm{f} - nb} - an^2\left(\frac{1}{V_\mathrm{f}} - \frac{1}{V_\mathrm{i}}\right) \end{aligned}$$

With the data given and using the values of the van der Waals constants for CO_2 from the *Resource section* the work of expansion evaluates as

$$\begin{aligned} w &= nRT\ln\frac{V_\mathrm{i} - nb}{V_\mathrm{f} - nb} - an^2\left(\frac{1}{V_\mathrm{f}} - \frac{1}{V_\mathrm{i}}\right) \\ &= (1.0\ \mathrm{mol}) \times (8.2057 \times 10^{-2}\ \mathrm{dm^3\ atm\ K^{-1}\ mol^{-1}}) \times (298\ \mathrm{K}) \times \\ &\quad \ln\frac{(1.0\ \mathrm{dm^3}) - (1\ \mathrm{mol}) \times (0.0429\ \mathrm{dm^3\ mol^{-1}})}{(3.0\ \mathrm{dm^3}) - (1\ \mathrm{mol}) \times (0.0429\ \mathrm{dm^3\ mol^{-1}})} \\ &\quad - (3.610\ \mathrm{dm^6\ atm\ mol^{-2}}) \times (1.0\ \mathrm{mol})^2 \times \left(\frac{1}{3.0\ \mathrm{dm^3}} - \frac{1}{1.0\ \mathrm{dm^3}}\right) \\ &= -25.1...\ \mathrm{dm^3\ atm} \end{aligned}$$

The work is converted into the more usual units of joules by changing the pressure and volume to SI units

$$w = (-25.1...\ \mathrm{dm^3\ atm}) \times \frac{10^{-3}\ \mathrm{m^3}}{1\ \mathrm{dm^3}} \times \frac{1.01325 \times 10^5\ \mathrm{Pa}}{1\ \mathrm{atm}} = \boxed{-2.6\ \mathrm{kJ}}$$

where $1\ \mathrm{Pa\ m^3} = 1\ \mathrm{N\ m^{-2}\ m^3} = 1\ \mathrm{N\ m} = 1\ \mathrm{J}$ is used.

For a perfect gas the work is $w = -nRT\ln V_\mathrm{f}/V_\mathrm{i} = -2.7\ \mathrm{kJ}$. The non-ideality of the gas results in a significant change in the work.

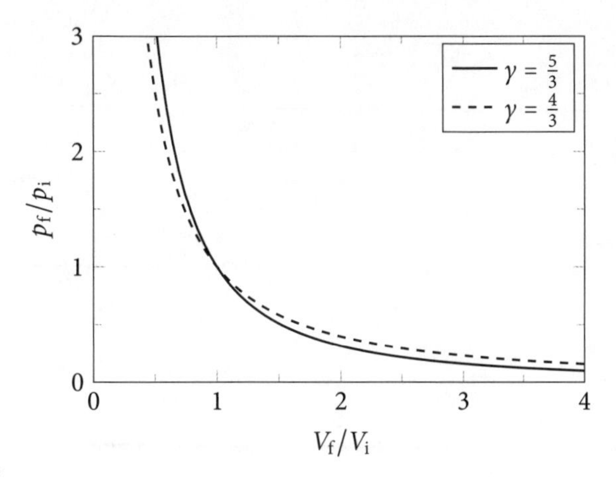

Figure 2.4

(b) For a reversible adiabatic expansion of a perfect gas the initial and final pressure and volume are related by [2E.3–68], $(p_f/p_i) = (V_i/V_f)^\gamma$, where $\gamma = C_{p,m}/C_{V,m}$. For a monatomic gas $\gamma = 5/3$ and for a gas of polyatomic molecules in which only translational and rotational modes contribute to the heat capacity, $\gamma = 4/3$.

To plot how pressure depends on volume it is sufficient to take $p_i = 1$ and $V_i = 1$, in whatever units are convenient, and then to plot $p_f/(1) = (1/V_f)^\gamma$. Such a plot is shown in Fig. 2.4.

As expected the adiabats cross when $V_f/V_i = 1$. The pressure–volume dependence, as evidenced by the slope of the adiabat, becomes weaker as the volume is increased. For the range of values that the parameter γ can reasonably take, it has a modest effect of the shape of the adiabat. Note that which of the two adiabats has the greater slope crosses over at $V_f/V_i = 1$.

The second and third laws

3A Entropy

Answers to discussion questions

D3A.1 The Second Law of thermodynamics states only that the total entropy of both the system (here, the molecules organizing themselves into cells) and the surroundings (here, the medium) must increase in a naturally occurring process. It does not state that entropy must increase in a portion of the universe that interacts with its surroundings. In this case, the cells grow by using chemical energy from their surroundings (the medium) and in the process the increase in the entropy of the medium outweighs the decrease in entropy of the system. Hence, the Second Law is not violated.

D3A.3 For a thorough discussion of the relationship between the various formulations of the Second Law, see Topic 3A. In summary, there are two equivalent statements of the Second Law that are based on directly observable processes:

(a) **The Kelvin statement**: No process is possible in which the sole result is the absorption of heat from a reservoir and its complete conversion into work.

(b) **The Clausius statement**: No process is possible in which the sole result is the transfer of energy from a cooler to a hotter body.

It can be shown that these statements are equivalent and that they both lead to the existence of a state function of the system called the entropy,S, and defined through the relation $\mathrm{d}S = \mathrm{d}q_{\mathrm{rev}}/T$. It can be shown that $\mathrm{d}S$ is an exact differential, that is, $\oint \mathrm{d}S = 0$. Hence, S is a property of all systems.

The definition given above leads to the Clausius inequality, that is $\mathrm{d}S \geq \mathrm{d}q/T$, where $\mathrm{d}q$ represents the actual heat associated with a real, necessarily irreversible, process. If the system is isolated from its surroundings, so that $\mathrm{d}q = 0$, the Clausius inequality then implies that $\mathrm{d}S \geq 0$, and it follows that in an isolated system the entropy cannot decrease when a spontaneous change occurs. Since the universe as a whole can be considered an isolated system, this implies that $\Delta S_{\mathrm{tot}} = \Delta S_{\mathrm{sys}} + \Delta S_{\mathrm{surr}} \geq 0$, which is another version of the Second Law.

Solutions to exercises

E3A.1(a) For the process to be spontaneous it must be irreversible and obey the Clausius inequality [3A.12–86] implying that $\Delta S_{\mathrm{tot}} = \Delta S + \Delta S_{\mathrm{sur}} > 0$. In this case,

$\Delta S_{tot} = 125\,\mathrm{J\,K^{-1}} + (-125\,\mathrm{J\,K^{-1}}) = 0$, thus the process is $\boxed{\text{not spontaneous}}$ in either direction and is at equilibrium.

E3A.2(a) The thermodynamic definition of entropy is [3A.1a–*80*], $\mathrm{d}S = \mathrm{d}q_{rev}/T$ or for a finite change at constant temperature $\Delta S = q_{rev}/T$. The transfer of heat is specified as being reversible, which can often be assumed for a large enough metal block, therefore $q_{rev} = 100$ kJ.

(i)

$$\Delta S = \frac{q_{rev}}{T} = \frac{100\ \mathrm{kJ}}{273.15\ \mathrm{K}} = 0.366\ \mathrm{kJ} = \boxed{+366\ \mathrm{J}}$$

(ii)

$$\Delta S = \frac{q_{rev}}{T} = \frac{100\ \mathrm{kJ}}{(273.15\ \mathrm{K} + 50\ \mathrm{K})} = 0.309\ \mathrm{kJ} = \boxed{+309\ \mathrm{J}}$$

E3A.3(a) As explained in Section 3A.2(a) on page 80 the change in entropy for an isothermal expansion of a gas is calculated using

$$\begin{aligned}\Delta S &= nR\ln\left(\frac{V_f}{V_i}\right) = \frac{m}{M}R\ln\left(\frac{V_f}{V_i}\right)\\ &= \left(\frac{15\ \mathrm{g}}{44.01\ \mathrm{g\,mol^{-1}}}\right)\times(8.3145\,\mathrm{J\,K^{-1}\,mol^{-1}})\times\ln\left(\frac{3.0\ \mathrm{dm^3}}{1.0\ \mathrm{dm^3}}\right)\\ &= \boxed{+3.1\,\mathrm{J\,K^{-1}}}.\end{aligned}$$

E3A.4(a) The change in entropy for an isothermal expansion of a gas is $\Delta S = nR\ln(V_f/V_i)$ as explained in Section 3A.2(a) on page 80. For a doubling of the volume $V_f/V_i = 2$.

(i) Isothermal reversible expansion

$$\Delta S = \left(\frac{14\ \mathrm{g}}{28.02\ \mathrm{g\,mol^{-1}}}\right)\times(8.3145\,\mathrm{J\,K^{-1}\,mol^{-1}})\times\ln(2) = \boxed{+2.9\,\mathrm{J\,K^{-1}}}.$$

Because the process is reversible $\Delta S_{tot} = \boxed{0}$.
Because $\Delta S_{tot} = \Delta S + \Delta S_{sur}$

$$\Delta S_{sur} = \Delta S_{tot} - \Delta S = \boxed{-2.9\,\mathrm{J\,K^{-1}}}.$$

(ii) Isothermal irreversible expansion against $p_{ex} = 0$ Because entropy is a state function and the initial and final states of the system are the same as in (a), ΔS is the same.

$$\Delta S = \boxed{+2.9\,\mathrm{J\,K^{-1}}}.$$

Expansion against an external pressure of 0 does no work, and for an isothermal process of an ideal gas $\Delta U = 0$. From the First Law if follows that $q = 0$ and therefore $\Delta S_{sur} = \boxed{0}$.

$$\Delta S_{tot} = \Delta S + \Delta S_{sur} = \boxed{+2.9\,\mathrm{J\,K^{-1}}}.$$

(iii) Adiabatic reversible expansion For an adiabatic expansion there is no heat flowing to or from the surroundings, thus $\Delta S_{sur} = \boxed{0}$. For a reversible process $\Delta S_{tot} = \boxed{0}$, therefore it follows that $\Delta S = \boxed{0}$ as well.

E3A.5(a) The efficiency is defined in [3A.7–84], $\eta = |w|/|q_h|$, and for a Carnot cycle efficiency is given by [3A.9–84], $\eta = 1 - (T_c/T_h)$. These two are combined and rearranged into an expression for the temperature of the cold sink

$$|w|/|q_h| = 1 - (T_c/T_h)$$

$$\text{hence} \quad T_c = \left(1 - \frac{|w|}{|q_h|}\right) \times T_h$$

$$= \left(1 - \frac{|3.00\ \text{kJ}|}{|-10.00\ \text{kJ}|}\right) \times (273\ \text{K}) = \boxed{191\ \text{K}}.$$

E3A.6(a) The efficiency of a Carnot cycle is given by [3A.9–84], $\eta = 1 - (T_c/T_h)$. Thus

$$\eta = 1 - \frac{T_c}{T_h} = 1 - \frac{(273.15\ \text{K} + 10\ \text{K})}{(273.15\ \text{K} + 100\ \text{K})} = 0.241 = \boxed{24.1\%}.$$

Note that the temperatures must be in kelvins.

Solutions to problems

P3A.1 (a) Isothermal reversible expansion

The work of a reversible isothermal expansion of an ideal gas is given by [2A.9–41], $w = -nRT\ln(V_f/V_i)$. Because at fixed temperature $p \propto (1/V)$ as given by Boyle's law, an equivalent expression is

$$w = -nRT\ln\left(\frac{p_i}{p_f}\right)$$

$$= -(1.00\ \text{mol}) \times (8.3145\ \text{J K}^{-1}\ \text{mol}^{-1})$$

$$\times (273.15\ \text{K} + 27\ \text{K}) \times \ln\left(\frac{3.00\ \text{atm}}{1.00\ \text{atm}}\right)$$

$$= -2.74 \times 10^3 ...\ \text{J} = \boxed{-2.74\ \text{kJ}}.$$

For an isothermal process of a perfect gas $\Delta U = \boxed{0}$ and $\Delta H = \boxed{0}$. The First Law is defined in [2A.2–38], $\Delta U = q + w$, hence

$$q = \Delta U - w = 0 - (-2.74...\ \text{kJ}) = \boxed{+2.74\ \text{kJ}}.$$

The heat transfer is reversible, therefore $q_{rev} = q$.

$$\Delta S = \frac{q_{rev}}{T} = \frac{2.74 \times 10^3 ...\ \text{J}}{273.15\ \text{K} + 27\ \text{K}}$$

$$= +9.13...\ \text{J K}^{-1} = \boxed{+9.13\ \text{J K}^{-1}}.$$

The process is reversible, therefore $\Delta S_{tot} = \boxed{0}$. Finally because $\Delta S_{tot} = \Delta S + \Delta S_{sur}$

$$\Delta S_{sur} = \Delta S_{tot} - \Delta S = 0 - (+9.13...\ \mathrm{J\,K^{-1}}) = \boxed{-9.13\ \mathrm{J\,K^{-1}}}.$$

(b) Isothermal expansion against $p_{ex} = 1.00$ atm The expansion work against a constant external pressure is given by [2A.6–40], $w = -p_{ex}(V_f - V_i)$. The volumes are written in terms of pressures by using the perfect gas law [1A.4–8], $pV = nRT$.

$$\begin{aligned} w &= -p_{ex}(V_f - V_i) \\ &= p_{ex}\left(\frac{nRT}{p_f} - \frac{nRT}{p_i}\right) = -nRT \times \left(\frac{p_{ex}}{p_f} - \frac{p_{ex}}{p_i}\right) \\ &= -(1.00\ \mathrm{mol}) \times (8.3145\ \mathrm{J\,K^{-1}\,mol^{-1}}) \\ &\quad \times (273.15\ \mathrm{K} + 27\ \mathrm{K}) \times \left(\frac{1.00\ \mathrm{atm}}{1.00\ \mathrm{atm}} - \frac{1.00\ \mathrm{atm}}{3.00\ \mathrm{atm}}\right) \\ &= -1.66... \times 10^3\ \mathrm{J} = \boxed{-1.66\ \mathrm{kJ}}. \end{aligned}$$

For an isothermal process in perfect gas $\Delta U = \boxed{0}$ and $\Delta H = \boxed{0}$. Using the First Law

$$q = \Delta U - w = 0 - (-1.66\ \mathrm{kJ}) = \boxed{+1.66\ \mathrm{kJ}}.$$

Because entropy is a state function and the initial and final states of the system are the same, the entropy change of the system is as in (a), $\Delta S = \boxed{+9.13\ \mathrm{J\,K^{-1}}}$.

The entropy change of the surroundings in terms of the heat of the surroundings, q_{sur}, is given by [3A.2b–81], $\Delta S_{sur} = q_{sur}/T$. This heat is simply the opposite of the heat of the system: $q_{sur} = -q$, therefore

$$\begin{aligned} \Delta S_{sur} &= \frac{q_{sur}}{T} = \frac{-q}{T} \\ &= \frac{-1.66... \times 10^3\ \mathrm{J}}{(273.15\ \mathrm{K} + 27\ \mathrm{K})} = -5.54...\ \mathrm{J\,K^{-1}} = \boxed{-5.54\ \mathrm{J\,K^{-1}}}. \end{aligned}$$

$$\begin{aligned} \Delta S_{tot} &= \Delta S + \Delta S_{sur} \\ &= (+9.13...\ \mathrm{J\,K^{-1}}) + (-5.53...\ \mathrm{J\,K^{-1}}) = \boxed{+3.59\ \mathrm{J\,K^{-1}}}. \end{aligned}$$

P3A.3 (a) After Stage 1 the volume doubles, thus $V_B = 2 \times V_A = 2 \times (1.00\ \mathrm{dm^3}) = \boxed{2.00\ \mathrm{dm^3}}$. Assuming $VT^{3/2}$ = constant for the adiabatic stages, the volume after Stage 2 is

$$\begin{aligned} V_C &= V_B \times \left(\frac{T_h}{T_c}\right)^{3/2} = (2.00\ \mathrm{dm^3}) \times \left(\frac{373\ \mathrm{K}}{273\ \mathrm{K}}\right)^{3/2} \\ &= 3.19...\ \mathrm{dm^3} = \boxed{3.19\ \mathrm{dm^3}}. \end{aligned}$$

(b) Again assuming $VT^{3/2}$ = constant for the adiabatic stage, the volume after Stage 3 can be related to the initial volume

$$V_D = V_A \times \left(\frac{T_h}{T_c}\right)^{3/2} = (1.00\ \text{dm}^3) \times \left(\frac{373\ \text{K}}{273\ \text{K}}\right)^{3/2}$$
$$= 1.59...\ \text{dm}^3 = \boxed{1.60\ \text{dm}^3}.$$

(c) As shown in Section 3A.3(a) on page 82 the heat transferred reversibly during an isothermal gas expansion is $q_{rev} = nRT \ln(V_f/V_i)$, thus the heats for Stage 1 and Stage 3 are, respectively

$$q_1 = q_h = nRT_h \ln\left(\frac{V_B}{V_A}\right)$$
$$= (0.100\ \text{mol}) \times (8.3145\ \text{J K}^{-1}\ \text{mol}^{-1}) \times (373\ \text{K}) \times \ln(2)$$
$$= +2.14... \times 10^2\ \text{J} = \boxed{+215\ \text{J}}.$$

$$q_3 = q_c = nRT_c \ln\left(\frac{V_D}{V_C}\right)$$
$$= (0.100\ \text{mol}) \times (8.3145\ \text{J K}^{-1}\ \text{mol}^{-1}) \times (273\ \text{K}) \times \ln\left(\frac{1.59...\ \text{dm}^3}{3.19...\ \text{dm}^3}\right)$$
$$= -1.57... \times 10^2\ \text{J} = \boxed{-157\ \text{J}}.$$

Because there is no heat exchange during adiabatic processes, rhe heat transfer for Stages 2 and 4 are $q_2 = \boxed{0}$ and $q_4 = \boxed{0}$, respectively.

(d) At the beginning and end of the cycle the temperature is the same. Because the working substance is a perfect gas, $\Delta U = 0$ over the cycle. The First Law [2A.2–*38*], $\Delta U = w + q$, therefore implies that $w = -q$, that is, the net heat over the cycle is converted to work. This net heat is the difference between that extracted from the hot source and deposited into the cold sink.

(e) The efficiency is defined in [3A.7–*84*], $\eta = |w|/|q_h|$. As has been explained, $|w|$ is the net heat.

$$|w| = |q_h| - |q_c| = |+2.14... \times 10^2\ \text{J}| - |-1.57... \times 10^2\ \text{J}|$$
$$= +5.7... \times 10^1\ \text{J} = \boxed{+58\ \text{J}}.$$

hence $$\eta = \frac{|w|}{|q_h|} = \frac{|+5.7... \times 10^1\ \text{J}|}{|+2.14... \times 10^2\ \text{J}|} = 0.268... = \boxed{27\%}.$$

(f) The Carnot efficiency is given by [3A.9–*84*],

$$\eta = 1 - \frac{T_c}{T_h} = 1 - \frac{273\ \text{K}}{373\ \text{K}} = 0.268 = \boxed{26.8\%}.$$

the result is the same as the above (the difference is due to the use of fewer significant figures in the previous calculation).

Using the values of the heat transfer calculated above in equation [3A.6–84] gives

$$\frac{q_c}{T_h} + \frac{q_c}{T_c} = \frac{214...\,\text{J}}{373\,\text{K}} + \frac{-157...\,\text{J}}{273\,\text{K}}$$
$$= \boxed{0.0\,\text{J}}.$$

the result is zero, as expected from a Carnot cycle.

P3A.5 (a) Consider a process in which heat dq_c is extracted from the cold source at temperature T_c, and heat dq_h is discarded into the hot sink at temperature T_h. The overall entropy change of such process is

$$dS = \frac{dq_c}{T_c} + \frac{dq_h}{T_h}$$

Assume that $dq_c = -dq$ and $dq_h = +dq$, where dq is a positive quantity. It follows that

$$dS = \frac{+dq}{T_h} + \frac{-dq}{T_c} = dq \times \left(\frac{1}{T_h} - \frac{1}{T_c}\right)$$

Because $T_h > T_c$, the term in parentheses is negative, therefore dS is negative. The process is therefore not spontaneous and not allowed by the Second Law. If work is done on the engine, $|dq_h|$ will become greater than $|dq_c|$ and eventually dS will be greater than zero.

(b) Assuming $q_c = -|q|$ and $q_h = |q| + |w|$ the overall change in entropy is

$$\Delta S = \frac{-|q|}{T_c} + \frac{|q| + |w|}{T_h}$$

For the process to be permissible by the Second Law the Clausius inequality defined in [3A.12–86], $dS \geq 0$, must be satisfied. Therefore

$$\frac{-|q|}{T_c} + \frac{|q| + |w|}{T_h} \geq 0$$

which implies

$$|w| \geq |q| \times \left(\frac{T_h}{T_c} - 1\right) = \boxed{|q| \times \left(\frac{T_h}{T_c} - 1\right)}.$$

P3A.7 Suppose two adiabatic paths intersect at point A as shown in the figure. Two remote points corresponding to the same temperature on each adiabat, A and B, are then connected by an isothermal path forming a cycle.

Consider energy changes for each Stage of the cycle. Stage 1 ($A \rightarrow B$) is adiabatic and, thus, no heat exchange takes place $q_1 = 0$. Therefore, the total change in internal energy is $\Delta U_1 = w_1 + q_1 = w_1$. Stage 2 ($B \rightarrow C$) is an isothermal change and assuming that the system energy is a function of temperature only (e.g.

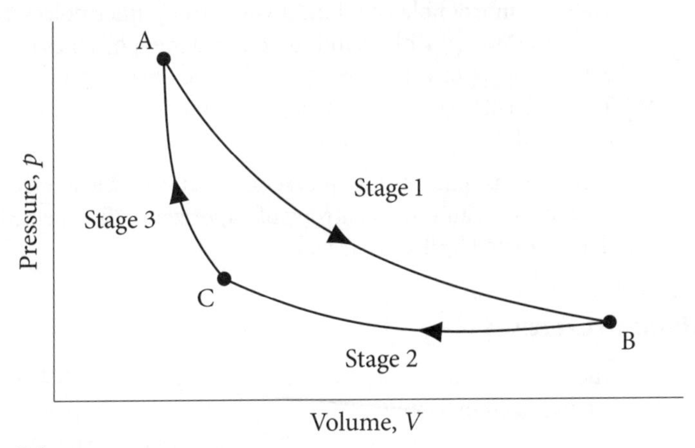

Figure 3.1

ideal gas): $\Delta U_2 = w_2 + q_2 = 0$. Stage 3 ($C \rightarrow A$) is again adiabatic, $q_3 = 0$, with $\Delta U_3 = w_3 + q_3 = w_3$. Because the system energy is a function of temperature only, $U_B = U_C$ and, thus

$$\Delta U_3 = U_A - U_C = U_A - U_B = -\Delta U_1$$

This implies that $w_1 = -w_3$.

Because internal energy is a state function and the cycle is closed:

$$\begin{aligned} U_{\text{cycle}} &= w_{\text{cycle}} + q_{\text{cycle}} = 0 \\ &= \Delta U_1 + \Delta U_2 + \Delta U_3 \end{aligned}$$

Finally, analyse the net work done, $w_{\text{cycle}} = w_1 + w_2 + w_3 = w_2$, and the net heat absorbed, $q_{\text{cycle}} = q_1 + q_2 + q_3 = q_2$, over the cycle. It is apparent that the sole result of the process is the absorption of heat q_2 and its convertion to work w_2, which directly contradicts the statement of the Second Law by Kelvin, unless the $q_2 = w_2 = 0$, i.e. points B and C are the same and correspond to the same path. Therefore, no two such adiabatic paths exist.

3B Entropy changes accompanying specific processes

Answer to discussion question

D3B.1 The explanation of Trouton's rule is that a comparable change in volume is expected whenever any unstructured liquid forms a vapour; accompanying this will be a comparable change in the number of accessible microstates. Hence, all unstructured liquids can be expected to have similar entropies of vaporization. Liquids that show significant deviations from Trouton's rule do so on account of strong molecular interactions that restrict molecular motion. As a result there is a greater dispersal of matter and energy when such liquids vaporize.

Water is an example of a liquid with strong intermolecular interactions (hydrogen bonding) which tend to organize the molecules in the liquid, hence its entropy of vaporization is expected to be greater than the value predicted by Trouton's rule. The same is true for ethanol, which is also hydrogen bonded in the liquid.

Mercury has quite strong interactions between the atoms, as evidenced by its cohesiveness, and so its entropy of vaporization is expected to be greater than that predicted by Trouton's rule.

Solutions to exercises

E3B.1(a) The entropy change of a phase transition is given by [3B.4–89], $\Delta_{trs}S = \Delta_{trs}H/T_{trs}$. As discussed in Section 3B.2 on page 89 because there is no hydrogen bonding in liquid benzene it is safe to apply Trouton's rule. That is $\Delta_{vap}S^{\ominus} = +85\,\mathrm{J\,K^{-1}\,mol^{-1}}$. It follows that

$$\begin{aligned}\Delta_{vap}H^{\ominus} &= T_b \times \Delta_{vap}S^{\ominus}\\ &= (273.15\,\mathrm{K} + 80.1\,\mathrm{K}) \times (+85\,\mathrm{J\,K^{-1}\,mol^{-1}})\\ &= 3.00... \times 10^4\,\mathrm{J\,mol^{-1}} = \boxed{+30\,\mathrm{kJ\,mol^{-1}}}.\end{aligned}$$

E3B.2(a) (i) The entropy change of a phase transition is given by [3B.4–89], $\Delta_{trs}S = \Delta_{trs}H/T_{trs}$. For vaporisation this becomes

$$\begin{aligned}\Delta_{vap}S^{\ominus} &= \frac{\Delta_{vap}H^{\ominus}}{T_b} = \frac{+29.4\times10^3\,\mathrm{J\,mol^{-1}}}{334.88\,\mathrm{K}}\\ &= \boxed{+87.8\,\mathrm{J\,K^{-1}\,mol^{-1}}}.\end{aligned}$$

(ii) Because the system at the transition temperature is at equilibrium, $\Delta S_{tot} = 0$, thus

$$\Delta S_{sur} = -\Delta_{vap}S^{\ominus} = \boxed{-87.8\,\mathrm{J\,K^{-1}\,mol^{-1}}}.$$

E3B.3(a) The change in entropy with temperature is given by [3B.6–90],

$$\Delta S = S(T_f) - S(T_i) = \int_{T_i}^{T_f} C_p \frac{dT}{T}$$

Assuming that C_p is constant in the temperature range T_i to T_f, this becomes $\Delta S = C_p \ln(T_f/T_i)$ as detailed in Section 3B.3 on page 90. Thus, the increase in the molar entropy of oxygen gas is

$$\begin{aligned}\Delta S_m &= S_m(348\,\mathrm{K}) - S_m(298\,\mathrm{K}) = (29.355\,\mathrm{J\,K^{-1}\,mol^{-1}}) \times \ln\left(\frac{348\,\mathrm{K}}{298\,\mathrm{K}}\right)\\ &= \boxed{+4.55\,\mathrm{J\,K^{-1}\,mol^{-1}}}.\end{aligned}$$

E3B.4(a) As explained in Section 3B.3 on page 90 the temperature variation of the entropy at constant volume is given by

$$\Delta S = S(T_f) - S(T_i) = \int_{T_i}^{T_f} C_V \frac{dT}{T}$$

Assuming that $C_V = \frac{3}{2}R$, the ideal gas limit, for the temperature range of interest, the molar entropy at 500 K is given by

$$\begin{aligned} S_m(500\ \mathrm{K}) &= S_m(298\ \mathrm{K}) + \int_{298\ \mathrm{K}}^{500\ \mathrm{K}} \tfrac{3}{2}R\frac{dT}{T} \\ &= S_m(298\ \mathrm{K}) + \tfrac{3}{2}R \times \ln\left(\frac{500\ \mathrm{K}}{298\ \mathrm{K}}\right) \\ &= (146.22\ \mathrm{J\,K^{-1}\,mol^{-1}}) \\ &\quad + (\tfrac{3}{2} \times 8.3145\ \mathrm{J\,K^{-1}\,mol^{-1}}) \times \ln\left(\frac{500\ \mathrm{K}}{298\ \mathrm{K}}\right) \\ &= \boxed{153\ \mathrm{J\,K^{-1}\,mol^{-1}}}. \end{aligned}$$

E3B.5(a) Two identical blocks must come to their average temperature. Therefore the final temperature is

$$T_f = \tfrac{1}{2}(T_1 + T_2) = \tfrac{1}{2} \times (50\ ^\circ\mathrm{C} + 0\ ^\circ\mathrm{C}) = 25\ ^\circ\mathrm{C} = \boxed{298\ \mathrm{K}}.$$

Although the above result may seem self-evident, the more detailed explaination is as follows. The heat capacity at constant volume is defined in [2A.14–43], $C_V = (\partial U/\partial T)_V$. As shown in Section 2A.4(b) on page 43, if the heat capacity is constant, the internal energy changes linearly with the change in temperature. That is $\Delta U = C_V\Delta T = C_V(T_f - T_i)$. For the two blocks at the initial temperatures of T_1 and T_2, the change in internal energy to reach the final temperature T_f is $\Delta U_1 = C_{V,1}(T_f - T_1)$ and $\Delta U_2 = C_{V,2}(T_f - T_2)$, respectively. The blocks of metal are made of the same substance and are of the same size, therefore $C_{V,1} = C_{V,2} = C_V$. Because the system is isolated the total change in internal energy is $\Delta U = \Delta U_1 + \Delta U_2 = 0$. This means that $\Delta U = C_V((T_f - T_1) - (T_f - T_2)) = C_V \times (2T_f - (T_1 + T_2)) = 0$, which implies that the final temperature is $T_f = \frac{1}{2}(T_1 + T_2)$, as stated above.

The temperature variation of the entropy at constant volume is given by [3B.7–90], $\Delta S = C_V \ln(T_f/T_i)$, with C_p replaced by C_V. Expressed with the specific heat $C_{V,s} = C_V/m$ it becomes

$$\Delta S = mC_{V,s}\ln\left(\frac{T_f}{T_i}\right).$$

Note that for a solid the internal energy does not change significantly with the volume or pressure, thus it can be assumed that $C_V = C_p = C$. The entropy change for each block is found using this expression

$$\begin{aligned}\Delta S_1 &= mC_{V,s}\ln\left(\frac{T_f}{T_1}\right)\\ &= (1.00\times10^3\ \text{g})\times(0.385\ \text{J K}^{-1}\ \text{g}^{-1})\times\ln\left(\frac{298\ \text{K}}{50\ \text{K}+273.15\ \text{K}}\right)\\ &= -31.0...\ \text{J K}^{-1} = \boxed{-31.0\ \text{J K}^{-1}}.\\ \Delta S_2 &= mC_{V,s}\ln\left(\frac{T_f}{T_2}\right)\\ &= (1.00\times10^3\ \text{g})\times(0.385\ \text{J K}^{-1}\ \text{g}^{-1})\times\ln\left(\frac{298\ \text{K}}{273.15\ \text{K}}\right)\\ &= 33.7...\ \text{J K}^{-1} = \boxed{+33.7\ \text{J K}^{-1}}.\end{aligned}$$

The total change in entropy is

$$\begin{aligned}\Delta S_{\text{tot}} &= \Delta S_1 + \Delta S_2 = (-31.0...\ \text{J K}^{-1}) + (33.7...\ \text{J K}^{-1})\\ &= 27.2...\ \text{J K}^{-1} = \boxed{+2.7\ \text{J K}^{-1}}.\end{aligned}$$

Because $\Delta S_{\text{tot}} > 0$ the process is spontaneous, in accord with experience.

E3B.6(a) Because entropy is a state function, ΔS between the initial and final states is the same irrespective of the path taken. Thus the overall process can be broken down into steps that are easier to evaluate. First consider heating the initial system at constant pressure to the final temperature. The variation of entropy with temperature at constant pressure is given by [3B.7–90], $S(T_f) = S(T_i) + C_p \ln(T_f/T_i)$. Thus the change in entropy, $\Delta S = S(T_f) - S(T_i)$, of this step is

$$\Delta S_1 = C_p \ln\left(\frac{T_f}{T_i}\right) = nC_{p,m}\ln\left(\frac{T_f}{T_i}\right)$$

Next consider an isothermal change in pressure. As explained in Section 3A.2(a) on page 80 the change in entropy of an isothermal expansion of an ideal gas is given by $\Delta S = nR\ln(V_f/V_i)$. Because for a fixed amount of gas at fixed temperature $p \propto (1/V)$ an equivalent expression for this entropy change is

$$\Delta S_2 = nR\ln\left(\frac{p_i}{p_f}\right)$$

Therefore the overall entropy change for the system is

$$\begin{aligned}\Delta S &= \Delta S_1 + \Delta S_2 = nC_{p,m}\ln\left(\frac{T_f}{T_i}\right) + nR\ln\left(\frac{p_i}{p_f}\right)\\ &= (3.00\ \text{mol})\times\left(\tfrac{5}{2}\times 8.3145\ \text{J K}^{-1}\ \text{mol}^{-1}\right)\times\ln\left(\frac{273.15\ \text{K}+125\ \text{K}}{273.15\ \text{K}+25\ \text{K}}\right)\\ &\quad + (3.00\ \text{mol})\times(8.3145\ \text{J K}^{-1}\ \text{mol}^{-1})\times\ln\left(\frac{1.00\ \text{atm}}{5.00\ \text{atm}}\right)\\ &= (+18.0...\ \text{J K}^{-1}) + (-40.1...\ \text{J K}^{-1}) = \boxed{-22.1\ \text{J K}^{-1}}.\end{aligned}$$

E3B.7(a) Because entropy is a state function, ΔS between the initial and final states is the same irrespective of the path taken. Thus the overall process can be broken down into steps that are easier to evaluate. First consider heating the ice at constant pressure from the initial temperature to the melting point, T_m. The variation of entropy with temperature at constant pressure is given by [3B.7–90], $S(T_f) = S(T_i) + C_p \ln(T_f/T_i)$. Thus the change in entropy, $\Delta S = S(T_f) - S(T_i)$, for this step is

$$\Delta S_1 = C_p \ln\left(\frac{T_m}{T_i}\right) = nC_{p,m}(H_2O(s)) \ln\left(\frac{T_m}{T_i}\right)$$

Next consider the phase transition from solid to liquid at the melting temperature. The entropy change of a phase transition is given by [3B.4–89], $\Delta_{trs}S = \Delta_{trs}H/T_{trs}$, thus

$$\Delta S_2 = n\frac{\Delta_{fus}H_m^{\ominus}}{T_m}$$

Then the liquid is heated to the boiling temperature, T_b. In analogy to the first step

$$\Delta S_3 = nC_{p,m}(H_2O(l)) \ln\left(\frac{T_b}{T_m}\right)$$

The next phase transition is from liquid to gas

$$\Delta S_4 = n\frac{\Delta_{vap}H_m^{\ominus}}{T_b}$$

Finally, the vapour is heated from T_b to T_f

$$\Delta S_5 = nC_{p,m}(H_2O(g)) \ln\left(\frac{T_f}{T_b}\right)$$

Therefore the overall entropy change for the system is

$$\Delta S/n = \Delta S_1 + \Delta S_2 + \Delta S_3 + \Delta S_4 + \Delta S_5$$

$$= C_{p,m}(H_2O(s)) \ln\left(\frac{T_m}{T_i}\right) + \frac{\Delta_{fus}H_m^\ominus}{T_m} + C_{p,m}(H_2O(l)) \ln\left(\frac{T_b}{T_m}\right)$$

$$+ \frac{\Delta_{vap}H_m^\ominus}{T_b} + C_{p,m}(H_2O(g)) \ln\left(\frac{T_f}{T_b}\right)$$

$$= (37.6\,J\,K^{-1}\,mol^{-1}) \times \ln\left(\frac{273.15\,K}{273.15\,K - 10.0\,K}\right)$$

$$+ \frac{6.01 \times 10^3\,J\,mol^{-1}}{273.15\,K}$$

$$+ (75.3\,J\,K^{-1}\,mol^{-1}) \times \ln\left(\frac{273.15\,K + 100.0\,K}{273.15\,K}\right)$$

$$+ \frac{40.7 \times 10^3\,J\,mol^{-1}}{273.15\,K + 100.0\,K}$$

$$+ (33.6\,J\,K^{-1}\,mol^{-1}) \times \ln\left(\frac{273.15\,K + 115.0\,K}{273.15\,K + 100.0\,K}\right)$$

$$= (+1.40...\,J\,K^{-1}\,mol^{-1}) + (+22.0...\,J\,K^{-1}\,mol^{-1})$$

$$+ (+23.4...\,J\,K^{-1}\,mol^{-1}) + (+1.09... \times 10^2\,J\,K^{-1}\,mol^{-1})$$

$$+ (+1.32...\,J\,K^{-1}\,mol^{-1})$$

$$= +1.57... \times 10^2\,J\,K^{-1}\,mol^{-1}$$

Hence

$$\Delta S = \frac{10.0\,g}{18.02\,g\,mol^{-1}} \times (+1.57... \times 10^2\,J\,K^{-1}) = \boxed{+87.3\,J\,K^{-1}}.$$

Solutions to problems

P3B.1 Because entropy is a state function, ΔS between the initial and final states is the same irrespective of the path taken. Thus the overall process can be broken down into steps that are easier to evaluate.

First consider heating the water at constant pressure from the initial temperature T to the melting point. The variation of the entropy with temperature at constant pressure is given by [3B.7–*90*], $S(T_f) = S(T_i) + C_p \ln(T_f/T_i)$. Thus the change in entropy for this step is

$$\Delta S_1 = C_p \ln\left(\frac{T_m}{T}\right) = nC_{p,m}(H_2O(l)) \ln\left(\frac{T_m}{T}\right)$$

Next consider the phase transition from liquid to solid at the melting temperature; note that freezing is just the opposite of fusion, thus $\Delta H_2 = n(-\Delta_{fus}H_m^\ominus)$. The entropy change of a phase transition is given by [3B.4–*89*], $\Delta_{trs}S = \Delta_{trs}H/T_{trs}$, thus

$$\Delta S_2 = \frac{\Delta H_2}{T_m} = n\frac{-\Delta_{fus}H_m^\ominus}{T_m}$$

The ice is then cooled to the final temperature, T. Similarly to the first step

$$\Delta S_3 = nC_{p,m}(H_2O(s))\ln\left(\frac{T}{T_m}\right)$$

Therefore the overall entropy change for the system is

$$\begin{aligned}\Delta S &= \Delta S_1 + \Delta S_2 + \Delta S_3\\ &= nC_{p,m}(H_2O(l))\ln\left(\frac{T_m}{T}\right) + n\frac{-\Delta_{fus}H_m^{\ominus}}{T_m} + nC_{p,m}(H_2O(s))\ln\left(\frac{T}{T_m}\right)\\ &= (1.00\ \text{mol}) \times (75.3\ \text{J K}^{-1}\ \text{mol}^{-1}) \times \ln\left(\frac{273.15\ \text{K}}{273.15\ \text{K} - 5.00\ \text{K}}\right)\\ &\quad + (1.00\ \text{mol}) \times \frac{-6.01 \times 10^3\ \text{J mol}^{-1}}{273.15\ \text{K}}\\ &\quad + (1.00\ \text{mol}) \times (37.6\ \text{J K}^{-1}\ \text{mol}^{-1}) \times \ln\left(\frac{273.15\ \text{K} - 5.00\ \text{K}}{273.15\ \text{K}}\right)\\ &= (+1.39...\ \text{J K}^{-1}) + (-22.0...\ \text{J K}^{-1}) + (-0.694...\ \text{J K}^{-1})\\ &= -21.3...\ \text{J K}^{-1} = \boxed{-21.3\ \text{J K}^{-1}}.\end{aligned}$$

Consider enthalphy change for the same path. The variation of enthalpy with temperature at constant pressure is given by [2B.6b–49], $\Delta H = C_p\Delta T$. Thus for the first and third steps, respectively

$$\Delta H_1 = nC_{p,m}(H_2O(l))(T_m - T) \quad \text{and} \quad \Delta H_3 = nC_{p,m}(H_2O(s))(T - T_m)$$

Therefore the overall enthalpy change for the system is

$$\begin{aligned}\Delta H &= \Delta H_1 + \Delta H_2 + \Delta H_3\\ &= nC_{p,m}(H_2O(l))(T_m - T) + n(-\Delta_{fus}H_m^{\ominus}) + nC_{p,m}(H_2O(s))(T - T_m)\\ &= (1.00\ \text{mol}) \times (75.3\ \text{J K}^{-1}\ \text{mol}^{-1}) \times (+5.00\ \text{K})\\ &\quad + (1.00\ \text{mol}) \times (-6.01 \times 10^3\ \text{J mol}^{-1})\\ &\quad + (1.00\ \text{mol}) \times (37.6\ \text{J K}^{-1}\ \text{mol}^{-1}) \times (-5.00\ \text{K})\\ &= (+3.76... \times 10^2\ \text{J}) + (-6.01... \times 10^3\ \text{J}) + (-1.88... \times 10^2\ \text{J})\\ &= -5.82... \times 10^3\ \text{J}\end{aligned}$$

At constant pressure the heat released by the system is the enthalpy change of the system, $q = \Delta H$. Because $q_{sur} = -q$, the entropy change of the surroundings is

$$\begin{aligned}\Delta S_{sur} &= \frac{-q}{T} = \frac{-(-5.82... \times 10^3\ \text{J})}{273.15\ \text{K} - 5.00\ \text{K}}\\ &= +21.7...\ \text{J K}^{-1} = \boxed{+21.7\ \text{J K}^{-1}}.\end{aligned}$$

Therefore the total entropy change is

$$\begin{aligned}\Delta S_{tot} &= \Delta S + \Delta S_{sur} = (-21.3...\ \text{J K}^{-1}) + (+21.7...\ \text{J K}^{-1})\\ &= +0.403...\ \text{J K}^{-1} = \boxed{+0.4\ \text{J K}^{-1}}.\end{aligned}$$

Because the total entropy change is positive, the Second Law implies that the process is $\boxed{\text{spontaneous}}$.

A similar method is used to find the entropy change when the liquid evaporates at T_2. Consider heating the liquid to the boiling temperature T_b, then the phase transition taking place, followed by cooling of the gas back to the temperature T_2. The entropy changes are calculated in an analagous way

$$\begin{aligned}
\Delta S &= \Delta S_1 + \Delta S_2 + \Delta S_3 \\
&= nC_{p,m}(H_2O(l)) \ln\left(\frac{T_b}{T_2}\right) + n\frac{\Delta_{vap}H_m^{\ominus}}{T_b} + nC_{p,m}(H_2O(g)) \ln\left(\frac{T_2}{T_b}\right) \\
&= (1.00\ \text{mol}) \times (75.3\ \text{J K}^{-1}\ \text{mol}^{-1}) \ln\left(\frac{273.15\ \text{K} + 100\ \text{K}}{273.15\ \text{K} + 95.0\ \text{K}}\right) \\
&\quad + (1.00\ \text{mol}) \times \frac{4.07 \times 10^4\ \text{J mol}^{-1}}{273.15\ \text{K} + 100\ \text{K}} \\
&\quad + (1.00\ \text{mol}) \times (33.6\ \text{J K}^{-1}\ \text{mol}^{-1}) \ln\left(\frac{273.15\ \text{K} + 95.0\ \text{K}}{273.15\ \text{K} + 100\ \text{K}}\right) \\
&= (+1.01...\ \text{J K}^{-1}) + (+1.09... \times 10^2\ \text{J K}^{-1}) + (-0.453...\ \text{J K}^{-1}) \\
&= +1.09... \times 10^2\ \text{J K}^{-1} = \boxed{+110\ \text{J K}^{-1}}.
\end{aligned}$$

$$\begin{aligned}
\Delta S_{sur} &= \frac{-\Delta H}{T_2} = -\frac{1}{T_2} \times (\Delta H_1 + \Delta H_2 + \Delta H_3) \\
&= -\left(nC_{p,m}(H_2O(l))\frac{T_b - T_2}{T_2} + n\frac{-\Delta_{vap}H_m^{\ominus}}{T_2} + nC_{p,m}(H_2O(g))\frac{T_2 - T_b}{T_2}\right) \\
&= -(1.00\ \text{mol}) \times (75.3\ \text{J K}^{-1}\ \text{mol}^{-1})\frac{5.00\ \text{K}}{273.15\ \text{K} + 95.0\ \text{K}} \\
&\quad - (1.00\ \text{mol}) \times \frac{4.07 \times 10^4\ \text{J mol}^{-1}}{273.15\ \text{K} + 95.0\ \text{K}} \\
&\quad - (1.00\ \text{mol}) \times (33.6\ \text{J K}^{-1}\ \text{mol}^{-1}) \times \frac{-5.00\ \text{K}}{273.15\ \text{K} + 95.0\ \text{K}} \\
&= -(+1.02...\ \text{J K}^{-1}) - (+1.10... \times 10^2\ \text{J K}^{-1}) - (-0.456...\ \text{J K}^{-1}) \\
&= -1.11... \times 10^2\ \text{J K}^{-1} = \boxed{-111\ \text{J K}^{-1}}.
\end{aligned}$$

Therefore the total entropy change is

$$\begin{aligned}
\Delta S_{tot} &= \Delta S + \Delta S_{sur} = (+1.09... \times 10^2\ \text{J K}^{-1}) + (-1.11... \times 10^2\ \text{J K}^{-1}) \\
&= -1.48...\ \text{J K}^{-1} = \boxed{-1.5\ \text{J K}^{-1}}.
\end{aligned}$$

Because the change in the entropy is negative, the Second Law implies that the process is $\boxed{\text{not spontaneous}}$.

P3B.3 Consider heating trichloromethane at constant pressure from the initial to final temperatures. The variation of the entropy with temperature is given by

[3B.6–90], $S(T_f) = S(T_i) + \int_{T_i}^{T_f}(C_p/T)\mathrm{d}T$. The contant-pressure molar heat capacity is given as a function of temperature of a form $C_{p,m} = a + bT$, with $a = +91.47\,\mathrm{J\,K^{-1}\,mol^{-1}}$ and $b = +7.5\times10^{-2}\,\mathrm{J\,K^{-2}\,mol^{-1}}$. Thus the change in molar entropy, $\Delta S_m = S_m(T_f) - S_m(T_i)$, of this process is

$$\begin{aligned}\Delta S_m &= \int_{T_i}^{T_f}(C_{p,m}/T)\mathrm{d}T = \int_{T_i}^{T_f}\frac{a+bT}{T}\mathrm{d}T\\ &= a\times\int_{T_i}^{T_f}\frac{1}{T}\mathrm{d}T + b\times\int_{T_i}^{T_f}\mathrm{d}T\\ &= a\times\ln\left(\frac{T_f}{T_i}\right) + b\times(T_f - T_i)\\ &= (+91.47\,\mathrm{J\,K^{-1}\,mol^{-1}})\times\ln\left(\frac{300\,\mathrm{K}}{273\,\mathrm{K}}\right)\\ &\quad + (+7.5\times10^{-2}\,\mathrm{J\,K^{-2}\,mol^{-1}})\times(300\,\mathrm{K} - 273\,\mathrm{K})\\ &= (+8.62...\,\mathrm{J\,K^{-1}\,mol^{-1}}) + (+2.02...\,\mathrm{J\,K^{-1}\,mol^{-1}})\\ &= \boxed{+10.7\,\mathrm{J\,K^{-1}\,mol^{-1}}}.\end{aligned}$$

P3B.5 Two identical blocks must come to their average temperature. Therefore the final temperature is

$$T = \tfrac{1}{2}(T_c + T_h)$$

Although the above result may seem self-evident, the more detailed explaination is as follows. The heat capacity at constant volume is defined in [2A.14–43], $C_V = (\partial U/\partial T)_V$. As shown in Section 2A.4(b) on page 43, if the heat capacity is constant, the internal energy changes linearly with the change in temperature. That is $\Delta U = C_V\Delta T = C_V(T_f - T_i)$. For the two blocks at the initial temperatures of T_c and T_h, the change in internal energy to reach the final temperature T is $\Delta U_c = C_{V,c}(T - T_c)$ and $\Delta U_h = C_{V,h}(T - T_h)$, respectively. The blocks of metal are made of the same substance and are of the same size, therefore $C_{V,c} = C_{V,h} = C_V$. Note that for a given solid the internal energy does not change significantly on the volume or pressure, thus it can be assumed that $C_V = C_p$. Assuming the system is isolated the total change in internal energy is $\Delta U = \Delta U_c + \Delta U_h = 0$. This means that $\Delta U = C_p\left((T - T_c) - (T - T_h)\right) = C_p\times(2T - (T_1 + T_2)) = 0$, which implies that the final temperature is $T = \frac{1}{2}(T_c + T_h)$, as stated above.

At constant pressure the temperature dependence of the entropy is given by [3B.7–90],

$$\Delta S = nC_{p,m}\ln\left(\frac{T_f}{T_i}\right)$$

Therefore for the two blocks

$$\Delta S_c = nC_{p,m}\ln\left(\frac{T}{T_c}\right) \quad\text{and}\quad \Delta S_h = nC_{p,m}\ln\left(\frac{T}{T_h}\right)$$

The total change in entropy is

$$\begin{aligned}
\Delta S_{\text{tot}} &= \Delta S_c + \Delta S_h \\
&= nC_{p,m} \ln\left(\frac{T}{T_c}\right) + nC_{p,m} \ln\left(\frac{T}{T_h}\right) \\
&= nC_{p,m} \times \ln\left(\frac{T^2}{T_c \times T_h}\right) \\
&= \frac{m}{M} C_{p,m} \times \ln\left(\frac{\left[\frac{1}{2}(T_c + T_h)\right]^2}{T_c \times T_h}\right) \\
&= \boxed{\frac{m}{M} C_{p,m} \ln\left(\frac{(T_c + T_h)^2}{4(T_c \times T_h)}\right)}.
\end{aligned}$$

where m is the mass of the block and M is the molar mass.

In the case given

$$\begin{aligned}
\Delta S_{\text{tot}} &= \frac{500\text{ g}}{63.55\text{ g mol}^{-1}} \times (24.4\text{ J K}^{-1}\text{ mol}^{-1}) \times \ln\left(\frac{(250\text{ K} + 500\text{ K})^2}{4 \times (250\text{ K} \times 500\text{ K})}\right) \\
&= \boxed{+22.6\text{ J K}^{-1}}.
\end{aligned}$$

P3B.7 The heat produced by the resistor over a time period Δt is q = power $\times\ \Delta t = IV\Delta t = I^2R\Delta t$, where the last expression was obtained using Ohm's law, $V = IR$. Note that care is needed handling the units. From the inside of the front cover of the textbook use $(1\text{ A}) \equiv (1\text{ Cs}^{-1})$ and $(1\text{ V}) \equiv (1\text{ JC}^{-1})$, so that $(1\ \Omega) \equiv (1\text{ JsC}^{-2})$. Therefore the units of the final expression for the heat are as expected

$$\text{A}^2 \times \Omega \times \text{s} \equiv (\text{C}^2\text{s}^{-2}) \times (\text{JsC}^{-2}) \times (\text{s}) \equiv \text{J}$$

Assuming that all the heat is absorbed by the large metal block at constant pressure, this heat is the change of enthalpy of the system, $\Delta H = q$. The enthalpy change on heating is given by [2B.6b–49], $\Delta H = C_p\Delta T$. This is rearranged to give an expression for a temperature change

$$\Delta T = \frac{\Delta H}{C_p} = \frac{q}{C_p} = \frac{I^2R\Delta t}{C_p} = \frac{I^2R\Delta t}{(m/M)C_{p,m}}$$

where m is the mass, M the molar mass and $C_{p,m}$ the molar heat capacity. Thus the final temperature of the metal block is

$$\begin{aligned}
T_f &= T_i + \Delta T = T_i + \frac{I^2R\Delta t}{(m/M)C_{p,m}} \\
&= (293\text{ K}) + \frac{(1.00\text{ A})^2 \times (1.00 \times 10^3\ \Omega) \times (15.0\text{ s})}{[(500\text{ g})/(63.55\text{ g mol}^{-1})] \times (24.4\text{ J K}^{-1}\text{ mol}^{-1})} \\
&= (293\text{ K}) + 78.1...\text{ K} = 3.71... \times 10^2\text{ K}.
\end{aligned}$$

The variation of entropy with temperature at constant pressure is given by [3B.7–90], $S(T_f) = S(T_i) + C_p \ln(T_f/T_i)$. Therefore the change in entropy is

$$\begin{aligned}\Delta S = S(T_f) - S(T_i) &= C_p \ln\left(\frac{T_f}{T_i}\right) = \left(\frac{m}{M}\right) C_{p,m} \ln\left(\frac{T_f}{T_i}\right) \\ &= \left(\frac{500\ \text{g}}{63.55\ \text{g mol}^{-1}}\right) \times (24.4\ \text{J K}^{-1}\ \text{mol}^{-1}) \times \ln\left(\frac{3.71... \times 10^2\ \text{K}}{293\ \text{K}}\right) \\ &= \boxed{+45.4\ \text{J K}^{-1}}.\end{aligned}$$

For the second experiment, the initial and final states of the metal block is the same, therefore $\boxed{\Delta S = 0}$. All the heat is released into surroundings, that is water bath, which can be assumed to be large enough to retain constant temperature. Thus

$$\begin{aligned}\Delta S_{\text{sur}} &= \frac{q}{T_{\text{sur}}} = \frac{I^2 R \Delta t}{T_{\text{sur}}} \\ &= \frac{(1.00\ \text{A})^2 \times (1.00 \times 10^3\ \Omega) \times (15.0\ \text{s})}{293\ \text{K}} = \boxed{+51.2\ \text{J K}^{-1}}.\end{aligned}$$

P3B.9 As suggested in the hint, first consider heating the folded protein at constant pressure to from the initial temperature T to that of the transition, T_{trs}. The variation of entropy with temperature at constant pressure is given by [3B.7–90], $S(T_f) = S(T_i) + C_p \ln(T_f/T_i)$. Thus the change in molar entropy, $\Delta S_m = S_m(T_f) - S_m(T_i)$, of this step is

$$\Delta S_{1,m} = C_{p,m}(\text{folded}) \ln\left(\frac{T_{\text{trs}}}{T}\right)$$

Next consider the unfolding step. The entropy change of such a transition is given by [3B.4–89], $\Delta_{\text{trs}} S = \Delta_{\text{trs}} H / T_{\text{trs}}$, thus

$$\Delta S_{2,m} = \frac{\Delta_{\text{trs}} H_m^{\ominus}}{T_{\text{trs}}}$$

The final step is cooling the unfolded protein to the initial temperature

$$\Delta S_{3,m} = C_{p,m}(\text{unfolded}) \ln\left(\frac{T}{T_{\text{trs}}}\right) = -C_{p,m}(\text{unfolded}) \ln\left(\frac{T_{\text{trs}}}{T}\right).$$

The overall entropy change is the sum of above steps

$$\begin{aligned}\Delta S_m &= \Delta S_{1,m} + \Delta S_{2,m} + \Delta S_{3,m} \\ &= C_{p,m}(\text{folded}) \ln\left(\frac{T_{\text{trs}}}{T}\right) + \frac{\Delta_{\text{trs}} H_m^{\ominus}}{T_{\text{trs}}} - C_{p,m}(\text{unfolded}) \ln\left(\frac{T_{\text{trs}}}{T}\right) \\ &= \frac{\Delta_{\text{trs}} H_m^{\ominus}}{T_{\text{trs}}} + \left[C_{p,m}(\text{folded}) - C_{p,m}(\text{unfolded})\right] \times \ln\left(\frac{T_{\text{trs}}}{T}\right)\end{aligned}$$

Given that $C_{p,\mathrm{m}}(\text{unfolded}) - C_{p,\mathrm{m}}(\text{folded}) = 6.28 \times 10^3\ \mathrm{J\,K^{-1}\,mol^{-1}}$, the molar entropy of unfolding at 25.0 °C is thus

$$\begin{aligned}\Delta S_\mathrm{m} &= \frac{5.09 \times 10^5\ \mathrm{J\,mol^{-1}}}{273.15\ \mathrm{K} + 75.5\ \mathrm{K}} \\ &\quad + (-6.28 \times 10^3\ \mathrm{J\,K^{-1}\,mol^{-1}}) \times \ln\left(\frac{273.15\ \mathrm{K} + 75.5\ \mathrm{K}}{273.15\ \mathrm{K} + 25.0\ \mathrm{K}}\right) \\ &= (1.45... \times 10^3\ \mathrm{J\,K^{-1}\,mol^{-1}}) + (-9.82 \times 10^2\ \mathrm{J\,K^{-1}\,mol^{-1}}) \\ &= +4.77... \times 10^2\ \mathrm{J\,K^{-1}\,mol^{-1}} = \boxed{+477\ \mathrm{J\,K^{-1}\,mol^{-1}}}.\end{aligned}$$

P3B.11 (a) Consider a process in which heat $|\mathrm{d}q|$ is extracted from the cold source at temperature T_c, and heat $q_\mathrm{h} = |\mathrm{d}q| + |\mathrm{d}w|$ is discarded into the hot sink at temperature T_h. The overall entropy change of such process is

$$\mathrm{d}S = \frac{-|\mathrm{d}q|}{T_\mathrm{c}} + \frac{|\mathrm{d}q| + |\mathrm{d}w|}{T_\mathrm{h}}$$

For the process to be permissible by the Second Law, the Clausius inequality defined in [3A.12–86], $\mathrm{d}S \geq 0$, must be satisfied. Therefore

$$\frac{-|\mathrm{d}q|}{T_\mathrm{c}} + \frac{|\mathrm{d}q| + |\mathrm{d}w|}{T_\mathrm{h}} \geq 0$$

the equality implies the minimum amount of work for which the process is permissible. Hence it follows that

$$\boxed{\frac{|\mathrm{d}q|}{T_\mathrm{c}} = \frac{|\mathrm{d}q| + |\mathrm{d}w|}{T_\mathrm{h}}}.$$

(b) The expression in (a) is rearranged to find $|dw|$ and the given relation, $dq = C\mathrm{d}T_\mathrm{c}$, is used to give

$$|\mathrm{d}w| = T_\mathrm{h}\frac{|\mathrm{d}q|}{T_\mathrm{c}} - |\mathrm{d}q|$$
$$|\mathrm{d}w| = CT_\mathrm{h}\left|\frac{\mathrm{d}T_\mathrm{c}}{T_\mathrm{c}}\right| - C|\mathrm{d}T_\mathrm{c}|$$

Integration of both sides between the appropriate limits gives

$$\int_0^w |\mathrm{d}w'| = CT_\mathrm{h}\int_{T_\mathrm{i}}^{T_\mathrm{f}}\left|\frac{\mathrm{d}T_\mathrm{c}}{T_\mathrm{c}}\right| - C\int_{T_\mathrm{i}}^{T_\mathrm{f}}|\mathrm{d}T_\mathrm{c}|$$

which evaluates to

$$\boxed{|w| = CT_\mathrm{h}\left|\ln\left(\frac{T_\mathrm{f}}{T_\mathrm{i}}\right)\right| - C|T_\mathrm{f} - T_\mathrm{i}|}.$$

(c) Using $C = (m/M)C_{p,\mathrm{m}}$, the work needed is

$$|w| = \frac{250\ \mathrm{g}}{18.02\ \mathrm{g\,mol^{-1}}} \times (75.3\ \mathrm{J\,K^{-1}\,mol^{-1}}) \times (293\ \mathrm{K}) \times \left|\ln\left(\frac{273\ \mathrm{K}}{293\ \mathrm{K}}\right)\right|$$
$$- \frac{250\ \mathrm{g}}{18.02\ \mathrm{g\,mol^{-1}}} \times (75.3\ \mathrm{J\,K^{-1}\,mol^{-1}}) \times |273\ \mathrm{K} - 293\ \mathrm{K}|$$
$$= |-2.16... \times 10^4|\ \mathrm{J} - |-2.08... \times 10^4|\ \mathrm{J}$$
$$= +7.47... \times 10^2\ \mathrm{J} = \boxed{+7.5 \times 10^2\ \mathrm{J}}.$$

(d) Assuming constant temperature, for finite amounts of heat and work, the expression derrived in (a) becomes

$$\frac{|q|}{T_\mathrm{c}} = \frac{|q| + |w|}{T_\mathrm{h}}$$

This is rearranged to give the work as

$$|w| = \left(\frac{T_\mathrm{h}}{T_\mathrm{c}} - 1\right) \times |q|$$

The heat transferred during freezing is equal to the enthalpy of the transition, which is the opposite of fusion, $q = \Delta_\mathrm{trs}H = (m/M)(-\Delta_\mathrm{fus}H^\ominus)$. Therefore the work needed is

$$|w| = \left(\frac{293\ \mathrm{K}}{273\ \mathrm{K}} - 1\right) \times \left|\frac{250\ \mathrm{g}}{18.02\ \mathrm{g\,mol^{-1}}} \times (-6.01 \times 10^3\ \mathrm{J\,K^{-1}\,mol^{-1}})\right|$$
$$= 6.10... \times 10^3\ \mathrm{J} = \boxed{6.11 \times 10^3\ \mathrm{J}}.$$

(e) The total work is the sum of the two steps described in (c) and (d). Therefore

$$w_\mathrm{tot} = (+7.47... \times 10^2\ \mathrm{J}) + (6.10... \times 10^3\ \mathrm{J})$$
$$= +6.85... \times 10^3\ \mathrm{J} = \boxed{+6.86\ \mathrm{kJ}}.$$

(f) Assuming no energy losses, power is the total work divided by the time interval over which the work is done, $P = w_\mathrm{tot}/\Delta t$, hence

$$\Delta t = \frac{w_\mathrm{tot}}{P} = \frac{6.85... \times 10^3\ \mathrm{J}}{100\ \mathrm{W}} = \boxed{68.6\ \mathrm{s}}.$$

3C The measurement of entropy

Answer to discussion question

D3C.1 Because solutions of cations cannot be prepared in the absence of anions, the standard molar entropies of ions in solution are reported on a scale in which, by convention, the standard entropy of H^+ ions in water is taken as zero at all temperatures: $S^\ominus(H^+, aq) = 0$.

Because the entropies of ions in water are values relative to the hydrogen ion in water, they may be either positive or negative. A positive entropy means that an ion has a higher molar entropy than H^+ in water, and a negative entropy means that the ion has a lower molar entropy than H^+ in water. An ion with zero entropy in fact has that *same* entropy as H^+.

Solutions to exercises

E3C.1(a) Assuming that the Debye extrapolation is valid, the constant-pressure molar heat capacity is $C_{p,m}(T) = aT^3$. The temperature dependence of the entropy is given by [3C.1a–92], $S(T_2) = S(T_1) + \int_{T_1}^{T_2}(C_{p,m}/T)dT$. For a given temperature T the change in molar entropy from zero temperature is therefore

$$S_m(T) - S_m(0) = \int_0^T \frac{C_{p,m}}{T'}\,dT' = \int_0^T \frac{aT'^3}{T'}\,dT'$$
$$= a\int_0^T T'^2\,dT' = \frac{aT^3}{3} = \frac{C_{p,m}(T)}{3}$$

Hence

$$S_m(4.2\text{ K}) - S_m(0) = \frac{C_{p,m}(4.2\text{ K})}{3} = \frac{0.0145\text{ J K}^{-1}\text{ mol}^{-1}}{3}$$
$$= \boxed{4.8\times10^{-3}\text{ J K}^{-1}\text{ mol}^{-1}}.$$

E3C.2(a) The standard reaction entropy is given by [3C.3b–94], $\Delta_r S^\ominus = \sum_J \nu_J S_m^\ominus(J)$, where ν_J are the signed stoichiometric numbers.

(i)

$$\Delta_r S^\ominus = 2S_m^\ominus(CH_3COOH,(l)) - 2S_m^\ominus(CH_3CHO,(g)) - S_m^\ominus(O_2,(g))$$
$$= 2\times(159.8\text{ J K}^{-1}\text{ mol}^{-1}) - 2\times(250.3\text{ J K}^{-1}\text{ mol}^{-1})$$
$$-(205.138\text{ J K}^{-1}\text{ mol}^{-1})$$
$$= \boxed{-386.1\text{ J K}^{-1}\text{ mol}^{-1}}.$$

(ii)

$$\Delta_r S^\ominus = 2S_m^\ominus(AgBr,(s)) + S_m^\ominus(Cl_2,(g)) - 2S_m^\ominus(AgCl,(s)) - S_m^\ominus(Br_2,(l))$$
$$= 2\times(107.1\text{ J K}^{-1}\text{ mol}^{-1}) + (223.07\text{ J K}^{-1}\text{ mol}^{-1})$$
$$-2\times(96.2\text{ J K}^{-1}\text{ mol}^{-1}) - (152.23\text{ J K}^{-1}\text{ mol}^{-1})$$
$$= \boxed{+92.6\text{ J K}^{-1}\text{ mol}^{-1}}.$$

(iii)

$$\begin{aligned}\Delta_r S^{\ominus} &= S_m^{\ominus}(\mathrm{HgCl_2,(s)}) - S_m^{\ominus}(\mathrm{Hg,(l)}) - S_m^{\ominus}(\mathrm{Cl_2,(g)})\\ &= (146.0\,\mathrm{J\,K^{-1}\,mol^{-1}}) - (76.02\,\mathrm{J\,K^{-1}\,mol^{-1}})\\ &\quad - (223.07\,\mathrm{J\,K^{-1}\,mol^{-1}})\\ &= \boxed{-153.1\,\mathrm{J\,K^{-1}\,mol^{-1}}}.\end{aligned}$$

E3C.3(a) Consider the chemical equation

$$\tfrac{1}{2}\mathrm{N_2(g)} + \tfrac{3}{2}\mathrm{H_2(g)} \longrightarrow \mathrm{NH_3(g)}$$

The standard reaction entropy is given by [3C.3b–94], $\Delta_r S^{\ominus} = \sum_J \nu_J S_m^{\ominus}(J)$, where ν_J are singed stoichiometric coefficients for a given reaction equation. Therefore, using data from the *Resource section*

$$\begin{aligned}\Delta_r S^{\ominus} &= nS_m^{\ominus}(\mathrm{NH_3,(g)}) - \tfrac{3}{2}nS_m^{\ominus}(\mathrm{H_2,(g)}) - \tfrac{1}{2}nS_m^{\ominus}(\mathrm{N_2,(g)})\\ &= (1.00\,\mathrm{mol}) \times (192.45\,\mathrm{J\,K^{-1}\,mol^{-1}})\\ &\quad - \left(\tfrac{3}{2} \times 1.00\,\mathrm{mol}\right) \times (130.684\,\mathrm{J\,K^{-1}\,mol^{-1}})\\ &\quad - \left(\tfrac{1}{2} \times 1.00\,\mathrm{mol}\right) \times (191.61\,\mathrm{J\,K^{-1}\,mol^{-1}})\\ &= \boxed{-99.38\,\mathrm{J\,K^{-1}}}.\end{aligned}$$

Solutions to problems

P3C.1 Consider the process of determining the calorimetric entropy from zero to the temperature of interest. Assuming that the Debye extrapolation is valid, the constant-pressure molar heat capacity at the lowest temperatures is of a form $C_{p,\mathrm{m}}(T) = aT^3$. The temperature dependence of the entropy is given by [3C.1a–92], $S(T_2) = S(T_1) = \int_{T_1}^{T_2} (C_{p,\mathrm{m}}/T)\mathrm{d}T$. Thus for a given (low) temperature T the change in molar entropy from zero is

$$\begin{aligned}S_\mathrm{m}(T) - S_\mathrm{m}(0) &= \int_0^T \frac{C_{p,\mathrm{m}}}{T'}\,\mathrm{d}T' = \int_0^T \frac{aT'^3}{T'}\,\mathrm{d}T'\\ &= a\int_0^T T'^2\,\mathrm{d}T' = \frac{a}{3}T^3 = \tfrac{1}{3}C_{p,\mathrm{m}}(T)\end{aligned}$$

Hence

$$S_m^{\ominus}(10\,\mathrm{K}) - S_m^{\ominus}(0) = \tfrac{1}{3} \times (4.64\,\mathrm{J\,K^{-1}\,mol^{-1}}) = 1.54...\,\mathrm{J\,K^{-1}\,mol^{-1}}$$

The increase in entropy on raising the temperature to the melting point is $S_m^{\ominus}(234.4\,\mathrm{K}) - S_m^{\ominus}(10\,\mathrm{K}) = 57.74\,\mathrm{J\,K^{-1}\,mol^{-1}}$. The entropy change of a phase transition is given by [3C.1b–92], $\Delta_\mathrm{trs}S(T_\mathrm{trs}) = \Delta_\mathrm{trs}H(T_\mathrm{trs})/T_\mathrm{trs}$. Thus

$$\Delta_\mathrm{fus}S_m^{\ominus}(234.4\,\mathrm{K}) = \frac{2322\,\mathrm{J\,mol^{-1}}}{234.4\,\mathrm{K}} = 9.90...\,\mathrm{J\,K^{-1}\,mol^{-1}}$$

Further raising the temperature to 298.0 K gives an increase in the entropy of $S_m^{\ominus}(298\text{ K}) - S_m^{\ominus}(234.4\text{ K}) = 6.85\text{ J K}^{-1}\text{ mol}^{-1}$.

The Third-Law standard molar entropy at 298 K is the sum of the above contributions.

$$\begin{aligned} S_m^{\ominus}(298\text{ K}) - S_m^{\ominus}(0) &= (S_m^{\ominus}(10\text{ K}) - S_m^{\ominus}(0)) + (S_m^{\ominus}(234.4\text{ K}) - S_m^{\ominus}(10\text{ K})) \\ &\quad + \Delta_{\text{fus}}S_m^{\ominus}(234.4\text{ K}) + (S_m^{\ominus}(298\text{ K}) - S_m^{\ominus}(234.4\text{ K})) \\ &= (1.54...\text{ J K}^{-1}\text{ mol}^{-1}) + (57.74\text{ J K}^{-1}\text{ mol}^{-1}) \\ &\quad + (9.90...\text{ J K}^{-1}\text{ mol}^{-1}) + (6.85\text{ J K}^{-1}\text{ mol}^{-1}) \\ &= \boxed{76.04\text{ J K}^{-1}\text{ mol}^{-1}}. \end{aligned}$$

P3C.3 (a) Assuming that the Debye extrapolation is valid, the constant pressure molar heat capacity is of a form $C_{p,m}(T) = aT^3$. The temperature dependence of the entropy is given by [3C.1a–92], $S(T_2) = S(T_1) = \int_{T_1}^{T_2}(C_{p,m}/T)\mathrm{d}T$. Thus for a given (low) temperature T the change in the molar entropy from zero is

$$\begin{aligned} S_m(T) - S_m(0) &= \int_0^T \frac{C_{p,m}}{T'}\mathrm{d}T' = \int_0^T \frac{aT'^3}{T'}\mathrm{d}T' \\ &= a\int_0^T T'^2\mathrm{d}T' = \frac{a}{3}T^3 = \tfrac{1}{3}C_{p,m}(T) \end{aligned}$$

Hence

$$\begin{aligned} S_m^{\ominus}(10\text{ K}) - S_m^{\ominus}(0) &= \tfrac{1}{3} \times (2.8\text{ J K}^{-1}\text{ mol}^{-1}) = 0.933...\text{ J K}^{-1}\text{ mol}^{-1} \\ &= \boxed{0.93\text{ J K}^{-1}\text{ mol}^{-1}}. \end{aligned}$$

(b) The change in entropy is determined calorimetrically by measuring the area under a plot of $(C_{p,m}/T)$ against T, as shown in Fig. 3.2.

The plot is rather irregular and is best fitted by two polynomials of order 3: one in the range 10 K to 30 K and the other in the range 30 K to 298 K. Define $y = (C_{p,m}/T)/(\text{J K}^{-2}\text{ mol}^{-1})$ and $x = T/\text{K}$, so that the fitted function is expressed

$$y = c_3x^3 + c_2x^2 + c_1x + c_0$$

where the best fitted coefficients c_i for the respective temperature ranges are

c_i	10 K to 30 K	30 K to 298 K
c_3	$+5.0222 \times 10^{-5}$	-5.2881×10^{-8}
c_2	-4.3010×10^{-3}	$+3.5425 \times 10^{-5}$
c_1	$+1.2025 \times 10^{-1}$	-8.1107×10^{-3}
c_0	-5.4187×10^{-1}	$+7.5533 \times 10^{-1}$

T/K	$C_{p,\mathrm{m}}/(\mathrm{J\,K^{-1}\,mol^{-1}})$	$(C_{p,\mathrm{m}}/T)/(\mathrm{J\,K^{-2}\,mol^{-1}})$
10	2.8	0.2800
15	7.0	0.4667
20	10.8	0.5400
25	14.1	0.5640
30	16.5	0.5500
50	21.4	0.4280
70	23.3	0.3329
100	24.5	0.2450
150	25.3	0.1687
200	25.8	0.1290
250	26.2	0.1048
298	26.6	0.0893

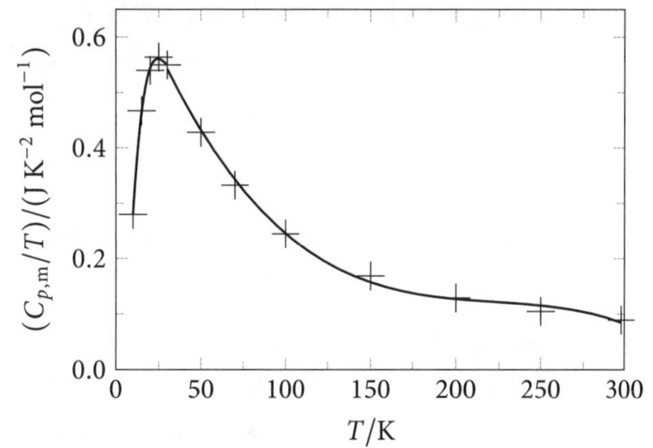

Figure 3.2

The integral of the fitted functions over the range x_i to x_f is

$$I = \int_{x_\mathrm{i}}^{x_\mathrm{f}} c_3x^3 + c_2x^2 + c_1x + c_0\,\mathrm{d}x$$
$$= \frac{c_3}{4}\left(x_\mathrm{f}^4 - x_\mathrm{i}^4\right) + \frac{c_2}{3}\left(x_\mathrm{f}^3 - x_\mathrm{i}^3\right) + \frac{c_1}{2}\left(x_\mathrm{f}^2 - x_\mathrm{i}^2\right) + c_0\left(x_\mathrm{f} - x_\mathrm{i}\right)$$

Using the appropriate coefficients and limits the integrals are evaluated to give the respective changes in entropy

$$S_\mathrm{m}^\ominus(30\,\mathrm{K}) - S_\mathrm{m}^\ominus(10\,\mathrm{K}) = 10.0...\,\mathrm{J\,K^{-1}\,mol^{-1}}$$
$$S_\mathrm{m}^\ominus(298\,\mathrm{K}) - S_\mathrm{m}^\ominus(30\,\mathrm{K}) = 53.8...\,\mathrm{J\,K^{-1}\,mol^{-1}}$$

The total entropy change is the sum of the two integrals. Therefore

$$S_\mathrm{m}^\ominus(298\,\mathrm{K}) - S_\mathrm{m}^\ominus(10\,\mathrm{K}) = (10.0...\,\mathrm{J\,K^{-1}\,mol^{-1}}) + (53.8...\,\mathrm{J\,K^{-1}\,mol^{-1}})$$
$$= 63.9...\,\mathrm{J\,K^{-1}\,mol^{-1}} = \boxed{63.9\,\mathrm{J\,K^{-1}\,mol^{-1}}}.$$

(c) The standard Third-Law entropy at 298 K is the sum of the above calculated contributions. Thus

$$\begin{aligned}S_m^\ominus(298\,\text{K}) - S_m^\ominus(0) &= (S_m^\ominus(298\,\text{K}) - S_m^\ominus(10\,\text{K}))\\&\quad + (S_m^\ominus(10\,\text{K}) - S_m^\ominus(0))\\&= (63.9...\,\text{J K}^{-1}\,\text{mol}^{-1}) + (0.933...\,\text{J K}^{-1}\,\text{mol}^{-1})\\&= \boxed{64.8\,\text{J K}^{-1}\,\text{mol}^{-1}}.\end{aligned}$$

For the standard Third-Law entropy at 273 K, the second integral in part (b) needs to be repeated with $T_f = 273$ K. Therefore

$$S_m^\ominus(273\,\text{K}) - S_m^\ominus(30\,\text{K}) = 51.4...\,\text{J K}^{-1}\,\text{mol}^{-1}$$

The other contributions are the same, hence

$$\begin{aligned}S_m^\ominus(273\,\text{K}) - S_m^\ominus(0) &= (10.0...\,\text{J K}^{-1}\,\text{mol}^{-1} + 51.4...\,\text{J K}^{-1}\,\text{mol}^{-1})\\&\quad + (0.933...\,\text{J K}^{-1}\,\text{mol}^{-1})\\&= \boxed{62.4\,\text{J K}^{-1}\,\text{mol}^{-1}}.\end{aligned}$$

P3C.5 The standard reaction entropy is given by [3C.3b–94], $\Delta_r S^\ominus = \sum_J \nu_J S_m^\ominus(J)$, where ν_J are the signed stoichiometric numbers.

$$\begin{aligned}\Delta_r S^\ominus(298\,\text{K}) &= S_m^\ominus(\text{CO},(\text{g})) + S_m^\ominus(\text{H}_2\text{O},(\text{g}))\\&\quad - S_m^\ominus(\text{CO}_2,(\text{g})) - S_m^\ominus(\text{H}_2,(\text{g}))\\&= (197.67\,\text{J K}^{-1}\,\text{mol}^{-1}) + (188.83\,\text{J K}^{-1}\,\text{mol}^{-1})\\&\quad - (213.74\,\text{J K}^{-1}\,\text{mol}^{-1}) - (130.684\,\text{J K}^{-1}\,\text{mol}^{-1})\\&= +42.07...\,\text{J K}^{-1}\,\text{mol}^{-1} = \boxed{+42.08\,\text{J K}^{-1}\,\text{mol}^{-1}}.\end{aligned}$$

Similarly, the standard reaction enthalpy is given by [2C.5b–55], $\Delta_r H^\ominus = \sum_J \nu_J \Delta_f H^\ominus(J)$.

$$\begin{aligned}\Delta_r H^\ominus(298\,\text{K}) &= \Delta_f H^\ominus(\text{CO},(\text{g})) + \Delta_f H^\ominus(\text{H}_2\text{O},(\text{g}))\\&\quad - \Delta_f H^\ominus(\text{CO}_2,(\text{g})) - \Delta_f H^\ominus(\text{H}_2,(\text{g}))\\&= (-110.53\,\text{kJ mol}^{-1}) + (-241.82\,\text{kJ mol}^{-1})\\&\quad - (-393.51\,\text{kJ mol}^{-1}) - 0\\&= \boxed{+41.16\,\text{kJ mol}^{-1}}.\end{aligned}$$

The temperature dependence of the reaction entropy is given by [3C.5a–95], $\Delta_r S^\ominus(T_2) = \Delta_r S^\ominus(T_1) + \int_{T_1}^{T_2}(\Delta_r C_p^\ominus/T)\text{d}T$. Similarly, the enthalpy dependence on temperature is given by Kirchhoff's law [2C.7a–55], $\Delta_r H^\ominus(T_2) = \Delta_r H^\ominus(T_1) + \int_{T_1}^{T_2}\Delta_r C_p^\ominus\text{d}T$. The quantity $\Delta_r C_p^\ominus$ is defined in [3C.5b–95], $\Delta_r C_p^\ominus =$

$\sum_J \nu_J C^{\ominus}_{p,\mathrm{m}}(\mathrm{J})$. For the reaction at 298 K

$$\begin{aligned}\Delta_r C^{\ominus}_p &= C^{\ominus}_{p,\mathrm{m}}(\mathrm{CO},(\mathrm{g})) + C^{\ominus}_{p,\mathrm{m}}(\mathrm{H_2O},(\mathrm{g}))\\ &\quad - C^{\ominus}_{p,\mathrm{m}}(\mathrm{CO_2},(\mathrm{g})) - C^{\ominus}_{p,\mathrm{m}}(\mathrm{H_2},(\mathrm{g}))\\ &= (29.14\ \mathrm{J\,K^{-1}\,mol^{-1}}) + (33.58\ \mathrm{J\,K^{-1}\,mol^{-1}})\\ &\quad - (37.11\ \mathrm{J\,K^{-1}\,mol^{-1}}) - (28.824\ \mathrm{J\,K^{-1}\,mol^{-1}})\\ &= -3.21...\ \mathrm{J\,K^{-1}\,mol^{-1}}\end{aligned}$$

Assuming that $\Delta_r C^{\ominus}_p$ is constant over the temperature range involved, the standard entropy and enthalpy changes of the reaction is given by, respectively, [3C.5b–95], $\Delta_r S^{\ominus}(T_2) = \Delta_r S^{\ominus}(T_1)+\Delta_r C^{\ominus}_p \ln(T_2/T_1)$, and [2C.7d–56], $\Delta_r H^{\ominus}(T_2) = \Delta_r H^{\ominus}(T_1) + \Delta_r C^{\ominus}_p(T_2 - T_1)$.

$$\begin{aligned}\Delta_r S^{\ominus}(398\ \mathrm{K}) &= \Delta_r S^{\ominus}(298\ \mathrm{K}) + \Delta_r C^{\ominus}_p \times \ln\left(\frac{398\ \mathrm{K}}{298\ \mathrm{K}}\right)\\ &= (+42.0...\ \mathrm{J\,K^{-1}\,mol^{-1}})\\ &\quad + (-3.21...\ \mathrm{J\,K^{-1}\,mol^{-1}}) \times \ln\left(\frac{398}{298}\right)\\ &= \boxed{+41.15\ \mathrm{J\,K^{-1}\,mol^{-1}}}.\end{aligned}$$

$$\begin{aligned}\Delta_r H^{\ominus}(398\ \mathrm{K}) &= \Delta_r H^{\ominus}(298\ \mathrm{K}) + \Delta_r C^{\ominus}_p \times (398\ \mathrm{K} - 298\ \mathrm{K})\\ &= (+41.1... \times 10^3\ \mathrm{J\,mol^{-1}})\\ &\quad + (-3.21...\ \mathrm{J\,K^{-1}\,mol^{-1}}) \times (100\ \mathrm{K})\\ &= +40.8... \times 10^3\ \mathrm{J\,mol^{-1}} = \boxed{+40.8\ \mathrm{kJ\,mol^{-1}}}.\end{aligned}$$

P3C.7 Assuming that the Debye extrapolation is valid, the constant-pressure molar heat capacity is of a form $C_{p,\mathrm{m}}(T) = aT^3$. The temperature dependence of the entropy is given by [3C.1a–92], $S(T_2) = S(T_1) = \int_{T_1}^{T_2}(C_{p,\mathrm{m}}/T)\mathrm{d}T$. Thus for a given (low) temperature T the change in the molar entropy from zero is

$$\begin{aligned}S_\mathrm{m}(T) - S_\mathrm{m}(0) &= \int_0^T \frac{C_{p,\mathrm{m}}}{T'}\,\mathrm{d}T' = \int_0^T \frac{aT'^3}{T'}\,\mathrm{d}T'\\ &= a\int_0^T T'^2\mathrm{d}T' = \frac{a}{3}T^3 = \tfrac{1}{3}C_{p,\mathrm{m}}(T)\end{aligned}$$

Hence

$$S^{\ominus}_\mathrm{m}(14.14\ \mathrm{K}) - S^{\ominus}_\mathrm{m}(0) = \tfrac{1}{3} \times (9.492\ \mathrm{J\,K^{-1}\,mol^{-1}}) = 3.16...\ \mathrm{J\,K^{-1}\,mol^{-1}}$$

The change in entropy is determined calorimetrically by measuring the area under a plot of $(C_{p,\mathrm{m}}/T)$ against T.

T/K	$C_{p,\mathrm{m}}/(\mathrm{J\,K^{-1}\,mol^{-1}})$	$(C_{p,\mathrm{m}}/T)/(\mathrm{J\,K^{-2}\,mol^{-1}})$
14.14	9.492	0.671 29
16.33	12.70	0.777 71
20.03	18.18	0.907 64
31.15	32.54	1.044 62
44.08	46.86	1.063 07
64.81	66.36	1.023 92
100.90	95.05	0.942 02
140.86	121.3	0.861 14
183.59	144.4	0.786 54
225.10	163.7	0.727 23
262.99	180.2	0.685 20
298.06	196.4	0.658 93

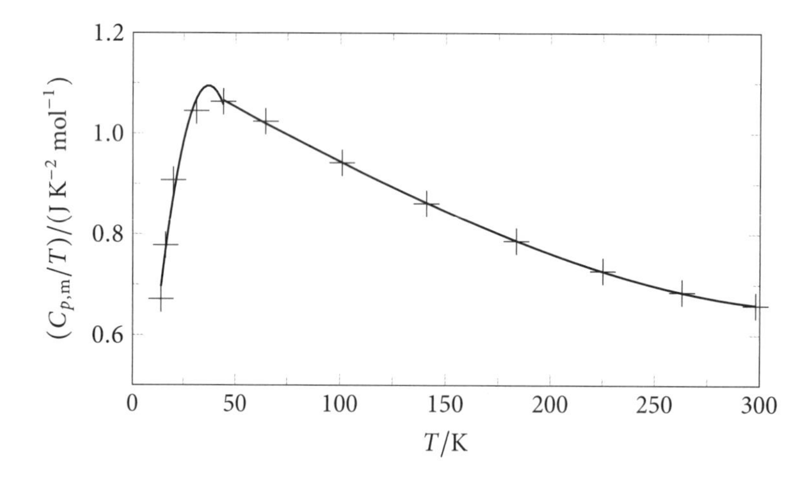

The plot is rather irregular and is best fitted by two polynomials of order 2 and 3, respectively, in the ranges 14.14 K to 44.08 K and 44.08 K to 298.06 K. Define $y = (C_{p,\mathrm{m}}/T)/(\mathrm{J\,K^{-2}\,mol^{-1}})$ and $x = T/\mathrm{K}$, so that the fitted function is expressed

$$y = c_3x^3 + c_2x^2 + c_1x + c_0$$

where the best fitted coefficients c_i for the respective temperature ranges are

c_i	14.14 K to 44.08 K	44.08 K to 298.06 K
c_3	0	$+7.1979 \times 10^{-9}$
c_2	-7.6119×10^{-4}	-3.0830×10^{-7}
c_1	$+5.6367 \times 10^{-2}$	-2.2415×10^{-3}
c_0	$+5.1090 \times 10^{-2}$	$+1.1644 \times 10^{0}$

The integral of the fitted functions over the range x_i to x_f is

$$I = \int_{x_i}^{x_f} c_3x^3 + c_2x^2 + c_1x + c_0 \, dx$$
$$= \frac{c_3}{4}\left(x_f^{\ 4} - x_i^{\ 4}\right) + \frac{c_2}{3}\left(x_f^{\ 3} - x_i^{\ 3}\right) + \frac{c_1}{2}\left(x_f^{\ 2} - x_i^{\ 2}\right) + c_0\left(x_f - x_i\right)$$

The temperatures of interest are beyond 44.08 K. Thus the contribution corresponding to the integral of the quadratic is included in the estimates of the entropy. Using the appropriate coefficients and limits the integral gives

$$S_m^{\ominus}(44.08\text{ K}) - S_m^{\ominus}(14.14\text{ K}) = 29.6...\text{ J K}^{-1}\text{ mol}^{-1}$$

Finally, the remaining contribution is found by estimating the integral of the cubic polynomial to the temperature of interest. Therefore for $T = 100$ K, 200 K and, after small extrapolation, 300 K

$$S_m^{\ominus}(100\text{ K}) - S_m^{\ominus}(44.08\text{ K}) = 56.1...\text{ J K}^{-1}\text{ mol}^{-1}$$
$$S_m^{\ominus}(200\text{ K}) - S_m^{\ominus}(44.08\text{ K}) = 1.41... \times 10^2\text{ J K}^{-1}\text{ mol}^{-1}$$
$$S_m^{\ominus}(300\text{ K}) - S_m^{\ominus}(44.08\text{ K}) = 2.11... \times 10^2\text{ J K}^{-1}\text{ mol}^{-1}$$

The standard Third-Law molar entropy at the temperatures of interests is the sum of all the contributions up to that point. Thus, the entropy at three temperatures are

$$\begin{aligned} S_m^{\ominus}(100\text{ K}) - S_m^{\ominus}(0) &= S_m^{\ominus}(14.14\text{ K}) - S_m^{\ominus}(0) \\ &\quad + (S_m^{\ominus}(44.08\text{ K}) - S_m^{\ominus}(14.14\text{ K})) \\ &\quad + (S_m^{\ominus}(100\text{ K}) - S_m^{\ominus}(44.08\text{ K})) \\ &= (3.16...\text{ J K}^{-1}\text{ mol}^{-1}) + (29.6...\text{ J K}^{-1}\text{ mol}^{-1}) \\ &\quad + (56.1...\text{ J K}^{-1}\text{ mol}^{-1}) \\ &= \boxed{89.0\text{ J K}^{-1}\text{ mol}^{-1}}. \end{aligned}$$

Similarly it is found for 200 K and 300 K, respectively

$$\begin{aligned} S_m^{\ominus}(200\text{ K}) - S_m^{\ominus}(0) &= (3.16...\text{ J K}^{-1}\text{ mol}^{-1}) + (29.6...\text{ J K}^{-1}\text{ mol}^{-1}) \\ &\quad + (1.41... \times 10^2\text{ J K}^{-1}\text{ mol}^{-1}) \\ &= \boxed{173.8\text{ J K}^{-1}\text{ mol}^{-1}}. \end{aligned}$$

$$\begin{aligned} S_m^{\ominus}(300\text{ K}) - S_m^{\ominus}(0) &= (3.16...\text{ J K}^{-1}\text{ mol}^{-1}) + (29.6...\text{ J K}^{-1}\text{ mol}^{-1}) \\ &\quad + (2.11... \times 10^2\text{ J K}^{-1}\text{ mol}^{-1}) \\ &= \boxed{243.9\text{ J K}^{-1}\text{ mol}^{-1}}. \end{aligned}$$

P3C.9 (a) Given the expression for the constant-pressure molar heat capacity, $C_{p,\mathrm{m}}(T) = aT^3 + bT$, consider $C_{p,\mathrm{m}}/T$.

$$\frac{C_{p,\mathrm{m}}}{T} = \frac{aT^3 + bT}{T} = aT^2 + b$$

This expression is of the form of a straight line, $y = (\text{slope}) \times x + (\text{intercept})$, if $y = C_{p,\mathrm{m}}(T)$ and $x = T^2$. It follows that (slope) = a and (intercept) = b.

(b) The data below are plotted in Fig. 3.3.

T/K	$C_{p,\mathrm{m}}/(\mathrm{J\,K^{-1}\,mol^{-1}})$	$T^2/(\mathrm{K}^2)$	$(C_{p,\mathrm{m}}/T)/(\mathrm{J\,K^{-2}\,mol^{-1}})$
0.20	0.437	0.040	2.1850
0.25	0.560	0.063	2.2100
0.30	0.693	0.090	2.3100
0.35	0.838	0.123	2.3943
0.40	0.996	0.160	2.4900
0.45	1.170	0.203	2.6000
0.50	1.361	0.250	2.7220
0.55	1.572	0.303	2.8582

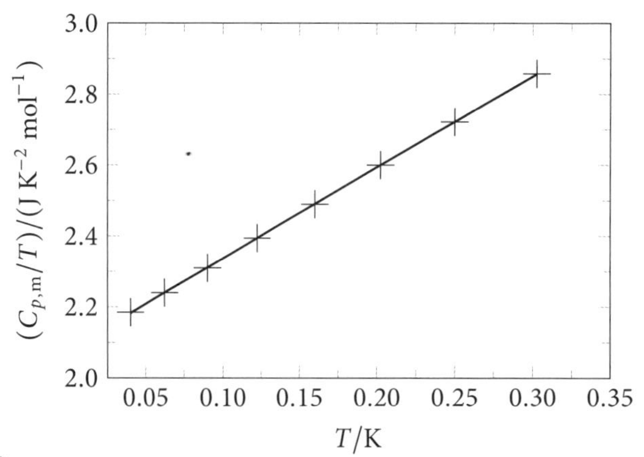

Figure 3.3

The data lie on a good straight line, the equation of which is

$$(C_{p,\mathrm{m}}/T)/(\mathrm{J\,K^{-2}\,mol^{-1}}) = 2.569 \times (T/\mathrm{K})^2 + 2.080$$

Thus $a = \boxed{2.569\ \mathrm{JK^{-4}mol^{-1}}}$ and $b = \boxed{2.080\ \mathrm{JK^{-2}mol^{-1}}}$.

(c) The dependence of the entropy on temperature is given by [3C.1a–92], $S(T_2) = S(T_1) + \int_{T_1}^{T_2} (C_{p,\mathrm{m}}/T)\mathrm{d}T$. Thus for a given (low) temperature T

the molar entropy change from the zero temperature is

$$S_m(T) = S_m(0) + \int_0^T \frac{C_{p,m}}{T'}\,dT' = S_m(0) + \int_0^T (aT'^2 + b)\,dT'$$
$$= \boxed{S_m(0) + \frac{a}{3}T^3 + bT}.$$

(d) Assuming that the expression derived above can be extrapolated to 2.0 K

$$S_m(2.0\text{ K}) - S_m(0) = \frac{(2.569\text{ JK}^{-4}\text{mol}^{-1})}{3} \times (2.0\text{ K})^3 + (2.080\text{ JK}^{-2}\text{mol}^{-1}) \times (2.0\text{ K})$$
$$= \boxed{11.01\text{ J K}^{-1}\text{ mol}^{-1}}.$$

3D Concentrating on the system

Answers to discussion questions

D3D.1 These criteria for spontaneity are obtained from a combination of the First Law of thermodynamics, $dU = dq + dw$, with the Second Law in the form of the Clausius inequality, $dS \geq dq/T$.

First imagine a process at constant volume which can therefore do no work of expansion. From the First Law it follows that $dU = dq$. The Clausius inequality is rewriten as $dS - dq/T \geq 0$, from which it follows that $dS - dU/T \geq 0$ and hence $TdS \geq dU$.

The Helmholtz energy is defined at $A = U - TS$. For a general change at constant temperature it follows that $dA = dU - TdS$. It is already established that, at constant volume, $TdS \geq dU$ so it follows that $dA \leq 0$. This is the criterion for spontaneity for a process at constant volume and temperature; the inequality specifies a spontaneous process, and the equality specifies an equilibrium process.

Now consider a process at constant pressure: for such a process the heat is equal to the enthalpy change, so $dH = dq$. Following a similar line of argument to that above leads to $TdS \geq dH$.

The Gibbs energy is defined at $G = H - TS$. For a general change at constant temperature it follows that $dG = dH - TdS$. It is already established that, at constant pressure, $TdS \geq dH$ so it follows that $dG \leq 0$. This is the criterion for spontaneity for a process at constant pressure and temperature.

Solutions to exercises

E3D.1(a) The standard reaction Gibbs energy is given by [3D.9–*100*], $\Delta_r G^\ominus = \Delta_r H^\ominus - T\Delta_r S^\ominus$. The standard reaction enthalpy is given in terms of the enthalpies of formation by [2C.5b–*55*], $\Delta_r H^\ominus = \sum_J \nu_J \Delta_f H^\ominus(J)$, where ν_J are the signed stoichiometric numbers.

(i)

$$\begin{aligned}\Delta_r H^\ominus &= 2\Delta_f H^\ominus(\text{CH}_3\text{COOH},(\text{l})) - 2\Delta_f H^\ominus(\text{CH}_3\text{CHO},(\text{g})) \\ &\quad - \Delta_f H^\ominus(\text{O}_2,(\text{g})) \\ &= 2\times(-484.5\ \text{kJ mol}^{-1}) - 2\times(-166.19\ \text{kJ mol}^{-1}) - 0 \\ &= -636.62\ \text{kJ mol}^{-1} = \boxed{-636.6\ \text{kJ mol}^{-1}}.\end{aligned}$$

Given the result for the previous execise, $\Delta_r S^\ominus = -386.1\ \text{J K}^{-1}\ \text{mol}^{-1}$

$$\begin{aligned}\Delta_r G^\ominus &= (-636.62\ \text{kJ mol}^{-1}) - (298.15\ \text{K})\times(-0.3861\ \text{kJ K}^{-1}\ \text{mol}^{-1}) \\ &= \boxed{-521.5\ \text{kJ mol}^{-1}}.\end{aligned}$$

(ii)

$$\begin{aligned}\Delta_r H^\ominus &= 2\Delta_f H^\ominus(\text{AgBr},(\text{s})) + \Delta_f H^\ominus(\text{Cl}_2,(\text{g})) - 2\Delta_f H^\ominus(\text{AgCl},(\text{s})) \\ &\quad - \Delta_f H^\ominus(\text{Br}_2,(\text{l})) \\ &= 2\times(-100.37\ \text{kJ mol}^{-1}) + 0 - 2\times(-127.07\ \text{kJ mol}^{-1}) - 0 \\ &= \boxed{+53.40\ \text{kJ mol}^{-1}}.\end{aligned}$$

Given the result for the previous execise, $\Delta_r S^\ominus = +92.6\ \text{J K}^{-1}\ \text{mol}^{-1}$

$$\begin{aligned}\Delta_r G^\ominus &= (+53.40\ \text{kJ mol}^{-1}) - (298.15\ \text{K})\times(+0.0926\ \text{kJ K}^{-1}\ \text{mol}^{-1}) \\ &= \boxed{+25.8\ \text{kJ mol}^{-1}}.\end{aligned}$$

(iii)

$$\begin{aligned}\Delta_r H^\ominus &= \Delta_f H^\ominus(\text{HgCl}_2,(\text{s})) - \Delta_f H^\ominus(\text{Hg},(\text{l})) - \Delta_f H^\ominus(\text{Cl}_2,(\text{g})) \\ &= (-224.3\ \text{kJ mol}^{-1}) - 0 - 0 = \boxed{-224.3\ \text{kJ mol}^{-1}}.\end{aligned}$$

Given the result for the previous execise, $\Delta_r S^\ominus = -153.1\ \text{J K}^{-1}\ \text{mol}^{-1}$.

$$\begin{aligned}\Delta_r G^\ominus &= (-224.3\ \text{kJ mol}^{-1}) - (298.15\ \text{K})\times(-0.1531\ \text{kJ K}^{-1}\ \text{mol}^{-1}) \\ &= \boxed{-178.7\ \text{kJ mol}^{-1}}.\end{aligned}$$

E3D.2(a) The standard reaction entropy is given by [3C.3b–94], $\Delta_r S^\ominus = \sum_J \nu_J S_m^\ominus(J)$, where ν_J are the signed stoichiometric numbers.

$$\begin{aligned}\Delta_r S^\ominus &= 2S_m^\ominus(\text{I}_2,(\text{s})) + 2S_m^\ominus(\text{H}_2\text{O},(\text{l})) - 4S_m^\ominus(\text{HI},(\text{g})) - S_m^\ominus(\text{O}_2,(\text{g})) \\ &= 2\times(116.135\ \text{J K}^{-1}\ \text{mol}^{-1}) + 2\times(69.91\ \text{J K}^{-1}\ \text{mol}^{-1}) \\ &\quad - 4\times(206.59\ \text{J K}^{-1}\ \text{mol}^{-1}) - (205.138\ \text{J K}^{-1}\ \text{mol}^{-1}) \\ &= -659.40...\ \text{J K}^{-1}\ \text{mol}^{-1}\end{aligned}$$

The standard reaction enthalpy is given by [2C.5b–*55*], $\Delta_r H^\ominus = \sum_J \nu_J \Delta_f H^\ominus(J)$.

$$\begin{aligned}\Delta_r H^\ominus &= 2\Delta_f H^\ominus(I_2,(s)) + 2\Delta_f H^\ominus(H_2O,(l)) - 4\Delta_f H^\ominus(HI,(g))\\ &\quad - \Delta_f H^\ominus(O_2,(g))\\ &= 2\times 0 + 2\times(-285.83\ \text{kJ mol}^{-1}) - 4\times(+26.48\ \text{kJ mol}^{-1}) - 0\\ &= -677.58\ \text{kJ mol}^{-1}\end{aligned}$$

The standard reaction Gibbs energy is given by [3D.9–*100*], $\Delta_r G^\ominus = \Delta_r H^\ominus - T\Delta_r S^\ominus$.

$$\begin{aligned}\Delta_r G^\ominus &= (-677.58\ \text{kJ mol}^{-1}) - (298.15\ \text{K}) \times (-0.65940...\ \text{kJ K}^{-1}\ \text{mol}^{-1})\\ &= \boxed{-480.98\ \text{kJ mol}^{-1}}.\end{aligned}$$

E3D.3(a) The maximum non-expansion work is equal to the Gibbs free energy as explained in Section 3D.1(e) on page 100. The standard reaction Gibbs energy is given by [3D.10b–*101*], $\Delta_r G^\ominus = \sum_J \nu_J \Delta_f G^\ominus(J)$, where ν_J are the signed stoichiometric numbers. For the reaction $CH_4(g) + 3O_2(g) \longrightarrow CO_2(g) + 2H_2O(l)$

$$\begin{aligned}\Delta_r G^\ominus &= \Delta_f G^\ominus(CO_2,(g)) + 2\Delta_f G^\ominus(H_2O,(l)) - \Delta_f H^\ominus(CH_4,(g))\\ &\quad - 3\Delta_f H^\ominus(O_2,(g))\\ &= (-394.36\ \text{kJ mol}^{-1}) + 2\times(-237.13\ \text{kJ mol}^{-1})\\ &\quad - (-50.72\ \text{kJ mol}^{-1}) - 3\times 0\\ &= -817.90\ \text{kJ mol}^{-1}.\end{aligned}$$

Therefore, $|w_{\text{add,max}}| = |\Delta_r G^\ominus| = \boxed{817.90\ \text{kJ mol}^{-1}}$.

E3D.4(a) The standard reaction Gibbs energy is given by [3D.10b–*101*], $\Delta_r G^\ominus = \sum_J \nu_J \Delta_f G^\ominus(J)$, where ν_J are the signed stoichiometric numbers.

(i)

$$\begin{aligned}\Delta_r G^\ominus &= 2\Delta_f G^\ominus(CH_3COOH,(l)) - 2\Delta_f G^\ominus(CH_3CHO,(g)) - \Delta_f G^\ominus(O_2,(g))\\ &= 2\times(-389.9\ \text{kJ mol}^{-1}) - 2\times(-128.86\ \text{kJ mol}^{-1}) - 0\\ &= \boxed{-522.1\ \text{kJ mol}^{-1}}.\end{aligned}$$

(ii)

$$\begin{aligned}\Delta_r G^\ominus &= 2\Delta_f G^\ominus(AgBr,(s)) + \Delta_f G^\ominus(Cl_2,(g)) - 2\Delta_f G^\ominus(AgCl,(s))\\ &\quad - \Delta_f G^\ominus(Br_2,(l))\\ &= 2\times(-96.90\ \text{kJ mol}^{-1}) + 0 - 2\times(-109.79\ \text{kJ mol}^{-1}) - 0\\ &= \boxed{+25.78\ \text{kJ mol}^{-1}}.\end{aligned}$$

(iii)

$$\Delta_r G^\ominus = \Delta_f G^\ominus(\mathrm{HgCl_2},(\mathrm{s})) - \Delta_f G^\ominus(\mathrm{Hg},(\mathrm{l})) - \Delta_f G^\ominus(\mathrm{Cl_2},(\mathrm{g}))$$
$$= (-178.6\ \mathrm{kJ\,mol^{-1}}) - 0 - 0 = \boxed{-178.6\ \mathrm{kJ\,mol^{-1}}}.$$

E3D.5(a) Consider the reaction

$$\mathrm{CH_3COOC_2H_5(l) + 5O_2(g) \longrightarrow 4CO_2(g) + 4H_2O(l)}$$

The standard reaction enthalpy is [2C.5b–55], $\Delta_r H^\ominus = \sum_J \nu_J \Delta_f H^\ominus(J)$, where ν_J are the signed stoichiometric numbers.

$$\Delta_r H^\ominus = 4\Delta_f H^\ominus(\mathrm{CO_2},(\mathrm{g})) + 4\Delta_f H^\ominus(\mathrm{H_2O},(\mathrm{l})) - \Delta_f H^\ominus(\mathrm{CH_3COOC_2H_5},(\mathrm{l})) - 5\Delta_f H^\ominus(\mathrm{O_2},(\mathrm{g}))$$

when rearranged this gives

$$\Delta_f H^\ominus(\mathrm{CH_3COOC_2H_5},(\mathrm{l})) = 4\Delta_f H^\ominus(\mathrm{CO_2},(\mathrm{g})) + 4\Delta_f H^\ominus(\mathrm{H_2O},(\mathrm{l})) - 5\Delta_f H^\ominus(\mathrm{O_2},(\mathrm{g})) - \Delta_r H^\ominus$$
$$= 4 \times (-393.51\ \mathrm{kJ}) + 4 \times (-285.83\ \mathrm{kJ}) - 5 \times 0 - (-2231\ \mathrm{kJ\,mol^{-1}})$$
$$= -486.36\ \mathrm{kJ\,mol^{-1}}$$

The standard reaction entropy is given by [3C.3b–94], $\Delta_r S^\ominus = \sum_J \nu_J S_m^\ominus(J)$. Therefore, for the formation of the compound

$$\Delta_f S^\ominus(\mathrm{CH_3COOC_2H_5},(\mathrm{l})) = S_m^\ominus(\mathrm{CH_3COOC_2H_5},(\mathrm{l})) - 4S_m^\ominus(\mathrm{C},(\mathrm{s})) - 4S_m^\ominus(\mathrm{H_2},(\mathrm{g})) - S_m^\ominus(\mathrm{O_2},(\mathrm{g}))$$
$$= (259.4\ \mathrm{J\,K^{-1}\,mol^{-1}}) - 4 \times (5.740\ \mathrm{J\,K^{-1}\,mol^{-1}}) - 4 \times (130.684\ \mathrm{J\,K^{-1}\,mol^{-1}}) - (205.138\ \mathrm{J\,K^{-1}\,mol^{-1}})$$
$$= -4.91... \times 10^2\ \mathrm{J\,K^{-1}\,mol^{-1}}$$

The standard reaction Gibbs energy is defined in [3D.9–100], $\Delta_r G^\ominus = \Delta_r H^\ominus - T\Delta_r S^\ominus$, thus

$$\Delta_f G^\ominus(\mathrm{CH_3COOC_2H_5},(\mathrm{l})) = (-486.36\ \mathrm{kJ\,mol^{-1}}) - (298.15\ \mathrm{K})(-0.491...\ \mathrm{kJ\,K^{-1}\,mol^{-1}})$$
$$= \boxed{-340\ \mathrm{kJ\,mol^{-1}}}.$$

Solutions to problems

P3D.1 (a) From the perfect gas law, $pV = nRT$, the final pressure is

$$p = \frac{n_B R T_B}{V_{B,f}} = \frac{(2.00\ \mathrm{mol}) \times (8.3145 \times 10^{-2}\ \mathrm{dm^3\,bar\,K^{-1}\,mol^{-1}}) \times (300\ \mathrm{K})}{1.00\ \mathrm{dm^3}}$$
$$= 49.8...\ \mathrm{bar} = \boxed{49.9\ \mathrm{bar}}.$$

The final volume of Section A is $V_{A,f} = (V_A + V_B) - V_{B,f} = 3.00\ \text{dm}^3$. Therefore the final temperature of Section A is

$$T_{A,f} = \frac{pV_{A,f}}{nR} = \frac{(49.8...\ \text{bar}) \times (3.00\ \text{dm}^3)}{(2.00\ \text{mol}) \times (8.3145 \times 10^{-2}\ \text{dm}^3\ \text{bar}\ \text{K}^{-1}\ \text{mol}^{-1})}$$
$$= \boxed{900\ \text{K}}.$$

(b) Taking the hint, first consider the heating at constant volume. The entropy dependence on temperature at constant volume is given by [3B.7–90], $\Delta S = nC_{V,m} \ln(T_f/T_i)$, with C_p replaced by $nC_{V,m}$. The volume is then allowed to expand to the final. The entropy change for the isothermal expansion of a perfect gas is given by [3B.2–88], $\Delta S = nR\ln(V_f/V_i)$. Therefore the total change in the entropy for the gas in Section A is

$$\Delta S_A = nC_{V,m} \ln\left(\frac{T_{A,f}}{T_A}\right) + nR\ln\left(\frac{V_{A,f}}{V_A}\right)$$
$$= (2.00\ \text{mol}) \times (20.0\ \text{J}\,\text{K}^{-1}\,\text{mol}^{-1}) \times \ln\left(\frac{900\ \text{K}}{300\ \text{K}}\right)$$
$$+ (2.00\ \text{mol}) \times (8.3145\,\text{J}\,\text{K}^{-1}\,\text{mol}^{-1}) \times \ln\left(\frac{3.00\ \text{dm}^3}{2.00\ \text{dm}^3}\right)$$
$$= +50.6...\ \text{J}\,\text{K}^{-1} = \boxed{+50.7\ \text{J}\,\text{K}^{-1}}.$$

(c) Section B is kept at the constant temperature throughout the process, thus only the change in the volume needs to be considered

$$\Delta S_B = nR\ln\left(\frac{V_{B,f}}{V_B}\right)$$
$$= (2.00\ \text{mol}) \times (8.3145\,\text{J}\,\text{K}^{-1}\,\text{mol}^{-1}) \times \ln\left(\frac{1.00\ \text{dm}^3}{2.00\ \text{dm}^3}\right)$$
$$= -11.5...\ \text{J}\,\text{K}^{-1} = \boxed{-11.5\ \text{J}\,\text{K}^{-1}}.$$

(d) The change in internal energy as a result of a change in temperature assuming constant heat capacity is given by [2A.15b–45], $\Delta U = C_V \Delta T$. Because the internal energy of a perfect gas depends only on the temperature

$$\Delta U_A = (2.00\ \text{mol}) \times (20.0\ \text{J}\,\text{K}^{-1}\,\text{mol}^{-1}) \times (900\ \text{K} - 300\ \text{K})$$
$$= +2.40 \times 10^4\ \text{J} = \boxed{+24.0\ \text{kJ}}.$$
$$\Delta U_B = \boxed{0}.$$

(e) The Helmholtz energy is defined in [3D.4a–97], $A = U - TS$. For the finite changes it becomes $\Delta A = \Delta U - \Delta(TS) = \Delta U - T\Delta S - S\Delta T$. Because Section B is kept at constant temperature $\Delta T = 0$ and so

$$\Delta A_B = \Delta U_B - T_B \Delta S_B$$
$$= 0 - (300\ \text{K}) \times (-11.5...\ \text{J}\,\text{K}^{-1}) = \boxed{+3.46 \times 10^3\ \text{J}}.$$

Because ΔT is not zero and S is not given for the Section A, the equivalent expression cannot be evaluated.

(f) Because the process is reversible, the total $\Delta A = \Delta A_{\mathrm{A}} + \Delta A_{\mathrm{B}} = \boxed{0}$. This implies $\Delta A_{\mathrm{A}} = -\Delta A_{\mathrm{B}}$.

P3D.3 Consider the thermodynamic cycle shown in Fig. 3.4.

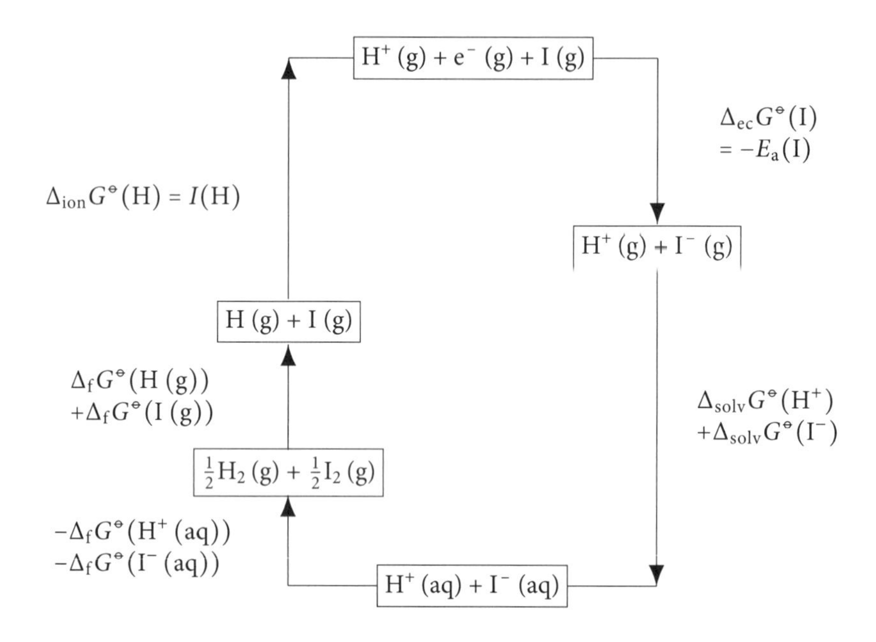

Figure 3.4

The sum of the Gibbs energies for all the steps around a closed cycle is zero.

$$0 = -\left[\Delta_f G^\ominus(\mathrm{H^+\,(aq)}) + \Delta_f G^\ominus(\mathrm{I^-\,(aq)})\right] + \Delta_f G^\ominus(\mathrm{H\,(g)}) + \Delta_f G^\ominus(\mathrm{I\,(g)})$$
$$I(\mathrm{H}) + (-E_a(\mathrm{I})) + \Delta_{\mathrm{solv}} G^\ominus(\mathrm{H^+}) + \Delta_{\mathrm{solv}} G^\ominus(\mathrm{I^-})$$

Because $\Delta_f G^\ominus(\mathrm{H^+\,(aq)} = 0$, by convention, the Gibbs energy of formation for the $\mathrm{I^-}$ (aq) is

$$\begin{aligned}\Delta_f G^\ominus(\mathrm{I^-\,(aq)}) &= \Delta_f G^\ominus(\mathrm{H\,(g)}) + \Delta_f G^\ominus(\mathrm{I\,(g)}) + I(\mathrm{H}) \\ &\quad - E_a(\mathrm{I}) + \Delta_{\mathrm{solv}} G^\ominus(\mathrm{H^+}) + \Delta_{\mathrm{solv}} G^\ominus(\mathrm{I^-}) \\ &= (203.25\ \mathrm{kJ\,mol^{-1}}) + (70.25\ \mathrm{kJ\,mol^{-1}}) + (1312.0\ \mathrm{kJ\,mol^{-1}}) \\ &\quad - (295.3\ \mathrm{kJ\,mol^{-1}}) + (-1090\ \mathrm{kJ\,mol^{-1}}) + (-247\ \mathrm{kJ\,mol^{-1}}) \\ &= \boxed{-47\ \mathrm{kJ\,mol^{-1}}}.\end{aligned}$$

P3D.5 The standard reaction Gibbs energy is given by [3D.9–*100*], $\Delta_r G^\ominus = \Delta_r H^\ominus - T\Delta_r S^\ominus$. The standard reaction entropy is [2C.5b–55], $\Delta_r S^\ominus = \sum_J \nu_J S_m^\ominus(\mathrm{J})$, where

ν_J are the signed stoichiometric numbers. Therefore

$$\begin{aligned}\Delta_r S_1^\ominus &= S_m^\ominus(\text{Li}^+,(\text{g})) + S_m^\ominus(\text{F}^-,(\text{g})) - S_m^\ominus(\text{LiF},(\text{s}))\\ &= (133\ \text{J K}^{-1}\ \text{mol}^{-1}) + (145\ \text{J K}^{-1}\ \text{mol}^{-1}) - (35.6\ \text{J K}^{-1}\ \text{mol}^{-1})\\ &= 242.4\ \text{J K}^{-1}\ \text{mol}^{-1}\end{aligned}$$

And so

$$\begin{aligned}\Delta_r G_1^\ominus &= (1037\ \text{kJ mol}^{-1}) - (298\ \text{K}) \times (0.2424\ \text{kJ K}^{-1}\ \text{mol}^{-1})\\ &= +9.64... \times 10^2\ \text{kJ mol}^{-1} = \boxed{+965\ \text{kJ mol}^{-1}}.\end{aligned}$$

For the second step $\Delta_r G_2^\ominus = \Delta_{\text{solv}} G^\ominus(\text{Li}^+) + \Delta_{\text{solv}} G^\ominus(\text{F}^-)$. The Gibbs energy of solvation in water is given by Born equation [3D.12b–*103*], $\Delta_{\text{solv}} G^\ominus = -(z_i^2/[r_i/\text{pm}]) \times 6.86 \times 10^4\ \text{kJ mol}^{-1}$, thus

$$\begin{aligned}\Delta_r G_2^\ominus &= -\left(\frac{(+1)^2}{[r(\text{Li}^+)/\text{pm}]} + \frac{(-1)^2}{[r(\text{F}^-)/\text{pm}]}\right) \times 6.86 \times 10^4\ \text{kJ mol}^{-1}\\ &= -\left(\frac{1}{127} + \frac{1}{163}\right) \times 6.86 \times 10^4\ \text{kJ mol}^{-1} = -9.61... \times 10^2\ \text{kJ mol}^{-1}\\ &= \boxed{-961\ \text{kJ mol}^{-1}}\end{aligned}$$

Therefore the total Gibbs energy change of the process

$$\begin{aligned}\Delta_r G^\ominus &= \Delta_r G_1^\ominus + \Delta_r G_2^\ominus = (+9.64... \times 10^2\ \text{kJ mol}^{-1}) + (-9.61... \times 10^2\ \text{kJ mol}^{-1})\\ &= +3.74...\ \text{kJ mol}^{-1} = \boxed{+4\ \text{kJ mol}^{-1}}.\end{aligned}$$

The change in positive implying that the reverse process is spontaneous.

3E Combining the First and Second Laws

Answer to discussion questions

D3E.1 The relation $(\partial G/\partial T)_p = -S$, combined with the fact that the entropy is always positive, shows that the Gibbs function of a system decreases as the temperature increases (at constant pressure).

Solutions to exercises

E3E.1(a) The Gibbs energy dependence on temperature for a perfect gas is given by [3E.14–*109*], $G_m(p_f) = G_m(p_i) + RT\ln(p_f/p_i)$. From the perfect gas law $p \propto (1/V)$. This allows rewriting the previous equation for the change in Gibbs energy due to isothermal gas expansion

$$\Delta G = nRT \ln\left(\frac{V_\mathrm{i}}{V_\mathrm{f}}\right)$$
$$= (2.5\times10^{-3}\ \mathrm{mol})\times(8.3145\ \mathrm{J\,K^{-1}\,mol^{-1}})\times(300\ \mathrm{K})\times\ln\left(\frac{42\ \mathrm{cm^3}}{600\ \mathrm{cm^3}}\right)$$
$$= \boxed{-17\ \mathrm{J}}.$$

E3E.2(a) The variation of the Gibbs energy with pressure is given by [3E.8–*107*], $(\partial G/\partial T)_p = -S$. The change in entropy is thus

$$\Delta S = S_\mathrm{f} - S_\mathrm{i} = -\left(\frac{\partial G_\mathrm{f}}{\partial T}\right)_p + \left(\frac{\partial G_\mathrm{i}}{\partial T}\right)_p = -\left(\frac{\partial (G_\mathrm{f}-G_\mathrm{i})}{\partial T}\right)_p$$
$$= -\left(\frac{\partial \Delta G}{\partial T}\right)_p = -\left(\frac{\partial[(-85.40\ \mathrm{J}) + T\times(36.5\ \mathrm{J\,K^{-1}})]}{\partial T}\right)_p$$
$$= -(36.5\ \mathrm{J\,K^{-1}}) = \boxed{-36.5\ \mathrm{J\,K^{-1}}}.$$

E3E.3(a) The Gibbs-Helmholtz relation for the change in Gibbs energy is given by [3E.11–*108*], $(\partial[\Delta G/T]/\partial T)_p = -\Delta H/T^2$. Expressing for the change in enthalpy gives

$$\Delta H = -T^2\left(\frac{\partial[\Delta G/T]}{\partial T}\right)_p = -T^2\left(\frac{\partial[(-85.40\ \mathrm{J})/T + (36.5\ \mathrm{J\,K^{-1}})]}{\partial T}\right)_p$$
$$= T^2\left(\frac{-85.40\ \mathrm{J}}{T^2}\right) = \boxed{-85.40\ \mathrm{J}}.$$

E3E.4(a) The molar Gibbs energy dependence on pressure for an incompressible substance is given by [3E.13–*108*], $G_\mathrm{m}(p_\mathrm{f}) = G_\mathrm{m}(p_\mathrm{i}) + (p_\mathrm{f}-p_\mathrm{i})V_\mathrm{m}$. Assuming that the volume of liquid octane changes little over the range of pressures considered

$$\Delta G = n[G_\mathrm{m}(p_\mathrm{f}) - G_\mathrm{m}(p_\mathrm{i})] = (p_\mathrm{f}-p_\mathrm{i})nV_\mathrm{m} = (p_\mathrm{f}-p_\mathrm{i})V$$
$$= (100\ \mathrm{atm} - 1.0\ \mathrm{atm})\times\frac{1.01325\times10^5\ \mathrm{Pa}}{1\ \mathrm{atm}}\times(1.0\times10^{-3}\ \mathrm{m^3})$$
$$= +1.00...\times10^4\ \mathrm{J} = \boxed{+10\ \mathrm{kJ}}.$$

For the molar Gibbs energy

$$\Delta G_\mathrm{m} = \frac{\Delta G}{n} = \frac{\Delta G}{m/M} = \frac{M\Delta G}{\rho V}$$
$$= \frac{(114.23\ \mathrm{g\,mol^{-1}})\times(+10.0...\ \mathrm{kJ})}{(0.703\ \mathrm{g\,cm^{-3}})\times(1.0\times10^3\ \mathrm{cm^3})} = \boxed{+1.6\ \mathrm{kJ\,mol^{-1}}}.$$

E3E.5(a) As explained in Section 3E.2(c) on page 108, the change in Gibbs energy of a phase transition varies with pressure as $\Delta_{trs}G_m(p_f) = \Delta_{trs}G_m(p_i)\int_{p_i}^{p_f}\Delta_{trs}V_m\,dp$. Assuming that $\Delta_{trs}V_m$ changes little over the range of pressures considered

$$\begin{aligned}\Delta G_m &= \Delta_{trs}G_m(p_f) - \Delta_{trs}G_m(p_i) = (p_f - p_i)\Delta_{trs}V_m \\ &= [(1000\times10^5\ \text{Pa}) - (1\times10^5\ \text{Pa})]\times(-1.6\times10^{-6}\ \text{m}^3\,\text{mol}^{-1}) \\ &= \boxed{-1.6\times10^2\ \text{J mol}^{-1}}.\end{aligned}$$

E3E.6(a) The Gibbs energy dependence on pressure for a perfect gas is given by [3E.14–*109*], $G_m(p_f) = G_m(p_i) + RT\ln(p_f/p_i)$, thus

$$\begin{aligned}\Delta G_m &= RT\ln\left(\frac{p_f}{p_i}\right) \\ &= (8.3145\,\text{J K}^{-1}\,\text{mol}^{-1})\times(298\ \text{K})\times\ln\left(\frac{100.0\ \text{atm}}{1.0\ \text{atm}}\right) \\ &= +11.4...\times10^3\ \text{J mol}^{-1} = \boxed{+11\ \text{kJ mol}^{-1}}.\end{aligned}$$

Solutions to problems

P3E.1 (a) The Gibbs-Helmholtz relation for the change in Gibbs energy is given by [3E.11–*108*], $(\partial[\Delta G/T]/\partial T)_p = -\Delta H/T^2$. Integrating the equation between the temperatures T_1 and T_2 and assuming that ΔH is temperature independent gives

$$\int_{T_1}^{T_2}\left(\frac{\partial}{\partial T}\frac{\Delta G(T)}{T}\right)_p dT = \Delta H\int_{T_1}^{T_2}-\frac{1}{T^2}\,dT$$

$$\frac{\Delta G(T_2)}{T_2} - \frac{\Delta G(T_1)}{T_1} = \Delta H\left(\frac{1}{T_2}-\frac{1}{T_1}\right)$$

Therefore

$$\frac{\Delta G(T_2)}{T_2} = \frac{\Delta G(T_1)}{T_1} + \Delta H\left(\frac{1}{T_2}-\frac{1}{T_1}\right)$$

(b) The standard reaction entropy is given by [3C.3b–*94*], $\Delta_r G^\ominus = \sum_J \nu_J\Delta_f G^\ominus(J)$, where ν_J are the signed stoichiometric numbers.

$$\begin{aligned}\Delta_r G^\ominus(298\ \text{K}) &= 2\Delta_f G^\ominus(\text{CO}_2\ (\text{g})) - 2\Delta_f G^\ominus(\text{CO}\ (\text{g})) - \Delta_f G^\ominus(\text{O}_2\ (\text{g})) \\ &= 2\times(-394.36\ \text{kJ mol}^{-1}) - 2\times(-137.17\ \text{kJ mol}^{-1}) - 0 \\ &= \boxed{-514.38\ \text{kJ mol}^{-1}}.\end{aligned}$$

Similarly, the standard reaction enthalpy is given by [2C.5b–*55*], $\Delta_r H^\ominus = \sum_J \nu_J\Delta_f H^\ominus(J)$.

$$\begin{aligned}\Delta_r H^\ominus(298\ \text{K}) &= 2\Delta_f H^\ominus(\text{CO}_2\ (\text{g})) - 2\Delta_f H^\ominus(\text{CO}\ (\text{g})) - \Delta_f H^\ominus(\text{O}_2\ (\text{g})) \\ &= 2\times(-393.51\ \text{kJ mol}^{-1}) - 2\times(-110.53\ \text{kJ mol}^{-1}) - 0 \\ &= \boxed{-565.96\ \text{kJ mol}^{-1}}.\end{aligned}$$

(c) The above derived expression is rearranged to give

$$\Delta G(T_2) = \Delta G(T_1)\frac{T_2}{T_1} + \Delta H\left(1 - \frac{T_2}{T_1}\right)$$

Hence

$$\begin{aligned}\Delta G(375\text{ K}) &= (-514.38\text{ kJ mol}^{-1})\frac{375\text{ K}}{298\text{ K}} \\ &\quad + (-565.96\text{ kJ mol}^{-1})\left(1 - \frac{375\text{ K}}{298\text{ K}}\right) \\ &= (-6.47\times10^{2}\text{ kJ mol}^{-1}) + (+1.46\times10^{2}\text{ kJ mol}^{-1}) \\ &= \boxed{-501\text{ kJ mol}^{-1}}.\end{aligned}$$

P3E.3 The given expression for the reaction Gibbs energy dependence on temperature is rearranged for $\Delta G(T_2)$ and becomes

$$\Delta_r G^{\ominus}(T_2) = \Delta_r G^{\ominus}(T_1)\frac{T_2}{T_1} + \Delta_r H^{\ominus}\left(1 - \frac{T_2}{T_1}\right)$$

Hence at 37 °C = 310 K

$$\begin{aligned}\Delta_r G^{\ominus}(310\text{ K}) &= (-6333\text{ kJ mol}^{-1})\frac{273.15\text{ K} + 37\text{ K}}{298\text{ K}} \\ &\quad + (-5797\text{ kJ mol}^{-1})\left(1 - \frac{273.15\text{ K} + 37\text{ K}}{298\text{ K}}\right) \\ &= -6355\text{ kJ mol}^{-1}\end{aligned}$$

The extra non-expansion work that is obtained by raising the temperature is the difference

$$\begin{aligned}\Delta_r G^{\ominus}(310\text{ K}) - \Delta_r G^{\ominus}(298\text{ K}) &= (-6355\text{ kJ mol}^{-1}) - (-6333\text{ kJ mol}^{-1}) \\ &= -22\text{ kJ mol}^{-1}\end{aligned}$$

Therefore the result is an extra $\boxed{22\text{ kJ mol}^{-1}}$ of energy that is available for non-expansion work.

P3E.5 Consider the exact differential of the Enthalpy, $H = U + pV$

$$dH = dU + d(pV) = dU + Vdp + pdV$$

The exact differential of the internal energy is given by the fundamental equation [3E.1–*104*], $dU = TdS - pdV$, hence

$$dH = TdS - pdV + Vdp + pdV = TdS + Vdp$$

Because dH is the exact differential this implies

$$\left(\frac{\partial H}{\partial S}\right)_p = T \quad \text{and} \quad \left(\frac{\partial H}{\partial p}\right)_S = V$$

The mixed partial derivatives are equal irrespective of the order

$$\left(\frac{\partial}{\partial p}\left(\frac{\partial H}{\partial S}\right)_p\right)_S = \left(\frac{\partial}{\partial S}\left(\frac{\partial H}{\partial p}\right)_S\right)_p$$

Therefore

$$\boxed{\left(\frac{\partial T}{\partial p}\right)_S = \left(\frac{\partial V}{\partial S}\right)_p}$$

Similarly consider the exact differentials of the Helmholtz energy, $A = U - TS$, and the Gibbs energy, $G = H - TS$. Starting with the Helmholtz energy

$$\begin{aligned} \mathrm{d}A &= \mathrm{d}U - \mathrm{d}(TS) = \mathrm{d}U - T\mathrm{d}S - S\mathrm{d}T \\ &= T\mathrm{d}S - p\mathrm{d}V - T\mathrm{d}S - S\mathrm{d}T = -p\mathrm{d}V - S\mathrm{d}T \end{aligned}$$

It follows that

$$\boxed{\left(\frac{\partial p}{\partial T}\right)_V = \left(\frac{\partial S}{\partial V}\right)_T}$$

For the Gibbs energy, the above derived result for $\mathrm{d}H$ is used

$$\begin{aligned} \mathrm{d}G &= \mathrm{d}H - \mathrm{d}(TS) = T\mathrm{d}S + V\mathrm{d}p - T\mathrm{d}S - S\mathrm{d}T \\ &= V\mathrm{d}p - S\mathrm{d}T \end{aligned}$$

It follows that

$$\boxed{\left(\frac{\partial V}{\partial T}\right)_p = -\left(\frac{\partial S}{\partial p}\right)_T}.$$

P3E.7 (a) Assuming that $a = 0$ and $b \neq 0$, the van der Waals equation becomes $p = RT/(V_\mathrm{m} - b)$. The molar volume is thus

$$V_\mathrm{m} = \frac{RT}{p} + b$$

Consider the exact differential of the molar Gibbs energy at constant temperature, $\mathrm{d}G_\mathrm{m} = (\partial G_\mathrm{m}/\partial p)_T\,\mathrm{d}p$. Integrating this gives

$$\int_{G_\mathrm{m,i}}^{G_\mathrm{m,f}} \mathrm{d}G_\mathrm{m} = \int_{p_\mathrm{i}}^{p_\mathrm{f}} \left(\frac{\partial G_\mathrm{m}}{\partial p}\right)_T \mathrm{d}p = \int_{p_\mathrm{i}}^{p_\mathrm{f}} V_\mathrm{m}\,\mathrm{d}p = \int_{p_\mathrm{i}}^{p_\mathrm{f}} \left(\frac{RT}{p} + b\right)\mathrm{d}p$$

Therefore

$$\boxed{G_\mathrm{m}(p_\mathrm{f}) = G_\mathrm{m}(p_\mathrm{i}) + RT\ln\left(\frac{p_\mathrm{f}}{p_\mathrm{i}}\right) + b(p_\mathrm{f} - p_\mathrm{i})}$$

The change in Gibbs energy energy increases more rapidly with pressure than the perfect gas due to the last term originating from the repulsion.

(b) Assuming that $a \neq 0$ and $b = 0$, the van der Waals equation becomes $p = RT/V_m) + a/V_m^2$. This is rearranged into a quadratic equation in V_m

$$pV_m^2 - RTV_m + a = 0$$

The solutions for V_m are

$$V_m = \frac{-(-RT) \pm \sqrt{(-RT)^2 - 4pa}}{2p} = \frac{RT}{2p} \pm \frac{RT}{2p}\sqrt{1 - \frac{4pa}{R^2T^2}}$$

Because the van der Waals equation is a correction to the ideal gas, the result should be approximately similar. Considering $4pa/(RT)^2 \ll 1$, it is obvious that only a positive root reproduces the perfect gas and hence is physically relevant. This solution is used further to apply the suggested approximate expansion

$$V_m = \frac{RT}{2p} + \frac{RT}{2p}\sqrt{1 - \frac{4pa}{R^2T^2}} \approx \frac{RT}{2p} + \frac{RT}{2p}\left(1 - \tfrac{1}{2}\left[\frac{4pa}{R^2T^2}\right]\right) = \boxed{\frac{RT}{p} - \frac{a}{pRT}}.$$

Integrating this as before gives the Gibbs energy dependence on pressure

$$\int_{G_{m,i}}^{G_{m,f}} dG_m = \int_{p_i}^{p_f} V_m\,dp = \int_{p_i}^{p_f} \frac{RT}{p} - \frac{a}{pRT}dp$$

Therefore

$$\boxed{G_m(p_f) = G_m(p_i) + RT\ln\left(\frac{p_f}{p_i}\right) - \frac{a}{RT}\ln\left(\frac{p_f}{p_i}\right)}$$

The change in Gibbs energy energy decreases more rapidly with pressure than the perfect gas due to the last term originating from the attractive interaction between the molecules term.

(c) Using the given data the change in molar Gibbs energy, $\Delta G_m(p) = G_m(p) - G_m(p^\ominus)$, is plotted against $(p/p^\ominus)$ at 298 K using the requested units (Fig. 3.5). A zoomed version of the same plot is shown in Fig. 3.6.

Answers to integrated activities

I3.1 The Joule–Thomson coefficient is defined as [2D.12–63], $\mu = (\partial T/\partial p)_H$. In a Joule–Thomson expansion the enthalpy is constant, therefore the temperature change is computed simply as

$$\Delta T = \mu\Delta p = (0.21\ \text{K atm}^{-1}) \times (100\ \text{atm} - 1.00\ \text{atm}) = -20.7...\ \text{K} = \boxed{-20.8\ \text{K}}$$

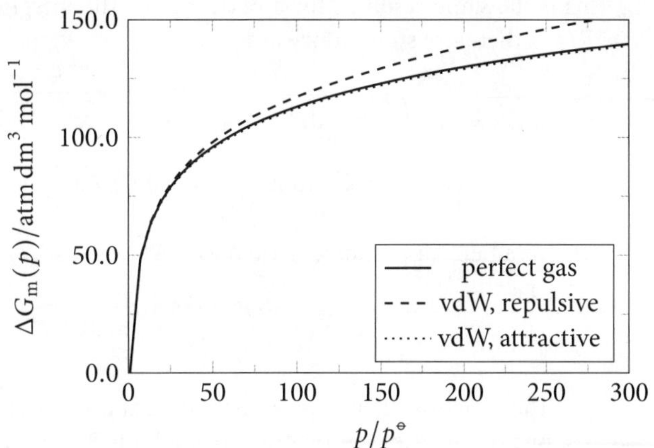

Figure 3.5

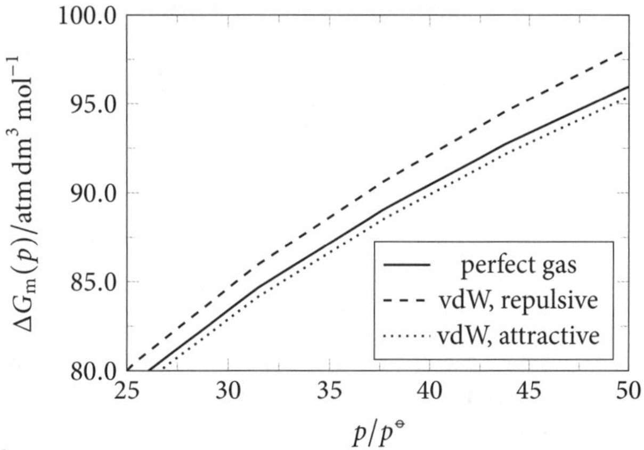

Figure 3.6

The temperature after the expansion is therefore 373 − 20.7... = $3.52... \times 10^2$ K.

The entropy change accompanying this expansion is separated into a constant temperature expansion and then a second step involving cooling at constant pressure. For the first step the entropy change is given by [3B.7–*90*]

$$\Delta S_{m,1} = C_{p,m}\ln(T_f/T_i) = \tfrac{5}{2} \times (8.3145\,\text{J K}^{-1}\,\text{mol}^{-1}) \times \ln\frac{3.52... \times 10^2\ \text{K}}{373\ \text{K}}$$
$$= -1.19...\ \text{J K}^{-1}\,\text{mol}^{-1}$$

For the second step the variation of entropy with pressure is given by one of the Maxwell relations from Table 3E.1 on page 105, $(\partial S/\partial p)_T = -(\partial V/\partial T)_p$. In this case $pV_m = RT(1+Bp)$, and the virial coefficient is temperature dependent and of the form $B = \alpha/T$. Therefore $V_m = RT/p + RTB = RT/p + \alpha R$. It follows that

$$\left(\frac{\partial V_m}{\partial T}\right)_p = R/p$$

This is the same result as for a perfect gas. The integration of $(\partial S_m/\partial p)_T = -R/p$ is therefore straightforward

$$\Delta S_{m,2} = -R\int_{p_i}^{p_f} p^{-1}\,dp = R\ln(p_i/p_f)$$
$$= (8.3145\,J\,K^{-1}\,mol^{-1}) \times \ln(100/1.00) = +38.2...\,J\,K^{-1}\,mol^{-1}$$

The overall entropy change is therefore $\Delta S_{m,1} + \Delta S_{m,2} = -1.19... + 38.2... = \boxed{+37.1\,J\,K^{-1}\,mol^{-1}}$.

I3.3 (a) The variation of entropy with volume at constant temperature is given by one of the Maxwell relations from Table 3E.1 on page 105, $(\partial S/\partial V)_T = (\partial p/\partial T)_V$. Working with molar quantities, the van der Waals equation of state is $p = RT/(V_m - b) - a/V_m^2$, therefore $(\partial p/\partial T)_V = R/(V_m - b)$. The integration is then straightforward

$$\Delta S_m = \int_{V_{m,i}}^{V_{m,f}} \frac{R}{V_m - b}\,dV_m = R\ln\frac{V_{m,f} - b}{V_{m,i} - b}$$
$$= (8.3145\,J\,K^{-1}\,mol^{-1})$$
$$\times \ln\frac{(10.0\,dm^3\,mol^{-1}) - (4.29\times10^{-2}\,dm^3\,mol^{-1})}{(1.00\,dm^3\,mol^{-1}) - (4.29\times10^{-2}\,dm^3\,mol^{-1})}$$
$$= \boxed{+19.5\,J\,K^{-1}\,mol^{-1}}$$

where the value of b is taken from the tables in the *Resource section*; note the conversion of the molar volumes to $dm^3\,mol^{-1}$ so as to match the units of b.

(b) The variation of entropy with temperature at constant volume and pressure are given by

$$\Delta S_m = C_{V,m}\ln(T_2/T_1) \quad \text{and} \quad \Delta S_m = C_{p,m}\ln(T_2/T_1)$$

respectively; both relationships assume that the heat capacities do not change in the temperature interval.

The equipartition theorem, *The chemist's toolkit* 7 in Topic 2A, is used to estimate the value of $C_{V,m}$; for a perfect gas $C_{p,m} = C_{V,m} + R$. For atoms there are just three translational degrees of freedom therefore $C_{V,m} = \frac{3}{2}R$ and $C_{p,m} = \frac{5}{2}R$. For linear rotors there are in addition two rotational degrees of freedom, therefore $C_{V,m} = \frac{5}{2}R$ and $C_{p,m} = \frac{7}{2}R$. For non-linear rotors there are three rotational degrees of freedom, $C_{V,m} = 3R$ and $C_{p,m} = 4R$.

Figures 3.7 and 3.8 show plots of $\Delta S_m/R$ against $\ln(T_2/T_1)$ for the constant volume and constant pressure cases, respectively.

(c) The change in entropy as a function of temperature is given by [3B.6–90];

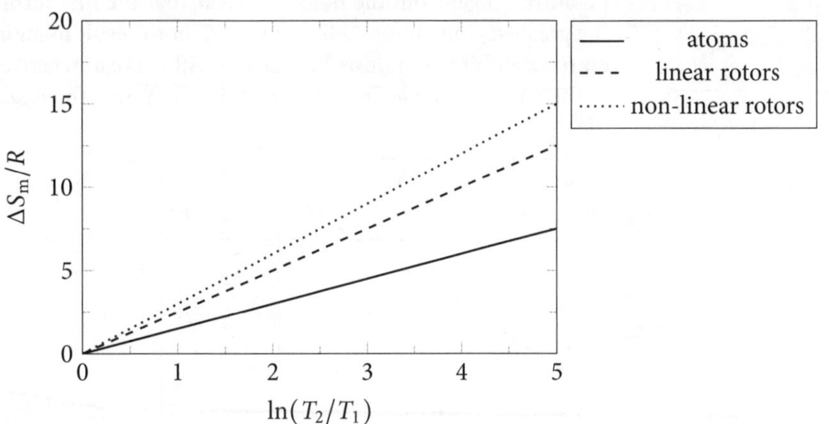

Figure 3.7

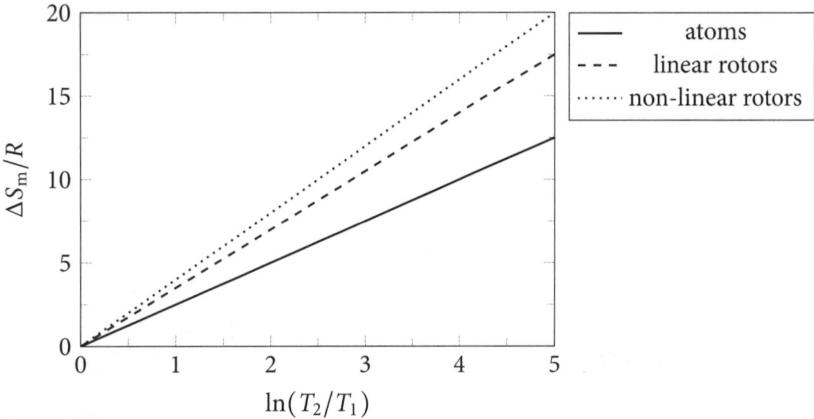

Figure 3.8

this is integrated for the particular form of the heat capacity suggested

$$\Delta S = \int_{T_i}^{T_f} \frac{C}{T}\,dT = \int_{T_i}^{T_f} \left(\frac{a}{T} + b + \frac{c}{T^3}\right) dT$$

$$= \underbrace{a \ln \frac{T_f}{T_i}}_{\text{term 1}} + \underbrace{b(T_f - T_f)}_{\text{term 2}} \underbrace{-\tfrac{1}{2}c\left(\frac{1}{T_f^2} - \frac{1}{T_i^2}\right)}_{\text{term 3}}$$

A convenient way of exploring this result is to choose a specific temperature range, say from 273 K to 473 K, and then plot the contribution of each of the three terms as a function of the relevant parameter, a, b, or c. Referring to the data in the *Resource section* it is seen that the ranges of these parameters are: for a, between 15 J K^{-1} mol^{-1} and 80 J K^{-1} mol^{-1}; for b between 0 and 50×10^{-3} J K^{-2} mol^{-1}; and for c between -10×10^5 J K mol^{-1} and $+2.0 \times 10^5$ J K mol^{-1}.

Figure 3.9 compares the contributions made by three terms over this temperature range; from the plots it is clear that the first term makes by far the greatest contribution. Terms 1 and 2 both result in an increase in the entropy with temperature, but term 3 will make a negative contribution to the entropy change if $c < 0$, which is commonly the case.

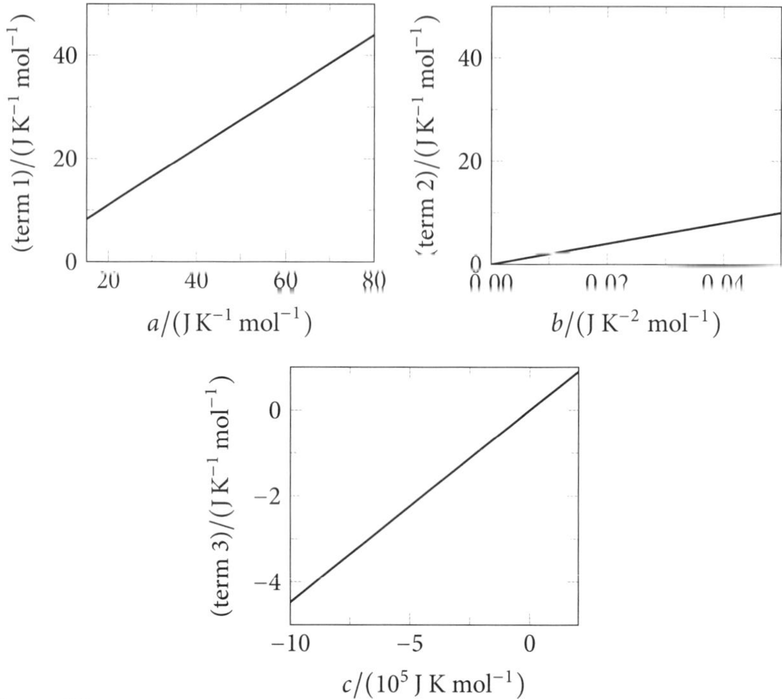

Figure 3.9

(d) The variation of G with p at constant T is given by [3E.8–*107*], $(\partial G/\partial p)_T = V$. The physical significance of the derivative is therefore that it is equal to the volume of the system. For a perfect gas, $V = nRT/p$, which makes the integration straightforward to give $\Delta G = nRT\ln(p_f/p_i)$ ([3E.14–*109*]). Figure 3.10 shows a plot of $\Delta G/nRT$ as a function of p_f/p_i. The Gibbs energy increases with pressure at constant temperature.

(e) The fugacity coefficient is given in terms of the compression factor Z by

$$\ln\phi = \int_0^p \frac{Z-1}{p}\,\mathrm{d}p \qquad Z = \frac{pV_m}{RT}$$

The pressure, volume and temperature can be expressed in terms of the reduced variables p_r, V_r, and T_r, given by

$$p_r = p/p_c \quad V_r = V_m/V_c \quad T_r = T/T_c$$

where the critical values of p, V, and T are given in terms of the van der Waals parameters by

$$p_c = a/27b^2 \quad V_c = 3b \quad T_c = 8a/27bR$$

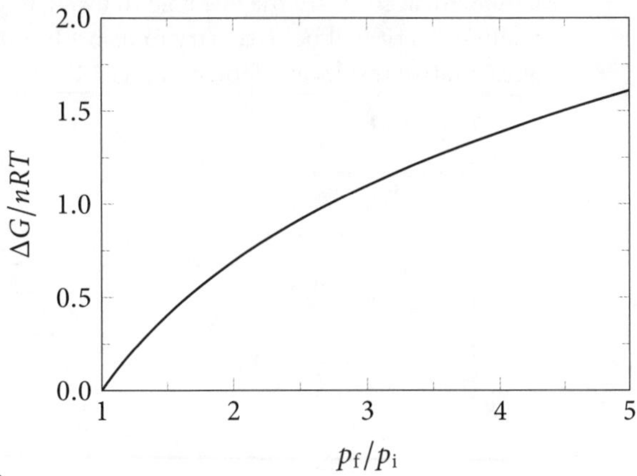

Figure 3.10

The compression factor can therefore be written

$$Z = \frac{pV_m}{RT} = \frac{p_c V_c}{RT_c}\frac{p_r V_r}{T_r} = \frac{1}{R}\frac{a}{27b^2} \times 3b \times \frac{27bR}{8a} \times \frac{p_r V_r}{T_r} = \frac{3}{8}\frac{p_r V_r}{T_r}$$

The aim is to write Z in terms of just V_r and T_r, therefore p_r is substituted by

$$p_r = \frac{8T_r}{3V_r - 1} - \frac{3}{V_r^2}$$

to give

$$Z = \frac{3}{8}\frac{V_r}{T_r}\left(\frac{8T_r}{3V_r - 1} - \frac{3}{V_r^2}\right) = \frac{3V_r}{3V_r - 1} - \frac{9}{8T_r V_r}$$

The variable of integration is p, and it is desired to change this to V_r, for which the following derivative is required

$$\frac{dp}{dV_r} = p_c \frac{dp_r}{dV_r} = p_c \frac{d}{dV_r}\left(\frac{8T_r}{3V_r - 1} - \frac{3}{V_r^2}\right)$$
$$= p_c\left(\frac{-24T_r}{(3V_r - 1)^2} + \frac{6}{V_r^3}\right)$$

The lower limit of the integral is $p = 0$, which corresponds to $V_r = \infty$. The integral therefore becomes

$$\ln\phi = \int_0^p \frac{Z - 1}{p}\,dp$$
$$= \int_\infty^{V_r} \frac{Z - 1}{p_c p_r} p_c\left(\frac{-24T_r}{(3V_r - 1)^2} + \frac{6}{V_r^3}\right) dV_r$$
$$= \int_\infty^{V_r} \frac{Z - 1}{p_r}\left(\frac{-24T_r}{(3V_r - 1)^2} + \frac{6}{V_r^3}\right) dV_r$$
$$= \int_\infty^{V_r} \left(\frac{3V_r}{3V_r - 1} - \frac{9}{8T_r V_r} - 1\right)\left(\frac{8T_r}{3V_r - 1} - \frac{3}{V_r^2}\right)^{-1}\left(\frac{-24T_r}{(3V_r - 1)^2} + \frac{6}{V_r^3}\right) dV_r$$

Mathematical software may be able to evaluate this integral analytically, or failing that it will be necessary to resort to numerical methods. Some representative results are shown in Fig. 3.11.

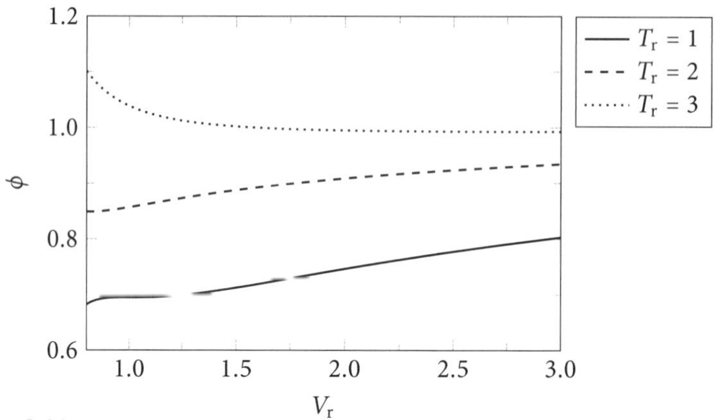

Figure 3.11

4 Physical transformations of pure substances

4A Phase diagrams of pure substances

Answers to discussion questions

D4A.1 Chemical potential is the single function that governs phase stability. The phase whose chemical potential is least under a set of given conditions is the most stable. Conditions under which two or more phases have equal chemical potentials are conditions under which those phases are in equilibrium. Understanding how chemical potential varies with physical conditions such as temperature, pressure, and composition makes it possible to compute chemical potentials for various phases and to map out the conditions for stability of those phases and for equilibrium between them.

D4A.3 For two phases to be in equilibrium, the chemical potentials of each component must be equal in the two phases. In a one-component system, this means that the chemical potential of that one component must be the same in all phases that are in equilibrium. The chemical potential is a function of two variables, say p and T (and not of composition in a one-component system). Thus, if there are four phases α, β, γ, and δ in equilibrium the chemical potentials would need to satisfy

$$\mu_{\alpha}(p,T) = \mu_{\beta}(p,T) = \mu_{\gamma}(p,T) = \mu_{\delta}(p,T)$$

This is a set of three independent equations in only two variables (p and T), which are not compatible.

Solutions to exercises

E4A.1(a) In a phase diagram, a single phase is represented by an area, while a line represents a phase boundary where two phases coexist in equilibrium. Point a lies within an area and therefore only $\boxed{\text{one phase}}$ is present. Points b and d each lie on the boundary between two areas, and therefore in each case $\boxed{\text{two phases}}$ are present. Point c lies at the intersection of three phase boundaries, where $\boxed{\text{three phases}}$ are present in equilibrium.

E4A.2(a) The change in Gibbs energy when an infinitesimal amount dn of substance is moved from location 1 to location 2 is given by (Section 4A.1(c) on page 121)

$$\mathrm{d}G = (\mu_2 - \mu_1)\mathrm{d}n$$

Assuming that 0.1 mmol is a sufficiently small amount to be regarded as infinitesimal, the Gibbs energy change in this case is

$$\Delta G = (\mu_2 - \mu_1)\Delta n = (7.1 \times 10^3\ \mathrm{J\,mol^{-1}}) \times (0.1 \times 10^{-3}\ \mathrm{mol}) = \boxed{0.71\ \mathrm{J}}$$

E4A.3(a) Use the phase rule [4A.1–*124*], $F = C - P + 2$, with $C = 2$ (for two components). Rearranging for the number of phases gives

$$P = C - F + 2 = 2 - F + 2 = 4 - F$$

The number of variables that can be changed arbitrarily, F, cannot be smaller than zero so the maximum number of phases in this case is $\boxed{4}$.

E4A.4(a) Use the phase rule [4A.1–*124*], $F = C - P + 2$, with $C = 1$ (one component). Inserting $P = 1$ gives $F = 1 - 1 + 2 = 2$. The condition $P = 1$ therefore represents an $\boxed{\text{area}}$. An area has $F = 2$ because it is possible to vary pressure and temperature independently (within limits) and stay within the area. $P = 1$ indicates that a single phase is present, so this result confirms that a single phase is represented by an area in a phase diagram.

E4A.5(a)

(i) 200 K and 2.5 atm lies on the boundary between solid and gas phases. $\boxed{\text{Two phases}}$, solid and gas, would therefore be present in equilibrium under these conditions.

(ii) 300 K and 4 atm lies in the vapour region, so only $\boxed{\text{one phase}}$, vapour, will be present.

(iii) 310 K is greater than the critical temperature, which means that there is no distinction between gas and liquid. Therefore only $\boxed{\text{one phase}}$ (a supercritical fluid) will be present at all pressures.

Solutions to problems

P4A.1

(a) 100 K and 1 atm lies in the solid region of the phase diagram, so initially only solid carbon dioxide (dry ice) will be present. When the temperature reaches 194.7 K, the sublimation point of CO_2 at 1 atm, solid and gas phases will be present in equilibrium. Above this temperature only gaseous CO_2 is present.

(b) 100 K and 70 atm lies in the solid region of the phase diagram, so again CO_2 will initially be a solid. On heating, a point is reached at which the solid melts; at this temperature solid and liquid phases are both present in equilibrium. Above this temperature only a liquid phase is present until the boiling temperature is reached, at which point liquid and gas will be in equilibrium. Above this temperature, only the gas phase will be present.

P4A.3 A schematic phase diagram is shown in Fig 4.1. Note that in reality the phase boundaries may be curved rather than straight. There are two triple points which are marked with dots.

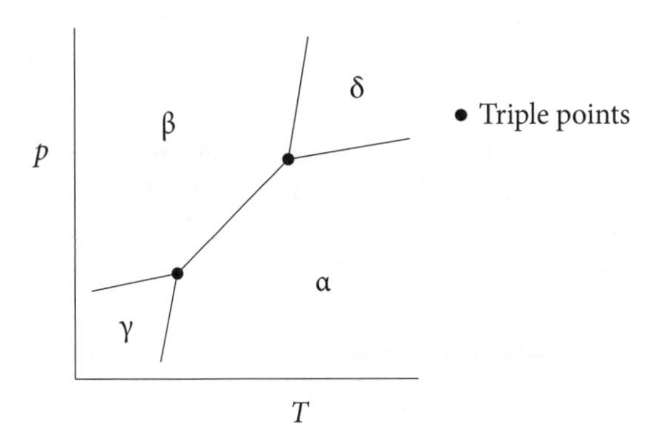

Figure 4.1

4B Thermodynamic aspects of phase transitions

Answers to discussion questions

D4B.1 Formally, the temperature derivative of the chemical potential is $(\partial\mu/\partial T)_p = -S_m$. Because the molar entropy is always positive for all pure substances, the slope of the change in chemical potential with respect to change in temperature is negative: that is, the chemical potential decreases with increasing temperature.

D4B.3 The operation of a DSC is described in Section 2C.4 on page 56. A phase transition involves an enthalpy change and the associated heat, as well as the temperature at which the transition occurs, may be measured using a DSC.

Solutions to exercises

E4B.1(a) The variation of chemical potential with temperature is given by [4B.1a–*128*], $(\partial\mu/\partial T)_p = -S_m$. For a finite change this gives $\Delta\mu = -S_m\Delta T$.

$$\Delta\mu(\text{liquid}) = -(65\ \text{J K}^{-1}\,\text{mol}^{-1}) \times (1\ \text{K}) = \boxed{-65\ \text{J mol}^{-1}}$$
$$\Delta\mu(\text{solid}) = -(43\ \text{J K}^{-1}\,\text{mol}^{-1}) \times (1\ \text{K}) = \boxed{-43\ \text{J mol}^{-1}}$$

The chemical potentials of both solid and liquid are decreased at the higher temperature, but the chemical potential of the liquid is decreased by a greater amount. As they were at equilibrium before it follows that the liquid is the more stable phase at the higher temperature, so melting will be spontaneous.

E4B.2(a) The variation of chemical potential with temperature is given by [4B.1a–*128*], $(\partial\mu/\partial T)_p = -S_\mathrm{m}$. For a finite change this gives $\Delta\mu = -S_\mathrm{m}\Delta T$, assuming that S_m is constant over the temperature range.

$$\Delta\mu = -(69.9\,\mathrm{J\,K^{-1}\,mol^{-1}}) \times ([35-25]\,\mathrm{K}) = \boxed{-699\,\mathrm{J\,mol^{-1}}}$$

E4B.3(a) The variation of chemical potential with pressure is given by [4B.1b–*128*], $(\partial\mu/\partial p)_T = V_\mathrm{m}$. For a finite change, and assuming that V_m is constant over the pressure range, this gives $\Delta\mu = V_\mathrm{m}\Delta p$. The molar volume V_m is given by M/ρ where M is the molar mass of copper and ρ is the mass density.

$$\begin{aligned}\Delta\mu &= V_\mathrm{m}\Delta p = (M/\rho)\Delta p\\ &= \frac{63.55\times10^{-3}\,\mathrm{kg\,mol^{-1}}}{8960\,\mathrm{kg\,m^{-3}}} \times ([10\times10^6 - 100\times10^3]\,\mathrm{Pa}) = \boxed{+70\,\mathrm{J\,mol^{-1}}}\end{aligned}$$

Note that $1\,\mathrm{Pa\,m^3} = 1\,\mathrm{J}$.

E4B.4(a) The variation of vapour pressure with applied pressure is given by [4B.2–*130*], $p = p^* \mathrm{e}^{V_\mathrm{m}(\mathrm{l})\Delta p/RT}$.

$$\begin{aligned}p &= (2.34\times10^3\,\mathrm{Pa})\times\exp\left(\frac{(18.1\times10^{-6}\,\mathrm{m^3\,mol^{-1}})\times([20\times10^6 - 1\times10^5]\,\mathrm{Pa})}{(8.3145\,\mathrm{J\,K^{-1}\,mol^{-1}})\times([20+273.15]\,\mathrm{K})}\right)\\ &= 2710\,\mathrm{Pa} = \boxed{2.71\,\mathrm{kPa}}\end{aligned}$$

E4B.5(a) The relationship between pressure and temperature along the solid–liquid boundary is given by [4B.7–*132*], $p = p^* + (\Delta_\mathrm{fus}H/T^*\Delta_\mathrm{fus}V)(T-T^*)$, which is rearranged to give $\Delta_\mathrm{fus}H$. In this case $p^* = 1.00\,\mathrm{atm}$, $T^* = 350.75\,\mathrm{K}$, $p = 100\,\mathrm{atm}$ and $T = 351.26\,\mathrm{K}$.

$$\begin{aligned}\Delta_\mathrm{fus}H &= \frac{p-p^*}{T-T^*}T^*\Delta_\mathrm{fus}V\\ &= \frac{([100-1]\,\mathrm{atm})\times(1.01325\times10^5\,\mathrm{Pa/1\,atm})}{([351.26-350.75]\,\mathrm{K})}\times(350.75\,\mathrm{K})\\ &\quad\times([163.3-161.0]\times10^{-6}\,\mathrm{m^3\,mol^{-1}})\\ &= 1.58...\times10^4\,\mathrm{J\,mol^{-1}} = \boxed{15.9\,\mathrm{kJ\,mol^{-1}}}\end{aligned}$$

The entropy of transition is given by [3B.4–*89*], $\Delta_\mathrm{fus}S = \Delta_\mathrm{fus}H/T$, where T is the transition temperature. At the melting temperature the entropy of fusion is

$$\Delta_\mathrm{fus}S = \frac{1.58...\times10^4\,\mathrm{J\,mol^{-1}}}{350.75\,\mathrm{K}} = \boxed{45.2\,\mathrm{J\,K^{-1}\,mol^{-1}}}$$

E4B.6(a) The integrated version of the Clausius–Clapeyron equation [4B.10–133] is given by $\ln(p/p^*) = -(\Delta_{\text{vap}}H/R)(1/T - 1/T^*)$. Rearranging for T gives

$$T = \left(\frac{1}{T^*} - \frac{R}{\Delta_{\text{vap}}H}\ln\frac{p}{p^*}\right)^{-1}$$

$$= \left(\frac{1}{[24.1+273.15]\,\text{K}} - \frac{8.3145\,\text{J K}^{-1}\,\text{mol}^{-1}}{28.7\times10^3\,\text{J mol}^{-1}} \times \ln\frac{70.0\,\text{kPa}}{53.3\,\text{kPa}}\right)^{-1}$$

$$= \boxed{304\,\text{K}} \text{ or } \boxed{31.2\,^\circ\text{C}}$$

E4B.7(a) The Clausius–Clapeyron equation [4B.9–133] is $\text{d}\ln p/\text{d}T = \Delta_{\text{vap}}H/RT^2$. This equation is rearranged for $\Delta_{\text{vap}}H$, and the expression for $\ln p$ is differentiated. It does not matter that the pressure is given in units of Torr because only the slope of $\ln p$ is required.

$$\Delta_{\text{vap}}H = RT^2\frac{\text{d}\ln p}{\text{d}T} = RT^2\frac{\text{d}}{\text{d}T}\left(16.255 - \frac{2501.8\,\text{K}}{T}\right) = RT^2\left(\frac{2501.8\,\text{K}}{T^2}\right)$$

$$= (2501.8\,\text{K})R = (2501.8\,\text{K}) \times (8.3145\,\text{J K}^{-1}\,\text{mol}^{-1}) = \boxed{20.801\,\text{kJ mol}^{-1}}$$

E4B.8(a) (i) The Clausius–Clapeyron equation [4B.9–133] is $\text{d}\ln p/\text{d}T = \Delta_{\text{vap}}H/RT^2$. This equation is rearranged for $\Delta_{\text{vap}}H$ and the expression for $\ln p$ is differentiated, noting from inside the front cover that $\ln x = (\ln 10)\log x$. It does not matter that the pressure is given in units of Torr because only the slope of $\ln p$ is required.

$$\Delta_{\text{vap}}H = RT^2\frac{\text{d}\ln p}{\text{d}T} = RT^2\ln 10\frac{\text{d}\log p}{\text{d}T} = RT^2\ln 10\frac{\text{d}}{\text{d}T}\left(7.960 - \frac{1780\,\text{K}}{T}\right)$$

$$= RT^2\ln 10\left(\frac{1780\,\text{K}}{T^2}\right) = (1780\,\text{K})R\ln 10$$

$$= (1780\,\text{K}) \times (8.3145\,\text{J K}^{-1}\,\text{mol}^{-1}) \times \ln 10 = \boxed{34.08\,\text{kJ mol}^{-1}}$$

(ii) The normal boiling point refers to the temperature at which the vapour pressure is 1 atm which is 760 Torr. The given expression, $\log(p/\text{Torr}) = 7.960 - (1780\text{K})/T$, is rearranged for T and a pressure of 760 Torr is substituted into it to give

$$T = \frac{1780\,\text{K}}{7.960 - \log(p/\text{Torr})} = \frac{1780\,\text{K}}{7.960 - \log 760} = \boxed{350.4\,\text{K}} \text{ or } \boxed{77.30\,^\circ\text{C}}$$

Note that this temperature lies outside the range 10 °C to 30 °C for which the expression for $\log(p/\text{Torr})$ is known to be valid, and is therefore an estimate.

E4B.9(a) The relationship between pressure and temperature along the solid–liquid boundary is given by [4B.7–132], $p = p^* + (\Delta_{\text{fus}}H/T^*\Delta_{\text{fus}}V)(T - T^*)$. The value of

$\Delta_{\text{fus}}V$ is found by using $V_{\text{m}} = M/\rho$ where M is the molar mass and ρ is the mass density:

$$\Delta_{\text{fus}}V = V_{\text{m}}(\text{l}) - V_{\text{m}}(\text{s}) = \frac{M}{\rho(\text{l})} - \frac{M}{\rho(\text{s})}$$

$$= \frac{78.1074\,\text{g}\,\text{mol}^{-1}}{0.879\times10^{6}\,\text{g}\,\text{m}^{-3}} - \frac{78.1074\,\text{g}\,\text{mol}^{-1}}{0.891\times10^{6}\,\text{g}\,\text{m}^{-3}} = 1.19...\times10^{-6}\,\text{m}^{3}\,\text{mol}^{-1}$$

Equation [4B.7–*132*] is then rearranged to find T:

$$T = T^{*} + (p-p^{*})\frac{T^{*}\Delta_{\text{fus}}V}{\Delta_{\text{fus}}H}$$

$$= ([5.5+273.15]\,\text{K}) + \left(([1000-1]\,\text{atm}) \times \frac{1.01325\times10^{5}\,\text{Pa}}{1\,\text{atm}}\right)$$

$$\times \frac{([5.5+273.15]\,\text{K})\times(1.19...\times10^{-6}\,\text{m}^{3})}{10.59\times10^{3}\,\text{J}\,\text{mol}^{-1}} = \boxed{2.8\times10^{3}\,\text{K}} \text{ or } \boxed{8.7\,°\text{C}}$$

E4B.10(a) The relationship between pressure and temperature along the solid–liquid boundary is given by [4B.7–*132*], $p = p^{*} + (\Delta_{\text{fus}}H/T^{*}\Delta_{\text{fus}}V)(T-T^{*})$. In this case $p^{*} = 1$ atm (corresponding to the normal melting point, $T^{*} = 273.15$ K) and $p = 1$ bar (corresponding to the standard melting point). Rearranging for $(T-T^{*})$, the difference in melting points, gives

$$(T-T^{*}) = (p-p^{*})\frac{T^{*}\Delta_{\text{fus}}V}{\Delta_{\text{fus}}H}$$

$$= \left(1\times10^{5}\,\text{Pa} - 1\,\text{atm}\times\frac{1.01325\times10^{5}\,\text{Pa}}{1\,\text{atm}}\right)$$

$$\times \frac{(273.15\,\text{K})\times(-1.6\times10^{-6}\,\text{m}^{3}\,\text{mol}^{-1})}{6.008\times10^{3}\,\text{J}\,\text{mol}^{-1}} = \boxed{9.6\times10^{-5}\,\text{K}}$$

This result shows that the standard melting point of ice is slightly higher than the normal melting point, but the difference is negligibly small for most purposes.

E4B.11(a) Since $1\,\text{W} = 1\,\text{J}\,\text{s}^{-1}$, the rate at which energy is absorbed is $(1.2\,\text{kW}\,\text{m}^{-2}) \times (50\,\text{m}^{2}) = 60\,\text{kJ}\,\text{s}^{-1}$. The rate of vaporization is then

$$\frac{\text{rate of energy absorption}}{\Delta_{\text{vap}}H} = \frac{60\,\text{kJ}\,\text{s}^{-1}}{44\,\text{kJ}\,\text{mol}^{-1}} = 1.36...\,\text{mol}\,\text{s}^{-1}$$

Multiplication by the molar mass of water gives the rate of loss of water as $(1.36...\,\text{mol}\,\text{s}^{-1}) \times (18.0158\,\text{g}\,\text{mol}^{-1}) = \boxed{25\,\text{g}\,\text{s}^{-1}}$.

E4B.12(a) The perfect gas equation [1A.4–*8*], $pV = nRT$, is used to calculate the amount as $n = pV/RT$. V is the volume of the laboratory (75 m^{3}) and p is the vapour pressure. The mass is found from $m = nM$, where M is the molar mass; hence $m = pVM/RT$

$$\text{Water:}\quad m = \frac{pVM}{RT} = \frac{(3.2\times10^3\,\text{Pa})\times(75\,\text{m}^3)\times(18.0158\,\text{g mol}^{-1})}{(8.3145\,\text{J K}^{-1}\,\text{mol}^{-1})\times([25+273.15]\,\text{K})} = \boxed{1.7\,\text{kg}}$$

$$\text{Benzene:}\quad m = \frac{pVM}{RT} = \frac{(13.1\times10^3\,\text{Pa})\times(75\,\text{m}^3)\times(78.1074\,\text{g mol}^{-1})}{(8.3145\,\text{J K}^{-1}\,\text{mol}^{-1})\times([25+273.15]\,\text{K})} = \boxed{31\,\text{kg}}$$

$$\text{Mercury:}\quad m = \frac{pVM}{RT} = \frac{(0.23\,\text{Pa})\times(75\,\text{m}^3)\times(200.59\,\text{g mol}^{-1})}{(8.3145\,\text{J K}^{-1}\,\text{mol}^{-1})\times([25+273.15]\,\text{K})} = \boxed{1.4\,\text{g}}$$

$1\,\text{Pa} = 1\,\text{kg m}^{-1}\,\text{s}^{-2}$ and $1\,\text{J} = 1\,\text{kg m}^2\,\text{s}^{-2}$ have been used. Note that an typically sized bottle of benzene (containing less than 31 kg of benzene) would evaporate completely before saturating the air of the laboratory with benzene vapour.

E4B.13(a) (i) The integrated form of the Clausius–Clapeyron equation [4B.10–*133*] is

$$\ln\frac{p}{p^*} = -\frac{\Delta_{\text{vap}}H}{R}\left(\frac{1}{T}-\frac{1}{T^*}\right)$$

Rearranging for $\Delta_{\text{vap}}H$ and substituting in the numbers, taking p^*, T^* at 85.8 °C and p, T at 119.3 °C, gives

$$\Delta_{\text{vap}}H = -R\left(\frac{1}{T}-\frac{1}{T^*}\right)^{-1}\ln\frac{p}{p^*}$$

$$= -(8.3145\,\text{J K}^{-1}\,\text{mol}^{-1})\times\left(\frac{1}{[119.3+273.15]\,\text{K}}-\frac{1}{[85.5+273.15]\,\text{K}}\right)^{-1}$$

$$\times\ln\left(\frac{5.3\,\text{kPa}}{1.3\,\text{kPa}}\right) = 4.86...\times10^4\,\text{J mol}^{-1} = \boxed{49\,\text{kJ mol}^{-1}}$$

(ii) The integrated form of the Clausius–Clapeyron equation is now rearranged for T. Substituting in $p = 1\,\text{atm}$, or $1.01325\times10^5\,\text{Pa}$, corresponding to the normal boiling point, together with the value of $\Delta_{\text{vap}}H$ from above and the same values for p^*, T^* as before, gives

$$T = \left(\frac{1}{T^*}-\frac{R}{\Delta_{\text{vap}}H}\ln\frac{p}{p^*}\right)^{-1}$$

$$= \left(\frac{1}{[85.5+273.15]\,\text{K}}-\frac{8.3145\,\text{J K}^{-1}\,\text{mol}^{-1}}{4.86...\times10^4}\times\ln\frac{1.01325\times10^5\,\text{Pa}}{1.3\times10^3\,\text{Pa}}\right)^{-1}$$

$$= 4.89...\times10^2\,\text{K} = \boxed{4.9\times10^2\,\text{K}}\ \text{or}\ \boxed{2.2\times10^2\,°\text{C}}$$

(iii) To find $\Delta_{\text{vap}}S$ at the boiling temperature, use [3B.4–*89*]:

$$\Delta_{\text{vap}}S = \frac{\Delta_{\text{vap}}H}{T} = \frac{4.86...\times10^4\,\text{J mol}^{-1}}{4.89...\times10^2\,\text{K}} = \boxed{99\,\text{J K}^{-1}\,\text{mol}^{-1}}$$

E4B.14(a) The relationship between pressure and temperature along the solid–liquid boundary is given by [4B.7–132], $p = p^* + (\Delta_{fus}H/T^*\Delta_{fus}V)(T - T^*)$. The molar volume is $V_m = M/\rho$ where M is the molar mass and ρ is the mass density

$$\Delta_{fus}V = V_m(l) - V_m(s) = M/\rho(l) - M/\rho(s)$$

This expression is inserted into [4B.7–132], which is then rearranged for T. T^*, p^*, and $\Delta_{vap}H$ are taken as the values corresponding to the normal melting point of ice, that is, 0 °C (273.15 K) and 1 atm (101.325 kPa). It is assumed that $\Delta_{vap}H$ is constant over the temperature range of interest.

$$T = T^* + (p - p^*)\frac{T^*}{\Delta_{fus}H}\left(\frac{M}{\rho(l)} - \frac{M}{\rho(s)}\right)$$

$$= (273.15\,\text{K}) + ([50 \times 10^5 - 1.01325 \times 10^5]\,\text{Pa}) \times \frac{273.15\,\text{K}}{6.008 \times 10^3\,\text{J mol}^{-1}}$$

$$\times \left(\frac{18.0158\,\text{g mol}^{-1}}{1.00 \times 10^6\,\text{g m}^{-3}} - \frac{18.0158\,\text{g mol}^{-1}}{0.92 \times 10^6\,\text{g m}^{-3}}\right) = \boxed{273\,\text{K}} \text{ or } \boxed{-0.35\,°\text{C}}$$

Solutions to problems

P4B.1 The work done in expanding against a constant external pressure is given by equation [2A.6–40], $w = -p_{ex}\Delta V$. Because the molar volume of a gas is so much greater the molar volume of a liquid, $\Delta_{vap}V \approx V_m(g)$. In addition, if the gas behaves perfectly, $V_m = RT/p$ (from the perfect gas law, [1A.4–8]) with $p = p_{ex}$ as the gas expands against constant external pressure. The work of expansion is therefore

$$w = -p_{ex} \times \frac{RT}{p_{ex}} = -RT = -(8.3145\,\text{J K}^{-1}\,\text{mol}^{-1}) \times ([100 + 273.15]\,\text{K})$$

$$= -3.10... \times 10^3\,\text{J mol}^{-1} = \boxed{-3.10\,\text{kJ mol}^{-1}}$$

The negative sign indicates that the system has done work on the surroundings, so the internal energy of the system falls. The fraction of the enthalpy of vaporization spent on expanding the vapour is

$$\frac{3.10...\,\text{kJ mol}^{-1}}{40.7\,\text{kJ mol}^{-1}} \times 100\,\% = \boxed{7.62\,\%}$$

P4B.3 The variation of vapour pressure with temperature is given by [4B.10–133], $p = p^* \exp[(-\Delta_{vap}H/R)(1/T - 1/T^*)]$. The values of T^* and p^* corresponding to the normal boiling point are used

$$p = p^* \exp\left(-\frac{\Delta_{vap}H}{R}\left(\frac{1}{T} - \frac{1}{T^*}\right)\right)$$

$$= (1\,\text{atm})$$

$$\times \exp\left(-\frac{20.25 \times 10^3\,\text{J mol}^{-1}}{8.3145\,\text{J K}^{-1}\,\text{mol}^{-1}} \times \left(\frac{1}{(40 + 273.15)\,\text{K}} - \frac{1}{(-29.2 + 273.15)\,\text{K}}\right)\right)$$

$$= \boxed{9.08\,\text{atm}} \text{ or } \boxed{920\,\text{kPa}}$$

P4B.5 (a) From the variation of chemical potential with temperature (at constant pressure) [4B.1a–*128*], $(\partial\mu/\partial T)_p = -S_m$, the slope of the chemical potential against temperature is equal to the negative of the molar entropy. The difference in slope on either side of the normal freezing point of water is therefore

$$\left(\frac{\partial\mu(l)}{\partial T}\right)_p - \left(\frac{\partial\mu(s)}{\partial T}\right)_p = -S_m(l) - (-S_m(s))$$
$$= -\Delta_{fus}S = \boxed{-22.0\,J\,K^{-1}\,mol^{-1}}$$

(b) In a similar way, the difference in slope on either side of the normal boiling point of water is

$$\left(\frac{\partial\mu(g)}{\partial T}\right)_p - \left(\frac{\partial\mu(l)}{\partial T}\right)_p = -S_m(g) - (-S_m(l))$$
$$= -\Delta_{vap}S = \boxed{-109.9\,J\,K^{-1}\,mol^{-1}}$$

(c) From part (a)

$$\left(\frac{\partial\mu(l)}{\partial T}\right)_p - \left(\frac{\partial\mu(s)}{\partial T}\right)_p = -\Delta_{fus}S \quad \text{hence} \quad \left(\frac{\partial[\mu(l)-\mu(s)]}{\partial T}\right)_p = -\Delta_{fus}S$$

For a finite change $\Delta[\mu(l) - \mu(s)] = -\Delta_{fus}S \times \Delta T$. For a 5 °C drop in temperature:

$$\Delta[\mu(l) - \mu(s)] = -(22.0\,J\,K^{-1}\,mol^{-1}) \times (-5\,K) = +110\,J\,mol^{-1}$$

Therefore, since water and ice are in equilibrium ($\mu(l) - \mu(s) = 0$) at 0 °C it follows that the chemical potential of liquid water exceeds that of ice by $\boxed{+110\,J\,mol^{-1}}$ at −5 °C. The fact that $\mu(l) > \mu(s)$ indicates that supercooled water at −5, °C has a tendency to freeze to ice.

P4B.7 The total pressure at the bottom of the column is

$$p = \rho g d + 1\,atm$$
$$= \left(13.6\,g\,cm^{-3} \times \frac{10^{-3}\,kg}{1\,g} \times \frac{10^6\,cm^3}{1\,m^3}\right) \times (9.81\,m\,s^{-2}) \times (10\,m)$$
$$+ (1.01325 \times 10^5\,Pa) = 1.44... \times 10^6\,Pa$$

To find the freezing point, use [4B.7–*132*], $p = p^* + (\Delta_{fus}H/T^*\Delta_{fus}V)(T-T^*)$. Rearranging for T gives

$$T = T^* + \frac{T^*\Delta_{fus}V}{\Delta_{fus}H}(p - p^*)$$
$$= (234.3\,K) + \frac{(234.3\,K) \times (0.517 \times 10^{-6}\,m^3\,mol^{-1})}{2.292 \times 10^3\,J\,mol^{-1}}$$
$$\times ([1.44... \times 10^6 - 1.01325 \times 10^5]\,Pa) = \boxed{234.4\,K}$$

Note that this is not a very large difference from the normal freezing point, reflecting the fact that the slope of the solid-liquid boundary is generally very steep compared to the liquid-vapour boundary. Large changes in pressure are therefore needed to bring about significant changes in freezing point.

P4B.9 The integrated form of the Clausius–Clapeyron equation [4B.10–*133*], $\ln(p/p^*) = -(\Delta_{\text{vap}}H/R)(1/T - 1/T^*)$, is rewritten

$$\ln\frac{p}{p^*} = -\frac{\Delta_{\text{vap}}H}{R}\frac{1}{T} + \frac{\Delta_{\text{vap}}H}{RT^*}$$

This implies that a plot of $\ln(p/p^*)$ against $1/T$ should be a straight line of slope $-\Delta_{\text{vap}}H/R$ and intercept $\Delta_{\text{vap}}H/RT^*$; such a plot is shown in Fig. 4.2. If p^* is taken to be 1 atm, or 101.325 kPa, then T^* corresponds to the normal boiling point which can then be obtained from the intercept.

$\theta/°\text{C}$	p/kPa	T^{-1}/K^{-1}	$\ln(p/p^*)$
0	1.92	0.003 66	−3.966
20	6.38	0.003 41	−2.765
40	17.70	0.003 19	−1.745
50	27.70	0.003 09	−1.297
70	62.30	0.002 91	−0.486
80	89.30	0.002 83	−0.126
90	124.90	0.002 75	0.209
100	170.90	0.002 68	0.523

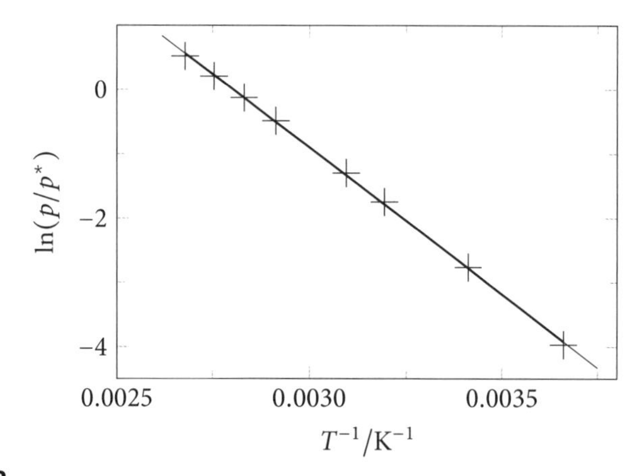

Figure 4.2

The data fall on a good straight line, the equation of which is

$$\ln(p/p^*) = (-4.570 \times 10^3) \times (T^{-1}/\text{K}^{-1}) + 12.81$$

The values of $\Delta_{\text{vap}}H$ and T^* are obtained from the slope and intercept respectively:

$$\Delta_{\text{vap}}H = -\text{slope} \times R = -(-4.570 \times 10^3\ \text{K}) \times (8.3145\ \text{J}\ \text{K}^{-1}\ \text{mol}^{-1})$$
$$= 3.79... \times 10^4\ \text{J}\ \text{mol}^{-1} = \boxed{38.0\ \text{kJ}\ \text{mol}^{-1}}$$
$$T^* = \frac{\Delta_{\text{vap}}H}{R \times \text{intercept}} = \frac{3.79... \times 10^4\ \text{J}\ \text{mol}^{-1}}{(8.3145\ \text{J}\ \text{K}^{-1}\ \text{mol}^{-1}) \times 12.81} = \boxed{357\ \text{K}} \text{ or } \boxed{84\ °\text{C}}$$

P4B.11 (a) The Clapeyron equation [4B.4a–*131*] is $\text{d}p/\text{d}T = \Delta_{\text{trs}}S/\Delta_{\text{trs}}V$. For sublimation, and with $\Delta_{\text{trs}}S = \Delta_{\text{trs}}H/T$ this becomes

$$\frac{\text{d}p}{\text{d}T} = \frac{\Delta_{\text{sub}}H}{T\Delta_{\text{sub}}V}$$

Since the molar volume of a gas is much greater than that of a solid, $\Delta_{\text{sub}}V$ can be approximated as $\Delta_{\text{sub}}V = V_{\text{m}}(\text{g}) - V_{\text{m}}(\text{s}) \approx V_{\text{m}}(\text{g})$, and if the gas behaves perfectly, $V_{\text{m}} = RT/p$. Substituting these into the above equation gives

$$\frac{\text{d}p}{\text{d}T} = \frac{\Delta_{\text{sub}}H}{T(RT/p)} = \frac{p\Delta_{\text{sub}}H}{RT^2}$$

Using $\text{d}x/x = \text{d}\ln x$ this becomes written as $\boxed{\text{d}\ln p/\text{d}T = \Delta_{\text{sub}}H/RT^2}$

(b) Integration of the equation derived in (a) under the assumption that $\Delta_{\text{vap}}H$ is independent of T gives

$$\ln p = -\frac{\Delta_{\text{sub}}H}{R}\frac{1}{T} + \text{constant}$$

This implies that a plot of $\ln p$ against $1/T$ should be a straight line of slope $-\Delta_{\text{sub}}H/R$; such a plot is shown in Fig. 4.3.

T/K	p/Pa	T^{-1}/K^{-1}	$\ln(p/\text{Pa})$
145.94	13.07	0.006 852	2.570
147.96	18.49	0.006 759	2.917
149.93	25.99	0.006 670	3.258
151.94	36.76	0.006 582	3.604
153.97	50.86	0.006 495	3.929
154.94	59.56	0.006 454	4.087

The data fall on a good straight line, the equation of which is

$$\ln(p/\text{Pa}) = (-3.816 \times 10^3) \times (T^{-1}/\text{K}^{-1}) + 28.71$$

The slope is equal to $-\Delta_{\text{vap}}H/R$, so:

$$\Delta_{\text{vap}}H = -\text{slope} \times R = -(-3.816 \times 10^3\ \text{K}) \times (8.3145\ \text{J}\ \text{K}^{-1}\ \text{mol}^{-1})$$
$$= \boxed{31.7\ \text{kJ}\ \text{mol}^{-1}}$$

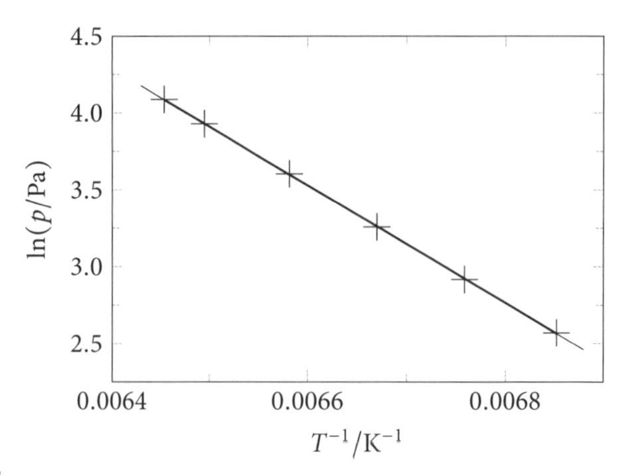

Figure 4.3

P4B.13 (a) If the mass of the liquid decreases by m, then the amount in moles of vapour formed is $n_{vap} = m/M$. The amount in moles of the input gas is given by $n_{gas} = PV/RT$ (from the perfect gas equation) so the mole fraction of the vapour is

$$x_{vap} = \frac{n_{vap}}{n_{tot}} = \frac{n_{vap}}{n_{vap} + n_{gas}} = \frac{m/M}{PV/RT + m/M} = \frac{mRT}{MPV + mRT}$$

(b) If the total pressure remains at P, the partial pressure of the vapour is

$$p = x_{vap} \times P = \frac{mRT}{MPV + mRT} \times P = \frac{mRTP}{MPV + mRT}$$

(c) Dividing top and bottom of this expression by MPV gives

$$p = \frac{AmP}{1 + Am} \quad \text{where} \quad A = \frac{RT}{MPV}$$

(d) For geraniol, noting that $P = 760$ Torr $= 1.01325 \times 10^5$ Pa,

$$A = \frac{(8.3145\,\mathrm{J\,K^{-1}\,mol^{-1}}) \times ([110 + 273.15]\,\mathrm{K})}{(154.2\,\mathrm{g\,mol^{-1}}) \times (1.01325 \times 10^5\,\mathrm{Pa}) \times (5.00 \times 10^{-3}\,\mathrm{m^3})} = 0.0407...\,\mathrm{g^{-1}}$$

hence

$$p = \frac{AmP}{1 + Am} = \frac{(0.0407...\,\mathrm{g^{-1}}) \times (0.32\,\mathrm{g}) \times (1.01325 \times 10^5\,\mathrm{Pa})}{1 + (0.0407...\,\mathrm{g^{-1}}) \times (0.32\,\mathrm{g})} = \boxed{1.31\,\mathrm{kPa}}$$

P4B.15 The integrated form of the Clausius–Claypeyron equation [4B.10–*133*] is

$$\ln \frac{p}{p^*} = -\frac{\Delta_{vap}H}{R}\left(\frac{1}{T} - \frac{1}{T^*}\right)$$

Taking p^* and T^* as corresponding to the pressure and boiling point at sea level, p_0 and T_0, and inserting $p/p_0 = e^{-a/H}$ from the barometric formula gives

$$\ln\left(e^{-a/H}\right) = -\frac{\Delta_{\text{vap}}H}{R}\left(\frac{1}{T} - \frac{1}{T_0}\right) \quad \text{hence} \quad \boxed{T = \left(\frac{1}{T_0} + \frac{R}{\Delta_{\text{vap}}H}\frac{a}{H}\right)^{-1}}$$

For water at 3 km, $a = 3000$ m, the boiling point is

$$T = \left(\frac{1}{373.15\,\text{K}} + \frac{8.3145\,\text{J K}^{-1}\,\text{mol}^{-1}}{40.7\times10^{3}\,\text{J mol}^{-1}} \times \frac{3\,\text{km}}{8\,\text{km}}\right)^{-1} = \boxed{363\,\text{K}} \text{ or } \boxed{89.6\,°\text{C}}$$

Solutions to integrated activities

I4.1 The relationship between p and T along the *solid-liquid boundary* is given by equation [4B.7–*132*]:

$$p = p^* + \frac{\Delta_{\text{fus}}H}{T^*\Delta_{\text{fus}}V}(T - T^*)$$

Using $V_m = M/\rho$, $\Delta_{\text{fus}}V$ is calculated as

$$\Delta_{\text{fus}}V = V_m(\text{l}) - V_m(\text{s}) = \frac{M}{\rho(\text{l})} - \frac{M}{\rho(\text{s})} = \frac{78.1074\,\text{g mol}^{-1}}{0.879\times10^{6}\,\text{g m}^{-3}} - \frac{78.1074\,\text{g mol}^{-1}}{0.891\times10^{6}\,\text{g m}^{-3}}$$

$$= 1.19...\times10^{-6}\,\text{m}^3\,\text{mol}^{-1}$$

This is used in equation [4B.7–*132*] together with the value of $\Delta_{\text{fus}}H$ quoted. Taking p^* and T^* as corresponding to the triple point, $p^* = 36$ Torr $= 4.80$ kPa and $T^* = 5.50\,°\text{C} = 278.65$ K gives the equation of the solid-liquid boundary as

$$p = (4.80\times10^{3}\,\text{Pa}) + \frac{10.6\times10^{3}\,\text{J mol}^{-1}}{(278.65\,\text{K})\times(1.19...\times10^{-6}\,\text{m}^3\,\text{mol}^{-1})}(T - 278.65\,\text{K})$$

$$= (4.80\times10^{3}\,\text{Pa}) + (3.18\times10^{7}\,\text{Pa K}^{-1})\times(T - 278.65\,\text{K})$$

$$\text{so that} \quad \boxed{(p/\text{kPa}) = 4.80 + (3.18\times10^{4})\times[(T/\text{K}) - 278.65]}$$

This takes the form of a steep straight line with a positive gradient extending upwards from the triple point. This is plotted in Fig. 4.4. The line is only drawn for $T \geq T^*$ ($p \geq p^*$) because the liquid does not exist below the triple point.

The relationship between p and T along the *liquid-vapour* boundary is given by equation [4B.10–*133*]; p^* and T^* are again taken as corresponding to the triple point.

$$p = p^*\exp\left(-\frac{\Delta_{\text{vap}}H}{R}\left(\frac{1}{T} - \frac{1}{T^*}\right)\right)$$

$$= (4.80\times10^{3}\,\text{Pa})\times\exp\left(-\frac{30.8\times10^{3}\,\text{J mol}^{-1}}{8.3145\,\text{J K}^{-1}\,\text{mol}^{-1}}\left(\frac{1}{T} - \frac{1}{278.65\,\text{K}}\right)\right)$$

$$\text{or} \quad \boxed{(p/\text{kPa}) = 4.80\times\exp\left[-3.70\times10^{3}\left(\frac{1}{T/\text{K}} - \frac{1}{278.65}\right)\right]}$$

This equation is also plotted in Fig. 4.4, again only for values of T in the range $T \geq 278.65$ K since the liquid does not exist below this temperature.

The relationship between p and T along the *solid-vapour* boundary is given by an equation that is analogous to the liquid-vapour one except that $\Delta_{vap}H$ is replaced by $\Delta_{sub}H$. $\Delta_{sub}H = \Delta_{fus}H + \Delta_{vap}H$ so the required equation is

$$p = p^* \exp\left(-\frac{\Delta_{sub}H}{R}\left(\frac{1}{T}-\frac{1}{T^*}\right)\right)$$

$$= (4.80\times10^3\,\text{Pa})\times\exp\left(-\frac{[10.6+30.8]\times10^3\,\text{J mol}^{-1}}{8.3145\,\text{J K}^{-1}\,\text{mol}^{-1}}\left(\frac{1}{T}-\frac{1}{278.65\,\text{K}}\right)\right)$$

or $$\boxed{(p/\text{kPa}) = 4.80\times\exp\left[-4.98\times10^3\left(\frac{1}{T/\text{K}}-\frac{1}{278.65}\right)\right]}$$

This equation is plotted Fig. 4.4 for values in the range $T \leq 278.65$ K since the solid and vapour phases are only in equilibrium at the triple point and below.

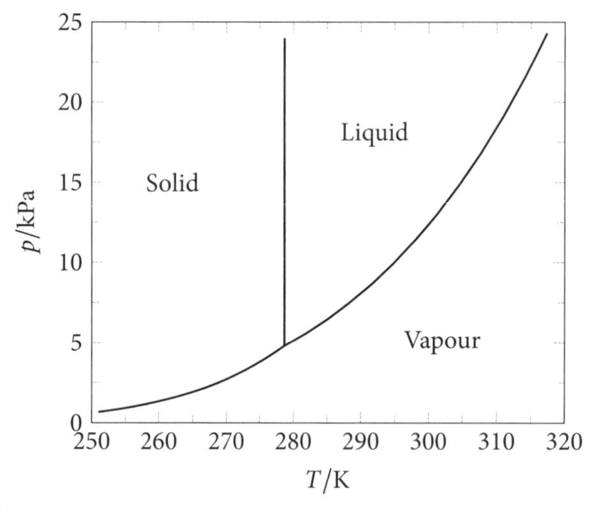

Figure 4.4

I4.3 (a) As outlined in the question, $N-4$ hydrogen bonds are formed when an α-helix forms. If each hydrogen bond has a bond enthalpy of $\Delta_{hb}H$ (that is, the enthalpy change when a hydrogen bond is broken is $\Delta_{hb}H$) then the total enthalpy change on breaking all the hydrogen bonds when the protein unfolds is $\Delta_{unfold}H = (N-4)\Delta_{hb}H$.

The question also explains that $N-2$ of the amino acids form the compact helix with restricted motion. When the helix unfolds, these $N-2$ residues are released and become free to move. If $\Delta_{hb}S$ is the entropy change associated with releasing one amino acid from the helix then the total entropy change when the helix unfolds is $\Delta_{unfold}S = (N-2)\Delta_{hb}S$.

The Gibbs energy change is therefore

$$\Delta_{unfold}G = \Delta_{unfold}H - T\Delta_{unfold}S = (N-4)\Delta_{hb}H - (N-2)T\Delta_{hb}S$$

(b) At the melting temperature, $T = T_m$ and $\Delta_{unfold}G = 0$ (because the folded and unfolded states are in equilibrium). Therefore

$$0 = (N-4)\Delta_{hb}H - (N-2)T_m\Delta_{hb}S,$$

which on rearrangement gives

$$T_m = \frac{(N-4)\Delta_{hb}H}{(N-2)\Delta_{hb}S}$$

(c) A plot of $T_m/(\Delta_{hb}H/\Delta_{hb}S)$ against N is shown in Fig. 4.5.

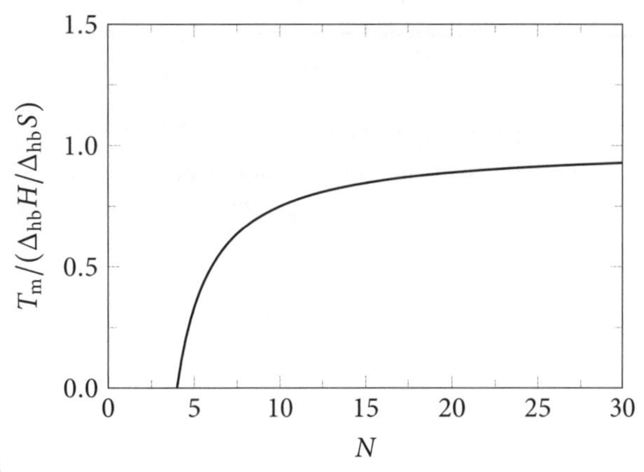

Figure 4.5

The fractional change in T_m when N increases by one is given by

$$\frac{T_m(N+1)}{T_m(N)} = \frac{\left(\frac{(N-3)\Delta_{hb}H}{(N-1)\Delta_{hb}S}\right)}{\left(\frac{(N-4)\Delta_{hb}H}{(N-2)\Delta_{hb}S}\right)} = \frac{(N-3)(N-2)}{(N-1)(N-4)}$$

If the increase is to be less than 1% then

$$\frac{(N-3)(N-2)}{(N-1)(N-4)} < 1.01 \quad \text{which can be rearranged to} \quad N^2 - 5N - 196 > 0$$

Solving this quadratic inequality yields $N > 16.7$ or $N < -11.7$. Discarding the negative solution it follows that the smallest value of N that will give an increase in T_m of less than 1% when N increases by 1 is $\boxed{N = 17}$.

I4.5 (a) The variation of G with pressure is given by [3E.8–*107*], $(\partial G/\partial p)_T = V$. This equation implies that

$$\left(\frac{\partial \Delta_r G}{\partial p}\right)_T = \Delta_r V = V_{m,d} - V_{m,gr}$$

(b) Differentiating the result from above, and noting that the definition of κ_T can be rearranged to $(\partial V/\partial p)_T = -\kappa_T V$, gives

$$\left(\frac{\partial^2 \Delta_r G}{\partial p^2}\right)_T = \left(\frac{\partial}{\partial p}\right)_T \left[V_{m,d} - V_{m,gr}\right] = \left(\frac{\partial V_{m,d}}{\partial p}\right)_T - \left(\frac{\partial V_{m,gr}}{\partial p}\right)_T$$
$$= -\kappa_{T,d} V_{m,d} + \kappa_{T,gr} V_{m,gr}$$

(c) Substituting the results from (a) and (b) into the Taylor expansion gives

$$\Delta_r G = \Delta_r G(p^\ominus) + \left(\frac{\partial \Delta_r G}{\partial p}\right)_{p=p^\ominus} (p-p^\ominus) + \frac{1}{2}\left(\frac{\partial^2 \Delta_r G}{\partial p^2}\right)_{p=p^\ominus} (p-p^\ominus)^2$$
$$= \Delta_r G(p^\ominus) + (V^\ominus_{m,d} - V^\ominus_{m,gr})(p-p^\ominus)$$
$$+ \frac{1}{2}(-\kappa_{T,d} V^\ominus_{m,d} + \kappa_{T,gr} V^\ominus_{m,gr})(p-p^\ominus)^2$$

where $V^\ominus_{m,d}$ and $V^\ominus_{m,gr}$ are the molar volumes under standard conditions.

(d) The transformation is spontaneous when $\Delta_r G < 0$. This is substituted into the above result together with the given data, noting that $V_m = V_s M$ where M is the molar mass.

$$(2.8678 \times 10^3 \text{ J mol}^{-1}) + A(p-p^\ominus) + \frac{1}{2}B(p-p^\ominus)^2 < 0$$

where

$$A = V^\ominus_{m,d} - V^\ominus_{m,gr}$$
$$= ([0.284 - 0.444] \text{ cm}^3 \text{ g}^{-1}) \times (12.01 \text{ g mol}^{-1}) \times \left(\frac{10^{-6} \text{ m}^3}{1 \text{ cm}^3}\right)$$
$$= -1.9216 \times 10^{-6} \text{ m}^3 \text{ mol}^{-1}$$

$$B = -\kappa_{T,d} V^\ominus_{m,d} + \kappa_{T,gr} V^\ominus_{m,gr}$$
$$= \big[-(0.187 \times 10^{-8} \text{ kPa}^{-1}) \times (0.284 \text{ cm}^3 \text{ g}^{-1}) + (3.04 \times 10^{-8} \text{ kPa}^{-1})$$
$$\times (0.444 \text{ cm}^3 \text{ g}^{-1})\big] \times (12.01 \text{ g mol}^{-1}) \times \left(\frac{10^{-6} \text{ m}^3}{1 \text{ cm}^3}\right) \times \left(\frac{10^{-3} \text{ Pa}^{-1}}{1 \text{ kPa}^{-1}}\right)$$
$$= 1.55... \times 10^{-16} \text{ m}^3 \text{ mol}^{-1} \text{ Pa}^{-1}$$

The resulting inequality is

$$\frac{1}{2}(1.55... \times 10^{-16} \text{ m}^3 \text{ mol}^{-1} \text{ Pa}^{-1})(p-p^\ominus)^2$$
$$+ (-1.9216 \times 10^{-6} \text{ m}^3 \text{ mol}^{-1})(p-p^\ominus) + (2.8678 \times 10^3 \text{ J mol}^{-1}) < 0$$

hence, on dividing through by 1 J mol^{-1}

$$(7.78... \times 10^{-17})\left(\frac{[p-p^\ominus]}{\text{Pa}}\right)^2 - (1.9216 \times 10^{-6})\left(\frac{[p-p^\ominus]}{\text{Pa}}\right) + (2.8678 \times 10^{-3}) < 0$$

where $1\,\text{J} = 1\,\text{Pa}\,\text{m}^3$ has been used. Solving this inequality gives

$$1.60 \times 10^9 < (p - p^\ominus)/\text{Pa} < 2.31 \times 10^{10}$$

and hence, using $p^\ominus = 1\,\text{bar} = 10^5\,\text{Pa}$

$$1.60 \times 10^4\,\text{bar} < p < 2.31 \times 10^5\,\text{bar}$$

The conversion of graphite to diamond at 298 K is therefore predicted to become spontaneous at $\boxed{1.60 \times 10^4\,\text{bar}}$. The result also predicts that the conversion will cease to be spontaneous at pressures above 2.31×10^5 bar. This can be attributed to graphite having a greater compressibility than diamond. Its molar volume will therefore decrease more quickly with pressure and this will eventually lead to graphite becoming the more stable phase again. Note, however, that this analysis neglects higher order terms in the Taylor expansion which may become significant at high pressures.

5 Simple mixtures

5A The thermodynamic description of mixtures

Answers to discussion questions

D5A.1 A partial molar quantity X_J is defined in analogy to the definition of partial molar volume, [5A.1–*143*], as the partial derivative of a state function with respect to the amount of a component J, holding all other variables, including the amounts of other species present, constant

$$X_J = \left(\frac{\partial X}{\partial n_J}\right)_{p,T,n'}$$

The partial molar quantity X_J can be interpreted as the change in property X when one mole of substance J is added to a very large amount of a mixture such that the addition does not change the composition; temperature and pressure being held constant.

The Gibbs–Duhem equation, [5A.12b–*146*], shows that partial molar quantities cannot change in an arbitrary way

$$\sum_J n_J \, dX_J = 0$$

It therefore follows that if the partial molar quantity of a solute changes, that of the solute must necessarily change in such a way as to satisfy the Gibbs–Duhem equation.

At a molecular level this can be rationalised by noting that if the amount of solute is changed then it follows that the solvent molecules will experience a different environment, and so their thermodynamic properties must change.

D5A.3 *Perfect* gases spontaneously mix in all proportions. There are, however, conceivable circumstances under which two *real* gases might not mix spontaneously. Consider allowing two gases initially at the same pressure p to mix (so that mixing them would not change the pressure) under conditions of constant temperature. Mixing is spontaneous if $\Delta_{mix}G < 0$, and this Gibbs energy change has an entropic and an enthalpic contribution

$$\Delta_{mix}G = \Delta_{mix}H - T\Delta_{mix}S$$

The entropy change, $\Delta_{mix}S$, is always positive, so mixing is always favoured entropically. The only circumstances under which mixing might not be spontaneous would be if $\Delta_{mix}H > T\Delta_{mix}S$, that is if the change in enthalpy on mixing was so unfavourable as to outweigh the entropic term.

For perfect gases, $\Delta_{\text{mix}}H = 0$, so mixing always occurs. However, there are *liquids* for which unfavourable interactions prevent mixing at least in some proportions and at some temperatures. If two such species were taken above their critical temperatures and held at a pressure high enough to make their densities more typical of liquids than gases, then it is possible to imagine that mixing might not occur. Because the temperature is above the critical temperatures the species are technically gases, although the term supercritical fluid might be more appropriate. In conclusion, there might be examples of immiscibility among supercritical fluids.

D5A.5 Raoult's law, [5A.22–*151*] defines the behaviour of ideal solutions. Like perfect gases, what makes the behaviour ideal can be expressed in terms of intermolecular interactions. Unlike perfect gases, however, the interactions in an ideal solution cannot be neglected. Instead, ideal behaviour amounts to having the same interactions between molecules of the different components of the mixture as there are between molecules of the same type.

In short, ideal behaviour consists of A–B interactions being the same as A–A and B–B interactions. If that is the case, then the cohesive forces that would keep a molecule in the liquid phase would be the same in the solution as in a pure liquid, and the vapour pressure of a component will differ from that of a pure liquid only in proportion to its abundance (mole fraction). Thus, Raoult's law is expected to be valid for mixtures of components that have very similar chemical structures. Similar structures imply both similar intermolecular interactions and similar sizes.

In an ideal dilute solution, on the other hand, Raoult's law holds for the solvent in the limit as x_A approaches 1, not because A–B interactions are like A–A interactions, but because there are so many more A–A interactions than A–B interactions that A–A interactions dominate the behaviour of the solvent. For the solute, on the other hand, there are many more A–B interactions than B–B interactions in the limit as x_B approaches zero. Thus, only one kind of interaction (A–B) is important in determining the affinity of the solute for the solution.

Solutions to exercises

E5A.1(a) The partial molar volume of B is defined from [5A.1–*143*] as

$$V_B = \left(\frac{\partial V}{\partial n_B}\right)_{p,T,n'}$$

The polynomial given relates v to x, and so from this it is possible to compute the derivative dv/dx. This required derivative is dV/dn_B (where the partials are dropped for simplicity), which is related to dv/dx in the following way

$$\left(\frac{dV}{dn_B}\right) = \left(\frac{dV}{dv}\right)\left(\frac{dv}{dx}\right)\left(\frac{dx}{dn_B}\right)$$

Because $x = n_B/\text{mol}$, $dx/dn_B = \text{mol}^{-1}$, and because $v = V/\text{cm}^3$, $dv/dV = \text{cm}^{-3}$ and so $dV/dv = \text{cm}^3$. Hence

$$\left(\frac{dV}{dn_B}\right) = \left(\frac{dV}{dv}\right)\left(\frac{dv}{dx}\right)\left(\frac{dx}{dn_B}\right) = \left(\frac{dv}{dx}\right) \text{cm}^3\,\text{mol}^{-1}$$

The required derivative is

$$\left(\frac{dv}{dx}\right) = 35.6774 - 0.91846\,x + 0.051975\,x^2$$

hence

$$\boxed{V_B = \left(35.6774 - 0.91846\,x + 0.051975\,x^2\right)\,\text{cm}^3\,\text{mol}^{-1}}$$

E5A.2(a) The partial molar volume of solute B (here NaCl) is defined from [5A.1–*143*] as

$$V_B = \left(\frac{\partial V}{\partial n_B}\right)_{p,T,n'}$$

The total volume is given as a function of the molality, but this volume is described as that arising from adding the solute to 1 kg of solvent. The molality of a solute is defined as (amount in moles of solute)/(mass of solvent in kg), therefore because in this case the mass of solvent is 1 kg, the molality is numerically equal to the amount in moles, n_B.

The polynomial given relates v to x, and so from this it is possible to compute the derivative dv/dx. This required derivative is dV/dn_B (where the partials are dropped for simplicity), which is related to dv/dx in the following way

$$\left(\frac{dV}{dn_B}\right) = \left(\frac{dV}{dv}\right)\left(\frac{dv}{dx}\right)\left(\frac{dx}{dn_B}\right)$$

The quantity x is defined as $b/b^\ominus$, but it has already been argued that the molality can be expressed as $n_B/(1\text{ kg})$, hence $x = n_B/(\text{mol})$ and therefore $dx/dn_B = \text{mol}^{-1}$. Because $v = V/\text{cm}^3$, $dv/dV = \text{cm}^{-3}$ and so $dV/dv = \text{cm}^3$. Hence

$$\left(\frac{dV}{dn_B}\right) = \left(\frac{dV}{dv}\right)\left(\frac{dv}{dx}\right)\left(\frac{dx}{dn_B}\right) = \left(\frac{dv}{dx}\right) \text{cm}^3\,\text{mol}^{-1}$$

The required derivative is

$$\left(\frac{dv}{dx}\right) = 16.62 + 2.655\,x^{1/2} + 0.24\,x$$

Hence the expression for the partial molar volume of B (NaCl) is

$$V_B = \left(16.62 + 2.655\,x^{1/2} + 0.24\,x\right)\,\text{cm}^3\,\text{mol}^{-1}$$

The partial molar volume when $b/b^\ominus = 0.1$ is given by

$$\begin{aligned} V_B/(\text{cm}^3\,\text{mol}^{-1}) &= (16.62 + 2.655\,x^{1/2} + 0.24\,x) \\ &= (16.62 + 2.655(0.100)^{1/2} + 0.24 \times 0.100) = 17.4... \end{aligned}$$

Therefore $\boxed{V_B = 17.5 \text{ cm}^3 \text{ mol}^{-1}}$.

The total volume is calculated from the partial molar volumes of the two components, [5A.3–*144*], $V = n_A V_A + n_B V_B$. In this case V and V_B are known, so V_A, the partial molar volume of the solvent water, can be found from $V_A = (V - n_B V_B)/n_A$.

The total volume when $b/b^{\ominus} = 0.1$ is given by

$$V = 1003 + 16.62 \times 0.100 + 1.77 \times 0.100^{3/2} + 0.12 \times 0.100^2 = 1004.7... \text{ cm}^3$$

The amount in moles of 1 kg of water is $(1000 \text{ g})/[(16.00+2\times1.0079) \text{ g mol}^{-1}] =$ 55.5... mol, hence

$$V_A = \frac{V - n_B V_B}{n_A} = \frac{(1004.7... \text{ cm}^3) - (0.100 \text{ mol}) \times (17.4... \text{ cm}^3 \text{ mol}^{-1})}{55.5... \text{ mol}}$$
$$= \boxed{18.1 \text{ cm}^3 \text{ mol}^{-1}}$$

where, as before, for this solution a molality of 0.100 mol kg^{-1} corresponds to $n_B = 0.100$ mol.

E5A.3(a) For a binary mixture the Gibbs–Duhem equation, [5A.12b–*146*], relates changes in the chemical potentials of A and B

$$n_A d\mu_A + n_B d\mu_B = 0$$

If it is assumed that the differential can be replaced by the small change

$$(0.1\, n_B) \times (+12 \text{ J mol}^{-1}) + n_B \delta\mu_B = 0$$

$$\text{hence} \quad \delta\mu_B = -\frac{(0.1\, n_B)}{n_B}(+12 \text{ J mol}^{-1}) = \boxed{-1.2 \text{ J mol}^{-1}}$$

E5A.4(a) Because the gases are assumed to be perfect and are at the same temperature and pressure when they are separated, the pressure and temperature will not change upon mixing. Therefore [5A.18–*149*], $\Delta_{\text{mix}} S = -nR(x_A \ln x_A + x_B \ln x_B)$, applies. The amount in moles is computed from the total volume, pressure and temperature using the perfect gas equation: $n = pV/RT$. Because the separate volumes are equal, and at the same pressure and temperature, each compartment contains the same amount of gas, so the mole fractions of each gas in the mixture are equal at 0.5.

$$\Delta_{\text{mix}} S = -nR(x_A \ln x_A + x_B \ln x_B) = -(pV/T)(x_A \ln x_A + x_B \ln x_B)$$
$$= -\frac{(1.01325 \times 10^5 \text{ Pa}) \times (5.0 \times 10^{-3} \text{ m}^3)}{298.15 \text{ K}}(0.5 \ln 0.5 + 0.5 \ln 0.5)$$
$$= \boxed{+1.2 \text{ J K}^{-1}}$$

Note that the pressure in expressed in Pa and the volume in m^3; the units of the result are therefore $(\text{N m}^{-2}) \times (\text{m}^3) \times (\text{K}^{-1}) = \text{N m K}^{-1} = \text{J K}^{-1}$.

Under these conditions the Gibbs energy of mixing is given by [5A.17–*148*], $\Delta_{\text{mix}}G = nRT(x_A \ln x_A + x_B \ln x_B)$; as before $n = pV/RT$.

$$\begin{aligned}\Delta_{\text{mix}}G &= nRT(x_A \ln x_A + x_B \ln x_B) = (pV)(x_A \ln x_A + x_B \ln x_B)\\ &= [(1.01325\times10^5\ \text{Pa})\times(5.0\times10^{-3}\ \text{m}^3)](0.5\ln 0.5 + 0.5\ln 0.5)\\ &= \boxed{-3.5\times10^2\ \text{J}}\end{aligned}$$

The units of the result are $(\text{N m}^{-2})\times(\text{m}^3) = \text{N m} = \text{J}$. As expected, the entropy of mixing is positive and the Gibbs energy of mixing is negative.

E5A.5(a) The partial pressure of gas A, p_A above a liquid mixture is given by Raoult's Law, [5A.22–*151*], $p_A = x_A p_A^*$, where x_A is the mole fraction of A in the liquid and p_A^* is the vapour pressure over pure A. The total pressure over a mixture of A and B is $p_A + p_B$.

The first step is to calculate the mole fractions. If the molar mass of A is M_A and the mass of A is m, then the amount in moles of A is m/M_A, and likewise because the mass of B is the same, the amount of B is m/M_B. The mole fraction of A is therefore

$$x_A = \frac{m/M_A}{m/M_A + m/M_B} = \frac{1/M_A}{1/M_A + 1/M_B} \quad \text{likewise} \quad x_B = \frac{1/M_B}{1/M_A + 1/M_B}$$

These mole fractions are used with Raoult's law to give the total pressure

$$p = x_A p_A^* + x_B p_B^* = \frac{1/M_A}{1/M_A + 1/M_B} p_A^* + \frac{1/M_B}{1/M_A + 1/M_B} p_B^*$$

If A is benzene, $M_A = 6\times 12.01\ \text{g mol}^{-1} + 6\times 1.0079\ \text{g mol}^{-1} = 78.1074\ \text{g mol}^{-1}$, and if B is methylbenzene $M_B = 7\times 12.01\ \text{g mol}^{-1} + 8\times 1.0079\ \text{g mol}^{-1} = 92.1332\ \text{g mol}^{-1}$.

$$\begin{aligned}p &= \frac{1/M_A}{1/M_A + 1/M_B} p_A^* + \frac{1/M_B}{1/M_A + 1/M_B} p_B^*\\ &= \frac{1/(78.1074\ \text{g mol}^{-1})}{1/(78.1074\ \text{g mol}^{-1}) + 1/(92.1332\ \text{g mol}^{-1})}\times(10\ \text{kPa})\\ &\quad + \frac{1/(92.1332\ \text{g mol}^{-1})}{1/(78.1074\ \text{g mol}^{-1}) + 1/(92.1332\ \text{g mol}^{-1})}\times(2.8\ \text{kPa})\\ &= 5.41...\ \text{kPa} + 1.28...\ \text{kPa} = \boxed{6.7\ \text{kPa}}\end{aligned}$$

E5A.6(a) The total volume is calculated from the partial molar volumes of the two components using [5A.3–*144*], $V = n_A V_A + n_B V_B$. The task is therefore to find the amount in moles, n_A and n_B, of A and B in a given mass m of solution. If the molar masses of A and B are M_A and M_B then it follows that

$$m = n_A M_A + n_B M_B$$

The mole fraction of A is defined as $x_A = n_A/(n_A + n_B)$, hence $n_A = x_A(n_A + n_B)$ and likewise for B. With these substitutions for n_A and n_B the previous equation becomes

$$m = x_A M_A(n_A + n_B) + x_B M_B(n_A + n_B) \text{ hence } (n_A + n_B) = \frac{m}{x_A M_A + x_B M_B}$$

This latter expression for the total amount in moles, $(n_A + n_B)$, is used with $n_A = x_A(n_A + n_B)$ to give

$$n_A = x_A(n_A + n_B) = \frac{m x_A}{x_A M_A + x_B M_B}$$

and likewise

$$n_B = \frac{m x_B}{x_A M_A + x_B M_B}$$

With these expressions for n_A and n_B the total volume is computed from the partial molar volumes

$$\begin{aligned} V = n_A V_A + n_B V_B &= \frac{m x_A V_A}{x_A M_A + x_B M_B} + \frac{m x_B V_B}{x_A M_A + x_B M_B} \\ &= \frac{m}{x_A M_A + x_B M_B}[x_A V_A + x_B V_B] \\ &= \frac{m}{x_A M_A + (1 - x_A) M_B}[x_A V_A + (1 - x_A) V_B] \end{aligned}$$

where on the last line $x_B = (1 - x_A)$ is used.

Taking A as trichloromethane and B as propanone the molar masses are $M_A = 12.01 + 1.0079 + 3 \times 35.45 = 119.3679 \text{ g mol}^{-1}$ and $M_B = 3 \times 12.01 + 6 \times 1.0079 + 16.00 = 58.0774 \text{ g mol}^{-1}$. With these values, the expression for the volume of 1.000 kg evaluates as

$$\begin{aligned} V &= \frac{1000 \text{ g}}{0.4693 \times (119.3679 \text{ g mol}^{-1}) + (1 - 0.4693) \times (58.0774 \text{ g mol}^{-1})} \\ &\quad \times \left[0.4693 \times (80.235 \text{ cm}^3 \text{ mol}^{-1}) + (1 - 0.4693) \times (74.166 \text{ cm}^3 \text{ mol}^{-1})\right] \\ &= \boxed{886.8 \text{ cm}^3} \end{aligned}$$

E5A.7(a) Consider a solution of A and B in which the fraction (by mass) of A is α (here $\alpha = \frac{1}{2}$). The total volume of a solution of A and B is calculated from the partial molar volumes of the two components using [5A.3–*144*], $V = n_A V_A + n_B V_B$. In this exercise V and V_A are known, so the task is therefore to find the amount in moles, n_A and n_B, of A and B in the solution of known mass density ρ.

The mass of a volume V of the solution is ρV, so the mass of A is $\alpha \rho V$. If the molar mass of A is M_A, then the amount in moles of A is $n_A = \alpha \rho V / M_A$. Similarly, $n_B = (1 - \alpha)\rho V / M_B$. The volume is expressed using these quantities as

$$V = n_A V_A + n_B V_B = \frac{\alpha \rho V V_A}{M_A} + \frac{(1 - \alpha)\rho V V_B}{M_B}$$

The term V cancels between the first and third terms to give

$$1 = \frac{\alpha\rho V_A}{M_A} + \frac{(1-\alpha)\rho V_B}{M_B}$$

This equation is rearranged to give an expression for V_A

$$V_A = \frac{M_A}{\alpha\rho}\left(1 - \frac{(1-\alpha)V_B\rho}{M_B}\right)$$

In this exercise let B be H_2O and A be ethanol, and as the mixture is 50% by mass, $\alpha = \frac{1}{2}$. The molar mass of B (H_2O) is $M_B = 16.00 + 2 \times 1.0079 = 18.0158\ \text{g mol}^{-1}$ and the molar mass of A (ethanol) is $M_A = 2 \times 12.01 + 16.00 + 6 \times 1.0079 = 46.0674\ \text{g mol}^{-1}$. The above expression for V_A evaluates as

$$\begin{aligned} V_A &= \frac{M_A}{\alpha\rho}\left(1 - \frac{(1-\alpha)V_B\rho}{M_B}\right) \\ &= \frac{(46.0674\ \text{g mol}^{-1})}{0.5 \times (0.914\ \text{g cm}^{-3})} \\ &\quad \times \left(1 - \frac{(1-0.5) \times (17.4\ \text{cm}^3\ \text{mol}^{-1}) \times (0.914\ \text{g cm}^{-3})}{18.0158\ \text{g mol}^{-1}}\right) \\ &= \boxed{56.3\ \text{cm}^3\ \text{mol}^{-1}} \end{aligned}$$

E5A.8(a) Henry's law gives the partial vapour pressure of a solute B as $p_B = K_B x_B$, [5A.24–*152*]. A test of this law is to make a plot of p_B against x_B which is expected to be a straight line with slope K_B; such a plot is shown in Fig. 5.1.

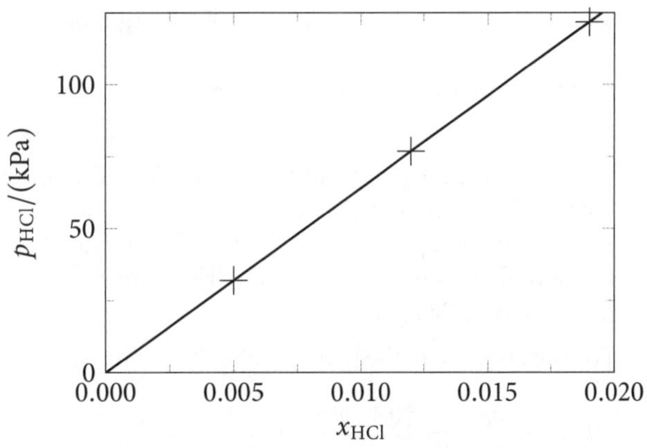

Figure 5.1

The data fall on a good straight line, the equation of which is

$$p_{HCl}/(\text{kPa}) = 6.41 \times 10^3 \times (x_{HCl}) - 0.071$$

If Henry's law is obeyed the pressure should go to zero as x_{HCl} goes to zero, and the graph shows that this is almost achieved. Overall the conclusion is that these data obey Henry's law quite closely. The Henry's law constant K_{HCl} is computed from the slope as $\boxed{6.4 \times 10^3 \text{ kPa}}$.

E5A.9(a) In Section 5A.3(b) on page 152 it is explained that for practical applications Henry's law is often expressed as $p_B = K_B b_B$, where b_B is the molality of the solute, usually expressed in mol kg^{-1}. The molality is therefore calculated from the partial pressure as $b_B = p_B/K_B$.

Molality is the amount of solute per kg of solvent. The mass m of a volume V of solvent is given by $m = \rho V$, where ρ is the mass density of the solvent. If the amount of solute in volume V is n_B, the molar concentration c_B is related to the molality by

$$c_B = \frac{n_B}{V} = \frac{n_B}{m/\rho} = \rho \overbrace{\frac{n_B}{m}}^{b_B} = \rho b_B$$

Using Henry's law the concentration is therefore given by

$$c_B = \rho b_B = \rho \overbrace{\frac{p_B}{K_B}}^{b_B} = \frac{\rho x_B p}{K_B}$$

where the partial pressure p_B is expressed in terms of the mole fraction and the total pressure p, $p_B = x_B p$.

The mole fraction of N_2 in air is 0.780, the Henry's law constant for N_2 in benzene is 1.87×10^4 kPa kg mol^{-1} and the density of benzene is 0.879 g cm^{-3}. If it is assumed that the total pressure is 1 atm then

$$c_{N_2} = \frac{\rho x_{N_2} p}{K_{N_2}} = \frac{(0.879 \times 10^3 \text{ kg m}^{-3}) \times (0.780) \times (101.325 \text{ kPa})}{1.87 \times 10^4 \text{ kPa kg mol}^{-1}} = 3.71... \text{ mol m}^{-3}$$

The molar concentration is therefore $\boxed{3.7 \times 10^{-3} \text{ mol dm}^{-3}}$.

E5A.10(a) In Section 5A.3(b) on page 152 it is explained that for practical applications Henry's law is often expressed as $p_B = K_B b_B$, where b_B is the molality of the solute, usually expressed in mol kg^{-1}. The molality is therefore calculated from the partial pressure as $b_B = p_B/K_B$. The Henry's law constant for CO_2 in water is 3.01×10^3 kPa kg mol^{-1}.

For the case where the pressure of CO_2 is 0.10 atm

$$b_{CO_2} = \frac{p_{CO_2}}{K_{CO_2}} = \frac{(0.10 \text{ atm}) \times (101.325 \text{ kPa}/1 \text{ atm})}{3.01 \times 10^3 \text{ kPa kg mol}^{-1}} = \boxed{3.4 \times 10^{-3} \text{ mol kg}^{-1}}$$

When the pressure is ten times greater at 1.00 atm the solubility is increased by the same factor to $\boxed{3.37 \times 10^{-2} \text{ mol kg}^{-1}}$

E5A.11(a) As explained in Exercise E5A.9(a) the concentration of a solute is estimated as $c_B = \rho p_B/K_B$ where ρ is the mass density of the solvent. The Henry's law constant for CO_2 in water is 3.01×10^3 kPa kg mol^{-1} and the density of water is $0.997\ \mathrm{g\,cm^{-3}}$ or $997\ \mathrm{kg\,m^{-3}}$.

$$c_{CO_2} = \frac{\rho p_{CO_2}}{K_{CO_2}} = \frac{(997\ \mathrm{kg\,m^{-3}}) \times (5.0\ \mathrm{atm}) \times (101.325\ \mathrm{kPa}/1\ \mathrm{atm})}{3.01 \times 10^3\ \mathrm{kPa\,kg\,mol^{-1}}}$$
$$= 1.67... \times 10^2\ \mathrm{mol\,m^{-3}}$$

The molar concentration is therefore $\boxed{0.17\ \mathrm{mol\,dm^{-3}}}$.

Solutions to problems

P5A.1 This problem is similar to the *Example* given in Section 5A.1(d) on page 146. The Gibbs–Duhem equation [5A.12b–*146*], expressed in terms of partial molar volumes is $n_A dV_A + n_B dV_B = 0$ which is rearranged to

$$dV_A = -\frac{n_B}{n_A} dV_B$$

If the variation of the solute partial molar volume V_B with concentration is described by a known function, then integration of this equation gives an expression for how the solvent partial molar volume V_A varies.

The range of integration of V_A is from pure A, at which the partial molar volume is equal to the molar volume of the pure solvent V_A^*, up to some arbitrary concentration. The corresponding range for V_B is from 0, the molar volume of B in the limit of of no B being present (that is pure A), up to some arbitrary concentration.

$$\int_{V_A^*}^{V_A} dV_A = -\int_0^{V_B} \frac{n_B}{n_A} dV_B$$

The expression for V_B is given as a function of the molality, which is the amount in moles divided by the mass of the solvent in kg. In 1 kg of solvent the amount in moles is $n_A = (1\ \mathrm{kg})/M_A$, where M_A is the molar mass of the solvent A. With this expression the ratio n_B/n_A is rewritten

$$\frac{n_B}{n_A} = \frac{n_B}{(1\ \mathrm{kg})/M_A} = \frac{n_B M_A}{(1\ \mathrm{kg})}$$

The quantity $n_B/(1\ \mathrm{kg})$ is recognised as the molality b of solute B, hence $n_B/n_A = bM_A$. The expression for V_B is given in terms of $x = b/b^\ominus$, thus $b = b^\ominus x$ and hence $n_B/n_A = M_A b^\ominus x$, With this, the integral to be evaluated becomes

$$\int_{V_A^*}^{V_A} dV_A = -\int_0^{V_B} M_A b^\ominus x\, dV_B$$

The partial molar volumes V_J are replaced throughout by the dimensionless quantities $v_J = V_J/(\mathrm{cm^3\,mol^{-1}})$ to give

$$\int_{v_A^*}^{v_A} dv_A = -\int_0^{v_B} M_A b^\ominus x\, dv_B$$

The next step is to change the variable of integration on the right from v_B to x; this is done by differentiating the relationship between these two quantities

$$v_B = 5.117 + 19.121\,x^{1/2} \quad \text{hence} \quad dv_B = 9.5605\,x^{-1/2}\,dx$$

The integral is then

$$\int_{v_A^*}^{v_A} dv_A = -M_A b^{\ominus} \int_0^x x(9.5605\,x^{-1/2})\,dx = -M_A b^{\ominus} \int_0^x 9.5605\,x^{1/2}\,dx$$

Evaluating the integrals gives

$$v_A - v_A^* = -M_A b^{\ominus} \times \tfrac{2}{3} \times 9.5605\,x^{3/2}$$

The molar mass of the solvent H_2O is 18.0158 g mol^{-1}; for compatibility with the units of molality this needs to be expressed as 1.80158×10^{-2} kg mol^{-1}. The value of v_A^* is given as 18.079; with these values the expression for v_A becomes

$$v_A = 18.079 - 0.11483\,x^{3/2}$$

P5A.3 The required molar masses are: N_2 28.02 g mol^{-1}; O_2 32.00 g mol^{-1}; Ar 39.95 g mol^{-1}; CO_2 44.01 g mol^{-1}.

Consider 100 g of the mixture. Of this 75.5 g is N_2 so the amount in moles of this gas is $n_{N_2} = (75.5\text{ g})/(28.02\text{ g mol}^{-1}) = 2.69...$ mol. Similar calculations are made for the other cases to give the results shown below in the table. The total amount in moles n is found by summing these individual contributions and this is then used to compute the mole fractions from $x_J = n_J/n$: the resulting values are also shown in the table.

gas	N_2	O_2	Ar	CO_2	total
mass %	75.5	23.2	1.3		
n_J/mol in 100 g	2.69...	0.725	0.0325...		3.45...
x_J	0.780...	0.210...	$9.42... \times 10^{-3}$		
mass %	75.52	23.15	1.28	0.046	
n_J/mol in 100 g	2.69...	0.723...	0.0320...	$1.04... \times 10^{-3}$	3.45...
x_J	0.780...	0.209...	$9.28... \times 10^{-3}$	$3.02... \times 10^{-4}$	

The entropy of mixing (at constant pressure and temperature) is given by a generalisation of [5A.18–*149*]

$$\Delta_{\text{mix}} S = -nR \sum_J x_J \ln x_J$$

The entropy of mixing per mole is $(\Delta_{\text{mix}} S)/n$ is given by

$$(\Delta_{\text{mix}} S)/n = -R \sum_J x_J \ln x_J$$

This expression is used togther with the values given in the table to compute the entropy of mixing for the first set of data as $\boxed{+4.70\ \mathrm{J\,K^{-1}\,mol^{-1}}}$ and for the second set of data as $\boxed{+4.711\ \mathrm{J\,K^{-1}\,mol^{-1}}}$. The difference is of the order of $\boxed{0.01\ \mathrm{J\,K^{-1}\,mol^{-1}}}$.

P5A.5 The definition of the partial molar volume V_B is

$$V_B = \left(\frac{\partial V}{\partial n_B}\right)_{n_A}$$

which is interpreted as the slope of a plot of V against n_B, at constant n_A. Let B be the solute $CuSO_4$ and A be the solvent H_2O.

The task is therefore to calculate the volume of a solution with a *fixed* amount of A as a function of the amount of B. The data given refer to a particular mass of the solution, whereas what is required is data for a particular mass of solvent, so some manipulation is required. Imagine a solution created from a fixed mass $m_A/(\mathrm{g})$ of solvent and which contains a mass $m_B/(\mathrm{g})$ of solute; the total mass is therefore $m_A/(\mathrm{g}) + m_B/(\mathrm{g})$. From the data supplied 100 g of solution contains $m/(\mathrm{g})$ of $CuSO_4$, so it follows that

$$\overbrace{\frac{m_A/(\mathrm{g}) + m_B/(\mathrm{g})}{100}}^{\text{multiples of 100 g}} \times m/(\mathrm{g}) = m_B/(\mathrm{g})$$

This equation is rearranged to give an expression for $m_B/(\mathrm{g})$

$$m_B/(\mathrm{g}) = \frac{m_A/(\mathrm{g}) \times m/(\mathrm{g})}{100 - m/(\mathrm{g})}$$

The amount in moles of B is found using m_B/M_B, where M_B is the molar mass of B, which in this case is 159.61 $\mathrm{g\,mol^{-1}}$.

The volume of this solution is computed from the mass density as $(m_A + m_B)/\rho$. The following table of data is drawn up using $m_A = 1000$ g as the fixed mass of solvent, and using this a plot of V against n_B is made, as shown in Fig. 5.2. Note that $m_{tot}/(\mathrm{g}) = 1000 + m_B/(\mathrm{g})$.

$m(CuSO_4)/\mathrm{g}$	$\rho/\mathrm{g\,cm^{-3}}$	m_B/g	n_B/mol	m_{tot}/g	$V/\mathrm{cm^3}$
5	1.051	52.6	0.330	1 053	1 001.6
10	1.107	111.1	0.696	1 111	1 003.7
15	1.167	176.5	1.106	1 176	1 008.1
20	1.230	250.0	1.566	1 250	1 016.3

The data fit well to the polynomial (shown as the smooth curve on the plot)

$$V/(\mathrm{cm^3}) = 7.2249(n_B/\mathrm{mol})^2 - 1.8512(n_B/\mathrm{mol}) + 1001.4$$

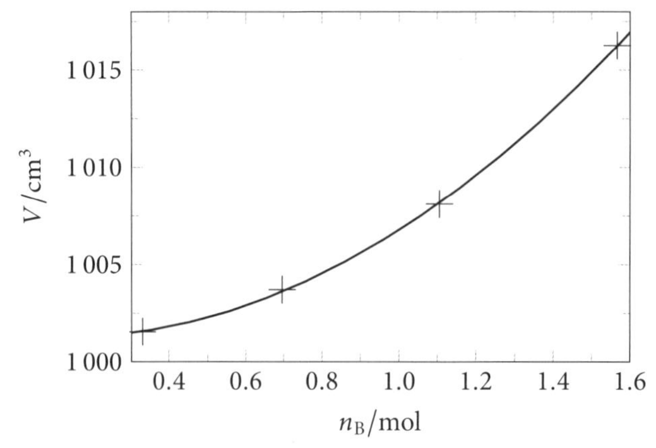

Figure 5.2

The partial molar volume is the slope of this curve which is the derivative with respect to n_B

$$V_B/(\text{cm}^3\,\text{mol}^{-1}) = 14.450(n_B/\text{mol}) - 1.8512$$

The following table gives values of V_B for each of the data points. These are plotted in Fig. 5.3; the line is the function above.

$m(\text{CuSO}_4)$/g	ρ/g cm^{-3}	n_B/mol	V_B/cm^3 mol^{-1}
5	1.051	0.330	2.91
10	1.107	0.696	8.21
15	1.167	1.106	14.13
20	1.230	1.566	20.78

P5A.7 In *Example* 5A.1 on page 144 the partial molar volume of ethanol is found to be given by

$$v = 54.6664 - 0.72788\,z + 0.084768\,z^2$$

where $v = V_E/(\text{cm}^3\,\text{mol}^{-1})$ and $z = n_E/\text{mol}$. The value of z at which v is a minimum or maximum is found by setting the derivative $\text{d}v/\text{d}z = 0$

$$\frac{\text{d}v}{\text{d}z} = -0.72788 + 0.169536\,z = 0 \quad \text{hence} \quad z = \frac{0.72788}{0.169536} = 4.2934$$

This value of z corresponds to 4.2934 mol in 1.000 kg of solvent water (specified in the *Example*). The molality is the amount in moles divided by the mass of the solvent in kg, thus the corresponding molality is $\boxed{4.2934\ \text{mol kg}^{-1}}$. The plot in the text confirms that this is indeed the position of the minimum in the partial molar volume.

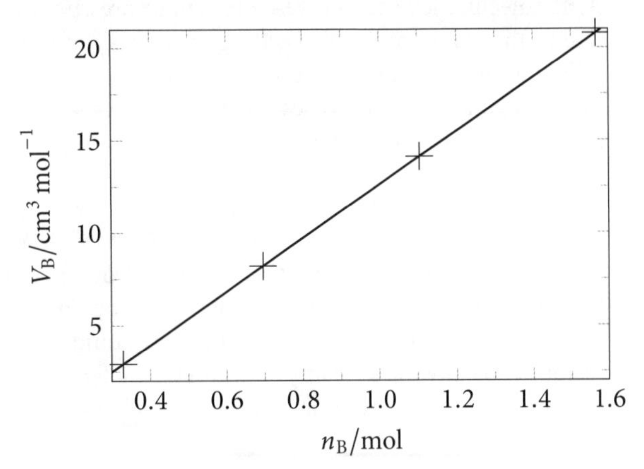

Figure 5.3

5B The properties of solutions

Answers to discussion question

D5B.1 A regular solution has excess entropy S^E of zero, but an excess enthalpy H^E that is non-zero. A regular solution of A and B can be thought of as one in which the different molecules of A and B are distributed randomly, as in an ideal solution, but where the energy of A–A, B–B, and A–B interactions are different.

In real solutions both S^E and H^E are non-zero, and in general both are likely to vary with composition. The non-zero value for S^E is interpreted as arising from the non-random distribution of molecules. This is exemplified by ionic solutions, in which ions of one charge are more likely to be surrounded by ions of the opposite charge than of the same charge (Topic 5F).

D5B.3 All of the colligative properties result from the lowering of the chemical potential of the *solvent* due to the presence of the *solute*. For an ideal solution, this reduction is predicted by $\mu_A = \mu_A^* + RT\ln x_A$. The relationship shows that as the amount of *solute* increases, the mole fraction of the *solvent* x_A decreases and hence the chemical potential the solvent A decreases.

If the chemical potential of the solvent is lowered, then the chemical potential of the vapour in equilibrium with the solvent is also lowered because at equilibrium these two chemical potentials must be equal. The chemical potential of a perfect gas is given by $\mu_A = \mu_A^\ominus + RT\ln p_A$, so a lowering of the chemical potential results in a reduction in the pressure.

The overall result is that addition of a solute reduces the vapour pressure of the solvent, and therefore the temperature at which the solvent boils is raised because a greater increase in temperature is needed to make the vapour pressure equal to the external pressure. Similarly, the freezing point of the solvent is decreased because the chemical potential of the solid will equal that of the solvent at a lower temperature.

At a molecular level the decrease in vapour pressure can be thought of as being due to the solute molecules getting in the way of the solvent molecules, thus reducing their tendency to escape. Another way of looking at this is that the presence of a solute increases the 'randomness', and hence the entropy, of the solution, thus reducing the tendency for the formation of the (pure) vapour or solid.

D5B.5 The boiling-point constant is given by [5B.9b–*160*], $K = RT^{*2}/\Delta_{vap}H$, where T^* is the boiling point of the pure liquid and $\Delta_{vap}H$ is its enthalpy of vaporisation, and the freezing-point constant is given by [5B.11–*161*], $K_f = RT_f^{*2}/\Delta_{fus}H$, where T_f^* is the freezing point of the pure liquid and $\Delta_{fus}H$ is its enthalpy of fusion. Typically the enthalpy of fusion is smaller than that of vaporisation, and this accounts for the freezing-point constant being larger than the boiling-point constant.

Another way of viewing this is to refer to Fig. 5B.6 on page 158 and note that the chemical potential of the solid changes more slowly with temperature than does that of the gas on account of the entropy of the solid being smaller than that of a gas. When the line showing how the chemical potential of the liquid changes with temperature is shifted down, which is what happens when solute is added, the intersection of this line with the line for the solid (the freezing point) changes by more than does the intersection with the line for the gas (the boiling point).

D5B.7 Colligative properties depend on the solvent and on the concentration, but not the identity, of the solute. Thus osmometry (and other colligative properties, for that matter) can be used to determine the *molar* concentration of a solute in a given solvent. If the mass of the solute in the solution and the volume of the solution is known, then it is possible compute the molar mass from the measured concentration.

Solutions to exercises

E5B.1(a) In *Exercise* E5A.8(a) it is found that the vapour pressure obeys

$$p_{HCl}/(\text{kPa}) = 6.41 \times 10^3 \times (x_{HCl}) - 0.071 \qquad (5.1)$$

The task is to work out the mole fraction that corresponds to the given molality. The molality of HCl is defined as $b_{HCl} = n_{HCl}/m_{GeCl_4}$, where n_{HCl} is the amount in moles of HCl and m_{GeCl_4} is the mass in kg of solvent $GeCl_4$. The mole fraction of HCl is $n_{HCl}/(n_{HCl}+n_{GeCl_4})$, where n_{GeCl_4} is the amount in moles of $GeCl_4$, which is given by $n_{GeCl_4} = m_{GeCl_4}/M_{GeCl_4}$, where M_{GeCl_4} is the molar mass of $GeCl_4$. These relationships allow the mole fraction to be rewritten as follows

$$x_{HCl} = \frac{n_{HCl}}{n_{HCl}+n_{GeCl_4}} = \frac{n_{HCl}}{n_{HCl}+m_{GeCl_4}/M_{GeCl_4}}$$

The amount in moles of HCl is written is $n_{HCl} = b_{HCl}m_{GeCl_4}$; using this the above expression for the mole fraction becomes

$$x_{HCl} = \frac{n_{HCl}}{n_{HCl}+m_{GeCl_4}/M_{GeCl_4}} = \frac{b_{HCl}m_{GeCl_4}}{b_{HCl}m_{GeCl_4}+m_{GeCl_4}/M_{GeCl_4}} = \frac{b}{b+1/M_{GeCl_4}}$$

The molar mass of $GeCl_4$ is 214.44 g mol^{-1}, therefore the mole fraction corresponding to $b = 0.10$ mol kg^{-1} is

$$x_{HCl} = \frac{b}{b+1/M} = \frac{(0.10\text{ mol kg}^{-1})}{(0.10\text{ mol kg}^{-1}) + 1/(214.44\times10^{-3}\text{ kg mol}^{-1})} = 0.0209...$$

The pressure is found by inserting this value into eqn 5.1

$$p_{HCl}/(\text{kPa}) = 6.41\times10^3\times(0.0209...) - 0.071 = 1.34...\times10^2$$

The vapour pressure of HCl is therefore $\boxed{1.3\times10^2\text{ kPa}}$.

E5B.2(a) Raoult's law, [5A.22–*151*], $p_A = x_A p_A^*$ relates the vapour pressure to the mole fraction of A, therefore from the given data is it possible to compute x_A. The task is to relate the mole fraction of A to the masses of A (the solvent) and B (the solute), and to do this the molar masses M_A and M_B are introduced. With these $n_A = m_A/M_A$, where m_A is the mass of A, and similarly for n_B. It follows that

$$x_A = \frac{n_A}{n_A+n_B} = \frac{m_A/M_A}{m_A/M_A + m_B/M_B} = \frac{M_B m_A}{M_B m_A + M_A m_B}$$

The final form of this expression for x_A is rearranged to given an expression for M_B, which is the desired quantity; then x_A is replaced by p_A/p_A^*

$$M_B = \frac{x_A M_A m_B}{m_A(1-x_A)} = \frac{(p_A/p_A^*)M_A m_B}{m_A[1-(p_A/p_A^*)]}$$

The molar mass of the solvent benzene, A, is 78.1074 g mol^{-1}, hence

$$\begin{aligned} M_B &= \frac{(p_A/p_A^*)M_A m_B}{m_A[1-(p_A/p_A^*)]} \\ &= \frac{[(51.5\text{ kPa})/(53.3\text{ kPa})]\times(78.1074\text{ g mol}^{-1})\times(19.0\text{ g})}{(500\text{ g})\times[1-(51.5\text{ kPa})/(53.3\text{ kPa})]} \\ &= \boxed{84.9\text{ g mol}^{-1}} \end{aligned}$$

E5B.3(a) The freezing point depression ΔT_f is related to the molality of the solute B, b_B, by [5B.12–*161*], $\Delta T_f = K_f b_B$, where K_f is the freezing-point constant. From the data and the known value of K_f it is possible to calculate b_B. The task is then to relate this to the given masses and the desired molar mass of the solute, M_B.

The molality of B is defined as $b_B = n_B/m_A$, where m_A is the mass of the solvent A in kg. It follows that

$$b_B = \frac{n_B}{m_A} = \frac{m_B/M_B}{m_A}$$

where m_B is the mass of solute B. From the freezing point data $b_B = \Delta T_f/K_f$, therefore

$$\frac{\Delta T_f}{K_f} = \frac{m_B/M_B}{m_A} \quad \text{hence} \quad M_B = \frac{m_B K_f}{m_A \Delta T_f}$$

With the data given and the value of the freezing-point constant from the *Resource section*

$$M_B = \frac{(100\,\text{g}) \times (30\,\text{K kg mol}^{-1})}{(0.750\,\text{kg}) \times (10.5\,\text{K})} = \boxed{381\,\text{g mol}^{-1}}$$

Note that because molality is defined as (amount in moles)/(mass of solvent in kg), the mass of solvent m_A is used as 0.750 kg.

E5B.4(a) The freezing point depression ΔT_f is related to the molality of the solute B, b_B, by [5B.12–*161*], $\Delta T_f = K_f b_B$, where K_f is the freezing-point constant. The molality of the solute B is defined as $b_B = n_B/m_A$, where n_B is the amount in moles of B and m_A is the mass in kg of solvent A. The amount is related to the mass of B, m_B, using the molar mass M_B: $n_B = m_B/M_B$. It therefore follows that

$$\Delta T_f = K_f b_B = \frac{K_f m_B}{M_B m_A}$$

The molar mass of sucrose $C_{12}H_{22}O_{11}$ is 342.2938 g mol^{-1}. A volume 200 cm^3 of water has mass 200 g to a good approximation. Using these values with the data given and the value of the freezing-point constant from the *Resource section* gives the freezing point depression as

$$\Delta T_f = \frac{K_f m_B}{M_B m_A} = \frac{(1.86\,\text{K kg mol}^{-1}) \times (2.5\,\text{g})}{(342.2938\,\text{g mol}^{-1}) \times (0.200\,\text{kg})} = 0.0679...\,\text{K}$$

Note that because molality is defined as (amount in moles)/(mass of solvent in kg), the mass of solvent m_A is used as 0.200 kg. The new freezing point is therefore 273.15 K – 0.0679... K = $\boxed{273.08\,\text{K}}$

E5B.5(a) The osmotic pressure Π is related to the molar concentration of solute B, [B], by [5B.16–*163*], $\Pi = [\text{B}]RT$. The freezing point depression ΔT_f is related to the molality of B, b_B, by [5B.12–*161*], $\Delta T_f = K_f b_B$, where K_f is the freezing-point constant. The task is to relate [B] to b_B so that these two relationships can be used together.

The molar concentration [B] is given by $[\text{B}] = n_B/V$, where n_B is the amount in moles of B and V is the volume of the solvent A. This volume is related to the mass of A, m_A, using the mass density ρ: $V = m_A/\rho$. It therefore follows that

$$[\text{B}] = \frac{n_B}{V} = \frac{n_B}{m_A/\rho} = \overbrace{\frac{n_B}{m_A}}^{b_B}\rho = b_B\rho$$

With this the osmotic pressure is related to the molality

$$[\text{B}] = \frac{\Pi}{RT} \quad \text{hence} \quad b_B\rho = \frac{\Pi}{RT} \quad \text{and so} \quad b_B = \frac{\Pi}{\rho RT}$$

The freezing point depression for a solution exerting this osmotic pressure is therefore

$$\Delta T_f = K_f b_B = \frac{K_f \Pi}{\rho RT}$$

Note that because molality is defined as (amount in moles)/(mass of solvent in kg), the mass of solvent m_A must be in kg and therefore the mass density must be used in kg volume^{-1}.

With the data given, the value of the freezing-point constant from the *Resource section*, and taking the mass density of water as $1\,\text{g cm}^{-3} = 1000\,\text{kg m}^{-3}$ gives the freezing point depression as

$$\Delta T_f = \frac{K_f \Pi}{\rho RT} = \frac{(1.86\,\text{K kg mol}^{-1}) \times (120 \times 10^3\,\text{Pa})}{(1000\,\text{kg m}^{-3}) \times (8.3145\,\text{J K}^{-1}\,\text{mol}^{-1}) \times (300\,\text{K})}$$
$$= 0.0894...\,\text{K}$$

In this expression all of the quantities are in SI units therefore the temperature is expected to be in K, which is verified as follows

$$\frac{(\text{K kg mol}^{-1}) \times (\text{Pa})}{(\text{kg m}^{-3}) \times (\text{J K}^{-1}\,\text{mol}^{-1}) \times (\text{K})} = \frac{\text{Pa}}{\text{J} \times \text{m}^{-3} \times \text{K}^{-1}}$$
$$= \frac{\text{kg m}^{-1}\,\text{s}^{-2}}{(\text{kg m}^2\,\text{s}^{-2}) \times \text{m}^{-3} \times \text{K}^{-1}} = \text{K}$$

The freezing point is therefore 273.15 K − 0.0894... K = $\boxed{273.06\,\text{K}}$

E5B.6(a) The Gibbs energy of mixing is given by [5B.3–*155*], $\Delta_{\text{mix}}G = nRT(x_A \ln x_A + x_B \ln x_B)$, the entropy of mixing by [5B.4–*155*], $\Delta_{\text{mix}}S = -nR(x_A \ln x_A + x_B \ln x_B)$. $\Delta_{\text{mix}}H$ for an ideal solution is $\boxed{\text{zero}}$.

The total amount in moles is 0.50 mol + 2.00 mol = 2.50 mol. With A as hexane and B as heptane the thermodynamic quantities are calculated as

$$\Delta_{\text{mix}}G = nRT(x_A \ln x_A + x_B \ln x_B)$$
$$= (2.50\,\text{mol}) \times (8.3145\,\text{J K}^{-1}\,\text{mol}^{-1}) \times (298\,\text{K}) \times \left(\frac{0.50}{2.50}\ln\frac{0.50}{2.50} + \frac{2.00}{2.50}\ln\frac{2.00}{2.50}\right)$$
$$= \boxed{-3.10 \times 10^3\,\text{J}}$$
$$\Delta_{\text{mix}}S = -nR(x_A \ln x_A + x_B \ln x_B)$$
$$= -(2.50\,\text{mol}) \times (8.3145\,\text{J K}^{-1}\,\text{mol}^{-1}) \times \left(\frac{0.50}{2.50}\ln\frac{0.50}{2.50} + \frac{2.00}{2.50}\ln\frac{2.00}{2.50}\right)$$
$$= \boxed{+10.4\,\text{J K}^{-1}}$$

E5B.7(a) The entropy of mixing is given by [5B.4–*155*], $\Delta_{\text{mix}}S = -nR(x_A \ln x_A + x_B \ln x_B)$, and is a maximum when $x_A = x_B = \boxed{\tfrac{1}{2}}$. This is evident from Fig. 5B.2 on page 156.

The task is to relate the mole fraction of A (heptane) to the masses of A and B (hexane), and to do this the molar masses M_J are introduced. With these $n_J = m_J/M_J$, where m_J is the mass of J. It follows that

$$x_A = \frac{n_A}{n_A + n_B} = \frac{m_A/M_A}{m_A/M_A + m_B/M_B} = \frac{M_B m_A}{M_B m_A + M_A m_B}$$

This is rearranged to give an expression for m_B/m_A

$$x_A = \frac{M_B m_A}{M_B m_A + M_A m_B} = \frac{M_B}{M_B + M_A(m_B/m_A)} \quad \text{hence} \quad \frac{m_B}{m_A} = \frac{M_B}{M_A}\left(\frac{1}{x_A} - 1\right)$$

The molar mass of A (heptane) is 100.1964 g mol^{-1}, and that of B (hexane) is 86.1706 g mol^{-1}. With these values and $x_A = \frac{1}{2}$

$$\frac{m_B}{m_A} = \frac{M_B}{M_A}\left(\frac{1}{x_A} - 1\right) = \frac{86.1706\ \text{g mol}^{-1}}{100.1964\ \text{g mol}^{-1}}\left(\frac{1}{1/2} - 1\right) = \boxed{0.8600}$$

More simply, if equal amounts in moles of A and B are required, the ratio of the corresponding masses of A and B must be equal to the ratio of their molar masses: $m_B/m_A = M_B/M_A$.

E5B.8(a) The ideal solubility of solute B at temperature T is given by [5B.14–162], $\ln x_B = (\Delta_{fus}H/R)(1/T_f - 1/T)$, where $\Delta_{fus}H$ is the enthalpy of fusion of the solute, and T_f is the freezing point of the pure solute.

$$\ln x_B = \frac{\Delta_{fus}H}{R}\left(\frac{1}{T_f} - \frac{1}{T}\right)$$

$$= \frac{28.8 \times 10^3\ \text{J mol}^{-1}}{8.3145\ \text{J K}^{-1}\text{mol}^{-1}}\left(\frac{1}{(217 + 273.15)\ \text{K}} - \frac{1}{(25 + 273.15)\ \text{K}}\right) = -4.55...$$

hence $x_B = 0.0105....$

The mole fraction is expressed in terms of the molality, $b_B = n_B/m_A$, where m_A is the mass of the solvent in kg, in the following way

$$x_B = \frac{n_B}{n_A + n_B} = \frac{n_B}{m_A/M_A + n_B} = \frac{n_B/m_A}{1/M_A + n_B/m_A} = \frac{b_B}{1/M_A + b_B}$$

$$\text{hence} \quad b_B = \frac{x_B}{(1 - x_B)M_A}$$

where M_A is the molar mass of A, expressed in kg mol^{-1}. The molar mass of solvent benzene is 78.1074 g mol^{-1} or $78.1074 \times 10^{-3}\ \text{kg mol}^{-1}$, therefore

$$b_B = \frac{x_B}{(1 - x_B)M_A} = \frac{0.0105...}{(1 - 0.0105...) \times (78.1074 \times 10^{-3}\ \text{kg mol}^{-1})} = 0.136...\ \text{mol kg}^{-1}$$

The molality of the solution is therefore $\boxed{0.137\ \text{mol kg}^{-1}}$. The molar mass of anthracene ($C_{14}H_{10}$) is 178.219 g mol^{-1}, so the mass of anthracene which is dissolved in 1 kg of solvent is $(0.136...\ \text{mol kg}^{-1}) \times (1\ \text{kg}) \times (178.219\ \text{g mol}^{-1}) = \boxed{24.3\ \text{g}}$.

E5B.9(a) Let the solvent CCl_4 be A and the solute Br_2 be B. The vapour pressure of the solute in an ideal dilute solution obeys Henry's law, [5A.24–152], $p_B = K_B x_B$, and the vapour pressure of the solvent obeys Raoult's law, [5A.22–151], $p_A = p_A^* x_A$.

$$p_B = K_B x_B = (122.36\ \text{Torr}) \times 0.050 = 6.11...\ \text{Torr}$$
$$p_A = p_A^* x_A = (33.85\ \text{Torr}) \times (1 - 0.050) = 32.1...\ \text{Torr}$$
$$p_{tot} = p_A + p_A = (6.11...\ \text{Torr}) + (32.1...\ \text{Torr}) = 38.2...\ \text{Torr}$$

Therefore the pressure are $\boxed{p_B = 6.1\ \text{Torr}}$, $\boxed{p_A = 32\ \text{Torr}}$, and $\boxed{p_{tot} = 38\ \text{Torr}}$.

The partial pressure of the gas is given by $p_A = y_A p_{tot}$, where y_A is the mole fraction in the vapour

$$y_A = \frac{p_A}{p_{tot}} = \frac{32.1...\ \text{Torr}}{38.2...\ \text{Torr}} = \boxed{0.84}$$
$$y_B = \frac{p_B}{p_{tot}} = \frac{6.11...\ \text{Torr}}{38.2...\ \text{Torr}} = \boxed{0.16}$$

E5B.10(a) Let methylbenzene be A and 1,2-dimethylbenzene be B. If the solution is ideal the vapour pressure obeys Raoult's law, [5A.22–151], $p_J = p_J^* x_J$. The mixture will boil when the sum of the partial vapour pressures of A and B equal the external pressure, here 0.50 atm.

$$p_{ext} = p_A + p_B = x_A p_A^* + x_B p_B^* = x_A p_A^* + (1 - x_A) p_B^*$$
$$\text{hence } x_A = \frac{p_{ext} - p_B^*}{p_A^* - p_B^*} \quad \text{and by analogy } x_B = \frac{p_{ext} - p_A^*}{p_B^* - p_A^*}$$

$$x_A = \frac{(0.50\ \text{atm}) \times [(101.325\ \text{kPa})/(1\ \text{atm})] - (20.0\ \text{kPa})}{(53.3\ \text{kPa}) - (20.0\ \text{kPa})} = 0.920... = \boxed{0.92}$$
$$x_B = \frac{(0.50\ \text{atm}) \times [(101.325\ \text{kPa})/(1\ \text{atm})] - (53.3\ \text{kPa})}{(20.0\ \text{kPa}) - (53.3\ \text{kPa})} = 0.0792... = \boxed{0.08}$$

The partial pressure of the gas is given by $p_J = y_J p_{ext}$, where y_J is the mole fraction in the gas, and p_J is given by $p_J = p_J^* x_J$, hence $y_J = x_J p_J^* / p_{ext}$

$$y_A = \frac{x_A p_A^*}{p_{ext}} = \frac{(0.920...) \times (53.3\ \text{kPa})}{(0.50\ \text{atm}) \times [(101.325\ \text{kPa})/(1\ \text{atm})]} = \boxed{0.97}$$
$$y_B = \frac{x_B p_B^*}{p_{ext}} = \frac{(0.0792...) \times (20.0\ \text{kPa})}{(0.50\ \text{atm}) \times [(101.325\ \text{kPa})/(1\ \text{atm})]} = \boxed{0.03}$$

E5B.11(a) The vapour pressure of component J in the solution obeys Raoult's law, [5A.22–151], $p_J = p_J^* x_J$, where x_J is the mole fraction in the solution. In the gas the partial pressure is $p_J = y_J p_{tot}$, where y_J is the mole fraction in the vapour.

These relationships give rise to four equations

$$p_A = p_A^* x_A \quad p_B = p_B^*(1 - x_A) \quad p_A = p_{tot} y_A \quad p_B = p_{tot}(1 - y_A)$$

where $x_A + x_B = 1$ is used and likewise for the gas. In these equations x_A and p_{tot} are the unknowns to be found. The expressions for p_A are set equal, as are those for p_B, to give

$$p_A^* x_A = p_{tot} y_A \text{ hence } p_{tot} = \frac{p_A^* x_A}{y_A}$$

$$p_B^*(1 - x_A) = p_{tot}(1 - y_A) \text{ hence } p_{tot} = \frac{p_B^*(1 - x_A)}{1 - y_A}$$

These two expressions for p_{tot} are set equal and the resulting equation rearranged to find x_A

$$\frac{p_A^* x_A}{y_A} = \frac{p_B^*(1 - x_A)}{1 - y_A} \quad \text{hence} \quad x_A = \frac{p_B^* y_A}{p_A^*(1 - y_A) + p_B^* y_A}$$

With the data given

$$x_A = \frac{p_B^* y_A}{p_A^*(1 - y_A) + p_B^* y_A} = \frac{(52.0 \text{ kPa}) \times (0.350)}{(76.7 \text{ kPa})*(1 - 0.350) + (52.0 \text{ kPa}) \times (0.350)}$$
$$= 0.267... \quad \text{and} \quad x_B = 1 - 0.267... = 0.732...$$

The composition of the liquid is therefore $\boxed{x_A = 0.267}$ and $\boxed{x_B = 0.733}$.

The total pressure is computed from $p_A = p_{tot} y_A$ and $p_A = p_A^* x_A$ to give $p_{tot} = x_A p_A^* / y_A$

$$p_{tot} = \frac{x_A p_A^*}{y_A} = \frac{(0.267...) \times (76.7 \text{ kPa})}{0.350} = \boxed{58.6 \text{ kPa}}$$

E5B.12(a) If the solution is ideal, the vapour pressure of component J in the solution obeys Raoult's law, [5A.22–151], $p_J = p_J^* x_J$, where x_J is the mole fraction in the solution. In the gas the partial pressure is $p_J = y_J p_{tot}$, where y_J is the mole fraction in the vapour.

Assuming ideality, the total pressure is computed as

$$p_{tot} = p_A + p_B = p_A^* x_A + p_B^*(1 - x_A)$$
$$= (127.6 \text{ kPa}) \times (0.6589) + (50.60 \text{ kPa}) \times (1 - 0.6589) = 101 \text{ kPa}$$

The normal boiling point is when the total pressure is 1 atm, and this is exactly the pressure found by assuming Raoult's law applies. The solution is therefore $\boxed{\text{ideal}}$.

The composition of the vapour is computed from $p_A = p_{tot} y_A$ and $p_A = p_A^* x_A$ hence

$$y_A = \frac{p_A}{p_{tot}} = \frac{p_A^* x_A}{p_{tot}} = \frac{(127.6 \text{ kPa}) \times (0.6589)}{101.325 \text{ kPa}} = 0.829...$$

It follows that $y_B = 1 - 0.829... = 0.170...$. The composition of the vapour is therefore $\boxed{y_A = 0.830}$ and $\boxed{y_B = 0.170}$.

Solutions to problems

P5B.1 The freezing point depression in terms of mole fraction is predicted by [5B.11–161]

$$\Delta T = x_B K' \qquad K' = \frac{RT^{*2}}{\Delta_{fus}H}$$

With the data given

$$K' = \frac{RT^{*2}}{\Delta_{fus}H} = \frac{(8.3145\,\mathrm{J\,K^{-1}\,mol^{-1}}) \times (290\,\mathrm{K})^2}{11.4 \times 10^3\,\mathrm{J\,mol^{-1}}} = 61.3...\,\mathrm{K}$$

The data are given in terms of molality, which is n_B/m_A, where n_B is the amount in moles of solute and m_A is the mass of solvent in kg. The mole fraction x_B is related to the molality by using the molar mass of the solvent, M_A

$$x_B = \frac{n_B}{n_A + n_B} = \frac{n_B}{m_A/M_A + n_B} = \frac{n_B/m_A}{1/M_A + n_B/m_A} = \frac{b_B}{1/M_A + b_B}$$

The molar mass of ethanoic acid CH_3COOH is $60.0516\,\mathrm{g\,mol^{-1}}$. Because m_A must be in kg the molar mass must be expressed in kg volume^{-1}, $M_A = 60.0516 \times 10^{-3}\,\mathrm{kg\,mol^{-1}}$. For the data given $1/M_A \gg b_B$ therefore the expression for the mole fraction is well approximated by $x_B = b_B M_A$. With this, the freezing point depression is given by

$$\Delta T = \overbrace{b_B M_A}^{x_B} K' \quad \text{hence} \quad b_B = \Delta T/M_A K'$$

The table below gives values of b_B calculated from the given ΔT and this expression; to distinguish these values for the experimental values of b_B, the calculated values are termed apparent molalities, $b_{B,app}$

$b_B/(\mathrm{mol\,kg^{-1}})$	$\Delta T/\mathrm{K}$	$b_{B,app}/(\mathrm{mol\,kg^{-1}})$	$b_{B,app}/b_B$	$M_{B,app}/(\mathrm{g\,mol^{-1}})$
0.015	0.115	0.031	2.081	27.9
0.037	0.295	0.080	2.165	26.8
0.077	0.470	0.128	1.657	35.1
0.295	1.381	0.375	1.271	45.7
0.602	2.67	0.725	1.204	48.3

The apparent molar mass of B, $M_{B,app}$, is computed using

$$\frac{M_{B,app}}{M_B} = \frac{b_B}{b_{B,app}}$$

with $M_B = 58.01\,\mathrm{g\,mol^{-1}}$, the molar mass of KF. The argument that leads to this is that the greater the apparent molality the smaller the molar mass: $M_{B,app} \propto 1/b_{B,app}$.

The data in the table show that the molality predicted from the experimental freezing point depression using the value of the freezing-point constant determined by [5B.11–*161*] is always *greater* than the molality know from the way the solution was prepared. Presumably this latter molality is based on adding a know mass of KF to a known mass of solvent, and assuming that the molar mass of KF is 58.1 g mol^{-1}. The fact that the apparent molality is *higher* than the molality of the prepared solution implies that the number of solute species is greater than expected.

The freezing point depression depends on the mole fraction of the solute, regardless of its identity. Therefore if the added KF were to dissociate completely on dissolution in ethanoic acid the mole fraction of the solute would be twice as large as expected on the basis of the amount of added KF, and in turn this would mean that the apparent molality (based on the freezing-point depression) is twice as large as expected.

The data in the table can be interpreted as indicating that there is dissociation of the KF, and that this dissociation is greater at lower molalities. However, this only part of the picture as it does no explain why $b_{B,app}/b_B$ is greater than 2 at some molalities.

P5B.3 Let the two components of the mixture be labelled 1 (propionic acid) and 2 (THP). The definition of the partial molar volume of 1, V_1, is

$$V_1 = \left(\frac{\partial V}{\partial n_1}\right)_{n_2}$$

To use this definition an expression for V as a function of the n_J is required.

The excess volume V^E is defined as $V^E = \Delta V - \Delta V^{\text{ideal}}$, where ΔV is the volume of mixing and ΔV^{ideal} is the volume of mixing of the ideal solution, which is zero. Therefore $V^E = \Delta V$.

The volume of mixing ΔV is written $\Delta V = V - V_{\text{separated}}$, which from the above is also written $V^E = V - V_{\text{separated}}$. The expression given in the problem for V^E is per mole, so for mixing n_1 moles of 1 is mixed with n_2 moles of 2 the excess volume is in fact $(n_1 + n_2)V^E$.

The volume of the separated components is computed from their mass densities: n_1 moles corresponds to a mass $n_1 M_1$, where M_1 is the molar mass, which has volume $n_1 M_1/\rho_1$, where ρ_1 is the mass density. It follows that

$$(n_1 + n_2)V^E = V - \frac{n_1 M_1}{\rho_1} - \frac{n_2 M_2}{\rho_2} \quad \text{hence} \quad V = (n_1 + n_2)V^E + \frac{n_1 M_1}{\rho_1} + \frac{n_2 M_2}{\rho_2}$$

The second equation above is the required expression for V as a function of n_1 and n_2. The expression given in the problem for V^E is a function of the mole fractions, which are easily written in terms of the amounts.

To compute the partial molar volume it is necessary to compute the derivative of V with respect to n_1, keeping in mind that V^E is a function of n_1

$$V = (n_1 + n_2)V^E + \frac{n_1 M_1}{\rho_1} + \frac{n_2 M_2}{\rho_2}$$

$$\text{hence } V_1 = \left(\frac{\partial V}{\partial n_1}\right)_{n_2} = (n_1 + n_2)\left(\frac{\partial V^E}{\partial n_1}\right)_{n_2} + V^E + \frac{M_1}{\rho_1} \qquad (5.2)$$

To compute the derivative it has been recognised that $(n_1 + n_2)V^E$ is a product of two functions of n_1.

The first step is to compute $(\partial V^E/\partial n_1)_{n_2}$, and this requires rewriting the mole fractions in terms of the n_i

$$\begin{aligned}
V^E &= x_1x_2a_0 + x_1^2x_2a_1 - x_1x_2^2a_1\\
&= \frac{n_1n_2a_0}{(n_1+n_2)^2} + \frac{n_1^2n_2a_1}{(n_1+n_2)^3} - \frac{n_1n_2^2a_1}{(n_1+n_2)^3}\\
\left(\frac{\partial V^E}{\partial n_1}\right)_{n_2} &= \frac{n_2a_0}{(n_1+n_2)^2} - \frac{2n_1n_2a_0}{(n_1+n_2)^3} + \frac{2n_1n_2a_1}{(n_1+n_2)^3}\\
&\quad - \frac{3n_1^2n_2a_1}{(n_1+n_2)^4} - \frac{n_2^2a_1}{(n_1+n_2)^3} + \frac{3n_1n_2^2a_1}{(n_1+n_2)^4}
\end{aligned}$$

The quantity required is $(n_1+n_2)(\partial V^E/\partial n_1)_{n_2}$, so the above expression is multiplied by (n_1+n_2). This cancels a term (n_1+n_2) in each of the denominators and allows the expression to be rewritten in terms of the mole fractions

$$(n_1+n_2)(\partial V^E/\partial n_1)_{n_2} = x_2a_0 - 2x_1x_2a_0 + 2x_1x_2a_1 - 3x_1^2x_2a_1 - x_2^2a_1 + 3x_1x_2^2a_1$$

The parts of eqn 5.2 are now assembled

$$\begin{aligned}
V_1 &= (n_1+n_2)\left(\frac{\partial V^E}{\partial n_1}\right)_{n_2} + V^E + \frac{M_1}{\rho_1}\\
&= (x_2a_0 - 2x_1x_2a_0 + 2x_1x_2a_1 - 3x_1^2x_2a_1 - x_2^2a_1 + 3x_1x_2^2a_1)\\
&\quad + (x_1x_2a_0 + x_1^2x_2a_1 - x_1x_2^2a_1) + M_1/\rho_1\\
&= a_0x_2(1-x_1) + a_1x_2(2x_1 - 2x_1^2 - x_2 + 2x_1x_2) + M_1/\rho_1\\
&= a_0x_2(x_2) + a_1x_2(2x_1[1-x_1] - x_2 + 2x_1x_2) + M_1/\rho_1\\
&= a_0x_2^2 + a_1x_2(2x_1x_2 - x_2 + 2x_1x_2) + M_1/\rho_1\\
&= a_0x_2^2 + a_1x_2(4x_1x_2 - x_2) + M_1/\rho_1\\
&= a_0x_2^2 + a_1x_2^2(4x_1 - 1) + M_1/\rho_1
\end{aligned}$$

The last four lines involve repeated use of $x_1 + x_2 = 1$ in order to simplify the expression. A similar process is used to find an expression for V_2. In principle all that is required is to swap the indices 1 and 2, however when this is done for the expression for V^E the result is

$$V^E = x_2x_1a_0 + x_2^2x_1a_1 - x_2x_1a_1$$

which, when compared with the original expression, shows that the sign of the term in a_1 is reversed: this change needs to be carried through to the end. In summary

$$\boxed{V_1 = a_0x_2^2 + a_1x_2^2(4x_1 - 1) + M_1/\rho_1}$$

$$\boxed{V_2 = a_0x_1^2 - a_1x_1^2(4x_2 - 1) + M_2/\rho_2}$$

The molar mass of propionic acid CH_3CH_2COOH is $M_1 = 74.0774\ \mathrm{g\,mol^{-1}}$ and that of THP $C_5H_{10}O$ is $M_1 = 86.129\ \mathrm{g\,mol^{-1}}$. For an equimolar mixture $x_1 = x_2 = \frac{1}{2}$ and therefore

$$\begin{aligned}
V_1 &= a_0x_2^2 + a_1x_2^2(4x_1 - 1) + M_1/\rho_1 = \tfrac{1}{4}a_0 + \tfrac{1}{4}a_1 + M_1/\rho_1 \\
&= 0.25 \times (-2.4697\ \mathrm{cm^3\,mol^{-1}}) + 0.25 \times (0.0608\ \mathrm{cm^3\,mol^{-1}}) \\
&\quad + (74.0774\ \mathrm{g\,mol^{-1}})/(0.97174\ \mathrm{g\,cm^{-3}}) = \boxed{75.6\ \mathrm{cm^3\,mol^{-1}}} \\
V_2 &= a_0x_1^2 - a_1x_1^2(4x_2 - 1) + M_2/\rho_2 = \tfrac{1}{4}a_0 - \tfrac{1}{4}a_1 + M_2/\rho_2 \\
&= 0.25 \times (-2.4697\ \mathrm{cm^3\,mol^{-1}}) - 0.25 \times (0.0608\ \mathrm{cm^3\,mol^{-1}}) \\
&\quad + (86.129\ \mathrm{g\,mol^{-1}})/(0.86398\ \mathrm{g\,cm^{-3}}) = \boxed{99.1\ \mathrm{cm^3\,mol^{-1}}}
\end{aligned}$$

P5B.5 The excess Gibbs energy G^E is defined in [5B.5–*156*], $G^E = \Delta_{mix}G - \Delta_{mix}G^{ideal}$. The ideal Gibbs energy of mixing (per mole) is given by [5B.3–*155*], $\Delta_{mix}G^{ideal} = RT(x_A \ln x_A + x_B \ln x_B)$. Let A by MCH and B be THF. The Gibbs energy of mixing of n_A moles of A with n_B moles of B is therefore given by

$$\begin{aligned}
\Delta_{mix}G &= \Delta_{mix}G^{ideal} + G^E \\
&= (n_A + n_B)RT(x_A \ln x_A + x_B \ln x_B) \\
&\quad + (n_A + n_B)RTx_A(1 - x_A)[0.4857 - 0.1077(2x_A - 1) \\
&\quad + 0.0191(2x_A - 1)^2]
\end{aligned}$$

With the values given $(n_A + n_B) = 4$ mol, $x_A = \frac{1}{4}$, and $x_B = \frac{3}{4}$

$$\begin{aligned}
\Delta_{mix}G &= (4.00\ \mathrm{mol}) \times (8.3145\ \mathrm{J\,K^{-1}\,mol^{-1}}) \times (303.15\ \mathrm{K}) \times (\tfrac{1}{4}\ln\tfrac{1}{4} + \tfrac{3}{4}\ln\tfrac{3}{4}) \\
&\quad + (4.00\ \mathrm{mol}) \times (8.3145\ \mathrm{J\,K^{-1}\,mol^{-1}}) \times (303.15\ \mathrm{K}) \times \tfrac{1}{4} \times (1 - \tfrac{1}{4}) \\
&\quad \times [0.4857 - 0.1077(2 \times \tfrac{1}{4} - 1) + 0.0191(2 \times \tfrac{1}{4} - 1)^2] \\
&= \boxed{-4.64\ \mathrm{kJ}}
\end{aligned}$$

P5B.7 The osmotic pressure Π is related to the molar concentration [J] through a virial-type equation, [5B.18–*163*], $\Pi = [\mathrm{J}]RT(1 + B[\mathrm{J}])$. The data are given in terms of the mass concentration, so the first task is to relate this to the molar concentration. If the amount in moles of solute dissolved in volume V is n_J and the molar mass is M_J it follows that

$$[\mathrm{J}] = \frac{n_J}{V} = \frac{m_J/M_J}{V} = \overbrace{\frac{m_J}{V}}^{c_J}\frac{1}{M_J} = \frac{c_J}{M_J}$$

where c_J is the mass concentration. With this the virial equation is rewritten

$$\Pi = [J]RT(1 + B[J]) = \frac{c_J}{M_J}RT\left(1 + B\frac{c_J}{M_J}\right)$$

Division of both sides by c_J gives an equation of a straight line

$$\frac{\Pi}{c_J} = \frac{RT}{M_J} + \frac{BRT}{M_J^2}c_J$$

A plot of Π/c_J against c_J will have intercept RT/M_J when $c_J = 0$, and from this it is possible to determine the molar mass.

The pressure is given by $\Pi = h\rho g$; for the pressure to be in Pa the height needs to be in m and ρ in kg m^{-3}; for the present case $\rho = 1\ \text{g cm}^{-3} = 1000\ \text{kg m}^{-3}$. The data are plotted in Fig. 5.4.

$c/(\text{mg cm}^{-3})$	$h/(\text{cm})$	Π/Pa	$(\Pi/c)/(\text{Pa mg}^{-1}\text{ cm}^3)$
3.221	5.746	563.7	175.002 4
4.618	8.238	808.1	174.999 5
5.112	9.119	894.6	174.994 9
6.722	11.990	1 176.2	174.980 5

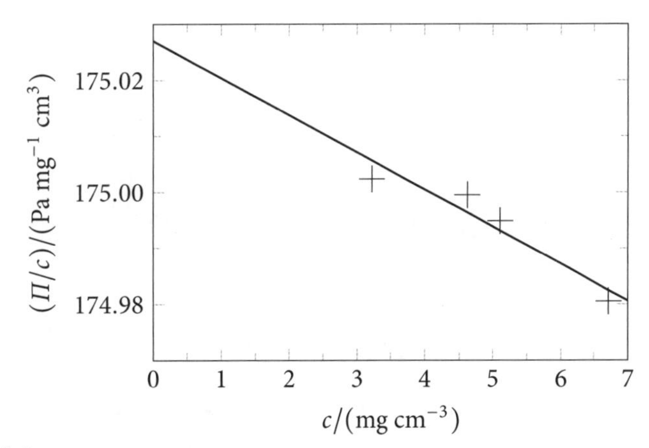

Figure 5.4

The data are a modest fit to a straight line, the equation of which is

$$(\Pi/c)/(\text{Pa mg}^{-1}\text{ cm}^3) = -6.628 \times 10^{-3} \times c/(\text{mg cm}^{-3}) + 175.03$$

The intercept is RT/M_J; before using the intercept in this expression it is best to convert it from $(\text{Pa mg}^{-1}\text{ cm}^3)$ to SI units, $(\text{Pa kg}^{-1}\text{ m}^3)$

$$(175.03\ \text{Pa mg}^{-1}\text{ cm}^3) \times \frac{1\ \text{m}^3}{10^6\ \text{cm}^3} \times \frac{1\ \text{mg}}{10^{-6}\ \text{kg}} = 175.03\ \text{Pa kg}^{-1}\text{ m}^3$$

$$M_J = \frac{RT}{\text{intercept}} = \frac{(8.3145\,\text{J K}^{-1}\,\text{mol}^{-1}) \times (293.15\,\text{K})}{175.03\,\text{Pa kg}^{-1}\,\text{m}^3} = 13.92...\,\text{kg mol}^{-1}$$

The molar mass of the protein is therefore $\boxed{1.39 \times 10^4\,\text{g mol}^{-1}}$.

P5B.9 The osmotic pressure Π is related to the molar concentration [J] through a virial-type equation, [5B.18–163], $\Pi = [\text{J}]RT(1+B[\text{J}])$. As is shown in *Problem* P5B.7 this equation can be rewritten in terms of the mass concentration c_J and the molar mass of J, M_J

$$\frac{\Pi}{c_J} = \frac{RT}{M_J} + \frac{BRT}{M_J^2}c_J$$

A plot of Π/c_J against c_J will have intercept RT/M_J when $c_J = 0$: from this it is possible to determine the molar mass. The second virial coefficient is obtained from the slope. The plot is shown in Fig. 5.5.

$c/(\text{mg cm}^{-3})$	Π/Pa	$(\Pi/c)/(\text{Pa mg}^{-1}\,\text{cm}^3)$
1.33	30	22.6
2.10	51	24.3
4.52	132	29.2
7.18	246	34.3
9.87	390	39.5

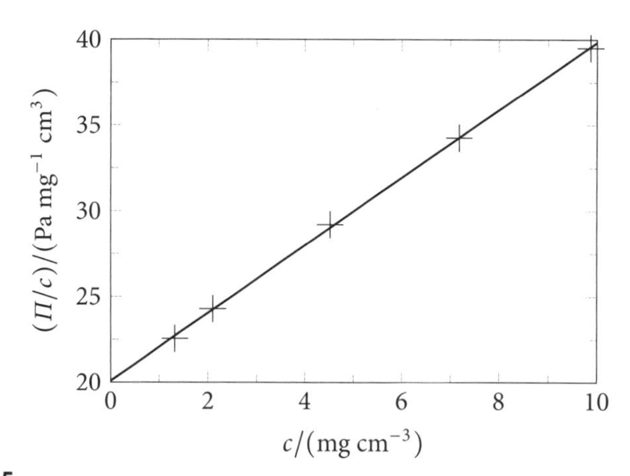

Figure 5.5

The data are a good fit to a straight line, the equation of which is

$$(\Pi/c)/(\text{Pa mg}^{-1}\,\text{cm}^3) = 1.975 \times (c/(\text{mg cm}^{-3})) + 20.09$$

The intercept is RT/M_J; before using the intercept in this expression it is best to convert it from $(\text{Pa mg}^{-1}\,\text{cm}^3)$ to SI units, $(\text{Pa kg}^{-1}\,\text{m}^3)$

$$(20.09\,\text{Pa mg}^{-1}\,\text{cm}^3) \times \frac{1\,\text{m}^3}{10^6\,\text{cm}^3} \times \frac{1\,\text{mg}}{10^{-6}\,\text{kg}} = 20.09\,\text{Pa kg}^{-1}\,\text{m}^3$$

$$M_J = \frac{RT}{\text{intercept}} = \frac{(8.3145\,\text{J K}^{-1}\,\text{mol}^{-1}) \times (303.15\,\text{K})}{20.09\,\text{Pa kg}^{-1}\,\text{m}^3} = 1.25... \times 10^2\,\text{kg mol}^{-1}$$

The molar mass of the polymer is therefore $\boxed{1.25 \times 10^5\,\text{g mol}^{-1}}$.

The second virial coefficient is found by taking the ratio (slope)/(intercept)

$$\frac{\text{slope}}{\text{intercept}} = \frac{B}{M_J} \quad \text{hence} \quad B = M_J \times \frac{\text{slope}}{\text{intercept}}$$

$$B = (1.25... \times 10^2\,\text{kg mol}^{-1}) \times \frac{1.975\,\text{Pa mg}^{-2}\,\text{cm}^6}{20.09\,\text{Pa mg}^{-1}\,\text{cm}^3}$$

$$= (1.25... \times 10^8\,\text{mg mol}^{-1}) \times \frac{1.975\,\text{Pa mg}^{-2}\,\text{cm}^6}{20.09\,\text{Pa mg}^{-1}\,\text{cm}^3} = 1.23... \times 10^7\,\text{cm}^3\,\text{mol}^{-1}$$

Given that [J] is usually in mol dm^{-3} it is convenient to quote the value of the second virial coefficient as $\boxed{B = 1.23 \times 10^4\,\text{mol}^{-1}\,\text{dm}^3}$.

P5B.11 The excess enthalpy of mixing for this particular regular solution is given by [5B.6–157], $H^{\text{E}} = nRT\xi x_{\text{A}}x_{\text{B}}$. The plot in Fig. 5.6 shows $H^{\text{E}}/(nRT)$ as a function of x_{A} for different values of ξ; recall that $x_{\text{A}} + x_{\text{B}} = 1$, so $x_{\text{A}}x_{\text{B}} = x_{\text{A}}(1 - x_{\text{A}})$.

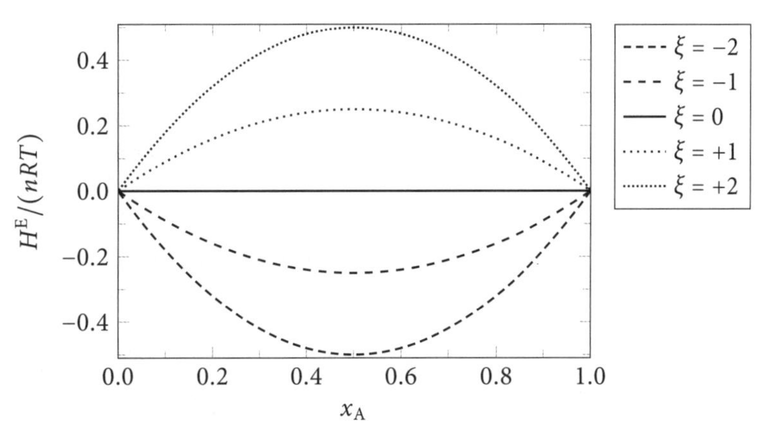

Figure 5.6

If ξ is fixed, the temperature dependence is explored by plotting $H^{\text{E}}/(nR\xi)$ as a function of x_{A}: $H^{\text{E}}/(nR\xi) = Tx_{\text{A}}x_{\text{B}} = Tx_{\text{A}}(1 - x_{\text{A}})$. This is shown in Fig. 5.7. Evidently the strongest temperature dependence is once more at $x_{\text{A}} = \frac{1}{2}$.

P5B.13 The osmotic pressures Π is related to the molar concentration [J] through a virial-type equation, [5B.18–163], $\Pi = [\text{J}]RT(1 + B[\text{J}])$. As is shown in *Problem* P5B.7 this equation can be rewritten in terms of the mass concentration c_{J} and the molar mass of J, M_{J}

$$\frac{\Pi}{c_{\text{J}}} = \frac{RT}{M_{\text{J}}} + \frac{BRT}{M_{\text{J}}^2}c_{\text{J}}$$

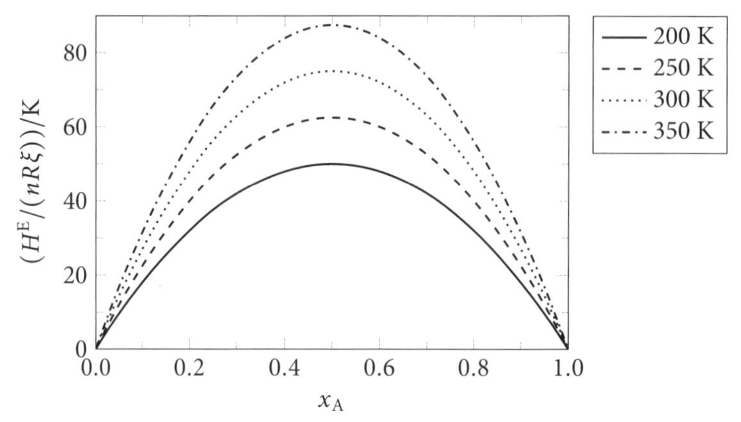

Figure 5.7

A plot of Π/c_J against c_J will have intercept RT/M_J when $c_J = 0$: from this it is possible to determine the molar mass. The second virial coefficient is obtained from the slope. Such a plot is shown in Fig. 5.8.

$c/(\text{g dm}^{-3})$	Π/Pa	$(\Pi/c)/(\text{Pa g}^{-1}\text{ dm}^3)$
1.00	27	27.0
2.00	70	35.0
4.00	197	49.3
7.00	500	71.4
9.00	785	87.2

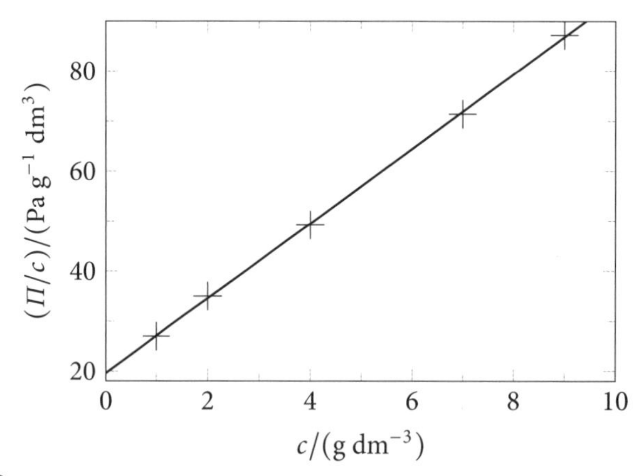

Figure 5.8

The data are a good fit to a straight line, the equation of which is

$$(\Pi/c)/(\text{Pa g}^{-1}\text{ dm}^3) = 7.466 \times (c/(\text{g dm}^{-3})) + 19.64$$

The intercept is RT/M_J; before using the intercept in this expression it is best to convert it from (Pa g^{-1} dm^3) to SI units, (Pa kg^{-1} m^3)

$$(19.64\ \text{Pa g}^{-1}\ \text{dm}^3) \times \frac{1\ \text{m}^3}{10^3\ \text{dm}^3} \times \frac{1\ \text{g}}{10^{-3}\ \text{kg}} = 19.64\ \text{Pa kg}^{-1}\ \text{m}^3$$

$$M_J = \frac{RT}{\text{intercept}} = \frac{(8.3145\ \text{J K}^{-1}\ \text{mol}^{-1}) \times (298\ \text{K})}{19.64\ \text{Pa kg}^{-1}\ \text{m}^3} = 1.26... \times 10^2\ \text{kg mol}^{-1}$$

The molar mass of the polymer is therefore $\boxed{1.26 \times 10^5\ \text{g mol}^{-1}}$.

The second virial coefficient is found by taking the ratio (slope)/(intercept)

$$\frac{\text{slope}}{\text{intercept}} = \frac{B}{M_J} \quad \text{hence} \quad B = M_J \times \frac{\text{slope}}{\text{intercept}}$$

$$B = (1.26... \times 10^2\ \text{kg mol}^{-1}) \times \frac{7.466\ \text{Pa g}^{-2}\ \text{dm}^6}{19.64\ \text{Pa g}^{-1}\ \text{dm}^3}$$

$$= (1.26... \times 10^5\ \text{g mol}^{-1}) \times \frac{7.466\ \text{Pa g}^{-2}\ \text{dm}^6}{19.64\ \text{Pa g}^{-1}\ \text{dm}^3} = 4.79... \times 10^4\ \text{dm}^3\ \text{mol}^{-1}$$

The second virial coefficient is therefore $\boxed{B = 4.80 \times 10^4\ \text{mol}^{-1}\ \text{dm}^3}$.

5C Phase diagrams of binary systems: liquids

Answers to discussion questions

D5C.1 The temperature-composition phase diagram is shown in Fig. 5.9; the diagram shows a high-boiling azeotrope.

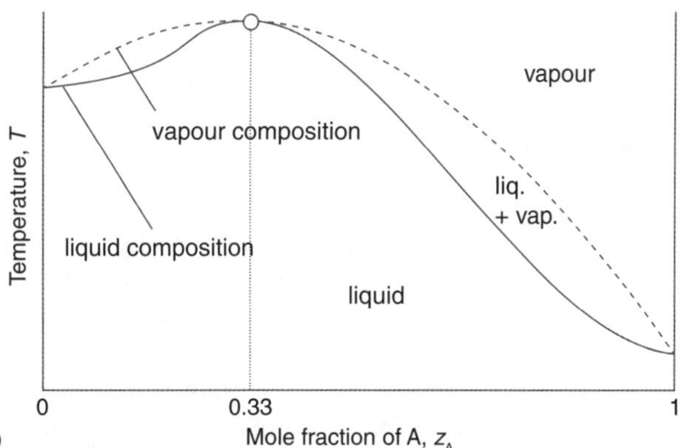

Figure 5.9

D5C.3 The principal factor is the shape of the two-phase liquid-vapour region in the phase diagram (usually a temperature-composition diagram). The closer the liquid and vapour lines are to each other, the more steps of the sort illustrated in Fig. 5C.9 on page 171 are needed to move from a given mixed composition to an acceptable enrichment in one of the components. However, the presence of an azeotrope could prevent the desired degree of separation from being achieved.

Solutions to exercises

E5C.1(a) The temperature–composition phase diagram is a plot of the boiling point against (1) composition of the liquid, x_M and (2) composition of the vapour, y_M. The horizontal axis is labelled z_M, which is interpreted as x_M or y_M according to which set of data are being plotted. In addition to the data in the table, the boiling points of the pure liquids are added. The plot is shown in Fig 5.10; in this plot, the lines are best-fit polynomials of order 3.

$\theta/°C$	x_M	y_M	$\theta/°C$	x_M	y_M
110.6	1	1	117.3	0.408	0.527
110.9	0.908	0.923	119.0	0.300	0.410
112.0	0.795	0.836	121.1	0.203	0.297
114.0	0.617	0.698	123.0	0.097	0.164
115.8	0.527	0.624	125.6	0	0

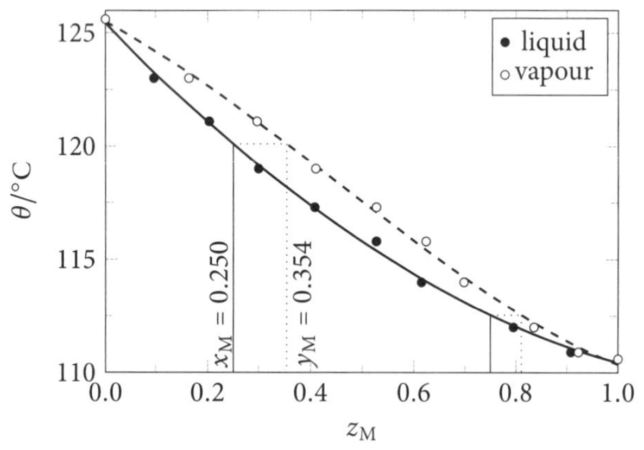

Figure 5.10

(i) The vapour composition corresponding to a liquid composition of x_M = 0.250 is found by taking the vertical line at this composition up to the intersection with the liquid curve, and then moving across horizontally to the intersection with the vapour curve; occurs at $\boxed{y_M = 0.354}$, which gives the composition of the vapour. The exact points of intersection can be found either from the graph or by using the fitted functions.

(ii) A composition x_O = 0.250 corresponds to x_M = 0.750; from the graph this corresponds to a vapour composition $\boxed{y_M = 0.811}$.

E5C.2(a) At the lowest temperature shown in the diagram the mixture is in the two-phase region, and the two phases have composition of approximately x_B = 0.88 and x_B = 0.05. The level rule shows that there is about 9 times more of the B-rich than of the B-poor phase. As the temperature is raised the B-rich phase becomes slightly less rich in B, and the other phase becomes richer in B. The lever rule implies that the proportion of the B-rich phase increases as the temperature rises.

At temperature T_1 the vertical line intersects the phase boundary. At this point the B-poor phase disappears and only one phase, with $x_B = 0.8$, is present.

E5C.3(a) The molar masses of phenol and water are $94.1074\,\mathrm{g\,mol^{-1}}$ and $18.0158\,\mathrm{g\,mol^{-1}}$, respectively. The mole fraction of phenol (P) is

$$x_P = \frac{(7.32\,\mathrm{g})/(94.1074\,\mathrm{g\,mol^{-1}})}{(7.32\,\mathrm{g})/(94.1074\,\mathrm{g\,mol^{-1}}) + (7.95\,\mathrm{g})/(18.0158\,\mathrm{g\,mol^{-1}})} = 0.149...$$

Hence $\boxed{x_P = 0.150}$. Let the two phases be α ($x_P = 0.042$) and β ($x_P = 0.161$). The proportions of these two phases, n_β/n_α is given by the level rule, [5C.6–*170*]

$$\frac{n_\beta}{n_\alpha} = \frac{l_\alpha}{l_\beta} = \frac{0.149... - 0.042}{0.161 - 0.149...} = \boxed{9.68}$$

The phenol-rich phase is more abundant by a factor of almost 10.

E5C.4(a) An approximate phase diagram is shown in Fig. 5.11; the given data points are shown with dots and the curve is a quadratic which is a modest fit to these points. The shape conforms to the expected phase diagram for such a system.

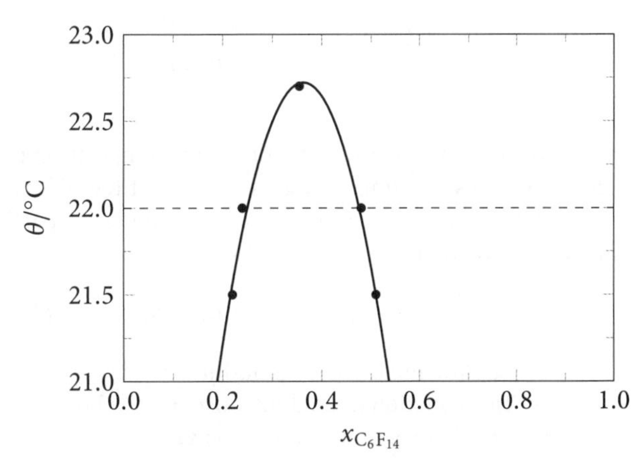

Figure 5.11

(i) A temperature of 23 °C is above the highest temperature at which partial miscibility occurs, and therefore the expectation is that hexane and perfluorohexane mix in all proportions to give a single phase.

(ii) At 22 °C the possibility of phase separation exists; as the mole fraction of perfluorohexane increases the phase diagram is traversed along the dashed line. When the mole fraction of perfluorohexane is low a single phase forms, but as the mole fraction goes beyond 0.24 phase separation occurs. Initially, according to the lever rule, the proportion of the perfluorohexane-rich phase is very small, but as more and more perfluorohexane is added the proportion of this phase increases. When the mole fraction is just under 0.48, there is very little of the perfluorohexane-poor phase present, and as the mole fraction increases further a one-phase zone is reached in which there is complete miscibility.

Solutions to problems

P5C.1 If it is assumed that Raoult's law applies, [5A.22–*151*], the partial vapour pressures of benzene (B) and methylbenzene (M) are

$$p_B = x_B p_B^* \qquad p_M = x_M p_M^*$$

where x_J are the mole fractions and p_J^* are the vapour pressures over the pure liquids. The total pressure is taken to be $p_{tot} = p_B + p_M$.

The mole fraction in the vapour, y_J, is related to the total pressure by $p_J = y_J p_{tot}$, so it follows that

$$y_J = \frac{p_J}{p_{tot}} = \frac{x_J p_J^*}{p_{tot}}$$

Therefore

$$y_B = \frac{x_B p_B^*}{p_{tot}} = \frac{0.75 \times (75\ \text{Torr})}{0.75 \times (75\ \text{Torr}) + 0.25 \times (21\ \text{Torr})} = \boxed{0.91}$$

$$y_M = \frac{x_M p_M^*}{p_{tot}} = \frac{0.25 \times (21\ \text{Torr})}{0.75 \times (75\ \text{Torr}) + 0.25 \times (21\ \text{Torr})} = \boxed{0.085}$$

P5C.3 It is convenient to construct a pressure–composition phase diagram in order to answer this question. If it is assumed that Raoult's law applies, [5A.22–*151*], the total pressure is computed from the sum of the partial vapour pressures of benzene (B) and methylbenzene (M)

$$p_{tot} = p_B + p_M = x_B p_B^* + x_M p_M^*$$

where x_J are the mole fractions and p_J^* are the vapour pressures over the pure liquids. This equation is used to construct the liquid line on the graph shown in Fig. 5.12, where z_B is interpreted as x_B.

The mole fraction in the vapour, y_J, is related to the total pressure by $p_J = y_J p_{tot}$. Using this it can be shown that the total pressure in terms of the mole fraction in the vapour in given by [5C.5–*167*],

$$p_{tot} = \frac{p_B^* p_M^*}{p_B^* + (p_M^* - p_B^*) y_B}$$

This equation is used to construct the vapour line on the phase diagram, where z_B is interpreted as y_B.

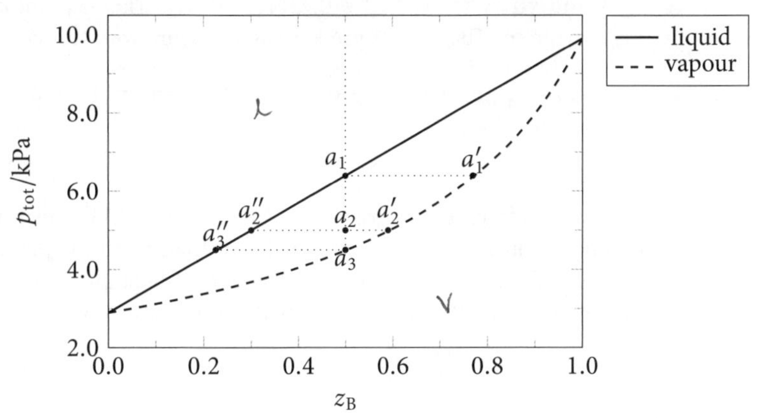

Figure 5.12

(a) A mixture with equal amounts of B and M has mole fractions $x_{\text{B}} = x_{\text{M}} = \frac{1}{2}$. The total pressure is therefore

$$p_{\text{tot}} = x_{\text{B}}p_{\text{B}}^* + x_{\text{M}}p_{\text{M}}^* = \tfrac{1}{2}(9.9\text{ kPa}) + \tfrac{1}{2}(2.9\text{ kPa}) = \boxed{6.4\text{ kPa}}$$

This is the pressure at which boiling first occurs, point a_1 in the diagram.

(b) The composition of the vapour is given by

$$y_{\text{B}} = \frac{p_{\text{B}}}{p_{\text{tot}}} = \frac{x_{\text{B}}p_{\text{B}}^*}{p_{\text{tot}}} = \frac{\frac{1}{2}(9.9\text{ kPa})}{6.4\text{ kPa}} = 0.773... = \boxed{0.77}$$

and therefore $y_{\text{M}} = 1 - 0.773... = \boxed{0.23}$. This is point a_1' on the diagram: the lever-rule also indicates that the fraction of the phase with composition a_1' (the vapour) is very small.

(c) As the pressure is reduced further, say to point a_2, the tie line indicates that the liquid will have composition a_2'' and the vapour will have composition a_2', the latter being richer in the more volatile component B. The level rule indicates that the proportion of the vapour phase is now significant.

The process continues until point a_3 is reached. At this pressure the composition of the liquid is given by point a_3'', and the level rule indicates that the proportion of the liquid phase is very small. It is also evident from the diagram that at point a_3 the vapour composition is $y_{\text{B}} = y_{\text{M}} = \frac{1}{2}$, therefore $p_{\text{B}} = \frac{1}{2}p_{\text{tot}}$ and $p_{\text{M}} = \frac{1}{2}p_{\text{tot}}$. Raoult's law gives the partial vapour pressures of B and M are $p_{\text{B}} = p_{\text{B}}^* x_{\text{B}}$ and $p_{\text{M}} = p_{\text{M}}^* x_{\text{M}}$. It follows that

$$\tfrac{1}{2}p_{\text{tot}} = x_{\text{B}}p_{\text{B}}^* \quad \text{and} \quad \tfrac{1}{2}p_{\text{tot}} = x_{\text{M}}p_{\text{M}}^* = (1 - x_{\text{B}})p_{\text{M}}^*$$

These two equations are combined to give

$$x_{\text{B}} = \frac{p_{\text{M}}^*}{p_{\text{B}}^* + p_{\text{M}}^*} = \frac{(2.9\text{ kPa})}{(9.9\text{ kPa}) + (2.9\text{ kPa})} = 0.226... = \boxed{0.23}$$

and $x_M = 1 - x_B = 1 - 0.226... = \boxed{0.77}$. This is point a_3'' on the phase diagram. The vapour pressure of a mixture with this composition is

$$p_{\text{tot}} = x_B p_B^* + x_M p_M^* = 0.226... \times (9.9\ \text{kPa}) + (1 - 0.226...) \times (2.9\ \text{kPa})$$
$$= \boxed{4.5\ \text{kPa}}$$

P5C.5 The annotated phase diagrams are shown in Fig. 5.13. Given that the normal boiling point of hexane is certainly lower than that of heptane the horizontal scale should presumably be mole fraction of heptane; however, the solution provided follows the labelling of the diagram in the text.

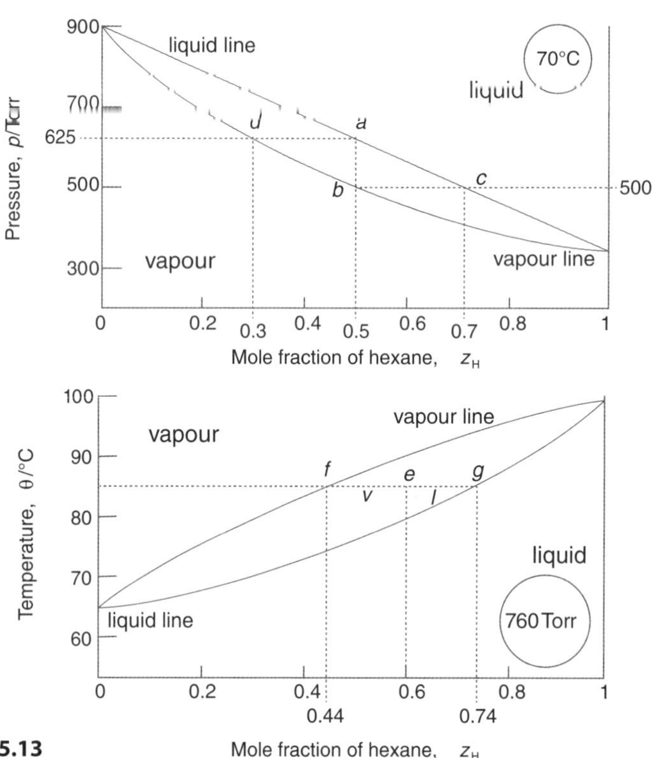

Figure 5.13

(a) The phases present are indicated on the diagrams above.

(b) For an equimolar mixture $x_H = 0.5$; the vertical line at this composition intersects the liquid line at a, and reading across the pressure is $\boxed{625\ \text{Torr}}$.

(c) At this pressure the composition of the vapour is given by the intersection of the horizontal line with the vapour line, which occurs at point d. The vapour is less rich in hexane than the liquid. As the solution continues to evapourate the composition of the liquid moves along the liquid line to point c. This is at the pressure at which the composition of the vapour matches the original composition of the liquid ($x_H = 0.5$, point b). The composition of the liquid is read off the scale as $\boxed{x_H = 0.7}$, and the pressure is $\boxed{500\ \text{Torr}}$.

(d) From part (b) the composition of the liquid is $\boxed{x_H = 0.5}$, and the composition of the vapour is read off from where the horizontal line at 625 Torr intersects the vapour curve, point *d*, which is at $\boxed{y_H = 0.3}$.

(e) From part (c) the composition of the vapour is $\boxed{y_H = 0.5}$ and the composition of the liquid can be read off from where the horizontal line at 500 Torr intersects the liquid curve, point *c*, which is at $\boxed{x_H = 0.7}$.

(f) Refer to the temperature–composition phase diagram; the stated composition is $z_{\text{heptane}} = 0.40$ which corresponds to $z_H = 0.60$. The vertical line at $z_H = 0.60$ intersects the horizontal line at 85 °C at point *e*. From the tie line the composition of the vapour is read off from point *f*, $y_H = 0.44$; the composition of the liquid is read off from point *g*, $y_H = 0.74$. From the lever rule

$$\frac{n_l}{n_v} = \frac{v}{l} = \frac{0.60 - 0.44}{0.74 - 0.60} = \boxed{1.1}$$

The two phases are roughly equally abundant.

P5C.7 The relationship between y_A and x_A is given in [5C.4–*166*]

$$y_A = \frac{x_A p_A^*}{p_B^* + (p_A^* - p_B^*)x_A} = \frac{x_A(p_A^*/p_B^*)}{1 + (p_A^*/p_B^* - 1)x_A}$$

The form of the function on the right gives y_A as a function of x_A and the ratio (p_A^*/p_B^*) as required. The plot if shown in Fig. 5.14

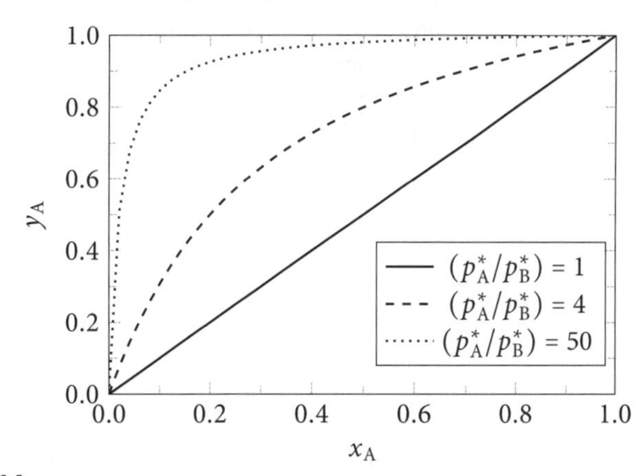

Figure 5.14

P5C.9 If the excess enthalpy is modelled as $H^E = \xi RT x_A^2 x_B^2$ then, by anaolgy with [5B.7–*157*], the expression for for Gibbs energy of mixing is

$$\Delta_{\text{mix}}G = nRT\left(x_A \ln x_A + x_B \ln x_B + \xi x_A^2 x_B^2\right)$$

The minima and maxima in this function are located by setting the derivative with respect to x_A to zero; it is convenient to take the derivative of $\Delta_{mix}G/nRT$ and before doing this x_B is replaced by $(1 - x_A)$

$$\begin{aligned}\Delta_{mix}G/nRT &= x_A \ln x_A + (1 - x_A)\ln(1 - x_A) + \xi x_A^2(1 - x_A)^2 \\ &= x_A \ln x_A + \ln(1 - x_A) - x_A \ln(1 - x_A) + \xi x_A^2(1 - x_A)^2 \\ \frac{d(\Delta_{mix}G/nRT)}{dx_A} &= 1 + \ln x_A - \frac{1}{1 - x_A} + \frac{x_A}{1 - x_A} - \ln(1 - x_A) \\ &\quad + 2\xi x_A(1 - x_A)^2 - 2\xi x_A^2(1 - x_A) \\ &= \frac{1 - x_A - 1 + x_A}{1 - x_A} + \ln \frac{x_A}{1 - x_A} + 2\xi x_A(1 - x_A)(1 - 2x_A) \\ &= \ln \frac{x_A}{1 - x_A} + 2\xi x_A(1 - x_A)(1 - 2x_A)\end{aligned}$$

As before, the derivative is zero at $x_A = 0.5$ for all values of ξ; this corresponds either to a minimum when ξ is small, or to a maximum when ξ is sufficiently large. Qualitatively the behaviour is similar to that shown in Fig. 5B.5.

Apart from this solution at $x_A = 0.5$, there are no analytical solutions for when this derivative is zero. However, solutions can be found by graphically by looking for the intersection between $\ln(x_A/[1 - x_A])$ and $-2\xi x_A(1-x_A)(1-2x_A)$. This is done with the aid of Fig. 5.3b. From the graph it is evident that for $\xi = 1$ there are no values of x_A at which the curves intersect, and so no minima, but at sufficiently high values of ξ (such as $\xi = 6$) such intersections do occur and lead to two minima.

Overall, as is seen in Fig. 5.15, the behaviour is qualitatively similar to that for $H^E = \xi RT x_A x_B$. If ξ is below some particular positive value $\Delta_{mix}G$ is always negative and shows a single minimum at $x_A = 0.5$. Above some critical value, $\Delta_{mix}G$ may become positive for some values of x_A, and the plot shows a maximum at $x_A = 0.5$, flanked symmetically by two minima. This indicates that phase separation will occur.

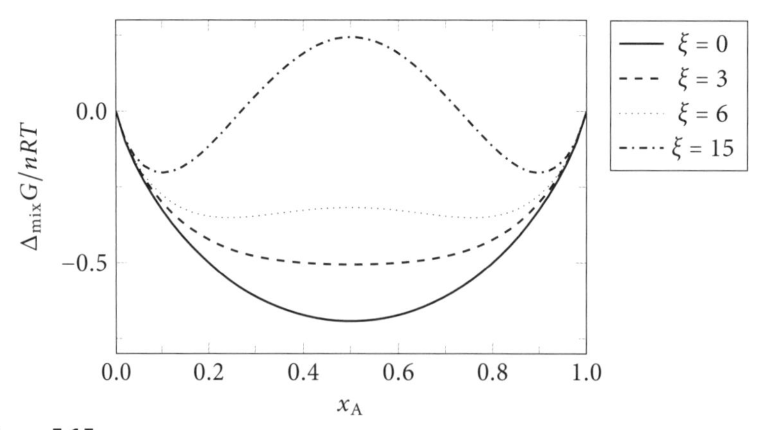

Figure 5.15

5D Phase diagrams of binary systems: solids

Answers to discussion questions

D5D.1 The schematic phase diagram is shown in Fig. 5.16. Congruent melting means that the compound AB also occurs in the liquid phase.

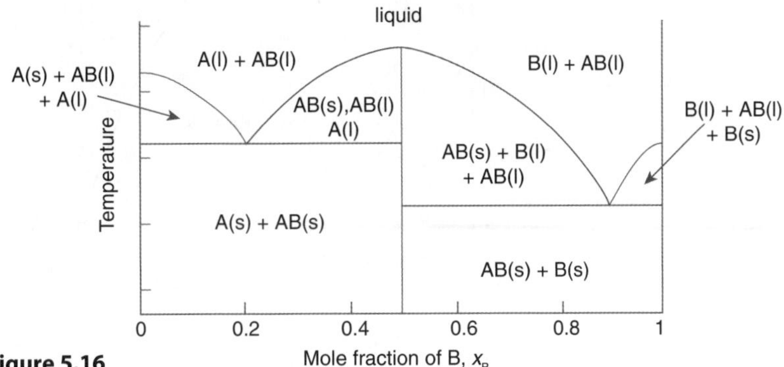

Figure 5.16

Solutions to exercises

E5D.1(a) The schematic phase diagram is shown in Fig 5.17. The solid points are the data given in the *Exercise*, and the lines are simply plausible connections between these points; it is assumed that the compound in 1:1. Note that to the right of $x_B = 0.5$ the solids are AB and B, whereas to the left of this composition the solids are AB and A.

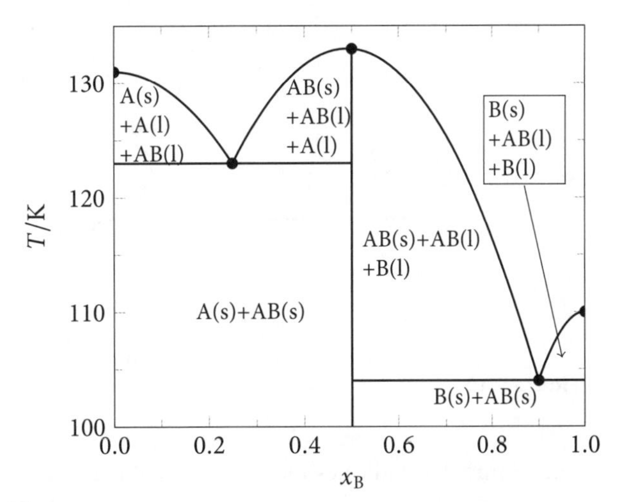

Figure 5.17

E5D.2(a) The schematic phase diagram is shown in Fig 5.18. The solid points are the data given in the *Exercise*, and lines are simply plausible connections between these points. (The dash-dotted lines are referred in to *Exercise* E5D.3(a).)

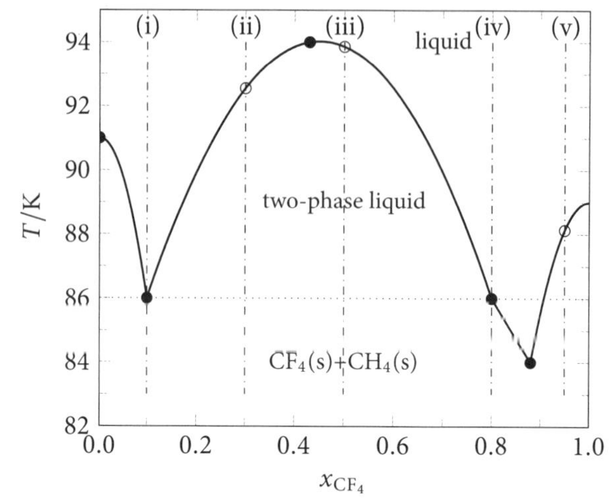

Figure 5.18

E5D.3(a) The compositions at which the cooling curves are plotted are indicated by the vertical dash-dotted lines on the phase diagram for *Exercise* E5D.2(a), Fig. 5.18. The cooling curves are shown in Fig 5.19. The break points, where solid phases start to form are shown by the short horizontal lines, and the dotted lines indicate the temperatures of the two eutectics (86 K and 84 K). The horizontal segments correspond to solidification of a eutectic.

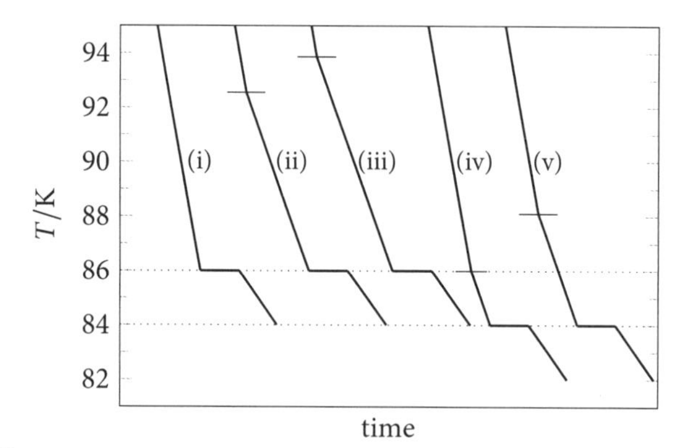

Figure 5.19

E5D.4(a) The feature that indicates incongruent melting is the intersection of the two liquid curves at around $x_B = 0.6$. The incongruent melting point is marked as $T_1 \approx 350\,°C$. The composition of the eutectic is $\boxed{x_B \approx 0.25}$ and its melting point is labelled $\boxed{T_2 \approx 190\,°C}$.

E5D.5(a) The cooling curves are shown in Fig 5.20; the break points are shown by the short horizontal lines. For isopleth a the first break point is at 380 °C where the isopleth crosses the liquid curve, there is a second break point where the isopleth crosses the boundary at T_1; there is then a eutectic halt at 190 °C. For isopleth b the first break point is at 450 °C where the isopleth crosses the liquid curve, there is a second break point where the isopleth crosses the boundary at T_1, and then a eutectic halt at 190 °C.

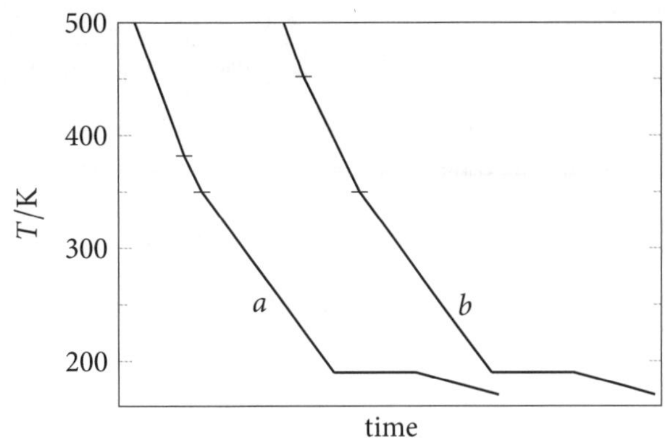

Figure 5.20

E5D.6(a) Figure 5.21 shows the relevant phase diagram to which dotted horizontal lines have been added at the relevant temperatures.

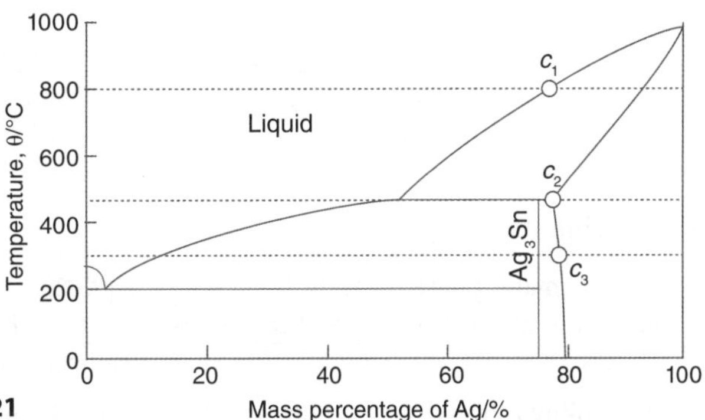

Figure 5.21

(i) The solubility of silver in tin at 800 °C is determined by the point c_1. (At higher proportions of silver, the system separates into two phases, a liquid and a solid phase rich in silver.) The point c_1 corresponds to 76% silver by mass.

(ii) The compound Ag_3Sn decomposes at this temperature. Three phases are in equilibrium here: a liquid containing atomic Ag and Sn about 52% Ag by mass; a solid solution of Ag_3Sn in Ag; and solid Ag_3Sn. See point c_2.

(iii) At point c_3, two phases coexist: solid Ag_3Sn and a solid solution of the compound and metallic silver. Because this point is close to the Ag_3Sn composition, the solid solution is mainly Ag_3Sn, at least when measured in mass terms. The composition of the solid solution is expressed as a ratio of moles of compound (n_c) to moles of atomic silver (n_a). These quantities are related to the silver mass fraction c_{Ag} by employing the definition of mass fraction, namely the mass of silver (from the compound and from atomic silver) over the total sample mass

$$c_{Ag} = \frac{m_{Ag}}{m_{Ag} + m_{Sn}} = \frac{(3n_c + n_a)M_{Ag}}{(3n_c + n_a)M_{Ag} + n_c M_{Sn}}$$

This relationship is rearranged, collecting terms in n_c on one side and n_a on the other

$$n_c\left[3M_{Ag}(c_{Ag} - 1) + M_{Sn}c_{Ag}\right] = n_a M_{Ag}(1 - c_{Ag})$$

The mole ratio of compound to atomic silver is given by

$$\frac{n_c}{n_a} = \frac{M_{Ag}(1 - c_{Ag})}{3M_{Ag}(c_{Ag} - 1) + M_{Sn}c_{Ag}}$$

At 460 °C, $c_{Ag} = 0.78$ (point c_3 on the coexistence curve), so

$$\frac{n_c}{n_a} = \frac{(107.9\,\mathrm{g\,mol^{-1}}) \times (1 - 0.78)}{3 \times (107.9\,\mathrm{g\,mol^{-1}}) \times (0.78 - 1) + (118.7\,\mathrm{g\,mol^{-1}}) \times 0.78} = \boxed{1.11}$$

At 300 °C, $c_{Ag} = 0.77$ (point c_2 on the coexistence curve), so

$$\frac{n_c}{n_a} = \frac{(107.9\,\mathrm{g\,mol^{-1}}) \times (1 - 0.77)}{3 \times (107.9\,\mathrm{g\,mol^{-1}}) \times (0.77 - 1) + (118.7\,\mathrm{g\,mol^{-1}}) \times 0.77} = \boxed{1.46}$$

Solutions to problems

P5D.1 The schematic phase diagram is shown in Fig 5.22. The solid points are the data given in the *Exercise*, and in the absence of any further information and because there are so few points, these have just been joined by straight lines.

On cooling a liquid with composition $x_{ZrF_4} = 0.4$ solid first starts to appear when the isopleth intersects the liquid line (at about 870 °C). The composition of the small amount of solid that forms is given by the left-hand open circle in the diagram (about $x = 0.29$; containing less ZrF_4 than the liquid). As the temperature drops further more solid is formed and its composition moves along the solid line to the right becoming richer in ZrF_4 until it reaches the point where the isopleth crosses the solid line. At this point what remains of the liquid has composition given by the right-hand open circle (about $x = 0.48$). A further drop in temperature results in complete solidification.

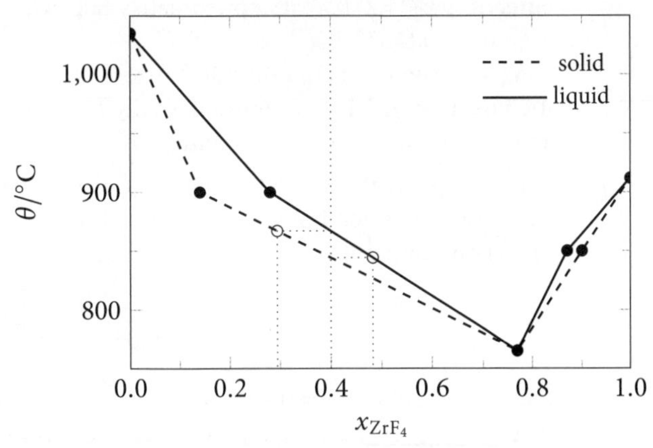

Figure 5.22

P5D.3 The phase diagram is shown in Fig. 5.23, along with the relevant cooling curves. The fact that there is a phase boundary indicated by the vertical line at $x_B = 0.67$ is taken to indicate the formation of compound AB_2 which has $x_B = \frac{2}{3}$. By analogy with the phase diagram shown in Fig. 5D.5 on page 179, the form of the given phase diagram indicates that AB_2 does not exist in the liquid phase.

The number of distinct chemical species (as opposed to components) and phases present at the indicated points are, respectively

$$b(3,2), d(2,2), e(4,3), f(4,3), g(4,3), k(2,2)$$

Liquid A and solid A are considered to be distinct species.

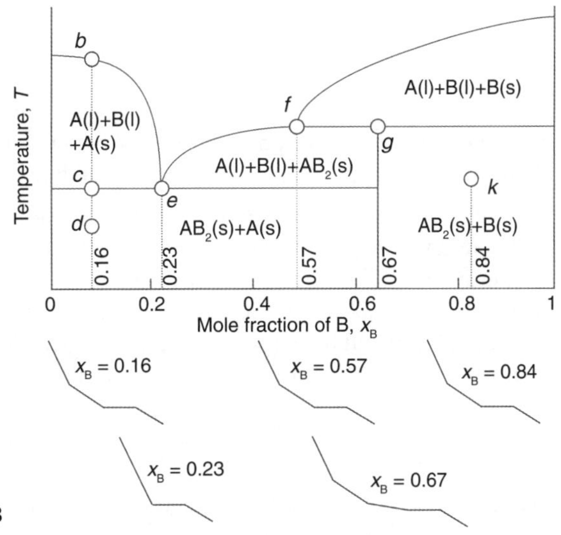

Figure 5.23

P5D.5 (a) Note that, as indicated on the diagram, Ca_2Si, CaSi, $CaSi_2$ appear at mole fractions of Si $\frac{1}{3}$, $\frac{1}{2}$ and $\frac{2}{3}$, as expected.

eutectics: $x_{\text{Si}} = 0.056$ at approximately 800 °C, $x_{\text{Si}} = 0.402$ at 1268 °C, $x_{\text{Si}} = 0.694$ at 1030 °C
congruent melting compounds: Ca_2Si $T_f = 1314$ °C, CaSi $T_f = 1324$ °C
incongruent melting compound: $CaSi_2$ $T_f = 1040$ °C (melts into CaSi(s) and Si-rich liquid with x_{Si} around 0.69)

(b) For an isopleth at $x_{\text{Si}} = 0.2$ and at 1000 °C the phases in equilibrium are $CaSi_2$ and a Ca-rich liquid ($x_{\text{Si}} = 0.11$). The lever rule, [5C.6–*170*], gives the relative amounts:

$$\frac{n_{\text{Ca}_2\text{Si}}}{n_{\text{liq}}} = \frac{l_{\text{liq}}}{l_{\text{Ca}_2\text{Si}}} = \frac{0.2 - 0.11}{0.333 - 0.2} = \boxed{0.7}$$

(c) For an isopleth at $x_{\text{Si}} = 0.8$ Si(s) begins to appear at about 1300 °C. Further cooling causes more Si(s) to freeze out of the melt so that the melt becomes more concentrated in Ca. There is a eutectic at $x_{\text{Si}} = 0.694$ and 1030 °C.

At a temperature just above the eutectic point the liquid has composition $x_{\text{Si}} = 0.694$ and the lever rule gives that the relative amounts of the Si(s) and liquid phases as:

$$\frac{n_{\text{Si}}}{n_{\text{liq}}} = \frac{l_{\text{liq}}}{l_{\text{Si}}} = \frac{0.80 - 0.694}{1.0 - 0.80} = \boxed{0.53}$$

At the eutectic temperature a third phase appears, $CaSi_2(s)$. As the melt cools at this temperature, both Si(s) and $CaSi_2(s)$ freeze out of the melt while the composition of the melt remains constant. At a temperature slightly below the eutectic point all the melt will have frozen to Si(s) and $CaSi_2(s)$ with the relative amounts:

$$\frac{n_{\text{Si}}}{n_{\text{CaSi}_2}} = \frac{l_{\text{CaSi}_2}}{l_{\text{Si}}} = \frac{0.80 - 0.667}{1.0 - 0.80} = \boxed{0.67}$$

P5D.7 The data are plotted as the phase diagram shown in Fig 5.24; the filled and open circles are the data points and the solid/dashed line is a best-fit cubic function.

A mixture of 0.750 mol of *N,N*-dimethylacetamide with 0.250 mol of heptane has mole fraction of the former of $x_1 = 0.750/(0.750 + 0.250) = 0.750$. The tie line at 296.0 °C is shown on the diagram, and this intersects with the two curves at $x_1 = 0.167$ and $x_2 = 0.805$ (determined from the best-fit polynomial - an alternative would be to use the data points given for this temperature). The lever rule, [5C.6–*170*] gives the proportion of the two phases as

$$\frac{n_{x=0.805}}{n_{x=0.167}} = \frac{0.750 - 0.167}{0.805 - 0.750} = \boxed{10.6}$$

The *N,N*-dimethylacetamide-rich phase is therefore more than ten times more abundant than the other phase.

A mixture of this composition will become a single phase at the temperature at which the $x_1 = 0.750$ isopleth intersects the right-hand phase boundary. Using the fitted function, this intersection is at $\boxed{302.5\ ^\circ\text{C}}$.

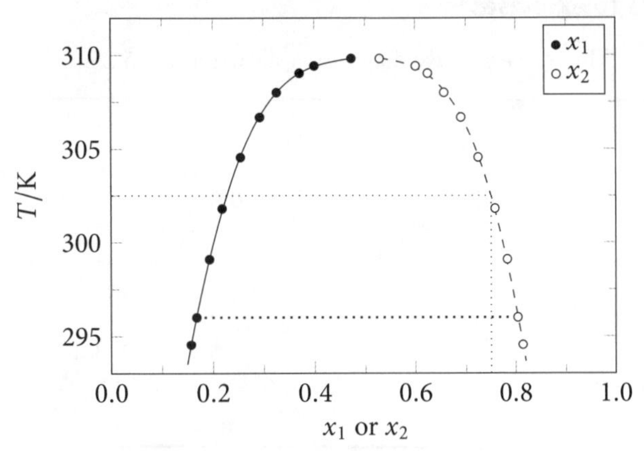

Figure 5.24

5E Phase diagrams of ternary systems

Answers to discussion questions

D5E.1 The phase rule [4A.1–*124*], $F = C - P + 2$, for three components ($C = 3$) implies that the degrees of freedom are $F = 5 - P$. If two of these degrees of freedom are used to fix the temperature and pressure, then there remain $3 - P$ degrees of freedom. The maximum number of phases in equilibrium at any given temperature and pressure, therefore, is $\boxed{3}$. A number of phases greater than this is not possible as there are would be no remaining degrees of freedom to describe their composition.

D5E.3 The phase rule [4A.1–*124*], $F = C - P + 2$, for four components ($C = 4$) implies that the degrees of freedom $F = 6 - P$. If two of those degrees of freedom are used to fix the temperature and pressure, then there are $4 - P$ remaining degrees of freedom. Those four degrees of freedom would be the proportions of the four components.

A regular tetrahedron would seem to be the object that could depict four mole fractions constrained to sum to one. It has four vertices, each of which could represent one of the four pure components (analogous to the three vertices of an equilateral triangle). It has four faces, each of which is an equilateral triangle. Points on any face would represent a three-component system (the component represented by the opposite vertex being missing); the faces, then, are exactly the ternary phase diagrams discussed in Topic 5E.

What is not immediately obvious, however, is that any point in the interior of the tetrahedron can represent compositions of four components constrained to sum to a constant (that is, four mole fractions constrained to sum to 1). It is not obvious, but it is true that for any interior point of a regular tetrahedron, the sum of its distances from the four faces is a constant: $d_1+d_2+d_3+d_4$ = constant, where the d_i represent the distances of the same point from the four sides of the tetrahedron.

Solutions to exercises

E5E.1(a) The ternary phase diagram is shown in Fig 5.25.

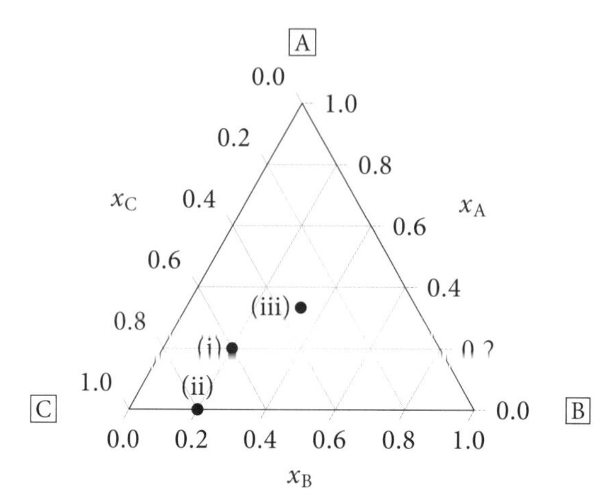

Figure 5.25

E5E.2(a) The composition by mass needs to be converted to mole fractions, which requires the molar masses: $M_{NaCl} = 58.44\,g\,mol^{-1}$, $M_{H_2O} = 18.016\,g\,mol^{-1}$, and $M_{Na_2SO_4\cdot 10\,H_2O} = 322.20\,g\,mol^{-1}$. Imagine that the solution contains 25 g NaCl, 25 g $Na_2SO_4 \cdot 10\,H_2O$ and hence $(100-25-25) = 50$ g H_2O. The mole fraction of NaCl is

$$x_{NaCl} = \frac{m_{NaCl}/M_{NaCl}}{m_{NaCl}/M_{NaCl} + m_{Na_2SO_4\cdot 10\,H_2O}/M_{Na_2SO_4\cdot 10\,H_2O} + m_{H_2O}/M_{H_2O}}$$

$$= \frac{(25\,g)/(58.44\,g\,mol^{-1})}{(25\,g)/(58.44\,g\,mol^{-1}) + (25\,g)/(322.20\,g\,mol^{-1}) + (50\,g)/(18.016\,g\,mol^{-1})}$$

$$= 0.13$$

Likewise, $x_{Na_2SO_4\cdot 10\,H_2O} = 0.024$ and $x_{H_2O} = 0.85$; this point is plotted in the ternary phase diagram shown in Fig 5.26.

The line with varying amounts of water but the same relative amounts of the two salts (in this case, equal by mass), passes through this point and the vertex corresponding to $x_{H_2O} = 1$. This line intersects the NaCl axis at a mole fraction corresponding to a 50:50 mixture (by mass) of the two salts

$$x_{NaCl} = \frac{(50\,g)/(58.44\,g\,mol^{-1})}{(50\,g)/(58.44\,g\,mol^{-1}) + (50\,g)/(322.20\,g\,mol^{-1})} = 0.85$$

The line is shown on the diagram.

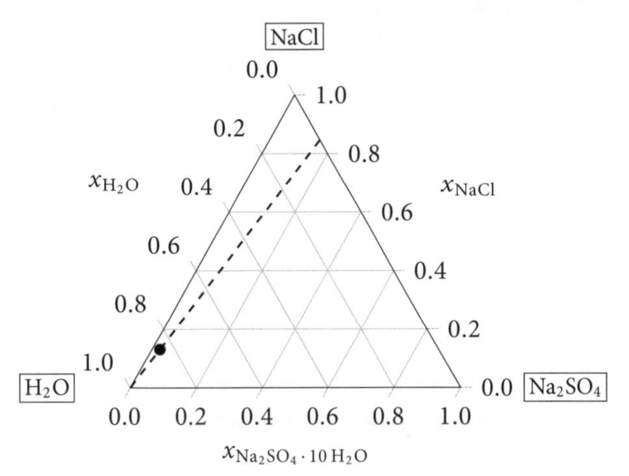

Figure 5.26

E5E.3(a) The composition by mass needs to be converted to mole fractions, which requires the molar masses: $M_{CHCl_3} = 119.37\ \text{g mol}^{-1}$, $M_{H_2O} = 18.016\ \text{g mol}^{-1}$, and $M_{CH_3COOH} = 60.052\ \text{g mol}^{-1}$. The mole fraction of $CHCl_3$ is

$$x_{CHCl_3} = \frac{m_{CHCl_3}/M_{CHCl_3}}{m_{CHCl_3}/M_{CHCl_3} + m_{CH_3COOH}/M_{CH_3COOH} + m_{H_2O}/M_{H_2O}}$$

$$= \frac{\dfrac{9.2\ \text{g}}{119.37\ \text{g mol}^{-1}}}{\dfrac{9.2\ \text{g}}{119.37\ \text{g mol}^{-1}} + \dfrac{3.1\ \text{g}}{60.052\ \text{g mol}^{-1}} + \dfrac{2.3\ \text{g}}{18.016\ \text{g mol}^{-1}}} = 0.30$$

Likewise, $x_{CH_3COOH} = 0.20$ and $x_{H_2O} = 0.50$. This point in marked with the open circle on the phase diagram shown in Fig. 5.27; it falls clearly in the two-phase region. The point almost lies on the tie line a_2'–a_2'', and using this as a guide the lever rule indicates that the phase with composition a_2', ($x_W = 0.57$, $x_T = 0.20$, $x_E = 0.23$), is approximately 5 times more abundant that the phase with composition a_2'', ($x_W = 0.06$, $x_T = 0.82$, $x_E = 0.12$).

(i) When water is added to the mixture the composition moves along the dashed line to the lower-left corner. The system will pass from the two-phase to the one-phase region when the line crosses the phase boundary, which is at approximately ($x_W = 0.75$, $x_T = 0.14$, $x_E = 0.10$).

(ii) When ethanoic acid is added to the mixture the composition moves along the dashed line to the vertex. The system will pass from the two-phase to the one-phase region when the line crosses the phase boundary, which is at approximately ($x_W = 0.44$, $x_T = 0.26$, $x_E = 0.30$).

E5E.4(a) The points corresponding to the given compositions are marked with letters on the phase diagram shown in Fig. 5.28. Composition (i) is in a two-phase region, (ii) is in a three-phase region, (iii) is in a region where there is only one phase. Composition (iv) corresponds to the point at which the phase boundaries meet and all of the phases are in equilibrium

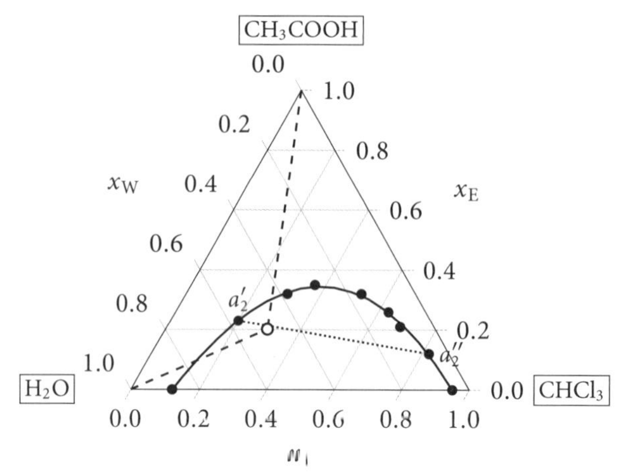

Figure 5.27

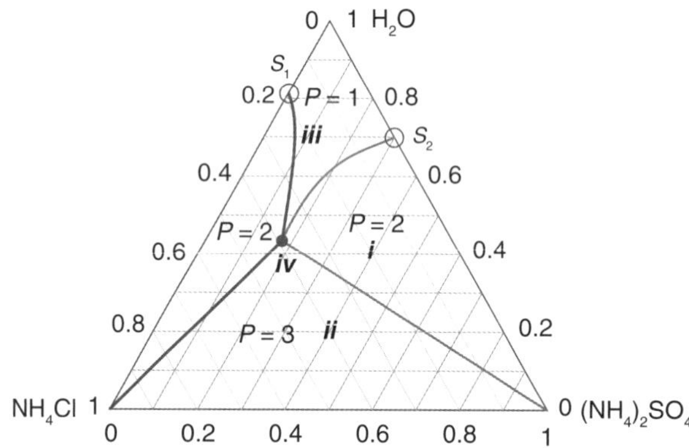

Figure 5.28

E5E.5(a) (i) The phase equilibrium between NH_4Cl and H_2O is indicated by the left-hand edge of the phase diagram shown in Fig. 5.28. At the point S_1 the system moves from two phases to one – in other words from a system of solid NH_4Cl in equilibrium with a solution of NH_4Cl in H_2O, to a system in which there is just one phase, a solution of NH_4Cl in H_2O. This point therefore marks the solubility of NH_4Cl, and from the diagram is occurs at $x_{NH_4Cl} = 0.19$.

The task is to convert this mole fraction to a molar concentration. Imagine a solution made from a mass m of H_2O: the amount in moles of H_2O is m/M, where M is the molar mass of H_2O. The mole fraction of NH_4Cl is therefore

$$x_{NH_4Cl} = \frac{n_{NH_4Cl}}{n_{NH_4Cl} + n_{H_2O}} = \frac{n_{NH_4Cl}}{n_{NH_4Cl} + m/M}$$

This rearranges to $n_{NH_4Cl} = x_{NH_4Cl}(m/M)/(1 - x_{NH_4Cl})$. If it is assumed that the mass density of the solution is approximately the same as the mass density of water, ρ, the volume of this solution is given by $V = m/\rho$. With this, the molar concentration of NH_4Cl is computed as

$$[NH_4Cl] = \frac{n_{NH_4Cl}}{V} = \frac{1}{V}\frac{x_{NH_4Cl}(m/M)}{1-x_{NH_4Cl}} = \frac{\rho}{m}\frac{x_{NH_4Cl}(m/M)}{1-x_{NH_4Cl}}$$
$$= \frac{x_{NH_4Cl}\rho}{M(1-x_{NH_4Cl})}$$

With the data from the diagram and assuming $\rho = 1000\ \mathrm{g\,dm^{-3}}$

$$[NH_4Cl] = \frac{x_{NH_4Cl}\rho}{M(1-x_{NH_4Cl})} = \frac{0.19 \times (1000\ \mathrm{g\,dm^{-3}})}{(18.0158\ \mathrm{g\,mol^{-1}}) \times (1-0.19)}$$
$$= \boxed{13\ \mathrm{mol\,dm^{-3}}}$$

This high concentration rather casts doubt on assuming the density of the solution is the same as that of water.

(ii) The solubility of $(NH_4)_2SO_4$ is indicated by point S_2 on the right-hand edge, at $x_{(NH_4)_2SO_4} = 0.3$. An analogous calculation to that in part (a) gives the concentration as $[(NH_4)_2SO_4] = \boxed{24\ \mathrm{mol\,dm^{-3}}}$.

Solutions to problems

P5E.1 The given points are shown by filled dots in the phase diagram shown in Fig. 5.29, and they are connected by a straight line, as indicated in the problem. Beneath this line lies a one-phase region because CO_2 and nitrobenzene are miscible in all proportions. Addition of I_2 eventually causes phase separation into a two-phase region for all compositions about the line.

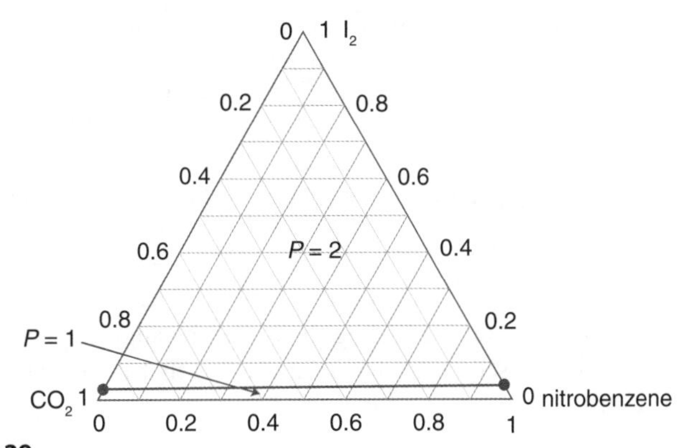

Figure 5.29

P5E.3 Consider the construction shown in Fig. 5.30. The line is interest is AW, where as indicated the position of W is determined by the mole fractions of B and C in the binary mixture; to avoid confusing these particular mole fractions with others, they are denoted y_B and y_C. The point P lies on this line and its

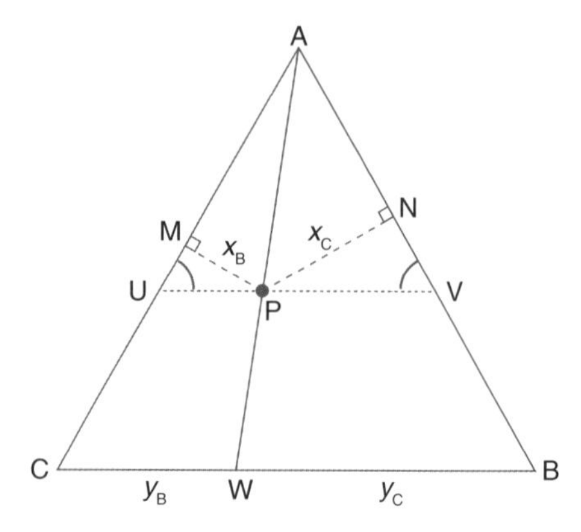

Figure 5.30

perpendicular distance from each of the edges of the triangle gives the mole fractions of B and C, x_B and x_C.

Construct the line UV passing through P and parallel to the base of the triangle. It follows that AWB and APV are *similar* triangles, therefore there exists the following relationship between the ratios of the sides.

$$\frac{\mathrm{WB}}{\mathrm{AW}} = \frac{\mathrm{PV}}{\mathrm{AP}} \tag{5.3}$$

The indicated angle is 60°, so it follows that $x_C = \mathrm{PV}\sin 60°$.

Now consider the triangles AWC and APU, which are also similar and therefore

$$\frac{\mathrm{WC}}{\mathrm{AW}} = \frac{\mathrm{PU}}{\mathrm{AP}}, \tag{5.4}$$

and as before $x_B = \mathrm{PU}\sin 60°$.

Dividing eqn 5.3 by eqn 5.4 gives

$$\frac{\mathrm{WB}}{\mathrm{WC}} = \frac{\mathrm{PV}}{\mathrm{PU}} \quad \text{hence} \quad \frac{y_C}{y_B} = \frac{\mathrm{PV}}{\mathrm{PU}}$$

Dividing $x_C = \mathrm{PV}\sin 60°$ by $x_B = \mathrm{PU}\sin 60°$ gives $x_C/x_B = \mathrm{PV}/\mathrm{PU}$. Equating the two expressions for PV/PU gives the required result

$$\frac{y_C}{y_B} = \frac{x_C}{x_B}$$

Alternatively (and avoiding the use of the angle explicitly) PNV and PMU are also recognised as similar triangles, so that it follows directly that $x_C/\mathrm{PV} = x_B/\mathrm{PU}$.

5F Activities

Answers to discussion questions

D5F.1 If a solvent or solute has a certain chemical potential, then this is related to the activity of the solvent or solute through [5F.1–*183*] and [5F.9–*184*]. At first sight it seems odd to think of the chemical potential determining the activity, but this approach is in fact logical as the chemical potential is the experimentally measurable quantity. For example, measurements of cell potentials or the vapour pressure of liquids provide (slightly indirect, it must be admitted) ways of measuring the chemical potential.

For ideal systems expressions are available which relate the chemical potential to the concentration (in the form of its various measures, such as mole fraction or molality). It therefore follows that the difference between the measured chemical potential and that predicted from idealised models is attributable to factors not taken into account in the ideal systems. Such factors are collectively described as non-ideal interactions; the difference between activity and concentration is therefore ascribed to the presence of such interactions

Non-ideal interactions are no different from the usual interactions between molecular species. For example, they may include interactions between charged or polar species, hydrogen bonding or more specific interactions.

D5F.3 The main way of measuring activities described in this Topic is from measurements of partial vapour pressures, as given in [5F.2–*183*] for the solvent and in [5F.10–*184*] for the solute activity. Other measurements from which the value of the chemical potential can be inferred (for example, cell potentials) are used to determine activities via the general relationship $\mu_J = \mu_J^\ominus + RT \ln a_J$.

D5F.5 The Debye–Hückel theory of electrolyte solutions formulates deviations from ideal behaviour (essentially, deviations due to electrostatic interactions between the ions) in terms of the work of charging the ions. The assumption is that the solute particles would behave ideally if they were not charged, and the difference in chemical potential between real and ideal behaviour amounts to the work of putting electrical charges onto the ions.

To find the work of charging, the distribution of ions must be found, and that is done using the shielded Coulomb potential which takes into account the ionic strength of the solution and the dielectric constant of the solvent. The Debye–Hückel limiting law, [5F.27–*188*], relates the mean ionic activity coefficient to the charges of the ions involved, the ionic strength of the solution, and depends on a constant that takes into account solvent properties and temperature.

Solutions to exercises

E5F.1(a) The activity in terms of the vapour pressure p is given by [5F.2–*183*], $a = p/p^*$, where p^* is the vapour pressure of the pure solvent. With the data given $a = p/p^* = (1.381\ \text{kPa})/(2.3393\ \text{kPa}) = \boxed{0.5903}$.

E5F.2(a) On the basis of Raoult's law, the activity in terms of the vapour pressure p_A is given by [5F.2–*183*], $a_A = p_A/p_A^*$, where p_A^* is the vapour pressure of the pure solvent. With the data given $a_A = p_A/p_A^* = (250\ \text{Torr})/(300\ \text{Torr}) =$

$0.833... = \boxed{0.833}$. The activity coefficient is defined through [5F.4–*183*], $a_A = \gamma_A x_A$, therefore $\gamma_A = a_A/x_A = 0.833.../0.900 = \boxed{0.926}$.

For the solute, Henry's law is used as the basis and the activity is given by [5F.10–*184*], $a_B = p_B/K_B$, where K_B is the Henry's law constant expressed in terms of mole fraction. In this case $a_B = p_B/K_B = (25\,\text{Torr}/200\,\text{Torr}) = \boxed{0.125}$.

E5F.3(a) On the basis of Raoult's law, the activity in terms of the vapour pressure p_J is given by [5F.2–*183*], $a_J = p_J/p_J^*$, where p_J^* is the vapour pressure of the pure solvent. The partial vapour pressure of component J in the gas is given by $p_J = y_J p_{\text{tot}}$. In this case

$$a_P = \frac{p_P}{p_P^*} = \frac{y_P p_{\text{tot}}}{p_P^*} = \frac{0.516 \times (1.00\,\text{atm}) \times [(101.325\,\text{kPa})/(1\,\text{atm})]}{105\,\text{kPa}} = 0.497...$$

The activity of propanone is therefore $a_P = \boxed{0.498}$. The activity coefficient is defined through [5F.4–*183*], $a_J = \gamma_J x_J$, therefore $\gamma_P = a_P/x_P = 0.497.../0.400 = \boxed{1.24}$.

For the other component the mole fractions are $y_M = 1 - y_P = 0.484$ and $x_M = 1 - x_P = 0.600$. The rest of the calculation follows as before

$$a_M = \frac{p_M}{p_M^*} = \frac{y_M p_{\text{tot}}}{p_M^*} = \frac{0.484 \times (1.00\,\text{atm}) \times [(101.325\,\text{kPa})/(1\,\text{atm})]}{73.5\,\text{kPa}} = 0.667...$$

The activity of methanol is therefore $a_M = \boxed{0.667}$ and its activity coefficient is given by $\gamma_M = a_M/x_M = 0.667.../0.600 = \boxed{1.11}$.

E5F.4(a) For this model of non-ideal solutions the vapour pressures are given by [5F.18–*186*], $p_A = p_A^* x_A \exp(\xi[1-x_A]^2)$ and likewise for p_B; the total pressure is given by $p_{\text{tot}} = p_A + p_B$. The vapour pressures are plotted in Fig. 5.31.

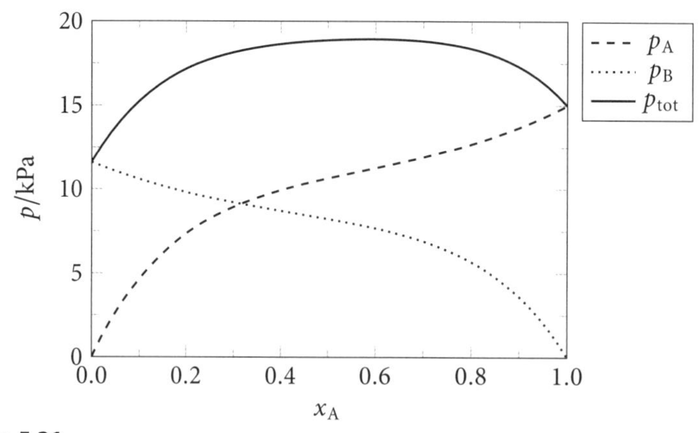

Figure 5.31

E5F.5(a) Ionic strength is defined in [5F.28–*188*]

$$I = \tfrac{1}{2}\sum_i z_i^2(b_i/b^{\ominus})$$

where the sum runs over all the ions in the solution, z_i is the charge number on ion i, and b_i is its molality. For KCl the molality of K^+ and Cl^- are both 0.10 mol kg^{-1}; $z_{K^+} = +1$ and $z_{Cl^-} = -1$. For $CuSO_4$ the molality of Cu^{2+} and $SO_4{}^{2-}$ are both 0.20 mol kg^{-1}; $z_{Cu^{2+}} = +2$ and $z_{SO_4{}^{2-}} = -2$. The ionic strength is therefore

$$I = \tfrac{1}{2}[1/(1\ \text{mol kg}^{-1})]\big[(+1)^2 \times (0.10\ \text{mol kg}^{-1}) + (-1)^2 \times (0.10\ \text{mol kg}^{-1}) \\ +(+2)^2 \times (0.20\ \text{mol kg}^{-1}) + (-2)^2 \times (0.20\ \text{mol kg}^{-1})\big] = \boxed{0.9}$$

E5F.6(a) Ionic strength is defined in [5F.28–*188*]

$$I = \tfrac{1}{2}\sum_i z_i^2(b_i/b^{\ominus})$$

where the sum runs over all the ions in the solution, z_i is the charge number on ion i, and b_i is its molality.

(i) The aim here is to increase the ionic strength by $0.250 - 0.150 = 0.100$; the task is therefore to compute the mass m of $Ca(NO_3)_2$ which, when added to a mass m_w of water, gives this increase in the ionic strength.
A solution of $Ca(NO_3)_2$ of molality b contributes Ca^{2+} at molality b and $NO_3{}^-$ at molality $2b$. The contribution to the ionic strength is therefore $\tfrac{1}{2}[(+2)^2 \times b + (-1)^2 \times 2b]/b^{\ominus} = 3b/b^{\ominus}$.
The molality arising from dissolving mass m of $Ca(NO_3)_2$ in a mass m_w of solvent is $(m/M)/m_w$, where M is the molar mass. It therefore follows that to achieve the desired increase in ionic strength

$$3 \times \frac{m}{Mm_w} \times \frac{1}{b^{\ominus}} = 0.100 \quad \text{hence} \quad m = \tfrac{1}{3} \times 0.100 \times Mm_w b^{\ominus}$$

The molar mass of $Ca(NO_3)_2$ is 164.10 g mol^{-1}; using this, and recalling that the molality is expressed in mol kg^{-1}, gives

$$m = \tfrac{1}{3} \times 0.100 \times (164.10 \times 10^{-3}\ \text{kg mol}^{-1}) \times (0.500\ \text{kg}) \times (1\ \text{mol kg}^{-1}) \\ = 2.73... \times 10^{-3}\ \text{kg}$$

Hence the $\boxed{2.74\ \text{g}}$ of $Ca(NO_3)_2$ needs to be added to achieve the desired ionic strength.

(ii) The argument is as in (a) except that the added solute is now NaCl which contributes singly-charged ions at the same molality as the solute so the contribution to the ionic strength is simply $b/b^{\ominus}$. It therefore follows that to achieve the desired increase in ionic strength

$$\frac{m}{Mm_w} \times \frac{1}{b^{\ominus}} = 0.100 \quad \text{hence} \quad m = 0.100 \times Mm_w b^{\ominus}$$

The molar mass of NaCl is 58.44 g mol^{-1}, hence

$$\begin{aligned} m &= 0.100 \times (58.44 \times 10^{-3}\ \text{kg mol}^{-1}) \times (0.500\ \text{kg}) \times (1\ \text{mol kg}^{-1}) \\ &= 2.92... \times 10^{-3}\ \text{kg} \end{aligned}$$

Hence the $\boxed{2.92\ \text{g}}$ of NaCl needs to be added to achieve the desired ionic strength.

E5F.7(a) The Debye–Hückel limiting law, [5F.27–*188*], is used to estimate the mean activity coefficient, $\gamma_{\pm}$, at 25 °C in water

$$\log \gamma_{\pm} = -0.509\,|z_+z_-|\,I^{1/2} \qquad I = \tfrac{1}{2}\sum_i z_i^2(b_i/b^{\ominus})$$

where $z_{\pm}$ are the charge numbers on the ions from the salt of interest and I is the ionic strength, defined in [5F.28–*188*]. In the definition of I the sum runs over all the ions in the solution, z_i is the charge number on ion i, and b_i is its molality.

A solution of $CaCl_2$ of molality b contributes Ca^{2+} at molality b and Cl^- at molality $2b$. The contribution to the ionic strength is therefore $\frac{1}{2}[(+2)^2 \times b + (-1)^2 \times 2b]/b^{\ominus} = 3b/b^{\ominus}$. A solution of NaF of molality b' contributes Na^+ at molality b' and F^- at molality b'. The contribution to the ionic strength is therefore $\frac{1}{2}[(+1)^2 \times b' + (-1)^2 \times b']/b^{\ominus} = b'/b^{\ominus}$. The ionic strength of the solution is therefore

$$(3b+b')/b^{\ominus} = [3\times(0.010\ \text{mol kg}^{-1})+1\times(0.030\ \text{mol kg}^{-1})]/(1\ \text{mol kg}^{-1}) = 0.060$$

For solute $CaCl_2$ $z_+ = +2$ and $z_- = -1$ so the limiting law evaluates as

$$\log \gamma_{\pm} = -0.509\,|z_+z_-|\,I^{1/2} = -0.509\,|(+2)\times(-1)|\,(0.060)^{1/2} = -0.249...$$

The mean activity coefficient is therefore $\gamma_{\pm} = 10^{-0.249...} = 0.563... = \boxed{0.56}$.

E5F.8(a) The Davies equation is given in [5F.30b–*189*]

$$\log \gamma_{\pm} = \frac{-A\,|z_+z_-|\,I^{1/2}}{1 + BI^{1/2}} + CI$$

Because the electrolyte is 1:1 with univalent ions, the ionic strength is simply $I = b_{HBr}/b^{\ominus}$. There is no obvious straight-line plot using which the data can be tested against the Davies equation, therefore a non-linear fit is made using mathematical software and assuming that $A = 0.509$; remember that the molalities must be expressed in mol kg^{-1}. The best-fit values are $\boxed{B = 1.96}$ and $C = 0.0183$; With these values the predicted activity coefficients are 0.930, 0.907 and 0.879, which is very good agreement.

Solutions to problems

P5F.1 In the Raoult's law basis the activity is given by [5F.2–183], $a_J = p_J/p_J^*$, and the activity coefficient by [5F.4–183], $a_J = \gamma_J x_J$. The partial pressure of the vapour is given by $p_J = y_J p$, where y_J is the mole fraction of J in the vapour and p the total pressure.

It follows that $\gamma_J = a_J/x_J = y_J p/x_J p_J^*$. In this binary system the activity coefficient for 1,2-epoxybutane (E) is found using the given data for trichloromethane (T) by using and $x_E = 1 - x_T$, and likewise for y_E. It follows that $\gamma_E = (1 - y_T)p/(1 - x_T)p_E^*$. The resulting activity coefficients are shown in the table.

p/kPa	x_T	y_T	γ_T	γ_E
23.40	0	0		1
21.75	0.129	0.065	0.417	0.998
20.25	0.228	0.145	0.490	0.958
18.75	0.353	0.285	0.576	0.885
18.15	0.511	0.535	0.723	0.738
20.25	0.700	0.805	0.885	0.563
22.50	0.810	0.915	0.966	0.430
26.30	1	1	1	

P5F.3 The Debye–Hückel limiting law, [5F.27–188], is

$$\log \gamma_\pm = -0.509\,|z_+ z_-|\, I^{1/2} \qquad I = \tfrac{1}{2}\sum_i z_i^2 (b_i/b^\ominus)$$

where $z_\pm$ are the charge numbers on the ions from the salt of interest and I is the ionic strength, defined in [5F.28–188]. For a 1:1 electrolyte of univalent ions at molality b, $I = b/b^\ominus$ (recall that $b^\ominus = 1$ mol kg^{-1} so b must also be in units of mol kg^{-1}). A test of this equation is to plot $\log \gamma_\pm$ against $I^{1/2}$, such a plot is shown in Fig. 5.32.

$b/(\text{mmol kg}^{-1})$	$\gamma_\pm$	$\log \gamma_\pm$	$I^{1/2}$
1.00	0.9649	−0.01552	0.0316
2.00	0.9519	−0.02141	0.0447
5.00	0.9275	−0.03269	0.0707
10.0	0.9024	−0.04460	0.1000
20.0	0.8712	−0.05988	0.1414

The data fit to quite a good straight line, the equation of which is

$$\log \gamma_\pm = -0.404 \times I^{1/2} - 3.395 \times 10^{-3}$$

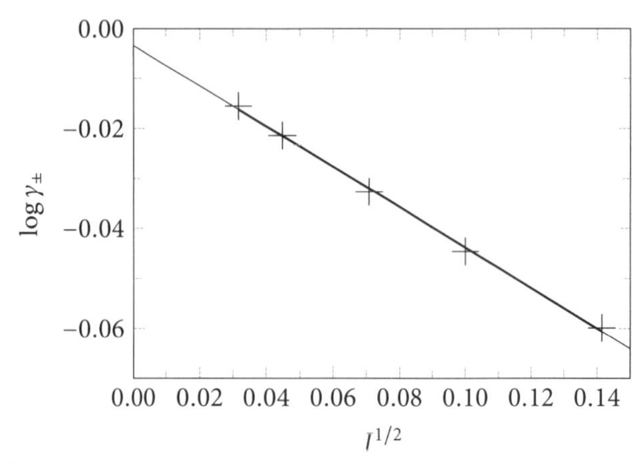

Figure 5.32

The Debye–Hückel limiting law is verified to the extent that $\log \gamma_\pm$ is linear in $I^{1/2}$, however the slope is -0.404 rather than the expected value of -0.509. According to the limiting law the intercept at $I^{1/2} = 0$ should be zero, which is not the case. Overall, the conclusion is that the limiting law predicts the correct functional dependence of $\gamma_\pm$ on ionic strength but fails to predict the correct values.

The Davies equation is given in [5F.30b–*189*]

$$\log \gamma_\pm = \frac{-A|z_+z_-|I^{1/2}}{1 + BI^{1/2}} + CI$$

The data can be fitted to this equation using mathematical software to implement a non-linear fit; the value of A is fixed as 0.509 for the fit. The best-fit parameters are $B = 1.2975$, $C = -0.0470$. Using these, a table is drawn up comparing the experimental values of $\gamma_\pm$ with those predicted by the limiting law and by the Davies equation.

$b/(\text{mmol kg}^{-1})$	$\gamma_\pm$	$\gamma_\pm$(DH)	error (%)	$\gamma_\pm$(Davies)	error (%)
1.00	0.9649	0.9636	−0.13	0.9651	0.02
2.00	0.9519	0.9489	−0.31	0.9519	0.00
5.00	0.9275	0.9205	−0.76	0.9274	−0.01
10.00	0.9024	0.8894	−1.44	0.9024	0.00
20.00	0.8712	0.8473	−2.75	0.8712	0.00

The table shows that the values predicted by the limiting law are increasingly in error as the ionic strength increases, whereas the Davies equation reproduces most of the experimental data to within the stated precision. However, it must be kept in mind that the B and C parameters in the Davies equation have been adjusted specifically to fit these data.

Answers to integrated activities

I5.1 (a) On the basis of Raoult's law, the activity in terms of the vapour pressure p_J is given by [5F.2–183], $a_J = p_J/p_J^*$, where p_J^* is the vapour pressure of the pure solvent. The activity coefficient is defined through [5F.4–183], $a_J = \gamma_J x_J$, therefore $\gamma_J = p_J/p_J^* x_J$. From the data given $p_E^* = 37.38$ kPa (when $x_I = 0$) and $p_I^* = 47.12$ kPa; recall that $x_E = 1 - x_I$. The table shows the computed values of the activity coefficients.

x_I	p_I/kPa	p_E/kPa	γ_I	γ_E	γ_I(Henry)
0	0	37.38		1	
0.0579	3.73	35.48	1.367	1.008	1.003
0.1095	7.03	33.64	1.362	1.011	1.000
0.1918	11.70	30.85	1.295	1.021	0.950
0.2353	14.05	29.44	1.267	1.030	0.930
0.3718	20.72	25.05	1.183	1.067	0.868
0.5478	28.44	19.23	1.102	1.138	0.809
0.6350	31.88	16.39	1.065	1.201	0.782
0.8253	39.58	8.88	1.018	1.360	0.747
0.9093	43.00	5.09	1.004	1.501	0.737
1	47.12	0	1		0.734

(b) On the basis of Henry's law, the activity in terms of the vapour pressure p_J is given by [5F.10–184], $a_J = p_J/K_J$, where K_J is the Henry's law constant for J as a solute. The activity coefficient is defined as before, $a_J = \gamma_J x_J$, and therefore $\gamma_J = p_J/L_J x_J$.

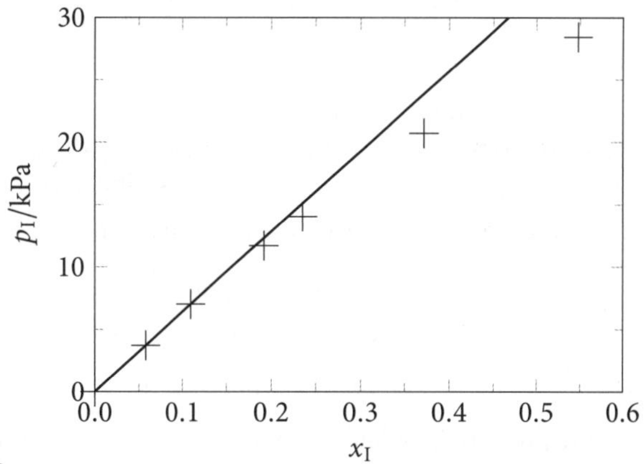

Figure 5.33

To find the Henry's law constant, p_I is plotted as a function of x_I and the limiting slope taken; such a plot is shown in Fig. 5.33. The limiting slope, taken from the first three data points is 64.20 and so $K_I = 64.20$ kPa. This value is used to compute the activity coefficients for I based on Henry's law, and the results are shown in right-most column of the table above.

I5.3 On the basis of Henry's law, the activity in terms of the vapour pressure p_J is given by [5F.10–*184*], $a_J = p_J/K_J$, where K_J is the Henry's law constant for J as a solute. The activity coefficient is defined as, $a_J = \gamma_J x_J$, and therefore $\gamma_J = p_J/K_J x_J$.

The data given are expressed in terms of the mole fractions of cyclohexanol in the liquid and gas, x_{cyc} and y_{cyc}, but what is required are the corresponding mole fractions for CO_2: these are computed as $x_C = 1 - x_{cyc}$ and $y_C = 1 - y_{cyc}$. The partial pressure of CO_2 is then computed as $p_C = y_C p_{tot}$. In summary

$$\gamma_C = \frac{p_C}{K_C x_C} = \frac{(1 - y_{cyc})p_{tot}}{K_C(1 - x_{cyc})}$$

To find the Henry's law constant for CO_2, the limiting slope of a plot of p_C against x_C is taken; the data set out in the table below are plotted in Fig. 5.34. The limiting slope taken from the first three data points is 371 and so $\boxed{K_C = 371 \text{ bar}}$. This value is used to compute the activity coefficients for CO_2, γ_C, shown in the table.

p_{tot}/bar	y_{cyc}	x_{cyc}	p_C/bar	x_C	γ_C
10.0	0.026 7	0.974 1	9.7	0.025 9	1.01
20.0	0.014 9	0.946 4	19.7	0.053 6	0.99
30.0	0.011 2	0.920 4	29.7	0.079 6	1.00
40.0	0.009 47	0.892 0	39.6	0.108 0	0.99
60.0	0.008 35	0.836 0	59.5	0.164 0	0.98
80.0	0.009 21	0.773 0	79.3	0.227 0	0.94

I5.5 If the concentration, expressed as (mass of solute)/(mass of solvent), dissolved gas c is given by $c = Kp$, then the mass of gas dissolved in mass m_S of solvent is $m_G = Kpm_S$. The partial pressure of N_2 in the atmosphere is $x_{N_2} \times p_{tot}$.

Hence at a total pressure of 4 atm the mass of N_2 dissolved in 100 g of water is

$$\begin{aligned} m_{N_2} &= Kp_{N_2}m_S = Kx_{N_2}p_{tot}m_S \\ &= (0.18\ \mu g/(g_{H_2O}\ atm)) \times 0.7808 \times (4\ atm) \times (100\ g_{H_2O}) \\ &= 56.2...\ \mu g = \boxed{56\ \mu g} \end{aligned}$$

If the pressure is reduced to 1 atm, the mass of dissolved gas is reduced by a factor of four to $\boxed{14\ \mu g}$.

If N_2 is four times more soluble in fatty tissues than in water, the increase in dissolved gas on going from 1 atm to 4 atm is

$$\Delta m_{N_2} = 4 \times [(56.2...\ \mu g) - (56.2...\ \mu g)/4] = \boxed{1.7 \times 10^2\ \mu g}$$

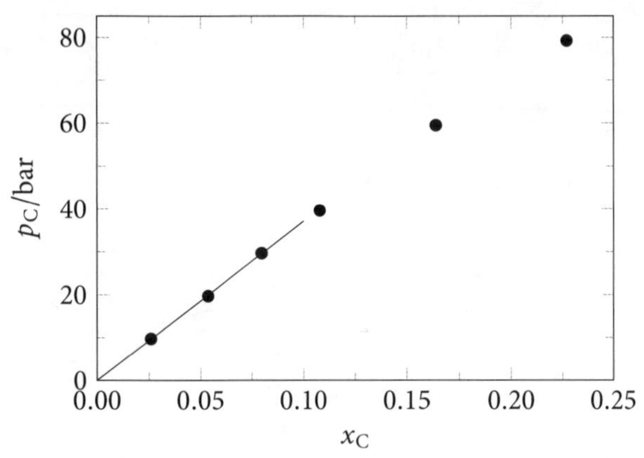

Figure 5.34

I5.7 (a) In this case each of the four terms in the sum is the same, with the same equilibrium constant, so the function to be plotted is

$$\frac{\nu c^\ominus}{[A]_{out}} = \frac{4 \times K}{1 + K[A]_{out}/c^\ominus} = \frac{4 \times (1.0 \times 10^7)}{1 + (1.0 \times 10^7) \times (10^{-6} \times [A]_{out}/\mu M)}$$

where 1 μM = 1 μmol dm^{-3} and $[A]_{out}$ is in units of μM.

(b) In this case there are six terms, four with binding constant 1×10^5 and two with binding constant 2×10^6. The function to be plotted is

$$\begin{aligned}\frac{\nu c^\ominus}{[A]_{out}} &= \frac{4 \times K}{1 + K[A]_{out}/c^\ominus} + \frac{2 \times K'}{1 + K'[A]_{out}/c^\ominus} \\ &= \frac{4 \times (1.0 \times 10^5)}{1 + (1.0 \times 10^5) \times (10^{-6} \times [A]_{out}/\mu M)} \\ &\quad + \frac{2 \times (2.0 \times 10^6)}{1 + (2.0 \times 10^6) \times (10^{-6} \times [A]_{out}/\mu M)}\end{aligned}$$

These two equations are plotted in Fig. 5.35

I5.9 Start with the Gibbs–Duhem equation, [5A.12b–*146*], $n_A d\mu_A + n_B d\mu_B = 0$ and divide both sides by $(n_A + n_B)$ to give $x_A d\mu_A + x_B d\mu_B = 0$. Next introduce the general dependence of the chemical potential on activity, $\mu_A = \mu_A^\ominus + RT \ln a_A$, from which it follows that $d\mu_A = RT d\ln a_A$. Introducing this into the Gibbs–Duhem equation, and dividing both sides by RT gives

$$x_A d\ln a_A + x_B d\ln a_B = 0$$

The aim is an expression for $\ln a_B$, so rearrange this equation to give an expression for $d\ln a_B$, and then introduce $r = x_B/x_A$ to give

$$d\ln a_B = -(x_A/x_B) d\ln a_A = -(1/r) d\ln a_A \tag{5.5}$$

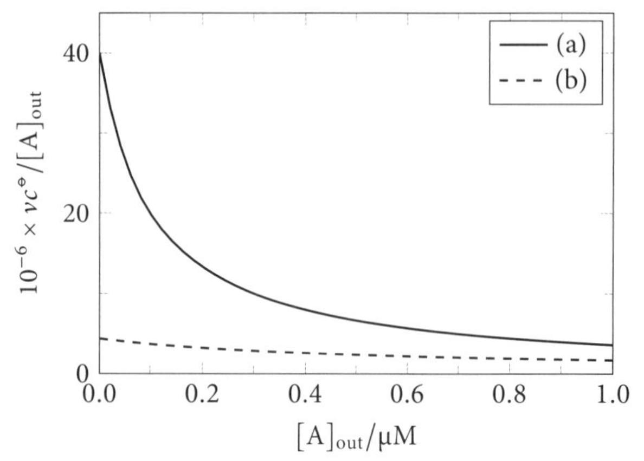

Figure 5.35

The osmotic function ϕ is defined as $\phi = -(x_A/x_B)\ln a_A = -(1/r)\ln a_A$, therefore $\ln a_A = -\phi r$. Taking the differential of both sides (using the product rule for the right-hand side) gives

$$\mathrm{d}\ln a_A = -\mathrm{d}(\phi r) = -r\mathrm{d}\phi - \phi\mathrm{d}r$$

This expression for $\mathrm{d}\ln a_A$ is substituted into the right-hand side of eqn 5.5

$$\mathrm{d}\ln a_B = -(1/r)\mathrm{d}\ln a_A = (-1/r)(-r\mathrm{d}\phi - \phi\mathrm{d}r) = \mathrm{d}\phi + (\phi/r)\,\mathrm{d}r = \mathrm{d}\phi + \phi\,\mathrm{d}\ln r$$

where $(1/r)\mathrm{d}r = \mathrm{d}\ln r$ is used. The final expression involves $\ln a_B/r$, so taking this as a hint, subtract $\mathrm{d}\ln r$ from both sides of the previous equation to give

$$\mathrm{d}\ln a_B - \mathrm{d}\ln r = \mathrm{d}\phi + \phi\,\mathrm{d}\ln r - \mathrm{d}\ln r = \mathrm{d}\phi + (\phi - 1)\mathrm{d}\ln r$$

$$\text{hence }\ \mathrm{d}\ln\frac{a_B}{r} = \mathrm{d}\phi + \frac{\phi - 1}{r}\mathrm{d}r$$

where $\mathrm{d}\ln r = (1/r)\mathrm{d}r$ is used. Both sides are now integrated from pure A up to the composition of interest

$$\ln\frac{a_B}{r}\bigg|_{\text{pure A}}^{a_B} = \phi\big|_{\text{pure A}}^{a_B} + \int_{\text{pure A}}^{a_B}\frac{\phi - 1}{r}\mathrm{d}r$$

The lower limits can all be evaluated: consider first the term on the left, which is developed in general as

$$\ln\frac{a_B}{r} = \ln\frac{\gamma_B x_B}{x_B/x_A} = \ln\gamma_B x_A$$

In the limit of pure A, $x_A \to 1$, and the activity coefficient of B also goes to 1; therefore, because $\ln 1 = 0$, this term goes to zero. The lower limit of the integral over ϕ is simply written as $\phi(0)$, and the integral over r goes from 0 for pure A to some value of r that corresponds to the composition of interest. Putting all this together gives the required result

$$\ln\frac{a_B}{r} = \phi - \phi(0) + \int_0^r\frac{\phi - 1}{r}\mathrm{d}r$$

I5.11 In Section 5B.2(a) on page 158 the derivation of the expression for the freezing point depression starts by equating the chemical potential of A as a pure solid with that of A in a solution of mole fraction x_A containing solute B: $\mu_A^*(s) = \mu_A(l, x_A)$. The latter chemical potential is written, for an ideal solution, as $\mu_A(l, x_A) = \mu_A^*(l) + RT \ln x_A$. If the solution is not ideal, then the mole fraction is replaced by the activity a_A to give

$$\mu_A^*(s) = \mu_A^*(l) + RT \ln a_A$$

The derivation then proceeds as before. First, $\mu_A^*(s) - \mu_A^*(l)$ is identified as $-\Delta_{fus}G$ to give

$$\ln a_A = \frac{\mu_A^*(s) - \mu_A^*(l)}{RT} = \frac{-\Delta_{fus}G}{RT}$$

Next, both sides are differentiated with respect to T and the Gibbs–Helmholtz equation, $d(G/T)/dT = -H/T^2$, is applied to the right-hand side

$$\frac{d \ln a_A}{dT} = \frac{-1}{R}\frac{d}{dT}\left(\frac{\Delta_{fus}G}{T}\right) = \frac{\Delta_{fus}H}{RT^2}$$

The freezing point depression ΔT is defined as $\Delta T = T^* - T$, where T is the freezing point of the solution and T^* is the freezing point of the pure solvent. It follows that $d\Delta T = -dT$. Finally, because the freezing point depression is small, T on the right-hand side of the previous equation can be replaced by T^* to give

$$\frac{d \ln a_A}{d\Delta T} = -\frac{\Delta_{fus}H}{RT^{*2}} \qquad (5.6)$$

The empirical freezing-point constant K_f is introduced in [5B.12–*161*], $\Delta T = K_f b_B$, where b_B is the molality of the solvent. This expression is developed by writing $b_B = n_B/m_A$, where m_A is the mass of solvent A (in kg) and then $m_A = n_A M$, where M is the molar mass of the solvent. It follows that $\Delta T = K_f n_B/(n_A M)$. For dilute solutions $x_B \approx n_B/n_A$ so $\Delta T = K_f x_B/M$.

The expression for the freezing point depression given in [5B.11–*161*] is

$$\Delta T = \frac{RT^{*2}}{\Delta_{fus}H} x_B$$

Comparison of this with the expression just derived, $\Delta T = K_f x_B/M$, gives

$$\frac{K_f}{M} = \frac{RT^{*2}}{\Delta_{fus}H}$$

Using this expression for the right-hand side in eqn 5.6 gives the required form

$$\frac{d \ln a_A}{d\Delta T} = -\frac{M}{K_f} \qquad (5.7)$$

Start with the Gibbs–Duhem equation, [5A.12b–*146*], $n_A d\mu_A + n_B d\mu_B = 0$ and divide both sides by $(n_A + n_B)$ to give $x_A d\mu_A + x_B d\mu_B = 0$. Next introduce the general dependence of the chemical potential on activity, $\mu_A = \mu_A^{\ominus} + RT \ln a_A$, from which it follows that $\mathrm{d}\mu_A = RT\mathrm{d}\ln a_A$. Introducing this into the Gibbs–Duhem equation, and dividing both sides by RT gives

$$x_A\mathrm{d}\ln a_A + x_B\mathrm{d}\ln a_B = 0 \quad \text{hence} \quad \mathrm{d}\ln a_A = -\frac{x_B}{x_A}\mathrm{d}\ln a_B \tag{5.8}$$

It follows that

$$\frac{\mathrm{d}\ln a_A}{\mathrm{d}\Delta T} = -\frac{x_B}{x_A}\frac{\mathrm{d}\ln a_B}{\mathrm{d}\Delta T}$$

This expression for $(\mathrm{d}\ln a_A)/\mathrm{d}\Delta T$ is substituted into eqn 5.7

$$-\frac{x_B}{x_A}\frac{\mathrm{d}\ln a_B}{\mathrm{d}\Delta T} = -\frac{M}{K_f}$$

This expression is developed further by using the approximations $x_A \approx 1$ and $x_B \approx n_B/n_A$ which are appropriate for dilute solutions

$$\begin{aligned}\frac{\mathrm{d}\ln a_B}{\mathrm{d}\Delta T} &= \frac{x_A}{x_B}\frac{M}{K_f}\\ &= \frac{1}{n_B/n_A}\frac{M}{K_f} = \frac{1}{n_B}\frac{n_A M}{K_f} = \frac{1}{n_B}\frac{m_A}{K_f}\\ &= \frac{1}{b_B K_f}\end{aligned}$$

On the penultimate line $m_A = n_A M$, where m_A is the mass of solvent A, is used, and to go to the final line the molality of B, $b_B = n_B/m_A$ is introduced.

Recall the definition of the osmotic coefficient $\phi = -(x_A/x_B)\ln a_A$ and the result from eqn 5.8 that $\mathrm{d}\ln a_A = -(x_B/x_A)\mathrm{d}\ln a_B$. It follows that

$$\int \mathrm{d}\ln a_A = \int -\frac{x_B}{x_A}\mathrm{d}\ln a_B \quad \text{hence} \quad \ln a_A = \int -\frac{x_B}{x_A}\mathrm{d}\ln a_B$$

where the integration is from pure A to some arbitrary composition. This result is used with the definition of ϕ to give

$$\phi = -\frac{x_A}{x_B}\ln a_A = -\frac{x_A}{x_B}\int -\frac{x_B}{x_A}\mathrm{d}\ln a_B$$

The terms in x_A and x_B cannot be cancelled because those inside the integral are functions of the variable of integration. For dilute dilute solutions $x_A \approx 1$ and $x_B \approx n_B/n_A = n_B/(m_A/M) = M(n_B/m_A) = Mb_B$, where M is the molar mass of A, m_A is the mass of the solvent, and b_B is the molality of B. The integral for ϕ therefore becomes

$$\phi = -\frac{x_A}{x_B}\ln a_A = \frac{1}{b_B}\int b_B\,\mathrm{d}\ln a_B \tag{5.9}$$

For a 1:1 univalent electrolyte the Debye–Hückel limiting law [5F.27–*188*], $\log\gamma_\pm = -A|z_+z_-|I^{1/2}$, becomes $\log\gamma_\pm = -A(b_B/b^\ominus)^{1/2}$, and changing to naural logarithms this becomes $\ln\gamma_\pm = -A'(b_B/b^\ominus)^{1/2}$, with $A' = 2.303 \times A$. Using this, the activity and its derivative are developed as

$$\ln a_B = \ln(b_B/b^\ominus) + \ln\gamma = \ln(b_B/b^\ominus) - A'(b_B/b^\ominus)^{1/2}$$

$$\text{hence}\quad \mathrm{d}\ln a_B = \left[\frac{1}{b_B} - \tfrac{1}{2}A'\left(\frac{1}{b_B b^\ominus}\right)^{1/2}\right]\mathrm{d}b_B$$

This expression for $\mathrm{d}\ln a_B$ is used in eqn 5.9 and the integral is then evaluated

$$\begin{aligned}\phi &= \frac{1}{b_B}\int b_B\,\mathrm{d}\ln a_B \\ &= \frac{1}{b_B}\int b_B\left[\frac{1}{b_B} - \tfrac{1}{2}A'\left(\frac{1}{b_B b^\ominus}\right)^{1/2}\right]\mathrm{d}b_B \\ &= \frac{1}{b_B}\int\left[1 - \tfrac{1}{2}A'\left(\frac{b_B}{b^\ominus}\right)^{1/2}\right]\mathrm{d}b_B \\ &= \frac{1}{b_B}\left[b_B - \tfrac{1}{2}\times\tfrac{2}{3}\times A'\left(\frac{1}{b^\ominus}\right)^{1/2} b_B^{3/2}\right]_{b_B=0}^{b_B} \\ &= 1 - \tfrac{1}{3}A'\left(\frac{b_B}{b^\ominus}\right)^{1/2}\end{aligned}$$

6 Chemical equilibrium

6A The equilibrium constant

Answers to discussion questions

D6A.1 Without the contribution to the Gibbs energy resulting from the physical mixing of the reactants and products there would be no position of equilibrium: the reaction would simply go to unmixed products or unmixed reactants, whichever had the lower Gibs energy. For further discussion see Section 6A.2(a) on page 205.

Solutions to exercises

E6A.1(a) In general if the extent of a reaction changes by an amount $\Delta\xi$ then the amount of a component J changes by $\nu_J\Delta\xi$ where ν_J is the stoichiometric number for species J (positive for products, negative for reactants). In this case $\nu_A = -1$ and $\nu_B = +2$.

$$n_A = n_{A,0} + \Delta n_A = n_{A,0} + \nu_A\Delta\xi = (1.50\text{ mol}) + (-1)\times(0.60\text{ mol}) = \boxed{0.90\text{ mol}}$$

$$n_B = n_{B,0} + \Delta n_B = n_{B,0} + \nu_B\Delta\xi = 0 + 2\times(0.60\text{ mol}) = \boxed{1.20\text{ mol}}$$

E6A.2(a) The reaction Gibbs energy $\Delta_r G$ is defined by [6A.1–*204*], $\Delta_r G = (\partial G/\partial\xi)_{p,T}$. Approximating the derivative by finite changes gives

$$\Delta_r G = \left(\frac{\partial G}{\partial \xi}\right)_{p,T} \approx \frac{\Delta G}{\Delta\xi} = \frac{-6.4\text{ kJ}}{+0.1\text{ mol}} = \boxed{-64\text{ kJ mol}^{-1}}$$

E6A.3(a) A reaction is exergonic if $\Delta_r G < 0$ and endergonic if $\Delta_r G > 0$. From the *Resource section* the standard Gibbs energy change for the formation of methane from its elements in their reference states at 298 K is $\Delta_f G^\ominus = -50.72\text{ kJ mol}^{-1}$. This is negative so the reaction is $\boxed{\text{exergonic}}$.

E6A.4(a) The reaction quotient is defined by [6A.10–*207*], $Q = \prod_J a_J^{\nu_J}$. For the reaction $A + 2B \rightarrow 3C$, $\nu_A = -1$, $\nu_B = -2$, and $\nu_C = +3$. The reaction quotient is then

$$Q = a_A^{-1}a_B^{-2}a_C^{3} = \frac{a_C^3}{a_A a_B^2}$$

E6A.5(a) The equilibrium constant is defined by [6A.14–207], $K = \left(\prod_J a_J^{\nu_J}\right)_{\text{equilibrium}}$. The 'equilibrium' subscript indicates that the activities are those at equilibrium rather than at an arbitrary stage in the reaction; however this subscript is not usually written explicitly. In this case

$$K = a_{P_4(s)}^{-1} a_{H_2(g)}^{-6} a_{PH_3(g)}^{4} = \frac{a_{PH_3(g)}^4}{a_{P_4(s)} a_{H_2(g)}^6}$$

The activity of $P_4(s)$ is 1, because it is a pure solid. Furthermore if the gases are treated as perfect then their activities are replaced by $a_J = p_J/p^\ominus$. The equilibrium constant becomes

$$K = \frac{(p_{PH_3}/p^\ominus)^4}{(p_{H_2}/p^\ominus)^6} = \frac{p_{PH_3}^4 p^{\ominus 2}}{p_{H_2}^6}$$

E6A.6(a) The standard reaction Gibbs energy is given by [6A.13a–207]

$$\Delta_r G^\ominus = \sum_{\text{Products}} \nu \Delta_f G^\ominus - \sum_{\text{Reactants}} \nu \Delta_f G^\ominus$$

The relationship between $\Delta_r G^\ominus$ and K, [6A.15–208], $\Delta_r G^\ominus = -RT \ln K$, is then used to calculate the equilibrium constant.

(i) For the oxidation of ethanal

$$\begin{aligned}\Delta_r G^\ominus &= 2\Delta_f G^\ominus(CH_3COOH, l) - \{2\Delta_f G^\ominus(CH_3CHO, g) + \Delta_f G^\ominus(O_2, g)\} \\ &= 2\Delta_f G^\ominus(CH_3COOH, l) - 2\Delta_f G^\ominus(CH_3CHO, g) \\ &= 2(-389.9\ \text{kJ mol}^{-1}) - 2(-128.86\ \text{kJ mol}^{-1}) = -5.22... \times 10^5\ \text{J mol}^{-1}\end{aligned}$$

Then

$$K = e^{-\Delta_r G^\ominus/RT} = \exp\left(-\frac{-5.22... \times 10^5\ \text{J mol}^{-1}}{(8.3145\ \text{J K}^{-1}\ \text{mol}^{-1}) \times (298\ \text{K})}\right) = \boxed{3.24 \times 10^{91}}$$

(ii) For the reaction of AgCl(s) with $Br_2(l)$

$$\begin{aligned}\Delta_r G^\ominus &= 2\Delta_f G^\ominus(AgBr, s) + \Delta_f G^\ominus(Cl_2, g) \\ &\quad - \{2\Delta_f G^\ominus(AgCl, s) + \Delta_f G^\ominus(Br_2, l)\} \\ &= 2\Delta_f G^\ominus(AgBr, s) - 2\Delta_f G^\ominus(AgCl, s) \\ &= 2(-96.90\ \text{kJ mol}^{-1}) - 2(-109.79\ \text{kJ mol}^{-1}) = +25.7...\ \text{kJ mol}^{-1}\end{aligned}$$

Then

$$K = e^{-\Delta_r G^\ominus/RT} = \exp\left(-\frac{25.7... \times 10^3\ \text{J mol}^{-1}}{(8.3145\ \text{J K}^{-1}\ \text{mol}^{-1}) \times (298\ \text{K})}\right) = \boxed{3.03 \times 10^{-5}}$$

Of these two reactions, the first has $K > 1$ at 298 K.

E6A.7(a) The relationship between $\Delta_r G^\ominus$ and the equilibrium constant is given by [6A.15–*208*], $\Delta_r G^\ominus = -RT \ln K$. The ratio of the equilibrium constants for the two reactions is

$$\frac{K_1}{K_2} = \frac{e^{-\Delta_r G_1^\ominus/RT}}{e^{-\Delta_r G_2^\ominus/RT}} = \exp\left(-\frac{\Delta_r G_1^\ominus - \Delta_r G_2^\ominus}{RT}\right)$$

$$= \exp\left(-\frac{(-320 \times 10^3 \text{ J mol}^{-1}) - (-55 \times 10^3 \text{ J mol}^{-1})}{(8.3145 \text{ J K}^{-1} \text{ mol}^{-1}) \times (300 \text{ K})}\right) = \boxed{1.4 \times 10^{46}}$$

E6A.8(a) The reaction Gibbs energy at an arbitrary stage is given by [6A.11–*207*], $\Delta_r G = \Delta_r G^\ominus + RT \ln Q$. In this case $\Delta_r G^\ominus = -32.9 \text{ kJ mol}^{-1}$. The values of $\Delta_r G$ for at value of Q are:

(i) At $Q = 0.010$

$$\Delta_r G = (-32.9 \times 10^3 \text{ J mol}^{-1}) + (8.3145 \text{ J K}^{-1} \text{ mol}^{-1}) \times (298 \text{ K}) \times \ln(0.010)$$
$$= -4.43... \times 10^4 \text{ J mol}^{-1} = \boxed{-44 \text{ kJ mol}^{-1}}$$

(ii) At $Q = 1.0$

$$\Delta_r G = (-32.9 \times 10^3 \text{ J mol}^{-1}) + (8.3145 \text{ J K}^{-1} \text{ mol}^{-1}) \times (298 \text{ K}) \times \ln(1.0)$$
$$= -3.29... \times 10^4 \text{ J mol}^{-1} = \boxed{-33 \text{ kJ mol}^{-1}} \quad (= \Delta_r G^\ominus)$$

(iii) At $Q = 10$

$$\Delta_r G = (-32.9 \times 10^3 \text{ J mol}^{-1}) + (8.3145 \text{ J K}^{-1} \text{ mol}^{-1}) \times (298 \text{ K}) \times \ln(10)$$
$$= -2.71... \times 10^4 \text{ J mol}^{-1} = \boxed{-27 \text{ kJ mol}^{-1}}$$

(iv) At $Q = 10^5$

$$\Delta_r G = (-32.9 \times 10^3 \text{ J mol}^{-1}) + (8.3145 \text{ J K}^{-1} \text{ mol}^{-1}) \times (298 \text{ K}) \times \ln(10^5)$$
$$= -4.37... \times 10^3 \text{ J mol}^{-1} = \boxed{-4.4 \text{ kJ mol}^{-1}}$$

(v) At $Q = 10^6$

$$\Delta_r G = (-32.9 \times 10^3 \text{ J mol}^{-1}) + (8.3145 \text{ J K}^{-1} \text{ mol}^{-1}) \times (298 \text{ K}) \times \ln(10^6)$$
$$= +1.33... \times 10^3 \text{ J mol}^{-1} = \boxed{+1.3 \text{ kJ mol}^{-1}}$$

The equilibrium constant K is the value of Q for which $\Delta_r G = 0$. From the above values, K will therefore be somewhere between 10^5 and 10^6. To find exactly where by linear interpolation, note that according to $\Delta_r G = \Delta_r G^\ominus + RT \ln Q$, a plot of $\Delta_r G$ against $\ln Q$ should be a straight line. Consider the two points on either side of zero, that is, (iv) and (v). The point $\Delta_r G = 0$ occurs a fraction $(4.37...)/(1.33... + 4.37...) = 0.766...$ of the way between points (iv) and (v), so is at

$$\ln K = \ln 10^5 + (0.766...) \times (\ln 10^6 - \ln 10^5) = 13.2...$$

Hence $K = e^{13.2...} = \boxed{5.84 \times 10^5}$

The value is calculated directly by setting $\Delta_r G = 0$ and $Q = K$ in $\Delta_r G = \Delta_r G^\ominus + RT \ln Q$ and rearranging for K

$$K = e^{-\Delta_r G^\ominus/RT} = \exp\left(-\frac{-32.9 \times 10^3 \text{ J mol}^{-1}}{(8.3145 \text{ J K}^{-1} \text{ mol}^{-1}) \times (298 \text{ K})}\right) = \boxed{5.84 \times 10^5}$$

which is the same result as obtained from the linear interpolation.

E6A.9(a) For the reaction $2H_2O(g) \rightleftharpoons 2H_2(g) + O_2(g)$ the following table is drawn up by supposing that there are n moles of H_2O initially and that at equilibrium a fraction α has dissociated.

	$2H_2O$	$\rightleftharpoons$	$2H_2$	+	O_2
Initial amount	n		0		0
Change to reach equilibrium	$-\alpha n$		$+\alpha n$		$+\frac{1}{2}\alpha n$
Amount at equilibrium	$(1-\alpha)n$		αn		$\frac{1}{2}\alpha n$
Mole fraction, x_J	$\frac{1-\alpha}{1+\frac{1}{2}\alpha}$		$\frac{\alpha}{1+\frac{1}{2}\alpha}$		$\frac{\frac{1}{2}\alpha}{1+\frac{1}{2}\alpha}$
Partial pressure, p_J	$\frac{(1-\alpha)p}{1+\frac{1}{2}\alpha}$		$\frac{\alpha p}{1+\frac{1}{2}\alpha}$		$\frac{\frac{1}{2}\alpha p}{1+\frac{1}{2}\alpha}$

The total amount in moles is $n_{tot} = (1-\alpha)n + \alpha n + \frac{1}{2}\alpha n = (1+\frac{1}{2}\alpha)n$. This value is used to find the mole fractions. In the last line, $p_J = x_J p$ has been used. Treating all species as perfect gases so that $a_J = (p_J/p^\ominus)$, the equilibrium constant is

$$K = \frac{a_{H_2}^2 a_{O_2}}{a_{H_2O}^2} = \frac{(p_{H_2}/p^\ominus)^2(p_{O_2}/p^\ominus)}{(p_{H_2O}/p^\ominus)^2} = \frac{p_{H_2}^2 p_{O_2}}{p_{H_2O}^2 p^\ominus} = \frac{\left(\frac{\alpha p}{1+\frac{1}{2}\alpha}\right)^2 \left(\frac{\frac{1}{2}\alpha p}{1+\frac{1}{2}\alpha}\right)}{\left(\frac{(1-\alpha)p}{1+\frac{1}{2}\alpha}\right)^2 p^\ominus}$$

$$= \frac{\frac{1}{2}\alpha^3 p^3 (1+\frac{1}{2}\alpha)^2}{(1-\alpha)^2 p^2 (1+\frac{1}{2}\alpha)^3 p^\ominus} = \frac{\alpha^3}{(1-\alpha)^2(2+\alpha)} \frac{p}{p^\ominus}$$

In this case $\alpha = 1.77\%$ ($= 0.0177$) and $p = 1.00$ bar; recall that $p^\ominus = 1$ bar.

$$K = \frac{0.0177^3}{(1-0.0177)^2 \times (2+0.0177)} \times \frac{1.00 \text{ bar}}{1 \text{ bar}} = \boxed{2.85 \times 10^{-6}}$$

E6A.10(a) The relationship between K and K_c is [6A.18b–209], $K = K_c \times (c^\ominus RT/p^\ominus)^{\Delta\nu}$. For the reaction $H_2CO(g) \rightleftharpoons CO(g) + H_2(g)$

$$\Delta\nu = \nu_{CO} + \nu_{H_2O} - \nu_{H_2CO} = 1 + 1 - 1 = +1 \quad \text{hence} \quad \boxed{K = K_c \times (c^\ominus RT/p^\ominus)}$$

$p^\ominus/c^\ominus R$ evaluates to 12.03 K so the relationship can alternatively be written as $K = K_c \times (T/\text{K})/(12.03)$.

E6A.11(a) The following table is drawn up:

	2A	+	B	$\rightleftharpoons$	3C	+	2D
Initial amount, $n_{\mathrm{J},0}/\mathrm{mol}$	1.00		2.00		0		1.00
Change, $\Delta n_{\mathrm{J}}/\mathrm{mol}$	−0.60		−0.30		+0.90		+0.60
Equilibrium amount, $n_{\mathrm{J}}/\mathrm{mol}$	0.40		1.70		0.90		1.60
Mole fraction, x_{J}	0.0869...		0.369...		0.195...		0.347...
Partial pressure, p_{J}	$(0.0869...)p$		$(0.369...)p$		$(0.195...)p$		$(0.347...)p$

To go to the second line, the fact that 0.90 mol of C has been produced is used to deduce the changes in the other species given the stoichiometry of the reaction. For example, 2 mol of A is consumed for every 3 mol of C produced so $\Delta\nu_{\mathrm{A}} = -\frac{2}{3}\Delta\nu_{\mathrm{C}} = \frac{2}{3}\times +0.90\ \mathrm{mol} = -0.60\ \mathrm{mol}$. The total amount in moles is $(0.40\ \mathrm{mol}) + (1.70\ \mathrm{mol}) + (0.90\ \mathrm{mol}) + (1.60\ \mathrm{mol}) = 4.6\ \mathrm{mol}$. This value has been used to find the mole fractions. In the last line, $p_{\mathrm{J}} = x_{\mathrm{J}}p$ has been used.

(i) The mole fractions are given in the above table.

(ii) Treating all species as perfect gases so that $a_{\mathrm{J}} = p_{\mathrm{J}}/p^{\ominus}$ the equilibrium constant is

$$K = \frac{a_{\mathrm{C}}^3 a_{\mathrm{D}}^2}{a_{\mathrm{A}}^2 a_{\mathrm{B}}} = \frac{(p_{\mathrm{C}}/p^{\ominus})^3 (p_{\mathrm{D}}/p^{\ominus})^2}{(p_{\mathrm{A}}/p^{\ominus})^2 (p_{\mathrm{B}}/p^{\ominus})} = \frac{p_{\mathrm{C}}^3 p_{\mathrm{D}}^2}{p_{\mathrm{A}}^2 p_{\mathrm{B}} p^{\ominus 2}} = \frac{x_{\mathrm{C}}^3 x_{\mathrm{D}}^2}{x_{\mathrm{A}}^2 x_{\mathrm{B}}}\frac{p^2}{p^{\ominus 2}}$$

$$= \frac{(0.195...)^3 (0.347...)^2}{(0.0869...)^2 (0.369...)} \times \frac{(1.00\ \mathrm{bar})^2}{(1\ \mathrm{bar})^2} = 0.324... = \boxed{0.32}$$

(iii) The relationship between $\Delta_{\mathrm{r}}G^{\ominus}$ and K [6A.15–*208*], $\Delta_{\mathrm{r}}G^{\ominus} = -RT\ln K$, is used to calculate $\Delta_{\mathrm{r}}G^{\ominus}$:

$$\Delta_{\mathrm{r}}G^{\ominus} = -(8.3145\ \mathrm{J\,K^{-1}\,mol^{-1}}) \times ([25 + 273.15]\ \mathrm{K}) \times \ln 0.324...$$

$$= \boxed{+2.8\ \mathrm{kJ\,mol^{-1}}}$$

E6A.12(a) The reaction Gibbs energy for an arbitrary reaction quotient is given by [6A.11–*207*], $\Delta_{\mathrm{r}}G = \Delta_{\mathrm{r}}G^{\ominus} + RT\ln Q$. Treating borneol and isoborneol as perfect gases so that $a_{\mathrm{J}} = p_{\mathrm{J}}/p^{\ominus}$, the reaction quotient Q is

$$Q = \frac{a_{\mathrm{isoborneol}}}{a_{\mathrm{borneol}}} = \frac{p_{\mathrm{isoborneol}}/p^{\ominus}}{p_{\mathrm{borneol}}/p^{\ominus}} = \frac{p_{\mathrm{isoborneol}}}{p_{\mathrm{borneol}}}$$

Because $p_{\mathrm{J}} = x_{\mathrm{J}}p = (n_{\mathrm{J}}/n)p \propto n_{\mathrm{J}}$ it follows that

$$Q = \frac{n_{\mathrm{isoborneol}}}{n_{\mathrm{borneol}}} = \frac{0.30\ \mathrm{mol}}{0.15\ \mathrm{mol}} = 2.0$$

Hence

$$\begin{aligned}\Delta_r G &= \Delta_r G^\ominus + RT \ln Q \\ &= (+9.4 \times 10^3\ \mathrm{J\,mol^{-1}}) + (8.3145\ \mathrm{J\,K^{-1}\,mol^{-1}}) \times (503\ \mathrm{K}) \times \ln 2.0 \\ &= \boxed{+12\ \mathrm{kJ\,mol^{-1}}}\end{aligned}$$

E6A.13(a) The reaction corresponding to the standard Gibbs energy change of formation of NH_3 is

$$\tfrac{1}{2}\mathrm{N_2(g)} + \tfrac{3}{2}\mathrm{H_2(g)} \rightleftharpoons \mathrm{NH_3(g)}$$

This is the reaction in question. The reaction Gibbs energy for an arbitrary reaction quotient is given by [6A.11–207], $\Delta_r G = \Delta_r G^\circ + RT \ln Q$. All species are treated as perfect gases so that $a_J = p_J/p^\ominus$. Therefore the reaction quotient Q is

$$\begin{aligned}Q &= \frac{a_{\mathrm{NH_3}}}{a_{\mathrm{N_2}}^{1/2} a_{\mathrm{H_2}}^{3/2}} = \frac{(p_{\mathrm{NH_3}}/p^\ominus)}{(p_{\mathrm{N_2}}/p^\ominus)^{1/2} \times (p_{\mathrm{H_2}}/p^\ominus)^{3/2}} = \frac{p_{\mathrm{NH_3}}\, p^\ominus}{p_{\mathrm{N_2}}^{1/2}\, p_{\mathrm{H_2}}^{3/2}} \\ &= \frac{(4.0\ \mathrm{bar}) \times (1\ \mathrm{bar})}{(3.0\ \mathrm{bar})^{1/2} \times (1.0\ \mathrm{bar})^{3/2}} = 2.30...\end{aligned}$$

Hence

$$\begin{aligned}\Delta_r G &= \Delta_r G^\ominus + RT \ln Q \\ &= (-16.5 \times 10^3\ \mathrm{J\,mol^{-1}}) + (8.3145\ \mathrm{J\,K^{-1}\,mol^{-1}}) \times (298\ \mathrm{K}) \times \ln(2.30...) \\ &= \boxed{-14\ \mathrm{kJ\,mol^{-1}}}\end{aligned}$$

Because $\Delta_r G < 0$ the spontaneous direction of the reaction under these conditions is from left to right.

E6A.14(a) The standard Gibbs energy change for the reaction is given in terms of the standard Gibbs energies of formation by [6A.13a–*207*]:

$$\Delta_r G^\ominus = \Delta_f G^\ominus(\mathrm{CaF_2, aq}) - \Delta_f G^\ominus(\mathrm{CaF_2, s})$$

This is rearranged for $\Delta_f G^\ominus(\mathrm{CaF_2, aq})$ and $\Delta_r G^\ominus$ is replaced by $-RT \ln K$ [6A.15–*208*] to give

$$\begin{aligned}\Delta_f G^\ominus(\mathrm{CaF_2, aq}) &= \Delta_r G^\ominus + \Delta_f G^\ominus(\mathrm{CaF_2, s}) = -RT \ln K + \Delta_f G^\ominus(\mathrm{CaF_2, s}) \\ &= -(8.3145\ \mathrm{J\,K^{-1}\,mol^{-1}}) \times ([25 + 273.15]\ \mathrm{K}) \times \ln(3.9 \times 10^{-11}) \\ &\quad + (-1167 \times 10^3\ \mathrm{J\,mol^{-1}}) = \boxed{-1.1 \times 10^3\ \mathrm{kJ\,mol^{-1}}}\end{aligned}$$

Solutions to problems

P6A.1 (a) The relationship between the equilibrium constant and $\Delta_r G^\ominus$ is [6A.15–208], $\Delta_r G^\ominus = -RT\ln K$.

$$\Delta_r G^\ominus = -(8.3145\,\text{J}\,\text{K}^{-1}\,\text{mol}^{-1}) \times (298.15\,\text{K}) \times \ln 0.164 = \boxed{+4.48\,\text{kJ}\,\text{mol}^{-1}}$$

(b) The following table is drawn up. Iodine is not included in the calculations as it is a solid.

	$I_2(s)$	+	$Br_2(g)$	$\rightleftharpoons$	$2IBr(g)$
Initial amount	—		n		0
Change to reach equilibrium	—		$-\alpha n$		$+2\alpha n$
Amount at equilibrium	—		$(1-\alpha)n$		$2\alpha n$
Mole fraction, x_J	—		$\dfrac{1-\alpha}{1+\alpha}$		$\dfrac{2\alpha}{1+\alpha}$
Partial pressure, p_J	—		$\dfrac{(1-\alpha)p}{1+\alpha}$		$\dfrac{2\alpha p}{1+\alpha}$

The total amount in moles is $n_{\text{tot}} = (1-\alpha)n + 2\alpha n = (1+\alpha)n$. This value is used to find the mole fractions. Treating $Br_2(g)$ and $IBr(g)$ as perfect gases, so that $a_J = p_J/p^\ominus$, and I_2 as a pure solid, so that $a_{I_2} = 1$, the equilibrium constant is:

$$K = \frac{a_{\text{IBr}}^2}{a_{I_2} a_{\text{Br}_2}} = \frac{(p_{\text{IBr}}/p^\ominus)^2}{1 \times (p_{\text{Br}_2}/p^\ominus)} = \frac{p_{\text{IBr}}^2}{p_{\text{Br}_2} p^\ominus} = \frac{\left(\frac{2\alpha p}{1+\alpha}\right)^2}{\frac{(1-\alpha)p}{1+\alpha} p^\ominus} = \frac{4\alpha^2}{(1-\alpha)(1+\alpha)} \frac{p}{p^\ominus}$$

Note that $(1-\alpha)(1+\alpha) = 1-\alpha^2$. With this, an expression for α is found by straightforward algebra.

$$\begin{aligned}\alpha &= \left(\frac{Kp^\ominus}{4p + Kp^\ominus}\right)^{\frac{1}{2}} \\ &= \left(\frac{0.164 \times (1\,\text{bar})}{4 \times (0.164\,\text{atm}) \times (1.01325\,\text{bar})/(1\,\text{atm}) + 0.164 \times (1\,\text{bar})}\right)^{\frac{1}{2}} \\ &= 0.444...\end{aligned}$$

Hence

$$p_{\text{IBr}} = \frac{2\alpha p}{1+\alpha} = \frac{2 \times 0.444... \times (0.164\,\text{atm})}{1 + 0.444...} = \boxed{0.101\,\text{atm}} \text{ or } \boxed{0.102\,\text{bar}}$$

(c) The issue here is that the reaction under discussion is that with $I_2(s)$. If the partial pressure of I_2 is not zero then p_{Br_2} and p_{IBr} no longer sum to the total pressure p but rather to $p - p_{I_2}$ where p_{I_2} is the partial pressure of iodine. The partial pressures in the last line of the above table therefore become

$$p_{\text{Br}_2} = \frac{1-\alpha}{1+\alpha}(p - p_{I_2}) \quad \text{and} \quad p_{\text{IBr}} = \frac{2\alpha}{1+\alpha}(p - p_{I_2})$$

The equilibrium constant K for the $I_2(s) + Br_2(g) \rightleftharpoons 2IBr(g)$ is still $K = p_{IBr}^2/p_{Br_2}p^\ominus$ but now with the new partial pressures:

$$K = \frac{\left[\left(\frac{2\alpha}{1+\alpha}\right)(p - p_{I_2})\right]^2}{\left(\frac{1-\alpha}{1+\alpha}\right)(p - p_{I_2})p^\ominus} = \frac{4\alpha^2}{(1+\alpha)(1-\alpha)}\left(\frac{p - p_{I_2}}{p^\ominus}\right)$$

Given the partial pressure of I_2 this equation can be solved for α, and p_{IBr} calculated as before.

P6A.3 The following table is drawn up for the reaction, assuming that to reach equilibrium the reaction proceeds by an amount z in the direction of the products.

	$H_2(s)$	+	$I_2(g)$	$\rightleftharpoons$	$2HI(g)$
Initial amount	$n_{H_2,0}$		$n_{I_2,0}$		$n_{HI,0}$
Change to reach equilibrium	$-z$		$-z$		$+2z$
Amount at equilibrium	$n_{H_2,0} - z$		$n_{I_2,0} - z$		$n_{HI,0} + 2z$
Mole fraction, x_J	$\frac{n_{H_2,0} - z}{n_{tot}}$		$\frac{n_{I_2,0} - z}{n_{tot}}$		$\frac{n_{HI,0} + 2z}{n_{tot}}$
Partial pressure, p_J	$\frac{(n_{H_2,0} - z)p}{n_{tot}}$		$\frac{(n_{I_2,0} - z)p}{n_{tot}}$		$\frac{(n_{HI,0} + 2z)p}{n_{tot}}$

where $n_{tot} = n_{H_2,0} + n_{I_2,0} + n_{HI,0}$. Treating all species as perfect gases, so that $a_J = p_J/p^\ominus$, the equilibrium constant is

$$K = \frac{a_{HI}^2}{a_{H_2}a_{I_2}} = \frac{(p_{HI}/p^\ominus)^2}{(p_{H_2}/p^\ominus)(p_{I_2}/p^\ominus)} = \frac{p_{HI}^2}{p_{H_2}p_{I_2}} = \frac{(n_{HI,0} + 2z)^2}{(n_{H_2,0} - z)(n_{I_2,0} - z)}$$

Rearranging gives

$$K(n_{H_2,0} - z)(n_{I_2,0} - z) = (n_{HI,0} + 2z)^2$$

Hence $(K - 4)z^2 - ([n_{H_2,0} + n_{I_2,0}]K + 4n_{HI,0})z + (n_{H_2,0}n_{I_2,0}K - n_{HI}^2) = 0$

Substituting in the values for n_J and K, dividing through by mol^2 and writing $x = z/\text{mol}$ yields the quadratic

$$866x^2 - 609.8x + 104.36 = 0$$

which has solutions $x = 0.410...$ and $x = 0.293...$ implying $z = (0.410...\ \text{mol})$ or $z = (0.293...\ \text{mol})$. The solution $z = (0.410...\ \text{mol})$ is rejected because z cannot be larger than $n_{H_2,0}$ or $n_{I_2,0}$. The amounts of each substance present at equilibrium are therefore

$$n_{H_2} = n_{H_2,0} - z = (0.300\ \text{mol}) - (0.293...\ \text{mol}) = \boxed{6.67 \times 10^{-3}\ \text{mol}}$$

$$n_{I_2} = n_{I_2,0} - z = (0.400\ \text{mol}) - (0.293...\ \text{mol}) = \boxed{0.107\ \text{mol}}$$

$$n_{HI} = n_{HI,0} + 2z = (0.200\ \text{mol}) + 2 \times (0.293...\ \text{mol}) = \boxed{0.787\ \text{mol}}$$

P6A.5 If the extent of reaction at equilibrium is ξ, then from the stoichiometry of the reaction the amounts of A and B that have reacted are ξ and 3ξ respectively and the amount of C that has been formed is 2ξ. If the initial amounts of A, B and C are n, $3n$ and 0, the following table is drawn up.

	A	+	3B	$\rightleftharpoons$	2C
Initial amount	n		$3n$		0
Change to reach equilibrium	$-\xi$		-3ξ		$+2\xi$
Amount at equilibrium	$n-\xi$		$3(n-\xi)$		2ξ
Mole fraction, x_J	$\dfrac{n-\xi}{4n-2\xi}$		$\dfrac{3(n-\xi)}{4n-2\xi}$		$\dfrac{2\xi}{4n-2\xi}$
Partial pressure, p_J	$\dfrac{(n-\xi)p}{4n-2\xi}$		$\dfrac{3(n-\xi)p}{4n-2\xi}$		$\dfrac{2\xi p}{4n-2\xi}$

The total amount in moles is $n_{\text{tot}} = (n-\xi)+3(n-\xi)+2\xi = 4n-2\xi$. This value is used to find the mole fractions. Treating all species as perfect gases, so that $a_J = p_J/p^{\ominus}$, the equilibrium constant is

$$K = \frac{a_C^2}{a_A a_B^3} = \frac{(p_C/p^{\ominus})^2}{(p_A/p^{\ominus})(p_B/p^{\ominus})^3} = \frac{p_C^2 p^{\ominus 2}}{p_A p_B^3} = \frac{\left(\frac{2\xi p}{4n-2\xi}\right)^2 p^{\ominus 2}}{\left(\frac{(n-\xi)p}{4n-2\xi}\right)\left(\frac{3(n-\xi)p}{4n-2\xi}\right)^3}$$

$$= \frac{16\xi^2(2n-\xi)^2}{27(n-\xi)^4}\left(\frac{p^{\ominus}}{p}\right)^2$$

Rearranging and then taking the square root gives

$$\frac{\xi^2(2n-\xi)^2}{(n-\xi)^4} = \frac{27Kp^2}{16p^{\ominus 2}} \quad \text{hence} \quad \frac{\xi(2n-\xi)}{(n-\xi)^2} = \tfrac{1}{4}\sqrt{27K}(p/p^{\ominus})$$

The negative square root is rejected because $0 \le \xi \le n$. This requirement arises because if $\xi < 0$ this would imply a negative amount of C, while if $\xi > n$ this would imply negative amounts of A and B. Because $0 \le \xi \le n$ the left hand side of the square rooted expression is always ≥ 0. Because $p/p^{\ominus}$ cannot be negative either it follows that the positive square root is required.

Rearranging further yields the quadratic

$$(\xi/n)^2 - 2(\xi/n) + \frac{\tfrac{1}{4}\sqrt{27K}(p/p^{\ominus})}{1+\tfrac{1}{4}\sqrt{27K}(p/p^{\ominus})} = 0$$

which is solved to give

$$(\xi/n) = 1 - \left(\frac{1}{1+\tfrac{1}{4}\sqrt{27K}(p/p^{\ominus})}\right)^{\frac{1}{2}}$$

The positive square root is rejected in order to ensure that $0 \le (\xi/n) \le 1$.

Inspection of this expression shows that $\xi \to 0$ as $p \to 0$, indicating that the reactants are favoured at low pressures. On the other hand $(\xi/n) \to 1$ as $p \to \infty$ indicating that the products are favoured at high pressure. (ξ/n) is plotted against $p/p^\ominus$ in the graph shown in Fig. 6.1, using three different values of K.

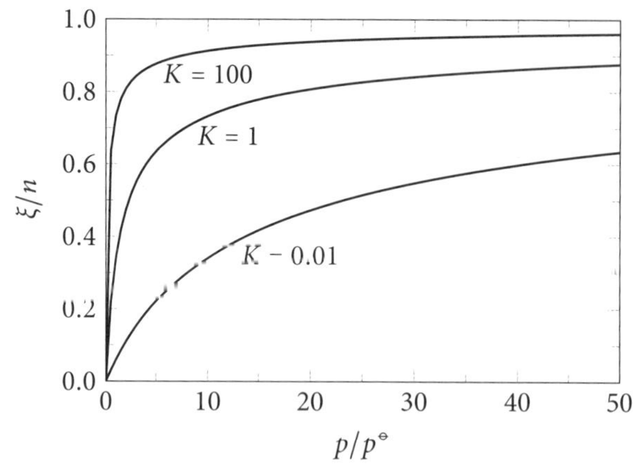

Figure 6.1

6B The response of equilibria to the conditions

Answer to discussion question

D6B.1 The thermodynamic equilibrium constant is that expressed in terms of activities. In the limit of low pressures the activity of a gaseous species is well approximated in terms of its partial pressure, $a_i = p_i/p^\ominus$: under these circumstances the equilibrium constant expressed in terms of partial pressures is the same as the thermodynamic equilibrium constant. At higher pressures the presence of significant molecular interactions means that the activities are no longer well-approximated in terms of partial pressures, and therefore the equilibrium 'constant' in terms of partial pressures will no longer be the same as the (true) thermodynamic equilibrium constant.

D6B.3 This is discussed in Section 6B.2(a) on page 213.

Solutions to exercises

E6B.1(a) For the reaction $N_2O_4(g) \rightleftharpoons 2NO_2(g)$ the following table is drawn up by supposing that there are n moles of N_2O_4 initially and that at equilibrium a fraction α has dissociated.

	N_2O_4	$\rightleftharpoons$ $2NO_2$
Initial amount	n	0
Change to reach equilibrium	$-\alpha n$	$+2\alpha n$
Amount at equilibrium	$(1-\alpha)n$	$2\alpha n$
Mole fraction, x_J	$\dfrac{1-\alpha}{1+\alpha}$	$\dfrac{2\alpha}{1+\alpha}$
Partial pressure, p_J	$\dfrac{(1-\alpha)p}{1+\alpha}$	$\dfrac{2\alpha p}{1+\alpha}$

The total amount in moles is $n_{tot} = (1-\alpha)n + 2\alpha n = (1+\alpha)n$. This value is used to find the mole fractions. In the last line, $p_J = x_J p$ [1A.6–9] has been used. Treating all species as perfect gases so that $a_J = (p_J/p^\ominus)$, the equilibrium constant is

$$K = \frac{a_{NO_2}^2}{a_{N_2O_4}} = \frac{(p_{NO_2}/p^\ominus)^2}{(p_{N_2O_4}/p^\ominus)} = \frac{p_{NO_2}^2}{p_{N_2O_4}p^\ominus} = \frac{\left(\frac{2\alpha p}{1+\alpha}\right)^2}{\left(\frac{(1-\alpha)p}{1+\alpha}\right)p^\ominus} = \frac{4\alpha^2}{(1-\alpha)(1+\alpha)}\frac{p}{p^\ominus}$$

In this case $\alpha = 0.1846$ and $p = 1.00$ bar; recall that $p^\ominus = 1$ bar.

$$K = \frac{4\times 0.1846^2}{(1-0.1846)\times(1+0.1846)} \times \frac{1.00\text{ bar}}{1\text{ bar}} = 0.141... = \boxed{0.141}$$

The temperature dependence of K is given by [6B.4–*215*],

$$\ln K_2 - \ln K_1 = -\frac{\Delta_r H^\ominus}{R}\left(\frac{1}{T_2} - \frac{1}{T_1}\right)$$

assuming that $\Delta_r H^\ominus$ is constant over the temperature range of interest. Taking $T_1 = 25\,°C$ (= 298.15 K) and $T_2 = 100\,°C$ (= 373.15 K) gives

$$\ln K_2 = \ln(0.141...) - \frac{56.2\times 10^3\text{ J mol}^{-1}}{8.3145\text{ J K}^{-1}\text{ mol}^{-1}}\left(\frac{1}{373.15\text{ K}} - \frac{1}{298.15\text{ K}}\right) = 2.59...$$

That is, $K_2 = \boxed{13.4}$, a larger value than at 25 °C, as expected for this endothermic reaction.

E6B.2(a) The data in the *Resource section* is used to calculate $\Delta_r G^\ominus$ and $\Delta_r H^\ominus$ at 298 K

$$\begin{aligned}\Delta_r G^\ominus &= \Delta_f G^\ominus(CO_2,g) - \Delta_f G^\ominus(PbO,s,red) - \Delta_f G^\ominus(CO,g)\\ &= (-394.36\text{ kJ mol}^{-1}) - (-188.93\text{ kJ mol}^{-1}) - (-137.17\text{ kJ mol}^{-1})\\ &= -68.26\text{ kJ mol}^{-1}\\ \Delta_r H^\ominus &= \Delta_f H^\ominus(CO_2,g) - \Delta_f H^\ominus(PbO,s,red) - \Delta_f H^\ominus(CO,g)\\ &= (-393.51\text{ kJ mol}^{-1}) - (-218.99\text{ kJ mol}^{-1}) - (-110.53\text{ kJ mol}^{-1})\\ &= -63.99\text{ kJ mol}^{-1}\end{aligned}$$

The equilibrium constant at 298 K is calculated from $\Delta_r G^\ominus$ using [6A.15–*208*], $\Delta_r G^\ominus = -RT \ln K$

$$\ln K = -\frac{\Delta_r G^\ominus}{RT} = -\frac{-68.26 \times 10^3 \text{ J mol}^{-1}}{(8.3145 \text{ J K}^{-1} \text{mol}^{-1}) \times (298 \text{ K})} = 27.5...$$

$$\text{hence} \quad K = e^{27.5...} = 9.21... \times 10^{11} = \boxed{9.22 \times 10^{11}}$$

The temperature dependence of K is given by [6B.4–*215*],

$$\ln K_2 - \ln K_1 = -\frac{\Delta_r H^\ominus}{R}\left(\frac{1}{T_2} - \frac{1}{T_1}\right)$$

assuming that $\Delta_r H^\ominus$ is constant over the temperature range of interest. This is used to calculate the equilibrium constant at 400 K

$$\ln K_2 = \ln(9.21... \times 10^{11}) - \frac{63.99 \times 10^3 \text{ J mol}^{-1}}{8.3145 \text{ J K}^{-1} \text{mol}^{-1}}\left(\frac{1}{400 \text{ K}} - \frac{1}{298 \text{ K}}\right) = 20.9...$$

That is, $K_2 = \boxed{1.27 \times 10^9}$, a smaller value than at 298 K, as expected for this exothermic reaction.

E6B.3(a) Assuming that $\Delta_r H^\ominus$ is constant over the temperature range of interest, the temperature dependence of K is given by [6B.4–*215*],

$$\ln K_2 - \ln K_1 = -\frac{\Delta_r H^\ominus}{R}\left(\frac{1}{T_2} - \frac{1}{T_1}\right)$$

Using $\Delta_r G^\ominus = -RT \ln K$ to substitute for K_1 and setting $\ln K_2 = \ln 1 = 0$ (the crossover point) gives

$$\frac{\Delta_r G^\ominus(T_1)}{RT_1} = -\frac{\Delta_r H^\ominus}{R}\left(\frac{1}{T_2} - \frac{1}{T_1}\right)$$

Rearranging for T_2 gives

$$T_2 = \frac{T_1 \Delta_r H^\ominus}{\Delta_r H^\ominus - \Delta_r G^\ominus(T_1)} = \frac{(1280 \text{ K}) \times (+224 \text{ kJ mol}^{-1})}{(+224 \text{ kJ mol}^{-1}) - (+33 \text{ kJ mol}^{-1})} = \boxed{1.5 \times 10^3 \text{ K}}$$

Note that this temperature is outside the range over which $\Delta_r H^\ominus$ is known to be constant and is therefore an estimate.

E6B.4(a) The van 't Hoff equation [6B.2–*214*], $\mathrm{d}\ln K/\mathrm{d}T = \Delta_r H^\ominus/RT^2$, is rearranged to obtain an expression for $\Delta_r H^\ominus$

$$\begin{aligned}
\Delta_r H^\ominus &= RT^2 \frac{\mathrm{d}\ln K}{\mathrm{d}T} \\
&= RT^2 \frac{\mathrm{d}}{\mathrm{d}T}\left(A + \frac{B}{T} + \frac{C}{T^2}\right) = RT^2\left(-\frac{B}{T^2} - \frac{2C}{T^3}\right) = -R\left(B + \frac{2C}{T}\right) \\
&= -(8.3145 \text{ J K}^{-1} \text{mol}^{-1}) \times \left((-1088 \text{ K}) + \frac{2 \times (1.51 \times 10^5 \text{ K}^2)}{400 \text{ K}}\right) \\
&= +2.76... \times 10^3 \text{ J mol}^{-1} = \boxed{+2.77 \text{ kJ mol}^{-1}}
\end{aligned}$$

The standard reaction entropy is obtained by first finding an expression for $\Delta_r G^\ominus$ using [6A.15–*208*]

$$\Delta_r G^\ominus = -RT\ln K = -RT\left(A + \frac{B}{T} + \frac{C}{T^2}\right) = -R\left(AT + B + \frac{C}{T}\right)$$

The equation $\Delta_r G^\ominus = \Delta_r H^\ominus - T\Delta_r S^\ominus$ [3D.9–*100*] is then rearranged to find $\Delta_r S^\ominus$

$$\Delta_r S^\ominus = \frac{\Delta_r H^\ominus - \Delta_r G^\ominus}{T} = \frac{1}{T}\left(\underbrace{-R\left(B + \frac{2C}{T}\right)}_{\Delta_r H^\ominus} + \underbrace{R\left(AT + B + \frac{C}{T}\right)}_{-\Delta_r G^\ominus}\right) = R\left(A - \frac{C}{T^2}\right)$$

$$= (8.3145\,\mathrm{J\,K^{-1}\,mol^{-1}}) \times \left(-1.04 - \frac{1.51\times10^5\,\mathrm{K}^2}{(400\,\mathrm{K})^2}\right) = \boxed{-16.5\,\mathrm{J\,K^{-1}\,mol^{-1}}}$$

An alternative approach to finding $\Delta_r S^\ominus$ is to use the variation of G with T which is given by [3E.8–*107*], $(\partial G/\partial T)_p = -S$. This implies that $\mathrm{d}\Delta_r G^\ominus/\mathrm{d}T = -\Delta_r S^\ominus$ where the derivative is complete (not partial) because $\Delta_r G^\ominus$ is independent of pressure. Using the expression for $\Delta_r G^\ominus$ from above it follows that

$$\Delta_r S^\ominus = -\frac{\mathrm{d}\Delta_r G^\ominus}{\mathrm{d}T} = -\frac{\mathrm{d}}{\mathrm{d}T}\left(\underbrace{-R\left(AT + B + \frac{C}{T}\right)}_{\Delta_r G^\ominus}\right) = R\left(A - \frac{C}{T^2}\right)$$

which is the same expression obtained above.

E6B.5(a) Treating all species as perfect gases so that $a_J = p_J/p^\ominus$, the equilibrium constant for the reaction $H_2CO(g) \rightleftharpoons CO(g) + H_2(g)$ is

$$K = \frac{a_{H_2}a_{CO}}{a_{H_2CO}} = \frac{(p_{H_2}/p^\ominus)(p_{CO}/p^\ominus)}{(p_{H_2CO}/p^\ominus)} = \frac{p_{H_2}p_{CO}}{p_{H_2CO}p^\ominus} = \frac{(x_{H_2}p)(x_{CO}p)}{(x_{H_2CO}p)p^\ominus}$$
$$= \frac{x_{H_2}x_{CO}p}{x_{H_2CO}p^\ominus} = K_x \times \frac{p}{p^\ominus}$$

where K_x is the part of the equilibrium constant expression that contains the equilibrium mole fractions of reactants and products. Because K is independent of pressure, if p doubles K_x must halve in order to preserve the value of K. In other words, K_x is reduced by $\boxed{50\%}$.

E6B.6(a) The following table is drawn up for the borneol $\rightleftharpoons$ isoborneol reaction, denoting the initial amounts of borneol and isoborneol by $n_{b,0}$ and $n_{iso,0}$ and supposing that in order to reach equilibrium an amount z of borneol has converted to isoborneol.

	Borneol	⇌	Isoborneol
Initial amount	$n_{b,0}$		$n_{iso,0}$
Change to reach equilibrium	$-z$		$+z$
Amount at equilibrium	$n_{b,0}-z$		$n_{iso,0}+z$
Mole fraction, x_J	$\dfrac{n_{b,0}-z}{n_{b,0}+n_{iso,0}}$		$\dfrac{n_{iso,0}+z}{n_{b,0}+n_{iso,0}}$
Partial pressure, p_J	$\dfrac{(n_{b,0}-z)p}{n_{b,0}+n_{iso,0}}$		$\dfrac{(n_{iso,0}+z)p}{n_{b,0}+n_{iso,0}}$

The total amount in moles is $(n_{b,0}-z)+n_{iso,0}+z = n_{b,0}+n_{iso,0}$. This value is used to find the mole fractions. Treating both species as perfect gases so that $a_J = p_J/p^{\ominus}$ the equilibrium constant is

$$K = \frac{a_{\text{borneol}}}{a_{\text{isoborneol}}} = \frac{p_{\text{borneol}}}{p_{\text{isoborneol}}} = \frac{n_{iso,0}+z}{n_{b,0}-z}$$

Rearranging for z gives $z = (Kn_{b,0}-n_{iso,0})/(1+K)$. Noting that $n = m/M$ where $M = 154.2422\ \text{g mol}^{-1}$ is the molar mass of borneol and isoborneol, gives

$$n_{b,0} = (7.50\ \text{g})/(154.2422\ \text{g mol}^{-1}) = 0.0486...\ \text{mol}$$

$$n_{iso,0} = (14.0\ \text{g})/(154.2422\ \text{g mol}^{-1}) = 0.0907...\ \text{mol}$$

$$z = \frac{Kn_{b,0}-n_{iso,0}}{1+K} = \frac{0.106\times(0.0486...\ \text{mol})-(0.0907...\ \text{mol})}{1+0.106} = -0.0774...\ \text{mol}$$

The negative value of z indicates that in order to reach equilibrium there is a net conversion of isoborneol to borneol. Using this value of z, and the expressions for x_{borneol} and $x_{\text{isoborneol}}$ in the above table, the mole fractions at equilibrium are calculated as

$$x_{\text{borneol}} = \frac{n_{b,0}-z}{n_{b,0}+n_{iso,0}} = \frac{(0.0486...\ \text{mol})-(-0.0774...\ \text{mol})}{(0.0486...\ \text{mol})+(0.0907...\ \text{mol})} = 0.904... = \boxed{0.904}$$

Then $x_{\text{isoborneol}} = 1 - x_{\text{borneol}} = 1 - 0.904... = \boxed{0.096}$

E6B.7(a) The temperature dependence of K is given by [6B.4–*215*]

$$\ln K_2 - \ln K_1 = -\frac{\Delta_r H^{\ominus}}{R}\left(\frac{1}{T_2}-\frac{1}{T_1}\right) \quad \text{hence} \quad \Delta_r H^{\ominus} = -\frac{R\ln(K_2/K_1)}{(1/T_2)-(1/T_1)}$$

(i) If the equilibrium constant is doubled then $K_2/K_1 = 2$

$$\Delta_r H^{\ominus} = -\frac{(8.3145\ \text{J K}^{-1}\ \text{mol}^{-1})\times\ln 2}{[1/(308\ \text{K})]-(1/[298\ \text{K}])} = \boxed{+52.9\ \text{kJ mol}^{-1}}$$

(ii) If the equilibrium constant is halved then $K_2/K_1 = 1/2$

$$\Delta_r H^\ominus = -\frac{(8.3145\,\mathrm{J\,K^{-1}\,mol^{-1}}) \times \ln(1/2)}{[1/(308\,\mathrm{K})] - (1/[298\,\mathrm{K}])} = \boxed{-52.9\,\mathrm{kJ\,mol^{-1}}}$$

E6B.8(a) The relationship between $\Delta_r G^\ominus$ and K is given by [6A.15–*208*], $\Delta_r G^\ominus = -RT\ln K$. Hence if $K = 1$, $\Delta_r G^\ominus = -RT\ln 1 = 0$. Furthermore $\Delta_r G^\ominus$ is related to $\Delta_r H^\ominus$ and $\Delta_r S^\ominus$ by [3D.9–*100*], $\Delta_r G^\ominus = \Delta_r H^\ominus - T\Delta_r S^\ominus$, so if $K = 1$

$$\Delta_r H^\ominus - T\Delta_r S^\ominus = 0 \quad \text{hence} \quad T = \frac{\Delta_r H^\ominus}{\Delta_r S^\ominus}$$

Values of $\Delta_r H^\ominus$ and $\Delta_r S^\ominus$ at 298 K are calculated using data from the *Resource section*.

$$\begin{aligned}
\Delta_r H^\ominus &= \Delta_f H^\ominus(\mathrm{CaO,s}) + \Delta_f H^\ominus(\mathrm{CO_2,g}) - \Delta_f H^\ominus(\mathrm{CaCO_3,s,calcite}) \\
&= (-635.09\,\mathrm{kJ\,mol^{-1}}) + (-393.51\,\mathrm{kJ\,mol^{-1}}) - (-1206.9\,\mathrm{kJ\,mol^{-1}}) \\
&= +178.3\,\mathrm{kJ\,mol^{-1}} \\
\Delta_r S^\ominus &= S_m^\ominus(\mathrm{CaO,s}) + S_m^\ominus(\mathrm{CO_2,g}) - S_m^\ominus(\mathrm{CaCO_3,s,calcite}) \\
&= (39.75\,\mathrm{J\,K^{-1}\,mol^{-1}}) + (213.74\,\mathrm{J\,K^{-1}\,mol^{-1}}) - (92.9\,\mathrm{J\,K^{-1}\,mol^{-1}}) \\
&= 160.59\,\mathrm{J\,K^{-1}\,mol^{-1}}
\end{aligned}$$

Substituting these values into the equation found above, assuming that $\Delta_r H^\ominus$ and $\Delta_r S^\ominus$ do not vary significantly with temperature over the range of interest, gives:

$$T = \frac{\Delta_r H^\ominus}{\Delta_r S^\ominus} = \frac{178.13 \times 10^3\,\mathrm{J\,mol^{-1}}}{160.59\,\mathrm{J\,K^{-1}\,mol^{-1}}} = \boxed{1109\,\mathrm{K}}$$

E6B.9(a) Treating the vapour as a perfect gas, so that $a_J = p_J/p^\ominus$, and noting that $a_{A_2B} = 1$ because it is a pure solid, the equilibrium constant for the dissociation $A_2B(s) \rightleftharpoons A_2(g) + B(g)$ is

$$K = \frac{a_{A_2,g}a_{B,g}}{a_{A_2B,s}} = \frac{(p_{A_2}/p^\ominus)(p_B/p^\ominus)}{1} = \frac{p_{A_2}p_{B_2}}{p^{\ominus 2}}$$

Furthermore, because A_2 and B are formed in a 1 : 1 ratio, they each have a mole fraction of 1/2 and the partial pressure of each is half the total pressure: $p_{A_2} = p_B = \frac{1}{2}p$. The equilibrium constant is therefore

$$K = \frac{(\frac{1}{2}p)(\frac{1}{2}p)}{p^{\ominus 2}} = \frac{p^2}{4p^{\ominus 2}}$$

The variation of K with temperature, assuming that $\Delta_r H^\ominus$ does not vary with T over the temperature range of interest, is given by [6B.4–*215*]:

$$\ln K_2 - \ln K_1 = -\frac{\Delta_r H^\ominus}{R}\left(\frac{1}{T_2} - \frac{1}{T_1}\right) \quad \text{hence} \quad \Delta_r H^\ominus = -\frac{R\ln(K_2/K_1)}{(1/T_2) - (1/T_1)}$$

Noting that the above expression for K implies that $\ln(K_2/K_1) = \ln(p_2^2/p_1^2)$, $\Delta_r H^\ominus$ is calculated as

$$\Delta_r H^\ominus = -\frac{(8.3145\,\mathrm{J\,K^{-1}\,mol^{-1}}) \times \ln\left((547\,\mathrm{kPa})^2/(208\,\mathrm{kPa})^2\right)}{[1/(477+273.15)\,\mathrm{K}] - [1/(367+273.15)\,\mathrm{K}]}$$

$$= 7.01\ldots \times 10^4\,\mathrm{J\,mol^{-1}} = \boxed{70.2\,\mathrm{kJ\,mol^{-1}}}$$

The standard entropy of reaction, $\Delta_r S^\ominus$ is found by rearranging $\Delta_r G^\ominus = \Delta_r H^\ominus - T\Delta_r S^\ominus$ [3D.9–*100*] and replacing $\Delta_r G^\ominus$ by $\Delta_r G^\ominus = -RT\ln K$ [6A.15–*208*]:

$$\Delta_r S^\ominus = \frac{\Delta_r H^\ominus - \Delta_r G^\ominus}{T} = \frac{\Delta_r H^\ominus + RT\ln K}{T} = \frac{\Delta_r H^\ominus}{T} + R\ln\left(\frac{p^2}{4p^\ominus}\right)$$

Using the data for 367 °C (both temperatures give the same result) gives:

$$\Delta_r S^\ominus = \frac{7.01\ldots \times 10^4\,\mathrm{J\,mol^{-1}}}{(367+273.15)\,\mathrm{K}} + (8.3145\,\mathrm{J\,K^{-1}\,mol^{-1}}) \times \ln\left(\frac{(208\,\mathrm{kPa})^2}{4\times(100\,\mathrm{kPa})^2}\right)$$

$$= 1.10\ldots \times 10^2\,\mathrm{J\,K^{-1}\,mol^{-1}} = \boxed{110\,\mathrm{J\,K^{-1}\,mol^{-1}}}$$

An alternative (but equivalent) approach to finding $\Delta_r H^\ominus$ and $\Delta_r S^\ominus$ is to first calculate $\Delta_r G^\ominus$ at both temperatures and hence obtain two equations of the form $\Delta_r G^\ominus = \Delta_r H^\ominus - T\Delta_r S^\ominus$. These can then be solved simultaneously to find the two unknowns $\Delta_r H^\ominus$ and $\Delta_r S^\ominus$, assuming them to be constant.

The values of $\Delta_r H^\ominus$ and $\Delta_r S^\ominus$ are then used with $\Delta_r G^\ominus = \Delta_r H^\ominus - T\Delta_r S^\ominus$ and $\Delta_r G^\ominus = -RT\ln K$ to calculate $\Delta_r G^\ominus$ and K at the temperature of interest, 422 °C or 695.15 K. In making this calculation it is again assumed that $\Delta_r H^\ominus$ and $\Delta_r S^\ominus$ do not vary with temperature.

$$\Delta_r G^\ominus = \Delta_r H^\ominus - T\Delta_r S^\ominus$$

$$= (7.01\ldots \times 10^4\,\mathrm{J\,mol^{-1}}) - (695.15\,\mathrm{K}) \times (1.10\ldots \times 10^2\,\mathrm{J\,K^{-1}\,mol^{-1}})$$

$$= -6.48\ldots \times 10^3\,\mathrm{J\,mol^{-1}} = \boxed{-6.48\,\mathrm{kJ\,mol^{-1}}}$$

$$K = \mathrm{e}^{-\Delta_r G^\ominus/RT} = \exp\left(-\frac{-6.48\ldots \times 10^3}{(8.3145\,\mathrm{J\,K^{-1}\,mol^{-1}}) \times (695.15\,\mathrm{K})}\right) = \boxed{3.07}$$

Solutions to problems

P6B.1 Assuming $\Delta_r H^\ominus$ to be constant over the temperature range of interest, the temperature dependence of K is given by [6B.4–*215*]

$$\ln K_2 - \ln K_1 = -\frac{\Delta_r H^\ominus}{R}\left(\frac{1}{T_2} - \frac{1}{T_1}\right) \quad \text{hence} \quad \Delta_r H^\ominus = -\frac{R\ln(K_2/K_1)}{(1/T_2) - (1/T_1)}$$

Therefore

$$\Delta_r H^\ominus = -\frac{(8.3145\,\mathrm{J\,K^{-1}\,mol^{-1}}) \times \ln\left[(1.75\times 10^5)/(2.13\times 10^6)\right]}{[1/(308\,\mathrm{K})] - [1/(288\,\mathrm{K})]}$$

$$= \boxed{-92.2\,\mathrm{kJ\,mol^{-1}}}$$

P6B.3 The reaction for which $\Delta_r H^\ominus$ is the standard enthalpy of formation of UH_3 is:

$$U(s) + \tfrac{3}{2}H_2(g) \rightleftharpoons UH_3(s)$$

Treating H_2 as a perfect gas (so that $a_{H_2} = p_{H_2}/p^\ominus$) and noting that pure solids have $a_J = 1$, the equilibrium constant for this reaction is written

$$K = \frac{a_{UH_3}}{a_{H_2}^{3/2}a_U} = \frac{1}{(p/p^\ominus)^{3/2} \times 1} = \left(\frac{p^\ominus}{p}\right)^{3/2} = \left(\frac{p}{p^\ominus}\right)^{-3/2}$$

where p is the pressure of H_2. The standard reaction enthalpy, which corresponds to $\Delta_f H^\ominus(UH_3, s)$, is obtained by rearranging the van 't Hoff equation [6B.2–*214*], $\mathrm{d}\ln K/\mathrm{d}T = \Delta_r H^\ominus/RT^2$, for $\Delta_r H^\ominus$

$$\begin{aligned}\Delta_f H^\ominus(UH_3, s) &= RT^2\frac{\mathrm{d}\ln K}{\mathrm{d}T} = RT^2\frac{\mathrm{d}}{\mathrm{d}T}\ln\left(\frac{p}{p^\ominus}\right)^{-3/2} = -\tfrac{3}{2}RT^2\frac{\mathrm{d}}{\mathrm{d}T}\ln\left(\frac{p}{p^\ominus}\right)\\ &= -\tfrac{3}{2}RT^2\frac{\mathrm{d}}{\mathrm{d}T}\left(\ln p - \ln p^\ominus\right) = -\tfrac{3}{2}RT^2\frac{\mathrm{d}}{\mathrm{d}T}\ln p\\ &= -\tfrac{3}{2}RT^2\frac{\mathrm{d}}{\mathrm{d}T}\left(A + \frac{B}{T} + C\ln T\right) = -\tfrac{3}{2}RT^2\left(-\frac{B}{T^2} + \frac{C}{T}\right)\\ &= \boxed{-\tfrac{3}{2}R(CT - B)}\end{aligned}$$

Heat capacity at constant pressure is defined by [2B.5–*48*], $C_p = (\partial H/\partial T)_p$, which implies that $\Delta_f C_p^\ominus = \mathrm{d}\Delta_f H^\ominus/\mathrm{d}T$ where the derivative is complete (not partial) because $\Delta_f H^\ominus$ does not depend on pressure. Therefore

$$\begin{aligned}\Delta_f C_p^\ominus &= \frac{\mathrm{d}}{\mathrm{d}T}\left(-\tfrac{3}{2}R[-B + CT]\right) = -\tfrac{3}{2}RC\\ &= -\tfrac{3}{2} \times (8.3145\,\mathrm{J\,K^{-1}\,mol^{-1}}) \times (-5.65) = \boxed{+70.5\,\mathrm{J\,K^{-1}\,mol^{-1}}}\end{aligned}$$

P6B.5 The van 't Hoff equation [6B.2–*214*] is:

$$\frac{\mathrm{d}\ln K}{\mathrm{d}T} = \frac{\Delta_r H^\ominus}{RT^2} \quad \text{which can also be written} \quad -\frac{\mathrm{d}\ln K}{\mathrm{d}(1/T)} = \frac{\Delta_r H^\ominus}{R}$$

The second form implies that a graph of $-\ln K$ against $1/T$ should be a straight line of slope $\Delta_r H^\ominus/R$. It is first necessary to relate K to α. To do this, the following table is drawn up for the $CO_2(g) \rightleftharpoons CO(g) + \tfrac{1}{2}O_2(g)$ equilibrium

	$CO_2(g)$	$\rightleftharpoons$ $CO(g)$	+ $\tfrac{1}{2}O_2(g)$
Initial amount	n	0	0
Change to reach equilibrium	$-\alpha n$	$+\alpha n$	$+\tfrac{1}{2}\alpha n$
Amount at equilibrium	$(1-\alpha)n$	αn	$\tfrac{1}{2}\alpha n$
Mole fraction, x_J	$\dfrac{1-\alpha}{1+\frac{1}{2}\alpha}$	$\dfrac{\alpha}{1+\frac{1}{2}\alpha}$	$\dfrac{\frac{1}{2}\alpha}{1+\frac{1}{2}\alpha}$
Partial pressure, p_J	$\dfrac{(1-\alpha)p}{1+\frac{1}{2}\alpha}$	$\dfrac{\alpha}{1+\frac{1}{2}\alpha}$	$\dfrac{\frac{1}{2}\alpha}{1+\frac{1}{2}\alpha}$

The total amount in moles is $n_{tot} = (1-\alpha)n + \alpha n + \frac{1}{2}\alpha n = (1+\frac{1}{2}\alpha)n$. This value is used to find the mole fractions. Treating all species as perfect gases (so that $a_J = p_J/p^\ominus$) the equilibrium constant is

$$K = \frac{a_{CO}a_{O_2}^{1/2}}{a_{CO_2}} = \frac{(p_{CO}/p^\ominus)(p_{O_2}/p^\ominus)^{1/2}}{(p_{CO_2}/p^\ominus)} = \frac{p_{CO}p_{O_2}^{1/2}}{p_{CO_2}p^{\ominus 1/2}} = \frac{\left(\frac{\alpha p}{1+\frac{1}{2}\alpha}\right)\left(\frac{\frac{1}{2}\alpha p}{1+\frac{1}{2}\alpha}\right)^{1/2}}{\left(\frac{(1-\alpha)p}{1+\frac{1}{2}\alpha}\right)p^\ominus}$$

$$= \frac{\left(\frac{1}{2}\alpha^3\right)^{1/2}}{(1-\alpha)(1+\frac{1}{2}\alpha)^{1/2}}\left(\frac{p}{p^\ominus}\right)^{1/2}$$

Using this expression, with $p = 1$ bar, K is calculated at each temperature and $-\ln K$ is plotted against $1/(T/\mathrm{K})$ as described above; the plot is shown in Fig. 6.2.

T/K	α	K	$10^4/(T/K)$	$-\ln K$
1 395	1.44×10^{-4}	1.222×10^{-6}	7.168	13.62
1 443	2.50×10^{-4}	2.794×10^{-6}	6.930	12.79
1 498	4.71×10^{-4}	7.224×10^{-6}	6.676	11.84

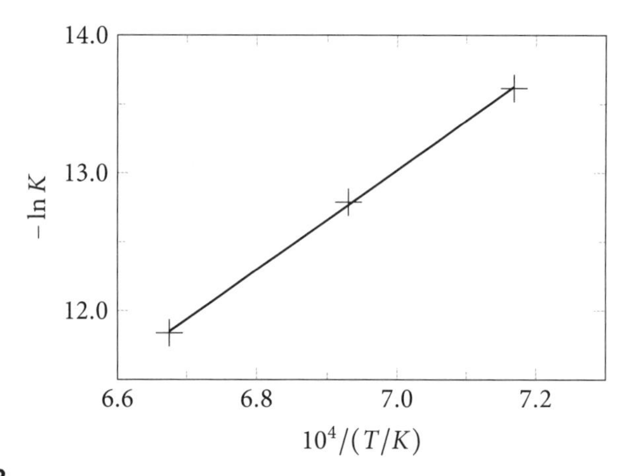

Figure 6.2

The data fall on a good straight line, the equation of which is

$$\ln K = 3.607 \times 10^4/(T/K) - 12.23$$

$\Delta_r H^\ominus/R$ is determined from the slope

$$\Delta_r H^\ominus = R \times (slope \times \mathrm{K}) = (8.3145\,\mathrm{J\,K^{-1}\,mol^{-1}}) \times (3.607 \times 10^4\,\mathrm{K})$$

$$= +2.99... \times 10^5\,\mathrm{J\,mol^{-1}} = \boxed{+3.00 \times 10^2\,\mathrm{kJ\,mol^{-1}}}$$

The equilibrium constant K has already been calculated; from the table above the value of K at 1443 K is $\boxed{2.79 \times 10^{-6}}$.

The standard reaction Gibbs energy is then calculated using $\Delta_r G^\ominus = -RT \ln K$, [6A.15–*208*], and the standard reaction entropy from $\Delta_r G^\ominus = \Delta_r H^\ominus - T\Delta_r S^\ominus$ [3D.9–*100*].

$$\Delta_r G^\ominus = -RT \ln K = -(8.3145\,\mathrm{J\,K^{-1}\,mol^{-1}}) \times (1443\,\mathrm{K}) \times \ln(2.79... \times 10^{-6})$$

$$= +1.53... \times 10^5\,\mathrm{J\,mol^{-1}} = \boxed{+153\,\mathrm{kJ\,mol^{-1}}}$$

$$\Delta_r S^\ominus = \frac{\Delta_r H^\ominus - \Delta_r G^\ominus}{T} = \frac{(2.99... \times 10^5\,\mathrm{J\,mol^{-1}}) - (1.53... \times 10^5\,\mathrm{J\,mol^{-1}})}{1443\,\mathrm{K}}$$

$$= \boxed{+102\,\mathrm{J\,K^{-1}\,mol^{-1}}}$$

P6B.7 The equilibrium is $2CH_3COOH(g) \rightleftharpoons (CH_3COOH)_2(g)$, the dimer being held together by hydrogen bonds. The following table is drawn up, assuming that the initial amount in moles of ethanoic acid is n and that at equilibrium a fraction α of the ethanoic acid has dimerised.

	$2CH_3COOH$ $\rightleftharpoons$	$(CH_3COOH)_2$
Initial amount	n	0
Change to reach equilibrium	$-\alpha n$	$+\frac{1}{2}\alpha n$
Amount at equilibrium	$(1-\alpha)n$	$\frac{1}{2}\alpha n$
Mole fraction, x_J	$\dfrac{1-\alpha}{1-\frac{1}{2}\alpha}$	$\dfrac{\frac{1}{2}\alpha}{1-\frac{1}{2}\alpha}$
Partial pressure, p_J	$\dfrac{(1-\alpha)p}{1-\frac{1}{2}\alpha}$	$\dfrac{\frac{1}{2}\alpha p}{1-\frac{1}{2}\alpha}$

The total amount in moles is $n_{tot} = (1-\alpha)n + \frac{1}{2}\alpha n = (1-\frac{1}{2}\alpha)n$. This value is used to find the mole fractions.

The total amount in moles present at equilibrium is found from the pressure by using the perfect gas law [1A.4–*8*], $pV = n_{tot}RT$

$$pV = (1-\tfrac{1}{2}\alpha)nRT \quad \text{hence} \quad \alpha = 2 - \frac{2pV}{nRT} = 2 - \frac{2pV}{(m/M)RT}$$

where $M = 60.0516\,\mathrm{g\,mol^{-1}}$ is the molar mass of ethanoic acid. The value of α is then used to calculate K. Assuming that both species present are perfect gases (so that $a_J = p_J/p^\ominus$) and using the expressions for p_J from the above table, the equilibrium constant is

$$K = \frac{a_{(CH_3COOH)_2}}{a^2_{CH_3COOH}} = \frac{(p_{(CH_3COOH)_2}/p^{\ominus})}{(p_{CH_3COOH}/p^{\ominus})^2} = \frac{p_{(CH_3COOH)_2}p^{\ominus}}{p^2_{CH_3COOH}} = \frac{\left(\frac{\frac{1}{2}\alpha p}{1-\frac{1}{2}\alpha}\right)p^{\ominus}}{\left(\frac{(1-\alpha)p}{1-\frac{1}{2}\alpha}\right)^2}$$

$$= \frac{\frac{1}{2}\alpha(1-\frac{1}{2}\alpha)p^{\ominus}}{(1-\alpha)^2 p}$$

The values of α and K at the two temperatures are then calculated using these formulae as

At 437 K

$$\alpha = 2 - \frac{2\times(101.9\times10^3\ \mathrm{Pa})\times(21.45\times10^{-6}\ \mathrm{m^3})}{(0.0519\ \mathrm{g}/60.0516\ \mathrm{g\,mol^{-1}})\times(8.3145\ \mathrm{J\,K^{-1}\,mol^{-1}})\times(437\ \mathrm{K})} = 0.607...$$

$$K = \frac{\frac{1}{2}\alpha(1-\frac{1}{2}\alpha)p^{\ominus}}{(1-\alpha)^2 p} = \frac{\frac{1}{2}\times(0.607...)\times(1-\frac{1}{2}\times0.607...)\times(100\ \mathrm{kPa})}{(1-0.607...)^2\times(101.9\ \mathrm{kPa})} = 1.35...$$

$$= \boxed{1.35}$$

At 471 K

$$\alpha = 2 - \frac{2\times(101.9\times10^3\ \mathrm{Pa})\times(21.45\times10^{-6}\ \mathrm{m^3})}{(0.0380\ \mathrm{g}/60.0516\ \mathrm{g\,mol^{-1}})\times(8.3145\ \mathrm{J\,K^{-1}\,mol^{-1}})\times(471\ \mathrm{K})} = 0.235...$$

$$K = \frac{\frac{1}{2}\alpha(1-\frac{1}{2}\alpha)p^{\ominus}}{(1-\alpha)^2 p} = \frac{\frac{1}{2}\times(0.235...)\times(1-\frac{1}{2}\times0.235...)\times(100\ \mathrm{kPa})}{(1-0.235...)^2\times(101.9\ \mathrm{kPa})} = 0.174...$$

$$= \boxed{0.175}$$

The standard enthalpy of the dimerization reaction is found using the temperature dependence of K [6B.4–*215*]

$$\ln K_2 - \ln K_1 = -\frac{\Delta_r H^{\ominus}}{R}\left(\frac{1}{T_2}-\frac{1}{T_1}\right) \quad \text{hence} \quad \Delta_r H^{\ominus} = -\frac{R\ln(K_2/K_1)}{(1/T_2)-(1/T_1)}$$

Taking $T_1 = 437$ K and $T_2 = 471$ K gives

$$\Delta_r H^{\ominus} = -\frac{R\ln(0.174.../1.35...)}{[1/(471\ \mathrm{K})]-[1/(437\ \mathrm{K})]} = \boxed{-103\ \mathrm{kJ\,mol^{-1}}}$$

P6B.9 The relationship between K and $\Delta_r G^{\ominus}$ [6A.15–*208*], $\Delta_r G^{\ominus} = -RT\ln K$, implies that

$$K = e^{-\Delta_r G^{\ominus}/RT} = e^{-(\Delta_r H^{\ominus} - T\Delta_r S^{\ominus})/RT} = e^{-\Delta_r H^{\ominus}/RT}e^{\Delta_r S^{\ominus}/R}$$

The ratio of the values of K that would be obtained using the lowest and highest values of $\Delta_r H^{\ominus}$ is

$$\frac{K_{\mathrm{lowH}}}{K_{\mathrm{highH}}} = \frac{e^{-\Delta_r H^{\ominus}_{\mathrm{low}}/RT}e^{\Delta_r S^{\ominus}/R}}{e^{-\Delta_r H^{\ominus}_{\mathrm{high}}/RT}e^{\Delta_r S^{\ominus}/R}} = \exp\left(-\frac{\Delta_r H^{\ominus}_{\mathrm{low}} - \Delta_r H^{\ominus}_{\mathrm{high}}}{RT}\right)$$

For the given data, the value of this factor is

At 298 K:

$$\frac{K_{\text{lowH}}}{K_{\text{highH}}} = \exp\left(-\frac{(243-289)\times 10^3\ \text{J mol}^{-1}}{(8.3145\ \text{J K}^{-1}\ \text{mol}^{-1})\times(298\ \text{K})}\right) = \boxed{1.2\times 10^8}$$

At 700 K:

$$\frac{K_{\text{lowH}}}{K_{\text{highH}}} = \exp\left(-\frac{(243-289)\times 10^3\ \text{J mol}^{-1}}{(8.3145\ \text{J K}^{-1}\ \text{mol}^{-1})\times(700\ \text{K})}\right) = \boxed{2.7\times 10^3}$$

P6B.11 The standard reaction Gibbs energy is related to the standard reaction enthalpy and entropy according to [3D.9–100], $\Delta_r G^\ominus = \Delta_r H^\ominus - T\Delta_r S^\ominus$. However, $\Delta_r H^\ominus$ and $\Delta_r S^\ominus$ themselves vary with temperature according to Kirchhoff's law [2C.7a–55] for $\Delta_r H^\ominus$ and the analogous equation [3C.5a–95] for $\Delta_r S^\ominus$

$$\Delta_r H^\ominus(T_2) = \Delta_r H^\ominus(T_1) + \int_{T_1}^{T_2} \Delta_r C_p^\ominus\,\mathrm{d}T$$

$$\Delta_r S^\ominus(T_2) = \Delta_r S^\ominus(T_1) + \int_{T_1}^{T_2} \frac{\Delta_r C_p^\ominus}{T}\,\mathrm{d}T$$

$\Delta_r C_p^\ominus$ is defined by [2C.7b–55], $\Delta_r C_p^\ominus = \sum_{\text{products}} \nu C_{p,\text{m}}^\ominus - \sum_{\text{reactants}} \nu C_{p,\text{m}}^\ominus$. Because each species has $C_{p,\text{m}}$ given by $C_{p,\text{m}} = a + bT + c/T^2$ it is convenient to write $\Delta_r C_p^\ominus = A + BT + C/T^2$ where A is defined by $A = \sum_{\text{products}} \nu a - \sum_{\text{reactants}} \nu a$ and similarly for B and C. Expressions for $\Delta_r H^\ominus$ and $\Delta_r S^\ominus$ at temperature T_2 are then obtained by performing the integrations

$$\begin{aligned}\Delta_r H^\ominus(T_2) &= \Delta_r H^\ominus(T_1) + \int_{T_1}^{T_2} A + BT + C/T^2\,\mathrm{d}T \\ &= \Delta_r H^\ominus(T_1) + A(T_2 - T_1) + \tfrac{1}{2}B\left(T_2^2 - T_1^2\right) - C\left(\frac{1}{T_2} - \frac{1}{T_1}\right)\end{aligned}$$

$$\begin{aligned}\Delta_r S^\ominus(T_2) &= \Delta_r S^\ominus(T_1) + \int_{T_1}^{T_2} \frac{A + BT + C/T^2}{T}\,\mathrm{d}T \\ &= \Delta_r S^\ominus(T_1) + \int_{T_1}^{T_2} \frac{A}{T} + B + \frac{C}{T^3}\,\mathrm{d}T \\ &= \Delta_r S^\ominus(T_1) + A\ln\frac{T_2}{T_1} + B(T_2 - T_1) - \tfrac{1}{2}C\left(\frac{1}{T_2^2} - \frac{1}{T_1^2}\right)\end{aligned}$$

The standard reaction Gibbs energy at temperature T_2 is then given by

$$\begin{aligned}\Delta_r G^\ominus(T_2) &= \Delta_r H^\ominus(T_2) - T_2\Delta_r S^\ominus(T_2)\\ &= \left[\Delta_r H^\ominus(T_1) + A(T_2-T_1) + \tfrac{1}{2}B\left(T_2^2-T_1^2\right) - C\left(\frac{1}{T_2}-\frac{1}{T_1}\right)\right]\\ &\quad - T_2\left[\Delta_r S^\ominus(T_1) + A\ln\frac{T_2}{T_1} + B(T_2-T_1) - \tfrac{1}{2}C\left(\frac{1}{T_2^2}-\frac{1}{T_1^2}\right)\right]\\ &= \Delta_r H^\ominus(T_1) - T_2\Delta_r S(T_1) + A\left[T_2 - T_1 - T_2\ln\left(\frac{T_2}{T_1}\right)\right]\\ &\quad + B\left[\tfrac{1}{2}\left(T_2^2-T_1^2\right) - T_2\left(T_2-T_1\right)\right]\\ &\quad + C\left[\frac{T_2}{2}\left(\frac{1}{T_2^2}-\frac{1}{T_1^2}\right) - \left(\frac{1}{T_2}\ \frac{1}{T_1}\right)\right]\end{aligned}$$

In order to obtain an expression that contains $\Delta_r G^\ominus(T_1)$, it is necessary to write the first part of the above expression as $\underbrace{\Delta_r H^\ominus(T_1) - T_1\Delta_r S^\ominus(T_1)}_{\Delta_r G^\ominus(T_1)} - (T_2 - T_1)\Delta_r S^\ominus(T_1)$ so that:

$$\begin{aligned}\Delta_r G^\ominus(T_2) &= \Delta_r G^\ominus(T_1) - (T_2-T_1)\Delta_r S^\ominus(T_1) + A\left[T_2 - T_1 - T_2\ln\left(\frac{T_2}{T_1}\right)\right]\\ &\quad + B\left[\tfrac{1}{2}\left(T_2^2-T_1^2\right) - T_2\left(T_2-T_1\right)\right]\\ &\quad + C\left[\frac{T_2}{2}\left(\frac{1}{T_2^2}-\frac{1}{T_1^2}\right) - \left(\frac{1}{T_2}-\frac{1}{T_1}\right)\right]\end{aligned}$$

The standard Gibbs energy for the formation of $H_2O(l)$ is $\Delta_r H^\ominus$ for the reaction $H_2(g) + \frac{1}{2}O_2(g) \rightarrow H_2O(l)$. From the data in the *Resource section*, at 298 K, $\Delta_f H^\ominus(H_2O,l) = -285.83\ \mathrm{kJ\ mol^{-1}}$ and $\Delta_f S^\ominus(H_2O,l)$ is calculated as

$$\begin{aligned}\Delta_f S^\ominus &= S_m^\ominus(H_2O,l) - S_m^\ominus(H_2,g) - \tfrac{1}{2}S_m^\ominus(O_2,g)\\ &= (69.91 - 130.684 - \tfrac{1}{2}\times 205.138)\ \mathrm{J\,K^{-1}\,mol^{-1}} = -163.343\ \mathrm{J\,K^{-1}\,mol^{-1}}\end{aligned}$$

The quantities A, B, and C are also calculated using the data from the *Resource section*:

$$\begin{aligned}A &= a_{H_2O,l} - a_{H_2,g} - \tfrac{1}{2}a_{O_2,g}\\ &= \left(75.29 - 27.28 - \tfrac{1}{2}\times 29.96\right)\ \mathrm{J\,K^{-1}\,mol^{-1}} = 33.03\ \mathrm{J\,K^{-1}\,mol^{-1}}\end{aligned}$$

$$\begin{aligned}B &= b_{H_2O,l} - b_{H_2,g} - \tfrac{1}{2}b_{O_2,g}\\ &= \left(0 - 3.26 - \tfrac{1}{2}\times 4.18\right)\times 10^{-3}\ \mathrm{J\,K^{-2}\,mol^{-1}} = -5.35\times 10^{-3}\ \mathrm{J\,K^{-2}\,mol^{-1}}\end{aligned}$$

$$\begin{aligned}C &= c_{H_2O,l} - c_{H_2,g} - \tfrac{1}{2}c_{O_2,g}\\ &= \left(0 - 0.50 - \tfrac{1}{2}\times(-1.67)\right)\times 10^{5}\ \mathrm{J\,K\,mol^{-1}} = 0.335\times 10^{5}\ \mathrm{J\,K\,mol^{-1}}\end{aligned}$$

Hence, using the expressions derived above with $T_1 = 298$ K and $T_2 = 372$ K:

$$\begin{aligned}\Delta_f H^\ominus(372\,\text{K}) &= \overbrace{(-285.83\times10^3\ \text{J mol}^{-1})}^{\Delta_f H^\ominus(298\,\text{K})}\\ &\quad+(33.03\ \text{J K}^{-1}\,\text{mol}^{-1})\times([372-298]\ \text{K})\\ &\quad+\tfrac{1}{2}(-5.35\times10^{-3}\ \text{J K}^{-2}\,\text{mol}^{-1})\times[(372\ \text{K})^2-(298\ \text{K})^2]\\ &\quad-(0.335\times10^5\ \text{J K mol}^{-1})\times\left(\frac{1}{372\ \text{K}}-\frac{1}{298\ \text{K}}\right)\\ &=-283.49...\ \text{kJ mol}^{-1}\end{aligned}$$

$$\begin{aligned}\Delta_f S^\ominus(372\,\text{K}) &= \overbrace{(-163.343\ \text{J K}^{-1}\,\text{mol}^{-1})}^{\Delta_f S^\ominus(298\,\text{K})}+(33.03\ \text{J K}^{-1}\,\text{mol}^{-1})\times\ln\left(\frac{372\ \text{K}}{298\ \text{K}}\right)\\ &\quad+(-5.35\times10^{-3}\ \text{J K}^{-2}\,\text{mol}^{-1})\times[(372\ \text{K})-(298\ \text{K})]\\ &\quad-\tfrac{1}{2}(0.335\times10^5\ \text{J K mol}^{-1})\times\left(\frac{1}{(372\ \text{K})^2}-\frac{1}{(298\ \text{K})^2}\right)\\ &=-156.34...\ \text{J K}^{-1}\,\text{mol}^{-1}\end{aligned}$$

$$\begin{aligned}\Delta_f G^\ominus(372\,\text{K}) &= \Delta_f H^\ominus(372\,\text{K})-(372\ \text{K})\times\Delta_f S^\ominus(372\,\text{K})\\ &=(-283.49...\times10^3\ \text{J mol}^{-1})\\ &\quad-(372\ \text{K})\times(-156.34...\ \text{J K}^{-1}\,\text{mol}^{-1})\\ &=\boxed{-225.34\ \text{kJ mol}^{-1}}\end{aligned}$$

This compares to -237.13 kJ mol^{-1} at 298 K (from the *Resource section*). Note that $\Delta_f H^\ominus$ and $\Delta_r S^\ominus$ do not change very much between 298 K and 372 K in this case. In fact, assuming that they are constant gives almost the same value of $\Delta_f G^\ominus(372\ \text{K})$, as is seen by calculating $\Delta_f G^\ominus$ at 372 K using the values of $\Delta_f H^\ominus$ and $\Delta_f S^\ominus$ at 298 K:

$$\begin{aligned}\Delta_f G^\ominus(372\,\text{K}) &\approx \Delta_f H^\ominus(298\,\text{K})+(372\ \text{K})\times\Delta_f S^\ominus(298\,\text{K})\\ &=(-285.83\times10^3\ \text{J mol}^{-1})-(372\ \text{K})\times(-163.343\ \text{J K}^{-1}\,\text{mol}^{-1})\\ &=-225.07\ \text{kJ mol}^{-1}\end{aligned}$$

which differs from the value obtained above by less than 0.3 kJ mol^{-1}.

6C Electrochemical cells

Answers to discussion questions

D6C.1 Any overall reaction that can be achieved by the combination of two half-cell reactions can in principle be used to generate a current. However, the reactions in the two half cells are necessarily redox reactions. In a concentration cell the left- and right-hand half cells only differ by concentrations of the species involved. Such a cell may generate a current, even though the cell reaction is essentially null. The reactions at the electrodes are still redox reactions.

D6C.3 The role of a salt bridge is to minimise the liquid junction potential which would otherwise occur as a result of the contact between the electrolytes in the two half cells. For a cell to generate a potential these solutions must be in electrical contact: the salt bridge achieves this without involving a physical contact between the two solutions.

D6C.5 When a current is being drawn from an electrochemical cell, the cell potential is altered by the formation of charge double layers at the surface of electrodes and by the formation of solution chemical potential gradients (concentration gradients). Resistive heating of the cell circuits may occur and junction potentials between dissimilar materials both external and external to the cell may change.

Solutions to exercises

E6C.1(a) (i) The reduction half-reactions for the cell $Zn(s)|ZnSO_4(aq)||AgNO_3|Ag(s)$, together with their standard electrode potentials from the *Resource section*, are

$$\begin{array}{lll} R: & Ag^+(aq) + e^- \rightarrow Ag(s) & E^{\ominus}(R) = +0.80\ V \\ L: & Zn^{2+}(aq) + 2e^- \rightarrow Zn(s) & E^{\ominus}(L) = -0.76\ V \end{array}$$

The cell reaction is obtained by subtracting the left-hand reduction half-reaction from the right-hand reduction half-reaction, after first multiplying the right-hand half-reaction by two so that the numbers of electrons in both half-reactions are the same

$$2Ag^+(aq) + Zn(s) \rightarrow 2Ag(s) + Zn^{2+}(aq)$$

The standard cell potential is calculated as the difference of the two standard electrode potentials, [6D.3–*224*], $E^{\ominus}_{cell} = E^{\ominus}(R) - E^{\ominus}(L)$

$$E^{\ominus}_{cell} = (+0.80\ V) - (-0.76\ V) = \boxed{+1.56\ V}$$

(ii) Following the same approach as part (i), and noting that the Pt(s) is an 'inert metal' that is only present to act as a source or sink of electrons, the half-reactions for the $Cd(s)|CdCl_2(aq)||HNO_3(aq)|H_2(g)|Pt(s)$ cell and their electrode potentials are

$$\begin{array}{lll} R: & 2H^+(aq) + 2e^- \rightarrow H_2(g) & E^{\ominus}(R) = 0\ \text{(by definition)} \\ L: & Cd^{2+}(aq) + 2e^- \rightarrow Cd(s) & E^{\ominus}(L) = -0.40\ V \end{array}$$

The cell reaction (R – L) is therefore

$$2H^+(aq) + Cd(s) \rightarrow H_2(g) + Cd^{2+}(aq)$$

and the standard cell potential is

$$E^{\ominus}_{cell} = 0 - (-0.40\ V) = \boxed{+0.40\ V}$$

(iii) For the $Pt(s)|K_3[Fe(CN)_6](aq),K_4[Fe(CN)_6](aq)||CrCl_3(aq)|Cr(s)$ cell the reduction half-reactions are:

$$\begin{array}{lll} \text{R:} & Cr^{3+}(aq) + 3e^- \rightarrow Cr(s) & E^\ominus(\text{R}) = -0.74\text{ V} \\ \text{L:} & [Fe(CN)_6]^{3-}(aq) + e^- \rightarrow [Fe(CN)_6]^{4-}(aq) & E^\ominus(\text{L}) = +0.36\text{ V} \end{array}$$

The cell reaction is obtained by subtracting the left-hand half-reaction from the right-hand half-reaction, after first multiplying the right-hand half reaction by three so that both half-reactions involve the same number of electrons.

$$Cr^{3+}(aq) + 3[Fe(CN)_6]^{4-}(aq) \rightarrow Cr(s) + 3[Fe(CN)_6]^{3-}(aq)$$

The standard cell potential is

$$E^\ominus_{\text{cell}} = (-0.74\text{ V}) - (+0.36\text{ V}) = \boxed{-1.10\text{ V}}$$

E6C.2(a) (i) The required reduction half-reactions are

$$\begin{array}{lll} \text{R:} & Cu^{2+}(aq) + 2e^- \rightarrow Cu(s) & E^\ominus(\text{R}) = +0.34\text{ V} \\ \text{L:} & Zn^{2+}(aq) + 2e^- \rightarrow Zn(s) & E^\ominus(\text{L}) = -0.76\text{ V} \end{array}$$

The cell reaction (R – L) generated from these reduction half-reactions is $Zn(s)+Cu^{2+}(aq) \rightarrow Zn^{2+}(aq)+Cu(s)$ which is equivalent to the required reaction. The cell required is

$$Zn(s)|ZnSO_4(aq)||CuSO_4(aq)|Cu(s)$$

and the standard cell potential is

$$E^\ominus_{\text{cell}} = E^\ominus(\text{R}) - E^\ominus(\text{L}) = (+0.34\text{ V}) - (-0.76\text{ V}) = \boxed{+1.10\text{ V}}$$

(ii) The required reduction half-reactions are:

$$\begin{array}{lll} \text{R:} & 2AgCl(s) + 2e^- \rightarrow 2Ag(s) + 2Cl^-(aq) & E^\ominus(\text{R}) = +0.22\text{ V} \\ \text{L:} & 2H^+(aq) + 2e^- \rightarrow H_2(g) & E^\ominus(\text{L}) = 0\text{ (by definition)} \end{array}$$

The cell reaction (R – L) generated from these reduction half-reactions is $2AgCl(s) + H_2(g) \rightarrow 2Ag(s) + 2Cl^-(aq) + 2H^+(aq)$ which is equivalent to the required reaction. The required cell is:

$$Pt(s)|H_2(g)|HCl(aq)|AgCl(s)|Ag(s)$$

The Pt(s) electrode is an 'inert metal' that acts as an electron source or sink. Note that there is no interface between the two half cells because both electrodes have a common electrolyte (HCl). The standard cell potential is

$$E^\ominus_{\text{cell}} = E^\ominus(\text{R}) - E^\ominus(\text{L}) = \boxed{+0.22\text{ V}}$$

(iii) The required reduction half reactions are

$$\begin{array}{lll} \text{R:} & O_2(g) + 4H^+(aq) + 4e^- \rightarrow 2H_2O(l) & E^\ominus(\text{R}) = +1.23\text{ V} \\ \text{L:} & 4H^+(aq) + 4e^- \rightarrow 2H_2(g) & E^\ominus(\text{L}) = 0\text{ (by definition)} \end{array}$$

The cell reaction (R – L) generated from these reduction half-reactions is the required reaction, $O_2(g) + 2H_2(g) \rightarrow 2H_2O(l)$. The required cell is

$$\mathrm{Pt(s)|H_2(g)|HCl(aq)|O_2(g)|Pt(s)}$$

As in part (ii) the platinum electrode is an 'inert metal' and there is no interface between the half cells because they have a common electrolyte. The standard cell potential is

$$E^{\ominus}_{\text{cell}} = E^{\ominus}(\text{R}) - E^{\ominus}(\text{L}) = \boxed{+1.23\ \text{V}}$$

An alternative combination of reduction half-reactions is

$$\begin{array}{lll} \text{R:} & O_2(g) + 2H_2O(l) + 4e^- \rightarrow 4OH^-(aq) & E^{\ominus}(\text{R}) = +0.40\ \text{V} \\ \text{L:} & 4H_2O(l) + 4e^- \rightarrow 4OH^-(aq) + 2H_2(g) & E^{\ominus}(\text{L}) = -0.83\ \text{V} \end{array}$$

which uses alkaline instead of acidic conditions. The cell required is

$$\mathrm{Pt(s)|H_2(g)|NaOH(aq)|O_2(g)|Pt(s)}$$

The overall cell reaction (R – L) is the same and so, therefore, is the standard cell potential:

$$E^{\ominus}_{\text{cell}} = (+0.40\ \text{V}) - (-0.83\text{V}) = +1.23\ \text{V}$$

E6C.3(a) The reduction half-reactions for the cell in question are:

$$\begin{array}{lll} \text{R:} & Cd^{2+}(aq) + 2e^- \rightarrow Cd(s) & E^{\ominus}(\text{R}) = -0.40\ \text{V} \\ \text{L:} & AgBr(s) + e^- \rightarrow Ag(s) + Br^-(aq) & E^{\ominus}(\text{L}) = +0.0713\ \text{V} \end{array}$$

The cell reaction is obtained by subtracting the left-hand half-reaction from the right-hand half-reaction, after first multiplying the left-hand half-reaction by two so that both half-reactions involve the same number of electrons.

$$Cd^{2+}(aq) + 2Ag(s) + 2Br^-(aq) \rightarrow Cd(s) + 2AgBr(s)$$

The cell potential is given by the Nernst equation [6C.4–*221*]

$$E_{\text{cell}} = E^{\ominus}_{\text{cell}} - (RT/\nu F)\ln Q$$

In this case $\nu = 2$ and the reaction quotient Q is

$$Q = \frac{\overbrace{a_{\text{Cd(s)}}}^{1}\ \overbrace{a^2_{\text{AgBr(s)}}}^{1}}{a_{\text{Cd}^{2+}\text{(aq)}}\ \underbrace{a^2_{\text{Ag(s)}}}_{1}\ a^2_{\text{Br}^-\text{(aq)}}} = \frac{1}{a_{\text{Cd}^{2+}\text{(s)}}a^2_{\text{Br}^-\text{(aq)}}}$$

where $a_J = 1$ for pure solids has been used. For ions in solution the activity is written as $a = \gamma_{\pm} b/b^{\ominus}$, where $\gamma_{\pm}$ is the mean activity coefficient, as established in Section 5F.4 on page 187. The Debye–Hückel limiting law [5F.27–*188*], which applies at low molalities, is

$$\log \gamma_\pm = -A|z_+z_-|I^{1/2}$$

where $A = 0.502$ for an aqueous solution at 298 K, z_+ and z_- are the charges on the ions, and I is the dimensionless ionic strength of the solution which for a solution containing two types of ion at molality b_+ and b_- is given by [5F.29–*188*], $I = \frac{1}{2}(b_+z_+^2 + b_-z_-^2)/b^\ominus$.

For the cell in question, the right-hand electrode contains a solution of $Cd(NO_3)_2$ of molality $b_R = 0.010\ \text{mol kg}^{-1}$. In this case $z_+ = +2$ (for Cd^{2+}), $z_- = -2$ (for NO_3^-), $b_+ = b_{Cd^{2+}} = b_R$ and $b_- = b_{NO_3^-} = 2b_R$. The ionic strength is

$$I_R = \tfrac{1}{2}\left(2^2 \times b_R + (-1)^2 \times (2b_R)\right)/b^\ominus = 3b_R/b^\ominus$$

and the mean activity coefficient for the right-hand electrode is therefore given by

$$\log \gamma_{\pm,R} = -A|z_+z_-|I_R^{1/2} = -A\left|(+2)(-1)\right|\left(\frac{3b_R}{b^\ominus}\right)^{\frac{1}{2}} = -2A\left(\frac{3b_R}{b^\ominus}\right)^{\frac{1}{2}}$$

$$= -2 \times 0.509 \times \left(\frac{3 \times (0.010\ \text{mol kg}^{-1})}{1\ \text{mol kg}^{-1}}\right)^{\frac{1}{2}} = -0.176...$$

Hence $\gamma_{\pm,R} = 10^{-0.176...} = 0.666...$, and so

$$a_{Cd^{2+}} = \gamma_{\pm,R}\frac{b_{Cd^{2+}}}{b^\ominus} = (0.666...) \times \frac{0.010\ \text{mol kg}^{-1}}{1\ \text{mol kg}^{-1}} = 6.66... \times 10^{-3}$$

In a similar way, the left-hand electrode contains a solution of KBr of molality $b_L = 0.050\ \text{mol kg}^{-1}$, so that $z_+ = +1$ (for K^+), $z_- = -1$ (for Br^-), and $b_+ = b_- = b_L$. It follows that

$$I_L = \tfrac{1}{2}(b_+z_+^2 + b_-z_-^2)/b^\ominus = \tfrac{1}{2}\left(1^2 \times b_L + (-1)^2 \times b_L\right)/b^\ominus = b_L/b^\ominus$$

and therefore

$$\log \gamma_{\pm,L} = -A|z_+z_-|I_L^{1/2} = -A \times \left|(+1) \times (-1)\right| \times \left(\frac{b_L}{b^\ominus}\right)^{\frac{1}{2}} = -A\left(\frac{b_L}{b^\ominus}\right)^{\frac{1}{2}}$$

$$= -0.509 \times \left(\frac{0.050\ \text{mol kg}^{-1}}{1\ \text{mol kg}^{-1}}\right)^{\frac{1}{2}} = -0.113...$$

Hence $\gamma_{\pm,L} = 10^{-0.113...} = 0.769...$, and so

$$a_{Br^-} = \gamma_{\pm,L}\frac{b_{Br^-}}{b^\ominus} = (0.769...) \times \frac{0.050\ \text{mol kg}^{-1}}{1\ \text{mol kg}^{-1}} = 0.0384...$$

Putting these activities into the Nernst equation for the cell with $v = 2$ and the expression for Q obtained above gives

$$E_{\text{cell}} = E^{\ominus}_{\text{cell}} - \frac{RT}{vF}\ln Q = \underbrace{[E^{\ominus}(\text{R}) - E^{\ominus}(\text{L})]}_{E^{\ominus}_{\text{cell}}} - \frac{RT}{2F}\ln\left(\frac{1}{a_{\text{Cd}^{2+}}a^2_{\text{Br}^-}}\right)$$

$$= [(-0.40\text{ V}) - (+0.0713\text{ V})] - \frac{(8.3145\text{ J K}^{-1}\text{ mol}^{-1}) \times (298\text{ K})}{2 \times (96485\text{ C mol}^{-1})}$$

$$\times \ln\left(\frac{1}{(6.66... \times 10^{-3}) \times (0.0384...)^2}\right) = \boxed{-0.619\text{ V}}$$

E6C.4(a) The reduction half-reactions for the cell in question are

$$\begin{aligned} &\text{R:} \quad \text{Cu}^{2+}(\text{aq}) + 2\text{e}^- \rightarrow \text{Cu(s)} \\ &\text{L:} \quad \text{Zn}^{2+}(\text{aq}) + 2\text{e}^- \rightarrow \text{Zn(s)} \end{aligned}$$

which reveal that $v = 2$ for the given cell reaction. The relationship between $\Delta_r G^{\ominus}$ and $E^{\ominus}_{\text{cell}}$ is given by [6C.2–*217*], $\Delta_r G^{\ominus} = -vFE^{\ominus}_{\text{cell}}$

$$\Delta_r G^{\ominus} = -2 \times (96485\text{ C mol}^{-1}) \times (+1.10\text{ V}) = \boxed{-212\text{ kJ mol}^{-1}}$$

where 1 C V = 1 J is used.

E6C.5(a) The Nernst equation [6C.4–*221*] is $E_{\text{cell}} = E^{\ominus}_{\text{cell}} - (RT/vF)\ln Q$. If Q changes from Q_1 to Q_2 then the change in cell potential is given by

$$E_{\text{cell},1} - E_{\text{cell},2} = \left[E^{\ominus}_{\text{cell}} - \frac{RT}{vF}\ln Q_2\right] - \left[E^{\ominus}_{\text{cell}} - \frac{RT}{vF}\ln Q_1\right] = -\frac{RT}{vF}\ln\left(\frac{Q_2}{Q_1}\right)$$

For $v = 2$ and $Q_2/Q_1 = 1/10$ the change in cell potential is

$$E_{\text{cell},1} - E_{\text{cell},2} = -\frac{(8.3145\text{ J K}^{-1}\text{ mol}^{-1}) \times (298\text{ K})}{2 \times (96485\text{ C mol}^{-1})} \times \ln\tfrac{1}{10} = \boxed{+0.030\text{ V}}$$

where $1\text{ J C}^{-1} = 1\text{ V}$ is used.

Solutions to problems

P6C.1 (a) The reaction of hydrogen and oxygen, $2\text{H}_2(\text{g}) + \text{O}_2(\text{g}) \rightarrow 2\text{H}_2\text{O(l)}$, can be broken down into the reduction half-reactions

$$\begin{aligned} &\text{R:} \quad \text{O}_2(\text{g}) + 4\text{H}^+(\text{aq}) + 4\text{e}^- \rightarrow 2\text{H}_2\text{O(l)} \quad && E^{\ominus}(\text{R}) = +1.23\text{ V} \\ &\text{L:} \quad 4\text{H}^+(\text{aq}) + 4\text{e}^- \rightarrow 2\text{H}_2(\text{g}) && E^{\ominus}(\text{L}) = 0 \text{ (by definition)} \end{aligned}$$

The standard cell potential is given by

$$E^{\ominus}_{\text{cell}} = E^{\ominus}(\text{R}) - E^{\ominus}(\text{L}) = \boxed{+1.23\text{ V}}$$

(b) The balanced chemical equation for the combustion of butane, $C_4H_{10}(g) + \frac{13}{2}O_2(g) \rightarrow 4CO_2(g) + 5H_2O(g)$ can be broken down into the reduction half-reactions

$$\begin{aligned}
&\text{R:}\quad \tfrac{13}{2}O_2(g) + 26H^+(aq) + 26e^- \rightarrow 13H_2O(l)\\
&\text{L:}\quad 4CO_2(g) + 26H^+(aq) + 26e^- \rightarrow C_4H_{10}(g) + 8H_2O(l)
\end{aligned}$$

The standard electrode potential for the left-hand reduction half-reaction is not in the *Resource section*, so $E^\ominus_{\text{cell}}$ cannot be calculated from $E^\ominus(\text{R}) - E^\ominus(\text{L})$ as in part (a). Instead $\Delta_r G^\ominus$ is calculated for the cell reaction by first using standard Gibbs energies of formation and then using $E^\ominus_{\text{cell}} = -\Delta_r G^\ominus/\nu F$ [6C.3–*221*] to calculate $E^\ominus_{\text{cell}}$. Note from the above half-reactions that $\nu = 26$.

$$\begin{aligned}
\Delta_r G^\ominus &= 4\Delta_f G^\ominus(CO_2, g) + 5\Delta_f G^\ominus(H_2O, l) - \Delta_f G^\ominus(C_4H_{10}, g)\\
&= 4 \times (-394.36\ \text{kJ mol}^{-1}) + 5 \times (-237.13\ \text{kJ mol}^{-1})\\
&\quad - (-17.03\ \text{kJ mol}^{-1}) = -2746.06\ \text{kJ mol}^{-1}
\end{aligned}$$

Hence

$$E^\ominus_{\text{cell}} = -\frac{\Delta_r G^\ominus}{\nu F} = -\frac{-2746.06 \times 10^3\ \text{J mol}^{-1}}{26 \times (96485\ \text{C mol}^{-1})} = \boxed{+1.09\ \text{V}}$$

P6C.3 The reduction half-reactions for the cell are

$$\begin{aligned}
&\text{R:}\quad Q(aq) + 2H^+(aq) + 2e^- \rightarrow QH_2(aq) \qquad E^\ominus(\text{R}) = +0.6994\ \text{V}\\
&\text{L:}\quad Hg_2Cl_2(s) + 2e^- \rightarrow 2Hg(l) + 2Cl^-(aq) \quad E^\ominus(\text{L}) = +0.27\ \text{V}
\end{aligned}$$

for which $\nu = 2$. The value of $E^\ominus(\text{L})$ is taken from the *Resource section*. The cell reaction (R – L) is

$$\begin{aligned}
&Q(aq) + 2H^+(aq) + 2Hg(l) + 2Cl^-(aq) \rightarrow QH_2(aq) + Hg_2Cl_2(s)\\
&E^\ominus_{\text{cell}} = E^\ominus(\text{R}) - E^\ominus(\text{L}) = (+0.6994\ \text{V}) - (+0.27\ \text{V}) = +0.4294\ \text{V}
\end{aligned}$$

Noting that $\nu = 2$ and that $a_J = 1$ for pure solids and liquids, the Nernst equation is

$$E_{\text{cell}} = E^\ominus_{\text{cell}} - \frac{RT}{2F}\ln\left(\frac{a_{QH_2}}{a_Q a^2_{H^+} a^2_{Cl^-}}\right)$$

Taking $a_{QH_2} = a_Q$, because Q and QH_2 are present at the same concentration, and $a_{H^+} = a_{Cl^-}$ gives:

$$\begin{aligned}
E_{\text{cell}} &= E^\ominus_{\text{cell}} - \frac{RT}{2F}\ln\left(\frac{1}{a^4_{H^+}}\right) = E^\ominus_{\text{cell}} + \frac{2RT}{F}\ln a_{H^+} = E^\ominus_{\text{cell}} + \frac{2RT}{F}\ln 10 \times \log a_{H^+}\\
&= E^\ominus_{\text{cell}} - \frac{2RT\ln 10}{F} \times \text{pH}
\end{aligned}$$

where $\mathrm{pH} = -\log a_{\mathrm{H}^+}$ and $\ln x = \ln 10 \times \log x$ (from inside the front cover) have been used. Rearranging for pH gives:

$$\mathrm{pH} = \frac{F}{2RT\ln 10}\left(E^{\ominus}_{\mathrm{cell}} - E_{\mathrm{cell}}\right)$$
$$= \frac{96485\,\mathrm{C\,mol^{-1}}}{2 \times (8.3145\,\mathrm{J\,K^{-1}\,mol^{-1}}) \times (298\,\mathrm{K}) \times \ln 10} \times [(+0.4294\,\mathrm{V}) - (+0.190\,\mathrm{V})] = \boxed{2.0}$$

6D Electrode potentials

Answer to discussion questions

D6D.1 This is described in Section 6D.1(a) on page 225.

Solutions to exercises

E6D.1(a) (i) The following electrodes are combined

$$\begin{array}{lll} \mathrm{R:} & \mathrm{Sn^{4+}(aq) + 2e^- \rightarrow Sn^{2+}(aq)} & E^{\ominus}(\mathrm{R}) = +0.15\,\mathrm{V} \\ \mathrm{L:} & \mathrm{Sn^{2+}(aq) + 2e^- \rightarrow Sn(s)} & E^{\ominus}(\mathrm{L}) = -0.14\,\mathrm{V} \end{array}$$

The overall cell reaction (R–L) is therefore $\mathrm{Sn^{4+}(aq)+Sn(s) \rightarrow 2Sn^{2+}(aq)}$, which is the required reaction, and has $\nu = 2$. The standard cell potential is given by [6D.3–*224*], $E^{\ominus}_{\mathrm{cell}} = E^{\ominus}(\mathrm{R}) - E^{\ominus}(\mathrm{L})$

$$E^{\ominus}_{\mathrm{cell}} = (+0.15\,\mathrm{V}) - (-0.14\,\mathrm{V}) = +0.29\mathrm{V}$$

The relationship between the equilibrium constant and the standard cell potential is given by [6C.5–*221*], $E^{\ominus}_{\mathrm{cell}} = (RT/\nu F)\ln K$. Rearranging gives

$$\ln K = \frac{\nu F}{RT}E^{\ominus}_{\mathrm{cell}} = \frac{2 \times (96485\,\mathrm{C\,mol^{-1}})}{(8.3145\,\mathrm{J\,K^{-1}\,mol^{-1}}) \times (298\,\mathrm{K})} \times (+0.29\,\mathrm{V}) = 22.5...$$

where $1\,\mathrm{V} = 1\,\mathrm{J\,C^{-1}}$ is used. Hence $K = \boxed{6.4 \times 10^9}$.

(ii) The following electrodes are combined

$$\begin{array}{lll} \mathrm{R:} & \mathrm{2AgCl(s) + 2e^- \rightarrow 2Ag(s) + 2Cl^-(aq)} & E^{\ominus}(\mathrm{R}) = +0.22\,\mathrm{V} \\ \mathrm{L:} & \mathrm{Sn^{2+}(aq) + 2e^- \rightarrow Sn(s)} & E^{\ominus}(\mathrm{L}) = -0.14\,\mathrm{V} \end{array}$$

The cell reaction is $\mathrm{2AgCl(s) + Sn(s) \rightarrow 2Ag(s) + 2Cl^-(aq) + Sn^{2+}(aq)}$ which is equivalent to the required reaction, and has $\nu = 2$. Therefore, using the same equations as in part (i)

$$E^{\ominus}_{\mathrm{cell}} = E^{\ominus}(\mathrm{R}) - E^{\ominus}(\mathrm{L}) = (+0.22\,\mathrm{V}) - (-0.14\,\mathrm{V}) = +0.36\,\mathrm{V}$$

$$\ln K = \frac{\nu F}{RT}E^{\ominus}_{\mathrm{cell}} = \frac{2 \times (96485\,\mathrm{C\,mol^{-1}})}{(8.3145\,\mathrm{J\,K^{-1}\,mol^{-1}}) \times (298\,\mathrm{K})} \times (+0.36\,\mathrm{V}) = 28.0...$$

Hence $K = \boxed{1.5 \times 10^{12}}$

E6D.2(a) The reduction half-reactions for the given cell are

$$\text{R:}\quad \mathrm{Ag^+(aq) + e^- \rightarrow Ag(s)}$$
$$\text{L:}\quad \mathrm{AgI(s) + e^- \rightarrow Ag(s) + I^-(aq)}$$

The cell reaction (R – L) is $\mathrm{Ag^+(aq) + I^-(aq) \rightarrow AgI(s)}$, with $\nu = 1$. The equilibrium constant for this reaction is calculated from $E^\ominus_{\text{cell}}$ using [6C.5–*221*], $E^\ominus_{\text{cell}} = (RT/\nu F)\ln K$. Rearranging gives

$$\ln K = \frac{\nu F}{RT}E^\ominus_{\text{cell}} = \frac{1 \times (96485\,\mathrm{C\,mol^{-1}})}{(8.3145\,\mathrm{J\,K^{-1}\,mol^{-1}}) \times (298.15\,\mathrm{K})} \times (0.9509\,\mathrm{V}) = 37.0...$$

where $1\,\mathrm{V} = 1\,\mathrm{J\,C^{-1}}$ is used. Hence $K = 1.18... \times 10^{16}$.

The dissolution reaction, $\mathrm{AgI(s) \rightarrow Ag^+(aq) + I^-(aq)}$, corresponds to the reverse of the cell reaction as written above. The required equilibrium constant is therefore the reciprocal of the one just calculated

$$K_{\text{diss}} = \frac{1}{1.18... \times 10^{16}} = \boxed{8.445 \times 10^{-17}}$$

E6D.3(a) (i) The reduction half-reactions for the specified cell and their corresponding electrode potentials from the *Resource section* are

$$\text{R:}\quad \mathrm{Cu^{2+}(aq) + 2e^- \rightarrow Cu(s)} \qquad E^\ominus(\text{R}) = +0.34\,\mathrm{V}$$
$$\text{L:}\quad \mathrm{2Ag^+(aq) + 2e^- \rightarrow 2Ag(s)} \qquad E^\ominus(\text{L}) = +0.80\,\mathrm{V}$$

The overall cell reaction is

$$\mathrm{Cu^{2+}(aq) + 2Ag(s) \rightarrow Cu(s) + 2Ag^+(aq)} \qquad \nu = 2$$

The standard cell potential is

$$E^\ominus_{\text{cell}} = E^\ominus(\text{R}) - E^\ominus(\text{L}) = (+0.34\,\mathrm{V}) - (+0.80\,\mathrm{V}) = -0.46\,\mathrm{V}$$

The reaction Gibbs energy is related to the cell potential according to [6C.3–*221*], $\Delta_r G^\ominus = -\nu F E^\ominus_{\text{cell}}$. Therefore

$$\begin{aligned}\Delta_r G^\ominus &= -\nu F E^\ominus_{\text{cell}} = -2 \times (96485\,\mathrm{C\,mol^{-1}}) \times (-0.46\,\mathrm{V}) = 88.7...\,\mathrm{kJ\,mol^{-1}}\\ &= \boxed{+89\,\mathrm{kJ\,mol^{-1}}}\end{aligned}$$

The standard reaction enthalpy is calculated using standard enthalpies of formation from the *Resource section*, noting that elements in their reference states have $\Delta_f H^\ominus = 0$.

$$\begin{aligned}\Delta_r H^\ominus &= 2\Delta_f H^\ominus(\mathrm{Ag^+, aq}) - \Delta_f H^\ominus(\mathrm{Cu^{2+}, aq})\\ &= 2 \times (+105.58\,\mathrm{kJ\,mol^{-1}}) - (+64.77\,\mathrm{kJ\,mol^{-1}}) = \boxed{+146.39\,\mathrm{kJ\,mol^{-1}}}\end{aligned}$$

(ii) The standard entropy change of reaction is obtained from $\Delta_r G^\ominus$ and $\Delta_r H^\ominus$ using [3D.9–*100*], $\Delta_r G^\ominus = \Delta_r H^\ominus - T\Delta_r S^\ominus$. Rearranging for $\Delta_r S^\ominus$ gives

$$\Delta_r S^\ominus = \frac{\Delta_r H^\ominus - \Delta_r G^\ominus}{T} = \frac{(146.39\times10^3\ \mathrm{J\,mol^{-1}}) - (88.7...\times10^3\ \mathrm{J\,mol^{-1}})}{298\ \mathrm{K}}$$
$$= +1.93...\times10^2\ \mathrm{J\,K^{-1}\,mol^{-1}}$$

The value of $\Delta_r G^\ominus$ at 308 K is then calculated using [3D.9–*100*], $\Delta_r G^\ominus = \Delta_r H^\ominus - T\Delta_r S^\ominus$, assuming that $\Delta_r H^\ominus$ and $\Delta_r S^\ominus$ do not vary significantly with temperature over this range

$$\Delta_r G^\ominus = (+146.39\times10^3\ \mathrm{J\,mol^{-1}}) - (308\ \mathrm{K})\times(+1.93...\times10^2\ \mathrm{J\,K^{-1}\,mol^{-1}})$$
$$= \boxed{+87\ \mathrm{kJ\,mol^{-1}}}$$

E6D.4(a) Assuming that the mercury forms $Hg_2SO_4(s)$ in the reaction, the required reduction half-equations and the corresponding standard electrode potentials are

R: $Zn^{2+}(aq) + 2e^- \rightarrow Zn(s)$ $\quad E^\ominus(R) = -0.76\ \mathrm{V}$
L: $Hg_2SO_4(s) + 2e^- \rightarrow 2Hg(l) + SO_4^{2-}(aq)$ $\quad E^\ominus(L) = +0.62\ \mathrm{V}$

The cell reaction is $Zn^{2+}(aq) + SO_4^{2-}(aq) + 2Hg(l) \rightarrow Zn(s) + Hg_2SO_4(s)$, and the standard cell potential is

$$E^\ominus_{cell} = E^\ominus(R) - E^\ominus(L) = (-0.76\ \mathrm{V}) - (+0.62\ \mathrm{V}) = -1.38\ \mathrm{V}$$

The negative value of $E^\ominus_{cell}$ indicates that the cell reaction as written will not be spontaneous. This means that $\boxed{\text{no}}$, mercury cannot produce zinc metal from aqueous zinc sulfate under standard conditions.

Solutions to problems

P6D.1 The given reaction can be broken down into the following reduction half equations

R: $2Fe^{3+}(aq) + 2e^- \rightarrow 2Fe^{2+}(aq)$
L: $Ag_2CrO_4(s) + 2e^- \rightarrow 2Ag(s) + CrO_4^{2-}(s)$

where the K^+ and Cl^- spectator ions have been ignored. These half-equations show that $\nu = 2$ for the given reaction.

(a) The standard potential is calculated from the standard reaction Gibbs energy using [6C.3–*221*], $E^\ominus_{cell} = -\Delta_r G^\ominus/\nu F$.

$$E^\ominus_{cell} = -\frac{-\Delta_r G^\ominus}{\nu F} = -\frac{-62.5\times10^3\ \mathrm{J\,mol^{-1}}}{2\times(96485\ \mathrm{C\,mol^{-1}})} = 0.323...\ \mathrm{V} = \boxed{+0.324\ \mathrm{V}}$$

(b) The standard potential of the $Ag_2CrO_4/Ag,CrO_4^{2-}$ couple, equal to $E^\ominus(L)$ of the cell considered above, is calculated from $E^\ominus_{cell}$ and the known value of $E^\ominus(R)$ using [6D.3–*224*], $E^\ominus_{cell} = E^\ominus(R) - E^\ominus(L)$. The value of $E^\ominus(R)$, in this case $E^\ominus(Fe^{3+}/Fe^{2+})$, is taken from the *Resource section.*

$$E^\ominus(L) = E^\ominus_{cell} - E^\ominus(R) = (+0.77\,V) - (+0.323...\,V) = \boxed{+0.45\,V}$$

P6D.3 (a) The equilibrium $HCO_3^-(aq) \rightleftharpoons CO_3^{2-}(aq) + H^+(aq)$ is broken down into the following reduction half-reactions

$$\begin{array}{ll} \text{R:} & HCO_3^-(aq) + e^- \rightarrow \tfrac{1}{2}H_2(g) + CO_3^{2-}(aq) \\ \text{L:} & H^+(aq) + e^- \rightarrow \tfrac{1}{2}H_2(g) \end{array}$$

The standard cell potential for this cell is given by

$$E^\ominus_{cell} = E^\ominus(R) - E^\ominus(L) = E^\ominus(HCO_3^-/CO_3^{2-}, H_2) - E^\ominus(H^+/H_2)$$

The standard electrode potential of the H^+/H_2 electrode is zero by definition, so it follows that $E^\ominus(HCO_3^-/CO_3^{2-}, H_2) = E^\ominus_{cell}$. The value of $E^\ominus_{cell}$ is calculated using [6C.3–*221*], $E^\ominus_{cell} = -\Delta_r G^\ominus/\nu F$, noting that $\nu = 1$. The value of $\Delta_r G^\ominus$ is calculated using the data in the question, noting from Section 3D.2(a) on page 101 that $\Delta_f G^\ominus(H^+, aq) = 0$.

$$\begin{aligned} \Delta_r G^\ominus &= \Delta_f G^\ominus(CO_3^{2-}, aq) - \Delta_f G^\ominus(HCO_3^-, aq) \\ &= (-527.81\,kJ\,mol^{-1}) - (-586.77\,kJ\,mol^{-1}) = +58.96\,kJ\,mol^{-1} \end{aligned}$$

Hence

$$E^\ominus_{cell} = -\frac{\Delta_r G^\ominus}{\nu F} = -\frac{58.96 \times 10^3\,J\,mol^{-1}}{1 \times (96485\,C\,mol^{-1})} = \boxed{-0.6111\,V}$$

As shown above, this is equal to $E^\ominus(HCO_3^-/CO_3^{2-}, H_2)$.

(b) The reaction $Na_2CO_3(aq) + H_2O(l) \rightarrow NaHCO_3(aq) + NaOH(aq)$ is broken down into the following reduction half-equations, in which the Na^+ counterions are ignored because they play no part in the reaction. The value of $E^\ominus(L)$ is as calculated in part (a), and $E^\ominus(R)$ is taken from the *Resource section.*

$$\begin{array}{lll} \text{R:} & H_2O(l) + e^- \rightarrow \tfrac{1}{2}H_2(g) + OH^-(aq) & E^\ominus(R) = -0.83\,V \\ \text{L:} & HCO_3^-(aq) + e^- \rightarrow \tfrac{1}{2}H_2(g) + CO_3^{2-}(aq) & E^\ominus(L) = -0.611...\,V \end{array}$$

The standard cell potential is given by

$$E^\ominus_{cell} = E^\ominus(R) - E^\ominus(L) = (-0.83\,V) - (-0.611...\,V) = -0.218...\,V = \boxed{-0.22\,V}$$

(c) The cell reaction for the cell considered in part (b) is

$$CO_3^{2-}(aq) + H_2O(l) \rightarrow HCO_3^-(aq) + OH^-(aq) \qquad \nu = 1$$

It is assumed that $a_{H_2O} = 1$ because solvent water is close to being in its standard state. Therefore the Nernst equation is

$$E_{cell} = E^\ominus_{cell} - \frac{RT}{F}\ln\left(\frac{a_{HCO_3^-}a_{OH^-}}{a_{CO_3^{2-}}}\right)$$

(d) The standard cell potential corresponds to all species involved in the cell reaction, which includes OH^-, being present at unit activity. This means that the pH will need to be approximately 14, in order to give $a_{OH^-} = 1$. At pH 7.0, the concentration of OH^- will be lower than at pH 14, which will mean that the cell reaction as written above will have a greater tendency to move in the forward direction. As a result E_{cell} is predicted to be larger at pH 7.0 than when $a_{OH^-} = 1$.

Assuming that the activities of all other species remain the same, the change in cell potential on going from $a_{OH^-} = 1$ to pH 7 is

$$\Delta E_{cell} = \underbrace{\left[E^{\ominus}_{cell} - \frac{RT}{F}\ln\left(\frac{a_{HCO_3^-}a_{OH^-}}{a_{CO_3^{2-}}}\right)\right]}_{E_{cell}\text{ at pH}=7} - \underbrace{\left[E^{\ominus}_{cell} - \frac{RT}{F}\ln\left(\frac{a_{HCO_3^-}\times 1}{a_{CO_3^{2-}}}\right)\right]}_{E_{cell}\text{ at }a_{OH^-}=1}$$

$$= -\frac{RT}{F}\ln a_{OH^-} = -\frac{RT\ln 10}{F}\log a_{OH^-}$$

where $\ln x = \ln 10 \times \log x$ from inside the front cover is used. To relate a_{OH^-} to the pH, use the relation $K_w = a_{H^+}a_{OH^-}$ so that

$$a_{OH^-} = \frac{K_w}{a_{H^+}} \quad \text{hence} \quad \log a_{OH^-} = \log K_w - \log a_{H^+} = -pK_w + pH$$

where $pH = -\log a_{H^+}$ and $pK_w = -\log K_w$. Taking $K_w = 1.00 \times 10^{-14}$, or $pK_w = 14.0$, the change in cell potential when the pH is changed to 7 is therefore

$$\Delta E_{cell} = -\frac{RT\ln 10}{F}(pH - pK_w)$$

$$= -\frac{(8.3145\,J\,K^{-1}\,mol^{-1}) \times (298\,K) \times \ln 10}{(96485\,C\,mol^{-1})} \times (-14.0 + 7.0)$$

$$= \boxed{+0.4139\,V}$$

Therefore the cell potential has increased on going from $a_{OH^-} = 1$ to pH 7.

P6D.5 The relationship between $\Delta_r S^{\ominus}$ and $E^{\ominus}_{cell}$ is given by [6C.6–222], $dE^{\ominus}_{cell}/dT = \Delta_r S^{\ominus}/\nu F$. If it is assumed that $\Delta_r S^{\ominus}$ is independent of temperature over the range of interest, integration of $dE^{\ominus}_{cell} = (\Delta_r S^{\ominus}/\nu F)dT$ between T_1 and T_2 gives

$$E^{\ominus}_{cell}(T_2) - E^{\ominus}_{cell}(T_1) = -\frac{\Delta_r S^{\ominus}}{\nu F}(T_2 - T_1)$$

where $E^{\ominus}_{cell}(T)$ is the potential at temperature T. This equation is conveniently written as $\Delta_r S^{\ominus} = \nu F \Delta E^{\ominus}_{cell}/\Delta T$.

$$\Delta_r S^{\ominus} = \nu F \times \frac{\Delta E^{\ominus}_{cell}}{\Delta T}$$

$$= 4\times(96485\,C\,mol^{-1})\times\frac{(+1.2251\,V)-(+1.2335\,V)}{(303\,K)-(293\,K)} = \boxed{-324\,J\,K^{-1}\,mol^{-1}}$$

The standard reaction enthalpy is then calculated from [3D.9–*100*], $\Delta_r G^\ominus = \Delta_r H^\ominus - T\Delta_r S^\ominus$, with $\Delta_r G^\ominus$ being given by [6C.3–*221*], $\Delta_r G^\ominus = -\nu F E^\ominus_{cell}$.

$$\begin{aligned}\Delta_r H^\ominus &= \Delta_r G^\ominus + T\Delta_r S^\ominus = -\nu F E^\ominus_{cell} + T\Delta_r S^\ominus \\ &= -4 \times (96485\,\text{C mol}^{-1}) \times (+1.2335\,\text{V}) \\ &\quad + (293\,\text{K}) \times (-3.24... \times 10^2\,\text{J K}^{-1}\,\text{mol}^{-1}) = \boxed{-571\,\text{kJ mol}^{-1}}\end{aligned}$$

In calculating $\Delta_r H^\ominus$, the value of $E^\ominus_{cell}$ at 293 K has been used. However, because $\Delta_r S^\ominus$ has been assumed to be constant over the temperature range, the data at 303 K will give the same value for $\Delta_r H^\ominus$.

Solutions to integrated activities

I6.1 The relationship between the equilibrium constant and $\Delta_r G^\ominus$ is given by [6A.15–*208*], $\Delta_r G^\ominus = -RT\ln K$. Combining this with [3D.9–*100*], $\Delta_r G^\ominus = \Delta_r H^\ominus - T\Delta_r S^\ominus$ and rearranging for $\Delta_r H^\ominus$ gives

$$\Delta_r H^\ominus = \Delta_r G^\ominus + T\Delta_r S^\ominus = -RT\ln K + T\Delta_r S^\ominus$$

The standard reaction enthalpy for the reaction $Cl_2O(g) + H_2O(g) \rightarrow 2HOCl(g)$ is therefore

$$\begin{aligned}\Delta_r H^\ominus &= -(8.3145\,\text{J K}^{-1}\,\text{mol}^{-1}) \times (298\,\text{K}) \times \ln\left(8.2 \times 10^{-2}\right) \\ &\quad + (298\,\text{K}) \times (+16.38\,\text{J K}^{-1}\,\text{mol}^{-1}) = +11.0...\,\text{kJ mol}^{-1}\end{aligned}$$

The standard reaction enthalpy is also expressed in terms of standard enthalpies of formation:

$$\Delta_r H^\ominus = 2\Delta_f H^\ominus(\text{HOCl, g}) - \Delta_f H^\ominus(\text{Cl}_2\text{O, g}) - \Delta_f H^\ominus(\text{H}_2\text{O, g})$$

Hence, using the given value for $\Delta_f H^\ominus(\text{Cl}_2\text{O, g})$ and taking $\Delta_f H^\ominus(\text{H}_2\text{O, g})$ from the *Resource section*, the standard enthalpy of formation of HOCl(g) is

$$\begin{aligned}\Delta_f H^\ominus(\text{HOCl, g}) &= \tfrac{1}{2}\left[\Delta_r H^\ominus + \Delta_f H^\ominus(\text{Cl}_2\text{O, g}) + \Delta_f H^\ominus(\text{H}_2\text{O, g})\right] \\ &= \tfrac{1}{2}\left[(+11.0...\,\text{kJ mol}^{-1}) + (+77.2\,\text{kJ mol}^{-1}) \right. \\ &\quad \left. + (-241.82\,\text{kJ mol}^{-1})\right] = \boxed{-77\,\text{kJ mol}^{-1}}\end{aligned}$$

I6.3 (a) The cell reaction is described by the reduction half-reactions

$$\begin{aligned}&\text{R:}\quad \text{Hg}_2\text{Cl}_2(\text{s}) + 2\text{e}^- \rightarrow 2\text{Hg(l)} + 2\text{Cl}^-(\text{aq}) \qquad E^\ominus(\text{R}) = +0.2676\,\text{V} \\ &\text{L:}\quad \text{Zn}^{2+}(\text{aq}) + 2\text{e}^- \rightarrow \text{Zn(s)} \qquad E^\ominus(\text{L}) = -0.7628\,\text{V}\end{aligned}$$

which show that $\nu = 2$ for this reaction. The Nernst equation is

$$E_{cell} = E^\ominus_{cell} - \frac{RT}{2F}\ln\left(a_{\text{Zn}^{2+}}a^2_{\text{Cl}^-}\right)$$

Note that $a_J = 1$ for pure solids and liquids. The activities of the ions are then replaced by $a = \gamma_\pm(b/b^\ominus)$ to give

$$E_{cell} = E^\ominus_{cell} - \frac{RT}{2F}\ln\left(\frac{\gamma_\pm b_{Zn^{2+}}}{b^\ominus} \times \left(\frac{\gamma_\pm b_{Cl^-}}{b^\ominus}\right)^2\right)$$

Using $b_{Zn^{2+}} = b$ and $b_{Cl^-} = 2b$ where b is the molality of $ZnCl_2$ gives

$$E_{cell} = E^\ominus_{cell} - \frac{RT}{2F}\ln\left(\frac{\gamma_\pm b}{b^\ominus} \times \left(\frac{\gamma_\pm \times 2b}{b^\ominus}\right)^2\right) = E^\ominus_{cell} - \frac{RT}{2F}\ln\left(\frac{4\gamma_\pm^3 b^3}{b^{\ominus 3}}\right)$$

(b) The standard cell potential is calculated from the standard electrode potentials using [6D.3–*224*], $E^\ominus_{cell} = E^\ominus(R) - E^\ominus(L)$

$$E^\ominus_{cell} = E^\ominus(R) - E^\ominus(L) = (+0.2676\ V) - (-0.7628\ V) = \boxed{+1.0304\ V}$$

(c) The relationship between $\Delta_r G$ and E_{cell} is given by [6C.2–*217*], $\Delta_r G = -\nu F E_{cell}$.

$$\Delta_r G = -\nu F E_{cell} = -2\times(96485\ C\,mol^{-1})\times(+1.2272\ V) = \boxed{-236.81\ kJ\,mol^{-1}}$$

The standard reaction Gibbs energy is calculated in the same way using $E^\ominus_{cell}$ in place of E_{cell}

$$\Delta_r G^\ominus = -\nu F E^\ominus_{cell} = -2\times(96485\ C\,mol^{-1})\times(+1.0304\ V) = \boxed{-198.84\ kJ\,mol^{-1}}$$

The relationship between $\Delta_r G^\ominus$ and the equilibrium constant is given by [6A.15–*208*], $\Delta_r G^\ominus = -RT\ln K$. Rearranging for K gives

$$K = e^{-\Delta_r G^\ominus/RT} = \exp\left(-\frac{-1.98... \times 10^5\ J\,mol^{-1}}{(8.3145\ J\,K^{-1}\,mol^{-1}) \times (298\ K)}\right) = \boxed{7.11 \times 10^{34}}$$

(d) The mean ionic activity coefficient $\gamma_\pm$ of $ZnCl_2$ is calculated from the measured cell potential by rearranging the Nernst equation found in (a):

$$\begin{aligned} E_{cell} &= E^\ominus_{cell} - \frac{RT}{2F}\ln\left(\frac{4\gamma_\pm^3 b^3}{b^{\ominus 3}}\right) \\ &= E^\ominus_{cell} - \frac{3RT}{2F}\ln\gamma_\pm - \frac{3RT}{2F}\ln\left(\frac{b}{b^\ominus}\right) - \frac{RT}{2F}\ln 4 \end{aligned}$$

Hence

$$\begin{aligned} \ln\gamma_\pm &= \frac{2F}{3RT}(E^\ominus_{cell} - E_{cell}) - \ln\left(\frac{b}{b^\ominus}\right) - \tfrac{1}{3}\ln 4 \\ &= \frac{2 \times (96485\ C\,mol^{-1})}{3 \times (8.3145\ J\,K^{-1}\,mol^{-1}) \times (298\ K)} \\ &\quad \times \left[(+1.0304\ V) - (+1.2272\ V)\right] - \ln\left(\frac{0.0050\ mol\,kg^{-1}}{1\ mol\,kg^{-1}}\right) - \tfrac{1}{3}\ln 4 \\ &= -0.272... \end{aligned}$$

Therefore $\gamma_\pm = e^{-0.272...} = \boxed{0.761}$

(e) According to the Debye–Hückel limiting law the mean ionic activity coefficient is given by [5F.27–*188*], $\log\gamma_\pm = -A|z_+z_-|I^{1/2}$ where $A = 0.509$ for an aqueous solution at 298 K, z_+ and z_- are the charges on the ions, and I is the ionic strength. The ionic strength for a solution containing only two types of ion is given by [5F.29–*188*], $I = \frac{1}{2}\left(b_+z_+^2 + b_-z_-^2\right)/b^\ominus$. For a solution of $ZnCl_2$ of molality b, $z_+ = +2$, $z_- = -1$, $b_- = 2b$ and $b_+ = b$, which gives

$$I = \tfrac{1}{2}\left(b_+z_+^2 + b_-z_-^2\right)/b^\ominus = \tfrac{1}{2}\left(b\times 2^2 + (2b)\times(-1)^2\right)/b^\ominus = 3b/b^\ominus$$

$$\text{Hence } \log\gamma_\pm = -A|z_+z_-|I^{1/2} = -A|z_+z_-|\left(\frac{3b}{b^\ominus}\right)^{1/2}$$

$$= -0.509\times\left|2\times(-1)\right|\times\left(\frac{3\times(0.0050\ \text{mol kg}^{-1})}{1\ \text{mol kg}^{-1}}\right)^{1/2}$$

$$= -0.124...$$

This gives $\gamma_\pm = 10^{-0.124...} = \boxed{0.750}$, which is in reasonable agreement with the value obtained from the measured cell potential in part (d).

(f) Combining [3E.8–*107*], $(\partial G/\partial T)_p = -S$, and [6C.2–*217*], $\Delta_r G = -\nu F E_{\text{cell}}$, gives

$$\left(\frac{\partial\Delta_r G}{\partial T}\right)_p = -\Delta_r S \quad\text{hence}\quad \left(\frac{\partial(-\nu F E_{\text{cell}})}{\partial T}\right)_p = -\Delta_r S$$

$$\text{hence}\quad \nu F\left(\frac{\partial E_{\text{cell}}}{\partial T}\right)_p = \Delta_r S$$

which is equivalent to [6C.6–*222*], $\mathrm{d}E^\ominus_{\text{cell}}/\mathrm{d}T = -\Delta_r S^\ominus/\nu F$ except that the partial derivative is required because E_{cell}, unlike $E^\ominus_{\text{cell}}$, in general depends on pressure. For this cell

$$\Delta_r S = \nu F\left(\frac{\partial E_{\text{cell}}}{\partial T}\right)_p = 2\times(96485\ \text{C mol}^{-1})\times(-4.52\times10^{-4}\ \text{V K}^{-1})$$

$$= \boxed{-87.2\ \text{J K}^{-1}\ \text{mol}^{-1}}$$

The reaction enthalpy is then calculated using [3D.9–*100*], $\Delta_r G = \Delta_r H - T\Delta_r S$, with the value of $\Delta_r G$ from part (c):

$$\Delta_r H = \Delta_r G + T\Delta_r S$$

$$= (-2.36...\times10^5\ \text{J mol}^{-1}) + (298\ \text{K})\times(-87.2...\ \text{J K}^{-1}\ \text{mol}^{-1})$$

$$= \boxed{-263\ \text{kJ mol}^{-1}}$$

I6.5 The specified cell consists of two cells connected in series with a common central electrode. The cell on the left is Ag(s),AgCl(s)|LiCl(b_1)|Li(amal), where Li(amal) denotes an amalgam. The reduction half-reactions are

$$\begin{aligned}\text{R:}\quad & Li^+(b_1)+e^- \rightarrow Li(amal)\\ \text{L:}\quad & AgCl(s)+e^- \rightarrow Ag(s)+Cl^-(b_1)\end{aligned}$$

The cell reaction (R – L) and Nernst equation are

$$Li^+(b_1)+Ag(s)+Cl^-(b_1)\rightarrow Li(amal)+AgCl(s) \qquad \nu=1$$

$$E_{cell,1}=E^{\ominus}_{cell,1}-\frac{RT}{F}\ln\left(\frac{1}{a_{Li^+,b_1}a_{Cl^-,b_1}}\right)$$

The cell on the right is essentially the same, but in the opposite orientation and with different concentrations: $Li_xHg|LiCl(b_2)|AgCl(s),Ag(s)$. The cell reaction and Nernst equation are

$$Li(amal)+AgCl(s)\rightarrow Li^+(b_2)+Ag(s)+Cl^-(b_2)$$

$$E_{cell,2}=E^{\ominus}_{cell,2}-\frac{RT}{F}\ln\left(a_{Li^+,b_2}a_{Cl^-,b_2}\right)$$

Because the two cells are connected in series, the overall cell potential is the sum of the individual cell potentials:

$$E_{cell}=E_{cell,1}+E_{cell,2}=E^{\ominus}_{cell,1}+E^{\ominus}_{cell,2}-\frac{RT}{F}\ln\left(\frac{a_{Li^+,b_2}a_{Cl^-,b_2}}{a_{Li^+,b_1}a_{Cl^-,b_1}}\right)$$

Because both cells have the same cell reaction but in opposite directions, their standard cell potentials are equal and opposite, $E^{\ominus}_{cell,1}=-E^{\ominus}_{cell,2}$, so the Nernst equation becomes

$$E_{cell}=-\frac{RT}{F}\ln\left(\frac{a_{Li^+,b_2}a_{Cl^-,b_2}}{a_{Li^+,b_1}a_{Cl^-,b_1}}\right)$$

The activities are replaced by $a=\gamma_\pm(b/b^{\ominus})$ to give

$$\begin{aligned}E_{cell}&=-\frac{RT}{F}\ln\left(\frac{(\gamma_{\pm,2}b_2/b^{\ominus})(\gamma_{\pm,2}b_2/b^{\ominus})}{(\gamma_{\pm,1}b_1/b^{\ominus})(\gamma_{\pm,1}b_1/b^{\ominus})}\right)=-\frac{RT}{F}\ln\left(\frac{\gamma_{\pm,2}^2b_2^2}{\gamma_{\pm,1}^2b_1^2}\right)\\&=-\frac{2RT}{F}\ln\left(\frac{\gamma_{\pm,2}b_2}{\gamma_{\pm,1}b_1}\right)=-\frac{2RT\ln 10}{F}\log\left(\frac{\gamma_{\pm,2}b_2}{\gamma_{\pm,1}b_1}\right)\end{aligned}$$

where in the last step the relationship $\ln x=\ln 10\times\log x$ from inside the front cover is used. The task is to use the Davies equation with the given parameters to calculate $\gamma_{\pm,1}$ when $b_1=1.350\ \text{mol kg}^{-1}$, and then use this value with the value of E_{cell} at this b_1 to find $\gamma_{\pm,2}$. Then because $\gamma_{\pm,2}$ is constant for all the measurements it is used with the other values of E_{cell} to calculate $\gamma_{\pm,1}$ at the other values of b_1.

The Davies equation [5F.30b–*189*] is

$$\log\gamma_\pm=-\frac{A|z_+z_-|I^{1/2}}{1+BI^{1/2}}+CI$$

where A, B and C have the values given in the question, z_+ and z_- are the charges on the ions, and I is the ionic strength which as specified in the question is taken as $I = b/b^\ominus$. For an LiCl electrolyte, the ions have charges +1 and −1, so noting that $b/b^\ominus = (1.350\ \text{mol kg}^{-1})/(1\ \text{mol kg}^{-1}) = 1.350$ for the concentration marked *, the value of $\log \gamma_{\pm,1}$ for this concentration is

$$\log \gamma_{\pm,1} = -\frac{A|z_+z_-|(b/b^\ominus)^{1/2}}{1+B(b/b^\ominus)^{1/2}} + C(b/b^\ominus)$$

$$= -\frac{1.461 \times \left|(+1)\times(-1)\right| \times (1.350)^{1/2}}{1+1.70\times(1.350)^{1/2}} + 0.20\times(1.350) = -0.300...$$

Hence $\gamma_{\pm,1} = 10^{-0.300...} = \boxed{0.501}$. This value of $\gamma_{\pm,1}$ is used with the Nernst equation derived above,

$$E_{\text{cell}} = -\frac{2RT\ln 10}{F}\log\left(\frac{\gamma_{\pm,2}b_2}{\gamma_{\pm,1}b_1}\right),$$

to find $\gamma_{\pm,2}$ from the data for $b_1 = 1.350\ \text{mol kg}^{-1}$. Rearranging for $\log \gamma_{\pm,2}$ gives

$$\log \gamma_{\pm,2} = \log \gamma_{\pm,1} + \log\left(\frac{b_1}{b_2}\right) - \frac{F\times E_{\text{cell}}}{2RT\ln 10}$$

$$= (-0.300...) + \log\left(\frac{1.350\ \text{mol kg}^{-1}}{0.09141\ \text{mol kg}^{-1}}\right)$$

$$- \frac{(96485\ \text{C mol}^{-1})\times(+0.1336\ \text{V})}{2\times(8.3145\ \text{J K}^{-1}\ \text{mol}^{-1})\times(298.15\ \text{K})\times\ln 10} = -0.260...$$

where 1 C V = 1 J is used. Hence $\gamma_{\pm,2} = 10^{-0.260...} = \boxed{0.549}$. This value of $\gamma_{\pm,2}$ is then used with the remaining values of E_{cell} to calculate the remaining values of $\gamma_{\pm,1}$. Rearranging the Nernst equation from above gives

$$\log \gamma_{\pm,1} = \log \gamma_{\pm,2} - \log\left(\frac{b_1}{b_2}\right) + \frac{F\times E_{\text{cell}}}{2RT\ln 10}$$

which gives the results in the table below

$b_1/\text{mol kg}^{-1}$	E_{cell}/V	$\log \gamma_{\pm,1}$	$\gamma_{\pm,1}$
0.0555	−0.0220	−0.2296	0.589
0.0914	0.0000	−0.2604	0.549
0.1652	0.0263	−0.2951	0.507
0.2171	0.0379	−0.3157	0.483
1.0400	0.1156	−0.3394	0.458
1.3500	0.1336	−0.3006	0.501

I6.7 (a) From *Impact* 9 the reaction for the hydrolysis of ATP to ADP and inorganic phosphate P_i^- is

$$\text{ATP(aq)} + \text{H}_2\text{O(l)} \rightarrow \text{ADP(aq)} + \text{P}_i^-\text{(aq)} + \text{H}_3\text{O}^+\text{(aq)}$$

This reaction produces three moles of dissolved solute from one mole of solute and one mole of liquid. The greater number of chemical species present in solution increases the disorder of the system by increasing the number of translational degrees of freedom. The number of translational energy levels available to the system is increased and therefore there is an increase in entropy as explained in Section 3A.2(a) on page 80.

(b) At physiological pH, the phosphate groups in ATP are deprotonated, resulting in the negatively charged molecule ATP^{4-}. The electrostatic repulsion between the negatively charged oxygen atoms of ATP^{4-} is expected to give it an exergonic hydrolysis Gibbs energy by making the hydrolysis $\boxed{\text{enthalpy}}$ negative.

Protonated ATP, H_4ATP, does not have negatively charged oxygen atoms and therefore the repulsions are not present. The observation that H_4ATP has a less exergonic Gibbs energy of hydrolysis is therefore consistent with the repulsion hypothesis. The same is true for $MgATP^{2-}$ because the Mg^{2+} ion lies between two negatively charged oxygen atoms, thereby reducing repulsions.

I6.9 (a) *Impact* 9 gives standard Gibbs energy for the complete oxidation of glucose as $-2880\ \text{kJ mol}^{-1}$. The reaction corresponding to this is

$$C_6H_{12}O_6(aq) + 6O_2(g) \rightarrow 6H_2O(l) + 6CO_2(g)$$

Because this reaction does not involve H^+ the standard reaction Gibbs energy is independent of pH and therefore the reaction Gibbs energy under standard biological conditions of pH = 7 is the same as the quoted value.

Impact 9 also gives the Gibbs energy of hydrolysis of ATP under biological standard conditions as $-31\ \text{kJ mol}^{-1}$. If 38 molecules of ATP are formed by oxidation of 1 mole of glucose then the efficiency of aerobic respiration under biological standard conditions is therefore

$$\text{Efficiency} = \frac{38 \times (-31\ \text{kJ mol}^{-1})}{-2880\ \text{kJ mol}^{-1}} \times 100\,\% = \boxed{41\,\%}$$

(b) The reaction Gibbs energies for both the oxidation of glucose and the hydrolysis of ATP must now be evaluated under the specified conditions using [6A.11–*207*], $\Delta_r G = \Delta_r G^\ominus + RT \ln Q$. For the oxidation of glucose this gives

$$\Delta_r G = \Delta_r G^\ominus + RT \ln\left(\frac{a^6_{H_2O,l}\, a^6_{CO_2,g}}{a_{C_6H_{12}O_6,aq}\, a^6_{O_2,g}}\right)$$

Treating O_2 and CO_2 as perfect gases so that $a_J = p_J/p^\ominus$ and glucose as an ideal solute so that $a_{C_6H_{12}O_6} = [C_6H_{12}O_6]/c^\ominus$, and noting that $a_{H_2O} = 1$

because water is a pure liquid gives

$$\Delta_r G = \Delta_r G^\ominus + RT\ln\left(\frac{1\times(p_{CO_2}/p^\ominus)^6}{([C_6H_{12}O_6]/c^\ominus)\times(p_{O_2}/p^\ominus)^6}\right)$$

$$= \Delta_r G^\ominus + RT\ln\left(\frac{p_{CO_2}^6 \times c^\ominus}{[C_6H_{12}O_6]\times p_{O_2}^6}\right)$$

$$= (-2880\times10^3\ \mathrm{J\,mol^{-1}}) + (8.3145\ \mathrm{J\,K^{-1}\,mol^{-1}})\times(310\ \mathrm{K})$$

$$\times\ln\left(\frac{(5.3\times10^{-2}\ \mathrm{atm})^6\times(1\ \mathrm{mol\,dm^{-3}})}{(5.6\times10^{-12}\ \mathrm{mol\,dm^{-3}})\times(0.132\ \mathrm{atm})^6}\right)$$

$$= -2.82...\times10^6\ \mathrm{J\,mol^{-1}}$$

For ATP hydrolysis, the reaction given in *Impact* 9 is

$$\mathrm{ATP(aq) + H_2O(l) \rightarrow ADP(aq) + P_i^-(aq) + H_3O^+(aq)}$$

The reaction Gibbs energy under standard biological conditions, that is, pH = 7, is given in *Impact* 9 as $\Delta_r G^\oplus = -31\ \mathrm{kJ\,mol^{-1}}$. This is converted to a value under the conditions specified in the question using

$$\Delta_r G = \Delta_r G^\oplus + RT\ln Q^\oplus$$

where $Q^\oplus$ is the reaction quotient calculated relative to the biological standard state. Because pH is defined by pH $= -\log a_{H_3O^+}$, pH 7 corresponds to $a_{H_3O^+} = 10^{-7}$ so that when computing $Q^\oplus$ the activity of H_3O^+ is measured relative to an activity of 10^{-7} rather than an activity of 1 as is usually the case. In practice this means that $a_{H_3O^+}$ is replaced by $(a_{H_3O^+}/10^{-7})$ in the expression for $Q^\oplus$.
For the ATP hydrolysis reaction this gives

$$\Delta_r G = \Delta_r G^\oplus + RT\ln\left(\frac{a_{ADP}\times a_{P_i^-}\times(a_{H_3O^+}/10^{-7})}{a_{ATP}\times a_{H_2O}}\right)$$

Water is a pure liquid so $a_{H_2O} = 1$, and for the environment specified in the question, pH = 7.4 so $a_{H_3O^+} = 10^{-7.4}$. For the other species activities are approximated by concentrations according to $a_J = [J]/c^\ominus$ where $c^\ominus =$ 1 mol dm^{-3}.

$$\Delta_r G = \Delta_r G^\oplus + RT\ln\left(\frac{([ADP]/c^\ominus)([P_i^-]/c^\ominus)(a_{H_3O^+}/10^{-7})}{([ATP]/c^\ominus)}\right)$$

$$= \Delta_r G^\oplus + RT\ln\left(\frac{[ADP][P_i^-](a_{H_3O^+}/10^{-7})}{[ATP]c^\ominus}\right)$$

$$= (-31\times10^3\ \mathrm{J\,mol^{-1}}) + (8.3145\ \mathrm{J\,K^{-1}\,mol^{-1}})\times(310\ \mathrm{K})$$

$$\times\ln\left(\frac{(0.1\times10^{-3}\ \mathrm{mol\,dm^{-3}})\times(0.1\times10^{-3}\ \mathrm{mol\,dm^{-3}})\times(10^{-7.4}/10^{-7})}{(0.1\times10^{-3}\ \mathrm{mol\,dm^{-3}})\times(1\ \mathrm{mol\,dm^{-3}})}\right)$$

$$= -5.71...\times10^4\ \mathrm{J\,mol^{-1}}$$

The efficiency of aerobic respiration under these conditions is therefore

$$\text{Efficiency} = \frac{38 \times (-5.71... \times 10^4 \text{ J mol}^{-1})}{-2.82... \times 10^6 \text{ J mol}^{-1}} \times 100\,\% = \boxed{77\,\%}$$

(c) The efficiency of the diesel engine operating at 75 % of the theoretical limit of $1 - T_c/T_h$ is

$$\text{Efficiency} = \left(1 - \frac{873 \text{ K}}{1923 \text{ K}}\right) \times 75\,\% = \boxed{41\,\%}$$

The efficiency of the diesel engine, or any heat engine, is limited by the fact that in order to convert heat into work, some of the heat has to be discarded into the surroundings in order that the process is allowed by the Second Law.

For biological processes the change in Gibbs energy of one process is simply used to drive another process. The change in Gibbs energies of the two processes must match up, but there is no requirement for any of the Gibbs energy to be "discarded" in a way analogous to heat engines. Higher efficiency is therefore possible.

I6.11 (a) The half-reactions for the overall equation $\text{cyt}_{\text{ox}} + \text{D}_{\text{red}} \rightarrow \text{cyt}_{\text{red}} + \text{D}_{\text{ox}}$, which is given to be a one-electron transfer, are:

$$\text{R:} \quad \text{cyt}_{\text{ox}} + \text{e}^- \rightarrow \text{cyt}_{\text{red}}$$
$$\text{L:} \quad \text{D}_{\text{ox}} + \text{e}^- \rightarrow \text{D}_{\text{red}}$$

The relationship between the the standard cell potential and the equilibrium constant is given by [6C.5–*221*], $E^{\ominus}_{\text{cell}} = (RT/\nu F)\ln K$. In this case, $\nu = 1$ and $E^{\ominus}_{\text{cell}} = E^{\ominus}(\text{R}) - E^{\ominus}(\text{L}) = E^{\ominus}_{\text{cyt}} - E^{\ominus}_{\text{D}}$.

$$E^{\ominus}_{\text{cell}} = \frac{RT}{\nu F}\ln K \quad \text{hence} \quad E^{\ominus}_{\text{cyt}} - E^{\ominus}_{\text{D}} = \frac{RT}{F}\ln\left(\frac{a_{\text{cyt}_{\text{red}}} a_{\text{D}_{\text{ox}}}}{a_{\text{cyt}_{\text{ox}}} a_{\text{D}_{\text{red}}}}\right)$$

Replaced by $a_{\text{J}} = [\text{J}]_{\text{eq}}/c^{\ominus}$, where $[\text{J}]_{\text{eq}}$ are the equilibrium molar concentrations, gives

$$\begin{aligned} E^{\ominus}_{\text{cyt}} - E^{\ominus}_{\text{D}} &= \frac{RT}{F}\ln\left(\frac{([\text{cyt}_{\text{red}}]_{\text{eq}}/c^{\ominus})([\text{D}_{\text{ox}}]_{\text{eq}}/c^{\ominus})}{([\text{cyt}_{\text{ox}}]_{\text{eq}}/c^{\ominus})([\text{D}_{\text{red}}]_{\text{eq}}/c^{\ominus})}\right) \\ &= \frac{RT}{F}\ln\left(\frac{[\text{cyt}_{\text{red}}]_{\text{eq}}[\text{D}_{\text{ox}}]_{\text{eq}}}{[\text{cyt}_{\text{ox}}]_{\text{eq}}[\text{D}_{\text{red}}]_{\text{eq}}}\right) \end{aligned}$$

Hence

$$\begin{aligned} \frac{F(E^{\ominus}_{\text{cyt}} - E^{\ominus}_{\text{D}})}{RT} &= \ln\left(\frac{[\text{cyt}_{\text{red}}]_{\text{eq}}[\text{D}_{\text{ox}}]_{\text{eq}}}{[\text{cyt}_{\text{ox}}]_{\text{eq}}[\text{D}_{\text{red}}]_{\text{eq}}}\right) = \ln\left(\frac{[\text{cyt}_{\text{red}}]_{\text{eq}}}{[\text{cyt}_{\text{ox}}]_{\text{eq}}}\right) + \ln\left(\frac{[\text{D}_{\text{ox}}]_{\text{eq}}}{[\text{D}_{\text{red}}]_{\text{eq}}}\right) \\ &= -\ln\left(\frac{[\text{cyt}_{\text{ox}}]_{\text{eq}}}{[\text{cyt}_{\text{red}}]_{\text{eq}}}\right) + \ln\left(\frac{[\text{D}_{\text{ox}}]_{\text{eq}}}{[\text{D}_{\text{red}}]_{\text{eq}}}\right) \end{aligned}$$

Hence

$$\underbrace{\ln\left([\mathrm{D_{ox}}]_{\mathrm{eq}}/[\mathrm{D_{red}}]_{\mathrm{eq}}\right)}_{y} = \underbrace{1}_{\text{slope}} \times \underbrace{\ln\left([\mathrm{cyt_{ox}}]_{\mathrm{eq}}/[\mathrm{cyt_{red}}]_{\mathrm{eq}}\right)}_{x} + \underbrace{F(E^{\ominus}_{\mathrm{cyt}} - E^{\ominus}_{\mathrm{D}})/RT}_{\text{intercept}}$$

Hence a plot of $\ln\left([\mathrm{D_{ox}}]_{\mathrm{eq}}/[\mathrm{D_{red}}]_{\mathrm{eq}}\right)$ against $\ln\left([\mathrm{cyt_{ox}}]_{\mathrm{eq}}/[\mathrm{cyt_{red}}]_{\mathrm{eq}}\right)$ should be a straight line with slope 1 and y-intercept $F(E^{\ominus}_{\mathrm{cyt}} - E^{\ominus}_{\mathrm{D}})/RT$.

(b) The data are tabulated below, and plotted in the graph shown in Fig. 6.3.

$\dfrac{[\mathrm{cyt_{ox}}]_{\mathrm{eq}}}{[\mathrm{cyt_{red}}]_{\mathrm{eq}}}$	$\dfrac{[\mathrm{D_{ox}}]_{\mathrm{eq}}}{[\mathrm{D_{red}}]_{\mathrm{eq}}}$	$\ln\dfrac{[\mathrm{cyt_{ox}}]_{\mathrm{eq}}}{[\mathrm{cyt_{red}}]_{\mathrm{eq}}}$	$\ln\dfrac{[\mathrm{D_{ox}}]_{\mathrm{eq}}}{[\mathrm{D_{red}}]_{\mathrm{eq}}}$
1.06×10^{-2}	2.79×10^{-3}	−4.547	−5.882
2.30×10^{-2}	8.43×10^{-3}	−3.772	−4.776
8.94×10^{-2}	2.57×10^{-2}	−2.415	−3.661
0.197	4.97×10^{-2}	−1.625	−3.002
0.335	7.48×10^{-2}	−1.094	−2.593
0.809	0.238	−0.212	−1.435
1.39	0.534	0.329	−0.627

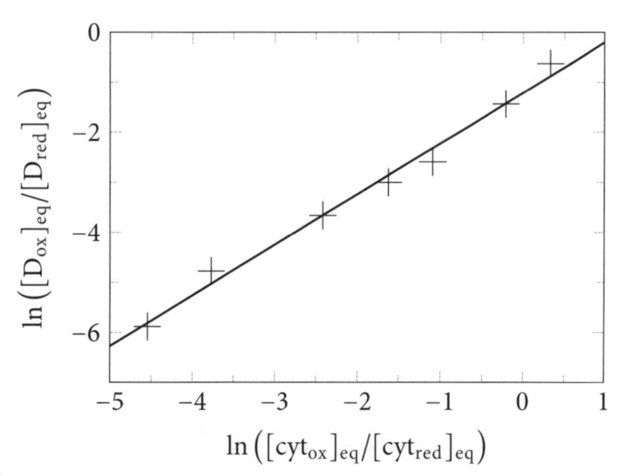

Figure 6.3

The data fall on a reasonable straight line, and the slope is quite close to 1. The equation of the line is

$$\ln\left([\mathrm{D_{ox}}]_{\mathrm{eq}}/[\mathrm{D_{red}}]_{\mathrm{eq}}\right) = 1.012 \times \ln\left([\mathrm{cyt_{ox}}]_{\mathrm{eq}}/[\mathrm{cyt_{red}}]_{\mathrm{eq}}\right) - 1.212$$

Equating the intercept to $F(E^{\ominus}_{\mathrm{cyt}} - E^{\ominus}_{\mathrm{D}})/RT$ and rearranging for $E^{\ominus}_{\mathrm{cyt}}$ gives

$$\begin{aligned} E^{\ominus}_{\mathrm{cyt}} &= \frac{RT}{F} \times \textit{intercept} + E^{\ominus}_{\mathrm{D}} \\ &= \frac{(8.3145\,\mathrm{J\,K^{-1}\,mol^{-1}}) \times (298\,\mathrm{K})}{(96485\,\mathrm{C\,mol^{-1}})} \times (-1.212) + (+0.237\,\mathrm{V}) \\ &= \boxed{+0.206\,\mathrm{V}} \end{aligned}$$

Quantum theory

7A The origins of quantum mechanics

Answers to discussion question

D7A.1 At the end of the nineteenth century and the beginning of the twentieth, there were many experimental results on the properties of matter and radiation that could not be explained on the basis of established physical principles and theories. Amongst these the most significant were:

(a) The form of black-body radiation. The radiation emitted by a black body was known to be a maximum at a particular wavelength which depended on the temperate, and then to fall off at shorter wavelengths. This distribution was inexplicable with the then current theories which predicted that the radiation would simply increase without limit as the wavelength became shorter (the 'ultraviolet catastrophe').

(b) Heat capacities. The heat capacities of solids were found to vary with temperature in a way which appeared to be inconsistent with the then established theory of the equipartition of energy.

(c) The spectra of atoms and molecules. Such spectra were known to consist of emissions and absorptions at discrete frequencies ('spectral lines') which were highly characteristic of the particular species being studied.

(d) Wave-particle duality. The accepted theories indicated that electromagnetic radiation should be considered as a wave and electrons as particles. However it appeared only to be possible to understand the photoelectric effect, in which electrons are ejected from a metal as a result of the bombardment with light, by imagining that the light consisted of particles. In addition, the phenomenon of diffraction, associated at that time with waves, was observed using electrons which were considered to be particles.

D7A.3 The heat capacities of solids are thought to arise from the energy associated with the oscillation of atoms about their equilibrium positions. In the classical theory each oscillator has an average energy of $3kT$, leading to a molar heat capacity of $3R$, independent of temperature. Einstein assumed that each atom vibrating with frequency ν has energy $E = nh\nu$ where n is 0, 1, 2, . . .; that is, the energy of the oscillation is quantized. He further assumed that the population of the states with increasing values of n, that is with higher energies, is governed by the Boltzmann distribution. Taken together, these assumptions result in the heat capacity becoming temperature dependent and in particular that the heat capacity decreases with temperature as the higher energy states of the oscillators are populated less and less.

Solutions to exercises

E7A.1(a) Wien's law [7A.1–*238*], $\lambda_{\text{max}}T = 2.9 \times 10^{-3}$ m K, is rearranged to give the wavelength at which intensity is maximised

$$\lambda_{\text{max}} = (2.9 \times 10^{-3}\ \text{m K})/T = (2.9 \times 10^{-3}\ \text{m K})/(298\ \text{K}) = \boxed{9.7 \times 10^{-6}\ \text{m}}$$

E7A.2(a) Assuming that the object is a black body is equivalent to assuming that Wien's law [7A.1–*238*], $\lambda_{\text{max}}T = 2.9 \times 10^{-3}$ m K, holds. Using $\tilde{\nu} = \lambda^{-1}$, Wien's law is expressed in terms of the wavenumber of maximum intensity ($\tilde{\nu}_{\text{max}}$)

$$T/\tilde{\nu}_{\text{max}} = 2.9 \times 10^{-3}\ \text{m K}$$

This is rearranged to give the temperature

$$\begin{aligned} T &= (2.9 \times 10^{-3}\ \text{m K}) \times \tilde{\nu}_{\text{max}} \\ &= (2.9 \times 10^{-3}\ \text{m K}) \times (2000 \times 10^{2}\ \text{m}^{-1}) = \boxed{580\ \text{K}} \end{aligned}$$

E7A.3(a) Molar heat capacities of monatomic non-metallic solids obey the Einstein relation [7A.8a–*241*],

$$C_{V,\text{m}}(T) = 3Rf_{\text{E}}(T), \quad f_{\text{E}}(T) = \left(\frac{\theta_{\text{E}}}{T}\right)^2 \left(\frac{e^{\theta_{\text{E}}/2T}}{e^{\theta_{\text{E}}/T} - 1}\right)^2$$

where the solid is at temperature T and is characterized by an Einstein temperature θ_{E}. Thus, for a solid at 298 K with an Einstein temperature of 2000 K

$$f_{\text{E}}(298\ \text{K}) = \left(\frac{2000\ \text{K}}{298\ \text{K}}\right)^2 \left(\frac{e^{(2000\ \text{K})/2(298\ \text{K})}}{e^{(2000\ \text{K})/298\ \text{K}} - 1}\right)^2 = 5.49... \times 10^{-2}$$

Hence, $C_{V,\text{m}}(298\ \text{K}) = \boxed{(5.49 \times 10^{-2}) \times 3R}$

E7A.4(a) The energy of the quantum is given by the Bohr frequency condition [7A.9–*241*], $\Delta E = h\nu$, and the frequency is $\nu = 1/T$. The energy per mole is $\Delta E_{\text{m}} = N_{\text{A}}\Delta E$.

(i) For $T = 1.0$ fs

$$\Delta E = (6.6261 \times 10^{-34}\ \text{J s})/(1.0 \times 10^{-15}\ \text{s}) = \boxed{6.6 \times 10^{-19}\ \text{J}}$$

$$\Delta E_{\text{m}} = (6.6... \times 10^{-19}\ \text{J}) \times (6.0221 \times 10^{23}\ \text{mol}^{-1}) = \boxed{4.0 \times 10^{2}\ \text{kJ mol}^{-1}}$$

(ii) For T = 10 fs

$$\Delta E = (6.6261 \times 10^{-34}\,\mathrm{J\,s})/(10 \times 10^{-15}\,\mathrm{s}) = \boxed{6.6 \times 10^{-20}\,\mathrm{J}}$$

$$\Delta E_\mathrm{m} = (6.6... \times 10^{-20}\,\mathrm{J}) \times (6.0221 \times 10^{23}\,\mathrm{mol^{-1}}) = \boxed{40\,\mathrm{kJ\,mol^{-1}}}$$

(iii) For T = 1.0 s

$$\Delta E = (6.6261 \times 10^{-34}\,\mathrm{J\,s})/(1.0\,\mathrm{s}) = \boxed{6.6 \times 10^{-34}\,\mathrm{J}}$$

$$\Delta E_\mathrm{m} = (6.6... \times 10^{-34}\,\mathrm{J}) \times (6.0221 \times 10^{23}\,\mathrm{mol^{-1}}) = \boxed{4.0 \times 10^{-13}\,\mathrm{kJ\,mol^{-1}}}$$

E7A.5(a) The energy of a photon with wavelength λ is given by

$$E = h\nu = hc/\lambda = (6.6261 \times 10^{-34}\,\mathrm{J\,s}) \times (2.9979 \times 10^{8}\,\mathrm{m\,s^{-1}})/\lambda$$
$$= (1.9825 \times 10^{-25}\,\mathrm{J})/(\lambda/\mathrm{m})$$

The energy per mole is then given by

$$E_\mathrm{m} = N_\mathrm{A}E = (0.11939\,\mathrm{J\,mol^{-1}})/(\lambda/\mathrm{m})$$

Hence, the following table is drawn up

	λ/nm	E/zJ	E_m/kJ mol^{-1}
(a)	600	330	199
(b)	550	360	217
(c)	400	496	298

E7A.6(a) When a photon is absorbed by a free hydrogen atom, the law of conservation of energy requires that the kinetic energy acquired by the atom is E_k, the energy of the absorbed photon. Assuming relativistic corrections are negligible the kinetic energy is $E_\mathrm{k} = E_\mathrm{photon} = \frac{1}{2}m_\mathrm{H}v^2$. The atom is accelerated to the speed

$$v = \left(\frac{2E_\mathrm{photon}}{m_\mathrm{H}}\right)^{1/2} = \left(\frac{2N_\mathrm{A}E_\mathrm{photon}}{M_\mathrm{H}}\right)^{1/2}$$
$$= \left(\frac{2 \times (6.0221 \times 10^{23}\,\mathrm{mol^{-1}}) \times E_\mathrm{photon}}{(1.0079 \times 10^{-3}\,\mathrm{kg\,mol^{-1}})}\right)^{1/2}$$
$$= (3.45... \times 10^{13}\,\mathrm{m\,s^{-1}}) \times (E_\mathrm{photon}/\mathrm{J})^{1/2}$$

The photon energies have been calculated in *Exercise* E7A.5(a), and thus the following table is drawn up

	λ/nm	E/zJ	v/km s^{-1}
(a)	600	330	19.9
(b)	550	360	20.8
(c)	400	496	24.4

E7A.7(a) The energy emitted from a lamp at (constant) power P in a time interval Δt is $P\Delta t$. The energy of a single photon of wavelength λ is $E = hc/\lambda$. Hence, the total number of photons emitted in this time interval is the total energy emitted divided by the energy per photon, $N = P\Delta t/E_{\text{photon}} = P\Delta t\lambda/hc$. Thus, for a time interval of 1 s and a wavelength of 550 nm

(i) $P = 1$ W

$$N = \frac{(1\text{ W}) \times (1\text{ s})(550 \times 10^{-9}\text{ m})}{(6.6261 \times 10^{-34}\text{ J s}) \times (2.9979 \times 10^{8}\text{ m s}^{-1})} = \boxed{2.77 \times 10^{18}}$$

(ii) $P = 100$ W

$$N = \frac{(100\text{ W}) \times (1\text{ s})(550 \times 10^{-9}\text{ m})}{(6.6261 \times 10^{-34}\text{ J s}) \times (2.9979 \times 10^{8}\text{ m s}^{-1})} = \boxed{2.77 \times 10^{20}}$$

E7A.8(a) As described in Section 7A.2 on page 242, photoejection can only occur if the energy of the incident photon is greater than or equal to the work function of the metal ϕ. If this condition is fulfilled, the energy of the emitted photon is given by [7A.10–*243*], $E_k = h\nu - \Phi = hc/\lambda - \Phi$. To convert the work function to Joules, multiply through by the elementary charge, as described in Section 7A.2 on page 242,

$$\Phi = (2.14\text{ eV}) \times e = (2.14\text{ eV}) \times (1.602 \times 10^{-19}\text{ J eV}^{-1}) = 3.42... \times 10^{-19}\text{ J}$$

and since $E_k = 1/2m_e v^2$, $v = \sqrt{2E_k/m_e}$

(i) For $\lambda = 700$ nm

$$E_{\text{photon}} = \frac{hc}{\lambda} = \frac{(6.6261 \times 10^{-34}\text{ J s}) \times (2.9979 \times 10^{8}\text{ m s}^{-1})}{700 \times 10^{-9}\text{ m}}$$
$$= 2.84... \times 10^{-19}\text{ J}$$

This is less than the threshold energy, hence $\boxed{\text{no electron ejection}}$ occurs.

(ii) For $\lambda = 300$ nm

$$E_{\text{photon}} = \frac{hc}{\lambda} = \frac{(6.6261 \times 10^{-34}\text{ J s}) \times (2.9979 \times 10^{8}\text{ m s}^{-1})}{300 \times 10^{-9}\text{ m}}$$
$$= 6.62... \times 10^{-19}\text{ J}$$

This is greater than the the threshold frequency, and so photoejection can occur. The kinetic energy of the electron is,

$$E_k = hc/\lambda - \Phi = 6.62... \times 10^{-19}\text{ J} - 3.42... \times 10^{-19}\text{ J} = \boxed{3.19 \times 10^{-19}\text{ J}}$$

$$v = \sqrt{2E_{\text{photon}}/m_e}$$
$$= \sqrt{2 \times (3.19 \times 10^{-19}\text{ J})/(9.109 \times 10^{-31}\text{ kg})} = \boxed{837\text{ km s}^{-1}}$$

E7A.9(a) If the power, P, is constant, the total energy emitted in time Δt is $P\Delta t$. The energy of each emitted photon is $E_{\text{photon}} = h\nu = hc/\lambda$. The total number of photons emitted in this time period is therefore the total energy emitted divided by the energy per photon

$$N = P\Delta t/E_{\text{photon}} = P\Delta t\lambda/hc$$

The conservation of linear momentum requires that the loss of a photon must impart an equivalent momentum in the opposite direction to the glow-worm, hence the total momentum p imparted to the glow-worm in time Δt is

$$p = Np_{\text{photon}} = Nh/c = (P\Delta t\lambda/hc) \times (h/\lambda) = P\Delta t/c$$

Because $p = (mv)_{\text{glow-worm}}$, the final speed of the glow-worm is

$$v = P\Delta t/cm_{\text{glow-worm}}$$
$$= \frac{(0.10\ \text{W}) \times (10\ \text{y}) \times (3.1536 \times 10^7\ \text{s y}^{-1})}{(2.9979 \times 10^8\ \text{m s}^{-1}) \times (0.0050\ \text{kg})} = \boxed{21\ \text{m s}^{-1}}$$

Noting that the number of seconds in one year is

$$365 \times 24 \times 60 \times 60 = 3.1536 \times 10^7$$

E7A.10(a) The de Broglie relation is [7A.11–*244*], $\lambda = h/p = h/(mv)$. Therefore,

$$v = \frac{h}{m_e\lambda} = \frac{6.6261 \times 10^{-34}\ \text{J s}}{(9.1094 \times 10^{-31}\ \text{kg}) \times (100 \times 10^{-12}\ \text{m})} = \boxed{7.27 \times 10^6\ \text{m s}^{-1}}$$

The kinetic energy acquired by an electron accelerated through a potential $\mathcal{E}$ is $e\mathcal{E}$: $E_k = \frac{1}{2}m_e v^2 = e\mathcal{E}$. Solving for the potential gives

$$\mathcal{E} = \frac{m_e v^2}{2e} = \frac{(9.1094 \times 10^{-31}\ \text{kg}) \times (7.27 \times 10^6\ \text{m s}^{-1})^2}{2 \times (1.6022 \times 10^{-19}\ \text{C})} = \boxed{150\ \text{V}}$$

E7A.11(a) The de Broglie relation is [7A.11–*244*] $\lambda = h/p = h/(mv)$.

$$v = \frac{h}{m_e\lambda} = \frac{6.6261 \times 10^{-34}\ \text{J s}}{(9.1094 \times 10^{-31}\ \text{kg}) \times (3 \times 10^{-2}\ \text{m})} = \boxed{2.4 \times 10^{-2}\ \text{m s}^{-1}}$$

E7A.12(a) The de Broglie relation is [7A.11–*244*] $\lambda = h/p = h/(mv)$. Therefore

$$\lambda = \frac{h}{m_e\alpha c} = \frac{\alpha^{-1}h}{m_e c} = \frac{137 \times (6.6261 \times 10^{-34}\ \text{J s})}{(9.1094 \times 10^{-31}\ \text{kg}) \times (2.9979 \times 10^8\ \text{m s}^{-1})} = \boxed{332\ \text{pm}}$$

E7A.13(a) The de Broglie wavelength is [7A.11–*244*], $\lambda = h/p = h/(mv)$

(i)

$$\lambda = \frac{6.6261 \times 10^{-34}\,\text{J s}}{(1.0 \times 10^{-3}\,\text{kg}) \times (1.0 \times 10^{-2}\,\text{m s}^{-1})} = \boxed{6.6 \times 10^{-29}\,\text{m}}$$

(ii)

$$\lambda = \frac{6.6261 \times 10^{-34}\,\text{J s}}{(1.0 \times 10^{-3}\,\text{kg}) \times (100 \times 10^{3}\,\text{m s}^{-1})} = \boxed{6.6 \times 10^{-36}\,\text{m}}$$

(iii)

$$\lambda = \frac{6.6261 \times 10^{-34}\,\text{J s}}{((4.00 \times 10^{-3}\,\text{kg mol}^{-1})/(6.0221 \times 10^{23}\,\text{mol}^{-1})) \times (1.00 \times 10^{3}\,\text{m s}^{-1})}$$
$$= \boxed{99.8\,\text{pm}}$$

Solutions to problems

P7A.1 A cavity approximates an ideal black body, hence the Planck distribution [7A.6a–*239*], applies

$$\rho(\lambda, T) = \frac{8\pi hc}{\lambda^5 \left(e^{hc/\lambda kT} - 1\right)}$$

Because the wavelength range is small (5 nm), the energy density is approximated by

$$\Delta E(T) = \rho(\lambda, T)\Delta\lambda$$

Taking $\lambda = 652.2$ nm gives

$$\frac{hc}{\lambda k} = \frac{(6.6261 \times 10^{-34}\,\text{J s}) \times (2.9979 \times 10^{8}\,\text{m s}^{-1})}{(652.5 \times 10^{-9}\,\text{m}) \times (1.3806 \times 10^{-23}\,\text{J K}^{-1})} = 2.20... \times 10^{4}\,\text{K}$$

and

$$\frac{8\pi hc}{\lambda^5} = \frac{8\pi \times (6.6261 \times 10^{-34}\,\text{J s}) \times (2.9979 \times 10^{8}\,\text{m s}^{-1})}{(652.5 \times 10^{-9}\,\text{m})^5} = 4.22... \times 10^{7}\,\text{J m}^{-4}$$

It follows that

$$\Delta E(T) = (4.22... \times 10^{7}\,\text{J m}^{-4}) \times \frac{1}{e^{(2.20...\times 10^4\,\text{K})/T} - 1} \times (5 \times 10^{-9}\,\text{m})$$
$$= \frac{0.211...\,\text{J m}^{-3}}{e^{(2.20...\times 10^4\,\text{K})/T} - 1}$$

(a)

$$\Delta E(298\,\text{K}) = \frac{0.211...\,\text{J m}^{-3}}{e^{(2.20...\times 10^4\,\text{K})/(298\,\text{K})} - 1} = \boxed{1.54 \times 10^{-33}\,\text{J m}^{-3}}$$

(b)

$$\Delta E(3273\,\text{K}) = \frac{0.211...\,\text{J m}^{-3}}{e^{(2.20...\times 10^4\,\text{K})/(3273\,\text{K})} - 1} = \boxed{2.51 \times 10^{-4}\,\text{J m}^{-3}}$$

P7A.3 As λ increases, $hc/\lambda kT$ decreases, and at very long wavelengths $hc/\lambda kT \ll 1$. Hence, the exponential can be expanded in a power series. Let $x = hc/\lambda kT$, then $e^x = 1 + x + \frac{1}{2!}x^2 + \frac{1}{3!}x^3...$, and the Planck distribution becomes

$$\rho = \frac{8\pi hc}{\lambda^5\left(1 + x + \frac{1}{2!}x^2 + \frac{1}{3!}x^3... - 1\right)}$$

When $x << 1$ the second and higher order terms in x become negligibly small compared to x. Consequently

$$\lim_{\lambda\to\infty} \rho = \frac{8\pi hc}{\lambda^5 x} = \frac{8\pi hc}{\lambda^5}\left(\frac{1}{hc/\lambda kT}\right) = \frac{8\pi kT}{\lambda^4}$$

This is the Rayleigh–Jeans law

P7A.5 Each data point is used to find a value for $\lambda_{\max}T$, and then the mean of these is used with $\lambda_{\max}T = hc/(4.965k)$ to find a value of h, $h = (4.965k/c)\,(\lambda_{\max}T)_{\text{mean}}$. The following table is drawn up

$\theta/\,°\text{C}$	1000	1500	2000	2500	3000	3500
$T/\,\text{K}$	1273	1773	2273	2773	3273	3773
$\lambda_{\max}/\,\text{nm}$	2181	1600	1240	1035	878	763
$\lambda_{\max}T/(10^{-3}\,\text{m K})$	2.776	2.837	2.819	2.870	2.874	2.879

The mean is $2.84... \times 10^{-3}$ m K with a standard deviation of $4.01... \times 10^{-5}$ m K. Therefore

$$h = \frac{4.965 \times (1.3806 \times 10^{-23}\,\text{J K}^{-1}) \times (2.84... \times 10^{-3}\,\text{m K})}{2.9979 \times 10^{8}\,\text{m s}^{-1}} = \boxed{6.54 \times 10^{-34}\,\text{J s}}$$

P7A.7 The total energy density is given by [7A.7–*240*],

$$E(T) = \int_0^\infty \rho(\lambda, T)\,d\lambda = 8\pi hc \int_0^\infty \frac{1}{\lambda^5}\frac{1}{e^{hc/\lambda kT} - 1}\,d\lambda$$

Let $x = hc/\lambda kT$, or $\lambda = hc/xkT$. Then, $d\lambda = -hc/x^2kT\,dx$

$$E(T) = 8\pi hc \int_0^\infty \frac{1}{(hc/xkT)^5}\frac{1}{e^x - 1}\frac{hc}{x^2kT}\,dx = 8\pi\frac{(kT)^4}{(hc)^3}\int_0^\infty \frac{x^3}{e^x - 1}\,dx$$
$$= 8\pi(kT)^4/(hc)^3 \times \pi^4/15 = 8\pi^5k^4T^4/15(hc)^5$$

This is the Stefan–Boltzmannn law, that the total energy density is proportional to T^4, with a constant of proportionality of $8\pi^5k^4/15(hc)^5$.

P7A.9 Assuming the Sun approximates an ideal black body, Wien's law, $\lambda_{\max}T = 2.9 \times 10^{-3}$ m K, applies. Hence, $\lambda_{\max} = (2.9 \times 10^{-3}\,\text{m K})/(5800\,\text{K}) = \boxed{500\,\text{nm}}$ which corresponds to $\boxed{\text{blue-green}}$.

7B Wavefunctions

Answers to discussion questions

D7B.1 The wavefunction itself allows the calculation of the probability density of the system: $P(x) = \psi(x)^*\psi(x)$, in one dimension. Such a probability density would describe, for example, the position of the object to which the wavefunction refers. Other dynamical properties of the system are found by calculating the expectation value of the operator which represents that property: this is discussed in Topic 7C.

D7B.3 This is discussed in Section 7B.2(b) on page 249.

Solutions to exercises

E7B.1(a) The task is to find N such that $\psi = N\sin(2\pi x/L)$ satisfies the normalization condition [7B.4c–248], $\int \psi^*\psi\,d\tau = 1$. In this case the integration is over x and the range is 0 to L; the function is real, so $\psi = \psi^*$.

$$N^2\int_0^L \sin^2(2\pi x/L)dx = N^2[L/2 - (L/8\pi)\overbrace{\sin(4\pi L/L)}^{=\sin 4\pi=0}] = N^2(L/2)$$

The integral is of the form of Integral T.2 with $a = L$ and $k = 2\pi/L$. For the wavefunction to be normalized, the integral must be 1 and therefore the normalizing factor is $\boxed{N = (2/L)^{1/2}}$.

E7B.2(a) The task is to find N such that $\psi = N\exp(-ax^2)$ satisfies the normalization condition [7B.4c–248], $\int \psi^*\psi\,d\tau = 1$. In this case the integration is over x and the range is $-\infty$ to ∞; the function is real, so $\psi = \psi^*$ and the integral is therefore $N^2\int_{-\infty}^{\infty}\exp(-2ax^2)\,dx$. The integrand is even, meaning that it has the same value at $+x$ and $-x$, so the integral between $-\infty$ and $+\infty$ is simply twice that between 0 and $+\infty$. The integral is then of the form of Integral G.1 with $k = 2a$

$$2N^2\int_0^\infty \exp(-2ax^2)dx = 2N^2[\tfrac{1}{2}(\pi/2a)^{1/2}]$$

Setting this equal to 1 gives $\boxed{N = (2a/\pi)^{1/4}}$.

E7B.3(a)

(i) The function $\psi = \exp(-ax^2)$ $\boxed{\text{can be normalized}}$ as $\psi^*\psi = \exp(-2ax^2)$ goes to 0 at $x = \pm\infty$ so that the integral over all space of $\psi^*\psi$ is finite.

(ii) The function $\psi = \exp(-ax)$ $\boxed{\text{cannot be normalized}}$ as it goes to ∞ as x goes to $-\infty$, therefore the integral over all space of $\psi^*\psi$ is infinite. If the wavefunction were limited to a finite region of space it could, however, be normalized.

A function is an acceptable wavefunction if it: (1) is not infinite over a finite region; (2) is single-valued; (3) is continuous; (4) has a continuous first derivative. Both functions satisfy all of these conditions and so are acceptable wavefunctions.

E7B.4(a) The probability of finding an electron in an infinitesimal region dx around x is $P(x)dx = \psi^*(x)\psi(x)dx$, provided that $\psi(x)$ is a normalized wavefunction. The wavefunction is real so that $\psi^* = \psi$, hence the probability is given by

$$P(x)dx = \left[(2/L)^{1/2}\sin(2\pi x/L)\right]^2 dx = (2/L)\sin^2(2\pi x/L)dx$$

Hence, at $x = L/2$, $P(L/2)dx = (2/L)\sin(\pi)dx = \boxed{0}$.

E7B.5(a) The normalized wavefunction is $\psi(x) = (2/L)^{1/2}\sin(2\pi x/L)$. The probability of finding the electron between $x = L/4$ and $x = L/2$ is, using Integral T.2

$$\int_{L/4}^{L/2}(2/L)\sin^2(2\pi x/L)\,dx = (2/L)\left[x/2 - (L/8\pi)\sin(4\pi x/L)|_{L/4}^{L/2}\right]$$

$$= (2/L)([L/4 - (L/8\pi)\sin(2\pi)] - [L/8 - (L/8\pi)\sin(\pi)])$$

$$= (2/L)[L/4 - L/8] = \boxed{1/4}$$

E7B.6(a) In two dimensions $|\psi(x,y)|^2dxdy$ is the probability of finding the particle in the infinitesimal area $dxdy$ at the position (x,y). The probability is a dimensionless quantity, and the area $dxdy$ has units of $(\text{length})^2$. Hence, $|\psi(x,y)|^2$ must have units of $(\text{length})^{-2}$, implying that ψ has units of $\boxed{\text{length}^{-1}}$.

E7B.7(a)

(i) The function $\psi = x^2$ $\boxed{\text{cannot be normalized}}$ as the area under $\psi^*\psi = x^4$ is infinite when the integral is evaluated over all space. However, if the function were restricted to a finite region of space, it could be normalized.

(ii) The function $\psi = 1/x$ $\boxed{\text{cannot be normalized}}$ as it goes to ∞ as x goes to 0 so the integral of $\psi^*\psi$ over all space is infinite. However, if the function were restricted to a finite region of space which did not include $x = 0$ it could be normalized.

(iii) The function $\psi = \exp(-x^2)$ $\boxed{\text{can be normalized}}$ as $\psi^*\psi = \exp(-2x^2)$ goes to 0 at $x = \pm\infty$ so that the integral of $\psi^*\psi$ over all space is finite.

A function is an acceptable wavefunction if it: (1) is not infinite over a finite region; (2) is single-valued; (3) is continuous; (4) has a continuous first derivative. The functions $\psi = x^2$ and $\psi = \exp(-x^2)$ satisfy all these conditions. The function $\psi = 1/x$ satisfies these conditions provided that the wavefunction is restricted to a region which excludes $x = 0$.

E7B.8(a) The normalized wavefunction is $\psi(x) = (2/L)^{1/2}\sin(2\pi x/L)$, and so the probability density is $P(x) = |\psi(x)|^2 = (2/L)\sin^2(2\pi x/L)$. This is maximized when $\sin^2(2\pi x/L) = 1$, and so when $\sin(2\pi x/L) = \pm 1$. These values occur when $2\pi x/L = \pi/2, 3\pi/2$ and hence $\boxed{x = L/4, 3L/4}$.

Nodes occur when the wavefunction goes through zero: $\sin(2\pi x/L) = 0$. The sine function is zero when $2\pi x/L = \pi$, hence $\boxed{x = L/2}$. The wavefunction goes to zero at $x = 0$ and $x = L$, but these do not count as nodes as the wavefunction does not pass through zero.

Solutions to problems

P7B.1 (a) The task is to find N such that $\psi = N\mathrm{e}^{\mathrm{i}\phi}$ satisfies the normalization condition [7B.4c–248], $\int \psi^*\psi\,\mathrm{d}\tau = 1$. In this case the integration is over ϕ and the range is 0 to 2π. The function is complex so $\psi^* = N\exp(-\mathrm{i}\phi)$; note that the normalization factor N is always real. The integrand is therefore

$$\psi^*\psi = \mathrm{e}^{-\mathrm{i}\phi}\mathrm{e}^{\mathrm{i}\phi} = \mathrm{e}^{-\mathrm{i}\phi+\mathrm{i}\phi} = \mathrm{e}^0 = 1$$

The integral to evaluate is therefore

$$N^2\int_0^{2\pi} 1\,\mathrm{d}\phi = N^2\,\phi|_0^{2\pi} = 2\pi N^2$$

Setting this equal to 1 gives $\boxed{N = (2\pi)^{-1/2}}$.

(b) For the function $\mathrm{e}^{\mathrm{i}m_l\phi}$ exactly the same argument applies and so the normalization factor is the same.

P7B.3 (a) The task is to find N such that $\psi = N\sin(\pi x/L_x)\sin(\pi y/L_y)$ satisfies the normalization condition [7B.4c–248], $\int \psi^*\psi\,\mathrm{d}\tau = 1$. The integration is over x from 0 to L_x, and y from 0 to L_y. The wavefunction is real, and so the integral to evaluate is

$$N^2\int_0^{L_x}\int_0^{L_y} \sin^2(\pi x/L_x)\sin^2(\pi x/L_y)\mathrm{d}x\mathrm{d}y$$

This separates into the product of two integrals

$$N^2\left(\int_0^{L_x}\sin^2(\pi x/L_x)\mathrm{d}x\right)\left(\int_0^{L_y}\sin^2(\pi y/L_y)\mathrm{d}y\right)$$

Each of these is evaluated using Integral T.2

$$\int_0^{L_x}\sin^2(\pi x/L_x)\mathrm{d}x = L_x/2 - (L_x/4\pi)\sin(2\pi L/L) = L_x/2$$

and similarly the integral over y is equal to $L_y/2$. Hence the integral is evaluated as $N^2(L_x/2)(L_y/2)$. Setting this equal to 1 gives $\boxed{N = 2/\sqrt{L_xL_y}}$.

(b) In the case when $L_x = L_y = L$ this reduces to $\boxed{N = 2/L}$.

P7B.5 The task is to find N such that $\psi(x) = N\mathrm{e}^{-ax}$, satisfies the normalization condition [7B.4c–248], $\int \psi^*\psi\,\mathrm{d}\tau = 1$. The integration is over x ranging from 0 to ∞.

$$N^2\int_0^{\infty}\mathrm{e}^{-2ax}\,\mathrm{d}x = -(N^2/2a)\,\mathrm{e}^{-2ax}\Big|_0^{\infty} = -(N^2/2a)(\mathrm{e}^{-\infty} - \mathrm{e}^0) = N^2/2a$$

where $\mathrm{e}^{-\infty} = 0$ and $\mathrm{e}^0 = 1$ are used. Setting this equal to 1 gives $N = (2a)^{1/2}$. Hence, the total probability of finding the particle at a distance $x \geq x_0$ is

$$\int_{x_0}^{\infty}|\psi(x)|^2\,\mathrm{d}x = 2a\int_{x_0}^{\infty}\mathrm{e}^{-2ax}\,\mathrm{d}x = -(2a/2a)\,\mathrm{e}^{-2ax}\Big|_{x_0}^{\infty} = \mathrm{e}^{-2ax_0}$$

With $a = 2\ \mathrm{m}^{-1}$ and $x_0 = 1$ m, the probability is $\exp(-2\times(2\ \mathrm{m}^{-1})\times(1\ \mathrm{m}))$ $= \exp(-4) = \boxed{0.0183}$

P7B.7 The probability of finding the particle in the range $x = a$ to $x = b$ is

$$P(a \to b) = \int_a^b |\psi(x)|^2 \, dx = \int_a^b (2/L)\sin^2(\pi x/L)\, dx$$

This is evaluated using Integral T.2 to to give

$$P(a \to b) = (2/L)\left[x/2 - (L/4\pi)\sin(2\pi x/L)|_a^b\right]$$
$$= \frac{b-a}{L} - \frac{1}{2\pi}\left[\sin\left(\frac{2\pi b}{L}\right) - \sin\left(\frac{2\pi a}{L}\right)\right]$$

Hence,

(a) $L = 10.00$ nm, $a = 4.95$ nm, $b = 5.05$ nm, $P(a \to b) = \boxed{2.00 \times 10^{-2}}$
(b) $L = 10.00$ nm, $a = 1.95$ nm, $b = 2.05$ nm, $P(a \to b) = \boxed{6.91 \times 10^{-3}}$
(c) $L = 10.00$ nm, $a = 9.90$ nm, $b = 10.00$ nm, $P(a \to b) = \boxed{6.58 \times 10^{-6}}$
(d) $L = 10.00$ nm, $a = 5.00$ nm, $b = 10.00$ nm, $P(a \to b) = \boxed{0.5}$. This is expected as for this wavefunction the probability density is symmetric around $x = 5.00$ nm.

P7B.9 The probability of being in a small volume δV at distance r from the nucleus is $\psi(r)^2\delta V$. The probability of the electron being in a small sphere of radius r' centred at r from the nucleus is

$$P(r) = \left[(\pi a_0^3)^{-1/2} e^{-r/a_0}\right]^2 \times \left[(4/3)\pi(r')^3\right] = \left[4(r')^3/3a_0^3\right] e^{-2r/a_0}$$

where $a_0 = 53$ pm is the Bohr radius. If the sphere has radius 1.0 pm then the expression evaluates to

$$\frac{4 \times (1.0\ \text{pm})^3}{3 \times (53\ \text{pm})^3} \times e^{-2r/a_0} = 8.95... \times 10^{-6} \times e^{-2r/a_0}$$

(a) If the sphere is centered at the nucleus $r = 0$, $P(0) = \boxed{8.95 \times 10^{-6}}$
(b) If the sphere is centered at $r = a_0$, $P(a_0) = 8.95...\times 10^{-6} \times e^{-2} = \boxed{1.21 \times 10^{-6}}$

P7B.11 The probability finding the atom between x and $x + dx$, where dx is an infinitesimal displacement, is given by

$$P(x)dx = \psi(x)^2 dx = N^2 x^2 e^{-x^2/a^2} dx$$

The task is to find the position of the maximum in $P(x)$, which is done by setting $dP(x)/dx = 0$. To compute the derivative it is necessary to use the product rule.

$$\frac{dP(x)}{dx} = \frac{d}{dx}\left(N^2x^2e^{-x^2/a^2}\right) = N^2\left(\frac{dx^2}{dx} \times e^{-x^2/a^2} + x^2 \times \frac{de^{-x^2/a^2}}{dx}\right)$$
$$= N^2\left(2xe^{-x^2/a^2} - x^2 \times \frac{2x}{a^2}e^{-x^2/a^2}\right) = 2xN^2e^{-x^2/a^2}\left[1 - \left(\frac{x}{a}\right)^2\right]$$

The derivative goes to zero at $x = 0, \pm a, \pm\infty$. Inspection of the form of $P(x)$ shows that $x = 0$ is a minimum ($P(x) = 0$), and at $x = \pm\infty$, $P(x)$ also goes asymptotically to zero. The maxima are therefore at $\boxed{x = \pm a}$.

7C Operators and observables

Answers to discussion questions

D7C.1 The operator which represents kinetic energy is $\hat{T} = (-\hbar^2/2m)\mathrm{d}^2/\mathrm{d}x^2$. The expectation value of the kinetic energy is therefore related to the average value of $\mathrm{d}^2\psi/\mathrm{d}x^2$, that is the average of the second derivative or curvature of the wavefunction. Sharply curved regions of the wavefunction will make a larger contribution to the kinetic energy than less sharply curved regions. The expectation value of the kinetic energy will have contributions from all parts of the wavefunction.

D7C.3 This is discussed in Section 7C.3 on page 257.

Solutions to exercises

E7C.1(a) To construct the potential energy operator, replace the position x in the classical expression by the operator for position $\hat{x}$ to give $\hat{V} = \frac{1}{2}k_\mathrm{f}\hat{x}^2$. However, because $\hat{x} = x\times$ the potential energy operator is $\boxed{\hat{V} = \frac{1}{2}k_\mathrm{f}x^2}$.

E7C.2(a) A function ψ is an eigenfunction of an operator $\hat{\Omega}$ if $\hat{\Omega}\psi = \omega\psi$ where ω is a constant called the eigenvalue.

(i) $(\mathrm{d}/\mathrm{d}x)\cos(kx) = -k\sin(kx)$. Hence $\cos kx$ is $\boxed{\text{not}}$ an eigenfunction of the operator $\mathrm{d}/\mathrm{d}x$.

(ii) $(\mathrm{d}/\mathrm{d}x)\mathrm{e}^{ikx} = ik\mathrm{e}^{ikx}$. Hence e^{ikx} $\boxed{\text{is}}$ an eigenfunction of the operator $\mathrm{d}/\mathrm{d}x$, with eigenvalue $\boxed{ik}$.

(iii) $(\mathrm{d}/\mathrm{d}x)kx = k$. Hence kx is $\boxed{\text{not}}$ an eigenfunction of the operator $\mathrm{d}/\mathrm{d}x$.

(iv) $(\mathrm{d}/\mathrm{d}x)\mathrm{e}^{-ax^2} = -2ax\mathrm{e}^{-ax^2}$. Hence e^{-ax^2} is $\boxed{\text{not}}$ an eigenfunction of the operator $\mathrm{d}/\mathrm{d}x$.

E7C.3(a) Wavefunctions ψ_1 and ψ_2 are orthogonal if $\int \psi_1^*\psi_2\,\mathrm{d}\tau = 0$, [7C.8–254]. Here $\psi_1(x) = \sin(\pi x/L)$, $\psi_2(x) = \sin(2\pi x/L)$, and the region is $0 \le x \le L$. The integral is evaluated using Integral T.5

$$\int \psi_1^*\psi_2\mathrm{d}\tau = \int_0^L \sin(\pi x/L)\sin(2\pi x/L)\,\mathrm{d}x$$
$$= [(-L/2\pi)\sin(-\pi) - (L/6\pi)\sin(3\pi)] = 0$$

where $\sin(n\pi) = 0$ for integer n is used. Thus, the two wavefunctions are orthogonal.

E7C.4(a) Wavefunctions ψ_1 and ψ_2 are orthogonal if $\int \psi_1^* \psi_2 \, d\tau = 0$, [7C.8–*254*]. Here $\psi_1(x) = \cos(\pi x/L)$, $\psi_2(x) = \cos(3\pi x/L)$, and the region is $-L/2 \le x \le L/2$. The integral is evaluated using Integral T.6

$$\begin{aligned}\int \psi_1^* \psi_2 d\tau &= \int_{-L/2}^{L/2} \cos(\pi x/L)\cos(3\pi x/L)\, dx \\ &= (-L/4\pi)\sin(-2\pi x/L) + (L/8\pi)\sin(4\pi x/L)|_{-L/2}^{L/2} \\ &= [(-L/4\pi)\sin(-\pi) + (L/8\pi)\sin(2\pi)] \\ &\quad - [(-L/4\pi)\sin(\pi) + (L/8\pi)\sin(-2\pi)] = 0\end{aligned}$$

where $\sin(n\pi) = 0$ for integer n is used. Thus, the two wavefunctions are orthogonal.

E7C.5(a) Two wavefunctions ψ_i and ψ_j are orthogonal if $\int \psi_i^* \psi_j \, d\tau = 0$, [7C.8–*254*]. In this case the integration is from $\phi = 0$ to $\phi = 2\pi$. Let $\psi_i = \exp(i\phi)$, the wavefunction with $m_l = +1$, and $\psi_j = \exp(2i\phi)$, the wavefunction with $m_l = +2$. Note that the functions are complex, so $\psi_i^* = \exp(-i\phi)$. The integrand is therefore

$$\psi_i^* \psi_j = \exp(-i\phi)\exp(2i\phi) = \exp(i\phi)$$

and the integral evaluates as

$$\int_0^{2\pi} \exp(i\phi)\, d\phi = (1/i)\exp(i\phi)|_0^{2\pi} = (1/i)[\overbrace{\exp(i\times 2\pi)}^{=1} - \overbrace{\exp(i\times 0)}^{=1}] = 0$$

The identity $\exp(ix) = \cos x + i \sin x$ (*The chemist's toolkit* 16 in Topic 7C on page 256) is used to evaluate $\exp(i2\pi) = \cos(2\pi) + i\sin(2\pi) = 1 + 0 = 1$. The integral is zero, so the functions are indeed orthogonal.

E7C.6(a) The normalized wavefunction is $\psi(x) = (2/L)^{1/2}\sin(2\pi x/L)$. The operator for position is $\hat{x} = x$, therefore the expectation value of the position of the electron is [7C.11–*256*]

$$\langle x \rangle = \int \psi^* \hat{x} \psi \, d\tau = (2/L)\int_0^L x \sin^2(2\pi x/L)\, dx$$

This integral is of the form of Integral T.11 with $k = 2\pi/L$ and $a = L$

$$\begin{aligned}&= \frac{2}{L}\left[\frac{L^2}{4} - \frac{L}{4\times 2\pi/L}\overbrace{\sin\left(\frac{4\pi L}{L}\right)}^{=\sin 4\pi=0} - \frac{1}{8\times(2\pi/L)^2}\left\{\overbrace{\cos\left(\frac{4\pi L}{L}\right)}^{=\cos 4\pi=1} - 1\right\}\right] \\ &= \frac{2}{L}\times\frac{L^2}{4} = \boxed{L/2}\end{aligned}$$

Because the probability density $|\psi(x)|^2$ is symmetric about $x = L/2$, the expected result is $\langle x \rangle = L/2$.

E7C.7(a) The normalized wavefunction is $\psi(x) = (2/L)^{1/2}\sin(2\pi x/L)$. The expectation value of the momentum is $\int \psi^* \hat{p}_x \psi \, dx$, and the momentum operator is $\hat{p}_x = (\hbar/i)d/dx$, therefore

$$\langle p_x \rangle = (2/L)\int_0^L \sin(2\pi x/L)\hat{p}_x \sin(2\pi x/L)\,dx$$
$$= (2\hbar/iL)\int_0^L \sin(2\pi x/L)(d/dx)\sin(2\pi x/L)\,dx$$

Using $(d/dx)\sin(2\pi x/L) = (2\pi/L)\cos(2\pi x/L)$ gives

$$\langle p_x \rangle = (4\pi\hbar/iL^2)\int_0^L \sin(2\pi x/L)\cos(2\pi x/L)\,dx$$

The integral is of the form of Integral T.7 with $a = L$, $k = 2\pi/L$

$$\langle p_x \rangle = \frac{4\pi\hbar}{iL^2} \times \frac{1}{2\times 2\pi/L}\sin^2\left(\frac{2\pi L}{L}\right) = \boxed{0}$$

This result is interpreted as meaning that there are equal probabilities of having momentum in the positive and negative x directions.

E7C.8(a) For the case when $m_l = +1$ the normalized wavefunction is $\psi_{+1}(\phi) = (2\pi)^{-1/2}e^{i\phi}$. This is complex, and so $\psi_{+1}^* = (2\pi)^{-1/2}e^{-i\phi}$. The expectation value of the position, specified by the operator ϕ, is

$$\langle \phi \rangle = \int_0^{2\pi} \psi_{+1}^*\phi\psi_{+1}\,d\phi = (1/2\pi)\int_0^{2\pi} e^{-i\phi}\phi\, e^{i\phi}\,d\phi$$
$$= (1/2\pi)\int_0^{2\pi}\phi\,d\phi = (1/2\pi)\times(\phi^2/2)\Big|_0^{2\pi} = \boxed{\pi}$$

where $e^{-i\phi}e^{i\phi} = 1$ is used.

For the general case the integral is

$$(1/2\pi)\int_0^{2\pi} e^{-im_l\phi}\phi e^{im_l\phi}\,d\phi$$

which evaluates to give the same result, $\langle\phi\rangle_{m_l} = \pi$. This result is interpreted by noting that the probability density around the ring is $(1/2\pi)e^{-im_l\phi}e^{im_l\phi} = 1/2\pi$, which is constant. Therefore the average position is half way round the ring at $\phi = \pi$.

E7C.9(a) The uncertainty in the momentum is given by $\Delta p = m\Delta v$, where m is the mass and Δv is the uncertainty in the velocity. The uncertainties in position (Δq) and momentum (Δp) must obey the Heisenberg uncertainty principle [7C.13a–258], $\Delta p_q \Delta q \geq (\hbar/2)$, which in this case is expressed as $m\Delta v\Delta q \geq (\hbar/2)$. This is rearranged to give the uncertainty in the velocity, $\Delta v \geq \hbar/(2m\Delta q)$. The minimum uncertainty in the speed is therefore $\Delta v_{min} = \hbar/(2m\Delta q)$, which is evaluated as

$$\frac{1.0546\times10^{-34}\,\mathrm{J\,s}}{2\times(500\times10^{-3}\,\mathrm{kg})\times(1.0\times10^{-6}\,\mathrm{m})} = \boxed{1.05\times10^{-28}\,\mathrm{m\,s^{-1}}}$$

The uncertainty principle can be rearranged for the uncertainty in the position $\Delta q \geq \hbar/2m\Delta v$, and so the minimum uncertainty is $\hbar/(2m\Delta v)$. The uncertainty in the position of the bullet is 1×10^{-5} m s^{-1}, and hence

$$\Delta q_{\min} = \frac{1.0546 \times 10^{-34}\,\text{J s}}{2 \times (5.0 \times 10^{-3}\,\text{kg}) \times (1 \times 10^{-5}\,\text{m s}^{-1})} = \boxed{1.05 \times 10^{-27}\,\text{m}}$$

E7C.10(a) The desired uncertainty in the momentum is

$$\begin{aligned}\Delta p &= 0.0100 \times 10^{-2} p = 1.00 \times 10^{-4} m_{\text{p}} v \\ &= 1.00 \times 10^{-4} \times (1.6726 \times 10^{-27}\,\text{kg}) \times (0.45 \times 10^{6}\,\text{m s}^{-1}) \\ &= 7.52... \times 10^{-26}\,\text{kg m s}^{-1}\end{aligned}$$

The Heisenberg uncertainty principle, [7C.13a–*258*] is rearranged to give the uncertainty in the position as $\Delta q \geq \hbar/(2\Delta p)$, which gives a minimum uncertainty of $\Delta q_{\min} = \hbar/(2\Delta p)$. This is evaluated as

$$\frac{1.0546 \times 10^{-34}\,\text{J s}}{2 \times (7.52... \times 10^{-26}\,\text{kg m s}^{-1})} = \boxed{7.01 \times 10^{-10}\,\text{m}}$$

Solutions to problems

P7C.1 Operate on each function f with $\hat{i}$, the inversion operator, which has the effect of making the replacement $x \rightarrow -x$. If the result of the operation is f multiplied by a constant, f is an eigenfunction of $\hat{i}$ and the constant is the eigenvalue.

(a) $\hat{i}(x^3 - kx) = (-x)^3 - k(-x) = -(x^3 - kx)$. $\boxed{\text{Yes}}$, f is an eigenfunction, eigenvalue $\boxed{-1}$.

(b) $\hat{i}(\cos(kx)) = \cos(k(-x)) = \cos(kx)$. $\boxed{\text{Yes}}$, f is an eigenfunction, eigenvalue $\boxed{+1}$.

(c) $\hat{i}(x^2 + 3x - 1) = (-x)^2 + 3(-x) - 1 = x^2 - 3x - 1$. $\boxed{\text{No}}$, f is not an eigenfunction.

P7C.3 An operator $\hat{\Omega}$ is hermitian if $\int \psi_i^* \hat{\Omega} \psi_j \, \text{d}\tau = \left[\int \psi_j^* \hat{\Omega} \psi_i \, \text{d}\tau\right]^*$, [7C.7–*253*]. For the kinetic energy operator, it is necessary to integrate by parts (*The chemist's toolkit* 15 in Topic 7C on page 254) twice

$$\int \psi_i^* \left(-\frac{\hbar^2}{2m}\frac{\text{d}^2}{\text{d}x^2}\right)\psi_j \, \text{d}x = -\frac{\hbar^2}{2m}\int_{-\infty}^{\infty} \psi_i^* \frac{\text{d}^2\psi_j}{\text{d}x^2}\,\text{d}x$$

$$= -\frac{\hbar^2}{2m}\left(\underbrace{\psi_i^* \frac{\text{d}\psi_j}{\text{d}x}\bigg|_{-\infty}^{\infty}}_{\text{A}} - \int_{-\infty}^{\infty} \frac{\text{d}\psi_i^*}{\text{d}x}\frac{\text{d}\psi_j}{\text{d}x}\,\text{d}x\right) = \frac{\hbar^2}{2m}\int_{-\infty}^{\infty} \frac{\text{d}\psi_i^*}{\text{d}x}\frac{\text{d}\psi_j}{\text{d}x}\,\text{d}x$$

Term A evaluates to zero because, as stated in the problem, the wavefunction and its derivative can be assumed to go to zero at $x = \pm\infty$. Then integrate the remaining term by parts once again,

$$\frac{\hbar^2}{2m}\int_{-\infty}^{\infty}\frac{d\psi_i^*}{dx}\frac{d\psi_j}{dx}\,dx = \frac{\hbar^2}{2m}\left(\underbrace{\frac{d\psi_i^*}{dx}\psi_j\bigg|_{-\infty}^{\infty}}_{\text{B}} - \int_{-\infty}^{\infty}\frac{d^2\psi_i^*}{dx^2}\psi_j\,dx\right)$$

As before, term B evaluates to zero, so the final result is

$$\int \psi_i^*\left(-\frac{\hbar^2}{2m}\frac{d^2}{dx^2}\right)\psi_j\,dx = \int_{-\infty}^{\infty}\psi_j\left(-\frac{\hbar^2}{2m}\frac{d^2}{dx^2}\right)\psi_i^*\,dx$$

The term on the right can be rewritten as a complex conjugate to give

$$\int \psi_i^*\left(-\frac{\hbar^2}{2m}\frac{d^2}{dx^2}\right)\psi_j\,dx = \left[\int \psi_j^*\left(-\frac{\hbar^2}{2m}\frac{d^2}{dx^2}\right)\psi_i\,dx\right]^*$$

Note that because the complex conjugate of the whole term is taken, to compensate for this the complex conjugate of the terms inside the bracket need to be taken too $\psi = [\psi^*]^*$. This final equation is consistent with [7C.7–*253*] and so demonstrates that the kinetic energy operator is hermitian.

P7C.5

(a) Consider the sum of two arbitrary hermitian operators $\hat{A}$ and $\hat{B}$,

$$\int \psi_i^*(\hat{A}+\hat{B})\psi_j\,d\tau = \int \psi_i^*\hat{A}\psi_j\,d\tau + \int \psi_i^*\hat{B}\psi_j\,d\tau$$

As $\hat{A}$ is hermitian, $\int \psi_i^*\hat{A}\psi_j\,d\tau = \left[\int \psi_j^*\hat{A}\psi_i\,d\tau\right]^*$ and similarly for $\hat{B}$.

$$\begin{aligned}\int \psi_i^*(\hat{A}+\hat{B})\psi_j\,d\tau &= \left[\int \psi_j^*\hat{A}\psi_i\,d\tau\right]^* + \left[\int \psi_j^*\hat{B}\psi_i\,d\tau\right]^* \\ &= \left[\int \psi_j^*(\hat{A}+\hat{B})\psi_i\,d\tau\right]^*\end{aligned}$$

which demonstrates that the sum of two hermitian operators is also hermitian.

(b) As $\hat{\Omega}\psi_j = \psi_k$ is also a function, the integral can be rewritten

$$\int \psi_i^*\hat{\Omega}\hat{\Omega}\psi_j\,d\tau = \int \psi_i^*\hat{\Omega}\psi_k\,d\tau$$

As $\hat{\Omega}$ is a hermitian operator it follows that

$$\int \psi_i^*\hat{\Omega}\psi_k\,d\tau = \left[\int \psi_k^*\hat{\Omega}\psi_i\,d\tau\right]^*$$

Next realise that $\hat{\Omega}\psi_i$ is a function which will be called ψ_l: $\hat{\Omega}\psi_i = \psi_l$. With this substitution

$$\int \psi_i^* \hat{\Omega}\psi_k \, d\tau = \left[\int \psi_k^* \psi_l \, d\tau\right]^* = \int \psi_k \psi_l^* \, d\tau$$
$$= \int \psi_l^* \psi_k \, d\tau$$
$$= \int \psi_l^* \hat{\Omega}\psi_j \, d\tau$$

To go to the second line the two functions have been re-ordered, which is always permitted, and to go to the third line the definition $\hat{\Omega}\psi_j = \psi_k$ has been used. Because $\hat{\Omega}$ is hermitian the final term can be rewritten as $[\int \psi_j^* \hat{\Omega}\psi_l \, d\tau]^*$ and so

$$\int \psi_i^* \hat{\Omega}\psi_k \, d\tau = \left[\int \psi_j^* \hat{\Omega}\psi_l \, d\tau\right]^*$$

In the term on the right the definition $\hat{\Omega}\psi_i = \psi_l$ is used, and in the term on the left $\hat{\Omega}\psi_j = \psi_k$ is used to give

$$\int \psi_i^* \hat{\Omega}\hat{\Omega}\psi_j \, d\tau = \left[\int \psi_j^* \hat{\Omega}\hat{\Omega}\psi_i \, d\tau\right]^*$$

This is exactly the property which proves that $\hat{\Omega}\hat{\Omega}$ is hermetian.

P7C.7 The operator for position is multiplication by x, so it follows that the expectation value of the position operator is

$$\langle x \rangle = \int \psi^* \hat{x} \psi \mathrm{d}\tau = a \int_0^\infty x\mathrm{e}^{-ax} \mathrm{d}x = a \times \frac{1!}{a^2} = \boxed{1/a}$$

The integral is of the form of Integral E.3 with $n = 1$ and $k = a$.

P7C.9 (a) The expectation value is given by [7C.11–256], $\langle \Omega \rangle = \int \psi^* \hat{\Omega}\psi \, d\tau$, where ψ is normalized. A hermitian operator has the property $\int \psi_i^* \hat{\Omega}\psi_j \, d\tau = [\int \psi_j^* \hat{\Omega}\psi_i \, d\tau]^*$, which for the case $\psi_i = \psi_j = \psi$ becomes

$$\overbrace{\int \psi^* \hat{\Omega}\psi \, d\tau}^{A} = \left[\overbrace{\int \psi^* \hat{\Omega}\psi \, d\tau}^{B}\right]^*$$

The quantities A and B and the same, so it follows that $A = B^* = A^*$, which is only true if A is real. The quantity A is the expectation value, so it follows that the expectation value of a hermitian operator is real.

(b) The expectation value of the square of the hermitian operator $\hat{\Omega}\hat{\Omega}$ is

$$\langle \Omega\Omega \rangle = \int \psi^* \hat{\Omega}\hat{\Omega}\psi \, d\tau$$

Recognise that $\hat{\Omega}\psi$ is a function, which will be written ϕ: $\hat{\Omega}\psi = \phi$; with this, and using the fact that $\hat{\Omega}$ is hermitian it follows that

$$\int \psi^* \hat{\Omega}\hat{\Omega}\psi \,\mathrm{d}\tau = \int \psi^* \hat{\Omega}\phi \,\mathrm{d}\tau = \left[\int \phi^* \hat{\Omega}\psi \,\mathrm{d}\tau\right]^*$$

Once again use $\hat{\Omega}\psi = \phi$ in the term on the right to give

$$\int \psi^* \hat{\Omega}\hat{\Omega}\psi \,\mathrm{d}\tau = \left[\int \phi^* \phi \,\mathrm{d}\tau\right]^*$$

The integral $\int \phi^* \phi \,\mathrm{d}\tau$ is the sum over all space of the square magnitude of a wavefunction. Using the Born interpretation, this is proportional to the total probability and therefore must be a real positive number. It is therefore shown that the expectation value of the square of a hermitian operator is real and positive.

P7C.11 (a) The normalized wavefunction is $\psi(x) = (2a/\pi)^{1/4}\mathrm{e}^{-ax^2}$, leading to a probability density of $|\psi(x)|^2 = (2a/\pi)^{1/2}\mathrm{e}^{-2ax^2}$. The expectation value of the position is therefore

$$\langle x\rangle = (2a/\pi)^{1/2}\int_{-\infty}^{\infty} x\mathrm{e}^{-2ax^2}\,\mathrm{d}x = \boxed{0}$$

as the integrand is odd and the integration is over a symmetric range. For $\langle x^2\rangle$ the required integral is

$$\langle x^2\rangle = (2a/\pi)^{1/2}\int_{-\infty}^{\infty} x^2\mathrm{e}^{-2ax^2}\,\mathrm{d}x = 2(2a/\pi)^{1/2}\int_{0}^{\infty} x^2\mathrm{e}^{-2ax^2}\,\mathrm{d}x$$

The integrand is an even function, and so its integral between $-\infty$ and ∞ is twice that from 0 to ∞. This integral is of the form of Integral G.3 with $k = 2a$

$$\langle x^2\rangle = 2(2a/\pi)^{1/2} \times \tfrac{1}{4}(\pi/(2a)^3)^{1/2} = \boxed{1/4a}$$

The effect of the momentum operator, $(\hbar/\mathrm{i})(\mathrm{d}/\mathrm{d}x)$, on the wavefunction is $(\hbar/\mathrm{i})(\mathrm{d}/\mathrm{d}x)(2a/\pi)^{1/4}\mathrm{e}^{-ax^2} = (\hbar/\mathrm{i})(2a/\pi)^{1/4} \times (-2ax\mathrm{e}^{-ax^2})$. Hence the expectation value of the momentum is

$$\langle p_x\rangle = \int \psi^* \hat{p}_x\psi \,\mathrm{d}\tau = \frac{\hbar}{\mathrm{i}}\left(\frac{2a}{\pi}\right)^{1/2}\int_{-\infty}^{\infty} -2ax\mathrm{e}^{-2ax^2}\,\mathrm{d}x = \boxed{0}$$

as the integrand is odd.

To evaluate the effect of the operator $\hat{p}_x^2 = -\hbar^2(\mathrm{d}^2/\mathrm{d}x^2)$ on the wavefunction the product rule is used

$$\hat{p}_x^2\psi = \hat{p}_x(\hat{p}_x)\psi = \frac{\hbar}{\mathrm{i}}\frac{\mathrm{d}}{\mathrm{d}x}\left((\hbar/\mathrm{i})(2a/\pi)^{1/4} \times [-2ax\mathrm{e}^{-ax^2}]\right)$$

This is equal to

$$-\hbar^2\,(2a/\pi)^{1/4}\,(4a^2x^2\mathrm{e}^{-ax^2} - 2a\mathrm{e}^{-ax^2})$$

which gives

$$\begin{aligned}\langle p_x^2\rangle &= \int \psi^* \hat{p}_x^2 \psi \,\mathrm{d}\tau \\ &= -\hbar^2 (2a/\pi)^{1/2} \int_{-\infty}^{+\infty} (4a^2x^2\mathrm{e}^{-2ax^2} - 2a\mathrm{e}^{-2ax^2}) \\ &= -\hbar^2 \left(\frac{2a}{\pi}\right)^{1/2} \times \left[2a^2\left(\frac{\pi}{(2a)^3}\right)^{1/2} - 2a\left(\frac{\pi}{2a}\right)^{1/2}\right] \\ &= \boxed{\hbar^2 a}\end{aligned}$$

where Integrals G.3 and G.1 are used.

(b) The uncertainty in the position is given by

$$\Delta x = \left[\langle x^2\rangle - \langle x\rangle^2\right]^{1/2} = (1/4a - 0^2)^{1/2} = (4a)^{-1/2}$$

and the uncertainty in the momentum is given by

$$\Delta p_x = \left[\langle p_x^2\rangle - \langle p_x\rangle^2\right]^{1/2} = (\hbar^2 a - 0^2)^{1/2} = \hbar\sqrt{a}$$

The product of the uncertainties is

$$\Delta x \Delta p_x = (4a)^{-1/2} \times \hbar\sqrt{a} = \hbar/2$$

This uncertainty is $\geq \hbar/2$ and so is consistent with the uncertainty principle; in fact it is the smallest possible uncertainty.

P7C.13 To evaluate the commutator of the operators consider the effect of the commutator on an arbitrary function ψ; use the product rule to evaluate the derivatives of products as necessary.

(a)

$$\begin{aligned}\left[\frac{\mathrm{d}}{\mathrm{d}x}, \frac{1}{x}\right]\psi &= \frac{\mathrm{d}}{\mathrm{d}x}\left(\frac{\psi}{x}\right) - \frac{1}{x}\frac{\mathrm{d}\psi}{\mathrm{d}x} = \frac{1}{x}\frac{\mathrm{d}\psi}{\mathrm{d}x} + \psi\frac{\mathrm{d}x^{-1}}{\mathrm{d}x} - \frac{1}{x}\frac{\mathrm{d}\psi}{\mathrm{d}x} \\ &= -\frac{1}{x^2}\psi\end{aligned}$$

The commutator is therefore identified as the term multiplying ψ, $\boxed{-1/x^2}$.

(b)

$$\begin{aligned}\left[\frac{\mathrm{d}}{\mathrm{d}x}, x^2\right]\psi &= \frac{\mathrm{d}(x^2\psi)}{\mathrm{d}x} - x^2\frac{\mathrm{d}\psi}{\mathrm{d}x} = x^2\frac{\mathrm{d}\psi}{\mathrm{d}x} + \psi\frac{\mathrm{d}x^2}{\mathrm{d}x} - x^2\frac{\mathrm{d}\psi}{\mathrm{d}x} \\ &= 2x\,\psi\end{aligned}$$

The commutator is therefore identified as the term multiplying ψ, $\boxed{2x}$.

P7C.15 To complete this problem it is useful to establish some general properties of commutators. First, an operator $\hat{A}$ commutes with powers of itself: for example, $[\hat{A}, \hat{A}^2] = 0$. This is proved by simply multiplying out the commutator:

$$[\hat{A}, \hat{A}^2] = \hat{A}(\hat{A}\hat{A}) - (\hat{A}\hat{A})\hat{A} = 0$$

Note that multiplying either of the terms by a constant does not alter this property. Second, any operator commutes with a constant: $[\hat{A}, c] = 0$, which follows trivially on multiplying out the commutator. Finally, the useful property that $[\hat{A} + \hat{B}, \hat{C}] = [\hat{A}, \hat{C}] + [\hat{B}, \hat{C}]$. Again, this follows by multiplying out the commutator

$$[\hat{A} + \hat{B}, \hat{C}] = (\hat{A} + \hat{B})\hat{C} - \hat{C}(\hat{A} + \hat{B}) = \hat{A}\hat{C} - \hat{C}\hat{A} + \hat{B}\hat{C} - \hat{C}\hat{B} = [\hat{A}, \hat{C}] + [\hat{B}, \hat{C}]$$

(a) For the case $\hat{V} = V_0$:

$$[\hat{H}, \hat{p}_x] = [\hat{p}_x^2/2m + V_0, \hat{p}_x] = \overbrace{[\hat{p}_x^2/2m, \hat{p}_x]}^{\text{A}} + \overbrace{[V_0, \hat{p}_x]}^{\text{B}} \boxed{= 0}$$

Term A is zero because it is a commutator of an operator with a power of itself, and term B is zero because it is a commutator with a constant.
For the case $\hat{V} = \frac{1}{2}k_f x^2$:

$$[\hat{H}, \hat{p}_x] = [\hat{p}_x^2/2m + \tfrac{1}{2}k_f x^2, \hat{p}_x] = [\hat{p}_x^2/2m, \hat{p}_x] + [\tfrac{1}{2}k_f x^2, \hat{p}_x]$$

The first commutator is zero. To evaluate the second commutator consider its effect on an arbitrary wavefunction ψ; to simplify the calculation set aside the constant multiplying term $\frac{1}{2}k_f$ and re-introduce it at the end

$$\begin{aligned}[x^2, \hat{p}_x]\psi &= (x^2\hat{p}_x - \hat{p}_x x^2)\psi \\ &= \frac{\hbar}{\mathrm{i}}\left(x^2\frac{\mathrm{d}\psi}{\mathrm{d}x} - \frac{\mathrm{d}(x^2\psi)}{\mathrm{d}x}\right) = \frac{\hbar}{\mathrm{i}}\left(x^2\frac{\mathrm{d}\psi}{\mathrm{d}x} - x^2\frac{\mathrm{d}\psi}{\mathrm{d}x} - \psi \times 2x\right) \\ &= -2x(\hbar/\mathrm{i})\psi\end{aligned}$$

Reintroducing the constants gives the result $\boxed{[\hat{H}, \hat{p}_x] = -(k_f\hbar/\mathrm{i})x}$.

(b) For the case $\hat{V} = V_0$:

$$[\hat{H}, \hat{x}] = [\hat{p}_x^2/2m + V_0, \hat{x}] = [\hat{p}_x^2/2m, \hat{x}] + [V_0, \hat{x}]$$

The second term is zero because it is a commutator with a constant. The first term is evaluated in the usual way, first setting aside the constants

$$\begin{aligned}[\hat{p}_x^2, \hat{x}]\psi &= (\hat{p}_x^2 x - x\hat{p}_x^2)\psi \\ &= -\hbar^2\left[\frac{\mathrm{d}^2(x\psi)}{\mathrm{d}x^2} - x\frac{\mathrm{d}^2\psi}{\mathrm{d}x^2}\right] = -\hbar^2\left[\frac{\mathrm{d}}{\mathrm{d}x}(\psi + x\frac{\mathrm{d}\psi}{\mathrm{d}x}) - x\frac{\mathrm{d}^2\psi}{\mathrm{d}x^2}\right] \\ &= -\hbar^2\left[\frac{\mathrm{d}\psi}{\mathrm{d}x} + x\frac{\mathrm{d}^2\psi}{\mathrm{d}x^2} + \frac{\mathrm{d}\psi}{\mathrm{d}x} - x\frac{\mathrm{d}^2\psi}{\mathrm{d}x^2}\right] = -2\hbar^2\frac{\mathrm{d}\psi}{\mathrm{d}x}\end{aligned}$$

Reintroducing the constants gives the result $\boxed{[\hat{H},\hat{x}] = -(\hbar^2/m)(\mathrm{d}/\mathrm{d}x)}$

For the case $\hat{V} = \frac{1}{2}k_\mathrm{f}x^2$:

$$[\hat{H},\hat{x}] = [\hat{p}_x^2/2m + \tfrac{1}{2}k_\mathrm{f}x^2, x] = [\hat{p}_x^2/2m, x] + [\tfrac{1}{2}k_\mathrm{f}x^2, x]$$

The second commutator is zero. The first has already been evaluated in the preceding calculation hence $\boxed{[\hat{H},\hat{x}] = -(\hbar^2/m)(\mathrm{d}/\mathrm{d}x)}$

7D Translational motion

Answers to discussion questions

D7D.1 The Schrödinger equation for a free particle, that is one experiencing zero potential energy, is solved by functions of the form $\psi = \sin kx$ for any value of k. The corresponding energy is $k^2\hbar^2/2m$. This solution applies inside the box but, because the potential goes to infinity outside the box, in these regions the wavefunction must be zero. At the walls of the box the wavefunction must be continuous, and therefore as the wavefunction inside the box goes to the walls it must go to zero. Assuming that the walls are at $x = 0$ and $x = L$ this condition is only satisfied if $k = n\pi/L$, where $n = 1, 2, \ldots$. The energy is therefore quantized with values $n^2\hbar^2\pi^2/2mL^2$.

D7D.3 The physical origin of tunnelling is related to the probability density of the particle, which according to the Born interpretation is the square of the wavefunction that represents the particle. This interpretation requires that the wavefunction of the system be everywhere continuous, even at barriers. Therefore, if the wavefunction is non zero on one side of a barrier it must be non zero on the other side of the barrier and this implies that the particle has tunnelled into the barrier. The transmission probability depends upon the mass of the particle: the greater the mass the smaller the probability of tunnelling. Electrons and protons have small masses, molecular groups large masses; therefore, tunnelling effects are more observable in process involving electrons and protons.

Solutions to exercises

E7D.1(a) The linear momentum of a free electron is given by

$$p = \hbar k = (1.0546\times10^{-34}\,\mathrm{J\,s})\times(3\times10^{9}\,\mathrm{m^{-1}}) = \boxed{3\times10^{-25}\,\mathrm{kg\,m\,s^{-1}}}$$

where $1\,\mathrm{J} = 1\,\mathrm{kg\,m^2\,s^{-2}}$ is used; note that $1\,\mathrm{nm^{-1}} = 1\times10^{9}\,\mathrm{m^{-1}}$. The kinetic energy is given by [7D.2–*261*]

$$E_\mathrm{k} = \frac{(\hbar k)^2}{2m} = \frac{((1.0546\times10^{-34}\,\mathrm{J\,s})\times(3\times10^{9}\,\mathrm{m^{-1}}))^2}{2\times(9.1094\times10^{-31}\,\mathrm{kg})} = \boxed{5\times10^{-20}\,\mathrm{J}}$$

E7D.2(a) The electron is travelling in the negative x direction hence the momentum, and the value of k, are both negative. The kinetic energy is related to k through [7D.2–261], $E_k = k^2\hbar^2/2m$, which is rearranged to give k

$$k = -\left(\frac{2mE_k}{\hbar^2}\right)^{1/2} = -\left(\frac{2\times(2.0\times10^{-3}\,\text{kg})\times(20\,\text{J})}{(1.0546\times10^{-34}\,\text{J s})^2}\right)^{1/2} = -2.7\times10^{33}\,\text{m}^{-1}$$

The wavefunction is then, with x is measured in metres,

$$\psi(x) = e^{ikx} = \boxed{e^{-i(2.7\times10^{33}\,\text{m}^{-1})x}}$$

E7D.3(a) The energy levels of a particle in a box are given by [7D.6–263], $E_n = n^2h^2/8mL^2$, where n is the quantum number. With the mass equal to that of the electron and the length as 1.0 nm, the energies are

$$E_n = \frac{n^2h^2}{8mL^2} = n^2\times\frac{(6.6261\times10^{-34}\,\text{J s})^2}{8\times(9.1094\times10^{-31}\,\text{kg})\times(1.0\times10^{-9}\,\text{m})^2}$$
$$= n^2\times(6.02...\times10^{-20}\,\text{J})$$

To convert to kJ mol^{-1}, multiply through by Avogadro's constant and divide by 1000.

$$= n^2\times(6.02...\times10^{-20}\,\text{J})\times\frac{6.0221\times10^{23}\,\text{mol}^{-1}}{1000} = n^2\times(36.2...\,\text{kJ mol}^{-1})$$

To convert to electronvolts divide through by the elementary charge

$$= n^2\times(6.02...\times10^{-20}\,\text{J})\times\frac{1}{1.6022\times10^{-19}\,\text{C}} = n^2\times(0.376...\,\text{eV})$$

To convert to reciprocal centimetres, divide by hc, with c in cm s^{-1}

$$= n^2\times\frac{(6.02...\times10^{-20}\,\text{J})}{(6.6261\times10^{-34}\,\text{J s})\times(2.9979\times10^{10}\,\text{cm s}^{-1})} = n^2\times(3.03...\times10^3\,\text{cm}^{-1})$$

The energy separation between two levels with quantum numbers n_1 and n_2 is

$$\Delta E(n_1,n_2) = E_{n_2} - E_{n_1} = (n_2^2 - n_1^2)\times(6.02...\times10^{-20}\,\text{J})$$

The values in the other units are found by using the appropriate value of the constant, computed above.

(i) $\Delta E(1,2) = \boxed{1.8\times10^{-19}\,\text{J}}, \boxed{1.1\times10^2\,\text{kJ mol}^{-1}}, \boxed{1.1\,\text{eV}}$, or $\boxed{9.1\times10^3\,\text{cm}^{-1}}$

(ii) $\Delta E(5,6) = \boxed{6.6\times10^{-19}\,\text{J}}, \boxed{4.0\times10^2\,\text{kJ mol}^{-1}}, \boxed{4.1\,\text{eV}}$, or $\boxed{3.3\times10^4\,\text{cm}^{-1}}$

E7D.4(a) The wavefunctions are $\psi_1(x) = (2/L)^{1/2}\sin(\pi x/L)$ and $\psi_2(x) = (2/L)^{1/2} \times \sin(2\pi x/L)$. They are orthogonal if $\int \psi_1^*\psi_2\,d\tau = 0$. In this case the integral is taken from $x = 0$ to $x = L$ as outside this range the wavefunctions are zero. The required integral is of the form of Integral T.5, with $A = \pi/L, B = 2\pi/L$, and $a = L$

$$\frac{2}{L}\int_0^L \sin\left(\frac{\pi x}{L}\right)\sin\left(\frac{2\pi x}{L}\right)dx = \frac{2}{L}\left[\frac{\sin(-\pi L/L)}{-2\pi/L} - \frac{\sin(3\pi L/L)}{6\pi/L}\right]$$
$$= \frac{2}{L}\left[\frac{\sin(-\pi)}{-2\pi/L} - \frac{\sin(3\pi)}{6\pi/L}\right] = 0$$

where $\sin n\pi = 0$ for integer n is used. The two wavefunctions are orthogonal.

E7D.5(a) The particle in a box wavefunction with quantum number n is given by $\psi_n(x) = (2/L)^{1/2}\sin(n\pi x/L)$ The probability of finding the electron in a small region of space δx centred on position x is approximated as $\psi^2(x)\delta x$. For this *Exercise* $x = 0.50L, \delta x = 0.02L$.

For the case where $n = 1$

$$\psi_1(0.50L)^2 \times 0.02L = \left[\sqrt{2/L}\sin\left(\pi(0.50L)/L\right)\right]^2 \times 0.02L$$
$$= (2/L)\sin^2(0.50\pi) \times 0.02L = \boxed{0.04}$$

For the case where $n = 2$

$$\psi_2(0.50L)^2 \times 0.02L = \left[\sqrt{2/L}\sin\left(2\times\pi(0.50L)/L\right)\right]^2 \times 0.02L$$
$$= (2/L)\sin^2(1.00\pi) \times 0.02L = \boxed{0}$$

E7D.6(a) The wavefunction of the lowest energy state for a particle in a box has $n = 1$ and is $\psi_1(x) = \sqrt{2/L}\sin(\pi x/L)$, which leads to a probability density $P_1(x) = |\psi_1(x)|^2 = (2/L)\sin^2(\pi x/L)$. Graphs of these functions are shown in Fig 7.1.

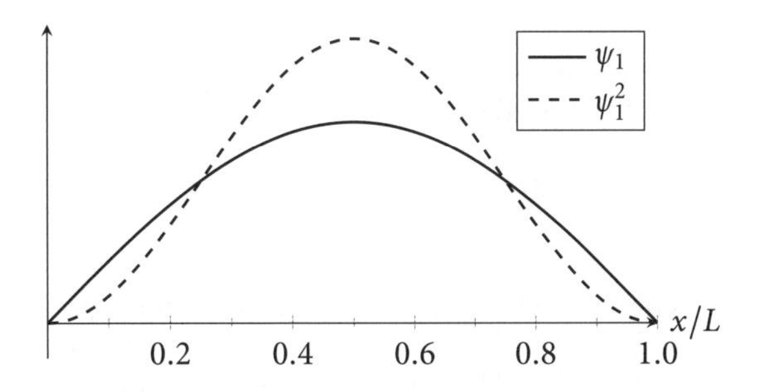

Figure 7.1

The probability density is symmetric about $x = L/2$. Therefore, there is an equal probability of observing the particle at an arbitrary position x' and at $L - x'$, so it follows that the average position of the particle must be at $L/2$.

E7D.7(a) The wavefunction for the state with $n = 1$ is $\psi_1(x) = \sqrt{2/L}\sin(\pi x/L)$, which leads to a probability density $P_1(x) = |\psi_1(x)|^2 = (2/L)\sin^2(\pi x/L)$; these are plotted in Fig 7.1.

The plotted probability density gives the probability of the particle being found in an interval at a particular value of x – that is, it is the probability density of x. It is *not* the probability density of x^2, so there is no simple way of inferring from the plot the value of $\langle x^2\rangle$. The fact that the probability density of x is symmetric about $L/2$ does not imply that the probability density of x^2 has the same property. There is therefore no reason to assume that $\langle x^2\rangle = (L/2)^2$.

E7D.8(a) For an electron in a square well of side L the energy of the state characterized by quantum numbers n_1 and n_2 is given by [7D.12b–*267*], $E_{n_x,n_y} = h^2(n_x^2 + n_y^2)/8m_eL^2$, where n_x, n_y are integers greater than or equal to one. Hence, the minimum or zero-point energy has $n_x = n_y = 1$,

$$E_{1,1} = h^2(1^2 + 1^2)/8m_eL^2 = h^2/4m_eL^2$$

Setting this equal to the rest energy m_ec^2, and rearranging for the length, L

$$h^2/4m_eL^2 = m_ec^2$$

$$L = h/2m_ec = \boxed{\lambda_C/2}$$

where $\lambda_C = h/(m_ec)$ is the Compton wavelength.

E7D.9(a) The wavefunction for a particle in a one-dimensional box with length L and in the state with $n = 3$ is $\psi(x) = \sqrt{2/L}\sin(3\pi x/L)$ giving a probability density of $P(x) = |\psi(x)|^2 = (2/L)\sin^2(3\pi x/L)$.

As this is a trigonometric function, the function is maximized when $\sin^2(3\pi x/L) = 1$, which is when $\sin(3\pi x/L) = \pm 1$; these values occur when the argument of the sine function is an odd multiple of $\pi/2$

$$3\pi x/L = \pi/2, 3\pi/2, 5\pi/2... \quad \text{hence} \quad x = \boxed{L/6, L/2, 5L/6}$$

The probability density is zero when the wavefunction is zero, which is when the argument of the sine function is a multiple of π

$$3\pi x/L = 0, \pi, 2\pi, ... \quad \text{hence} \quad x = \boxed{0, L/3, 2L/3, L}$$

E7D.10(a) The energy levels of a particle in a box with length L are given by [7D.6–*263*], $E_n = h^2n^2/8mL^2$ and when the length is increased to $1.1 \times L$ the energies become $E_n' = h^2n^2/8m(1.1 \times L)^2 = h^2n^2/8m(1.21 \times L^2)$ giving a fractional change of

$$\frac{E_n' - E_n}{E_n} = \frac{[h^2n^2/8m(1.21 \times L^2)] - [h^2n^2/8mL^2]}{h^2n^2/8mL^2} = \frac{1}{1.21} - 1 = \boxed{-0.174}$$

E7D.11(a) The energy levels of a particle in a box of length L are given by [7D.6–263], $E_n = h^2n^2/8mL^2$. Hence the spacing between adjacent levels (n and $n+1$) is given by

$$\Delta E(n) = E_{n+1} - E_n = \frac{h^2(n+1)^2}{8mL^2} - \frac{h^2n^2}{8mL^2} = \frac{h^2}{8mL^2}\left((n+1)^2 - n^2\right)$$
$$= \frac{h^2}{8mL^2}(n^2 + 2n + 1 - n^2) = \frac{h^2}{8mL^2}(2n+1)$$

Setting $\Delta E(n)$ to the average thermal energy gives $h^2(2n+1)/8mL^2 = kT/2$, and so $(2n+1) = 4mkTL^2/h^2$, leading to

$$\boxed{n = \frac{2mkTL^2}{h^2} - \frac{1}{2}}$$

For a helium atom, mass 4.00 m_u in a box for length 1 cm,

$$n = \Big[2(4.00 \times 1.6605 \times 10^{-27}\,\text{kg}) \times (1.3806 \times 10^{-23}\,\text{J K}^{-1}) \times (298\,\text{K}) \times \frac{(1 \times 10^{-2}\,\text{m})^2}{(6.6261 \times 10^{-34}\,\text{J s})^2}\Big] - \tfrac{1}{2} = \boxed{1.24 \times 10^{16}}$$

E7D.12(a) The wavefunction of a particle in a square box of side length L with quantum numbers $n_1 = 2, n_2 = 2$ is $\psi_{2,2}(x,y) = (2/L)\sin(2\pi x/L)\sin(2\pi y/L)$. The probability density is $P_{2,2}(x,y) = (2/L)^2 \sin^2(2\pi x/L)\sin^2(2\pi y/L)$. The probability density is maximized when $\sin^2(2\pi x/L) \times \sin^2(2\pi y/L) = 1$ which occurs only when each sin term is equal to ± 1,

$$\sin(2\pi x/L) = \pm 1$$
$$2\pi x/L = \pi/2, 3\pi/2 \quad \text{hence} \quad x = L/4, 3L/4$$

and similarly for y. Hence the maxima in probability density occurs at $(x,y) =$ $\boxed{(L/4, L/4), (L/4, 3L/4), (3L/4, L/4), (3L/4, 3L/4)}$

Nodes occur when the wavefunction passes through zero, which is when either of the sin terms are zero, excluding the boundaries at $x = 0, L$ $y = 0, L$ because at these points the wavefunction does not pass through zero. There is thus a node when $\sin(2\pi x/L) = 0$, corresponding to $x = L/2$ and any value of y. This node is therefore a line at $\boxed{x = L/2 \text{ and parallel to the } y \text{ axis}}$. Similarly there is another nodal line at $\boxed{y = L/2 \text{ and parallel to the } x \text{ axis}}$.

E7D.13(a) The energy levels for a 2D rectangular box, side lengths L_1, L_2 are

$$E_{n_1,n_2} = \frac{h^2}{8m}\left(\frac{n_1^2}{L_1^2} + \frac{n_2^2}{L_2^2}\right)$$

where n_1 and n_2 are integers greater than or equal to 1.

For the specific case where $L_1 = L, L_2 = 2L$,

$$E_{n_1,n_2} = \frac{h^2}{8m}\left(\frac{n_1^2}{L^2} + \frac{n_2^2}{(2L)^2}\right) = \frac{h^2}{8mL^2}\left(n_1^2 + \frac{n_2^2}{4}\right)$$

The energy of the state with $n_1 = n_2 = 2$ is then

$$E_{2,2} = \frac{h^2}{8mL^2}\left(2^2 + \frac{2^2}{4}\right) = \frac{h^2}{8mL^2}(5)$$

This is degenerate with the state with $n_1 = 1, n_2 = 4$

$$E_{(1,4)} = \frac{h^2}{8mL^2}\left(1^2 + \frac{4^2}{4}\right) = \frac{h^2}{8mL^2}(5)$$

The question notes that degeneracy frequently accompanies symmetry, and suggests that one might be surprised to find degeneracy in a box with unequal lengths. Symmetry is a matter of degree. This box is less symmetric than a square box, but it is more symmetric than boxes whose sides have a non-integer or irrational ratio. Every state of a square box except those with $n_1 = n_2$ is degenerate (with the state that has n_1 and n_2 reversed). Only a few states in this rectangular box are degenerate. In this system, a state (n_1, n_2) is degenerate with a state $(n_2/2, 2n_1)$ as long as the latter state (a) exists (that is, $n_2/2$ must be an integer) and (b) is distinct from (n_1, n_2). A box with incommensurable sides, say, L and $\sqrt{2}L$, would have no degenerate levels.

E7D.14(a) The energy levels of a cubic box are given by [7D.13b–*267*]

$$E_{n_1,n_2,n_3} = \frac{h^2}{8mL^2}(n_1^2 + n_2^2 + n_3^2)$$

where n_1, n_2, n_3 are integers greater than or equal to 1. Hence the lowest energy state is that with $n_1 = n_2 = n_3 = 1$, with energy

$$E_{(1,1,1)} = \frac{h^2}{8mL^2}(1^2 + 1^2 + 1^2) = \frac{h^2}{8mL^2}(3)$$

and so the energy of the level with energy three times that of the lowest is $9h^2/8mL^2$, which will be produced by states for which $n_1^2 + n_2^2 + n_3^2 = 9$. There are three states with this energy, $(n_1, n_2, n_3) = (1,2,2), (2,1,2), (2,2,1)$, and so the degeneracy is $\boxed{3}$.

E7D.15(a) The transmission probability [7D.20a–*269*] depends on the energy of the tunnelling particle relative to the barrier height ($\varepsilon = E/V_0 = (1.5\text{ eV})/(2.0\text{ eV}) = 0.75$), the width of the barrier, ($L = 100$ pm), and the decay parameter of the wavefunction inside the barrier (κ)

$$\begin{aligned}\kappa &= \frac{(2m(E - V_0))^{1/2}}{\hbar} \\ &= \frac{(2 \times (9.1094 \times 10^{-31}\text{ kg}) \times [(2.0 - 1.5)\text{ eV} \times 1.6022 \times 10^{-19}\text{ J eV}^{-1}])^{1/2}}{1.0546 \times 10^{-34}\text{ J s}} \\ &= 3.68... \times 10^{9}\text{ m}^{-1}\end{aligned}$$

The product $\kappa L = (3.68... \times 10^9\ \text{m}^{-1}) \times (100 \times 10^{-12}\ \text{m}) = 0.368...$, and so the transmission probability is given by

$$T = \left[1 + \frac{(e^{\kappa L} - e^{-\kappa L})^2}{16\varepsilon(1-\varepsilon)}\right]^{-1} = \left[1 + \frac{(e^{0.368...} - e^{-0.368...})^2}{16 \times 0.75 \times (1-0.75)}\right]^{-1} = \boxed{0.84}$$

Solutions to problems

P7D.1 The energy levels of a particle in a one dimensional box are $E_n = h^2n^2/8mL^2$. The mass of an O_2 molecule is $2 \times 16.00\ \text{amu} \times 1.6605 \times 10^{-27}\ \text{kg} = 5.3136 \times 10^{-26}\ \text{kg}$. To find the value of n that makes the energy equal to the average thermal energy, set $E_n = \frac{1}{2}kT$ and solve for n

$$n = \frac{2L\sqrt{mkT}}{h}$$

$$= \frac{2\times(5.0 \times 10^{-2}\ \text{m})\times\left[(5.3136 \times 10^{-26}\ \text{kg})\times(300\ \text{K})\times(1.3806 \times 10^{-23}\ \text{J K}^{-1})\right]^{1/2}}{6.6261 \times 10^{-34}\ \text{J s}}$$

$$= 2.23... \times 10^9 = \boxed{2.2 \times 10^9}$$

The separation of level n from the one immediately below it, $n-1$, is $E_n - E_{n-1}$

$$E_n - E_{n-1} = \frac{h^2n^2}{8mL^2} - \frac{h^2(n-1)^2}{8mL^2} = \frac{h^2(2n-1)}{8mL^2}$$

$$= \frac{(6.6261 \times 10^{-34}\ \text{J s})^2 \times (2.23... \times 10^9 - 1)}{8 \times (5.3136 \times 10^{-26}\ \text{kg}) \times (5.0 \times 10^{-2}\ \text{m})^2}$$

$$= \boxed{1.8 \times 10^{-30}\ \text{J}}$$

P7D.3 The normalized wavefunction with $n = 1$ is $\psi_1(x) = (2/L)^{1/2}\sin(\pi x/L)$, and this is used in [7C.11–256], $\langle\Omega\rangle = \int \psi^*\hat{\Omega}\psi\,d\tau$, to compute the expectation value for the cases $\hat{\Omega} = x$ and $\hat{\Omega} = x^2$.

$$\langle x\rangle = \int_0^L \psi_1^* x \psi_1\,dx = (2/L)\int_0^L x\sin^2(\pi x/L)\,dx$$

The integral is of the form of Integral T.11 with $a = L$ and $k = \pi/L$

$$= (2/L)\left[L^2/4 - (L^2/4\pi)\overbrace{\sin(2\pi L/L)}^{=\sin 2\pi = 0} - (L^2/8\pi^2)\{\overbrace{\cos(2\pi L/L)}^{=\cos(2\pi)=1} - 1\}\right]$$

$$= \boxed{L/2}$$

$$\langle x^2\rangle = \int_0^L \psi_1^* x^2 \psi_1\,dx = (2/L)\int_0^L x^2\sin^2(\pi x/L)\,dx$$

The integral is of the form of Integral T.12 with $a = L$ and $k = \pi/L$

$$= (2/L)\left[L^3/6 - (L^3/4\pi - L^3/8\pi^3)\sin(2\pi L/L) - (L^3/4\pi^2)\cos(2\pi L/L)\right]$$

$$= \boxed{L^2/3 - 1/2\pi^2}$$

P7D.5 (a) Assuming that the system can be modelled as a 1D particle in a box, the energy levels of the system are given by [7D.6–263], $E_n = h^2n^2/8mL^2$. Hence, the separation between the levels n and $n+1$ is

$$\Delta E = E_{n+1} - E_n = h^2[(n+1)^2 - n^2]/8mL^2 = h^2(2n+1)/8mL^2$$

The conjugated system consists of 12 atoms, and so there are 11 bonds, therefore the length of the box is 11 times the average internuclear distance $L = 11 \times 140\ \text{pm} = 1.54\ \text{nm}$. Hence the separation between the levels with $n = 6$ and $n = 7$ is

$$\Delta E = \frac{(6.6261 \times 10^{-34}\ \text{J s})^2 \times (2 \times 6 + 1)}{8 \times (9.1094 \times 10^{-31}\ \text{kg}) \times (1.54 \times 10^{-9}\ \text{m})^2} = \boxed{3.30 \times 10^{-19}\ \text{J}}$$

(b) The energy and frequency of a transition are related by [7A.9–241], $\Delta E = h\nu$, rearranging for the frequency gives

$$\nu = \Delta E/h = (3.30... \times 10^{-19}\ \text{J})/(6.6261 \times 10^{-34}\ \text{J s}) = \boxed{4.98 \times 10^{14}\ \text{Hz}}$$

(c) The terms in the energy expression that change with the number of conjugated atoms N are the quantum number of the highest energy occupied state, n, and the length of the box, L. The energy, and hence frequency, of the transition is directly proportional to $2n+1$ and inversely proportional to L^2. Because the there are N electrons which occupy the states in pairs, the quantum number of the highest occupied state is $n = N/2$ (rounded up if N is odd). The length L is proportional to the number of bonds in the molecule, which is $N-1$. Hence, the frequency of the transition is proportional to $(2(N/2)+1)/(N-1)^2 = (N+1)/(N-1)^2 \approx N^{-1}$, and so the absorption spectrum of a linear polyene shifts to a $\boxed{\text{lower}}$ frequency as the number of conjugated atoms $\boxed{\text{increases}}$.

P7D.7 The expectation values are $\langle x \rangle = L/2$, $\langle x^2 \rangle = L^2(1/3 - 1/2n^2\pi^2)$, $\langle p_x \rangle = 0$, $\langle p_x^2 \rangle = n^2h^2/4L^2$.

(a) The uncertainty in the position is

$$\Delta x = [\langle x^2 \rangle - \langle x \rangle^2]^{1/2} = [L^2(1/3 - 1/2n^2\pi^2) - (L/2)^2]^{1/2}$$
$$= \boxed{L(1/12 - 1/2n^2\pi^2)^{1/2}}$$

and in the momentum

$$\Delta p_x = [\langle p_x^2 \rangle - \langle p_x \rangle^2]^{1/2} = [n^2h^2/4L^2 - 0^2]^{1/2} = \boxed{nh/2L}$$

(b) The product of these is then

$$\Delta x \Delta p_x = L(1/12 - 1/2n^2\pi^2)^{1/2} \times nh/2L$$
$$= \boxed{(nh/2)(1/12 - 1/2n^2\pi^2)^{1/2}}$$

(c) The Heisenberg uncertainty principle states that $\Delta x \Delta p_x \geq \hbar/2$. Rewriting the expression in (b) in terms of $\hbar$ gives $\Delta x \Delta p_x = (\hbar/2) \times 2\pi n(1/12 - 1/2n^2\pi^2)^{1/2}$. For $n = 1$

$$\Delta x \Delta p_x = (\hbar/2) \times 2\pi(1/12 - 1/2\pi^2)^{1/2} = 1.13... \times (\hbar/2)$$

and for $n = 2$,

$$\Delta x \Delta p_x = (\hbar/2) \times 4\pi(1/12 - 1/8\pi^2)^{1/2} = 3.34... \times (\hbar/2)$$

In both cases $\Delta x \Delta p_x \geq (\hbar/2)$, which is consistent with the uncertainty principle. It is evident that $\Delta x \Delta p_x$ is an increasing function of n and therefore it can be inferred that the uncertainly principle is obeyed for all n.

P7D.9 (a) In 3D the Schrödinger equation with a potential energy function $V(x, y, z)$ is given by

$$-\frac{\hbar^2}{2m}\left(\frac{\partial^2}{\partial x^2} + \frac{\partial^2}{\partial y^2} + \frac{\partial^2}{\partial z^2}\right)\psi + V(x, y, z)\psi = E\psi$$

In this case $V(x, y, z) = 0$ inside the box and infinite elsewhere. It follows that the wavefunction is zero outside the box. Inside the box the Schrödinger equation is

$$-\frac{\hbar^2}{2m}\left(\frac{\partial^2}{\partial x^2} + \frac{\partial^2}{\partial y^2} + \frac{\partial^2}{\partial z^2}\right)\psi = E\psi$$

Assume that the solution is a product of three functions of a single variable, that is let $\psi = X(x)Y(y)Z(z)$. Substituting this into the Schrödinger equation gives

$$-\frac{\hbar^2}{2m}\left(YZ\frac{\mathrm{d}^2X}{\mathrm{d}x^2} + XZ\frac{\mathrm{d}^2Y}{\mathrm{d}y^2} + YZ\frac{\mathrm{d}^2Z}{\mathrm{d}z^2}\right) = EXYZ$$

Dividing through by the function XYZ gives

$$-\frac{\hbar^2}{2m}\left(\frac{1}{X}\frac{\mathrm{d}^2X}{\mathrm{d}x^2} + \frac{1}{Y}\frac{\mathrm{d}^2Y}{\mathrm{d}y^2} + \frac{1}{Z}\frac{\mathrm{d}^2Z}{\mathrm{d}z^2}\right) = E$$

Now rearrange such that the terms that depend on x only are on the left hand side

$$-\frac{\hbar^2}{2mX}\frac{\mathrm{d}^2X}{\mathrm{d}x^2} = E + \frac{\hbar^2}{2m}\left(\frac{1}{Y}\frac{\mathrm{d}^2Y}{\mathrm{d}y^2} + \frac{1}{Z}\frac{\mathrm{d}^2Z}{\mathrm{d}z^2}\right)$$

The left hand side depends only on x, whilst the right side depends only on y and z. The only way that this can be true for all x, y, z is if both sides are equal to a constant, arbitrarily called E_x, such that

$$-\frac{\hbar^2}{2mX}\frac{\mathrm{d}^2X}{\mathrm{d}x^2} = E_x \quad E_x = E + \frac{\hbar^2}{2m}\left(\frac{1}{Y}\frac{\mathrm{d}^2Y}{\mathrm{d}y^2} + \frac{1}{Z}\frac{\mathrm{d}^2Z}{\mathrm{d}z^2}\right)$$

The same procedure is followed with the functions of y and z to give

$$-\frac{\hbar^2}{2mY}\frac{d^2Y}{dy^2} = E_y \qquad -\frac{\hbar^2}{2mZ}\frac{d^2Z}{dz^2} = E_z$$

where $E = E_x + E_y + E_z$. The assumption that the wavefunction can be written as a product of single-variable functions is a valid one, as ordinary differential equations can be found for the assumed factors. That is what it means for a partial differential equation to be separable.

(b) The differential equation involving the function X is recognized as being the same as the Schrödinger equation for a particle in a one-dimensional box, whose solutions are already known: $X(x) = (2/L)^{1/2}\sin(n\pi x/L)$ with energies $E_n = n^2h^2/8mL^2$. The product $X(x)Y(y)Z(z)$ which is the solution to this three-dimensional problem will therefore be a product with three such one-dimensional functions along x, y and z, each with the appropriate length L_i and quantum number n_i

$$\psi(x,y,z) = (2/L_1)^{1/2}\sin(n_x\pi x/L_1) \times (2/L_2)^{1/2}\sin(n_y\pi y/L_2) \times (2/L_3)^{1/2}\sin(n_z\pi z/L_3)$$

Likewise, the total energy will be the sum of energies along each direction

$$E = E_x + E_y + E_z = \frac{h^2}{8m}\left(\frac{n_x^2}{L_1^2} + \frac{n_y^2}{L_2^2} + \frac{n_z^2}{L_3^2}\right)$$

(c) For a cubic box, $L_1 = L_2 = L_3 = L$ and so

$$E = h^2(n_x^2 + n_y^2 + n_z^2)/8mL^2$$

Generally states with all three quantum numbers the same are non degenerate, if two quantum numbers are the same they are triply degenerate, and if all three quantum numbers are the different they are six-fold degenerate. In addition, there may be 'accidental' degeneracies, such as that between $(n_x, n_y, n_z) = (3,3,3)$ and $(5,1,1)$. The energy level diagram is shown in Fig. 7.2; in the diagram the states are labelled (n_x, n_y, n_z).

(d) Comparing the 3D and 1D energy diagrams, the one dimensional case is more sparse than the three dimensional case, as the fifteenth level is not reached until $E_n/(h^2/8ml^2) = 225$. In addition, none of the levels in the one-dimensional case are degenerate.

P7D.11 The rate of tunnelling is proportional to the transmission probability, given by [7D.20b–*270*], $T = 16\varepsilon(1-\varepsilon)e^{-2\kappa L}$. Therefore the ratio of tunnelling rates is equal to the corresponding ratio of transmission probabilities. The desired factor is T_1/T_2 where the subscripts denote different tunnelling distances.

$$\frac{T_1}{T_2} = \frac{16\varepsilon(1-\varepsilon)e^{-2\kappa L_1}}{16\varepsilon(1-\varepsilon)e^{-2\kappa L_2}} = e^{-2\kappa(L_1-L_2)}$$

With $\kappa = 7\ \text{nm}^{-1}$, $L_1 = 1$ nm and $L_1 = 2$ nm

$$\frac{T_1}{T_2} = e^{-2\kappa(L_1-L_2)} = e^{-2(7\ \text{nm}^{-1})[(1\ \text{nm})-(2\ \text{nm})]} = \boxed{1.20\times10^6}$$

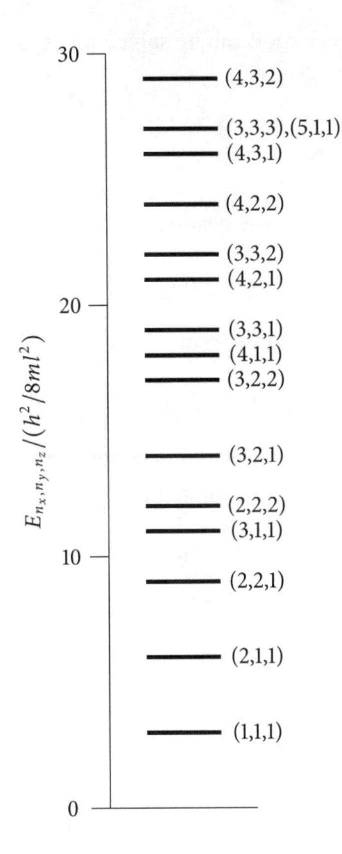

Figure 7.2

P7D.13 (a) The wavefunctions in each region are $\psi_1(x) = e^{ik_1x} + B_1e^{ik_1x}$, $\psi_2(x) = A_2e^{k_2x} + B_2e^{-k_2x}$, $\psi_3(x) = A_3e^{ik_3x}$. With this choice of $A_1 = 1$, the transmission probability is simply $T = |A_3|^3$.

The wavefunction coefficients are determined by the criteria that both the wavefunction and its first derivative with respect to x are continuous at the potential boundaries, $\psi_1(0) = \psi_2(0)$, $d\psi_1(0)/dx = d\psi_2(0)/dx$, $\psi_2(W) = \psi_3(W)$, $d\psi_3(W)/dx = d\psi_2(W)/dx$. These establish the algebraic relationships

(a) $1 + B_1 = A_2 + B_2$

(b) $ik_1 - ik_1B_1 = A_2k_2 - B_2k_2$

(c) $A_2e^{k_2W} + B_2e^{-k_2W} = A_3e^{ik_3W}$

(d) $A_2k_2e^{k_2W} - B_2k_2e^{-k_2W} = ik_3A_3e^{ik_3W}$

These equations need to be solved for A_3 and hence B_1, A_2, B_2 need to be eliminated. A_3 appears in (c) and (d) only. Solving these equations for $A_3e^{ik_3W}$ and setting the results equal to each other yields

$$A_2e^{k_2W} + B_2e^{-k_2W} = (k_2/ik_3)(A_2k_2e^{k_2W} - B_2e^{-k_2W})$$

This equation can be solved for B_2 in terms of A_2

$$B_2 = \frac{A_2 e^{2k_2 W}(k_2/ik_3 - 1)}{k_2/ik_3 + 1} = \frac{A_2 e^{2k_2 W}(k_2 - ik_3)}{k_2 + ik_3}$$

Note that B_1 appears only in (a) and (b). Solving these equations for B_1 and setting the results equal to each other yields

$$B_1 = A_2 + B_2 - 1 = 1 - (1/ik_1)(A_2 k_2 - B_2 k_2)$$

Substituting the expression for B_2 into this equation yields

$$A_2\left[1 + \frac{e^{2k_2 W}(k_2 - ik_3)}{k_2 + ik_3}\right] = 2ik_1 - A_2 k_2\left[1 - \frac{e^{2k_2 W}(k_2 - ik_3)}{k_2 + ik_3}\right]$$

This can be rearranged to give

$$A_2 = \left[\frac{1}{2} + \frac{e^{2k_2 W}(k_2 - ik_3)}{2(k_2 + ik_3)} + \frac{k_2(k_2 - ik_3)}{2ik_1(k_2 + ik_3)} - \frac{k_2 e^{2k_2 W}(k_2 - ik_3)}{2ik_1(k_2 + ik_3)}\right]^{-1}$$

Hence,

$$B_2 = \frac{A_2 e^{2k_2 W}(k_2 - ik_3)}{k_2 + ik_3} = A_2\left[\frac{k_2 + ik_3}{e^{2k_2 W}(k_2 - ik_3)}\right]^{-1}$$

$$= \left[\frac{k_2 + ik_3}{2e^{2k_2 W}(k_2 - ik_3)} + \frac{1}{2} + \frac{k_2}{2ik_1 e^{2k_2 W}} - \frac{k_2}{2ik_1}\right]^{-1}$$

This gives

$$A_3 = e^{-ik_3 W}(A_2 e^{k_2 W} + B_2 e^{-k_2 W}) = \frac{4k_1 k_2 e^{ik_2 W}}{(ia + b)e^{k_2 W} - (ia - b)e^{-k_2 W}}$$

Since $\sinh z = \frac{1}{2}(e^z - e^{-z})$, substitute $e^{k_2 L} = 2\sinh k_2 L + e^{k_2 L}$

$$A_3 = \frac{2k_1 k_2 e^{ik_2 L}}{(ia + b)\sinh(k_2 W) + be^{-k_2 W}}$$

The transmission coefficient is then

$$T = |A_3|^2 = \frac{4k_1^2 k_2^2}{(a^2 + b^2)\sinh^2(k_2 L) + b^2}$$

where $a^2 + b^2 = (k_1^2 + k_2^2)(k_2^2 + k_3^2)$, and $b^2 = k_2^2(k_1 + k_3)^2$.

(b) In the special case, for which $V_1 = V_3 = 0$, $k_1 = k_3$. Additionally, $k_1/k_2 = E/(V_2 - E) = \varepsilon/(1 - \varepsilon)$, where $\varepsilon = E/V_2$. Also

$$a^2 + b^2 = (k_1^2 + k_2^2)^2 = k_2^4[1 + (k_1/k_2)^2]^2$$

$$b^2 = 4k_1^2 k_2^2$$

Hence

$$T = \frac{b^2}{b^2 + (a^2 + b^2)\sinh^2(k_2 W)} = \left(1 + \frac{a^2 + b^2}{b^2}\sinh^2 k_2 W\right)^{-1}$$
$$= \left(1 + \frac{\sinh^2(k_2 W)}{4\varepsilon(1-\varepsilon)}\right)^{-1} = \left(1 + \frac{[\frac{1}{2}(e^{k_2 W} - e^{-k_2 W})]^2}{4\varepsilon(1-\varepsilon)}\right)^{-1}$$
$$= \left(1 + \frac{(e^{k_2 W} - e^{-k_2 W})^2}{16\varepsilon(1-\varepsilon)}\right)^{-1} \quad (7.1)$$

Which is eqn 7D.20a as expected. In the case of a high, wide barrier, $k_2 W \gg 1$. This implies that $e^{-k_2 W}$ is negligibly small in comparison to $e^{k_2 W}$, and that 1 is negligibly small compared to $e^{2k_2 W}/16\varepsilon(1-\varepsilon)$, such that

$$T \approx \left(e^{2k_2 W}/16\varepsilon(1-\varepsilon)\right)^{-1} = 16\varepsilon(1-\varepsilon)e^{-2k_2 W}$$

(c) The function plotted in Fig. 7.3 is eqn 7.1 with the parameters given and

$$k_2 = [2m(V_2 - E)]^{1/2}/\hbar \quad \varepsilon = E/V_2$$

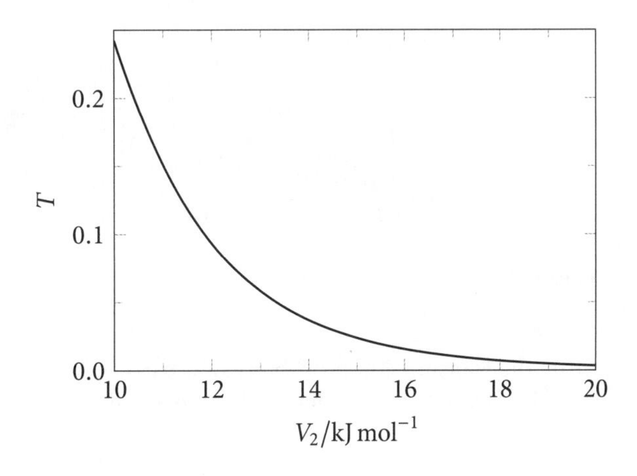

Figure 7.3

P7D.15 (a) The particle in a box wavefunctions are given by $\psi_n(x) = \sqrt{2/L}\sin(n\pi x/L)$, giving a probability density of $P_n(x) = |\psi_n(x)|^2 = (2/L)\sin^2(n\pi x/L)$. This probability density is plotted in Fig. 7.4 for $n = 1 \ldots 5$, and for $n = 50$ in Fig. 7.5.

The correspondence principle is that as the quantum number becomes large the quantum result must converge with the classical result. For a particle in a box, the classical result is an even probability. As n increases the oscillations become more rapid leading to a more uniform distribution especially if it is averaged over short distance which is a small fraction of the width of the box.

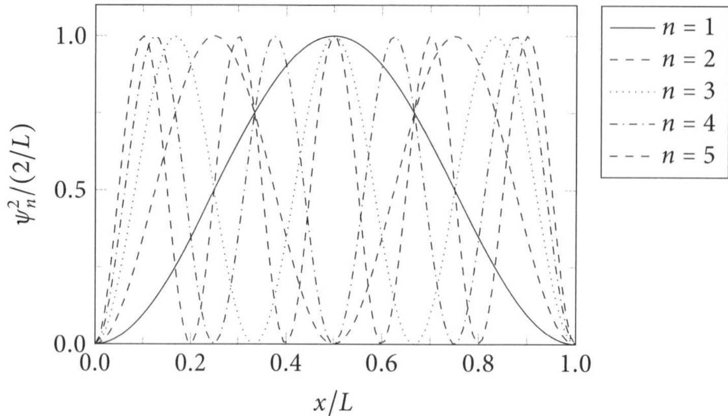

Figure 7.4

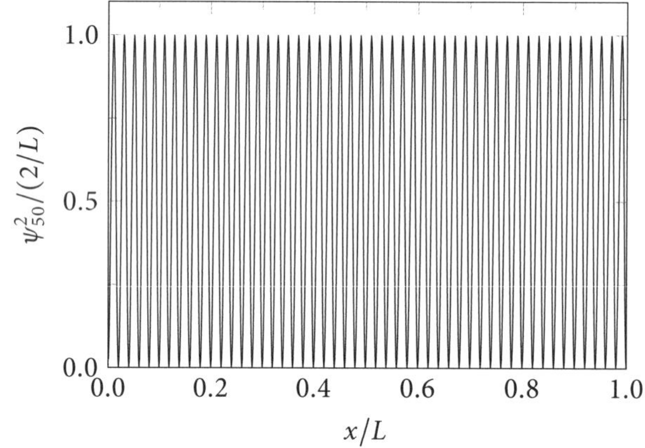

Figure 7.5

(b) The transmission probability of a particle of mass m passing through a barrier of height V_0, length W is given by, [7D.20a–*269*]

$$T = \left(1 + \frac{(e^{\kappa W} - e^{-\kappa W})^2}{16\varepsilon(1-\varepsilon)}\right)^{-1}$$

where $\varepsilon = E/V_0$ and $\kappa = \sqrt{2m(V_0 - E)}/\hbar = \sqrt{2mV_0(1-\varepsilon)}/\hbar$. A plot of T versus ε is shown in Fig. 7.6 for the passage by a H_2 molecule, a proton, and an electron.

The curves in the figure differ in the value of $W(mV_0)^{1/2}/\hbar$, a measure of the obstruction that the barrier represents taking into account its height, width and the type of particle. The curves can be thought of as having the same value of W and V_0, but differing only in m. The values of W and V_0 were chosen such that the proton and hydrogen molecule exhibit 'typical' tunnelling behaviour: if the incident energy is small enough, there is practically no transmission, and if the incident energy is high enough, transmission is virtually complete. A barrier through which a proton and

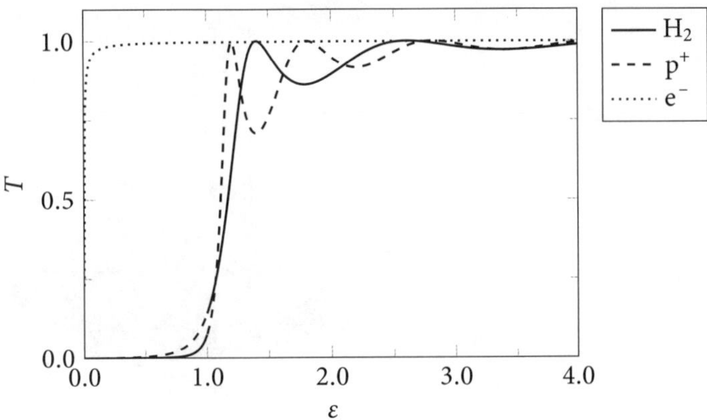

Figure 7.6

hydrogen molecule exhibit tunnelling behaviour is practically no barrier for an electron on account of the much smaller mass of the latter. On this plot, T for the electron is essentially 1.

(c) The wavefunction of a 2D particle in a box with quantum numbers n_1 and n_2 is given by [7D.12a–*267*]

$$\psi_{n_1,n_2}(x,y) = (2/L_x)^{1/2}(2/L_y)^{1/2}\sin(n_1\pi x/L_x)\sin(n_2\pi y/L_y)$$

This is plotted for $n_1 = 1, n_2 = 1$, and for $n_1 = 1, n_2 = 2$ in Fig. 7.7; the wavefunctions with $n_1 = 2, n_2 = 1$, and with $n_1 = 2, n_2 = 2$ are plotted in Fig. 7.8.

As can be seen from the plots, the function $\sin(n_1\pi x/L_x)$ has n_1-1 nodes where the function passes through zero; the boundaries do not count as nodes because the wavefunction does not pass through zero at these points. Similarly the function $\sin(n_2\pi y/L_y)$ has $n_2 - 1$ nodes. In two dimensions a node in the wavefunction along x or y leads to a nodal line, and the total number of these is $(n_1 - 1) + (n_2 - 1) = \boxed{n_1 + n_2 - 2}$.

7E Vibrational motion

Answers to discussion questions

D7E.1 The energy levels of a harmonic oscillator are given by $E_\upsilon = (\upsilon + \frac{1}{2})\hbar\omega$, where $\omega = (k_f/m)^{1/2}$; in this expression k_f is the force constant and m is the mass. The energy levels are therefore evenly spaced, with spacing $\hbar\omega$; the spacing is proportional to ω which goes as $k_f^{1/2}$ and as $m^{-1/2}$.

D7E.3 If the harmonic oscillator were to have zero energy the particle would be at rest and located at the bottom of the potential energy well ($x = 0$). Such an arrangement has no uncertainty in either the position or the momentum, which is not in accord with the uncertainty principle. By giving the oscillator

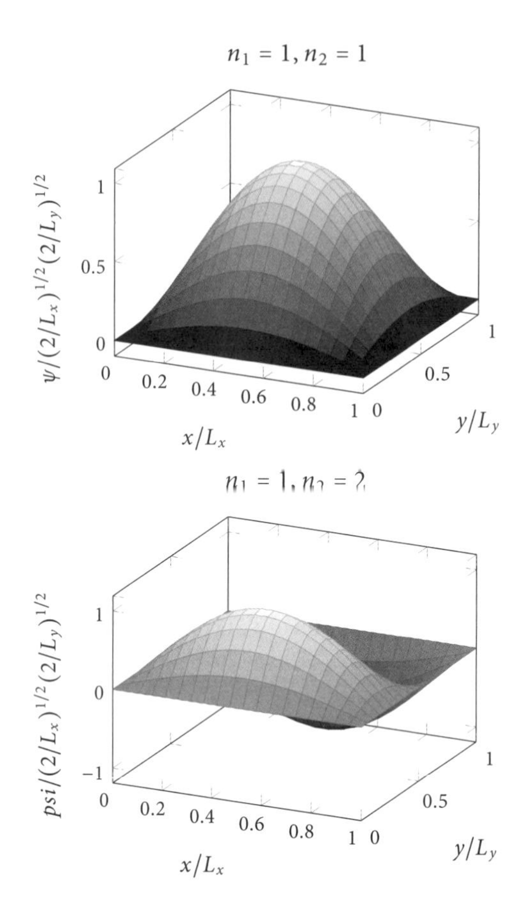

Figure 7.7

zero-point energy, there is some uncertainty in the position as the particle oscillates back and forth, and then it is possible for there to be the appropriate uncertainty in the momentum such that the uncertainty principle is satisfied. The existence of zero-point energy is thus traced to the need to satisfy the uncertainty principle.

Solutions to exercises

E7E.1(a) The zero-point energy of a harmonic oscillator is given by [7E.5–*274*], $E_0 = \frac{1}{2}\hbar\omega$, where the frequency ω is given by [7E.3–*274*], $\omega = (k_f/m)^{1/2}$. For this system,

$$E_0 = \tfrac{1}{2} \times (1.0546 \times 10^{-34}\,\mathrm{J\,s}) \times \left[(155\,\mathrm{N\,m^{-1}})/(2.33 \times 10^{-26}\,\mathrm{kg})\right]^{1/2}$$
$$= \boxed{4.30 \times 10^{-21}\,\mathrm{J}}$$

E7E.2(a) The separation between adjacent energy levels of a harmonic oscillator is [7E.4–*274*], $\Delta E = \hbar\omega$, where the frequency, ω is given by [7E.3–*274*], $\omega = (k_f/m)^{1/2}$.

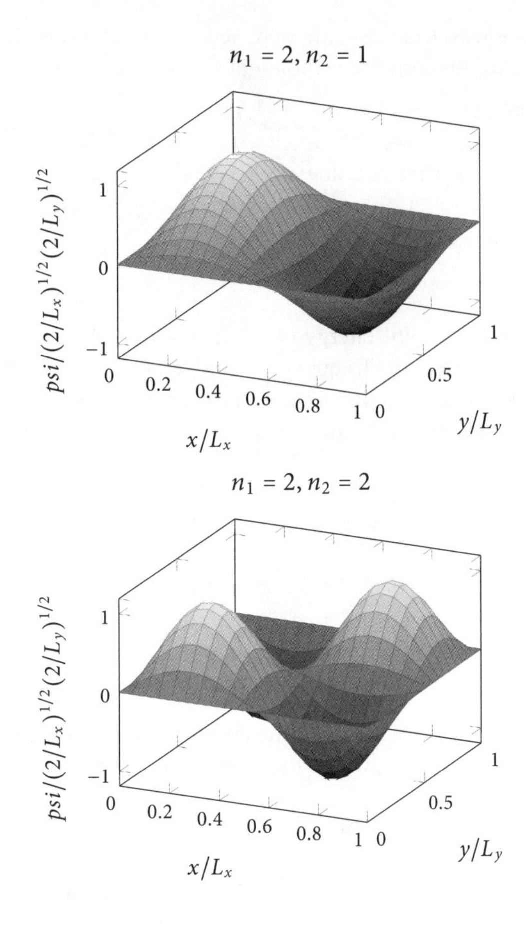

Figure 7.8

This is rearranged for the force constant as $k_f = m(\Delta E/\hbar)^2$. Evaluating this gives

$$k_f = (1.33 \times 10^{-25}\,\text{kg}) \times \left[(4.82 \times 10^{-21}\,\text{J})/(1.0546 \times 10^{-34}\,\text{J s})\right]^2 = \boxed{278\,\text{N m}^{-1}}$$

E7E.3(a) The separation between adjacent energy levels of a harmonic oscillator is [7E.4–*274*], $\Delta E = \hbar\omega$, where the frequency, ω is given by [7E.3–*274*], $\omega = (k_f/m)^{1/2}$. The Bohr frequency condition [7A.9–*241*], $\Delta E = h\nu$, can be rewritten in terms of the wavelength as $\Delta E = hc/\lambda$. The wavelength of the photon corresponding to a transition between adjacent energy levels is therefore given by $\hbar\omega = hc/\lambda$, or $\hbar(k_f/m)^{1/2} = hc/\lambda$. Solving for λ gives $\lambda = 2\pi c/(k_f/m)^{1/2}$; with the data given

$$\lambda = \frac{2\pi \times (2.9979 \times 10^8\,\text{m s}^{-1})}{\left[(855\,\text{N m}^{-1})/(1.0078 \times 1.6605 \times 10^{-27}\,\text{kg})\right]^{1/2}}$$

$$= \boxed{2.64 \times 10^{-6}\,\text{m}}$$

E7E.4(a) The wavefunctions are depicted in Fig. 7E.6 on page 276; they are real. Two wavefunctions are orthogonal if $\int \psi_i^* \psi_j \, d\tau = 0$. In this case the wavefunctions are $\psi_0(y) = N_0 e^{-y^2/2}$ and $\psi_1(y) = N_1 y e^{-y^2/2}$, and the integration is from $y = -\infty$ to $+\infty$. The integrand $\psi_0\psi_1$ is $N_0 N_1 y e^{-y^2}$, which is an odd function, meaning that its value at $-y$ is the negative of its value at $+y$. The integral of an odd function over a symmetric range is zero, hence these wavefunctions are orthogonal.

E7E.5(a) The zero-point energy of a harmonic oscillator is given by [7E.5–*274*], $E_0 = \frac{1}{2}\hbar\omega$, where the frequency ω is given by [7E.3–*274*], $\omega = (k_f/\mu)^{1/2}$. The effective mass μ of a diatomic AB is given by [7E.6–*274*], $\mu = (m_A m_B)/(m_A + m_B)$. In the case of a homonuclear diatomic A_2 this reduces to $\mu = m_A/2$. With the data given

$$\begin{aligned} E_0 &= \tfrac{1}{2} \times (1.0546 \times 10^{-34}\,\mathrm{J\,s}) \\ &\quad \times \left[(329\,\mathrm{N\,m^{-1}})/(\tfrac{1}{2} \times 34.9688 \times 1.6605 \times 10^{-27}\,\mathrm{kg})\right]^{1/2} \\ &= \boxed{5.61 \times 10^{-21}\,\mathrm{J}} \end{aligned}$$

E7E.6(a) For the state with $v = 0$ the energy of a harmonic oscillator is given by [7E.5–*274*], $E_0 = \frac{1}{2}\hbar\omega$, where the frequency ω is given by [7E.3–*274*], $\omega = (k_f/m)^{1/2}$.

(i) For the system with $k_f = 1000\,\mathrm{N\,m^{-1}}$ the energy of the state with $v = 0$ is

$$\begin{aligned} E_0 &= \tfrac{1}{2} \times (1.0546 \times 10^{-34}\,\mathrm{J\,s}) \times \left[(1000\,\mathrm{N\,m^{-1}})/(1 \times 1.6605 \times 10^{-27}\,\mathrm{kg})\right]^{1/2} \\ &= 4.09... \times 10^{-20}\,\mathrm{J} = \boxed{4.09 \times 10^{-20}\,\mathrm{J}} \end{aligned}$$

The classical turning points of this state occur when $E_0 = \frac{1}{2}k_f x_{tp}^2$. Solving this for x_{tp} leads to $x_{tp} = \pm\sqrt{2E_0/k_f}$, giving a separation of

$$2\sqrt{2E_0/k_f} = 2\sqrt{2 \times (4.09... \times 10^{-20}\,\mathrm{J})/(1000\,\mathrm{N\,m^{-1}})} = \boxed{18.1\,\mathrm{pm}}$$

(ii) For the system with $k_f = 100\,\mathrm{N\,m^{-1}}$ the energy of the state with $v = 0$ is

$$\begin{aligned} E_0 &= \tfrac{1}{2} \times (1.0546 \times 10^{-34}\,\mathrm{J\,s}) \times \left[(100\,\mathrm{N\,m^{-1}})/(1 \times 1.6605 \times 10^{-27}\,\mathrm{kg})\right]^{1/2} \\ &= \boxed{1.29 \times 10^{-20}\,\mathrm{J}} \end{aligned}$$

The classical turning points of this state occur when $E_0 = \frac{1}{2}k_f x_{tp}^2$. Solving this for x_{tp} leads to $x_{tp} = \pm\sqrt{2E_0/k_f}$, giving a separation of

$$2\sqrt{2E_0/k_f} = 2\sqrt{2 \times (1.29... \times 10^{-20}\,\mathrm{J})/(100\,\mathrm{N\,m^{-1}})} = \boxed{32.2\,\mathrm{pm}}$$

E7E.7(a) The wavefunctions are depicted in Fig. 7E.6 on page 276. The general form of the harmonic oscillator wavefunctions is $\psi_v = N_v H_v(y)e^{-y^2/2}$, where N_v is a normalization constant and $H_v(y)$ is the Hermite polynomial of order v in the reduced position variable y. Nodes occur when the wavefunction passes through zero. The wavefunction asymptotically approaches zero at $y = \pm\infty$, but as the function does not pass through zero these limits do not count as nodes. The nodes in the wavefunction therefore correspond to the solutions of $H_v(y) = 0$. $H_v(y)$ is a polynomial of order v, meaning that the highest power of y that occurs is y^v; such polynomials in general have v solutions and hence there are v nodes. Therefore (a) the wavefunction with $v = 3$ has $\boxed{3}$ nodes; (b) the wavefunction with $v = 4$ has $\boxed{4}$ nodes.

E7E.8(a) The wavefunction with $v = 2$ is $\psi_2(y) = N_2(y^2 - 1)e^{-y^2/2}$. Nodes occur when the wavefunction passes through zero; the wavefunction approaches zero at $y = \pm\infty$, but these do not count as nodes as the wavefunction does not pass through zero. It is evident that the nodes occur when $y^2 - 1 = 0$, which solves to give nodes at $\boxed{y = -1, +1}$.

E7E.9(a) The wavefunction with $v = 1$ is $\psi_1(y) = N_1 ye^{-y^2/2}$, which gives a probability density of $P(y) = |\psi_2(y)|^2 = N_1^2 y^2 e^{-y^2}$. The extrema are located by differentiating the wavefunction, setting the result to 0 and solving for y. The differential is evaluated using the product rule

$$\frac{d\psi_1(y)}{dy} = N_1^2\left(\frac{dy^2}{dy}e^{-y^2} + y^2\frac{de^{-y^2}}{dy}\right) = N_1^2[2ye^{-y^2} + y^2 \times (-2ye^{-y^2})]$$
$$= N_1^2 2y(1 - y^2)e^{-y^2}$$

This function is equal to zero when $y = 0$, or $1 - y^2 = 0$ or when $e^{-y^2} = 0$. These correspond to extrema at $y = 0, \pm1, \pm\infty$.

To identify the maxima, consider the form of the probability density. This goes to zero at $y = 0, \pm\infty$, and so these must be minima, implying that $\boxed{y = \pm1}$ correspond to maxima.

Solutions to problems

P7E.1 The intermolecular potential is electrostatic in origin and arises from the interaction between the protons and electrons, the number and distribution of which remain the same on isotopic substitution and so therefore does the potential. The force constant, which reflects the shape of the intermolecular potential, is therefore unaffected by such a change.

(a) The force constants for the molecules are the same therefore $\omega_{A'B}/\omega_{AB} = (k_f/\mu_{A'B})^{1/2}/(k_f/\mu_{AB})^{1/2} = (\mu_{AB}/\mu_{A'B})^{1/2}$; multiplying through by ω_{AB}, gives $\omega_{A'B} = \omega_{AB}(\mu_{AB}/\mu_{A'B})^{1/2}$. The vibrational frequency in Hz, ν, is related to ω by $\nu = \omega/2\pi$, hence $\nu_{A'B} = \nu_{AB}(\mu_{AB}/\mu_{A'B})^{1/2}$.

(b) The effective mass of $^1H^{35}Cl$ is $\mu_1 = (1 \times 35)/(35+1) = 0.972...\ m_u$, that of $^2H^{35}Cl$ is $\mu_2 = (2 \times 35)/(35+2) = 1.89...\ m_u$, and that of $^1H^{37}Cl$ is $\mu_3 = (1 \times 37)/(1+37) = 0.973...\ m_u$.
The vibrational frequency of $^2H^{35}Cl$ is

$$\nu_2 = \nu_1(\mu_1/\mu_2)^{1/2} = (5.63 \times 10^{14}\ \text{Hz})[(0.972...\ m_u)/(1.89...\ m_u)]^{1/2} = \boxed{4.04 \times 10^{14}\ \text{Hz}}$$

and for $^1H^{37}Cl$

$$\nu_3 = \nu_1(\mu_1/\mu_3)^{1/2} = (5.63 \times 10^{14}\ \text{Hz})[(0.972...\ m_u)/(0.973...\ m_u)]^{1/2} = \boxed{5.63 \times 10^{14}\ \text{Hz}}$$

P7E.3 Assuming that the force constant is the same for all the molecules, then the ratio of frequencies is $\omega_2/\omega_1 = \nu_2/\nu_1 = \sqrt{k_f/\mu_2}/\sqrt{k_f/\mu_1} = \sqrt{\mu_1/\mu_2}$. For a homonuclear diatomic molecule A_2 the effective mass is $\mu_{A_2} = m_A^2/2m_A = m_A/2$. Hence, $\mu_{^1H_2} = 1/2 = 0.5\ m_u$, $\mu_{^2H_2} = 2/2 = 1\ m_u$, $\mu_{^3H_2} = 3/2 = 1.5\ m_u$.

$$\nu_{^2H_2} = \nu_{^1H_2}\sqrt{\mu_{^1H_2}/\mu_{^2H_2}} = 131.9\ \text{THz} \times \sqrt{0.5\ m_u/1\ m_u} = \boxed{93.27\ \text{THz}}$$

$$\nu_{^3H_2} = \nu_{^1H_2}\sqrt{\mu_{^1H_2}/\mu_{^3H_2}} = 131.9\ \text{THz} \times \sqrt{0.5\ m_u/1.5\ m_u} = \boxed{76.15\ \text{THz}}$$

P7E.5 (a) The wavenumber, $\tilde{\nu}$ is given by $\tilde{\nu} = \nu/c$, where c is the speed of light (see *The chemist's toolkit* 13). The frequency in Hz is related to the angular frequency by $\nu = \omega/2\pi$, hence $\boxed{\tilde{\nu} = \omega/2\pi c}$.

(b) The vibrational frequency of $^1H^{35}Cl$ is $5.63 \times 10^{14}\ s^{-1}$, which gives a wavenumber of

$$\tilde{\nu} = (5.63 \times 10^{14}\ \text{s}^{-1})/[2\pi \times (2.9979 \times 10^{10}\ \text{cm s}^{-1})] = \boxed{2.99 \times 10^3\ \text{cm}^{-1}}$$

Note that in order to express the wavenumber in cm^{-1}, the speed of light is used in $cm\ s^{-1}$.

(c) The frequency ω is given by [7E.3–*274*], $\omega = (k_f/\mu)^{1/2}$, and so the vibrational frequency expressed as a wavenumber is $\tilde{\nu} = (1/2\pi c)(k_f/\mu)^{1/2}$. This rearranges to give $\boxed{k_f = \mu(2\pi\tilde{\nu}c)^2}$.

(d) The effective mass of $^{12}C^{16}O$ is $\mu = (12 \times 16)/(12+16) = 6.85...\ m_u$. Using the previous result, and converting the mass to kg, gives the force constant

$$k_f = [6.85... \times (1.6605 \times 10^{-27}\ \text{kg})] \times [2\pi \times (2170\ \text{cm}^{-1}) \times (2.9979 \times 10^{10}\ \text{cm s}^{-1})]^2 = \boxed{1902\ \text{N m}^{-1}}$$

For an isotopic substitution the vibrational frequencies are related by the effective masses (*Problem* P7E.1) $\tilde{\nu}_2/\tilde{\nu}_1 = \sqrt{\mu_1/\mu_2}$. The effective mass of

$^{14}C^{16}O$ is $\mu_{^{14}C^{16}O} = (14 \times 16)/(14+16) = 7.46...\ m_u$. This gives the wavenumber for $^{14}C^{16}O$ as

$$\begin{aligned}\tilde{\nu}_{^{14}C^{16}O} &= \tilde{\nu}_{^{12}C^{16}O}\sqrt{\mu_{^{12}C^{16}O}/\mu_{^{14}C^{16}O}}\\ &= (2170\ \mathrm{cm^{-1}}) \times \sqrt{6.85...\ m_u/7.46\ m_u} = \boxed{2080\ \mathrm{cm^{-1}}}\end{aligned}$$

P7E.7 If the C atom is immobilized, only the O will be moving, and the force constant will be the same as that in the CO molecule. The wavenumber of the CO vibration is 2170 cm^{-1}, as given in *Problem* P7E.6, and as the force constant is the same this is linked to the wavenumber of the vibration when bound as $\tilde{\nu}_2 = \tilde{\nu}_1\sqrt{\mu_1/\mu_2}$, where 1 refers to the free molecule and 2 that bound to the haem group. The effective mass of $^{12}C^{16}O$ is $\mu_1 = (12\times16)/(12+16) = 6.85...\ m_u$, and in the bound case the effective mass is simply the mass of the O atom, 16 m_u, as this is the only atom which is moving. Hence

$$\tilde{\nu}_2 = (2170\ \mathrm{cm^{-1}}) \times \sqrt{(6.85...\ m_u)/(16\ m_u)} = \boxed{1420\ \mathrm{cm^{-1}}}$$

Expressed as a frequency this is $\nu = 4.259 \times 10^{13}$ Hz.

P7E.9 (a) The second derivative of the function $\psi = e^{-gx^2}$ is evaluated using the chain rule

$$\begin{aligned}\frac{d^2\psi}{dx^2} &= \frac{d^2}{dx^2}\left(e^{-gx^2}\right) = \frac{d}{dx}\left(-2gxe^{-gx^2}\right) = -2g\left(\frac{dx}{dx}e^{-gx^2} + x\frac{de^{-gx^2}}{dx}\right)\\ &= -2g\left(e^{-gx^2} + x \times -2gxe^{-gx^2}\right) = 2g(2gx^2-1)e^{-gx^2}\end{aligned}$$

(b) The Hamiltonian for the harmonic oscillator is $\hat{H} = -(\hbar^2/2m)d^2/dx^2 + \frac{1}{2}k_f x^2$. To find the condition for e^{-gx^2} to be an eigenfunction of $\hat{H}$, first allow the operator to act on the function

$$\begin{aligned}\hat{H}e^{-gx^2} &= \left[-(\hbar^2/2m)d^2/dx^2 + \tfrac{1}{2}k_f x^2\right]e^{-gx^2}\\ &= -(\hbar^2/2m)2g(2gx^2-1)e^{-gx^2} + \tfrac{1}{2}k_f x^2 e^{-gx^2} \qquad (7.2)\end{aligned}$$

Evidently e^{-gx^2} is not an eigenfunction of $\hat{H}$ because when this operator acts on e^{-gx^2} the result is not a constant times that function. However, this condition would be satisfied if the terms in x^2 were to cancel one another, that is

$$-(\hbar^2/2m)2g(2gx^2) + \tfrac{1}{2}k_f x^2 = 0$$

Solving this for g gives $\boxed{g = (mk_f)^{1/2}/2\hbar}$; this is the value of g needed for e^{-gx^2} to be an eigenfunction of the hamiltonian. The resulting wavefunction is $\psi = e^{-(mk_f)^{1/2}x^2/2\hbar}$, which is in agreement with that quoted in [7E.8b–*275*].

(c) Returning to eqn 7.2, if the terms in x^2 cancel the remaining terms are

$$-(\hbar^2/2m)2g(-1)e^{-gx^2}$$

which is of the form of a constant times the original function. The constant is the eigenvalue, which in this case is the energy

$$E = (\hbar^2/2m)2g = (\hbar^2/2m) \times 2 \times (mk_f)^{1/2}/2\hbar = \boxed{\tfrac{1}{2}\hbar(k_f/m)^{1/2}}$$

The energy is indeed the same as that of the lowest level of the harmonic oscillator as described in the text.

P7E.11 (a) The variable y is defined as $y = x/\alpha$ where $\alpha = (\hbar^2/mk_f)^{1/4}$, and the kinetic energy operator in terms of x is $\hat{T} = -(\hbar^2/2m)d^2/dx^2$. Substituting in y leads to

$$\hat{T} = -\frac{\hbar^2}{2m}\frac{d^2}{d(\alpha y)^2} = -\frac{\hbar^2}{2\alpha^2 m}\frac{d^2}{dy^2}$$
$$= -\frac{\hbar^2}{2m}\left(\frac{mk_f}{\hbar^2}\right)^{1/2}\frac{d^2}{dy^2} = -\tfrac{1}{2}\hbar\left(\frac{k_f}{m}\right)^{1/2}\frac{d^2}{dy^2} = -\tfrac{1}{2}\hbar\omega\frac{d^2}{dy^2}$$

where for the last line the definition of the frequency ω is used, $\omega = (k_f/m)^{1/2}$.

(b) The expectation value of the kinetic energy of the state with quantum number v is given by

$$\langle T\rangle_v = -\tfrac{1}{2}\hbar\omega N_v^2\int_{-\infty}^{\infty} H_v e^{-y^2/2}\frac{d^2}{dy^2}H_v e^{-y^2/2}\,dy$$

To evaluate the derivative, the product rule is employed,

$$\frac{d^2}{dy^2}H_v e^{-y^2/2} = \frac{d}{dy}(H_v' e^{-y^2/2} + H_v\frac{de^{-y^2/2}}{dy}) = \frac{d}{dy}(H_v' - yH_v)e^{-y^2/2}$$

The product rule is applied once more

$$\frac{d^2}{dy^2}H_v e^{-y^2/2} = \frac{d(H_v' - yH_v)}{dy}e^{-y^2/2} + (H_v' - yH_v)\frac{de^{-y^2/2}}{dy}$$
$$= (H_v'' - yH_v' - H_v)e^{-y^2/2} - y(H_v' - yH_v)e^{-y^2/2}$$
$$= [H_v'' - 2yH_v' + (y^2 - 1)H_v]e^{-y^2/2}$$

From Table 7E.1 on page 275 one of the properties of the Hermite polynomials is $H_v'' - 2yH_v' + 2vH_v = 0$, so it follows that $H_v'' - 2yH_v' = -2vH_v$. Using this in the last line above gives

$$\frac{d^2}{dy^2}H_v e^{-y^2/2} = -(2v+1)H_v e^{-y^2/2} + y^2 H_v e^{-y^2/2}$$

(c) A further property of the Hermite polynomials is $H_{v+1} - 2yH_v + 2vH_{v-1} = 0$, which rearranges to $yH_v = vH_{v-1} + \frac{1}{2}H_{v+1}$. Using this, the term $y^2 H_v$ is rewritten

$$y^2 H_v = y(yH_v) = y(vH_{v-1} + \tfrac{1}{2}H_{v+1}) = yvH_{v-1} + \tfrac{1}{2}yH_{v+1}$$

The same relationship is used to substitute for the two terms on the right, but rewritten firstly by making the substitution $v \to v-1$ to give $yH_{v-1} = (v-1)H_{v-2} + \frac{1}{2}H_v$, and secondly with with the substitution $v \to v+1$ to give $yH_{v+1} = (v+1)H_v + \frac{1}{2}H_{v+2}$. With these substitutions

$$\begin{aligned} yvH_{v-1} + \tfrac{1}{2}yH_{v+1} &= v[(v-1)H_{v-2} + \tfrac{1}{2}H_v] + \tfrac{1}{2}[(v+1)H_v + \tfrac{1}{2}H_{v+2}] \\ &= v(v-1)H_{v-2} + (v + \tfrac{1}{2})H_v + \tfrac{1}{4}H_{v+2} \end{aligned}$$

Hence, the second derivative is given by

$$\begin{aligned} \frac{d^2}{dy^2} H_v e^{-y^2/2} &= \left[-(2v+1)H_v + y^2 H_v\right] e^{-y^2/2} \\ &= \left[-(2v+1)H_v + v(v-1)H_{v-2} + (v + \tfrac{1}{2})H_v + \tfrac{1}{4}H_{v+2}\right] e^{-y^2/2} \\ &= \left[v(v-1)H_{v-2} - (v + \tfrac{1}{2})H_v + \tfrac{1}{4}H_{v+2}\right] e^{-y^2/2} \end{aligned}$$

(d) The expectation value of the kinetic energy is therefore given by

$$\begin{aligned} \langle T \rangle_v = -\tfrac{1}{2}\hbar\omega N_v^2 \int_{-\infty}^{\infty} & H_v e^{-y^2/2} [v(v-1)H_{v-2} \\ & - (v + \tfrac{1}{2})H_v + \tfrac{1}{4}H_{v+2}] e^{-y^2/2}\, dy \\ = -\tfrac{1}{2}\hbar\omega N_v^2 \Big[v(v-1) & \int_{-\infty}^{\infty} H_v H_{v-2}\, e^{-y^2}\, dy + \tfrac{1}{4} \int_{-\infty}^{\infty} H_v H_{v+2}\, e^{-y^2}\, dy \\ & - (v + \tfrac{1}{2}) \int_{-\infty}^{\infty} H_v H_v e^{-y^2}\, dy \Big] \end{aligned}$$

The first two integrals are zero due to orthogonality of the Hermite polynomials, leaving

$$\langle T \rangle_v = \tfrac{1}{2}(v + \tfrac{1}{2})\hbar\omega N_v^2 \int_{-\infty}^{\infty} H_v^2 e^{-y^2}\, dy \qquad (7.3)$$

The next step to to evaluate the normalization constant N_v which is found from

$$N_v^2 \int_{-\infty}^{\infty} H_v^2 e^{-y^2}\, dy = 1$$

A property of the Hermite polynomials is (Table 7E.1 on page 275)

$$\int_{-\infty}^{\infty} H_v^2 e^{-y^2}\, dy = \pi^{1/2} 2^v v!$$

If follows that $N_v^2 = 1/(\pi^{1/2} 2^v v!)$. However, by the same property of the Hermite polynomials the integral in eqn 7.3 is also equal to $\pi^{1/2} 2^v v!$, therefore this term cancels with N_v^2 to leave

$$\langle T \rangle_v = \tfrac{1}{2}(v + \tfrac{1}{2})\hbar\omega = \tfrac{1}{2}E_v$$

P7E.13 As is shown in *Example* 7E.3 on page 279, in terms of the dimensionless variable y the classical turning points are at $y_{tp} = \pm(2\upsilon+1)^{1/2}$, where υ is the quantum number of the state. For the first excited state $\upsilon = 1$, and so $y_{tp} = \pm\sqrt{3}$.

The wavefunction is $\psi_1 = N_1 y e^{-y^2/2}$, which is normalized by solving

$$N_1^2 \int_{-\infty}^{+\infty} y^2 e^{-y^2}\, dy = 2N_1^2 \int_0^{+\infty} y^2 e^{-y^2}\, dy = 1$$

The integral is of the form of Integral G.3 with $k = 1$ and evaluates to $\frac{1}{4}\pi^{1/2}$, thus $N_1^2 = 2\pi^{-1/2}$. The probability of finding the particle outside the range of the turning points is then

$$P = 2 \times 2\pi^{-1/2} \int_{\sqrt{3}}^{\infty} y^2 e^{-y^2}\, dy = \frac{4}{\pi^{1/2}} \int_{\sqrt{3}}^{\infty} y^2 e^{-y^2}\, dy$$

Evaluating this integral numerically gives $\boxed{P = 0.112}$.

P7E.15 The integral to be evaluated is $I = \int_{-\infty}^{\infty} \psi_{\upsilon'}^* x \psi_\upsilon\, dx$. This can be written in terms of the dimensionless variable y, defined as $x = \alpha y$ where $\alpha = (\hbar^2/mk_f)^{1/4}$, as $I = \int_{-\infty}^{\infty} \psi_{\upsilon'}^*(\alpha y)\psi_\upsilon\, d(\alpha y) = \alpha^2 \int_{-\infty}^{\infty} \psi_{\upsilon'}^* y \psi_\upsilon\, dy$. The wavefunctions are $\psi_\upsilon = N_\upsilon H_\upsilon e^{-y^2/2}$, where N_υ is a normalization constant, and H_υ is a Hermite polynomial of order υ. With these substitutions I is

$$I = \alpha^2 \int_{-\infty}^{\infty} (N_{\upsilon'} H_{\upsilon'} e^{-y^2/2}) y (N_\upsilon H_\upsilon e^{-y^2/2})\, dy$$

$$= \alpha^2 N_\upsilon N_{\upsilon'} \int_{-\infty}^{\infty} H_{\upsilon'}(yH_\upsilon) e^{-y^2}\, dy$$

Using the properties of the Hermite polynomials given in Table 7E.1 on page 275, the term yH_υ can be rewritten as $yH_\upsilon = \frac{1}{2}H_{\upsilon+1} + \upsilon H_{\upsilon-1}$, and so the integral can be rewritten as

$$I = \alpha^2 N_\upsilon N_{\upsilon'} \left[\frac{1}{2}\int_{-\infty}^{\infty} H_{\upsilon'} H_{\upsilon+1}\, e^{-y^2}\, dy + \upsilon \int_{-\infty}^{\infty} H_{\upsilon'} H_{\upsilon-1}\, e^{-y^2}\, dy\right] \quad (7.4)$$

Due to orthogonality of the Hermite polynomials, the integral $\int_{-\infty}^{\infty} H_{\upsilon'} H_\upsilon e^{-y^2}\, dy$ is only non-zero when $\upsilon' = \upsilon$. Hence, the integral I is only non-zero if $\upsilon' = \upsilon+1$ or $\upsilon' = \upsilon - 1$. Hence, the only transitions with intensities not equal to zero are those with $\boxed{\upsilon' = \upsilon \pm 1}$.

The normalization constant N_υ is found from

$$N_\upsilon^2 \int_{-\infty}^{\infty} H_\upsilon^2 e^{-y^2}\, dy = 1$$

It is a property of the Hermite polynomials that $\int_{-\infty}^{\infty} H_\upsilon^2 e^{-y^2}\, dy = \pi^{1/2} 2^\upsilon \upsilon!$ so it follows that $N_\upsilon = (\pi^{1/2} 2^\upsilon \upsilon!)^{-1/2}$.

For $\upsilon' = \upsilon+1$ only the first integral in eqn 7.4 is non-zero. From the properties of Hermite polynomials it follows that $\int_{-\infty}^{\infty} H_{\upsilon+1}^2 e^{-y^2}\, dy = [\pi^{1/2} 2^{\upsilon+1}(\upsilon+1)!]^{1/2}$.

Using this, and the normalization factors already quoted, gives

$$I = \alpha^2 N_v N_{v+1} \tfrac{1}{2} \int_{-\infty}^{\infty} H_{v+1}^2 \mathrm{e}^{-y^2} \mathrm{d}y$$

$$= \tfrac{1}{2}\alpha^2 \frac{1}{(\pi^{1/2}2^v v!)^{1/2}} \frac{1}{[\pi^{1/2}2^{v+1}(v+1)!]^{1/2}} (\pi^{1/2}2^{v+1}(v+1)!) = \frac{\alpha^2}{2^{1/2}}(v+1)^{1/2}$$

In the case that $v' = v-1$ only the second integral in eqn 7.4 is non-zero. Using the same procedure as before with $\int_{-\infty}^{\infty} H_{v-1}^2 \mathrm{e}^{-y^2} \mathrm{d}y = [\pi^{1/2}2^{v-1}(v-1)!]^{1/2}$, gives

$$I = \alpha^2 N_v N_{v-1} v \int_{-\infty}^{\infty} H_{v-1}^2 \mathrm{e}^{-y^2} \mathrm{d}y$$

$$= v\alpha^2 \frac{1}{[\pi^{1/2}2^v v!]^{1/2}} \frac{1}{(\pi^{1/2}2^{v-1}(v-1)!)^{1/2}} [\pi^{1/2}2^{v-1}(v-1)!] = \frac{\alpha^2}{2^{1/2}}\sqrt{v}$$

P7E.17

(a) The harmonic oscillator potential energy is $V(x) = \tfrac{1}{2}k_f x^2$ which is symmetric about $x = 0$. It therefore follows that the probability density must also be symmetric about $x = 0$, for all wavefunctions. Hence the probability of finding the particle at $x = x'$ is the same as finding it at $x = -x'$ and so, for all v, $\boxed{\langle x \rangle_v = 0}$.

(b) As the potential is symmetric about $x = 0$ the probability of finding the particle moving to the right with a particular positive momentum is equal to that of finding the particle moving to the left with same momentum, but with opposite sign. It therefore follows that $\boxed{\langle p_x \rangle_v = 0}$.

(c) From the virial theorem, $\langle E_\mathrm{k} \rangle_v = \tfrac{1}{2}E_v = \tfrac{1}{2}(v+\tfrac{1}{2})\hbar\omega$. As the kinetic energy is $E_\mathrm{k} = p_x^2/2m$, where m is the mass of the particle, it follows that $\langle p_x^2 \rangle_v = 2m\langle E_\mathrm{k}\rangle_v = 2m \times \tfrac{1}{2}(v+\tfrac{1}{2})\hbar\omega = \boxed{m(v+\tfrac{1}{2})\hbar\omega}$.

(d) The value of $\langle x^2 \rangle_v$ is given in [7E.12b–*278*] as $\langle x^2 \rangle_v = (v+\tfrac{1}{2})\hbar/(mk_f)^{1/2}$.

$$\Delta x_v = [\langle x^2 \rangle_v - \langle x \rangle_v^2]^{1/2} = [(v+\tfrac{1}{2})\hbar/(mk_f)^{1/2} - 0^2]^{1/2}$$
$$= (v+\tfrac{1}{2})^{1/2}(\hbar^2/mk_f)^{1/4}$$
$$\Delta p_v = [\langle p_x^2 \rangle_v - \langle p_x \rangle_v^2]^{1/2} = [m(v+\tfrac{1}{2})\hbar\omega - 0^2]^{1/2}$$
$$= (v+\tfrac{1}{2})^{1/2}(\hbar m\omega)^{1/2}$$

(e) The product of the uncertainties is

$$\Delta x_v \Delta p_v = (v+\tfrac{1}{2})(\hbar^2/mk_f)^{1/4} \times (\hbar m\omega)^{1/2} = (v+\tfrac{1}{2})(\hbar^4 m\omega^2/k_f)^{1/4}$$

This can be simplified by noting that $\omega = (k_f/m)^{1/2}$; using this in the expression above gives $\Delta x_v \Delta p_v = \boxed{(v+\tfrac{1}{2})\hbar}$. This function increases linearly with v and its minimum value is when $v = 0$: $\Delta x_0 \Delta p_0 = \hbar/2$. This value satisfies the uncertainty principle, $\Delta x \Delta p \geq \hbar/2$, as do wavefunctions with higher values of v.

(f) The minimum product of the uncertainties is for $\boxed{v = 0}$.

7F Rotational motion

Answers to discussion questions

D7F.1 This is discussed in Section 7F.1(a) on page 283 and in particular in *How is that done?* 7F.1 on page 283.

D7F.3 The vector model of angular momentum is described in Section 7F.2(c) on page 288. The model captures the key idea that the magnitude and the z-component of the angular momentum can be known simultaneously and precisely. However, if this is so there is no knowledge of the other components of the angular momentum other than that the sum of their squares must be equal to a particular value.

Solutions to exercises

E7F.1(a) The magnitude of the angular momentum associated with a wavefunction with angular momentum quantum number l is given by [7F.11–*288*], magnitude = $\hbar[l(l+1)]^{1/2}$. Hence for $l = 1$ the magnitude is $\hbar[1(1+1)]^{1/2} = \boxed{2^{1/2}\hbar}$.

The projection of the angular momentum onto the z-axis is given by [7F.6–*284*], $\hbar m_l$, where m_l is a quantum number that takes values between $-l$ and $+l$ in integer steps, $m_l = -l, -l+1, \ldots +l$. Hence the possible projections onto the z-axis are $\boxed{-\hbar, 0, \hbar}$.

E7F.2(a) The wavefunction of a particle on a ring, with quantum number m_l is $\psi_{m_l} = e^{im_l\phi} = \cos(m_l\phi) + i\sin(m_l\phi)$ in the range $0 \leq \phi \leq 2\pi$. The real and imaginary parts of the wavefunction are therefore $\cos(m_l\phi)$ and $\sin(m_l\phi)$ respectively.

Nodes occur when the function passes through zero, which for trigonometric functions are the same points at which the function is zero. Hence in the real part, nodes occur when $\cos(m_l\phi) = 0$, and so when $m_l\phi = (2n+1)\pi/2$ for integer n, which gives $\phi = (2n+1)\pi/2m_l$. In the imaginary part, nodes occur when $\sin(m_l\phi) = 0$ and so when $m_l\phi = n\pi$ for an integer n, which gives $\phi = n\pi/m_l$.

(i) With $m_l = 0$ the real part is a constant and has $\boxed{\text{no nodes}}$; the imaginary part is zero everywhere.

(ii) With $m_l = 3$, nodes in the real part occur at $\boxed{\pi/6, \pi/2, 5\pi/6, 7\pi/6, 3\pi/2, 11\pi/6}$. In the imaginary part nodes occur at $\boxed{0, \pi/3, 2\pi/3, \pi, 4\pi/3, 5\pi/3}$. There are $\boxed{6}$ nodes in each of the parts.

E7F.3(a) The normalization condition is $\int \psi^*_{m_l}\psi_{m_l}\,d\tau = 1$. In this case the integral is over ϕ in the range $0 \leq \phi \leq 2\pi$, and the wavefunction is $\psi_{m_l} = Ne^{im_l\phi}$. Hence, noting that the wavefunction is complex

$$N^2\int \psi^*_{m_l}\psi_{m_l}\,d\tau = N^2\int_0^{2\pi} e^{-im_l\phi}e^{im_l\phi}\,d\phi = N^2\int_0^{2\pi} d\phi = 2\pi N^2$$

where $e^{-i\theta}e^{+i\theta} = e^0 = 1$ is used. Setting this integral to 1 gives $\boxed{N = (2\pi)^{-1/2}}$.

E7F.4(a) The integral to evaluate is $\int_0^{2\pi} \psi^*_{m'_l}\psi_{m_l}\,d\phi$, with $\psi_{m_l} = e^{im_l\phi}$. This wavefunction is complex and so $\psi^*_{m'_l} = e^{-im'_l\phi}$. Note that both m_l and m'_l are integers and therefore $m_l \pm m'_l$ is also an integer. The integral is then

$$\begin{aligned}\int_0^{2\pi} e^{-im'_l\phi}e^{im_l\phi}d\phi &= \int_0^{2\pi} e^{i(m_l-m'_l)\phi}d\phi \\ &= (i[m_l - m'_l])^{-1}(e^{2\pi i(m_l-m'_l)} - e^0) \\ &= (i[m_l - m'_l])^{-1}(1-1) = \boxed{0}\end{aligned}$$

where $e^{2\pi in} = 1$ for any integer n is used. Hence, wavefunctions for a particle on a ring with different quantum numbers are orthogonal.

E7F.5(a) The energy levels of a particle on a ring are [7F.4–283], $E_{m_l} = m_l^2\hbar^2/2I$ where I is the momentum of inertia of the system, $I = mr^2$, see *The chemist's toolkit* 20 in Topic 7F on page 282. The minimum excitation is therefore $\Delta E = E_1 - E_0 = (\hbar^2/2I)(1^2 - 0^2) = \hbar^2/2I$ Evaluating this gives

$$\Delta E = \frac{(1.0546 \times 10^{-34}\,\mathrm{J\,s})^2}{2 \times (1.6726 \times 10^{-27}\,\mathrm{kg}) \times (100 \times 10^{-12}\,\mathrm{m})^2} = \boxed{3.32 \times 10^{-22}\,\mathrm{J}}$$

E7F.6(a) The energy levels are [7F.10–287], $E_l = \hbar^2 l(l+1)/2I$, where I is the moment of inertia. The minimum energy to start it rotating is the minimum excitation energy, the energy to take it from the motionless $l = 0$ to the rotating $l = 1$ state, $\Delta E = E_1 - E_0 = (\hbar^2/2I)(1(1+1) - 0(0+1)] = \hbar^2/I$. Evaluating this gives

$$\Delta E = (1.0546 \times 10^{-34}\,\mathrm{J\,s})^2/(5.27 \times 10^{-47}\,\mathrm{kg\,m^2}) = \boxed{2.11 \times 10^{-22}\,\mathrm{J}}$$

E7F.7(a) The energy levels are [7F.10–287], $E_l = \hbar^2 l(l+1)/2I$, where I is the moment of inertia. So that the excitation energy is $\Delta E = E_2 - E_1 = (\hbar^2/2I)[2(2+1) - 1(1+1)] = 2\hbar^2/I$. Evaluating this gives

$$\Delta E = 2(1.0546 \times 10^{-34}\,\mathrm{J\,s})^2/(5.27 \times 10^{-47}\,\mathrm{kg\,m^2}) = \boxed{4.22 \times 10^{-22}\,\mathrm{J}}$$

E7F.8(a) The energy levels are [7F.10–287], $E_l = \hbar^2 l(l+1)/2I$, where I is the moment of inertia. The corresponding angular momentum is $\langle l^2\rangle^{1/2} = \hbar\sqrt{l(l+1)}$. Hence, the minimum energy allowed is 0, through this corresponds to zero angular momentum, and so rest and not motion. So the minimum energy of rotation occurs for the state that has $l = 1$. The angular momentum of that state is $\langle l^2\rangle_1^{1/2} = \hbar\sqrt{1(1+1} = \sqrt{2}\hbar = \sqrt{2} \times (1.0546 \times 10^{-34}\,\mathrm{J\,s}) = \boxed{1.49 \times 10^{-34}\,\mathrm{J\,s}}$.

E7F.9(a) The diagrams shown in Fig. 7.9 are drawn by forming a vector of length $[l(l+1)]^{1/2}$ and with a projection m_l on the z-axis. For $l = 1$, $m_l = +1$ the vector is of length $\sqrt{2}$ and has projection $+1$ on the z-axis; for $l = 2$, $m_l = 0$ the vector is of length $\sqrt{6}$ and has projection 0 on the z-axis. Each vector may lie anywhere on a cone described by rotating the vector about the z-axis.

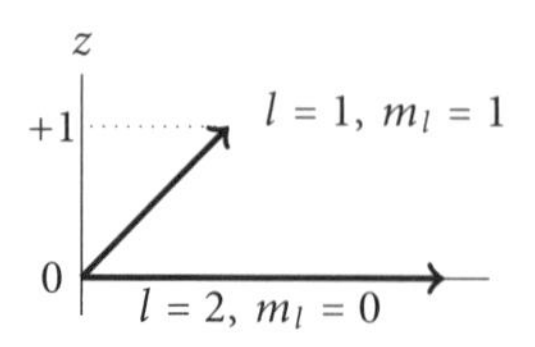

Figure 7.9

E7F.10(a) The spherical harmonic is

$$Y_{3,0} = \tfrac{1}{4}(7/\pi)^{1/2}(5\cos^3\theta - 3\cos\theta) = \tfrac{1}{4}\times(7/\pi)^{1/2}\cos\theta(5\cos^2\theta - 3)$$

The nodes occur when $Y_{3,0}$ passes through zero, which happens when either $\cos\theta = 0$ or $5\cos^2\theta - 3 = 0$, that is when $\cos\theta = \pm\sqrt{3/5}$, recall that θ is in the range 0 to π. Hence, the nodes are $\boxed{\theta = \pi/2,\ 0.684,\ 2.46}$. There are $\boxed{3}$ angular nodes.

E7F.11(a) The real part of the spherical harmonic $Y_{1,+1}$ is $-\frac{1}{2}\sqrt{3/\pi}\sin\theta\cos\phi$. When ϕ varies, an angular node therefore occurs when $\cos\phi = 0$, i.e. at $\boxed{\phi = \pi/2, 3\pi/2}$. This plane corresponds to the plane $x = 0$, i.e. the $\boxed{yz}$ plane.

The imaginary part of the same spherical harmonic is $-\frac{1}{2}\sqrt{3/\pi}\sin\theta\sin\phi$. When ϕ varies, an angular node therefore occurs when $\sin\phi = 0$, i.e. at $\phi = 0, \pi$. This plane corresponds to the plane $y = 0$, i.e. the $\boxed{xz}$ plane.

E7F.12(a) The rotational energy depends only on the quantum number l [7F.10–*287*], but there are distinct states for every allowed value of m_l, which can range from $-l$ to $+l$ in integer steps. There are $2l + 1$ such states, as there are l of these with $m_l > 0$, l of these with $m_l < 0$ and $m_l = 0$. Hence $l = 3$ has a degeneracy of $\boxed{7}$.

E7F.13(a) The diagrams shown in Fig. 7.10 are drawn by forming a vector of length $[l(l+1)]^{1/2}$ and with a projection m_l on the z-axis. For $l = 1$ the vector is of length $\sqrt{2}$ and has projection $-1, 0, +1$ on the z-axis. For $l = 2$ the vector is of length $\sqrt{6}$ and has projection $-2, \ldots +2$ in integer steps on the z-axis. Each vector may lie anywhere on a cone described by rotating the vector about the z-axis.

E7F.14(a) The angle in question is that between the z-axis and the vector representing the angular momentum. The projection of the vector onto the z-axis is $m_l\hbar$, and the length of the vector is $\hbar\sqrt{l(l+1)}$. Therefore the angle θ that the vector makes to the z-axis is given by $\cos\theta = m_l/\sqrt{l(l+1)}$.

When $m_l = l$, $\cos\theta = l/\sqrt{l(l+1)}$, which for $l = 1$ gives $\cos\theta = 1/\sqrt{2}$, and so $\boxed{\theta = \pi/4}$, and for $l = 5$ gives $\cos\theta = 5/\sqrt{30}$, and so $\boxed{\theta = 0.420}$.

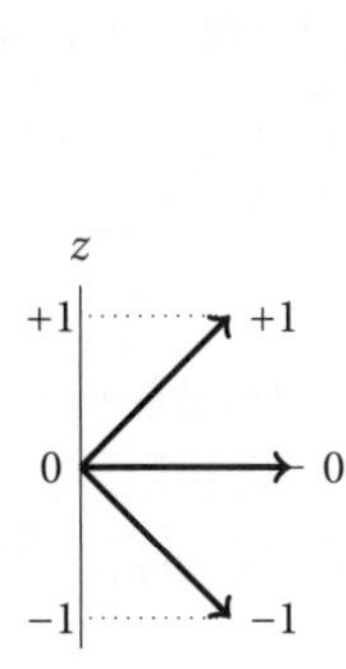

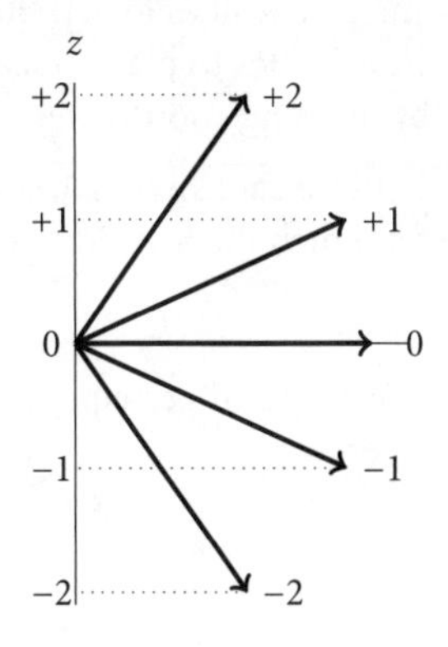

Figure 7.10

Solutions to problems

P7F.1 The angular momentum states have quantum numbers $m_l = 0, \pm1, \pm2....$ The energy levels for a particle on a ring are given by [7F.4–*283*], $E_{m_l} = m_l^2\hbar^2/2I$, and have angular momentum [7F.6–*284*], $J_z = m_l\hbar$. The moment of inertia for an electron on this ring is $I = mr^2 = (9.1094 \times 10^{-31}\,\text{kg}) \times (440 \times 10^{-12}\,\text{m})^2 = 1.76... \times 10^{-49}\text{kg m}^2$

(a) If there are 22 electrons in the system the highest occupied state will be the degenerate levels $m_l = \pm5$. These states have an energy of $E_{\pm5} = (\pm5)^2(1.0546 \times 10^{-34}\,\text{J s})^2/2(1.76... \times 10^{-49}\text{kg m}^2) = \boxed{7.88 \times 10^{-19}\,\text{J}}$, and angular momenta of

$$J_z = \pm5\hbar = \pm5 \times (1.0546 \times 10^{-34}\,\text{J s}) = \boxed{5.273 \times 10^{-34}\,\text{J s}}$$

(b) The lowest unoccupied levels are those with $m_l = \pm6$, and so the difference in energy between the highest occupied and lowest unoccupied levels is

$$\Delta E = E_{\pm6} - E_{\pm5} = (\hbar^2/2I)(6^2 - 5^2)$$
$$= 11 \times 1.0546 \times 10^{-34}\,\text{J s})^2/2(1.76... \times 10^{-49}\text{kg m}^2) = 3.46... \times 10^{-19}\text{J}$$

The Bohr frequency condition, [7A.9–*241*] states that the frequency of radiation that will excite such a transition is

$$\nu = \Delta E/h = \frac{3.46... \times 10^{-19}\,\text{J}}{6.6261 \times 10^{-34}\,\text{J s}} = \boxed{5.23 \times 10^{14}\,\text{Hz}}$$

P7F.3 In Cartesian coordinates, the equation defining the ellipse is $x^2/a^2 + y^2/b^2 = 1$. An appropriate change of variables can transform this ellipse into a circle. That change of variable is most conveniently described in terms of new Cartesian coordinates (X, Y) where $X = x$ and $Y = ay/b$. In these coordinates, the equation for the ellipse can be rewritten as $X^2 + Y^2 = a^2$, which is the equation of a circle radius a centered on the origin. The text found the eigenfunctions and eigenvalues for a particle on a circular ring by transforming from Cartesian coordinates to plane polar coordinates. A similar transformation can be made by defining coordinates (R, Φ) such that $X = R\cos\Phi$, $Y = R\sin\Phi$.

In these coordinates, this is simply a particle on a ring, as described in the text, for which the Schrödinger equation $\boxed{\text{is separable}}$.

P7F.5 The Schrödinger equation for a particle on a sphere is $-(\hbar^2/2I)\hat{\Lambda}^2\psi = E\psi$, where

$$\hat{\Lambda}^2 = \frac{1}{\sin^2\theta}\frac{\partial^2}{\partial\phi^2} + \frac{1}{\sin\theta}\frac{\partial}{\partial\theta}\sin\theta\frac{\partial}{\partial\theta}$$

(a) $Y_{0,0} = (1/2)\pi^{-1/2}$, which is a constant, and so its derivatives with respect to all θ and ϕ are zero, so $\hat{\Lambda}^2 Y_{0,0} = 0$ implying that $E = 0$ and $\hat{l}_z Y_{0,0} = 0$, so that $\boxed{J_z = 0}$.

(b) $Y_{2,-1} = N\sin\theta\cos\theta e^{-i\phi}$, thus

$$\partial Y_{2,-1}/\partial\theta = Ne^{-i\phi}(\cos^2\theta - \sin^2\theta) \quad \partial Y_{2,-1}/\partial\phi = -iN\sin\theta\cos\theta e^{-i\phi} = -iY_{2,-1}$$

In addition, $\partial^2 Y_{2,-1}/\partial\phi^2 = N\sin\theta\cos\theta e^{-i\phi}$

$$\begin{aligned}\hat{\Lambda}^2 Y_{2,-1} &= \frac{1}{\sin^2\theta}\frac{\partial^2 Y_{2,-1}}{\partial\phi^2} + \frac{1}{\sin\theta}\frac{\partial}{\partial\theta}\sin\theta\frac{\partial Y_{2,-1}}{\partial\theta} \\ &= \frac{N\cos\theta e^{-i\phi}}{\sin\theta} + \frac{Ne^{-i\phi}}{\sin\theta}\frac{\partial}{\partial\theta}\sin\theta(\cos^2\theta - \sin^2\theta)\end{aligned}$$

The derivative is evaluated using the product rule

$$\begin{aligned}&= \frac{N\cos\theta e^{-i\phi}}{\sin\theta} \\ &\qquad + \frac{Ne^{-i\phi}}{\sin\theta}\left[\sin\theta(-4\cos\theta\sin\theta) + \cos\theta(\cos^2\theta - \sin^2\theta)\right]\end{aligned}$$

as $\cos^3\theta = \cos\theta\cos^2\theta = \cos\theta(1 - \sin^2\theta)$

$$\begin{aligned}&= \frac{N\cos\theta e^{-i\phi}}{\sin\theta} + Ne^{-i\phi}\left(-6\sin\theta\cos\theta + \frac{\cos\theta}{\sin\theta}\right) \\ &= -6Ne^{-i\phi}\sin\theta\cos\theta\end{aligned}$$

This has an eigenvalue of -6, giving an energy eigenvalue of $\boxed{6\hbar^2/2I}$. For angular momentum, $\hat{l}_z Y_{2,-1} = (\hbar/i)\times -iY_{2,-1} = \hbar Y_{2,-1}$, giving an angular momentum eigenvalue of $\boxed{J_z = -\hbar}$.

(c) $Y_{3,+3} = Ne^{3i\phi}\sin^3\theta$, thus

$$\partial Y_{3,+3}/\partial\theta = 3Ne^{3i\phi}\sin^2\theta\cos\theta \quad \partial Y_{3,+3}/\partial\phi = 3iNe^{3i\phi}\sin^3\theta = 3iY_{3,+3}$$

In addition, $\partial^2 Y_{3,+3}/\partial\phi^2 = -9Ne^{3i\phi}\sin^3\theta$. Hence,

$$\begin{aligned}\hat{\Lambda}^2 Y_{3,+3} &= \frac{1}{\sin^2\theta}\frac{\partial^2 Y_{3,+3}}{\partial\phi^2} + \frac{1}{\sin\theta}\frac{\partial}{\partial\theta}\sin\theta\frac{\partial Y_{3,+3}}{\partial\theta} \\ &= \frac{-9Ne^{3i\phi}\sin^3\theta}{\sin^2\theta} + \frac{1}{\sin\theta}\frac{\partial}{\partial\phi}3Ne^{3i\phi}\sin^3\theta\cos\theta\end{aligned}$$

The derivative is evaluated using the product rule

$$\begin{aligned}&= -9Ne^{3i\phi}\sin\theta + \frac{3Ne^{3i\phi}}{\sin\theta}(3\sin^2\theta\cos^2\theta - \sin^4\theta) \\ &= -9Ne^{3i\phi}\sin\theta + 3Ne^{3i\phi}(3\sin\theta\cos^2\theta - \sin^3\theta) \\ &= -12Ne^{3i\phi}\sin^3\theta = -12Y_{3,+3}\end{aligned}$$

This has an eigenvalue of -12, giving an energy eigenvalue of $\boxed{12\hbar^2/2I}$. For angular momentum, $\hat{l}_z Y_{2,-1} = (\hbar/i)\times 3iY_{2,-1} = 3\hbar Y_{2,-1}$, giving an angular momentum eigenvalue of $\boxed{J_z = 3\hbar}$.

The energies are given by [7F.10–287], $E_l = \hbar^2 l(l+1)/2I$, and therefore $E_0 = 0$, $E_2 = 2(3)\hbar^2/2I = 6\hbar^2/2I$ and $E_3 = 4(3)\hbar^2/2I = 12\hbar^2/2I$, all of which $\boxed{\text{are consistent}}$ with the calculated eigenvalues.

P7F.7 The function $Y_{1,+1} = -\frac{1}{2}\sqrt{3/2\pi}\sin\theta e^{i\phi}$ and $Y_{1,0} = \frac{1}{2}\sqrt{3/\pi}\cos\theta$. The integral to evaluate is

$$\int_{\theta=0}^{\pi}\int_{\phi=0}^{2\pi} Y_{1,0}^{*}Y_{1,+1}\sin\theta\,d\theta\,d\phi$$

$Y_{1,0}$ is real and so the integrand is

$$-\tfrac{1}{2}\sqrt{3/2\pi}\sin\theta e^{i\phi}\times\tfrac{1}{2}\sqrt{3/\pi}\cos\theta\times\sin\theta = -(3/4\pi\sqrt{2})\sin^2\theta\cos\theta e^{i\phi}$$

This gives the integral as

$$I = \int_{\theta=0}^{\pi}\int_{\phi=0}^{2\pi} Y_{1,0}^{*}Y_{1,+1}\sin\theta\,d\theta\,d\phi = -\frac{3}{4\pi\sqrt{2}}\int_{\theta=0}^{\pi}\int_{\phi=0}^{2\pi}\sin^2\theta\cos\theta e^{i\phi}\,d\theta\,d\phi$$

The integrand is the product of separate functions of θ and ϕ, and so the integral can be separated

$$I = -\frac{3}{4\pi\sqrt{2}}\int_0^{2\pi} e^{i\phi}\,d\phi\int_0^{\pi}\sin^2\theta\cos\theta\,d\theta$$

Evaluating the first integral gives

$$\int_0^{2\pi} e^{i\phi}\,d\phi = (1/i)\,e^{i\phi}\Big|_0^{2\pi} = (1/i)[e^{2\pi i} - e^0] = (1/i)(1-1) = 0$$

Hence

$$\int_{\theta=0}^{\pi}\int_{\phi=0}^{2\pi} Y_{1,0}^{*}Y_{1,+1}\sin\theta\,d\theta\,d\phi = 0$$

so the two functions are $\boxed{\text{orthogonal}}$.

P7F.9 (a) Multiplying out the brackets, noting that the derivatives in the left brackets act on the whole term in the right brackets, e.g. $x\partial f/\partial x$

$$\hat{l}_x\hat{l}_y f = -\hbar^2\left(y\frac{\partial}{\partial z} - z\frac{\partial}{\partial y}\right)\left(z\frac{\partial f}{\partial x} - x\frac{\partial f}{\partial z}\right)$$

$$= -\hbar^2\left[y\frac{\partial}{\partial z}\left(z\frac{\partial f}{\partial x}\right) - y\frac{\partial}{\partial z}\left(x\frac{\partial f}{\partial z}\right) - z\frac{\partial}{\partial y}\left(z\frac{\partial f}{\partial x}\right) + z\frac{\partial}{\partial y}\left(x\frac{\partial f}{\partial z}\right)\right]$$

To evaluate the first term in this, the product rule is used

$$= -\hbar^2\left[y\frac{\partial z}{\partial z}\frac{\partial f}{\partial x} + yz\frac{\partial^2 f}{\partial z\partial x} - xy\frac{\partial^2 f}{\partial z^2} - z^2\frac{\partial^2 f}{\partial y\partial x} + zx\frac{\partial^2 f}{\partial y\partial z}\right]$$

$$= -\hbar^2\left[y\frac{\partial f}{\partial x} + yz\frac{\partial^2 f}{\partial z\partial x} - xy\frac{\partial^2 f}{\partial z^2} - z^2\frac{\partial^2 f}{\partial y\partial x} + zx\frac{\partial^2 f}{\partial y\partial z}\right]$$

(b) Similarly,

$$\hat{l}_y\hat{l}_x f = -\hbar^2\left(z\frac{\partial}{\partial x} - x\frac{\partial}{\partial z}\right)\left(y\frac{\partial f}{\partial z} - z\frac{\partial f}{\partial y}\right)$$

$$= -\hbar^2\left[z\frac{\partial}{\partial x}\left(y\frac{\partial f}{\partial z}\right) - z\frac{\partial}{\partial x}\left(z\frac{\partial f}{\partial y}\right) - x\frac{\partial}{\partial z}\left(y\frac{\partial f}{\partial z}\right) + x\frac{\partial}{\partial z}\left(z\frac{\partial f}{\partial y}\right)\right]$$

To evaluate the final term in this, the product rule must be used

$$= -\hbar^2\left[yz\frac{\partial^2 f}{\partial x\partial z} - z^2\frac{\partial^2 f}{\partial x\partial y} - xy\frac{\partial^2 f}{\partial z^2} + x\frac{\partial z}{\partial z}\frac{\partial f}{\partial y} + xz\frac{\partial^2 f}{\partial z\partial y}\right]$$

$$= -\hbar^2\left[yz\frac{\partial^2 f}{\partial x\partial z} - z^2\frac{\partial^2 f}{\partial x\partial y} - xy\frac{\partial^2 f}{\partial z^2} + x\frac{\partial f}{\partial y} + xz\frac{\partial^2 f}{\partial z\partial y}\right]$$

(c) Due to the symmetry of mixed partial derivatives, the only terms that are not repeated in both of these terms are the first derivatives. Hence,

$$\hat{l}_x\hat{l}_y f - \hat{l}_y\hat{l}_x f = \hbar^2\left(x\frac{\partial f}{\partial y} - y\frac{\partial f}{\partial x}\right) = \mathrm{i}\hbar \times \frac{\hbar}{\mathrm{i}}\left(x\frac{\partial f}{\partial y} - y\frac{\partial f}{\partial x}\right) = \mathrm{i}\hbar\hat{l}_z f$$

where the definition of the $\hat{l}_z$ operator given in [7F.13–*289*] is used. It follow that $[\hat{l}_x, \hat{l}_y] = \hat{l}_z$.

(d) Applying cyclic permutation to $\hat{l}_x$ gives $(\hbar/\mathrm{i})(z\partial/\partial x - x\partial/\partial z) = \hat{l}_y$. Likewise $\hat{l}_y$ gives $(\hbar/\mathrm{i})(x\partial/\partial y - y\partial/\partial x) = \hat{l}_z$, and $\hat{l}_z$ gives $(\hbar/\mathrm{i})(y\partial/\partial z - z\partial/\partial y) = \hat{l}_x$.

(e) Applying this permutation to the expression $[\hat{l}_x, \hat{l}_y] = \mathrm{i}\hbar\hat{l}_z$ gives $[\hat{l}_y, \hat{l}_z] = \mathrm{i}\hbar\hat{l}_x$. Permuting this expression gives $[\hat{l}_z, \hat{l}_x] = \mathrm{i}\hbar\hat{l}_y$, and permuting that expression gives $[\hat{l}_x, \hat{l}_y] = \mathrm{i}\hbar\hat{l}_z$.

P7F.11 The Cartesian coordinates expressed in terms of the spherical polar coordinates are $x = r\sin\theta\cos\phi$, $y = r\sin\theta\sin\phi$, $z = r\cos\theta$, see *The chemist's toolkit* 21 in Topic 7F on page 286. The chain rule is therefore used to express $\partial/\partial\phi$ in terms of derivatives of x, y and z.

$$\frac{\partial}{\partial\phi} = \frac{\partial x}{\partial\phi}\frac{\partial}{\partial x} + \frac{\partial y}{\partial\phi}\frac{\partial}{\partial y} + \frac{\partial z}{\partial\phi}\frac{\partial}{\partial z}$$

Evaluating the derivatives gives

$$= -r\sin\theta\sin\phi\frac{\partial}{\partial x} + r\sin\theta\cos\phi\frac{\partial}{\partial y} + 0\frac{\partial}{\partial z} = -y\frac{\partial}{\partial x} + x\frac{\partial}{\partial y}$$

where it has been noted that the factors multiplying the derivatives are Cartesian coordinates. Hence, $\hat{l}_z = (\hbar/\mathrm{i})(x\partial/\partial y - y\partial/\partial x) = (\hbar/\mathrm{i})\partial/\partial\phi$

Answers to integrated activities

I7.1 The reaction considered is $CH_4(g) \rightarrow C(graphite) + 2\,H_2(g)$.

(a) The reverse of this reaction is the formation of $CH_4(g)$ from elements in their standard states, and so at standard temperature, 298 K

$$\Delta_r G^\ominus = -\Delta_f G^\ominus = -(-50.72\ \mathrm{kJ\,mol^{-1}}) = +50.72\ \mathrm{kJ\,mol^{-1}}\ \text{at 298 K}$$

$$\Delta_r H^\ominus = -\Delta_f H^\ominus = -(-74.81\ \mathrm{kJ\,mol^{-1}}) = \boxed{+74.81\ \mathrm{kJ\,mol^{-1}}}\ \text{at 298 K}$$

At a fixed temperature, $\Delta G = \Delta H - T\Delta S$, and hence, at 298 K,

$$\Delta_r S^\ominus = (\Delta_r H^\ominus - \Delta_r G^\ominus)/T = ([74.81 - 50.72]\ \mathrm{kJ\,mol^{-1}})/(298\ \mathrm{K})$$
$$= \boxed{+80.8...\ \mathrm{J\,K^{-1}\,mol^{-1}}}$$

To convert the enthalpy change to an arbitrary temperature [2C.7d–56] is used, $\Delta_r H^\ominus(T_2) = \Delta_r H^\ominus(T_1) + \Delta_r C_p^\ominus(T_2 - T_1)$, and similarly [3C.5c–95] is used for the entropy change, $\Delta_r S^\ominus(T_2) = \Delta_r S^\ominus(T_1) + \Delta_r C_p^\ominus \ln(T_2/T_1)$; in theses expressions

$$\Delta_r C_p^\ominus = 2C_p^\ominus(H_2(g)) + C_p^\ominus(C(graphite)) - C_p^\ominus(CH_4(g))$$

and it is assumed that $\Delta_r C_p^\ominus$ is independent of temperature. For this reaction $\Delta_r C_p^\ominus = [8.527{+}2(28.824){-}35.31]\ \mathrm{J\,K^{-1}\,mol^{-1}} = +30.865\ \mathrm{J\,K^{-1}\,mol^{-1}}$.

$$\Delta_r H^\ominus(T) = (74.81\ \mathrm{kJ\,mol^{-1}}) + (30.865\ \mathrm{J\,K^{-1}\,mol^{-1}}) \times (T - 298\ \mathrm{K})$$
$$= (6.56... \times 10^4\ \mathrm{J\,mol^{-1}}) + T(30.865\ \mathrm{J\,K^{-1}\,mol^{-1}})$$
$$\Delta_r H^\ominus(T)/(\mathrm{J\,mol^{-1}}) = 6.56... \times 10^4 + (T/\mathrm{K}) \times (30.865)$$
$$\Delta_r S^\ominus(T) = (80.8...\ \mathrm{J\,K^{-1}\,mol^{-1}}) + (30.865\ \mathrm{J\,K^{-1}\,mol^{-1}}) \times \ln[T/(298\ \mathrm{K})]$$
$$= -95.0...\ \mathrm{J\,K^{-1}\,mol^{-1}} + (30.865\ \mathrm{J\,K^{-1}\,mol^{-1}})\ln(T/\mathrm{K})$$
$$\Delta_r S^\ominus(T)/(\mathrm{J\,K^{-1}\,mol^{-1}}) = -95.0... + 30.865\ln(T/\mathrm{K})$$

(b) Using the expressions for $\Delta_r H^\ominus(T)$ and $\Delta_r S^\ominus(T)$ from part (a) and expression for $\Delta_r G^\ominus(T)$ is found

$$\Delta_r G^\ominus(T) = \Delta_r H^\ominus(T) - T\Delta_r S^\ominus(T)$$

$$\begin{aligned}\Delta_r G^\ominus(T)/(\mathrm{J\,mol^{-1}}) &= 6.56... \times 10^4 + (T/\mathrm{K}) \times (30.865) \\ &\quad - (T/\mathrm{K}) \times [-95.0... + 30.865 \ln(T/\mathrm{K})] \\ &= (6.56... \times 10^4) + (1.26... \times 10^2) \times (T/\mathrm{K}) \\ &\quad - (30.865) \times (T/\mathrm{K}) \ln(T/\mathrm{K})\end{aligned}$$

The reaction is endothermic so the products will be increasingly favoured as the temperature is increased. One way to proceed is to find the temperature at which $\Delta_r G^\ominus(T) = 0$: at this temperature the equilibrium constant is = 1, and so above this temperature the products will be increasingly favoured. Using mathematical software it is found that $\Delta_r G^\ominus(T) = 0$ at $\boxed{T = 812\ \mathrm{K}}$. It can be concluded that above $T = 812$ K the equilibrium for the reaction CH_4(g)→ C(graphite) + 2 H_2(g) will favour the elements.

(c) If the star behaves as a black body emitter, then the wavelength at which the radiation from it is a maximum is given by Wien's law [7A.1–*238*], $\lambda_{max} T = 2.9 \times 10^{-3}$ m K, and therefore at 1000 K

$$\lambda_{max} = (2.9 \times 10^{-3}\ \mathrm{m\,K})/(1000\ \mathrm{K}) = \boxed{2.9 \times 10^{-6}\ \mathrm{m}}$$

(d) The fraction of the total energy density is found by integration of the Planck distribution over the visible range of radiation (between about 700 nm (red) and 420 nm (violet)) followed by division by the total energy density. Hence, the energy density in the visible range is

$$E_{vis} = \int_{420\ \mathrm{nm}}^{700\ \mathrm{nm}} \rho(\lambda, T)\,\mathrm{d}\lambda = 8\pi hc \int_{420\ \mathrm{nm}}^{700\ \mathrm{nm}} \lambda^{-5}(\mathrm{e}^{hc/\lambda kT} - 1)^{-1}\,\mathrm{d}\lambda \quad (7.5)$$

where the expression for the Planck distribution, [7A.6a–*239*], is used. Mathematical software is used to evaluate this integral numerically giving the result $E_{vis} = \boxed{1.39 \times 10^{-9}\ \mathrm{J\,m^{-3}}}$.

This energy density is compared with the total energy density given by the Stefan–Boltzmann law, $E_{tot} = (7.567 \times 10^{-16}\ \mathrm{J\,m^{-3}\,K^{-4}})T^4$, which gives

$$\text{fraction} = \frac{E_{vis}}{E_{tot}} = \frac{1.39 \times 10^{-9}\ \mathrm{J\,m^{-3}}}{(7.567 \times 10^{-16}\ \mathrm{J\,m^{-3}\,K^{-4}}) \times (1000\ \mathrm{K})^4} = \boxed{1.84 \times 10^{-6}}$$

Very little of the radiation from the brown dwarf radiation is in the visible region.

If it is assumed that the integral of eqn 7.5 can be approximated as $E_{vis} = \rho(\lambda, T)\Delta\lambda$, where ρ is evaluated at the midpoint of the integration range, 560 nm, and $\Delta\lambda = 700 - 420 = 280$ nm, the energy density is found to be $1.76 \times 10^{-10}\ \mathrm{J\,m^{-3}}$. This is quite different from the result obtained by numerical integration: the approximation of the integral in this way is rather poor.

I7.3 The particle in a box and the harmonic oscillator are useful models because they are sufficiently simple that there are analytic forms for the wavefunctions and energy levels, yet at the same time the models capture the essence of some interesting chemical and physical systems. The particle in a box is the starting point for modelling systems in which an electron is held in a confined space, for example the π electrons in a linear conjugated polyene (see *Example* 7D.1 on page 265 in the text) or the electrons in a nanostructure, such as a quantum dot. The harmonic oscillator is the starting point for modelling the motion of atoms which are confined by a potential well, such as in a solid material or in a chemical bond. The results from such a model are used to interpret spectroscopic data (infrared spectroscopy) and physical properties (heat capacities).

I7.5 (a) In *Problem* P7D.6 and *Problem* P7D.7 it is shown that for a particle in a box in a state with quantum number n

$$\Delta x = L(1/12 - 1/2n^2\pi^2)^{1/2} \quad \text{and} \quad \Delta p_x = nh/2L$$

Hence for $n = 1$

$$\Delta x\Delta p_x = L(1/12 - 1/2\pi^2)^{1/2} \times h/2L = (h/2)(1/12 - 1/2\pi^2)^{1/2} \approx 0.57\hbar$$

The Heisenberg uncertainty principle is satisfied.

(b) In *Problem* P7E.17 it is shown that for a harmonic oscillator in a state with quantum number υ

$$\Delta x_\upsilon \Delta p_\upsilon = (\upsilon + \tfrac{1}{2})\hbar$$

Therefore, for the ground state with $\upsilon = 0$, $\Delta x\Delta p = \frac{1}{2}\hbar$: the Heisenberg uncertainty principle is satisfied with the smallest possible uncertainty.

8 Atomic structure and spectra

8A Hydrogenic Atoms

Answers to discussion questions

D8A.1 The separation of variables method as applied to the hydrogen atom is described in *A deeper look* 3 on the website for the main text.

D8A.3 (i) A boundary surface for a hydrogenic orbital is drawn so as to contain most (say 90%) of the probability density of an electron in that orbital. Its shape varies from orbital to orbital because the electron density distribution is different for different orbitals.

(ii) The radial distribution function $P(r)$ gives the probability density that the electron will be found at a distance r from the nucleus. It is defined such that $P(r)\,dr$ is the probability of finding the electron in a shell of radius r and thickness dr. Because the radial distribution function gives the total density, summed over all angles, it has no angular dependence and, as a result, perhaps gives a clearer indication of how the electron density varies with distance from the nucleus.

Solutions to exercises

E8A.1(a) The energy of the level of the H atom with quantum number n is given by [8A.8–*306*], $E_n = -hc\tilde{R}_H/n^2$. As described in Section 8A.2(d) on page 309, the degeneracy of a state with quantum number n is n^2.

The state with $E = -hc\tilde{R}_H$ has $n = 1$ and degeneracy $(1)^2 = \boxed{1}$; that with $E = -hc\tilde{R}_H/9$ has $n = 3$ and degeneracy $(3)^2 = \boxed{9}$; and that with $E = -hc\tilde{R}_H/25$ has $n = 5$ and degeneracy $(5)^2 = \boxed{25}$.

E8A.2(a) The task is to find the value of N such that the integral $\int \psi^*\psi\,d\tau = 1$, where $\psi = Ne^{-r/a_0}$. The integration is over the range $r = 0$ to ∞, $\theta = 0$ to π, and $\phi = 0$ to 2π; the volume element is $r^2 \sin\theta\,dr\,d\theta\,d\phi$. The required integral is therefore

$$N^2 \int_0^\infty \int_0^\pi \int_0^{2\pi} r^2 e^{-2r/a_0} \sin\theta\,dr\,d\theta\,d\phi$$

The integrand is a product of functions of each of the variables, and so the integral separates into three

$$N^2\int_0^\infty r^2 e^{-2r/a_0}\,dr\int_0^\pi \sin\theta\,d\theta\int_0^{2\pi}d\phi$$
$$= N^2[2!/(2/a_0)^3]\times(-\cos\theta)|_0^\pi\times\phi|_0^{2\pi}$$
$$= N^2[2!/(2/a_0)^3]\times 2\times 2\pi = N^2 a_0^3\pi$$

The integral over r is evaluated using Integral E.3 with $n = 2$ and $k = 2/a_0$. Setting the full integral equal to 1 gives $\boxed{N = (a_0^3\pi)^{-1/2}}$.

E8A.3(a) The wavefunction is given by [8A.12–307], $\psi_{n,l,m_l} = Y_{l,m_l}(\theta,\phi)R_{n,l}(r)$; for the state with $n = 2$, $l = 0$, $m_l = 0$ this is

$$\psi_{2,0,0} = Y_{0,0}(\theta,\phi)R_{1,0}(r) = (4\pi)^{-1/2}(Z/2a_0)^{3/2}(2-\rho)e^{-\rho/2}$$

where the radial wavefunction is taken from Table 8A.1 on page 306, the angular wavefunction (the spherical harmonic) is taken from Table 7F.1 on page 286, and $\rho = 2Zr/na_0$. The probability density is therefore

$$P_{2,0,0} = |\psi_{2,0,0}|^2 = (4\pi)^{-1}(Z/2a_0)^3(2-\rho)^2 e^{-\rho}$$

The probability density at the nucleus, $\rho = 0$, is then $(1/4\pi)Z^3/(8a_0^3)(2-0)^2 e^0 = \boxed{Z^3/(8\pi a_0^3)}$.

E8A.4(a) The radial wavefunction of a 2s orbital is taken from Table 8A.1 on page 306, $R_{2,0}(r) = N(2-\rho)e^{-\rho/2}$, where $\rho = 2Zr/na_0$; for $n = 2$, $\rho = Zr/a_0$. The extrema are located by finding the values of ρ for which $dR_{2,0}/d\rho = 0$; the product rule is required

$$\frac{dR_{2,0}}{d\rho} = N\frac{d(2-\rho)}{d\rho}e^{-\rho/2} + N(2-\rho)\frac{de^{-\rho/2}}{d\rho}$$
$$= N(-1)e^{-\rho/2} + N(2-\rho)(-\tfrac{1}{2}e^{-\rho/2}) = N(\rho/2-2)e^{-\rho/2}$$

The derivative is zero when $\rho = 4$, which corresponds to $\boxed{r = 4a_0/Z}$. The wavefunction is positive at $\rho = 0$, negative at $\rho = 4$, and asymptotically approaches zero as $\rho\to\infty$; $\rho = 4$ must therefore correspond to a minimum.

E8A.5(a) Assuming that the electron is in the ground state, the wavefunction is $\psi = Ne^{-r/a_0}$, and so the probability density is $P(r) = \psi^2 = N^2e^{-2r/a_0}$. $P(r)$ is a maximum at $r = 0$ and then simply falls off as r increases; it falls to $\frac{1}{2}$ its initial value when $P(r')/P(0) = \frac{1}{2}$

$$P(r')/P(0) = e^{-2r'/a_0} = \tfrac{1}{2}$$

Hence $r' = -\frac{1}{2}\ln\frac{1}{2}a_0 = \boxed{0.347a_0}$.

E8A.6(a) The radial wavefunction of a 3s orbital is given in Table 8A.1 on page 306 as $R_{3,0} = N(6 - 6\rho + \rho^2)e^{-\rho/2}$, where $\rho = 2Zr/3a_0$. Radial nodes occur when the wavefunction passes through 0, which is when $6 - 6\rho + \rho^2 = 0$. The roots of this quadratic equation are at $\rho = 3 \pm \sqrt{3}$ and hence the nodes are at

$$\boxed{r = (3 \pm \sqrt{3})(3a_0/2Z)}$$

The wavefunction goes to zero as $\rho \to \infty$, but this does not count as a node as the wavefunction does not pass through zero.

E8A.7(a) Angular nodes occur when $\cos\theta \sin\theta \cos\phi = 0$, which occurs when any of $\cos\theta$, $\sin\theta$, or $\cos\phi$ is equal to zero; recall that the range of θ is $0 \to \pi$ and of ϕ is $0 \to 2\pi$.

Although the function is zero for $\theta = 0$ this does not describe a plane, and so is discounted. The function is zero for $\boxed{\theta = \pi/2}$ with any value of ϕ: this is the xy plane. The function is also zero for $\boxed{\phi = \pi/2}$ with any value of θ: this is the yz plane. There are two nodal planes, as expected for a d orbital.

E8A.8(a) The radial distribution function is defined in [8A.17b–*312*], $P(r) = r^2R(r)^2$. For the 2s orbital $R(r)$ is given in Table 8A.1 on page 306 as $R_{2,0} = N(2 - \rho)e^{-\rho/2}$ where $\rho = 2Zr/na_0$, which for $n = 2$ is $\rho = Zr/a_0$. With the substitution $r^2 = \rho^2(a_0/Z)^2$, the radial distribution function is therefore $P(\rho) = N^2(a_0/Z)^2\rho^2(2-\rho)^2e^{-\rho}$.

Mathematical software is used to find the values of ρ for which $dP(\rho)/d\rho = 0$, giving the results $\rho = 0,\ 2,\ 3 \pm \sqrt{5}$. The simplest way to identify which of these is a maximum is to plot $P(\rho)$ against ρ, from which it is evident that $\rho = 2$ is a minimum and $\rho = 3 \pm \sqrt{5}$ are both maxima, with the principal maximum being at $\rho = 3 + \sqrt{5}$. The maximum in the radial distribution function is therefore at $r = \boxed{(3 + \sqrt{5})(a_0/Z)}$.

E8A.9(a) The radius at which the electron is most likely to be found is that at which the radial distribution function is a maximum. The radial distribution function is defined in [8A.17b–*312*], $P(r) = r^2R(r)^2$. For the 2p orbital $R(r)$ is given in Table 8A.1 on page 306 as $R_{2,1} = N\rho e^{-\rho/2}$ where $\rho = 2Zr/na_0$, which for $n = 2$ is $\rho = Zr/a_0$. With the substitution $r^2 = \rho^2(a_0/Z)^2$, the radial distribution function is therefore $P(\rho) = N^2(a_0/Z)^2\rho^4e^{-\rho}$.

To find the maximum in this function the derivative is set to zero; the multiplying constants can be discarded for the purposes of this calculation

$$\frac{d}{dr}\rho^4e^{-\rho} = 4\rho^3e^{-\rho} - \rho^4e^{-\rho}$$

Setting this derivative to zero gives the solutions $\rho = 0$ and $\rho = 4$. $P(\rho)$ is zero for $\rho = 0$ and as $\rho \to \infty$, therefore $\rho = 4$ must be a maximum. This occurs at $r = \boxed{4a_0/Z}$.

E8A.10(a) The M shell has $n = 3$. The possible values of l (subshells) are 0, corresponding to the s orbital, $l = 1$ corresponding to the p orbitals, and $l = 2$ corresponding to the d orbital; there are therefore $\boxed{\text{3 subshells}}$. As there is one s orbital, 3 p orbitals and 5 d orbitals, there are $\boxed{\text{9 orbitals}}$ in total.

E8A.11(a) The magnitude of the orbital angular momentum of an orbital with quantum number l is $\sqrt{l(l+1)}\hbar$. The total number of nodes for an orbital with quantum number n is $n - 1$, l of these are angular and so the number of radial nodes is $n - l - 1$.

orbital	n	l	ang. mom.	angular nodes	radial nodes
1s	1	0	0	0	0
3s	3	0	0	0	2
3d	3	2	$\sqrt{6}\hbar$	2	0

E8A.12(a) All the 2p orbitals have the same value of n and l, and hence have the same radial function, which is given in Table 8A.1 on page 306 as $R_{2,1} = N\rho e^{-\rho/2}$ where $\rho = 2Zr/na_0$, which for $n = 2$ is $\rho = Zr/a_0$. Radial nodes occur when the wavefunction passes through zero. The function goes to zero at $\rho = 0$ and as $\rho \to \infty$, but it does not pass through zero at these points so they are not nodes. The number of radial nodes is therefore 0.

Solutions to problems

P8A.1 The 2p orbitals only differ in the axes along which they are directed. Therefore, the distance from the origin to the position of maximum probability density will be the same for each.

The radial function for the 2p orbitals is $R_{2,1} = N\rho e^{-\rho/2}$, where $\rho = Zr/a_0$. The probability density is the square of the radial function, $R_{2,1}^2 = N^2\rho^2 e^{-\rho}$, and the maximum in this is found by setting $dR^2/d\rho = 0$

$$\frac{dR_{2,1}^2}{d\rho} = 2N^2\rho e^{-\rho} - N^2\rho^2 e^{-\rho} = 0$$

Turning points occur at $\rho = 0,\ 2$, and it is evident from a plot of $R(\rho)$ that $\rho = 2$ is the maximum. This corresponds to $r = 2a_0/Z$. For the $2p_z$ orbital, for which the angular part goes as $\cos\theta$, the maximum will be at $\theta = 0$, which corresponds to $x = y = 0$. The position of maximum probability density is therefore at $\boxed{x = 0,\ y = 0,\ z = 2a_0/Z}$. The corresponding positions for the other 2p orbitals are found by permuting the x, y and z coordinates.

P8A.3 The energy levels of a hydrogenic atom with atomic number Z are given by [8A.13–308], $E_n = -hcZ^2\tilde{R}_N/n^2$, where the Rydberg constant for the atom is given by [8A.14–308], $\tilde{R}_N = (\mu/m_e)\tilde{R}_\infty$; μ is the reduced mass of the atom.

For the D atom, with nuclear mass m_D, the reduced mass is

$$\mu = \frac{m_D m_e}{m_D + m_e}$$

and therefore the Rydberg constant for D is

$$\begin{aligned}\tilde{R}_D &= \frac{\mu}{m_e}\tilde{R}_\infty = \frac{m_D m_e}{m_e(m_D + m_e)}\tilde{R}_\infty = \frac{m_D}{m_D + m_e}\tilde{R}_\infty \\ &= \frac{2.01355 \times (1.660\,539 \times 10^{-27}\,\text{kg})}{[2.01355 \times (1.660\,539 \times 10^{-27}\,\text{kg})] + (9.109\,382 \times 10^{-31}\,\text{kg})} \\ &\quad \times (109\,737\,\text{cm}^{-1}) = 109\,707\,\text{cm}^{-1}\end{aligned}$$

where the constants have been used to sufficient precision to match the data. The energy of the ground state is

$$\begin{aligned}E_1 = -hc\tilde{R}_D &= -(6.626\,070 \times 10^{-34}\,\text{J s}) \times (2.997\,925 \times 10^{10}\,\text{cm s}^{-1}) \\ &\times (109\,707\,\text{cm}^{-1}) = \boxed{-2.179\,27 \times 10^{-18}\,\text{J}}\end{aligned}$$

Expressed as a molar quantity this is -1312.39 kJ mol^{-1}.

P8A.5 (a) By analogy with [8A.21–314], the wavefunction for a $3p_x$ orbital is $\psi_{3p_x} = R_{3,1}(r) \times [Y_{1,+1}(\theta,\phi) - Y_{1,-1}(\theta,\phi)](2)^{-1/2}$. The required integral to verify normalization is

$$\int_0^\infty R_{3,1}^2 r^2\,dr \int_0^\pi \int_0^{2\pi} 2^{-1}\left|Y_{1,+1}(\theta,\phi) - Y_{1,-1}(\theta,\phi)\right|^2 \sin\theta\,d\theta\,d\phi$$

Consider first the integral over r. From Table 8A.1 on page 306 $R_{3,1}(\rho) = (486)^{-1/2}(Z/a_0)^{3/2}(4-\rho)\rho e^{-\rho/2}$, where $\rho = 2Zr/na_0$ which is this case is $2Zr/3a_0$. It is convenient to calculate the integral over ρ, noting that $r = \rho(3a_0/2Z)$ so that $r^2 = \rho^2(3a_0/2Z)^2$ and $dr = d\rho\,(3a_0/2Z)$. The integral becomes

$$\begin{aligned}&\frac{1}{486}\frac{Z^3}{a_0^3}\int_0^\infty (4-\rho)^2\rho^2(3a_0/2Z)^2\rho^2 e^{-\rho}\,(3a_0/2Z)\,d\rho \\ &= \frac{1}{486}\frac{3^3}{2^3}\int_0^\infty (4-\rho)^2\rho^4 e^{-\rho}\,d\rho = \frac{1}{144}\int_0^\infty (16\rho^4 - 8\rho^5 + \rho^6)e^{-\rho}\,d\rho \\ &= \frac{1}{144}\left[16(4!/1^5) - 8(5!/1^6) + (6!/1^7)\right] = 1\end{aligned}$$

where Integral E.3 is used, with $k = 1$ and the appropriate value of n. The radial part of the function is therefore normalized.
The angular function is found using the explicit form of the spherical harmonics listed in Table 7F.1 on page 286

$$\begin{aligned}Y_x &= \frac{1}{2^{1/2}}(Y_{1,+1} - Y_{1,-1}) = -\left(\frac{3}{8\pi}\right)^{1/2}\frac{1}{2^{1/2}}\left(\sin\theta\,e^{i\phi} + \sin\theta\,e^{-i\phi}\right) \\ &= -\left(\frac{3}{8\pi}\right)^{1/2}\frac{1}{2^{1/2}} \times 2\sin\theta\,\cos\phi = -\left(\frac{3}{4\pi}\right)^{1/2}\sin\theta\,\cos\phi\end{aligned}$$

The normalization integral over the angles becomes

$$\left(\frac{3}{4\pi}\right)\int_0^{\pi}\int_0^{2\pi}\sin^3\theta\cos^2\phi\,d\theta\,d\phi$$

Using Integral T.3 with $a = \pi$ and $k = 1$ gives the integral over θ as $4/3$. The term $\cos^2\phi$ is written as $1 - \sin^2\phi$. The integral $\int_0^{2\pi} 1\,d\phi = 2\pi$ and, using Integral T.2 with $a = 2\pi$ and $k = 1$, gives $\int_0^{2\pi}\sin^2\phi\,d\phi = \pi$. Hence the integral over ϕ is $2\pi - \pi = \pi$. The overall integral over the angles is therefore $(3/4\pi) \times (4/3) \times (\pi) = 1$; the angular part is also normalized, and as a result the complete wavefunction is normalized.

The next task is to show that ψ_{3p_x} and $\psi_{3d_{xy}}$ are mutually orthogonal; taking the hint from question, attention is focused on the angular parts, because if these are orthogonal the overall wavefunctions will also be orthogonal. Setting aside all of the normalization factors, which will not be relevant to orthogonality, the angular part of the wavefunction for p_x is $(Y_{1,+1} - Y_{1,-1})$. In a similar way, it is expected that the angular parts of a d orbital can be constructed from spherical harmonics with $l = 2$.

To find the combination that represents d_{xy} recall that this function is of the form $xyf(r)$ and that in spherical polar coordinates $x = r\sin\theta\cos\phi$ and $y = r\sin\theta\sin\phi$, therefore

$$\begin{aligned}xyf(r) &= (r\sin\theta\cos\phi)(r\sin\theta\sin\phi)f(r) = r^2 f(r)\sin^2\theta\cos\phi\sin\phi\\ &= \tfrac{1}{2}r^2 f(r)\sin^2\theta\sin 2\phi\end{aligned}$$

The spherical harmonics $Y_{2,\pm2}$ have the form (again, omitting the normalization factors) $\sin^2\theta\,e^{\pm 2i\phi}$. The angular part of d_{xy} is therefore obtained by the combination

$$\begin{aligned}Y_{2,+2} - Y_{2,-2} &= \sin^2\theta\left[e^{2i\phi} - e^{-2i\phi}\right]\\ &= \sin^2\theta\left[\cos 2\phi + i\sin 2\phi - \cos 2\phi + i\sin 2\phi\right] = 2i\sin^2\theta\sin 2\phi\end{aligned}$$

It therefore follows that, to within some numerical factors, the angular part of d_{xy} is given by $Y_{2,+2} - Y_{2,-2}$. The orthogonality of the angular parts of d_{xy} and p_x therefore involves the following integral

$$\int_0^{\pi}\int_0^{2\pi}(Y_{2,+2} - Y_{2,-2})^*(Y_{1,+1} - Y_{1,-1})\sin\theta\,d\theta\,d\phi \qquad (8.1)$$

Concentrating on the integral over ϕ, this will involve terms such as

$$\int_0^{2\pi} Y_{2,+2}^* Y_{1,+1}\,d\phi = \int_0^{2\pi}\sin^2\theta\,e^{-2i\phi}\sin\theta\,e^{i\phi}\,d\phi = \int_0^{2\pi}\sin^3\theta\,e^{-i\phi}\,d\phi$$

This integral is zero because $\int_0^{2\pi} e^{ni\phi}\,d\phi$ is zero for integer n. All of the terms in eqn 8.1 follow this pattern and therefore the overall integral is zero; the orbitals are therefore orthogonal.

(b) The radial nodes for the 3s, 3p and 3d orbitals are found by examining the radial wavefunctions, which are listed in Table 8A.1 on page 306, expressed as functions of $\rho = 2Zr/3a_0$. These functions all go to zero as

$\rho \to \infty$ and, in some cases they are also zero at $\rho = 0$; these do not count as radial nodes as the wavefunction does not pass through zero at these points. The nodes are located by finding the values of ρ at which the polynomial part of the radial function is zero.

The positions of the nodes in the 3s orbital are given by the solutions to $6-6\rho+\rho^2 = 0$, which are at $\rho = 3 \pm \sqrt{3}$. In terms of r these nodes occur at $\boxed{r = (3a_0/2Z)(3 \pm \sqrt{3})}$. The positions of the nodes in the 3p orbital are given by the solutions to $(4-\rho)\rho = 0$; there is just one node t $\rho = 4$ which corresponds to $\boxed{r = 6a_0/Z}$. For the 3d orbital the polynomial is simply ρ^2, which does not lead to any nodes.

The 3s orbital has $\boxed{\text{no angular nodes}}$, as it has no angular variation. The $3p_x$ orbital has an angular node when $x = 0$, that is the $\boxed{yz}$ plane. The $3d_{xy}$ orbital has angular nodes when $x = 0$ or $y = 0$, corresponding to the $\boxed{yz}$ and $\boxed{xz}$ planes.

(c) The mean radius is calculated as

$$\langle r \rangle = \int \psi_{3s}^* r \psi_{3s} \, d\tau = \int_0^\infty \int_0^\pi \int_0^{2\pi} \psi_{3s}^* r \psi_{3s} r^2 \sin\theta \, dr \, d\theta \, d\phi$$

The wavefunction is written in terms of its radial and angular parts: $\psi_{3s} = R_{3,0}(r) Y_{0,0}(\theta, \phi)$. The angular part, the spherical harmonic $Y_{0,0}(\theta, \phi)$, is normalized with respect to integration over the angles

$$\int_0^\pi \int_0^{2\pi} Y_{0,0}(\theta, \phi)^* Y_{0,0}(\theta, \phi) \sin\theta \, d\theta \, d\phi = 1$$

All that remains is to compute the integral over r

$$\langle r \rangle = \int_0^\infty R_{3,0}(r)^2 r^3 \, dr$$

The form of $R(\rho)$ is given in Table 8A.1 on page 306, where $\rho = 2Zr/3a_0$. It is convenient to compute the integral over ρ using $r^3 = \rho^3(3a_0/2Z)^3$ and $dr = (3a_0/2Z)d\rho$

$$\begin{aligned}
\langle r \rangle &= \int_0^\infty R_{3,0}(r)^2 r^3 \, dr = (3a_0/2Z)^4 \int_0^\infty R_{3,0}(\rho)^2 \rho^3 \, d\rho \\
&= \frac{1}{243}\left(\frac{3a_0}{2Z}\right)^4 \left(\frac{Z}{a_0}\right)^3 \int_0^\infty (6-6\rho+\rho^2)^2 \rho^3 e^{-\rho} \, d\rho \\
&= \frac{1}{243}\frac{3^4}{2^4}\frac{a_0}{Z} \int_0^\infty (6-6\rho+\rho^2)^2 \rho^3 e^{-\rho} \, d\rho \\
&= \frac{1}{48}\frac{a_0}{Z} \int_0^\infty (\rho^7 - 12\rho^6 + 48\rho^5 - 72\rho^4 + 36\rho^3) e^{-\rho} \, d\rho \\
&= \frac{1}{48}\frac{a_0}{Z} [7! - 12(6!) + 48(5!) - 72(4!) + 36(3!)] = \frac{1}{48}\frac{a_0}{Z} \times 648 \\
&= \boxed{(27a_0)/(2Z)}
\end{aligned}$$

where the integrals are evaluated using Integral E.3 with $k = 1$ and the appropriate value of n.

(d) The radial distribution function is defined in [8A.17b–312], $P(r) = r^2 R(r)^2$. It is convenient to express this in terms of $\rho = 2Zr/3a_0$ using $R(\rho)$ from Table 8A.1 on page 306, and with $r^2 = \rho^2(3a_0/2Z)^2$

$$P_{3s} = \frac{1}{243}\left(\frac{Z}{a_0}\right)^3\left(\frac{3a_0}{2Z}\right)^2(6-6\rho+\rho^2)^2\rho^2 e^{-\rho}$$
$$= \frac{1}{108}\frac{Z}{a_0}\left(\rho^6 - 12\rho^5 + 48\rho^4 - 72\rho^3 + 36\rho^2\right)e^{-\rho}$$
$$P_{3p} = \frac{1}{486}\left(\frac{Z}{a_0}\right)^3\left(\frac{3a_0}{2Z}\right)^2(4-\rho)^2\rho^2\rho^2 e^{-\rho}$$
$$= \frac{1}{216}\frac{Z}{a_0}\left(\rho^6 - 8\rho^5 + 16\rho^4\right)e^{-\rho}$$
$$P_{3d} = \frac{1}{2430}\left(\frac{Z}{a_0}\right)^3\left(\frac{3a_0}{2Z}\right)^2(\rho^2)^2\rho^2 e^{-\rho}$$
$$= \frac{1}{1080}\frac{Z}{a_0}\rho^6 e^{-\rho}$$

Plots of these three functions are shown in Fig. 8.1

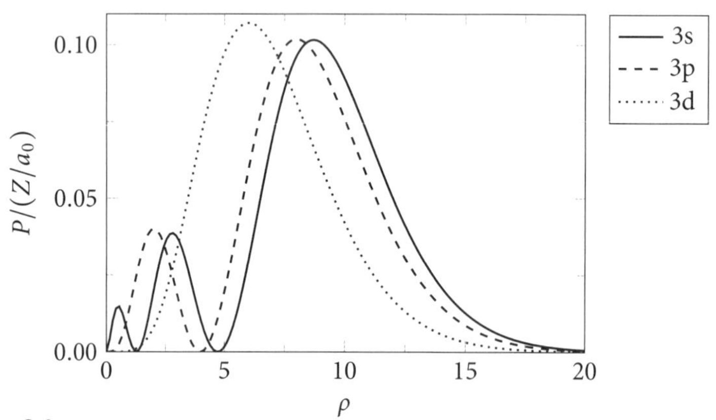

Figure 8.1

The radial distribution function for the 3s orbital has two subsidiary maxima which lie close in to the nucleus, and that for 3p has one such maximum. In multi-electron atoms this density close to the nucleus results in the energies of the 3s, 3p and 3d orbitals no longer being equal: see Section 8B.3 on page 319.

P8A.7 The probability of finding an electron within a sphere of radius σ is found by integrating the probability density over all angles and from $r = 0$ to $r = \sigma$

$$P(\sigma) = \int_0^{\sigma}\int_0^{\pi}\int_0^{2\pi}|\psi(r,\theta,\phi)|^2 r^2 \sin\theta\, dr\, d\theta\, d\phi$$

The ground state of the H atom is the 1s orbital for which the wavefunction is $\psi_{1s} = (\pi a_0^3)^{-1/2}e^{-r/a_0}$ therefore $P(r) = (\pi a_0^3)^{-1}e^{-2r/a_0}$. Because $P(r)$ does not depend on the angles, the integral over the angles can be evaluated separately to give 4π. The expression for $P(\sigma)$ therefore becomes

$$P(\sigma) = (4/a_0^3)\int_0^\sigma r^2 e^{-2r/a_0}\,dr$$

The integral is evaluated using integration by parts

$$\begin{aligned}\int_0^\sigma r^2 e^{-2r/a_0}\,dr &= -(a_0/2)\, r^2 e^{-2r/a_0}\Big|_0^\sigma + a_0\int_0^\sigma r e^{-2r/a_0}\,dr\\ &= -(a_0\sigma^2/2)e^{-2\sigma/a_0}\\ &\quad + a_0\left[-(a_0/2)\, r e^{-2r/a_0}\Big|_0^\sigma + (a_0/2)\int_0^\sigma e^{-2r/a_0}\,dr\right]\\ &= -(a_0\sigma^2/2)e^{-2\sigma/a_0} + a_0\left[-(a_0\sigma/2)e^{-2\sigma/a_0} - (a_0^2/4)\, e^{-2r/a_0}\Big|_0^\sigma\right]\\ &= a_0^3/4 - e^{-2\sigma/a_0}\left[(a_0\sigma^2/2) + (a_0^2\sigma/2) + a_0^3/4\right]\end{aligned}$$

hence

$$P(\sigma) = 1 - e^{-2\sigma/a_0}[2(\sigma/a_0)^2 + 2(\sigma/a_0) + 1]$$

To find the radius at which $P(\sigma) = 0.9$ needs the solution to the equation $0.9 = 1 - e^{-2\sigma/a_0}[2(\sigma/a_0)^2 + 2(\sigma/a_0) + 1]$, which is found numerically to be $\boxed{\sigma = 2.66a_0}$. Figure 8.2 is a plot of $P(\sigma)$ as a function of σ; it is a sigmoid curve and shows, as expected, that the the radius of the sphere increases as the total enclose probability increases.

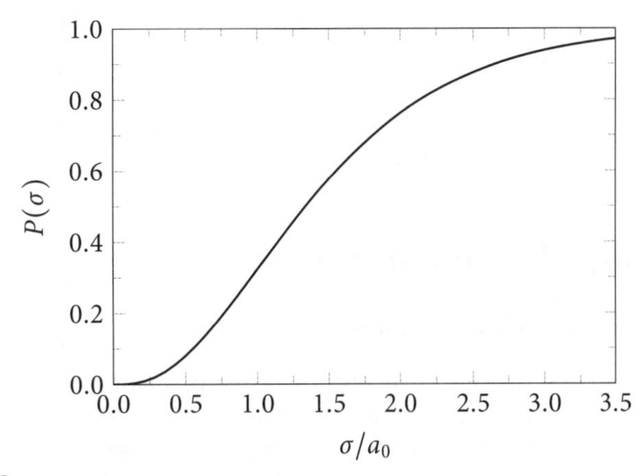

Figure 8.2

P8A.9 The repulsive centrifugal force of an electron travelling with an angular momentum J in a circle radius r is $J^2/m_e r^3$. Bohr postulated that $J = n\hbar$ where n can only take integer values, making the centrifugal force $(n\hbar)^2/m_e r^3$. The atom is in a stationary state when repulsive force is balanced by the attractive

Coulombic force $Ze^2/4\pi\varepsilon_0 r^2$, that is $Ze^2/4\pi\varepsilon_0 r^2 = (n\hbar)^2/m_e r^3$. This relationship is rearranged to give an expression for the radius r_n of the orbit for an electron in state n, $r_n = 4\pi(n\hbar)^2\varepsilon_0/Ze^2 m_e$.

The total energy of the state with an electron orbiting at radius r_n is the sum of the kinetic and potential energies. The kinetic energy is written in terms of the angular momentum as $J^2/2I = J^2/2m_e r_n^2$, with I the moment of inertia and $J = n\hbar$. The potential energy depends only on r_n

$$\begin{aligned} E_n = E_k + V &= \frac{J^2}{2m_e r_n^2} - \frac{Ze^2}{4\pi\varepsilon_0 r_n} \\ &= \frac{(n\hbar)^2}{2m_e[4\pi(n\hbar)^2\varepsilon_0/Ze^2 m_e]^2} - \frac{Ze^2}{4\pi\varepsilon_0[4\pi(n\hbar)^2\varepsilon_0/Ze^2 m_e]} \\ &= -\boxed{\frac{Z^2 e^4 m_e}{32\pi^2\varepsilon_0^2\hbar^2}\times\frac{1}{n^2}} \end{aligned}$$

P8A.11 The Bohr radius a_0 is given by [8A.9–*306*] and the Hartree is defined as $E_h = 2hc\tilde{R}_\infty$, where the Rydberg constant is given by [8A.14–*308*]

$$a_{0,H} = \frac{4\pi\varepsilon_0\hbar^2}{m_e e^2} \qquad E_{h,H} = \frac{m_e e^4}{4\varepsilon_0^2 h^2}$$

These constants are based on the approximation that the nucleus is infinitely heavy. If this is not the case, then the mass of the electron m_e must be replaced by the reduced mass of the atom, $\mu = m_e m_N/(m_e + m_N)$, where m_N is the mass of the nucleus.

In the case of positronium the 'nucleus' has the same mass as the electron, so $\mu = m_e/2$ and hence

$$a_{0,pos} = \frac{4\pi\varepsilon_0\hbar^2}{e^2(m_e/2)} = \boxed{2a_{0,H}} \qquad E_{h,pos} = \frac{(m_e/2)e^4}{4\varepsilon_0^2 h^2} = \boxed{\tfrac{1}{2}E_{h,H}}$$

8B Many-electron atoms

Answers to discussion questions

D8B.1 In the crudest form of the orbital approximation, the many-electron wavefunctions for atoms are represented as a simple product of one-electron wavefunctions. At a somewhat more sophisticated level, the many electron wavefunctions are written as linear combinations of such simple product functions that explicitly satisfy the Pauli exclusion principle. Relatively good one-electron functions are generated by the Hartree–Fock self-consistent field method described in Section 8B.4 on page 325.

If no restrictions are placed on the form of the one-electron functions, the Hartree–Fock limit is reached which gives the best value of the calculated energy within the orbital approximation. The orbital approximation is based on the disregard of significant portions of the electron-electron interaction terms in the many-electron Hamiltonian, so it cannot be expected that it will be quantitatively accurate. By abandoning the orbital approximation, in principle it is possible to obtain essentially exact energies; however, there are significant conceptual advantages to retaining the orbital approximation. Increased accuracy can be obtained by reintroducing the neglected electron-electron interaction terms and including their effects on the energies of the atom.

D8B.3 In Period 2, the first ionization energies increase markedly from Li to Be, decrease slightly from Be to B, again increase markedly from B to N, again decrease slightly from N to O, and finally increase markedly from O to Ne. The general trend is an overall increase in the ionization energy with atomic number across the period. This is to be expected since the principal quantum number of the outer electron remains the same, while the nuclear charge increases. The slight decrease from Be to B is a reflection of the outer electron in B being in an orbital with $l = 1$, whereas that in Be has $l = 0$. The slight decrease from N to O is due to the half-filled subshell effect, by which half-filled subshells have increased stability. Oxygen has one electron outside of the half-filled p subshell and that electron must pair with another resulting in strong electron-electron repulsions between them. The same kind of variation is expected for the elements of Period 3 because in both periods the outer shell electrons are only s and p.

Solutions to exercises

E8B.1(a) Hydrogenic orbitals are written in the form [8A.12–*307*], $R_{n,l}(r)Y_{l,m_l}(\theta,\phi)$, where the appropriate radial function $R_{n,l}$ is selected from Table 8A.1 on page 306 and the appropriate angular function Y_{l,m_l} is selected from Table 7F.1 on page 286. Using $Z = 2$ for the 1s and $Z = 1$ for the 2s gives

$$\psi_{1s}(r) = R_{1,0}Y_{0,0} = 2(2/a_0)^{3/2}e^{-2r/a_0} \times (4\pi)^{-1/2}$$

$$\psi_{2s}(r) = R_{2,0}Y_{0,0} = (8)^{-1/2}(1/a_0)^{3/2}[2-(r/a_0)]e^{-r/2a_0} \times (4\pi)^{-1/2}$$

The overall wavefunction is simply the product of the orbital wavefunctions

$$\Psi(r_1,r_2) = \psi_{1s}(r_1)\psi_{2s}(r_2)$$

E8B.2(a) For a subshell with angular momentum quantum number l there are $2l+1$ values of m_l, each of which corresponds to a separate orbital. Each orbital can accommodate two electrons, therefore the total number of electrons is $2\times(2l+1)$. The subshell with $l = 3$ can therefore accommodate $2(6+1) = \boxed{14}$ electrons.

E8B.3(a) All configurations have the [Ar] core.

Sc	Ti	V	Cr	Mn
$4s^23d^1$	$4s^23d^2$	$4s^23d^3$	$4s^13d^5$	$4s^23d^5$
Fe	Co	Ni	Cu	Zn
$4s^23d^6$	$4s^23d^7$	$4s^23d^8$	$4s^13d^{10}$	$4s^23d^{10}$

E8B.4(a) [Ar] $3d^8$

E8B.5(a) Across the period the energy of the orbitals generally decreases as a result of the increasing nuclear charge. Therefore $\boxed{\text{Li}}$ is expected to have the lowest ionization energy as its outer electron has the highest orbital energy.

Solutions to problems

P8B.1 The radial distribution function for a 1s orbital is given by [8A.18–*312*], $P(r) = (4Z^3/a_0^3)r^2e^{-2Zr/a_0}$. This gives the probability density of finding the electron in a shell of radius r. The most probable radius is found by finding the maximum in $P(r)$, when $dP(r)/dr = 0$. In finding this maximum the multiplying constants are not relevant and can be discarded

$$\frac{d}{dr}r^2e^{-2Zr/a_0} = [2r - (2Z/a_0)r^2]\,e^{-2Zr/a_0} = 0$$

It follows that $r_{max} = a_0/Z$; that this is a maximum is most easily seen by plotting $P(r)$. For $Z = 126$ the most probable radius will be $\boxed{a_0/126}$.

P8B.3 Toward the middle of the first transition series (Cr, Mn, and Fe) elements exhibit the widest ranges of oxidation states. This is due to the large number of electrons in the 3d and 4s subshells that have similar energies, and as the 3d electrons that are generally removed provide very little shielding to the 4s orbitals, the effective nuclear charge does not increase significantly between adjacent oxidation states, meaning that the ionization energies of these levels are close, and as these are the outermost electrons the ionization levels are relatively small, meaning that large numbers of reactions will release enough energy to lose many electrons.

However, it should be noted that the higher oxidation states of the middle transition metals do not exist as cations, but only in compounds or compound ions where there is a significant stabilization of this ion by electron rich atoms, for example the Mn^{VII} state only exists in the MnO_4^- ion where there is large electron donation from bonding with four O^{2-} ions, as here the effective nuclear charge has increased a lot over the neutral atom.

This phenomenon is related to the availability of both electrons and orbitals favourable for bonding. Elements to the left (Sc and Ti) of the series have few electrons and relatively low effective nuclear charge leaves d orbitals at high energies that are relatively unsuitable for bonding. To the far right (Cu and

Zn) effective nuclear charge may be higher but there are few, if any, orbitals available for bonding. Consequently, it is more difficult to produce a range of compounds that promote a wide range of oxidation states for elements at either end of the series. At the middle and right of the series the +2 oxidation state is very commonly observed because normal reactions can provide the requisite ionization energies for the removal of 4s electrons.

P8B.5 The first, second and third ionization energies for the group 13 elements are plotted in Fig. 8.3

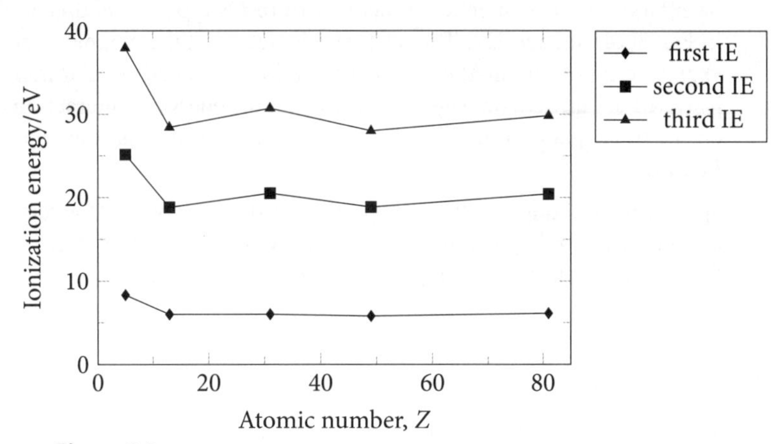

Figure 8.3

The following trends are identified.

(a) In all cases, $I_1 < I_2 < I_3$ because of decreased nuclear shielding as each successive electron is removed.

(b) The ionization energies of boron are much larger than those of the remaining group elements because the valence shell of boron is very small and compact with little nuclear shielding. The boron atom is much smaller than the aluminum atom.

(c) The ionization energies of Al, Ga, In, and Tl are comparable even though successive valence shells are further from the nucleus because the ionization energy decrease expected from large atomic radii is balanced by an increase in effective nuclear charge.

8C Atomic spectra

Answers to discussion questions

D8C.1 When electronically excited hydrogen atoms drop down to a lower level they emit photons of characteristic energies which depend on the quantum number of the upper and lower levels. A series of lines, such as the Lyman series, involves transitions from a range of excited states all to the *same* lower state, in this case $n = 1$; similarly the Balmer series involves transitions in which the lower state is always $n = 2$.

The lines of the Lyman series are seen in the ultraviolet, those of the Balmer series in the visible, and those of the Paschen series in the infra-red.

D8C.3 An electron has a magnetic moment due to its orbital angular momentum (specified by l) and a magnetic moment which arises from its spin angular momentum (specified by s). These two magnetic moments interact and hence affect the energy of the state; the energy of the interaction depends on the orientation of the magnetic moments with respect to one another, as depicted in Fig. 8C.4 on page 329. The relative orientation of the magnetic moments is expressed by specifying the vector sum of the associated angular momenta, illustrated in Fig. 8C.5 on page 330. As a result the energy of interaction depends on the three quantum numbers l, s and j, where j specifies the vector sum of l and s.

Spin-orbit coupling results in a splitting of the energy levels associated with atomic terms, and in turn these splittings give rise to fine structure in the spectra – for example, the splitting of of the sodium D lines.

Solutions to exercises

E8C.1(a) The spectral lines of a hydrogen atom are given by [8A.1–*304*], $\tilde{\nu} = \tilde{R}_{\text{H}}(n_1^{-2} - n_2^{-2})$, where $\tilde{R}_{\text{H}}$ is the Rydberg constant and $\tilde{\nu}$ is the wavenumber of the transition.

The Lyman series corresponds to $n_1 = 1$. The lowest energy transition, which would involve a photon with the longest wavelength, is to the next highest energy level which has $\boxed{n_2 = 2}$. Transitions to higher energy levels involve more an more energy, and the limit of this is the transition to $\boxed{n_2 = \infty}$ which involves the greatest possible energy change and hence the shortest wavelength.

E8C.2(a) The energy levels of a hydrogenic atom are $E_n = -hcZ^2\tilde{R}_{\text{N}}n^{-2}$, where Z is the atomic number; for all but the most precise work it is sufficient to approximate $\tilde{R}_{\text{N}}$ by $\tilde{R}_{\infty}$. The wavenumber of the transition between states with quantum numbers n_1 and n_2 in the He^+ ion is given by a modified version of [8A.1–*304*], $\tilde{\nu} = Z^2\tilde{R}_{\infty}(n_1^{-2} - n_2^{-2})$. For the $2 \rightarrow 1$ transition and with $Z = 2$

$$\tilde{\nu} = 2^2 \times (1.0974 \times 10^5\ \text{cm}^{-1}) \times (1^{-2} - 2^{-2}) = \boxed{3.29 \times 10^5\ \text{cm}^{-1}}$$

$$\lambda = \tilde{\nu}^{-1} = 1/[2^2 \times (1.0974 \times 10^5\ \text{cm}^{-1}) \times (1^{-2} - 2^{-2})]$$

$$= 3.03... \times 10^{-6}\ \text{cm} = \boxed{30.4\ \text{nm}}$$

$$\nu = c/\lambda = (2.9979 \times 10^8\ \text{m s}^{-1})/(3.03... \times 10^{-8}\ \text{m}) = \boxed{9.87\ \text{PHz}}$$

E8C.3(a) The selection rules for a many-electron atom are given in [8C.8–*335*]. For a single electron these reduce to $\Delta l = \pm 1$; there is no restriction on changes in n.

(i) 2s ($n = 2, l = 0$) → 1s ($n = 1, l = 0$) has $\Delta l = 0$, and so is $\boxed{\text{forbidden}}$.

(ii) 2p ($n = 2, l = 1$) → 1s ($n = 1, l = 0$) has $\Delta l = -1$, and so is $\boxed{\text{allowed}}$.

(iii) 3d ($n = 3, l = 2$) → 2p ($n = 2, l = 1$) has $\Delta l = -1$, and so is $\boxed{\text{allowed}}$.

E8C.4(a) The single electron in a p orbital has $l = 1$ and hence $L = 1$, and $s = \frac{1}{2}$ hence $S = \frac{1}{2}$. The spin multiplicity is $2S + 1 = 2$. Using the Clebsh–Gordon series, [8C.5–*332*], the possible values of J are $J = L + S,\ L + S - 1, \ldots |L - S| = \frac{3}{2}, \frac{1}{2}$. Hence, the term symbols for the levels are $\boxed{{}^2\text{P}_{1/2}, {}^2\text{P}_{3/2}}$.

E8C.5(a) For a d electron $l = 2$ and $s = \frac{1}{2}$. Using the Clebsh–Gordon series, [8C.5–*332*], the possible values of j are $l + s,\ l + s - 1, \ldots |l - s|$, which in this case are $\boxed{j = \frac{5}{2}, \frac{3}{2}}$.

For an f electron $l = 3$ and $s = \frac{1}{2}$ hence $\boxed{j = \frac{7}{2}, \frac{5}{2}}$.

E8C.6(a) The Clebsch–Gordan series [8C.5–*332*], in the form $j = l+s,\ l+s-1, \ldots |l-s|$, with $s = \frac{1}{2}$ implies that there are two possible values of j, $j = l \pm \frac{1}{2}$. Hence, given that $j = \frac{3}{2}, \frac{1}{2}$ it follows that $\boxed{l = 1}$.

E8C.7(a) The symbol D implies that the total orbital angular momentum $\boxed{L = 2}$, the superscript 1 implies that the multiplicity $2S + 1 = 1$, so that the total spin angular momentum $\boxed{S = 0}$. The subscript 2 implies that the total angular momentum $\boxed{J = 2}$.

E8C.8(a) The Clebsch–Gordan series, [8C.5–*332*], is used to combine two spin angular momenta s_1 and s_2 to give $S = s_1 + s_2,\ s_1 + s_2 - 1 \ldots,\ |s_1 - s_2|$. (i) For two electrons, each with $s = \frac{1}{2}$, $\boxed{S = 1, 0}$ with multiplicities, $2S = 1$, of $\boxed{3, 1}$. (ii) Three electrons are treated by first combining the angular momenta of two of them to give $S' = 0, 1$ and then combining each value of S' with $s_3 = \frac{1}{2}$ for the third spin. Therefore, for $S' = 1$, $S = 1 + \frac{1}{2}, |1 - \frac{1}{2}| = \frac{3}{2}, \frac{1}{2}$. Combining $S' = 0$ with $s = \frac{1}{2}$ simply results in $S = \frac{1}{2}$. The overall result is $\boxed{S = \frac{3}{2}, \frac{1}{2}}$ with corresponding multiplicities $\boxed{4, 1}$.

E8C.9(a) The valence electron configuration of the Ni^{2+} is [Ar] $3d^8$. In principle the same process could be adopted as in *Exercise* E8C.8(a), in which the spin angular momenta of all eight electrons are coupled together in successive steps to find the overall spin angular momentum. Such an approach would be rather tedious and would also run the risk of generating values of S which come from arrangements of electrons which violate the Pauli principle. A quicker method, and one which ensures that the Pauli principle is not violated, is to consider combinations of the quantum number m_s which gives the z-component of the spin angular momentum and which takes values $\pm\frac{1}{2}$. The total z-component of the spin angular momentum is found by simply adding together the m_s values: $M_S = m_{s_1} + m_{s_2} + \ldots$.

With 8 electrons in the 5 d orbitals, 6 of these electrons must doubly occupy three of the orbitals, and the Pauli principle requires that the two electrons in each orbital are spin paired: one has $m_s = +\frac{1}{2}$ and one has $m_s = -\frac{1}{2}$. These six electrons therefore make no net contribution to M_S, in the sense that the sum of the individual m_s values is 0. The remaining two electrons can either occupy the same orbital with spins paired, giving $M_S = +\frac{1}{2} - \frac{1}{2} = 0$, or they can occupy different orbitals with either their spins paired, giving $M_S = 0$ once more, or with their spins parallel, giving $M_S = +\frac{1}{2} + \frac{1}{2} = +1$ or $M_S = -\frac{1}{2} - \frac{1}{2} = -1$. Recall that a total spin S gives M_S values of S, $(S-1) \ldots -S$. Therefore the first arrangement with just $\boxed{M_S = 0}$ is interpreted as arising from $\boxed{S = 0}$, and the second arrangement with $\boxed{M_S = 0, \pm 1}$ is interpreted as arising from $\boxed{S = 1}$.

E8C.10(a) These electrons are not equivalent, as they are in different subshells, hence all the terms that arise from the vector model and the Clebsch–Gordan series are allowed. The orbital angular momentum of the s and d electrons are $l_1 = 0$ and $l_2 = 2$ respectively, and these are combined using $L = l_1 + l_2,\ l_1 + l_2 - 1,\ \ldots |l_1 - l_2|$ which in this case gives $L = 2$ only. The spin angular momenta of each electron is $s_1 = s_2 = \frac{1}{2}$, and these combine in the same way to give $S = 1,\ 0$; these values of S have spin multiplicities of $2S+1 = 3,\ 1$. The terms which arise are therefore ^{3}D and ^{1}D.

The possible values of J are given by $J = L+S,\ L+S-1,\ \ldots,\ |L-S|$, and hence for $S = 1$, $L = 2$ the values of J are 3, 2, and 1. For $S = 0$, $L = 2$ only $J = 2$ is possible. The term symbols are therefore $\boxed{^3\text{D}_3,\ ^3\text{D}_2,\ ^3\text{D}_1,\text{ and }^1\text{D}_2}$. From Hund's rules, described in Section 8C.2(d) on page 335, the lowest energy state is the one with the greatest spin and then, because the shell is less than half full, the smallest J. This is $\boxed{^3\text{D}_1}$.

E8C.11(a) (i) ^{1}S has $L = 0$, $S = 0$ and so $\boxed{J = 0}$ only; there are $2J + 1$ values of M_J, which for $J = 0$ is just $\boxed{1}$ state. (ii) ^{2}P has $L = 1$, $S = \frac{1}{2}$, and so $\boxed{J = \frac{3}{2},\ \frac{1}{2}}$; the former has $\boxed{4}$ states and the latter has $\boxed{2}$ states. (iii) ^{3}P has $L = 1$, $S = 1$, and so $\boxed{J = 2,\ 1,\ 0}$, with $\boxed{5,\ 3,\ 1}$ states, respectively.

E8C.12(a) Closed shells have total spin and orbital angular momenta of zero, and so do not contribute to the overall values of S and L. (i) For the configuration $2s^1$ there is just one electron to consider with $l = 0$ and $s = \frac{1}{2}$, so $L = 0$, $S = \frac{1}{2}$, and $J = \frac{1}{2}$. The term symbol is $\boxed{^2\text{S}_{1/2}}$. (ii) For the configuration $2p^1$ there is just one electron to consider with $l = 1$ and $s = \frac{1}{2}$, so $L = 1$, $S = \frac{1}{2}$, and $J = \frac{3}{2},\ \frac{1}{2}$. The term symbols are therefore $\boxed{^2\text{P}_{3/2}}$ and $\boxed{^2\text{P}_{1/2}}$.

E8C.13(a) The two terms arising from a d^1 configuration are $^2\text{D}_{3/2}$, $^2\text{D}_{5/2}$, which have $S = \frac{1}{2}$, $L = 2$ and $J = \frac{3}{2}, \frac{5}{2}$. The energy shift due to spin-orbit coupling is given by [8C.4–331], $E_{L,S,J} = \frac{1}{2}hc\tilde{A}[J(J+1) - L(L+1) - S(S+1)]$, where $\tilde{A}$ is the spin-orbit coupling constant. Hence, $E_{2,1/2,3/2} = \boxed{-(3/2)hc\tilde{A}}$, and $E_{2,1/2,5/2} = \boxed{+hc\tilde{A}}$.

E8C.14(a) The selection rules for a many-electron atom are given in [8C.8–335].

(i) $^3D_2$ $(S=1, L=2, J=2) \rightarrow {}^3P_1$ $(S=1, L=1, J=1)$ has $\Delta S = 0$, $\Delta L = -1$, $\Delta J = -1$ and so is $\boxed{\text{allowed}}$.

(ii) $^3P_2$ $(S=1, L=1, J=2) \rightarrow {}^1S_0$ $(S=0, L=0, J=0)$ has $\Delta S = -1$, $\Delta L = -1$, $\Delta J = -2$ and so is $\boxed{\text{forbidden}}$ by the S and J selection rules.

(iii) $^3F_4$ $(S=1, L=3, J=4) \rightarrow {}^3D_3$ $(S=1, L=2, J=3)$ has $\Delta S = 0$, $\Delta L = -1$, $\Delta J = -1$ and so is $\boxed{\text{allowed}}$.

Solutions to problems

P8C.1 The wavenumbers of the spectral lines of the H atom for the $n_2 \rightarrow n_1$ transition are given by [8A.1–304], $\tilde{\nu} = \tilde{R}_H(n_1^{-2} - n_2^{-2})$, where $\tilde{R}_H$ is the Rydberg constant for Hydrogen, $\tilde{R}_H = 109677\ \text{cm}^{-1}$. Hence, the wavelength of this transition is $\lambda = \tilde{\nu}^{-1} = \tilde{R}_H^{-1}(n_1^{-2} - n_2^{-2})^{-1}$.

The lowest energy, and therefore the longest wavelength transition (the one at $\lambda_{max} = 12368\ \text{nm} = 1.2368 \times 10^{-3}\ \text{cm}$) corresponds to the transition from $n_1 + 1 \rightarrow n_1$, therefore

$$\frac{1}{\lambda_{max}\tilde{R}_H} = \frac{1}{n_1^2} - \frac{1}{(n_1+1)^2} = \frac{(n_1+1)^2 - n_1^2}{n_1^2(n_1+1)^2} = \frac{2n_1+1}{n_1^2(n_1+1)^2}$$

From the given data $(\lambda_{max}\tilde{R}_H)^{-1} = [(1.2368 \times 10^{-3}\ \text{cm}) \times (109677\ \text{cm}^{-1})]^{-1} = (135.6...)^{-1}$. The value of n_1 is found by seeking an integer value of n_1 for which $n_1^2(n_1+1)^2/(2n_1+1) = 135.6....$ For $n_1 = 6$ the fraction on the left is $6^2 \times 7^2/13 = 135.6....$ Therefore, the Humphreys series is that with $\boxed{n_1 = 6}$.

The wavelengths of the transitions in the Humphreys series are therefore given by $\lambda = (109677\ \text{cm}^{-1})^{-1} \times (6^{-2} - n_2^{-2})^{-1}$ for $n_2 = 7, 8, ...$ The next few lines, with $n_2 = 8, 9,$ and 10 are at 7502.5 nm, 5908.3 nm, 5128.7 nm, respectively. The convergence limit, corresponding to $n_2 = \infty$ is 3282.4 nm, as given in the data.

P8C.3 The wavenumbers of transitions between energy levels in hydrogenic atoms are given by a modified version of [8A.1–304]

$$\tilde{\nu} = Z^2\tilde{R}_N\left(n_1^{-2} - n_2^{-2}\right) \tag{8.2}$$

where Z is the nuclear charge and $\tilde{R}_N$ is the Rydberg constant for the nucleus in question. In turn this is given by [8A.14–308]

$$\tilde{R}_N = \frac{\mu}{m_e}\tilde{R}_\infty \qquad \mu = \frac{m_e m_N}{m_e + m_N}$$

where m_N is the mass of the nucleus. The spectra of $^4He^+$ and $^3He^+$ differ because $\tilde{R}_N$ is different for the two atoms. However, this difference is very small because the value of the reduced mass μ is dominated by the mass of the electron ($m_e \ll m_N$). It is therefore necessary to work at high precision.

The first step is to compute $\tilde{R}_N$ for each nucleus, using $m_{^4He} = 4.002\,602 m_u$ and $m_{^3He} = 3.016\,029 m_u$.

$$\tilde{R}_{^4He} = \frac{\mu}{m_e}\tilde{R}_\infty = \frac{m_N}{m_e + m_N}\tilde{R}_\infty$$
$$= \frac{4.002\,602 \times (1.660\,539 \times 10^{-27}\ \text{kg})}{(9.109\,383 \times 10^{-31}\ \text{kg}) + 4.002\,602 \times (1.660\,539 \times 10^{-27}\ \text{kg})}$$
$$\times (1.097\,373 \times 10^5\ \text{cm}^{-1}) = 1.097\,223 \times 10^5\ \text{cm}^{-1}$$

A similar calculation gives $\tilde{R}_{^3He} = 1.097\,173 \times 10^5\ \text{cm}^{-1}$. With these values of the Rydberg constant the wavenumber of the relevant transitions is computed using eqn 8.2; the results are given in the table.

	$\tilde{\nu}_{3\to2}/\text{cm}^{-1}$	$\tilde{\nu}_{2\to1}/\text{cm}^{-1}$
$^4He^+$	60 956.8	329 167
$^3He^+$	60 954.1	329 152
difference	2.8	15

If the spectrometer has sufficient resolution these differences are detectable; the greatest difference is for the higher wavenumber transition.

P8C.5 The three transitions originate from the same level, the ^2P, with energy $E_{^2P}$, and if it is assumed that the $nd\ ^2D$ states are hydrogenic their energies may be written $E_n = -A/n^2$, where A is some constant. It follows that the wavenumber of the transitions can be written

$$\tilde{\nu}_n = E_{^2P}/hc - (A/hc)/n^2$$

A plot of $\tilde{\nu}_n$ against $1/n^2$ is therefore expected to a straight line with y-intercept (at $1/n^2 = 0$) $E_{^2P}/hc$. The data are tabulated below and the graph is given in Fig. 8.4.

n	$1/n^2$	λ/nm	$\tilde{\nu}_n/\text{cm}^{-1}$
3	0.111	610.36	16 384
4	0.063	460.29	21 725
5	0.040	413.23	24 200

The data fall on a good straight line which has y-intercept $\tilde{\nu}_\infty = E_{^2P}/hc = 28\,595\ \text{cm}^{-1}$. The transition from the ^2S ground state to the ^2P state is at a wavelength of 670.78 nm, which corresponds to a wavenumber of 14 908 cm^{-1}.

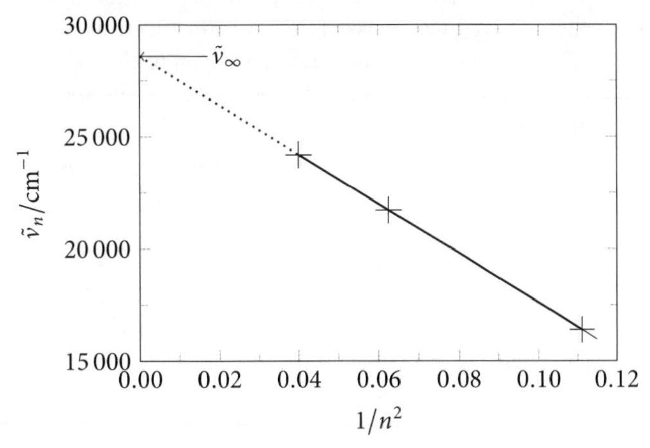

Figure 8.4

Therefore the transition from ^{2}S to the ionization limit of the ^{2}P–^{2}D series will be at wavenumber

$$14\,908\ \mathrm{cm}^{-1} + \tilde{\nu}_\infty = 14\,908\ \mathrm{cm}^{-1} + 28\,595\ \mathrm{cm}^{-1} = 43\,503\ \mathrm{cm}^{-1}$$

This corresponds to the ionization energy of the ground state, which can be expressed in eV as $\boxed{5.39\ \mathrm{eV}}$. Although the data are given to high precision, quite a long extrapolation is needed to find the energy of the ^{2}P state and it also has been assumed that the constant A is independent of n, which may not be the case. As a result, the ionization energy is quoted to more modest precision.

P8C.7 The outer electron in K can occupy an s, p or d orbital and such configurations gives rise to $^2\mathrm{S}_{1/2}$, $^2\mathrm{P}_{3/2,1/2}$, and $^2\mathrm{D}_{5/2,3/2}$ states, respectively. Taking into account the selection rules and the effect of spin-orbit coupling, two closely spaced lines are expected as a result of the transitions $^2\mathrm{S}_{1/2} \rightarrow {}^2\mathrm{P}_{3/2}$ and $^2\mathrm{S}_{1/2} \rightarrow {}^2\mathrm{P}_{1/2}$. The separation of the two lines will reflect the separation of the $^2\mathrm{P}_{3/2}$ and $^2\mathrm{P}_{1/2}$ levels, which is computed using [8C.4–*331*]; the terms in L and S cancel as they take the same value for the two states

$$\begin{aligned}\Delta E &= E_{1,1/2,3/2} - E_{1,1/2,1/2}\\ &= \tfrac{1}{2}hc\tilde{A}[\tfrac{3}{2}(\tfrac{3}{2}+1) - L(L+1) - S(S+1)]\\ &\quad - \tfrac{1}{2}hc\tilde{A}[\tfrac{1}{2}(\tfrac{1}{2}+1) - L(L+1) - S(S+1)] = \tfrac{3}{2}hc\tilde{A}\end{aligned}$$

The wavenumber of the separation between the two lines is therefore $\frac{3}{2}\tilde{A}$, hence

$$\begin{aligned}\tilde{A} &= \tfrac{2}{3}\left[(766.70\times10^{-7}\ \mathrm{cm})^{-1} - (770.11\times10^{-7}\ \mathrm{cm})^{-1}\right]\\ &= \tfrac{2}{3}(57.7...\ \mathrm{cm}^{-1}) = \boxed{38.5\ \mathrm{cm}^{-1}}\end{aligned}$$

P8C.9 The Rydberg constant for positronium is [8A.14–*308*] $\tilde{R}_{Ps} = \tilde{R}_\infty \times (\mu_{Ps}/m_e)$, where the reduced mass of the positron–electron system is $\mu_{Ps} = m_e^2/(2m_e) = m_e/2$, as the mass of the nucleus is equal to that of the electron. Hence, $\tilde{R}_{Ps} = \tilde{R}_\infty/2 = (109737\ \text{cm}^{-1})/2 = 54868.5\ \text{cm}^{-1}$. The spectral lines of the positronium atom are given by $\tilde{\nu} = \tilde{R}_{Ps}(n_1^{-2} - n_2^{-2})$.

The Balmer series are those lines with $n_1 = 2$, and so the wavenumbers of these are

$$\tilde{\nu} = \tilde{R}_{Ps}(2^{-2} - n_2^{-2}) = (54868.5\ \text{cm}^{-1}) \times (2^{-2} - n_2^{-2}) \quad n_2 = 3, 4 \ldots$$

The first three lines have $n_2 = 3, 4, 5$ and are at $\boxed{7621\ \text{cm}^{-1}}$, $\boxed{10288\ \text{cm}^{-1}}$, and $\boxed{11522\ \text{cm}^{-1}}$, respectively. The ionization energy is simply the binding energy of the ground state, which is $hc\tilde{R}_{Ps}$. Hence $I = hc\tilde{R}_{Ps}/e = \boxed{6.803\ \text{eV}}$

P8C.11 The derivation follows the method used in *How is that done?* 8C.1 on page 327. For a transition to be allowed the transition dipole moment $\boldsymbol{\mu}_{fi}$ must be non-zero. It is convenient to explore this condition by examining the x-, y-, and z-components of the moment: if any of these are non-zero, the overall moment will also be non-zero. The x- and y-components are given by $\mu_{x,fi} = -e\int \psi_f^* x\psi_i\,d\tau$ and $\mu_{y,fi} = -e\int \psi_f^* y\psi_i\,d\tau$, respectively. The limits of integration are $r = 0$ to ∞, $\theta = 0$ to π, and $\phi = 0$ to 2π, and the volume element is $d\tau = r^2 \sin\theta\,dr\,d\theta\,d\phi$.

The first step is to express the Cartesian co-ordinates in spherical polar coordinates and then in terms of the spherical harmonics. From *The chemist's toolkit* 21 in Topic 7F on page 286 it is seen that $x = r\sin\theta\cos\phi$, and $y = r\sin\theta\sin\phi$. From Table 7F.1 on page 286 the spherical harmonics $Y_{1,\pm1}$ are $\mp N\sin\theta e^{\pm i\phi}$, where N is the normalization constant. Using the identity $e^{\pm i\phi} = \cos\phi \pm i\sin\phi$ it follows that

$$\begin{aligned} Y_{1,+1} + Y_{1,-1} &= -N\sin\theta(\cos\phi + i\sin\phi) + N\sin\theta(\cos\phi - i\sin\phi) \\ &= -2iN\sin\theta\sin\phi \end{aligned}$$

Similarly

$$\begin{aligned} Y_{1,+1} - Y_{1,-1} &= -N\sin\theta(\cos\phi + i\sin\phi) - N\sin\theta(\cos\phi - i\sin\phi) \\ &= -2N\sin\theta\cos\phi \end{aligned}$$

With these relationships x and y are expressed as

$$\begin{aligned} x &= r\sin\theta\cos\phi = -r(Y_{1,+1} - Y_{1,-1})/2N \\ y &= r\sin\theta\sin\phi = -r(Y_{1,+1} + Y_{1,-1})/2iN \end{aligned}$$

The wavefunctions of the atomic orbitals are expressed in terms of a radial and an angular part: $\psi_{n,l,m_l} = R_{n,l}(r)Y_{l,m_l}(\theta,\phi)$. As is seen in the calculation in the text, the selection rule is derived by considering only the integral over the angles. Focusing just on this, and setting aside all the normalization and other factors, the integral to consider for the x-component of the transition moment is

$$\mu_{x,\mathrm{fi}} \propto \int_0^{\pi}\int_0^{2\pi} Y^*_{l_\mathrm{f},m_{l,\mathrm{f}}}(Y_{1,+1}-Y_{1,-1})Y_{l_\mathrm{i},m_{l,\mathrm{i}}}\sin\theta\,\mathrm{d}\theta\,\mathrm{d}\phi$$

It is a property of spherical harmonics that the 'triple integral'

$$\int_0^{\pi}\int_0^{2\pi} Y^*_{l_\mathrm{f},m_{l,\mathrm{f}}}Y_{l,m}Y_{l_\mathrm{i},m_{l,\mathrm{i}}}\sin\theta\,\mathrm{d}\theta\,\mathrm{d}\phi$$

vanishes unless $l_\mathrm{f} = l_\mathrm{i} \pm l$ and $m_{l,\mathrm{f}} = m_{l,\mathrm{i}} \pm m$. In this case the integrals of interest have $l = 1$ and $m = \pm1$, therefore they are non-zero only if $l_\mathrm{f} = l_\mathrm{i} \pm 1$ and $m_{l,\mathrm{f}} = m_{l,\mathrm{i}} \pm 1$. It follows that the integral on which the x-component of the transition depends is only non-zero if these conditions are satisfied. A similar argument applies to the y-component. The selection rules are therefore $\boxed{\Delta l = \pm1, \Delta m_l = \pm1}$.

In the text it is seen that the selection rule deriving from the z-component of the transition moment is $\Delta l = \pm1$, $\Delta m_l = 0$; when the x- and y- components are considered as well, transitions with $\Delta m_l = \pm1$ are also allowed.

Answers to integrated activities

I8.1

(a) The ground state of the He^+ ion is $1s^1$ with $S = \frac{1}{2}$, $L = 0$ and hence $J = \frac{1}{2}$. The term symbol is therefore ${}^2S_{1/2}$. The excited state configuration is $4p^1$ which has $S = \frac{1}{2}$, $L = 1$ and hence $J = \frac{3}{2}$ or $\frac{1}{2}$; the term symbols are ${}^2P_{3/2}$ and ${}^2P_{1/2}$, the lowest of which is that with $J = \frac{1}{2}$. According to the selection rules the transitions from ${}^2S_{1/2}$ to both ${}^2P$ states are allowed: hence, the transitions are $\boxed{{}^2S_{1/2} \rightarrow {}^2P_{1/2}}$ and $\boxed{{}^2S_{1/2} \rightarrow {}^2P_{3/2}}$.

(b) The wavenumber the spectral line corresponding to the $n_1 \rightarrow n_2$ transition is given by a modified version of [8A.1–*304*] which takes into account the nuclear charge Z: $\tilde{\nu} = Z^2\tilde{R}_\mathrm{H}(n_1^{-2} - n_2^{-2})$, where $\tilde{R}_\mathrm{H}$ is the Rydberg constant for hydrogen, $109\,677\ \mathrm{cm^{-1}}$; $Z = 2$ for He^+. In principle the Rydberg constant is different for He, but the change is so small that it can safely be ignored. Hence for a transition for $n = 1 \rightarrow 4$, the wavenumber is $\tilde{\nu} = 4 \times (109\,677\ \mathrm{cm^{-1}}) \times (1^{-2} - 4^{-2}) = \boxed{411\,289\ \mathrm{cm^{-1}}}$.
This corresponds to a wavelength of $\lambda = \tilde{\nu}^{-1} = (411\,289\ \mathrm{cm^{-1}})^{-1} = 2.43... \times 10^{-6}\ \mathrm{cm} = \boxed{24.313\,8\ \mathrm{nm}}$. The corresponding frequency is

$$\nu = c\lambda^{-1} = c\tilde{\nu} = (2.997\,925 \times 10^{10}\ \mathrm{cm\,s^{-1}}) \times (411\,289\ \mathrm{cm^{-1}})$$
$$= \boxed{1.233\,01 \times 10^{16}\ \mathrm{Hz}}$$

(c) The mean radius of a hydrogenic orbital, characterized by quantum numbers, n, l, m_l is given by

$$\langle r\rangle_{n,l,m_l} = \frac{n^2a_0}{Z}\left[1 + \tfrac{1}{2}\left(1 - \frac{l(l+1)}{n^2}\right)\right]$$

For the ground state orbital, with $Z = 2$, $n = 1$ and $l = 0$ in He^+

$$\langle r \rangle_{1,0,0} = \frac{(1^2)a_0}{2}\left[1 + \tfrac{1}{2}\left(1 - \frac{0(0+1)}{1^2}\right)\right] = \frac{3a_0}{4}$$

For the upper state with $n = 4$ and $l = 1$

$$\langle r \rangle_{4,1,0} = \frac{(4)^2 a_0}{2}\left[1 + \tfrac{1}{2}\left(1 - \frac{1(1+1)}{4^2}\right)\right] = \frac{23a_0}{2}$$

Hence, the mean radius of the atom increases by $23a_0/2 - 3a_0/4 = \boxed{43a_0/4}$.

I8.3 Because the beam splits into two, with deflections $\pm(\mu_B L^2/4E_k)\mathrm{d}\mathcal{B}/\mathrm{d}z$, a splitting between the two beams of Δx is achieved by satisfying the condition $\Delta x = (\mu_B L^2/2E_k)\mathrm{d}\mathcal{B}/\mathrm{d}z$, which is rearranged to give an expression for the field gradient $\mathrm{d}\mathcal{B}/\mathrm{d}z = 2E_k\Delta x/\mu_B L^2$. A reasonable estimate for the mean kinetic energy is to take the equipartition value of $\frac{3}{2}kT$.

$$\begin{aligned}\frac{\mathrm{d}\mathcal{B}}{\mathrm{d}z} &= \frac{2E_k\Delta x}{\mu_B L^2} = \frac{3kT\Delta x}{\mu_B L^2}\\ &= \frac{3 \times (1.3806 \times 10^{-23}\,\mathrm{J\,K^{-1}}) \times (1000\,\mathrm{K}) \times (1.00 \times 10^{-3}\,\mathrm{m})}{(9.2740 \times 10^{-24}\,\mathrm{J\,T^{-1}}) \times (50 \times 10^{-2}\,\mathrm{m})^2}\\ &= \boxed{17.9\,\mathrm{T\,m^{-1}}}\end{aligned}$$

Molecular Structure

9A Valence-bond theory

Answers to discussion questions

D9A.1 The Born–Oppenheimer approximation treats the nuclei of the multi-particle system of electrons and nuclei as if they were fixed. The dependence of energy on nuclear positions is then obtained by solving the Schrödinger equation at many different (fixed) nuclear geometries. Molecular potential energy curves and surfaces are plots of the total molecular energy (computed under the Born–Oppenheimer approximation) as a function of nuclear coordinates.

D9A.3 See Section 9A.3(b) on page 347 for details on hybridization applied to simple carbon compounds. The carbon atoms in alkanes are sp^3 hybridized. This explains the nearly tetrahedral bond angles about the carbon atoms in such molecules.

The double-bonded carbon atoms in alkenes are sp^2 hybridized. This explains the bond angles of approximately 120° about these atoms. The simultaneous overlap of sp^2 hybridized orbitals and unhybridized p orbitals in C=C double bonds explains the resistance of such bonds to torsion and the co-planarity of the atoms attached to those atoms.

The triple-bonded carbon atoms in alkynes are sp hybridized, which explains the 180° bond angles about these atoms. The central carbon atom in allene is also sp hybridized. Each of its C=C double bonds involves one of its sp hybrids and one unhybridized p orbital. The two resulting π orbitals are oriented perpendicular to one another, which is why the two CH_2 groups are rotated by 90° relative to one another. This arrangement of orbitals also accounts for the resistance to the two CH_2 groups being rotated relative to one another about the long axis.

D9A.5 Resonance refers to the superposition of the wave functions representing different electron distributions in the same nuclear framework. The wavefunction resulting from the superposition is called a resonance hybrid.

Resonance allows for a more refined description of the electron distribution, and hence bonding, than is given by a single valence bond wavefunction. Different valence bond structures are allowed to contribute to different extents, meaning that the overall wavefunction is built up from contributions from different valence-bond wavefunctions.

This approach makes it possible to describe polar bonds as a combination of a purely covalent and a purely ionic structure, and delocalized bonding in terms of combinations of valence-bond structures in which, for example, a double bond is located in different parts of a molecule. Resonance is a device for calculating an improved wavefunction: it does not imply that wavefunction flickers between those for the different structures.

Solutions to exercises

E9A.1(a) Using [9A.2–*344*] and assuming that the valence-bond in HF is formed between the H1s and F2p_z atomic orbitals, the spatial part of the valence-bond wavefunction is written as $\Psi(1,2) = \psi_{\text{F2p}_z}(1)\psi_{\text{H1s}}(2) + \psi_{\text{F2p}_z}(2)\psi_{\text{H1s}}(1)$. The overall wavefunction must be antisymmetric to satisfy the Pauli principle, therefore the symmetric spatial part has to be combined with the antisymmetric two-electron spin wavefunction given by [8B.3–319], $\sigma_-(1,2)$. The (unnormalized) complete two-electron wavefunction is therefore

$$\Psi(1,2) = [\psi_{\text{F2p}_z}(1)\psi_{\text{H1s}}(2) + \psi_{\text{F2p}_z}(2)\psi_{\text{H1s}}(1)] \times [\alpha(1)\beta(2) - \beta(1)\alpha(2)]$$

E9A.2(a) The resonance hybrid wavefunction constructed from one two-electron wavefunction corresponding to the purely covalent form of the bond and one two-electron wavefunction corresponding to the ionic form of the bond is given in [9A.3–*346*] as $\Psi = \Psi_{\text{covalent}} + \lambda\Psi_{\text{ionic}}$. Therefore the (unnormalized) resonance hybrid wavefunction of HF with two ionic structures is written as $\Psi_{\text{HF}} = \Psi_{\text{H-F}} + \lambda\Psi_{\text{H}^+\text{F}^-} + \kappa\Psi_{\text{H}^-\text{F}^+}$. $\Psi_{\text{H-F}}$ is written as in *Exercise* E9A.1(a), $\Psi_{\text{H-F}} = [\psi_{\text{F2p}_z}(1)\psi_{\text{H1s}}(2) + \psi_{\text{F2p}_z}(2)\psi_{\text{H1s}}(1)] \times \sigma_-(1,2)$. The wavefunction $\Psi_{\text{H}^+\text{F}^-}$ describes the electron distribution when both electrons reside on the F2p_z orbital. The spatial part of this wavefunction is given by $\psi_{\text{F2p}_z(1)}\psi_{\text{F2p}_z(2)}$, which is symmetric, therefore it has to be combined with the antisymmetric spin wavefunction resulting in $\Psi_{\text{H}^+\text{F}^-} = [\psi_{\text{F2p}_z(1)}\psi_{\text{F2p}_z(2)}] \times \sigma_-(1,2)$. Similarly, the other ionic structure has $\Psi_{\text{H}^-\text{F}^+} = [\psi_{\text{H1s}(1)}\psi_{\text{H1s}(2)}] \times \sigma_-(1,2)$.

E9A.3(a) Both phosphorus and nitrogen are in Group 15, therefore the valence bond description of the bonding in P_2 is similar to that of N_2. There is a triple bond between the two sp hybridized phosphorus atoms. A σ bond is formed by the overlap of two sp hybrid atomic orbitals projecting towards each other along the internuclear axis. The two π bonds are the result of the side-by-side overlap of $3p_x$ with $3p_x$ and $3p_y$ with $3p_y$ orbitals. There is one lone pair on each phosphorus atom, contained in the sp orbital projecting outwards along the internuclear axis.

Consider the equilibrium $P_4 \rightleftharpoons 2\,P_2$. In the tetrahedral P_4 there are six σ bonds, whereas in two molecules of P_2 there are two σ and four π bonds overall. π bonds are generally weaker than σ bonds, therefore the equilibrium favors P_4.

E9A.4(a) The ammonium ion is iso-electronic with methane, therefore the two species are expected to have the same description of bonding. Four sp^3 hybrid atomic orbitals are formed from the 2s and the three 2p orbitals of the nitrogen atom; each hybrid then forms a σ bond by overlapping with a hydrogen 1s orbital.

E9A.5(a) All the carbon atoms in 1,3-butadiene are sp^2 hybridized. The σ framework of the molecule consists of C–H and C–C σ bonds. Each C–H σ bond is formed by the overlap of an sp^2 hybrid atomic orbital on a carbon atom with a 1s atomic orbital on a neighbouring hydrogen atom. Similarly, C–C σ bonds are formed by the overlap of sp^2 hybrid atomic orbitals on neighbouring carbon atoms. The two π bonds are formed by the side-by-side overlap of unhybridized 2p orbitals on carbon atoms C1 and C2, and likewise between C3 and C4.

E9A.6(a) The carbon and nitrogen atoms in methylamine are sp^3 hybridized. The C–N bond is formed by the overlap of an sp^3 orbital on carbon with an sp^3 orbital on nitrogen. The C–H bonds are formed by the overlap of a carbon sp^3 hybrid atomic orbital with a hydrogen 1s atomic orbital. Similarly, the N–H bonds are formed by the overlap of a nitrogen sp^3 hybrid atomic orbital with a hydrogen 1s atomic orbital. The lone pair on nitrogen resides on an sp^3 hybrid atomic orbital.

E9A.7(a) The condition of orthogonality is given by [7C.8–*254*], $\int \Psi_i^* \Psi_j \,d\tau = 0$ for $i \neq j$. The atomic orbitals are all real, therefore $\Psi_i^* = \Psi_i$. The orthogonality condition becomes

$$\int h_1^* h_2 \,d\tau = \int (s + p_x + p_y + p_z)(s - p_x - p_y + p_z)\,d\tau$$

$$= \overbrace{\int s^2\,d\tau}^{1} - \overbrace{\int sp_x\,d\tau}^{0} - \overbrace{\int sp_y\,d\tau}^{0} + \overbrace{\int sp_z\,d\tau}^{0}$$

$$+ \ldots - \overbrace{\int p_x^2\,d\tau}^{1} + \ldots - \overbrace{\int p_y^2\,d\tau}^{1} + \ldots + \overbrace{\int p_z^2\,d\tau}^{1}$$

$$= 1 - 1 - 1 + 1 = 0$$

All the integrals of the form $\int sp_i\,d\tau$ are zero because the s and p orbitals are orthogonal, and all the integrals of the form $\int s^2\,d\tau$ and $\int p_i^2\,d\tau$ are 1 because the orbitals are normalized. The condition for the orthogonality of h_1 and h_2 is satisfied.

E9A.8(a) A normalized wavefunction satisfies [7B.4c–*248*], $\int \Psi^* \Psi\,d\tau = 1$. The wavefunction is normalized by finding the value of N for which $h = N(s + 2^{1/2}p)$ satisfies this condition. The orbital wavefunctions s and p are real as is N, therefore

$$\int h^* h\,d\tau = N^2 \int (s + 2^{1/2}p)^2\,d\tau$$

$$= N^2 \left[\overbrace{\int s^2\,d\tau}^{1} + 2\overbrace{\int p^2\,d\tau}^{1} + 2^{3/2}\overbrace{\int sp\,d\tau}^{0} \right] = 3N^2$$

The integral $\int sp\,d\tau$ is zero because the s and p orbitals are orthogonal, and the integrals $\int s^2\,d\tau$ and $\int p^2\,d\tau$ are 1 because the orbitals are normalized. From the normalization condition it follows that $3N^2 = 1$ and hence $N = 1/3^{1/2}$.

Solutions to problems

P9A.1 The wavefunction in terms of the polar coordinates of each electron is given in *Brief illustration* 9A.1 on page 344 as

$$\Psi(1,2) = \frac{1}{\pi a_0^3}\left[\mathrm{e}^{-(r_{\mathrm{A}1}+r_{\mathrm{B}2})/a_0} + \mathrm{e}^{-(r_{\mathrm{A}2}+r_{\mathrm{B}1})/a_0}\right]$$

Given that the internuclear separation along the z-axis is R, in Cartesian coordinates $r_{\mathrm{A}i}$ and $r_{\mathrm{B}i}$ becomes

$$r_{\mathrm{A}i} = (x_i^2 + y_i^2 + z_i^2)^{1/2} \quad \text{and} \quad r_{\mathrm{B}i} = (x_i^2 + y_i^2 + (z_i - R)^2)^{1/2}$$

Therefore the wavefunction is

$$\Psi(1,2) = \frac{1}{\pi a_0^3}$$

$$\times\left[\mathrm{e}^{-[(x_1^2+y_1^2+z_1^2)^{1/2}+(x_2^2+y_2^2+(z_2-R)^2)^{1/2}]/a_0} + \mathrm{e}^{-[(x_2^2+y_2^2+z_2^2)^{1/2}+(x_1^2+y_1^2+(z_1-R)^2)^{1/2}]/a_0}\right]$$

P9A.3 For the purposes of this problem, the p_x and p_y orbitals are represented by unit vectors along the x-and y-axes, respectively. The given hybrid atomic orbitals are created by the linear combination of the s, p_x and p_y orbitals. The s orbital is spherically symmetric about the origin, therefore it does not modify the directions in which the hybrids point. The vector representations of the hybrid atomic orbitals are

$$\mathbf{h}_1 = \sqrt{2}\,\mathbf{j} \quad \mathbf{h}_2 = \sqrt{3/2}\,\mathbf{i} - \sqrt{1/2}\,\mathbf{j} \quad \mathbf{h}_3 = -\sqrt{3/2}\,\mathbf{i} - \sqrt{1/2}\,\mathbf{j}$$

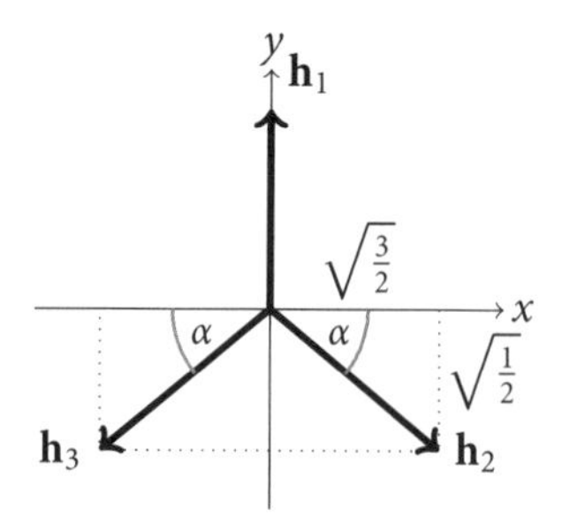

From the diagram it is evident that $\alpha = \tan^{-1}\left(1/\sqrt{3}\right) = 30°$. It follows that the angle between adjacent hybrids is 120°.

9B Molecular orbital theory: the hydrogen molecule-ion

Answer to discussion questions

D9B.1 The Born–Oppenheimer approximation treats the nuclei of the multi-particle system of electrons and nuclei as if they were fixed. The dependence of energy on nuclear positions is then obtained by solving the Schrödinger equation at many different (fixed) nuclear geometries. Molecular potential energy curves and surfaces are plots of molecular energy (computed under the Born–Oppenheimer approximation) as a function of nuclear coordinates.

D9B.3 The lowest energy arrangement is obtained by filling the lowest energy molecular orbital first, then the next lowest and so on. It is a consequence of the Pauli principle, the requirement that the overall wavefunction is antisymmetric with respect the interchange of the labels of the electrons, that a maximum of two electrons can be accommodated in each orbital and that these two electrons must have an antisymmetric spin wavefunction. Such a wavefunction in only achieved by pairing the spins.

Solutions to exercises

E9B.1(a) The normalization condition is given by [7B.4c–*248*], $\int \psi^*\psi\,\mathrm{d}\tau = 1$. The wavefunction is normalized by finding N such that $\psi = N(\psi_A + \lambda\psi_B)$ satisfies this condition. The wavefunctions ψ_A and ψ_B are real, as is N, therefore

$$\begin{aligned}\int \psi^*\psi\,\mathrm{d}\tau &= N^2\int(\psi_A+\lambda\psi_B)^2\,\mathrm{d}\tau\\ &= N^2\left[\overbrace{\int\psi_A^2\,\mathrm{d}\tau}^{1}+\lambda^2\overbrace{\int\psi_B^2\,\mathrm{d}\tau}^{1}+2\lambda\overbrace{\int\psi_A\psi_B\,\mathrm{d}\tau}^{S}\right]\\ &= N^2(1+\lambda^2+2\lambda S)\end{aligned}$$

The integrals $\int\psi_A^2\,\mathrm{d}\tau$ and $\int\psi_B^2\,\mathrm{d}\tau$ are 1 because the wavefunctions ψ_A and ψ_B are normalized. It follows that $\boxed{N = 1/(1+\lambda^2+2\lambda S)^{1/2}}$.

E9B.2(a) The condition of orthogonality is given by [7C.8–*254*], $\int\psi_i^*\psi_j\,\mathrm{d}\tau = 0$ for $i \neq j$. The given molecular orbital, $\psi_i = 0.145A + 0.844B$ is real, therefore $\psi_i^* = \psi_i$. The new linear combination for A and B, which is orthogonal to ψ_i must have the form of $\psi_j = A + \beta B$, where the coefficient of wavefunction A is chosen to be 1 for simplicity. Substitution of these wavefunctions in the condition of orthogonality gives

$$\begin{aligned}\int\psi_i^*\psi_j\,\mathrm{d}\tau &= \int(0.145A+0.844B)\times(A+\beta B)\,\mathrm{d}\tau\\ &= 0.145\overbrace{\int A^2\,\mathrm{d}\tau}^{1}+0.844\beta\overbrace{\int B^2\,\mathrm{d}\tau}^{1}+(0.145\beta+0.844)\overbrace{\int AB\,\mathrm{d}\tau}^{S}\\ &= 0.145+0.844\beta+(0.145\beta+0.844)S\end{aligned}$$

Using $S = 0.250$ the value of the integral becomes $0.356 + 0.88025\beta$. This value must be zero for the two wavefunctions to be orthogonal, therefore $\beta = -0.404$ and hence $\psi_j = A - 0.404B$.

Normalization of ψ_i follows the same logic as in *Exercise* E9B.1(a). First the wavefunction is written as $\psi_i = N(0.145A + 0.844B)$ and then the normalization constant N is found such that $\int \psi^* \psi \, d\tau = 1$.

$$\int \psi_i^* \psi_i \, d\tau = \int [N(0.145A + 0.844B)]^2 \, d\tau$$

$$= N^2\Big(0.145^2 \overbrace{\int A^2 \, d\tau}^{1} + 0.844^2 \beta \overbrace{\int B^2 \, d\tau}^{1} + (2 \times 0.145 \times 0.844) \overbrace{\int AB \, d\tau}^{S}\Big)$$

$$= N^2(0.733 + 0.245S)$$

Using $S = 0.250$ gives a value of $0.794N^2$ for the integral, therefore $N = 1/\sqrt{0.794} =$ 1.12. Therefore the normalized wavefunction is

$$\psi_i = 1.12 \times (0.145A + 0.844B) = \boxed{0.163A + 0.947B}$$

Normalization of ψ_j follows a similar procedure as for ψ_i, giving $N = 1.02$ and therefore $\boxed{\psi_j = 1.02A - 0.412B}$.

E9B.3(a) The energy of the σ bonding orbital in H_2^+ is given by [9B.4–*353*], $E_\sigma = E_{H1s} + j_0/R - (j + k)/(1 + S)$. Molecular potential energy curves are usually plotted with respect to the energy of the separated atoms, therefore the energies to be plotted are $E_\sigma - E_{H1s} = j_0/R - (j + k)/(1 + S)$.

Using [9B.5d–*353*], $j_0/a_0 = 27.21$ eV $= 1\,E_h$ the energy for $R/a_0 = 1$ is computed as

$$E_\sigma - E_{H1s} = \frac{(1\,E_h)}{1} - \frac{(0.729\,E_h) + (0.736\,E_h)}{(1 + 0.858)} = +0.211\,E_h$$

Similar calculations give the following energies

R/a_0	1	2	3	4
$(E_\sigma - E_{H1s})/E_h$	+0.211	-5.32×10^{-2}	-5.88×10^{-2}	-3.76×10^{-2}

These data are plotted in Fig. 9.1; with so few data points it is difficult to locate the minimum. The data are fitted well by the following cubic

$$(E_\sigma - E_{H1s})/E_h = -0.0386(R/a_0)^3 + 0.3611(R/a_0)^2 - 1.0771(R/a_0) + 0.9656$$

Note that this cubic equation has no physical meaning, it is only used to draw the line on the plot above and to locate the minimum by setting the derivative to zero; in particular the maximum close to the final data point has no physical basis. This minimum is found to be at $\boxed{R = 2.5\,a_0}$, which corresponds to the predicted equilibrium bond length. The depth of the potential energy well at this distance is about $-0.073\,E_h$ which is $\boxed{2.0 \text{ eV}}$.

E9B.4(a) A sketch of the bonding and the antibonding molecular orbitals resulting from the side-by-side overlap of two p orbitals is shown in Fig. 9C.5 on page 359. The bonding molecular orbital is antisymmetric with respect to inversion, therefore it is denoted as a π_u orbital. The antibonding molecular orbital is symmetric with respect to inversion, therefore it is a π_g orbital.

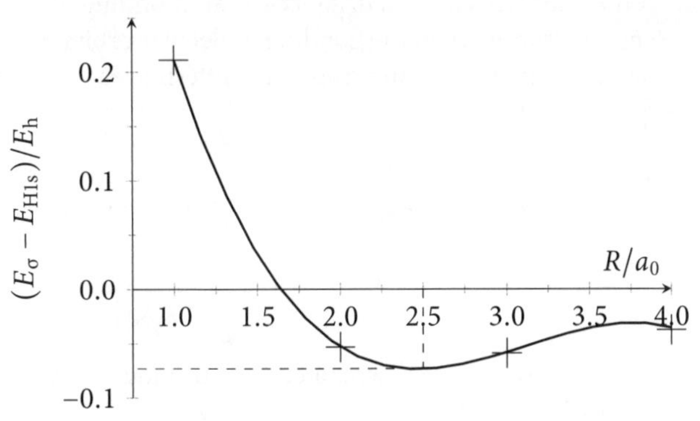

Figure 9.1

Solutions to problems

P9B.1 Inspection of [9B.1–*351*] reveals that the repulsion energy between two hydrogen nuclei is given by $e^2/4\pi\varepsilon_0 R$, where R is the internuclear separation. In molar quantities, the repulsion energy is $N_{\rm A}e^2/4\pi\varepsilon_0 R$, which, at an equilibrium separation of $R = 74.1$ pm becomes

$$\frac{(6.0221\times10^{23}\,\text{mol}^{-1})\times(1.6022\times10^{-19}\,\text{C})^2}{4\pi\times(8.8542\times10^{-12}\,\text{J}^{-1}\,\text{C}^2\,\text{m}^{-1})\times(74.1\times10^{-12}\,\text{m})} = \boxed{1.87\times10^{6}\,\text{J mol}^{-1}}$$

The molar gravitational potential energy between two hydrogen nuclei is

$$\frac{N_{\rm A}Gm_{\rm p}^2}{R}$$
$$= \frac{(6.0221\times10^{23}\,\text{mol}^{-1})\times(6.6738\times10^{-11}\,\text{N m}^2\,\text{kg}^{-2})\times(1.6726\times10^{-27}\,\text{kg})^2}{(74.1\times10^{-12}\,\text{m})}$$
$$= \boxed{1.52\times10^{-30}\,\text{J mol}^{-1}}$$

Therefore the gravitational attraction is entirely negligible compared to the electrostatic repulsion between the two nuclei.

P9B.3 Refer to the data presented in *Exercise* E9B.3(a). The energy of the σ bonding orbital in H_2^+ is given by [9B.4–*353*], $E_\sigma = E_{\rm H1s} + j_0/R - (j+k)/(1+S)$. This energy in usually measured with respect to the energy of the separated atoms, therefore the energy is $E_\sigma - E_{\rm H1s} = j_0/R - (j+k)/(1+S)$. Likewise for the σ^* antibonding orbital the energy is given by [9B.7–*355*], $E_{\sigma^*} = E_{\rm H1s} + j_0/R - (j-k)/(1-S)$. Relative to the separated atoms, the energy is $E_{\sigma^*} - E_{\rm H1s} = j_0/R - (j-k)/(1-S)$. With the data given, and using $j_0/a_0 = 27.21$ eV $= 1\,E_{\rm h}$, the energies of these molecular orbitals are

R/a_0	1	2	3	4
$(E_\sigma - E_{\rm H1s})/E_{\rm h}$	+0.211	-5.32×10^{-2}	-5.88×10^{-2}	-3.76×10^{-2}
$(E_\sigma^* - E_{\rm H1s})/E_{\rm h}$	+1.05	+0.340	+0.132	$+5.52\times10^{-2}$

It is evident that at each distance the antibonding molecular orbital is raised in energy by more than the bonding molecular orbital is lowered. This appears to be generally true for any reasonable internuclear separation.

P9B.5 The bonding and antibonding MOl wavefunctions are $\psi_\pm = N_\pm(\psi_A \pm \psi_B)$, where $N_\pm$ is the normalizing factor, given by (*Example* 9B.1 on page 352)

$$N_\pm = \frac{1}{[(2(1 \pm S)]^{1/2}}$$

where for two 1s AOs separated by a distance R the overlap integral is given by [9B.5a–*353*], $S = \left(1 + R/a_0 + \frac{1}{3}(R/a_0)^2\right) e^{-R/a_0}$. The form of ψ_A and ψ_B are given in *Brief illustration* 9B.1 on page 352

$$\psi_A = (1/\pi a_0^3)^{1/2} e^{-r_{A1}/a_0} \qquad \psi_B = (1/\pi a_0^3)^{1/2} e^{-r_{B1}/a_0}$$

Without loss of generality, it is assumed that atom A is located at $z_{A1} = 0$ and atom B at $z_{B1} = R$, the internuclear separation. The requirement is to plot the wavefunction along the z-axis, so $x_{A1} = y_{A1} = 0$, and likewise for orbital B. With all of these conditions imposed the function to plotted is

$$\psi_\pm = \frac{1}{[(2(1 \pm S)]^{1/2}} \frac{1}{(\pi a_0^3)^{1/2}} \left(e^{-|z|/a_0} \pm e^{-|(z-R)|/a_0}\right)$$

The modulus signs are needed because the argument of the exponential is the distrance from the nucleus, which is always positive.

Figure 9.2 shows plots of (a) the bonding and antibonding wavefunctions, and (b) the squares of these functions (the probability density) for the case $R = 2a_0$. The quantity plotted in (a) is $(a_0)^{3/2}\psi_\pm$, and in (b) it is $(a_0)^3\psi_\pm^2$; the same scale is used for each orbital.

The antibonding orbital has a node at the mid-point of the bond, whereas the bonding orbital has significant electron density at this point. From the diagrams it appears that the antibonding orbital has greater overall probability, but this is not in fact the case – both orbitals are normalized. It is just that when the functions are plotted along the z-axis there appears to be such a difference due to the different distribution of electron density elsewhere in the orbitals.

The difference density is the difference in electron density between the molecular orbital and two non-interacting 1s orbitals, one on each atom. It is a measure of the way in which the electron density is changed when the molecular orbitals are compared to non-interacting atomic orbitals. The difference density is given by

$$\psi_\pm^2 - \tfrac{1}{2}\left(\psi_A^2 + \psi_B^2\right)$$

Also shown in Fig. 9.2 is this difference density for (c) the bonding and (d) the antibonding molecular orbital; the same scale is used for each plot. If the difference density is positive the implication is that the electron density is greater than for two non-interacting atomic orbitals, whereas if the difference density is negative the implication is that there is a reduction in electron density. It is evident from the plots that in the bonding molecular orbital there is an increase in the electron density in the internuclear region, whereas for the antibonding orbital the density in this region is reduced, but the density is increased further away. These observations account (partially, at least) for the fact that occupying the bonding molecular orbital promotes bond formation. The apparent difference in size between the difference densities between (c) and (d) is a result of simply plotting the function along the z-axis.

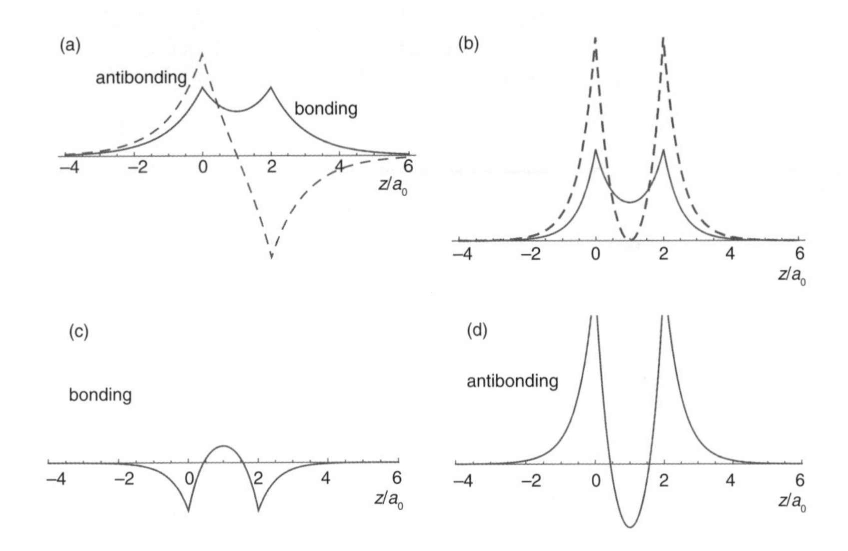

Figure 9.2

9C Molecular orbital theory: homonuclear diatomic molecules

Answer to discussion questions

D9C.1 Both σ and π interactions are possible, as shown below; for the antibonding combination simply reverse the sign of the p orbital.

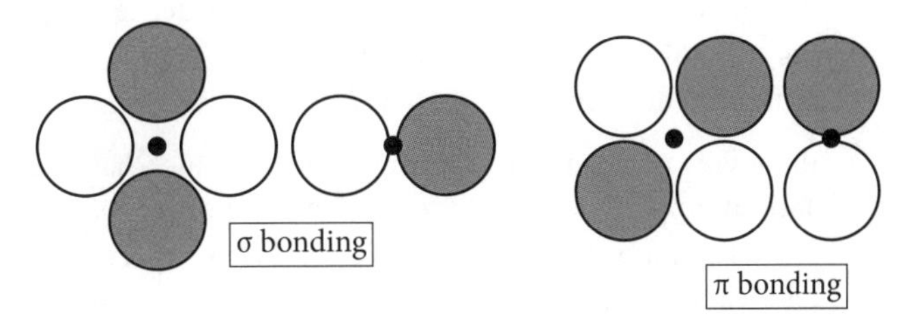

D9C.3 Molecular orbitals are made up of linear combinations of atomic orbitals of similar energy and appropriate symmetry. The s and p atomic orbitals have distinctly different energies, so the molecular orbitals that result from linear combinations primarily of s orbitals have very little character of the higher-energy p orbitals and vice versa. In addition, only a p orbital directed along the internuclear axis has the right symmetry to combine with an s orbital.

Solutions to exercises

E9C.1(a) The molecular orbital diagram for the homonuclear diatomic molecules Li_2, Be_2, and C_2 is shown in Fig. 9C.12 on page 361. According to the Pauli principle, up to two valence electrons can be placed in each of the molecular orbitals. First the lowest energy orbital is filled up, then the next lowest and so on, until all the valence electrons are used up.

(i) Li_2 has $1+1=2$ valence electrons (VE) overall, therefore the ground-state electron configuration is $1\sigma_g^2$. The bond order is defined in [9C.4–361] as $b=\frac{1}{2}(N-N^*)$, therefore $b=\frac{1}{2}(2-0)=1$.

(ii) Be_2: $2+2=4$ VE; $1\sigma_g^2\,1\sigma_u^{*2}$; $b=\frac{1}{2}(2-2)=0$.

(iii) C_2: $4+4=8$ VE; $1\sigma_g^2\,1\sigma_u^{*2}\,1\pi_u^4$; $b=\frac{1}{2}(6-2)=2$.

E9C.2(a) The molecule with the greater bond order is expected to have the larger dissociation energy. Qualitatively B_2 and C_2 share the same molecular orbital energy level diagram, shown in Fig. 9C.12 on page 361. B_2 has $3+3=6$ valence electrons overall, therefore its ground-state electron configuration is $1\sigma_g^2\,1\sigma_u^{*2}\,1\pi_u^2$. The bond order is defined in [9C.4–*361*] as $b=\frac{1}{2}(N-N^*)$, therefore $b=\frac{1}{2}(4-2)=1$.

C_2 has $4+4=8$ valence electrons, its configuration is $1\sigma_g^2\,1\sigma_u^{*2}\,1\pi_u^4$, and the bond order is $b=\frac{1}{2}(6-2)=2$. C_2 has greater bond order than B_2, therefore C_2 is expected to have the larger bond dissociation energy.

E9C.3(a) The molecule with the greater bond order is expected to have the larger dissociation energy. The molecular orbital energy level diagram of F_2 and F_2^+ is shown in Fig. 9C.11 on page 361. F_2 has $7+7=14$ valence electrons overall, therefore the ground-state electron configuration is $1\sigma_g^2\,1\sigma_u^{*2}\,2\sigma_g^2\,1\pi_u^4\,1\pi_g^{*4}$. The bond order is defined in [9C.4–*361*] as $b=\frac{1}{2}(N-N^*)$, therefore $b=\frac{1}{2}(8-6)=1$.

Removing one electron from F_2 gives F_2^+, which has one fewer electron in the antibonding π_g^* orbital, therefore the bond order is $b=\frac{1}{2}(8-5)=\frac{3}{2}$. F_2^+ has greater bond order than F_2, therefore F_2^+ is expected to have the larger bond dissociation energy.

E9C.4(a) The molecular orbital energy level diagram for Li_2, Be_2, B_2, C_2 and N_2 is shown in Fig. 9C.12 on page 361, and for O_2, F_2 and Ne_2 in Fig. 9C.11 on page 361. Following the same logic as in *Exercise* E9C.1(a) and *Exercise* E9C.2(a) gives

Li_2 $1+1=2$ VE $1\sigma_g^2$ $b=\frac{1}{2}(2-0)=1$

Be_2 $2+2=4$ VE $1\sigma_g^2\,1\sigma_u^{*2}$ $b=\frac{1}{2}(2-2)=0$

B_2 $3+3=6$ VE $1\sigma_g^2\,1\sigma_u^{*2}\,1\pi_u^2$ $b=\frac{1}{2}(4-2)=1$

C_2 $4+4=8$ VE $1\sigma_g^2\,1\sigma_u^{*2}\,1\pi_u^4$ $b=\frac{1}{2}(6-2)=2$

N_2 $5+5=10$ VE $1\sigma_g^2\,1\sigma_u^{*2}\,1\pi_u^4\,2\sigma_g^2$ $b=\frac{1}{2}(8-2)=3$

O_2 $6+6=12$ VE $1\sigma_g^2\,1\sigma_u^{*2}\,2\sigma_g^2\,1\pi_u^4\,1\pi_g^{*2}$ $b=\frac{1}{2}(8-4)=2$

F_2 $7+7=14$ VE $1\sigma_g^2\,1\sigma_u^{*2}\,2\sigma_g^2\,1\pi_u^4\,1\pi_g^{*4}$ $b=\frac{1}{2}(8-6)=1$

Ne_2 $8+8=16$ VE $1\sigma_g^2\,1\sigma_u^{*2}\,2\sigma_g^2\,1\pi_u^4\,1\pi_g^{*4}\,2\sigma_u^{*2}$ $b=\frac{1}{2}(8-8)=0$

E9C.5(a) The molecular orbital energy level diagram for Li_2, Be_2, B_2, C_2, N_2 and their ions is shown in Fig. 9C.12 on page 361, and for O_2, F_2, Ne_2 and their ions in Fig. 9C.11 on page 361. The highest occupied molecular orbital (HOMO) is the molecular orbital which is the highest in energy and is at least singly occupied. The HOMO of each of the listed ions is indicated by a box around it.

${Li_2}^+$ $1+1-1=1$ VE $\boxed{1\sigma_g^1}$

${Be_2}^+$ $2+2-1=3$ VE $1\sigma_g^2\,\boxed{1\sigma_u^{*1}}$

${B_2}^+$ $3+3-1=5$ VE $1\sigma_g^2\,1\sigma_u^{*2}\,\boxed{1\pi_u^1}$

${C_2}^+$ $4+4-1=7$ VE $1\sigma_g^2\,1\sigma_u^{*2}\,\boxed{1\pi_u^3}$

${N_2}^+$ $5+5-1=9$ VE $1\sigma_g^2\,1\sigma_u^{*2}\,1\pi_u^4\,\boxed{2\sigma_g^1}$

${O_2}^+$ $6+6-1=11$ VE $1\sigma_g^2\,1\sigma_u^{*2}\,2\sigma_g^2\,1\pi_u^4\,\boxed{1\pi_g^{*1}}$

${F_2}^+$ $7+7-1=13$ VE $1\sigma_g^2\,1\sigma_u^{*2}\,2\sigma_g^2\,1\pi_u^4\,\boxed{1\pi_g^{*3}}$

${Ne_2}^+$ $8+8-1=15$ VE $1\sigma_g^2\,1\sigma_u^{*2}\,2\sigma_g^2\,1\pi_u^4\,1\pi_g^{*4}\,\boxed{2\sigma_u^{*1}}$

${Li_2}^-$ $1+1+1=3$ VE $1\sigma_g^2\,\boxed{1\sigma_u^{*1}}$

${Be_2}^-$ $2+2+1=5$ VE $1\sigma_g^2\,1\sigma_u^{*2}\,\boxed{1\pi_u^1}$

${B_2}^-$ $3+3+1=7$ VE $1\sigma_g^2\,1\sigma_u^{*2}\,\boxed{1\pi_u^3}$

${C_2}^-$ $4+4+1=9$ VE $1\sigma_g^2\,1\sigma_u^{*2}\,1\pi_u^4\,\boxed{2\sigma_g^1}$

${N_2}^-$ $5+5+1=11$ VE $1\sigma_g^2\,1\sigma_u^{*2}\,1\pi_u^4\,2\sigma_g^2\,\boxed{1\pi_g^{*1}}$

${O_2}^-$ $6+6+1=13$ VE $1\sigma_g^2\,1\sigma_u^{*2}\,2\sigma_g^2\,1\pi_u^4\,\boxed{1\pi_g^{*3}}$

${F_2}^-$ $7+7+1=15$ VE $1\sigma_g^2\,1\sigma_u^{*2}\,2\sigma_g^2\,1\pi_u^4\,1\pi_g^{*4}\,\boxed{2\sigma_u^{*1}}$

${Ne_2}^-$ $8+8+1=17$ VE $1\sigma_g^2\,1\sigma_u^{*2}\,2\sigma_g^2\,1\pi_u^4\,1\pi_g^{*4}\,2\sigma_u^{*2}\,\boxed{3\sigma_g^1}$

Note that the extra electron in Ne_2 is accommodated on a bonding molecular orbital resulting from the overlap of the 3s atomic orbitals.

E9C.6(a) The energy of the incident photon must equal the sum of the ionization energy of the orbital and the kinetic energy of the ejected photoelectron, [9C.5–362], $h\nu = I + \frac{1}{2}m_e v^2$. The energy of the incident photon is given by $h\nu = hc/\lambda = (6.6261 \times 10^{-34}\,\text{J s}) \times (2.9979 \times 10^8\,\text{m s}^{-1})/(100 \times 10^{-9}\,\text{m}) = 1.98... \times 10^{-18}\,\text{J}$. Rearranging the equation to give the speed of the ejected electron gives

$$\begin{aligned} v &= \left[\frac{2}{m_e}\left(\frac{hc}{\lambda} - I\right)\right]^{1/2} \\ &= \left[\frac{2}{(9.1094 \times 10^{-31}\,\text{kg})} \times [(1.98... \times 10^{-18}\,\text{J}) \right. \\ &\quad \left. - (12.0\,\text{eV}) \times (1.6022 \times 10^{-19}\,\text{J eV}^{-1})]\right]^{1/2} = \boxed{3.70 \times 10^5\,\text{m s}^{-1}} \end{aligned}$$

Solutions to problems

P9C.1 (a) Figure 9.3 shows a plot of the overlap integral ($Z = 1$ is assumed)

$$S(2\text{s}, 2\text{s}) = \left[1 + \frac{1}{2}\left(\frac{R}{a_0}\right) + \frac{1}{12}\left(\frac{R}{a_0}\right)^2 + \frac{1}{240}\left(\frac{R}{a_0}\right)^4\right]\text{e}^{-R/2a_0}$$

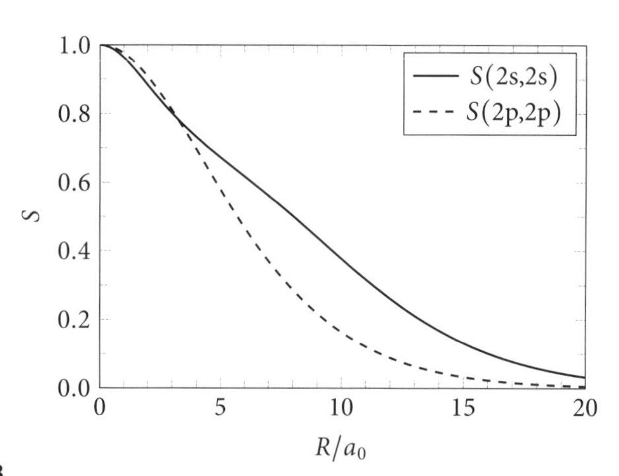

Figure 9.3

(b) The overlap integral $S(2\text{s}, 2\text{s})$ reaches a value of 0.50 at $\boxed{R/a_0 = 8.03}$; this value can be read off a graph or found by using mathematical software.

(c) Figure 9.3 shows a plot of the overlap integral ($Z = 1$ is assumed)

$$S(2\text{p}, 2\text{p}) = \left[1 + \frac{1}{2}\left(\frac{R}{a_0}\right) + \frac{1}{10}\left(\frac{R}{a_0}\right)^2 + \frac{1}{120}\left(\frac{R}{a_0}\right)^3\right]\text{e}^{-R/2a_0}$$

(d) The value of the overlap integral at $R/a_0 = 8.03$ is

$$S(2\text{p}, 2\text{p}) = \left[1 + \tfrac{1}{2}\times 8.03 + \tfrac{1}{10}\times(8.03)^2 + \tfrac{1}{120}\times(8.03)^3\right]\times\text{e}^{-8.03/2} = \boxed{0.29}$$

P9C.3 Figure 9.4 shows contour plots of the bonding and antibonding 2pσ and 2pπ molecular orbitals for a representative internuclear distance of $R = 6a_0$; negative amplitude is indicated by dashed contours, and the locations of the nuclei are shown by the black dots. Density plots, in which the intensity of the shading is proportional to the square of the wavefucntion, are shown in Fig. 9.5.

These plots are useful as they identify aspects of the symmetry of the wavefunctions and the positions of nodal planes. In addition, they illustrate that for the bonding orbitals electron density is accumulating in the internuclear region.

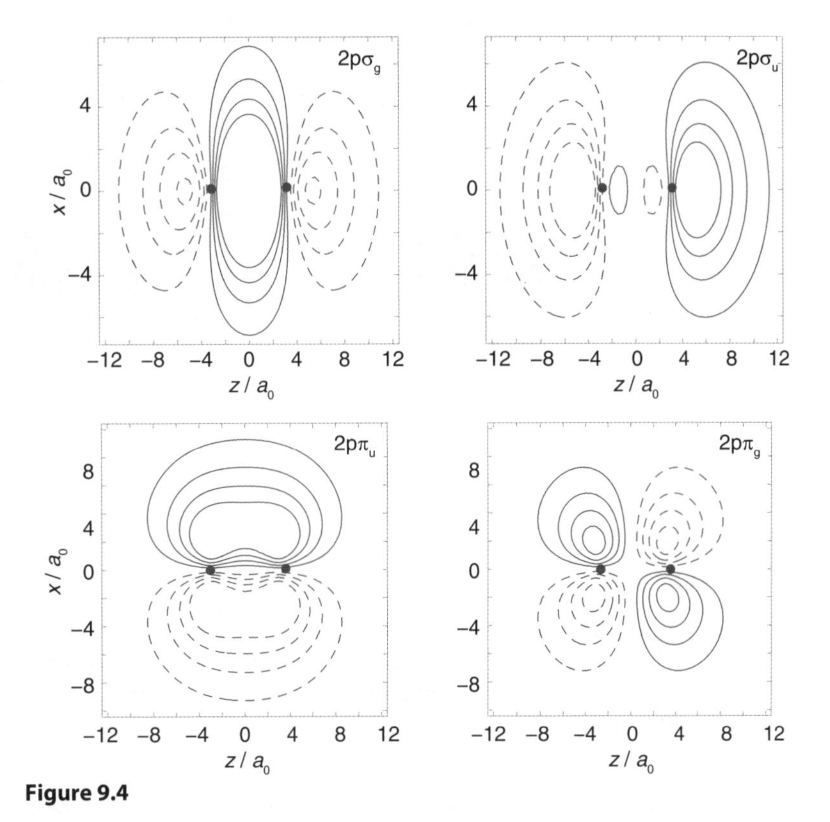

Figure 9.4

9D Molecular orbital theory: heteronuclear diatomic molecules

Answer to discussion questions

D9D.1 Both the Pauling and Mulliken methods for measuring the attracting power of atoms for electrons seem to make good chemical sense. Referring to [9D.2–365], the definition of the Pauling scale, it is seen that if $\tilde{D}_0(AB)$ were equal to $\frac{1}{2}[\tilde{D}_0(\text{AA})+\tilde{D}_0(\text{BB})]$ the calculated electronegativity difference would be zero, as expected for completely non-polar bonds. Hence, any increased strength of the A–B bond over the average of the A–A and B–B bonds can reasonably

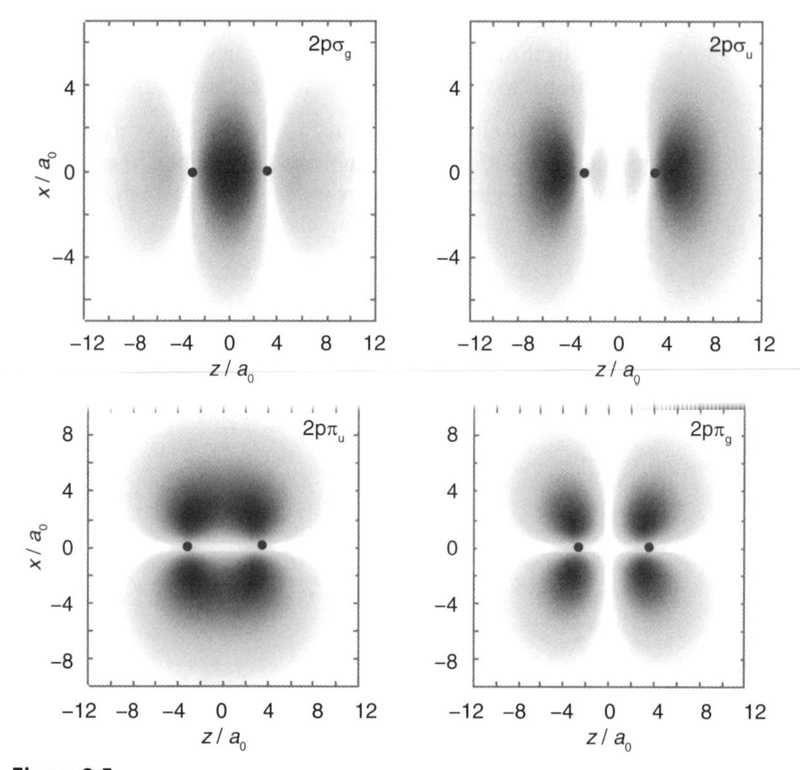

Figure 9.5

be thought of as being due to the polarity of the A–B bond, which in turn is due to the difference in electronegativity of the atoms involved. Therefore, this difference in bond strengths can be used as a measure of electronegativity difference.

The Mulliken scale, [9D.3–*366*], may be more intuitive than the Pauling scale because it is common to think of ionization energies and electron affinities as measures of the electron attracting powers of atoms.

D9D.3 The variation principle is that the energy of a system which is calculated from an arbitrary wavefunction is never less than the true ground-state energy; the only constraint is that the trail wavefunction must satisfy any relevant boundary conditions.

If the trial wavefunction is expressed in terms of certain parameters, the variation principle makes it possible to find the best values of these parameters simply by altering them until the lowest energy is achieved. The principle therefore provides a method to refining a trial wavefunction.

Solutions to exercises

E9D.1(a) The molecular orbital energy level diagram for a heteronuclear diatomic AB is similar to that for a homonuclear diatomic A_2 (Fig. 9C.11 on page 361 or Fig. 9C.12 on page 361) except that the atomic orbitals on A and B are no longer at the same energies. As a result the molecular orbitals no longer have equal contributions from the orbitals on A and B; furthermore, it is more likely that the 2s and 2p orbitals will mix. From simple considerations it it not possible to predict the exact ordering of the resulting molecular orbitals, so the diagram shown in Fig. 9.6 is simply one possibility. Note that because the heteronuclear diatomic no longer has a centre of symmetry the g/u labels are not applicable.

The electronic configurations are: (i) CO (10 valence electrons) $1\sigma^2\,2\sigma^2\,3\sigma^2\,1\pi^4$; (ii) NO (11 valence electrons) $1\sigma^2\,2\sigma^2\,3\sigma^2\,1\pi^4\,2\pi^1$; (iii) CN^- is isoelectronic with CO and therefore has the same configuration.

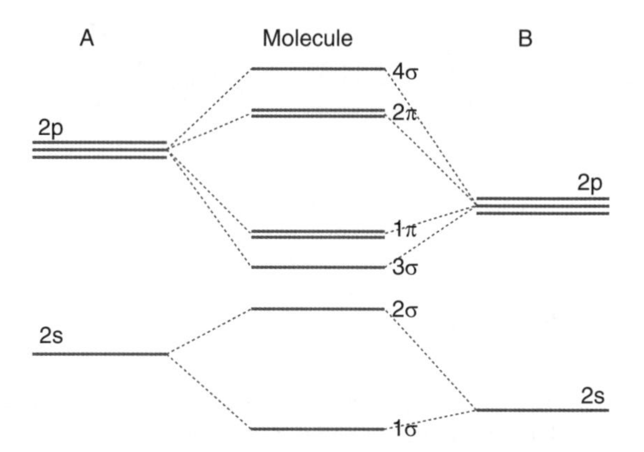

Figure 9.6

E9D.2(a) The molecular orbital energy level diagram of the heteronuclear diatomic molecule XeF is similar to the one shown in the solution to *Exercise* E9D.1(a) except that the orbitals on atom A are 5s and 5p. It is not possible to predict the precise energy ordering of the orbitals from simple considerations, so this diagram is simply a plausible suggestion.

XeF has $8+7=15$ valence electrons, therefore the ground state electron configuration is $1\sigma^2\,2\sigma^2\,3\sigma^2\,1\pi^4\,2\pi^4\,4\sigma^1$. The configuration of XeF^+ is the same except that, as there is one fewer electrons, the antibonding 4σ orbital is not occupied. This means that the bond order in XeF^+ ($b=1$) is greater than the bond order in XeF ($b=\frac{1}{2}$), therefore XeF^+ is likely to have a shorter bond length than XeF.

E9D.3(a) A suitable MO diagram in shown in the solution to *Exercise* E9D.1(a). The ion with the greater bond order is expected to have the shorter bond length. NO^+ has $5+6-1=10$ valence electrons, just enough to completely fill up all the bonding molecular orbitals, leading to a ground state electron configuration of $1\sigma^2\,2\sigma^2\,3\sigma^2\,1\pi^4$. NO^- has two more electrons, both accommodated in the antibonding 2π orbital. It follows that NO^+ has a greater bond order than NO^-, therefore NO^+ is expected to have the shorter bond length.

E9D.4(a) The relationship between the Pauling and Mulliken electronegativities is given by [9D.4–*366*], $\chi_{\text{Pauling}} = 1.35\chi_{\text{Mulliken}}^{1/2} - 1.37$. A plot of the Pauling electronegativities of Period 2 atoms against the square root of their Mulliken electronegativities is shown in Fig. 9.7.

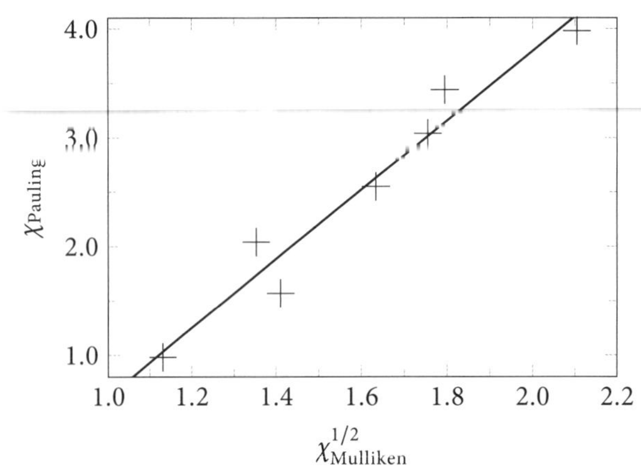

Figure 9.7

The equation of the best fit line is $\chi_{\text{Pauling}} = 3.18\chi_{\text{Mulliken}}^{1/2} - 2.57$, which is very far from the expected relationship.

E9D.5(a) The orbital energy of an atomic orbital in a given atom is estimated using the procedure outlined in *Brief illustration* 9D.2 on page 369, and using data from the *Resource section*. The orbital energy of hydrogen is

$$\alpha_{\text{H}} = -\tfrac{1}{2}[I + E_{\text{ea}}]$$
$$= -\tfrac{1}{2}\times[(1312.0\ \text{kJ mol}^{-1}) + (72.8\ \text{kJ mol}^{-1})]\times\frac{(1\ \text{eV})}{(96.485\ \text{kJ mol}^{-1})} = \boxed{-7.18\ \text{eV}}$$

The conversion factor between kJ mol^{-1} and eV is taken from inside the front cover . Similarly for chlorine

$$\alpha_{\text{Cl}} = -\tfrac{1}{2}[I + E_{\text{ea}}]$$
$$= -\tfrac{1}{2}\times[(1251.1\ \text{kJ mol}^{-1}) + (348.7\ \text{kJ mol}^{-1})]\times\frac{(1\ \text{eV})}{(96.485\ \text{kJ mol}^{-1})} = \boxed{-8.29\ \text{eV}}$$

E9D.6(a) The orbital energies of hydrogen ($\alpha_{\text{H}} = -7.18$ eV) and chlorine ($\alpha_{\text{Cl}} = -8.29$ eV) are calculated in *Exercise* E9D.5(a). Taking $\beta = -1.0$ eV as a typical value and setting $S = 0$ for simplicity, substitution into [9D.9c–368] gives

$$E_\pm = \tfrac{1}{2}(\alpha_\mathrm{H} + \alpha_\mathrm{Cl}) \pm \tfrac{1}{2}(\alpha_\mathrm{H} - \alpha_\mathrm{Cl})\left[1 + \left(\frac{2\beta}{\alpha_\mathrm{H} - \alpha_\mathrm{Cl}}\right)^2\right]^{1/2}$$

$$= \tfrac{1}{2}[(-7.18\ \mathrm{eV}) + (-8.29\ \mathrm{eV})]$$

$$\pm \tfrac{1}{2}[(-7.18\ \mathrm{eV}) - (-8.29\ \mathrm{eV})]\left[1 + \left(\frac{(-2.0\ \mathrm{eV})}{(-7.18\ \mathrm{eV}) - (-8.29\ \mathrm{eV})}\right)^2\right]^{1/2}$$

$$= (-7.73...\ \mathrm{eV}) \pm (1.14...\ \mathrm{eV})$$

Therefore the energy of the bonding molecular orbital is $E_- = (-7.73...\ \mathrm{eV}) - (1.14...\ \mathrm{eV}) = \boxed{-8.88\ \mathrm{eV}}$, and the antibonding orbital is at an energy level of $E_+ = (-7.73...\ \mathrm{eV}) + (1.14...\ \mathrm{eV}) = \boxed{-6.59\ \mathrm{eV}}$.

E9D.7(a) The orbital energies of hydrogen ($\alpha_\mathrm{H} = -7.18$ eV) and chlorine ($\alpha_\mathrm{Cl} = -8.29$ eV) are calculated in *Exercise* E9D.5(a). Taking $\beta = -1.0$ eV as a typical value, and setting $S = 0.2$, substitution into [9D.9a–368] gives

$$E_\pm = \frac{\alpha_\mathrm{H} + \alpha_\mathrm{Cl} - 2\beta S \pm [(2\beta S - (\alpha_\mathrm{H} + \alpha_\mathrm{Cl}))^2 - 4(1 - S^2)(\alpha_\mathrm{H}\alpha_\mathrm{Cl} - \beta^2)]^{1/2}}{2(1 - S^2)}$$

$$= \frac{(-15.0...\ \mathrm{eV}) \pm (1.54...\ \mathrm{eV})}{(1.92)} = (-7.84...\ \mathrm{eV}) \pm (0.803...\ \mathrm{eV})$$

Therefore the energy of the bonding molecular orbital is $E_- = (-7.48...\ \mathrm{eV}) - (0.803...\ \mathrm{eV}) = \boxed{-8.65\ \mathrm{eV}}$, and the antibonding orbital is at an energy level of $E_+ = (-7.48...\ \mathrm{eV}) + (0.803...\ \mathrm{eV}) = \boxed{-7.05\ \mathrm{eV}}$.

Solutions to problems

P9D.1 (a) A normalized wavefunction satisfies the condition given by [7B.4c–248], $\int \psi\psi^*\,\mathrm{d}\tau = 1$. The given wavefunction is real, therefore $\psi = \psi^*$.

$$\int \psi\psi^*\,\mathrm{d}\tau = \int (\psi_\mathrm{A}\cos\theta + \psi_\mathrm{B}\sin\theta)^2\,\mathrm{d}\tau$$

$$= \cos^2\theta\overbrace{\int \psi_\mathrm{A}^2\,\mathrm{d}\tau}^{1} + \sin^2\theta\overbrace{\int \psi_\mathrm{B}^2\,\mathrm{d}\tau}^{1} + 2\cos\theta\sin\theta\overbrace{\int \psi_\mathrm{A}\psi_\mathrm{B}\,\mathrm{d}\tau}^{0}$$

$$= \cos^2\theta + \sin^2\theta = 1$$

The values of the integrals come from the fact that ψ_A and ψ_B are orthonormal.

(b) The wavefunction which describes the bonding molecular orbital is formed by the in-phase interference of the atomic orbitals ψ_A and ψ_B, therefore the coefficients of ψ_A and ψ_B must have the same sign. Similarly, the antibonding orbital is the result of the out-of-phase interference of the basis atomic orbitals, therefore the corresponding coefficients must have

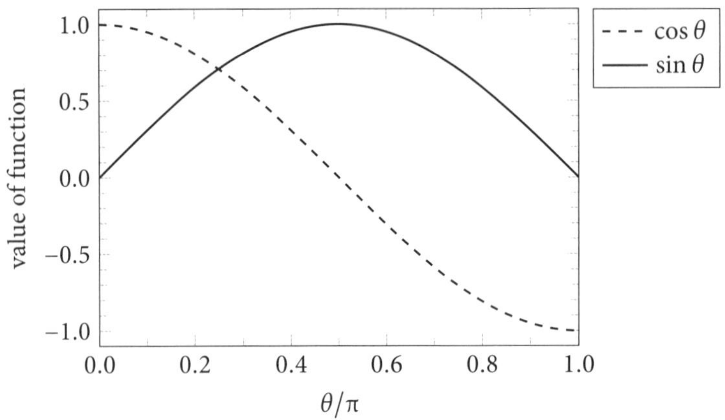

Figure 9.8

opposite signs. A plot of the coefficients, $\cos\theta$ and $\sin\theta$ as a function of θ is shown in Fig. 9.8.

Therefore ψ describes a bonding molecular orbital for $0 < \theta < \pi/2$, and an antibonding molecular orbital for $\pi/2 < \theta < \pi$.

P9D.3 The energy of ψ_A and ψ_C are kept constant in the following, but the energy of ψ_B is progressively lowered.

(a) Taking the energy of the orbital ψ_B to be -12.0 eV, the secular determinant becomes

$$\begin{vmatrix} (-7.2\text{ eV})-E & (-1.0\text{ eV}) & (-0.8\text{ eV}) \\ (-1.0\text{ eV}) & (-12.0\text{ eV})-E & 0 \\ (-0.8\text{ eV}) & 0 & (-8.4\text{ eV})-E \end{vmatrix}$$
$$= -E^3-(27.6\text{ eV})E^2-(246.04\text{ eV}^2)E-(709.68\text{ eV}^3)$$

Setting the polynominal to zero and solving the cubic gives the following energies: $\boxed{E_1=-12.2\text{ eV}}$, $\boxed{E_2=-8.75\text{ eV}}$ and $\boxed{E_3=-6.65\text{ eV}}$. The matrix which diagonalizes the hamiltonian matrix is

$$\begin{pmatrix} 0.202 & 0.394 & 0.897 \\ 0.978 & -0.121 & -0.168 \\ 0.0425 & 0.911 & -0.409 \end{pmatrix}$$

The entries in each column of the matrix above give the coefficients of the atomic orbitals for the corresponding molecular orbital. Note that E_1 is close to the energy of ψ_B, and that in the first column the orbital with the largest coefficient by far is ψ_B.

(b) If the energy of ψ_B is lowered further to -15.0 eV, the secular determinant becomes

$$\begin{vmatrix} (-7.2\text{ eV})-E & (-1.0\text{ eV}) & (-0.8\text{ eV}) \\ (-1.0\text{ eV}) & (-15.0\text{ eV})-E & 0 \\ (-0.8\text{ eV}) & 0 & (-8.4\text{ eV})-E \end{vmatrix}$$
$$= -E^3-(30.6\text{ eV})E^2-(292.84\text{ eV}^2)E-(889.2\text{ eV}^3)$$

The energies are $\boxed{E_1 = -15.1 \text{ eV}}$, $\boxed{E_2 = -8.77 \text{ eV}}$ and $\boxed{E_3 = -6.70 \text{ eV}}$, and the matrix which diagonalizes the hamiltonian is

$$\begin{pmatrix} 0.127 & 0.419 & 0.899 \\ 0.992 & -0.0672 & -0.108 \\ 0.0151 & 0.906 & -0.424 \end{pmatrix}$$

As before E_1 is close in energy to the energy of ψ_B, and in the first column the coefficient of that atomic orbital is close to 1. Furthermore, in columns 2 and 3 the other molecular orbitals are seen to have only small contributions from ψ_B.

The interpretation is that as the energy of ψ_B becomes more and more separate from the other orbitals, one of the molecular orbitals becomes very much like ψ_B and would be classed as non-bonding, and the other two molecular orbitals have little contribution from ψ_B.

9E Molecular orbital theory: polyatomic molecules

Answer to discussion questions

D9E.1 The Hückel method parametrizes, rather than calculates, the energy integrals that arise in molecular orbital theory. In the simplest version of the method, the overlap integral is also parametrized. The energy integrals, α and β, are always considered to be adjustable parameters; their numerical values emerge only at the end of the calculation by comparison to experimental energies. The simple form of the method has three other rather drastic approximations, listed in Section 9E.1(a) on page 371 , which eliminate many terms from the secular determinant and make it easier to solve. Ease of solution was important in the early days of quantum chemistry before the advent of computers. Without the use of these approximations, calculations on polyatomic molecules would have been difficult to accomplish. The simple Hückel method is usually applied only to the calculation of π-electron energies in conjugated organic systems. It is based on the assumption of the separability of the σ- and π-electron systems in the molecule. This is a very crude approximation and works best when the energy level pattern is determined largely by the symmetry of the molecule.

D9E.3 See Section 9E.3 on page 377.

D9E.5 In *ab initio* methods an attempt is made to evaluate all integrals that appear in the secular determinant. Approximations are still employed, but these are mainly associated with the construction of the wavefunctions involved in the integrals. In semi-empirical methods, many of the integrals are expressed in terms of spectroscopic data or physical properties. Semi-empirical methods exist at several levels. At some levels, in order to simplify the calculations, many of the integrals are set equal to zero.

Density functional theory (DFT) is different from the Hartree-Fock (HF) self-consistent field (HF-SCF) methods, the *ab initio* methods, in that DFT focuses on the electron density while HF-SCF methods focus on the wavefunction. However, they both attempt to evaluate integrals from first principles, so DFT methods are in that sense *ab initio* methods: both are iterative self-consistent methods in that the calculations are repeated until the energy and wavefunctions (HF-SCF) or energy and electron density (DFT) are unchanged to within some acceptable tolerance.

Solutions to exercises

E9E.1(a) (i) Without making the Hückel approximations, the secular determinant of the H_3 molecule is written as

$$\begin{vmatrix} \alpha_1 - E & \beta_{12} - S_{12}E & \beta_{13} - S_{13}E \\ \beta_{21} - S_{21}E & \alpha_2 - E & \beta_{23} - S_{23}E \\ \beta_{31} - S_{31}E & \beta_{32} - S_{32}E & \alpha_3 - E \end{vmatrix}$$

where α_n is the Coulomb integral of the orbital on atom n, β_{nm} is the resonance integral accounting for the interaction between the orbitals on atoms n and m, E is the energy of the molecular orbital and S_{nm} is the overlap integral between the orbtials on atoms n and m.

Within the Hückel approximations the energy of the basis atomic orbitals is taken to be independent of the position of the corresponding atoms in the molecule, therefore all Coulomb integrals are set equal to α (given that there is only one type of basis atomic orbital and only one type of atom is involved in the problem). Interaction between orbitals on non-neighbouring atoms is neglected, that is $\beta_{nm} = 0$ if atoms n and m are not neighbouring. All other resonance integrals are set equal to β. The overlap between atomic orbitals is also neglected, therefore all overlap integrals S_{nm} with $n \neq m$ are set to zero. Hence the secular determinant for linear H_3 is

$$\begin{vmatrix} \alpha - E & \beta & 0 \\ \beta & \alpha - E & \beta \\ 0 & \beta & \alpha - E \end{vmatrix}$$

(ii) In this case hydrogen atoms 1 and 3 are neighbours, therefore $\beta_{13} = \beta$, and the secular determinant is

$$\begin{vmatrix} \alpha - E & \beta & \beta \\ \beta & \alpha - E & \beta \\ \beta & \beta & \alpha - E \end{vmatrix}$$

E9E.2(a) The energies of the molecular orbitals in benzene are given by [9E.13–*377*] as $E = \alpha \pm 2\beta, \alpha \pm \beta, \alpha \pm \beta$. Note that α and β are negative quantities, therefore the a_{2u} molecular orbital is the lowest in energy with energy of $\alpha + 2\beta$, as shown in Fig. 9E.4 on page 377.

(i) The benzene anion has $6 + 1 = 7$ electrons in its π system, its electronic configuration is $a_{2u}^2\, e_{1g}^4\, e_{2u}^1$ and the total π-electron binding energy is $E_\pi = 2(\alpha + 2\beta) + 4(\alpha + \beta) + (\alpha - \beta) = \boxed{7\alpha + 7\beta}$.

(ii) The benzene cation has $6 - 1 = 5$ electrons in its π system, its electronic configuration is $a_{2u}^2 e_{1g}^3$ and the total π-electron binding energy is $E_\pi = 2(\alpha + 2\beta) + 3(\alpha + \beta) = \boxed{5\alpha + 7\beta}$.

E9E.3(a) The delocalization energy is the energy difference between the π-electron binding energy E_π in the given species and the hypothetical π-electron binding energy if the given species had isolated π bonds. Therefore the delocalization energy is given by $E_{\text{deloc}} = E_\pi - N_\pi(\alpha + \beta)$, where N_π is the number of π electrons. The π-bond formation energy is defined in [9E.12–*376*] as $E_{\text{bf}} = E_\pi - N_\pi\alpha$.

(i) The benzene anion has 7 π electrons and its π-electron binding energy is calculated in *Exercise* E9E.2(a) as $E_\pi = 7\alpha + 7\beta$. Therefore $E_{\text{deloc}} = (7\alpha + 7\beta) - 7(\alpha + \beta) = \boxed{0}$ and $E_{\text{bf}} = (7\alpha + 7\beta) - 7\alpha = \boxed{7\beta}$.

(ii) The benzene cation has 5 π electrons and its π-electron binding energy is calculated in *Exercise* E9E.2(a) as $E_\pi = 5\alpha + 7\beta$. Therefore $E_{\text{deloc}} = (5\alpha + 7\beta) - 5(\alpha + \beta) = \boxed{2\beta}$ and $E_{\text{bf}} = (5\alpha + 7\beta) - 5\alpha = \boxed{7\beta}$.

E9E.4(a) (i) Following the same logic as in *Exercise* E9E.1(a) and applying the Hückel approximations as explained there the secular determinant for anthracene is written as (the numbers in bold refer to the numbering of the carbon atoms in the molecule)

	1	**2**	**3**	**4**	**5**	**6**	**7**	**8**	**9**	**10**	**11**	**12**	**13**	**14**
1	$\alpha-E$	β	0	0	0	0	0	0	0	0	0	0	0	β
2	β	$\alpha-E$	β	0	0	0	0	0	0	0	0	0	0	0
3	0	β	$\alpha-E$	β	0	0	0	0	0	0	0	0	0	0
4	0	0	β	$\alpha-E$	β	0	0	0	0	0	0	0	0	0
5	0	0	0	β	$\alpha-E$	β	0	0	0	0	0	0	0	β
6	0	0	0	0	β	$\alpha-E$	β	0	0	0	0	0	0	0
7	0	0	0	0	0	β	$\alpha-E$	β	0	0	0	β	0	0
8	0	0	0	0	0	0	β	$\alpha-E$	β	0	0	0	0	0
9	0	0	0	0	0	0	0	β	$\alpha-E$	β	0	0	0	0
10	0	0	0	0	0	0	0	0	β	$\alpha-E$	β	0	0	0
11	0	0	0	0	0	0	0	0	0	β	$\alpha-E$	β	0	0
12	0	0	0	0	0	0	β	0	0	0	β	$\alpha-E$	β	0
13	0	0	0	0	0	0	0	0	0	0	0	β	$\alpha-E$	β
14	β	0	0	0	β	0	0	0	0	0	0	0	β	$\alpha-E$

(ii) Similarly for phenanthrene

	1	2	3	4	5	6	7	8	9	10	11	12	13	14
1	$\alpha-E$	β	0	0	0	0	0	0	0	0	0	0	0	β
2	β	$\alpha-E$	β	0	0	0	0	0	0	0	0	0	0	0
3	0	β	$\alpha-E$	β	0	0	0	0	0	0	0	0	0	0
4	0	0	β	$\alpha-E$	β	0	0	0	0	0	0	0	0	0
5	0	0	0	β	$\alpha-E$	β	0	0	0	0	0	0	0	β
6	0	0	0	0	β	$\alpha-E$	β	0	0	0	0	0	0	0
7	0	0	0	0	0	β	$\alpha-E$	β	0	0	0	0	0	0
8	0	0	0	0	0	0	β	$\alpha-E$	β	0	0	0	β	0
9	0	0	0	0	0	0	0	β	$\alpha-E$	β	0	0	0	0
10	0	0	0	0	0	0	0	0	β	$\alpha-E$	β	0	0	0
11	0	0	0	0	0	0	0	0	0	β	$\alpha-E$	β	0	0
12	0	0	0	0	0	0	0	0	0	0	β	$\alpha-E$	β	0
13	0	0	0	0	0	0	0	β	0	0	0	β	$\alpha-E$	β
14	β	0	0	0	β	0	0	0	0	0	0	0	β	$\alpha-E$

E9E.5(a) To calculate the π-electron binding energy of the given systems, it is necessary to calculate the energies of the occupied molecular orbitals. This is done by diagonalising the hamiltonian matrix: the diagonal elements of the resulting matrix are the energies of the molecular orbitals. The hamiltonian matrix has the same form as the secular matrix except that the diagonal elements are α instead of $\alpha - E$. Alternatively, the energies can be found my finding the eigenvalues of the hamiltonian matrix, or by multiplying out the secular determinant, setting the resulting polynomial in E to zero and then finding the roots. Mathematical software is needed for all of these approaches.

The secular determinants are derived in *Exercise* E9E.4(a), and from these the form of the hamiltonain matrix is easily found.

(i) The orbital energies for anthracene are $E = \alpha + 2.41\beta$, $\alpha + 2\beta$, $\alpha + 1.41\beta$ (doubly degenerate), $\alpha + \beta$ (doubly degenerate), $\alpha + 0.414\beta$, $\alpha - 0.414\beta$, $\alpha - \beta$ (doubly degenerate), $\alpha - 1.41\beta$ (doubly degenerate), $\alpha - 2\beta$, $\alpha - 2.41\beta$. The π system of anthracene accommodates 14 electrons, therefore the 7 lowest energy π molecular orbitals are filled. The π-electron binding energy is therefore $E_\pi = 2(\alpha + 2.41\beta) + 2(\alpha + 2\beta) + 4(\alpha + 1.41\beta) + 4(\alpha + \beta) + 2(\alpha + 0.414\beta) = \boxed{14\alpha + 19.3\beta}$.

(ii) The orbital energies for phenanthrene are $E = \alpha + 2.43\beta$, $\alpha + 1.95\beta$, $\alpha + 1.52\beta$, $\alpha + 1.31\beta$, $\alpha + 1.14\beta$, $\alpha + 0.769\beta$, $\alpha + 0.605\beta$, $\alpha - 0.605\beta$, $\alpha - 0.769\beta$, $\alpha - 1.14\beta$, $\alpha - 1.31\beta$, $\alpha - 1.52\beta$, $\alpha - 1.95\beta$, $\alpha - 2.43\beta$. The π system of anthracene accommodates 14 electrons, therefore the 7 lowest energy π molecular orbitals are filled. The π-electron binding energy is therefore $E_\pi = 2(\alpha + 2.43\beta) + 2(\alpha + 1.95\beta) + 2(\alpha + 1.52\beta) + 2(\alpha + 1.31\beta) + 2(\alpha + 1.14\beta) + 2(\alpha + 0.769\beta) + 2(\alpha + 0.605\beta) = \boxed{14\alpha + 19.5\beta}$

E9E.6(a) The hamiltonian for a single electron in H_2^+ is given by [9B.1–*351*]. It has a kinetic energy term, $\hat{T} = -(\hbar^2/2m_e)\nabla_1^2$, and a potential energy term, $\hat{V}$. The species HeH^+ has two electrons, therefore the kinetic energy term is written as

$$\hat{T} = -\frac{\hbar^2}{2m_e}\nabla_1^2 - \frac{\hbar^2}{2m_e}\nabla_2^2$$

The energy of interaction between an electron and a nucleus with charge number Z at distance r is given by $-Ze^2/4\pi\varepsilon_0 r$. The potential energy operator consists of terms for each electron interacting with the H nuclues ($Z = 1$) and the He nucleus ($Z = 2$)

$$\hat{V} = -\frac{e^2}{4\pi\varepsilon_0}\left(\frac{1}{r_{1\mathrm{H}}} + \frac{1}{r_{2\mathrm{H}}} + \frac{2}{r_{1\mathrm{He}}} + \frac{2}{r_{2\mathrm{He}}} - \frac{1}{r_{12}}\right)$$

The first term represent the interaction between electron 1 and the H nucleus, and the second is for electron 2 with the same nucleus. The third and fourth terms represent the interactions of the two electrons with the He nucleus. The last term accounts for the repulsion between the two electrons. The complete electronic hamiltonian is $\hat{H}_{\mathrm{elec}} = \hat{T} + \hat{V}$. Because only the electronic hamiltonian is required, the repulsion between the two nuclei is not included.

Solutions to problems

P9E.1 Number the oxygen atoms from 1 to 3 and the carbon atom as number 4. Taking the Hückel approximations the secular determinant of the carbonate ion is written as

$$\begin{vmatrix} \alpha_\mathrm{O}-E & 0 & 0 & \beta \\ 0 & \alpha_\mathrm{O}-E & 0 & \beta \\ 0 & 0 & \alpha_\mathrm{O}-E & \beta \\ \beta & \beta & \beta & \alpha_\mathrm{C}-E \end{vmatrix} = (\alpha_\mathrm{O} - E)^2[(\alpha_\mathrm{O} - E)(\alpha_\mathrm{C} - E) - 3\beta^2]$$

Hence the energies of the molecular orbitals of the carbonate ion are the solutions of the equation $(\alpha_\mathrm{O} - E)^2[(\alpha_\mathrm{O} - E)(\alpha_\mathrm{C} - E) - 3\beta^2] = 0$. The solution $E = \alpha_\mathrm{O}$ occurs twice and represents a degenerate pair of orbitals. The other two roots are the solutions of the equation $(\alpha_\mathrm{O} - E)(\alpha_\mathrm{C} - E) - 3\beta^2 = 0$, which gives the energies

$$E_\pm = \tfrac{1}{2}[\alpha_\mathrm{O} + \alpha_\mathrm{C} \pm \sqrt{(\alpha_\mathrm{O} - \alpha_\mathrm{C})^2 + 12\beta^2}]$$

where $E_- < E_+$. Because oxygen is the more electronegative element it is likely that the three lowest energy molecular orbitlas will be those with $E = \alpha_\mathrm{O}$ and $E = E_-$. The carbonate ion has 6 π electrons which will occupy these orbitals and hence give a π-electron binding energy of

$$E_\pi = (\alpha_\mathrm{O}+\alpha_\mathrm{C}-\sqrt{(\alpha_\mathrm{O} - \alpha_\mathrm{C})^2 + 12\beta^2})+4\alpha_\mathrm{O} = 5\alpha_\mathrm{O}+\alpha_\mathrm{C}-\sqrt{(\alpha_\mathrm{O} - \alpha_\mathrm{C})^2 + 12\beta^2}$$

In a localized description of the bonding two electrons occupy a π-bonding molecular orbital between C and O, and four electrons will be localized on oxygen atoms in orbitals with energy α_O. Using the orbital energies from [9D.9c–*368*], the total energy of the two electrons in the bonding molecular orbital is

$\alpha_O + \alpha_C + \sqrt{(\alpha_O - \alpha_C)^2 + 4\beta^2}$. Hence, in a localized structure the π-electron binding energy is

$$E_{loc} = 5\alpha_O + \alpha_C + \sqrt{(\alpha_O - \alpha_C)^2 + 4\beta^2}$$

The delocalization energy is given by $E_\pi - E_{loc}$

$$\begin{aligned} E_{deloc} &= E_\pi - E_{loc} \\ &= [5\alpha_O + \alpha_C + \sqrt{(\alpha_O - \alpha_C)^2 + 12\beta^2}] - [5\alpha_O + \alpha_C + \sqrt{(\alpha_O - \alpha_C)^2 + 4\beta^2}] \\ &= \sqrt{(\alpha_O - \alpha_C)^2 + 12\beta^2} - \sqrt{(\alpha_O - \alpha_C)^2 + 4\beta^2} \end{aligned}$$

P9E.3 The secular equations are

$$\begin{aligned} (\alpha_A - E)c_A + (\beta_{AB} - S_{AB}E)c_B + (\beta_{AC} - S_{AC}E)c_C = 0 \\ (\beta_{BA} - S_{BA}E)c_A + (\alpha_B - E)c_B + (\beta_{BC} - S_{BC}E)c_C = 0 \\ (\beta_{CA} - S_{CA}E)c_A + (\beta_{CB} - S_{CB}E)c_B + (\alpha_C - E)c_C = 0 \end{aligned}$$

In this case, orbitals B and C are on the same atom. It follows that the resonance integral β_{BC} and the overlap integral S_{BC} are zero, as the atomic orbitals on one atom are orthogonal to each other. Therefore the secular equations simplify to

$$\begin{aligned} (\alpha_A - E)c_A + (\beta_{AB} - S_{AB}E)c_B + (\beta_{AC} - S_{AC}E)c_C = 0 \\ (\beta_{BA} - S_{BA}E)c_A + (\alpha_B - E)c_B = 0 \\ (\beta_{CA} - S_{CA}E)c_A + (\alpha_C - E)c_C = 0 \end{aligned}$$

and hence the secular determinant is

$$\begin{vmatrix} \alpha_A - E & \beta_{AB} - S_{AB}E & \beta_{AC} - S_{AC}E \\ \beta_{BA} - S_{BA}E & \alpha_B - E & 0 \\ \beta_{CA} - S_{CA}E & 0 & \alpha_C - E \end{vmatrix}$$

P9E.5 Within the Hückel approximations, the secular determinant of cyclobutadiene is

$$\begin{vmatrix} \alpha - E & \beta & 0 & \beta \\ \beta & \alpha - E & \beta & 0 \\ 0 & \beta & \alpha - E & \beta \\ \beta & 0 & \beta & \alpha - E \end{vmatrix}$$

The hamiltonian matrix is of the same form, but with diagonal elements α

$$H = \begin{pmatrix} \alpha & \beta & 0 & \beta \\ \beta & \alpha & \beta & 0 \\ 0 & \beta & \alpha & \beta \\ \beta & 0 & \beta & \alpha \end{pmatrix} \qquad H = \alpha\mathbf{1} + \beta\begin{pmatrix} 0 & 1 & 0 & 1 \\ 1 & 0 & 1 & 0 \\ 0 & 1 & 0 & 1 \\ 1 & 0 & 1 & 0 \end{pmatrix}$$

As explained in the text, in order to find the energies it is sufficient to diagonalize the matrix on the right, and this is convenient because most mathematical software packages are only able to diagonalize numerical matrices. The diagonal elements in the resulting diagonalized matrix are +2, 0, 0, −2 giving the energies as $\alpha + 2\beta$, α (doubly degenerate), and $\alpha - 2\beta$.

Similarly, the secular determinant and hamiltonian matrix for benzene is

$$\begin{vmatrix} \alpha-E & \beta & 0 & 0 & 0 & \beta \\ \beta & \alpha-E & \beta & 0 & 0 & 0 \\ 0 & \beta & \alpha-E & \beta & 0 & 0 \\ 0 & 0 & \beta & \alpha-E & \beta & 0 \\ 0 & 0 & 0 & \beta & \alpha-E & \beta \\ \beta & 0 & 0 & 0 & \beta & \alpha-E \end{vmatrix} \qquad H = \alpha\mathbf{1} + \beta\begin{pmatrix} 0&1&0&0&0&1 \\ 1&0&1&0&0&0 \\ 0&1&0&1&0&0 \\ 0&0&1&0&1&0 \\ 0&0&0&1&0&1 \\ 1&0&0&0&1&0 \end{pmatrix}$$

The diagonal elements in the resulting diagonalized matrix are +2, +1, +1, −1, −1, and −2, giving the energies $\alpha + 2\beta$, $\alpha + \beta$ (doubly degenerate), $\alpha - \beta$ (doubly degenerate), and $\alpha - 2\beta$.

The secular determinant and hamiltonian matrix for cyclooctatetraene is

$$\begin{vmatrix} \alpha-E & \beta & 0 & 0 & 0 & 0 & 0 & \beta \\ \beta & \alpha-E & \beta & 0 & 0 & 0 & 0 & 0 \\ 0 & \beta & \alpha-E & \beta & 0 & 0 & 0 & 0 \\ 0 & 0 & \beta & \alpha-E & \beta & 0 & 0 & 0 \\ 0 & 0 & 0 & \beta & \alpha-E & \beta & 0 & 0 \\ 0 & 0 & 0 & 0 & \beta & \alpha-E & \beta & 0 \\ 0 & 0 & 0 & 0 & 0 & \beta & \alpha-E & \beta \\ \beta & 0 & 0 & 0 & 0 & 0 & \beta & \alpha-E \end{vmatrix} \qquad H = \alpha\mathbf{1} + \beta\begin{pmatrix} 0&1&0&0&0&0&0&1 \\ 1&0&1&0&0&0&0&0 \\ 0&1&0&1&0&0&0&0 \\ 0&0&1&0&1&0&0&0 \\ 0&0&0&1&0&1&0&0 \\ 0&0&0&0&1&0&1&0 \\ 0&0&0&0&0&1&0&1 \\ 1&0&0&0&0&0&1&0 \end{pmatrix}$$

The diagonal elements in the resulting diagonalized matrix are +2.00, +1.41, +1.41, 0, 0, −1.41, −1.41, and −2, giving energies $\alpha + 2\beta$, $\alpha + 1.41\beta$ (doubly degenerate), α (doubly degenerate), $\alpha - 1.41\beta$ (doubly degenerate), and $\alpha - 2\beta$.

For all three molecules the lowest and highest energy molecular orbital is not degenerate, and all the other orbitals occur as degenerate pairs.

P9E.7 (a) Within the Hückel approximations, the secular determinant of the triangular species H_3 is

$$\begin{vmatrix} \alpha-E & \beta & \beta \\ \beta & \alpha-E & \beta \\ \beta & \beta & \alpha-E \end{vmatrix}$$

The hamiltonian matrix is of the same form, but with diagonal elements α

$$H = \begin{pmatrix} \alpha & \beta & \beta \\ \beta & \alpha & \beta \\ \beta & \beta & \alpha \end{pmatrix} \qquad H = \alpha\mathbf{1} + \beta\begin{pmatrix} 0 & 1 & 1 \\ 1 & 0 & 1 \\ 1 & 1 & 0 \end{pmatrix}$$

As explained in the text, in order to find the energies it is sufficient to diagonalize the matrix on the right, and this is convenient because most mathematical software packages are only able to diagonalize numerical matrices. The diagonal elements in the resulting diagonalized matrix are +2, −1, and −1 giving the energies as $\alpha+2\beta$, and $\alpha-\beta$ (doubly degenerate). The molecular orbital energy level diagram is shown below.

energy

$E = \alpha - \beta$

$E = \alpha + 2\beta$

The number of valence electrons (VE) and electron binding energies of each species are

$$\begin{array}{lll} H_3^+ & 2\,\text{VE} & E_{\text{tot}} = 2\alpha + 4\beta \\ H_3 & 3\,\text{VE} & E_{\text{tot}} = 3\alpha + 3\beta \\ H_3^- & 4\,\text{VE} & E_{\text{tot}} = 4\alpha + 2\beta \end{array}$$

(b) Consider the following set of equations

$$\begin{array}{ll} H_3^+(g) \longrightarrow 2\,H(g) + H^+(g) & \Delta_r U^\ominus(1) = +849\ \text{kJ mol}^{-1} \\ H_2(g) \longrightarrow 2\,H(g) & \Delta_r U^\ominus(2) = +432.1\ \text{kJ mol}^{-1} \\ H^+(g) + H_2(g) \longrightarrow H_3^+(g) & \Delta_r U^\ominus(3) \end{array}$$

Reaction(3) is reaction(2) – reaction(1), therefore

$$\begin{aligned} \Delta_r U^\ominus(3) &= \Delta_r U^\ominus(2) - \Delta_r U^\ominus(1) = (+432.1\ \text{kJ mol}^{-1}) - (+849\ \text{kJ mol}^{-1}) \\ &= -4.16... \times 10^2\ \text{kJ mol}^{-1} = \boxed{-417\ \text{kJ mol}^{-1}} \end{aligned}$$

(c) The change in the total electron binding energy directly gives the change in the internal energy in the reaction $H^+(g)+H_2(g) \longrightarrow H_3^+(g)$. Therefore $\Delta_r U^\ominus$ is expressed in terms of the resonance and Coulomb integrals as

$$\begin{aligned} \Delta_r U^\ominus &= E_{\text{tot}}(\text{products}) - E_{\text{tot}}(\text{reactants}) \\ &= \overbrace{(2\alpha + 4\beta)}^{H_3^+} - \overbrace{0}^{H^+} - \overbrace{2(\alpha + \beta)}^{H_2} = 2\beta \end{aligned}$$

Therefore $\beta = (-4.16... \times 10^2\ \text{kJ mol}^{-1})/2 = \boxed{-208\ \text{kJ mol}^{-1}}$. The electron binding energies are $2\alpha - 834\ \text{kJ mol}^{-1}$ for H_3^+, $3\alpha - 625\ \text{kJ mol}^{-1}$ for H_3 and $4\alpha - 416\ \text{kJ mol}^{-1}$ for H_3^-.

P9E.9 For H_2 the 6-31G* basis set is equivalent to the 6-31G basis set because the star indicates that the basis set adds d-type polarization functions for each atom other than hydrogen. Consequently, the basis sets (a) 6-31G* and (b) 6-311+G** were chosen. Since the calculated energy is with respect to the energy of widely separated stationary electrons and nuclei, the experimental ground electronic energy of dihydrogen is calculated as $D_e + 2I$.

H_2			
	6-31G*	6-311+G**	experimental
R/pm	73.0	73.5	74.1
ground-state energy, E_0/eV	−30.6626	−30.8167	−32.06
F_2			
R/pm	134.5	132.9	141.8
ground-state energy, E_0/eV	−5406.30	−5407.92	

Both computational basis sets give satisfactory bond length agreement with the experimental value for H_2. However the 6-31G* basis set is not as accurate as the larger basis set as illustrated by consideration of both its higher ground-state energy and the variation principle that the energy of a trial wavefunction is never less than the true energy. That is, the energy provided by the 6-311+G** basis set is closer to the true energy.

P9E.11 (a) Linear conjugated polyenes do not have degenerate energy levels, so each value of k corresponds to a molecular orbital which can be occupied by up to two electrons. Therefore for ethene, with 2 electrons, the HOMO has $k = 1$, and the LUMO has $k = 2$. The HOMO–LUMO energy gap is

$$\Delta E = \left(\alpha + 2\beta\cos\frac{2\pi}{3}\right) - \left(\alpha + 2\beta\cos\frac{\pi}{3}\right) = (\alpha - \beta) - (\alpha + \beta) = -2\beta$$

This energy gap corresponds to 61500 cm^{-1}, therefore $\beta = -3.07... \times 10^4\ \text{cm}^{-1}$.

Butadiene has 4 electrons, the HOMO has $k = 2$ and the LUMO has $k = 3$; the HOMO–LUMO energy gap is given by

$$\Delta E = \left(\alpha + 2\beta\cos\frac{3\pi}{5}\right) - \left(\alpha + 2\beta\cos\frac{2\pi}{5}\right) = -1.23\beta$$

ΔE corresponds to 46080 cm^{-1}, therefore $\beta = -3.72... \times 10^4\ \text{cm}^{-1}$.

Hexatriene has 6 electrons, the HOMO has $k = 3$ and the LUMO has $k = 4$; the HOMO–LUMO energy gap is given by

$$\Delta E = \left(\alpha + 2\beta\cos\frac{4\pi}{7}\right) - \left(\alpha + 2\beta\cos\frac{3\pi}{7}\right) = -0.890\beta$$

ΔE corresponds to 39750 cm^{-1}, therefore $\beta = -4.46... \times 10^4\ \text{cm}^{-1}$.

Octatetraene has 8 electrons, the HOMO has $k = 4$ and the LUMO has $k = 5$; the HOMO–LUMO energy gap is given by

$$\Delta E = \left(\alpha + 2\beta\cos\frac{5\pi}{9}\right) - \left(\alpha + 2\beta\cos\frac{4\pi}{9}\right) = -0.695\beta$$

ΔE corresponds to 32900 cm^{-1}, therefore $\beta = -4.73... \times 10^4\ \text{cm}^{-1}$.

The average value of β is $-4.00... \times 10^4\ \text{cm}^{-1}$, which in electronvolts is

$$\beta = \frac{(-4.00... \times 10^4\ \text{cm}^{-1}) \times (6.6261 \times 10^{-34}\ \text{J s}) \times (2.9979 \times 10^{10}\ \text{cm s}^{-1})}{(1.6022 \times 10^{-19}\ \text{J eV}^{-1})}$$

$$= \boxed{-4.96\ \text{eV}}$$

(b) Octatetraene has 8 π electrons, therefore the orbitals with quantum numbers $k = 1, 2, 3$ and 4 are all fully occupied. Hence the π-electron binding energy is

$$E_\pi = 2\left(\alpha + 2\beta\cos\frac{\pi}{9}\right) + 2\left(\alpha + 2\beta\cos\frac{2\pi}{9}\right) + 2\left(\alpha + 2\beta\cos\frac{3\pi}{9}\right) + 2\left(\alpha + 2\beta\cos\frac{4\pi}{9}\right) = 8\alpha + 9.52\beta$$

Therefore the delocalization energy is $E_{\text{deloc}} = 8\alpha + 9.52\beta - 8(\alpha + \beta) = \boxed{1.52\beta}$.

(c) The energies and the orbital coefficients are calculated according to the given formulae and presented in the following tabe; k is the index for the molecular orbital.

k	E_k	$c_{k,1}$	$c_{k,2}$	$c_{k,3}$	$c_{k,4}$	$c_{k,5}$	$c_{k,6}$
1	$\alpha + 1.80\beta$	0.232	0.418	0.521	0.521	0.418	0.232
2	$\alpha + 1.25\beta$	0.418	0.521	0.232	−0.232	−0.521	−0.418
3	$\alpha + 0.445\beta$	0.521	0.232	−0.418	−0.418	0.232	0.521
4	$\alpha - 0.445\beta$	0.521	−0.232	−0.418	0.418	0.232	−0.521
5	$\alpha - 1.25\beta$	0.418	−0.521	0.232	0.232	−0.521	0.418
6	$\alpha - 1.80\beta$	0.232	−0.418	0.521	−0.521	0.418	−0.232

For the lowest energy molecular orbital ($k = 1$) all the coefficients of the atomic orbitals are positive, therefore this molecular orbital is bonding between all pairs of carbon atoms. As the energy of the molecular orbitals increase, the number of nodes increases as indicated by the number of sign changes of the coefficients of neighbouring atomic orbitals. In the highest energy molecular orbital ($k = 6$) the sign of the neighbouring coefficients alternates, hence it is a fully antibonding molecular orbital.

Integrated activities

I9.1 The carbon atoms in ethene are sp^2 hybridized. Each carbon atom has three sp^2 hybrids which are used to form the C–C and H–H σ bonds in the plane of the molecule. Both carbon atoms have a $2p_z$ atomic orbital perpendicular to the plane of the molecule, the overlap of which form the π system. An approximate molecular orbital energy level diagram is shown in Fig. 9.9.

I9.3 (a) The calculated and the measured values of the standard enthalpy of formation ($\Delta_f H^\ominus/\text{kJ mol}^{-1}$) of ethene, butadiene, hexatriene and octatetraene are shown in the table below, together with the relative error in the calculated values.

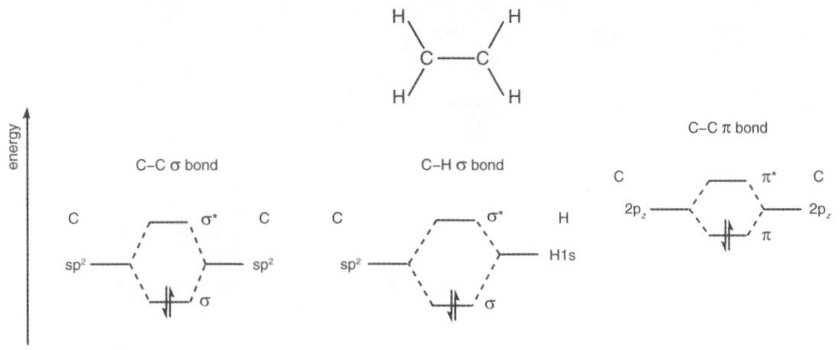

Figure 9.9

molecule	computed	experimental	% error
C_2H_4	69.58	52.46694	32.6
C_4H_6	129.8	108.8 ± 0.79	19.3
C_6H_8	188.5	168 ± 3	12.2
C_8H_{10}	246.9	295.9	16.6

The experimental values are taken from *webbook.nist.gov/chemistry/* and book *Thermodynamic Data of Organic Compounds* by Pedley, Naylor and Kirby.

(b) The % errors shown in the table above are calculated using the expression

$$\% \text{ error} = \frac{|\Delta_f H^{\ominus}(\text{calc}) - \Delta_f H^{\ominus}(\text{expt})|}{\Delta_f H^{\ominus}(\text{expt})}$$

(c) For all of the molecules, the computed enthalpies of formation deviate from the experimental values by much more than the uncertainty in the experimental value. It is clear that molecular modeling software is not a substitute for experimentation when it comes to quantitative measures.

I9.5 (a) The energies of the LUMOs of the given molecules are calculated using the DF/B3LYP/6-31G* method. The results along with the standard reduction potentials are listed in the table below.

molecule	$E_{\text{LUMO}}/\text{eV}$	$E^{\ominus}/\text{V}$
A	−3.54	0.078
B	−3.39	0.023
C	−3.24	−0.067
D	−3.11	−0.165
E	−3.01	−0.260

The plot of E_{LUMO} against $E^{\ominus}$ is shown in Fig. 9.10
The data points are a moderate fit to a straight line, the equation of which is

$$E_{\text{LUMO}}/\text{eV} = (-1.53) \times E^{\ominus}/\text{V} - 3.38$$

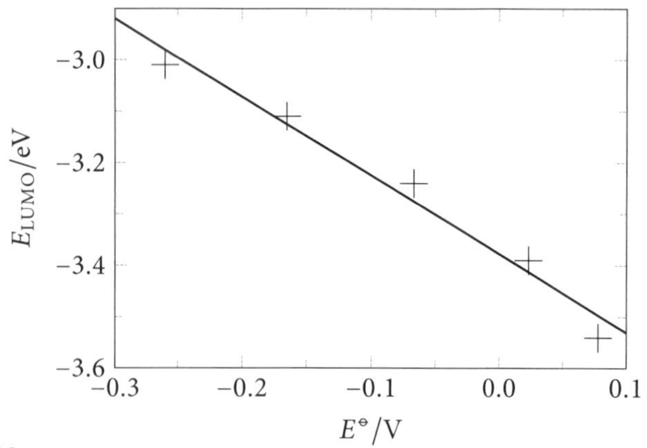

Figure 9.10

(b) The energy of the LUMO of this molecule is calculated using the same method as above as −2.99 eV. Hence the predicted reduction potential is

$$E^{\ominus}/\mathrm{V} = \frac{E_{\mathrm{LUMO}}/\mathrm{eV} + (3.38...)}{(-1.53...)} = \frac{(-2.99) + (3.38...)}{(-1.53...)} = \boxed{-0.25}$$

(c) The energy of the LUMO of the given molecule is calculated as −3.11 eV. Hence the predicted reduction potential is

$$E^{\ominus}/\mathrm{V} = \frac{E_{\mathrm{LUMO}}/\mathrm{eV} + (3.38...)}{(-1.53...)}\mathrm{V} = \frac{(-3.11) + (3.38...)}{(-1.53...)} = \boxed{-0.18}$$

Plastoquinone has less negative reduction potential, therefore it is the better oxidizing agent.

I9.7 The energy of a normalized trial wavefunction Ψ_{trial} is $E = \int \Psi^*_{\mathrm{trial}} \hat{H} \Psi_{\mathrm{trial}}\,\mathrm{d}\tau$. The hamiltonian operator for the hydrogen atom can be inferred from [8A.4–305] as

$$\hat{H} = -\frac{\hbar^2}{2\mu}\nabla^2 - \frac{e^2}{4\pi\varepsilon_0 r}$$

The Laplacian operator ∇^2 is given in Section 7F.2(a) on page 285 as

$$\nabla^2 = \frac{1}{r}\frac{\partial^2}{\partial r^2}r + \frac{1}{r^2}\Lambda^2$$

but an entirely equivalent form, which is more convenient here, is

$$\nabla^2 = \frac{\partial^2}{\partial r^2} + \frac{2}{r}\frac{\partial}{\partial r} + \frac{1}{r^2}\Lambda^2$$

The legendrian operator Λ^2 contains derivatives with respect to angles only. As the given trial wavefunction is independent of angles, $\Lambda^2\Psi = 0$ and therefore the laplacian operating on the trial wavefunction gives

$$\nabla^2\Psi_{\mathrm{trial}} = \frac{\partial^2}{\partial r^2}\left[N\mathrm{e}^{-\alpha r^2}\right] + \frac{2}{r}\frac{\partial}{\partial r}\left[N\mathrm{e}^{-\alpha r^2}\right] + 0 = 4N\alpha^2 r^2\mathrm{e}^{-\alpha r^2} - 6N\alpha\mathrm{e}^{-\alpha r^2}$$

The wavefunction Ψ_{trial} is real, therefore $\Psi^*_{\text{trial}} = \Psi_{\text{trial}}$. Therefore the hamiltonian operating on the trial wavefunction gives

$$\hat{H}\Psi_{\text{trial}} = -\frac{N\alpha\hbar^2}{\mu}\left[2\alpha r^2 e^{-\alpha r^2} - 3e^{-\alpha r^2}\right] - \frac{Ne^2}{4\pi\varepsilon_0 r}e^{-\alpha r^2}$$

Hence the energy of the trial wavefunction is given by

$$\begin{aligned} E &= \int \Psi^*_{\text{trial}}\hat{H}\Psi_{\text{trial}}\,d\tau \\ &= \int_0^\infty -\frac{N^2\alpha\hbar^2}{\mu}\left[2\alpha r^2 e^{-2\alpha r^2} - 3e^{-2\alpha r^2}\right] - \frac{N^2e^2}{4\pi\varepsilon_0 r}e^{-2\alpha r^2}\,d\tau \end{aligned}$$

The volume element $d\tau$ in polar coordinates is $r^2 \sin\theta\, d\theta\, d\phi\, dr$. There is no angular dependence in the integrand, hence integrating over all angles gives 4π. Thus the integral becomes

$$E = 4\pi N^2 \int_0^\infty -\frac{\alpha\hbar^2}{\mu}\left[2\alpha r^4 e^{-2\alpha r^2} - 3r^2 e^{-2\alpha r^2}\right] - \frac{e^2 r}{4\pi\varepsilon_0}e^{-2\alpha r^2}\,dr$$

The integral is best evaluated term by term. To evaluate the first term, Integral G.5 is used from the *Resource section* to give

$$\int_0^\infty -\frac{2\alpha^2\hbar^2}{\mu}r^4 e^{-2\alpha r^2}\,dr = -\frac{3\hbar^2}{16\mu}\left(\frac{\pi}{2\alpha}\right)^{1/2}$$

Using Integral G.3, the second term gives

$$\int_0^\infty \frac{3\alpha\hbar^2}{\mu}r^2 e^{-2\alpha r^2}\,dr = \frac{3\hbar^2}{8\mu}\left(\frac{\pi}{2\alpha}\right)^{1/2}$$

The last term is evaluated using Integral G.2

$$\int_0^\infty -\frac{e^2 r}{4\pi\varepsilon_0}e^{-2\alpha r^2}\,dr = -\frac{e^2}{16\pi\varepsilon\alpha}$$

Therefore the energy is given by

$$E = N^2\left[\frac{3\hbar^2\pi}{4\mu}\left(\frac{\pi}{2\alpha}\right)^{1/2} - \frac{e^2}{4\varepsilon_0\alpha}\right]$$

The value of N^2 is found using the normalization condition given by [7B.4c–*248*], $\int \Psi^*_{\text{trial}}\Psi_{\text{trial}}\,d\tau = 1$. Again, note that in polar coordinates the volume element $d\tau$ is given by $r^2 \sin\theta\, d\theta\, d\phi\, dr$, and because the wavefunction is spherically symmetric, integration over all angles gives 4π. Therefore the normalization condition becomes

$$4\pi N^2 \int_0^\infty r^2 e^{-2\alpha r^2}\,dr = 1$$

The integral is evaluated using Integral G.3 to give

$$4\pi N^2 \frac{1}{4}\left(\frac{\pi}{8\alpha^3}\right)^{1/2} = 1 \quad \text{and therefore} \quad N^2 = \frac{2\alpha}{\pi}\left(\frac{2\alpha}{\pi}\right)^{1/2}$$

Hence the energy corresponding to the trial wavefunction is

$$E = \frac{3\hbar^2\alpha}{2\mu} - \frac{e^2}{2^{1/2}\pi^{3/2}\varepsilon_0}\alpha^{1/2}$$

According to the variation principle the minimum energy is obtained by taking the derivative of the trial energy with respect to the adjustable parameter, which is in this case α. Setting the derivative equal to zero and then solving the equation for α gives the value of α which minimises the energy of the trial wavefunction.

$$\frac{\mathrm{d}E}{\mathrm{d}\alpha} = \frac{3\hbar^2}{2\mu} - \frac{e^2}{2^{3/2}\pi^{3/2}\varepsilon_0}\frac{1}{\alpha^{1/2}} = 0$$

Hence α is given by

$$\alpha = \frac{\mu^2 e^4}{18\hbar^4\pi^3\varepsilon_0^2}$$

Substituting this back into the energy expression yields the minimum energy for this trial wavefunction.

$$E = \frac{\mu e^4}{12\pi^3\hbar^2\varepsilon_0^2} - \frac{\mu e^4}{6\pi^3\hbar^2\varepsilon_0^2} = -\frac{\mu e^4}{12\pi^3\hbar^2\varepsilon_0^2}$$

10 Molecular symmetry

10A Shape and symmetry

Answers to discussion questions

D10A.1 This is described in Section 10A.2 on page 390.

D10A.3 This is described in Section 10A.3(a) on page 394.

Solutions to exercises

E10A.1(a) From the table in Section 10A.2(b) on page 392 a molecule belonging to the point group C_{3v}, such as chloromethane, possesses

- the $\boxed{\text{identity } E}$
- a $\boxed{C_3 \text{ axis}}$ passing through the chlorine and carbon atoms
- three $\boxed{\text{vertical mirror planes } \sigma_v}$, each containing the chlorine, carbon and one of the hydrogen atoms

The C_3 axis and one of the σ_v mirror planes is shown in Fig. 10.1

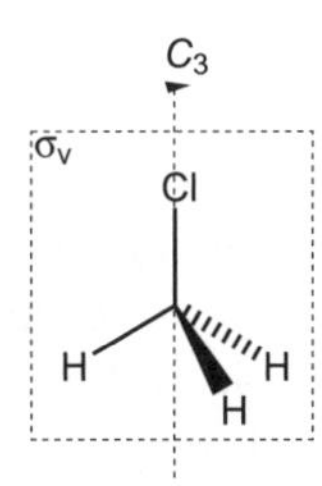

Figure 10.1

E10A.2(a) The point group of the naphthalene molecule is identified by following the flow diagram in Fig. 10A.7 on page 391. Firstly, naphthalene is not linear, which leads to the question "*Two or more* C_n, $n > 2$?" Naphthalene has three two-fold axes of rotation but no higher order axes, so the answer to this question is 'No'. However the fact that it does have three C_2 axes means that the answer to the next question "C_n?" is 'Yes'.

This leads to the question "*Select C_n with the highest n; then, are there nC_2 perpendicular to C_n?*" As already identified there are three mutually perpendicular C_2 axes and therefore no one axis with highest n, that is, no principal axis. As explained in Section 10A.1 on page 388 in the case of a planar molecule such as naphthalene with more than one axis competing for the title of principal axis, it is common to choose the one perpendicular to the plane of the molecule. However, because all three axes are mutually perpendicular, whichever of the three is selected it still has two other C_2 axes perpendicular to it, so the answer to this question is necessarily 'Yes'.

This leads to the question "σ_h?" to which the answer is 'Yes' because, whichever C_2 axis is selected, there is a mirror plane perpendicular to it. In the case of the C_2 axis perpendicular to the plane of the molecule, the σ_h mirror plane lies in the plane of the molecule. Answering 'Yes' to this question leads to the result D_{nh}, and because the C_n axis with highest n in this molecule is C_2, it follows that the point group is $\boxed{D_{2h}}$.

Alternatively, the point group may be identified from the table in Fig. 10A.8 on page 391 by drawing an analogy between the planar naphthalene molecule and the rectangle which belongs to D_{2h}.

From the table in Section 10A.2(c) on page 393 a molecule belonging to the point group D_{2h} possesses

- the $\boxed{\text{identity } E}$
- a $\boxed{C_2 \text{ axis}}$, which in the case of a planar molecule such as naphthalene is commonly taken to be the axis perpendicular to the plane of the molecule. This axis passes through the mid-point of the central bond joining the two rings.
- two $\boxed{C_2' \text{ axes}}$, perpendicular to the C_2 axis and lying in the plane of the molecule. One of these passes along the line of the central bond joining the two rings, while the other bisects this bond.
- a $\boxed{\text{horizontal mirror plane } \sigma_h}$ in the plane of the molecule

In addition

- The presence of the C_2 axis and the two C_2' axes jointly imply the presence of two $\boxed{\text{vertical mirror planes } \sigma_v}$, both containing the principal axis. One of these σ_v planes also contains the central bond joining the two rings, while the other bisects this bond.
- The C_2 and σ_h elements jointly imply a $\boxed{\text{centre of inversion } i}$, which lies at the midpoint of the central bond.

These symmetry elements are shown in Fig. 10.2.

E10A.3(a) The objects to be assigned are shown in Fig. 10.3. For clarity not all symmetry elements are shown.

(i) As explained in Section 10A.2(f) on page 394 a sphere possesses an infinite number of rotation axes with all possible values of n, and belongs to the full rotation group $\boxed{R_3}$.

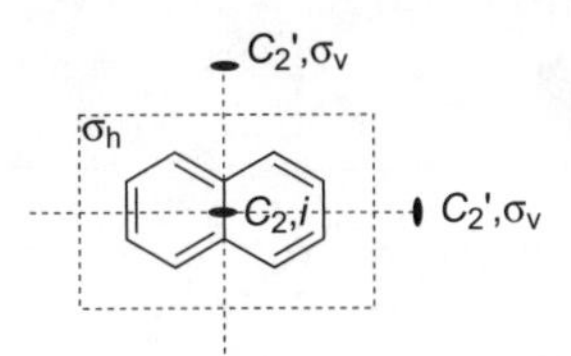

Figure 10.2

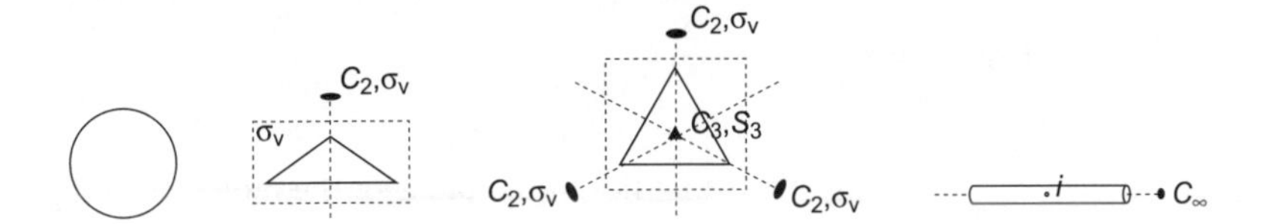

Figure 10.3

(ii) The point group of an isosceles triangle is identified using the flow diagram in Fig. 10A.7 on page 391. It has a C_2 axis lying in the plane of the paper on which the shape is drawn which bisects the non-equal side and passes through the vertex opposite this side. There are no other C_n axes, and there is not a σ_h mirror plane perpendicular to the C_2 axis. There are, however, two σ_v mirror planes which contain the C_2 axis, one in the plane of the paper and the other perpendicular to it. These considerations establish the point group as $\boxed{C_{2v}}$.

(iii) The point group of an equilateral triangle is readily identified as $\boxed{D_{3h}}$ by reference to the table of shapes in Fig. 10A.8 on page 391.

(iv) Modelling the unsharpened cylindrical pencil as a cylinder (Fig. 10.3) and assuming no lettering or other pattern on it, its point group is determined using the flow diagram in Fig. 10A.7 on page 391. It is linear, in the sense that rotation by any angle around the long axis of the pencil is a symmetry operation so that the pencil possesses a C_∞ axis. Because the pencil has not been sharpened, both ends are the same and this means that the pencil possesses a centre of inversion i. Using the flow diagram this establishes the point group as $\boxed{D_{\infty h}}$.

E10A.4(a) The molecules to be assigned are shown in Fig. 10.4. For clarity not all symmetry elements are shown.

In each case the point group is identified using the flow diagram in Fig. 10A.7 on page 391 having identified the symmetry elements, or by drawing an analogy with one of the shapes in the summary table in Fig. 10A.8 on page 391.

(i) Nitrogen dioxide, NO_2, is a V-shaped molecule shape with a bond angle of approximately 134°. It possesses

- the $\boxed{\text{identity, } E}$

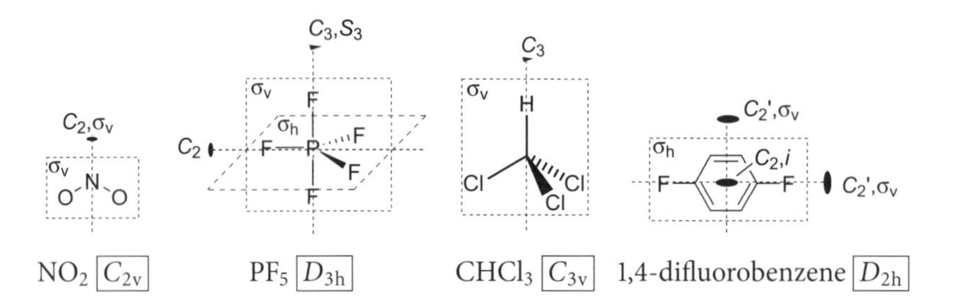

Figure 10.4

- a C_2 axis that lies in the plane of the molecule and passes through the nitrogen atom.
- two different vertical mirror planes σ_v, both containing the C_2 axis but with one lying in the plane of the molecule and the other perpendicular to it.

The point group is C_{2v}.

(ii) PF_5 has a trigonal bipyramidal shape with two axial and three equatorial fluorine atoms. It possesses

- the identity, E
- a C_3 axis passing through the phosphorus and the two axial chlorine atoms
- a horizontal mirror plane σ_h perpendicular to the C_3 axis and containing the phosphorus and the three equatorial fluorine atoms
- three C_2' axes perpendicular to the C_3 axis, each passing through the phosphorus and one of the equatorial fluorine atoms. Only one of the C_2' axes is shown in Fig. 10.4.
- a S_3 axis coincident with the C_3 axis
- three vertical mirror planes σ_v, each containing the phosphorus, the two axial fluorine atoms, and one of the three equatorial fluorine atoms. Only one of the σ_v mirror planes is shown in Fig. 10.4.

The point group is D_{3h}

(iii) $CHCl_3$, chloroform, possesses

- the identity E
- a C_3 axis passing through the hydrogen and the carbon atom
- three vertical mirror planes σ_v, each containing the hydrogen, carbon, and one of the chlorine atoms; only one of these is shown in Fig. 10.4.

The point group is C_{3v}

(iv) 1,4-difluorobenzene possesses

- the identity, E

- a $\boxed{C_2 \text{ axis}}$ perpendicular to the plane of the molecule and passing through the centre of the benzene ring
- two different $\boxed{C_2'}$ axes, both perpendicular to the C_2 axis and lying in the plane of the molecule. One C_2' axis passes through the two fluorine atoms, while the other perpendicular to this.
- a $\boxed{\text{horizontal mirror plane } \sigma_h}$ lying in the plane of the molecule
- two different $\boxed{\text{vertical mirror planes } \sigma_v}$, both containing the C_2 axis and perpendicular to the plane of the molecule. One contains the two fluorine atoms while the other is perpendicular to the line joining the two fluorine atoms.
- a $\boxed{\text{centre of inversion } i}$, which lies at the centre of the benzene ring

The point group is $\boxed{D_{2h}}$

E10A.5(a) The *cis* and *trans* isomers are shown in Fig. 10.5.

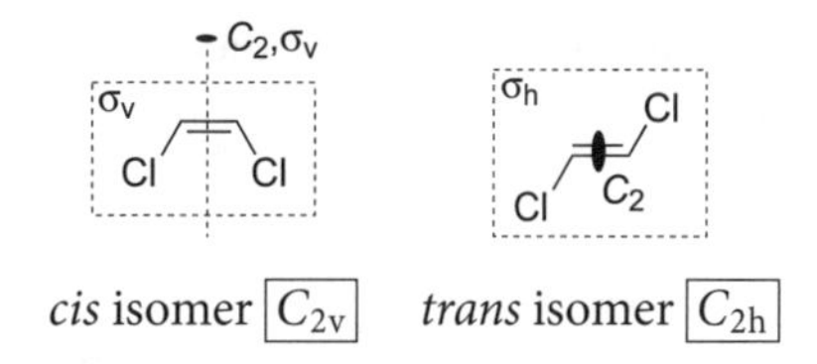

cis isomer $\boxed{C_{2v}}$ *trans* isomer $\boxed{C_{2h}}$

Figure 10.5

(i) *cis*-dichloroethene possesses a C_2 axis lying in the plane of the molecule and bisecting the C=C double bond. It also has two σ_v mirror planes containing this axis, one lying in the plane of the molecule and the other perpendicular to it. The flow diagram in Fig. 10A.7 on page 391 is then used to establish that the point group is $\boxed{C_{2v}}$.

(ii) *trans*-dichloroethene possesses a C_2 axis that is perpendicular to the plane of the molecule and passes through the centre of the C=C double bond. It also has a horizontal mirror plane σ_h perpendicular to this axis and lying in the plane of the molecule, and also has a centre of inversion i. The flow diagram in Fig. 10A.7 on page 391 is then used to establish that the point group is $\boxed{C_{2h}}$.

E10A.6(a) The molecules are shown in Fig. 10.6 along with their point groups. For clarity not all symmetry elements are shown.

As explained in Section 10A.3(a) on page 394, only molecules belonging to the groups C_n, C_{nv}, and C_s may be polar.

(i) Pyridine has point group C_{2v}, so it $\boxed{\text{may be polar}}$. The dipole must lie along the C_2 axis, which passes through the nitrogen and the carbon atom opposite it in the ring.

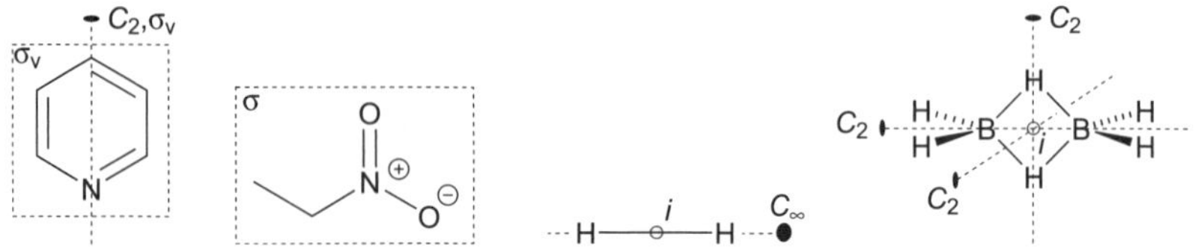

C_2,σ_v σ_v σ C_2 C_∞ i

Pyridine, C_{2v} Nitroethane, C_s BeH_2, $D_{\infty h}$ (Be not shown for clarity) B_2H_6, D_{2h}

Figure 10.6

(ii) Nitroethane, $CH_3CH_2NO_2$, belongs to point group C_s, so it may be polar.

(iii) Linear BeH_2 belongs to point group $D_{\infty h}$, so it may not be polar.

(iv) Diborane, B_2H_6, belongs to point group D_{2h}, so it may not be polar.

E10A.7(a) There are 10 distinct isomers of dichloronaphthalene, shown in Fig. 10.7 together with their point groups. All isomers have a mirror plane in the plane of the paper; additional symmetry elements are marked.

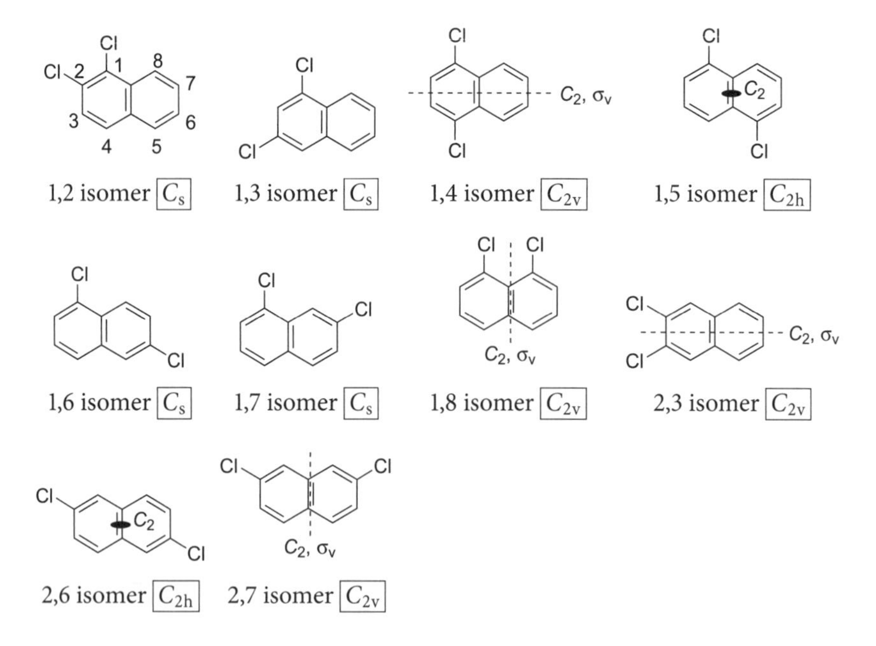

Figure 10.7

E10A.8(a) As explained in Section 10A.3(b) on page 395, to be chiral a molecule must not possess an axis of improper rotation, S_n. This includes mirror planes, which are the same as S_1, and centres of inversion, which are the same as S_2. The table in Section 10A.2(c) on page 393 shows that a molecule belonging to D_{2h} possesses three mirror planes, so such a molecule $\boxed{\text{may not be chiral}}$. Similarly the table in Section 10A.2(b) on page 392 shows that a molecule belonging to C_{3h} possesses a σ_h mirror plane, so such a molecule $\boxed{\text{may not be chiral}}$.

Solutions to problems

P10A.1 The molecules concerned are shown in Fig. 10.8. For clarity not every symmetry element is shown in each diagram.

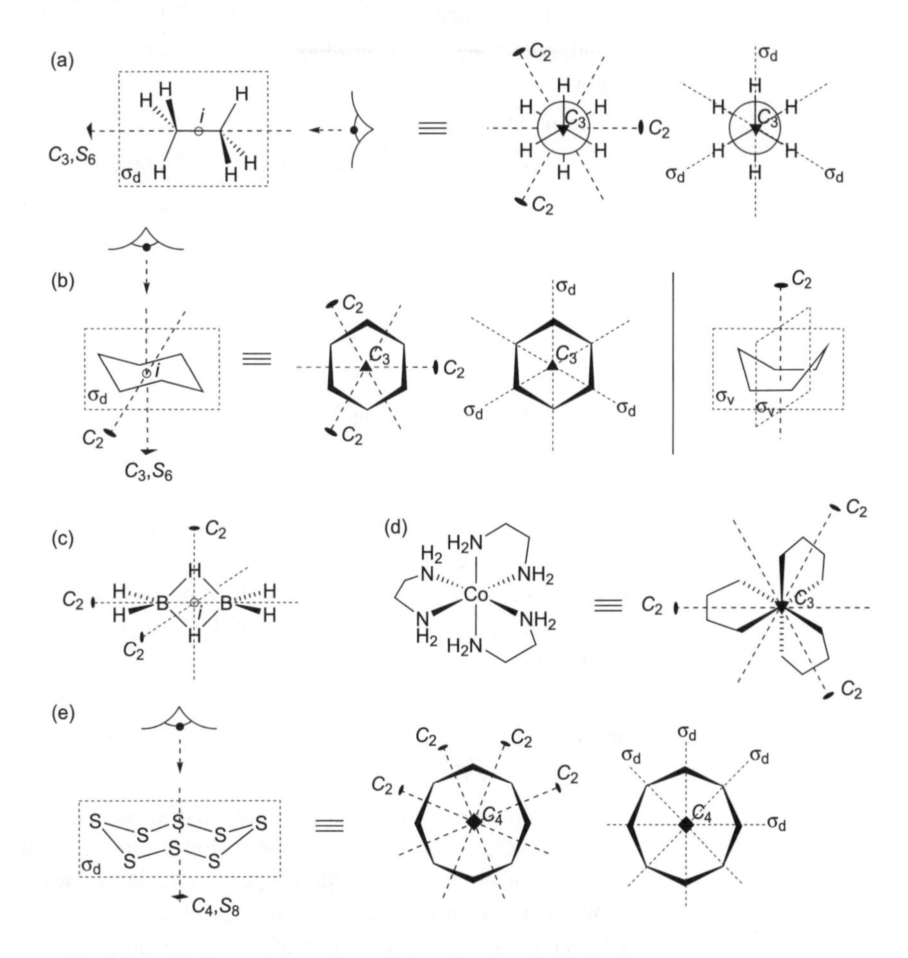

Figure 10.8

(a) The staggered conformation of ethane possesses the following symmetry elements

- The $\boxed{\text{identity } E}$

- A C_3 axis along the the C–C bond
- Three C_2 axes perpendicular to the C_3 axis. These are best seen in the Newman projection also shown in Fig. 10.8(a), that is, the view along the C_3 axis
- Three dihedral mirror planes σ_d, each containing the C_3 principal axis and two hydrogen atoms. These mirror planes are 'dihedral' rather than 'vertical' because they bisect the angles between the C_2 axes. The three σ_d mirror planes are seen most easily in the Newman projection; for clarity only one σ_d plane is shown
- An S_6 axis coincident with the C_3 axis
- A centre of inversion i at the midpoint of the C–C bond.

Using the flow diagram in Fig. 10A.7 on page 391 the point group is D_{3d}.

(b) The chair conformation of cyclohexane possesses the following symmetry elements

- The identity E
- A C_3 axis perpendicular to the approximate plane of the molecule
- Three C_2 axis perpendicular to the C_3 axis, each passing through the midpoint of two opposite C–C bonds
- Three dihedral mirror planes σ_d, each containing the C_3 axis as well as two opposite carbon atoms
- An S_6 axis coincident with the C_3 axis
- A centre of inversion i at the intersection of the C_3 and C_2 axes.

Using the flow diagram in Fig. 10A.7 on page 391 the point group is D_{3d}.
The boat conformation of cyclohexane possesses

- The identity E
- A C_2 axis passing vertically through the centre of the boat where the mast would be
- Two vertical mirror planes σ_v, both containing the C_2 axis but one also containing the bow and stern of the boat and the other perpendicular to this

Using the flow diagram in Fig. 10A.7 on page 391 the point group is C_{2v}.

(c) Diborane, B_2H_6, possesses

- The identity E
- Three different C_2 axes, all perpendicular to each other. One passes through the two bridging hydrogen atoms, one passes through the two boron atoms, and the third is perpendicular to the plane defined by the boron and bridging hydrogen atoms.
- Three different mirror planes σ, one perpendicular to each of the three C_2 axes. One mirror plane contains the boron atoms and the four terminal hydrogen atoms. The second contains the two bridging hydrogen atoms and forms the perpendicular bisector of a line joining the two boron atoms. The third contains the boron atoms and the two bridging hydrogen atoms. For clarity these mirror planes are not shown in Fig. 10.8(c).

- A centre of inversion i at the intersection of the C_2 axes.

Using the flow diagram in Fig. 10A.7 on page 391 the point group is D_{2h}.

(d) Ignoring the detailed structure of the en ligands, $[Co(en)_3]^{3+}$ can be drawn in a way that resembles a propeller. It possesses the following symmetry elements

- The identity E
- A C_3 axis corresponding to the axis of the propeller
- Three C_2 axes, each passing through the cobalt and the centre of one of the en ligands

Using the flow diagram in Fig. 10A.7 on page 391 the point group is D_3.

(e) Crown-shaped S_8 possesses

- The identity E
- A C_4 axis perpendicular to the approximate plane of the molecule
- Four C_2 axes perpendicular to the C_4 axis, analogous to the C_2 axes in the chair conformation of cyclohexane
- Four dihedral mirror planes σ_d, again analogous to the σ_d planes in the chair conformation of cyclohexane
- An S_8 axis coincident with the C_4 axis

Using the flow diagram in Fig. 10A.7 on page 391 the point group is D_{4d}.

(i) As explained in Section 10A.3(a) on page 394, only molecules belonging to the groups C_n, C_{nv}, or C_s may have a permanent electric dipole moment. Therefore of the molecules considered only boat cyclohexane (C_{2v}) can be polar.

(ii) As explained in Section 10A.3(b) on page 395, a molecule may be chiral only if it does not possess an axis of improper rotation S_n. This includes mirror planes, which are equivalent to S_1, and a centre of inversion, which is equivalent to S_2. Therefore of the molecules considered only $[Co(en)_3]^{3+}$ is chiral.

P10A.3 The molecules are shown in Fig. 10.9 along with some of their key symmetry elements.

(a) In addition to the identity element, ethene possesses three C_2 axes, three mirror planes, and a centre of inversion i. As explained in Section 10A.1 on page 388, in the case of a planar molecule with several C_n axes competing for the title of principal axis it is common to choose the axis perpendicular to plane of the molecule to be the principal axis. This axis is labelled C_2 in Fig. 10.9, while the other C_2 axes are labelled C_2'. The mirror plane that lies in the plane of the molecule is denoted σ_h while the other two are denoted σ_v. Using the flow diagram in Fig. 10A.7 on page 391 the point group is established as D_{2h}.

In the case of allene there is a C_2 axis, coincident with an S_4 axis, that passes through all three carbon atoms. There are also two other C_2 axes,

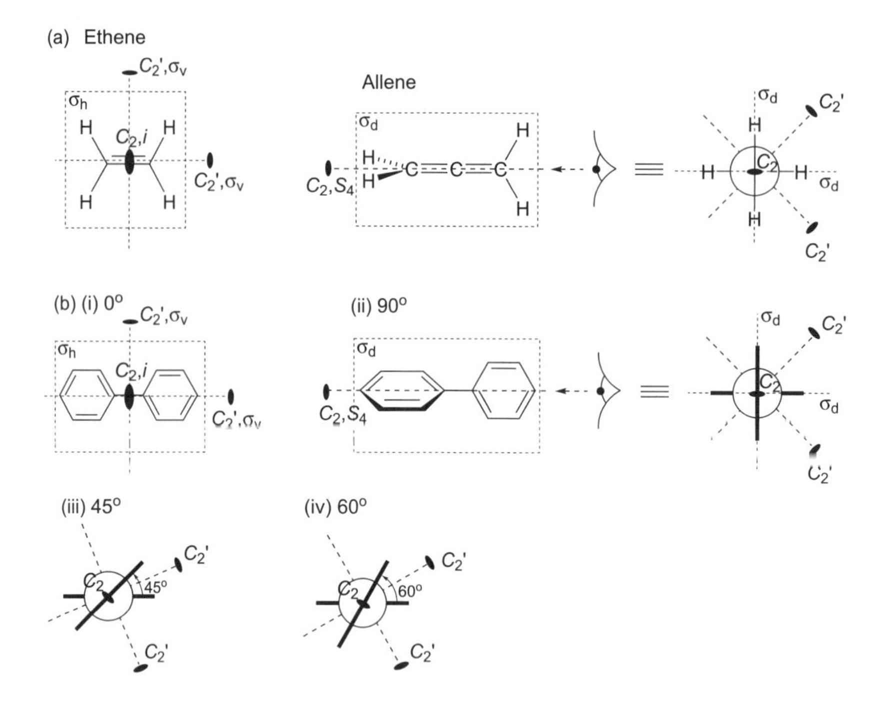

Figure 10.9

denoted C_2', that are perpendicular to C_2 and which pass through the central carbon; these are most clearly seen in the Newman projection shown in Fig. 10.9(a). Finally there are two dihedral mirror planes σ_d which bisect the angle between the two C_2' axes and which each pass through all three carbon atoms and two of the hydrogen atoms. Using the flow diagram in Fig. 10A.7 on page 391 these symmetry elements establish the point group as $\boxed{D_{2d}}$.

(b) (i) The biphenyl molecule with a dihedral angle of 0°, that is, with both phenyl groups in the same plane, has a shape analogous to that of the ethene molecule from part (a). Like ethene it belongs to the point group $\boxed{D_{2h}}$ and possesses the symmetry elements E, C_2, $2C_2'$, $2\sigma_v$, σ_h and i.

(ii) The biphenyl molecule with a dihedral angle of 90°, that is, with the two rings perpendicular, has a shape analogous to that of allene. Like allene it belongs to the point group $\boxed{D_{2d}}$ and possesses the symmetry elements E, C_2, $2C_2'$, $2\sigma_d$, and S_4.

(iii) If the dihedral angle is changed from 90° to 45° the C_2 and C_2' axes are retained but the σ_d mirror planes and the S_4 axis are no longer present. Using the flow diagram in Fig. 10A.7 on page 391 the symmetry elements establish the point group as $\boxed{D_2}$.

(iv) The biphenyl molecule with a dihedral angle of 60° possesses the same set of symmetry elements as when the dihedral angle is 45°

and therefore belongs to the same point group of $\boxed{D_2}$.

P10A.5 The ion is shown in Fig. 10.10. In each diagram only the mirror plane referred to in the question is shown; any other mirror planes are omitted for clarity.

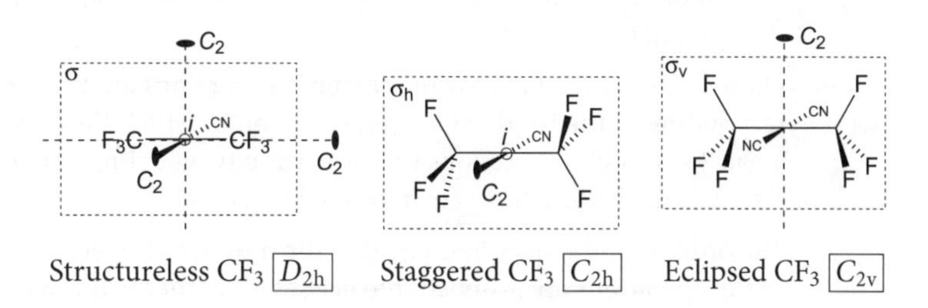

Figure 10.10

(a) If the CF_3 groups are treated as structureless ligands then in addition to the identity E the ion possesses three C_2 axes, a centre of inversion, and three mirror planes, only one of which is shown in Fig. 10.10. The point group is $\boxed{D_{2d}}$.

(b) Fig. 10.10 shows the staggered and eclipsed conformations of the CF_3 groups. In the staggered arrangement the centre of inversion i is retained, as is the C_2 axis that lies along the CN–Ag–CN bond and the mirror plane perpendicular to this, but the other C_2 axes and mirror planes are lost. The point group is $\boxed{C_{2h}}$. In the eclipsed arrangement, the C_2 axis perpendicular to the plane of the Ag and the four ligands is retained, as are the two mirror planes containing this axis, only one of which is shown in Fig. 10.10. The other C_2 axes, the other mirror plane, and the centre of inversion are lost. The point group is $\boxed{C_{2v}}$.

10B Group theory

Answer to discussion questions

D10B.1 Within the context of quantum theory and molecular symmetry a group is a collection of transformations $R, S, T \ldots$ that satisfy the criteria: (1) One of the transformations is the identity (E). (2) For every transformation R, the inverse transformation R^{-1} is included in the collection so that the combination RR^{-1} is equivalent to the identity: $RR^{-1} = E$. (3) The product RS is equivalent to a single member of the collection of transformations. (4) Multiple transformations obey the associative rule: $R(ST) = (RS)T$.

D10B.3 The top row of the table lists the operations of the group; operations in the same class are grouped together. Subsequent rows list the characters of each of the irreducible representations (symmetry species), the symbols for which are shown if the far left column. Operations in the same class have the same character, which is why they can conveniently be grouped together.

One-dimensional irreducible representations are labelled A or B, and have character 1 under the operation E. Two- and three-dimensional irreducible representations are labelled E and T, respectively, and have characters 2 and 3 under the operation E.

On the right of the table various cartesian functions are listed on the row corresponding to the irreducible representation as which they transform. Where a pair (or more) of functions transform as a two (or higher) dimensional representation, the functions are bracketed together.

The order of the group, h, is equal to the number of operations. Alternatively, as the operations are grouped into classes, the order is the sum of the number of operations belonging to each class.

The number of irreducible representations equals the number of classes. Also, apart from the groups C_n with $n > 2$, the sum of the squares of the dimensions of each irreducible representation is equal to the order.

D10B.5 The letters and subscripts of a symmetry species provide information about the symmetry behaviour of the species. An A or a B is used to denote a one-dimensional representation; A is used if the character under the principal rotation is +1 (symmetric behaviour), and B is used if the character is −1 (antisymmetric behaviour). Subscripts are used to distinguish the irreducible representations if there is more than one of the same type: A_1 is reserved for the representation with the character 1 under all operations.

E denotes a two-dimensional irreducible representation, and T a three dimensional irreducible representation. For groups with an inversion centre, a subscript g (gerade) indicates symmetric behaviour under the inversion operation; a subscript u (ungerade) indicates antisymmetric behaviour. If a horizontal mirror plane is present, ′ or ″ superscripts are added if the behaviour is symmetric or antisymmetric, respectively, under the σ_h operation.

Solutions to exercises

E10B.1(a) BF_3 belongs to the point group D_{3h}. The σ_h operation corresponds to reflection in the plane of the molecule.

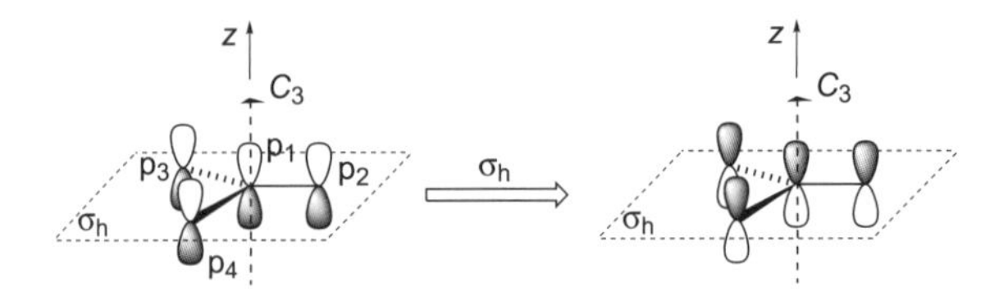

Figure 10.11

As shown in Fig. 10.11 the effect of this operation is to change the sign of all four p orbitals. This effect is written as $(-p_1\ -p_2\ -p_3\ -p_4) \leftarrow (p_1\ p_2\ p_3\ p_4)$ which is expressed using matrix multiplication as

$$(-p_1\ -p_2\ -p_3\ -p_4) = (p_1\ p_2\ p_3\ p_4)\overbrace{\begin{pmatrix} -1 & 0 & 0 & 0 \\ 0 & -1 & 0 & 0 \\ 0 & 0 & -1 & 0 \\ 0 & 0 & 0 & -1 \end{pmatrix}}^{D(\sigma_h)} = (p_1\ p_2\ p_3\ p_4)D(\sigma_h)$$

The representative of this operation is therefore

$$\boxed{D(\sigma_h) = \begin{pmatrix} -1 & 0 & 0 & 0 \\ 0 & -1 & 0 & 0 \\ 0 & 0 & -1 & 0 \\ 0 & 0 & 0 & -1 \end{pmatrix}}$$

E10B.2(a) BF_3 belongs to the point group D_{3h}. The σ_h operation corresponds to a reflection in the plane of the molecule, while the C_3 operation corresponds to rotation by 120° around the C_3 axis which passes through the boron atom and is perpendicular to the plane of the molecule (Fig. 10.12).

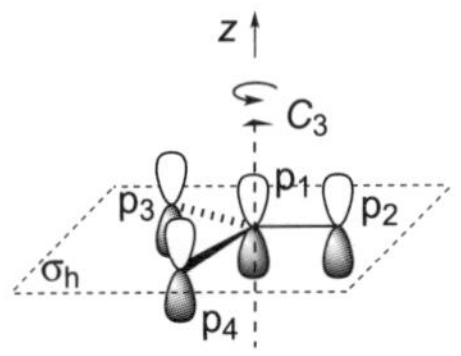

Figure 10.12

Using the orbital numbering shown in Fig. 10.12, the matrix representatives for the σ_h and C_3 operations were found in *Exercise* E10B.1(a) and *Exercise* E10B.1(b) to be

$$D(\sigma_h) = \begin{pmatrix} -1 & 0 & 0 & 0 \\ 0 & -1 & 0 & 0 \\ 0 & 0 & -1 & 0 \\ 0 & 0 & 0 & -1 \end{pmatrix} \quad \text{and} \quad D(C_3) = \begin{pmatrix} 1 & 0 & 0 & 0 \\ 0 & 0 & 0 & 1 \\ 0 & 1 & 0 & 0 \\ 0 & 0 & 1 & 0 \end{pmatrix}$$

The matrix representative of the operation $\sigma_h C_3$ is found by multiplying the matrix representatives of σ_h and C_3. Basic information about how to handle matrices is given in *The chemist's toolkit* 24 in Topic 9E on page 373.

$$D(\sigma_h)D(C_3) = \begin{pmatrix} -1 & 0 & 0 & 0 \\ 0 & -1 & 0 & 0 \\ 0 & 0 & -1 & 0 \\ 0 & 0 & 0 & -1 \end{pmatrix}\begin{pmatrix} 1 & 0 & 0 & 0 \\ 0 & 0 & 0 & 1 \\ 0 & 1 & 0 & 0 \\ 0 & 0 & 1 & 0 \end{pmatrix} = \boxed{\begin{pmatrix} -1 & 0 & 0 & 0 \\ 0 & 0 & 0 & -1 \\ 0 & -1 & 0 & 0 \\ 0 & 0 & -1 & 0 \end{pmatrix}}$$

The operation corresponding to this representative is found by considering its effect on the starting basis

$$(\, p_1\ p_2\ p_3\ p_4\,)\begin{pmatrix} -1 & 0 & 0 & 0 \\ 0 & 0 & 0 & -1 \\ 0 & -1 & 0 & 0 \\ 0 & 0 & -1 & 0 \end{pmatrix} = (\, -p_1\ -p_3\ -p_4\ -p_2\,)$$

The operation $\sigma_h C_3$ therefore changes the sign of p_1, and converts p_2, p_3 and p_4 into $-p_3$, $-p_4$, and $-p_2$ respectively. This is precisely the same outcome as achieved by the S_3 operation, that is, a 120° rotation around the C_3 axis followed by a reflection in the σ_h plane. Thus, $\boldsymbol{D}(\sigma_h)\boldsymbol{D}(C_3) = \boldsymbol{D}(S_3)$, as expected from the fact that by definition the S_3 operation corresponds to a C_3 rotation followed by a reflection in the plane perpendicular to the C_3 axis.

E10B.3(a) Fig. 10.13 shows BF_3, an example of a molecule with D_{3h} symmetry. The three C_2 axes are labelled C_2, C_2' and C_2'', and likewise for the σ_v mirror planes.

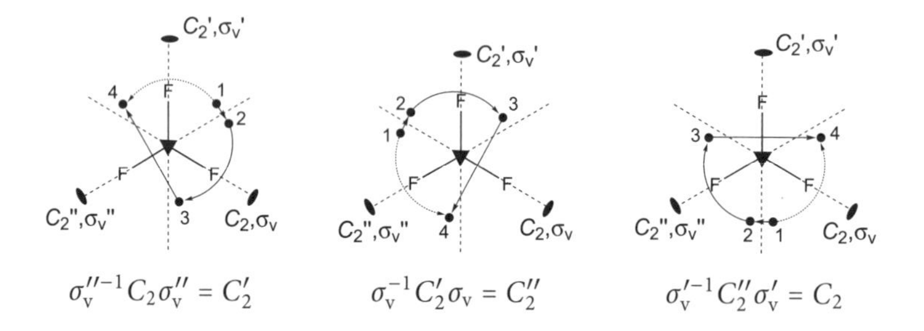

Figure 10.13

The criteria for two operations R and R' to be in the same class is given by [10B.1–398], $R' = S^{-1}RS$ where S is another operation in the group. The task is therefore to find an operation S in D_{3h} such that this equation is satisfied when R and R' are two of the C_2 axes.

Referring to the first diagram in Fig. 10.13, to show that C_2 and C_2' are in the same class consider the operation $\sigma_v''^{-1}C_2\sigma_v''$. Start at the arbitrary point 1, and recall that the operations are applied starting from the right. The operation σ_v'' moves the point to 2, and then C_2 moves the point to 3. The inverse of a reflection is itself, $\sigma_v''^{-1} = \sigma_v''$, so the effect of $\sigma_v''^{-1}$ is to move the point to 4. From the diagram it can be seen that 4 can be reached by applying C_2' to point 1, thus demonstrating that $\sigma_v''^{-1}C_2\sigma_v'' = C_2'$ and hence that C_2 and C_2' belong to the same class.

In a similar way the second diagram in Fig. 10.13 shows that $\sigma_v^{-1}C_2'\sigma_v = C_2''$ and hence that C_2' and C_2'' belong to the same class, while the third diagram shows that C_2'' and C_2 belong to the same class.

E10B.4(a) The orthonormality of irreducible representations is defined by [10B.7–402],

$$\frac{1}{h}\sum_C N(C)\chi^{\Gamma^{(i)}}(C)\chi^{\Gamma^{(j)}}(C) = \begin{cases} 0 \text{ for } i \neq j \\ 1 \text{ for } i = j \end{cases}$$

where the sum is over all classes of the group, $N(C)$ is the number of operations in class C, and h is the number of operations in the group (its order). $\chi^{\Gamma^{(i)}}(C)$ is the character of class C in irreducible representation $\Gamma^{(i)}$, and similarly for $\chi^{\Gamma^{(j)}}(C)$.

The character table for the point group C_{2h} is available in the *Online resource centre*. The operations of the group are $\{E, C_2, i, \sigma_h\}$, so $h = 4$. There are four classes and in this group each class has just one member, so all $N(C) = 1$. The irreducible representations have the following characters

$$\begin{array}{llll} A_g & \{1,1,1,1\} & B_g & \{1,-1,1,-1\} \\ A_u & \{1,1,-1,-1\} & B_u & \{1,-1,-1,1\} \end{array}$$

To prove orthogonality, [10B.7–402] is used with each pair of irreducible representations; the result is zero in each case which shows that each pair are orthogonal

$$\begin{array}{ll}
A_g \text{ and } B_g & \tfrac{1}{4}\{1\times1\times1 + 1\times1\times(-1) + 1\times1\times1 + 1\times1\times(-1)\} = 0 \\
A_g \text{ and } A_u & \tfrac{1}{4}\{1\times1\times1 + 1\times1\times1 + 1\times1\times(-1) + 1\times1\times(-1)\} = 0 \\
A_g \text{ and } B_u & \tfrac{1}{4}\{1\times1\times1 + 1\times1\times(-1) + 1\times1\times(-1) + 1\times1\times1\} = 0 \\
B_g \text{ and } A_u & \tfrac{1}{4}\{1\times1\times1 + 1\times(-1)\times1 + 1\times1\times(-1) + 1\times(-1)\times(-1)\} = 0 \\
B_g \text{ and } B_u & \tfrac{1}{4}\{1\times1\times1 + 1\times(-1)\times(-1) + 1\times1\times(-1) + 1\times(-1)\times1\} = 0 \\
A_u \text{ and } B_u & \tfrac{1}{4}\{1\times1\times1 + 1\times1\times(-1) + 1\times(-1)\times(-1) + 1\times(-1)\times1\} = 0
\end{array}$$

To prove that each irreducible representation is normalised, [10B.7–402] is used with $i = j$ for each irreducible representation $\Gamma^{(i)}$; the result is 1 in each case.

$$\begin{array}{ll}
A_g \text{ and } A_g & \tfrac{1}{4}\{1\times1\times1 + 1\times1\times1 + 1\times1\times1 + 1\times1\times1\} = 1 \\
B_g \text{ and } B_g & \tfrac{1}{4}\{1\times1\times1 + 1\times(-1)\times(-1) + 1\times1\times1 + 1\times(-1)\times(-1)\} = 1 \\
A_u \text{ and } A_u & \tfrac{1}{4}\{1\times1\times1 + 1\times1\times1 + 1\times(-1)\times(-1) + 1\times(-1)\times(-1)\} = 1 \\
B_u \text{ and } B_u & \tfrac{1}{4}\{1\times1\times1 + 1\times(-1)\times(-1) + 1\times(-1)\times(-1) + 1\times1\times1\} = 1
\end{array}$$

E10B.5(a) The D_{3h} character table is given in the *Resource section*. As explained in Section 10B.3(a) on page 402, the symmetry species of s, p and d orbitals on a central atom are indicated at the right hand side of the character table. The position of z in the D_{3h} character table shows that p_z, which is proportional to $zf(r)$, has symmetry species $\boxed{A_2''}$. Similarly the positions of x and y in the character table shows that p_x and p_y, which are proportional to $xf(r)$ and $yf(r)$ respectively, jointly span the irreducible representation of symmetry species $\boxed{E'}$.

In the same way the positions of z^2, $x^2 - y^2$, xy, yz and zx in the character table show that d_{z^2} has symmetry species $\boxed{A_1'}$, $d_{x^2-y^2}$ and d_{xy} jointly span $\boxed{E'}$, and d_{yz} and d_{zx} jointly span $\boxed{E'}$.

E10B.6(a) As explained in Section 10B.3(c) on page 404, the highest dimensionality of irreducible representations in a group is equal to the maximum degree of degeneracy in the group. The highest dimensionality is found by noting the maximum value of $\chi(E)$ in the character table. An octahedral hole in a crystal has O_h symmetry, and from the O_h character table in the *Resource section* the maximum value of $\chi(E)$ is three, corresponding to the T irreducible representations. Therefore the maximum degeneracy of a particle in an octahedral hole is $\boxed{\text{three}}$.

E10B.7(a) As explained in Section 10B.3(c) on page 404, the highest dimensionality of any irreducible representation in a group is equal to the maximum degree of degeneracy in the group. The highest dimensionality is found by noting the maximum value of $\chi(E)$ in the character table. Benzene has D_{6h} symmetry, the character table for which is available in the *Online resource centre*. The maximum value of $\chi(E)$ is two, corresponding to the E irreducible representations. Therefore the maximum degeneracy of the orbitals in benzene is $\boxed{\text{two}}$.

Solutions to problems

P10B.1 Fig. 10.14 shows *trans*-difluoroethene, an example of a molecule with C_{2h} symmetry.

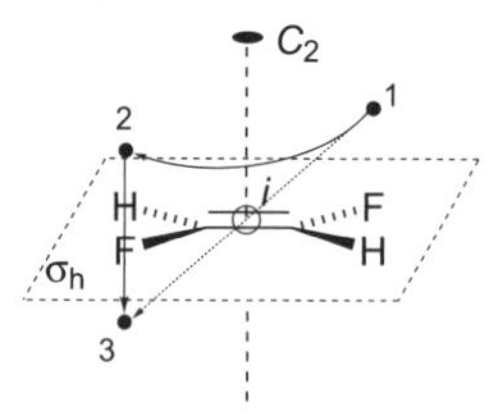

Figure 10.14

The group multiplication table is constructed by considering the effect of successive transformations. For example, Fig. 10.14 shows the effect on an arbitrary point of a C_2 rotation followed by a σ_h reflection. The C_2 operation moves the point from 1 to 2, and the σ_h reflection moves the point to 3. The overall effect, 1 → 3, is equivalent to carrying out the i operation. Thus, $\sigma_h C_2 = i$. Considering all other pairs of operations in the same way gives the following table.

$R \downarrow R' \rightarrow$	E	C_2	σ_h	i
E	E	C_2	σ_h	i
C_2	C_2	E	i	σ_h
σ_h	σ_h	i	E	C_2
i	i	σ_h	C_2	E

P10B.3 Fig. 10.15 shows that a water molecule together with the specified axis system and the six basis orbitals, which are labelled s_A, s_B, s_O, p_x, p_y, and p_z. The σ_v plane is the xz plane, and the σ_v' plane is the yz plane.

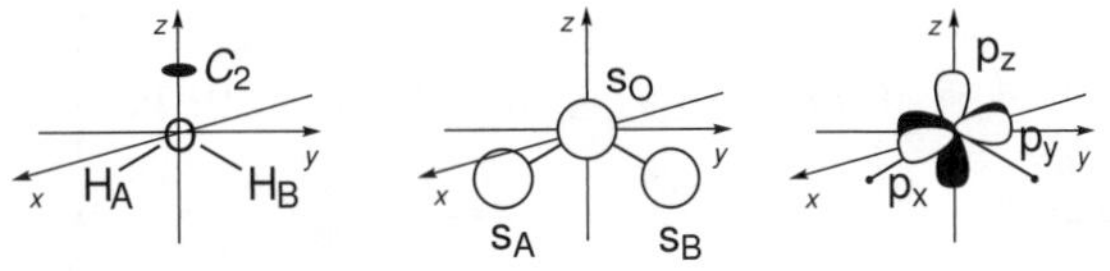

Figure 10.15

The matrix representatives are obtained using the approach described in Section 10B.2(a) on page 398. Beginning with C_2, this operation exchanges s_A and s_B, leaves s_O and p_z unchanged, and changes the sign of p_x and of p_y. Its effect is written $(s_B\ s_A\ s_O\ -p_x\ -p_y\ p_z) \leftarrow (s_A\ s_B\ s_O\ p_x\ p_y\ p_z)$ which is expressed using matrix multiplication as

$$(s_B\ s_A\ s_O\ -p_x\ -p_y\ p_z) = (s_A\ s_B\ s_O\ p_x\ p_y\ p_z)\overbrace{\begin{pmatrix} 0 & 1 & 0 & 0 & 0 & 0 \\ 1 & 0 & 0 & 0 & 0 & 0 \\ 0 & 0 & 1 & 0 & 0 & 0 \\ 0 & 0 & 0 & -1 & 0 & 0 \\ 0 & 0 & 0 & 0 & -1 & 0 \\ 0 & 0 & 0 & 0 & 0 & 1 \end{pmatrix}}^{D(C_2)}$$

Similarly, the σ_v operation exchanges s_A and s_B, leaves s_O, p_x and p_z unchanged, and changes the sign of p_y: $(s_B\ s_A\ s_O\ p_x\ -p_y\ p_z) \leftarrow (s_A\ s_B\ s_O\ p_x\ p_y\ p_z)$ This is expressed using matrix multiplication as

$$(s_B\ s_A\ s_O\ p_x\ -p_y\ p_z) = (s_A\ s_B\ s_O\ p_x\ p_y\ p_z)\overbrace{\begin{pmatrix} 0 & 1 & 0 & 0 & 0 & 0 \\ 1 & 0 & 0 & 0 & 0 & 0 \\ 0 & 0 & 1 & 0 & 0 & 0 \\ 0 & 0 & 0 & 1 & 0 & 0 \\ 0 & 0 & 0 & 0 & -1 & 0 \\ 0 & 0 & 0 & 0 & 0 & 1 \end{pmatrix}}^{D(\sigma_v)}$$

The σ_v' operation leaves all orbitals unchanged except p_x, which changes sign.

$$(s_A\ s_B\ s_O\ -p_x\ p_y\ p_z) = (s_A\ s_B\ s_O\ p_x\ p_y\ p_z)\overbrace{\begin{pmatrix} 1 & 0 & 0 & 0 & 0 & 0 \\ 0 & 1 & 0 & 0 & 0 & 0 \\ 0 & 0 & 1 & 0 & 0 & 0 \\ 0 & 0 & 0 & -1 & 0 & 0 \\ 0 & 0 & 0 & 0 & 1 & 0 \\ 0 & 0 & 0 & 0 & 0 & 1 \end{pmatrix}}^{D(\sigma_v')}$$

Finally, the E operation, 'do nothing', leaves all orbitals unchanged.

$$(s_A\ s_B\ s_O\ p_x\ p_y\ p_z) = (s_A\ s_B\ s_O\ p_x\ p_y\ p_z)\overbrace{\begin{pmatrix} 1&0&0&0&0&0\\0&1&0&0&0&0\\0&0&1&0&0&0\\0&0&0&1&0&0\\0&0&0&0&1&0\\0&0&0&0&0&1 \end{pmatrix}}^{D(E)}$$

(a) The representative of the operation $C_2\sigma_v$ is found by multiplying together the matricies $\boldsymbol{D}(C_2)$ and $\boldsymbol{D}(\sigma_v)$ found above. Basic information about how to handle matrices is given in *The chemist's toolkit* 24 in Topic 9E on page 373.

$$\boldsymbol{D}(C_2)\boldsymbol{D}(\sigma_v) = \overbrace{\begin{pmatrix} 0&1&0&0&0&0\\1&0&0&0&0&0\\0&0&1&0&0&0\\0&0&0&-1&0&0\\0&0&0&0&-1&0\\0&0&0&0&0&1 \end{pmatrix}}^{D(C_2)}\overbrace{\begin{pmatrix} 0&1&0&0&0&0\\1&0&0&0&0&0\\0&0&1&0&0&0\\0&0&0&1&0&0\\0&0&0&0&-1&0\\0&0&0&0&0&1 \end{pmatrix}}^{D(\sigma_v)} = \overbrace{\begin{pmatrix} 1&0&0&0&0&0\\0&1&0&0&0&0\\0&0&1&0&0&0\\0&0&0&-1&0&0\\0&0&0&0&1&0\\0&0&0&0&0&1 \end{pmatrix}}^{D(\sigma_v')}$$

$$= \boldsymbol{D}(\sigma_v')$$

This establishes that $C_2\sigma_v = \sigma_v'$. Similarly, the representative of the operation $\sigma_v\sigma_v'$ is found by representative matricies of σ_v and σ_v'; the result is the representative of C_2, establishing that $\sigma_v\sigma_v' = C_2$.

$$\boldsymbol{D}(\sigma_v)\boldsymbol{D}(\sigma_v') = \overbrace{\begin{pmatrix} 0&1&0&0&0&0\\1&0&0&0&0&0\\0&0&1&0&0&0\\0&0&0&1&0&0\\0&0&0&0&-1&0\\0&0&0&0&0&1 \end{pmatrix}}^{D(\sigma_v)}\overbrace{\begin{pmatrix} 1&0&0&0&0&0\\0&1&0&0&0&0\\0&0&1&0&0&0\\0&0&0&-1&0&0\\0&0&0&0&1&0\\0&0&0&0&0&1 \end{pmatrix}}^{D(\sigma_v')} = \overbrace{\begin{pmatrix} 0&1&0&0&0&0\\1&0&0&0&0&0\\0&0&1&0&0&0\\0&0&0&-1&0&0\\0&0&0&0&-1&0\\0&0&0&0&0&1 \end{pmatrix}}^{D(C_2)}$$

$$= \boldsymbol{D}(C_2)$$

(b) The representation is reduced using the method in Section 10B.2(c) on page 400. Inspection of the matrix representatives reveales that they are all of block-diagonal format

$$\boldsymbol{D} = \begin{pmatrix} \blacksquare&\blacksquare&0&0&0&0\\ \blacksquare&\blacksquare&0&0&0&0\\ 0&0&\blacksquare&0&0&0\\ 0&0&0&\blacksquare&0&0\\ 0&0&0&0&\blacksquare&0\\ 0&0&0&0&0&\blacksquare \end{pmatrix}$$

This implies that the symmetry operations of C_{2v} never mix the s_O, p_x, p_y and p_z orbitals together, nor do they mix these orbitals with s_A and s_B, but s_A and s_B are mixed together by the operations of the group. Consequently the basis can be cut into five parts: four one-dimensional bases for the individual orbitals on the oxygen and a two-dimensional basis for the two hydrogen s orbitals. The representations corresponding to the four one-dimensional bases are

$$\begin{array}{lllll}\text{For } s_O: & \boldsymbol{D}(E)=1 & \boldsymbol{D}(C_2)=1 & \boldsymbol{D}(\sigma_v)=1 & \boldsymbol{D}(\sigma_v')=1 \\ \text{For } p_x: & \boldsymbol{D}(E)=1 & \boldsymbol{D}(C_2)=-1 & \boldsymbol{D}(\sigma_v)=1 & \boldsymbol{D}(\sigma_v')=-1 \\ \text{For } p_y: & \boldsymbol{D}(E)=1 & \boldsymbol{D}(C_2)=-1 & \boldsymbol{D}(\sigma_v)=-1 & \boldsymbol{D}(\sigma_v')=1 \\ \text{For } p_z: & \boldsymbol{D}(E)=1 & \boldsymbol{D}(C_2)=1 & \boldsymbol{D}(\sigma_v)=1 & \boldsymbol{D}(\sigma_v')=1\end{array}$$

The characters of one-dimensional representatives are just the representatives themselves. Therefore, inspection of the C_{2v} character table from the *Resource section* indicates that the one-dimensional representations for these orbitals correspond to A_1, B_1, B_2, and A_1 respectively. The oxygen-based orbitals therefore span $2A_1 + B_1 + B_2$.

The remaining two orbitals, s_A and s_B, are a basis for a two-dimensional representation

$$\boldsymbol{D}(E)=\begin{pmatrix}1&0\\0&1\end{pmatrix} \quad \boldsymbol{D}(C_2)=\begin{pmatrix}0&1\\1&0\end{pmatrix} \quad \boldsymbol{D}(\sigma_v)=\begin{pmatrix}0&1\\1&0\end{pmatrix} \quad \boldsymbol{D}(\sigma_v')=\begin{pmatrix}1&0\\0&1\end{pmatrix}$$

where the matrices correspond to the top left-hand corners of the 6 × 6 matricies found above. That this two-dimensional representation is reducible is demonstrated by considering the linear combinations $s_1 = s_A + s_B$ and $s_2 = s_A - s_B$. The C_2 and σ_v operations both exchange s_A and s_B: $(\,s_B\ s_A\,) \leftarrow (\,s_A\ s_B\,)$ which means that $(s_B + s_A) \leftarrow (s_A + s_B)$, corresponding to $(s_1) \leftarrow (s_1)$. Similarly, $(s_B - s_A) \leftarrow (s_A - s_B)$, corresponding to $(-s_2) \leftarrow (s_2)$. On the other hand, the E and σ_v' operations leave s_A and s_B unchanged, which leads to the results $(s_1) \leftarrow (s_1)$ and $(s_2) \leftarrow (s_2)$. The representation in the basis $(\,s_1\ s_2\,)$ is therefore

$$\boldsymbol{D}(E)=\begin{pmatrix}1&0\\0&1\end{pmatrix} \quad \boldsymbol{D}(C_2)=\begin{pmatrix}1&0\\0&-1\end{pmatrix} \quad \boldsymbol{D}(\sigma_v)=\begin{pmatrix}1&0\\0&-1\end{pmatrix} \quad \boldsymbol{D}(\sigma_v')=\begin{pmatrix}1&0\\0&1\end{pmatrix}$$

The new representatives are all in block diagonal format, in this case in the form $\begin{pmatrix}\blacksquare&0\\0&\blacksquare\end{pmatrix}$, which means that the two combinations are not mixed with each other by any operation of the group. The two dimensional basis $(\,s_1\ s_2\,)$ can therefore be cut into two one-dimensional bases corresponding to s_1 and s_2. The representations for these bases are

$$\begin{array}{lllll}\text{For } s_1: & \boldsymbol{D}(E)=1 & \boldsymbol{D}(C_2)=1 & \boldsymbol{D}(\sigma_v)=1 & \boldsymbol{D}(\sigma_v')=1 \\ \text{For } s_2: & \boldsymbol{D}(E)=1 & \boldsymbol{D}(C_2)=-1 & \boldsymbol{D}(\sigma_v)=-1 & \boldsymbol{D}(\sigma_v')=1\end{array}$$

Because the characters of one-dimensional representatives are just the representatives themselves, inspection of the C_{2v} character table immediately indicates that these one-dimensional representations correspond to A_1 and B_2 respectively. Combining these results with those from the oxygen orbitals found earlier, which span $2A_1 + B_1 + B_2$, the original six-dimensional representation therefore spans $3A_1 + B_1 + 2B_2$.

P10B.5 The ethene molecule and axis system are shown in Fig. 10.16.

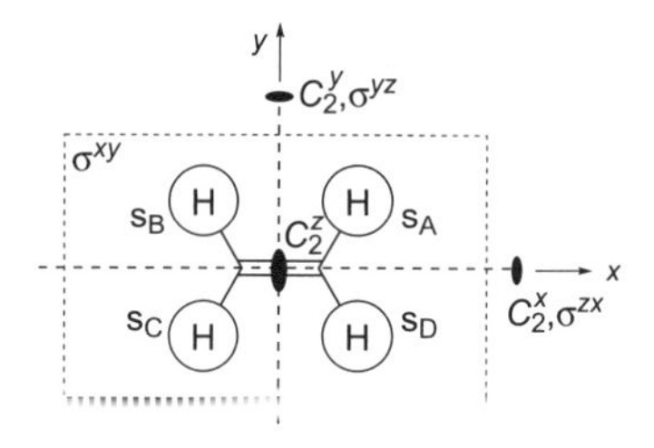

Figure 10.16

The C_2^x operation exchanges s_A with s_D, and s_B with s_C:

$$(s_D\ s_C\ s_B\ s_A) \leftarrow (s_A\ s_B\ s_C\ s_D)$$

This is written using matrix multiplication as

$$(s_D\ s_C\ s_B\ s_A) = (s_A\ s_B\ s_C\ s_D)\overbrace{\begin{pmatrix} 0 & 0 & 0 & 1 \\ 0 & 0 & 1 & 0 \\ 0 & 1 & 0 & 0 \\ 1 & 0 & 0 & 0 \end{pmatrix}}^{D(C_2^x)}$$

The matrix $D(C_2^x)$ is the representative of the C_2^x operation in this basis. The representatives of the other seven operations are obtained in the same way.

$$\text{For } C_2^y\text{:}\quad (s_B\ s_A\ s_D\ s_C) = (s_A\ s_B\ s_C\ s_D)\overbrace{\begin{pmatrix} 0 & 1 & 0 & 0 \\ 1 & 0 & 0 & 0 \\ 0 & 0 & 0 & 1 \\ 0 & 0 & 1 & 0 \end{pmatrix}}^{D(C_2^y)}$$

$$\text{For } C_2^z\text{:}\quad (s_C\ s_D\ s_A\ s_B) = (s_A\ s_B\ s_C\ s_D)\overbrace{\begin{pmatrix} 0 & 0 & 1 & 0 \\ 0 & 0 & 0 & 1 \\ 1 & 0 & 0 & 0 \\ 0 & 1 & 0 & 0 \end{pmatrix}}^{D(C_2^z)}$$

$$\text{For } i\text{:}\quad (s_C\ s_D\ s_A\ s_B) = (s_A\ s_B\ s_C\ s_D)\overbrace{\begin{pmatrix} 0 & 0 & 1 & 0 \\ 0 & 0 & 0 & 1 \\ 1 & 0 & 0 & 0 \\ 0 & 1 & 0 & 0 \end{pmatrix}}^{D(i)}$$

$$\text{For } \sigma^{xy}\text{:}\quad (s_A\ s_B\ s_C\ s_D) = (s_A\ s_B\ s_C\ s_D)\overbrace{\begin{pmatrix} 1 & 0 & 0 & 0 \\ 0 & 1 & 0 & 0 \\ 0 & 0 & 1 & 0 \\ 0 & 0 & 0 & 1 \end{pmatrix}}^{D(\sigma^{xy})}$$

$$\text{For } \sigma^{yz}\text{:}\quad (s_B\ s_A\ s_D\ s_C) = (s_A\ s_B\ s_C\ s_D)\overbrace{\begin{pmatrix} 0 & 1 & 0 & 0 \\ 1 & 0 & 0 & 0 \\ 0 & 0 & 0 & 1 \\ 0 & 0 & 1 & 0 \end{pmatrix}}^{D(\sigma^{yz})}$$

$$\text{For } \sigma^{zx}\text{:}\quad (s_D\ s_C\ s_B\ s_A) = (s_A\ s_B\ s_C\ s_D)\overbrace{\begin{pmatrix} 0 & 0 & 0 & 1 \\ 0 & 0 & 1 & 0 \\ 0 & 1 & 0 & 0 \\ 1 & 0 & 0 & 0 \end{pmatrix}}^{D(\sigma^{zx})}$$

$$\text{For } E\text{:}\quad (s_C\ s_D\ s_A\ s_B) = (s_A\ s_B\ s_C\ s_D)\overbrace{\begin{pmatrix} 0 & 0 & 1 & 0 \\ 0 & 0 & 0 & 1 \\ 1 & 0 & 0 & 0 \\ 0 & 1 & 0 & 0 \end{pmatrix}}^{D(E)}$$

P10B.7 For one-dimensional representatives, the characters as just the representatives themselves. Thus for the first representation, $\boldsymbol{D}(C_3) = 1$ and $\boldsymbol{D}(C_2) = 1$, the characters are $\chi(C_3) = 1$ and $\chi(C_2) = 1$. Inspection of the C_{6v} character table in the *Resource section* shows that this combination of characters corresponds to either A_1 or A_2. Both of these irreducible representations have $\chi(C_6) = +1$, as expected from the matrix multiplication $\boldsymbol{D}(C_6) = \boldsymbol{D}(C_3)\boldsymbol{D}(C_2) = 1 \times 1 = 1$ which implies that $\chi(C_6) = +1$. The character table gives the characters of σ_v and σ_d as both $+1$ for A_1 or both -1 for A_2, implying that $\boxed{\boldsymbol{D}(\sigma_v) = +1}$, $\boxed{\boldsymbol{D}(\sigma_d) = +1}$ in the case of A_1, or $\boxed{\boldsymbol{D}(\sigma_v) = -1, \boldsymbol{D}(\sigma_d) = -1}$ in the case of A_2. For the second representation, $\boldsymbol{D}(C_3) = 1$ and $\boldsymbol{D}(C_2) = -1$, the characters are $\chi(C_3) = 1$ and $\chi(C_2) = -1$. Inspection of the character table shows that this corresponds to either B_1 or B_2; as expected these both have $\chi(C_6) = -1$. The characters of σ_v and σ_d are $+1$ and -1 in the case of B_1, or -1 and $+1$ in the case of B_2. Hence $\boxed{\boldsymbol{D}(\sigma_v) = +1, \boldsymbol{D}(\sigma_d) = -1}$ in the case of B_1, or $\boxed{\boldsymbol{D}(\sigma_v) = -1}$, $\boxed{\boldsymbol{D}(\sigma_d) = +1}$ in the case of B_2.

P10B.9 (a) The C_{2v} character table is given in the *Resource section.* As explained in the question, r^2 is invariant to all operations of the group, and furthermore the C_{2v} character table shows that both z and z^2 belong to the totally symmetric symmetric representation A_1 which means that they are also invariant to all operations of the group. Because all parts of the function $z(5z^2 - 3r^2)$ are invariant to all operations, it follows that the entire function and therefore the f orbital that it represents is invariant to all operations. Consequently this f orbital belongs to the totally symmetric representation A_1.

(b) The function $y(5y^2 - 3r^2)$ is considered as a product of the functions y and $(5y^2 - 3r^2)$. The C_{2v} character table shows that the function y^2 belongs to the totally symmetric representation and is therefore invariant to all operations of the group; because this is also true of r^2 it follows that the factor $(5y^2 - 3r^2)$ is similarly invariant to all operations. The character table also shows that the function y belongs to the B_2 representation, for which the characters are $+1$ under E and σ_v' and -1 under C_2 and σ_v. This indicates that the function y changes sign under C_2 and σ_v and, therefore, because $(5y^2 - 3r^2)$ is invariant to all operations, the product $y(5y^2 - 3r^2)$ behaves in the same way as y. The f orbital therefore belongs to the B_2 representation.

(c) In the same way, the function $x(5x^2 - 3r^2)$ behaves in the same way as x, because $(5x^2 - 3r^2)$ is invariant to all operations. The character table shows that x, and therefore this f orbital, belongs to the B_1 representation.

(d) The function $z(x^2 - y^2)$ belongs to the totally symmetric A_1 representation, because as shown in the character table each of z, x^2 and y^2 belongs to A_1 and are therefore invariant to all operations of the group. It follows that the product $z(x^2 - y^2)$ is also invariant to all operations and hence belongs to A_1.

(e) The function $y(x^2 - z^2)$ behaves in the same way as y, because the factor $(x^2 - z^2)$ is invariant to all operations of the group. Hence this f orbital belongs to B_2, the same as y.

(f) Similarly the function $x(z^2 - y^2)$ behaves the same way as x and therefore belongs to B_1.

(g) The function xyz is considered as the product $xy \times z$. The character table shows that z belongs to the totally symmetric representation A_1, so the function xyz behaves in the same way as xy which as shown in the character table belongs to A_2.

The seven f orbitals therefore span $2A_1 + A_2 + 2B_1 + 2B_2$.

10C Applications of symmetry

Answers to discussion questions

D10C.1 Character tables provide a way to: (a) assign symmetry symbols for orbitals; (b) know whether overlap integrals are nonzero; (c) determine what atomic orbitals can contribute to a LCAO-MO; (d) determine the maximum orbital degeneracy of a molecule; and (e) determine whether a transition is allowed.

Solutions to exercises

E10C.1(a) As explained in Section 10C.1 on page 406 an integral can only be non-zero if the integrand spans the totally symmetric irreducible representation, which in C_{2v} is A_1. From Section 10C.1(a) on page 407 the symmetry species spanned by the integrand $p_x z p_z$ is found by the forming the direct product of the symmetry species spanned by p_x, z, and p_z separately. These are read off the C_{2v} character table by looking for the appropriate Cartesian functions listed on the right of the table: x and hence p_x spans B_1, while z and p_z both span A_1. The direct product required is therefore $B_1 \times A_1 \times A_1$.

The order does not matter, so this is equal to $A_1 \times A_1 \times B_1$, which is equal to B_1 because, from the first simplifying feature described in Section 10C.1(a) on page 407, the direct product of the totally symmetric representation with any other representation is the latter representation itself: $A_1 \times \Gamma^{(i)} = \Gamma^{(i)}$. The integrand therefore spans B_1. This is not the totally symmetric irreducible representation, therefore the integral is $\boxed{\text{zero}}$.

E10C.2(a) As explained in Section 10C.3 on page 411, an electric dipole transition is forbidden if the electric transition dipole moment $\mu_{q,\text{fi}}$ is zero. The transition dipole moment is given by [10C.6–*411*], $\mu_{q,\text{fi}} = -e \int \psi_\text{f}^* q \psi_\text{i} \, d\tau$ where q is x, y, or z. The integral is only non-zero if the integrand contains the totally symmetric representation, which from the C_{3v} character table is A_1.

For a transition $A_1 \to A_2$, the symmetry species of the integrand is given by the direct product $A_2 \times \Gamma^{(q)} \times A_1$. The order does not matter so this is equal to $A_1 \times A_2 \times \Gamma^{(q)}$, which is simply equal to $A_2 \times \Gamma^{(q)}$ because, from the first simplifying feature listed in Section 10C.1(a) on page 407, the direct product of the totally symmetric irreducible representation with any other representation is the latter representation itself. Therefore $A_1 \times A_2 = A_2$.

The direct product $A_2 \times \Gamma^{(q)}$ contains the totally symmetric irreducible representation only if $\Gamma^{(q)}$ spans A_2 because, according to the second simplifying feature listed in Section 10C.1(a) on page 407, the direct product of two irreducible representations contains the totally symmetric irreducible representation only if the two irreducible representations are identical. The C_{3v} character table shows that none of x, y or z span A_2, so it follows that the integrand does not contain A_1 and hence the transition is $\boxed{\text{forbidden}}$.

E10C.3(a) The D_{2h} character table shows that xy spans B_{1g} in D_{2h}. To show this explicitly, the direct product is formed between the irreducible representations spanned by x and y individually. The character table shows that x spans B_{3u} and y spans B_{2u}; the direct product $B_{3u} \times B_{2u}$ is calculated using the method described in Section 10C.1(a) on page 407

	E	C_2^x	C_2^y	C_2^z	i	σ^{xy}	σ^{yz}	σ^{zx}
B_{3u}	1	−1	−1	1	−1	1	−1	1
B_{2u}	1	−1	1	−1	−1	1	1	−1
product	1	1	−1	−1	1	1	−1	−1

The characters in the product row are those of symmetry species B_{1g}, thus confirming that the function xy has symmetry species B_{1g} in D_{2h}.

E10C.4(a) As explained in Section 10C.2(a) on page 409, only orbitals of the same symmetry species may have a nonzero overlap. Inspection of the C_{2v} character table shows that the oxygen $\boxed{2s}$ and $\boxed{2p_z}$ orbitals both have A_1 symmetry, so they can interact with the A_1 combination of fluorine orbitals. The oxygen $\boxed{2p_y}$ orbital has B_2 symmetry, so it can interact with the B_2 combination of fluorine orbitals. The oxygen $2p_x$ orbital has B_1 symmetry, so cannot interact with either combination of fluorine orbitals and therefore remains nonbonding.

In SF_2 the same interactions with the sulfur s and p orbitals are possible but there is now the possibility of additional interactions involving the d orbitals. The C_{2v} character table shows that $\boxed{d_{z^2}}$ and $\boxed{d_{x^2-y^2}}$ have A_1 symmetry, so they can interact with the A_1 combination of fluorine orbitals. The $\boxed{d_{yz}}$ orbital has B_2 symmetry so can interact with the B_2 combination. The d_{xy} and d_{zx} orbitals have A_2 and B_1 symmetry respectively, so they cannot interact with either combination of fluorine orbitals and therefore remain nonbonding.

E10C.5(a) As explained in Section 10C.2(a) on page 409, only orbitals of the same symmetry species may have a nonzero overlap. Inspection of the C_{2v} character table shows that the nitrogen 2s, p_x, p_y, and p_z orbitals span A_1, B_1, B_2, and A_1 respectively. Because none of these orbitals have A_2 symmetry, $\boxed{\text{none of them}}$ can interact with the A_2 combination of oxygen orbitals.

The character table also shows that the d_{z^2}, $d_{x^2-y^2}$, d_{xy}, d_{yz}, and d_{zx} orbitals of the sulfur in SO_2 transform as A_1, A_1, A_2, B_2 and B_1 respectively. Therefore the $\boxed{d_{xy}}$ orbital has the correct symmetry to interact with the A_2 combination of oxygen p orbitals.

E10C.6(a) As explained in Section 10C.3 on page 411, a transition from a state with symmetry $\Gamma^{(i)}$ to one with symmetry $\Gamma^{(f)}$ is only allowed if the direct product $\Gamma^{(f)} \times \Gamma^{(q)} \times \Gamma^{(i)}$ contains the totally symmetric irreducible representation, which in C_{2v} is A_1. If the ground state is A_1, then the direct product becomes $\Gamma^{(f)} \times \Gamma^{(q)} \times A_1$. This is simply $\Gamma^{(f)} \times \Gamma^{(q)}$ because, from the first simplifying feature of direct products listed in Section 10C.1(a) on page 407, the direct product of the totally symmetric irreducible representation A_1 with any other representation is the latter representation itself.

If $\Gamma^{(f)} \times \Gamma^{(q)}$ is to be A_1, then $\Gamma^{(f)}$ must equal $\Gamma^{(q)}$ because, from the second simplifying feature of direct products listed in Section 10C.1(a) on page 407, the direct product of two irreducible representations only contains the totally symmetric irreducible representation if the two irreducible representations are identical. The C_{2v} character table shows that $\Gamma^{(q)} = B_1$ for x polarized light ($q = x$), B_2 for y polarised light, and A_1 for z polarised light. It follows that x, y and z polarised light can excite the molecule to $\boxed{B_1, B_2, \text{and } A_1}$ states respectively.

E10C.7(a) The number of times $n(\Gamma)$ that a given irreducible representation Γ occurs in a representation is given by [10C.3a–*408*], $n(\Gamma) = (1/h)\sum_C N(C)\chi^{(\Gamma)}(C)\chi(C)$, where h is the order of the group, $N(C)$ is the number of operations in class C, $\chi^{(\Gamma)}$ is the character of class C in the irreducible representation Γ, and $\chi(C)$ is the character of class C in the representation being reduced.

$$\begin{aligned}n(\mathrm{A_1}) = \tfrac{1}{8}\Big(&1\times\chi^{(\mathrm{A_1})}(E)\times\chi(E) + 1\times\chi^{(\mathrm{A_1})}(C_2)\times\chi(C_2) + 2\times\chi^{(\mathrm{A_1})}(C_4)\times\chi(C_4)\\ &+ 2\times\chi^{(\mathrm{A_1})}(\sigma_\mathrm{v})\times\chi(\sigma_\mathrm{v}) + 2\times\chi^{(\mathrm{A_1})}(\sigma_\mathrm{d})\times\chi(\sigma_\mathrm{d})\Big)\\ = \tfrac{1}{8}\Big(&1\times1\times5 + 1\times1\times1 + 2\times1\times1 + 2\times1\times3 + 2\times1\times1\Big) = 2\end{aligned}$$

Similarly

$$n(\mathrm{A_2}) = \tfrac{1}{8}\Big(1\times1\times5 + 1\times1\times1 + 2\times1\times1 + 2\times(-1)\times3 + 2\times(-1)\times1\Big) = 0$$

$$n(\mathrm{B_1}) = \tfrac{1}{8}\Big(1\times1\times5 + 1\times1\times1 + 2\times(-1)\times1 + 2\times1\times3 + 2\times(-1)\times1\Big) = 1$$

$$n(\mathrm{B_2}) = \tfrac{1}{8}\Big(1\times1\times5 + 1\times1\times1 + 2\times(-1)\times1 + 2\times(-1)\times3 + 2\times1\times1\Big) = 0$$

$$n(\mathrm{E}) = \tfrac{1}{8}\Big(1\times2\times5 + 1\times(-2)\times1 + 2\times0\times1 + 2\times0\times3 + 2\times0\times1\Big) = 1$$

The representation therefore spans $\boxed{2\mathrm{A_1} + \mathrm{B_1} + \mathrm{E}}$.

E10C.8(a) The number of times $n(\Gamma)$ that a given irreducible representation Γ occurs in a representation is given by [10C.3a–*408*], $n(\Gamma) = (1/h)\sum_C N(C)\chi^{(\Gamma)}(C)\chi(C)$, where h is the order of the group, $N(C)$ is the number of operations in class C, $\chi^{(\Gamma)}$ is the character of class C in the irreducible representation Γ, and $\chi(C)$ is the character of class C in the representation being reduced. In the case of $D_{4\mathrm{h}}$, $h = 16$.

Because the representation being reduced has characters of zero for all classes except E, C_2', σ_h, and σ_v, only these latter four classes make a non-zero contribution to the sum and therefore only these classes need be considered. The number of times that the irreducible representation $\mathrm{A_{1g}}$ occurs is therefore

$$\begin{aligned}n(\mathrm{A_1}) = \tfrac{1}{16}\big(&N(E)\times\chi^{(\mathrm{A_{1g}})}(E)\times\chi(E) + N(C_2')\times\chi^{(\mathrm{A_{1g}})}(C_2')\times\chi(C_2')\\ &+ N(\sigma_\mathrm{h})\times\chi^{(\mathrm{A_{1g}})}(\sigma_\mathrm{h})\times\chi(\sigma_\mathrm{h}) + N(\sigma_\mathrm{v})\times\chi^{(\mathrm{A_{1g}})}(\sigma_\mathrm{v})\times\chi(\sigma_\mathrm{v})\big)\\ = \tfrac{1}{16}\,&(1\times1\times4 + 2\times1\times2 + 1\times1\times4 + 2\times1\times2) = 1\end{aligned}$$

Similarly

$$n(A_{2g}) = \tfrac{1}{16}(1\times1\times4 + 2\times(-1)\times2 + 1\times1\times4 + 2\times(-1)\times2) = 0$$
$$n(B_{1g}) = \tfrac{1}{16}(1\times1\times4 + 2\times1\times2 + 1\times1\times4 + 2\times1\times2) = 1$$
$$n(B_{2g}) = \tfrac{1}{16}(1\times1\times4 + 2\times(-1)\times2 + 1\times1\times4 + 2\times(-1)\times2) = 0$$
$$n(E_{g}) = \tfrac{1}{16}(1\times2\times4 + 2\times0\times2 + 1\times(-2)\times4 + 2\times0\times2) = 0$$
$$n(A_{1u}) = \tfrac{1}{16}(1\times1\times4 + 2\times1\times2 + 1\times(-1)\times4 + 2\times(-1)\times2) = 0$$
$$n(A_{2u}) = \tfrac{1}{16}(1\times1\times4 + 2\times(-1)\times2 + 1\times(-1)\times4 + 2\times1\times2) = 0$$
$$n(B_{1u}) = \tfrac{1}{16}(1\times1\times4 + 2\times1\times2 + 1\times(-1)\times4 + 2\times(-1)\times2) = 0$$
$$n(B_{2u}) = \tfrac{1}{16}(1\times1\times4 + 2\times(-1)\times2 + 1\times(-1)\times4 + 2\times1\times2) = 0$$
$$n(E_{u}) = \tfrac{1}{16}(1\times2\times4 + 2\times0\times2 + 1\times0\times4 + 2\times1\times2) = 1$$

The representation therefore spans $\boxed{A_{1g} + B_{1g} + E_u}$.

E10C.9(a) As explained in Section 10C.3 on page 411, a transition from a state with symmetry $\Gamma^{(i)}$ to one with symmetry $\Gamma^{(f)}$ is only allowed if the direct product $\Gamma^{(f)} \times \Gamma^{(q)} \times \Gamma^{(i)}$ contains the totally symmetric irreducible representation, which for both molecules is A_{1g}. The ground state is totally symmetric, implying that it transforms as A_{1g}. Therefore the direct product becomes $\Gamma^{(f)} \times \Gamma^{(q)} \times A_{1g}$. This is simply $\Gamma^{(f)} \times \Gamma^{(q)}$ because, from the first simplifying feature of direct products listed in Section 10C.1(a) on page 407, the direct product of the totally symmetric representation A_{1g} with any other representation is the latter representation itself.

If $\Gamma^{(f)} \times \Gamma^{(q)}$ is to be A_{1g}, then $\Gamma^{(f)}$ must equal $\Gamma^{(q)}$ because, from the second simplifying feature of direct products listed in Section 10C.1(a) on page 407, the direct product of two irreducible representations only contains the totally symmetric irreducible representation if the two irreducible representations are identical.

(i) Benzene belongs to point group D_{6h}. The D_{6h} character table in the *Online resource centre* shows that shows that z transforms as A_{2u}, and x and y together transform as E_{1u}. Therefore light polarized along z can excite benzene to an $\boxed{A_{2u}}$ state, and x or y polarised light can excite it to an $\boxed{E_{1u}}$ state.

(ii) Naphthalene belongs to point group D_{2h}. The D_{2h} character table in the *Resource section* shows that x transforms as B_{3u}, y transforms as B_{2u}, and z transforms as B_{1u}. Therefore naphthalene can be excited to $\boxed{B_{3u}, B_{2u}, \text{ or } B_{1u}}$ states by x, y, and z polarised light respectively.

Solutions to problems

P10C.1 Methane belongs to point group T_d. The methane molecule and its H1s orbitals are shown in Fig. 10.17, along with one operation of each class. It is sufficient to

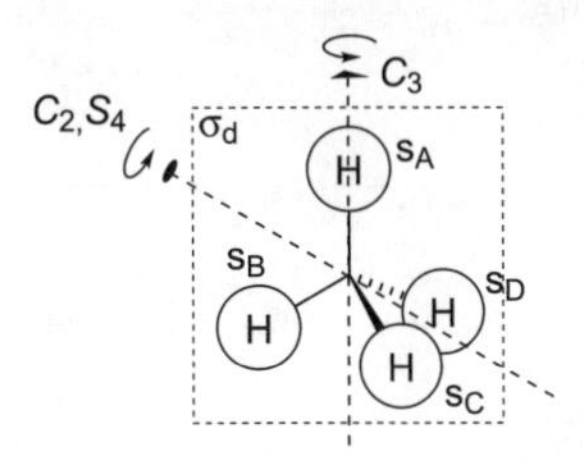

Figure 10.17

consider just one operation in each class because, by definition, all operations the same class have the same character.

The C_3 operation shown in Fig. 10.17 leaves s_A unchanged but converts s_B into s_C, s_C into s_D, and s_D into s_B: $(s_A\ s_C\ s_D\ s_B) \leftarrow (s_A\ s_B\ s_C\ s_D)$. This is written using matrix multiplication as

$$(s_A\ s_C\ s_D\ s_B) = (s_A\ s_B\ s_C\ s_D)\overbrace{\begin{pmatrix} 1 & 0 & 0 & 0 \\ 0 & 0 & 0 & 1 \\ 0 & 1 & 0 & 0 \\ 0 & 0 & 1 & 0 \end{pmatrix}}^{\boldsymbol{D}(C_3)}$$

The matrix $\boldsymbol{D}(C_3)$ is the representative of the C_3 operation in this basis. Similarly, the C_2 operation shown in Fig. 10.17 exchanges s_A and s_B, and also exchanges s_C and s_D. This is written using matrix multiplication as

$$(s_B\ s_A\ s_D\ s_C) = (s_A\ s_B\ s_C\ s_D)\overbrace{\begin{pmatrix} 0 & 1 & 0 & 0 \\ 1 & 0 & 0 & 0 \\ 0 & 0 & 0 & 1 \\ 0 & 0 & 1 & 0 \end{pmatrix}}^{\boldsymbol{D}(C_2)}$$

The σ_d operation shown in Fig. 10.17 leaves s_A and s_B unchanged and exchanges s_C and s_D; this gives

$$(s_A\ s_B\ s_D\ s_c) = (s_A\ s_B\ s_C\ s_D)\overbrace{\begin{pmatrix} 1 & 0 & 0 & 0 \\ 0 & 1 & 0 & 0 \\ 0 & 0 & 0 & 1 \\ 0 & 0 & 1 & 0 \end{pmatrix}}^{\boldsymbol{D}(\sigma_d)}$$

The S_4 operation shown in Fig. 10.17 converts s_A to s_D, s_B to s_C, s_C to s_A, and

s_D to s_B, giving

$$(s_D\ s_C\ s_A\ s_B) = (s_A\ s_B\ s_C\ s_D)\overbrace{\begin{pmatrix} 0 & 0 & 1 & 0 \\ 0 & 0 & 0 & 1 \\ 0 & 1 & 0 & 0 \\ 1 & 0 & 0 & 0 \end{pmatrix}}^{\boldsymbol{D}(S_4)}$$

Finally, the E operation leaves all orbitals unchanged, meaning that its representative is simply the identity matrix

$$(s_A\ s_B\ s_C\ s_D)\quad(s_A\ s_B\ s_C\ s_D)\overbrace{\begin{pmatrix} 1 & 0 & 0 & 0 \\ 0 & 1 & 0 & 0 \\ 0 & 0 & 1 & 0 \\ 0 & 0 & 0 & 1 \end{pmatrix}}^{\boldsymbol{D}(E)}$$

The representatives found above have the following characters

$$\chi(E) = 4 \quad \chi(C_3) = 1 \quad \chi(C_2) = 0 \quad \chi(\sigma_d) = 2 \quad \chi(S_4) = 0$$

This result can be arrived at much more quickly by noting that: (1) only the diagonal elements of the representative matrix contribute to the trace; (2) orbitals which are unmoved by an operation will result in a 1 on the diagonal; (3) orbitals which are moved to other positions by an operation will result in a 0 on the diagonal. The character is found simply by counting the number of orbitals which do not move. In the present case 4 are unmoved by E, 1 is unmoved by C_3, none are unmoved by C_2, 2 are unmoved by σ_d, and none are unmoved by S_4. The characters are thus $\{4, 1, 0, 2, 0\}$.

This representation is decomposed using the method described in Section 10C.1(b) on page 408. The number of times $n(\Gamma)$ that a given irreducible representation Γ occurs in a representation is given by [10C.3a–*408*],

$$n(\Gamma) = \frac{1}{h}\sum_C N(C)\chi^{(\Gamma)}(C)\chi(C)$$

where h is the order of the group, $N(C)$ is the number of operations in class C, $\chi^{(\Gamma)}$ is the character of class C in the irreducible representation Γ, and $\chi(C)$ is the character of class C in the representation being reduced. In the case of the group T_d, $h = 24$. The number of times that the irreducible representation A_1 occurs is

$$\begin{aligned} n(A_1) = \tfrac{1}{24}\big(&N(E)\times\chi^{(A_1)}(E)\times\chi(E) + N(C_3)\times\chi^{(A_1)}(C_3)\times\chi(C_3) \\ &+ N(C_2)\times\chi^{(A_1)}(C_2)\times\chi(C_2) + N(\sigma_d)\times\chi^{(A_1)}(\sigma_d)\times\chi(\sigma_d) \\ &+ N(S_4)\times\chi^{(A_1)}(S_4)\times\chi(S_4)\big) \\ = \tfrac{1}{24}&(1\times1\times4 + 8\times1\times1 + 3\times1\times0 + 6\times1\times2 + 6\times1\times0) = 1 \end{aligned}$$

Similarly

$$n(A_2) = \tfrac{1}{24}\left(1\times1\times4 + 8\times1\times1 + 3\times1\times0 + 6\times(-1)\times2 + 6\times(-1)\times0\right) = 0$$
$$n(E) = \tfrac{1}{24}\left(1\times2\times4 + 8\times(-1)\times1 + 3\times2\times0 + 6\times0\times2 + 6\times0\times0\right) = 0$$
$$n(T_1) = \tfrac{1}{24}\left(1\times3\times4 + 8\times0\times1 + 3\times(-1)\times0 + 6\times(-1)\times2 + 6\times1\times0\right) = 0$$
$$n(T_2) = \tfrac{1}{24}\left(1\times3\times4 + 8\times0\times1 + 3\times(-1)\times0 + 6\times1\times2 + 6\times(-1)\times0\right) = 1$$

The four H1s orbitals in methane therefore span $\boxed{A_1 + T_2}$.

As explained in Section 10C.2(a) on page 409, only orbitals of the same symmetry species may have a nonzero overlap. The carbon 2s orbital spans the totally symmetric irreducible representation A_1 as it is unchanged under all symmetry operations in the group. It can therefore form molecular orbitals with the A_1 combination of hydrogen orbitals. Inspection of the T_d character table shows that the carbon p_x, p_y, and p_z orbitals jointly span T_2, so they can form molecular orbitals with the T_2 combinations of hydrogen orbitals.

The character table also shows that the d_{z^2} and $d_{x^2-y^2}$ orbitals on the silicon in SiH_4 jointly span E; note that the d_{z^2} orbital appears as $(3z^2 - r^2)$ in the character table. These d orbitals therefore cannot form molecular orbitals with the A_1 or T_2 combinations of hydrogen orbitals. However, the d_{xy}, d_{yz}, and d_{zx} orbitals on the silicon span T_2 and can therefore form molecular orbitals with the T_2 combinations of hydrogen orbitals.

P10C.3 As explained in Section 10C.1 on page 406, an integral over a region will necessarily vanish unless the integrand, or part thereof, spans the totally symmetric irreducible representation of the point group of the region. In the case of the function $3x^2 - 1$, the -1 part will always span the totally symmetric representation because it is invariant to all symmetry operations. Therefore the integral will $\boxed{\text{not necessarily vanish}}$ in any of the ranges.

P10C.5 The p_1 symmetry adapted linear combination is shown in Fig. 10.18, along with some of the symmetry elements.

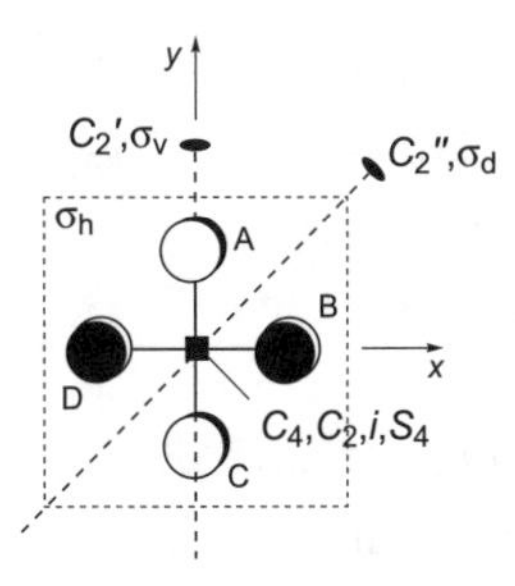

Figure 10.18

This SALC is unaffected by any of the operations E, C_2, C_2'', S_4, and σ_v, but changes sign under C_4, C_2', i, σ_h, and σ_d. The representatives, which are the same as the characters because the representatives are one-dimensional, are therefore

class C	E	C_4	C_2	C_2'	C_2''	i	S_4	σ_h	σ_v	σ_d
$D(C)$	1	−1	1	−1	1	−1	1	−1	1	−1

The D_{4h} character table shows that this representation corresponds to the B_{2u} irreducible representation. The s orbital on the xenon spans the totally symmetric representation A_{1g}, and the D_{4h} character table indicates that p_x and p_y jointly span E_u; p_z spans A_{2u}. It also shows that d_{yz} and d_{zx} jointly span E_g. The orbitals d_{z^2}, $d_{x^2-y^2}$ and d_{xy} span respectively A_{1g}, B_{1g} and B_{2g}. Because only orbitals of the same symmetry may have a non-zero overlap it follows that $\boxed{\text{none}}$ of the s, p and d orbitals on the xenon can form molecular orbitals with the B_{2u} orbital p_1.

P10C.7 The ethene molecule is shown in Fig. 10.19.

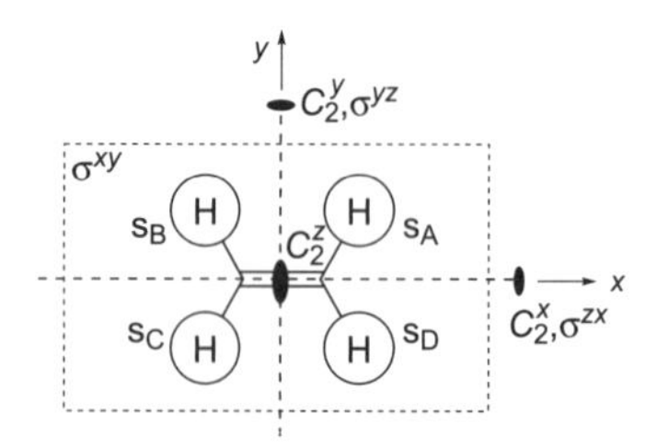

Figure 10.19

The SALCs are generated using the method described in Section 10C.2(b) on page 409, applying each operation to s_A. The results are given in the following table.

Row		E	C_2^z	C_2^y	C_2^x	i	σ^{xy}	σ^{yz}	σ^{zx}
1	effect on s_A	s_A	s_C	s_B	s_D	s_C	s_A	s_B	s_D
2	characters for A_g	1	1	1	1	1	1	1	1
3	product of rows 1 and 2	s_A	s_C	s_B	s_D	s_C	s_A	s_B	s_D
4	characters for B_{2u}	1	−1	1	−1	−1	1	1	−1
5	product of rows 1 and 4	s_A	$-s_C$	s_B	$-s_D$	$-s_C$	s_A	s_B	$-s_D$
6	characters for B_{3u}	1	−1	−1	1	−1	1	−1	1
7	product of rows 1 and 6	s_A	$-s_C$	$-s_B$	s_D	$-s_C$	s_A	$-s_B$	s_D
8	characters for B_{1g}	1	1	−1	−1	1	1	−1	−1
9	product of rows 1 and 8	s_A	s_C	$-s_B$	$-s_D$	s_C	s_A	$-s_B$	$-s_D$
10	characters for B_{1u}	1	1	−1	−1	−1	−1	1	1
11	product of rows 1 and 10	s_A	s_C	$-s_B$	$-s_D$	$-s_C$	$-s_A$	s_B	s_D

The SALCs are formed by summing rows 3, 5, 7, 9 and 11 and dividing each by the order of the group (h = 8).

Row 3: $\psi^{(A_{1g})} = \frac{1}{8}(s_A + s_C + s_B + s_D + s_C + s_A + s_B + s_D) = \boxed{\frac{1}{4}(s_A + s_B + s_C + s_D)}$

Row 5: $\psi^{(B_{2u})} = \frac{1}{8}(s_A - s_C + s_B - s_D - s_C + s_A + s_B - s_D) = \boxed{\frac{1}{4}(s_A + s_B - s_C - s_D)}$

Row 7: $\psi^{(B_{3u})} = \frac{1}{8}(s_A - s_C - s_B + s_D - s_C + s_A - s_B + s_D) = \boxed{\frac{1}{4}(s_A - s_B - s_C + s_D)}$

Row 9: $\psi^{(B_{1g})} = \frac{1}{8}(s_A + s_C - s_B - s_D + s_C + s_A - s_B - s_D) = \boxed{\frac{1}{4}(s_A - s_B + s_C - s_D)}$

Row 11: $\psi^{(B_{1u})} = \frac{1}{8}(s_A + s_C - s_B - s_D - s_C - s_A + s_B + s_D) = 0$

The results from row 11 show that attempting to project out a SALC with symmetry B_{1u} gives $\boxed{\text{zero}}$. This is because the four hydrogen 1s orbitals do not span B_{1u}.

Molecular Spectroscopy

11A General features of molecular spectroscopy

Answers to discussion questions

D11A.1 A selection rule arises because for the molecule to be able to interact with the electromagnetic field and absorb or create a photon of frequency ν, it must possess, at least transiently, a dipole oscillating at that frequency. A gross selection rule specifies the general features that a molecule must have if it is to have a spectrum of a given kind, and specific selection rules express the allowed transitions in terms of the changes in quantum numbers.

For both rotational (microwave) and infrared spectroscopy, the allowed transitions depend on the existence of an oscillating dipole moment which can stir the electromagnetic field into oscillation (and vice versa for absorption). For rotational spectroscopy, this implies that the molecule must have a permanent dipole moment, which is equivalent to an oscillating dipole when the molecule is rotating. In the case of vibrational spectroscopy, the physical basis of the gross selection rule is that the molecule must have a structure that results in an dipole moment that changes when the molecule vibrates.

For rotational Raman spectroscopy the gross selection rule is that the molecule must have an anisotropic polarizability: all molecules other than spherical rotors satisfy this condition. For vibrational Raman spectroscopy the gross selection rule is that the polarizability of the molecule must change as the molecule vibrates. All diatomic molecules satisfy this condition because as the molecule swells and contracts during a vibration, the control of the nuclei over the electrons varies, and the polarizability changes. In polyatomic molecules it can be quite difficult to judge by inspection whether or not the polarizability changes for a particular normal mode; a symmetry analysis using group theory is the best way determining the Raman activity of normal modes.

D11A.3 This is discussed in Section 11A.3 on page 425.

Solutions to exercises

E11A.1(a) The ratio A/B is given by [11A.6a–*420*], $A/B = 8\pi h\nu^3/c^3$; the frequency ν is related to the wavelength though $\nu = c/\lambda$, and to the wavenumber through $\nu = \tilde{\nu}c$.

(i) For X-rays with $\lambda = 70.8$ pm

$$\frac{A}{B} = \frac{8\pi h(c/\lambda)^3}{c^3} = \frac{8\pi h}{\lambda^3} = \frac{8\pi \times (6.6261 \times 10^{-34}\,\text{J s})}{(70.8 \times 10^{-12}\,\text{m})^3} = \boxed{0.0469\,\text{J s m}^{-3}}$$

(ii) For visible light with $\lambda = 500$ nm

$$\frac{A}{B} = \frac{8\pi h}{\lambda^3} = \frac{8\pi \times (6.6261 \times 10^{-34}\,\text{J s})}{(500 \times 10^{-9}\,\text{m})^3} = \boxed{1.33 \times 10^{-13}\,\text{J s m}^{-3}}$$

(iii) For infrared radiation with $\tilde{\nu} = 3000\,\text{cm}^{-1}$

$$\frac{A}{B} = \frac{8\pi h(c\tilde{\nu})^3}{c^3} = 8\pi h\tilde{\nu}^3$$

$$= 8\pi \times (6.6261 \times 10^{-34}\,\text{J s}) \times (3000 \times 10^2\,\text{m}^{-1})^3 = \boxed{4.50 \times 10^{-16}\,\text{J s m}^{-3}}$$

Note the conversion of the wavenumber from cm^{-1} to m^{-1}.

E11A.2(a) The Beer–Lambert law [11A.8–*421*], $I = I_0 10^{-\varepsilon[\text{J}]L}$ relates the intensity of the transmitted light I to that of the incident light I_0.

$$I/I_0 = 10^{-\varepsilon[\text{J}]L} = 10^{-(723\,\text{dm}^3\,\text{mol}^{-1}\,\text{cm}^{-1})\times(4.25\times10^{-3}\,\text{mol dm}^{-3})\times(0.250\,\text{cm})}$$
$$= 0.171...$$

Using this, the percentage reduction in intensity is calculated as $100(I_0 - I)/I_0 = 100(1 - I/I_0) = 100(1 - 0.171...) = \boxed{82.9\%}$. Note the conversion of L to cm and [J] to mol dm^{-3} in order to match the units of ε.

E11A.3(a) The Beer–Lambert law [11A.8–*421*], $I = I_0 10^{-\varepsilon[\text{J}]L}$ relates the intensity of the transmitted light I to that of the incident light I_0. If a fraction T of the incident light passes through the sample, $I = TI_0$ and hence $I/I_0 = T$; T is the transmittance. It follows that $\log T = -\varepsilon[\text{J}]L$ hence $\varepsilon = -(\log T)/[\text{J}]L$. If 18.1% of the light is transmitted, $T = 0.181$

$$\varepsilon = -(\log T)/[\text{J}]L = -[\log(0.181)]/(0.139 \times 10^{-3}\,\text{mol dm}^{-3}) \times (1.00\,\text{cm})$$
$$= \boxed{5.34 \times 10^3\,\text{dm}^3\,\text{mol}^{-1}\,\text{cm}^{-1}}$$

Note the use of L in cm and the conversion of [J] to mol dm^{-3} in order to give the usual units of ε.

E11A.4(a) The Beer–Lambert law [11A.8–*421*], $I = I_0 10^{-\varepsilon[\text{J}]L}$ relates the intensity of the transmitted light I to that of the incident light I_0. If a fraction α is absorbed, then a fraction $T = 1 - \alpha$ of the incident light passes through the sample, $I = TI_0$ and hence $I/I_0 = T$; T is the transmittance. It follows that $\log T = -\varepsilon[\text{J}]L$ hence $[\text{J}] = -(\log T)/\varepsilon L$. If 38.5% of the light is absorbed, $\alpha = 0.385$ and $T = 1 - 0.385 = 0.615$

$$[\text{J}] = -(\log T)/\varepsilon L = -[\log(0.615)]/(386\,\text{dm}^3\,\text{mol}^{-1}\,\text{cm}^{-1}) \times (0.500\,\text{cm})$$
$$= 1.09... \times 10^{-3}\,\text{mol dm}^{-3} = \boxed{1.09\,\text{mM}}$$

Note the use of L in cm.

E11A.5(a) The transmittance T is the ratio I/I_0, hence the Beer–Lambert law [11A.8–*421*] can be written $T = I/I_0 = 10^{-\varepsilon[\mathrm{J}]L}$. It follows that $\log T = -\varepsilon[\mathrm{J}]L$ and hence $\varepsilon = -(\log T)/[\mathrm{J}]L$. With this, the following table is drawn up, with $L = 0.20$ cm

$[\text{dye}]/(\text{mol dm}^{-3})$	0.0010	0.0050	0.0100	0.0500
$T/\%$	81.4	35.6	12.7	3.0×10^{-3}
T	0.814	0.356	0.127	3.0×10^{-5}
$\varepsilon/(\text{dm}^3\,\text{mol}^{-1}\,\text{cm}^{-1})$	447	449	448	452

The average value of ε from these measurements is $\boxed{449\ \text{dm}^3\,\text{mol}^{-1}\,\text{cm}^{-1}}$.

E11A.6(a) The transmittance T is the ratio I/I_0, hence the Beer–Lambert law [11A.8–*421*] can be written $T = I/I_0 = 10^{-\varepsilon[\mathrm{J}]L}$. It follows that $\log T = -\varepsilon[\mathrm{J}]L$ and hence $\varepsilon = -(\log T)/[\mathrm{J}]L$. With the given data, $T = 0.48$ and $L = 0.20$ cm, the molar absorption coefficient is calculated as

$$\varepsilon = -(\log 0.48)/[(0.010\,\text{mol dm}^{-3})\times(0.20\,\text{cm})] = 1.59...\times 10^2\,\text{dm}^3\,\text{mol}^{-1}\,\text{cm}^{-1}$$

The molar absorption coefficient is therefore $\boxed{\varepsilon = 1.6 \times 10^2\ \text{dm}^3\,\text{mol}^{-1}\,\text{cm}^{-1}}$. For a path length of 0.40 cm the transmittance is

$$T = 10^{-(1.59...\times 10^2\,\text{dm}^3\,\text{mol}^{-1}\,\text{cm}^{-1})\times(0.40\,\text{cm})\times(0.01\,\text{mol dm}^{-3})} = 0.230$$

Hence $\boxed{T = 23\%}$.

E11A.7(a) The ratio of the incident to the transmitted intensities of light after passing through a sample of length L, molar concentration $[\mathrm{H_2O}]$, and molar absorption coefficient ε is given by [11A.8–*421*], $T = I/I_0 = 10^{-\varepsilon[\mathrm{H_2O}]L}$. It follows that $\log T = -\varepsilon[\mathrm{H_2O}]L$, which rearranges to give $L = -(\log T)/\varepsilon[\mathrm{H_2O}]$.

The molar concentration of H_2O is calculated by noting that its mass density is $\rho = 1000\ \text{kg m}^{-3}$ and its molar mass is $M = 18.016\ \text{g mol}^{-1}$. The concentration is therefore $\rho/M = (1000\ \text{kg m}^{-3})/(18.016 \times 10^{-3}\ \text{kg mol}^{-1}) = 55.5... \times 10^3\ \text{mol m}^{-3} = 55.5...\ \text{mol dm}^{-3}$.

The light intensity is half that at the surface when $T = 0.5$, hence the depth is calculated as

$$\begin{aligned} L &= -(\log 0.5)/[(6.2 \times 10^{-5}\,\text{dm}^3\,\text{mol}^{-1}\,\text{cm}^{-1}) \times (55.5...\,\text{mol dm}^{-3})] \\ &= 87.4...\,\text{cm} = \boxed{0.875\ \text{m}} \end{aligned}$$

The light intensity reaches a tenth of at the surface when $T = 0.1$

$$\begin{aligned} L &= -(\log 0.1)/[(6.2 \times 10^{-5}\,\text{dm}^3\,\text{mol}^{-1}\,\text{cm}^{-1}) \times (55.5...\,\text{mol dm}^{-3})] \\ &= 290...\,\text{cm} = \boxed{2.90\ \text{m}} \end{aligned}$$

E11A.8(a) The integrated absorption coefficient is given by [11A.10–*423*], $\mathcal{A} = \int_{\text{band}} \varepsilon(\tilde{\nu})\, d\tilde{\nu}$, where the integration is over the band and $\tilde{\nu} = \lambda^{-1}$ is the wavenumber. The initial, peak, and final wavenumbers of the lineshape are given by $(220\times10^{-7}\text{ cm})^{-1} = 4.54...\times10^4\text{ cm}^{-1}$, $(270\times10^{-7}\text{ cm})^{-1} = 3.70...\times10^4\text{ cm}^{-1}$ and $(300\times10^{-7}\text{ cm})^{-1} = 3.33...\times10^4\text{ cm}^{-1}$.

Assuming that the lineshape is triangular the area under it is $\frac{1}{2}\times\text{base}\times\text{height}$

$$\begin{aligned}\mathcal{A} &= \tfrac{1}{2}\times[(4.54... - 3.33...)\times10^4\text{ cm}^{-1}]\times(2.21\times10^4\text{ dm}^3\text{ mol}^{-1}\text{ cm}^{-1})\\ &= 1.33...\times10^8\text{ dm}^3\text{ mol}^{-1}\text{ cm}^{-2}\end{aligned}$$

The integrated absorption coefficient is therefore $\boxed{1.34\times10^8\text{ dm}^3\text{ mol}^{-1}\text{ cm}^{-2}}$.

E11A.9(a) The Doppler linewidth is given by [11A.12a–*424*], $\delta\nu_{\text{obs}} = (2\nu_0/c)(2kT\ln 2/m)^{1/2}$. Because $\lambda = c/\nu$ this may be rewritten $\delta\nu_{\text{obs}} = (2/\lambda_0)(2kT\ln 2/m)^{1/2}$. Taking the mass of a hydrogen atom as $1\,m_u$ gives the linewidth as

$$\begin{aligned}\delta\nu_{\text{obs}} &= (2/\lambda_0)(2kT\ln 2/m)^{1/2}\\ &= [2/(821\times10^{-9}\text{ m})]\times\left(\frac{2\times(1.3806\times10^{-23}\text{ J K}^{-1})\times(300\text{ K})\times\ln 2}{1.6605\times10^{-27}\text{ kg}}\right)^{1/2}\\ &= 4.53...\times10^9\text{ Hz}\end{aligned}$$

where $1\text{ J} = 1\text{ kg m}^2\text{ s}^{-2}$ is used. Expressed as a wavenumber the linewidth is $(4.53...\times10^9\text{ Hz})/(2.9979\times10^{10}\text{ cm s}^{-1}) = \boxed{0.151\text{ cm}^{-1}}$.

E11A.10(a) If a light source of frequency ν_0 is approached at a speed s, the Doppler shifted frequency ν_a is [11A.11a–*423*],

$$\nu_a = \nu_0\left(\frac{1+s/c}{1-s/c}\right)^{1/2}$$

Writing the frequencies in terms of the wavelength as $\nu = c/\lambda$ and then inverting before sides gives

$$\lambda_a = \lambda_0\left(\frac{1-s/c}{1+s/c}\right)^{1/2}$$

At nonrelativistic speeds, $s \ll c$, this simplifies to $\lambda_a = \lambda_0\,(1-s/c)^{1/2}$. Hence

$$\begin{aligned}\lambda_a = (680\text{ nm})\times[1 - (60\text{ km h}^{-1})\times(1\text{ h}/3600\text{ s})\\ \times(1000\text{ m}/1\text{ km})/(2.9979\times10^8\text{ m s}^{-1})]^{1/2} = \boxed{680\text{ nm}}\end{aligned}$$

Within the precision of the data given, the Doppler shift is insignificant.

E11A.11(a) The uncertainty in the energy of a state with lifetime τ is $\delta E \approx \hbar/\tau$. Therefore a spectroscopic transition involving this state has an uncertainty in its frequency, and hence a linewidth, of the order of $\delta\nu = \delta E/h \approx (2\pi\tau)^{-1}$. This expression is rearranged to give the lifetime as $\tau = (2\pi\delta\nu)^{-1}$; expressing the linewidth as a wavenumber gives $\tau = (2\pi\delta\tilde{\nu}c)^{-1}$.

(i) For $\delta\tilde{\nu} = 0.20\ \mathrm{cm^{-1}}$

$$\tau = [2\pi\times(0.20\ \mathrm{cm^{-1}})\times(2.9979\times10^{10}\ \mathrm{cm\,s^{-1}})]^{-1} = 2.65...\times10^{-11}\ \mathrm{s} = \boxed{27\ \mathrm{ps}}$$

(ii) For $\delta\tilde{\nu} = 2.0\ \mathrm{cm^{-1}}$

$$\tau = [2\pi\times(2.0\ \mathrm{cm^{-1}})\times(2.9979\times10^{10}\ \mathrm{cm\,s^{-1}})]^{-1} = 2.65...\times10^{-12}\ \mathrm{s} = \boxed{2.7\ \mathrm{ps}}$$

E11A.12(a) The uncertainty in the energy of a state with lifetime τ is $\delta E \approx \hbar/\tau$. Therefore a spectroscopic transition involving this state has an uncertainty in its frequency, and hence a linewidth, of the order of $\delta\nu = \delta E/h \approx (2\pi\tau)^{-1}$. If the linewidth is expressed as a wavenumber the expression becomes $\delta\tilde{\nu} = \delta E/hc \approx (2\pi\tau c)^{-1}$.

If each collision deactivates the molecule, the lifetime is 1/(collision frequency), but if only 1 in N of the collisions deactivates the molecule, the lifetime is N/(collision frequency). Thus $\tau = N/z$, where z is the collision frequency. The linewidth is therefore $\delta\tilde{\nu} = (2\pi cN/z)^{-1}$.

(i) If each collision is effective at deactivation, $N = 1$ and with the data given

$$\delta\tilde{\nu} = [2\pi\times(2.9979\times10^{10}\ \mathrm{cm\,s^{-1}})\times1/(1.0\times10^{13}\ \mathrm{s^{-1}})]^{-1} = \boxed{53\ \mathrm{cm^{-1}}}$$

(ii) If only 1 in 100 collisions are effective at deactivation, $N = 100$

$$\delta\tilde{\nu} = [2\pi\times(2.9979\times10^{10}\ \mathrm{cm\,s^{-1}})\times100/(1.0\times10^{13}\ \mathrm{s^{-1}})]^{-1} = \boxed{0.53\ \mathrm{cm^{-1}}}$$

Solutions to problems

P11A.1 The fraction of the incident photons that reach the retina is $(1-0.30)\times(1-0.25)\times(1-0.09)\times(1-0.43) = 0.272....$ Hence the number of photons reaching the retina in 0.1 s is

$$(4.0\times10^{3}\ \mathrm{mm^{-2}\,s^{-1}})\times(40\ \mathrm{mm^{2}})\times(0.1\ \mathrm{s})\times0.272... = \boxed{4.4\times10^{3}}$$

P11A.3 The absorbance at λ_1 and λ_2 are A_1 and A_2, respectively

$$A_1 = \varepsilon_{A1}[A]L + \varepsilon_{B1}[B]L \qquad (11.1)$$
$$A_2 = \varepsilon_{A2}[A]L + \varepsilon_{B2}[B]L \qquad (11.2)$$

At each wavelength the absorbance depends on the concentration of each species and the relevant molar absorption coefficient. Equation 11.1 is multiplied by ε_{A2} and eqn 11.2 is multiplied by ε_{A1} to give

$$\varepsilon_{A2}A_1 = \varepsilon_{A2}\varepsilon_{A1}[A]L + \varepsilon_{A2}\varepsilon_{B1}[B]L$$
$$\varepsilon_{A1}A_2 = \varepsilon_{A1}\varepsilon_{A2}[A]L + \varepsilon_{A1}\varepsilon_{B2}[B]L$$

Subtracting the two equations eliminates [A], and rearrangement gives the required expression for [B]

$$\varepsilon_{A2}A_1 - \varepsilon_{A1}A_2 = \varepsilon_{A2}\varepsilon_{B1}[B]L - \varepsilon_{A1}\varepsilon_{B2}[B]L \qquad [B] = \frac{\varepsilon_{A2}A_1 - \varepsilon_{A1}A_2}{(\varepsilon_{A2}\varepsilon_{B1} - \varepsilon_{A1}\varepsilon_{B2})L}$$

Simply exchanging the labels A and B gives the corresponding expression for [A]

$$[A] = \frac{\varepsilon_{B2}A_1 - \varepsilon_{B1}A_2}{(\varepsilon_{B2}\varepsilon_{A1} - \varepsilon_{B1}\varepsilon_{A2})L}$$

P11A.5 Following the hint, a plot is made of $\ln \varepsilon$ against $\tilde{\nu}$; the data are shown in the following table and the plot is shown in Fig. 11.1.

λ/nm	$\varepsilon/(\mathrm{dm^3\ mol^{-1}\ cm^{-1}})$	$\ln[\varepsilon/(\mathrm{dm^3\ mol^{-1}\ cm^{-1}})]$	$\tilde{\nu}/(10^4\ \mathrm{cm^{-1}})$
292.0	1512	7.32	3.425
296.3	865	6.76	3.375
300.8	477	6.17	3.324
305.4	257	5.55	3.274
310.1	135.9	4.91	3.225
315.0	69.5	4.24	3.175
320.0	34.5	3.54	3.125

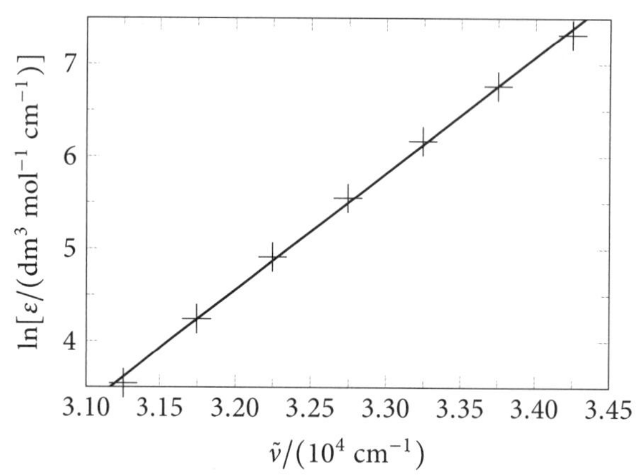

Figure 11.1

The data are quite a good fit to the line

$$\ln(\varepsilon/\mathrm{dm^3\ mol^{-1}\ cm^{-1}}) = 12.609 \times [\tilde{\nu}/(10^4\ \mathrm{cm^{-1}})] - 35.793$$

This can be expressed as $\ln(\varepsilon/\mathrm{dm^3\ mol^{-1}\ cm^{-1}}) = a(\tilde{\nu}/\mathrm{cm^{-1}}) + b$ with $a = 1.2609 \times 10^{-3}$ and $b = -35.793$. It follows that $\varepsilon = e^{a\tilde{\nu}}e^{b}$, where the units have been omitted for clarity. With this expression for ε, the integrated absorption coefficient is found by evaluating the integral

$$\begin{aligned}\mathcal{A} &= \int_{\tilde{\nu}_{\min}}^{\tilde{\nu}_{\max}} \varepsilon\, d\tilde{\nu} = \int_{\tilde{\nu}_{\min}}^{\tilde{\nu}_{\max}} e^{a\tilde{\nu}} e^{b}\, d\tilde{\nu} = e^{b}\ (1/a)e^{a\tilde{\nu}}\Big|_{\tilde{\nu}_{\min}}^{\tilde{\nu}_{\max}} \\ &= e^{b}(1/a)\left(e^{a\tilde{\nu}_{\max}} - e^{a\tilde{\nu}_{\min}}\right) \\ &= e^{-35.793}\frac{1}{1.2609\times 10^{-3}}\left(e^{(1.2609\times10^{-3})\times(3.425\times10^{4})} - e^{(1.2609\times10^{-3})\times(3.125\times10^{4})}\right) \\ &= \boxed{1.26\times10^{6}\ \mathrm{dm^{3}\ mol^{-1}\ cm^{-2}}}\end{aligned}$$

Again, units have been omitted for clarity.

P11A.7 (a) The area of a triangle is $\frac{1}{2}$ × base × height, so the integrated absorption coefficient is

$$\begin{aligned}\mathcal{A} &= \tfrac{1}{2}\times[(34483-31250)\ \mathrm{cm^{-1}}]\times(150\ \mathrm{dm^{3}\ mol^{-1}\ cm^{-1}}) \\ &= \boxed{2.42\times10^{5}\ \mathrm{dm^{3}\ mol^{-1}\ cm^{-2}}}\end{aligned}$$

(b) Assume that the equilibrium involved is $2\,\mathrm{M} \rightleftharpoons \mathrm{M_2}$, where M is CH_3I. The total pressure is known, and from this it is possible to compute the total concentration of M and M_2 together; the fraction of the total present as M_2 is also known. Using these data it is possible to find the concentration of M, and hence the absorbance.

Suppose that initially there are n_0 moles of M which then come to equilibrium by forming n moles of M_2: the amount in moles of M is then $n_{\mathrm{M}} = n_0 - 2n$, and the total amount in moles of all species is $n_{\mathrm{tot}} = n_0 - n$. Let the fraction that is present as dimer be α, $\alpha = n/n_{\mathrm{tot}}$.

The aim is to express $n_{\mathrm{M}} = n_0 - 2n$ in terms of the known quantities n_{tot} and α

$$n_{\mathrm{M}} = \overbrace{n_0}^{=n_{\mathrm{tot}}+n} - 2n = n_{\mathrm{tot}} - \overbrace{n}^{=\alpha n_{\mathrm{tot}}} = n_{\mathrm{tot}}(1-\alpha)$$

Assuming that the perfect gas law applies

$$c_{\mathrm{tot}} = \frac{n_{\mathrm{tot}}}{V} = \frac{p}{RT}$$

where c_{tot} is the total concentration of both M and M_2. It follows that the concentration of M is

$$[\mathrm{M}] = \frac{n_{\mathrm{M}}}{V} = \frac{n_{\mathrm{tot}}(1-\alpha)}{V} = c_{\mathrm{tot}}(1-\alpha) = \frac{p}{RT}(1-\alpha)$$

With the data given

$$\begin{aligned}[\mathrm{M}] &= \frac{(2.4\ \mathrm{Torr})\times[(1\ \mathrm{atm})/(760\ \mathrm{Torr})]\times[(1.01325\times10^{5}\ \mathrm{Pa})/(1\ \mathrm{atm})]}{(8.3145\ \mathrm{J\ K^{-1}\ mol^{-1}})\times(373\ \mathrm{K})} \\ &\quad\times(1-0.01) = 0.102...\ \mathrm{mol\ m^{-3}} = 1.02...\times10^{-4}\ \mathrm{mol\ dm^{-3}}\end{aligned}$$

The absorbance at the mid-point is

$$A = \varepsilon[\mathrm{M}]L$$
$$= (150\ \mathrm{dm^3\ mol^{-1}\ cm^{-1}}) \times (1.02... \times 10^{-4}\ \mathrm{mol\,dm^{-3}}) \times (12.0\ \mathrm{cm})$$
$$= \boxed{0.18}$$

(c) With the data at 100 Torr and 18% dimers, the concentration of the monomer is

$$[\mathrm{M}] = \frac{(100\ \mathrm{Torr})\times[(1\ \mathrm{atm})/(760\ \mathrm{Torr})]\times[(1.01325\times10^5\ \mathrm{Pa})/(1\ \mathrm{atm})]}{(8.3145\ \mathrm{J\,K^{-1}\,mol^{-1}})\times(373\ \mathrm{K})}$$
$$\times(1-0.18) = 3.52...\ \mathrm{mol\,m^{-3}} = 3.52... \times 10^{-3}\ \mathrm{mol\,dm^{-3}}$$

The absorbance at the mid-point is

$$A = \varepsilon[\mathrm{M}]L$$
$$= (150\ \mathrm{dm^3\ mol^{-1}\ cm^{-1}}) \times (3.52... \times 10^{-3}\ \mathrm{mol\,dm^{-3}}) \times (12.0\ \mathrm{cm})$$
$$= 6.34...$$

The absorbance at the mid-point is therefore $\boxed{A = 6.35}$. From this value the molar absorption coefficient would be inferred as $\varepsilon = A/c_{\mathrm{tot}}L$ and c_{tot} is computed as before from the pressure

$$c_{\mathrm{tot}} = \frac{(100\ \mathrm{Torr})\times[(1\ \mathrm{atm})/(760\ \mathrm{Torr})]\times[(1.01325\times10^5\ \mathrm{Pa})/(1\ \mathrm{atm})]}{(8.3145\ \mathrm{J\,K^{-1}\,mol^{-1}})\times(373\ \mathrm{K})}$$
$$= 4.29...\ \mathrm{mol\,m^{-3}} = 4.29... \times 10^{-3}\ \mathrm{mol\,dm^{-3}}$$

Hence

$$\varepsilon = A/c_{\mathrm{tot}}L = (6.34...)/[(4.29... \times 10^{-3}\ \mathrm{mol\,dm^{-3}}) \times (12.0\ \mathrm{cm})]$$
$$= \boxed{123\ \mathrm{dm^3\ mol^{-1}\ cm^{-1}}}$$

P11A.9 The line from the star is at longer wavelength, and hence lower frequency, than for the Earth-bound observation, therefore the object is receding. The Doppler shift is given by [11A.11a–*423*]

$$f = \nu_\mathrm{r}/\nu_0 = \left(\frac{1-(s/c)}{1+(s/c)}\right)^{1/2}$$

The ratio f is equal to $\lambda_0/\lambda_\mathrm{r}$ because the frequency is inversely proportional to the wavelength. Writing $x = s/c$ gives

$$f = \left(\frac{1-x}{1+x}\right)^{1/2} \quad \text{hence} \quad f^2(1+x) = (1-x) \quad \text{hence} \quad x = \frac{1-f^2}{1+f^2}$$

It follows that $s = c[1-(\lambda_0/\lambda_\mathrm{r})^2]/[1+(\lambda_0/\lambda_\mathrm{r})^2]$.

$$s = (2.9979\times10^8\ \mathrm{m\,s^{-1}})\times\frac{1-[(654.2\ \mathrm{nm})/(706.5\ \mathrm{nm})]^2}{1+[(654.2\ \mathrm{nm})/(706.5\ \mathrm{nm})]^2} = \boxed{2.301\times10^6\ \mathrm{m\,s^{-1}}}$$

The Doppler linewidth is given by [11A.12a–*424*], $\delta\nu/\nu_0 = (2/c)(2kT\ln 2/m)^{1/2}$. Provided that the linewidth is small compared to the absolute frequency of the line (which is the case here), $\delta\nu/\nu_0$ is well approximated by $\delta\lambda/\lambda_0$

$$\frac{\delta\lambda}{\lambda_0} = \frac{2}{c}\left(\frac{2kT\ln 2}{m}\right)^{1/2} \quad \text{hence} \quad T = \left(\frac{\delta\lambda}{\lambda_0}\right)^2 \frac{c^2 m}{8k\ln 2}$$

With the data given

$$T = \left(\frac{0.0618\ \text{nm}}{706.5\ \text{nm}}\right)^2 \frac{(2.9979\times 10^8\ \text{m s}^{-1})^2 \times 47.95 \times (1.6605\times 10^{-27}\ \text{kg})}{8\times(1.3806\times 10^{-23}\ \text{J K}^{-1})\times \ln 2}$$
$$= \boxed{7.15\times 10^5\ \text{K}}$$

P11A.11 If each collision is effective at changing the energy of a state, the lifetime is simply the inverse of the collision rate: $\boxed{\tau = 1/z}$.

The uncertainty in the energy of a state with lifetime τ is $\delta E \approx \hbar/\tau$. Therefore a spectroscopic transition involving this state has an uncertainty in its frequency, and hence a linewidth, of the order of $\delta\nu = \delta E/h \approx (2\pi\tau)^{-1}$. Using $\tau = 1/z$ and the given expression for z gives the linewidth as

$$\delta\nu = 1/2\pi\tau = z/2\pi = \frac{4\sigma}{2\pi}\left(\frac{kT}{\pi m}\right)^{1/2}\frac{p}{kT} = \left(\frac{4\sigma^2}{\pi^3 mkT}\right)^{1/2} p$$

With the given data and taking $m = 36\ m_\text{u}$

$$\delta\nu = \left(\frac{4\times(0.30\times 10^{-18}\ \text{m}^2)^2}{\pi^3\times(36)\times(1.6605\times 10^{-27}\ \text{kg})\times(1.3806\times 10^{-23}\ \text{J K}^{-1})\times(298\ \text{K})}\right)^{1/2}$$
$$\times (1.01325\times 10^5\ \text{Pa}) = \boxed{0.70\ \text{GHz}}$$

The Doppler linewidth is given by [11A.12a–*424*], $\delta\nu/\nu_0 = (2/c)(2kT\ln 2/m)^{1/2}$; with the data given

$$\delta\nu = \frac{2\nu_0}{c}\left(\frac{2kT\ln 2}{m}\right)^{1/2} = \frac{2\tilde{\nu}_0 c}{c}\left(\frac{2kT\ln 2}{m}\right)^{1/2} = 2\tilde{\nu}_0\left(\frac{2kT\ln 2}{m}\right)^{1/2}$$
$$= 2\times(6356\ \text{m}^{-1})\times\left(\frac{2\times(1.3806\times 10^{-23}\ \text{J K}^{-1})\times(298\ \text{K})\times\ln 2}{(36)\times(1.6605\times 10^{-27}\ \text{kg})}\right)^{1/2}$$
$$= 3.93\ \text{MHz}$$

Note that $\tilde{\nu}_0$ is used in m^{-1}. For the collisional broadening to be equal to the Doppler broadening the former must be reduced by a factor $700/3.93 = 178$; because the linewidth is proportional to the pressure, this means that the pressure must be reduced by this factor to $(1.01325\times 10^5\ \text{Pa})/178 = \boxed{569\ \text{Pa}}$ or $\boxed{4.27\ \text{Torr}}$.

P11A.13 The best way to approach this is to generate the interferogram in a numerical form, that is as a table of data points. As is seen in the previous *Problem* it is necessary to have at least two data points per cycle in order to represent the wavenumber correctly, which implies that the distance by which the mirror must be moved in one step is $\delta = 1/2\tilde{\nu}_{\text{max}}$, where $\tilde{\nu}_{\text{max}}$ is the highest wavenumber which will be represented correctly. The ith data point in the interferogram is constructed using the expression

$$I_i = \sum_j \{a_j[1 + \cos(2\pi\tilde{\nu}_j i\delta)]\}\ \overbrace{e^{-\alpha(i\delta)^2}}^{\text{apodization}}$$

where a_j and $\tilde{\nu}_j$ are the intensity and wavenumber, respectively, of the jth peak in the spectrum; i runs from 0 to N, the number of data points. The value of N is a matter of choice, but a sensible starting value might be 256; the reason for this apparently odd choice is that some numerical implementations of the Fourier transform require that the number of points be a power of 2 ($256 = 2^8$).

The apodization term is there in order to force the interferogram to go smoothly to zero (or at least near to zero) for the largest value of the pathlength difference $N\delta$. If this is not done, the peaks in the spectrum will have 'wiggles' around their bases, as seen in Fig. 11A.2 on page 421. In a practical spectrometer this term might not be required because with radiation passing through the interferometer covers a wide range of frequencies and interference between these will naturally drive the interferogram to zero. In this simulation, with only a few frequencies present, apodization is required. The parameter α is adjusted to achieve the desired smoothing of the envelope.

Figure 11.2 shows an interferogram computed using the following parameters; the data points have been joined up by a continuous line

$$N = 256 \quad \tilde{\nu}_{\text{max}} = 100\ \text{cm}^{-1} \quad \delta = 1/(2 \times 100\ \text{cm}^{-1}) = 0.005\ \text{cm} \quad \alpha = 2.5\ \text{cm}^{-2}$$

$$a_1 = 0.25\ \ \tilde{\nu}_1 = 5.0\ \text{cm}^{-1} \quad a_2 = 1.00\ \ \tilde{\nu}_2 = 15\ \text{cm}^{-1} \quad a_3 = 0.75\ \ \tilde{\nu}_3 = 50\ \text{cm}^{-1}$$

To find the spectrum it is necessary to compute the Fourier transform of the interferogram. There are many variants of the way this transform is implemented as a numerical procedure, and the one needed here is usually referred to as the discrete cosine Fourier transform. As can be seen from Fig. 11.2, the interferogram is always positive and decays to zero; this will give a large peak in the spectrum at a wavenumber of zero, in addition to the peaks corresponding to the wavenumbers of the oscillating terms that have been introduced. Figure 11.3 shows the spectrum obtained by Fourier transformation of the interferogram; the peak at zero wavenumber has been truncated.

11B Rotational spectroscopy

Answers to discussion questions

D11B.1 *Symmetric rotor*: The energy depends on J and K^2, hence each level except the $K = 0$ level is doubly degenerate. In addition, states of a given J have $(2J + 1)$

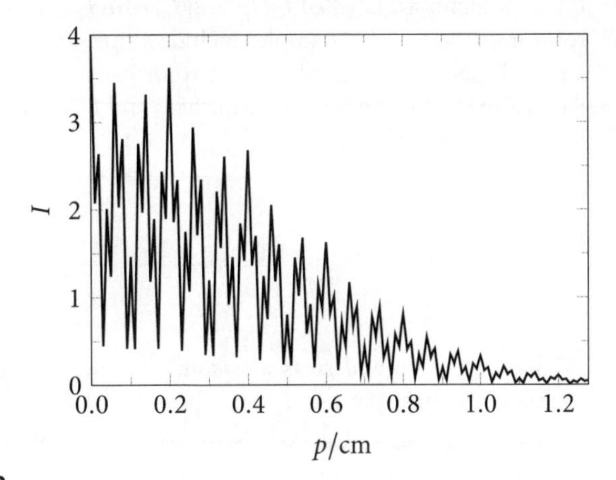

Figure 11.2

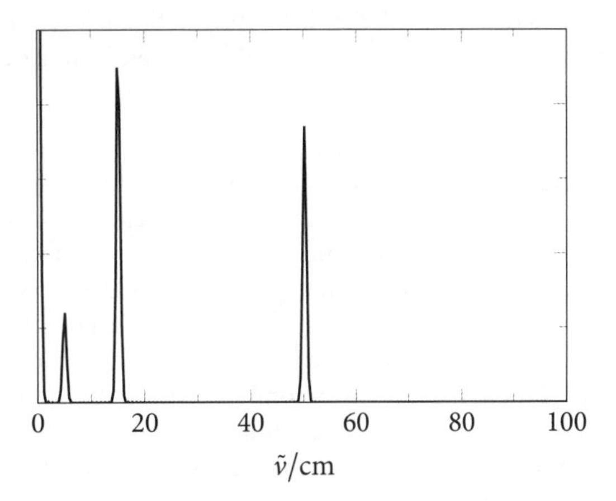

Figure 11.3

values of the component of their angular momentum along an external axis, characterized by the quantum number M_J. The energy is not affected by M_J, so there is a degeneracy of $2J + 1$ for each J. It follows that a symmetric rotor level is $2(2J + 1)$-fold degenerate for $K \neq 0$, and $2J + 1$ degenerate for $K = 0$.

Linear rotor: A linear rotor has K fixed at 0, but there are still $2J + 1$ values of M_J, so the degeneracy is $2J + 1$.

Spherical rotor: A spherical rotor can be regarded as a version of a symmetric rotor in which $A = B$; consequently the energy is independent of the $2J + 1$ values that K can assume. Hence, there is a degeneracy of $2J + 1$ associated with both K and M_J, resulting in a total degeneracy of $(2J + 1)^2$.

If a decrease in rigidity affects the symmetry of the molecule, the rotational degeneracy could be affected also.

D11B.3 A molecule has three principal moments of inertia about perpendicular axes:

these moments are labelled I_a, I_b, and I_c, with $I_c \geq I_b \geq I_a$. A prolate symmetric rotor has $I_a \neq I_b = I_c$; examples include a thin rod, any linear molecule, CH_3F and CH_3CN. An oblate symmetric rotor has $I_a = I_b \neq I_c$; examples include a flat disc, benzene and BF_3. In terms of $I_{||}$ and $I_{\perp}$, prolate rotors have $I_{||} < I_{\perp}$ and oblate tops have $I_{||} > I_{\perp}$.

D11B.5 This is discussed in Section 11B.3 on page 437.

D11B.7 ^{12}C has spin zero and so is a boson; ^{13}C and ^1H have spin half and so are fermions; ^2H has spin 1 and so is a boson. All the molecules are linear so the same considerations as described in Section 11B.4 on page 439 apply.

For $^1H^{12}C \equiv {}^{12}C^1H$ the ^{12}C have no effect as they are spin 0, so the rotational levels behave in just the same way as $^1H_2$: the (odd J)/(even J) statistical weight ratio is therefore 3/1. Similarly, the rotational levels of $^2H^{12}C \equiv {}^{12}C^2H$ behave in just the same way as $^2H_2$: the (odd J)/(even J) statistical weight ratio is therefore $I/(I+1) = 1/2$.

For $^1H^{13}C \equiv {}^{13}C^1H$ there are four nuclear spin wavefunctions arising from the ^1H nuclei, three symmetric and one antisymmetric with respect to exchange of the nuclei. In addition there are four nuclear spin wavefunctions arising from the ^{13}C nuclei, three symmetric and one antisymmetric. Overall, there are 16 nuclear spin wavefunctions. Of these, 9 arise from combining a symmetric wavefunction for $^1H_2$ and a symmetric wavefunction for $^{13}C_2$, giving overall symmetric wavefunctions. In addition there is one more overall symmetric wavefunction obtained by combining the antisymmetric wavefunction for $^{13}C_2$ with that for $^1H_2$. The total number of symmetric wavefunctions is therefore 10, and the remaining 6 are therefore antisymmetric.

The ratio of symmetric to antisymmetric nuclear spin functions is therefore 10/6, therefore the (odd J)/(even J) statistical weight ratio is 10/6.

Solutions to exercises

E11B.1(a) The moment of inertia I of a molecule about a specified axis is given by [11B.2–430], $I = \sum_i m_i r_i^2$ where the sum is over all the atoms, m_i is the mass of atom i and r_i is its perpendicular distance to the axis. For the calculation of the moment of inertia about the bisector, the central atom makes no contribution.

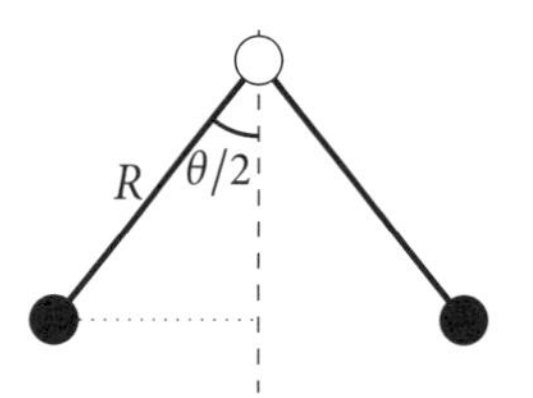

Each of the other atoms is at a perpendicular distance $R\sin(\theta/2)$, where θ is the bond angle and R the bond length. The moment of inertia is therefore

$$\begin{aligned} I &= 2\times m_{\mathrm{O}}R^2\sin^2(\theta/2) \\ &= 2\times(15.9949)\times(1.6605\times10^{-27}\,\mathrm{kg}) \\ &\qquad \times[(128\times10^{-12}\,\mathrm{m})\times\sin(117°/2)]^2 \\ &= 6.32...\times10^{-46}\,\mathrm{kg\,m^2} = \boxed{6.33\times10^{-46}\,\mathrm{kg\,m^2}} \end{aligned}$$

The corresponding rotational constant is given by [11B.7–*432*],

$$\begin{aligned} \tilde{B} &= \frac{\hbar}{4\pi cI} = \frac{1.0546\times10^{-34}\,\mathrm{J\,s}}{4\pi\times(2.9979\times10^{10}\,\mathrm{cm\,s^{-1}})\times(6.32...\times10^{-46}\,\mathrm{kg\,m^2})} \\ &= \boxed{0.442\,\mathrm{cm^{-1}}} \end{aligned}$$

E11B.2(a) The required expressions are the first listed under symmetric rotors in Table 11B.1 on page 431

$$\begin{aligned} I_\perp &= m_{\mathrm{A}}f_1(\theta)R^2 + \frac{m_{\mathrm{A}}(m_{\mathrm{B}}+m_{\mathrm{A}})}{m}f_2(\theta)R^2 \\ &\quad + \frac{m_{\mathrm{C}}}{m}\left\{(3m_{\mathrm{A}}+m_{\mathrm{B}})R' + 6m_{\mathrm{A}}R[\tfrac{1}{3}f_2(\theta)]^{1/2}\right\}R' \\ I_\parallel &= 2m_{\mathrm{A}}f_1(\theta)R^2 \end{aligned}$$

Note that the molecule described by these relationships is BA_3C, which becomes BA_4 by letting C=A; the question refers to a molecule AB_4, but for consistency with the main text the exercise will be continued with BA_4. Let $m_{\mathrm{C}} = m_{\mathrm{A}}$ and $R' = R$ to give

$$\begin{aligned} I_\perp &= m_{\mathrm{A}}f_1(\theta)R^2 + \frac{m_{\mathrm{A}}(m_{\mathrm{B}}+m_{\mathrm{A}})}{m}f_2(\theta)R^2 \\ &\quad + \frac{m_{\mathrm{A}}}{m}\left\{(3m_{\mathrm{A}}+m_{\mathrm{B}}) + 6m_{\mathrm{A}}[\tfrac{1}{3}f_2(\theta)]^{1/2}\right\}R^2 \end{aligned}$$

with $m = m_{\mathrm{B}} + 4m_{\mathrm{A}}$. To simplify the expression somewhat let $m_{\mathrm{B}} = \alpha m_{\mathrm{A}}$. This gives $m = \alpha m_{\mathrm{A}} + 4m_{\mathrm{A}} = (4+\alpha)m_{\mathrm{A}}$

$$\begin{aligned} I_\perp &= m_{\mathrm{A}}f_1(\theta)R^2 + \frac{m_{\mathrm{A}}^2(1+\alpha)}{(4+\alpha)m_{\mathrm{A}}}f_2(\theta)R^2 \\ &\quad + \frac{m_{\mathrm{A}}}{(4+\alpha)m_{\mathrm{A}}}\left\{m_{\mathrm{A}}(3+\alpha) + 6m_{\mathrm{A}}[\tfrac{1}{3}f_2(\theta)]^{1/2}\right\}R^2 \\ I_\perp/(m_{\mathrm{A}}R^2) &= f_1(\theta) + \frac{(1+\alpha)}{(4+\alpha)}f_2(\theta) + \frac{1}{(4+\alpha)}\left\{(3+\alpha) + 6[\tfrac{1}{3}f_2(\theta)]^{1/2}\right\} \\ I_\parallel/(m_{\mathrm{A}}R^2) &= 2f_1(\theta) \end{aligned}$$

The variation of the moments of inertia with θ are shown in Fig. 11.4; $I_\perp$ is shown for three representative values of α. Not surprisingly, $I_\perp$ and $I_\parallel$ converge onto the same value when θ is the tetrahedral angle (shown by the vertical dotted line). This is because in this limit the molecule becomes tetrahedral and is then a spherical rotor, for which all the moments of inertia are the same.

At the tetrahedral angle $\cos\theta_{\text{tet}} = -\frac{1}{3}$; hence $f_1(\theta_{\text{tet}}) = \frac{4}{3}$ and $f_2(\theta_{\text{tet}}) = \frac{1}{3}$

$$\begin{aligned} I_\perp/(m_A R^2) &= \tfrac{4}{3} + \frac{(1+\alpha)}{(4+\alpha)} \times \tfrac{1}{3} + \frac{1}{(4+\alpha)}\{(3+\alpha) + 6[\tfrac{1}{3}\times\tfrac{1}{3}]^{1/2}\} \\ &= \tfrac{4}{3} + \frac{(1+\alpha)}{(4+\alpha)} \times \tfrac{1}{3} + \frac{1}{(4+\alpha)}\{(3+\alpha)+2\} \\ &= \frac{4(4+\alpha)+(1+\alpha)+3(5+\alpha)}{3(4+\alpha)} = \frac{32+8\alpha}{3(4+\alpha)} = \tfrac{8}{3} \end{aligned}$$

The moment of inertia for a tetrahedral molecule is, from the table, $I/(m_A R^2) = \frac{8}{3}$, in agreement with the result just derived. In this limit the moment of inertia does not depend on the mass of B (the central atom), as the axes pass through this atom.

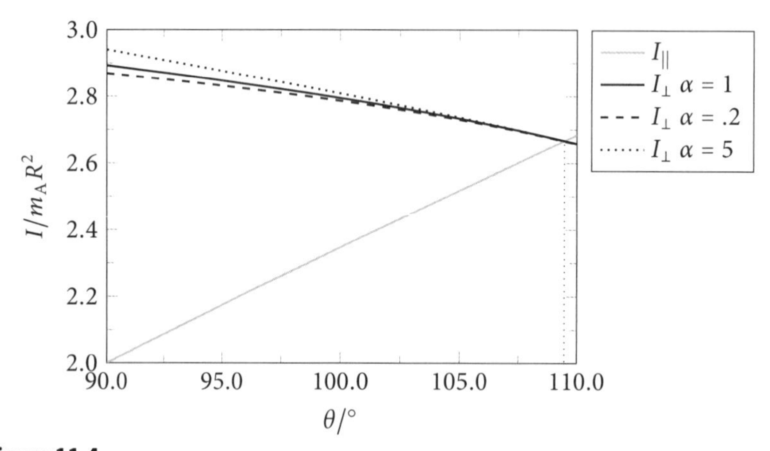

Figure 11.4

E11B.3(a) To be a symmetric rotor a molecule most possess an n-fold axis with $n > 2$. (i) O_3 is bent (like H_2O), it has a two-fold axis and so is an asymmetric rotor. (ii) CH_3CH_3 has a three-fold axis and so is a symmetric rotor. (iii) XeO_4 is tetrahedral, and so is a spherical rotor. (iv) Ferrocene has a five-fold axis and so is a symmetric rotor.

E11B.4(a) In order to determine two unknowns, data from two independent experiments are needed. In this exercise two values of B for two isotopologues of HCN are given; these are used to find two moments of inertia. The moment of inertia of a linear triatomic is given in Table 11B.1 on page 431, and if it is assumed that the bond lengths are unaffected by isotopic substitution, the expressions for the moment of inertia of the two isotopologues can be solved simultaneously to obtain the two bond lengths.

The rotational constant in wavenumber is given by [11B.7–432], $\tilde{B} = \hbar/4\pi cI$; multiplication by the speed of light gives the rotational constant in frequency units $B = \hbar/4\pi I$, which rearranges to $I = \hbar/4\pi B$

$$I_{\rm HCN} = (1.0546\times10^{-34}\,{\rm J\,s})/[4\pi\times(44.316\times10^{9}\,{\rm Hz})] = 1.89...\times10^{-46}\,{\rm kg\,m^2}$$
$$I_{\rm DCN} = (1.0546\times10^{-34}\,{\rm J\,s})/[4\pi\times(36.208\times10^{9}\,{\rm Hz})] = 2.31...\times10^{-46}\,{\rm kg\,m^2}$$

It is somewhat more convenient for the subsequent manipulations to express the moments of inertia in units of the atomic mass constant $m_{\rm u}$ and nm.

$$I_{\rm HCN} = (1.89...\times10^{-46}\,{\rm kg\,m^2})\times\left(\frac{10^9\,{\rm nm}}{1\,{\rm m}}\right)^2\times\frac{1\,m_{\rm u}}{1.6605\times10^{-27}\,{\rm kg}}$$
$$= 0.114...\,m_{\rm u}\,{\rm nm^2}$$
$$I_{\rm DCN} = (2.31...\times10^{-46}\,{\rm kg\,m^2})\times\left(\frac{10^9\,{\rm nm}}{1\,{\rm m}}\right)^2\times\frac{1\,m_{\rm u}}{1.6605\times10^{-27}\,{\rm kg}}$$
$$= 0.139...\,m_{\rm u}\,{\rm nm^2}$$

Using the expressions from Table 11B.1 on page 431, the moments of inertia are expressed in terms of the masses and bond lengths, where the former are expressed as multiples on $m_{\rm u}$. In this case A = ^{1}H or ^{2}H, B = ^{12}C and C = ^{14}N.

$$I_{\rm HCN} = m_{\rm H}R^2 + m_{\rm N}R'^2 - \frac{(m_{\rm H}R - m_{\rm N}R')^2}{m_{\rm H}+m_{\rm C}+m_{\rm N}}$$
$$= 1.0078R^2 + 14.0031R'^2 - \frac{(1.0078R - 14.0031R')^2}{1.0078+12.0000+14.0031}$$
$$= 1.0078R^2 + 14.0031R'^2 - \frac{(1.0078R - 14.0031R')^2}{27.0109}$$
$$I_{\rm DCN} = m_{\rm D}R^2 + m_{\rm N}R'^2 - \frac{(m_{\rm D}R - m_{\rm N}R')^2}{m_{\rm D}+m_{\rm C}+m_{\rm N}}$$
$$= 2.0141R^2 + 14.0031R'^2 - \frac{(2.0141R - 14.0031R')^2}{2.0141+12.0000+14.0031}$$
$$= 2.0141R^2 + 14.0031R'^2 - \frac{(2.0141R - 14.0031R')^2}{28.0172}$$

These two equations need to be solved simultaneously for R and R', but because they are quadratics this is a very laborious process by hand: it is best achieved using mathematical software. This gives the resulting bond lengths as $\boxed{R = R_{\rm CH} = 0.1062\,{\rm nm}}$ and $\boxed{R' = R_{\rm CN} = 0.1157\,{\rm nm}}$.

E11B.5(a) The centrifugal distortion constant is given by [11B.16–434], $\tilde{D}_J = 4\tilde{B}^3/\tilde{\nu}^2$. With the given data $\tilde{D}_J = 4(6.511\,{\rm cm^{-1}})^3/(2308\,{\rm cm^{-1}})^2 = \boxed{2.073\times10^{-4}\,{\rm cm^{-1}}}$.

The rotational constant is inversely proportional to the moment of inertia of the molecule, $I = m_{\rm eff}R^2$ where R is the bond length and $m_{\rm eff}$ is the effective mass. Assuming that isotopic substitution does not affect the bond length, it follows that $\tilde{B} \propto m_{\rm eff}^{-1}$. Assuming that isotopic substitution does not affect the force constant, the vibrational frequency is proportional to $m_{\text{eff}}^{-1/2}$. Thus $\tilde{D} \propto (m_{\text{eff}}^{-1})^3/(m_{\text{eff}}^{-1/2})^2 = m_{\text{eff}}^{-2}$. For this estimation it is sufficient to use integer masses, and because a ratio is involved these can be expressed as multiples of m_{u}.

$$\tilde{D}_{^2\text{HI}}/\tilde{D}_{^1\text{HI}} = (m_{\text{eff}\,^1\text{HI}}/m_{\text{eff}\,^2\text{HI}})^2 = \left(\frac{1 \times 127}{1 + 127} \times \frac{2 + 127}{2 \times 127}\right)^2 = \boxed{0.25}$$

E11B.6(a) For a molecule to show a pure rotational (microwave) absorption spectrum is must have a permanent dipole moment. Of the molecules given, the only ones to satisfy this requirement are $\boxed{\text{HCl, CH}_3\text{Cl and CH}_2\text{Cl}_2}$.

E11B.7(a) The wavenumbers of the lines in the rotational spectrum are given by [11B.20a–436], $\tilde{\nu}(J) = 2\tilde{B}(J + 1)$; the $J = 3 \leftarrow 2$ transition is therefore at $\tilde{\nu}(2) = 2\tilde{B}(2 + 1) = 6\tilde{B}$. The rotational constant is given by [11B.7–432], $\tilde{B} = \hbar/4\pi cI$, and the moment of inertia is given by $m_{\text{eff}}R^2$, where $m_{\text{eff}} = m_1m_2/(m_1 + m_2)$.

$$I = \frac{(14.0031 \times 15.9949)m_{\text{u}}^2}{(14.0031 + 15.9949)m_{\text{u}}} \times \frac{1.6605 \times 10^{-27}\text{ kg}}{1\ m_{\text{u}}} \times (115 \times 10^{-12}\text{ m})^2$$
$$= 1.63... \times 10^{-46}\text{ kg m}^2$$

$$\tilde{B} = \frac{\hbar}{4\pi cI} = \frac{1.0546 \times 10^{-34}\text{ J s}}{4\pi \times (2.9979 \times 10^{10}\text{ cm s}^{-1}) \times (1.63... \times 10^{-46}\text{ kg m}^2)}$$
$$= 1.70...\text{ cm}^{-1}$$

The transition occurs at $6\tilde{B} = 6 \times (1.70...\text{ cm}^{-1}) = \boxed{10.2\text{ cm}^{-1}}$. Expressed in frequency units this is $6c\tilde{B} = 6 \times (2.9979 \times 10^{10}\text{ cm s}^{-1}) \times (1.70...\text{ cm}^{-1}) = 3.07... \times 10^{11}\text{ Hz} = \boxed{307\text{ GHz}}$.

Centrifugal distortion will lower the frequency.

E11B.8(a) The wavenumbers of the lines in the rotational spectrum are given by [11B.20a–436], $\tilde{\nu}(J) = 2\tilde{B}(J + 1)$. The $J = 3 \leftarrow 2$ transition is therefore at $\tilde{\nu}(2) = 2\tilde{B}(2 + 1) = 6\tilde{B}$, hence $\tilde{B} = (63.56/6)\text{ cm}^{-1}$. The rotational constant is given by [11B.7–432], $\tilde{B} = \hbar/4\pi cI$, and the moment of inertia is given by $m_{\text{eff}}R^2$, where $m_{\text{eff}} = m_1m_2/(m_1 + m_2)$. It follows that $R = (\hbar/4\pi c m_{\text{eff}}\tilde{B})^{1/2}$.

$$m_{\text{eff}} = \frac{(1.0078 \times 34.9688)m_{\text{u}}^2}{(1.0078 + 34.9688)m_{\text{u}}} \times \frac{1.6605 \times 10^{-27}\text{ kg}}{1\ m_{\text{u}}} = 1.62\ldots \times 10^{-27}\text{ kg}$$

$$R = \left(\frac{1.0546 \times 10^{-34}\text{ J s}}{4\pi \times (2.9979 \times 10^{10}\text{ cm s}^{-1}) \times (1.62... \times 10^{-27}\text{ kg}) \times [(65.36/6)\text{cm}^{-1}]}\right)^{1/2}$$
$$= \boxed{125.7\text{ pm}}$$

E11B.9(a) The wavenumbers of the lines in the rotational spectrum are given by [11B.20a–*436*], $\tilde{\nu}(J) = 2\tilde{B}(J+1)$; the lines are therefore spaced by $2\tilde{B}$, it therefore follows that $\tilde{B} = (12.604/2)\ \text{cm}^{-1}$. The rotational constant is given by [11B.7–*432*], $\tilde{B} = \hbar/4\pi c I$, and the moment of inertia is given by $m_{\text{eff}}R^2$, where $m_{\text{eff}} = m_1 m_2/(m_1 + m_2)$. It follows that $I = \hbar/4\pi c\tilde{B}$ and $R = (I/m_{\text{eff}})^{1/2}$.

$$I = \hbar/4\pi c\tilde{B}$$

$$= \frac{1.0546 \times 10^{-34}\ \text{J s}}{4\pi \times (2.9979 \times 10^{10}\ \text{cm s}^{-1}) \times [(12.604/2]\ \text{cm}^{-1})} = \boxed{4.4420 \times 10^{-47}\ \text{kg m}^2}$$

$$m_{\text{eff}} = \frac{(1.0078 \times 26.9815)m_{\text{u}}^2}{(1.0078 + 26.9815)m_{\text{u}}} \times \frac{1.6605 \times 10^{-27}\ \text{kg}}{1\ m_{\text{u}}} = 1.61... \times 10^{-27}\ \text{kg}$$

$$R = (I/m_{\text{eff}})^{1/2} = [(4.44... \times 10^{-47}\ \text{kg m}^2)/(1.61... \times 10^{-27}\ \text{kg})]^{1/2}$$
$$= \boxed{165.9\ \text{pm}}$$

E11B.10(a) The most occupied J state is given by [11B.21–*437*], $J_{\text{max}} = (kT/2hc\tilde{B})^{1/2} - \frac{1}{2}$.

(i) At 25 °C, 298 K, this gives

$$J_{\text{max}} =$$
$$\left(\frac{(1.3806 \times 10^{-23}\ \text{J K}^{-1}) \times (298\ \text{K})}{2 \times (6.6261 \times 10^{-34}\ \text{J s}) \times (2.9979 \times 10^{10}\ \text{cm s}^{-1}) \times (0.244\ \text{cm}^{-1})}\right)^{1/2} - \frac{1}{2}$$
$$= \boxed{20}$$

(ii) At 100 °C, 373 K, this gives

$$J_{\text{max}} =$$
$$\left(\frac{(1.3806 \times 10^{-23}\ \text{J K}^{-1}) \times (373\ \text{K})}{2 \times (6.6261 \times 10^{-34}\ \text{J s}) \times (2.9979 \times 10^{10}\ \text{cm s}^{-1}) \times (0.244\ \text{cm}^{-1})}\right)^{1/2} - \frac{1}{2}$$
$$= \boxed{23}$$

E11B.11(a) For a molecule to show a pure rotational Raman spectrum it must have an anisotropic polarizability. With the exception of spherical rotors, all molecules satisfy this requirement. Therefore $\boxed{H_2, HCl, CH_3Cl}$ all give rotational Raman spectra.

E11B.12(a) The Stokes lines appear at wavenumbers given by [11B.24a–*438*], $\tilde{\nu}(J+2 \leftarrow J) = \tilde{\nu}_{\text{i}} - 2\tilde{B}(2J+3)$, where the wavenumber of the incident radiation is $\tilde{\nu}_{\text{i}}$, and J is the quantum number of the initial state. With the given data

$$\tilde{\nu}(2 \leftarrow 0) = 20\,487\ \text{cm}^{-1} - 2 \times (1.9987\ \text{cm}^{-1})(2 \times 0 + 3) = \boxed{20\,475\ \text{cm}^{-1}}$$

E11B.13(a) The Stokes lines appear at wavenumbers given by [11B.24a–438], $\tilde{\nu}(J+2 \leftarrow J) = \tilde{\nu}_i - 2\tilde{B}(2J+3)$, where the wavenumber of the incident radiation is $\tilde{\nu}_i$, and J is the quantum number of the initial state. It therefore follows that the separation between adjacent lines is $4\tilde{B}$, hence $\tilde{B} = (0.9752/4)\ \text{cm}^{-1}$.

The rotational constant is given by [11B.7–432], $\tilde{B} = \hbar/4\pi cI$, and the moment of inertia is given by $m_{\text{eff}}R^2$, where $m_{\text{eff}} = m_1 m_2/(m_1+m_2)$. It follows that $I = \hbar/4\pi c\tilde{B}$ and $R = (I/m_{\text{eff}})^{1/2}$.

$$I = \hbar/4\pi c\tilde{B}$$

$$= \frac{1.0546 \times 10^{-34}\ \text{J s}}{4\pi \times (2.9979 \times 10^{10}\ \text{cm s}^{-1}) \times [(0.9752/4)\ \text{cm}^{-1}]} = 1.14... \times 10^{-45}\ \text{kg m}^2$$

For a homonuclear diatomic the effective mass is simply $m_{\text{eff}} = \frac{1}{2}m$

$$R = (I/m_{\text{eff}})^{1/2} = \left(\frac{1.14... \times 10^{-45}\ \text{kg m}^2}{\frac{1}{2} \times 34.9688 \times (1.6605 \times 10^{-27}\ \text{kg})}\right)^{1/2}$$

$$= \boxed{198.9\ \text{pm}}$$

E11B.14(a) The ratio of the weights for (odd J)/(even J) is given by [11B.25–440]. For ^{35}Cl, $I = \frac{3}{2}$ and the nucleus is therefore a fermion. The ratio is (odd J)/(even J) $= (I+1)/I = (\frac{3}{2}+1)/(\frac{3}{2}) = \boxed{\frac{5}{3}}$.

Solutions to problems

P11B.1 Suppose that the bond length is R and that the centre of mass is at a distance x from mass m_1 and therefore $(R-x)$ from mass m_2. Balancing moments gives $m_1 x = m_2(R-x)$, hence $x = m_2 R/(m_1+m_2)$. Using this result it follows that $(R-x) = R - m_2R/(m_1+m_2) = m_1R/(m_1+m_2)$. The moment of inertia is therefore

$$I = m_1x^2 + m_2(R-x)^2 = \frac{m_1 m_2^2 R^2}{(m_1+m_2)^2} + \frac{m_2 m_1^2 R^2}{(m_1+m_2)^2}$$

$$= \frac{m_1 m_2(m_2+m_1)R^2}{(m_1+m_2)^2} = \frac{m_1 m_2 R^2}{(m_1+m_2)} = m_{\text{eff}}R^2$$

P11B.3 The rotational terms for a symmetric rotor are given by [11B.13a–433], $\tilde{F}(J,K) = \tilde{B}J(J+1) + (\tilde{A}-\tilde{B})K^2$. The selection rules are $\Delta J = \pm 1$ and $\Delta K = 0$, and therefore the term in K does not affect the wavenumber of the lines in the spectrum; the result is that the lines are at exactly the same wavenumbers as for a linear rotor, [11B.20a–436], $\tilde{\nu}(J) = 2\tilde{B}(J+1)$. The separation of the lines is $2\tilde{B}$.

In frequency units the spacing is $2B = 2 \times (298\ \text{GHz}) = \boxed{596\ \text{GHz}}$. Expressed as a wavenumber this spacing is $(596 \times 10^9\ \text{Hz})/(2.9979 \times 10^{10}\ \text{cm s}^{-1}) = \boxed{19.9\ \text{cm}^{-1}}$.

The rotational constant is given by [11B.7–*432*], $\tilde{B} = \hbar/4\pi c I$. Expressed in frequency units this is $B = \hbar/4\pi I$. It follows that $I = \hbar/4\pi B$

$$I = \hbar/4\pi B = \frac{1.0546 \times 10^{-34}\,\mathrm{J\,s}}{4\pi \times (298 \times 10^{9}\,\mathrm{Hz})} = 2.82 \times 10^{-47}\,\mathrm{kg\,m^2}$$

Expressions for the moment of inertia are given in Table 11B.1 on page 431; NH_3 is a symmetric rotor and the second entry under symmetric rotors is the required one. The moment of inertia corresponding to the rotational constant B is $I_\perp$. With the data given

$$I_\perp = m_\mathrm{H}(1 - \cos\theta)R^2 + \frac{m_\mathrm{H} m_\mathrm{N}}{m_\mathrm{N} + 3m_\mathrm{H}}(1 + 2\cos\theta)R^2$$

It is convenient to work with the masses as multiples of m_u and R in nm

$$\begin{aligned} I_\perp &= (0.1014\,\mathrm{nm})^2 \times \Big[(1.0078) \times (1 - \cos 106.78°) \\ &\quad + \frac{1.0078 \times 14.0031}{14.0031 + 3 \times 1.0078}(1 + 2\cos 106.78°)\Big] \times m_\mathrm{u} \\ &= 0.0169...\, m_\mathrm{u}\,\mathrm{nm}^2 \end{aligned}$$

Converting to the usual units gives

$$I = (0.0169...\, m_\mathrm{u}\,\mathrm{nm}^2) \times \frac{10^{-18}\,\mathrm{m}^2}{1\,\mathrm{nm}^2} \times \frac{1.6605 \times 10^{-27}\,\mathrm{kg}}{1\,m_\mathrm{u}} = 2.815 \times 10^{-47}\,\mathrm{kg\,m^2}$$

This value is consistent with the moment of inertia determined from the given rotational constant.

P11B.5 Bonding is essentially the result of electrostatic interactions so to a very good approximation it is expected that adding an uncharged neutron will have no effect on the bond length.

The wavenumbers of the lines expected for a diatomic are given by [11B.20a–*436*], $\tilde{\nu}(J) = 2\tilde{B}(J+1)$; the separation of the lines is $2\tilde{B}$. The rotational constant is inversely proportional to the effective mass, therefore if the bond length is unaffected by isotopic substitution the ratio of the rotational constants should be equal to the inverse ratio of the effective masses. With the data given

$$\tilde{B}_{^1\mathrm{H}^{35}\mathrm{Cl}}/\tilde{B}_{^2\mathrm{H}^{35}\mathrm{Cl}} = (20.8784\,\mathrm{cm}^{-1})(10.7840\,\mathrm{cm}^{-1}) = 1.93605$$

$$\begin{aligned} \frac{m_{\mathrm{eff},^2\mathrm{H}^{35}\mathrm{Cl}}}{m_{\mathrm{eff},^1\mathrm{H}^{35}\mathrm{Cl}}} &= \frac{m_{^1\mathrm{H}} m_{^{35}\mathrm{Cl}}}{m_{^1\mathrm{H}} + m_{^{35}\mathrm{Cl}}} \times \frac{m_{^2\mathrm{H}} + m_{^{35}\mathrm{Cl}}}{m_{^2\mathrm{H}} m_{^{35}\mathrm{Cl}}} \\ &= \frac{1.007\,825 + 34.968\,85}{1.007\,825 \times 34.968\,85} \times \frac{2.0140 \times 34.968\,85}{2.0140 + 34.968\,85} \\ &= 1.93440 \end{aligned}$$

These two quantities differ by less than 0.1% so the hypothesis that the bond length is invariant to isotopic substitution is confirmed to quite a high level of precision; with the accuracy of the data given there is, however, some perceptible change.

P11B.7 *Note*: there is an error in the problem; for the ^{34}S isotopologue the line at 47.462 40 GHz is for $J = 3$.

The wavenumbers of the lines expected for a linear rotor are given by [11B.20a–*436*], $\tilde{\nu}(J) = 2\tilde{B}(J+1)$; the separation of the lines is $2\tilde{B}$. For $OC^{32}S$ the average spacing of the lines is 12.16272 GHz, so the best estimate for the rotational constant is $B_{OCS} = \frac{1}{2} \times (12.16272\ \text{GHz}) = 6.08136\ \text{GHz}$. For $OC^{34}S$ there are just two lines, one for $J = 1$ and one for $J = 3$; these are separated by 23.73007 GHz, which is $4B$. The best estimate for the rotational constant is $B_{OCS'} = \frac{1}{4} \times (23.73007\ \text{GHz}) = 5.93252\ \text{GHz}$.

The rotational constant in wavenumber is given by [11B.7–*432*], $\tilde{B} = \hbar/4\pi cI$; multiplication by the speed of light gives the rotational constant in frequency units $B = \hbar/4\pi I$, hence $I = \hbar/4\pi B$

$$I_{OCS} = (1.0546 \times 10^{-34}\ \text{J s})/[4\pi \times (6.08136 \times 10^{9}\ \text{Hz})] = 1.37... \times 10^{-45}\ \text{kg m}^2$$

$$I_{OCS'} = (1.0546 \times 10^{-34}\ \text{J s})/[4\pi \times (5.93252 \times 10^{9}\ \text{Hz})] = 1.41... \times 10^{-45}\ \text{kg m}^2$$

where for short S implies ^{32}S and S′ implies ^{34}S. It is somewhat more convenient for the subsequent manipulations to express the moments of inertia in units of the atomic mass constant m_u and nm.

$$I_{OCS} = (1.37... \times 10^{-45}\ \text{kg m}^2) \times \left(\frac{10^9\ \text{nm}}{1\ \text{m}}\right)^2 \times \frac{1\ m_u}{1.6605 \times 10^{-27}\ \text{kg}}$$
$$= 0.831...\ m_u\ \text{nm}^2$$

$$I_{OCS'} = (1.41... \times 10^{-45}\ \text{kg m}^2) \times \left(\frac{10^9\ \text{nm}}{1\ \text{m}}\right)^2 \times \frac{1\ m_u}{1.6605 \times 10^{-27}\ \text{kg}}$$
$$= 0.851...\ m_u\ \text{nm}^2$$

Using the expressions from Table 11B.1 on page 431, the moments of inertia are expressed in terms of the masses and bond lengths, where the former are expressed as multiples on m_u. In this case A = ^{16}O, B = ^{12}C and C = ^{32}S or ^{32}S.

$$I_{OCS} = m_O R^2 + m_S R'^2 - \frac{(m_O R - m_S R')^2}{m_O + m_C + m_S}$$
$$= 15.9949R^2 + 31.9721R'^2 - \frac{(15.9949R - 31.9721R')^2}{15.9949 + 12.0000 + 31.9721}$$
$$= 15.9949R^2 + 31.9721R'^2 - \frac{(15.9949R - 31.9721R')^2}{59.967}$$

$$I_{OCS'} = m_O R^2 + m_{S'} R'^2 - \frac{(m_O R - m_{S'} R')^2}{m_O + m_C + m_{S'}}$$
$$= 15.9949R^2 + 33.9679R'^2 - \frac{(15.9949R - 33.9679R')^2}{15.9949 + 12.0000 + 33.9679}$$
$$= 15.9949R^2 + 33.9679R'^2 - \frac{(15.9949R - 33.9679R')^2}{61.9628}$$

These two equations need to be solved simultaneously for R and R', but because they are quadratics this is a very laborious process by hand: it is best achieved using mathematical software. This gives the resulting bond lengths as $\boxed{R = R_{OC} = 0.1167\ \text{nm}}$ and $\boxed{R' = R_{CS} = 0.1565\ \text{nm}}$.

P11B.9 The wavenumbers of the lines expected for a linear rotor are given by [11B.20a–*436*], $\tilde{\nu}(J) = 2\tilde{B}(J+1)$; the separation of the lines is $2\tilde{B}$. However, the separation between adjacent lines in the given data is not constant, but increases along the series. To account for this, the effects of centrifugal distortion are included, and in this case the frequencies of the lines are given by [11B.20b–*436*], $\nu(J) = 2B(J+1) - 4D_J(J+1)^3$ (written with the constants in frequency units). Division of both side of this expression by $2(J+1)$ indicates that a plot of $[\nu(J)]/2(J+1)$ against $(J+1)^2$ should be a straight line with slope $-2D_J$ and intercept B. The data are tabulated below; δ is the difference between successive lines. The plot is shown in Fig. 11.5.

J	$\nu(J)$/MHz	δ/MHz	$[\nu(J)/2(J+1)]$/MHz	$(J+1)^2$
24	214 777.7		4 295.6	625
25	223 379.0	8 601.3	4 295.8	676
26	231 981.2	8 602.2	4 296.0	729
27	240 584.4	8 603.2	4 296.2	784
28	249 188.5	8 604.1	4 296.4	841
29	257 793.5	8 605.0	4 296.6	900

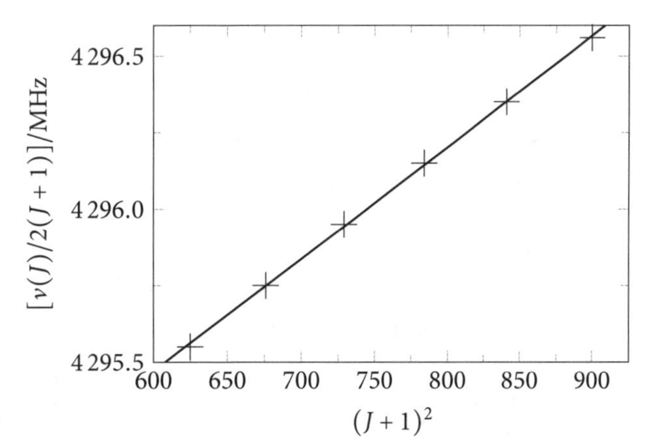

Figure 11.5

The data are a good fit to the line

$$\{\nu(J)]/2(J+1)\}/\text{MHz} = 3.652 \times 10^{-3} \times (J+1)^2 + 4293.28$$

The value of the rotational constant is found from the intercept: (B/MHz) = intercept. Some elementary statistics on the best-fit line indicates an error of about 0.03 MHz in the intercept, so the best estimate for the rotational constant is $\boxed{B = 4293.28 \pm 0.03\ \text{MHz}}$ or, expressed as a wavenumber, $\tilde{B} = 0.1432\ \text{cm}^{-1}$.

It is somewhat unusual that the centrifugal distortion constant appears to be negative.

The most occupied J state is given by [11B.15–434], $J_{\max} = (kT/2hc\tilde{B})^{1/2} - \frac{1}{2}$. At 298 K

$$J_{\max} = \left(\frac{(1.3806\times10^{-23}\,\text{J K}^{-1})\times(298\,\text{K})}{2\times(6.6261\times10^{-34}\,\text{J s})\times(2.9979\times10^{10}\,\text{cm s}^{-1})\times(0.1432\,\text{cm}^{-1})}\right)^{1/2} - \frac{1}{2} = \boxed{26}$$

A similar calculation at 100 K gives $\boxed{J_{\max} = 15}$.

P11B.11 The population of level J, N_J, is given by $N_J \propto g_J e^{-E_J/kT}$. In this expression g_J is the degeneracy of level J, $g_J = (2J+1)$, and E_J is the energy of that level, $E_J = hc\tilde{B}J(J+1)$. To find the level with the greatest population the derivative $\text{d}N_J/\text{d}J$ is computed and then set to zero; it is not necessary to know the constant of proportion, which will be written A. To compute the derivative requires the product rule and the chain rule

$$\frac{\text{d}}{\text{d}J}A(2J+1)e^{-hc\tilde{B}J(J+1)/kT}$$
$$= A\times2\times e^{-hc\tilde{B}J(J+1)/kT} - A(2J+1)\times(2J+1)\times(hc\tilde{B}/kT)e^{-hc\tilde{B}J(J+1)/kT}$$

setting the derivative to zero and gathering terms gives

$$0 = Ae^{-hc\tilde{B}J(J+1)/kT}\left[2-(2J+1)^2(hc\tilde{B}/kT)\right]$$

The exponential term goes to zero as $J\to\infty$, but this is not a maximum; rather, the maximum is when the term in square brackets is zero

$$0 = \left[2-(2J_{\max}+1)^2(hc\tilde{B}/kT)\right]$$
$$(2J_{\max}+1)^2 = 2kT/hc\tilde{B} \quad \text{hence} \quad (2J_{\max}+1) = (2kT/hc\tilde{B})^{1/2}$$
$$J_{\max} = \tfrac{1}{2}\times(2kT/hc\tilde{B})^{1/2} - \tfrac{1}{2}$$

The level with the greatest population is therefore $\boxed{J_{\max} = (kT/2hc\tilde{B})^{1/2} - \frac{1}{2}}$. With the given data

$$J_{\max} = \left(\frac{(1.3806\times10^{-23}\,\text{J K}^{-1})\times(298\,\text{K})}{2(6.6261\times10^{-34}\,\text{J s})\times(2.9979\times10^{10}\,\text{cm s}^{-1})\times(0.1142\,\text{cm}^{-1})}\right)^{1/2} - \frac{1}{2}$$
$$= \boxed{30}$$

For a spherical rotor the degeneracy of each level is $(2J+1)^2$. Finding the most populated level proceeds as before

$$\frac{\text{d}}{\text{d}J}A(2J+1)^2e^{-hc\tilde{B}J(J+1)/kT}$$
$$= A\times4(2J+1)\times e^{-hc\tilde{B}J(J+1)/kT}$$
$$- A(2J+1)^2\times(2J+1)\times(hc\tilde{B}/kT)e^{-hc\tilde{B}J(J+1)/kT}$$

setting the derivative to zero and gathering terms gives

$$0 = Ae^{-hc\tilde{B}J(J+1)/kT}(2J+1)\left[4-(2J+1)^2(hc\tilde{B}/kT)\right]$$

As before the maximum occurs when the term in square brackets is zero

$$0 = \left[4-(2J+1)^2(hc\tilde{B}/kT)\right]$$

$$(2J_{\text{max}}+1)^2 = 4kT/hc\tilde{B} \quad \text{hence} \quad (2J_{\text{max}}+1) = (4kT/hc\tilde{B})^{1/2}$$

$$J_{\text{max}} = \tfrac{1}{2}\times(4kT/hc\tilde{B})^{1/2} - \tfrac{1}{2}$$

The level with the greatest population is therefore $\boxed{J_{\text{max}} = (kT/hc\tilde{B})^{1/2} - \tfrac{1}{2}}$.

With the given data

$$J_{\text{max}} = \left(\frac{(1.3806\times10^{-23}\,\text{J K}^{-1})\times(298\,\text{K})}{(6.6261\times10^{-34}\,\text{J s})\times(2.9979\times10^{10}\,\text{cm s}^{-1})\times(5.24\,\text{cm}^{-1})}\right)^{1/2} - \frac{1}{2}$$

$$= \boxed{6}$$

In such calculations it may be helpful to use $kT/hc = 207.225\ \text{cm}^{-1}$ at 298 K (from inside the front cover).

P11B.13 *Temperature effects.* At extremely low temperatures (10 K) only the lowest rotational states are populated. No emission spectrum is expected for the CO in the cloud and star-light microwave absorptions by the CO in the cloud are from the lowest rotational states. At higher temperatures additional high-energy lines appear because higher energy rotational states are populated. Circumstellar clouds may exhibit infrared absorptions due to vibrational excitation as well as electronic transitions in the ultraviolet. Ultraviolet absorptions may indicate the photodissocation of carbon monoxide. High temperature clouds exhibit emissions.

Density effects. The density of an interstellar cloud may range from one particle to a billion particles per cm^3. This is still very much a vacuum compared to the laboratory high vacuum of a trillion particles per cm^3. Under such extreme vacuum conditions the half-life of any quantum state is expected to be extremely long and absorption lines should be very narrow. At the higher densities the vast size of nebulae obscures distant stars. High densities and high temperatures may create conditions in which emissions stimulate emissions of the same wavelength by molecules. A cascade of stimulated emissions greatly amplifies normally weak lines – the maser phenomena of microwave amplification by stimulated emission of radiation.

Particle velocity effects. Particle velocity can cause Doppler broadening of spectral lines. The effect is extremely small for interstellar clouds at 10 K but is appreciable for clouds near high temperature stars. Outflows of gas from pulsing stars exhibit a red Doppler shift when moving away at high speed and a blue shift when moving toward us.

There will be many more transitions observable in circumstellar gas than in interstellar gas, because many more rotational states will be accessible at the higher temperatures. Higher velocity and density of particles in circumstellar material can be expected to broaden spectral lines compared to those of interstellar material by shortening collisional lifetimes. (Doppler broadening is not likely to be significantly different between circumstellar and interstellar material in the same astronomical neighbourhood. The relativistic speeds involved are due to large-scale motions of the expanding universe, compared to which local thermal variations are insignificant.)

A temperature of 1000 K is not high enough to significantly populate electronically excited states of CO; such states would have different bond lengths, thereby producing transitions with different rotational constants. Excited vibrational states would be accessible, though, and ro-vibrational transitions with P and R branches as detailed in this following Topic would be observable in circumstellar but not interstellar material. The rotational constant for CO is $1.691\ \text{cm}^{-1}$. The first excited rotational energy level, $J = 1$, with energy $2hc\tilde{B}$, is thermally accessible at about 6 K (based on the rough equation of the rotational energy to thermal energy kT). In interstellar space, only two or three rotational lines would be observable; in circumstellar space (at about 1000 K) the number of transitions would be more like 20.

11C Vibrational spectroscopy of diatomic molecules

Answers to discussion questions

D11C.1 Harmonic oscillation results from a parabolic potential energy curve. For low vibrational energies, near the bottom of the potential well, the assumption of a parabolic potential energy is a good approximation for real molecules. However, molecular vibrations are always anharmonic to a greater or lesser extent. At higher excitation energies, the parabolic approximation is poor, and in particular it fails to predict dissociation. An advantage of the parabolic potential energy is that it allows for a relatively straightforward solution of the Schrödinger equation for the vibrational motion.

The Morse potential is a closer approximation to the true potential energy curve for molecular vibrations. It allows for the convergence of the energy levels at higher values of the quantum numbers and for dissociation at large displacements. However, although it has the same general form as typical potential energy curves, it fails to represent the detailed shape of these curves, especially at large internuclear distances. An advantage is that the Schrödinger equation can be solved exactly for the Morse potential.

D11C.3 This is discussed in Section 11C.4(b) on page 448.

Solutions to exercises

E11C.1(a) The vibrational frequency of a harmonic oscillator is given by [7E.3–274], $\omega = (k_f/m)^{1/2}$; ω is an angular frequency, so to convert to frequency in Hz, ν, use

$\omega = 2\pi\nu$. Therefore $2\pi\nu = (k_f/m)^{1/2}$. Rearranging this gives the force constant as $k_f = m(2\pi\nu)^2$

$$k_f = (0.100\ \text{kg}) \times (2\pi \times 2.0\ \text{Hz})^2 = \boxed{16\ \text{N m}^{-1}}$$

where $1\ \text{N} = 1\ \text{kg m s}^{-2}$ and $1\ \text{Hz} = 1\ \text{s}^{-1}$ are used.

E11C.2(a) The vibrational frequency, expressed as a wavenumber, of a harmonic oscillator is given by [11C.4b–*443*], $\tilde{\nu} = (1/2\pi c)(k_f/m_{\text{eff}})^{1/2}$, where m_{eff} is the effective mass, given by $m_{\text{eff}} = m_1 m_2/(m_1 + m_2)$. Assuming that the force constants of the two isotopologues are the same, $\tilde{\nu}$ simply scales as $(m_{\text{eff}})^{-1/2}$. The fractional change is therefore

$$\frac{\tilde{\nu}_{\text{Na}^{35}\text{Cl}} - \tilde{\nu}_{\text{Na}^{37}\text{Cl}}}{\tilde{\nu}_{\text{Na}^{35}\text{Cl}}} = 1 - \frac{\tilde{\nu}_{\text{Na}^{37}\text{Cl}}}{\tilde{\nu}_{\text{Na}^{35}\text{Cl}}} = 1 - \left(\frac{m_{\text{eff Na}^{35}\text{Cl}}}{m_{\text{eff Na}^{37}\text{Cl}}}\right)^{1/2}$$

$$= 1 - \left(\frac{22.9898 \times 34.9688}{22.9898 + 34.9688} \times \frac{22.9898 + 36.9651}{22.9898 \times 36.9651}\right)^{1/2} = 0.0107...$$

The fractional change, expressed as a percentage, is therefore $\boxed{1.077\%}$.

E11C.3(a) The wavenumber of the fundamental vibrational transition is simply equal to the vibrational frequency expressed as a wavenumber. This is given by [11C.4b–*443*], $\tilde{\nu} = (1/2\pi c)(k_f/m_{\text{eff}})^{1/2}$, where m_{eff} is the effective mass, given by $m_{\text{eff}} = m_1 m_2/(m_1 + m_2)$. It follows that $k_f = m_{\text{eff}}(2\pi c\tilde{\nu})^2$; for a homonuclear diatomic $m_{\text{eff}} = \frac{1}{2}m_1$. With the data given

$$k_f = (\tfrac{1}{2} \times [34.9688 \times (1.6605 \times 10^{-27}\ \text{kg})])$$
$$\times [2\pi \times (2.9979 \times 10^{10}\ \text{cm s}^{-1}) \times (564.9\ \text{cm}^{-1})]^2$$
$$= \boxed{328.7\ \text{N m}^{-1}}$$

Note the conversion of the mass to kg.

E11C.4(a) The wavenumber of the fundamental vibrational transition is simply equal to the vibrational frequency expressed as a wavenumber. This is given by [11C.4b–*443*], $\tilde{\nu} = (1/2\pi c)(k_f/m_{\text{eff}})^{1/2}$, where m_{eff} is the effective mass, given by $m_{\text{eff}} = m_1 m_2/(m_1 + m_2)$. It follows that $k_f = m_{\text{eff}}(2\pi c\tilde{\nu})^2$. With the data given the following table is drawn up.

	$^1\text{H}^{19}\text{F}$	$^1\text{H}^{35}\text{Cl}$	$^1\text{H}^{81}\text{Br}$	$^1\text{H}^{127}\text{I}$
$\tilde{\nu}/\text{cm}^{-1}$	4141.3	2988.9	2649.7	2309.5
m_{eff}/m_u	0.9570	0.9796	0.9954	0.9999
$k_f/\text{N m}^{-1}$	967.0	515.6	411.7	314.2

E11C.5(a) The terms (energies expressed as wavenumbers) of the harmonic oscillator are given by [11C.4b–*443*], $\tilde{G}(v) = (v + \frac{1}{2})\tilde{\nu}$; these are wavenumbers and so can be converted to energy by multiplying by hc to give $E(v) = (v + \frac{1}{2})hc\tilde{\nu}$. The ground state has $v = 0$, and the first excited state has $v = 1$. The relative population of these levels is therefore given by the Boltzmann distribution, $n_1/n_0 = e^{-(E_1-E_0)/kT}$. The energy difference $E_1 - E_0 = hc\tilde{\nu}$, and hence $n_1/n_0 = e^{-hc\tilde{\nu}/kT}$. It is convenient to compute the quantity $hc\tilde{\nu}/k$ first to give

$$hc\tilde{\nu}/k = \frac{(6.6261 \times 10^{-34}\ \text{J s}) \times (2.9979 \times 10^{10}\ \text{cm s}^{-1}) \times (559.7\ \text{cm}^{-1})}{1.3806 \times 10^{-23}\ \text{J K}^{-1}}$$

$$= 805.3...\ \text{K}$$

It follows that $n_1/n_0 = e^{-(805.3...\ \text{K})/T}$

(i) At 298 K, $n_1/n_0 = e^{-(805.3...\ \text{K})/(298\ \text{K})} = \boxed{0.0670}$

(ii) At 500 K, $n_1/n_0 = e^{-(805.3...\ \text{K})/(500\ \text{K})} = \boxed{0.200}$

As expected, the relative population of the upper level increases with temperature.

E11C.6(a) Taking $y_e = 0$ is equivalent to using the terms for the Morse oscillator, which are given in [11C.8–*444*], $\tilde{G}(v) = (v + \frac{1}{2})\tilde{\nu} - (v + \frac{1}{2})^2\tilde{\nu}x_e$. The transition $v \leftarrow 0$ has wavenumber

$$\begin{aligned}\Delta\tilde{G}(v) &= \tilde{G}(v) - \tilde{G}(0)\\ &= [(v + \tfrac{1}{2})\tilde{\nu} - (v + \tfrac{1}{2})^2\tilde{\nu}x_e] - [(0 + \tfrac{1}{2})\tilde{\nu} - (0 + \tfrac{1}{2})^2\tilde{\nu}x_e]\\ &= v\tilde{\nu} - v(v + 1)\tilde{\nu}x_e\end{aligned}$$

Data on three transitions are provided, but only two are needed to obtain values for $\tilde{\nu}$ and x_e. The $\Delta\tilde{G}(v)$ values for the first two transitions are

$$\begin{aligned}1 \leftarrow 0 \qquad \tilde{\nu} - 2\tilde{\nu}x_e &= 1556.22\ \text{cm}^{-1}\\ 2 \leftarrow 0 \qquad 2\tilde{\nu} - 6\tilde{\nu}x_e &= 3088.28\ \text{cm}^{-1}\end{aligned}$$

Multiplying the first expression by 3 and subtracting the second gives

$$3(\tilde{\nu} - 2\tilde{\nu}x_e) - (2\tilde{\nu} - 6\tilde{\nu}x_e) = \tilde{\nu}$$

$$\text{hence } \tilde{\nu} = 3 \times (1556.22\ \text{cm}^{-1}) - (3088.28\ \text{cm}^{-1}) = \boxed{1580.4\ \text{cm}^{-1}}$$

This value for $\tilde{\nu}$ is used in the first equation, which is then solved for x_e to give $x_e = \frac{1}{2} - (1556.22\ \text{cm}^{-1})/[2 \times (1580.4\ \text{cm}^{-1})] = \boxed{7.65 \times 10^{-3}}$.

E11C.7(a) Following the discussion in Section 11C.3(b) on page 445, $\tilde{D}_0$ is given by the area under a plot of $\Delta\tilde{G}_{v+1/2}$ against $(v + \frac{1}{2})$, where $\Delta\tilde{G}_{v+1/2} = \tilde{G}(v + 1) - \tilde{G}(v)$. The data are shown in the table and the plot in Fig. 11.6.

v	$\tilde{G}_v/\text{cm}^{-1}$	$\Delta\tilde{G}_{v+1/2}/\text{cm}^{-1}$	$v+\frac{1}{2}$
0	1 481.86	2 885.64	0.5
1	4 367.50	2 781.54	1.5
2	7 149.04	2 677.44	2.5
3	9 826.48	2 573.32	3.5
4	12 399.80		

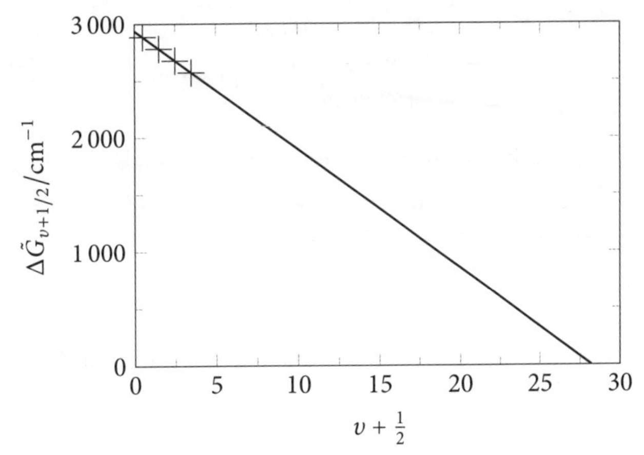

Figure 11.6

The data are a good fit to the line

$$\Delta\tilde{G}_{v+1/2}/\text{cm}^{-1} = -104.11 \times (v+\tfrac{1}{2}) + 2\,937.7$$

This line intercepts the horizontal axis when

$$0 = -104.11 \times (v+\tfrac{1}{2})_{\text{max}} + 2\,937.7 \quad \text{hence} \quad (v+\tfrac{1}{2})_{\text{max}} = 28.22$$

The area under the line is simply the area of a triangle, $\frac{1}{2} \times \text{base} \times \text{height}$, which in this case is $\frac{1}{2} \times (28.22) \times (2\,937.7) = 4.14 \times 10^4$. The dissociation energy is therefore $\boxed{\tilde{D}_0 = 4.14 \times 10^4 \text{ cm}^{-1}}$; only modest precision is quoted because a long extrapolation is made on the basis of few data points.

The fact that the data fall on a good straight line indicates that the Morse levels apply, in which case, according to [11C.9b–*445*], $\Delta\tilde{G}_{v+1/2} = \tilde{\nu} - 2(v+1)x_e\tilde{\nu}$. This expression is rewritten

$$\Delta\tilde{G}_{v+1/2} = \tilde{\nu} - 2(v+\tfrac{1}{2})x_e\tilde{\nu} - x_e\tilde{\nu}$$

which implies that a plot of $\Delta\tilde{G}_{v+1/2}$ against $(v+\frac{1}{2})$ will have slope $-2x_e\tilde{\nu}$ and intercept $(\tilde{\nu} - x_e\tilde{\nu})$. Hence, using the slope of the plot already made

$$x_e\tilde{\nu}/\text{cm}^{-1} = -\tfrac{1}{2}(-104.11) \quad \text{hence} \quad x_e\tilde{\nu} = 52.06 \text{ cm}^{-1}$$

and then using the intercept

$$(\tilde{\nu} - x_e\tilde{\nu})/\text{cm}^{-1} = 2\,937.7$$
$$\text{hence } \tilde{\nu} = (2\,937.7\ \text{cm}^{-1}) + (52.06\ \text{cm}^{-1}) = 2989.8\ \text{cm}^{-1}$$

The depth of the well, $\tilde{D}_e$ is then found using [11C.8–*444*], $x_e = \tilde{\nu}/4\tilde{D}_e$ rearranged to $\tilde{D}_e = \tilde{\nu}/4x_e = \tilde{\nu}^2/4\tilde{\nu}x_e$. The dissociation energy is $\tilde{D}_0 = \tilde{D}_e - \tilde{G}(0)$ (Fig. 11C.3 on page 444), hence

$$\tilde{D}_0 = \tilde{D}_e - \tilde{G}(0) = \frac{\tilde{\nu}^2}{4\tilde{\nu}x_e} - \tilde{G}(0)$$
$$= \frac{(2989.8\ \text{cm}^{-1})^2}{4 \times (52.06\ \text{cm}^{-1})} - (1481.86\ \text{cm}^{-1}) = \boxed{4.14 \times 10^4\ \text{cm}^{-1}}$$

So within the precision quoted, both methods give the same result.

To convert to eV, the conversion 1 eV = 8065.5 cm^{-1} from inside the front cover is used to give $\boxed{D_0 = 5.14\ \text{eV}}$.

E11C.8(a) The wavenumber of the transition arising from the rotational state J in the R branch ($\Delta J = +1$) of the fundamental transition ($v = 1 \leftarrow v = 0$) is given by [11C.13c–*447*], $\tilde{\nu}_R(J) = \tilde{\nu} + 2\tilde{B}(J+1)$. In this case $\tilde{\nu} = 2308.09\ \text{cm}^{-1}$ and $\tilde{B} = 6.511\ \text{cm}^{-1}$ hence

$$\tilde{\nu}_R(2) = (2308.09\ \text{cm}^{-1}) + 2 \times (6.511\ \text{cm}^{-1}) \times (2+1) = \boxed{2347.2\ \text{cm}^{-1}}$$

Solutions to problems

P11C.1 (a) Figure 11.7 shows plot of the total electronic energy (with respect to the free atoms) as a function of the bond length for each of the hydrogen halides. Calculations are performed with Spartan 10 using the MP2 method with the 6-311++G** basis set.

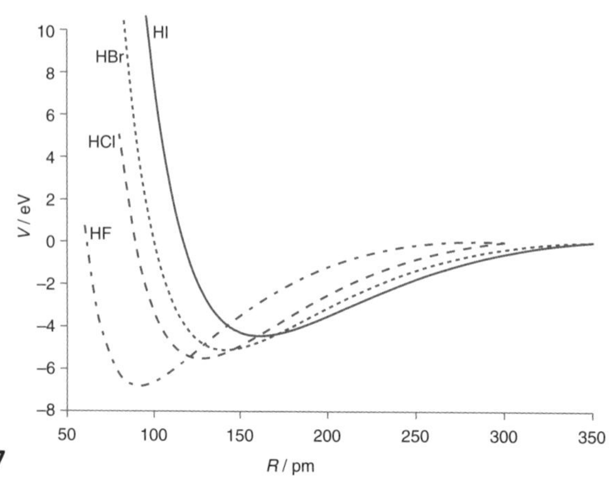

Figure 11.7

The plot clearly shows that in going down the halogen group from HF to HI the equilibrium bond length increases and the depth of the potential well decreases. The equilibrium properties of each molecule are summarized in the following table. The force constants are computed in the harmonic approximation using [11C.4b–443], $\tilde{\nu} = (1/2\pi c)(k_f/m_{eff})^{1/2}$, with $m_{eff} = m_1 m_2/(m_1 + m_2)$. It follows that $k_f = m_{eff}(2\pi c\tilde{\nu})^2$. The calculated bond lengths are in good agreement with the experimental values, but the vibrational frequencies do not agree very well at all.

property	$H^{19}F$	$H^{35}Cl$	$H^{81}Br$	$H^{127}I$
R_e/pm	91.7	127.3	141.3	161.2
$R_{e,\,expt}$/pm	91.680	127.45	141.44	160.92
$\tilde{\nu}/cm^{-1}$	4198.162	3086.560	2729.302	2412.609
$\tilde{\nu}_{expt}/cm^{-1}$	4138.29	2990.95	2648.98	2309.01
$k_f/(N\,m^{-1})$	993.7	549.8	436.7	342.9

(b) The force constants decrease steadily down the series, as expected.

P11C.3 Figure 11.8 shows a plot of $V(x)/V_0$ as a function of x/a; the minimum at $x = 0$ is clearly rather 'flat'.

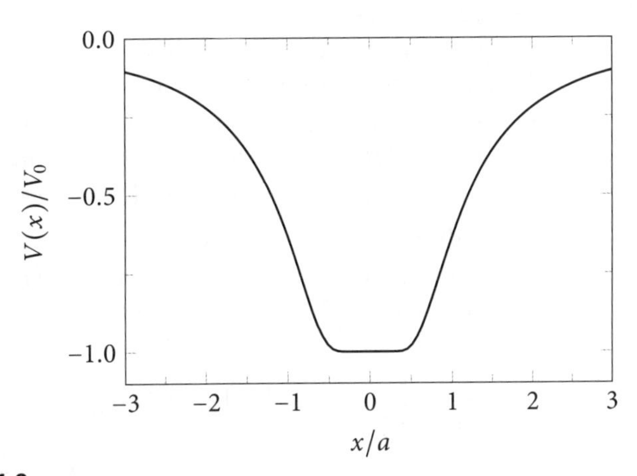

Figure 11.8

For a given potential the force constant is defined in [11C.2b–442] in terms of the second derivative as $k_f = (d^2V/dx^2)_0$.

$$\frac{d}{dx}V_0(e^{-a^2/x^2} - 1) = V_0\frac{2a^2}{x^3}e^{-a^2/x^2}$$

$$\frac{d^2}{dx^2}V(x) = \frac{d}{dx}V_0\frac{2a^2}{x^3}e^{-a^2/x^2} = \frac{2V_0a^2}{x^3}\left(-\frac{3}{x} + \frac{2a^2}{x^3}\right)e^{-a^2/x^2}$$

The second derivative, and hence the force constant, goes to zero at $x = 0$ on account of the argument of the exponential term going to $-\infty$; this dominates the other terms. Thus, for small displacements there is no restoring force and harmonic motion will not occur.

The potential is confining so it is expected that there will be quantized energy levels. By loose analogy with the harmonic case the ground state wavefunction is expected to have a maximum at $x = 0$ and then decay away to zero as $x \rightarrow \pm\infty$. The first excited state is likely to have a node at $x = 0$, increase to a maximum at some positive value of x and then decay away to zero. The function will be odd with respect to $x = 0$, and so will show a symmetrically placed minimum at a negative value of x.

P11C.5 (a) The dissociation energy is $\tilde{D}_0 = \tilde{D}_e - \tilde{G}(0)$ (Fig. 11C.3 on page 444), where $\tilde{G}(0)$ is the energy of the lowest vibrational term. For the Morse energy levels given by [11C.8–444], $\tilde{G}(\upsilon) = (\upsilon + \frac{1}{2})\tilde{\nu} - (\upsilon + \frac{1}{2})^2 x_e\tilde{\nu}$ it follows that $\tilde{G}(0) = \frac{1}{2}\tilde{\nu} - \frac{1}{4}x_e\tilde{\nu}$. The conversion between cm^{-1} and eV is achieved using 1 eV = 8065.5 cm^{-1} from inside the front cover. For $^1H^{35}Cl$

$$
\begin{aligned}
hc\tilde{D}_0 &= hc\tilde{D}_e - \tilde{G}(0) = hc\tilde{D}_e - (\tfrac{1}{2}\tilde{\nu} - \tfrac{1}{4}x_e\tilde{\nu}) \\
&= (5.33\ \text{eV}) - (\tfrac{1}{2} \times 2989.7 - \tfrac{1}{4} \times 52.05) \times [(1\ \text{eV})/(8065.5\ \text{cm}^{-1})] \\
&= \boxed{5.15\ \text{eV}}
\end{aligned}
$$

(b) The task is to calculate the values of $\tilde{\nu}$ and $x_e\tilde{\nu}$ for the isotopologue $^2H^{35}Cl$. The potential energy curve, and hence the value of the depth of the well $\tilde{D}_e$, is the same for the two isotopologues.

In the harmonic limit the vibrational frequency is given by [11C.4b–*443*], $\tilde{\nu} = (1/2\pi c)(k_f/m_{eff})^{1/2}$, with $m_{eff} = m_1 m_2/(m_1 + m_2)$. Assuming that the force constants of the two isotopologues are the same, $\tilde{\nu} \propto m_{eff}^{-1/2}$.

From [11C.8–*444*] it is seen that $x_e = \tilde{\nu}/4\tilde{D}_e$ which rearranges to $\tilde{D}_e = \tilde{\nu}/4x_e$. Because $\tilde{\nu} \propto m_{eff}^{-1/2}$ it follows that $x_e \propto m_{eff}^{-1/2}$ also in order for $\tilde{D}_e$ to be unaffected by isotopic substitution. Thus $x_e\tilde{\nu} \propto m_{eff}^{-1}$.

$$
\frac{\tilde{\nu}_{^2H^{35}Cl}}{\tilde{\nu}_{^1H^{35}Cl}} = \left(\frac{m_{eff,^1HX}}{m_{eff,^2H^{35}Cl}}\right)^{1/2} \text{ hence } \tilde{\nu}_{^2H^{35}Cl} = \tilde{\nu}_{^1H^{35}Cl} \times \left(\frac{m_{eff,^1H^{35}Cl}}{m_{eff,^2H^{35}Cl}}\right)^{1/2}
$$

$$
\begin{aligned}
\tilde{\nu}_{^2H^{35}Cl} &= (2989.7\ \text{cm}^{-1})\left(\frac{1.0078 \times 34.9688}{1.0078 + 34.9688} \times \frac{2.0140 + 34.9688}{2.02140 \times 34.9688}\right)^{1/2} \\
&= 2144.25\ \text{cm}^{-1}
\end{aligned}
$$

Similarly

$$
\begin{aligned}
x_e\tilde{\nu}_{^2H^{35}Cl} &= x_e\tilde{\nu}_{^1H^{35}Cl} \times \left(\frac{m_{eff,^1H^{35}Cl}}{m_{eff,^2H^{35}Cl}}\right) \\
&= (52.05\ \text{cm}^{-1})\left(\frac{1.0078 \times 34.9688}{1.0078 + 34.9688} \times \frac{2.0140 + 34.9688}{2.02140 \times 34.9688}\right) \\
&= 26.77\ \text{cm}^{-1}
\end{aligned}
$$

Hence for $^2H^{35}Cl$

$$hc\tilde{D}_0 = hc\tilde{D}_e - \tilde{G}(0) = hc\tilde{D}_e - (\tfrac{1}{2}\tilde{\nu} - \tfrac{1}{4}x_e\tilde{\nu})$$
$$= (5.33\ \text{eV}) - (\tfrac{1}{2} \times 2144.25 - \tfrac{1}{4} \times 26.77) \times [(1\ \text{eV})/(8065.5\ \text{cm}^{-1})]$$
$$= \boxed{5.20\ \text{eV}}$$

The term $\frac{1}{4}x_e\tilde{\nu}$ evaluates to 8.3×10^{-4} eV, so at the precision to which $hc\tilde{D}_e$ is quoted this term has no effect.

P11C.7 (a) The dissociation energy $\tilde{D}_0$ and the well depth $\tilde{D}_e$ are related by $\tilde{D}_e = \tilde{D}_0 + \tilde{G}(0)$ (Fig. 11C.3 on page 444), where $\tilde{G}(0)$ is the vibrational term of the ground vibrational state. In the harmonic approximation $\tilde{G}(0) = \frac{1}{2}\tilde{\nu}$, so it follows that

$$\tilde{\nu} = 2(\tilde{D}_e - \tilde{D}_0) = 2(D_e/hc - D_0/hc)$$

With the data given

$$\tilde{\nu} = 2[(1.51 \times 10^{-23}\ \text{J}) - (2 \times 10^{-26}\ \text{J})]$$
$$/[(6.6261 \times 10^{-34}\ \text{J s}) \times (2.9979 \times 10^{10}\ \text{cm s}^{-1})] = \boxed{1.5\ \text{cm}^{-1}}$$

In the harmonic limit the vibrational frequency is given by [11C.4b–*443*], $\tilde{\nu} = (1/2\pi c)(k_f/m_{\text{eff}})^{1/2}$, with $m_{\text{eff}} = m_1m_2/(m_1 + m_2)$. For a homonuclear diatomic $m_{\text{eff}} = \frac{1}{2}m$. It follows that $k_f = \frac{1}{2}m(2\pi c\tilde{\nu})^2$.

$$k_f = \tfrac{1}{2} \times (4.0026) \times (1.6605 \times 10^{-27}\ \text{kg})$$
$$\times [2\pi \times (2.9979 \times 10^{10}\ \text{cm s}^{-1}) \times (1.5\ \text{cm}^{-1})]^2$$
$$= \boxed{2.7 \times 10^{-4}\ \text{N m}^{-1}}$$

The moment of inertia is $I = m_{\text{eff}}R^2 = \frac{1}{2}mR^2$

$$I = \tfrac{1}{2} \times (4.0026) \times (1.6605 \times 10^{-27}\ \text{kg}) \times (297 \times 10^{-12}\ \text{m})^2$$
$$= \boxed{2.93 \times 10^{-46}\ \text{kg m}^2}$$

(b) If the Morse energy levels are assumed $G(0) = \frac{1}{2}\tilde{\nu} - \frac{1}{4}x_e\tilde{\nu}$, and from [11C.8–*444*] $x_e = \tilde{\nu}/4\tilde{D}_e$. It follows that

$$\tilde{D}_e = \tilde{D}_0 + \tfrac{1}{2}\tilde{\nu} - \tfrac{1}{4}x_e\tilde{\nu} = \tilde{D}_0 + \tfrac{1}{2}\tilde{\nu} - \tilde{\nu}^2/16\tilde{D}_e$$

The result is a quadratic in $\tilde{\nu}$ which is solved in the usual way

$$\tilde{\nu}^2/16\tilde{D}_e - \tfrac{1}{2}\tilde{\nu} + (\tilde{D}_e - \tilde{D}_0) = 0 \quad \text{hence} \quad \tilde{\nu} = \frac{\frac{1}{2} \pm [\frac{1}{4} - (\tilde{D}_e - \tilde{D}_0)/4\tilde{D}_e]^{1/2}}{1/8\tilde{D}_e}$$

With the data given

$$(\tilde{D}_e - \tilde{D}_0)/4\tilde{D}_e = [(1.51 \times 10^{-23}\ \text{J}) - (2 \times 10^{-26}\ \text{J})]/[4 \times (1.51 \times 10^{-23}\ \text{J})]$$
$$= 0.2497$$

$$\tilde{D}_e = (1.51 \times 10^{-23}\ \text{J})/[(6.6261 \times 10^{-34}\ \text{J s}) \times (2.9979 \times 10^{10}\ \text{cm s}^{-1})]$$
$$= 0.760\ \text{cm}^{-1}$$

hence

$$\tilde{\nu} = 8 \times (0.760\ \mathrm{cm^{-1}}) \times [\tfrac{1}{2} \pm (\tfrac{1}{4} - 0.2497)^{1/2}] = 2.93\ \mathrm{cm^{-1}}\ \text{or}\ 3.15\ \mathrm{cm^{-1}}$$

With these values the anharmonicity constant is computed using $x_e = \tilde{\nu}/4\tilde{D}_e$

$$x_e = \frac{2.93\ \mathrm{cm^{-1}}}{4 \times (0.760\ \mathrm{cm^{-1}})} = 0.96 \quad \text{or} \quad x_e = \frac{3.15\ \mathrm{cm^{-1}}}{4 \times (0.760\ \mathrm{cm^{-1}})} = 1.04$$

The anharmonicity constant is expected to be < 1, so the plausible values are $\boxed{\tilde{\nu} = 2.9\ \mathrm{cm^{-1}}}$ and $\boxed{x_e = 0.96}$. These values are very approximate given the data used to derive them.

P11C.9 The data are shown in the table and the plot in Fig. 11.9.

v	$\Delta\tilde{G}_{v+1/2}/\mathrm{cm^{-1}}$	$v+1$
0	2 143.1	1
1	2 116.1	2
2	2 088.9	3
3	2 061.3	4
4	2 033.5	5

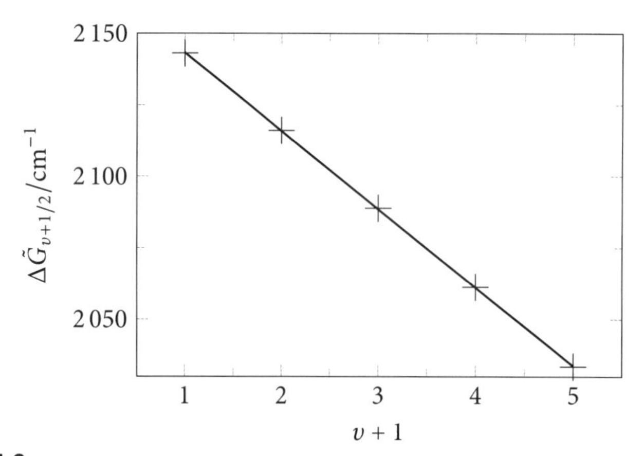

Figure 11.9

The data are a good fit to the line

$$\Delta\tilde{G}_{v+1/2}/\mathrm{cm^{-1}} = -27.40 \times (v+1) + 2\,170.7$$

From the slope

$$x_e\tilde{\nu}/\mathrm{cm^{-1}} = -\tfrac{1}{2} \times \text{slope} = -\tfrac{1}{2}(-27.40) \quad \text{hence} \quad \boxed{x_e\tilde{\nu} = 13.7\ \mathrm{cm^{-1}}}$$

and from the intercept

$$\tilde{\nu}/\mathrm{cm^{-1}} = \text{intercept} = 2\,170.70 \quad \text{hence} \quad \boxed{\tilde{\nu} = 2\,170.7\ \mathrm{cm^{-1}}}$$

P11C.11 The data provided allow the calculation of two independent moments of inertia. If it is assumed that the bond lengths are unaffected by isotopic substitution, then it is possible to set up two equations and solve them simultaneously for the CC and CH bond lengths.

Expressions for the wavenumbers of the lines in the P and R branches are given by [11C.13a–*447*] and [11C.13c–*447*]; from these it follows that the spacing between the lines is $2\tilde{B}$. The rotational constant is given by [11B.7–*432*], $\tilde{B} = \hbar/4\pi c I$; it follows that $I = \hbar/4\pi c\tilde{B}$.

$$\begin{aligned} I_H &= \hbar/4\pi c\tilde{B}_H \\ &= \frac{1.0546\times10^{-34}\,\text{J s}}{4\pi\times(2.9979\times10^{10}\,\text{cm s}^{-1})\times[(2.352/2)\,\text{cm}^{-1}]} = 2.38...\times10^{-46}\,\text{kg m}^2 \\ I_D &= \hbar/4\pi c\tilde{B}_D \\ &= \frac{1.0546\times10^{-34}\,\text{J s}}{4\pi\times(2.9979\times10^{10}\,\text{cm s}^{-1})\times[(1.696/2)\,\text{cm}^{-1}]} = 3.30...\times10^{-46}\,\text{kg m}^2 \end{aligned}$$

The moment of inertia is defined as $I = \sum_i m_i r_i^2$, where r_i is the perpendicular distance from the atom with mass m_i to the axis. In HCCH the axis passes through the mid-point of the CC bond and is perpendicular to the long axis of the molecule. Thus

$$\begin{aligned} I_H &= 2m_C(r_{CC}/2)^2 + 2m_H(r_{CH} + r_{CC}/2)^2 \\ I_D &= 2m_C(r_{CC}/2)^2 + 2m_D(r_{CH} + r_{CC}/2)^2 \end{aligned}$$

These are the two equations which need to be solved simultaneously. Finding the solution is much simplified by letting $p = (r_{CC}/2)^2$ and $q = (r_{CH} + r_{CC}/2)^2$ to give

$$I_H = 2m_C p + 2m_H q \qquad I_D = 2m_C p + 2m_D q$$

It follows that

$$\begin{aligned} q &= \frac{I_H - I_D}{2(m_H - m_D)} \\ &= \frac{(2.38...\times10^{-46}\,\text{kg m}^2) - (3.30...\times10^{-46}\,\text{kg m}^2)}{2(1.0078 - 2.0140)\times(1.6605\times10^{-27}\,\text{kg})} = 2.75...\times10^{-20}\,\text{m}^2 \\ p &= \frac{m_D I_H - m_H I_D}{2m_C(m_D - m_H)} \\ &= \frac{(2.0140)\times(2.38...\times10^{-46}\,\text{kg m}^2) - (1.0078)\times(3.30...\times10^{-46}\,\text{kg m}^2)}{2\times(12.0000)\times(2.0140 - 1.0078)\times(1.6605\times10^{-27}\,\text{kg})} \\ &= 3.65...\times10^{-20}\,\text{m}^2 \end{aligned}$$

If follows that $r_{CC} = 2\times p^{1/2} = \boxed{121.0\text{ pm}}$, and $r_{CH} = q^{1/2} - r_{CC}/2 = q^{1/2} - p^{1/2} = \boxed{105.5\text{ pm}}$.

P11C.13 The variable x is the displacement from the equilibrium separation R_e. The fact that the potential is symmetric about R_e means that $\langle R\rangle = R_e$ and $\langle x\rangle = 0$.

On the other hand $\langle x^2 \rangle$ is definitely non-zero, as was seen for the case of the harmonic oscillator in Topic 7E.

It follows straightforwardly that $\boxed{1/\langle R \rangle^2 = 1/R_e^2}$.

$\langle R^2 \rangle$ is found in the following way

$$\langle R^2 \rangle = \langle (R_e + x)^2 \rangle = \langle (R_e^2 + 2xR_e + x^2) \rangle$$
$$= \overbrace{\langle R_e^2 \rangle}^{A} + \overbrace{\langle 2xR_e \rangle}^{B} + \langle x^2 \rangle$$

Term A is simply the average of a constant term, which is equal to the term itself, in this case R_e^2. Term B is rewritten $2R_e\langle x \rangle$ by taking constant terms outside the averaging; this term is zero because $\langle x \rangle = 0$. Therefore $\langle R^2 \rangle = R_e^2 + \langle x^2 \rangle$.

Using this $1/\langle R^2 \rangle$ is found in the following way

$$\frac{1}{\langle R^2 \rangle} = \frac{1}{R_e^2 + \langle x^2 \rangle} = \frac{1}{R_e^2} \times \frac{1}{1 + \langle x^2 \rangle / R_e^2}$$
$$\approx \boxed{\frac{1}{R_e^2}\left(1 - \frac{\langle x^2 \rangle}{R_e^2}\right)}$$

where to go to the last line the expansion $(1+y)^{-1} \approx 1 - y$ is used. The resulting expression includes the lowest power of $\langle x^2 \rangle / R_e^2$, as required

$\langle 1/R^2 \rangle$ is found in the following way

$$\left\langle \frac{1}{R^2} \right\rangle = \left\langle \frac{1}{(R_e + x)^2} \right\rangle = \frac{1}{R_e^2}\left\langle \frac{1}{(1 + x/R_e)^2} \right\rangle$$
$$\approx \frac{1}{R_e^2}\langle 1 - 2x/R_e + 3x^2/R_e^2 \rangle = \frac{1}{R_e^2}(\langle 1 \rangle - (2/R_e)\langle x \rangle + (3/R_e^2)\langle x^2 \rangle)$$
$$= \boxed{\frac{1}{R_e^2}\left(1 + \frac{3\langle x^2 \rangle}{R_e^2}\right)}$$

On the penultimate line the expansion $(1+y)^{-2} \approx 1 - 2y + 3y^2$ is used. To go to the final line the fact that $\langle x \rangle = 0$ is used; the final expression has the lowest non-zero power of $\langle x^2 \rangle / R_e^2$, as required.

It is evident that none of the averages are the same and that

$$\left\langle \frac{1}{R^2} \right\rangle > \frac{1}{\langle R \rangle^2} > \frac{1}{\langle R^2 \rangle}$$

P11C.15 The rotational constant $\tilde{B}_0$ is computed from

$$\tilde{B}_0 = \tilde{B}_e - \tfrac{1}{2}a = (0.27971\ \text{cm}^{-1}) - \tfrac{1}{2}(0.187 \times 10^{-2}\ \text{cm}^{-1}) = \boxed{0.27877\ \text{cm}^{-1}}$$

and similarly for $\tilde{B}_1$

$$\tilde{B}_1 = \tilde{B}_e - \tfrac{3}{2}a = (0.27971\ \text{cm}^{-1}) - \tfrac{3}{2}(0.187 \times 10^{-2}\ \text{cm}^{-1}) = \boxed{0.27691\ \text{cm}^{-1}}$$

The wavenumber of the lines in the P and R branches are given by [11C.14–*448*]

$$\tilde{\nu}_P(J) = \tilde{\nu}_0 - (\tilde{B}_1 + \tilde{B}_0)J + (\tilde{B}_1 - \tilde{B}_0)J^2$$
$$\tilde{\nu}_R(J) = \tilde{\nu}_0 + (\tilde{B}_1 + \tilde{B}_0)(J+1) + (\tilde{B}_1 - \tilde{B}_0)(J+1)^2$$

In these expressions $\tilde{\nu}_0$ is the wavenumber of the pure vibrational transition. If the Morse levels are assumed, and if it is assumed that it is the $v = 1 \leftarrow v = 0$ transition which is being observed, the wavenumber is given by [11C.9b–*445*], $\tilde{\nu}_0 = \tilde{\nu} - 2x_e\tilde{\nu}$.

For the line in the P branch from $J = 3$

$$\tilde{\nu}_P(J)/\text{cm}^{-1} = 610.258 - 2 \times 3.141 - (0.27691 + 0.27877) \times 3 + (0.27691 - 0.27877) \times 3^2 = \boxed{602.292}$$

and for the corresponding line in the R branch

$$\tilde{\nu}_R(J)/\text{cm}^{-1} = 610.258 - 2 \times 3.141 + (0.27691 + 0.27877) \times 4 + (0.27691 - 0.27877) \times 4^2 = \boxed{606.170}$$

The depth of the well, $\tilde{D}_e$ is found using [11C.8–*444*], $x_e = \tilde{\nu}/4\tilde{D}_e$ rearranged to $\tilde{D}_e = \tilde{\nu}/4x_e = \tilde{\nu}^2/4\tilde{\nu}x_e$. The dissociation energy is $\tilde{D}_0 = \tilde{D}_e - \tilde{G}(0)$ (Fig. 11C.3 on page 444), and for the Morse oscillator $\tilde{G}(0) = \frac{1}{2}\tilde{\nu} - \frac{1}{4}\tilde{\nu}x_e$.

$$\begin{aligned}\tilde{D}_0 &= \frac{\tilde{\nu}^2}{4\tilde{\nu}x_e} - \tfrac{1}{2}\tilde{\nu} + \tfrac{1}{4}\tilde{\nu}x_e \\ &= \frac{(610.258\ \text{cm}^{-1})^2}{4 \times (3.141\ \text{cm}^{-1})} - \tfrac{1}{2}(610.258\ \text{cm}^{-1}) + \tfrac{1}{4}(3.141\ \text{cm}^{-1}) \\ &= \boxed{2.93 \times 10^4\ \text{cm}^{-1}}\end{aligned}$$

To convert to eV, the conversion 1 eV = 8065.5 cm^{-1} from inside the front cover is used to give $\boxed{D_0 = 3.64\ \text{eV}}$.

P11C.17 The features centred about 2143.26 cm^{-1} are the P and R branches. From [11C.13a–*447*] and [11C.13c–*447*] the first line in the R branch occurs at $\tilde{\nu} + 2\tilde{B}$, and the first line in the P branch is $\tilde{\nu} - 2\tilde{B}$. The separation of these two, 7.655 cm^{-1}, is therefore $4\tilde{B}$.

(a) The centre of the band is at the vibrational wavenumber, $\boxed{\tilde{\nu} = 2143.26\ \text{cm}^{-1}}$

(b) In the harmonic approximation the vibrational terms are $\tilde{G}(v) = (v+\frac{1}{2})\tilde{\nu}$, and so the lowest term is $\tilde{G}(0) = \frac{1}{2}\tilde{\nu}$. The molar zero-point energy is therefore $N_A \times hc \times \frac{1}{2}\tilde{\nu}$

$$E_{zpe} = (6.0221 \times 10^{23}\ \text{mol}^{-1}) \times (6.6261 \times 10^{-34}\ \text{J s}) \times (2.9979 \times 10^{10}\ \text{cm s}^{-1}) \times \tfrac{1}{2} \times (2143.26\ \text{cm}^{-1}) = \boxed{12.82\ \text{kJ mol}^{-1}}$$

(c) The harmonic frequency is given by [11C.4b–443], $\tilde{\nu} = (1/2\pi c)(k_f/m_{eff})^{1/2}$, with $m_{eff} = m_1 m_2/(m_1 + m_2)$. It follows that $k_f = m_{eff}(2\pi c\tilde{\nu})^2$. With the data given

$$k_f = \frac{12.0000 \times 15.9949}{12.000 + 15.9949} \times (1.6605 \times 10^{-27}\,\text{kg}) \times [2\pi \times (2.9979 \times 10^{10}\,\text{cm s}^{-1}) \times (2143.26\,\text{cm}^{-1})]^2 = \boxed{1856\,\text{N m}^{-1}}$$

(d) As noted at the start of the answer, $4\tilde{B} = 7.655\,\text{cm}^{-1}$, hence $\boxed{\tilde{B} = 1.914\,\text{cm}^{-1}}$.

(e) The rotational constant is given by [11B.7–432], $\tilde{B} = \hbar/4\pi c I$, and the moment of inertia is given by $m_{eff}R^2$. It follows that $R = (\hbar/4\pi c m_{eff}\tilde{B})^{1/2}$.

$$m_{eff} = \frac{12.0000 \times 15.9949}{12.0000 + 15.9949} \times (1.6605 \times 10^{-27}\,\text{kg}) = 1.13... \times 10^{-26}\,\text{kg}$$

$$R = (\hbar/4\pi c m_{eff}\tilde{B})^{1/2} = \left(\frac{1.0546 \times 10^{-34}\,\text{J s}}{4\pi(2.9979 \times 10^{10}\,\text{cm s}^{-1}) \times (1.13... \times 10^{-26}\,\text{kg}) \times (1.914\,\text{cm}^{-1})}\right)^{1/2} = \boxed{113.3\,\text{pm}}$$

Although the data are given to quite high precision the assumption that the harmonic oscillator/rigid rotor models apply means that the derived values of the bond length and so on are likely to have systematic errors which are higher than the apparent precision of the data.

P11C.19 The method of combination differences, described in Section 11C.4(b) on page 448, involves taking the difference between two transitions which share a common lower rotational level or a common upper rotational level. In the case of O and S branches, which correspond to $\Delta J = -2$ and $\Delta J = +2$, respectively, the two transitions which share a common lower level are $\tilde{\nu}_O(J)$ and $\tilde{\nu}_S(J)$: these are the transitions from J to $J-2$, and from J to $J+2$. As is evident from Fig. 11.10, the difference in wavenumber between these two transitions is the interval indicated by the dashed arrow which is simply $\tilde{G}(J+2) - \tilde{G}(J-2)$ for the upper vibrational state (assumed to be $v = 1$)

$$\tilde{G}(J+2) - \tilde{G}(J-2) = \tilde{B}_1(J+2)(J+3) - \tilde{B}_1(J-2)(J-1) = \tilde{B}_1(8J+4)$$

hence $\boxed{\tilde{\nu}_S(J) - \tilde{\nu}_O(J) = 8\tilde{B}_1(J+\tfrac{1}{2})}$

The two transitions sharing a common upper level are $\tilde{\nu}_O(J+2)$ and $\tilde{\nu}_S(J-2)$: these are the transitions from $J+2$ to J, and from $J-2$ to J. As is evident from Fig. 11.10, the difference in wavenumber between these two transitions is the interval indicated by the dotted arrow which is simply $\tilde{G}(J+2) - \tilde{G}(J-2)$ for the lower vibrational state (assumed to be $v = 0$). This is the same interval as above, with the exception that the rotational constant is $\tilde{B}_0$.

$$\tilde{G}(J+2) - \tilde{G}(J-2) = \tilde{B}_0(J+2)(J+3) - \tilde{B}_0(J-2)(J-1) = \tilde{B}_0(8J+4)$$

hence $\boxed{\tilde{\nu}_S(J-2) - \tilde{\nu}_O(J+2) = 8\tilde{B}_0(J+\tfrac{1}{2})}$

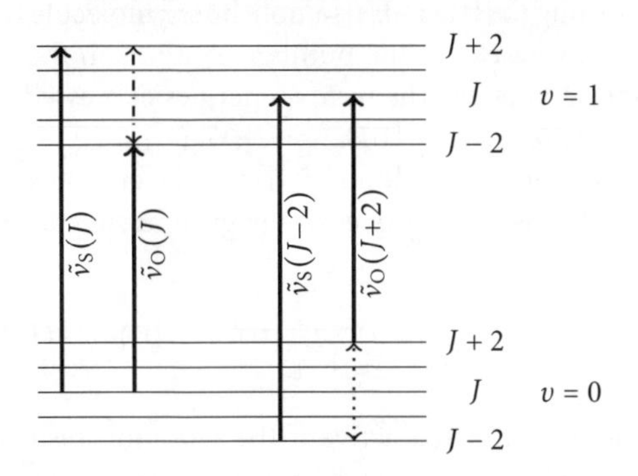

Figure 11.10

11D Vibrational spectroscopy of polyatomic molecules

Answers to discussion questions

D11D.1 The gross selection rule for infrared spectroscopy is that for a particular normal mode to be active the vibration must result in a change in the dipole moment as the molecule vibrates about the equilibrium position. The origin of this rule is that such an oscillating dipole is needed to stir the electromagnetic field into oscillation (and vice versa for absorption).

D11D.3 CO_2 is a centro-symmetric linear molecule. It gives rise to rotational Raman scattering because it has an anisotropic polarizability. Some of the normal modes may give rise to vibrational Raman scattering, but on account of the rule of mutual exclusion, modes which are infrared active are not Raman active, and vice versa. In fact for CO_2 only the symmetric stretch is Raman active.

Solutions to exercises

E11D.1(a) With the exception of homonuclear diatomics, all molecules have at least one infrared active normal mode. Of the molecules listed, $\boxed{\text{HCl, CO}_2\text{, and H}_2\text{O}}$ have infrared active modes.

E11D.2(a) According to [11D.1–*451*], a non-linear molecule has $3N-6$ vibrational normal modes, where N is the number of atoms in the molecule; a linear molecule has $3N-5$ normal modes. All of the molecules listed are non-linear.

(i) H_2O has $N=3$ and hence $\boxed{3}$ normal modes.

(ii) H_2O_2 has $N=4$ and hence $\boxed{6}$ normal modes.

(iii) C_2H_4 has $N=6$ and hence $\boxed{12}$ normal modes.

E11D.3(a) According to [11D.1–*451*], a linear molecule has $3N-5$ normal modes, where N is the number of atoms in the molecule. There are 44 atoms in this linear molecule, and so there are $3(44)-5=\boxed{127}$ normal modes.

E11D.4(a) According to [11D.1–*451*], a non-linear molecule has $3N-6$ vibrational normal modes, where N is the number of atoms in the molecule; therefore H_2O has 3 normal modes. The terms (energies expressed as wavenumbers) for normal mode q are given by [11D.2–*452*], $\tilde{G}_q(v) = (v_q + \frac{1}{2})\tilde{\nu}_q$, where v_q is the quantum number for that mode and $\tilde{\nu}_q$ is the wavenumber of the vibration of that mode. These terms are additive, so the ground state term corresponds to each mode having $v_q = 0$

$$\tilde{G}_1(0) + \tilde{G}_2(0) + \tilde{G}_3(0) = \boxed{\tfrac{1}{2}(\tilde{\nu}_1 + \tilde{\nu}_2 + \tilde{\nu}_3)}$$

E11D.5(a) A mode is infrared active if the vibration results in a dipole which changes as the molecule oscillates back and forth about the equilibrium geometry. A mode is Raman active if the vibration results in the polarizability changing as the molecule oscillates back and forth about the equilibrium geometry.

(i) The three normal modes of an angular AB_2 molecule are analogous to those of H_2O illustrated in Fig. 11D.3 on page 452; $\boxed{\text{all three}}$ modes are both infrared and Raman active.

(ii) The four normal modes of a linear AB_2 molecule are analogous to those of CO_2 illustrated in Fig. 11D.2 on page 452. Of these, $\boxed{\text{three}}$ are infrared active (the asymmetric stretch and the doubly degenerate bend), and $\boxed{\text{one}}$ (the symmetric stretch) is Raman active. The molecule has a centre of symmetry, so the rule of mutual exclusion applies and no mode is both Raman and infrared active.

E11D.6(a) The benzene molecule has a centre of symmetry, so the rule of mutual exclusion applies. The molecule has no permanent dipole moment and if the ring expands uniformly this situation does not change: such a vibration does not lead to a changing dipole and so the mode is $\boxed{\text{infrared inactive}}$.

This kind of 'breathing' vibration does lead to a change in the polarizability, so the mode is $\boxed{\text{Raman active}}$.

E11D.7(a) The exclusion rule applies only to molecules with a centre of symmetry. H_2O does not possess such symmetry, and so the exclusion rule $\boxed{\text{does not apply}}$.

Solutions to problems

P11D.1 Figure 11.11 shows a plot of $V(h)/V_0$ as a function of $hb^{1/4}$.

For a given potential the force constant is defined in [11C.2b–*442*] in terms of the second derivative of the potential with respect to the displacement from equilibrium; in this case h is the varaible which describes the displacement. Thus $k_f = (d^2V/dh^2)_0$.

$$\frac{d}{dh}V_0(1-e^{-bh^4}) = V_0 4bh^3 e^{-bh^4}$$

$$\frac{d^2}{dh^2}V(h) = \frac{d}{dh}V_0 4bh^3 e^{-bh^4} = 4V_0 b\left(3h^2 - 4bh^6\right)e^{-bh^4}$$

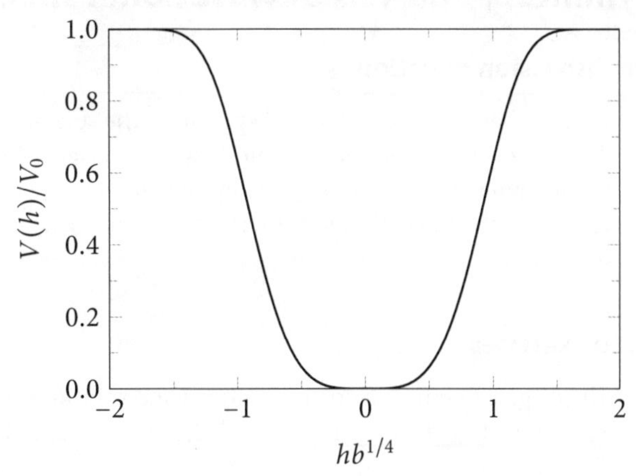

Figure 11.11

The second derivative, and hence the force constant, goes to zero at $h = 0$. Thus, for small displacements there is no restoring force.

The potential is confining so it is expected that there will be quantized vibrational energy levels. By loose analogy with the harmonic case the ground state wavefunction is expected to have a maximum at $h = 0$ and then decay away to zero as $h \to \pm\infty$.

P11D.3 (a) Calculations on SO_2 are performed with Spartan 10 using the MP2 method with the 6-311++G** basis set. The calculated equilibrium structure is shown in Fig. 11.12 where it is seen that the two S–O bonds have equal length, as expected. The calculated bond length and angle agrees quite well with the experimental values of 143.21 pm and 119.54°.

(b) The calculated values of the fundamental vibrational wavenumbers, and illustrations of the displacements involved in the normal modes, are also shown in Fig. 11.12. They correlate well with experimental values but are about 35–80 cm^{-1} lower. SCF calculations often yield systematically lower or higher values than experiment while approximately paralleling the experimental to within an additive constant.

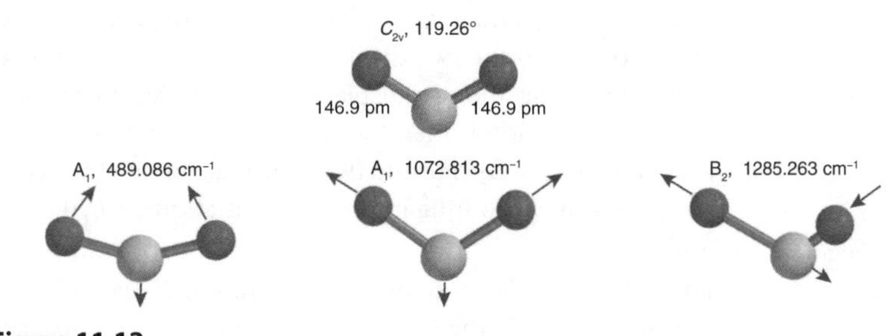

Figure 11.12

11E Symmetry analysis of vibrational spectroscopy

Answer to discussion question

D11E.1 Because benzene has a centre of symmetry the rule of mutual exclusion applies. Therefore a particular normal mode will be observed in either the infrared or in the Raman spectrum (or, possibly, in neither). The most complete characterization of the normal modes therefore requires the observation of both kinds of spectra.

Solutions to exercises

E11E.1(a) The displacements include translations and rotations. For the point group C_{2v}, x, y, and z transform as B_1, B_2, and A_1, respectively. The rotations R_x, R_y, and R_z transform as B_2, B_1, and A_2, respectively. Taking these symmetry species away leaves just the normal modes as $\boxed{4A_1 + A_2 + 2B_1 + 2B_2}$. These correspond to 9 normal modes, which is the number expected for CH_2Cl_2.

E11E.2(a) A mode is infrared active if it has the same symmetry species as one of the functions x, y, and z; in this point group these span $B_1 + B_2 + A_1$. A mode is Raman active if it has the same symmetry as a quadratic form; in this group such forms span $A_1 + A_2 + B_1 + B_2$. Therefore $\boxed{\text{all}}$ of the normal modes are both Raman and infrared active.

E11E.3(a) (i) H_2O belongs to the point group C_{2v}. Rather than considering all 9 displacement vectors together it is convenient to consider them in sub-sets of displacement vectors which are mapped onto one another by the operations of the group. The x, y, and z vectors on the oxygen are not mapped onto the displacements of the H atoms and so can be considered separately. In fact, because these displacement vectors are attached to the principal axis, they transform as the cartesian functions x, y, and z as listed in the character table: that is as $B_1 + B_2 + A_1$.

Assuming the same axis system as in Fig. 11E.1 on page 456, the two x displacements on the H atoms map onto one another, as do the two y displacements, as do the two z displacements: however, the x, y, and z displacements are not mixed with one another. For the two z displacements the operation E leaves both unaffected so the character is 2; the C_2 operation swaps the two displacements so the character is 0; the $\sigma_v(xz)$ operation swaps the two displacements so the character is 0; the $\sigma'_v(yz)$ operation leaves the two displacements unaffected so the character is 2. The representation is therefore $(2, 0, 0, 2)$, which is easily reduced by inspection to $A_1 + B_2$. For the two y displacements the argument is essentially the same, resulting in the representation $(2, 0, 0, 2)$, which reduces to $A_1 + B_2$.

For the two x displacements the operation E leaves both unaffected so the character is 2; the C_2 operation swaps the two displacements so the character is 0; the $\sigma_v(xz)$ operation swaps the two displacements so the character is 0; the $\sigma'_v(yz)$ operation leaves the two displacements in the same position by changes their direction, so the character is −2. The representation is therefore $(2, 0, 0, -2)$, which is easily reduced by inspection to $A_2 + B_1$.

The 9 displacements therefore transform as $3A_1 + A_2 + 2B_1 + 3B_2$.

The displacements include translations and rotations. For the point group C_{2v}, x, y, and z transform as B_1, B_2, and A_1, respectively. The rotations R_x, R_y, and R_z transform as B_2, B_1, and A_2, respectively. Taking these symmetry species away leaves just the normal modes as $\boxed{2A_1 + B_2}$.

A mode is infrared active if it has the same symmetry species as one of the functions x, y, and z; in this point group these span $B_1 + B_2 + A_1$. Therefore $\boxed{\text{all}}$ of the normal modes are infrared active.

(ii) H_2CO is a straightforward extension of the case of H_2O as both molecules belong to the point group C_{2v}. The H_2C portion lies in the same position as H_2O, with the carbonyl O atom lying on the z axis (the principal axis). The analysis therefore includes three more displacement vectors for the O, and as they are connected to the principal axis they transform as the cartesian functions x, y, and z, that is as $B_1 + B_2 + A_1$. The tally of normal modes is therefore those for H_2O plus these three in addition: $\boxed{3A_1 + B_1 + 2B_2}$. All these modes are infrared active.

Solutions to problems

P11E.1

(a) CH_3Cl has a C_3 axis (the principal axis) along the C–Cl bond, and three σ_v planes, one passing along each C–H bond and containing the principal axis. The point group is therefore $\boxed{C_{3v}}$.

(b) The molecule is non-linear and has $N = 5$, there are thus $3 \times 5 - 6 = \boxed{9}$ normal modes.

(c) The task is to find the symmetry species spanned by the set of (x, y, z) displacement vectors on each atom. The (x, y, z) displacement vectors on the Cl can be considered separately and, as these vectors are connected to the principal axis, their symmetry species is simply read from the character table as $E + A_1$. The (x, y, z) displacement vectors on the C behave in the same way and so transform as $E + A_1$.

Consider the set of 9 (x, y, z) displacement vectors on the H atoms. Under the operation E these are all unaffected so the character is 9; under the C_3 operation they are all moved to new positions so the character is 0. Next consider one of the σ_v planes: the (x, y, z) vectors on the H atoms which do not lie in this plane are all moved and so contribute 0 to the character. Now consider the H atom which lies in the plane, and arrange a local axis system such that z points along the C–H bond, y lies in the plane and x lies perpendicular to the plane. The effect of σ_v on (x, y, z) is to transform it to $(-x, +y, +z)$; two of the basis functions remain the same and one changes sign, so the character is $-1 + 1 + 1 = +1$. The representation formed from the 9 (x, y, z) displacement vectors on the

H atoms is thus $(9, 0, 1)$. This is reduced using the reduction formula, [10C.3a–*408*], to give $2A_1 + A_2 + 3E$.

In total, the displacements therefore transform as $4A_1 + A_2 + 5E$. Taking away the translations, $A_1 + E$, and the rotations, $A_2 + E$, leaves the symmetry species of the vibrations as $\boxed{3A_1 + 3E}$. As expected, there are 9 normal modes (recall that the E modes are doubly degenerate).

(d) A mode is infrared active if it has the same symmetry species as one of the functions x, y, and z; in this point group these span $A_1 + E$. $\boxed{\text{All}}$ the normal modes are infrared active.

(e) A mode is Raman active if it has the same symmetry as a quadratic form; in this point group these span $A_1 + E$. $\boxed{\text{All}}$ the normal modes are Raman active.

11F Electronic spectra

Answers to discussion questions

D11F.1 This is explained in Section 11F.1(a) on page 459.

D11F.3 The wavenumbers of the lines in the P, Q and R branches are given in [11F.7–*465*]. Consider first the P branch $\tilde{\nu}_P(J) = \tilde{\nu} - (\tilde{B}' + \tilde{B})J + (\tilde{B}' - \tilde{B})J^2$, and recall that $(\tilde{B}' + \tilde{B}) \gg |(\tilde{B}' - \tilde{B})|$. As J increases the lines move to lower wavenumber on account of the term $-(\tilde{B}' + \tilde{B})J$. However, as J becomes larger still the term in J^2 becomes proportionately more and more important. If $(\tilde{B}' - \tilde{B}) > 0$ this term contributes to an increase in the wavenumber of the lines, and for sufficiently large J it will overcome the $-(\tilde{B}' + \tilde{B})J$ term and cause the lines to start to move to higher wavenumber as J increases further. There will therefore be a lowest wavenumber at which any line appears: this is the band head. If $(\tilde{B}' - \tilde{B}) < 0$ the term in J^2 simply causes the lines to move to lower wavenumber and no band head is formed.

Next consider the R branch $\tilde{\nu}_R(J) = \tilde{\nu} + (\tilde{B}' + \tilde{B})(J + 1) + (\tilde{B}' - \tilde{B})(J + 1)^2$. As J increases the lines move to higher wavenumber on account of the term $+(\tilde{B}' + \tilde{B})(J + 1)$. However, as J becomes larger still the term in $(J + 1)^2$ becomes proportionately more and more important. If $(\tilde{B}' - \tilde{B}) < 0$ this term contributes to an decrease in the wavenumber of the lines, and for sufficiently large J it will overcome the $+(\tilde{B}' + \tilde{B})(J + 1)$ term and cause the lines to start to move to lower wavenumber as J increases further. There will therefore be a highest wavenumber at which any line appears: this is the band head. If $(\tilde{B}' - \tilde{B}) > 0$ the term in $(J + 1)^2$ simply causes the lines to move to higher wavenumber and no band head is formed.

The lines in the Q branch, $\tilde{\nu}_Q(J) = \tilde{\nu} + (\tilde{B}' - \tilde{B})J(J + 1)$, all appear at higher or lower wavenumber than $\tilde{\nu}$ depending on the sign of $(\tilde{B}' - \tilde{B})$; no band heads are formed.

If a parameter α is defined as $\alpha = \tilde{B}'/\tilde{B}$, so that $\tilde{B}' = \alpha\tilde{B}$, the wavenumbers of the lines in the P and R branches can be written

$$[\tilde{\nu}_P(J)-\tilde{\nu}]/\tilde{B} = -(1+\alpha)J-(1-\alpha)J^2 \quad [\tilde{\nu}_R(J)-\tilde{\nu}]/\tilde{B} = (1+\alpha)(J+1)-(1-\alpha)(J+1)^2$$

These functions are plotted in Fig. 11.13 for representative values of α. If $\alpha < 1$, meaning that $(\tilde{B}' - \tilde{B}) < 0$, the band head occurs in the R branch, but if $\alpha > 1$, meaning that $(\tilde{B}' - \tilde{B}) > 0$, the band head occurs in the P branch.

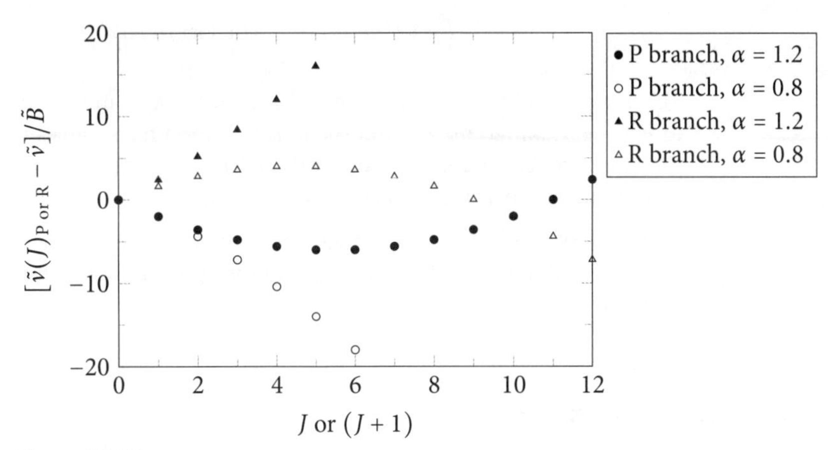

Figure 11.13

D11F.5 A simple model for the energy of the HOMO–LUMO transition in a polyene is discussed in *Example* 7D.1 on page 265. In this the energy levels of the π electrons in a polyene are modelled by those of a particle in a one-dimensional box of length L. If the polyene consists of n conjugated double bonds, the length L may be written as $L = nd$, where d is the length of a single conjugated bond. A molecule with n conjugated double bonds will have $2n$ π electrons which will occupy the energy levels pairwise. Therefore the HOMO is the level with quantum number n and the LUMO has quantum number $n + 1$. The energy of the HOMO–LUMO transition is therefore

$$E_{n+1} - E_n = [(n+1)^2 - n^2]\frac{h^2}{8mL^2} = \frac{(2n+1)h^2}{8mL^2}$$

For large n, $(E_{n+1} - E_n)$ goes as n. However, $L = nd$, therefore overall $(E_{n+1} - E_n)$ goes as $1/n$, and hence the wavelength of the transition goes as n. Thus, increasing the number of conjugated double bonds will increase the wavelength of the absorption; that is, shift it to the red.

The intensity of the transition will depend on the square of the transition dipole moment, given by

$$\int_0^L \psi_{n+1}\hat{\mu}_x\psi_n\,dx = (2/L)\int_0^L \sin[(n+1)\pi x/L]\,\hat{\mu}_x \sin[n\pi x/L]\,dx$$

where $\hat{\mu}_x$ is the operator for the dipole moment along x, and the normalized wavefunctions $\psi_n = (2/L)^{1/2}\sin(n\pi x/L)$ are used. The dipole moment operator is proportional to x, so setting aside the constants of proportion and using $(n+1) \approx n$ for large n, the required integral is

$$I = (2/L)\int_0^L x\sin^2(n\pi x/L)\,dx$$

As noted above, $L = nd$ therefore

$$I = (2/nd)\int_0^{nd} x\sin^2(\pi x/d)\,dx$$

The integral is of the form of Integral T.11 with $k = \pi/d$ and $a = nd$; it evaluates to $n^2d^2/4$. Taking into account the normalization factor gives the final result $nd/2$. The conclusion is therefore that the transition dipole moment increases with n, the number of conjugated double bonds.

In summary, the expectation is that increasing the number of conjugated double bonds will increase the wavelength of the absorption and also the intensity of the absorption.

Solutions to exercises

E11F.1(a) The electronic configuration of H_2 is σ_g^2. The two electrons are in the same orbital and so must be spin paired, hence $S = 0$, and $(2S+1) = 1$. Each σ electron has $\lambda = 0$, thus $\Lambda = 0 + 0 = 0$, which is represented by Σ. Two electrons with g symmetry have overall symmetry $g \times g = g$. The σ orbital is symmetric with respect to reflection in a plane containing the internuclear axis, therefore the two electrons in this orbital are also overall symmetric with respect to this mirror plane; this is indicated by a right-superscript +. The term symbol is therefore $\boxed{{}^1\Sigma_g^+}$.

E11F.2(a) The electronic configuration of Li_2^+ is $1\sigma_g^2\,1\sigma_u^2\,2\sigma_g^1$. The filled orbitals, $1\sigma_g^2$ and $1\sigma_u^2$, make no contribution to Λ and S, so can be ignored. Therefore, only the single electron in the $2\sigma_g$ orbital needs to be considered: it has $\lambda = 0$ and $s = \frac{1}{2}$, hence $\Lambda = 0$ and $S = \frac{1}{2}$ (giving a left superscript of $2S + 1 = 2$). The symmetry with respect to inversion is g and with respect to reflection is +. The term symbol is therefore $\boxed{{}^2\Sigma_g^+}$.

E11F.3(a) The electronic configuration given is $1\sigma_g^2\,1\sigma_u^2\,1\pi_u^3\,1\pi_g^1$. The filled orbitals, $1\sigma_g^2$ and $1\sigma_u^2$, make no contribution to Λ and S, and so can be ignored. With three electrons in a pair of degenerate π_u orbitals, two of the spins must be paired leaving one unpaired. There is another electron in a π_g orbital. These two electrons can be paired, giving $S = 0$ (a singlet, $2S + 1 = 1$), or parallel, giving $S = 1$ (a triplet, $2S + 1 = 3$).

The three electrons in the π_u have overall symmetry with respect to inversion (parity) $u \times u \times u = g \times u = u$. The remaining electron has g symmetry, so overall the state has symmetry $u \times g = u$.

In summary, the multiplicity is $\boxed{1}$ or $\boxed{3}$, and the parity is $\boxed{u}$.

E11F.4(a)

(i) Allowed

(ii) Allowed

(iii) No allowed, $\Delta\Sigma = 2$

(iv) Not allowed, $+ \nleftrightarrow -$

(v) Allowed

E11F.5(a) To evaluate the normalizing factor for the function $e^{-ax^2/2}$ requires the integral

$$\int_{-\infty}^{+\infty} e^{-ax^2}\, dx$$

which is of the form of Integral G.1 with $k = a$ and evaluates to $(\pi/a)^{1/2}$. The normalizing factor is therefore $N_0 = (a/\pi)^{1/4}$. The same factor applies to the function $e^{-a(x-x_0)^2/2}$ as this is simply a Gaussian shifted to x_0: the area under the square of this function is the same.

The Franck–Condon factor is given by [11F.5–*464*] and involves the square of integral of the product of the two wavefunctions

$$I = N_0^2 \int_{-\infty}^{+\infty} e^{-ax^2/2} e^{-a(x-x_0)^2/2}\, dx = (a/\pi)^{1/2} \int_{-\infty}^{+\infty} e^{-a[x^2/2+(x-x_0)^2/2]}\, dx$$

$$= (a/\pi)^{1/2} \int_{-\infty}^{+\infty} e^{-a[x^2/2+x^2/2-xx_0+x_0^2/2]}\, dx$$

$$= (a/\pi)^{1/2} \int_{-\infty}^{+\infty} e^{-a[x^2-xx_0+x_0^2/2]}\, dx = (a/\pi)^{1/2} \int_{-\infty}^{+\infty} e^{-a(x-x_0/2)^2} e^{-ax_0^2/4}\, dx$$

the final equality above is verified by expanding out the square and recombining the terms. Taking out the constant factors gives

$$I = (a/\pi)^{1/2} e^{-ax_0^2/4} \int_{-\infty}^{+\infty} e^{-a(x-x_0/2)^2}\, dx$$

as before, the integral is form of Integral G.1 with $k = a$ and evaluates to $(\pi/a)^{1/2}$

$$I = (a/\pi)^{1/2} e^{-ax_0^2/4} (\pi/a)^{1/2} = e^{-ax_0^2/4}$$

The Franck–Condon factor is therefore $\boxed{I^2 = e^{-ax_0^2/2}}$. As expected this factor is a maximum of 1 when $x_0 = 0$, that is when the two functions are aligned, and falls off towards zero as x_0 increases.

E11F.6(a) The Franck–Condon factor is given by [11F.5–464] and involves the square of integral of the product of the two wavefunctions. The region over which both wavefunctions are non-zero is from $L/4$ to L: this is the domain of integration

$$I = (2/L)\int_{L/4}^{L} \sin(\pi x/L)\sin(\pi[x - L/4]/L)\,dx$$

applying the identity $\sin A \sin B = \frac{1}{2}[\cos(A - B) - \cos(A + B)]$ gives

$$I = (1/L)\int_{L/4}^{L} \cos\left(\frac{\pi x}{L} - \frac{\pi(x - L/4)}{L}\right) - \cos\left(\frac{\pi x}{L} + \frac{\pi(x - L/4)}{L}\right) dx$$
$$= (1/L)\int_{L/4}^{L} \cos\left(\frac{\pi}{4}\right) - \cos\left(\frac{2\pi x}{L} - \frac{\pi}{4}\right) dx$$

Next the identity $\cos(A - B) = \cos A \cos B + \sin A \sin B$ is used

$$= (1/L)\int_{L/4}^{L} \cos\left(\frac{\pi}{4}\right) - \left[\cos\left(\frac{2\pi x}{L}\right)\cos\left(\frac{\pi}{4}\right) + \sin\left(\frac{2\pi x}{L}\right)\sin\left(\frac{\pi}{4}\right)\right] dx$$

Recognising that $\cos \pi/4 = \sin \pi/4 = (2)^{-1/2}$ allows this factor to be taken

$$= (1/L)2^{-1/2}\int_{L/4}^{L} 1 - \cos\left(\frac{2\pi x}{L}\right) - \sin\left(\frac{2\pi x}{L}\right) dx$$

Each term is now integrated and evaluated between the limits

$$= (1/L)2^{-1/2}\left| x - \frac{L}{2\pi}\sin\left(\frac{2\pi x}{L}\right) + \frac{L}{2\pi}\cos\left(\frac{2\pi x}{L}\right)\right|_{L/4}^{L}$$
$$= (1/L)2^{-1/2}\left[L - L/4 + \frac{L}{2\pi}\left\{-\sin\left(\frac{2\pi L}{L}\right) + \sin\left(\frac{2\pi(L/4)}{L}\right)\right.\right.$$
$$\left.\left. + \cos\left(\frac{2\pi L}{L}\right) - \cos\left(\frac{2\pi L/4}{L}\right)\right\}\right]$$
$$= (1/L)2^{-1/2}\left[3L/4 + \frac{L}{2\pi}\left\{-\sin(2\pi) + \sin(\pi/2) + \cos(2\pi) - \cos(\pi/2)\right\}\right]$$
$$= (1/L)2^{-1/2}(3L/4 + L/\pi) = 2^{-1/2}(3/4 + 1/\pi) = \frac{3 + 4/\pi}{4 \times 2^{1/2}}$$

The Franck–Condon factor is $\boxed{I^2 = (1/32)(3 + 4/\pi)^2}$; numerically this is 0.134.

E11F.7(a) The wavenumbers of the lines in the P branch are given in [11F.7–465], $\tilde{\nu}_P(J) = \tilde{\nu} - (\tilde{B}' + \tilde{B})J + (\tilde{B}' - \tilde{B})J^2$. The band head is located by finding the value of J which gives the smallest wavenumber, which can be inferred by solving $d\tilde{\nu}_P(J)/dJ = 0$.

$$\frac{d}{dJ}\left[\tilde{\nu} - (\tilde{B}' + \tilde{B})J + (\tilde{B}' - \tilde{B})J^2\right] = -(\tilde{B}' + \tilde{B}) + 2J(\tilde{B}' - \tilde{B})$$

Setting the derivative to zero and solving for J gives

$$J_{\text{head}} = \boxed{\frac{\tilde{B}' + \tilde{B}}{2(\tilde{B}' - \tilde{B})}}$$

A band head only occurs in the P branch if $\tilde{B}' > \tilde{B}$.

E11F.8(a) Because $\tilde{B}' < \tilde{B}$ a band head will occur in the R branch.
The wavenumbers of the lines in the R branch are given in [11F.7–*465*], $\tilde{\nu}_{\text{R}}(J) = \tilde{\nu} + (\tilde{B}' + \tilde{B})(J+1) + (\tilde{B}' - \tilde{B})(J+1)^2$. The band head is located by finding the value of J which gives the largest wavenumber, which can be inferred by solving $\text{d}\tilde{\nu}_{\text{R}}(J)/\text{d}J = 0$.

$$\frac{\text{d}}{\text{d}J}\left[\tilde{\nu} + (\tilde{B}' + \tilde{B})(J+1) + (\tilde{B}' - \tilde{B})(J+1)^2\right] = (\tilde{B}' + \tilde{B}) + 2(J+1)(\tilde{B}' - \tilde{B})$$

Setting the derivative to zero and solving for J gives

$$J_{\text{head}} = \frac{-(\tilde{B}' + \tilde{B})}{2(\tilde{B}' - \tilde{B})} - 1 = \frac{\tilde{B} - 3\tilde{B}'}{2(\tilde{B}' - \tilde{B})}$$

With the data given

$$J_{\text{head}} = \frac{0.3540 - 3 \times 0.3101}{2(0.3101 - 0.3540)} = 6.56$$

Assuming that it is satisfactory simply to round this to the nearest integer the band head occurs at $\boxed{J = 7}$.

E11F.9(a) The fact that a band head is seen in the R branch implies that $\tilde{B}' < \tilde{B}$. It is shown in *Exercise* E11F.7(b) that the band head in the R branch occurs at

$$J_{\text{head}} = \frac{\tilde{B} - 3\tilde{B}'}{2(\tilde{B}' - \tilde{B})} \qquad (11.3)$$

This rearranges to

$$\tilde{B}' = \tilde{B} \times \frac{2J+1}{2J+3} \qquad (11.4)$$

A band head at $J = 1$ might arise from a value of J determined from eqn 11.3 anywhere in the range 0.5 to 1.5, followed by subsequent rounding. Using these non-integer values of J in eqn 11.4 gives $\tilde{B}'$ in the range $\boxed{30\ \text{cm}^{-1} \text{ to } 40\ \text{cm}^{-1}}$.

The bond length in the upper state is $\boxed{\text{longer}}$ than that in the lower state (a longer bond means a larger moment of inertia and hence a smaller rotational constant).

E11F.10(a) Assuming that the transition corresponds to that between the two sets of d orbitals which are split as a result on the interaction with the ligands (Section 11F.2(a) on page 467), the energy of the transition is the value of Δ_o. Hence $\tilde{\Delta}_o = 1/(700 \times 10^{-7}\ \text{cm}) = \boxed{1.43 \times 10^4\ \text{cm}^{-1}}$ or $\boxed{1.77\ \text{eV}}$. This value is very approximate as it does not take into account the energy involved in rearranging the electron spins.

E11F.11(a) A rectangular wavefunction with value h between $x = 0$ and $x = a$ is normalized if the area under the square of the wavefunction is equal to 1: in this case $1 = ah^2$, hence $h = a^{-1/2}$. For the wavefunction which is non-zero between $x = a/2$ and $x = b$ the height is $h' = (b - a/2)^{-1/2}$. The region where the wavefunctions are both non-zero is $x = a/2$ to $x = a$ (because $b > a$); this is the domain of integration. The transition moment is

$$\begin{aligned}\int_{a/2}^{a} \psi_i x \psi_f \,dx &= \left(\frac{1}{a(b-a/2)}\right)^{1/2} \int_{a/2}^{a} x \,dx \\ &= \left(\frac{1}{a(b-a/2)}\right)^{1/2} \left|\tfrac{1}{2}x^2\right|_{a/2}^{a} = \left(\frac{1}{a(b-a/2)}\right)^{1/2} \tfrac{1}{2}(a^2 - a^2/4) \\ &= \left(\frac{1}{a(b-a/2)}\right)^{1/2} \frac{3a^2}{8} = \boxed{\frac{3}{8}\left(\frac{a^3}{b-a/2}\right)^{1/2}}\end{aligned}$$

E11F.12(a) The Gaussian functions are written $e^{-\alpha x^2/2}$ and $e^{-\alpha(x-a/2)^2/2}$, where the parameter α determines the width. To evaluate the normalizing factor for the function $e^{-\alpha x^2/2}$ requires the integral

$$\int_{-\infty}^{+\infty} e^{-\alpha x^2} \,dx$$

which is of the form of Integral G.1 with $k = \alpha$ and evaluates to $(\pi/\alpha)^{1/2}$. The normalizing factor is therefore $N_0 = (\alpha/\pi)^{1/4}$. The same factor applies to the other Gaussian function as this is simply the same Gaussian shifted to $a/2$: the area under the square of this function is the same.

The transition moment is given by the integral

$$\begin{aligned}I &= (\alpha/\pi)^{1/2} \int_{-\infty}^{+\infty} x e^{-\alpha x^2/2} e^{-\alpha(x-a/2)^2/2} \,dx \\ &= (\alpha/\pi)^{1/2} \int_{-\infty}^{+\infty} x e^{-\alpha[x^2/2 + (x-a/2)^2/2]} \,dx \\ &= (\alpha/\pi)^{1/2} \int_{-\infty}^{+\infty} x e^{-\alpha[x^2/2 + x^2/2 - xa/2 + a^2/8]} \,dx \\ &= (\alpha/\pi)^{1/2} \int_{-\infty}^{+\infty} x e^{-\alpha[x^2 - xa/2 + a^2/8]} \,dx \\ &= (\alpha/\pi)^{1/2} \int_{-\infty}^{+\infty} x e^{-\alpha(x-a/4)^2} e^{-\alpha a^2/16} \,dx\end{aligned}$$

The final equality above is verified by expanding out the square and recombining the terms. Taking out the constant factor and then writing x as $(x - a/4) + a/4$ gives

$$I = (\alpha/\pi)^{1/2} e^{-\alpha a^2/16} \int_{-\infty}^{+\infty} \left[(x - a/4)e^{-\alpha(x-a/4)^2} + (a/4)e^{-\alpha(x-a/4)^2}\right] dx$$

The first term in the integral is an odd function, and so evaluates to zero. The second term is simply a shifted Gaussian and, as before, the integral is form of Integral G.1 with $k = \alpha$ and evaluates to $(\pi/\alpha)^{1/2}$

$$I = (\alpha/\pi)^{1/2} e^{-\alpha a^2/16}(a/4)(\pi/\alpha)^{1/2} = (a/4)e^{-\alpha a^2/16}$$

The 'width' w of a Gaussian can be defined as the distance between the values of the coordinates $(\pm w/2)$ at which the height falls to half its maximum value. Because the maximum of the function is 1, the value of α for a given width is found by solving

$$\tfrac{1}{2} = e^{-\alpha(w/2)^2/2} \quad \text{hence} \quad \alpha = (8\ln 2)/w^2$$

The exercise specifies that the Gaussian should have width a, so $\alpha = (8\ln 2)/a^2$. With this the transition moment becomes

$$\begin{aligned}(a/4)e^{-\alpha a^2/16} &= (a/4)e^{-(8\ln 2)a^2/16a^2} = (a/4)e^{-(\ln 2)/2} \\ &= (a/4)e^{-(\ln 2^{1/2})} = (a/4)(1/2^{1/2}) = \boxed{a/(4 \times 2^{1/2})}\end{aligned}$$

E11F.13(a) A simple model for the wavelength of the HOMO–LUMO transition in a polyene is discussed in *Example* 7D.1 on page 265. In this model the energy levels of the π electrons in a polyene are modelled by those of a particle in a one-dimensional box of length L. If the polyene consists of n conjugated double bonds, the length L may be written as $L = nd$, where d is the length of a single conjugated bond. A molecule with n conjugated double bonds will have $2n$ π electrons which will occupy the energy levels pairwise. Therefore the HOMO is the level with quantum number n and the LUMO has quantum number $n+1$. The energy of the HOMO–LUMO transition is therefore

$$E_{n+1} - E_n = [(n+1)^2 - n^2]\frac{h^2}{8mL^2} = \frac{(2n+1)h^2}{8mL^2}$$

For large n, $E_{n+1} - E_n$ goes as n. However, $L = nd$, therefore overall $E_{n+1} - E_n$ goes as $1/n$, and hence the wavelength of the transition goes as n. Thus, increasing the number of conjugated double bonds will increase the wavelength of the absorption; that is, shift it to the red. The transition at 243 nm is therefore likely to be from **4**, and that at 192 nm is likely to be from **5**.

Solutions to problems

P11F.1 The first transition is not allowed because it violates the spin selection rule, $\Delta S = 0$; this transition has $\Delta S = 1$. The second transition is not allowed because it violates the selection rule for Λ, $\Delta\Lambda = 0, \pm 1$: this transition has $\Delta\Lambda = 2$.

P11F.3 Figure 11.14 is helpful in understanding the various quantities involved in this problem. The pure electronic energy of the ground electronic state, that is the energy at the equilibrium separation, is (as a wavenumber) $\tilde{T}_e(\text{X})$. Likewise, the pure electronic energy of the excited state is $\tilde{T}_e(\text{B})$. The vibrational terms of the ground state (in the harmonic approximation) are $\tilde{G}_\text{X}(v_\text{X}) = (v_\text{X} + \frac{1}{2})\tilde{\nu}_\text{X}$, with these terms measured from $\tilde{T}_e(\text{X})$. Likewise, those of the excited state are $\tilde{G}_\text{B}(v_\text{B}) = (v_\text{B} + \frac{1}{2})\tilde{\nu}_\text{B}$ measured from $\tilde{T}_e(\text{B})$.

The wavenumber of the 0–0 transition, $\tilde{\nu}_{00}$, is therefore

$$\begin{aligned}\tilde{\nu}_{00} &= \tilde{T}_e(\text{B}) - \tilde{T}_e(\text{X}) + \tilde{G}_\text{B}(0) - \tilde{G}_\text{X}(0)\\ &= \tilde{T}_e(\text{B}) - \tilde{T}_e(\text{X}) + \tfrac{1}{2}\tilde{\nu}_\text{B} - \tfrac{1}{2}\tilde{\nu}_\text{X}\end{aligned}$$

The difference $\tilde{T}_e(\text{B}) - \tilde{T}_e(\text{X})$ is the value quoted as 6.175 eV; this is converted to a wavenumber using the factor from inside the front cover

$$\begin{aligned}\tilde{\nu}_{00} &= (6.175\ \text{eV}) \times \frac{8065.5\ \text{cm}^{-1}}{1\ \text{eV}} + \tfrac{1}{2} \times (700\ \text{cm}^{-1}) - \tfrac{1}{2} \times (1580\ \text{cm}^{-1})\\ &= \boxed{4.936 \times 10^4\ \text{cm}^{-1}}\end{aligned}$$

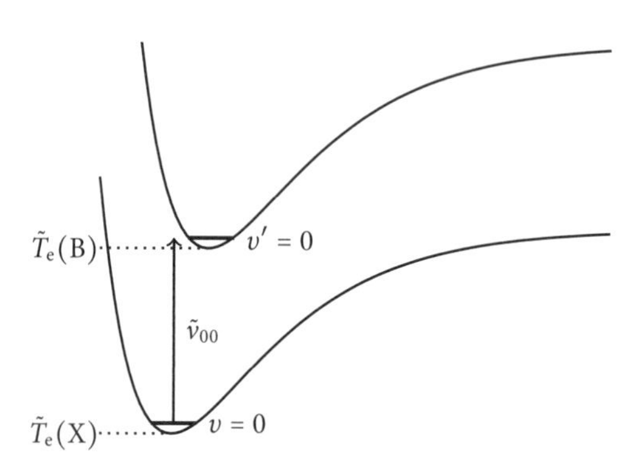

Figure 11.14

P11F.5 (a) The photoelectron spectrum involves a transition from the ground state of the molecule to an electronic state of the molecular ion. The energy needed for the transition is measured indirectly by measuring the energy of the ejected electron, but in all other respects the spectrum is interpreted in the same way as electronic absorption spectra.

In HBr only the ground vibrational state of the ground electronic state will be significantly populated, so only transitions from this level need be considered. In principle there can be transitions from this vibrational level to a range of different vibrational levels of the upper electronic state (a $v = 0$ progression) and the intensities of these transitions will be governed by the Franck–Condon factors. As described in Section 11F.1(c) on page 462, if the two electronic states have similar equilibrium bond lengths the $v = 0 \rightarrow v' = 0$ transition will be the strongest, and then the intensity will drop of quickly for higher values of v'. On the other hand, if the upper state is displaced to the left or right, several transitions will have significant Franck–Condon factors and several lines in the progression will be observed.

In the photoelectron spectrum of HBr the band centred at about 16 eV shows extensive structure which is interpreted as being due to several lines of a vibrational progression. This is consistent with this band being due to the removal of a bonding electron, resulting in the upper state (of the ion) having a significantly longer bond length than the ground state.

(b) The band at around 11.5 eV shows two peaks: these are not due to vibrational fine structure, but to spin-orbit coupling in the molecular ion, specifically such coupling associated with the Br atom. In the ion there are unpaired electrons, so spin-orbit coupling in manifested in the spectrum. Removal of a nonbonding electron from a lone pair on the Br is not expected to give a change in equilibrium bond length, so only the $v = 0 \rightarrow v' = 0$ transition has significant intensity: there is no vibrational progression.

P11F.7 The intensity of a transition depends on the transition dipole moment, in this case given by the integral

$$\int_0^L \psi_n \hat{\mu}_x \psi_1 \, dx = -e(1/L) \int_0^L \sin(n\pi x/L)\, x \sin(\pi x/L) \, dx$$

where the normalized wavefunctions $\psi_n = (2/L)^{1/2} \sin(n\pi x/L)$ are used, and $n = 2$ or 3. The integral is most easily evaluated by first using the given identity

$$I = -e(1/L) \int_0^L x \left[\cos\left(\frac{n\pi x}{L} - \frac{\pi x}{L} \right) - \cos\left(\frac{n\pi x}{L} + \frac{\pi x}{L} \right) \right] dx$$

$$= -e(1/L) \int_0^L x \left[\cos\left(\frac{(n-1)\pi x}{L} \right) - \cos\left(\frac{(n+1)\pi x}{L} \right) \right] dx$$

The integral is of the form of Integral T.13 with $a = L$

$$= -e(1/L) \left\{ \frac{L^2}{(n-1)^2\pi^2} [\cos(n-1)\pi - 1] + \frac{L^2}{(n-1)\pi} \sin(n-1)\pi \right.$$

$$\left. - \frac{L^2}{(n+1)^2\pi^2} [\cos(n+1)\pi - 1] - \frac{L^2}{(n+1)\pi} \sin(n+1)\pi \right\}$$

For the case $n = 2$

$$I_{1,2} = -e(1/L)\left\{\frac{L^2}{\pi^2}[-1-1] - \frac{L^2}{9\pi^2}[-1-1]\right\} = \frac{16Le}{9\pi^2}$$

For the case $n = 3$

$$I_{1,3} = -e(1/L)\left\{\frac{L^2}{4\pi^2}[1-1] + -\frac{L^2}{16\pi^2}[1-1]\right\} = 0$$

Thus, as was to be shown, the transition dipole is non-zero for the transition $1 \rightarrow 2$, but zero for $1 \rightarrow 3$.

This result is obtained much more simply by rewriting the integral using $x = (x - a/2) + a/2$

$$\begin{aligned} I &= -e(1/L)\int_0^L \sin(n\pi x/L)\,[(x-a/2)+a/2]\sin(\pi x/L)\,\mathrm{d}x \\ &= -e(1/L)\int_0^L \sin(n\pi x/L)\,(x-a/2)\sin(\pi x/L)\,\mathrm{d}x \\ &\quad - e(1/L)\int_0^L \sin(n\pi x/L)\,(a/2)\sin(\pi x/L)\,\mathrm{d}x \end{aligned}$$

The second integral is zero because the two eigenfunctions $\sin(n\pi x/L)$ and $\sin(\pi x/L)$ are orthogonal for $n \neq 1$. For the first integral the integrand is a product of three functions which can all be classified as odd or even with respect to $x = a/2$. For $n = 3$, $\sin(n\pi x/L)$ is even, $(x - a/2)$ is odd, and $\sin(\pi x/L)$ is even: the integrand is therefore odd overall, and hence when integrated over a symmetrical interval the result is necessarily zero.

For $n = 2$, $\sin(n\pi x/L)$ is odd, $(x-a/2)$ is odd, and $\sin(\pi x/L)$ is even: the integrand is therefore even overall, and hence when integrated over a symmetrical interval the result is not necessarily zero. What is not shown by this argument is that the integral is non-zero: however, a quick sketch of the integrand shows that it is negative everywhere, so the integral is non-zero.

P11F.9 The overlap integral for two 1s hydrogen ($Z = 1$) orbitals separated by a distance R is

$$S = \left[1 + \frac{R}{a_0} + \frac{1}{3}\left(\frac{R}{a_0}\right)^2\right]\mathrm{e}^{-R/a_0}$$

where a_0 is the Bohr radius. The transition moment, given as $-eRS$, is therefore

$$\mu = -eR\left[1 + \frac{R}{a_0} + \frac{1}{3}\left(\frac{R}{a_0}\right)^2\right]\mathrm{e}^{-R/a_0}$$

hence

$$-\mu/ea_0 = \frac{R}{a_0}\left[1 + \frac{R}{a_0} + \frac{1}{3}\left(\frac{R}{a_0}\right)^2\right]\mathrm{e}^{-R/a_0}$$

Figure 11.15 shows a plot of $-\mu/ea_0$ against R/a_0. The maximum occurs at $R/a_0 \approx 2.1$, and the transition moment tends to zero at large distances simply because the overlap also goes to zero in this limit.

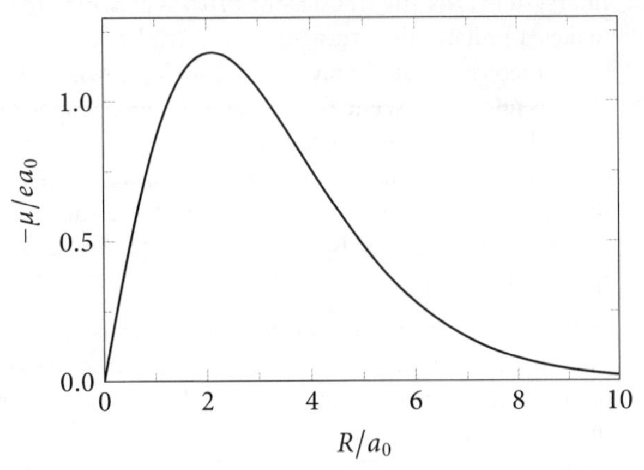

Figure 11.15

P11F.11 The adsorption at 189 nm is due to a $\pi^* \leftarrow \pi$ transition of the carbonyl group. The weaker adsorption at 280 nm is due to a $\pi^* \leftarrow$ n transition of the carbonyl group.

11G Decay of excited states

Answers to discussion questions

D11G.1 The overall process associated with fluorescence involves the following steps. Absorption of a photon promotes the molecule from the ground vibrational level of the ground electronic state to an excited vibrational level of an upper electronic state. Next, the molecule looses vibrational energy as a result of collisions (radiationless decay), eventually falling down to the ground vibrational level of the upper electronic state. Finally, spontaneous emission occurs to various vibrational levels of the ground electronic state: this is the fluorescent emission.

In the initial absorption step transitions are possible to different vibrational levels of the upper electronic state, and the intensity of these transitions will depend on the relevant Franck–Condon factors. The absorption spectrum therefore shows vibrational fine structure characteristic of the upper electronic state. Similarly, the emission spectrum (the fluorescence) will show vibrational structure characteristic of the ground electronic state.

Because of the loss of energy due to the radiationless transitions which precede emission, the fluorescence spectrum will appear at longer wavelengths than the absorption spectrum. If the potential energy curves of the two electronic states involved are similar they will have similar vibrational levels and hence the vibrational fine structure in the fluorescence spectrum will mirror that in the absorption spectrum. However, if the two electronic states are substantially different, for example in their bonding or equilibrium geometry, then the vibrational structure will be different and the spectra will not mirror one another.

D11G.3 Intersystem crossing (ISC) is the process by which the excited singlet state (S_1) makes a radiationless transition to a triplet state, T_1. Following this process spontaneous emission may occur from T_1 down to the ground electronic state – this is phosphorescence. Both ISC and the phosphorescent transition are spin forbidden and so only occur relatively slowly, and to the extent that the spin selection rule is broken. Spin-orbit coupling is one of the effects that leads to this rule being broken, and it is observed that the rates of these spin forbidden transitions is enhanced by the presence of heavy atoms which have significant spin orbit coupling.

In the present case, if the added iodide ion is able to interact with the chromophore for a significant period of time (a time comparable with ISC or phosphorescence) then the spin-orbit coupling in the transient species may increase the rate of ISC and/or of the phosphorescence, leading to an increase in the intensity of the latter.

D11G.5 This is described in Section 11G.3 on page 473.

Solutions to exercises

E11G.1(a) (i) The vibrational fine structure of the fluorescent transition is determined by the vibrational energy levels of the ground electronic state because the transitions observed are from the ground vibrational level of the upper electronic state to various vibrational levels of the ground electronic state.

(ii) No information is available about the vibrational levels of the upper electronic state because the spectrum only shows transitions from the ground vibrational level of this state.

E11G.2(a) This observed increase in the linewidth is a result of predissociation, as illustrated in Fig. 11G.8 on page 473. Where the dissociative $^5\Pi_u$ state crosses the bound upper electronic state the possibility exists that molecules in the upper electronic state will undergo radiationless transitions to the dissociative state leading to subsequent dissociation. This process reduces the lifetime of the excited states and so increases the linewidth of the associated transitions (lifetime broadening, see Section 11A.2(b) on page 425).

Solutions to problems

P11G.1 The anthracene fluorescence spectrum reflects the vibrational levels of the lower electronic state. The given wavelengths correspond to wavenumbers 22 730 cm^{-1}, 24 390 cm^{-1}, 25 640 cm^{-1}, and 27 030 cm^{-1}; these indicate spacings of the vibrational levels of $(24\,390\ cm^{-1}) - (22\,730\ cm^{-1}) = 1660\ cm^{-1}$, and, by similar calculations, 1250 cm^{-1}, and 1390 cm^{-1}.

The vibrational fine structure in the absorption spectrum reflects the vibrational levels of the upper electronic state. The wavenumbers of the absorption peaks are 27 800 cm^{-1}, 29 000 cm^{-1}, 30 300 cm^{-1}, and 32 800 cm^{-1}. The vibrational spacings are therefore 1200 cm^{-1}, 1300 cm^{-1}, and 2500 cm^{-1}.

The fact that the fluorescent transitions are all at longer wavelength (lower energy) that the absorption transitions is consistent with the loss of vibrational energy (by collision induced radiationless decay) after the initial excitation of the molecule. The vibrational fine structure in the absorption and fluorescence spectra do not mirror one another, suggesting that the bonding in the ground and excited electronic states are dissimilar.

P11G.3 (a) The resonant modes satisfy $n\lambda/2 = L$ therefore $\lambda = 2L/n$ and $\nu = nc/2L$. With the data given

$$\nu = \frac{n \times (2.9979 \times 10^8 \text{ m s}^{-1})}{2 \times (1.00 \text{ m})} = \boxed{n \times 150 \text{ MHz}}$$

(b) The spacing of the modes is therefore $\boxed{150 \text{ MHz}}$.

P11G.5 The peak power is energy/(duration of pulse)

$$P_{\text{peak}} = \frac{0.10 \text{ J}}{3.0 \times 10^{-9} \text{ s}} = \boxed{33 \text{ MW}}$$

where 1 W = 1 J s^{-1} is used. Consider a total time of 1 s: because the repetition frequency is 10 Hz, there will be 10 pulses in this time.

$$P_{\text{av}} = \frac{\text{total energy}}{\text{total time}} = \frac{10 \times (0.10 \text{ J})}{1 \text{ s}} = \boxed{1.0 \text{ W}}$$

The average power is very much less than the peak power.

Integrated activities

I11.1 (a) CH_4 belongs to the point group T_d which is a cubic group: the molecule is therefore a $\boxed{\text{spherical rotor}}$.

(b) CH_3CN belongs to the point group C_{3v} which includes a threefold axis: the molecule is therefore a $\boxed{\text{symmetric rotor}}$.

(c) CO_2 is self-evidently a $\boxed{\text{linear rotor}}$.

(d) If it is assumed that in CH_3OH the O–H bond either eclipses a C–H bond, or bisects the angle between two such bonds, the molecule belongs to the point group C_s. There are no rotation axes, so the molecule is therefore an $\boxed{\text{asymmetric rotor}}$.

(e) C_6H_6 belongs to the point group D_{6h} which includes a sixfold axis: the molecule is therefore a $\boxed{\text{symmetric rotor}}$.

(f) Pyridine belongs to the point group C_{2v} which includes a twofold axis: the molecule is therefore an $\boxed{\text{asymmetric rotor}}$.

I11.3 These calculations were performed with Spartan 06 using the both MP2 and DFT(B3LYP) methods both 6-31G* and 6-311G* basis sets. The tables summarize the data obtained, along with experimental values for comparison.

H_2O					
	MP2/6-31G*	MP2/6-311G*	DFT/6-31G*	DFT/6-311G*	exp.
basis funcs	19	24	19	24	
R/pm	96.9	95.7	96.8	96.3	95.8
E_0/eV	−2073.4	−2074.5	−2074.6	−2079.9	
angle/°	104.00	106.58	103.72	105.91	104.45
$\tilde{\nu}_1/\mathrm{cm}^{-1}$	3774.25	3858.00	3731.72	3764.70	3652
$\tilde{\nu}_2/\mathrm{cm}^{-1}$	1735.35	1730.88	1700.79	1705.47	1595
$\tilde{\nu}_3/\mathrm{cm}^{-1}$	3915.76	3994.30	3853.53	3877.60	3756
μ/D	n.s.	n.s.	2.0950	2.2621	1.854

CO_2					
	MP2/6-31G*	MP2/6-311G*	DFT/6-31G*	DFT/6-311G*	Exp.
basis funcs	45	54	45	54	
R/pm	118.0	116.9	116.9	116.0	116.3
E_0/eV	−5118.7	−5121.2	−5131.6	−5133.2	
angle/°	180.00	180.00	180.00	180.00	180
$\tilde{\nu}_1/\mathrm{cm}^{-1}$	1332.82	1341.46	1373.05	1376.55	1388
$\tilde{\nu}_2/\mathrm{cm}^{-1}$	636.22	657.60	641.47	666.39	667
$\tilde{\nu}_3/\mathrm{cm}^{-1}$	2446.78	2456.16	2438.17	2437.85	2349
μ/D / D	n.s.	n.s.	0.0000	0.0000	0

Except for the dipole moment of H_2O, all calculations are typically within a reasonable 1–3% of the experimental value. The dipole moment is very sensitive to the distribution of charge density. The significant difference between the dipole moment calculations and the experimental dipole moment may indicate that the computation methods do not adequately account for charge distribution in the very polar water molecule.

I11.5 The most populated rotational level for a linear rotor is given by [11B.12–*433*], $J_{\max} = (kT/2hc\tilde{B})^{1/2} - \frac{1}{2}$. The wavenumber of the lines in the O and S branches (Stokes scattering) are given by [11C.16–*449*] (note that the expression in the text for $\tilde{\nu}_{\mathrm{S}}(J)$ is incorrect)

$$\tilde{\nu}_{\mathrm{O}}(J) = \tilde{\nu}_{\mathrm{i}} - \tilde{\nu} + 4\tilde{B}(J - \tfrac{1}{2}) \quad \tilde{\nu}_{\mathrm{S}}(J) = \tilde{\nu}_{\mathrm{i}} - \tilde{\nu} - 4\tilde{B}(J + \tfrac{3}{2})$$

The separation between the lines in the O and S branch arising from the same value of J is

$$\tilde{\delta} = \tilde{\nu}_{\text{O}}(J) - \tilde{\nu}_{\text{S}}(J)$$
$$= 4\tilde{B}(J - \tfrac{1}{2}) - [-4\tilde{B}(J + \tfrac{3}{2})] = 8\tilde{B}(J + \tfrac{1}{2})$$

Using $J_{\text{max}} + \frac{1}{2} = (kT/2hc\tilde{B})^{1/2}$ gives the separation between the highest intensity lines in the two branches

$$\tilde{\delta} = 8\tilde{B} \times (kT/2hc\tilde{B})^{1/2} = (32\tilde{B}kT/hc)^{1/2}$$

which is the required expression.

Consider first the data for $Hg^{35}Cl_2$; the above expression for $\tilde{\delta}$ is rearranged to give $\tilde{B}$

$$\tilde{B} = \frac{hc\tilde{\delta}^2}{32kT}$$
$$= \frac{(6.6261 \times 10^{-34}\,\text{J s}) \times (2.9979 \times 10^{10}\,\text{cm s}^{-1}) \times (23.8\,\text{cm}^{-1})^2}{32 \times (1.3806 \times 10^{-23}\,\text{J K}^{-1}) \times [(282 + 273.15)\,\text{K}]}$$
$$= 0.0458...\,\text{cm}^{-1}$$

The moment of inertia is related to the rotational constant through [11B.7–*432*], $\tilde{B} = \hbar/4\pi cI$. It follows that $I = \hbar/4\pi c\tilde{B}$. The moment of inertia is $2m_{\text{Cl}}R^2$ thus $2m_{\text{Cl}}R^2 = \hbar/4\pi c\tilde{B}$. This rearranges to give the following expression for R

$$R = \left(\frac{\hbar}{8\pi m_{\text{Cl}}c\tilde{B}}\right)^{1/2}$$
$$= \left(\frac{1.0546 \times 10^{-34}\,\text{J s}}{8\pi(5.80... \times 10^{-26}\,\text{kg}) \times (2.9979 \times 10^{10}\,\text{cm s}^{-1}) \times (0.0458...\,\text{cm}^{-1})}\right)^{1/2}$$
$$= \boxed{229\,\text{pm}}$$

where the mass of ^{35}Cl is given by $(34.9688\,m_{\text{u}}) \times (1.6605 \times 10^{-27}\,\text{kg})/(1\,m_{\text{u}}) = 5.80... \times 10^{-26}\,\text{kg}$.

Similar calculations give: for $Hg^{79}Br_2$ $\tilde{B} = 0.0183...\,\text{cm}^{-1}$ and $\boxed{R = 241\,\text{pm}}$; and for $Hg^{127}I_2$ $\tilde{B} = 0.0103...\,\text{cm}^{-1}$ and $\boxed{R = 253\,\text{pm}}$.

I11.7 The notation $(v_1–v_0)$ is taken to imply a transition from vibrational level v_1 in the S_1 electronic state to vibrational level v_0 in the S_0 electronic state. The first step is to convert the wavelengths to wavenumbers.

transition	(0–0) centre	(1–1) centre	(0–0) band head	(0–1) band head
λ/nm	387.6	386.4	388.3	421.6
$\tilde{\nu}/\text{cm}^{-1}$	25 799.8	25 879.9	25 753.3	23 719.2

The separation of the (0–1) band head from the (0–0) band head is taken as a good approximation for the separation of the $v_0 = 1$ and $v_0 = 0$ vibrational levels of S_0, which in the harmonic approximation is simply $\tilde{\nu}_0$. This assumes that the separation of the band head from the band origin is the same in each case, which is equivalent to assuming that the rotational constants are the same for $v_0 = 1$ and $v_0 = 0$. It is likely that in an electronic spectrum the formation of a band head will be the result of significant differences in the rotational constants of the two electronic states, and that the variation in rotational constants between vibrational states is a minor factor.

$$\tilde{\nu}_0 = \text{(0,0) band head} - \text{(0–1) band head} = (25\,753.3\ \text{cm}^{-1}) - (23\,719.2\ \text{cm}^{-1})$$
$$= \boxed{2034.1\ \text{cm}^{-1}}$$

The wavenumber of the $(v_1 - v_0)$ transition is $\Delta\tilde{T}_e + \tilde{G}_1(v_1) - \tilde{G}_0(v_0)$, where $\tilde{G}_1(v_1)$ are the vibrational terms of S_1, $\tilde{G}_0(v_0)$ are the vibrational terms for S_0, and $\Delta\tilde{T}_e$ is the difference in the pure electronic energy between S_1 and S_0. Therefore, using the harmonic approximation for the vibrational terms,

$$\text{(0–0) centre} = \Delta\tilde{T}_e + \tfrac{1}{2}\tilde{\nu}_1 - \tfrac{1}{2}\tilde{\nu}_0 \qquad \text{(1–1) centre} = \Delta\tilde{T}_e + \tfrac{3}{2}\tilde{\nu}_1 - \tfrac{3}{2}\tilde{\nu}_0 \qquad (11.5)$$

from which it follows that

$$\text{(1–1) centre} - \text{(0–0) centre} = \tilde{\nu}_1 - \tilde{\nu}_0$$

with the data given and using $\tilde{\nu}_0 = 2034.1\ \text{cm}^{-1}$

$$25\,879.9 - 25\,799.8 = (\tilde{\nu}_1/\text{cm}^{-1}) - 2034.1 \quad \text{hence} \quad \boxed{\tilde{\nu}_1 = 2114.2\ \text{cm}^{-1}}$$

From eqn 11.5 it follows that

$$3 \times [\text{(0–0) centre}] - [\text{(1–1) centre}] = 2\Delta\tilde{T}_e$$
$$\text{hence } \Delta\tilde{T}_e = \tfrac{1}{2}[3 \times (25\,799.8\ \text{cm}^{-1}) - (25\,879.9\ \text{cm}^{-1})] = \boxed{25\,759.8\ \text{cm}^{-1}}$$

The difference $\tilde{\nu}_1 - \tilde{\nu}_0$ is, from eqn 11.5, (1–1) centre – (0–0) centre, which is $\boxed{80.1\ \text{cm}^{-1}}$. Given the various approximations made to derive these results it is unlikely that the accuracy reflects the quoted precision of the numbers.

The relative intensity of (1–1) and (0–0) reflects the relative populations of the $v_1 = 1$ and $v_1 = 0$ vibrational levels of the S_1 state. The ratio of these is given by $n_1/n_0 = e^{-\Delta\varepsilon/kT}$, where $\Delta\varepsilon = hc[\tilde{G}_1(1) - \tilde{G}_1(0)] = hc\tilde{\nu}_1$. It follows that

$$\ln n_1/n_0 = -\Delta\varepsilon/kT \quad \text{hence} \quad T = -hc\tilde{\nu}_1/(k \ln n_1/n_0)$$

With the data given

$$T = \frac{-(6.6261 \times 10^{-34}\ \text{J s}) \times (2.9979 \times 10^{10}\ \text{cm s}^{-1}) \times (2114.2\ \text{cm}^{-1})}{(1.3806 \times 10^{-23}\ \text{J K}^{-1}) \times (\ln 0.1)}$$
$$= \boxed{1.3 \times 10^3\ \text{K}}$$

If this were indeed the temperature, many more than 8 rotational levels of S_1 would be populated.

I11.9 (a) and (b) The calculations are perform with Spartan 10 using density functional theory (B3LYP) with the 6-31G* basis set. The relevant data for the *trans* and *cis* forms of **11** are summarized in the following table.

conformation	11-*trans*	11-*cis*
$E_{\text{LUMO}}/\text{eV}$	−5.78	−5.79
$E_{\text{HOMO}}/\text{eV}$	−7.91	−7.97
$\Delta E/\text{eV}$	2.13	2.18
λ/nm	582	569

(c) The lowest energy $\pi^* \leftarrow \pi$ transition is from the HOMO to the LUMO, and occurs in the visible (yellow) for both confirmations. The transition energy appears to be very slightly lower for the *trans* conformation and at a slightly longer wavelength. However, this difference is so small that it may be an artifact of the computation and not of physical significance.

I11.11 In *Problem* P11A.6 it is shown that the integrated absorption coefficient for a Gaussian line shape is $\mathcal{A} = 1.0645\,\varepsilon_{\text{max}}\Delta\tilde{\nu}_{1/2}$, where $\Delta\tilde{\nu}_{1/2}$ is the width at half height. Interpolating, by eye, a smooth curve across the band centred at about 280 nm, gives $\varepsilon_{\text{max}} = 250\ \text{mol}^{-1}\ \text{dm}^3\ \text{cm}^{-1}$. The molar absorbance drops to half this value at about 270 nm, which is $3.70\times10^4\ \text{cm}^{-1}$ and at about 310 nm, which is $3.23\times10^4\ \text{cm}^{-1}$, giving a width of $4700\ \text{cm}^{-1}$. The integrated absorption coefficient is therefore

$$\mathcal{A} = 1.0645 \times (250\ \text{mol}^{-1}\ \text{dm}^3\ \text{cm}^{-1}) \times (4700\ \text{cm}^{-1})$$
$$= \boxed{1.25\times10^6\ \text{mol}^{-1}\ \text{dm}^3\ \text{cm}^{-2}}$$

The transition moment is given by $\int \psi_{\text{f}}^*\hat{\mu}\psi_{\text{i}}\,\text{d}\tau$; $\hat{\mu}$ transforms as x, y, or z, which in this case is B_1, B_2, or A_1. The integral is only non-zero if the integrand transforms as the totally symmetric irreducible representation, which is determined by computing the direct product

$$\overbrace{\Gamma_{\text{f}}}^{\psi_{\text{f}}} \times \overbrace{\Gamma_{x,y,z}}^{\hat{\mu}_x} \times \overbrace{\text{A}_1}^{\psi_{\text{i}}} = \Gamma_{\text{f}} \times \Gamma_{x,y,z}$$

where that fact that the direct product with the totally symmetric irreducible representation has no effect is used. The only way this product can contain the totally symmetric irreducible representation is if the irreducible representation of the final state, Γ_{f}, is equal to the irreducible representation of $\Gamma_{x,y,z}$. Thus, transitions from the A_1 ground state to $\boxed{\text{A}_1}$, $\boxed{\text{B}_1}$, or $\boxed{\text{B}_2}$ excited states are allowed.

12 Magnetic resonance

12A General principles

Answers to discussion questions

D12A.1 The chemical shift, expressed in frequency units, scales directly with the applied magnetic field. Therefore, as the field increases the spectrum is spread over a wider range, making it easier to resolve complex, overlapping spectra. The sensitivity of the spectrum also increases with the applied field, making it possible to study smaller amounts of material in a shorter time. High fields are therefore very advantageous when it comes to studying the spectra of large molecules.

D12A.3 The Larmor frequency is the rate of precession of a magnetic moment (electron or nuclear) in a magnetic field. Resonance occurs when the frequency of the applied radiation matches the Larmor frequency.

Solutions to exercises

E12A.1(a) The nuclear g-factor g_I is given by [12A.4c–489], $g_I = \gamma_N \hbar / \mu_N$, where μ_N is the nuclear magneton ($5.051 \times 10^{-27}\ \mathrm{J\,T^{-1}}$) and γ_N is the nuclear magnetogyric ratio, the value of which depends on the identity of the nucleus. The units of $\hbar$ are J s and g_I is a dimensionless number, so the nuclear magnetogyric ratio γ_N has units $(\mathrm{J\,T^{-1}})/(\mathrm{J\,s}) = \boxed{\mathrm{T^{-1}\,s^{-1}}}$.

E12A.2(a) The magnitude of the angular momentum is given by $[I(I+1)]^{1/2}\hbar$ where I is the nuclear spin quantum number. For a proton, $I = \frac{1}{2}$, hence the magnitude of the angular momentum is $[\frac{1}{2}(\frac{1}{2}+1)]^{1/2}\hbar = \boxed{\sqrt{3}\hbar/2}$.

The component of the angular momentum along the z-axis is $m_I\hbar$ where $m_I = I, I-1, \ldots, -I$. For a proton, the components along the z-axis are $\boxed{\pm\frac{1}{2}\hbar}$ and the angle between angular momentum vector and the z-axis takes the values

$$\theta = \cos^{-1}\left(\frac{\pm\frac{1}{2}\hbar}{\frac{\sqrt{3}}{2}\hbar}\right) = \cos^{-1}\left(\frac{1}{\sqrt{3}}\right) = \boxed{\pm 0.9553\ \mathrm{rad} = \pm 54.74^\circ}$$

E12A.3(a) The NMR frequency is equal to the Larmor precession frequency, ν_L, which is given by [12A.7–489], $\nu_L = \gamma_N \mathcal{B}_0 / 2\pi$, where $\mathcal{B}_0$ is the magnitude of the magnetic field and γ_N is the nuclear magnetogyric ratio. Use Table 12A.2 on page 289 in the *Resource section* for the value of γ_N. Hence,

$$\nu_L = \frac{\gamma_N \mathcal{B}_0}{2\pi} = \frac{(26.752\times10^7\ \mathrm{T^{-1}\,s^{-1}})\times(13.5\ \mathrm{T})}{2\pi} = 5.75\times10^8\ \mathrm{Hz} = \boxed{575\ \mathrm{MHz}}$$

E12A.4(a) The energies of the nuclear spin states in a magnetic field are given by [12A.4d–*489*], $E_{m_I} = -g_I\mu_N\mathcal{B}_0 m_I$ where g_I is the nuclear g-factor, μ_N is the nuclear magneton, $\mathcal{B}_0$ is the magnitude of the magnetic field, and the component of the angular momentum on a specified axis is $m_I\hbar$ where $m_I = I, I-1, ..., -I$.

Therefore, since the possible values of m_I are $\pm\frac{3}{2}, \pm\frac{1}{2}$, the energies of nuclear spin states are

$$\begin{aligned} E_{m_I} &= -g_I\mu_N\mathcal{B}_0 m_I \\ &= -(0.4289)\times(5.0508\times10^{-27}\ \mathrm{J\,T^{-1}})\times(6.800\ \mathrm{T})\times m_I \\ &= (-1.473...\times10^{-26}\ \mathrm{J})\times m_I \end{aligned}$$

Hence $\boxed{E_{\pm3/2} = \mp2.210\times10^{-26}\ \mathrm{J}\ \text{and}\ E_{\pm1/2} = \mp7.365\times10^{-27}\ \mathrm{J}}$.

E12A.5(a) The energy level separation is $\Delta E = h\nu$ where $\nu = \gamma_N\mathcal{B}_0/2\pi$, [12A.6–*489*]. Hence, in megahertz, the frequency separation is

$$\nu = 10^{-6}\times\frac{\gamma_N\mathcal{B}_0}{2\pi} = 10^{-6}\times\frac{(6.73\times10^7\ \mathrm{T^{-1}\,s^{-1}})\times(15.4\ \mathrm{T})}{2\pi} = \boxed{165\ \mathrm{MHz}}$$

E12A.6(a) The energy level separation is $\Delta E = h\nu$ where $\nu = \gamma_N\mathcal{B}_0/2\pi$, [12A.6–*489*]. Hence, for a given magnetic field, $\Delta E \propto \gamma_N$. Using $\gamma_N(^{15}\mathrm{N}) = -2.712\times10^7\ \mathrm{T^{-1}\,s^{-1}}$ and $\gamma_N(^{31}\mathrm{P}) = 10.84\times10^7\ \mathrm{T^{-1}\,s^{-1}}$, it follows that $|\gamma_N(^{31}\mathrm{P})| > |\gamma_N(^{15}\mathrm{N})|$ and so the separation of energy levels is larger for ^{31}P than for ^{15}N.

E12A.7(a) The ground state has $m_I = +\frac{1}{2}$ (α spin) and population N_α, and the upper state has $m_I = -\frac{1}{2}$ (β spin) and population N_β. The total population N is $N = N_\alpha + N_\beta$, and the population difference is $N_\alpha - N_\beta$. The Boltzmann distribution gives $N_\beta/N_\alpha = e^{-\Delta E/kT}$, where ΔE is the energy difference between the two states: $\Delta E = \gamma_N\hbar\mathcal{B}_0$. It follows that $N_\beta = N_\alpha e^{-\Delta E/kT}$. With these results

$$\frac{N_\alpha - N_\beta}{N} = \frac{N_\alpha - N_\beta}{N_\alpha + N_\beta} = \frac{N_\alpha(1-e^{-\Delta E/kT})}{N_\alpha(1+e^{-\Delta E/kT})} = \frac{1-e^{-\Delta E/kT}}{1+e^{-\Delta E/kT}}$$

Because $\Delta E \ll kT$ the exponential $e^{-\Delta E/kT}$ is approximated as $1-\Delta E/kT$ to give

$$\frac{N_\alpha - N_\beta}{N} \approx \frac{1-(1-\Delta E/kT)}{1+(1-\Delta E/kT)} = \frac{\Delta E/kT}{2+\Delta E/kT} = \frac{\Delta E}{2kT} = \frac{\gamma_N\hbar\mathcal{B}_0}{2kT}$$

For a ^{1}H nucleus and at 298 K

$$\frac{N_\alpha - N_\beta}{N} = \frac{\gamma_N \hbar \mathcal{B}_0}{2kT} = \frac{(26.75\times10^{7}\ \mathrm{T^{-1}\,s^{-1}})\times(1.0546\times10^{-34}\ \mathrm{J\,s})\times\mathcal{B}_0}{2\times(1.3806\times10^{-23}\ \mathrm{J\,K^{-1}})\times(298\ \mathrm{K})}$$
$$= 3.42...\times10^{-6}\times(\mathcal{B}_0/\mathrm{T})$$

For $\mathcal{B}_0 = 0.30$ T, $(N_\alpha - N_\beta)/N = \boxed{1.0\times10^{-6}}$; for $\mathcal{B}_0 = 1.5$ T, the ratio is $\boxed{5.1\times10^{-6}}$; for $\mathcal{B}_0 = 10$ T, the ratio is $\boxed{3.4\times10^{-5}}$.

E12A.8(a) The population difference for a collection of N spin-$\frac{1}{2}$ nuclei is given by [12A.8b–*491*], $(N_\alpha - N_\beta) \approx N\gamma_N\hbar\mathcal{B}_0/2kT$, where N_α is the number of spins in the lower energy state and N_β is the number of spins in the higher energy state. At constant temperature, $(N_\alpha - N_\beta)/N \propto \mathcal{B}_0$. Hence, for the relative population difference to be increased by a factor of 5, the applied magnetic field must increase by a factor of $\boxed{5}$. This is independent of the type of nucleus.

E12A.9(a) The EPR resonance frequency ν is given by [12A.12b–*492*], $h\nu = g_e\mu_B\mathcal{B}_0$, where g_e is the magnetogyric ratio of the electron and $\mathcal{B}_0$ is the magnetic field strength. With $\nu = c/\lambda$ it follows that

$$\mathcal{B}_0 = \frac{h\nu}{g_e\mu_B} = \frac{hc}{g_e\mu_B\lambda}$$
$$= \frac{(6.6261\times10^{-34}\ \mathrm{J\,s})\times(2.9979\times10^{8}\ \mathrm{m\,s^{-1}})}{2.0023\times(9.2740\times10^{-24}\ \mathrm{J\,T^{-1}})\times(8\times10^{-3}\ \mathrm{m})} = \boxed{1.3\ \mathrm{T}}$$

Solutions to problems

P12A.1 The resonance condition in NMR is [12A.6–*489*], $h\nu = \gamma_N\hbar\mathcal{B}_0$, where ν is the resonance frequency, γ_N is the nuclear magnetogyric ratio, and $\mathcal{B}_0$ is the magnetic field strength. Hence, assuming the same magnetic field for both neutrons and ^{1}H,

$$\frac{h\nu_n}{h\nu_{^1\mathrm{H}}} = \frac{\gamma_N(n)\hbar\mathcal{B}_0}{\gamma_N(^1\mathrm{H})\hbar\mathcal{B}_0} = \frac{\gamma_N(n)}{\gamma_N(^1\mathrm{H})}$$

With the data given, and taking the modulus of the negative magnetogyric ratio of the neutron,

$$\nu_n = \nu_{^1\mathrm{H}}\times\frac{|\gamma_N(n)|}{\gamma_N(^1\mathrm{H})} = (300\times10^{6}\ \mathrm{Hz})\times\frac{(18.324\times10^{7}\ \mathrm{T^{-1}\,s^{-1}})}{(26.752\times10^{7}\ \mathrm{T^{-1}\,s^{-1}})}$$
$$= 2.1\times10^{8}\ \mathrm{Hz} = \boxed{210\ \mathrm{MHz}}$$

The energy of the state with quantum number m_I is $E_{m_I} = -m_I\gamma_N\hbar\mathcal{B}_0$. The neutron has spin $\frac{1}{2}$ and a negative magnetogyric ratio. Therefore the state with $\boxed{m_I = -\frac{1}{2}}$ (the β state) has the lower energy.

The relative population difference for spin-$\frac{1}{2}$ nuclei with positive γ_N is given by [12A.8b–*491*], $N_\alpha - N_\beta \approx N\gamma_N\hbar\mathcal{B}_0/2kT$. This applies equally well to negative

γ_N: the term on the right is then negative, implying that $(N_\alpha - N_\beta)$ is negative, which is expected because β is the lower energy, more populated, level.

If the ^{1}H resonance frequency is $\nu_{^1H}$, it follows from the resonance condition $h\nu_{^1H} = \gamma_N(^1H)\hbar\mathcal{B}_0$ that $\mathcal{B}_0 = 2\pi\nu_{^1H}/\gamma_N(^1H)$. The population difference is computed as $(N_\beta - N_\alpha)$, requiring a reversal of the sign on both sides

$$\frac{N_\beta - N_\alpha}{N} = \frac{-\gamma_N(n)\hbar\mathcal{B}_0}{2kT}$$

The above expression for $\mathcal{B}_0$ is used to give

$$\frac{N_\beta - N_\alpha}{N} = \frac{-\gamma_N(n)\hbar}{2kT} \times \frac{2\pi\nu_{^1H}}{\gamma_N(^1H)} = \frac{-\gamma_N(n)h\nu_{^1H}}{2kT\gamma_N(^1H)}$$

With the data given

$$\frac{N_\beta - N_\alpha}{N} = \frac{-(-18.324\times10^7\ \mathrm{T^{-1}\,s^{-1}})\times(6.6261\times10^{-34}\ \mathrm{J\,s})\times(300\times10^6\ \mathrm{Hz})}{2(1.3806\times10^{-23}\ \mathrm{J\,K^{-1}})\times(298\ \mathrm{K})\times(26.752\times10^7\ \mathrm{T^{-1}\,s^{-1}})}$$
$$= \boxed{1.65\times10^{-5}}$$

P12A.3 (a) Absorption intensity is given by [12A.8c–*491*], intensity $\propto N\gamma_N^2\mathcal{B}_0^2/T$. For the same N, $\mathcal{B}_0$ and T, the intensity is simply $\propto \gamma_N^2$. For the intensity to be equal for ^{13}C with natural abundance A, and ^{15}N with an enriched abundance A'

$$A'\times[\gamma_N(^{15}N)]^2 = A\times[\gamma_N(^{13}C)]^2$$
$$\text{hence } A' = \frac{A[\gamma_N(^{13}C)]^2}{[\gamma_N(^{15}N)]^2}$$
$$= \frac{1.108\%\times(6.7272\times10^7\ \mathrm{T^{-1}\,s^{-1}})^2}{(-2.7126\times10^7\ \mathrm{T^{-1}\,s^{-1}})^2}$$
$$= 6.81\%$$

Therefore, a 6.8 per cent enrichment in ^{15}N is needed.

(b) The intensity I relative to natural abundance ^{13}C, for 100 per cent enrichment of ^{17}O is

$$I = \frac{A'\times[\gamma_N(^{17}O)]^2}{A\times[\gamma_N(^{13}C)]^2}\times I_{^{13}C}$$
$$= \frac{100\%\times(-3.627\times10^7\ \mathrm{T^{-1}\,s^{-1}})^2}{1.108\%\times(6.7272\times10^7\ \mathrm{T^{-1}\,s^{-1}})^2}\times I_{^{13}C}$$
$$= 26.2\, I_{^{13}C}$$

12B Features of NMR spectra

Answers to discussion questions

D12B.1 For a chemically shifted nucleus the resonance frequency is given by [12B.3–*494*], $\nu_L = (\gamma_N \mathcal{B}_0/2\pi)(1-\sigma)$. For an NMR spectrometer operating at a fixed frequency, the external magnetic field required to fulfill the resonance condition is found by rearranging this to $\mathcal{B}_0 = 2\pi\nu_L/[\gamma_N(1-\sigma)]$.

Thus, a positive value of the shielding constant σ shifts the resonance field to 'high field', and a negative value of σ shifts the resonance field to 'low field'. Conversely, in a spectrometer at fixed external magnetic field, positive values of σ shift the resonance frequency to lower values, and negative values of σ shift it to higher values.

D12B.3 The reasons why no splitting is observed as a result of the coupling between equivalent nuclei is discussed in detail in Section 12B.3(d) on page 503. Coupling to a third spin, inequivalent to the first two, will result in a splitting for the same reason as the coupling of two inequivalent spins, as discussed in Section 12B.3(a) on page 499.

D12B.5 This is discussed in Section 12B.3(c) on page 502.

Solutions to exercises

E12B.1(a) The δ scale is defined by [12B.4a–*494*], $\delta = (\nu - \nu^\circ) \times 10^6/\nu^\circ$, where δ is the chemical shift of the peak, ν is the resonance frequency of the peak, and ν° is the resonance frequency of the standard. Hence

$$\delta = \frac{\nu - \nu^\circ}{\nu^\circ} \times 10^6 = \frac{(500.132500\ \text{MHz}) - (500.130000\ \text{MHz})}{(500.130000\ \text{MHz})} \times 10^6 = \boxed{5.0}$$

E12B.2(a) The δ scale is defined by [12B.4a–*494*], $\delta = (\nu - \nu^\circ) \times 10^6/\nu^\circ$, where δ is the chemical shift of the peak, ν is the resonance frequency of the peak, and ν° is the resonance frequency of the standard. Hence

$$\delta = \frac{\nu - \nu^\circ}{\nu^\circ} \times 10^6 = \frac{750\ \text{Hz}}{(500.130000 \times 10^6\ \text{Hz})} \times 10^6 = \boxed{1.5}$$

E12B.3(a) The resonance frequency ν is given by [12B.5–*494*], $\nu = \nu^\circ + (\nu_{\text{spect}}/10^6)\delta$, where ν° is the resonance frequency of the standard. The frequency separation of the two peaks, $\Delta\nu$, is

$$\Delta\nu = \nu_2 - \nu_1 = \left[\nu^\circ + \left(\frac{\nu_{\text{spect}}}{10^6}\right)\delta_2\right] - \left[\nu^\circ + \left(\frac{\nu_{\text{spect}}}{10^6}\right)\delta_1\right] = \left(\frac{\nu_{\text{spect}}}{10^6}\right)(\delta_2 - \delta_1)$$

$$= \left(\frac{400.130000 \times 10^6\ \text{Hz}}{10^6}\right)(9.80 - 2.20) = \boxed{3040\ \text{Hz}}$$

E12B.4(a) The resonance frequency ν is given by [12B.5–*494*], $\nu = \nu^\circ + (\nu_{\text{spect}}/10^6)\delta$, where ν° is the resonance frequency of the standard. The frequency separation of the two peaks, $\Delta\nu$, is

$$\Delta\nu = \nu_2 - \nu_1 = \left[\nu^\circ + \left(\frac{\nu_{\text{spect}}}{10^6}\right)\delta_2\right] - \left[\nu^\circ + \left(\frac{\nu_{\text{spect}}}{10^6}\right)\delta_1\right] = \left(\frac{\nu_{\text{spect}}}{10^6}\right)(\delta_2 - \delta_1)$$

This is rearranged to give $\Delta\delta$, the separation of the two peaks on the shift scale

$$\Delta\delta = \delta_2 - \delta_1 = \Delta\nu \times \left(\frac{10^6}{\nu_{\text{spec}}}\right) = (550\text{ Hz}) \times \frac{10^6}{400.130000 \times 10^6\text{ Hz}} = \boxed{1.37}$$

E12B.5(a) The combination of [12B.1–*494*], $\mathcal{B}_{\text{loc}} = \mathcal{B}_0 + \delta\mathcal{B}$, and [12B.2–*494*], $\delta\mathcal{B} = -\sigma\mathcal{B}_0$, gives the relationship $\mathcal{B}_{\text{loc}} = (1 - \sigma)\mathcal{B}_0$. For $|\sigma^\circ| \ll 1$, [12B.6–*495*] gives $\delta = (\sigma^\circ - \sigma) \times 10^6$. Hence,

$$\Delta\delta = \delta_2 - \delta_1 = (\sigma^\circ - \sigma_2) \times 10^6 - (\sigma^\circ - \sigma_1) \times 10^6 = -(\sigma_2 - \sigma_1) \times 10^6 = -(\Delta\sigma) \times 10^6$$

$$\Delta\mathcal{B}_{\text{loc}} = (1 - \sigma_2)\mathcal{B}_0 - (1 - \sigma_1)\mathcal{B}_0 = -(\sigma_2 - \sigma_1)\mathcal{B}_0 = -(\Delta\sigma)\mathcal{B}_0$$

Substituting for $\Delta\sigma$ gives $\Delta\mathcal{B}_{\text{loc}} = (\Delta\delta/10^6)\mathcal{B}_0$.

(i) For $\mathcal{B}_0 = 1.5\text{ T}$, $\Delta\mathcal{B}_{\text{loc}} = [(9.80 - 2.20)/10^6] \times (1.5\text{ T}) = 1.1 \times 10^{-5}\text{ T} = \boxed{11\ \mu\text{T}}$.

(ii) For $\mathcal{B}_0 = 15\text{ T}$, $\Delta\mathcal{B}_{\text{loc}} = [(9.80 - 2.20)/10^6] \times (15\text{ T}) = 1.1 \times 10^{-4}\text{ T} = \boxed{110\ \mu\text{T}}$.

E12B.6(a) The resonance frequency ν is given by [12B.5–*494*], $\nu = \nu^\circ + (\nu_{\text{spect}}/10^6)\delta$, where ν° is the resonance frequency of the standard. The resonance from TMS is taken as the origin so $\nu^\circ = 0$.

For ^{1}H nuclei, $I = \frac{1}{2}$ and so each signal is split into $(2nI + 1) = 2 \times 1 \times \frac{1}{2} + 1 = 2$ peaks, with a splitting equal to the coupling constant $J_{\text{AX}} = 10\text{ Hz}$. For a spectrometer operating at 250 MHz the resonance frequencies of the two spins are

$$\nu_{\text{A}} = \nu^\circ + \left(\frac{\nu_{\text{spect}}}{10^6}\right)\delta = 0 + \frac{250 \times 10^6\text{ Hz}}{10^6} \times 1.00 = 250\text{ Hz}$$

$$\nu_{\text{X}} = \nu^\circ + \left(\frac{\nu_{\text{spect}}}{10^6}\right)\delta = 0 + \frac{250 \times 10^6\text{ Hz}}{10^6} \times 2.00 = 500\text{ Hz}$$

There will therefore be peaks of equal height at 245 Hz, 255 Hz, 495 Hz, and 505 Hz.

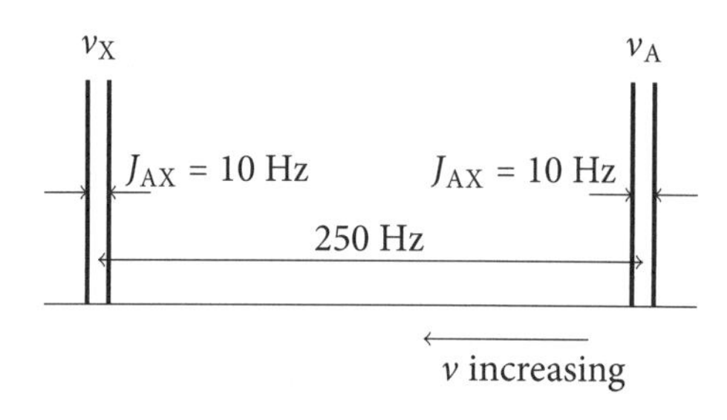

For a spectrometer operating at 800 MHz the resonance frequencies of the two spins are

$$\nu_A = \nu^\circ + \left(\frac{\nu_{spect}}{10^6}\right)\delta = 0 + \frac{800 \times 10^6\ \text{Hz}}{10^6} \times 1.00 = 800\ \text{Hz}$$

$$\nu_X = \nu^\circ + \left(\frac{\nu_{spect}}{10^6}\right)\delta = 0 + \frac{800 \times 10^6\ \text{Hz}}{10^6} \times 2.00 = 1600\ \text{Hz}$$

There will therefore be peaks of equal height at 795 Hz, 805 Hz, 1595 Hz, and 1605 Hz.

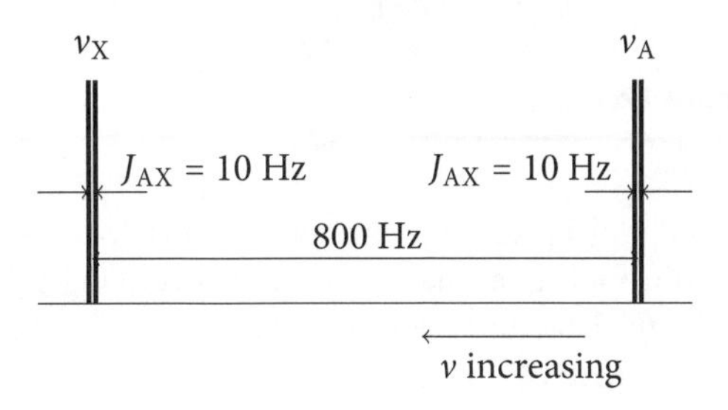

E12B.7(a) ^{19}F NMR: the four ^{19}F nuclei are equivalent and so all have the same shift. This resonance is split by coupling to the single ^{10}B nucleus into $2nI + 1$ lines; here $n = 1$ and $I = 3$ so there are 7 lines of equal intensity. The splitting between each is the B–F coupling constant, and the multiplet is centred at the shift of the fluorine.

^{10}B NMR: the ^{10}B nucleus is coupled to four equivalent ^{19}F nuclei, giving a multiplet with $2nI + 1$ lines; here $n = 4$ and $I = \frac{1}{2}$ so there are 5 lines. These lines have intensities in the ratio 1:4:6:4:1, each line is separated from the next by the B–F coupling constant, and the whole multiplet is centred at the shift of the boron.

E12B.8(a) The ^{31}P nucleus is coupled to six equivalent ^{19}F nuclei which have $I = \frac{1}{2}$. This splits the resonance into $2nI+1 = 2\times6\times\frac{1}{2}+1 = 7$ lines (a septet) with intensity ratio 1:6:15:20:15:6:1. The separation between adjacent lines in the multiplet is the P–F coupling constant.

E12B.9(a) The multiplet resulting from coupling to four equivalent spin-$\frac{1}{2}$ nuclei originates from the resonance of nucleus A being split into two by coupling with one X nucleus, and then each of those two lines being split into two by coupling to the second X nucleus. These lines are then each further split into two by coupling to the third X nucleus, and finally each of these lines is split into two by the fourth X nucleus.

As a result of the X nuclei being equivalent, each causes the same splitting and thus some lines are coincident and give rise to absorption lines of increased intensity. As shown in Fig. 12.1, the overall result is a $\boxed{\text{1:4:6:4:1 quintet}}$.

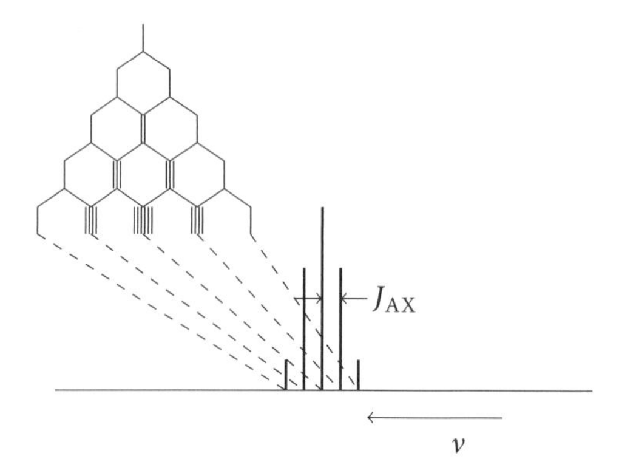

Figure 12.1

E12B.10(a) The multiplet resulting from coupling to two non-equivalent spin-$\frac{1}{2}$ nuclei originates from the resonance of nucleus A being split into two by coupling with the X_1 nucleus, and then each of these lines being split into two by coupling to the X_2 nucleus.

Because the coupling between A and X_1 is not the same as that between A and X_2, no lines are necessarily coincident. Coupling to two non-equivalent spin-$\frac{1}{2}$ nuclei thus results in a multiplet with four lines of equal intensity, in this context usually referred to as a doublet of doublets, as shown in Fig. 12.2.

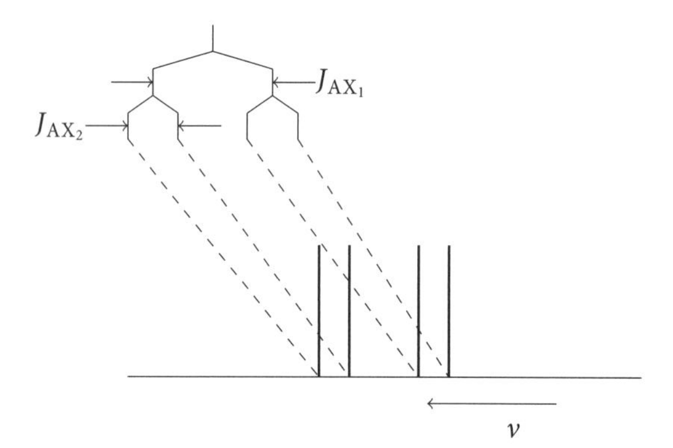

Figure 12.2

E12B.11(a) Coupling to one $I = \frac{5}{2}$ nucleus results in the original line being split into $(2I + 1) = 6$ lines. Each of these lines is then split into six in the same way by coupling to the second $I = \frac{5}{2}$ nucleus.

As a result of each nucleus being equivalent, the splittings are the same and thus some lines are coincident and give rise to absorption lines of increased intensity. The result is a 1:2:3:4:5:6:5:4:3:2:1 multiplet, as is shown in Fig. 12.3.

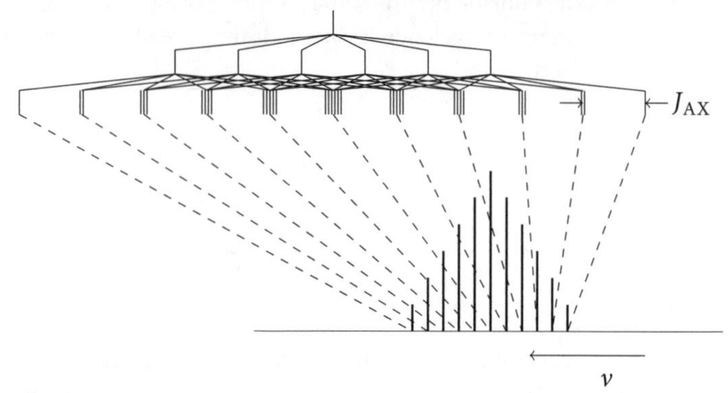

Figure 12.3

E12B.12(a) A group of nuclei are chemically equivalent if they are related by a symmetry operation of the molecule and therefore have the same chemical shift. Two chemically equivalent nuclei A and A′ are magnetically equivalent if the coupling between A and any other magnetic nucleus in the molecule, say Z, is the same as the coupling between A′ and Z. If Z itself is to be magnetically equivalent to Z′, then the couplings A–Z, A–Z′, A′–Z, and A′–Z′ must all be equal: that is, the couplings between any member of a group of equivalent spins with any member of another group of equivalent spins must be the same in order for the groups each to be magnetically equivalent.

In 1-chloro-4-bromobenzene, H_2 and H_6 are chemically equivalent because they are related by a 180° degree rotation about the C_1–C_4 axis. Similarly, H_3 and H_5 are chemically equivalent because they are related by the same operation.

The coupling between H_2 and H_3 is *not* the same as between H_2 and H_5, therefore the test for magnetic equivalence is not satisfied: H_2 and H_6 are *not* magnetically equivalent, nor are H_3 and H_5.

Cl
H_6 H_2
H_5 H_3
Br

E12B.13(a) The molecule PF_5 is trigonal bipyramidal. The three equatorial fluorine atoms, F_e, are related by a 120° rotation about the F_a–P–F_a axis: they are therefore chemically equivalent. The two axial fluorine atoms, F_a, are related by reflection in the plane formed by the three equatorial fluorine atoms: the axial fluorine atoms are therefore chemically equivalent.

Each axial fluorine has the same coupling to each equatorial fluorine so the test for magnetic equivalence is satisfied: the three F_e are magnetically equivalent, as are the two F_a.

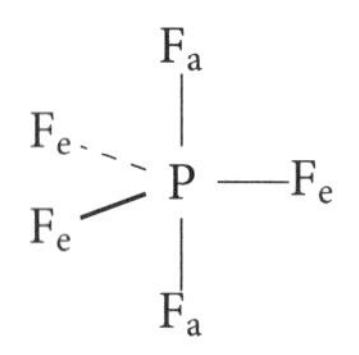

E12B.14(a) Coalescence of two NMR lines due to rapid nuclei exchange occurs when the condition given in [12B.16–505], $\tau = 2^{1/2}/\pi\delta\nu$, is satisfied; in this expression τ is the lifetime of an environment, and $\delta\nu$ is the difference between the Larmor frequencies of the two environments. Using [12B.5–494],

$$\delta\nu = \nu_2 - \nu_1 = \left[\nu^\circ + \left(\frac{\nu_{\text{spec}}}{10^6}\right)\delta_2\right] - \left[\nu^\circ + \left(\frac{\nu_{\text{spec}}}{10^6}\right)\delta_1\right] = \left(\frac{\nu_{\text{spec}}}{10^6}\right)(\delta_2 - \delta_1)$$

For a proton jumping between two sites with first-order rate constant k, $k = 1/\tau$

$$\delta\nu = \left(\frac{\nu_{\text{spec}}}{10^6}\right)(\delta_2 - \delta_1) = \left(\frac{550\times10^6}{10^6}\right)(4.8 - 2.7) = 1155\ \text{Hz}$$

$$k = \frac{1}{\tau} = \frac{\pi\delta\nu}{2^{1/2}} = \frac{\pi\times(1155\ \text{Hz})}{2^{1/2}} = \boxed{2.6\times10^3\ \text{s}^{-1}}$$

Solutions to problems

P12B.1 ^{129}Xe NMR spectrum: the signal is split by coupling to the ^{19}F atom (100% natural abundance, spin $\frac{1}{2}$) into a doublet with coupling constant $J = 7600$ Hz.

^{19}F NMR spectrum: this consists of a superposition of spectra from the different isotopes of Xe, and the intensity of these spectra reflects the abundance of each isotope. A doublet, with splitting 7600 Hz, arises from those 26% of molecules containing ^{129}Xe. The remaining 74% of molecules do not contain Xe nuclei which give rise to splitting, so a single line appears.

The observed spectrum is not a triplet, in the sense that it does not arise from coupling to two equivalent spin-$\frac{1}{2}$ nuclei. Rather it is a superposition of a singlet and a doublet.

P12B.3 Three possible structures for SF_4 are shown below.

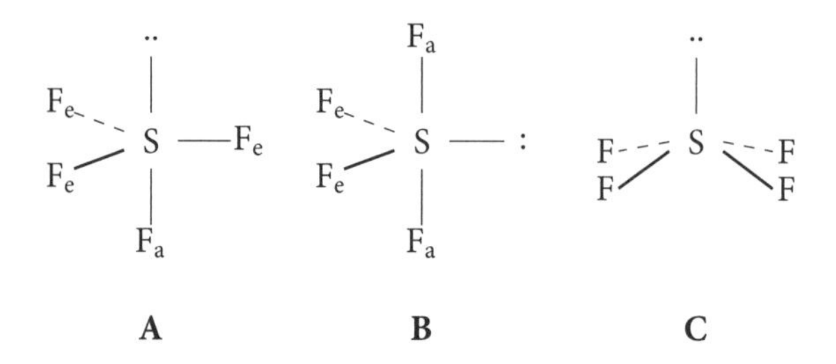

Structure **A**. There are two environments for fluorine atoms: the three F_e are equivalent as they are related by a threefold rotation about the S–F_a bond; F_a is unique. Each F_e has the same coupling to F_a, so the F_e are magnetically equivalent. The spectrum will therefore consist of a $\boxed{\text{doublet}}$ from F_e and a $\boxed{\text{quartet}}$ from F_a.

Structure **B**. There are two environments for fluorine atoms: the two F_e are equivalent as they are related by a twofold rotation about the S–lone pair axis; the two F_a atoms are equivalent because they are related by reflection in the mirror plane formed by the two F_e and the S. Each F_e has the same coupling to each F_a, so the F_e are magnetically equivalent, as are the F_a. The spectrum will therefore consist of a $\boxed{\text{triplet}}$ from F_e, and likewise a $\boxed{\text{triplet}}$ from F_a.

Structure **C**. All the fluorine atoms are equivalent as they are related by a fourfold rotation about the S–lone pair axis. A $\boxed{\text{single line}}$ will be seen in the spectrum, therefore.

P12B.5 Figure 12.4 shows the three possible staggered conformations of $XYCHCHR_3R_4$. When a staggered conformation is adopted, the angle between adjacent C–H bonds is $\phi = 60°$. Conformation *a* is trans ($\phi = 180°$, ${}^3J_{HH} = {}^3J_t$) and conformations *b* and *c* are gauche ($\phi = 60°$, ${}^3J_{HH} = {}^3J_g$).

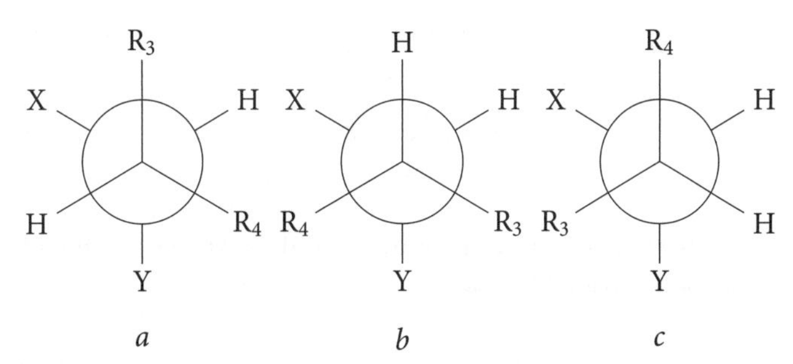

Figure 12.4

For $R_3 = R_4 = H$, all conformations have the same energy and each occurs with equal probability ($P = \frac{1}{3}$). Thus

$$ {}^3J_{HH} = \tfrac{1}{3}({}^3J_t + 2 \times {}^3J_g) = 7.3\ \text{Hz} \tag{12.1} $$

For $R_3 = CH_3$ and $R_4 = H$, the steric bulk of the methyl group makes conformation *c* very unfavourable. Thus, only equivalent conformations *a* and *b* are relevant for calculating ${}^3J_{HH}$: these have equal energy and hence equal probability ($P = \frac{1}{2}$). Thus

$$ {}^3J_{HH} = \tfrac{1}{2}({}^3J_t + {}^3J_g) = 8.0\ \text{Hz} \tag{12.2} $$

For $R_3 = R_4 = CH_3$, the steric bulk of the methyl group makes conformations *b* and *c* very unfavourable due to close proximity to the bulky X and Y groups. Therefore only conformation *a* is relevant for calculating ${}^3J_{HH}$. Thus

$$ {}^3J_{HH} = {}^3J_t = 11.2\ \text{Hz} \tag{12.3} $$

These three simultaneous equations, with two unknown variables ${}^3J_t$ and ${}^3J_g$, can be solved to estimate the values of the unknowns; note that with three equations and two unknowns the system is overdetermined.

Using eqn 12.1 and eqn 12.2, ${}^3J_g = 5.9$ Hz and ${}^3J_t = 10.1$ Hz.

Using eqn 12.1 and eqn 12.3, ${}^3J_g = 5.4$ Hz and ${}^3J_t = 11.2$ Hz.

Using eqn 12.2 and eqn 12.3, ${}^3J_g = 4.8$ Hz and ${}^3J_t = 11.2$ Hz.

The latter two results are used to give an average value ${}^3J_g = 5.1$ Hz and ${}^3J_t = 11.2$ Hz.

Using the original Karplus equation, ${}^3J_{HH} = A\cos^2\phi + B$, the gauche and trans couplings are calculated as

$$^3J_g = A\cos^2(60°) + B = \tfrac{1}{4}A + B \qquad ^3J_t = A\cos^2(180°) + B = A + B$$

With the values ${}^3J_g = 5.1$ Hz and ${}^3J_t = 11.2$ Hz these simultaneous equations yield $A = 8.1$ Hz and $B = 3.1$ Hz. The original Karplus equation is consistent with these data.

The modern form of the Karplus equation quoted in the text is [12B.14–*501*], ${}^3J_{HH} = A + B\cos\phi + C\cos 2\phi$. Using the values $A = 7$ Hz, $B = -1$ Hz, and $C = 5$ Hz, quoted in the text, gives

$$\begin{aligned} ^3J_g &= A + B\cos(60°) + C\cos(2\times 60°) = A + \tfrac{1}{2}B - \tfrac{1}{2}C \\ &= [7 + \tfrac{1}{2}(-1) - \tfrac{1}{2}(5)]\ \text{Hz} = 4\ \text{Hz} \\ ^3J_t &= A + B\cos(180°) + C\cos(2\times 180°) = A - B + C \\ &= [7 - (-1) + (5)]\ \text{Hz} = 13\ \text{Hz} \end{aligned}$$

There is poor agreement between these values and the ones determined from the experimental data.

P12B.7 The Karplus equation is given in [12B.14–*501*] as ${}^3J_{HH} = A + B\cos\phi + C\cos 2\phi$. From Fig. 12B.15 on page 502 it is evident that with typical values of the constants A, B, and C the coupling constant is a minimum around $\phi = 90°$.

Differentiating the Karplus equation with respect to ϕ, and then setting the derivative to zero allows the position of the minimum to be found.

$$\begin{aligned} \frac{\partial^3 J_{HH}}{\partial\phi} &= \frac{\partial}{\partial\phi}[A + B\cos\phi + C\cos 2\phi] \\ &= -B\sin\phi - 2C\sin 2\phi \end{aligned}$$

The trigonometric identity $\sin 2\phi = 2\sin\phi\cos\phi$ is then used to give

$$\begin{aligned} \frac{\partial\, ^3J_{HH}}{\partial\phi} &= -B\sin\phi - 4C\sin\phi\cos\phi \\ &= -(B + 4C\cos\phi)\sin\phi \end{aligned}$$

The derivative is zero when either $\sin\phi = 0$ or $B + 4C\cos\phi = 0$. The former of these solutions corresponds to $\phi = 0$ or π, and it is evident from the plot that these are maxima. For the other solution it follows that $\cos\phi = -B/4C$, which is the required result. Given that typically $C \approx 5|B|$, it follows that $|\cos\phi| \ll 1$, making ϕ close to 90°, as expected.

P12B.9 (a) The structure of cyclohexane, in its most stable chair conformation, is shown below. The H atoms occupy either 'axial' or 'equatorial' positions, and these two environments have different chemical shifts. However, at room temperature, the molecule undergoes a conformational change in which this chair form converts to another chair, but in the process the axial and equatorial H atoms are interchanged. The process is sufficiently fast at this temperature that the H atoms experience an average environment and only a single resonance is seen. In other words, the rate constant for the exchange is large compared to the frequency difference between the axial and equatorial environments (fast exchange).

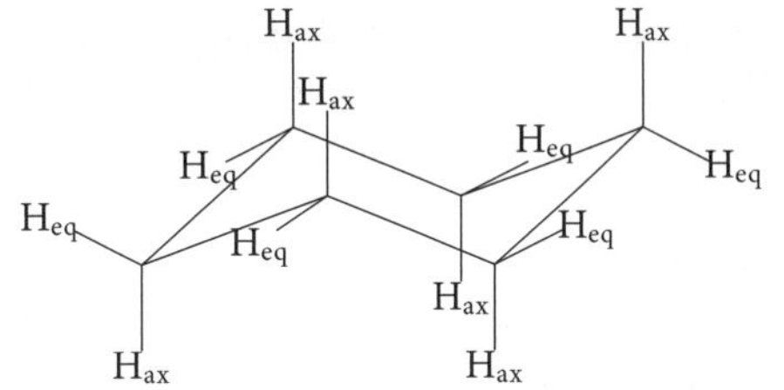

At lower temperatures, the rate constant for the interconversion of the two chair forms is reduced, and at sufficiently low temperatures separate resonances for the axial and equatorial H atoms are seen (slow exchange). Between the fast and slow exchange limits the system passes through an intermediate rate regime where the lines are broadened.

(b) The molecule PF_5 is trigonal bipyramidal, as shown below. The axial and equatorial F atoms are not equivalent, but there is a very fast process which interconverts them (the Berry pseudorotation), and even at the lowest accessible temperatures the rate constant for this process is sufficiently large that the two fluorine environments are in fast exchange. The two lines seen in the spectrum are due to coupling to ^{31}P.

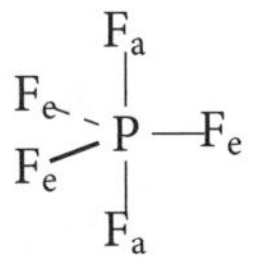

(c) Unless water is excluded rigorously from the sample there will exchange between the OH of the alcohol and the residual water in the solvent. This process leads to a collapse of the splitting due to any coupling to the OH proton because its spin state is scrambled as it leaves and returns to the molecule. The spectrum then consists of a triplet and a quartet, characteristic of a CH_3CH_2 group.

If water is excluded, the exchange is inhibited and then splittings due to coupling to the OH proton are seen, giving more complex multiplets.

12C Pulse techniques in NMR

Answers to discussion questions

D12C.1 The action of radiofrequency pulses is described in Section 12C.1(a) on page 510. The application of a radiofrequency field results in a small magnetic field $\mathcal{B}_1$ lying in the xy-plane and rotating about the z-axis at the radiofrequency. When viewed in a frame rotating about the z-axis at the same frequency, this field appears to be static, and it is easiest to understand its effect in such a frame. The radiofrequency needs to be close to the Larmor frequency, and for simplicity it is easiest to set it at exactly this frequency.

Because of the unequal populations of the α and β spins states, at equilibrium there will be a net magnetization of the sample along the direction of the applied field $\mathcal{B}_0$. This magnetization is represented by a vector, initially lying along the z-axis when the sample is at equilibrium. When a resonant radiofrequency field is applied the magnetization vector is rotated about the $\mathcal{B}_1$ field and towards the xy-plane (Fig. 12C.3 on page 510). The angle through which the magnetization is rotated is called the flip angle, and this angle is controlled by altering the duration for which the resonant field is applied (the pulse length).

If the length of the pulse is set such that the flip angle is 90° – a 90° pulse – the magnetization is rotated from the z-axis to the xy-plane. After such a pulse is complete, and the radiofrequency removed, the magnetization precesses in xy-plane, and it is this precession which results in the detected free-induction decay. If the length of the pulse is set such that the flip angle is 180° – a 180° pulse – the magnetization is rotated from the $+z$-axis to the $-z$-axis. Such a pulse does not result in a free-induction decay as no magnetization is generated in the xy-plane. However, the inversion of the magnetization corresponds to swapping the populations of the α and β spin states. A 180° pulses therefore generates a non-equilibrium state, whose relaxation back to equilibrium is studied in the inversion–recovery experiment.

When applied to a system in which there is magnetization in the xy-plane, a 180° pulse will flip the magnetization to a mirror image position with respect to the axis about which the pulse is applied (Fig. 12C.11 on page 515). Such a pulse is key in the formation of a spin echo.

D12C.3 Longitudinal relaxation is the result of fluctuating local magnetic fields, and for such fields to be effective at causing relaxation they must be fluctuating at frequencies near to the Larmor frequency. In liquids, such fluctuations arise from the random tumbling motion of the molecules. Typically, such motion result in fluctuations covering a range of frequenies. As the temperature increases the fluctuations spread over a wider range of frequencies and out to higher frequencies.

At the lowest temperatures there is little motion at the Larmor frequency and relaxation is slow. As the temperature increases the frequency range covered by the motion increases, resulting in more motion at the Larmor frequency, and hence faster relaxation. However, as the temperature is increased further still the motion continues to spread over a wider range of frequencies and the amount of motion at the Larmor frequency decreases, thus leading to slower relaxation. Thus, there is a temperature at which the relaxation is most rapid (and the relaxation time at a minimum, see Fig. 12C.9 on page 514).

Small molecules tumble rapidly in non-viscous solvents and at normal temperatures the resulting fluctuations are spread over a wide frequency range, well beyond the Larmor frequency. Increasing the temperature results in the fluctuations spreading out even more, such that the amount at the Larmor frequency is decreased, resulting in slower relaxation, and hence longer longitudinal relaxation times. In terms of the graph shown in Fig. 12C.9 on page 514 the molecule is to the right of the minimum.

In contrast, large molecules tumble much more slowly, and at normal temperatures the tumbling rate is such that the fluctuations are confined to lower frequencies. As the temperature increases the fluctuations spread to higher frequencies, therefore increasing the component at the Larmor frequency, leading to faster relaxation. In terms of the graph, the molecule is to the left of the minimum.

D12C.5 Homogeneous broadening is the contribution to the linewidth which results from transverse relaxation. It is the result of local fields modulated by random molecular motion, and its effects cannot be undone by the application of a spin echo. Inhomogeneous broadening is the contribution to the linewidth from the spatial inhomogeneity of the applied magnetic field. If the field varies from location to location in the sample, so does the Larmor frequency and therefore the spread of magnetic fields results in a contribution to the linewidth. This contribution arises from time-independent magnetic fields and, as a result, its effects can be reversed by the application of a spin echo.

Solutions to exercises

E12C.1(a) From the discussion in Section 12C.1(a) on page 510, the flip angle ϕ of a pulse of duration $\Delta\tau$ is given by $\phi = \gamma_N \mathcal{B}_1 \Delta\tau$, where $\mathcal{B}_1$ is the strength of the applied field. It follows that $\mathcal{B}_1 = \phi/\gamma_N \Delta\tau$. With the data given

$$\mathcal{B}_1 = \frac{\phi}{\gamma_N \Delta\tau} = \frac{\pi}{(26.752 \times 10^7\ \mathrm{T^{-1}\,s^{-1}}) \times (12.5 \times 10^{-6}\ \mathrm{s})} = \boxed{9.40 \times 10^{-4}\ \mathrm{T}}$$

The corresponding 90° pulse has half the flip angle so, for the same $\mathcal{B}_1$, $\Delta\tau$ is halved: $\Delta\tau = \frac{1}{2} \times (12.5\ \mu\mathrm{s}) = \boxed{6.25\ \mu\mathrm{s}}$.

E12C.2(a) The effective transverse relaxation time, T_2^*, is given by [12C.6–*514*], $T_2^* = 1/\pi\Delta\nu_{1/2}$, where $\Delta\nu_{1/2}$ is the width of the (assumed) Lorentzian signal measured at half the peak height. Hence

$$T_2^* = \frac{1}{\pi\Delta\nu_{1/2}} = \frac{1}{\pi \times (1.5\ \text{Hz})} = \boxed{0.21\ \text{s}}$$

E12C.3(a) The amplitude of the free induction decay, $S(t)$, is proportional to e^{-t/T_2}, where t is time and T_2 is the transverse relaxation time . If $S(t_2) = \frac{1}{2}S(t_1)$ then

$$\frac{S(t_2)}{S(t_1)} = \frac{1}{2} = \frac{e^{-t_2/T_2}}{e^{-t_1/T_2}} = e^{-(t_2-t_1)/T_2} = e^{-\Delta t/T_2}$$

taking logarithms gives

$$\ln\tfrac{1}{2} = -\Delta t/T_2 \quad \text{therefore} \quad T_2 = \frac{-\Delta t}{\ln\frac{1}{2}} = \frac{\Delta t}{\ln 2}$$

With the data given

$$T_2 = \frac{(1.0\ \text{s})}{\ln 2} = \boxed{1.4\ \text{s}}$$

E12C.4(a) Ethanoic acid has two carbon environments: a methyl carbon and a carboxylic acid carbon. The ^{13}C NMR spectrum shows a signal at $\delta = 21$, which is due to the methyl carbon. This C atom is adjacent to three protons and couples to them, giving rise to a quartet. The splitting of the quartet is the one-bond C–H coupling: $^1J = 130$ Hz.

The signal at $\delta = 178$ is due to the carboxylic acid carbon. This C atom also couples to the three methyl protons, but this time over two bonds which results in a much smaller coupling. As with the methyl carbon, a quartet is seen; the splitting is the much smaller two-bond C–H coupling.

When the ^{13}C NMR spectrum is recorded with proton decoupling, the splittings due to C–H couplings are collapsed, leaving two singlets.

E12C.5(a) The maximal NOE enhancement is given by [12C.8–517], $\eta = \gamma_X/2\gamma_A$, where γ_X and γ_A are the magnetogyric ratios of nuclei X and A, respectively.

$$\eta = \frac{\gamma_{^1\text{H}}}{2\gamma_{^{31}\text{P}}} = \frac{(26.752 \times 10^7\ \text{T}^{-1}\ \text{s}^{-1})}{2 \times (10.840 \times 10^7\ \text{T}^{-1}\ \text{s}^{-1})} = \boxed{1.234}$$

Solutions to problems

P12C.1 (a) The flip angle ϕ of the pulse is given by $\phi = \Delta\tau\gamma_N\mathcal{B}_1$ (Section 12C.1(a) on page 510). Therefore, increasing the pulse duration $\Delta\tau$ increases the flip angle. The strongest signal will be observed after application of a 90° pulse because such a pulse rotates the magnetization completely to the xy-plane. A 180° pulse rotates the magnetization from $+z$ to $-z$ and therefore creates no transverse magnetization, and hence no observable signal.

From the description of the experiment it is clear that pulses of duration 2.5 μs and 5.0 μs correspond to flip angles of less than 90°, and about 90°, respectively. The 7.5 μs pulse must correspond to a flip angle of more than 90° because the signal is weaker than for a shorter pulse. The 10.0 μs pulse gives no signal and so must correspond to a 180° pulse.

(b) The 10.0 μs pulse gives no signal and so corresponds to a flip angle of 180°. A 90° pulse has half the duration of a 180° pulse, that is $\boxed{\Delta\tau_{90} = 5.0\ \mu\text{s}}$.

(c) The magnetic field strength $\mathcal{B}_1$ is determined by rearranging $\phi = \Delta\tau\gamma_N\mathcal{B}_1$ to $\mathcal{B}_1 = \phi/\Delta\tau\gamma_N$. The length of the 180° pulse is normally used in the calculation of $\mathcal{B}_1$ because such a pulse gives a null signal, which is easier to determine than the maximum seen for a 90° pulse. If the duration of the 180° pulse is $\Delta\tau_{180}$ it follows that $\mathcal{B}_1 = \pi/\Delta\tau_{180}\gamma_N$. Substituting this expression into that for the $\mathcal{B}_1$ Larmor frequency gives

$$\nu_L' = \frac{\gamma_N\mathcal{B}_1}{2\pi} = \frac{\gamma_N}{2\pi} \times \frac{\pi}{\Delta\tau_{180}\gamma_N} = \frac{1}{2\Delta\tau_{180}}$$

$$= \frac{1}{2 \times (10.0 \times 10^{-6}\ \text{s})} = \boxed{5.00 \times 10^4\ \text{Hz}}$$

An alternative way at looking at this is to say that the magnetization completes a full rotation in 20.0 μs (a 360° pulse), so the period of the rotation is 20.0 μs. The corresponding frequency is 1/period $= 1/(20.0 \times 10^{-6}\ \text{s}) =$ 50.0 kHz. This is the $\mathcal{B}_1$ Larmor frequency, the frequency at which the magnetization rotates about $\mathcal{B}_1$.

P12C.3 To use the discrete Fourier transform the function needs to be generated as an array of points, regularly sampled in time. For the plots shown here 512 time-domain points were generated out to a maximum time of 3.0 s (some Fourier transform routines will only transform numbers of data points which are a power of 2, $2^9 = 512$); the interval between the points is therefore $(3.0\ \text{s})/512 = 5.86 \times 10^{-3}$ s. This means that in the spectrum the maximum frequency which is represented correctly is $1/(2 \times 5.86 \times 10^{-3}\ \text{s}) = 85$ Hz; this determines the scale on the spectra.

Figure 12.5 shows the free-induction decay and spectrum for the parameters given in the *Problem*. The peak at 10 Hz is higher than that at 50 Hz both because its intensity is greater and also because it is narrower on account of having a longer T_2; recall that the area under the peak is constant, so if it broadens the peak height decreases.

In Fig. 12.6 the high-frequency peak has been moved from 50 Hz to 60 Hz, and its intensity has been increased from 1 to 3; its peak height is still lower than the peak at 10 Hz on account of the higher-frequency resonance having a shorter T_2 and hence a broader line. In Fig. 12.7 the T_2 of the high-frequency peak is reduced from 0.5 s to 0.2 s, as is evident from the broadening of the line and the reduction in its peak height. In all three cases it is evident that the time-domain signal is changing, but only the frequency-domain signal, the spectrum, is readily interpretable.

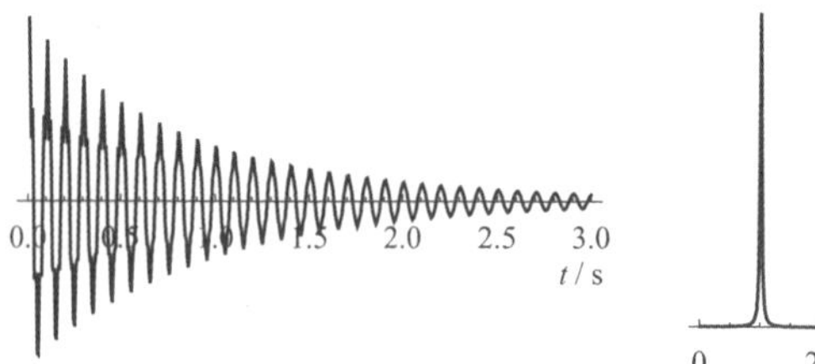

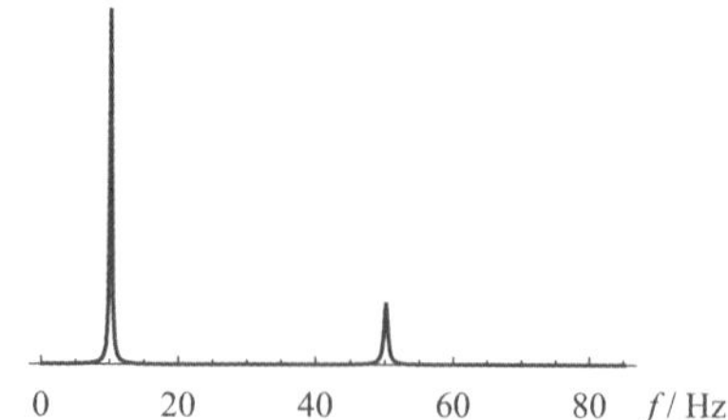

Figure 12.5

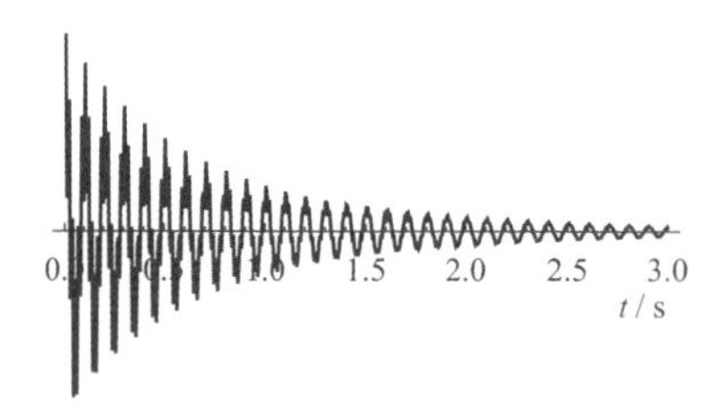

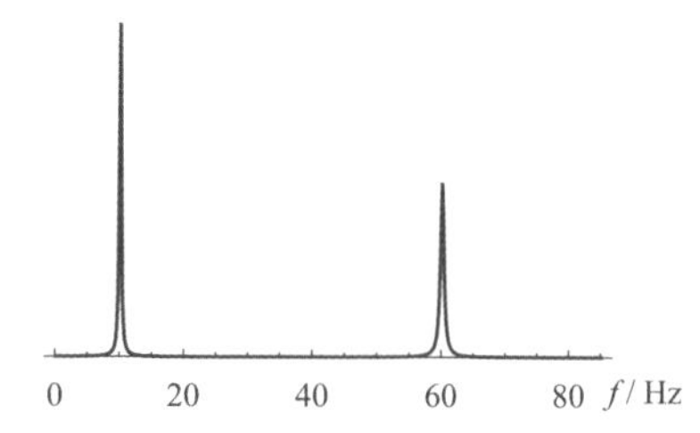

Figure 12.6

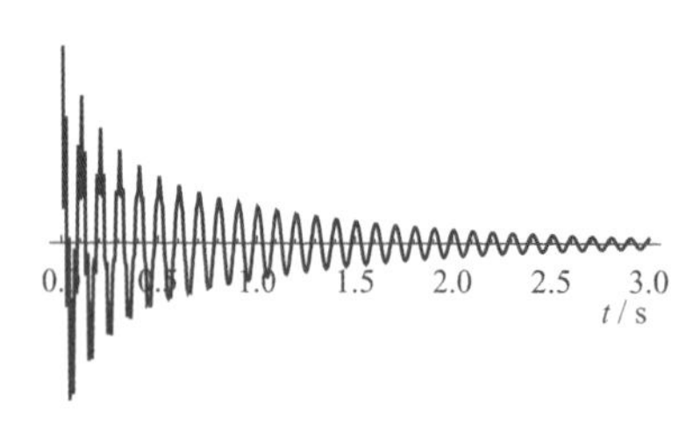

0 20 40 60 80 f / Hz

Figure 12.7

P12C.5 The shape of a spectral line, $I(\omega)$, is given by

$$I(\omega) = \text{Re}\left[\int_0^\infty G(t)\text{e}^{\text{i}\omega t}\,\text{d}t\right]$$

Using the suggested substitution, the free-induction decay $G(t) = a\text{e}^{-t/\tau}\cos\omega_0 t$ is rewritten

$$G(t) = \tfrac{1}{2}a\text{e}^{-t/\tau}\left(\text{e}^{-\text{i}\omega_0 t} + \text{e}^{\text{i}\omega_0 t}\right)$$

$$
\begin{aligned}
I(\omega) &= a\text{Re}\left[\int_0^\infty \tfrac{1}{2}\text{e}^{-t/\tau}\left(\text{e}^{-\text{i}\omega_0 t}+\text{e}^{\text{i}\omega_0 t}\right)\text{e}^{\text{i}\omega t}\,\text{d}t\right] \\
&= \tfrac{1}{2}a\text{Re}\left[\int_0^\infty \text{e}^{\text{i}t(\omega-\omega_0+\text{i}/\tau)}+\text{e}^{\text{i}t(\omega+\omega_0+\text{i}/\tau)}\,\text{d}t\right] \\
&= \tfrac{1}{2}a\text{Re}\left|\frac{1}{\text{i}(\omega-\omega_0+\text{i}/\tau)}\text{e}^{\text{i}t(\omega-\omega_0+\text{i}/\tau)}+\frac{1}{\text{i}(\omega+\omega_0+\text{i}/\tau)}\text{e}^{\text{i}t(\omega+\omega_0+\text{i}/\tau)}\right|_0^\infty \\
&= -\tfrac{1}{2}a\text{Re}\left[\frac{1}{\text{i}(\omega-\omega_0+\text{i}/\tau)}+\frac{1}{\text{i}(\omega+\omega_0+\text{i}/\tau)}\right] \\
&= \tfrac{1}{2}a\text{Re}\left[\frac{\text{i}}{(\omega-\omega_0+\text{i}/\tau)}+\frac{\text{i}}{(\omega+\omega_0+\text{i}/\tau)}\right] \\
&= \tfrac{1}{2}a\text{Re}\left[\frac{\text{i}(\omega-\omega_0-\text{i}/\tau)}{(\omega-\omega_0+\text{i}/\tau)(\omega-\omega_0-\text{i}/\tau)}+\frac{\text{i}(\omega+\omega_0-\text{i}/\tau)}{(\omega+\omega_0+\text{i}/\tau)(\omega+\omega_0-\text{i}/\tau)}\right] \\
&= \tfrac{1}{2}a\text{Re}\left[\frac{\text{i}(\omega-\omega_0)+1/\tau}{(\omega-\omega_0)^2+1/\tau^2}+\frac{\text{i}(\omega+\omega_0)+1/\tau}{(\omega+\omega_0)^2+1/\tau^2}\right] \\
&= \frac{a}{2\tau}\left[\frac{1}{(\omega-\omega_0)^2+1/\tau^2}+\frac{1}{(\omega+\omega_0)^2+1/\tau^2}\right] \\
&= \frac{a\tau}{2}\left[\frac{1}{\tau^2(\omega-\omega_0)^2+1}+\frac{1}{\tau^2(\omega+\omega_0)^2+1}\right]
\end{aligned}
$$

The Lorentzian lineshape is of the form $I_L = S_0T_2/[1+T_2^2(\omega-\omega_0)^2]$, and this lineshape has a width at half-height of $2/T_2$ in angular frequency units. Comparison with the result above shows that the spectrum contains two Lorentzian peaks at $\omega = \pm\omega_0$, with width $2/\tau$. Two peaks arise in the spectrum because the original free-induction decay, being of the form of a cosine, is invariant to the sign of ω_0.

P12C.7 Note that there is an error in the *Problem*: the initial condition should be $M_z(0) = -M_0$

(a) Separate the differential equation and integrate between $t = 0$ and $t = \tau$, and the corresponding limits $M_z(0) = -M_0$ and $M_z(\tau)$.

$$
\begin{aligned}
\int_{-M_0}^{M_z(\tau)} \frac{1}{M_z(t)-M_0}\,\text{d}M_z(t) &= -\int_0^\tau \frac{1}{T_1}\,\text{d}t \\
\left[\ln\left(M_z(t)-M_0\right)\right]_{-M_0}^{M_z(\tau)} &= \left[-\frac{t}{T_1}\right]_0^\tau \\
\ln\left(\frac{M_z(\tau)-M_0}{-2M_0}\right) &= -\frac{\tau}{T_1} \\
\text{hence }\ M_z(\tau) &= M_0(1-2\text{e}^{-\tau/T_1})
\end{aligned}
$$

(b) Figure 12.8 shows plots of $M_z(\tau)/M_0 = 1-2\text{e}^{-\tau/T_1}$ for different values of T_1. It is evident that increasing T_1 increases the time required for the z-component of magnetisation to relax back to its equilibrium value, M_0.

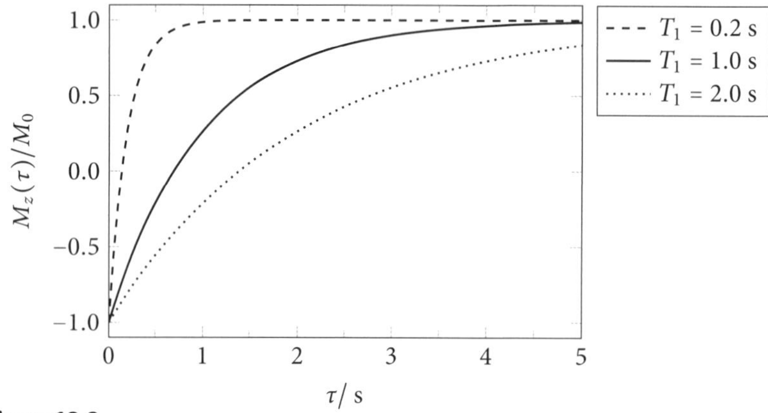

Figure 12.8

(c) From the result above, $M_z(\tau) = M_0(1-2e^{-\tau/T_1})$, it follows that $M_z(\tau)/M_0 = (1-2e^{-\tau/T_1})$, and hence $-\frac{1}{2}[M_z(\tau)/M_0-1] = e^{-\tau/T_1}$. Taking logarithms gives $\ln\{[M_0 - M_z(\tau)]/2M_0\} = -\tau/T_1$. It therefore follows that a plot of $\ln\{[M_0 - M_z(\tau)]/2M_0\}$ against τ is expected to be a straight line of slope $= 1/T_1$.

(d) The data are shown in the table below and the plot in Fig 12.9.

$\tau/$ s	$M_z(\tau)/M_0$	$\ln\{[M_0 - M_z(\tau)]/2M_0\}$
0.000	−1.000	0.000
0.100	−0.637	−0.200
0.200	−0.341	−0.400
0.300	−0.098	−0.600
0.400	0.101	−0.800
0.600	0.398	−1.201
0.800	0.596	−1.599
1.000	0.729	−1.999
1.200	0.819	−2.402

The data are a good fit to the equation

$$\ln\{[M_0 - M_z(\tau)]/2M_0\} = (-2.001) \times (\tau/\text{ s}) + 3.206 \times 10^{-4}$$

The value of T_1 is found from the slope

$$T_1 = -\frac{1}{(\text{slope})} = -\frac{1}{(-2.001\text{ s}^{-1})} = \boxed{0.500\text{ s}}$$

P12C.9 (a) Separate the differential equation and integrate between $t = 0$ and $t = \tau$,

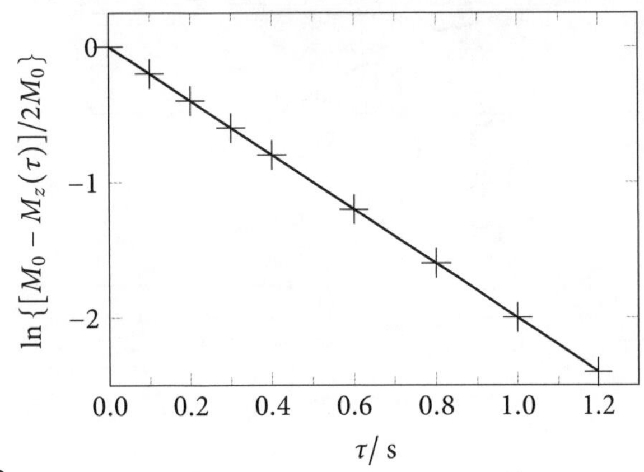

Figure 12.9

and the corresponding limits $M_{xy}(0)$ and $M_{xy}(\tau)$.

$$\int_{M_{xy}(0)}^{M_{xy}(\tau)} \frac{1}{M_{xy}(t)} \mathrm{d}M_{xy}(t) = -\int_0^\tau \frac{1}{T_2} \mathrm{d}t$$

$$\left|\ln M_{xy}(t)\right|_{M_{xy}(0)}^{M_{xy}(\tau)} = \left|-\frac{t}{T_2}\right|_0^\tau$$

$$\ln\left(\frac{M_{xy}(\tau)}{M_{xy}(0)}\right) = -\frac{\tau}{T_2}$$

$$\text{hence } \boxed{M_{xy}(\tau) = M_{xy}(0)\mathrm{e}^{-\tau/T_2}}$$

(b) Divide both sides of this expression by $M_{xy}(0)$ and take the natural logarithm to give

$$\ln\frac{M_{xy}(\tau)}{M_{xy}(0)} = -\frac{\tau}{T_2}$$

Hence a plot of $\ln[M_{xy}(\tau)/M_{xy}(0)]$ against τ is expected to be a straight line with slope $= -1/T_2$.

(c) The data are shown in the table below and the plot in Fig 12.10.

$\tau/$ ms	$M_{xy}(\tau)/M_0$	$\ln[M_{xy}(\tau)/M_{xy}(0)]$
10.0	0.819	−0.200
20.0	0.670	−0.400
30.0	0.549	−0.600
50.0	0.368	−1.000
70.0	0.247	−1.398
90.0	0.165	−1.802
110.0	0.111	−2.198
130.0	0.074	−2.604

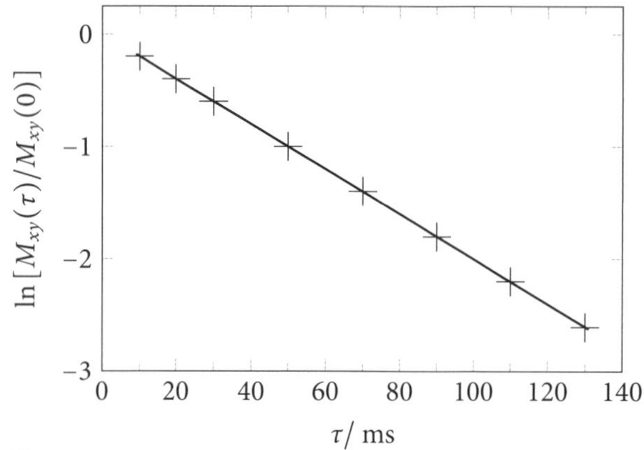

Figure 12.10

The data are a good fit to the equation

$$\ln[M_{xy}(\tau)/M_{xy}(0)] = (-0.0200) \times (\tau/\text{ ms}) + 8.1662 \times 10^{-4}$$

The value of T_2 is found from the slope,

$$T_2 = -\frac{1}{\text{slope}} = -\frac{1}{-0.0200\text{ ms}^{-1}} = \boxed{50.0\text{ ms}}$$

P12C.11 The magnitude of the local magnetic field is given by [12B.17a–*506*]

$$\mathcal{B}_{\text{loc}} = -\frac{\gamma_{\text{N}}\hbar\mu_0 m_I}{4\pi R^3}(1 - 3\cos^2\theta)$$

where $\mathcal{B}_{\text{loc}}$ is the local magnetic field, γ_{N} is the nuclear magnetogyric ratio, μ_0 is the vacuum permeability, $m_I\hbar$ is the z-component of the spin angular momentum, R is the internuclear separation, and θ is the angle between the direction of local field vector and the direction of the applied magnetic field. Rearrange this expression to give R, assume that $\theta = 0$ so that $(1 - 3\cos^2\theta) = -2$, and take $m_I = +\frac{1}{2}$

$$R = \left[-\frac{\gamma_{\text{N}}\hbar\mu_0 m_I}{4\pi\mathcal{B}_0}(1 - 3\cos^2\theta)\right]^{1/3}$$

$$= \left[\frac{\gamma_{\text{N}}\hbar\mu_0}{4\pi\mathcal{B}_0}\right]^{1/3}$$

$$= \left[\frac{(26.752 \times 10^7\text{ T}^{-1}\text{ s}^{-1}) \times (1.0546 \times 10^{-34}\text{ J s}) \times (1.257 \times 10^{-6}\text{ T}^2\text{ J}^{-1}\text{ m}^3)}{4\pi \times (0.715 \times 10^{-3}\text{ T})}\right]^{1/3}$$

$$= 1.58 \times 10^{-10}\text{ m} = \boxed{158\text{ pm}}$$

12D Electron paramagnetic resonance

Answers to discussion questions

D12D.1 This is described in Section 12D.2(c) on page 522.

Solutions to exercises

E12D.1(a) The EPR resonance condition is given by [12D.2–*519*], $h\nu = g\mu_B\mathcal{B}_0$, where ν is the spectrometer operating frequency, g is the g-value of the radical, μ_B is the Bohr magneton, and $\mathcal{B}_0$ is the magnetic field. Rearranging the expression gives

$$g = \frac{h\nu}{\mu_B\mathcal{B}_0} = \frac{(6.6261\times10^{-34}\,\mathrm{J\,s})\times(9.2231\times10^{9}\,\mathrm{Hz})}{(9.2740\times10^{-24}\,\mathrm{J\,T^{-1}})\times(329.12\times10^{-3}\,\mathrm{T})} = \boxed{2.0022}$$

E12D.2(a) As described in Section 12D.2 on page 520, the splitting between the lines in the multiplet is the hyperfine coupling constant. In this case the splitting is $(334.8 - 332.5)$ mT = 2.3 mT and $(332.5 - 330.2)$ mT = 2.3 mT, therefore $\boxed{a = 2.3\ \mathrm{mT}}$.

The central line is unaffected by coupling and so is used to compute the g-value using [12D.2–*519*], $h\nu = g\mu_B\mathcal{B}_0$

$$g = \frac{h\nu}{\mu_B\mathcal{B}_0} = \frac{(6.6261\times10^{-34}\,\mathrm{J\,s})\times(9.319\times10^{9}\,\mathrm{Hz})}{(9.2740\times10^{-24}\,\mathrm{J\,T^{-1}})\times(332.5\times10^{-3}\,\mathrm{T})} = \boxed{2.0025}$$

E12D.3(a) The total local magnetic field is given by [12D.3–*520*], $\mathcal{B}_{loc} = \mathcal{B}_0 + am_I$, where $m_I = \pm\frac{1}{2}$ and a is the hyperfine coupling constant. The centre of the spectrum is at 332.5 mT. Proton 1 splits the centre line into two components with separation $a_1 = 2.0$ mT.

$$\mathcal{B}_{loc} = \mathcal{B}_0 + a_1 m_I = (332.5\ \mathrm{mT}) \pm \tfrac{1}{2}\times(2.0\ \mathrm{mT}) = (332.5\pm1.0)\ \mathrm{mT}$$

Hence, $B_{loc,+1/2} = 331.5$ mT or $B_{loc,-1/2} = 333.5$ mT. Coupling to proton 2 splits each of these lines into two with $a_2 = 2.6$ mT, so the four lines occur at

$$\mathcal{B}_{loc} = B_{loc,+1/2}+a_2m_I = (331.5\,\mathrm{mT})\pm\tfrac{1}{2}\times(2.6\,\mathrm{mT}) = \boxed{330.2\ \mathrm{mT\ and\ }332.8\ \mathrm{mT}}$$

$$\mathcal{B}_{loc} = B_{loc,-1/2}+a_2m_I = (333.5\,\mathrm{mT})\pm\tfrac{1}{2}\times(2.6\,\mathrm{mT}) = \boxed{332.2\ \mathrm{mT\ and\ }334.8\ \mathrm{mT}}$$

The four lines have $\boxed{\text{equal intensity}}$.

E12D.4(a) (i) The three protons in the $\cdot CH_3$ radical are equivalent, each with spin quantum number $I = \frac{1}{2}$. If a radical contains N equivalent nuclei with spin I, then there are $2NI + 1$ hyperfine lines with an intensity distribution given by Pascal's triangle. Hence for the CH_3 radical, there are $(2\times3\times\frac{1}{2}+1) = 4$ hyperfine lines with intensity distribution $\boxed{1:3:3:1}$.

(ii) The three ^{2}H atoms in the $\cdot CD_3$ radical are equivalent, each with spin quantum number $I = 1$. The number of hyperfine lines is $(2\times3\times1+1) = 7$, and their intensity distribution is the same as that shown in Fig. 12D.5 on page 521, $\boxed{1:3:6:7:6:3:1}$.

E12D.5(a) The resonance frequency is related to the g-value by [12D.2–*519*], $h\nu = g\mu_B\mathcal{B}_0$. Rearranging this gives $\mathcal{B}_0 = h\nu/\mu_B$.

(i) For a spectrometer operating at 9.313 GHz,

$$\mathcal{B}_0 = \frac{h\nu}{g\mu_B} = \frac{(6.6261\times10^{-34}\,\mathrm{J\,s})\times(9.313\times10^{9}\,\mathrm{Hz})}{2.0025\times(9.2740\times10^{-24}\,\mathrm{J\,T^{-1}})} = \boxed{332.3\,\mathrm{mT}}$$

(ii) For a spectrometer operating at 33.80 GHz,

$$\mathcal{B}_0 = \frac{h\nu}{g\mu_B} = \frac{(6.6261\times10^{-34}\,\mathrm{J\,s})\times(33.80\times10^{9}\,\mathrm{Hz})}{2.0025\times(9.2740\times10^{-24}\,\mathrm{J\,T^{-1}})} = \boxed{1.206\,\mathrm{T}}$$

E12D.6(a) If a radical contains N equivalent nuclei with spin quantum number I, then there are $2NI + 1$ hyperfine lines. The radical contains a single magnetic nucleus, $N = 1$, and there are a total of four hyperfine lines. Therefore $2NI + 1 = 2\times1\times I + 1 = 4$ hence $\boxed{I = \tfrac{3}{2}}$.

Solutions to problems

P12D.1 The EPR resonance condition is given by [12D.1–*519*], $h\nu = g_e\mu_B\mathcal{B}_0$, where ν is the spectrometer operating frequency, g_e is the g-value of the electron, μ_B is the Bohr magneton and $\mathcal{B}_0$ is the magnetic field. The expression is rearranged to make ν the subject.

$$\nu = \frac{g_e\mu_B\mathcal{B}_0}{h} = \frac{2.0023\times(9.2740\times10^{-24}\,\mathrm{J\,T^{-1}})\times(1.0\times10^{3}\,\mathrm{T})}{6.6261\times10^{-34}\,\mathrm{J\,s}} = \boxed{2.8\times10^{13}\,\mathrm{Hz}}$$

Expressed as a wavenumber this is $\tilde{\nu} = 924\,\mathrm{cm^{-1}}$. This frequency is in the infrared region of the electromagnetic spectrum and hence is comparable to the separation between the energy levels associated with $\boxed{\text{molecular vibrations}}$. It is much greater than the separations of rotational levels, but much less than the separations of electronic energy levels.

P12D.3 (a) The hyperfine coupling is a result of an interaction between the magnetic moment of a nucleus and that of an electron. The magnetic moment of a nucleus is proportional to the magnetogyric ratio γ_N, so it is not unreasonable to expect the hyperfine coupling a to have the same dependence.

(b) Assuming that $a \propto \gamma_N$ it follows that

$$a_{\cdot CD_3} = \frac{\gamma_{N,D}}{\gamma_{N,H}} \times a_{\cdot CH_3} = \frac{4.1067 \times 10^7\ T^{-1}\ s^{-1}}{26.752 \times 10^7\ T^{-1}\ s^{-1}} \times (2.3\ mT) = \boxed{0.35\ mT}$$

The $\cdot CH_3$ spectrum consists of 4 lines (quartet, 1:3:3:1) and so has width $3a_H$. As shown in Fig. 12D.5, the $\cdot CD_3$ spectrum consists of 7 lines (septet, 1:3:6:7:6:3:1) and so has width $6a_D$. Hence the overall width of the $\cdot CH_3$ spectrum is $3a_H = 3 \times (2.3\ mT) = \boxed{6.9\ mT}$, and the overall width of the $\cdot CD_3$ spectrum is $6a_D = 6 \times (0.35\ mT) = \boxed{2.1\ mT}$.

P12D.5 The McConnell equation is given in [12D.5–*521*], $a = Q\rho$, where a is the hyperfine coupling constant observed for a H atom in a given molecule, Q is a constant, and ρ is the spin density on the adjacent carbon atom.

carbon 1 $\quad \rho = \frac{a}{Q} = \frac{0.274\ mT}{2.25\ mT} = \boxed{0.122}$

carbon 1 $\quad \rho = \frac{a}{Q} = \frac{0.151\ mT}{2.25\ mT} = \boxed{0.067}$

carbon 9 $\quad \rho = \frac{a}{Q} = \frac{0.534\ mT}{2.25\ mT} = \boxed{0.237}$

P12D.7 The proportion of time the unpaired electron spends in the N 2s orbital is given by the ratio of the hyperfine couplings: $P(N2s) = (5.7\ mT)/(55.2\ mT) = 0.103...$, that is $\boxed{10\%}$ of the time. The proportion of time the unpaired electron spends in the N 2s orbital is $P(N2p_z) = (1.3\ mT)/(3.4\ mT) = 0.382...$, that is $\boxed{38\%}$ of the time.

The total probability that the electron will be found on the N atom is $P(N) = 10\% + 38\% = \boxed{48\%}$. The total probability that the electron will be found on the O atoms is $P(O) = 100\% - 48\% = \boxed{52\%}$.

In *Problem* P9A.4 it is shown that the angle θ between two sp$^\lambda$ hybrids is related to the hybridisation ratio λ by $\lambda = \pm(-1/\cos\theta)^{1/2}$. The hybridization ratio is given by $\lambda^2 = (\text{p character})/(\text{s character})$, therefore in the present case $\lambda^2 = (38\%)/(10\%) = 3.8$, hence $\boxed{\lambda = 1.95}$. The angle between the hybrids is found by rearranging the above expression to $\cos\theta = -1/\lambda^2$, hence $\cos\theta = -1/3.8$ giving $\boxed{\theta = 105°}$. Such a hybrid is suitable for bonding an angular arrangement, and so supports the view that this is indeed the geometry for NO_2.

Answers to integrated activities

I12.1 (a) The first table below displays experimental ^{13}C chemical shifts and computed atomic charges (using semi-empirical, PM3 level, PC Spartan Pro) for the carbon *para* to the indicated substituent in a mono-substituted benzene. Two sets of charges are shown, one derived by fitting the electrostatic potential and the other by Mulliken population analysis.

substituent	CH_3	H	CF_3	CN	NO_2
δ	128.4	128.5	128.9	129.1	129.4
electrostatic charge/e	−0.1273	−0.0757	−0.0227	−0.0152	−0.0541
Mulliken charge/e	−0.1089	−0.1021	−0.0665	−0.0805	−0.0392

In the table below the net charges have been recalculated at a higher level than the semi-empirical PM3 level used above. The table shows both the experimental and calculated ^{13}C chemical shifts, and the computed atomic charges (using HF-SCF/6-311G* with Spartan 06) on the carbon atom of interest. Three sets of charges are shown, one derived by fitting the electrostatic potential, another by Mulliken population analysis, and the other by the method of 'natural' charges.

substituent	CH_3	H	CF_3	CN	NO_2
δ_{exp}	128.4	128.5	128.9	129.1	129.4
δ_{calc}	134.4	133.5	132.1	138.8	141.8
electrostatic charge/e	−0.240	−0.135	−0.138	−0.102	−0.116
Mulliken charge/e	−0.231	−0.217	−0.205	−0.199	−0.182
natural charge/e	−0.199	−0.183	−0.158	−0.148	−0.135

(b) None of the sets of charges correlates well to the chemical shifts, but there is a modest correlation for the Mulliken charges.

(c) The diamagnetic local contribution to shielding is roughly proportional to the electron density on the atom, so if this contribution is dominant it might be expected that there will be some correlation between the net charge and the chemical shift. That this is not seen in practice points either to inadequacies in the calculation of the charges or that there are other contributions to the shielding, such as the paramagnetic contribution. In fact the calculation of chemical shifts is a challenging test for computational chemistry and requires more sophisticated approaches than that used here.

I12.3 Coalescence of two NMR lines due to rapid exchange occurs when the condition given in [12B.16–*505*], $\tau = 2^{1/2}/\pi\delta\nu$, is satisfied. Here τ is the lifetime of an environment, and $\delta\nu$ is the difference between the Larmor frequencies of the two environments. The first-order rate constant for the process, k_{1st} is $1/\tau$. Using [12B.5–*494*] to convert between chemical shifts and frequencies

$$\delta\nu = \nu_2 - \nu_1 = \left[\nu^\circ + \left(\frac{\nu_{spec}}{10^6}\right)\delta_2\right] - \left[\nu^\circ + \left(\frac{\nu_{spec}}{10^6}\right)\delta_1\right] = \left(\frac{\nu_{spec}}{10^6}\right)(\delta_2 - \delta_1)$$

60 MHz spectrometer:

$$\delta\nu = \left(\frac{\nu_{spec}}{10^6}\right)(\delta_2 - \delta_1) = \left(\frac{60 \times 10^6 \text{ Hz}}{10^6}\right)(5.2 - 4.0) = 72 \text{ Hz}$$

$$k_{1st} = \frac{1}{\tau} = \frac{\pi\delta\nu}{2^{1/2}} = \frac{\pi \times (72\ \text{Hz})}{2^{1/2}} = \boxed{160\ \text{s}^{-1}}$$

300 MHz spectrometer:

$$\delta\nu = \left(\frac{\nu_{spec}}{10^6}\right)(\delta_2 - \delta_1) = \left(\frac{300 \times 10^6\ \text{Hz}}{10^6}\right)(5.2 - 4.0) = 360\ \text{Hz}$$

$$k_{1st} = \frac{1}{\tau} = \frac{\pi\delta\nu}{2^{1/2}} = \frac{\pi \times (360\ \text{Hz})}{2^{1/2}} = \boxed{800\ \text{s}^{-1}}$$

If it is assumed that the rate constant follows the Arrhenius law, $k_{1st}(T) = Ae^{-E_a/RT}$, it follows that

$$\ln\left(\frac{k_{1st}(T_2)}{k_{1st}(T_1)}\right) = \frac{-E_a}{R}\left(\frac{1}{T_2} - \frac{1}{T_1}\right)$$

$$\begin{aligned} E_a &= -R\ln\left(\frac{k_{1st}(T_2)}{k_{1st}(T_1)}\right) \times \frac{1}{1/T_2 - 1/T_1} \\ &= -(8.3145\ \text{J K}^{-1}\ \text{mol}^{-1}) \times \ln\left(\frac{800\ \text{s}^{-1}}{160\ \text{s}^{-1}}\right) \times \frac{1}{1/(300\ \text{K}) - 1/(280\ \text{K})} \\ &= 5.62... \times 10^4\ \text{J mol}^{-1} = \boxed{56\ \text{kJ mol}^{-1}} \end{aligned}$$

13 Statistical thermodynamics

13A The Boltzmann distribution

Answers to discussion questions

D13A.1 The *population* of a state is the number of molecules in a sample that are in that state. The *configuration* of a system is a list of populations in order of the energy of the corresponding states. For example, $\{N-3, 2, 1, 0, \ldots\}$ is a possible configuration of a system of N molecules in which all but three molecules are in the ground state, two are in the next highest state, and one in the state above that. The *weight* of a configuration is the number of ways a given configuration can be achieved, and is given by [13A.1–*533*]. When N is large (as it is for any macroscopic sample), the *most probable configuration* has a much greater weight, that is it is more probable, than any other configuration. Under such circumstances it can be assumed that the configuration adopted by the system is this most probable configuration.

D13A.3 In terms of molecular energy levels the thermodynamic temperature is the one quantity that determines the most probable populations of the states of the system at thermal equilibrium, as discussed in Section 13A.1(b) on page 533.

The equipartition theorem allows a connection to be made between the temperature as understood in statistical thermodynamics and the empirical concept of temperature which arises in classical thermodynamics. Temperature is a measure of the intensity of thermal energy, and is directly proportional to the mean energy for each quadratic contribution to the energy (provided that the temperature is sufficiently high).

Solutions to exercises

E13A.1(a) The weight of a configuration is given by [13A.1–*533*], $\mathcal{W} = N!/(N_0!N_1!N_2!\ldots)$, thus

$$\mathcal{W} = \frac{16!}{0! \times 1! \times 2! \times 3! \times 8! \times 0! \times 0! \times 0! \times 0! \times 2!} = \boxed{21\,621\,600}$$

E13A.2(a) (i) $8! = 8 \times 7 \times 6 \times 5 \times 4 \times 3 \times 2 \times 1 = \boxed{40\,320}$.

(ii) Stirling's approximation for $x \gg 1$ is given by [13A.2–533], $\ln(x!) \approx x\ln x - x$. This is rearranged to $x! \approx e^{(x\ln x - x)}$, thus

$$8! \approx e^{(8\times\ln 8-8)} = \boxed{5.63\times10^3}.$$

(iii) Using the more accurate version of Stirling's approximation

$$8! \approx (2\pi)^{(1/2)}\times 8^{(8+1/2)}\times e^{-8} = \boxed{3.99\times10^4}$$

E13A.3(a) The Boltzmann population ratio is given by [13A.13a–536], $N_i/N_j = e^{-\beta(\varepsilon_i-\varepsilon_j)}$, where $\beta = 1/(kT)$. At infinite temperature β becomes zero, therefore the relative populations of two levels $N_1/N_0 = e^{-0} = \boxed{1}$.

E13A.4(a) The Boltzmann population ratio is given by [13A.13a–536], $N_i/N_j = e^{-\beta(\varepsilon_i-\varepsilon_j)}$. This is rearranged to $\beta = -\ln(N_i/N_j)/\Delta\varepsilon$, where $\Delta\varepsilon = (\varepsilon_i - \varepsilon_j)$. Substituting $\beta = 1/(kT)$ and rearranging for T gives

$$T = -\frac{\Delta\varepsilon}{k\ln(N_1/N_0)} = -\frac{hc\tilde{\nu}}{k\ln(N_1/N_0)}$$
$$= -\frac{(6.6261\times10^{-34}\,\mathrm{J\,s})\times(2.9979\times10^{10}\,\mathrm{cm\,s^{-1}})\times(400\,\mathrm{cm^{-1}})}{(1.3806\times10^{-23}\,\mathrm{J\,K^{-1}})\times\ln(1/3)}$$
$$= \boxed{524\,\mathrm{K}}$$

E13A.5(a) The Boltzmann population ratio for degenerate energy levels is given by [13A.13b–536], $N_i/N_j = (g_i/g_j)e^{-\beta(\varepsilon_i-\varepsilon_j)}$. The rotational term of a linear rotor is given by [11B.14–434], $\tilde{F}(J) = \tilde{B}J(J+1)$ and, as explained in Section 11B.1(c) on page 434, its degeneracy is given as $g_J = 2J+1$. The rotational energy is related to the rotational term as $\varepsilon_J = hc\tilde{F}(J)$. Therefore

$$\frac{N_5}{N_0} = \frac{2\times5+1}{2\times0+1}\times e^{-hc\tilde{B}[5\times(5+1)-0\times(0+1)]/kT} = 11\times e^{-30\tilde{B}hc/kT}$$

using $kT/hc = 207.224\,\mathrm{cm^{-1}}$ at 298.15 K (from inside the front cover)

$$\frac{N_5}{N_0} = 11\times e^{-30\times(2.71\,\mathrm{cm^{-1}})/(207.224\,\mathrm{cm^{-1}})} = \boxed{7.43}$$

E13A.6(a) The Boltzmann population ratio is given by [13A.13a–536], $N_i/N_j = e^{-\beta(\varepsilon_i-\varepsilon_j)}$. This is rearranged to $\beta = -\ln(N_i/N_j)/\Delta\varepsilon$, where $\Delta\varepsilon = (\varepsilon_i - \varepsilon_j)$. Substituting $\beta = 1/(kT)$ and rearranging for T gives

$$T = -\frac{\Delta\varepsilon}{k\ln(N_1/N_0)} = -\frac{hc\tilde{\nu}}{k\ln(N_1/N_0)}$$
$$= -\frac{(6.6261\times10^{-34}\,\mathrm{J\,s})\times(2.9979\times10^{10}\,\mathrm{cm\,s^{-1}})\times(540\,\mathrm{cm^{-1}})}{(1.3806\times10^{-23}\,\mathrm{J\,K^{-1}})\times\ln(10\%/90\%)}$$
$$= \boxed{354\,\mathrm{K}}$$

Solutions to problems

P13A.1 (a) There is no configuration in which the molecules are distributed evenly over the states and which, at the same time, satisfies the constraint that the energy is 5ε.

(b) The energy of a configuration is $E/\varepsilon = N_1 + 2N_2 + 3N_3 \ldots$, and the weight of a configuration is given by [13A.1–533], $\mathcal{W} = N!/(N_0!N_1!N_2!\ldots)$. The configurations satisfying the total energy constraint $E = 5\varepsilon$ are

N_0	N_1	N_2	N_3	N_4	N_5	$\mathcal{W}$
4	0	0	0	0	1	5
3	1	0	0	1	0	20
3	0	1	1	0	0	20
2	2	0	1	0	0	30
2	1	2	0	0	0	30
1	3	1	0	0	0	20
0	5	0	0	0	0	1

Hence the most probable configurations are $\boxed{\{2,2,0,1,0,0\} \text{ and } \{2,1,2,0,0,0\}}$.

P13A.3 The energy of a configuration is $E/\varepsilon = N_1 + 2N_2 + 3N_3 \ldots$, and the weight of a configuration is given by [13A.1–533], $\mathcal{W} = N!/(N_0!N_1!N_2!\ldots)$. There are a very large number of possible configurations, but one way of selecting an interesting and diverse set of these is to consider configurations in which N_0 is 19, then 18, then 16 and so on.

N_0	N_1	N_2	N_3	N_4	N_5	N_6	N_7	N_8	N_9	N_{10}	$\mathcal{W}$
19	0	0	0	0	0	0	0	0	0	1	20
18	1	0	0	0	0	0	0	0	1	0	380
17	1	1	0	0	0	0	0	1	0	0	6 840
16	1	1	1	1	0	0	0	0	0	0	116 280
15	2	2	1	0	0	0	0	0	0	0	465 120
14	3	2	1	0	0	0	0	0	0	0	2 325 600
13	4	3	0	0	0	0	0	0	0	0	2 713 200
12	6	2	0	0	0	0	0	0	0	0	7 054 320
11	8	1	0	0	0	0	0	0	0	0	1 511 640
10	10	0	0	0	0	0	0	0	0	0	184 756

Of the ones listed in the table the configuration with the greatest weight is $\boxed{\{12,6,2,0,0,0,0,0,0,0,0\}}$.

The Boltzmann population ratio is given by [13A.13a–536], $N_i/N_0 = e^{-\beta\varepsilon_i}$, where it is assumed that $\varepsilon_0 = 0$. It follows that $\ln N_i/N_0 = -\beta\varepsilon_i = -i\beta\varepsilon$, and hence $\beta\varepsilon = (\ln N_i/N_0)/(-i)$. For $i = 1$, $\beta\varepsilon = (\ln 6/12)/(-1) = 0.693$; for

$i = 2$, $\beta\varepsilon = (\ln 2/12)/(-2) = 0.896$. Taking an average of these two values gives $\beta\varepsilon = 0.795$ and hence $\boxed{T = \varepsilon/(0.795k)}$.

With this value for the temperature the populations predicted by the Boltzmann distribution are

$$N_i/N_0 = e^{-\beta\varepsilon_i} = e^{-i\beta\varepsilon} = e^{-0.795\times i}$$

Therefore $N_1/N_0 = 0.452$, $N_2/N_0 = 0.204$, $N_3/N_0 = 0.092$. For the most probable distribution given above these ratios are $N_1/N_0 = 0.500$, $N_2/N_0 = 0.167$, $N_3/N_0 = 0$, which are roughly comparable.

P13A.5 The Boltzmann population ratio for degenerate energy levels is given by [13A.13b–536], $N_i/N_j = (g_i/g_j)e^{-\beta(\varepsilon_i-\varepsilon_j)}$. Taking logarithms gives

$$\ln(N_i/N_j) = \ln(g_i/g_j) - \beta(\varepsilon_i - \varepsilon_j) \quad \text{hence} \quad \beta = -\ln[(N_i/N_j)(g_j/g_i)]/\Delta\varepsilon$$

where $\Delta\varepsilon = (\varepsilon_i - \varepsilon_j)$. Substituting $\beta = 1/(kT)$ and rearranging for T gives

$$\begin{aligned} T &= -\frac{\Delta\varepsilon}{k\ln[(N_1/N_0)(g_0/g_1)]} = -\frac{hc\tilde{\nu}}{k\ln[(N_1/N_0)(g_0/g_1)]} \\ &= -\frac{(6.6261\times10^{-34}\,\text{J s})\times(2.9979\times10^{10}\,\text{cm s}^{-1})\times(450\,\text{cm}^{-1})}{(1.3806\times10^{-23}\,\text{J K}^{-1})\times\ln[(30\%/70\%)\times(2/4)]} \\ &= \boxed{420\,\text{K}} \end{aligned}$$

The populations of the electronic states do not correspond to the translational temperature; therefore the electronic states are $\boxed{\text{not}}$ in equilibrium with the translational states.

P13A.7 The Boltzmann population ratio is given by [13A.13a–536], $N_i/N_j = e^{-\beta(\varepsilon_i-\varepsilon_j)}$. The energy ε_i is interpreted as that at height h, and ε_j is interpreted as that at height 0, giving

$$N(h)/N(0) = e^{-\beta(mgh - mg\times0)} = e^{-\beta mgh}$$

From the perfect gas law, $pV = nRT$, and because $N \propto n$, it follows that $p \propto N$, and therefore $p(h)/p_0 = N(h)/N(0)$. Because $\beta = 1/(kT)$, $m = M/N_A$, and $k = R/N_A$ it follows that

$$\beta mgh = \frac{mgh}{kT} = \frac{(M/N_A)gh}{(R/N_A)T} = \frac{Mgh}{RT} = \frac{h}{H}$$

where $H = RT/Mg$ is used. Hence

$$p(h)/p_0 = e^{-h/H} \quad \text{and so} \quad p(h) = p_0 e^{-h/H}$$

From the perfect gas law $\mathcal{N} = N/V \propto p$, therefore for O_2

$$\begin{aligned} \mathcal{N}(h)/\mathcal{N}_0 &= \exp\left(-\frac{Mgh}{RT}\right) \\ &= \exp\left(-\frac{(32.00\times10^{-3}\,\text{kg mol}^{-1})\times(9.807\,\text{m s}^{-2})\times(8.0\times10^3\,\text{m})}{(8.3145\,\text{J K}^{-1}\,\text{mol}^{-1})\times(298\,\text{K})}\right) \\ &= \boxed{0.36} \end{aligned}$$

and for H_2O

$$\frac{\mathcal{N}(h)}{\mathcal{N}_0} = \exp\left(-\frac{Mgh}{RT}\right)$$
$$= \exp\left(-\frac{(18.02\times10^{-3}\,\text{kg mol}^{-1})\times(9.807\,\text{m s}^{-2})\times(8.0\times10^{3}\,\text{m})}{(8.3145\,\text{J K}^{-1}\,\text{mol}^{-1})\times(298\,\text{K})}\right)$$
$$= \boxed{0.57}$$

In these calculations the temperature is taken as 298 K and is assumed to be constant with height, which is in fact not the case.

13B Partition functions

Answer to discussion questions

D13B.1 The molecular partition is roughly equal to the number of physically distinct states that are thermally accessible to a molecule at a given temperature. At low temperatures, very little energy is available, so only the lowest-energy states of a molecule are accessible; therefore, as the temperature approaches absolute zero, the partition function approaches the degeneracy of the ground state of the molecule. The higher the temperature, the greater the Boltzmann weighting factor $e^{-\beta\varepsilon}$, and the more accessible a state of energy ε becomes. Thus the number of accessible states increases with temperature.

D13B.3 As discussed in Section 13B.2(b) on page 542, the symmetry number is the number of indistinguishable orientations that the molecule can be rotated into. This factor is needed in order to avoid counting contributions to the rotational partition function which are forbidden by symmetry considerations arising from the effects of nuclear spin (Section 11B.4 on page 439).

If the partition function is computed term-by-term then the symmetry number is not needed because those terms which are forbidden are simply omitted. However, in the high-temperature limit in which many terms are included, it is convenient to allow all terms to contribute to the sum and then compensate for those which should not have been included by division by the symmetry number.

Solutions to exercises

E13B.1(a) (i) The thermal wavelength is defined in [13B.7–541], $\Lambda = h/(2\pi mkT)^{1/2}$. Because the mass of a molecule m is $m = M/N_A$ and $k = R/N_A$ it follows that

$$\Lambda = \frac{h}{[2\pi(M/N_A)(R/N_A)T]^{1/2}} = \frac{hN_A}{(2\pi MRT)^{1/2}}$$

$$\Lambda(300\text{ K}) = \frac{(6.6261\times10^{-34}\text{ J s})\times(6.0221\times10^{23}\text{ mol}^{-1})}{[2\pi\times(0.150\text{ kg mol}^{-1})\times(8.3145\text{ J K}^{-1}\text{ mol}^{-1})\times(300\text{ K})]^{1/2}}$$
$$= 8.22...\times10^{-12}\text{ m} = \boxed{8.23\times10^{-12}\text{ m}}$$

Similarly, $\Lambda(3000\text{ K}) = \boxed{2.60\times10^{-12}\text{ m}}$

(ii) The translational partition function in three dimensions is given by [13B.10b–*541*], $q^{\text{T}} = V/\Lambda^3$.

$$q^{\text{T}}(300\text{ K}) = (1.00\times10^{-6}\text{ m}^3)/(8.22...\times10^{-12}\text{ m})^3 = \boxed{1.78\times10^{27}}$$
$$q^{\text{T}}(3000\text{ K}) = \boxed{5.67\times10^{28}}$$

E13B.2(a) The translational partition function in three dimensions is given by [13B.10b–*541*], $q^{\text{T}} = V/\Lambda^3$, where Λ is the thermal wavelength defined in [13B.7–*541*], $\Lambda = h/(2\pi mkT)^{1/2}$.

$$\frac{q^{\text{T}}_{\text{H}_2}}{q^{\text{T}}_{\text{He}}} = \frac{V/\Lambda_{\text{H}_2}{}^3}{V/\Lambda_{\text{He}}{}^3} = \left(\frac{\Lambda_{\text{He}}}{\Lambda_{\text{H}_2}}\right)^3 = \left(\frac{h/(2\pi m_{\text{He}}kT)^{1/2}}{h/(2\pi m_{\text{H}_2}kT)^{1/2}}\right)^3 = \left(\frac{m_{\text{H}_2}}{m_{\text{He}}}\right)^{3/2}$$

Because the mass of a molecule m is $m = M/N_{\text{A}}$ it follows that

$$\frac{q^{\text{T}}_{\text{H}_2}}{q^{\text{T}}_{\text{He}}} = \left(\frac{M_{\text{H}_2}}{M_{\text{He}}}\right)^{3/2} = \left(\frac{2\times1.0079\text{ g mol}^{-1}}{4.00\text{ g mol}^{-1}}\right)^{3/2} = \boxed{0.358}$$

E13B.3(a) The rotational partition function of a symmetric linear rotor is given by [13B.13a–*544*], $q^{\text{R}} = kT/(2hc\tilde{B})$, where the rotational constant is defined in [11B.7–*432*], $\tilde{B} = \hbar/(4\pi cI)$. The moment of inertia of a diatomic is $I = \mu R^2$, where R is the bond length and $\mu = m_{\text{A}}m_{\text{B}}/(m_{\text{A}}+m_{\text{B}})$. For a homonuclear diatomic $m_{\text{A}} = m_{\text{B}}$ so it follows that $\mu = m_{\text{B}}/2$. Using $m = M/N_{\text{A}}$, this becomes $\mu = M_{\text{B}}/2N_{\text{A}}$.

$$I = \mu R^2 = \frac{M_{\text{B}}R^2}{2N_{\text{A}}} = \frac{(0.01600\text{ kg mol}^{-1})\times(120.75\times10^{-12}\text{ m})^2}{2\times(6.0221\times10^{23}\text{ mol}^{-1})}$$
$$= 1.93...\times10^{-46}\text{ kg m}^2$$
$$q^{\text{R}} = \frac{kT}{2hc\tilde{B}} = \left(\frac{kT}{2hc}\right)\left(\frac{4\pi cI}{\hbar}\right) = \left(\frac{kT}{4\pi c\hbar}\right)\left(\frac{4\pi cI}{\hbar}\right) = \frac{kT}{\hbar^2}I$$
$$= \frac{(1.3806\times10^{-23}\text{ J K}^{-1})\times(300\text{ K})}{(1.0546\times10^{-34}\text{ J s})^2}\times(1.93...\times10^{-46}\text{ kg m}^2) = \boxed{72.1}$$

E13B.4(a) The rotational partition function of a non-linear rotor is given by [13B.14–*545*], $q^{\text{R}} = (1/\sigma)(kT/hc)^{3/2}(\pi/\tilde{A}\tilde{B}\tilde{C})^{1/2}$, where σ is the symmetry number. NOF is not centro-symmetric so $\sigma = 1$.

(i) At 25 °C which is 298.15 K

$$q^{\mathrm{R}} = \left(\frac{kT}{hc}\right)^{3/2}\left(\frac{\pi}{\tilde{A}\tilde{B}\tilde{C}}\right)^{1/2}$$
$$= \left(\frac{(1.3806\times10^{-23}\,\mathrm{J\,K^{-1}})\times(298.15\,\mathrm{K})}{(6.6261\times10^{-34}\,\mathrm{J\,s})\times(2.9979\times10^{10}\,\mathrm{cm\,s^{-1}})}\right)^{3/2}$$
$$\times\left(\frac{\pi}{(3.1752\,\mathrm{cm^{-1}})\times(0.3951\,\mathrm{cm^{-1}})\times(0.3505\,\mathrm{cm^{-1}})}\right)^{1/2}$$
$$= \boxed{7.97\times10^{3}}$$

(ii) At 100 °C which is 373.15 K

$$q^{\mathrm{R}} = \left(\frac{(1.3806\times10^{-23}\,\mathrm{J\,K^{-1}})\times(373.15\,\mathrm{K})}{(6.6261\times10^{-34}\,\mathrm{J\,s})\times(2.9979\times10^{10}\,\mathrm{cm\,s^{-1}})}\right)^{3/2}$$
$$\times\left(\frac{\pi}{(3.1752\,\mathrm{cm^{-1}})\times(0.3951\,\mathrm{cm^{-1}})\times(0.3505\,\mathrm{cm^{-1}})}\right)^{1/2}$$
$$= \boxed{1.12\times10^{4}}$$

E13B.5(a) The rotational partition function of a heteronuclear diatomic is given by [13B.11–*542*], $q^{\mathrm{R}} = \sum_J(2J+1)e^{-\beta hc\tilde{B}J(J+1)}$. This is evaluated explicitly by summing successive terms until they become too small to affect the result to a given level of precision. The partition function in the high-temperature limit is given by [13B.12a–*543*], $q^{\mathrm{R}} = kT/hc\tilde{B}$. For the data given it follows that

$$q^{\mathrm{R}} = \frac{k\times T}{hc\tilde{B}} = \frac{(1.3806\times10^{-23}\,\mathrm{J\,K^{-1}})\times T}{(6.6261\times10^{-34}\,\mathrm{J\,s})\times(2.9979\times10^{10}\,\mathrm{cm\,s^{-1}})\times(1.931\,\mathrm{cm^{-1}})}$$
$$= (0.359...\,\mathrm{K^{-1}})\times T$$

The values of q^{R} computed in these two different ways are compared in Fig. 13.1. The high temperature limit becomes accurate to within 5 % of the exact solution at around $\boxed{18\,\mathrm{K}}$.

E13B.6(a) The partition function is given by [13B.1b–*538*], $q^{\mathrm{R}} = \sum_J g_J e^{-\beta\varepsilon_J}$, where the degeneracy is given as $g_J = (2J+1)^2$, as explained in Section 11B.1(c) on page 434, and ε_J is given by [13B.1b–*538*], $\varepsilon_J = hc\tilde{B}J(J+1)$. This is evaluated explicitly by summing successive terms until they become too small to affect the result to a given level of precision.

The partition function in the high-temperature limit is given by [13B.12b–*544*], $q^{\mathrm{R}} = (kT/hc)^{3/2}(\pi/\tilde{A}\tilde{B}\tilde{C})^{1/2} = \pi^{1/2}(kT/hc\tilde{B})^{3/2}$, because for a spherical rotor $\tilde{B} = \tilde{A} = \tilde{C}$. Ignoring the role of the nuclear spin means that all J states are

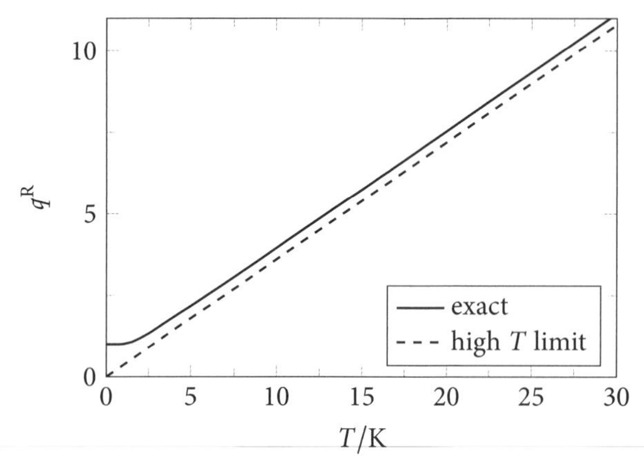

Figure 13.1

accessible and have equal weight. For the data given it follows that

$$q^{R} = \pi^{1/2}\left(\frac{k \times T}{hc\tilde{B}}\right)^{3/2}$$

$$= \pi^{1/2}\left(\frac{(1.3806 \times 10^{-23}\,\mathrm{J\,K^{-1}})}{(6.6261 \times 10^{-34}\,\mathrm{J\,s})\times(2.9979 \times 10^{10}\,\mathrm{cm\,s^{-1}})\times(5.241\,\mathrm{cm^{-1}})}\right)^{3/2} \times T^{3/2}$$

$$= (0.0855...\,\mathrm{K^{-3/2}}) \times T^{3/2}$$

The values of q^{R} computed in these two different ways are compared in Fig. 13.2. The high temperature limit becomes accurate to within 5 % of the exact solution at around $\boxed{37\,\mathrm{K}}$.

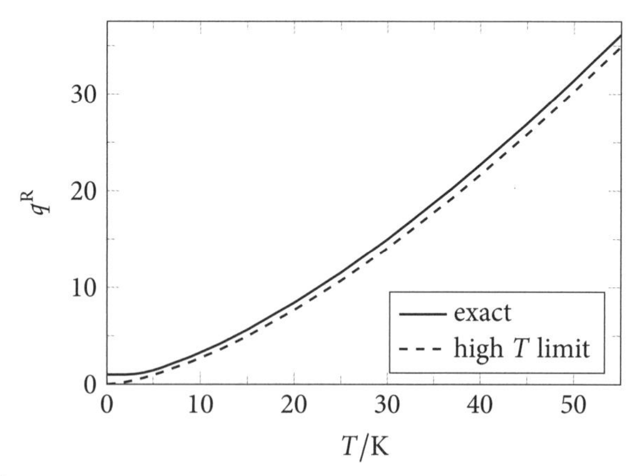

Figure 13.2

E13B.7(a) (i) CO is a heteronuclear diatomic and so has $\boxed{\sigma = 1}$.

(ii) O_2 is a homonuclear diatomic and so has $\boxed{\sigma = 2}$.

(iii) H_2S has a twofold rotational axis bisecting the H–S–H angle; rotation about this axis interchanges two identical hydrogen atoms, therefore $\boxed{\sigma = 2}$.

(iv) SiH_4 is tetrahedral and so has the same symmetry number as CH_4: $\boxed{\sigma = 12}$.

(v) $CHCl_3$ has a threefold rotational axis along the C–H bond; rotation about this axis interchanges three identical chlorine atoms, therefore $\boxed{\sigma = 3}$.

E13B.8(a) The rotational partition function of an asymmetric rotor is given by [13B.14–*545*], $q^{R} = (1/\sigma)(kT/hc)^{3/2}(\pi/\tilde{A}\tilde{B}\tilde{C})^{1/2}$, where σ is the symmetry number. For a molecule with high symmetry the simplest was to determine the symmetry number is to count the total number of rotational symmetry operations, C_n, listed in the character table of the point group to which the molecule belongs, which in this case is D_{2h}. For this group the rotational operations are (E, C_2^x, C_2^y, C_2^z), and therefore $\sigma = 4$.

$$q^{R} = \frac{1}{4}\left(\frac{kT}{hc}\right)^{3/2}\left(\frac{\pi}{\tilde{A}\tilde{B}\tilde{C}}\right)^{1/2}$$
$$= \frac{1}{4} \times \left(\frac{(1.3806 \times 10^{-23}\,\mathrm{J\,K^{-1}}) \times (298.15\,\mathrm{K})}{(6.6261 \times 10^{-34}\,\mathrm{J\,s}) \times (2.9979 \times 10^{10}\,\mathrm{cm\,s^{-1}})}\right)^{3/2}$$
$$\times \left(\frac{\pi}{(4.828\,\mathrm{cm^{-1}}) \times (1.0012\,\mathrm{cm^{-1}}) \times (0.8282\,\mathrm{cm^{-1}})}\right)^{1/2}$$
$$= \boxed{660.6}$$

E13B.9(a) The vibrational partition function is given by [13B.15–*546*], $q^{V} = 1/(1-e^{-\beta hc\tilde{\nu}})$, where $\beta = 1/kT$. The high-temperature approximation is given by [13B.16–*547*], $q^{V} \approx kT/hc\tilde{\nu}$.

$$\frac{k \times T}{hc\tilde{\nu}} = \frac{(1.3806 \times 10^{-23}\,\mathrm{J\,K^{-1}}) \times T}{(6.6261 \times 10^{-34}\,\mathrm{J\,s}) \times (2.9979 \times 10^{10}\,\mathrm{cm\,s^{-1}}) \times (323.2\,\mathrm{cm^{-1}})}$$
$$= (2.15... \times 10^{-3}\,\mathrm{K^{-1}}) \times T$$

The values of q^{V} computed using these two different expressions are compared in Fig. 13.3. The high temperature limit becomes accurate to within 5 % of the exact solution at $\boxed{4500\,\mathrm{K}}$.

E13B.10(a) The vibrational partition function for each mode is given by [13B.15–*546*], $q^{V} = 1/(1-e^{-\beta hc\tilde{\nu}})$, where $\beta = 1/kT$. The overall vibrational partition function is the product of the partition functions of the individual modes; the bend is included twice as it is doubly degenerate.

$$hc\beta = \frac{(6.6261 \times 10^{-34}\,\mathrm{J\,s}) \times (2.9979 \times 10^{10}\,\mathrm{cm\,s^{-1}})}{(1.3806 \times 10^{-23}\,\mathrm{J\,K^{-1}}) \times (500\,\mathrm{K})}$$
$$= 2.87... \times 10^{-3}\,\mathrm{cm}$$

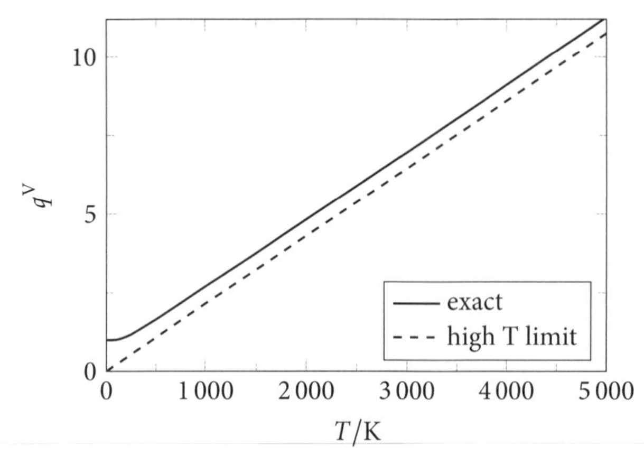

Figure 13.3

$$q_1^{\mathrm{V}} = \left(1 - \mathrm{e}^{-hc\beta\tilde{\nu}_1}\right)^{-1}$$
$$= \left(1 - \mathrm{e}^{-(2.87...\times10^{-3}\ \mathrm{cm})\times(658\ \mathrm{cm}^{-1})}\right)^{-1} = 1.17...$$

Similarly

$$q_2^{\mathrm{V}} = \left(1 - \mathrm{e}^{-(2.87...\times10^{-3}\ \mathrm{cm})\times(397\ \mathrm{cm}^{-1})}\right)^{-1} = 1.46...$$
$$q_3^{\mathrm{V}} = \left(1 - \mathrm{e}^{-(2.87...\times10^{-3}\ \mathrm{cm})\times(1535\ \mathrm{cm}^{-1})}\right)^{-1} = 1.01...$$

$$q^{\mathrm{V}} = q_1^{\mathrm{V}} \times (q_2^{\mathrm{V}})^2 \times q_3^{\mathrm{V}} = (1.17...) \times (1.46...)^2 \times (1.01...) = \boxed{2.57}$$

E13B.11(a) The vibrational partition function for each mode is given by [13B.15–*546*], $q^{\mathrm{V}} = 1/(1-\mathrm{e}^{-\beta hc\tilde{\nu}})$, where $\beta = 1/kT$. The overall vibrational partition function is the product of the partition functions of the individual modes, taking into account the stated degeneracies.

$$hc\beta = \frac{(6.6261\times10^{-34}\ \mathrm{J\,s})\times(2.9979\times10^{10}\ \mathrm{cm\,s^{-1}})}{(1.3806\times10^{-23}\ \mathrm{J\,K^{-1}})\times(500\ \mathrm{K})}$$
$$= 2.87...\times10^{-3}\ \mathrm{cm}$$

$$q_1^{\mathrm{V}} = \left(1 - \mathrm{e}^{-hc\beta\tilde{\nu}_1}\right)^{-1}$$
$$= \left(1 - \mathrm{e}^{-(2.87...\times10^{-3}\ \mathrm{cm})\times(459\ \mathrm{cm}^{-1})}\right)^{-1} = 1.36...$$

Similarly

$$q_2^{\mathrm{V}} = \left(1 - \mathrm{e}^{-(2.87...\times10^{-3}\ \mathrm{cm})\times(217\ \mathrm{cm}^{-1})}\right)^{-1} = 2.15...$$
$$q_3^{\mathrm{V}} = \left(1 - \mathrm{e}^{-(2.87...\times10^{-3}\ \mathrm{cm})\times(776\ \mathrm{cm}^{-1})}\right)^{-1} = 1.12...$$
$$q_4^{\mathrm{V}} = \left(1 - \mathrm{e}^{-(2.87...\times10^{-3}\ \mathrm{cm})\times(314\ \mathrm{cm}^{-1})}\right)^{-1} = 1.68...$$

$$q^{\mathrm{V}} = q_1^{\mathrm{V}} \times (q_2^{\mathrm{V}})^2 \times (q_3^{\mathrm{V}})^3 \times (q_4^{\mathrm{V}})^3$$
$$= (1.36...) \times (2.15...)^2 \times (1.12...)^3 \times (1.68...)^3 = \boxed{42.1}.$$

E13B.12(a) The partition function is given by [13B.1b–*538*], $q = \sum_i g_i e^{-\beta\varepsilon_i}$, where g_i is degeneracy and the corresponding energy is given as $\varepsilon_i = hc\tilde{\nu}_i$. At $T = 1900$ K

$$\beta hc = \frac{(6.6261 \times 10^{-34}\,\mathrm{J\,s}) \times (2.9979 \times 10^{10}\,\mathrm{cm\,s^{-1}})}{(1.3806 \times 10^{-23}\,\mathrm{J\,K^{-1}}) \times (1900\,\mathrm{K})} = 7.57... \times 10^{-4}\,\mathrm{cm}$$

Therefore the electronic partition function is

$$q^{\mathrm{E}} = g_0 + g_1 \times e^{-\beta\varepsilon_1} + g_2 \times e^{-\beta\varepsilon_2}$$
$$= 4 + 1 \times e^{-(7.57...\times10^{-4}\,\mathrm{cm})\times(2500\,\mathrm{cm^{-1}})} + 2 \times e^{-(7.57...\times10^{-4}\,\mathrm{cm})\times(3500\,\mathrm{cm^{-1}})}$$
$$= 4 + 0.150... + 0.141... = \boxed{4.291} \qquad (13.1)$$

The population of level i with degeneracy g_i is $N_i = (Ng_i/q)e^{-\beta\varepsilon_i}$, therefore the relative populations of the levels are proportional to $g_i e^{-\beta\varepsilon_i}$, which are the terms in eqn 13.1. Thus the populations, relative the ground state are

$$N_0/N_0 : N_1/N_0 : N_2/N_0 = 4/4 : (0.150.../4) : (0.141.../4)$$
$$= \boxed{1 : 0.0376 : 0.0353}$$

Solutions to problems

P13B.1 The partition function is given by [13B.1b–*538*], $q = \sum_i g_i e^{-\beta\varepsilon_i}$, where g_i is degeneracy and the corresponding energy is ε_i. Therefore the partition function is

$$q = 1 \times e^{-\beta 0} + 1 \times e^{-\beta\varepsilon} + 1 \times e^{-\beta(2\varepsilon)} = 1 + e^{-\varepsilon/kT} + e^{-2\varepsilon/kT}$$

This is plotted in Fig 13.4. As expected, the partition function rises from a value of 1 at low temperatures, where only the ground state is occupied, and approaches a value of 3 at high temperatures, when all three states are nearly equally populated.

P13B.3 As discussed in Section 11C.3(a) on page 444, the Morse oscillator has a finite number of bound levels between the ground level and the dissociation limit. The number of these is found by noting that E_v reaches a maximum value at the dissociation limit, and therefore this limit is found by solving $\mathrm{d}E_v/\mathrm{d}v = 0$

$$\frac{\mathrm{d}}{\mathrm{d}v}\left[(v + \tfrac{1}{2})hc\tilde{\nu} - (v + \tfrac{1}{2})^2 hcx_e\tilde{\nu}\right] = hc\tilde{\nu} - 2(v + \tfrac{1}{2})hcx_e\tilde{\nu}$$

$$\text{solving} \quad 0 = hc\tilde{\nu} - 2(v_{\max} + \tfrac{1}{2})hcx_e\tilde{\nu} \quad \text{gives} \quad v_{\max} = 1/2x_e - \tfrac{1}{2}$$

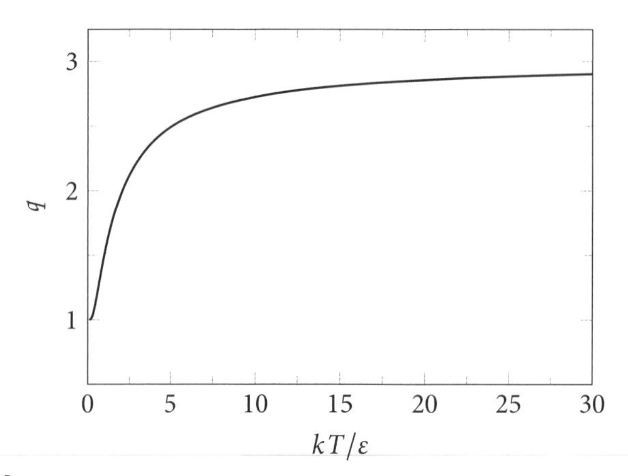

Figure 13.4

The energy of the lowest state is $E_0 = \frac{1}{2}hc\tilde{\nu} - \frac{1}{4}hcx_e\tilde{\nu}$, therefore the energies used to evaluate the partition function are

$$\begin{aligned}E'_v = E_v - E_0 &= \left[(v+\tfrac{1}{2})hc\tilde{\nu} - (v+\tfrac{1}{2})^2 hcx_e\tilde{\nu}\right] - \left[\tfrac{1}{2}hc\tilde{\nu} - \tfrac{1}{4}hcx_e\tilde{\nu}\right]\\ &= vhc\tilde{\nu} - (v^2+v)hcx_e\tilde{\nu}\end{aligned}$$

The partition function is evaluated from the sum

$$q_M^V = \sum_{v=0}^{v_{max}} e^{-(vhc\tilde{\nu}-(v^2+v)hcx_e\tilde{\nu})/kT}$$

Defining the characteristic vibrational temperature as $\theta^V = hc\tilde{\nu}/k$ gives

$$q_M^V = \sum_{v=0}^{v_{max}} e^{-(v-(v^2+v)x_e)\theta^V/T}$$

For the harmonic oscillator the partition is given by the exact expression [13B.15–546], $q_{HO}^V = (1-e^{-\theta^V/T})^{-1}$.

Figure 13.5 compares the partition functions for various values of x_e with that for the harmonic case. For the smallest value of x_e the partition function is initially larger than that for the harmonic oscillator. This can be attributed to more energy levels contributing at these temperatures as they are closer in energy for the Morse oscillator than for the harmonic case. However, at higher temperatures the partition function for the Morse oscillator starts to level off because there are a finite number of levels, whereas for the harmonic case the partition function continues without limit as there are an infinite number of levels.

This behaviour is even more pronounced for $x_e = 0.05$ and $x_e = 0.10$. In these two cases v_{max} is 10 and 5, respectively, and these values set the limiting high-temperature value of the partition function.

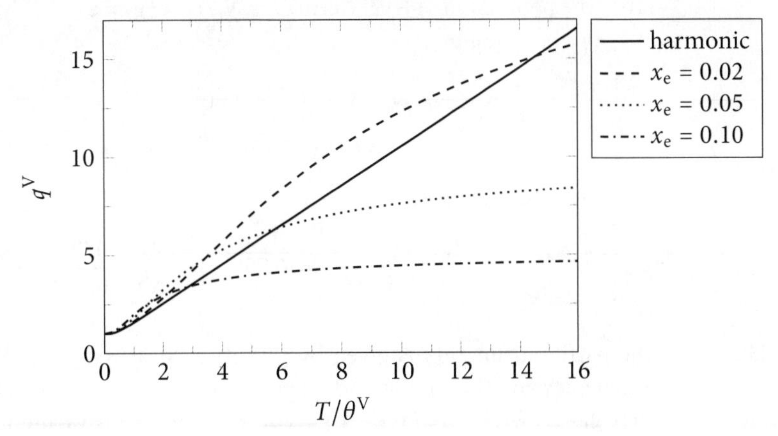

Figure 13.5

P13B.5 The partition function is given by [13B.1b–538], $q = \sum_i g_i e^{-\beta\varepsilon_i}$, where g_i is degeneracy and the corresponding energy is given as $\varepsilon_i = hc\tilde{\nu}_i$, and $\beta = 1/kT$. Therefore the electronic partition function is

$$q^E = g_0 + g_1 \times e^{-hc\tilde{\nu}_1/kT} + g_2 \times e^{-hc\tilde{\nu}_2/kT} + g_3 \times e^{-hc\tilde{\nu}_3/kT}$$

(a) (i) At T = 298 K

$$hc\beta = \frac{(6.6261 \times 10^{-34}\,\mathrm{J\,s}) \times (2.9979 \times 10^{10}\,\mathrm{cm\,s^{-1}})}{(1.3806 \times 10^{-23}\,\mathrm{J\,K^{-1}}) \times (298\,\mathrm{K})} = 4.82... \times 10^{-3}\,\mathrm{cm}$$

$$\begin{aligned} q^E &= 5 + 1 \times e^{-(4.82...\times 10^{-3}\,\mathrm{cm})\times(4707\,\mathrm{cm^{-1}})} \\ &+ 3 \times e^{-(4.82...\times 10^{-3}\,\mathrm{cm})\times(4751\,\mathrm{cm^{-1}})} \\ &+ 5 \times e^{-(4.82...\times 10^{-3}\,\mathrm{cm})\times(10559\,\mathrm{cm^{-1}})} \\ &= 5 + 1.34... \times 10^{-10} + 3.27... \times 10^{-10} + 3.61... \times 10^{-22} \\ &= \boxed{5.00} \end{aligned}$$

(ii) At T = 5000 K

$$\begin{aligned} hc\beta &= \frac{(6.6261 \times 10^{-34}\,\mathrm{J\,s}) \times (2.9979 \times 10^{10}\,\mathrm{cm\,s^{-1}})}{(1.3806 \times 10^{-23}\,\mathrm{J\,K^{-1}}) \times (5000\,\mathrm{K})} \\ &= 2.87... \times 10^{-4}\,\mathrm{cm} \end{aligned}$$

$$\begin{aligned} q^E &= 5 + 1 \times e^{-(2.87...\times 10^{-4}\,\mathrm{cm})\times(4707\,\mathrm{cm^{-1}})} \\ &+ 3 \times e^{-(2.87...\times 10^{-4}\,\mathrm{cm})\times(4751\,\mathrm{cm^{-1}})} \\ &+ 5 \times e^{-(2.87...\times 10^{-4}\,\mathrm{cm})\times(10559\,\mathrm{cm^{-1}})} \\ &= 5 + 0.258... + 0.764... + 0.239... = 6.262... = \boxed{6.262} \end{aligned}$$

(b) The population of level (term) i is $N_i/N = (g_i \times e^{-hc\tilde{\nu}_i/kT})/q^E$

(i) $T = 298$ K

$$N_0/N = 5/5.00... = \boxed{1.00}$$

$$N_2/N = (3.26... \times 10^{-10})/5.00... = \boxed{6.54 \times 10^{-11}}$$

(ii) $T = 5000$ K

$$\frac{N_0}{N} = \frac{5}{6.262...} = \boxed{0.798} \qquad \frac{N_2}{N} = \frac{0.764...}{6.262...} = \boxed{0.122}$$

P13B.7 The partition function is given by [13B.1b–538], $q = \sum_i g_i e^{-\beta\varepsilon_i}$, where g_i is degeneracy and the corresponding energy is given as $\varepsilon_i = hc\tilde{\nu}_i$, and $\beta = 1/kT$. Here $g_i = 2J+1$, where J is the right subscript in the term symbol. At $T = 298$ K

$$hc\beta = \frac{(6.6261 \times 10^{-34}\,\text{J s}) \times (2.9979 \times 10^{10}\,\text{cm s}^{-1})}{(1.3806 \times 10^{-23}\,\text{J K}^{-1}) \times (298\,\text{K})} = 4.82... \times 10^{-3}\,\text{cm}$$

Therefore the electronic partition function is

$$\begin{aligned} q^E &= 1 + 3 \times e^{-(4.82...\times10^{-3}\,\text{cm})\times(557.1\,\text{cm}^{-1})} \\ &\quad + 5 \times e^{-(4.82...\times10^{-3}\,\text{cm})\times(1410.0\,\text{cm}^{-1})} + 5 \times e^{-(4.82...\times10^{-3}\,\text{cm})\times(7125.3\,\text{cm}^{-1})} \\ &\quad + 1 \times e^{-(4.82...\times10^{-3}\,\text{cm})\times(16367.3\,\text{cm}^{-1})} \\ &= 1 + 0.203... + 5.52... \times 10^{-3} + 5.72... \times 10^{-15} + 4.78... \times 10^{-35} \\ &= \boxed{1.209} \end{aligned}$$

Similarly, at $T = 1000$ K

$$hc\beta = \frac{(6.6261 \times 10^{-34}\,\text{J s}) \times (2.9979 \times 10^{10}\,\text{cm s}^{-1})}{(1.3806 \times 10^{-23}\,\text{J K}^{-1}) \times (1000\,\text{K})} = 1.43... \times 10^{-3}\,\text{cm}$$

Therefore the electronic partition function is

$$\begin{aligned} q^E &= 1 + 3 \times e^{-(1.43...\times10^{-3}\,\text{cm})\times(557.1\,\text{cm}^{-1})} \\ &\quad + 5 \times e^{-(1.43...\times10^{-3}\,\text{cm})\times(1410.0\,\text{cm}^{-1})} + 5 \times e^{-(1.43...\times10^{-3}\,\text{cm})\times(7125.3\,\text{cm}^{-1})} \\ &\quad + 1 \times e^{-(1.43...\times10^{-3}\,\text{cm})\times(16367.3\,\text{cm}^{-1})} \\ &= 1 + 1.34... + 0.657... + 1.76... \times 10^{-4} + 5.92... \times 10^{-11} \\ &= \boxed{3.003} \end{aligned}$$

P13B.9 The partition function is given by [13B.1b–538], $q^R = \sum_i g_i e^{-\beta\varepsilon_i}$. Where g_i is degeneracy and $\varepsilon_i = hc\tilde{F}_i$. The rotational terms of a symmetric rotor is given by [11B.13a–433], $\tilde{F}(J,K) = \tilde{B}J(J+1) + (\tilde{A} - \tilde{B})K^2$, with $J = 0, 1, 2, ...$ and $K = 0, \pm1, ..., \pm J$. Thus, the partition function is

$$q^R = \sum_{J=0}^{\infty}(2J+1)e^{-hc\beta\tilde{B}J(J+1)}\left[\sum_{K=-J}^{+J} e^{-hc\beta(\tilde{A}-\tilde{B})K^2}\right]$$

Using mathematical software the terms are evaluated and summed until convergence is achieved to within the required precision.

In the high-temperature limit the partition function is given by [13B.12b–544], $q^{\mathrm{R}} = (kT/hc)^{3/2}(\pi/\tilde{A}\tilde{B}\tilde{C})^{1/2}$; for a symmetric rotor $\tilde{B} = \tilde{C}$, therefore $q^{\mathrm{R}} = (kT/hc)^{3/2}(\pi/\tilde{A}\tilde{B}^2)^{1/2}$. With the given data

$$q^{\mathrm{R}} = (kT/hc)^{3/2}(\pi/\tilde{A}\tilde{B}^2)^{1/2} = (1.02... \times 10\ \mathrm{K}) \times T^{3/2}$$

The two forms of the partition function are plotted in Fig. 13.6; the high temperature limit is accurate to within 5 % of the exact solution at $\boxed{4.5\ \mathrm{K}}$.

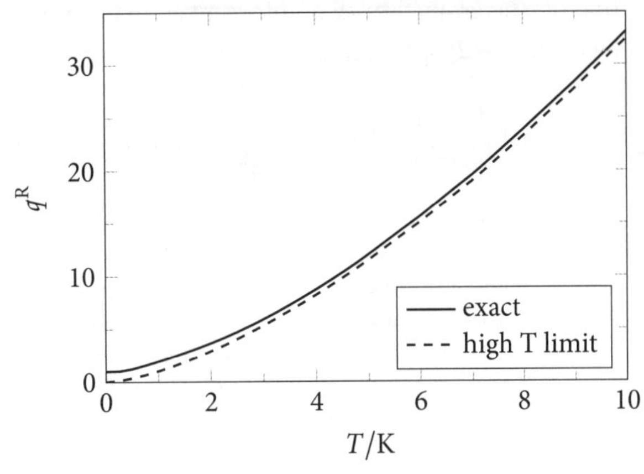

Figure 13.6

13C Molecular energies

Answers to discussion questions

D13C.1 The predictions of the equipartition theorem coincide with the energies computed using partition functions when the conditions are such that many energy levels have significant populations. Put another way, this means that many levels contribute to the partition function such that $q \gg 1$, or alternatively the spacing between the energy levels is small compared to kT.

Solutions to exercises

E13C.1(a) The mean energy of a molecule is given by [13C.2–549], $\langle\varepsilon\rangle = (1/q)\sum_i \varepsilon_i e^{-\beta\varepsilon_i}$, where $\varepsilon_i = hc\tilde{\nu}_i$, $\beta = 1/kT$, and q is the partition function given by [13A.11–

535], $q = \sum_i e^{-\beta\varepsilon_i}$. Therefore for the two-level system

$$\langle\varepsilon\rangle = \frac{0 + \varepsilon e^{-\beta\varepsilon}}{1 + e^{-\beta\varepsilon}} = \frac{\varepsilon}{e^{\beta\varepsilon} + 1} = \frac{hc\tilde{\nu}}{e^{hc\tilde{\nu}/kT} + 1}$$

$$= \frac{(6.6261 \times 10^{-34}\,\mathrm{J\,s}) \times (2.9979 \times 10^{10}\,\mathrm{cm\,s^{-1}}) \times (500\,\mathrm{cm^{-1}})}{e^{\frac{(6.6261\times10^{-34}\,\mathrm{J\,s})\times(2.9979\times10^{10}\,\mathrm{cm\,s^{-1}})\times(500\,\mathrm{cm^{-1}})}{(1.3806\times10^{-23}\,\mathrm{J\,K^{-1}})\times(298\,\mathrm{K})}} + 1}$$

$$= \boxed{8.15 \times 10^{-22}\,\mathrm{J}}$$

E13C.2(a) The mean molecular energy is given by [13C.4a–549], $\langle\varepsilon\rangle = -(1/q)(\partial q/\partial\beta)_V$, where $\beta = 1/kT$ and q is the partition function. The rotational partition function of a heteronuclear diatomic is given in terms of the rotational constant $\tilde{B}$ by [13B.11–542], $q^{\mathrm{R}} = \sum_J (2J + 1)e^{-\beta hc\tilde{B}J(J+1)}$.

$$\langle\varepsilon^{\mathrm{R}}\rangle = -\frac{1}{q^{\mathrm{R}}}\left(\frac{\partial q^{\mathrm{R}}}{\partial\beta}\right)_V = \frac{1}{q^{\mathrm{R}}}\sum_J hc\tilde{B}J(J+1)(2J+1)e^{-\beta hc\tilde{B}J(J+1)}$$

The terms of the sum above, and also of the sum needed to compute q^{R}, are evaluated and summed until the result has converged to the required precision. The equipartition value is $\langle\varepsilon^{\mathrm{R}}\rangle = kT$. These two expressions for the energy are plotted as a function of T in Fig. 13.7. The value from the equipartition theorem comes within 5 % of the exact value at $\boxed{19.6\,\mathrm{K}}$.

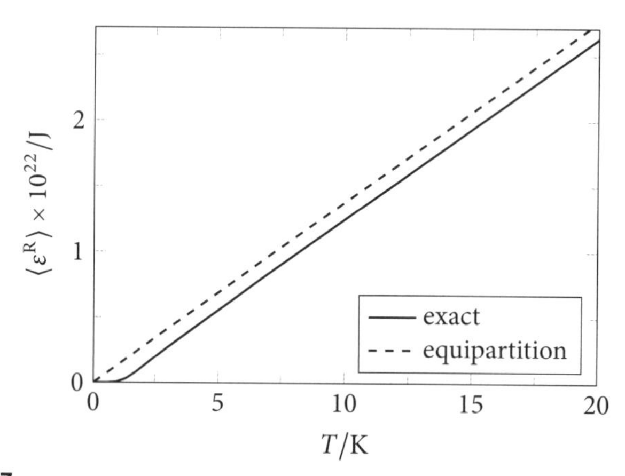

Figure 13.7

E13C.3(a) The mean molecular energy is given by [13C.4a–549], $\langle\varepsilon\rangle = -(1/q)(\partial q/\partial\beta)_V$, where $\beta = 1/kT$, and q is the partition function given by [13B.1b–538], $q^{\mathrm{R}} = \sum_J g_J e^{-\beta\varepsilon_J}$. The energy levels of a spherical rotor are given in [11B.8–432], $\varepsilon_J = hc\tilde{B}J(J+1)$ and, as is explained in Section 11B.1(c) on page 434, each has a degeneracy $g_J = (2J+1)^2$. It follows that

$$q^{\mathrm{R}} = \sum_J (2J+1)^2 e^{-\beta hc\tilde{B}J(J+1)}$$

$$\langle \varepsilon^{\mathrm{R}} \rangle = -\frac{1}{q^{\mathrm{R}}}\left(\frac{\partial q^{\mathrm{R}}}{\partial \beta}\right)_V = \frac{1}{q^{\mathrm{R}}}\sum_J hc\tilde{B}J(J+1)(2J+1)^2 e^{-\beta hc\tilde{B}J(J+1)}$$

The terms in the sum needed to compute q^{R} and $\langle \varepsilon^{\mathrm{R}} \rangle$ are evaluated and summed until the result has converged to the required precision. The equipartition value is $\langle \varepsilon^{\mathrm{R}} \rangle = \frac{3}{2}kT$, because for this non-linear molecule there are three rotational degrees of freedom. These two expressions for the energy are plotted as a function of T in Fig. 13.8. The value from the equipartition theorem comes within 5 % of the exact value at $\boxed{26.4\ \mathrm{K}}$.

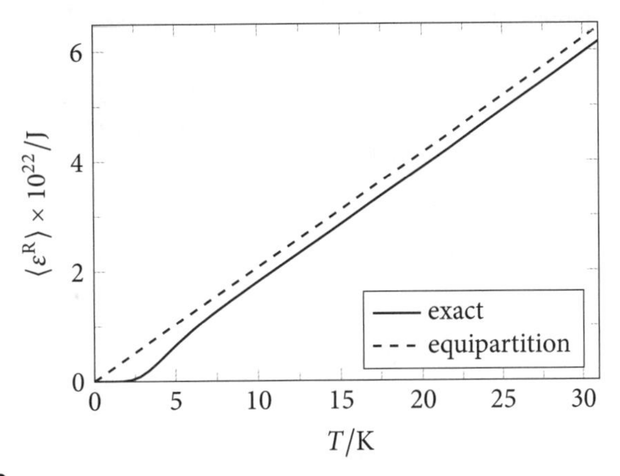

Figure 13.8

E13C.4(a) The mean vibrational energy is given by [13C.8–551], $\langle \varepsilon^{\mathrm{V}} \rangle = hc\tilde{\nu}/(e^{\beta hc\tilde{\nu}} - 1)$; this result is exact. The equipartition value is $\langle \varepsilon^{\mathrm{V}} \rangle = kT$, because there are two quadratic terms for a harmonic oscillator. These two expressions for the energy are plotted as a function of T in Fig. 13.9. The value from the equipartition theorem comes within 5 % of the exact value at $\boxed{4.80 \times 10^3\ \mathrm{K}}$.

E13C.5(a) The mean vibrational energy per vibrational mode is given by [13C.8–551], $\langle \varepsilon_i^{\mathrm{V}} \rangle = hc\tilde{\nu}_i/(e^{\beta hc\tilde{\nu}_i} - 1)$; this result is exact. The overall vibrational energy is the sum of the contributions from each normal mode, taking into account the degeneracy of each

$$\langle \varepsilon^{\mathrm{V}} \rangle = \langle \varepsilon_1^{\mathrm{V}} \rangle + 2 \times \langle \varepsilon_2^{\mathrm{V}} \rangle + \langle \varepsilon_3^{\mathrm{V}} \rangle$$

The equipartition value is $\langle \varepsilon^{\mathrm{V}} \rangle = 4kT$, because there are two quadratic terms for a harmonic oscillator, and four modes in total. These two expressions for the energy are plotted as a function of T in Fig. 13.10. The value from the equipartition theorem comes within 5% of the exact value at $\boxed{1.10 \times 10^4\ \mathrm{K}}$.

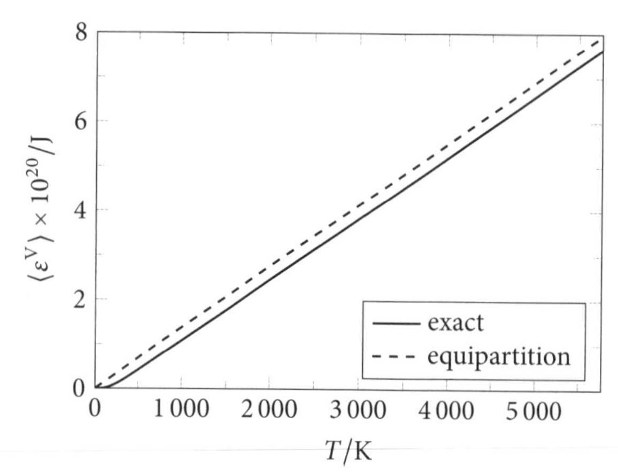

Figure 13.9

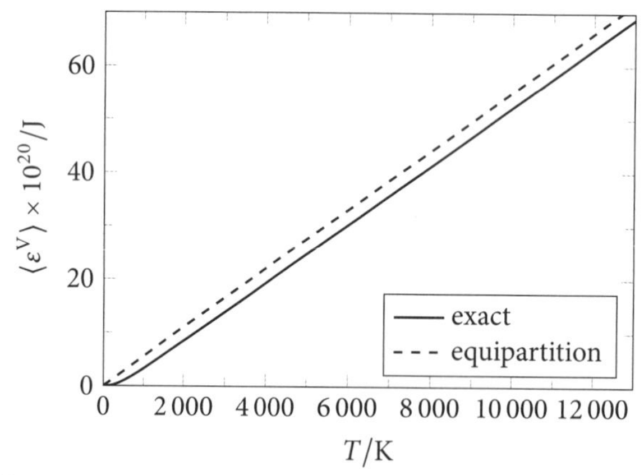

Figure 13.10

E13C.6(a) The mean vibrational energy per vibrational mode is given by [13C.8–551], $\langle \varepsilon_i^{\mathrm{V}} \rangle = hc\tilde{\nu}_i/(\mathrm{e}^{\beta hc\tilde{\nu}_i}-1)$; this result is exact. The overall vibrational energy is the sum of the contributions from each normal mode, taking into account the degeneracy of each

$$\langle \varepsilon^{\mathrm{V}} \rangle = \langle \varepsilon_1^{\mathrm{V}} \rangle + 2 \times \langle \varepsilon_2^{\mathrm{V}} \rangle + 3 \times \langle \varepsilon_3^{\mathrm{V}} \rangle + 3 \times \langle \varepsilon_4^{\mathrm{V}} \rangle$$

The equipartition value is $\langle \varepsilon^{\mathrm{V}} \rangle = 9kT$, because there are two quadratic terms for a harmonic oscillator, and nine modes in total. These two expressions for the energy are plotted as a function of T in Fig. 13.11. The value from the equipartition theorem comes within 5% of the exact value at $\boxed{6.85 \times 10^3\ \mathrm{K}}$.

E13C.7(a) The mean molecular energy is given by [13C.4a–549], $\langle \varepsilon \rangle = -(1/q)(\partial q/\partial \beta)_V$, where $\beta = 1/kT$, and q is the partition function given by [13B.1b–538], $q = \sum_i g_i \mathrm{e}^{-\beta \varepsilon_i}$, where g_i is degeneracy and the corresponding energy is given as

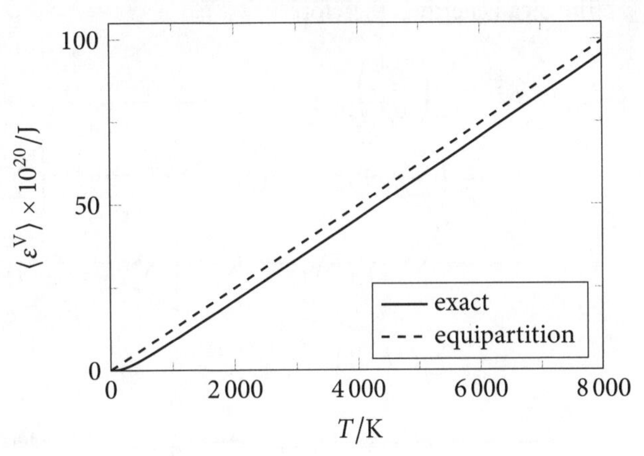

Figure 13.11

$\varepsilon_i = hc\tilde{\nu}_i$. At $T = 1900$ K

$$\beta hc = \frac{(6.6261 \times 10^{-34}\,\mathrm{J\,s}) \times (2.9979 \times 10^{10}\,\mathrm{cm\,s^{-1}})}{(1.3806 \times 10^{-23}\,\mathrm{J\,K^{-1}}) \times (1900\,\mathrm{K})} = 7.57... \times 10^{-4}\,\mathrm{cm}$$

Therefore the electronic partition function is

$$q^{\mathrm{E}} = g_0 + g_1 \mathrm{e}^{-\beta hc\tilde{\nu}_1} + g_2 \mathrm{e}^{-\beta hc\tilde{\nu}_2} = 4.29...$$

Therefore the mean energy is

$$\begin{aligned}\langle \varepsilon^{\mathrm{E}} \rangle &= -\frac{1}{q^{\mathrm{E}}}\left(\frac{\partial q^{\mathrm{E}}}{\partial \beta}\right)_V = \frac{hc}{q^{\mathrm{E}}}\left(g_1\tilde{\nu}_1 \mathrm{e}^{-\beta hc\tilde{\nu}_1} + g_2\tilde{\nu}_2 \mathrm{e}^{-\beta hc\tilde{\nu}_2}\right) \\ &= \frac{(6.6261 \times 10^{-34}\,\mathrm{J\,s}) \times (2.9979 \times 10^{10}\,\mathrm{cm\,s^{-1}})}{4.29...} \\ &\quad \times \Big[1 \times (2500\,\mathrm{cm^{-1}}) \times \mathrm{e}^{-(7.57...\times 10^{-4}\,\mathrm{cm})\times(2500\,\mathrm{cm^{-1}})} \\ &\quad\quad + 2 \times (3500\,\mathrm{cm^{-1}}) \times \mathrm{e}^{-(7.57...\times 10^{-4}\,\mathrm{cm})\times(3500\,\mathrm{cm^{-1}})}\Big] \\ &= \boxed{4.03 \times 10^{-21}\,\mathrm{J}}\end{aligned}$$

Solutions to problems

P13C.1 The mean molecular energy is given by [13C.4a–*549*], $\langle \varepsilon \rangle = -(1/q)(\partial q/\partial \beta)_V$, where $\beta = 1/kT$, and q is the partition function given by [13B.1b–*538*], $q = \sum_i g_i \mathrm{e}^{-\beta \varepsilon_i}$. For a symmetric rotor the rotational terms are given in [11B.13a–*433*], $\tilde{F}(J,K) = \tilde{B}J(J+1) + (\tilde{A} - \tilde{B})K^2$, with $J = 0, 1, 2, ...$ and $K = 0, \pm 1, ..., \pm J$; the corresponding energies are $hc\tilde{F}(J,K)$ and the degeneracy is $(2J+1)$. The rotational partition function is therefore given by

$$q^{\mathrm{R}} = \sum_{J=0}^{\infty}(2J+1)\mathrm{e}^{-\beta hc\tilde{B}J(J+1)}\left[\sum_{K=-J}^{+J} \mathrm{e}^{-\beta hc(\tilde{A}-\tilde{B})K^2}\right]$$

The mean energy is therefore

$$
\begin{aligned}
\langle \varepsilon^{\mathrm{R}} \rangle &= -\frac{1}{q^{\mathrm{R}}}\left(\frac{\partial q^{\mathrm{R}}}{\partial \beta}\right)_V \\
&= \frac{1}{q^{\mathrm{R}}}\left(\sum_{J=0}^{\infty} hc\tilde{B}J(J+1)(2J+1)\mathrm{e}^{-\beta hc\tilde{B}J(J+1)}\left[\sum_{K=-J}^{+J}\mathrm{e}^{-\beta hc(\tilde{A}-\tilde{B})K^2}\right]\right. \\
&\quad \left. + \sum_{J=0}^{\infty}(2J+1)\mathrm{e}^{-\beta hc\tilde{B}J(J+1)}\left[\sum_{K=-J}^{+J} hc(\tilde{A}-\tilde{B})K^2\mathrm{e}^{-\beta hc(\tilde{A}-\tilde{B})K^2}\right]\right) \\
&= \frac{1}{q^{\mathrm{R}}}\sum_{J=0}^{\infty}(2J+1)\mathrm{e}^{-\beta hc\tilde{B}J(J+1)} \\
&\quad \times \left[hc\sum_{K=-J}^{+J}\left(\tilde{B}J(J+1)+(\tilde{A}-\tilde{B})K^2\right)\mathrm{e}^{-\beta hc(\tilde{A}-\tilde{B})K^2}\right]
\end{aligned}
$$

The terms in the sum in this expression, and the terms in q^{R}, are evaluated and summed until the value converges to the required precision. The equipartition value of the energy is $\langle \varepsilon^{\mathrm{R}} \rangle = \frac{3}{2}kT$ because there are three rotational degrees of freedom. The two expressions for the energy are plotted as a function of temperature in Fig. 13.12. The equipartition value is within 5% of the exact solution at $\boxed{4.59\ \mathrm{K}}$.

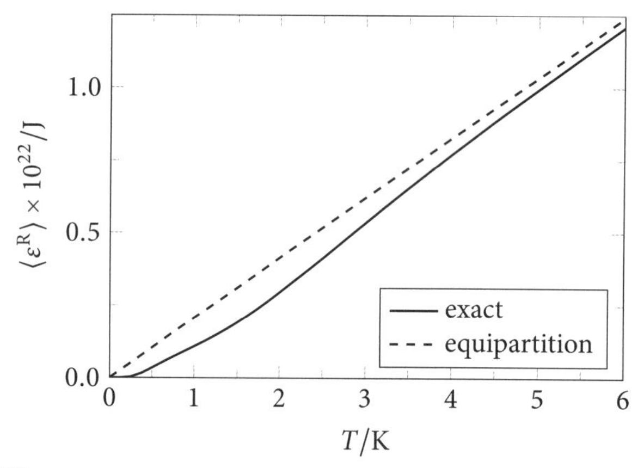

Figure 13.12

P13C.3 H_2O has three translational and three rotational degrees of freedom. It therefore follows from the equipartition principle that the molar internal energy is $U_{\mathrm{m}} = (3+3)\times\frac{1}{2}RT = 3RT$. The constant-volume molar heat capacity is therefore $C_{V,\mathrm{m}} = (\partial U_{\mathrm{m}}/\partial T)_V = 3R$.

The energy needed to raise the temperature of n mol of H_2O by ΔT is equal to the change in the internal energy which is $\Delta U = nC_{V,\mathrm{m}}\Delta T$. In this case $\Delta U = (1.0\ \mathrm{mol})\times(3\times 8.3145\ \mathrm{J\,K^{-1}\,mol^{-1}})\times(100\ \mathrm{K}) = \boxed{2.5\ \mathrm{kJ}}$.

P13C.5 The energy levels for a spin in a magnetic field are given by [12A.4d–489], $E_{m_I} = -g_I\mu_N\mathcal{B}_0 m_I$, where $m_I = +1, 0, -1$. These energies are conveniently written as $E_{m_I} = -m_I\delta$, with $\delta = g_I\mu_N\mathcal{B}_0$. If the energy of the lowest state is defined as the energy zero, then the three levels have energies $E_0 = 0$, $E_1 = \delta$, and $E_2 = 2\delta$. The partition function is

$$q = 1 + e^{-\beta\delta} + e^{-2\beta\delta}$$

The mean molecular energy is given by [13C.4a–549]

$$\begin{aligned}\langle\varepsilon\rangle &= \varepsilon_{gs} - \frac{1}{q}\left(\frac{\partial q}{\partial\beta}\right)_V \\ &= \varepsilon_{gs} - \frac{1}{1 + e^{-\beta\delta} + e^{-2\beta\delta}}\left(-\delta e^{-\beta\delta} - 2\delta e^{-2\beta\delta}\right) \\ &= \varepsilon_{gs} + \frac{\delta e^{-\beta\delta} + 2\delta e^{-2\beta\delta}}{1 + e^{-\beta\delta} + e^{-2\beta\delta}} = \boxed{-\delta + \frac{\delta e^{-\beta\delta} + 2\delta e^{-2\beta\delta}}{1 + e^{-\beta\delta} + e^{-2\beta\delta}}}\end{aligned}$$

In the last step the fact that the energy of ground state (the one with $m_I = +1$) is $-\delta$ is used. With the data given

$$\delta = g_I\mu_N\mathcal{B}_0 = 2.0 \times (5.0508 \times 10^{-27}\ \text{J T}^{-1}) \times (2.5\ \text{T}) = 2.52... \times 10^{-26}\ \text{J}$$

and assuming that $T = 298$ K

$$\begin{aligned}\beta\delta &= \delta/kT = (2.52... \times 10^{-26}\ \text{J})/[(1.3806 \times 10^{-23}\ \text{J K}^{-1}) \times (298\ \text{K})] \\ &= 6.13... \times 10^{-6}\end{aligned}$$

$$\begin{aligned}\langle\varepsilon\rangle &= (-2.52... \times 10^{-26}\ \text{J}) \\ &+ \frac{(2.52... \times 10^{-26}\ \text{J})e^{-6.13...\times10^{-6}} + 2 \times (2.52... \times 10^{-26}\ \text{J})e^{-2\times(6.13...\times10^{-6})}}{1 + e^{-6.13...\times10^{-6}} + e^{-2\times6.13...\times10^{-6}}} \\ &= -1.03 \times 10^{-31}\ \text{J}\end{aligned}$$

The separation of the energy levels is very much smaller than kT, therefore the three levels have almost equal populations giving a mean energy of very close to zero.

P13C.7 The partition function given by [13B.1b–538], $q = \sum_i g_i e^{-\beta\varepsilon_i}$, where g_i is degeneracy, $\varepsilon_i = hc\tilde{\nu}$ and $\beta = 1/kT$. Therefore

$$q^{E} = 2e^{0} + 2e^{-\beta hc\tilde{\nu}} = 2 + 2e^{-hc\tilde{\nu}/kT}$$

This function in plotted in Fig. 13.13.

(a) The ratio of the populations is given by [13A.13b–536]

$$N_i/N_j = (g_i/g_j)e^{-\beta(\varepsilon_i-\varepsilon_j)}$$

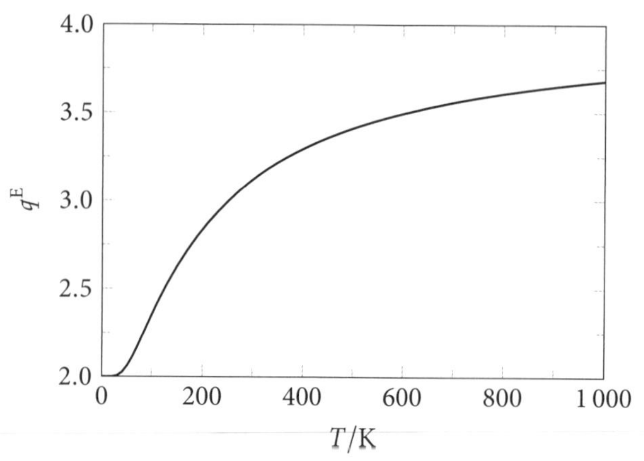

Figure 13.13

At 300 K this ratio is

$$N_1/N_0 = (g_1/g_0) \times e^{-\beta hc\tilde{\nu}} = (2/2) \times e^{-\beta hc\tilde{\nu}}$$

$$= e^{-\frac{(6.6261\times10^{-34}\,\mathrm{J\,s})\times(2.9979\times10^{10}\,\mathrm{cm\,s^{-1}})\times(121.1\,\mathrm{cm^{-1}})}{(1.3806\times10^{-23}\,\mathrm{J\,K^{-1}})\times(300\,\mathrm{K})}} = 0.559....$$

Thus the populations expressed as a fraction of N are

$$N_0/N = N_0/(N_0 + N_1) = 1/(1 + 0.559...) = \boxed{0.641}$$
$$N_1/N = N_1/(N_0 + N_1) = 0.559.../(1 + 0.559...) = \boxed{0.359}$$

(b) The mean molecular energy is given by [13C.4a–549], $\langle\varepsilon\rangle = -(1/q)(\partial q/\partial\beta)_V$. Thus

$$\langle\varepsilon^E\rangle = -\frac{1}{q^E}\frac{dq^E}{d\beta} = hc\tilde{\nu} \times \frac{e^{-hc\tilde{\nu}/kT}}{1 + e^{-hc\tilde{\nu}/kT}} = \frac{hc\tilde{\nu}}{e^{hc\tilde{\nu}/kT} + 1}$$

$$= \frac{(6.6261 \times 10^{-34}\,\mathrm{J\,s}) \times (2.9979 \times 10^{10}\,\mathrm{cm\,s^{-1}}) \times (121.1\,\mathrm{cm^{-1}})}{e^{\frac{(6.6261\times10^{-34}\,\mathrm{J\,s})\times(2.9979\times10^{10}\,\mathrm{cm\,s^{-1}})\times(121.1\,\mathrm{cm^{-1}})}{(1.3806\times10^{-23}\,\mathrm{J\,K^{-1}})\times(300\,\mathrm{K})}} + 1}$$

$$= \boxed{8.63 \times 10^{-22}\,\mathrm{J}}$$

P13C.9 Mean values of any observable are given by a sum of the observed value over all the possible states weighted by the probability of each state. Thus, the mean of the square of energy is given by $\langle\varepsilon^2\rangle = (1/q)\sum_j \varepsilon_j{}^2 e^{-\beta\varepsilon_j}$, where q is the partition function given by [13A.11–535], $q = \sum_i e^{-\beta\varepsilon_i}$.

It is useful to consider the first and second derivatives of the term $e^{-\beta\varepsilon_j}$ with respect to β. The first derivative is

$$\frac{d}{d\beta}e^{-\beta\varepsilon_j} = -\varepsilon_j e^{-\beta\varepsilon_j}$$

and hence the second is

$$\frac{d^2}{d\beta^2}e^{-\beta\varepsilon_j} = \frac{d}{d\beta}\left(-\varepsilon_j e^{-\beta\varepsilon_j}\right) = (-\varepsilon_j) \times \frac{d}{d\beta}e^{-\beta\varepsilon_j}$$
$$= (-\varepsilon_j) \times (-\varepsilon_j) \times e^{-\beta\varepsilon_j} = \varepsilon_j^{\,2} \times e^{-\beta\varepsilon_j}$$

This latter expression is used to rewrite the definition of $\langle\varepsilon^2\rangle$ as

$$\langle\varepsilon^2\rangle = \frac{1}{q}\sum_j \varepsilon_j^2 e^{-\beta\varepsilon_j}$$
$$= \frac{1}{q}\sum_j \frac{d^2}{d\beta^2}e^{-\beta\varepsilon_j} = \frac{1}{q}\frac{d^2}{d\beta^2}\left[\sum_j e^{-\beta\varepsilon_j}\right] = \frac{1}{q}\frac{d^2 q}{d\beta^2}$$

Therefore

$$\langle\varepsilon^2\rangle^{1/2} = \boxed{\left(\frac{1}{q}\frac{d^2 q}{d\beta^2}\right)^{1/2}}$$

These results are also used to find the root mean square of the deviation from the mean as

$$\left(\langle\varepsilon^2\rangle - \langle\varepsilon\rangle^2\right)^{1/2} = \left(\frac{1}{q}\frac{d^2 q}{d\beta^2} - \left(\frac{1}{q}\frac{dq}{d\beta}\right)^2\right)^{1/2} = \boxed{\frac{1}{q}\left(q\frac{d^2 q}{d\beta^2} - \left(\frac{dq}{d\beta}\right)^2\right)^{1/2}}$$

For a harmonic oscillator the partition function is given by [13B.15–*546*], $q = (1 - e^{-\beta hc\tilde{\nu}})^{-1}$. Therefore

$$\frac{dq}{d\beta} = \frac{-hc\tilde{\nu}\,e^{-\beta hc\tilde{\nu}}}{(1 - e^{-\beta hc\tilde{\nu}})^2} = -hc\tilde{\nu}\,e^{-\beta hc\tilde{\nu}}q^2$$
$$\frac{d^2 q}{d\beta^2} = -hc\tilde{\nu}\left[\frac{-hc\tilde{\nu}\,e^{-\beta hc\tilde{\nu}}}{(1 - e^{-\beta hc\tilde{\nu}})^2} + \frac{-2hc\tilde{\nu}\,e^{-\beta hc\tilde{\nu}}\,e^{-\beta hc\tilde{\nu}}}{(1 - e^{-\beta hc\tilde{\nu}})^3}\right]$$
$$= (hc\tilde{\nu})^2 e^{-\beta hc\tilde{\nu}}\left[\frac{1}{(1 - e^{-\beta hc\tilde{\nu}})^2} + \frac{2e^{-\beta hc\tilde{\nu}}}{(1 - e^{-\beta hc\tilde{\nu}})^3}\right]$$
$$= (hc\tilde{\nu})^2 e^{-\beta hc\tilde{\nu}}\left[q^2 + 2e^{-\beta hc\tilde{\nu}}q^3\right]$$
$$= q^2(hc\tilde{\nu})^2 e^{-\beta hc\tilde{\nu}}\left[1 + 2e^{-\beta hc\tilde{\nu}}q\right]$$

Using these results and making the substitution $x = hc\tilde{\nu}$

$$\langle\varepsilon^2\rangle - \langle\varepsilon\rangle^2 = \frac{1}{q}\frac{d^2 q}{d\beta^2} - \left(\frac{1}{q}\frac{dq}{d\beta}\right)^2$$
$$= qx^2 e^{-\beta x}\left[1 + 2e^{-\beta x}q\right] - \left[-x\,e^{-\beta x}q\right]^2$$
$$= qx^2 e^{-\beta x} + q^2 x^2 e^{-2\beta x} = \frac{x^2 e^{-\beta x}}{1 - e^{-\beta x}} + \frac{x^2 e^{-2\beta x}}{(1 - e^{-\beta x})^2}$$
$$= \frac{x^2 e^{-\beta x}(1 - e^{-\beta x}) + x^2 e^{-2\beta x}}{(1 - e^{-\beta x})^2} = \frac{x^2 e^{-\beta x}}{(1 - e^{-\beta x})^2}$$

hence

$$\left(\langle\varepsilon^2\rangle - \langle\varepsilon\rangle^2\right)^{1/2} = \boxed{\frac{hc\tilde{\nu}\,\mathrm{e}^{-\beta hc\tilde{\nu}/2}}{1-\mathrm{e}^{-\beta hc\tilde{\nu}}}}$$

In terms of the vibrational temperature this is

$$\frac{k\theta^{\mathrm{R}}\,\mathrm{e}^{-\theta^{\mathrm{R}}/2T}}{1-\mathrm{e}^{-\theta^{\mathrm{R}}/T}}$$

At high temperatures, $T \gg \theta^{\mathrm{R}}$, the exponential in the denominator is approximated as $1 - \theta^{\mathrm{R}}/T$, and the exponential in the numerator goes to 1, giving $(\langle\varepsilon^2\rangle - \langle\varepsilon\rangle^2)^{1/2} = kT$. At low temperatures, $T \ll \theta^{\mathrm{R}}$, $(\langle\varepsilon^2\rangle - \langle\varepsilon\rangle^2)^{1/2}$ tends to 0. This is as expected because if all the particles are in the ground state there is no uncertainty in their average energy.

13D The canonical ensemble

Answer to discussion questions

D13D.1 Ensembles are needed to treat systems of interacting particles (in contrast to molecular partition functions that apply to independent particles). An ensemble is a set of a large number of imaginary replications of an actual physical (thermodynamic) system. Ensembles are useful in statistical thermodynamics because it is mathematically more tractable to perform an ensemble average to determine time averaged thermodynamic properties than it is to perform an average over time to determine these properties. Macroscopic thermodynamic properties are averages over the time dependent properties of the particles that compose the macroscopic system.

The replications in an ensemble are identical in some, but not all, respects. In the canonical ensemble, all replications have the same number of particles, the same volume, and the same temperature, but they need not have the same energy. Because they have the same temperature, they can exchange energy. Thus the canonical ensemble corresponds to a closed physical system at thermal equilibrium: averages using canonical ensembles apply to closed physical systems at thermal equilibrium.

D13D.3 Identical particles can be regarded as distinguishable when they are localized as in a crystal lattice where a set of coordinates can be assigned to each particle. Strictly speaking, it is the lattice site that carries the set of coordinates, but as long as the particle is tethered to the site, it too can be considered distinguishable.

Solutions to exercises

E13D.1(a) It is essential to include the factor $1/N!$ when considering indistinguishable particles which are free to move. Thus, such a factor is always needed for gases.

In the solid state, particles are distinguished by their positions in the lattice and therefore the particles are regarded as distinguishable on the basis that their locations are distinguishable. For the cases mentioned, the factor $1/N!$ is needed for all but solid CO.

Solutions to problems

P13D.1

$$p = kT\left(\frac{\partial \ln Q}{\partial V}\right)_T = kT\left(\frac{\partial \ln(q^N/N!)}{\partial V}\right)_T = kT\left(\frac{\partial(N\ln q - \ln N!)}{\partial V}\right)_T$$

The $\ln N!$ term is volume independent and thus $p = NkT\left(\partial \ln q/\partial V\right)_T$. The molecular partition function, q, for the perfect gas is just the translational partition function given by [13B.10b–*541*], $q^{\mathrm{T}} = V/\Lambda^3$, where Λ is the thermal wavelength which is independent of volume. Therefore

$$p = NkT\left(\frac{\partial \ln(V/\Lambda^3)}{\partial V}\right)_T = NkT\left(\frac{\partial(\ln V - \ln \Lambda^3)}{\partial V}\right)_T$$

$$= NkT\left(\frac{\partial \ln V}{\partial V}\right)_T = \frac{NkT}{V} = \frac{nRT}{V}$$

where for the last step $N = nN_{\mathrm{A}}$ and $R = kN_{\mathrm{A}}$ are used.

13E The internal energy and entropy

Answer to discussion questions

D13E.1 The temperature is always high enough for the mean translational energy to be $\frac{3}{2}kT$, the equipartition value. Therefore, the molar constant-volume heat capacity for translation is $C_{V,\mathrm{m}}^{\mathrm{T}} = \frac{3}{2}R$.

When the temperature is high enough for the rotations of the molecules to be highly excited (when $T \gg \theta^{\mathrm{R}}$) the equipartition value kT for the mean rotational energy (for a linear rotor) can be used to obtain $C_{V,\mathrm{m}}^{\mathrm{R}} = R$. For nonlinear molecules, the mean rotational energy is $\frac{3}{2}kT$, so the molar rotational heat capacity rises to $\frac{3}{2}R$ when $T \gg \theta^{\mathrm{R}}$. At intermediate temperatures the total heat capacity takes a value between that due to translation, $\frac{3}{2}R$, and $\frac{5}{2}R$ (for a linear molecule) when both translation and rotation contribute fully.

Molecular vibrations contribute to the heat capacity, but only when the temperature is high enough for them to be significantly excited. For each vibrational mode, the equipartition mean energy is kT, so the maximum contribution to the molar heat capacity is R. However, it is unusual for the vibrations to be so highly excited that equipartition is valid, and in general the contribution to the heat capacity has to be calculated using [13E.3–*560*].

D13E.3 The statistical entropy is defined by Boltzmann's formula, $S = k \ln \mathcal{W}$, in terms of the number of configurations or microstates consistent with a given total energy. The thermodynamic entropy is defined by $\mathrm{d}S = \mathrm{d}q_{\mathrm{rev}}/T$ that is, in terms of reversible heat transfer.

The concept of the number of microstates makes quantitative the qualitative concepts of 'disorder' and 'dispersal of matter and energy' that are often used to introduce the concept of entropy: a more 'disorderly' distribution of energy and matter corresponds to a greater number of microstates consistent with the same total energy. The more molecules that can participate in the distribution of the energy, the more microstates there are for a given total energy and hence the greater the entropy.

The molecular interpretation of entropy embodied in the Boltzmann formula also suggests the thermodynamic definition. At high temperatures, where the molecules of a system can occupy a large number of energy levels, a small additional transfer of energy as heat will cause only a small change in the number of accessible energy levels, whereas at low temperatures the transfer of the same quantity of heat will increase the number of accessible energy levels and microstates significantly. Hence the change in entropy upon heating will be greater when the energy is transferred to a cold body than when it is transferred to a hot body, as required by the thermodynamic definition.

D13E.5 The entropy of a monatomic perfect gas is given by the Sackur–Tetrode equation [13E.9a–*563*]

$$S_{\mathrm{m}} = R \ln\left(\frac{V_{\mathrm{m}} \mathrm{e}^{5/2}}{N_{\mathrm{A}} \Lambda^3}\right) \quad \Lambda = h/(2\pi mkT)^{1/2} \quad V_{\mathrm{m}} = RT/p$$

Because the molar volume appears in the numerator, the molar entropy increases with the molar volume. In terms of the Boltzmann distribution, this relationship is expected: large containers have more closely spaced energy levels than do small ones, so more states are thermally accessible. Temperature appears in the numerator of the expression (through the denominator of Λ), so the molar entropy increases with the temperature. Again, this is consistent with the Boltzmann distribution, because more states are accessible at higher temperatures than at lower ones.

The fact that diatomic and polyatomic gases have rotational and vibrational modes of motion as well does not change the above arguments. The partition functions of those modes are independent of volume, so the volume dependence of the entropy is as described above. At most temperatures, rotational modes of motion are active and contribute to the entropy, as expressed in [13E.11a–*564*]; the contribution increases with temperature. Finally, most vibrational modes contribute little if at all to the entropy, but as with rotation the contribution increases with temperature.

Solutions to exercises

E13E.1(a) The equipartition value for $C_{V,\mathrm{m}}$ is expressed in [13E.6–560]: each translational or rotational mode contributes $\frac{1}{2}R$, and each active vibrational mode contributes R.

(i) I_2: three translational modes, two rotational modes (linear) and, because the vibrational frequency of the molecule is rather low, one vibrational mode: $C_{V,\mathrm{m}}/R = 3 \times \frac{1}{2} + 2 \times \frac{1}{2} + 1 = \boxed{\frac{7}{2}}$.

(ii) CH_4: three translational modes, three rotational modes (non-linear), and no active vibrational modes: $C_{V,\mathrm{m}}/R = 3 \times \frac{1}{2} + 3 \times \frac{1}{2} = \boxed{3}$.

(iii) C_6H_6: three translational modes, three rotational modes (non-linear), and no active vibrational modes: $C_{V,\mathrm{m}}/R = 3 \times \frac{1}{2} + 3 \times \frac{1}{2} = \boxed{3}$. There are four low-frequency normal modes which, if active, will contribute a further $4R$.

E13E.2(a) The equipartition value for $C_{V,\mathrm{m}}$ is expressed in [13E.6–560]: each translational or rotational mode contributes $\frac{1}{2}R$, and each active vibrational mode contributes R. The number of vibrational modes is $(3N-6)$, which is 6 for NH_3 and 9 for CH_4. For NH_3 there are three translational modes, three rotational modes (non-linear) giving $C_{V,\mathrm{m}} = 3R$; if the 6 vibrations are included, $C_{V,\mathrm{m}} = 9R$. For CH_4 there are three translational modes, three rotational modes (non-linear) giving $C_{V,\mathrm{m}} = 3R$; if the 9 vibrations are included, $C_{V,\mathrm{m}} = 12R$.

$$\gamma = C_{p,\mathrm{m}}/C_{V,\mathrm{m}} = (C_{V,\mathrm{m}} + R)/C_{V,\mathrm{m}} = 1 + R/C_{V,\mathrm{m}}$$

$$\gamma_{\mathrm{NH_3}} = 1 + R/3R = 1.33 \quad \text{no vibrational contribution}$$

$$\gamma_{\mathrm{NH_3}} = 1 + R/9R = 1.11 \quad \text{with vibrational contribution}$$

$$\gamma_{\mathrm{CH_4}} = 1 + R/3R = 1.33 \quad \text{no vibrational contribution}$$

$$\gamma_{\mathrm{CH_3}} = 1 + R/12R = 1.08 \quad \text{with vibrational contribution}$$

The experimental value for γ is 1.31 for both gases: evidently the vibrational modes are not active.

E13E.3(a) The partition function of this two-level system is

$$q = g_0 + g_1 \mathrm{e}^{-\beta hc\tilde{\nu}}$$

where g_0 and g_1 are the degeneracies of the ground and first excited state, respectively. The mean energy is given by [13C.4a–549], $\langle \varepsilon \rangle = -(1/q)(\partial q/\partial \beta)_V$

$$\langle \varepsilon \rangle = \frac{g_1 hc\tilde{\nu}\,\mathrm{e}^{-\beta hc\tilde{\nu}}}{g_0 + g_1 \mathrm{e}^{-\beta hc\tilde{\nu}}} = \frac{g_1 hc\tilde{\nu}}{g_0 \mathrm{e}^{\beta hc\tilde{\nu}} + g_1}$$

$$\text{hence } U_\mathrm{m} = N_\mathrm{A}\langle \varepsilon \rangle = \frac{N_\mathrm{A} g_1 hc\tilde{\nu}}{g_0 \mathrm{e}^{\beta hc\tilde{\nu}} + g_1}$$

By definition $C_{V,\mathrm{m}} = (\partial U_\mathrm{m}/\partial T)_V$, therefore

$$C_{V,\mathrm{m}} = \left(\frac{\partial U_\mathrm{m}}{\partial T}\right)_V = \left(\frac{\partial U_\mathrm{m}}{\partial \beta}\right)_V \frac{\mathrm{d}\beta}{\mathrm{d}T} = \left(\frac{\partial U_\mathrm{m}}{\partial \beta}\right)_V \times \frac{-1}{kT^2}$$

$$= \frac{1}{kT^2} \times N_\mathrm{A} g_1 hc\tilde{\nu} \frac{g_0 hc\tilde{\nu} \mathrm{e}^{\beta hc\tilde{\nu}}}{(g_0 \mathrm{e}^{\beta hc\tilde{\nu}} + g_1)^2}$$

$$= \frac{N_\mathrm{A}(hc\tilde{\nu})^2}{kT^2} \frac{g_0 g_1 \mathrm{e}^{\beta hc\tilde{\nu}}}{(g_0 \mathrm{e}^{\beta hc\tilde{\nu}} + g_1)^2}$$

For an electronic term the degeneracy is $2J + 1$, hence $g_0 = 2 \times \frac{3}{2} + 1 = 4$ and $g_1 = 2 \times \frac{1}{2} + 1 = 2$. With the data given

$$hc\tilde{\nu} = (6.6261 \times 10^{-34}\ \mathrm{J\,s}) \times (2.9979 \times 10^{10}\ \mathrm{cm\,s^{-1}}) \times (881\ \mathrm{cm^{-1}})$$
$$= 1.75... \times 10^{-20}\ \mathrm{J}$$

at 500 K $\beta hc\tilde{\nu} = hc\tilde{\nu}/kT$

$$= (1.75... \times 10^{-20}\ \mathrm{J})/[(1.3806 \times 10^{-23}\ \mathrm{J\,K^{-1}}) \times (500\ \mathrm{K})] = 2.53...$$

at 900 K $\beta hc\tilde{\nu} = hc\tilde{\nu}/kT$

$$= (1.75... \times 10^{-20}\ \mathrm{J})/[(1.3806 \times 10^{-23}\ \mathrm{J\,K^{-1}}) \times (900\ \mathrm{K})] = 1.40...$$

$$C_{V,\mathrm{m}}(500) = \frac{N_\mathrm{A}(hc\tilde{\nu})^2}{kT^2} \frac{8\,\mathrm{e}^{\beta hc\tilde{\nu}}}{(4\,\mathrm{e}^{\beta hc\tilde{\nu}} + 2)^2}$$

$$= \frac{(6.0221 \times 10^{23}\ \mathrm{mol^{-1}}) \times (1.75... \times 10^{-20}\ \mathrm{J})^2}{(1.3806 \times 10^{-23}\ \mathrm{J\,K^{-1}}) \times (500\ \mathrm{K})^2} \frac{8\,\mathrm{e}^{2.53...}}{(4\,\mathrm{e}^{2.53...} + 2)^2}$$

$$= \boxed{1.96\ \mathrm{J\,K^{-1}\,mol^{-1}}}$$

$$C_{V,\mathrm{m}}(900) = \frac{(6.0221 \times 10^{23}\ \mathrm{mol^{-1}}) \times (1.75... \times 10^{-20}\ \mathrm{J})^2}{(1.3806 \times 10^{-23}\ \mathrm{J\,K^{-1}}) \times (900\ \mathrm{K})^2} \frac{8\,\mathrm{e}^{1.40...}}{(4\,\mathrm{e}^{1.40...} + 2)^2}$$

$$= \boxed{1.60\ \mathrm{J\,K^{-1}\,mol^{-1}}}$$

E13E.4(a) The contribution of a collection of harmonic oscillators to the molar heat capacity is given by [13E.3–*560*]

$$C_{V,\mathrm{m}} = R\left(\frac{\theta^\mathrm{V}}{T}\right)^2 \left(\frac{\mathrm{e}^{-\theta^\mathrm{V}/2T}}{1 - \mathrm{e}^{-\theta^\mathrm{V}/T}}\right)^2 \qquad \theta^\mathrm{V} = hc\tilde{\nu}/k$$

This function is plotted in Fig 13.14.

The following table shows the vibrational temperatures and the contribution to the heat capacity for each of the normal modes

		298 K		500 K	
$\tilde{\nu}/\text{cm}^{-1}$	$\theta^{\text{V}}/\text{K}$	T/θ^{V}	$C_{V,\text{m}}/R$	T/θ^{V}	$C_{V,\text{m}}/R$
612	881	0.338	0.506	0.568	0.777
729	1049	0.284	0.390	0.477	0.702
1974	2840	0.1049	6.593×10^{-3}	0.1760	0.1109
3287	4729	0.06301	3.226×10^{-5}	0.1057	6.980×10^{-3}
3374	4855	0.06139	2.233×10^{-5}	0.1030	5.725×10^{-3}

The heat capacity is obtained by summing the contributions from each normal mode, taking into account the double degeneracy of the modes at 612 cm^{-1} and 729 cm^{-1} by counting each twice. Thus at 298 K $\boxed{C_{V,\text{m}} = 14.95\ \text{J K}^{-1}\ \text{mol}^{-1}}$ and at 500 K $\boxed{C_{V,\text{m}} = 25.62\ \text{J K}^{-1}\ \text{mol}^{-1}}$.

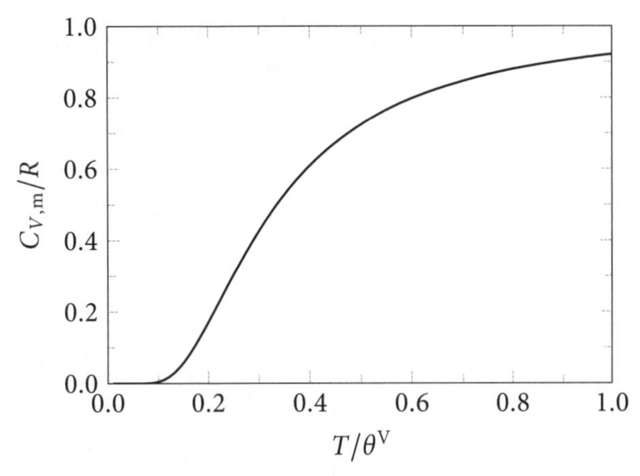

Figure 13.14

E13E.5(a) For atoms with filled shells the only contribution to the entropy is translational. The standard molar entropy of a monatomic perfect gas is given by the Sackur–Tetrode equation [13E.9b–*563*]

$$S_{\text{m}}^{\ominus} = R\ln\left(\frac{kT\text{e}^{5/2}}{p^{\ominus}\Lambda^3}\right) \qquad \Lambda = h/(2\pi mkT)^{1/2}$$

(i) Taking the mass of a He atom as 4.00 m_{u}

$$\Lambda = \frac{6.6261 \times 10^{-34}\ \text{J s}}{\left[2\pi(4.00\times 1.6605 \times 10^{-27}\ \text{kg})\times(1.3806 \times 10^{-23}\ \text{J K}^{-1})\times(298\ \text{K})\right]^{1/2}}$$
$$= 5.05... \times 10^{-11}\ \text{m}$$

$$S_{\text{m}}^{\ominus} = (8.3145\ \text{J K}^{-1}\ \text{mol}^{-1}) \times \ln\left[\frac{(1.3806 \times 10^{-23}\ \text{J K}^{-1}) \times (298\ \text{K}) \times \text{e}^{5/2}}{(10^5\ \text{N m}^{-2}) \times (5.05\ldots \times 10^{-11}\ \text{m})^3}\right]$$
$$= \boxed{126\ \text{J K}^{-1}\ \text{mol}^{-1}}$$

(ii) Taking the mass of a Xe atom as 131.29 m_u

$$\Lambda = \frac{6.6261\times10^{-34}\,\mathrm{J\,s}}{[2\pi(131.29\times1.6605\times10^{-27}\,\mathrm{kg})\times(1.3806\times10^{-23}\,\mathrm{J\,K^{-1}})\times(298\,\mathrm{K})]^{1/2}}$$
$$= 8.82...\times10^{-12}\,\mathrm{m}$$

$$S_m^{\ominus} = (8.3145\,\mathrm{J\,K^{-1}\,mol^{-1}})\times\ln\left[\frac{(1.3806\times10^{-23}\,\mathrm{J\,K^{-1}})\times(298\,\mathrm{K})\times e^{5/2}}{(10^5\,\mathrm{N\,m^{-2}})\times(8.82...\times10^{-12}\,\mathrm{m})^3}\right]$$
$$= \boxed{169.7\,\mathrm{J\,K^{-1}\,mol^{-1}}}$$

E13E.6(a) For atoms with filled shells the only contribution to the entropy is translational. The standard molar entropy of a monatomic perfect gas is given by the Sackur–Tetrode equation [13E.9b, 563]

$$S_m^{\ominus} = R\ln\left(\frac{kTe^{5/2}}{p^{\ominus}\Lambda^3}\right) \qquad \Lambda = h/(2\pi mkT)^{1/2}$$

It follows that $S_m^{\ominus} = A\ln(T^{5/2}m^{3/2})$, where A is a constant. Therefore

$$S_m^{\ominus}(\mathrm{He},T_1) - S_m^{\ominus}(\mathrm{Xe},T_2) = A\ln(T_1^{5/2}m_{\mathrm{He}}^{3/2}) - A\ln(T_2^{5/2}m_{\mathrm{Xe}}^{3/2})$$

If $S_m^{\ominus}(\mathrm{He},T_1) = S_m^{\ominus}(\mathrm{Xe},298)$

$$0 = A\ln[T_1^{5/2}m_{\mathrm{He}}^{3/2}] - A\ln[(298\,\mathrm{K})^{5/2}m_{\mathrm{Xe}}^{3/2}]$$
$$\text{hence } T_1^{5/2}m_{\mathrm{He}}^{3/2} = (298\,\mathrm{K})^{5/2}m_{\mathrm{Xe}}^{3/2}$$
$$T_1 = \left[\frac{(298\,\mathrm{K})^{5/2}\times(131.29)^{3/2}}{(4.00)^{3/2}}\right]^{2/5} = \boxed{2.42\times10^3\,\mathrm{K}}$$

E13E.7(a) The rotational partition function for a non-linear molecule is given by [13B.14–545]

$$q^R = \frac{1}{\sigma}\left(\frac{kT}{hc}\right)^{3/2}\left(\frac{\pi}{\tilde{A}\tilde{B}\tilde{C}}\right)^{1/2}$$

For H_2O the symmetry factor $\sigma = 2$. At 298 K

$$kT/hc = \frac{(1.3806\times10^{-23}\,\mathrm{J\,K^{-1}})\times(298\,\mathrm{K})}{(6.6261\times10^{-34}\,\mathrm{J\,s})\times(2.9979\times10^{10}\,\mathrm{cm\,s^{-1}})}$$
$$= 207.1...\,\mathrm{cm^{-1}}$$
$$q^R = \tfrac{1}{2}(207.1...\,\mathrm{cm^{-1}})^{3/2}\times$$
$$\left(\frac{\pi}{(27.878\,\mathrm{cm^{-1}})\times(14.509\,\mathrm{cm^{-1}})\times(9.827\,\mathrm{cm^{-1}})}\right)^{1/2}$$
$$= 43.1... = \boxed{43.1}$$

The entropy is given in terms of the partition function by [13E.8a–*562*]

$$S_m = [U_m(T) - U_m(0)]/T + R\ln q$$

This is the appropriate form for the rotational contribution; for the translational contribution the ln term is $\ln qe/N$. At 298 K $kT/hc = 207\ \mathrm{cm^{-1}}$ which is significantly greater than any of the rotational constants, therefore the equipartition theorem can be used to find $U_m(T)$: there are three rotational modes, therefore $U_m(T) - U_m(0) = \frac{3}{2}RT$.

$$\begin{aligned} S_m^R &= (\tfrac{3}{2}RT)/T + R\ln q^R = R(\tfrac{3}{2} + \ln q^R) \\ &= (8.3145\ \mathrm{J\,K^{-1}\,mol^{-1}}) \times [\tfrac{3}{2} + \ln(43.1...)] \\ &= \boxed{43.76\ \mathrm{J\,K^{-1}\,mol^{-1}}} \end{aligned}$$

E13E.8(a) Only the ground electronic state contributes to the electronic partition function, which is therefore simply the degeneracy of the ground state $q^E = g_0$. For a given term the degeneracy is given by the value of J, which is the right subscript: $g_0 = (2J+1) = (2 \times \frac{9}{2} + 1) = 10$. The entropy is given in terms of the partition function by [13E.8a–*562*]

$$S_m = [U_m(T) - U_m(0)]/T + R\ln q$$

This is the appropriate form for the electronic contribution; for the translational contribution the ln term is $\ln qe/N$. In this case $U_m(T) - U_m(0) = 0$ as only the ground state is considered

$$S_m = R\ln q = R\ln 10 = \boxed{19.14\ \mathrm{J\,K^{-1}\,mol^{-1}}}$$

E13E.9(a) The contribution of a collection of harmonic oscillators to the standard molar entropy is given by [13E.12b–*564*] (note that there is an error in the expression in the text: the argument of the exponential term in the ln should be negative)

$$S_m^V = R\left[\frac{\theta^V/T}{e^{\theta^V/T} - 1} - \ln(1 - e^{-\theta^V/T})\right] \qquad \theta^V = hc\tilde{\nu}/k$$

The following table shows the vibrational temperatures and the contribution to the molar entropy for each of the normal modes

		298 K		500 K	
$\tilde{\nu}/\mathrm{cm^{-1}}$	θ^V/K	θ^V/T	S_m^V/R	θ^V/T	S_m^V/R
625	899	3.02	0.205	1.80	0.538
638	918	3.08	0.195	1.84	0.522
1033	1486	4.988	0.04110	2.973	0.2128
1105	1590	5.335	0.03066	3.180	0.1805
1229	1768	5.934	0.01841	3.537	0.1356
1387	1996	6.697	9.515×10^{-3}	3.991	0.09378
1770	2547	8.546	1.855×10^{-3}	5.093	0.03761
2943	4234	14.21	1.026×10^{-5}	8.469	1.988×10^{-3}
3570	5137	17.24	5.957×10^{-7}	10.27	3.895×10^{-4}

The molar entropy is obtained by summing the contributions from each normal mode. Thus at 298 K $\boxed{S_m^V = 4.18\ \mathrm{J\,K^{-1}\,mol^{-1}}}$ and at 500 K $\boxed{S_m^V = 14.3\ \mathrm{J\,K^{-1}\,mol^{-1}}}$.

Solutions to problems

P13E.1 The electronic levels of NO form a two-level system, an expression for the heat capacity of which is derived in the solution to *Exercise* E13E.3(a).

$$C_{V,\mathrm{m}} = \frac{N_\mathrm{A}(hc\tilde{\nu})^2}{kT^2}\frac{g_0g_1\mathrm{e}^{\beta hc\tilde{\nu}}}{(g_0\mathrm{e}^{\beta hc\tilde{\nu}}+g_1)^2} = N_\mathrm{A}k(\beta hc\tilde{\nu})^2\frac{g_0g_1\mathrm{e}^{\beta hc\tilde{\nu}}}{(g_0\mathrm{e}^{\beta hc\tilde{\nu}}+g_1)^2}$$

where g_0 and g_1 are the degeneracies of the ground and first excited state, respectively. For NO $g_0 = 2$ and $g_1 = 2$. With the data given

$$\begin{aligned}\beta hc\tilde{\nu} &= hc\tilde{\nu}/k \times T^{-1}\\ &= \frac{(6.6261\times10^{-34}\ \mathrm{J\,s})\times(2.9979\times10^{10}\ \mathrm{cm\,s^{-1}})\times(121.1\ \mathrm{cm^{-1}})}{1.3806\times10^{-23}\ \mathrm{J\,K^{-1}}}\times T^{-1}\\ &= (174.2...\ \mathrm{K})\times T^{-1}\end{aligned}$$

A plot of $C_{V,\mathrm{m}}$ as a function of T is shown in Fig. 13.15.

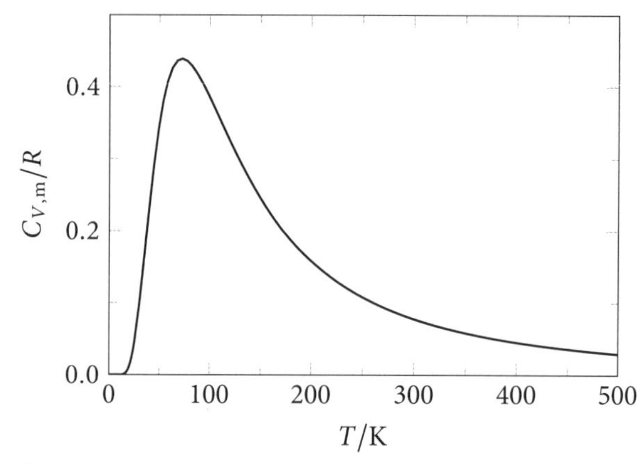

Figure 13.15

P13E.3 The energy levels for a particle on a ring are given by [7F.4–283], $E_m = m^2\hbar^2/2I$ where I is the moment of inertia and $m = 0, \pm1, \pm2, \dots$. In the high-temperature limit the partition function is well-approximated by an intergral

$$q^\mathrm{R} = \sum_m \mathrm{e}^{-\beta E_m} = \int_{-\infty}^{+\infty}\mathrm{e}^{-\beta m^2\hbar^2/2I}\,\mathrm{d}m$$

The integral is of the form of Integral G.1 with $k' = \beta\hbar^2/2I$; the value needed is twice that given for G.1 as that integral is from 0 to $+\infty$.

$$\boxed{q^\mathrm{R} = \left(\frac{2\pi I}{\beta\hbar^2}\right)^{1/2}}$$

These energy levels contribute one quadratic contribution to the energy, that is there is one rotational mode. In the high-temperature limit the equipartition theorem applies and hence the internal energy is $U_m = \frac{1}{2}RT$ and $\boxed{C^R_{V,m} = \frac{1}{2}R}$.

The entropy is given in terms of the partition function by [13E.8a–562]

$$S_m = [U_m(T) - U_m(0)]/T + R \ln q$$

This is the appropriate form for the rotational contribution; for the translational contribution the ln term is $\ln qe/N$. As has already been explained, $U_m(T) - U_m(0) = \frac{1}{2}RT$, therefore

$$S^R_m = (\tfrac{1}{2}RT)/T + R\ln q^R = R(\tfrac{1}{2} + \ln q^R)$$

With the data given

$$q^R = \left(\frac{2\pi I}{\beta\hbar^2}\right)^{1/2} = \left(\frac{2\pi kTI}{\hbar^2}\right)^{1/2}$$

$$= \left(\frac{2\pi(1.3806 \times 10^{-23}\,\mathrm{J\,K^{-1}}) \times (298\,\mathrm{K}) \times (5.341 \times 10^{-47}\,\mathrm{kg\,m^2})}{(1.0546 \times 10^{-34}\,\mathrm{J\,s})^2}\right)^{1/2}$$

$$= 11.1...$$

$$S^R_m = (8.3145\,\mathrm{J\,K^{-1}\,mol^{-1}}) \times [\tfrac{1}{2} + \ln(11.1...)] = \boxed{24.1\,\mathrm{J\,K^{-1}\,mol^{-1}}}$$

This calculation is for a particle on a ring. When used as a model for a rotating CH_3 group a symmetry factor of 3 is needed, so that q^R is one third of the value calculated here, giving $S^R_m = 15.1\,\mathrm{J\,K^{-1}\,mol^{-1}}$.

P13E.5 The characteristic vibrational temperature is defined as $\theta^V = hc\tilde{\nu}/k$. It follows that $\tilde{\nu} = k\theta^V/hc$, so the vibrational frequency for a characteristic temperature of 1000 K is

$$\tilde{\nu} = \frac{(1.3806 \times 10^{-23}\,\mathrm{J\,K^{-1}}) \times (1000\,\mathrm{K})}{(6.6261 \times 10^{-34}\,\mathrm{J\,s}) \times (2.9979 \times 10^{10}\,\mathrm{cm\,s^{-1}})} = 695\,\mathrm{cm^{-1}}$$

The following modes have vibrational frequencies of 695 $\mathrm{cm^{-1}}$ or less (the degeneracies are given in parentheses)

$$525(3)\ 578(3)\ 354(3)\ 345(4)\ 403(5)\ 525(5)\ 667(5)$$

Thus in total there are $\boxed{28}$ modes with characteristic vibrational temperatures of less than 1000 K.

The contributions to $C_{V,m}$ are three translational, three rotational and 28 vibrational modes, giving a heat capacity of $C_{V,m} = \frac{3}{2}R + \frac{3}{2}R + 28R = \boxed{31R}$

P13E.7 The partition function of this three-level system is

$$q = 1 + e^{-\beta\varepsilon} + e^{-2\beta\varepsilon}$$

The mean energy is given by [13C.4a–*549*], $E_{\text{mean}} = -(1/q)(\partial q/\partial\beta)_V$

$$E_{\text{mean}} = \frac{\varepsilon\,e^{-\beta\varepsilon} + 2\varepsilon\,e^{-2\beta\varepsilon}}{1 + e^{-\beta\varepsilon} + e^{-2\beta\varepsilon}} = \frac{\varepsilon\,e^{\beta\varepsilon} + 2\varepsilon}{e^{2\beta\varepsilon} + e^{\beta\varepsilon} + 1}$$

$$\text{hence } U_{\text{m}} = N_{\text{A}}E_{\text{mean}} = N_{\text{A}}\varepsilon\,\frac{e^{\beta\varepsilon} + 2}{e^{2\beta\varepsilon} + e^{\beta\varepsilon} + 1}$$

The entropy is given in terms of the partition function by [13E.8a–*562*]

$$\begin{aligned} S_{\text{m}} &= [U_{\text{m}}(T) - U_{\text{m}}(0)]/T + R\ln q \\ &= \frac{N_{\text{A}}\varepsilon}{T}\,\frac{e^{\beta\varepsilon} + 2}{e^{2\beta\varepsilon} + e^{\beta\varepsilon} + 1} + R\ln(1 + e^{-\beta\varepsilon} + e^{-2\beta\varepsilon}) \\ &= R\left(\frac{\beta\varepsilon(e^{\beta\varepsilon} + 2)}{e^{2\beta\varepsilon} + e^{\beta\varepsilon} + 1} + \ln(1 + e^{-\beta\varepsilon} + e^{-2\beta\varepsilon})\right) \end{aligned}$$

where to go to the last line $N_{\text{A}}/T = R\beta$ is used. At high temperatures, $\beta\varepsilon \to 0$ and S_{m} tends to $R\ln 3$. At low temperatures, $\beta\varepsilon \to \infty$ and S_{m} tends to 0, as expected.

P13E.9 The data in *Problem* P13B.8 fit very well to the terms $\tilde{F}(J) = \tilde{B}J(J+1)$ with $\tilde{B} = 10.593\ \text{cm}^{-1}$. The rotational contribution to the entropy is given by [13E.8a–*562*]

$$S_{\text{m}}^{\text{R}} = [U_{\text{m}}(T) - U_{\text{m}}(0)]/T + R\ln q^{\text{R}}$$

where the internal energy is given by [13E.2a–*559*]

$$U_{\text{m}}(T) - U_{\text{m}}(0) = -\frac{N_{\text{A}}}{q^{\text{R}}}\left(\frac{\partial q^{\text{R}}}{\partial\beta}\right)_V$$

and the partition function is

$$q^{\text{R}} = \sum_J (2J+1)e^{-hc\beta\tilde{B}J(J+1)}$$

In order to compute the entropy down to low temperatures it is necessary to evaluate the sums term by term rather than approximating them by an integral. The derivative of q^{R} is

$$\left(\frac{\partial q^{\text{R}}}{\partial\beta}\right)_V = -hc\tilde{B}\sum_J (2J+1)[J(J+1)]e^{-hc\beta\tilde{B}J(J+1)}$$

It is convenient to rewrite these expressions in terms of the characteristic vibrational temperature $\theta^{\text{R}} = hc\tilde{B}/k$: using this $hc\tilde{B} = k\theta^{\text{R}}$ and $hc\beta\tilde{B} = \theta^{\text{R}}/T$. With these, the expressions for q^{R} and its derivative become

$$q^{\text{R}} = \sum_J (2J+1)e^{-\theta^{\text{R}}J(J+1)/T} \qquad \left(\frac{\partial q^{\text{R}}}{\partial\beta}\right)_V = -k\theta^{\text{R}}\sum_J (2J+1)[J(J+1)]e^{-\theta^{\text{R}}J(J+1)/T}$$

The internal energy is therefore

$$U_m(T) - U_m(0) = \frac{R\theta^R}{q^R} \sum_J (2J+1)[J(J+1)]e^{-\theta^R J(J+1)/T}$$

The sums are best evaluated using mathematical software and the results are expressed in terms of the dimensionless parameter T/θ^R. The result of such a calculation is shown in Fig. 13.16

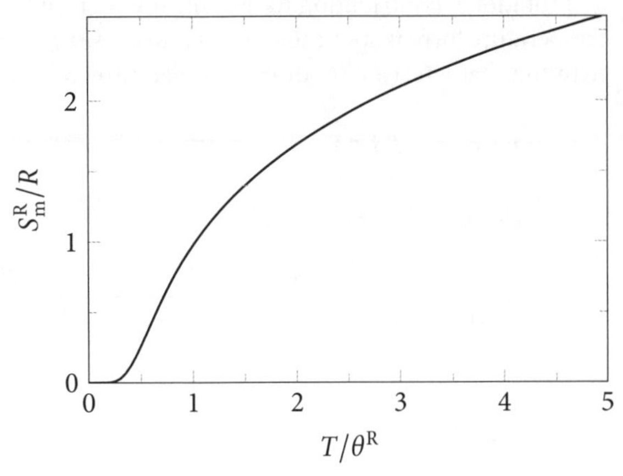

Figure 13.16

P13E.11 Contributions to the entropy from translation, rotation and vibration are expected. The molecule has a doubly-degenerate ground electronic state, so this will also contribute to the entropy. However, the excited electronic states are at energies very much greater than kT at 298 K (kT at 298 K is 0.026 eV), so their contribution is negligible.

The translational contribution to the standard molar entropy is given by the Sackur–Tetrode equation [13E.9b–*563*]

$$S_m^T = R\ln\left(\frac{kTe^{5/2}}{p^\ominus \Lambda^3}\right) \qquad \Lambda = h/(2\pi mkT)^{1/2}$$

Taking the mass of F_2^- as 38.00 m_u

$$\Lambda = \frac{6.6261\times10^{-34}\,\text{J s}}{\left[2\pi(38.00\times1.6605\times10^{-27}\,\text{kg})\times(1.3806\times10^{-23}\,\text{J K}^{-1})\times(298\,\text{K})\right]^{1/2}}$$
$$= 1.64...\times10^{-11}\,\text{m}$$

$$S_m^T = (8.3145\,\text{J K}^{-1}\,\text{mol}^{-1})\times\ln\left[\frac{(1.3806\times10^{-23}\,\text{J K}^{-1})\times(298\,\text{K})\times e^{5/2}}{(10^5\,\text{N m}^{-2})\times(1.64...\times10^{-11}\,\text{m})^3}\right]$$
$$= 1.54...\times10^2\,\text{J K}^{-1}\,\text{mol}^{-1}$$

The rotational constant is given by [11B.7–*432*], $\tilde{B} = \hbar/4\pi cI$, with $I = \mu R^2$ and $\mu = \frac{1}{2}m$ for a homonuclear diatomic.

$$\tilde{B} = \frac{\hbar}{4\pi cI} = \frac{\hbar}{2\pi cmR^2}$$

$$= \frac{1.0546 \times 10^{-34}\,\text{J s}}{2\pi(2.9979 \times 10^{10}\,\text{cm s}^{-1}) \times [19.00 \times 1.6605 \times 10^{-27}\,\text{kg})] \times (190.0 \times 10^{-12}\,\text{m})^2}$$

$$= 0.491...\,\text{cm}^{-1}$$

The rotational contribution to the entropy is given by [13E.11a–*564*]; this high-temperature form is applicable at 298 K because this temperature is much higher than the characteristic rotational temperature, $\theta^{\text{R}} = hc\tilde{B}/k = 0.707$ K.

$$S_{\text{m}}^{\text{R}} = R\left(1 + \ln\frac{kT}{\sigma hc\tilde{B}}\right)$$

$$= (8.3145\,\text{J K}^{-1}\,\text{mol}^{-1}) \times \left(1 + \ln\frac{(1.3806 \times 10^{-23}\,\text{J K}^{-1}) \times (298\,\text{K})}{2(6.6261 \times 10^{-34}\,\text{J s}) \times (2.9979 \times 10^{10}\,\text{cm s}^{-1}) \times (0.491...\,\text{cm}^{-1})}\right)$$

$$= 52.7...\,\text{J K}^{-1}\,\text{mol}^{-1}$$

The characteristic vibrational temperature is

$$\theta^{\text{V}} = hc\tilde{\nu}/k$$

$$= \frac{(6.6261 \times 10^{-34}\,\text{J s}) \times (2.9979 \times 10^{10}\,\text{cm s}^{-1}) \times (450.0\,\text{cm}^{-1})}{1.3806 \times 10^{-23}\,\text{J K}^{-1}}$$

$$= 647.4...\,\text{K}$$

The vibrational contribution to the standard molar entropy is given by [13E.12b–*564*] (note that there is an error in the expression in the text: the argument of the exponential term in the ln should be negative)

$$S_{\text{m}}^{\text{V}} = R\left[\frac{\theta^{\text{V}}/T}{\text{e}^{\theta^{\text{V}}/T} - 1} - \ln(1 - \text{e}^{-\theta^{\text{V}}/T})\right]$$

$$= (8.3145\,\text{J K}^{-1}\,\text{mol}^{-1}) \times \left[\frac{(647.4...\,\text{K})/(298\,K)}{\text{e}^{(647.4...\,\text{K})/(298\,\text{K})} - 1} - \ln[1 - \text{e}^{(-647.4...\,\text{K})/(298\,\text{K})}]\right]$$

$$= 3.32...\,\text{J K}^{-1}\,\text{mol}^{-1}$$

The electronic partition function is $q^{\text{E}} = g_0 = 2$, therefore the electronic contribution to the molar entropy is

$$S_{\text{m}}^{\text{E}} = R\ln q^{\text{E}} = (8.3145\,\text{J K}^{-1}\,\text{mol}^{-1}) \times \ln 2 = 5.76...\,\text{J K}^{-1}\,\text{mol}^{-1}$$

The molar entropy is therefore

$$S_{\text{m}}^{\ominus} = S_{\text{m}}^{\text{T}} + S_{\text{m}}^{\text{R}} + S_{\text{m}}^{\text{V}} + S_{\text{m}}^{\text{E}} = 1.54... \times 10^2 + 52.7... + 3.32... + 5.76... = \boxed{216.1\,\text{J K}^{-1}\,\text{mol}^{-1}}$$

P13E.13 It is convenient to calculate the entropy using the approach set out in *Problem* P13E.8 in which the two quantities q (the partition function) and $\dot{q}$ are defined

$$q = \sum_j e^{-\beta\varepsilon_j} \qquad \dot{q} = \sum_j \beta\varepsilon_j e^{-\beta\varepsilon_j}$$

It is shown in the solution to that problem that the entropy can be written in terms of these as

$$S_m = R\left(\frac{\dot{q}}{q} + \ln q\right) \tag{13.2}$$

As discussed in Section 11C.3(a) on page 444, the Morse oscillator has a finite number of bound levels between the ground level and the dissociation limit. The number of these is found by noting that E_v reaches a maximum value at the dissociation limit, and therefore this limit is found by solving $dE_v/dv = 0$

$$\frac{d}{dv}\left[(v+\tfrac{1}{2})hc\tilde{\nu} - (v+\tfrac{1}{2})^2 hcx_e\tilde{\nu}\right] = hc\tilde{\nu} - 2(v+\tfrac{1}{2})hcx_e\tilde{\nu}$$

$$\text{solving}\quad 0 = hc\tilde{\nu} - 2(v_{max}+\tfrac{1}{2})hcx_e\tilde{\nu} \quad\text{gives}\quad v_{max} = 1/2x_e - \tfrac{1}{2}$$

The energy of the lowest state is $E_0 = \frac{1}{2}hc\tilde{\nu} - \frac{1}{4}hcx_e\tilde{\nu}$, therefore the energies used to evaluate the partition function are

$$\begin{aligned} E'_v &= E_v - E_0 \\ &= \left[(v+\tfrac{1}{2})hc\tilde{\nu} - (v+\tfrac{1}{2})^2 hcx_e\tilde{\nu}\right] - \left[\tfrac{1}{2}hc\tilde{\nu} - \tfrac{1}{4}hcx_e\tilde{\nu}\right] \\ &= vhc\tilde{\nu} - (v^2+v)hcx_e\tilde{\nu} \end{aligned}$$

The partition function is evaluated from the sum

$$q = \sum_{v=0}^{v_{max}} e^{-(vhc\tilde{\nu}-(v^2+v)hcx_e\tilde{\nu})/kT}$$

Defining the characteristic vibrational temperature as $\theta^V = hc\tilde{\nu}/k$ gives

$$q = \sum_{v=0}^{v_{max}} e^{-(v-(v^2+v)x_e)\theta^V/T}$$

The sum needed to compute the quantity $\dot{q}$ is written, by analogy, as

$$\dot{q} = \sum_{v=0}^{v_{max}} (v-(v^2+v)x_e)\frac{\theta^V}{T} e^{-(v-(v^2+v)x_e)\theta^V/T}$$

These results are used with eqn 13.2 to compute the entropy. For comparison, for a harmonic oscillator the entropy is given by [13E.12b–*564*] (note that there is an error in the expression in the text: the argument of the exponential term in the ln should be negative)

$$S_m/R = \frac{\theta^V/T}{e^{\theta^V/T}-1} - \ln(1-e^{-\theta^V/T})$$

Figure 13.17 compares the entropy for various values of x_e with that for the harmonic case. For the smallest value of x_e the entropy is initially larger than that for the harmonic oscillator. This can be attributed to fact that the energy levels are more closely spaced for the Morse oscillator than for the harmonic oscillator. However, at higher temperatures the entropy for the Morse oscillator starts to level off because there are a finite number of levels, whereas for the harmonic case the entropy continues to increase without limit as there are an infinite number of levels. This behaviour is even more pronounced for $x_e = 0.05$ and $x_e = 0.10$, with the plateau at high temperatures being evident.

These plots are, however, somewhat unrealistic. For a typical molecule $\theta^V \approx 1000$ K, so at 298 K $T/\theta^V \approx 0.1$, and x_e is around 0.001. With these parameters the contribution to the entropy determined using the Morse levels is $5.0 \times 10^{-4} \times R$; the result obtained using the harmonic levels is the same. This is because there is very little contribution from excited vibrational states, so the small difference between these low-lying states for the Morse and harmonic oscillators has no significant effect on the partition function.

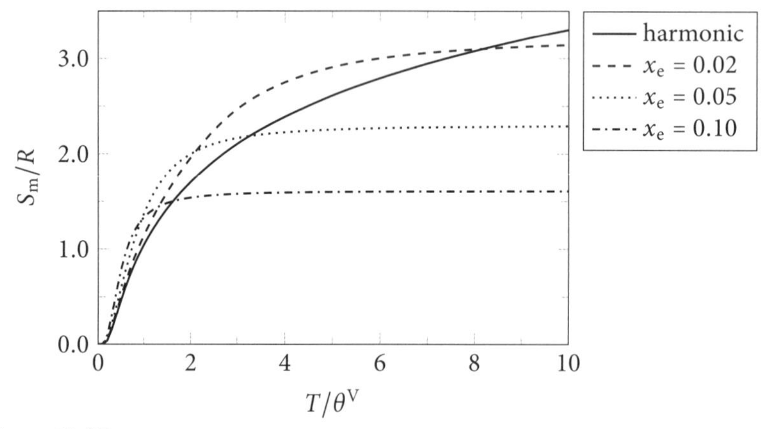

Figure 13.17

P13E.15 The partition function for a particle confined to a box of length X in one dimension is given by [13B.7–*541*]

$$q_X = X/\Lambda \quad \Lambda = h/(2\pi mkT)^{1/2}$$

Therefore the partition function for a particle confined to a two-dimensional box of dimensions X and Y is

$$q_{XY} = q_X q_Y = XY/\Lambda^2 = A/\Lambda^2$$

where A is the area. Because there are two translational modes, the internal energy of such a system is given by the equipartition theorem as $U = nRT$, where n is the amount in moles. The entropy of n moles is given by [13E.8b–

562]

$$
\begin{aligned}
S &= U/T + Nk\ln qe/N \\
&= nR + Nk\ln\frac{Ae}{\Lambda^2 N} = nR + nN_A k\ln\frac{Ae}{\Lambda^2 nN_A} \\
&= nR + nR\ln\frac{A_m e}{\Lambda^2 N_A}
\end{aligned}
$$

where in the last step the molar area, $A_m = A/n$, is introduced. The molar entropy is therefore

$$
\begin{aligned}
S_m^{2D} &= R + R\ln\frac{A_m e}{\Lambda^2 N_A} = R\ln e + R\ln\frac{A_m e}{\Lambda^2 N_A} \\
&= \boxed{R\ln\frac{A_m e^2}{\Lambda^2 N_A}}
\end{aligned}
$$

The translational molar entropy in three dimensions is given by [13E.9a–563]

$$
S_m^{3D} = R\ln\frac{V_m e^{5/2}}{\Lambda^3 N_A}
$$

Therefore the molar entropy of condensation is

$$
\begin{aligned}
\Delta S_{cond.} &= S_m^{2D} - S_m^{3D} \\
&= R\ln\frac{A_m e^2}{\Lambda^2 N_A} - R\ln\frac{V_m e^{5/2}}{\Lambda^3 N_A} \\
&= \boxed{R\ln\frac{A_m \Lambda}{V_m e^{1/2}}}
\end{aligned}
$$

P13E.17 If there are N binucleotides of four different kinds then $\mathcal{W} = 4^N$ and

$$
\begin{aligned}
S &= k\ln\mathcal{W} = Nk\ln 4 = (5\times10^8)\times(1.3806\times10^{-23}\,\mathrm{J\,K^{-1}})\times\ln 4 \\
&= \boxed{9.6\times10^{-15}\,\mathrm{J\,K^{-1}}}
\end{aligned}
$$

13F Derived functions

Answers to discussion questions

D13F.1 The relationship between pressure and the canonical partition function is given by [13F.3–567]

$$
p = kT\left(\frac{\partial \ln Q}{\partial V}\right)_T
$$

The derivative is a relationship between two non-negative quantities (Q and V). Furthermore, the derivative itself must be non-negative since the pressure is positive and so is kT. Therefore, the derivative describes a monotonic relationship between Q and V: Q can only increase or remain constant with increasing V. The relationship might almost be interpreted anthropomorphically: to the extent that increasing the volume makes more states available (increasing Q), the system will push harder (higher p) to bring about that volume increase. Ultimately, it is a reflection of the Second Law of Thermodynamics and the tendency toward increased disorder as quantified by the partition function.

D13F.3 This is discussed in Section 13F.2(c) on page 571.

Solutions to exercises

E13F.1(a) The Gibbs energy is computed from the partition function using [13F.8–*568*], $G(T) = G(0) - nRT\ln q/N$. As usual, the partition function is factored into separate contributions from translation, rotation and so on. The factor of $1/N$ is usually taken with the translational contribution, so that, for example, the rotational contribution to the Gibbs energy is $-nRT\ln q^{\mathrm{R}}$, or $-RT\ln q^{\mathrm{R}}$ for the molar quantity.

For a centro-symmetric linear molecule the rotational partition function in the high-temperature limit is given by [13B.13a–*544*]

$$\begin{aligned}
q^{\mathrm{R}} &= \frac{kT}{2hc\tilde{B}} \\
&= \frac{(1.3806\times10^{-23}\,\mathrm{J\,K^{-1}})\times(298\,\mathrm{K})}{2(6.6261\times10^{-34}\,\mathrm{J\,s})\times(2.9979\times10^{10}\,\mathrm{cm\,s^{-1}})\times(0.3902\,\mathrm{cm^{-1}})} \\
&= 265.3\ldots \\
G_{\mathrm{m}}^{\mathrm{R}} &= -RT\ln q^{\mathrm{R}} \\
&= -(8.3145\,\mathrm{J\,K^{-1}\,mol^{-1}})\times(298\,\mathrm{K})\times\ln(265.3\ldots) = \boxed{-13.83\,\mathrm{kJ\,mol^{-1}}}
\end{aligned}$$

The vibrational partition function for each mode is given by [13B.15–*546*], $q^{\mathrm{V}} = 1/(1-\mathrm{e}^{-\beta hc\tilde{\nu}})$, where $\beta = 1/kT$. The overall vibrational partition function is the product of the partition functions of the individual modes, taking into account any degeneracy. In this case the contribution from the mode at 667.4 $\mathrm{cm^{-1}}$ is included twice.

$$\begin{aligned}
hc\beta &= \frac{(6.6261\times10^{-34}\,\mathrm{J\,s})\times(2.9979\times10^{10}\,\mathrm{cm\,s^{-1}})}{(1.3806\times10^{-23}\,\mathrm{J\,K^{-1}})\times(298\,\mathrm{K})} \\
&= 4.82\ldots\times10^{-3}\,\mathrm{cm}
\end{aligned}$$

$$\begin{aligned}
q_1^{\mathrm{V}} &= \left(1-\mathrm{e}^{-hc\beta\tilde{\nu}_1}\right)^{-1} \\
&= \left(1-\mathrm{e}^{-(4.82\ldots\times10^{-3}\,\mathrm{cm})\times(1388.2\,\mathrm{cm^{-1}})}\right)^{-1} = 1.00\ldots \\
q_2^{\mathrm{V}} &= \left(1-\mathrm{e}^{-(4.82\ldots\times10^{-3}\,\mathrm{cm})\times(2349.2\,\mathrm{cm^{-1}})}\right)^{-1} = 1.00\ldots \\
q_3^{\mathrm{V}} &= \left(1-\mathrm{e}^{-(4.82\ldots\times10^{-3}\,\mathrm{cm})\times(667.4\,\mathrm{cm^{-1}})}\right)^{-1} = 1.04\ldots \\
q^{\mathrm{V}} &= q_1^{\mathrm{V}}\times q_2^{\mathrm{V}}\times(q_3^{\mathrm{V}})^2 = (1.00\ldots)\times(1.00\ldots)\times(1.04\ldots)^2 = 1.08\ldots
\end{aligned}$$

Hence

$$G_m^V = -RT \ln q^V$$
$$= -(8.3145\,\mathrm{J\,K^{-1}\,mol^{-1}}) \times (298\,\mathrm{K}) \times \ln(1.08...) = \boxed{-0.204\,\mathrm{kJ\,mol^{-1}}}$$

E13F.2(a) The Gibbs energy is computed from the partition function using [13F.8–*568*], $G(T) = G(0) - nRT \ln q/N$. As usual, the partition function is factored into separate contributions from translation, rotation and so on. The factor of $1/N$ is usually taken with the translational contribution, therefore the electronic contribution to the Gibbs energy is $-nRT \ln q^E$, or $-RT \ln q^E$ for the molar quantity.

The electronic partition function of this two-level system is

$$q^E = g_0 + g_1 e^{-\beta hc\tilde{\nu}}$$

where g_0 and g_1 are the degeneracies of the ground and first excited state, respectively. For an electronic term the degeneracy is $2J+1$, hence $g_0 = 2 \times \frac{3}{2} + 1 = 4$ and $g_1 = 2 \times \frac{1}{2} + 1 = 2$. With the data given

$$\beta hc\tilde{\nu} = hc\tilde{\nu}/k \times T^{-1}$$
$$= \frac{(6.6261 \times 10^{-34}\,\mathrm{J\,s}) \times (2.9979 \times 10^{10}\,\mathrm{cm\,s^{-1}}) \times (881\,\mathrm{cm^{-1}})}{1.3806 \times 10^{-23}\,\mathrm{J\,K^{-1}}} \times T^{-1}$$
$$= (1.26... \times 10^3\,\mathrm{K}) \times T^{-1}$$

at 500 K

$$q^E = 4 + 2e^{-(1.26... \times 10^3\,\mathrm{K})/(500\,\mathrm{K})} = 4.15...$$
$$G_m^E = -RT \ln q^E$$
$$= -(8.3145\,\mathrm{J\,K^{-1}\,mol^{-1}}) \times (500\,\mathrm{K}) \times \ln(4.15...) = \boxed{-5.92\,\mathrm{kJ\,mol^{-1}}}$$

at 900 K

$$q^E = 4 + 2e^{-(1.26... \times 10^3\,\mathrm{K})/(900\,\mathrm{K})} = 4.48...$$
$$G_m^E = -(8.3145\,\mathrm{J\,K^{-1}\,mol^{-1}}) \times (900\,\mathrm{K}) \times \ln(4.48...) = \boxed{-11.2\,\mathrm{kJ\,mol^{-1}}}$$

E13F.3(a) The equilibrium constant for this dissociation reaction is computed using [13F.12–*571*]

$$K = \frac{g_I^2 kT\Lambda_{I_2}^3}{g_{I_2} p^{\ominus} q_{I_2}^R q_{I_2}^V \Lambda_I^6} e^{-hc\tilde{D}_0/kT}$$

where $g_{\rm I} = 4$ and $g_{\rm I_2} = 1$. The various factors are computed separately. Λ is given by [13B.7–*541*]

$$\Lambda = h/(2\pi mkT)^{1/2}$$

$$\Lambda_{\rm I_2} = \frac{6.6261\times10^{-34}\,{\rm J\,s}}{[2\pi(253.8\times1.6605\times10^{-27}\,{\rm kg})\times(1.3806\times10^{-23}\,{\rm J\,K^{-1}})\times(1000\,{\rm K})]^{1/2}}$$
$$= 3.46...\times10^{-12}\,{\rm m}$$
$$\Lambda_{\rm I} = 4.90...\times10^{-12}\,{\rm m}$$

The rotational partition function for a homonuclear diatomic in the high-temperature limit is given by [13B.13a–*544*]

$$q^{\rm R} = \frac{kT}{2hc\tilde{B}}$$
$$= \frac{(1.3806\times10^{-23}\,{\rm J\,K^{-1}})\times(1000\,{\rm K})}{2(6.6261\times10^{-34}\,{\rm J\,s})\times(2.9979\times10^{10}\,{\rm cm\,s^{-1}})\times(0.0373\,{\rm cm^{-1}})} = 9.31...\times10^{3}$$

The vibrational partition function is given by [13B.15–*546*], $q^{\rm V} = (1-{\rm e}^{-hc\tilde{\nu}/kT})^{-1}$

$$hc\tilde{\nu}/kT = \frac{(6.6261\times10^{-34}\,{\rm J\,s})\times(2.9979\times10^{10}\,{\rm cm\,s^{-1}})\times(214.36\,{\rm cm^{-1}})}{(1.3806\times10^{-23}\,{\rm J\,K^{-1}})\times(1000\,{\rm K})}$$
$$= 0.308...$$
$$q^{\rm V} = (1-{\rm e}^{-0.308...})^{-1} = 3.76...$$

The dissociation energy is computed from the well depth (the conversion factor from eV to cm^{-1} from inside the front cover is used), using the energy of the ground state of the harmonic oscillator $\tilde{\varepsilon}_0 = \frac{1}{2}\tilde{\nu}$

$$\tilde{D}_0 = \tilde{D}_{\rm e} - \tilde{\varepsilon}_0$$
$$= (1.5422\,{\rm eV})\times\frac{8065.5\,{\rm cm^{-1}}}{1\,{\rm eV}} - \tfrac{1}{2}\times(214.36\,{\rm cm^{-1}})$$
$$= 1.23...\times10^{4}\,{\rm cm^{-1}}$$
$$\frac{hc\tilde{D}_0}{kT} = \frac{(6.6261\times10^{-34}\,{\rm J\,s})\times(2.9979\times10^{10}\,{\rm cm\,s^{-1}})\times(1.23...\times10^{4}\,{\rm cm^{-1}})}{(1.3806\times10^{-23}\,{\rm J\,K^{-1}})\times(1000\,{\rm K})}$$
$$= 17.7...$$
$${\rm e}^{-hc\tilde{D}_0/kT} = {\rm e}^{-17.7...} = 1.96...\times10^{-8}$$

With these results the equilibrium constant is computed as

$$K = \frac{4^2\times(1.3806\times10^{-23}\,{\rm J\,K^{-1}})\times(1000\,{\rm K})\times(3.46...\times10^{-12}\,{\rm m})^3}{1\times(10^5\,{\rm Pa})\times(9.31...\times10^3)\times(3.76...)\times(4.90...\times10^{-12}\,{\rm m})^6}$$
$$\times(1.96...\times10^{-8})$$
$$= \boxed{3.72\times10^{-3}}$$

Solutions to problems

P13F.1 The equilibrium constant for this reaction is given by [13F.10b–*570*]

$$K = \frac{q^{\ominus}_{\mathrm{CHD_3,m}} q^{\ominus}_{\mathrm{DCl,m}}}{q^{\ominus}_{\mathrm{CD_4,m}} q^{\ominus}_{\mathrm{HCl,m}}} e^{-\Delta_r E_0/RT}$$

It is convenient to consider the contribution of each mode to the fraction in the above expression separately.

The standard molar translational partition function is $q^{\ominus}_{\mathrm{m}} = V^{\ominus}_{\mathrm{m}}/\Lambda^3$, with $\Lambda = h/(2\pi mkT)^{1/2}$, therefore $q^{\ominus}_{\mathrm{m}}$ goes as $m^{3/2}$. In the fraction all of the other constants cancel to leave

$$\left(\frac{q^{\ominus}_{\mathrm{CHD_3,m}} q^{\ominus}_{\mathrm{DCl,m}}}{q^{\ominus}_{\mathrm{CD_4,m}} q^{\ominus}_{\mathrm{HCl,m}}}\right)_{\mathrm{trans}} = \left(\frac{m_{\mathrm{CHD_3}} m_{\mathrm{DCl}}}{m_{\mathrm{CD_4}} m_{\mathrm{HCl}}}\right)^{3/2}$$

$$= \left(\frac{19.06 \times 37.46}{20.07 \times 36.46}\right)^{3/2} = 0.964...$$

Assuming the high-temperature limit, the rotational partition function for a heteronuclear diatomic is given by [13B.13b–*544*], $q^{\mathrm{R}} = kT/hc\tilde{B}$, and for a nonlinear molecule by [13B.14–*545*], $q^{\mathrm{R}} = (1/\sigma)(kT/hc)^{3/2}(\pi/\tilde{A}\tilde{B}\tilde{C})^{1/2}$; for CD_4 $\tilde{A} = \tilde{C} = \tilde{B}$ and for CHD_3 $\tilde{C} = \tilde{B}$. The symmetry number is 12 for CD_4 and 3 for CHD_3. In the fraction the terms in kT/hc cancel to leave

$$\left(\frac{q_{\mathrm{CHD_3}} q_{\mathrm{DCl}}}{q_{\mathrm{CD_4}} q_{\mathrm{HCl}}}\right)_{\mathrm{rot}} = \frac{12}{3}\left(\frac{\tilde{B}^3_{\mathrm{CD_4}}}{\tilde{A}_{\mathrm{CHD_3}} \tilde{B}^2_{\mathrm{CHD_3}}}\right)^{1/2} \frac{\tilde{B}_{\mathrm{HCl}}}{\tilde{B}_{\mathrm{DCl}}}$$

$$= \frac{12}{3}\left(\frac{(2.63)^3}{(2.63) \times (3.28)^2}\right)^{1/2} \frac{10.59}{5.445} = 6.23...$$

The vibrational partition function is given by [13B.15–*546*] $q^{\mathrm{V}} = (1 - e^{-hc\tilde{\nu}/kT})^{-1}$, which is conveniently expressed as $q^{\mathrm{V}} = (1 - e^{-(1.4388\ \mathrm{cm\ K})\tilde{\nu}/T})^{-1}$. This term is temperature dependent and so needs to be re-evaluated at each temperature. The vibrational partition function for CHD_3 and CD_4 is the product of the partition function for each normal mode, raised to the power of its degeneracy. For example

$$q^{\mathrm{V}}_{\mathrm{CD_4}} = q^{\mathrm{V}}_{(2109\ \mathrm{cm^{-1}})} \times (q^{\mathrm{V}}_{(1092\ \mathrm{cm^{-1}})})^2 \times (q^{\mathrm{V}}_{(2259\ \mathrm{cm^{-1}})})^3 \times (q^{\mathrm{V}}_{(996\ \mathrm{cm^{-1}})})^3$$

The term $\Delta_r E_0$ is computed as

$$\Delta_r E_0 = E_0(\mathrm{CHD_3}) + E_0(\mathrm{DCl}) - E_0(\mathrm{CD_4}) - E_0(\mathrm{HCl})$$

To a good approximation it can be assumed that the pure electronic energy of a species is unaffected by isotopic substitution, however the vibrational zero point energy will be affected. For a harmonic oscillator the energy of the ground state is $\frac{1}{2}hc\tilde{\nu}$, therefore to compute the total vibrational zero point energy of

CHD_3 and CD_4 the contribution from each normal mode has to be taken into account; a mode with degeneracy g contributes $g \times \frac{1}{2}hc\tilde{\nu}$.

$$E_0(\mathrm{CHD_3})_{\mathrm{vib}} = \tfrac{1}{2}N_\mathrm{A}hc(2993 + 2142 + 3 \times 1003 + 2 \times 1291 + 2 \times 1036) = N_\mathrm{A}hc(6399\ \mathrm{cm^{-1}})$$

$$E_0(\mathrm{CD_4})_{\mathrm{vib}} = \tfrac{1}{2}N_\mathrm{A}hc(2109 + 2 \times 1092 + 3 \times 2259 + 3 \times 996) = N_\mathrm{A}hc(7029\ \mathrm{cm^{-1}})$$

$$E_0(\mathrm{HCl})_{\mathrm{vib}} = N_\mathrm{A}hc(1495.5\ \mathrm{cm^{-1}}) \qquad E_0(\mathrm{DCl})_{\mathrm{vib}} = N_\mathrm{A}hc(1072.5\ \mathrm{cm^{-1}})$$

$$\Delta_\mathrm{r}E_0 = N_\mathrm{A}hc(6399 + 1072.5 - 7029 - 1495.5) = N_\mathrm{A}hc(-1053\ \mathrm{cm^{-1}})$$

Thus the term $-\Delta_\mathrm{r}E_0/RT$ evaluates as

$$\frac{-\Delta_\mathrm{r}E_0}{RT} = \frac{-N_\mathrm{A}hc(-1053\ \mathrm{cm^{-1}})}{RT}$$
$$= -(6.0221 \times 10^{23}\ \mathrm{mol^{-1}}) \times (6.6261 \times 10^{-34}\ \mathrm{J\,s}) \times (2.9979 \times 10^{10}\ \mathrm{cm\,s^{-1}}) \times \frac{(-1053\ \mathrm{cm^{-1}})}{(8.3145\ \mathrm{J\,K^{-1}\,mol^{-1}}) \times T} = (1515\ \mathrm{K})/T$$

With these expressions the equilibrium constant is evaluated using mathematical software and the results are plotted as a function of temperature in Fig 13.18. At 300 K $K = 945$ and at 1000 K $K = 36.9$; the value of the equilibrium constant is dominated by the symmetry factors and the $e^{-\Delta_\mathrm{r}E_0/RT}$ term.

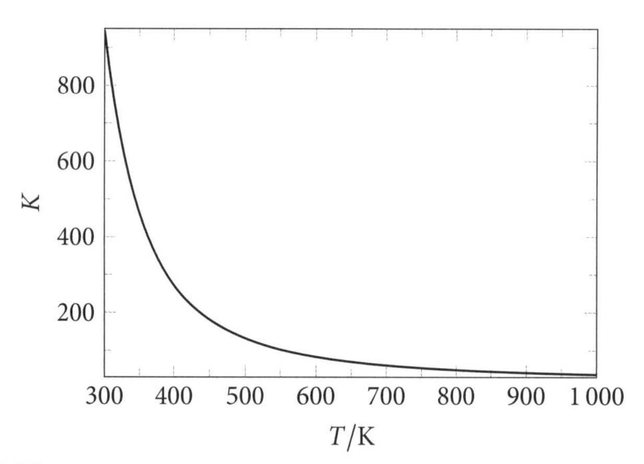

Figure 13.18

P13F.3 In the absence of a magnetic field the ground state of the I atom, with term symbol ${}^2\mathrm{P}_{3/2}$, has a degeneracy given by $(2J + 1) = (2 \times \frac{3}{2} + 1) = 4$. When a magnetic field is applied this level splits into four states characterised by $M_J = +\frac{3}{2}, +\frac{1}{2}, -\frac{1}{2}, -\frac{3}{2}$. The energy of these states is given, by analogy with [12A.11c–492], by $E_{M_J} = g\mu_\mathrm{B}\mathcal{B}M_J$, where the g-value is given as $\frac{4}{3}$, μ_B is the Bohr magneton, and $\mathcal{B}$ is the applied magnetic field. The energies are therefore $E_{\pm3/2} =$

$\pm 2\mu_B\mathcal{B}$, $E_{\pm 1/2} = \pm\frac{2}{3}\mu_B\mathcal{B}$, giving the partition function as

$$q^E = e^{2\mu_B\mathcal{B}/kT} + e^{(2/3)\mu_B\mathcal{B}/kT} + e^{-2\mu_B\mathcal{B}/kT} + e^{-(2/3)\mu_B\mathcal{B}/kT}$$

Because it is expected that $\mu_B\mathcal{B} \ll kT$ the exponentials are expanded to second order; letting $x = \mu_B\mathcal{B}/kT$ gives

$$\begin{aligned} q^E &= [1 + 2x + 2x^2] + [1 + (2/3)x + (2/9)x^2] \\ &\quad + [1 - 2x + 2x^2] + [1 - (2/3)x + (2/9)x^2] \\ &= 4 + (40/9)x^2 = 4\left[1 + (10/9)(\mu_B\mathcal{B}/kT)^2\right] \end{aligned}$$

As is seen, the linear terms cancel which is why it is necessary to expand to second order. In the absence of a magnetic field $q^E = 4$, and because the equilibrium is $I_2 \rightleftharpoons 2I$, the electronic partition function appears squared in the numerator of the expression for K. Therefore

$$\begin{aligned} \frac{K(\mathcal{B})}{K(0)} &= \left(\frac{4\left[1 + (10/9)(\mu_B\mathcal{B}/kT)^2\right]}{4}\right)^2 \\ &= \left[1 + (10/9)(\mu_B\mathcal{B}/kT)^2\right]^2 \approx 1 + (20/9)(\mu_B\mathcal{B}/kT)^2 \end{aligned}$$

where to go to the final expression only the squared term is retained.

For a change of 1%

$$\begin{aligned} (20/9)(\mu_B\mathcal{B}/kT)^2 &= 0.01 \\ \text{hence } \mathcal{B}^2 &= 0.01 \times (9/20) \times (kT/\mu_B)^2 \\ \mathcal{B} &= \left(\frac{9}{2000}\right)^{1/2} \times \frac{(1.3806 \times 10^{-23}\,\mathrm{J\,K^{-1}}) \times (1000\,\mathrm{K})}{9.2740 \times 10^{-24}\,\mathrm{J\,T^{-1}}} \\ &= \boxed{100\,\mathrm{T}} \end{aligned}$$

This is a very strong magnetic field which at present can only be generated by special techniques and only then for a very short times.

P13F.5 The standard molar Gibbs energy is computed from the partition function using [13F.9b–569], $G^\ominus_m(T) = G^\ominus_m(0) - RT\ln q^\ominus_m/N_A$. As usual, the partition function is factored into separate contributions from translation, rotation and so on. The factor of $1/N_A$ is usually taken with the translational contribution.

The standard molar translational partition function is given by $q^\ominus_m = V^\ominus_m/\Lambda^3 = RT/p^\ominus\Lambda^3$. Taking the mass of Cl_2O_2 as $2(35.45+16.00) = 102.9\,m_u$, Λ is given by [13B.7–*541*]

$$\Lambda = h/(2\pi mkT)^{1/2}$$

$$= \frac{6.6261\times10^{-34}\,\mathrm{J\,s}}{[2\pi(102.9\times1.6605\times10^{-27}\,\mathrm{kg})\times(1.3806\times10^{-23}\,\mathrm{J\,K^{-1}})\times(200\,\mathrm{K})]^{1/2}}$$

$$= 1.21...\times10^{-11}\,\mathrm{m}$$

$$q^{\ominus}_{\mathrm{m}}/N_{\mathrm{A}} = \frac{RT}{p^{\ominus}\Lambda^3 N_{\mathrm{A}}}$$

$$= \frac{(8.3145\,\mathrm{J\,K^{-1}\,mol^{-1}})\times(200\,\mathrm{K})}{(10^5\,\mathrm{N\,m^{-2}})\times(1.21...\times10^{-11}\,\mathrm{m})^3\times(6.0221\times10^{23}\,\mathrm{mol^{-1}})}$$

$$= 1.53...\times10^{7}$$

For a nonlinear molecule the rotational partition function is given by [13B.14–545], $q^{\mathrm{R}} = (1/\sigma)(kT/hc)^{3/2}(\pi/\tilde{A}\tilde{B}\tilde{C})^{1/2}$; here $\sigma = 2$. If the rotational constants are expressed in frequency units this expression becomes

$$q^{\mathrm{R}} = (1/\sigma)(kT/h)^{3/2}(\pi/ABC)^{1/2}$$

$$= \tfrac{1}{2}\left(\frac{(1.3806\times10^{-23}\,\mathrm{J\,K^{-1}})\times(200\,\mathrm{K})}{6.6261\times10^{-34}\,\mathrm{J\,s}}\right)^{3/2}$$

$$\times\left(\frac{\pi}{(1.31094\times10^{10}\,\mathrm{Hz})\times(2.4098\times10^{9}\,\mathrm{Hz})\times(2.1397\times10^{9}\,\mathrm{Hz})}\right)^{1/2}$$

$$= 2.89...\times10^{4}$$

The vibrational partition function for each mode is given by [13B.15–*546*], $q^{\mathrm{V}} = 1/(1-\mathrm{e}^{-\beta hc\tilde{\nu}})$, where $\beta = 1/kT$.

$$hc\beta = \frac{(6.6261\times10^{-34}\,\mathrm{J\,s})\times(2.9979\times10^{10}\,\mathrm{cm\,s^{-1}})}{(1.3806\times10^{-23}\,\mathrm{J\,K^{-1}})\times(200\,\mathrm{K})}$$

$$= 7.19...\times10^{-3}\,\mathrm{cm}$$

$$q^{\mathrm{V}}_1 = \left(1-\mathrm{e}^{-hc\beta\tilde{\nu}_1}\right)^{-1}$$

$$= \left(1-\mathrm{e}^{-(7.19...\times10^{-3}\,\mathrm{cm})\times(753\,\mathrm{cm^{-1}})}\right)^{-1} = 1.00...$$

The partition functions for the other normal modes evaluate to 1.02..., 1.12..., 1.66..., 1.00..., 1.05... in order of the given modes. The overall vibrational partition function is the product of these individual contributions: $q^{\mathrm{V}} = 2.03...$.

The overall partition function is the product of these contributions from the different modes, therefore

$$G^{\ominus}_{\mathrm{m}}(200) - G^{\ominus}_{\mathrm{m}}(0) = -RT\ln q^{\ominus}_{\mathrm{m}}/N_{\mathrm{A}}$$

$$= -(8.3145\,\mathrm{J\,K^{-1}\,mol^{-1}})\times(200\,\mathrm{K})$$

$$\times\ln\left[(1.53...\times10^{7})\times(2.89...\times10^{4})\times(2.03...)\right]$$

$$= \boxed{-45.8\,\mathrm{kJ\,mol^{-1}}}$$

Answers to integrated activities

I13.1 (a) In the high-temperature limit, the rotational partition function of an asymmetric rotor is given by [13B.14–545], $q^{\mathrm{R}} = (1/\sigma)(kT/hc)^{3/2}(\pi/\tilde{A}\tilde{B}\tilde{C})^{1/2}$, where σ is the symmetry number. The point group for ethene is $D_{2\mathrm{h}}$ which contains the rotational operations (E, C_2^x, C_2^y, C_2^z); therefore $\sigma = 4$.

$$\begin{aligned} q^{\mathrm{R}} &= \frac{1}{4}\left(\frac{kT}{hc}\right)^{3/2}\left(\frac{\pi}{\tilde{A}\tilde{B}\tilde{C}}\right)^{1/2} \\ &= \frac{1}{4} \times \left(\frac{(1.3806\times10^{-23}\,\mathrm{J\,K^{-1}})\times(298.15\,\mathrm{K})}{(6.6261\times10^{-34}\,\mathrm{J\,s})\times(2.9979\times10^{10}\,\mathrm{cm\,s^{-1}})}\right)^{3/2} \\ &\quad\times\left(\frac{\pi}{(4.828\,\mathrm{cm^{-1}})\times(1.0012\,\mathrm{cm^{-1}})\times(0.8282\,\mathrm{cm^{-1}})}\right)^{1/2} \\ &= \boxed{660.6} \end{aligned}$$

(b) Pyridine belongs to the point group $C_{2\mathrm{v}}$ which contains the rotational operations (E, C_2); therefore $\sigma = 2$.

$$\begin{aligned} q^{\mathrm{R}} &= \frac{1}{2}\left(\frac{kT}{hc}\right)^{3/2}\left(\frac{\pi}{\tilde{A}\tilde{B}\tilde{C}}\right)^{1/2} \\ &= \frac{1}{2} \times \left(\frac{(1.3806\times10^{-23}\,\mathrm{J\,K^{-1}})\times(298.15\,\mathrm{K})}{(6.6261\times10^{-34}\,\mathrm{J\,s})\times(2.9979\times10^{10}\,\mathrm{cm\,s^{-1}})}\right)^{3/2} \\ &\quad\times\left(\frac{\pi}{(0.2014\,\mathrm{cm^{-1}})\times(0.1936\,\mathrm{cm^{-1}})\times(0.0987\,\mathrm{cm^{-1}})}\right)^{1/2} \\ &= \boxed{4.26\times10^{4}} \end{aligned}$$

I13.3 The argument that leads to the expression in *Integrated activity* I13.2 is independent of the set of energy levels, and so can be adapted for the harmonic oscillator levels, $\varepsilon(v) = (v+\frac{1}{2})hc\tilde{\nu}$, all of which are singly degenerate, so that $g(v) = 0$. To make the notation more compact the energy levels will be written ε_v.

Note that there is an error in the question: the expression for $\xi(\beta)$ should include an additional factor of $g(J)$. The corresponding expression for the hamonic oscillator is therefore

$$C_V = \frac{Nk\beta^2}{2q^2}\sum_{v,v'}(\varepsilon_v - \varepsilon_{v'})^2 \mathrm{e}^{-\beta(\varepsilon_v+\varepsilon_{v'})}$$

For a harmonic oscillator (taking the ground state to have zero energy) $\beta\varepsilon_v = \beta hc\tilde{\nu}v = hc\tilde{\nu}v/kT = \theta^{\mathrm{V}}v/T$, where $\theta^{\mathrm{V}} = hc\tilde{\nu}/k$. For the molar quantity $N_{\mathrm{A}}k\beta^2 = N_{\mathrm{A}}k/k^2T^2 = R/k^2T^2$. The molar hear capacity is therefore given by

$$C_{V,\mathrm{m}}/R = \frac{1}{k^2T^2}\frac{1}{2q^2}(hc\tilde{\nu})^2 \sum_{\upsilon,\upsilon'}(\upsilon-\upsilon')^2\,\mathrm{e}^{-\theta^{\mathrm{V}}(\upsilon+\upsilon')/T}$$

$$= \left(\frac{\theta^{\mathrm{V}}}{T}\right)^2 \frac{1}{2q^2}\sum_{\upsilon,\upsilon'}(\upsilon-\upsilon')^2\,\mathrm{e}^{-\theta^{\mathrm{V}}(\upsilon+\upsilon')/T}$$

For the harmonic oscillator $q = (1-\mathrm{e}^{-\theta^{\mathrm{V}}/T})^{-1}$.

This expression is used to generate the curves in Fig. 13.19 for particular pairs of values of υ and υ', that is just one term from the double sum. However, the term for $\upsilon = 0, \upsilon' = 1$ is identical to that for $\upsilon = 1, \upsilon' = 0$, so the curves plotted in the figure are *twice* the value for the particular combination of υ and υ' indicated. This double sum is a poor way of computing the heat capacity because a closed form can be found, [13E.3–560]. This is used to compute the curve shown in Fig. 13.19.

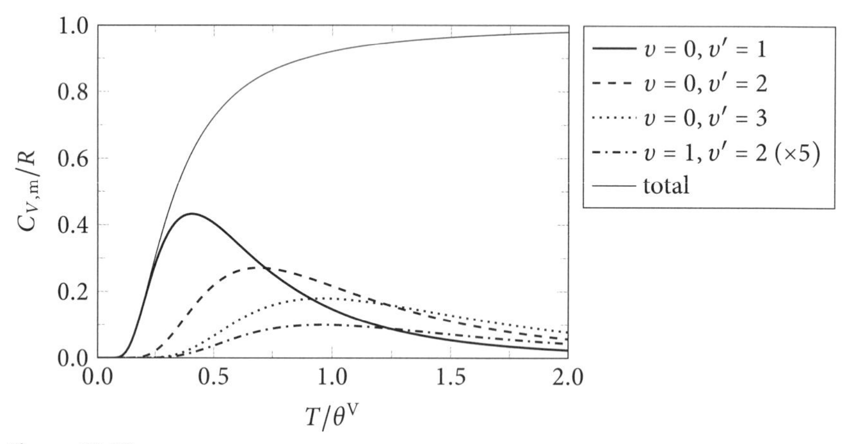

Figure 13.19

Molecular Interactions

14A Electric properties of molecules

Answers to discussion questions

D14A.1 A dipole moment arises due to the uneven distribution of charge density in a molecule; this charge density is usually described in terms of the different partial charges on the constituent atoms. These partial charges arise as a result of the details of the bonding in the molecule, but at a simple level they can be thought of as arising from the different electronegativity of the atoms which make up the molecule: the more electronegative atoms are likely to have the largest partial negative charges, and the least electronegative atoms are likely to have the largest partial positive charges. However, the presence of atoms with different partial charges does not always result in a dipole because the arrangement of the charges may be such that the net dipole is zero. For example, in CO_2 there is a partial negative charge on the oxygens and a partial positive charge on the carbon, but the symmetry of the molecule is such that there is no net dipole (see Section 14A.1 on page 585).

The application of an electric field can result in a distortion of the electron density which results in a change in the permanent dipole or the appearance of a dipole in molecules which have no permanent dipole. It is said that a dipole is induced in the molecule by the applied field, and the constant of proportionality between the field and the induced dipole is the polarizability. In quantum mechanical terms the induced dipole is a result of the electric field causing mixing between the ground and excited-state electronic wavefunctions. This mixing is greatest when the energy gap between the ground and excited states is smallest, and so molecules with the smallest HOMO-LUMO gaps are found to be the most polarizable.

D14A.3 Dipole moments are not measured directly, but are calculated from a measurement of the relative permittivity, $\varepsilon_r = \varepsilon/\varepsilon_0$ of the medium as measured by comparing the capacitance of a capacitor with and without the sample present using $\varepsilon_r = C/C_0$. From [14A.10–*590*] and [14A.11–*590*] it follows that the dipole moment can be determined from a measurement of ε_r as a function of temperature (see *Example* 14A.2 on page 591). An alternative method involves measuring the refractive index and then using [14A.13–*592*] to relate this to ε_r. It is also possible to infer a value of the dipole moment by measuring the splitting seen in the rotational (microwave) spectra of some molecules when a field is applied (the *Stark effect*).

Solutions to exercises

E14A.1(a) The molecules are shown in Fig. 14.1.

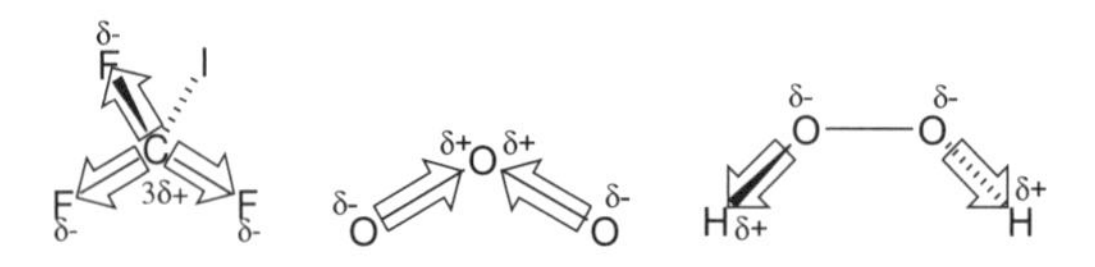

Figure 14.1

In none of the cases do the dipole moments associated with each bond cancel, so all three molecules are $\boxed{\text{polar}}$. Note that the O–O bonds in ozone are polar because the central atom is different from the other two. In the case of CIF_3 the dipole moment associated with the C–I bond is much less than that for the C–F bonds, which results in the dipole moments not cancelling.

An alternative approach is use a symmetry argument. As explained in Section 10A.3(a) on page 394, only molecules belonging to point groups C_n, C_{nv} or C_s may have a permanent electric dipole moment. CIF_3, O_3 and H_2O_2 belong to point groups C_{3v}, C_{2v} and C_2 respectively, so all three are polar.

E14A.2(a) The magnitude of the resultant dipole moment, μ_{res}, is given by [14A.3a–*586*], $\mu_{res} = (\mu_1^2 + \mu_2^2 + 2\mu_1\mu_2\cos\theta)^{1/2}$.

$$\mu_{res} = \left((1.5\text{ D})^2 + (0.8\text{ D})^2 + 2\times(1.5\text{ D})\times(0.8\text{ D})\times\cos 109.5°\right)^{1/2} = \boxed{1.4\text{ D}}$$

E14A.3(a) The arrangement of charges is shown on the left of Fig. 14.2

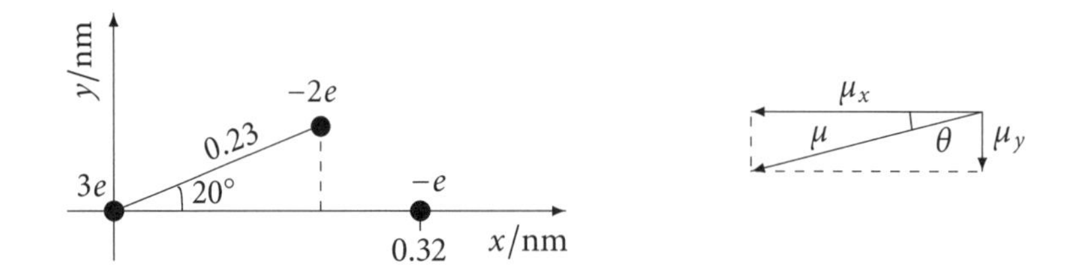

Figure 14.2

The x component of the dipole are given by [14A.4a–*586*], $\mu_x = \sum_J Q_J x_J$, and similarly for the y and z components; note that in this case $\mu_z = 0$ because all of the point charges have a z coordinate of zero. The components are then combined using [14A.4b–*586*], $\mu = (\mu_x^2 + \mu_y^2 + \mu_z^2)^{1/2}$, which is represented graphically on the right half of Fig. 14.2.

$$\mu_x = \sum_J Q_J x_J = \Big[-(1.6022\times10^{-19}\,\text{C})\times(0.32\times10^{-9}\,\text{m})$$
$$-2\times(1.6022\times10^{-19}\,\text{C})\times(0.23\times10^{-9}\,\text{m})\times\cos 20°\Big]$$
$$\times\frac{1\,\text{D}}{3.3356\times10^{-30}\,\text{C m}} = -36.1...\,\text{D}$$

where $1\,\text{D} = 3.3356\times10^{-30}\,\text{C m}$ is used from inside the front cover.

$$\mu_y = \sum_J Q_J y_J$$
$$= -2\times(1.6022\times10^{-19}\,\text{C})\times(0.23\times10^{-9}\,\text{m})\times\sin 20°$$
$$\times\frac{1\,\text{D}}{3.3356\times10^{-30}\,\text{C m}} = -7.55...\,\text{D}$$

The magnitude of the resultant is

$$\mu = \left(\mu_x^2+\mu_y^2\right)^{1/2} = \left[(-36.1...\,\text{D})^2+(-7.55...\,\text{D})^2\right] = \boxed{37\,\text{D}}$$

and, from the diagram on the right of Fig. 14.2, the direction is given by

$$\theta = \tan^{-1}\left(\frac{7.55...\,\text{D}}{31.6...\,\text{D}}\right) = \boxed{12°}$$

E14A.4(a) The relationship between the induced dipole moment μ^* and the electric field strength $\mathcal{E}$ is given by [14A.5a–*587*], $\mu^* = \alpha\mathcal{E}$, where α is the polarizability. The polarizability volume α' is related to the polarizability α by [14A.6–*587*], $\alpha' = \alpha/4\pi\varepsilon_0$. Combining these equations, rearranging for $\mathcal{E}$, and using $1\,\text{V} = 1\,\text{J C}^{-1}$ gives

$$\mathcal{E} = \frac{\mu^*}{4\pi\varepsilon_0\alpha'} = \frac{(1.0\times10^{-6}\,\text{D})\times[(3.3356\times10^{-30}\,\text{C m})/(1\,\text{D})]}{4\pi\times(8.8542\times10^{-12}\,\text{J}^{-1}\,\text{C}^2\,\text{m}^{-1})\times(2.6\times10^{-30}\,\text{m}^3)}$$
$$= \boxed{1.2\times10^4\,\text{V m}^{-1}}$$

E14A.5(a) The molar polarization P_m is defined by [14A.11–*590*], $P_\text{m} = (N_\text{A}/3\varepsilon_0)(\alpha + \mu^2/3kT)$, where α is the polarizability of the molecule and μ is its dipole moment. This equation is written as

$$P_\text{m} = \frac{N_\text{A}\alpha}{3\varepsilon_0} + \frac{N_\text{A}\mu^2}{9\varepsilon_0 k}\frac{1}{T}$$

which implies that a graph of P_m against $1/T$ should be a straight line with slope $N_\text{A}\mu^2/9\varepsilon_0 k$ and intercept $N_\text{A}\alpha/3\varepsilon_0$. However, as there are only two data points it is convenient to calculate the required quantities directly from the data. Writing the molar polarization at the two temperatures as $P_m(T_1)$ and $P_m(T_2)$ and considering $P_m(T_2) - P_m(T_1)$ gives

$$P_m(T_2) - P_m(T_1) = \frac{N_A\mu^2}{9\varepsilon_0 k}\left(\frac{1}{T_2} - \frac{1}{T_1}\right)$$

which is rearranged to give

$$\begin{aligned}\mu &= \left(\frac{9\varepsilon_0 k}{N_A} \times \frac{P_m(T_2) - P_m(T_1)}{1/T_2 - 1/T_1}\right)^{1/2} \\ &= \left(\frac{9 \times (8.8542 \times 10^{-12}\,\text{J}^{-1}\,\text{C}^2\,\text{m}^{-1}) \times (1.3806 \times 10^{-23}\,\text{J}\,\text{K}^{-1})}{6.0221 \times 10^{23}\,\text{mol}^{-1}} \right. \\ &\quad \left. \times \frac{(62.47 \times 10^{-6}\,\text{m}^3\,\text{mol}^{-1}) - (70.62 \times 10^{-6}\,\text{m}^3\,\text{mol}^{-1})}{1/(423.2\,\text{K}) - 1/(351.0\,\text{K})}\right)^{1/2} \\ &= 5.53... \times 10^{-30}\,\text{C}\,\text{m} = \boxed{1.659\,\text{D}}\end{aligned}$$

The value of α is found using this value of μ together with one of the data points; both give the same answer. Rearranging [14A.11–*590*], $P_m = (N_A/3\varepsilon_0)(\alpha + \mu^2/3kT)$, for α and using the data for 351.0 K gives

$$\begin{aligned}\alpha &= \frac{3\varepsilon_0 P_m}{N_A} - \frac{\mu^2}{3kT} \\ &= \frac{3 \times (8.8542 \times 10^{-12}\,\text{J}^{-1}\,\text{C}^2\,\text{m}^{-1}) \times (70.62 \times 10^{-6}\,\text{m}^3\,\text{mol}^{-1})}{6.0221 \times 10^{23}\,\text{mol}^{-1}} \\ &\quad - \frac{(5.53... \times 10^{-30}\,\text{C}\,\text{m})^2}{3 \times (1.3806 \times 10^{-23}\,\text{J}\,\text{K}^{-1}) \times (351.0\,\text{K})} = \boxed{1.008 \times 10^{-39}\,\text{C}^2\,\text{m}^2\,\text{J}^{-1}}\end{aligned}$$

E14A.6(a) The relationship between relative permittivity and molar polarization is given by the Debye equation, [14A.10–*590*], $(\varepsilon_r - 1)/(\varepsilon_r + 2) = \rho P_m/M$. Rearranging gives

$$\varepsilon_r - 1 = \frac{\rho P_m}{M}(\varepsilon_r + 2) \quad \text{hence} \quad \varepsilon_r - 1 = \varepsilon_r\left(\frac{\rho P_m}{M}\right) + \frac{2\rho P_m}{M}$$

$$\text{hence} \quad \varepsilon_r\left(1 - \frac{\rho P_m}{M}\right) = 1 + \frac{2\rho P_m}{M} \quad \text{hence} \quad \varepsilon_r = \frac{1 + 2\rho P_m/M}{1 - \rho P_m/M}$$

The molar mass of ClF_3 is $M = 92.45\,\text{g}\,\text{mol}^{-1}$, which gives

$$\begin{aligned}\varepsilon_r &= \frac{1 + 2\rho P_m/M}{1 - \rho P_m/M} \\ &= \frac{1 + 2 \times (1.89\,\text{g}\,\text{cm}^{-3}) \times (27.18\,\text{cm}^3\,\text{mol}^{-1})/(92.45\,\text{g}\,\text{mol}^{-1})}{1 - (1.89\,\text{g}\,\text{cm}^{-3}) \times (27.18\,\text{cm}^3\,\text{mol}^{-1})/(92.45\,\text{g}\,\text{mol}^{-1})} = \boxed{4.75}\end{aligned}$$

E14A.7(a) The relationship between the refractive index n_r at a specified wavelength and the relative permittivity ε_r at the same wavelength is given by [14A.13–592], $n_r = \varepsilon_r^{1/2}$, hence $\varepsilon_r = n_r^2$. In addition the relationship between relative permittivity and the polarizability α is given by the Clausius–Mossotti equation, [14A.12–590], $(\varepsilon_r - 1)/(\varepsilon_r + 2) = \rho N_A \alpha / 3M\varepsilon_0$. In using this equation it is assumed that there are no contributions from permanent electric dipole moments to the polarization, either because the molecules are nonpolar, which is not the case for CH_2I_2, or because the frequency of the applied field is so high that the molecules cannot orientate quickly enough to follow the change in direction of the field. Replacing ε_r by n_r^2 in the Clausius–Mossotti equation gives

$$\frac{n_r^2 - 1}{n_r^2 + 2} = \frac{\rho N_A \alpha}{3M\varepsilon_0} \quad \text{hence} \quad \alpha = \frac{3M\varepsilon_0}{\rho N_A} \times \frac{n_r^2 - 1}{n_r^2 + 2}$$

The molar mass of CH_2I_2 is $M = 267.8258\ \mathrm{g\,mol^{-1}}$ which gives

$$\alpha = \frac{3 \times (267.8258\ \mathrm{g\,mol^{-1}}) \times (8.8542 \times 10^{-12}\ \mathrm{J^{-1}\,C^2\,m^{-1}})}{(3.32 \times 10^6\ \mathrm{g\,m^{-3}}) \times (6.0221 \times 10^{23}\ \mathrm{mol^{-1}})} \times \frac{1.732^2 - 1}{1.732^2 + 2}$$
$$= \boxed{1.42 \times 10^{-39}\ \mathrm{C^2\,m^2\,J^{-1}}}$$

E14A.8(a) The relationship between the refractive index n_r at a specified wavelength and the relative permittivity ε_r at the same wavelength is given by [14A.13–592], $n_r = \varepsilon_r^{1/2}$. In order to find ε_r the Clausius–Mossotti equation, [14A.12–590], $(\varepsilon_r - 1)/(\varepsilon_r + 2) = \rho N_A \alpha / 3M\varepsilon_0$, is used, with the value of the polarizability α being determined from the polarizability volume α' using [14A.6–587], $\alpha' = \alpha / 4\pi\varepsilon_0$.

In using the Clausius–Mossotti equation it is assumed that there are no contributions from permanent electric dipole moments to the polarization, either because the molecules are nonpolar or because the frequency of the applied field is so high that the molecules cannot orientate quickly enough to follow the change in direction of the field.

The first step is to rearrange the Clausius–Mossotti equation for ε_r

$$\frac{\varepsilon_r - 1}{\varepsilon_r + 2} = \frac{\rho N_A \alpha}{3M\varepsilon_0} \quad \text{hence} \quad \varepsilon_r - 1 = \frac{\rho N_A \alpha}{3M\varepsilon_0}(\varepsilon_r + 2)$$

$$\text{hence} \quad \varepsilon_r \left(1 - \frac{\rho N_A \alpha}{3M\varepsilon_0}\right) = 1 + \frac{2\rho N_A \alpha}{3M\varepsilon_0} \quad \text{hence} \quad \varepsilon_r = \frac{1 + 2\rho N_A \alpha / 3M\varepsilon_0}{1 - \rho N_A \alpha / 3M\varepsilon_0}$$

Replacing α by $4\pi\varepsilon_0\alpha'$ and ε_r by n_r^2 gives

$$n_r^2 = \frac{1 + 2\rho N_A (4\pi\varepsilon_0\alpha')/3M\varepsilon_0}{1 - \rho N_A (4\pi\varepsilon_0\alpha')/3M\varepsilon_0} = \frac{1 + 8\pi\rho N_A \alpha'/3M}{1 - 4\pi\rho N_A \alpha'/3M} = \frac{1 + 2C}{1 - C}$$

where $C = 4\pi\rho N_A \alpha'/3M$. Taking the mass density of water from the *Resource section* as 0.997 g cm^{-3} and the molar mass as M = 18.0158 g mol^{-1}

$$C = 4\pi\rho N_A\alpha'/3M$$

$$= \frac{4\pi \times (0.997\,\text{g}\,\text{cm}^{-3}) \times (6.0221 \times 10^{23}\,\text{mol}^{-1}) \times (1.5 \times 10^{-24}\,\text{cm}^3)}{3 \times (18.0158\,\text{g}\,\text{mol}^{-1})}$$

$$= 0.209...$$

$$n_r = \left(\frac{1+2C}{1-C}\right)^{1/2} = \left(\frac{1+2\times 0.209...}{1-0.209...}\right)^{1/2} = \boxed{1.3}$$

E14A.9(a) The Debye equation, [14A.10–*590*], is $(\varepsilon_r-1)/(\varepsilon_r+2) = \rho P_m/M$, where P_m is the molar polarizability. The latter is defined by [14A.11–*590*], $P_m = (N_A/3\varepsilon_0)(\alpha + \mu^2/3kT)$, where μ is the dipole moment and α is the polarizability, which is related to the polarizability volume α' according to [14A.6–*587*], $\alpha' = \alpha/4\pi\varepsilon_0$. Replacing α in the expression for P_m by $4\pi\varepsilon_0\alpha'$ gives

$$P_m = \frac{N_A}{3\varepsilon_0}\left(4\pi\varepsilon_0\alpha' + \frac{\mu^2}{3kT}\right)$$

$$= \frac{6.0221 \times 10^{23}\,\text{mol}^{-1}}{3 \times (8.8542 \times 10^{-12}\,\text{J}^{-1}\,\text{C}^2\,\text{m}^{-1})} \times \left(4\pi \times (8.8542 \times 10^{-12}\,\text{J}^{-1}\,\text{C}^2\,\text{m}^{-1}) \times (1.23 \times 10^{-29}\,\text{m}^3) + \frac{[(1.57\,\text{D}) \times (3.3356 \times 10^{-30}\,\text{C m})/(1\,\text{D})]^2}{3 \times (1.3806 \times 10^{-23}\,\text{J}\,\text{K}^{-1}) \times ([25+273.15]\,\text{K})}\right)$$

$$= 8.13... \times 10^{-5}\,\text{m}^3\,\text{mol}^{-1}$$

The Debye equation is then rearranged for ε_r

$$\frac{\varepsilon_r - 1}{\varepsilon_r + 2} = \frac{\rho P_m}{M} \quad \text{hence} \quad \varepsilon_r = \frac{1 + 2\rho P_m/M}{1 - \rho P_m/M}$$

The molar mass of chlorobenzene, C_6H_5Cl is 112.5495 g mol^{-1}. Noting that ρ needs to be in g m^{-3} in order for the units to cancel appropriately, the relative permittivity is

$$\varepsilon_r = \frac{1 + 2\rho P_m/M}{1 - \rho P_m/M}$$

$$= \frac{1 + 2 \times (1.173 \times 10^6\,\text{g}\,\text{m}^{-3}) \times (8.13... \times 10^{-5}\,\text{m}^3\,\text{mol}^{-1})/(112.5495\,\text{g}\,\text{mol}^{-1})}{1 - (1.173 \times 10^6\,\text{g}\,\text{m}^{-3}) \times (8.13... \times 10^{-5}\,\text{m}^3\,\text{mol}^{-1})/(112.5495\,\text{g}\,\text{mol}^{-1})}$$

$$= \boxed{17.8}$$

Solutions to problems

P14A.1 The molecules are shown in Fig. 14.3.

The dipole moments of the three isomers of dimethylbenzene are estimated using the method described in Section 14A.1 on page 585. In each case the

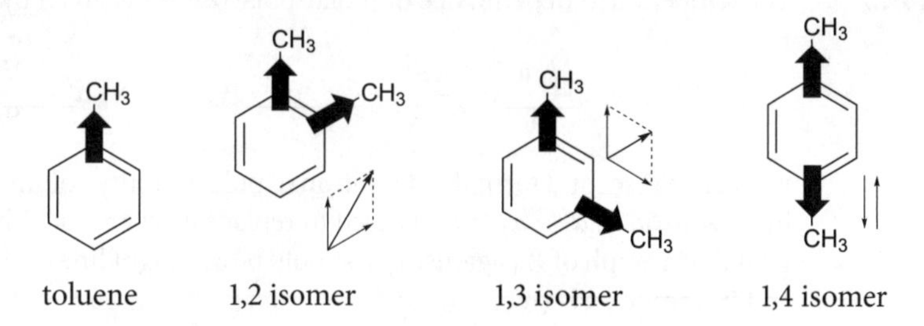

Figure 14.3

resultant dipole moment is assumed to be that of a vector sum of two toluene dipole moments μ_1 arranged at the appropriate angle to each other. In the case of two contributing dipoles of equal magnitude at angle Θ to each other, the size of the resultant dipole is given by [14A.3b–*586*], $\mu_{\text{res}} = 2\mu_1\cos(\frac{1}{2}\Theta)$. In this case $\mu_1 = 0.4$ D, the dipole moment in toluene, and the angle Θ is 60°, 120° or 180° for the 1,2-, 1,3- and 1,4- isomers respectively.

$$\text{1,2 isomer: } \mu_{\text{res}} = 2\mu_1\cos(\tfrac{1}{2}\Theta) = 2\times(0.4\text{ D})\times\cos(\tfrac{1}{2}\times 60°) = \boxed{0.7\text{ D}}$$
$$\text{1,3 isomer: } \mu_{\text{res}} = 2\mu_1\cos(\tfrac{1}{2}\Theta) = 2\times(0.4\text{ D})\times\cos(\tfrac{1}{2}\times 120°) = \boxed{0.4\text{ D}}$$
$$\text{1,4 isomer: } \mu_{\text{res}} = 2\mu_1\cos(\tfrac{1}{2}\Theta) = 2\times(0.4\text{ D})\times\cos(\tfrac{1}{2}\times 180°) = \boxed{0}$$

It is possible to be certain about the result for the 1,4 isomer, because the symmetry of the molecule means that the dipoles *must* cancel and so μ_{res} is necessarily zero.

P14A.3 The individual ethanoic acid molecules have a permanent dipole moment but the dimers do not because of their symmetry which cause the individual dipoles associated with polar bonds to cancel. As the temperature increases the dimer $\rightleftharpoons$ monomer equilibrium shifts in favour of the monomers because the process of breaking the hydrogen bonds between the molecules is endothermic. Consequently a greater proportion of the species present are monomers and therefore have a dipole, and consequently the apparent dipole moment per molecule appears to increase.

P14A.5 The relationship between induced dipole moment μ^* and electric field strength $\mathcal{E}$ is given by [14A.5a–*587*], $\mu^* = \alpha\mathcal{E}$ where α is the polarizability. The polarizability volume α' is related to α by [14A.6–*587*], $\alpha' = \alpha/4\pi\varepsilon_0$, so the induced dipole is

$$\begin{aligned}\mu^* &= \alpha\mathcal{E} = 4\pi\varepsilon_0\alpha'\mathcal{E}\\ &= 4\pi\times(8.8542\times10^{-12}\text{ J}^{-1}\text{ C}^2\text{ m}^{-1})\times(2.22\times10^{-30}\text{ m}^3)\times(15.0\times10^3\text{ V m}^{-1})\\ &= 3.70...\times10^{-36}\text{ C m} = 1.11...\times10^{-6}\text{ D} = \boxed{1.11\ \mu\text{D}}\end{aligned}$$

P14A.7 The temperature-dependence of molar polarization is given by [14A.11–*590*],

$$P_m = \frac{N_A\alpha}{3\varepsilon_0} + \frac{N_A\mu^2}{9\varepsilon_0 k} \times \frac{1}{T} \quad \text{or} \quad P_m = \tfrac{4}{3}\pi N_A\alpha' + \frac{N_A\mu^2}{9\varepsilon_0 k}$$

where in the second form the definition of polarizability volume, which is given by [14A.6–*587*], $\alpha' = \alpha/4\pi\varepsilon_0$, is used to replace α by $4\pi\varepsilon_0\alpha'$. This equation implies that a graph of P_m against $1/T$ should be a straight line of slope $N_A\mu^2/9\varepsilon_0 k$ and intercept $\frac{4}{3}\pi N_A\alpha'$.

The molar polarization is calculated using the Debye equation [14A.10–*590*], $(\varepsilon_r-1)/(\varepsilon_r+2) = \rho P_m/M$, which is rearranged to $P_m = (M/\rho)(\varepsilon_r-1)/(\varepsilon_r+2)$. Taking the molar mass of trichloromethane as $M = 119.3679\ \text{g mol}^{-1}$ the values of P_m are as shown in the table below and are plotted in Fig. 14.4.

$\theta/°\text{C}$	ε_r	$\rho/\text{g cm}^{-3}$	$1/(T/K)$	$P_m/\text{cm}^3\ \text{mol}^{-1}$
−80	3.1	1.65	0.005 18	29.79
−70	3.1	1.64	0.004 92	29.97
−60	7.0	1.64	0.004 69	48.52
−40	6.5	1.61	0.004 29	47.97
−20	6.0	1.57	0.003 95	47.52
0	5.5	1.53	0.003 66	46.81
20	5.0	1.50	0.003 41	45.47

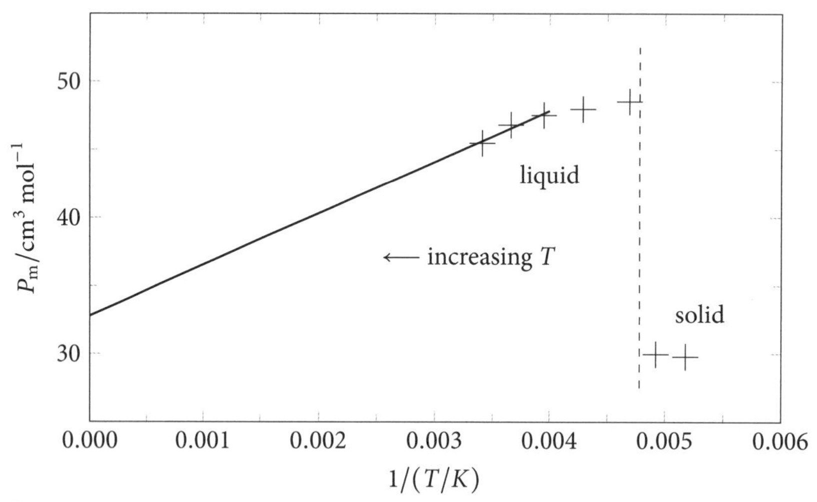

Figure 14.4

The first thing to note is that the molar polarization P_m at −80 °C and −70 °C is much lower than for the other temperatures. These are the temperatures for which trichloromethane (freezing point −64 °C) is a solid. The lower P_m is consistent with the molecules in the solid being fixed in position so that they cannot change orientation to respond to the electric field and therefore their permanent dipole moment makes no contribution to the polarization of the sample; that is, there is no orientation polarization. In the liquid state there is a contribution from orientation polarization and therefore the values of P_m are larger.

For the liquid state, the plot of P_m against $1/T$ should be linear with a slope of $N_A\mu^2/9\varepsilon_0 k$ and intercept $\frac{4}{3}\pi N_A\alpha'$ as deduced earlier. In fact, Fig. 14.4 shows that the data do not lie very close to a straight line. The line drawn on Fig. 14.4 uses only the points corresponding to the three highest temperatures and has equation

$$(P_m/\text{cm}^3\,\text{mol}^{-1}) = 3.76\times10^3 \times 1/(T/\text{K}) + 32.8$$

Identifying the slope with $N_A\mu^2/9\varepsilon_0 k$ gives

$$N_A\mu^2/9\varepsilon_0 k = (3.76\times10^3\ \text{cm}^3\,\text{mol}^{-1}\,\text{K}) = (3.76\times10^{-3}\text{m}^3\,\text{mol}^{-1}\,\text{K})$$

and hence

$$\mu = \left(\frac{9\varepsilon_0 k\times(3.76\times10^{-3}\ \text{m}^3\,\text{mol}^{-1}\,\text{K})}{N_A}\right)^{1/2}$$

$$= \left(\frac{9\times(8.8542\times10^{-12}\,\text{J}^{-1}\,\text{C}^2\,\text{m}^{-1})\times(3.76\times10^{-3}\ \text{m}^3\,\text{mol}^{-1}\,\text{K})}{6.0221\times10^{23}\,\text{mol}^{-1}}\right)^{1/2}$$

$$= 2.62...\times10^{-30}\ \text{C m} = \boxed{0.79\ \text{D}}$$

Similarly, identifying $\frac{4}{3}\pi N_A\alpha'$ with the intercept gives

$$\tfrac{4}{3}\pi N_A\alpha' = (32.8\ \text{cm}^3\,\text{mol}^{-1})$$

Rearranging gives

$$\alpha' = \frac{3}{4\pi N_A}\times(32.8\ \text{cm}^3\,\text{mol}^{-1})$$

$$= \frac{3}{4\pi\times(6.0221\times10^{23}\,\text{mol}^{-1})}\times(32.8\ \text{cm}^3\,\text{mol}^{-1}) = \boxed{1.3\times10^{-23}\ \text{cm}^3}$$

Note that the intercept of Fig. 14.4 corresponds approximately to P_m for the solid. This is interpreted as arising because the intercept gives P_m in the limit of high temperature, under which conditions the molecules then have so much thermal energy that they cannot be oriented by the applied electric field. Consequently there is no orientation polarization contribution to P_m, just as is the case in the solid where the molecules are prevented from rotating.

P14A.9 The temperature-dependence of molar polarization is given by [14A.11–*590*],

$$P_m = \frac{N_A\alpha}{3\varepsilon_0} + \frac{N_A\mu^2}{9\varepsilon_0 k}\times\frac{1}{T} \quad\text{or}\quad P_m = \tfrac{4}{3}\pi N_A\alpha' + \frac{N_A\mu^2}{9\varepsilon_0 k}\times\frac{1}{T}$$

where in the second form the definition of polarizability volume, which is given by [14A.6–*587*], $\alpha' = \alpha/4\pi\varepsilon_0$, is used to replace α by $4\pi\varepsilon_0\alpha'$. This equation implies that a graph of P_m against $1/T$ should be a straight line of slope $N_A\mu^2/9\varepsilon_0 k$ and intercept $\frac{4}{3}\pi N_A\alpha'$. The data are plotted in Fig. 14.5.

T/K	$1/(T/K)$	$P_m/\text{cm}^3\,\text{mol}^{-1}$
292.2	0.003 42	57.57
309.0	0.003 24	55.01
333.0	0.003 00	51.22
387.0	0.002 58	44.99
413.0	0.002 42	42.51
446.0	0.002 24	39.59

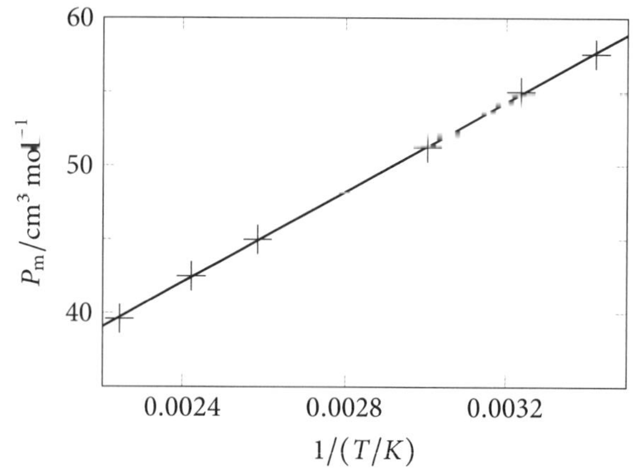

Figure 14.5

The data fall on a good straight line, the equation for which is

$$(P_m/\text{cm}^3\,\text{mol}^{-1}) = (1.5236\times 10^4)\times 1/(T/\text{K}) + 5.5432$$

Identifying the slope with $N_A\mu^2/9\varepsilon_0 k$ gives

$$N_A\mu^2/9\varepsilon_0 k = (1.5236\times 10^4\ \text{cm}^3\,\text{mol}^{-1}\,\text{K}) = (1.5236\times 10^{-2}\text{m}^3\,\text{mol}^{-1}\,\text{K})$$

and hence

$$\mu = \left(\frac{9\varepsilon_0 k}{N_A}\times(1.5236\times 10^{-2}\ \text{m}^3\,\text{mol}^{-1}\,\text{K})\right)^{1/2}$$

$$= \left(\frac{9\times(8.8542\times 10^{-12}\,\text{J}^{-1}\,\text{C}^2\,\text{m}^{-1})\times(1.3806\times 10^{-23}\,\text{J}\,\text{K}^{-1})}{6.0221\times 10^{23}\,\text{mol}^{-1}}\right.$$

$$\left.\times(1.5236\times 10^{-2}\ \text{m}^3\,\text{mol}^{-1}\,\text{K})\right)^{1/2} = 5.27...\times 10^{-30}\ \text{C}\,\text{m} = \boxed{1.582\ \text{D}}$$

Similarly, identifying $\frac{4}{3}\pi N_A\alpha'$ with the intercept gives

$$\alpha' = \frac{3}{4\pi N_A}\times(5.5432\ \text{cm}^3\,\text{mol}^{-1})$$

$$= \frac{3}{4\pi\times(6.0221\times 10^{23}\ \text{mol}^{-1})}\times(5.5432\ \text{cm}^3\,\text{mol}^{-1}) = \boxed{2.197\times 10^{-24}\ \text{cm}^3}$$

In the case of polarization measurements made with radiation of relatively high frequency, such as visible light, the permanent electric dipole moment of the molecules does not contribute to the polarization because the electric field oscillates too quickly for the molecules to orientate themselves to follow the field direction. The difference in molar polarization recorded under these conditions from the total molar polarization therefore corresponds to the permanent dipole contribution, that is, the term $N_A\mu^2/9\varepsilon_0 kT$ in the expression for total molar polarization.

The molar polarisation under the high frequency conditions is determined from the refractive index measurement using the Debye equation [14A.10–*590*], $(\varepsilon_r - 1)/(\varepsilon_r + 2) = \rho P_m/M$, together with the relationship between refractive index and relative permittivity which is given by [14A.13–*592*], $n_r = \varepsilon_r^{1/2}$. Combining these equations gives

$$\frac{n_r^2 - 1}{n_r^2 + 2} = \frac{\rho P_m}{M} \quad \text{hence} \quad P_m = \frac{M}{\rho}\left(\frac{n_r^2 - 1}{n_r^2 + 2}\right)$$

The mass density ρ is given by M/V_m, where M is the molar mass and V_m is the molar volume. Assuming perfect gas behaviour, $V_m = RT/p$ and hence

$$P_m = \frac{M}{\rho}\left(\frac{n_r^2 - 1}{n_r^2 + 2}\right) = \frac{M}{M/V_m}\left(\frac{n_r^2 - 1}{n_r^2 + 2}\right) = V_m\left(\frac{n_r^2 - 1}{n_r^2 + 2}\right) = \frac{RT}{p}\left(\frac{n_r^2 - 1}{n_r^2 + 2}\right)$$

Noting that $1\ \text{Pa} = 1\ \text{J m}^{-3}$, the high frequency molar polarization is therefore

$$P_m = \frac{RT}{p}\left(\frac{n_r^2 - 1}{n_r^2 + 2}\right) = \frac{(8.3145\ \text{J K}^{-1}\ \text{mol}^{-1}) \times (273\ \text{K})}{100 \times 10^3\ \text{Pa}} \times \left(\frac{1.000379^2 - 1}{1.000379^2 + 2}\right)$$

$$= 5.73... \times 10^{-6}\ \text{m}^3\ \text{mol}^{-1} = \boxed{5.73\ \text{cm}^3\ \text{mol}^{-1}}$$

This corresponds to the term $N_A\alpha/3\varepsilon_0$ in the expression

$$P_m = \frac{N_A\alpha}{3\varepsilon_0} + \frac{N_A\mu^2}{9\varepsilon_0 kT}$$

because as discussed above the term involving μ does not contribute at high frequency. Assuming the high frequency contribution to be the same at 292 K, the difference between this value and the static polarization $P_m = 57.57\ \text{cm}^3\ \text{mol}^{-1}$ at this temperature corresponds to the term $N_A\mu^2/9\varepsilon_0 kT$.

$$\frac{N_A\mu^2}{9\varepsilon_0 kT} = P_{m,\text{static}} - P_{m,\text{high frequency}}$$

Hence

$$\mu = \left(\frac{9\varepsilon_0 kT}{N_A}(P_{m,\text{static}} - P_{m,\text{high frequency}})\right)^{1/2}$$

$$= \left(\frac{9 \times (8.8542 \times 10^{-12}\ \text{J}^{-1}\ \text{C}^2\ \text{m}^{-1}) \times (1.3806 \times 10^{-23}\ \text{J K}^{-1}) \times (292\ \text{K})}{6.0221 \times 10^{23}\ \text{mol}^{-1}} \times \right.$$

$$\left. \left[(57.57 \times 10^{-6}\ \text{m}^3\ \text{mol}^{-1}) - (5.73... \times 10^{-6}\ \text{m}^3\ \text{mol}^{-1})\right]\right)^{1/2}$$

$$= 5.24... \times 10^{-30}\ \text{C m} = \boxed{1.57\ \text{D}}$$

The result is in good agreement with that obtained from the graph.

P14A.11 The molar polarization is given by [14A.11–*590*], $P_m = (N_A/3\varepsilon_0)(\alpha + \mu^2/3kT)$. However, assuming that the frequency of oscillation of the electric field is high, as it will be for visible light, there will be no contribution from orientation polarization because the molecules cannot reorientate themselves quickly enough to follow the field. Therefore the term $\mu^2/3kT$ in the expression for P_m does not contribute, and the equation reduces to $P_m = N_A\alpha/3\varepsilon_0$.

Taking the polarizability of methanol as $\alpha = 3.59 \times 10^{-40}\ \mathrm{J^{-1}\,C^2\,m^2}$ from Table 14A.1 on page 585 gives

$$P_m = \frac{N_A\alpha}{3\varepsilon_0} = \frac{(6.0221 \times 10^{23}\ \mathrm{mol^{-1}}) \times (3.59 \times 10^{-40}\ \mathrm{J^{-1}\,C^2\,m^2})}{3 \times (8.8542 \times 10^{-12}\ \mathrm{J^{-1}\,C^2\,m^{-1}})}$$

$$= 8.13... \times 10^{-6}\ \mathrm{m^3\,mol^{-1}} \quad \boxed{8.14\ \mathrm{cm^3\,mol^{-1}}}$$

The relationship between molar polarization and the relative permittivity is given by the Debye equation [14A.10–*590*], $(\varepsilon_r - 1)/(\varepsilon_r + 2) = \rho P_m/M$. Rearranging for ε_r and taking the molar mass of methanol as $M = 32.0416\ \mathrm{g\,mol^{-1}}$ gives

$$\varepsilon_r = \frac{1 + 2\rho P_m/M}{1 - \rho P_m/M}$$

$$= \frac{1 + 2 \times (0.7914\ \mathrm{g\,cm^{-3}}) \times (8.13...\ \mathrm{cm^3\,mol^{-1}})/(32.0416\ \mathrm{g\,mol^{-1}})}{1 - (0.7914\ \mathrm{g\,cm^{-3}}) \times (8.13...\ \mathrm{cm^3\,mol^{-1}})/(32.0416\ \mathrm{g\,mol^{-1}})}$$

$$= 1.75... = \boxed{1.75}$$

The refractive index n_r is related to the relative permittivity according to [14A.13–*592*], $n_r = \sqrt{\varepsilon_r}$. Hence $n_r = \sqrt{1.75...} = \boxed{1.32}$.

P14A.13 The monomer–dimer equilibrium is shown in Fig. 14.6.

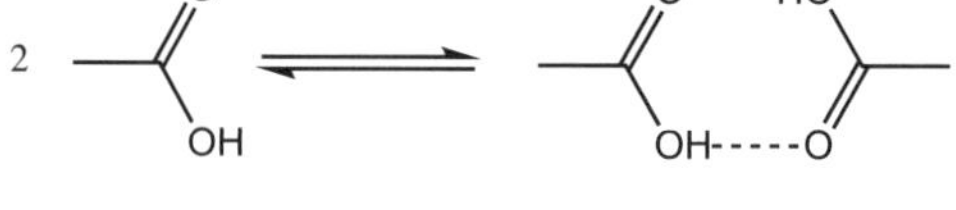

Figure 14.6

The relationship between relative permittivity and molar polarization is given by the Debye equation, [14A.10–*590*]

$$\frac{\varepsilon_r - 1}{\varepsilon_r + 2} = \frac{\rho P_m}{M} \quad \text{hence} \quad \varepsilon_r = \frac{1 + 2\rho P_m/M}{1 - \rho P_m/M}$$

This equation implies that if the molar polarization P_m increases then ε_r will increase, because the numerator in the expression for ε_r will increase while the denominator will decrease. The molar polarization in turn is given by [14A.11–590], $P_m = (N_A/3\varepsilon_0)(\alpha + \mu^2/3kT)$.

The individual molecules are polar but the dimers are not, because their symmetry means the individual dipoles associated with their polar bonds cancel out. At low temperatures, a significant proportion of molecules exist as the non-polar dimer and therefore do not contribute to the $\mu^2/3kT$ term in the expression for P_m. As the temperature is raised, the hydrogen bonds holding the dimers together are broken and the equilibrium shifts in favour of the monomers. Consequently more molecules contribute to the $\mu^2/3kT$ term in the expression for P_m and consequently P_m increases. In turn this leads to an increase in ε_r as shown above.

If the sample is diluted, then the equilibrium will also shift in favour of the monomers, according to Le Chatelier's principle as discussed in Section 6B.1 on page 212. Consequently the molar polarization and hence the relative permittivity will increase in the same way as when the temperature is increased.

14B Interactions between molecules

Answer to discussion questions

D14B.1 (a) V is the potential energy of interaction between a point dipole μ_1 and the point charge Q_2 at the separation r. The point charge lies on the axis of the dipole.

(b) V is the potential energy of interaction between a point dipole μ_1 and the point charge Q_2 at the separation r. The point charge lies at an angle θ to the axis of the dipole.

(c) V is the potential energy of interaction between the two point dipoles μ_1 and μ_2 at the separation r. The dipoles are parallel and the vector joining them is at an angle θ to the dipoles.

D14B.3 There are three van der Waals type interactions that depend upon distance as $1/r^6$; they are the Keesom interaction between rotating permanent dipoles, the permanent-dipole–induced-dipole interaction, and the induced-dipole–induced-dipole, or London dispersion, interaction. In each case, the distance dependence of the potential energy can be thought of as arising from the $1/r^3$ dependence of the field (and hence the magnitude of the induced dipole) and the $1/r^3$ dependence of the potential energy of interaction of the dipoles (either permanent or induced).

D14B.5 The monomer unit as shown is expected to be planar and rigid on account of the benzene rings and the amide linkage between them; the π system will be delocalised across both rings and the amide linkage, resulting in a significant barrier to rotation. Hydrogen bonds can be formed between the NH hydrogens on one polymer chain and the carbonyl groups of another, resulting in strong interactions between the chains.

The flatness of the Kevlar polymeric molecule makes it possible to process the material so that many molecules with parallel alignment form highly ordered, untangled crystal bundles. The alignment makes possible both considerable van der Waals attractions between adjacent molecules and for strong hydrogen bonding between the polar amide groups on adjacent molecules. These bonding forces create the high thermal stability and mechanical strength observed in Kevlar.

Solutions to exercises

E14B.1(a) The interaction between a point charge and a point dipole orientated directly away the charge is given by [14B.1–593], $V = -\mu_1 Q_2/4\pi\varepsilon_0 r^2$. In this case the lithium ion has a charge of $+e = +1.6022 \times 10^{-19}$ C.

$$V = \frac{\mu_1 Q_2}{4\pi\varepsilon_0 r^2} = \frac{(1.85\,\mathrm{D}) \times [(3.3356 \times 10^{-30}\,\mathrm{C\,m})/(1\,\mathrm{D})] \times (1.6022 \times 10^{-19}\,\mathrm{C})}{4\pi \times (8.8542 \times 10^{-12}\,\mathrm{J^{-1}\,C^2\,m^{-1}}) \times (100 \times 10^{-12}\,\mathrm{m})^2}$$
$$= -8.88... \times 10^{-19}\,\mathrm{J}$$

The potential energy if the dipole has the reverse orientation is $+8.88... \times 10^{-19}$ J so the energy required to reverse the direction is

$$\Delta V = (8.88... \times 10^{-19}\,\mathrm{J}) \times 2 = 1.77... \times 10^{-19}\,\mathrm{J} = \boxed{1.77 \times 10^{-18}\,\mathrm{J}}$$

The energy per mole is found by multiplying by N_A

$$\Delta V_m = (1.77... \times 10^{-21}\,\mathrm{kJ}) \times (6.0221 \times 10^{23}\,\mathrm{mol^{-1}}) = \boxed{1.07 \times 10^3\,\mathrm{kJ\,mol^{-1}}}$$

E14B.2(a) The potential energy interaction between two parallel point dipoles separated by distance r at angle Θ is given by [14B.3b–595], $V = \mu_1\mu_2(1-3\cos^2\Theta)/4\pi\varepsilon_0 r^3$.

$$V = \frac{\mu_1\mu_2(1-3\cos^2\Theta)}{4\pi\varepsilon_0 r^3}$$
$$= \frac{(2.7\,\mathrm{D}) \times (2.7\,\mathrm{D}) \times [(3.3356 \times 10^{-30}\,\mathrm{C\,m})/(1\,\mathrm{D})]^2 \times (1 - 3\cos^2 45°)}{4\pi \times (8.8542 \times 10^{-12}\,\mathrm{J^{-1}\,C^2\,m^{-1}}) \times (3.0 \times 10^{-9}\,\mathrm{m})^3}$$
$$= \boxed{-1.3 \times 10^{-23}\,\mathrm{J}} \quad \text{which corresponds to} \quad \boxed{-8.1\,\mathrm{J\,mol^{-1}}}$$

E14B.3(a) The shape of a linear quadrupole is given in Fig. 14A.2 on page 587; an example of such an arrangement is a CO_2 molecule which has negative charges on the oxygen atoms and a balancing positive charge on the central carbon. Two such quadrupoles are shown in Fig. 14.7, arranged so that they are collinear.

The interaction energy is derived in a similar way to that used in Section 14B.1(b) on page 594 for dipole–dipole interactions. The total interaction energy is the sum of nine pairwise terms, one for each combination of a point charge in one quadrupole with a point charge in the other quadrupole. Each term has the

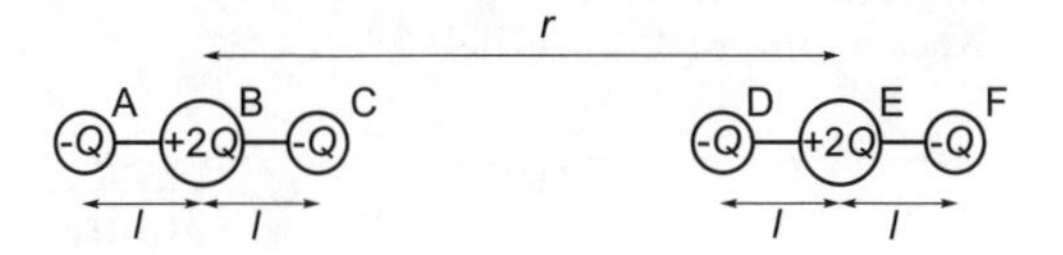

Figure 14.7

form $Q_1Q_2/4\pi\varepsilon_0 r_{12}$ where Q_1 and Q_2 are the charges being considered in that term and r_{12} is the distance between them.

$$\begin{aligned}V = -\frac{1}{4\pi\varepsilon_0}\Bigg(&\overbrace{\frac{(-Q)(-Q)}{r}}^{\text{A–D}}+\overbrace{\frac{(-Q)(2Q)}{r+l}}^{\text{A–E}}+\overbrace{\frac{(-Q)(-Q)}{r+2l}}^{\text{A–F}}+\overbrace{\frac{(2Q)(-Q)}{r-l}}^{\text{B–D}}\\&+\underbrace{\frac{(2Q)(2Q)}{r}}_{\text{B–E}}+\underbrace{\frac{(2Q)(-Q)}{r+l}}_{\text{B–F}}+\underbrace{\frac{(-Q)(-Q)}{r-2l}}_{\text{C–D}}+\underbrace{\frac{(-Q)(2Q)}{r-l}}_{\text{C–E}}+\underbrace{\frac{(-Q)(-Q)}{r}}_{\text{C–F}}\Bigg)\\=\frac{1}{4\pi\varepsilon_0}&\left(\frac{Q^2}{r}-\frac{2Q^2}{r+l}+\frac{Q^2}{r+2l}-\frac{2Q^2}{r-l}+\frac{4Q^2}{r}-\frac{2Q^2}{r+l}+\frac{Q^2}{r-2l}-\frac{2Q^2}{r-l}+\frac{Q^2}{r}\right)\\=\frac{1}{4\pi\varepsilon_0}&\left(\frac{6Q^2}{r}-\frac{4Q^2}{r+l}-\frac{4Q^2}{r-l}+\frac{Q^2}{r+2l}+\frac{Q^2}{r-2l}\right)\\=\frac{Q^2}{4\pi\varepsilon_0 r}&\left(6-\frac{4}{1+x}-\frac{4}{1-x}+\frac{1}{1+2x}+\frac{1}{1-2x}\right)\end{aligned}$$

where $x = l/r$. Assuming that $r \gg l$, so that $x \ll 1$, the fractions can be expanded using the series

$$\frac{1}{1+x} = 1-x+x^2-x^3+x^4-\ldots \quad \text{and} \quad \frac{1}{1+x} = 1+x+x^2+x^3+x^4+\ldots$$

from inside the front cover. The first term that survives is the one in x^4; higher order terms are much smaller and so are not included.

$$\begin{aligned}V = \frac{Q^2}{4\pi\varepsilon_0 r}\Big(&6-4\left(1-x+x^2-x^3+x^4-\ldots\right)-4\left(1+x+x^2+x^4+\ldots\right)\\&+\left(1-(2x)+(2x)^2-(2x)^3+(2x)^4-\ldots\right)\\&+\left(1+(2x)+(2x)^2+(2x)^3+(2x)^4+\ldots\right)\Big)\\\approx\frac{Q^2}{4\pi\varepsilon_0 r}\Big(&6-4-4+1+1+4x-4x-2x+2x-4x^2-4x^2+4x^2+4x^2\\&+4x^3-4x^3-8x^3+8x^3-4x^4-4x^4+16x^4+16x^4\Big)\\=\frac{Q^2}{4\pi\varepsilon_0 r}&\times 24x^4=\frac{Q^2}{4\pi\varepsilon_0 r}\times 24\left(\frac{l}{r}\right)^4=\boxed{\frac{6Q^2l^4}{\pi\varepsilon_0 r^5}}\end{aligned}$$

E14B.4(a) The average energy of interaction between rotating polar molecules is given by the Keesom interaction [14B.4–596].

$$\langle V\rangle = -\frac{C}{r^6} \quad C = \frac{2\mu_1^2\mu_2^2}{3(4\pi\varepsilon_0)^2kT}$$

In this case $\mu_1 = \mu_2$, so

$$C = \frac{2\mu_1^4}{3(4\pi\varepsilon_0)^2kT}$$
$$= \frac{2\times\left[(1\,\mathrm{D})\times(3.3356\times10^{-30}\,\mathrm{C\,m})/(1\,\mathrm{D})\right]^4}{3\times[4\pi\times(8.8542\times10^{-12}\,\mathrm{J^{-1}\,C^2\,m^{-1}})]^2\times(1.3806\times10^{-23}\,\mathrm{J\,K^{-1}})\times(298\,\mathrm{K})}$$
$$= 1.62...\times10^{-78}\,\mathrm{J\,m^6}$$

Hence $\langle V\rangle = -\frac{C}{r^6} = -\frac{1.62...\times10^{-78}\,\mathrm{J\,m^6}}{(0.5\times10^{-9}\,\mathrm{m})^6} = \boxed{-1.0\times10^{-22}\,\mathrm{J}}$. This energy corresponds, after multiplication by N_A, to $\boxed{-62\,\mathrm{J\,mol^{-1}}}$. This is very much smaller than the average molar kinetic energy of the molecules which, as explained in Section 2A.2(a) on page 37, is given by

$$\tfrac{3}{2}RT = \tfrac{3}{2}\times(8.3145\,\mathrm{J\,K^{-1}\,mol^{-1}})\times(298\,\mathrm{K}) = 3.7\,\mathrm{kJ\,mol^{-1}}$$

E14B.5(a) The dipole–induced dipole interaction between a polar molecule such as water and a polarizable molecule such as benzene is given by [14B.6–597], $V = -\mu_1^2\alpha_2'/4\pi\varepsilon_0 r^6$. From the data in the *Resource section* the dipole moment of water is 1.85 D and the polarizability volume of benzene is $10.4\times10^{-30}\,\mathrm{m^3}$.

$$V = -\frac{\mu_1^2\alpha_2'}{4\pi\varepsilon_0 r^6}$$
$$= -\frac{\left[(1.85\,\mathrm{D})\times(3.3356\times10^{-30}\,\mathrm{C\,m})/(1\,\mathrm{D})\right]^2\times(10.4\times10^{-30}\,\mathrm{m^3})}{4\pi\times(8.8542\times10^{-12}\,\mathrm{J^{-1}\,C^2\,m^{-1}})\times(1.0\times10^{-9}\,\mathrm{m})^6}$$
$$= -3.55...\times10^{-24}\,\mathrm{J}$$

This interaction energy corresponds, after multiplication by Avogadro's constant, to $\boxed{-2.1\,\mathrm{J\,mol^{-1}}}$.

E14B.6(a) The London formula for the energy of the dispersion interaction is given by [14B.7–598]

$$V = -\frac{C}{r^6} \quad C = \tfrac{3}{2}\alpha_1'\alpha_2'\frac{I_1I_2}{I_1+I_2}$$

In the case that the two interacting species are the same, with polarizability volume α' and ionisation energy I, this expression becomes

$$C = -\tfrac{3}{2}\alpha'^2\frac{I^2}{2I} = \tfrac{3}{4}\alpha'^2 I \quad \text{hence} \quad V = -\frac{3\alpha'^2 I}{4r^6}$$

Table 14A.1 on page 585 in the *Resource section* gives the polarizability volume of helium as $\alpha' = 0.20 \times 10^{-30}$ m^3 and Table 8B.4 on page 325 gives the first ionisation energy as $I = 2372.3$ kJ mol^{-1}, therefore

$$V = -\frac{3\alpha'^2 I}{4r^6} = -\frac{3 \times (0.20 \times 10^{-30}\ \mathrm{m}^3)^2 \times (2372.3 \times 10^3\ \mathrm{J\,mol^{-1}})}{4 \times (1.0 \times 10^{-9}\ \mathrm{m})^6} = \boxed{0.071\ \mathrm{J\,mol^{-1}}}$$

Solutions to problems

P14B.1 (a) In a vacuum the interaction energy is

$$V = \frac{Q_1 Q_2}{4\pi\varepsilon_0 r} = \frac{(-0.36 \times 1.6022 \times 10^{-19}\ \mathrm{C}) \times (+0.45 \times 1.6022 \times 10^{-19}\ \mathrm{C})}{4\pi \times (8.8542 \times 10^{-12}\ \mathrm{J^{-1}\,C^2\,m^{-1}}) \times (3.0 \times 10^{-9}\ \mathrm{m})}$$
$$= \boxed{-1.2 \times 10^{-20}\ \mathrm{J}}$$

This interaction energy corresponds, after multiplication by Avogadro's constant, to $\boxed{-7.5\ \mathrm{kJ\,mol^{-1}}}$.

(b) If instead the medium is bulk water, ε_0 should be replaced by $\varepsilon_0\varepsilon_r$ where ε_r is the relative permittivity of water.

$$V = \frac{Q_1 Q_2}{4\pi\varepsilon_0\varepsilon_r r} = \frac{(-0.36 \times 1.6022 \times 10^{-19}\ \mathrm{C}) \times (+0.45 \times 1.6022 \times 10^{-19}\ \mathrm{C})}{4\pi \times (8.8542 \times 10^{-12}\ \mathrm{J^{-1}\,C^2\,m^{-1}}) \times 80 \times (3.0 \times 10^{-9}\ \mathrm{m})}$$
$$= \boxed{-1.6 \times 10^{-22}\ \mathrm{J}}$$

This interaction energy corresponds, after multiplication by Avogadro's constant, to $\boxed{-94\ \mathrm{J\,mol^{-1}}}$. Note that the presence of an intervening medium significantly reduces the energy of the interaction in this case.

P14B.3 The energy of a dipole μ in an electric field $\mathcal{E}$ is given in *How is that done?* 14A.2 on page 589 by $E(\theta) = -\mu\mathcal{E}\cos\theta$, where θ is the angle between the direction of the dipole and the field. If the dipole is parallel to the field, $\theta = 0$, this energy is simply $-\mu\mathcal{E}$, while if it is at $\theta = 90°$ the energy is zero. The dipole–induced dipole interaction between a polar molecule such as water with dipole μ and a polarizable species such as an argon atom with polarizability volume α' is given by [14B.6–597], $V = -\mu^2\alpha'/4\pi\varepsilon_0 r^6$. The task is to find the distance r for which this energy is equal to that of the dipole aligned parallel to the field, that is, for which

$$-\mu\mathcal{E} = -\frac{\mu^2\alpha'}{4\pi\varepsilon_0 r^6}$$

Rearranging, and noting that 1 V = 1 J C^{-1}, gives

$$r = \left(\frac{\mu\alpha'}{4\pi\varepsilon_0\mathcal{E}}\right)^{1/6}$$
$$= \left(\frac{(1.85\ \mathrm{D}) \times (3.3356 \times 10^{-30}\ \mathrm{C\,m})/(1\ \mathrm{D}) \times (1.66 \times 10^{-30}\ \mathrm{m}^3)}{4\pi \times (8.8542 \times 10^{-12}\ \mathrm{J^{-1}\,C^2\,m^{-1}}) \times (1.0 \times 10^3\ \mathrm{V\,m^{-1}})}\right)^{1/6}$$
$$= 2.12... \times 10^{-9}\ \mathrm{m} = \boxed{2.1\ \mathrm{nm}}$$

For distances closer than this the arrangement with the dipole directed towards the argon atom is lower in energy than that with the dipole parallel to the field.

P14B.5 The dipole–induced dipole interaction between a dipole and a polarizable molecule is given by [14B.6–597], $V = -\mu_1^2\alpha_2'/4\pi\varepsilon_0 r^6$.

$$\begin{aligned} V &= -\frac{\mu_1^2\alpha_2'}{4\pi\varepsilon_0 r^6} \\ &= -\frac{\left[(2.7\,\mathrm{D}) \times (3.3356\times10^{-30}\,\mathrm{C\,m})/(1\,\mathrm{D})\right]^2 \times (1.04\times10^{-29}\,\mathrm{m^3})}{4\pi\times(8.8542\times10^{-12}\,\mathrm{J^{-1}\,C^2\,m^{-1}})\times(0.4\times10^{-9}\,\mathrm{m})^6} \\ &= -1.85...\times10^{-21}\,\mathrm{J} \end{aligned}$$

This interaction energy corresponds, after multiplication by Avogadro's constant, to $\boxed{-1.1\,\mathrm{kJ\,mol^{-1}}}$

P14B.7 The London formula for the energy of the dispersion interaction is given by [14B.7–598]

$$V = -\frac{C}{r^6} \qquad C = \tfrac{3}{2}\alpha_1'\alpha_2'\frac{I_1 I_2}{I_1+I_2}$$

Differentiation with respect to r gives

$$F = -\frac{\mathrm{d}V}{\mathrm{d}r} = -\frac{\mathrm{d}}{\mathrm{d}r}\left(-\frac{C}{r^6}\right) = -\frac{6C}{r^7} = \boxed{-9\alpha_1\alpha_2\frac{I_1 I_2}{I_1+I_2}\frac{1}{r^7}}$$

The negative sign indicates that the force is attractive.

P14B.9 One way of rewriting the Lennard-Jones potential with an exponential repulsive term (the exp-6 potential) is

$$4\varepsilon\left\{\left(\frac{r_0}{r}\right)^{12} - \left(\frac{r_0}{r}\right)^6\right\} \quad\rightarrow\quad A\varepsilon\left\{B\mathrm{e}^{-r/r_0} - \left(\frac{r_0}{r}\right)^6\right\}$$

where the constants A and B are introduced to scale the overall potential and to scale the repulsive term relative to the attractive term; their values can be found by introducing constraints to the potential. This modified potential immediately presents a problem. In the limit that $r \to 0$ the repulsive term, $B\mathrm{e}^{-r/r_0}$, tends to B, whereas the attractive term becomes more and more negative, without limit. This is not the required behaviour: a realistic potential is expected to be repulsive for small r.

The Lennard-Jones potential crosses zero at $r = r_0$. If this condition is applied to the modified potential the value of B is found by solving

$$A\varepsilon\{B\mathrm{e}^{-1} - 1\} = 0$$

which gives $B = \mathrm{e}$ and hence the potential becomes

$$V(r) = A\varepsilon\left\{\mathrm{e}^{1-r/r_0} - \left(\frac{r_0}{r}\right)^6\right\}$$

The function in the curly braces is positive for $r > r_0$ and negative for $r < r_0$. Therefore, for it to represent a realistic potential with the attractive part dominating at large distances, the constant A must be negative. Assuming this to be so, the potential shows a minimum at a distance somewhat beyond $r = r_0$; the minimum is located by solving $dV(r)/dr = 0$. There is no analytical solution to the resulting equation, but mathematical software is able to locate the minimum numerically at $r = 1.360\,r_0$.

If, like the Lennard-Jones potential, the depth of the well is to be ε at this point, then $A = -1.853$. A plot of the potential with this value of the constant is shown in Fig. 14.8. For comparison, the Lennard-Jones potential is also plotted.

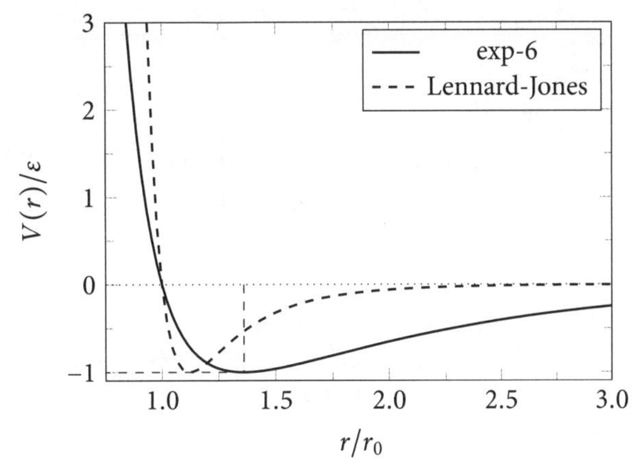

Figure 14.8

14C Liquids

Answers to discussion questions

D14C.1 This is discussed in Section 14C.1(b) on page 603.

Solutions to exercises

E14C.1(a) The vapour pressure of a liquid when it is dispersed as spherical droplets of radius r is given by the Kelvin equation [14C.15–611], $p = p^* e^{2\gamma V_m(l)/rRT}$, where p^* is the vapour pressure of bulk liquid to which no additional pressure has been applied. Because the mass density of a substance with molar volume V_m and molar mass M is given by $\rho = M/V_m$, it follows that $V_m = M/\rho$. Substituting this into the Kelvin equation gives $p = p^* e^{2\gamma(M/\rho)/rRT}$.

The surface tension γ of water at 20 °C is taken as 72.75 mN m^{-1}, which is equal to 72.75×10^{-3} J m^{-2} (Table 14C.1 on page 605), and the molar mass of water is

$M = 18.0158\,\mathrm{g\,mol^{-1}}$. Hence, taking p^* as 2.3 kPa,

$$p = p^* \exp\left(\frac{2\gamma M/\rho}{rRT}\right) = (2.3\,\mathrm{kPa})$$
$$\times \exp\left(\frac{2\times(72.75\times10^{-3}\,\mathrm{J\,m^{-2}})\times(18.0158\,\mathrm{g\,mol^{-1}})/(0.9982\times10^{6}\,\mathrm{g\,m^{-3}})}{(10\times10^{-9}\,\mathrm{m})\times(8.3145\,\mathrm{J\,K^{-1}\,mol^{-1}})\times([20+273.15]\,\mathrm{K})}\right)$$
$$= \boxed{2.6\,\mathrm{kPa}}$$

E14C.2(a) The height climbed by a liquid in a capillary tube of radius r is given by [14C.8–607], $h = 2\gamma/\rho g_{\mathrm{acc}} r$, assuming that the contact angle is zero. Rearranging for γ, taking $\rho = 0.9982\,\mathrm{g\,cm^{-3}} = 998.2\,\mathrm{kg\,m^{-3}}$, and noting that $1\,\mathrm{N} = 1\,\mathrm{kg\,m\,s^{-2}}$ gives

$$\gamma = \tfrac{1}{2}\rho g_{\mathrm{acc}} r h$$
$$= \tfrac{1}{2}\times(998.2\,\mathrm{kg\,m^{-3}})\times(9.807\,\mathrm{m\,s^{-2}})\times(0.300\times10^{-3}\,\mathrm{m})\times(4.96\times10^{-2}\,\mathrm{m})$$
$$= 0.0728...\,\mathrm{kg\,s^{-2}} = \boxed{72.8\,\mathrm{mN\,m^{-1}}}$$

E14C.3(a) The pressure difference between the inside and outside of a spherical droplet is given by the Laplace equation [14C.7–606], $p_{\mathrm{in}} = p_{\mathrm{out}} + 2\gamma/r$. Hence, taking the surface tension of water as $72.75\,\mathrm{mN\,m^{-1}}$ from Table 14C.1 on page 605, and noting that $1\,\mathrm{Pa} = 1\,\mathrm{N\,m^{-2}}$,

$$\Delta p = p_{\mathrm{in}} - p_{\mathrm{out}} = \frac{2\gamma}{r} = \frac{2\times(72.75\times10^{-3}\,\mathrm{N\,m^{-1}})}{(200\times10^{-9}\,\mathrm{m})} = \boxed{728\,\mathrm{kPa}}$$

E14C.4(a) The height climbed by a liquid in a capillary tube of radius r is given by [14C.8–607], $h = 2\gamma/\rho g_{\mathrm{acc}} r$, assuming that the contact angle is zero. Rearranging for γ, replacing r by $\frac{1}{2}d$ where d is the diameter of the tube, and noting that $1\,\mathrm{N} = 1\,\mathrm{kg\,m\,s^{-2}}$ gives

$$\gamma = \tfrac{1}{2}\rho g_{\mathrm{acc}} r h = \tfrac{1}{4}\rho g_{\mathrm{acc}} d h$$
$$= \tfrac{1}{4}\times(997.0\,\mathrm{kg\,m^{-3}})\times(9.807\,\mathrm{m\,s^{-2}})\times(0.500\times10^{-3}\,\mathrm{m})\times(5.89\times10^{-2}\,\mathrm{m})$$
$$= 0.0719...\,\mathrm{kg\,s^{-2}} = \boxed{72.0\,\mathrm{mN\,m^{-1}}}$$

Solutions to problems

P14C.1 (a) The function is plotted in Fig. 14.9

The plot resembles Fig. 14C.1 on page 603 in that the function oscillates for short values of r, corresponding to short-range order, but approaches 1 for large separations.

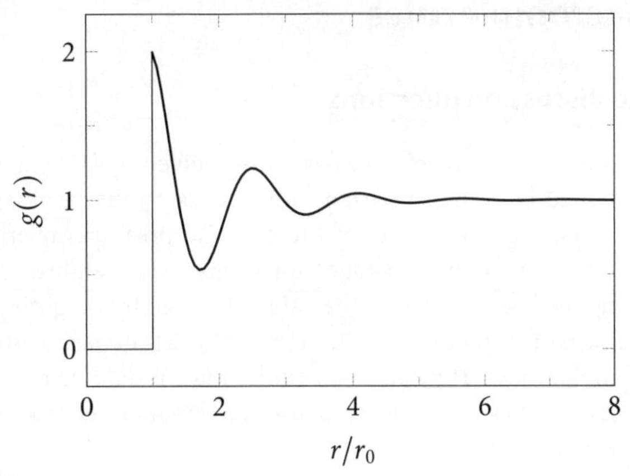

Figure 14.9

(b) The virial $v_2(r) = r(\mathrm{d}V/\mathrm{d}r)$ is obtained by differentiating the Lennard-Jones expression for V

$$\begin{aligned}
v_2(r) &= r\frac{\mathrm{d}}{\mathrm{d}r}4\varepsilon\left[\left(\frac{r_0}{r}\right)^{12} - \left(\frac{r_0}{r}\right)^{6}\right] = r\frac{\mathrm{d}}{\mathrm{d}r}4\varepsilon\left(r_0^{12}r^{-12} - r_0^{6}r^{-6}\right) \\
&= r \times 4\varepsilon\left(-12r_0^{12}r^{-13} + 6r_0^{6}r^{-7}\right) = 4\varepsilon\left(-12r_0^{12}r^{-12} + 6r_0r^{-6}\right) \\
&= -24\varepsilon\left[2\left(\frac{r_0}{r}\right)^{12} - \left(\frac{r_0}{r}\right)^{6}\right]
\end{aligned}$$

The quantity $v_2(r)/\varepsilon$ is plotted in Fig. 14.10; it falls away steeply to the left of the maximum.

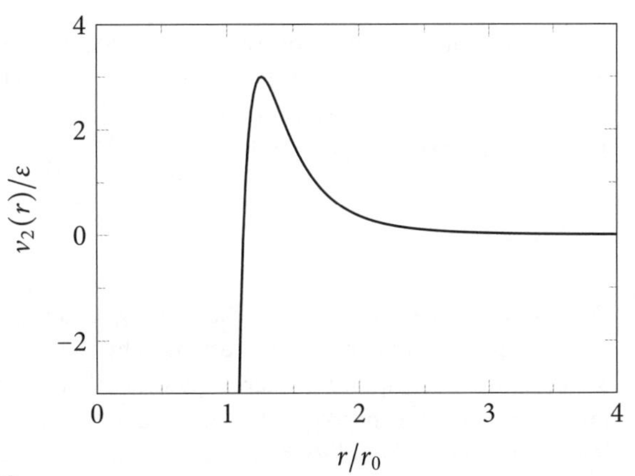

Figure 14.10

14D Macromolecules

Answers to discussion questions

D14D.1 The number-average molar mass is given by [14D.1a–*613*] and is the value obtained by weighting each molar mass by the number of molecules with that mass. Measurements of the osmotic pressures of macromolecular solutions yield the number-average molar mass. The weight-average molar mass is given by [14D.1b–*613*] and is the value obtained by weighting each molar mass by the mass of each one present. Light scattering experiments give the weight-average molar mass. The Z-average molar mass is defined through the formula given in *Example* 14D.1 on page 613; this mass is obtained from sedimentation equilibria experiments.

D14D.3 The freely jointed random coil model of a polymer chain of 'units' or 'residues' gives the simplest possibility for the conformation of the polymer that is not capable of forming hydrogen bonds or any other type of non-linkage bond. In this model, a bond that links adjacent units in the chain is free to make any angle with respect to the preceding one. The residues are assumed to occupy zero volume, so different parts of the chain can occupy the same region of space. It is also assumed in the derivation of the expression for the probability of the ends of the chain being a distance nl apart, that the chain is compact in the sense that $n \ll N$. This model is obviously an oversimplification because a bond is actually constrained to a cone of angles around a direction defined by its neighbour and it is impossible for one section of a chain to overlap with another. Constrained angles and self-avoidance tend to swell the coil, so it is better to regard the R_{rms} and R_{g} values of a random coil as lower bounds to the actual values.

The freely jointed chain is improved by constraining each successive individual bond to a single cone of angle θ relative to its neighbour. This constrained chain reduces R_{rms} and R_{g} values of a freely jointed random coil by a factor of $F = [(1 - \cos\theta)/(1 + \cos\theta)]^{1/2}$.

D14D.5 Polymer melting occurs at a specific melting temperature, T_{m}, above which the crystallinity of polymers can be destroyed by thermal motion. Higher melting temperatures correspond to increased strength and number of intermolecular interactions in the material. Polymers undergo a transition from a state of high to low chain mobility at the glass transition temperature, T_{g}. There is sufficient energy available to an elastomer at normal temperatures for limited bond rotation to occur and the flexible chains writhe. At lower temperatures, the amplitudes of the writhing motion decrease until a specific temperature, T_{g}, is reached at which motion is frozen completely and the sample forms a glass.

Solutions to exercises

E14D.1(a) The number-average molar mass is given by [14D.1a–613], $\overline{M}_{\rm n} = (1/N_{\rm total})\sum_i N_i M_i$. Denoting the polymers as 1 and 2 gives

$$\overline{M}_{\rm n} = \frac{1}{N_{\rm total}}\sum_i N_i M_i = \frac{1}{N_1+N_2}(N_1M_1+N_2M_2)$$

Because the two polymers are present in equal amounts (equal amounts in moles), $N_1 = N_2 = N$ and hence

$$\overline{M}_{\rm n} = \frac{1}{2N}(NM_1+NM_2) = \tfrac{1}{2}(M_1+M_2) = \tfrac{1}{2}\left[(62\,{\rm kg\,mol^{-1}})+(78\,{\rm kg\,mol^{-1}})\right]$$
$$= \boxed{70\,{\rm kg\,mol^{-1}}}$$

The weight-average molar mass is given by [14D.1b–613], $\overline{M}_{\rm W} = (1/m_{\rm total})\sum_i m_i M_i$, where m_i is the mass of polymer i present. Using $m_i = n_i M_i$ and $m_{\rm total} = \sum_i m_i$ gives

$$\overline{M}_{\rm W} = \frac{\sum_i m_i M_i}{\sum_i m_i} = \frac{\sum_i (n_i M_i) M_i}{\sum_i n_i M_i} = \frac{\sum_i n_i M_i^2}{\sum_i n_i M_i}$$

The two polymers are present in equal amounts $n_1 = n_2 = n$, therefore

$$\overline{M}_{\rm W} = \frac{nM_1^2+nM_2^2}{nM_1+nM_2} = \frac{M_1^2+M_2^2}{M_1+M_2} = \frac{(62\,{\rm kg\,mol^{-1}})^2+(78\,{\rm kg\,mol^{-1}})^2}{(62\,{\rm kg\,mol^{-1}})+(78\,{\rm kg\,mol^{-1}})}$$
$$= \boxed{71\,{\rm kg\,mol^{-1}}}$$

E14D.2(a) The root mean square separation of the ends of a freely jointed one-dimensional chain is given by [14D.6–617], $R_{\rm rms} = N^{1/2}l$, where N is the number of monomer units and l is the length of each unit. In this case

$$R_{\rm rms} = N^{1/2}l = 700^{1/2}\times(0.9\,{\rm nm}) = \boxed{24\,{\rm nm}}$$

E14D.3(a) The contour length $R_{\rm c}$ of a polymer is given by [14D.5–617], $R_{\rm c} = Nl$, and the root mean square separation of the ends of a freely jointed one-dimensional chain is given by [14D.6–617], $R_{\rm rms} = N^{1/2}l$. In both cases N is the number of monomer units and l is the length of each unit.

The monomer of polyethene $-(CH_2CH_2)_n-$ is taken to be CH_2CH_2. The number of monomers in the chain is given by

$$N = \frac{M_{\rm polymer}}{M_{\rm CH_2CH_2}} = \frac{280\times10^3\,{\rm g\,mol^{-1}}}{28.0516\,{\rm g\,mol^{-1}}} = 9.98...\times10^3$$

The length of each CH_2CH_2 unit is estimated as the length of a two C–C bonds: one C–C bond in the centre and half a bond length either side where the unit connects to carbons in adjacent units. From Table 9C.2 on page 362 in the

Resource section a C–C bond length is approximately 154 pm, so the monomer length l is taken as $2 \times (154\ \text{pm}) = 308\ \text{pm}$. The contour length and root mean square separation are then given by

$$R_c = Nl = (9.98... \times 10^3) \times (308\ \text{pm}) = 3.07... \times 10^6\ \text{pm} = \boxed{3.07\ \mu\text{m}}$$

$$R_{\text{rms}} = N^{1/2}l = (9.98... \times 10^3)^{1/2} \times (308\ \text{pm}) = 3.07... \times 10^4\ \text{pm} = \boxed{30.8\ \text{nm}}$$

E14D.4(a) The radius of gyration R_g of a one-dimensional random coil is given by [14D.7a–*618*], $R_g = N^{1/2}l$. Rearranging, and taking the length of each C–C link as 154 pm (Table 9C.2 on page 362) gives

$$N = \left(\frac{R_g}{l}\right)^2 = \left(\frac{7.3 \times 10^{-9}\ \text{m}}{154 \times 10^{-12}\ \text{m}}\right)^2 = \boxed{2.2 \times 10^3}$$

E14D.5(a) The probability that the ends of a one-dimensional random coil are a distance nl apart is given by [14D.3–*616*], $P = (2/\pi N)^{1/2}e^{-n^2/2N}$ where N is the total number of monomers in the chain and l is the length of each monomer unit. The monomer of polyethene $-[CH_2CH_2]_n-$ is taken to be CH_2CH_2, so the number of monomers in the chain is given by

$$N = \frac{M_{\text{polymer}}}{M_{CH_2CH_2}} = \frac{65 \times 10^3\ \text{g mol}^{-1}}{28.0516\ \text{g mol}^{-1}} = 2.31... \times 10^3$$

The length of each CH_2CH_2 unit is estimated as the length of a two C–C bonds: one C–C bond in the centre and half a bond length either side where the unit connects to carbons in adjacent units. From Table 9C.2 on page 362 a C–C bond length is approximately 154 pm, so the monomer length l is taken as $2 \times (154\ \text{pm}) = 308\ \text{pm}$.

If the end-to-end distance is d, then $d = nl$ and hence $n = d/l$. In this case $d = 10\ \text{nm}$ and $l = 308\ \text{pm}$ hence

$$n = \frac{d}{l} = \frac{10 \times 10^{-9}\ \text{m}}{308 \times 10^{-12}\ \text{m}} = 32.4...$$

The probability of the ends being this distance apart is therefore

$$\begin{aligned} P &= \left(\frac{2}{\pi N}\right)^{1/2} e^{-n^2/2N} \\ &= \left(\frac{2}{\pi \times (2.31... \times 10^3)}\right)^{1/2} \times e^{-(32.4...)^2/2\times(2.31...\times10^3)} = \boxed{0.013} \end{aligned}$$

E14D.6(a) The probability distribution function for a three-dimensional freely jointed chain is given by [14D.4–*616*]

$$f(r) = 4\pi\left(\frac{a}{\pi^{1/2}}\right)^3 r^2 e^{-a^2r^2} \qquad a = \left(\frac{3}{2Nl^2}\right)^{1/2}$$

where N is the number of monomers in the chain, l is the length of each monomer, and $f(r)\,dr$ is the probability that the ends of the chain are a distance between r and $r + dr$ apart.

The monomer of polyethene $-[CH_2CH_2]_n-$ is taken to be CH_2CH_2, so the number of monomers in the chain is given by

$$N = \frac{M_{\text{polymer}}}{M_{CH_2CH_2}} = \frac{65 \times 10^3\ \text{g mol}^{-1}}{28.0516\ \text{g mol}^{-1}} = 2.31... \times 10^3$$

The length of each CH_2CH_2 unit is estimated as the length of a two C–C bonds: one C–C bond in the centre and half a bond length either side where the unit connects to carbons in adjacent units. From Table 9C.2 on page 362 a C–C bond length is approximately 154 pm, so the monomer length l is taken as $2 \times (154\ \text{pm}) = 308\ \text{pm} = 0.308\ \text{nm}$. Therefore

$$a = \left(\frac{3}{2Nl^2}\right)^{1/2} = \left(\frac{3}{2 \times (2.31... \times 10^3) \times (0.308\ \text{nm})^2}\right)^{1/2} = 0.0826...\ \text{nm}^{-1}$$

$$\begin{aligned} f(10.0\ \text{nm}) &= 4\pi\left(\frac{a}{\pi^{1/2}}\right)^3 r^2 e^{-a^2r^2} \\ &= 4\pi\left(\frac{0.0826...\ \text{nm}^{-1}}{\pi^{1/2}}\right)^3 \times (10.0\ \text{nm})^2 \times e^{-(0.0826\ \text{nm}^{-1})^2 \times (10.0\ \text{nm})^2} \\ &= 0.0642...\ \text{nm}^{-1} \end{aligned}$$

The probability that the ends will be found in a narrow range of width $\delta r = 0.1\ \text{nm}$ at 10.0 nm is therefore

$$f(10.0\ \text{nm})\delta r = (0.0642...\ \text{nm}^{-1}) \times (0.1\ \text{nm}) = \boxed{6.4 \times 10^{-3}}$$

E14D.7(a) As explained in Section 14D.3(b) on page 618, the radius of gyration of a constrained chain is given by the value for a free chain multiplied by a factor F, where F is given by [14D.8–618], $F = [(1 - \cos\theta)/(1 + \cos\theta)]^{1/2}$. For $\theta = 109°$,

$$F = \left(\frac{1 - \cos\theta}{1 + \cos\theta}\right)^{1/2} = \left(\frac{1 - \cos 109°}{1 + \cos 109°}\right)^{1/2} = 1.40...$$

This corresponds to a percentage increase of $[(1.40...) - 1] \times 100\% = \boxed{+40.1\%}$. The volume is proportional to the cube of the radius, so the volume of the constrained chain is related to that of a free chain by a factor of $F^3 = (1.40...)^3 = 2.75....$ This corresponds to a percentage increase of $[(2.75...) - 1] \times 100\% = \boxed{+176\%}$.

E14D.8(a) As explained in Section 14D.3(c) on page 618, the root mean square separation of the ends of a partially rigid chain with persistence length l_p is given by the value for a free chain, $N^{1/2}l$, multiplied by a factor F, where F is given by

[14D.10–*619*], $F = (2l_p/l - 1)^{1/2}$. The contour length is given by [14D.5–*617*], $R_c = Nl$, so a persistence length of 5.0% of the contour length corresponds to $l_p = 0.050R_c = 0.050Nl$. Hence, for $N = 1000$,

$$F = \left(\frac{2l_p}{l} - 1\right)^{1/2} = \left(\frac{2 \times (0.050Nl)}{l} - 1\right)^{1/2} = (0.100N - 1)^{1/2}$$
$$= (0.100 \times 1000 - 1)^{1/2} = 9.94\ldots$$

This corresponds to a percentage increase of $[(9.94...) - 1] \times 100\% = \boxed{+895\%}$. The volume is proportional to the cube of the radius, so the volume of the partially rigid chain is related to that of a free chain by a factor of $F^3 = (9.94...)^3 = 9.85... \times 10^2$. This corresponds to a percentage increase of $[(9.85... \times 10^2) - 1] \times 100\% = \boxed{(9.84 \times 10^4)\%}$.

E14D.9(a) By analogy with [14D.10–*619*], the radius of gyration R_g of a partially rigid coil is related to that of a freely jointed chain according to $R_g = F \times R_{g,\text{free}}$ where $F = (2l_p/l - 1)^{1/2}$. The radius of gyration for a three-dimensional freely jointed chain is given by [14D.7b–*618*], $R_{g,\text{free}} = (N/6)^{1/2}l$, so

$$R_g = F \times R_{g,\text{free}} = \left(\frac{2l_p}{l} - 1\right)^{1/2} \times \left(\frac{N}{6}\right)^{1/2} l$$

Rearranging gives

$$R_g^2 = \left(\frac{2l_p}{l} - 1\right)\left(\frac{N}{6}\right) l^2 \quad \text{hence} \quad \frac{6R_g^2}{Nl^2} = \frac{2l_p}{l} - 1 \quad \text{hence} \quad l_p = \frac{l}{2}\left(\frac{6R_g^2}{Nl^2} + 1\right)$$

Therefore for the polymer in question, taking $l = 0.150$ nm,

$$l_p = \frac{l}{2}\left(\frac{6R_g^2}{Nl^2} + 1\right) = \frac{0.150\ \text{nm}}{2}\left(\frac{6 \times (2.1\ \text{nm})^2}{1000 \times (0.150\ \text{nm})^2} + 1\right) = \boxed{0.16\ \text{nm}}$$

E14D.10(a) Modelling the polyethene as a 1D random coil perfect elastomer, the restoring force is given by [14D.12a–*620*], $F = (kT/2l)\ln[(1 + \lambda)/(1 - \lambda)]$ where $\lambda = x/Nl$.

The monomer of polyethene $-[CH_2CH_2]_n-$ is taken to be CH_2CH_2, so the number of monomers in the chain is given by

$$N = \frac{M_{\text{polymer}}}{M_{CH_2CH_2}} = \frac{65 \times 10^3\ \text{g mol}^{-1}}{28.0516\ \text{g mol}^{-1}} = 2.31... \times 10^3$$

The length of each CH_2CH_2 unit is estimated as the length of a two C–C bonds: one C–C bond in the centre and half a bond length either side where the unit connects to carbons in adjacent units. From Table 9C.2 on page 362 a C–C bond length is approximately 154 pm, so the monomer length l is taken as 2 ×

$(154\text{ pm}) = 308\text{ pm} = 0.308\text{ nm}$. The value of λ corresponding to a 1.0 nm extension is therefore

$$\lambda = \frac{x}{Nl} = \frac{1.0\text{ nm}}{(2.31... \times 10^3) \times (0.308\text{ nm})} = 1.40... \times 10^{-3}$$

Because $\lambda \ll 1$ the simplified equation for the restoring force, [14D.12b–621], $F = (kT/Nl^2)x$ is used. Noting that $1\text{ J m}^{-1} = 1\text{ N}$ gives

$$F = \frac{kT}{Nl^2}x = \frac{(1.3806 \times 10^{-23}\text{ J K}^{-1}) \times ([20 + 273.15]\text{ K})}{(2.31... \times 10^3) \times (0.308 \times 10^{-9}\text{ m})^2} \times (1.0 \times 10^{-9}\text{ nm})$$
$$= \boxed{1.8 \times 10^{-14}\text{ N}}$$

E14D.11(a) The entropy change when a 1D random coil is stretched or compressed by a distance x is given by [14D.11–620], $\Delta S = -\frac{1}{2}kN\ln[(1+\lambda)^{(1+\lambda)}(1-\lambda)^{(1-\lambda)}]$ where $\lambda = x/R_c$. The contour length R_c is given by [14D.5–617], $R_c = Nl$, so it follows that $\lambda = x/Nl$.

The monomer of polyethene $-[CH_2CH_2]_n-$ is taken to be CH_2CH_2, so the number of monomers in the chain is given by

$$N = \frac{M_{\text{polymer}}}{M_{CH_2CH_2}} = \frac{65 \times 10^3\text{ g mol}^{-1}}{28.0516\text{ g mol}^{-1}} = 2.31... \times 10^3$$

The length of each CH_2CH_2 unit is estimated as the length of a two C–C bonds: one C–C bond in the centre and half a bond length either side where the unit connects to carbons in adjacent units. From Table 9C.2 on page 362 a C–C bond length is approximately 154 pm, so the monomer length l is taken as $2 \times (154\text{ pm}) = 308\text{ pm} = 0.308\text{ nm}$. The value of λ corresponding to a 1.0 nm extension is therefore

$$\lambda = \frac{x}{Nl} = \frac{1.0\text{ nm}}{(2.31... \times 10^3) \times (0.308\text{ nm})} = 0.00140...$$

The entropy change is therefore

$$\Delta S = \tfrac{1}{2}kN\ln\left[(1+\lambda)^{(1+\lambda)}(1-\lambda)^{(1-\lambda)}\right]$$
$$= \tfrac{1}{2} \times (1.3806 \times 10^{-23}\text{ J K}^{-1}) \times (2.31... \times 10^3)$$
$$\times \ln\left[(1 + 0.00140...)^{(1+0.00140...)} \times (1 - 0.00140...)^{(1-0.00140...)}\right]$$
$$= -3.14... \times 10^{-26}\text{ J K}^{-1}$$

The molar entropy change is obtained by multiplying by Avogadro's constant

$$\Delta S_m = (-3.14... \times 10^{-26}\text{ J K}^{-1}) \times (6.0221 \times 10^{23}\text{ mol}^{-1}) = \boxed{-0.019\text{ J K}^{-1}\text{ mol}^{-1}}$$

Solutions to problems

P14D.1 There is some lack of clarity in the text over the definition of the radius of gyration, R_g. For a polymer consisting of N identical monomer units, R_g is defined as

$$R_g^2 = (1/N)\sum_{i=1}^{N} r_i^2 \tag{14.1}$$

where r_i is the distance of monomer unit i from the centre of mass. In other words, the radius of gyration is the root-mean-square of the distance of the monomer units from the centre of mass.

A related quantity is the moment of inertia I about an axis, which is defined in the following way

$$I = \sum_{i=1}^{N} m d_i^2 \tag{14.2}$$

where m is the mass of the monomer unit and d_i is the *perpendicular distance* from the monomer to the axis. In general, the distance d_i is *not* the same as r_i: the first is the perpendicular distance to the axis, the second is the distance to the centre of mass.

A radius of gyration can be related to a moment of inertia by imagining a rigid rotor consisting of a mass m_{tot} equal to the total mass of the polymer held at a distance R_g from the origin; the moment of inertia of this rotor is $I = m_{tot}R_g^2$, and hence $R_g^2 = I/m_{tot}$. However, note that this radius of gyration is associated by the rotation about a particular axis.

(a) For a solid sphere the mass is distributed continuously rather than at discrete points as for the simple polymer. The definition in eqn 14.1 is adapted by imagining that the monomer unit at r_i is a volume element $\mathrm{d}V$ located at distance r from the centre of mass (the origin). The sum over r_i^2 becomes the integral of $r^2\mathrm{d}V$ over the sphere, and division by N becomes division by the volume of the sphere V_s. Hence $R_g^2 = (1/V_s)\int_{\text{sphere}} r^2\,\mathrm{d}V$. It is convenient to complete the calculation using spherical polar coordinates

$$\begin{aligned}
R_g^2 &= \frac{1}{V_s}\int_{r=0}^{r=a}\int_{\theta=0}^{\theta=\pi}\int_{\phi=0}^{\phi=2\pi} r^2 \times r^2 \sin\theta\,\mathrm{d}r\,\mathrm{d}\theta\,\mathrm{d}\phi \\
&= \frac{3}{4\pi a^3}\int_{r=0}^{r=a} r^4\,\mathrm{d}r \int_{\theta=0}^{\theta=\pi} \sin\theta\,\mathrm{d}\theta \int_{\phi=0}^{\phi=2\pi} \mathrm{d}\phi \\
&= \frac{3}{4\pi a^3} \times \frac{a^5}{5} \times 2 \times 2\pi \\
&= \tfrac{3}{5}a^2
\end{aligned}$$

Hence the radius of gyration is $\boxed{(3/5)^{1/2}a}$.

(b) For a solid rod it is more convenient to use eqn 14.2 for rotation about (i) the long axis of the rod, and (ii) an axis perpendicular to this and which passes through the centre of mass. The long axis of the rod defines the z-axis and the centre of mass is at $z = 0$; the rod therefore extends from

$-l/2$ to $+l/2$ along z. It is convenient to use cylindrical polar coordinates described in *The chemist's toolkit* 19 in Topic 7F on page 281. In such a coordinate system the volume element is $r\,dr\,d\phi\,dz$, and ϕ ranges from 0 to 2π.

Equation 14.2 is adapted for a solid object by replacing the mass by a volume element dV which has mass $\rho\,dV$, where ρ is the mass density; the summation becomes an integration over the relevant coordinates which describe the rod: $z = -l/2$ to $+l/2$, $\phi = 0$ to 2π, and $r = 0$ to a. To compute the moment of inertia about the long axis note that the perpendicular distance to the axis is r so the integral is

$$\begin{aligned} I_{\|} &= \int_{z=-l/2}^{z=+l/2}\int_{r=0}^{r=a}\int_{\phi=0}^{\phi=2\pi} r^2 \times \rho\,dV \\ &= \rho\int_{z=-l/2}^{z=+l/2}\int_{r=0}^{r=a}\int_{\phi=0}^{\phi=2\pi} r^2 \times r\,dz\,dr\,d\phi \\ &= \rho\int_{z=-l/2}^{z=+l/2} dz\int_{r=0}^{r=a} r^3\,dr\int_{\phi=0}^{\phi=2\pi} d\phi \\ &= \rho \times l \times \frac{a^4}{4} \times 2\pi = \rho l a^4\pi/2 = m_{tot}a^2/2 \end{aligned}$$

where on the last line the total mass is given by $m_{tot} = \rho \times \pi a^2 \times l$. A rigid rotor with the same total mass has moment of inertia $I = m_{tot}R^2_{g,\|}$, hence $\boxed{R_{g,\|} = (2)^{-1/2}a}$.

To compute the moment of inertia perpendicular to the long axis, say about the x-axis, it is necessary to know the perpendicular distance d between an arbitrary point (x, y, z) and that axis. This distance is that between the points (x, y, z) and $(x, 0, 0)$; by Pythagoras' theorem $d^2 = y^2 + z^2 = r^2 \sin^2\phi + z^2$. The moment is inertia is therefore found from the integral

$$I = \int_{cyl.} (r^2\sin^2\phi + z^2) \times \rho\,dV = \underbrace{\rho\int_{cyl.} r^2\sin^2\phi\,dV}_{A} + \underbrace{\rho\int_{cyl.} z^2\,dV}_{B}$$

where the integration is over the complete cylinder. The integrals A and B are conveniently evaluated separately.

$$\begin{aligned} A &= \rho\int_{z=-l/2}^{z=+l/2}\int_{r=0}^{r=a}\int_{\phi=0}^{\phi=2\pi} r^2\sin^2\phi \times r\,dz\,dr\,d\phi \\ &= \rho\int_{z=-l/2}^{z=+l/2} dz\int_{r=0}^{r=a} r^3\,dr\int_{\phi=0}^{\phi=2\pi} \sin^2\phi\,d\phi \\ &= \rho \times l \times \frac{a^4}{4} \times \pi = \rho l a^4\pi/4 = m_{tot}a^2/4 \end{aligned}$$

where the integral over ϕ is found using Integral T.2 with $k = 1$ and $a =$

2π.

$$\begin{aligned} B &= \rho \int_{z=-l/2}^{z=+l/2} \int_{r=0}^{r=a} \int_{\phi=0}^{\phi=2\pi} z^2 \times r \,\mathrm{d}z\,\mathrm{d}r\,\mathrm{d}\phi \\ &= \rho \int_{z=-l/2}^{z=+l/2} z^2 \,\mathrm{d}z \int_{r=0}^{r=a} r\,\mathrm{d}r \int_{\phi=0}^{\phi=2\pi} \mathrm{d}\phi \\ &= \rho \times \frac{l^3}{12} \times \frac{a^2}{2} \times 2\pi = \rho l^3 a^2 \pi/12 = m_{\text{tot}} l^2/12 \end{aligned}$$

The moment of inertia about the perpendicular axis is therefore

$$I_{\perp} = A + B = m_{\text{tot}} a^2/4 + m_{\text{tot}} l^2/12 = m_{\text{tot}}(a^2/4 + l^2/12)$$

A rigid rotor with the same total mass has moment of inertia $I = m_{\text{tot}} R_{\text{g},\perp}^2$, hence $\boxed{R_{\text{g},\perp} = (a^2/4 + l^2/12)^{1/2}}$.

(c) The specific volume is the volume divided by the mass, $v_{\text{s}} = V/m$, and the mass of an individual macromolecule is given by M/N_{A}, where M is the molar mass.

$$v_{\text{s}} = \frac{V}{m} = \frac{(4/3)\pi a^3}{M/N_{\text{A}}} \quad \text{hence} \quad a^3 = \frac{3v_{\text{s}}M}{4\pi N_{\text{A}}}$$

The radius of gyration is therefore

$$\begin{aligned} R_{\text{g}} &= \left(\frac{3}{5}\right)^{1/2} \left(\frac{3v_{\text{s}}M}{4\pi N_{\text{A}}}\right)^{1/3} \\ &= \left(\frac{3}{5}\right)^{1/2} \left(\frac{3}{4\pi N_{\text{A}}}\right)^{1/3} (v_{\text{s}}M)^{1/3} \\ R_{\text{g}}/\text{m} &= \left(\frac{3}{5}\right)^{1/2} \left(\frac{3}{4\pi \times (6.0221\times 10^{23}\ \text{mol}^{-1})}\right)^{1/3} \\ &\quad \times \left[(v_{\text{s}}/\text{cm}^3\ \text{g}^{-1})(M/\text{g mol}^{-1}) \times 10^{-6}\right]^{1/3} \\ &= 5.6902 \times 10^{-11} \times \left[(v_{\text{s}}/\text{cm}^3\ \text{g}^{-1})(M/\text{g mol}^{-1})\right]^{1/3} \\ R_{\text{g}}/\text{nm} &= 0.056902 \times \left[(v_{\text{s}}/\text{cm}^3\ \text{g}^{-1})(M/\text{g mol}^{-1})\right]^{1/3} \end{aligned}$$

The factor of 10^{-6} on the fourth line is there to convert cm^3 to m^3. For the given data

$$R_{\text{g}}/\text{nm} = 0.056902 \times \left[(0.750) \times (100 \times 10^3)\right]^{1/3} = 2.40$$

The radius of gyration is therefore $\boxed{2.40\ \text{nm}}$.

For a rod $R_{\text{g},\parallel} = (1/2)^{1/2} a = (1/2)^{1/2} \times (0.50\ \text{nm}) = \boxed{0.35\ \text{nm}}$. To find the radius of gyration about the perpendicular axis requires a knowledge of l and this is found from the specific volume in a similar way to the method used for a sphere

$$v_{\text{s}} = \frac{V}{m} = \frac{\pi a^2 l}{M/N_{\text{A}}} \quad \text{hence} \quad l = \frac{v_{\text{s}}M}{\pi a^2 N_{\text{A}}}$$

$$\begin{aligned} l &= \frac{v_s M}{\pi a^2 N_A} \\ &= \frac{(0.750 \times 10^{-6}\ \mathrm{m^3\ g^{-1}}) \times (100 \times 10^3\ \mathrm{g\ mol^{-1}})}{\pi \times (0.50 \times 10^{-9}\ \mathrm{m})^2 \times (6.0221 \times 10^{23}\ \mathrm{mol^{-1}})} \\ &= 1.585... \times 10^{-7}\ \mathrm{m} = 158.5...\ \mathrm{nm} \end{aligned}$$

Hence

$$R_{g,\perp} = (a^2/4 + l^2/12)^{1/2} = [(0.50\ \mathrm{nm})^2/4 + (158.5...\ \mathrm{nm})^2/12]^{1/2} = \boxed{46\ \mathrm{nm}}$$

Because $l \gg a$ the radius of gyration is dominated by the term in l and $R_{g,\perp} \approx (12)^{-1/2} l$.

P14D.3 The problem should be stated as $\langle r^2 \rangle = N l^2$. The probability distribution for the separation of the ends in a 3D random coil is given by [14D.4–*616*]

$$f(r) = 4\pi \left(\frac{a}{\pi^{1/2}}\right)^3 r^2 \mathrm{e}^{-a^2 r^2} \quad \text{where} \quad a = \left(\frac{3}{2Nl^2}\right)^{1/2}$$

The mean-square separation is calculated as

$$\begin{aligned} \langle r^2 \rangle &= \int r^2 f(r)\,\mathrm{d}r = \int_0^\infty r^2 \times 4\pi \left(\frac{a}{\pi^{1/2}}\right)^3 r^2 \mathrm{e}^{-a^2 r^2}\,\mathrm{d}r \\ &= 4\pi \left(\frac{a}{\pi^{1/2}}\right)^3 \underbrace{\int_0^\infty r^4 \mathrm{e}^{-a^2 r^2}\,\mathrm{d}r}_{\text{Integral G.5 with } k = a^2} = 4\pi \left(\frac{a}{\pi^{1/2}}\right)^3 \times \frac{3}{8(a^2)^2} \left(\frac{\pi}{a^2}\right)^{1/2} \\ &= \frac{3}{2a^2} = \frac{3}{2}\left(\frac{2Nl^2}{3}\right) = Nl^2 \end{aligned}$$

P14D.5 The walk is constructed by starting at the origin and taking steps of unit length in a direction specified by a randomly generated angle θ between 0 and 360°. Each step then involves incrementing the x coordinate by $\cos\theta$ and the y coordinate by $\sin\theta$. The final distance r reached at the end of the walk is found calculating $r = \sqrt{x^2 + y^2}$ where x and y are the final x and y coordinates.

Two such walks, of 50 and 100 steps, are shown in Fig. 14.11. The final values of r in these cases are 6.24 and 7.13.

To investigate whether the mean and most probable values of r vary as $N^{1/2}$, where N is the number of steps, a large number of random walks with varying numbers of steps are generated and the value of r found for each. The table shows the mean and most probable values of r estimated from samples of 100 random walks carried out with each of 10, 20, 30, 40, 50, 60, 70, 80, 90 and 100 steps; the most probable values have been estimated by constructing a histogram of the values of r, fitting a curve to the histogram, and finding the maximum of the curve.

If r_{mean} and $r_{\text{most probable}}$ vary as $N^{1/2}$ then plots of these values against $N^{1/2}$ should give a straight line passing through the origin. The data are plotted in

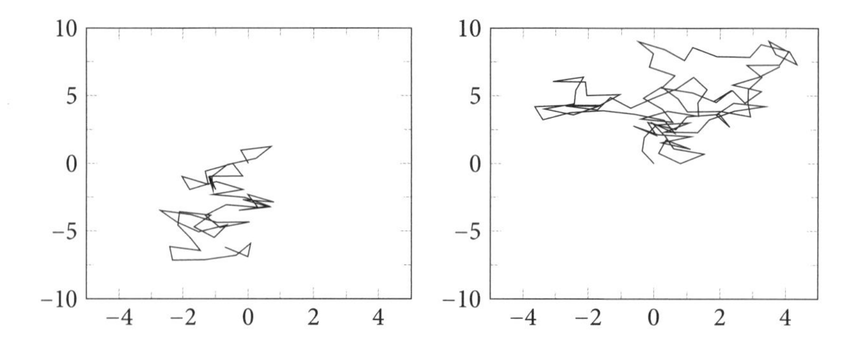

Figure 14.11

Fig. 14.12. In both cases the the data fall on a reasonable straight line that almost passes through the origin, thus indicating that r_{mean} and $r_{\text{most probable}}$ do indeed vary as $N^{1/2}$.

N	$N^{1/2}$	r_{mean}	$r_{\text{most probable}}$
10	3.16	2.49	3.44
20	4.47	3.75	4.81
30	5.48	4.54	7.03
40	6.32	6.32	8.94
50	7.07	6.30	8.81
60	7.75	7.05	8.77
70	8.37	7.10	8.94
80	8.94	7.83	10.99
90	9.49	8.40	11.80
100	10.00	8.70	12.94

P14D.7 The formula given in the text for the radius of gyration of a sphere of radius R is $R_g = (3/5)^{1/2}R$. If the molecules given are globular, that is, roughly spherical, their specific volume v_s should be given by $v_s = V/m = (4/3)\pi R^3/m$ where R is the radius of the sphere and m is the mass of one molecule. Replacing m by M/N_A and rearranging gives

$$v_s = \frac{\frac{4}{3}\pi R^3}{M/N_A} \quad \text{hence} \quad R = \left(\frac{3v_sM}{4\pi N_A}\right)^{1/3} \quad \text{hence} \quad R_g = \left(\frac{3}{5}\right)^{1/2}\left(\frac{3v_sM}{4\pi N_A}\right)^{1/3}$$

In the last step $R_g = (3/5)^{1/2}R$ is used. Using this expression, the value of R_g expected for each of the molecules if they are spherical is calculated from the v_s and M data, and compared to the experimental value of R_g. If the values are similar then there is evidence that the molecules are globular.

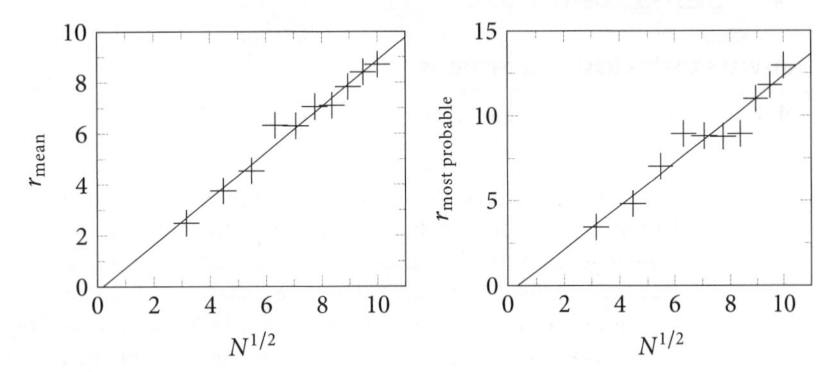

Figure 14.12

For serum albumin

$$R_g = \left(\frac{3}{5}\right)^{1/2}\left(\frac{3v_sM}{4\pi N_A}\right)^{1/3} = \left(\frac{3}{5}\right)^{1/2}\left(\frac{3\times(0.752\ \text{cm}^3\ \text{g}^{-1})\times(66\times10^3\ \text{g mol}^{-1})}{4\pi\times(N_A)}\right)^{1/3}$$
$$= 2.09...\times10^{-7}\ \text{cm} = 2.09\ \text{nm}$$

For bushy stunt virus

$$R_g = \left(\frac{3}{5}\right)^{1/2}\left(\frac{3v_sM}{4\pi N_A}\right)^{1/3} = \left(\frac{3}{5}\right)^{1/2}\left(\frac{3\times(0.741\ \text{cm}^3\ \text{g}^{-1})\times(10.6\times10^6\ \text{g mol}^{-1})}{4\pi\times(N_A)}\right)^{1/3}$$
$$= 1.13...\times10^{-6}\ \text{cm} = 11.3\ \text{nm}$$

For DNA

$$R_g = \left(\frac{3}{5}\right)^{1/2}\left(\frac{3v_sM}{4\pi N_A}\right)^{1/3} = \left(\frac{3}{5}\right)^{1/2}\left(\frac{3\times(0.556\ \text{cm}^3\ \text{g}^{-1})\times4\times10^6\ \text{g mol}^{-1})}{4\pi\times(N_A)}\right)^{1/3}$$
$$= 7.42...\times10^{-7}\ \text{cm} = 7.43\ \text{nm}$$

For serum albumin and bushy stunt virus the experimental radii of gyration (2.98 nm and 12.0 nm) are similar to the values that would be expected if these macromolecules were spherical, thus suggesting that they are globular. In the case of DNA the experimental radius of gyration (117.0 nm) is much greater than the value expected if it were spherical, suggesting that DNA is not globular and therefore more rod-like.

P14D.9 As explained in the *How is that done?* 14D.4 on page 620 the restoring force for an extended elastomer is given by $F = -T(\partial S/\partial x)_T$. This restoring force is equal to the tension t required to keep the sample at a particular length, hence $t = -T(\partial S/\partial x)_T$. The restoring force therefore depends on $(\partial S/\partial x)$. Extension of a polymer reduces the disorder and hence entropy of the chains, so there is a tendency to revert to the more disordered non-extended state.

14E Self-assembly

Answers to discussion questions

D14E.1 This is discussed in Section 14E.1(a) on page 623.

D14E.3 The formation of micelles is favoured by the interaction between hydrocarbon tails and is opposed by charge repulsion of the polar groups which are placed close together at the micelle surface. As the salt concentration is increased, the repulsion of head groups is reduced because their charges are partly shielded by the ions of the salt. This favours micelle formation causing the micelles to be formed at a lower concentration and hence reducing the critical micelle concentration.

D14E.5 Lipids with unsaturated chains 'freeze' at lower temperatures, so membranes can remain fluid at lower ambient temperatures.

Solutions to exercises

E14E.1(a) The isoelectric point of a protein is the pH at which the protein has no net charge and therefore is unaffected by an electric field. This is the pH at which the velocity is zero; solving for this gives

$$0 = 0.50 - 0.10(\mathrm{pH}) - (3.0\times10^{-3})(\mathrm{pH})^2 + (5.0\times10^{-4})(\mathrm{pH})^3$$

This equation is solved numerically to yield the solutions pH = −13.8, pH = 14.9, and pH = 4.9. The −13.8 and 14.9 solutions are rejected as they are outside the pH range 3.0–7.0 over which the expression is valid. Therefore the isoelectric point is $\boxed{4.9}$.

Solutions to problems

P14E.1 (a) The data show that π increases by 0.5 for every additional CH_2 group. The R group in question, $(CH_2)_6CH_3$, has two more CH_2 groups than $(CH_2)_4CH_3$ which has $\pi = 2.5$, so the predicted value for $(CH_2)_6CH_3$ is $2.5 + 2\times0.5 = \boxed{3.5}$.

(b) The data are plotted in Fig. 14.13. The points fall on a reasonable straight line, the equation of which is

$$\log K_{\mathrm{I}} = -1.49\pi - 1.95$$

The slope and intercept of the line are therefore $\boxed{-1.49}$ and $\boxed{-1.95}$.

(c) The definition of the hydrophobicity constant π is given by [14E.5–*627*], $\pi = \log[s(\mathrm{RX})/s(\mathrm{HX})]$. This definition implies that $\pi = \log 1 = 0$ for the case R = H. It follows that for this case

$$\log K_{\mathrm{I}} = -1.95 \quad \text{hence} \quad K_{\mathrm{I}} = \boxed{0.011}$$

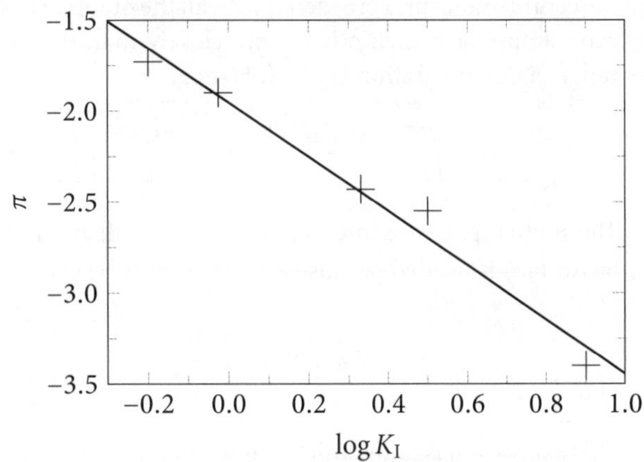

Figure 14.13

P14E.3 The equilibrium constant for the formation of micelles containing N monomers, M_N, is given by [14E.6b–*627*],

$$K = \frac{[M_N]}{([M]_{total} - N[M_N])^N}$$

where the factors of $1/c^{\ominus}$ are omitted for clarity. For the case that $K = 1$ and $N = 2$ this equation becomes

$$1 = \frac{[M_2]}{([M]_{total} - 2[M_2])^2} \quad \text{hence} \quad ([M]_{total} - 2[M_2])^2 = M_2$$

$$\text{hence} \quad 4[M_2]^2 - (1 + 4[M]_{total})[M_2] + [M]_{tot}^2 = 0$$

$$\text{hence} \quad [M_2] = \frac{(1 + 4[M]_{total}) \pm \sqrt{(1 + 4[M]_{total})^2 - 4 \times 4 \times [M]_{total}^2}}{8}$$

$$\text{hence} \quad [M_2] = \frac{1 + 4[M]_{total} - \sqrt{1 + 8[M]_{total}}}{8}$$

The positive square root is rejected because this predicts $[M_2] \neq 0$ when $[M]_{total} = 0$.

Solutions to integrated activities

I14.1 Consider a single molecule surrounded by $N - 1$ ($\approx N$) others in a container of volume V. The number of molecules in a spherical shell centred on this molecule with radius r and thickness dr is

$$dN = \text{volume of shell} \times \text{number of molecules per unit volume}$$

$$= 4\pi r^2 dr \times \frac{N}{V} = \frac{4\pi r^2 N}{V} dr$$

The interaction between the central molecule and the ones in the shell is $du = -C_6/r^6\, dN = -4\pi C_6 N/Vr^4\, dr$. Integration over r gives the total interaction of the central molecule considered with all the others. However, because molecules cannot approach each other more closely than the molecular diameter d the range of the integration is $r \geq d$. Hence

$$u = \int_d^\infty -\frac{4\pi C_6 N}{V}\frac{1}{r^4}\,\mathrm{d}r = \frac{4\pi C_6 N}{V}\left[\frac{1}{3}\frac{1}{r^3}\right]_d^\infty = -\frac{4\pi C_6 N}{3Vd^3}$$

The mutual pairwise interaction energy of all N molecules is $U = \frac{1}{2}Nu$; the factor of $\frac{1}{2}$ is needed because each pair must be counted once only

$$U = \tfrac{1}{2}N \times -\frac{4\pi C_6 N}{3d^3} = \boxed{-\frac{2\pi C_6 N^2}{3Vd^3}}$$

This expression is differentiated and compared to $n^2a/V^2 = (\partial U/\partial V)_T$ to give

$$\left(\frac{\partial U}{\partial V}\right)_T = \frac{2\pi C_6 N^2}{3V^2d^3} = \frac{n^2 a}{V^2} \quad \text{hence} \quad a = \frac{2\pi N^2}{3d^3n^2} = \boxed{\frac{2\pi N_A^2}{3d^3}}$$

In the last step, $N = nN_A$ is used.

I14.3 (a) The atomic charges on methyl adenine and methyl thymine are calculated with Spartan 10 using a Hartree–Fock procedure with a 6-31G* basis set. Some of the resulting charges are shown in Fig. 14.14.

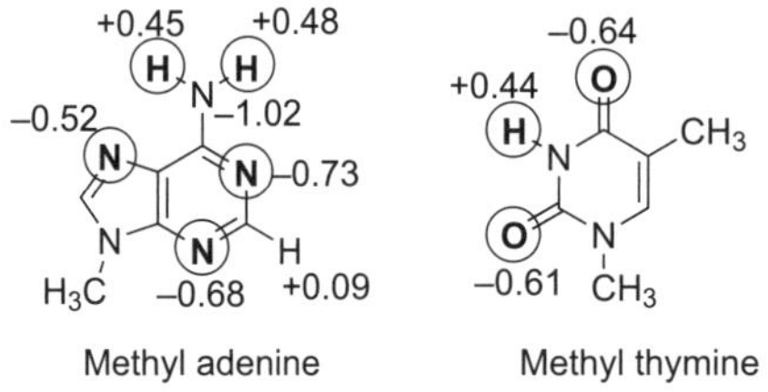

Figure 14.14

(b) Hydrogen bonds are most likely to be formed between hydrogen atoms carrying a substantial positive charge. It must also be possible for these atoms to approach each other without causing a steric clash. Suitable atoms are circled in Fig. 14.14.

(c) Two possible arrangements for a hydrogen-bonded dimer are shown in Fig. 14.15. In both arrangements two hydrogen bonds are formed between positive hydrogen atoms and negative nitrogen or oxygen atoms. In each case the two hydrogen bonds have approximately the same length and are linear, the arrangement which is the most favourable as explained in Section 14B.2 on page 598.

(d) *Impact* 21 shows that the first arrangement in Fig. 14.15 is the one that occurs in DNA. Reference to the atomic charges in Fig. 14.14 shows that this arrangement pairs the most positive hydrogen in methyl adenine

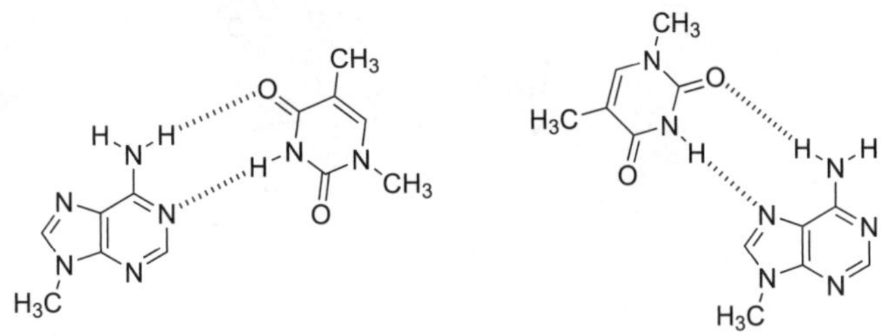

Figure 14.15

with the most negative oxygen in methyl thymine, and also the most positive hydrogen in methyl thymine with the most negative nitrogen in methyl adenine. This arrangement is therefore expected to be particularly favourable. This arrangement also positions the methyl groups of methyl adenine and methyl thymine at the correct positions corresponding to the sugars in DNA.

(e) A similar calculation for methyl guanine and methyl cytosine gives the charges shown in Fig. 14.16. Suitable atoms for forming hydrogen bonds are circled.

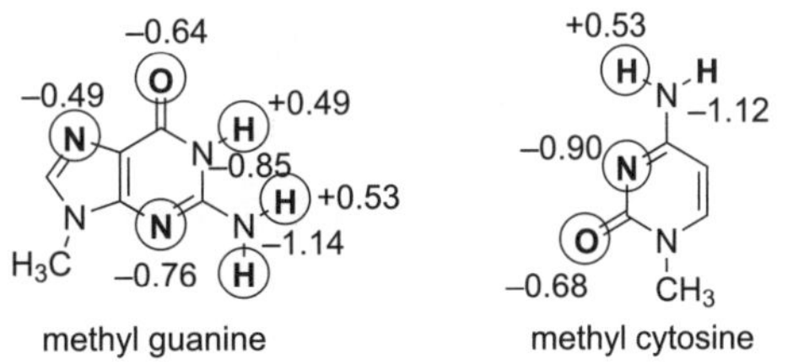

Figure 14.16

Several possible arrangements for a hydrogen-bonded dimer are shown in Fig. 14.17. Only one of these involves three linear hydrogen bonds, and from *Impact* 21 this is the arrangement that occurs in DNA. Comparison of the adenine–thymine and guanine–cytosine base pairs that occur in DNA shows that the methyl groups representing the attachment to the rest of the DNA molecule are in the same orientation for both base pairs, and also that all hydrogen bonds are a similar length. These are essential features in the formation of the DNA double helix.

I14.5

(a) Carrying out a multiple regression analysis using a spreadsheet or mathematical software gives the result

$$\log A = 3.59 + 0.957S + 0.362W$$

Hence $\boxed{b_0 = 3.59}$, $\boxed{b_1 = 0.957}$, and $\boxed{b_2 = 0.362}$.

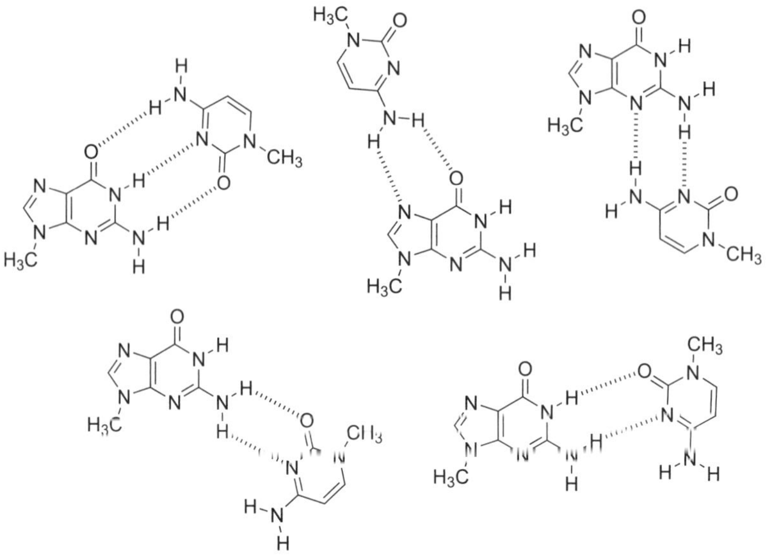

Figure 14.17

(b) The value of W for the drug is found by rearranging the equation obtained above

$$W = (\log A - 3.59 - 0.957S)/0.362 = (7.60 - 3.59 - 0.957 \times 4.84) = \boxed{-1.72}$$

I14.7 The three conformations are shown in Fig. 14.18. As explained in *Impact* 21, the ϕ angle describes the rotation around the N–C bond and is defined as the angle about which the bond must be rotated in order to bring the front CO group into an eclipsed conformation with the back CO group, a clockwise rotation being defined as positive. This is most clearly seen from the views down the N–C bond in Fig. 14.18. The ψ angle is similarly defined as the angle around which the C–CO bond must be rotated in order to bring the front NH group into an eclipsed conformation with the back NH group. This is shown by the views down the C–CO bond in Fig. 14.18.

Starting with these conformations, the final ϕ and ψ angles after optimizing the structures using Spartan 10 with the B3LYP/6-31G* density functional method are given below. The ΔE column gives the energies of each optimized structure, measured relative to that of the optimized geometry obtained starting with the $\phi = 180°$, $\psi = 180°$ structure of either the R = H or R = CH_3 molecule as appropriate.

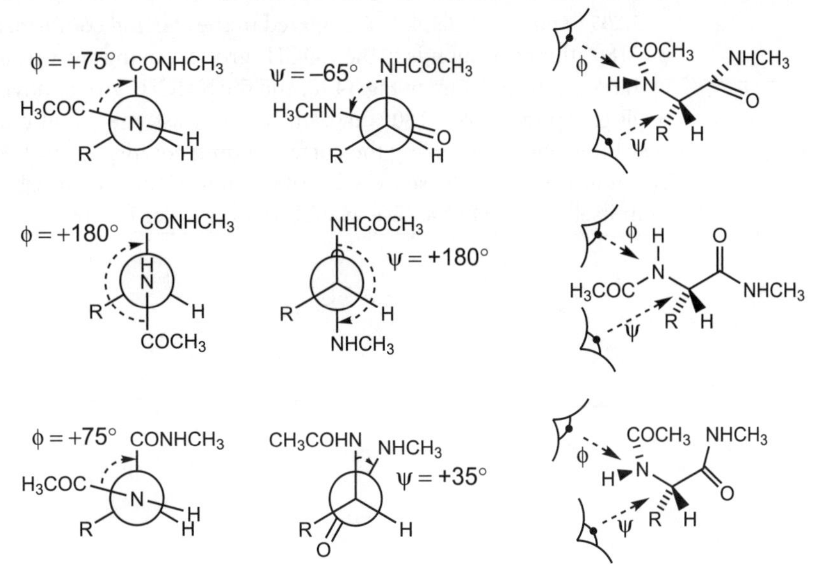

Figure 14.18

	Starting		Optimized		
	$\phi/°$	$\psi/°$	$\phi/°$	$\psi/°$	$\Delta E/10^{-3}$ au
(a) R = H	+75	−65	+81.99	−69.22	−1.377
	+180	+180	+178.09	+178.80	0.000
	+65	+35	+120.56	−20.12	+2.849
(b) R = CH_3	+75	−65	+73.78	−56.79	+1.933
	+180	+180	−157.97	+163.43	0.000
	+65	+35	+66.91	+28.01	+9.892

The results show that none of the starting structures converge to the same final conformation. This reflects the fact that there are several energy minima for this molecule, so that starting with a different structure leads to a different optimized geometry. It is possible that starting with other structures would lead to the discovery of further optimized geometries; in general an extensive conformational search beginning with a large number of starting structures is needed to find all possible conformers and hence to identify which is the lowest energy conformation, that is, the global minimum.

For the case that R = H, the lowest energy optimized conformer found has $\phi = +81.99°$ and $\psi = -69.22°$. The corresponding optimized conformer for the R = CH_3 case has slightly smaller ϕ and ψ, +73.78° and −56.79°, presumably as a result of the larger size of the R group. Furthermore this is not the lowest energy conformer found for the R = CH_3 case: the conformer obtained starting from the $\phi = \psi = 180°$ geometry lies lower.

In this latter geometry, the final optimized structure for R = CH_3 has ϕ = $-157.97°$ and $\psi = +163.43°$. Compared to the starting conformation with $\phi = \psi = 180°$ this corresponds to the $COCH_3$ group moving away from the R group in the $\phi = 180°$ diagram in Fig. 14.18, and the $NHCH_3$ group moving away from the R group in the $\psi = 180°$ diagram. By contrast, the ϕ and ψ angles change very little when optimizing the corresponding conformer for the R = H case. This is attributed to the steric effect of the methyl pushing nearby groups away from itself relative to the case when R is the smaller H atom.

15 Solids

15A Crystal structure

Answers to discussion questions

D15A.1 A space lattice is the three-dimensional structural pattern formed by lattice points representing the locations of motifs which may be atoms, molecules, or groups of atoms, molecules, or ions within a crystal. All points of the space lattice have identical environments and they define the crystal structure. The unit cell is an imaginary parallelepiped from which the entire crystal structure can be generated, without leaving gaps, using translations of the unit cell alone. Each unit cell is defined in terms of lattice points and the unit cell is commonly formed by joining neighbouring lattice points by straight lines. The smallest possible unit cell is called the primitive unit cell. Non-primitive unit cells may exhibit lattice points within the cell, at the cell centre, on cell faces, or on cell edges.

D15A.3 The three cubic lattices, P, I and F, are depicted in Fig. 15A.8 on page 643. As shown there, cubic P has one lattice point, cubic I has two, and cubic F has four. Only the cubic P lattice is primitive, therefore.

Solutions to exercises

E15A.1(a) The volume of an orthorhombic unit cell is given by $V = abc$, and the mass of the unit cell m is given by $m = \rho V$, where ρ is the mass density. Using the estimate of mass density $\rho = 3.9\ \mathrm{g\,cm^{-3}}$

$$m = abc\rho = [(634 \times 784 \times 516) \times 10^{-36}\ \mathrm{m^3}] \times (3.9 \times 10^6\ \mathrm{g\,m^{-3}})$$
$$= 1.00... \times 10^{-21}\ \mathrm{g}$$

The mass of a unit cell is also related to the molar mass by $m = nM = NM/N_\mathrm{A}$, where n is the amount in moles of $NiSO_4$ in a unit cell, M is the molar mass, and N is the number of formula units per unit cell.

$$N = \frac{mN_\mathrm{A}}{M} = \frac{(1.00... \times 10^{-21}\ \mathrm{g}) \times (6.0221 \times 10^{23}\ \mathrm{mol^{-1}})}{154.75\ \mathrm{g\,mol^{-1}}} = 3.89...$$

If it is assumed that there are no defects in the crystal lattice N is expected to be an integer and hence $\boxed{N = 4}$. With this value a more precise value of the mass density is calculated as

$$\rho = \frac{m}{V} = \frac{NM}{N_A V}$$

$$= \frac{4 \times (154.75\ \mathrm{g\,mol^{-1}})}{(6.0221 \times 10^{23}\ \mathrm{mol^{-1}}) \times [(634 \times 784 \times 516) \times 10^{-36}\ \mathrm{m^3}]} = \boxed{4.01\ \mathrm{g\,cm^{-3}}}$$

E15A.2(a) Miller indices are of the form (hkl) where h, k, and l are the reciprocals of the intersection distances along the a, b and c axes, respectively. If the reciprocal intersection distances are fractions then the Miller indices are achieved by multiplying through by the lowest common denominator.

intersect axes at	$(2a, 3b, 2c)$	$(2a, 2b, \infty c)$
remove cell dimensions	$(2,3,2)$	$(2,2,\infty)$
take reciprocals	$(\frac{1}{2}, \frac{1}{3}, \frac{1}{2})$	$(\frac{1}{2}, \frac{1}{2}, 0)$
Miller indices	(323)	(110)

E15A.3(a) The separation of (hkl) planes d_{hkl} of a cubic lattice is give by [15A.1a–645], $d_{hkl} = a/(h^2 + k^2 + l^2)^{1/2}$.

$$d_{112} = \frac{(562\ \mathrm{pm})}{(1^2 + 1^2 + 2^2)^{1/2}} = \boxed{229\ \mathrm{pm}} \qquad d_{110} = \frac{(562\ \mathrm{pm})}{(1^2 + 1^2 + 0^2)^{1/2}} = \boxed{397\ \mathrm{pm}}$$

$$d_{224} = \frac{(562\ \mathrm{pm})}{(2^2 + 2^2 + 4^2)^{1/2}} = \boxed{115\ \mathrm{pm}}$$

E15A.4(a) The separation of (hkl) planes d_{hkl} of an orthorhombic lattice is given by [15A.1b–645], $1/d_{hkl}^2 = h^2/a^2 + k^2/b^2 + l^2/c^2$. Therefore

$$d_{hkl} = (h^2/a^2 + k^2/b^2 + l^2/c^2)^{-1/2}$$

$$d_{321} = \left[\frac{3^2}{(812\ \mathrm{pm})^2} + \frac{2^2}{(947\ \mathrm{pm})^2} + \frac{1^2}{(637\ \mathrm{pm})^2}\right]^{-1/2}$$

$$= (2.05... \times 10^{19})^{-1/2}\ \mathrm{m} = \boxed{220\ \mathrm{pm}}$$

Solutions to problems

P15A.1 A face-centred cubic unit cell has lattice points at its 8 corners and also at the centres of its six faces. Therefore there are $(8 \times \frac{1}{8} + 6 \times \frac{1}{2}) = 4$ lattice points per unit cell. The mass density ρ is therefore $\rho = 4m/V$, where m is the mass per lattice point and the volume V is a^3, where a is the unit cell dimension. The molar mass M is calculated from $M = mN_A$

$$M = N_A \rho a^3/4$$

$$= (6.0221 \times 10^{23}\ \mathrm{mol^{-1}}) \times (1.287 \times 10^6\ \mathrm{g\,m^{-3}}) \times (12.3 \times 10^{-9}\ \mathrm{m})^3 \times \tfrac{1}{4}$$

$$= \boxed{3.61 \times 10^5\ \mathrm{g\,mol^{-1}}}$$

P15A.3 From Fig. 15A.8 on page 643 it is seen that the unit cell can be envisaged as a prism of height c whose base is rhombus with sides a and interior angle 120°, which is depicted below.

The area of the rhombus is $ax = a \times a\sin(60°)$, hence the volume is $V = cax = c \times [a \times a\sin(60°)] = \boxed{(\sqrt{3}/2)a^2c}$.

P15A.5 For a monoclinic unit cell, $V = abc\sin\beta$. From the information given, $a = 1.377b$ and $c = 1.436b$. Because there are two napthalene molecules within the unit cell it follows that the mass density is $\rho = 2m/V$, where m is the mass per molecule given by $m = M/N_A$, where M is the molar mass of napthalene (128.1... g mol^{-1}).

Using $\rho V = 2m$ and $V = abc\sin\beta$, it follows that $abc\sin\beta = 2m/\rho$ and hence $abc = 2m/(\rho\sin\beta) = 2M/(N_A\rho\sin\beta)$. The product $abc = 1.377 \times 1.436 \times b^3$ and so

$$b = \left[\frac{2M}{N_A\rho\sin\beta \times 1.377 \times 1.436}\right]^{1/3}$$

$$= \left[\frac{2 \times (128.1...\ \mathrm{g\,mol^{-1}})}{(6.0221 \times 10^{23}\ \mathrm{mol^{-1}}) \times (1.152 \times 10^6\ \mathrm{g\,m^{-3}}) \times \sin(122.82°) \times 1.377 \times 1.436}\right]^{1/3}$$

$$= \boxed{605.8\ \mathrm{pm}}$$

Thus $\boxed{a = 834.2\ \mathrm{pm}}$ and $\boxed{c = 870.0\ \mathrm{pm}}$

P15A.7 The mass of the unit cell m is given by $m = NM/N_A$, where N is the number of monomer units per unit cell and M is the molar mass of a monomer unit. The mass is also written in terms of the mass density ρ and the volume V as $m = \rho V$. Hence $NM/N_A = \rho V$ and so $N = \rho N_A V/M$. The molar mass is

$$M = 63.55 + 7 \times 12.01 + 13 \times 1.0079 + 5 \times 14.01 + 8 \times 16.00 + 32.06 = 390.8...\ \mathrm{g\,mol^{-1}}$$

For a monoclinic unit cell,

$$V = abc\sin\beta$$
$$= (1.0427\ \mathrm{nm}) \times (0.8876\ \mathrm{nm}) \times (1.3777\ \mathrm{nm}) \times \sin(93.254°)$$
$$= 1.27... \times 10^{-27}\ \mathrm{m^3}$$

hence

$$N = \frac{\rho N_A V}{M} = \frac{(2.024 \times 10^6\ \mathrm{g\,m^{-3}}) \times (6.0221 \times 10^{23}\ \mathrm{mol^{-1}}) \times (1.27... \times 10^{-27}\ \mathrm{m^3})}{(390.8...\ \mathrm{g\,mol^{-1}})}$$
$$= 3.97...$$

There are $\boxed{4}$ monomer units per unit cell.

P15A.9 Consider the two-dimensional lattice and planes shown in Fig. 15A.12 on page 644. The $(hk0)$ planes intersect the a, and b axes at distances a/h and b/k from the origin, respectively. Using trigonometry,

$$\sin\phi = \frac{d_{hk0}}{a/h} = \frac{d_{hk0}h}{a} \qquad \cos\phi = \frac{d_{hk0}}{b/k} = \frac{d_{hk0}k}{b}$$

Because $\sin^2\theta + \cos^2\theta = 1$ it follows that

$$\left(\frac{d_{hk0}h}{a}\right)^2 + \left(\frac{d_{hk0}k}{b}\right)^2 = 1$$

Rearranging gives

$$\frac{1}{d_{hk0}^2} = \frac{h^2}{a^2} + \frac{k^2}{b^2}$$

Because the third side of the cell is mutually perpendicular to the other two, the extension to three dimensions simply involves adding an additional term, as in the derivation for a cubic lattice

$$\boxed{\frac{1}{d_{hkl}^2} = \frac{h^2}{a^2} + \frac{k^2}{b^2} + \frac{l^2}{c^2}}$$

15B Diffraction techniques

Answers to discussion questions

D15B.1 A systematic absence is a reflection for which the structure factor happens to be zero. As a result the reflection has zero intensity and is therefore 'absent' from the diffraction pattern. The structure factor is zero when the contributions to it from various atoms combine in such a way as to cancel one another. Such absences therefore reflect both the lattice type and the identity of the atoms in the lattice; examples are given in Section 15B.1(d) on page 649.

For a given lattice, certain combinations of h, k and l lead to absent reflections, and therefore the observation of a particular pattern of such absences can be helpful in identifying the type of unit cell. Examples are seen in Fig. 15B.10 on page 652.

D15B.3 This is discussed in Section 15B.1(e) on page 652.

Solutions to exercises

E15B.1(a) Bragg's law [15B.1b–648], $\lambda = 2d\sin\theta$, describes the relationship between wavelength of the X-rays λ, the Bragg angle θ, and the plane separation d. Thus

$$\lambda = 2 \times (99.3\text{ pm}) \times \sin(20.85°) = \boxed{70.7\text{ pm}}$$

E15B.2(a) As shown in Fig. 15B.10 on page 652, for the cubic I lattice reflections from planes with $h + k + l$ = odd are absent from the diffraction pattern. Hence the first three possible reflections occur for planes (110), (200) and (211). Using the Bragg law [15B.1b–648], $\lambda = 2d_{hkl}\sin\theta$, and the expression for the spacing of the planes [15A.1a–645], $d_{hkl} = a/(h^2+k^2+l^2)^{1/2}$, the following table is drawn up

Miller indices	(110)	(200)	(211)
d_{hkl}	$a/(1^2+1^2)^{1/2}$	$a/(2^2)^{1/2}$	$a/(2^2+1^2+1^2)^{1/2}$
d_{hkl}/pm	205.7...	145.5	118.8...
$\sin\theta$	0.174...	0.247...	0.303...
$\theta/°$	10.1	14.3	17.6

E15B.3(a) The separation of the (hkl) planes of an orthorhombic lattice is given by [15A.1b–645], $1/d_{hkl}^2 = h^2/a^2 + k^2/b^2 + l^2/c^2$. This distance is used with [15B.1b–648] to compute the angle of reflection as $\theta = \sin^{-1}(\lambda/2d_{hkl})$.

Miller indices	(100)	(010)	(111)
$(1/d_{hkl}^2)/\text{pm}^{-2}$	$1^2/542^2$	$1^2/917^2$	$1^2/542^2 + 1^2/917^2 + 1^2/645^2$
d_{hkl}/pm	542	917	378.0...
$\theta/°$	8.17	4.82	11.8

E15B.4(a) The Bragg law [15B.1b–648], $\lambda = 2d\sin(\theta)$, is rearranged to give the glancing angle as $2\theta = 2\sin^{-1}(\lambda/2d)$, where d is the plane separation and λ is the wavelength of the X-rays. For the case where $\lambda = 154.433$ pm,

$$2\theta = 2 \times \sin^{-1}[(154.433\ \text{pm})/2 \times (77.8\ \text{pm})] = 165.9...°$$

For the case where $\lambda = 154.051$ pm,

$$2\theta = 2 \times \sin^{-1}[(154.051\ \text{pm})/2 \times (77.8\ \text{pm})] = 163.8...°$$

The difference in the glancing angles is $165.9...° - 163.8...° = \boxed{2.14°}$.

E15B.5(a) In Section 15B.1(c) on page 649 it is shown that the scattering factor in the forward direction, $f(0)$, is equal to the total number of electrons in the species, N_e. Thus for Br^- $\boxed{f(0) = 36}$.

E15B.6(a) The structure factor is given by [15B.3–650]

$$F_{hkl} = \sum_j f_j\, e^{i\phi_{hkl}(j)}$$

where f_j is the scattering factor of species j and $\phi_{hkl}(j) = 2\pi(hx_j + ky_j + lz_j)$ is the phase of the scattering from that species.

Species at the corners of the unit cell are shared between eight adjacent unit cells so they have weight of $\frac{1}{8}$ and so, if all the atoms are the same and have the same scattering factor f, the contribution from each is $\frac{1}{8}f$. The structure factor is

$$F_{hkl} = \sum_j f_j e^{i\phi_{hkl}(j)}$$

$$= \tfrac{1}{8}f\left[1+e^{2i\pi k}+e^{2i\pi l}+e^{2i\pi(k+l)}+e^{2i\pi h}+e^{2i\pi(h+k)}+e^{2i\pi(h+l)}+e^{2i\pi(h+k+l)}\right]$$

The indices h, k and l are all integers, and $e^{in\pi} = (-1)^n$ for integer n. All the exponents in the sum are even multiples of $i\pi$, so all the exponential terms are equal to $+1$. Hence $\boxed{F_{hkl} = f}$.

E15B.7(a) The orthorhombic C unit cell is shown in Fig. 15A.8 on page 643. The structure factor is given by [15B.3–650]

$$F_{hkl} = \sum_j f_j e^{i\phi_{hkl}(j)}$$

where f_j is the scattering factor of species j and $\phi_{hkl}(j) = 2\pi(hx_j + ky_j + lz_j)$ is the phase of the scattering from that species.

The ions at the corners of the unit cell are shared between eight adjacent unit cells so they have weight $\frac{1}{8}$ and therefore, if they all have the same scattering factor f, the contribution from each is $\frac{1}{8}f$. As is shown in *Exercise* E15B.6(a), these ions together contribute $+f$ to the structure factor.

The ions on the faces have positions $(\frac{1}{2}a, \frac{1}{2}a, 0)$ and $(\frac{1}{2}a, \frac{1}{2}a, a)$ and are shared between two adjacent unit cells. Each face ion thus contributes $\frac{1}{2} \times (2f) = f$, where $(2f)$ is the scattering factor for the face ions, given as twice that of the other ions. The contribution to the scattering factor from the face ions is

$$fe^{2i\pi(\frac{1}{2}h+\frac{1}{2}k)} + fe^{2i\pi(\frac{1}{2}h+\frac{1}{2}k+l)} = f\left(1+e^{2i\pi l}\right)e^{i\pi(h+k)} = 2f(-1)^{(h+k)}$$

The structure factor is therefore $F_{hkl} = f + 2f(-1)^{(h+k)}$. Therefore for $(h+k)$ odd, $\boxed{F_{hkl} = f - 2f = -f}$, and for $(h+k)$ even, $\boxed{F_{hkl} = f + 2f = 3f}$.

E15B.8(a) The electron density distribution $\rho(\boldsymbol{r})$ in the unit cell is given by [15B.4–651], $\rho(\boldsymbol{r}) = (1/V)\sum_{hkl} F_{hkl}e^{-2\pi i(hx+ky+lz)}$, where V is the volume of the unit cell. In this case the structure factors are only given for the x direction so the sum is just over the index h. Furthermore, because $F_h = F_{-h}$ the summation can be taken from $h = 0$ to $h = +\infty$

$$V\rho(x) = \sum_{h=-\infty}^{\infty} F_h e^{-2\pi ihx} = F_0 + \sum_{h=1}^{\infty}\left(F_h e^{-2\pi ihx} + F_{-h}e^{2\pi ihx}\right)$$

$$= F_0 + \sum_{h=1}^{\infty} F_h\left(e^{-2\pi ihx} + e^{2\pi ihx}\right) = F_0 + 2\sum_{h=1}^{\infty} F_h\cos(2\pi hx)$$

In this case there are a total of ten terms to include, $h = 0$ to 9. Figure 15.1 shows a plot of $V\rho(x)$ against x; the electron density is at a maximum of $110/V$ at $x = 0.5$, the centre of the unit cell.

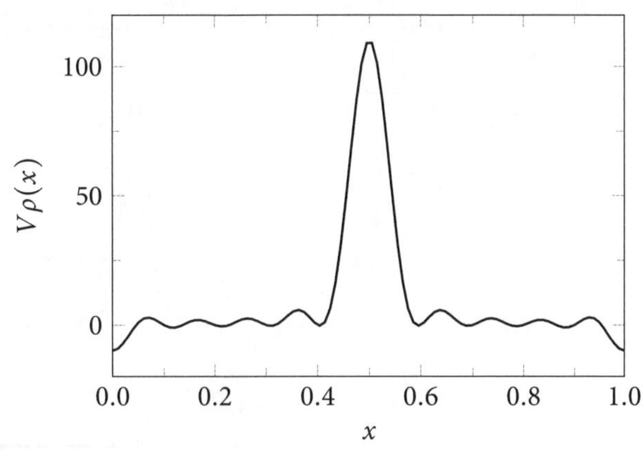

Figure 15.1

E15B.9(a) The Patterson synthesis is given by [15B.5–653],

$$P(\boldsymbol{r}) = \frac{1}{V}\sum_{hkl}|F_{hkl}|^2\,\mathrm{e}^{-2\pi\mathrm{i}(hx+ky+lz)}$$

In this case the structure factors are only given for the x direction so the sum is just over the index h. Furthermore, because $F_h = F_{-h}$ the summation can be taken from from $h = 0$ to $h = +\infty$. Using a similar line of argument to that in *Exercise* E15B.7(a), the Patterson synthesis is

$$VP(x) = F_0^2 + 2\sum_{h=1}^{\infty} F_h^2\cos(2\pi hx)$$

In this case there are a total of ten terms to include, $h = 0$ to 9. Figure 15.2 shows a plot of $VP(x)$ against x. As expected, there strong feature at the origin; this arises from the separation between each atom and itself. There is also a strong feature at $x = 1$ which indicate that atoms are separated by $1 \times a$ unit along the x-axis.

E15B.10(a) To constructor the Patterson map, choose the position of one atom to be the origin (here, the boron). Then add peaks to the map corresponding to vectors joining each pair of atoms (Fig. 15.3). Heavier atoms give more intense contributions than light atoms, so peaks arising from F and F separations are shown with greater diameter than those representing B and F separations. The vector between atom A and atom B has the same magnitude as that between B and A, but points in the opposite direction; the map therefore includes two symmetry related peaks on either side of the origin. The vectors between each atom and itself give a peak at the centre point of the Patterson map, and the many contributions at this position create an intense peak.

E15B.11(a) Using the de Broglie relation [7A.11–244], $\lambda = h/p = h/(mv)$, where p is the momentum, m is the mass of a neutron and v its speed, it follows that

$$v = \frac{h}{\lambda m} = \frac{6.6261\times10^{-34}\,\mathrm{J\,s}}{(65\times10^{-12}\,\mathrm{m})\times(1.6749\times10^{-27}\,\mathrm{kg})} = \boxed{6.1\,\mathrm{km\,s^{-1}}}$$

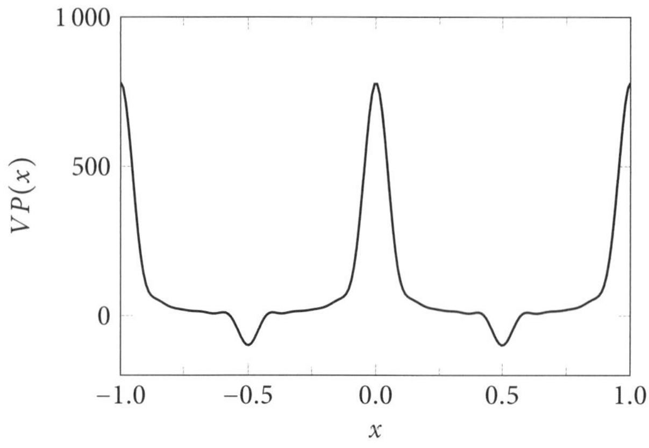

Figure 15.2

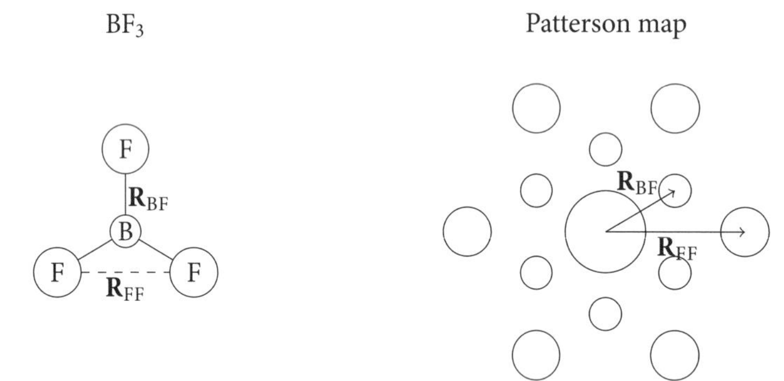

Figure 15.3

E15B.12(a) From the equipartition principle the kinetic energy is $E_k = \frac{1}{2}kT$. This energy can be written in terms of the momentum as $p^2/(2m)$ and hence $p = (mkT)^{1/2}$. The de Broglie relation [7A.11–*244*], $\lambda = h/p$, is then used to find the wavelength

$$\lambda = \frac{h}{(mkT)^{1/2}} = \frac{6.6261 \times 10^{-34}\,\text{J s}}{[(1.6749 \times 10^{-27}\,\text{kg}) \times (1.3806 \times 10^{-23}\,\text{J K}^{-1}) \times (350\,\text{K})]^{1/2}}$$
$$= \boxed{233\,\text{pm}}$$

Solutions to problems

P15B.1 The NaCl unit cell is depicted in Fig. 15B.9 on page 650. The unit cell is a cube with volume $V = a^3$ where a is the unit cell side length. There are eight Na^+ ions at vertices, six Na^+ ions on faces, one Cl^- ion at centre and 12 Cl^- ions at the edges of the unit cell. Thus in total there are four NaCl units per unit cell. The mass density ρ is $\rho = m/V$ where m is the total mass per unit cell. It follows that $m = 4m_{\text{NaCl}} = 4M/N_A$ where M is the molar mass of an NaCl unit ($M = 58.44\ \text{g mol}^{-1}$).

The spacing of the planes is given by [15A.1a–645], $d_{hkl} = a/(h^2 + k^2 + l^2)^{1/2}$; for the (100) reflection, this evaluates to $d_{100} = a$. The angle of refraction, the spacing and the wavelength are related by the Bragg law, [15B.1b–648], $\lambda = 2d \sin\theta$ which rearranges to $d_{100} = a = \lambda/(2\sin\theta)$. Using this, the density can be expressed as

$$\rho = \frac{4M}{N_A V} = \frac{4M}{N_A a^3} = \frac{32M \sin^3\theta}{N_A \lambda^3}$$

In turn this expression is rearranged to give the wavelength in terms of the known parameters

$$\begin{aligned}\lambda &= \left(\frac{32M}{N_A \rho}\right)^{1/3} \sin\theta \\ &= \left(\frac{32 \times (58.44\ \text{g mol}^{-1})}{(6.0221 \times 10^{23}\ \text{mol}^{-1}) \times (2.17 \times 10^6\ \text{g m}^{-3})}\right)^{1/3} \sin(6.0°) \\ &= \boxed{118\ \text{pm}}\end{aligned}$$

P15B.3 Combining Bragg's law [15B.1b–648], $\lambda = 2d \sin\theta$, with the expression for the the separation of planes for a cubic lattice [15A.1a–645], $d_{hkl} = a/(h^2 + k^2 + l^2)^{1/2}$, gives $\sin\theta = (\lambda/2a)(h^2 + k^2 + l^2)^{1/2}$.

The first three reflections for a cubic P lattice are (100), (110) and (200). Consider the ratio of $\sin\theta$ for the first two of these compared to ratio of $\sin\theta$ for the first two observed lines:

$$\frac{\sin\theta_{110}}{\sin\theta_{100}} = \frac{(1^2 + 1^2)^{1/2}}{(1^2)^{1/2}} = 1.41... \qquad \frac{\sin\theta_{1\text{st}}}{\sin\theta_{2\text{nd}}} = \frac{\sin 22.171°}{\sin 19.076°} = 1.15...$$

These do not match up, so the lattice is not cubic P. For cubic I the first three reflections are (110), (200) and (211); making the same comparison gives

$$\frac{\sin\theta_{200}}{\sin\theta_{110}} = \frac{(2^2)^{1/2}}{(1^2 + 1^2)^{1/2}} = 1.41... \qquad \frac{\sin\theta_{1\text{st}}}{\sin\theta_{2\text{nd}}} = \frac{\sin 22.171°}{\sin 19.076°} = 1.15...$$

These do not match up, so the lattice is not cubic I. For cubic F the first three reflections are (111), (200) and (220).

$$\frac{\sin\theta_{200}}{\sin\theta_{111}} = \frac{(2^2)^{1/2}}{(1^2 + 1^2 + 1^2)^{1/2}} = 1.15... \qquad \frac{\sin\theta_{1\text{st}}}{\sin\theta_{2\text{nd}}} = \frac{\sin 22.171°}{\sin 19.076°} = 1.15...$$

This matches well. The same procedure is used for the second and third reflections

$$\frac{\sin\theta_{220}}{\sin\theta_{200}} = \frac{(2^2 + 2^2)^{1/2}}{(2^2)^{1/2}} = 1.41... \qquad \frac{\sin\theta_{2\text{nd}}}{\sin\theta_{3\text{rd}}} = \frac{\sin 32.256°}{\sin 22.171°} = 1.41...$$

Again,there is a good match. Therefore silver adopts a $\boxed{\text{cubic F lattice}}$.

The lattice parameter is computed from $a = \lambda(h^2 + k^2 + l^2)^{1/2}/(2\sin\theta)$. With the data for the (111) reflection this gives

$$a = \frac{(154.18\ \text{pm})(1^2 + 1^2 + 1^2)^{1/2}}{2\sin(19.076°)} = \boxed{408.55\ \text{pm}}$$

Cubic F has four atoms per unit cell and so the mass density is $\rho = 4m/V = 4M/N_A V$, where M is the molar mass of silver.

$$\rho = \frac{4M}{N_A a^3} = \frac{4 \times (107.87\ \text{g mol}^{-1})}{(6.0221 \times 10^{23}\ \text{mol}^{-1}) \times (408.55 \times 10^{-12}\ \text{m})^3} = \boxed{10.51\ \text{g cm}^{-3}}$$

P15B.5 The scattering factor is given by [15B.2–*649*]

$$f(\theta) = 4\pi \int_0^\infty \rho(r)\frac{\sin kr}{kr} r^2\,\text{d}r \quad k = \frac{4\pi}{\lambda}\sin\theta$$

For $\rho(r) = 3Z/4\pi R^3$ where $0 \leq r \leq R$ and $\rho(r) = 0$ for $r \geq R$, the integral is evaluated by parts to give

$$f(\theta) = \frac{3Z}{kR^3}\int_0^R r\sin kr\,\text{d}r = \frac{3Z}{kR^3}\left.\frac{-r\cos kr}{k}\right|_0^R + \frac{3Z}{k^2R^3}\int_0^R \cos kr\,\text{d}r$$

$$= \frac{-3Z\cos kR}{k^2R^2} + \frac{3Z}{k^2R^3}\left.\frac{\sin kr}{k}\right|_0^R = \frac{3Z}{k^3R^3}(\sin kR - kR\cos kR)$$

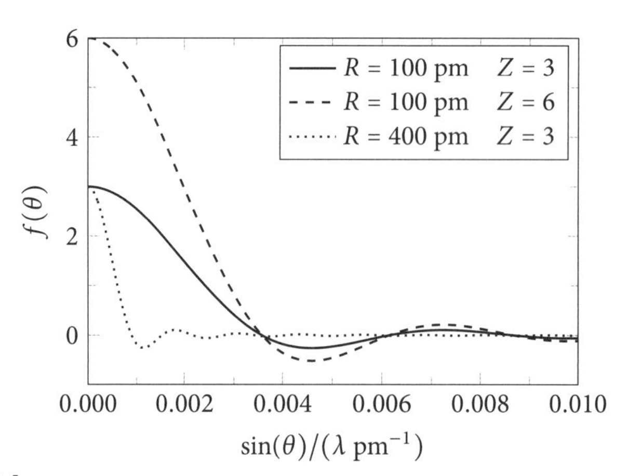

Figure 15.4

A plot of $f(\theta)$ against $(\sin\theta)/\lambda$ is shown in Fig. 15.4. In the forward direction, $\theta = 0$, the scattering factor is equal to Z, which is the expected result because Z is equal to the number of electrons. The scattering oscillates with increasing angle but superimposed on this is an overall decay that becomes faster as R increases. Thus, the larger the atom, the more concentrated is the scattering in the forward direction.

P15B.7 The structure factor is given by [15B.3–650], $F_{hkl} = \sum_j f_j e^{2\pi i(hx_j+ky_j+lz_j)}$. Each atom A is shared between 8 unit cells and therefore has weight $\frac{1}{8}$, whereas the B atom contributes to just one unit cell and so has weight 1.

$$\begin{aligned} F_{hkl} &= \tfrac{1}{8} f_A \left[1+e^{2\pi ik}+e^{2\pi il}+e^{2\pi i(k+l)}+e^{2\pi ih}+e^{2\pi i(h+k)}+e^{2\pi i(h+l)}+e^{2\pi i(h+k+l)}\right] \\ &\quad + f_B e^{\pi i(h+k+l)} \\ &= f_A + f_B(-1)^{(h+k+l)} \end{aligned}$$

where to go to the last line the relationship $e^{\pi in} = (-1)^n$ for integer n is used. The intensity of the diffraction pattern is directly proportional to the square modulus of the structure factor.

(a) For $f_A = f$, $f_B = 0$ then $F_{hkl} = f$ for all (hkl) planes so the diffraction pattern will display no systematic absences.

(b) For $f_B = \frac{1}{2} f_A$ then $F_{hkl} = f_A[1 + \frac{1}{2}(-1)^{(h+k+l)}]$. For all reflections with $(h+k+l)$ odd the intensity will be proportional to the square of $\frac{1}{2} f_A$, and for all reflections with $(h+k+l)$ even the intensity will be proportional to the square of $\frac{3}{2} f_A$.

(c) For $f_A = f_B = f$ then $F_{hkl} = f[1 + (-1)^{(h+k+l)}]$. For all reflections with $(h+k+l)$ odd the structure factor is zero, and for all reflections with $(h+k+l)$ even the structure factor is $2f$. The diffraction pattern will show systematic absences for lines where $(h+k+l)$ is odd.

15C Bonding in solids

Answers to discussion questions

D15C.1 The majority of metals crystallize in structures which can be interpreted as the closest packing arrangements of hard spheres. These are the cubic close-packed (ccp) and hexagonal close-packed (hcp) structures. In these models, 74% of the volume of the unit cell is occupied by the atoms (packing fraction = 0.74). Most of the remaining metallic elements crystallize in the body-centred cubic (bcc) arrangement which is not too much different from the close-packed structures in terms of the efficiency of the use of space (packing fraction 0.68 in the hard sphere model). If atoms were truly hard spheres, the expectation is that all metals would crystallize in either the ccp or hcp close-packed structures. The fact that a significant number crystallize in other structures indicates that a simple hard sphere model is an inaccurate representation of the interactions between the atoms.

Solutions to exercises

E15C.1(a) The densest packing arrangement possible for cylinders is the hexagonal packing shown in Fig. 15.5; the unit cell is the rhombus indicated, and the internal angles in this rhombus are 60° and 120°. The centre-to-centre spacing of the cylinders is $2R$, where R is the radius of one cylinder. The distance h is given by $h = 2R\sin 60° = R\sqrt{3}$, therefore the area of the rhombus is base × height $= 2R^2\sqrt{3}$.

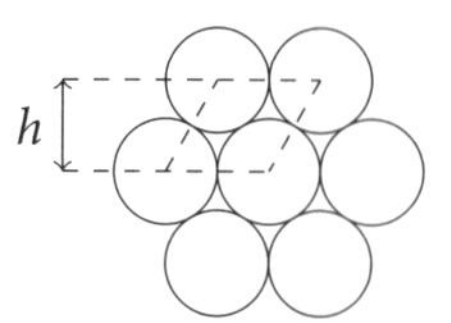

Figure 15.5

If the depth of the unit cell is z then the volume of the unit cell is the area of the rhombus times this depth, $V = 2\sqrt{3}R^2z$. Each cylinder occupies volume $\pi R^2 z$ and there is a total of one cylinder per unit cell. The packing density, f, is

$$f = \frac{\pi R^2 z}{2\sqrt{3}R^2 z} = \frac{\pi}{2\sqrt{3}} = \boxed{0.9069}$$

E15C.2(a) The packing fraction is $f = NV_a/V_c$ where N is the number of spheres per unit cell, $V_a = 4\pi R^3/3$ is the volume of each sphere of radius R, and V_c is the volume of the unit cell.

(i) For a primitive cubic unit cell the spheres touch along the edges of the cell, so the edges of the cube have length $2R$ and hence $V_c = (2R)^3$. There is one sphere per unit cell, $N = 1$, and therefore

$$f = \frac{4\pi R^3/3}{8R^3} = \frac{\pi}{6} = \boxed{0.5236}$$

(ii) For a bcc unit cell, the spheres touch along the body diagonal so the length of this diagonal is $4R$. Imagine a right-angle triangle in which the hypotenuse is the body diagonal, and the other two sides are an edge of the cube, length a, and a face diagonal, length $\sqrt{2}a$. It follows that $(4R)^2 = a^2 + 2a^2$ and hence $a = 4R/\sqrt{3}$. The volume is therefore $V_c = (4R/\sqrt{3})^3$, and as $N = 2$ it follows

It follows that side of the cube is $4R/\sqrt{3}$ so $N = 2$, $V_a = 4\pi R^3/3$ and $V_c = (4R/\sqrt{3})^3$. Thus

$$f = \frac{2 \times 4\pi R^3/3}{(4R/\sqrt{3})^3} = \frac{\sqrt{3}\pi}{8} = \boxed{0.6802}$$

(iii) For a fcc unit cell, the spheres touch along a face diagonal which therefore has length $4R$. If the edge of the cube has length a it follows, by considering a face, that $(4R)^2 = a^2 + a^2$ and hence $a = 2\sqrt{2}R$. The volume is therefore $V_c = (2\sqrt{2}R)^3$, and because $N = 4$ the packing fraction is

$$f = \frac{4 \times 4\pi R^3/3}{16\sqrt{2}R^3} = \frac{\pi}{3\sqrt{2}} = \boxed{0.7405}$$

E15C.3(a) The coordination number N of an ionic lattice depends on the radius ratio of the cation and anion of the lattice. The radius-ratio rule, which considers the maximum possible packing density of hard spheres of a given radius around a hard sphere of a different radius, provides a method to determine the structure type. The radius ratio is $\gamma = r_s/r_l$ where r_s is the radius of the smallest ion and r_l is the radius of the largest ion. If $\gamma \le (2^{1/2} - 1)$ then $N < 6$; for $(2^{1/2} - 1) < \gamma < (3^{1/2} - 1)$ then $N = 6$; for $\gamma \ge (3^{1/2} - 1)$ then $N = 8$.

The range for sixfold coordination is therefore $0.414 < \gamma < 0.732$, and hence $r_l \times 0.414 < r_s < r_l \times 0.732$. For the case of the Cl^- anion $(181\ \text{pm}) \times 0.414 = 75.0\ \text{pm}$ and $(181\ \text{pm}) \times 0.732 = 132.5\ \text{pm}$. Therefore for sixfold coordination the smallest radius for the cation is $\boxed{75.0\ \text{pm}}$, whilst for eightfold coordination the smallest radius is $\boxed{133\ \text{pm}}$.

E15C.4(a) The unit cell volume V is related to the packing density f and the atomic volume v by $fV = v$. Assuming the atoms can be approximated as spheres then $v = 4\pi R^3/3$ where R is the atomic radius. Using the packing densities calculated in *Exercise* E15C.2(a) and the given data

$$\frac{V_{\text{bcc}}}{V_{\text{hcp}}} = \frac{v_{\text{bcc}}}{v_{\text{hcp}}} \times \frac{f_{\text{hcp}}}{f_{\text{bcc}}} = \frac{(R_{\text{bcc}})^3 f_{\text{hcp}}}{(R_{\text{hcp}})^3 f_{\text{bcc}}} = \frac{142.5^3 \times 0.7405...}{145.8^3 \times 0.6802...} = 1.016...$$

Thus transformation from hcp to bcc causes cell volume to $\boxed{\text{expand by } 1.6\%}$

E15C.5(a) The lattice enthalpy ΔH_L is the change in standard molar enthalpy for the process $MX(s) \rightarrow M^+(g) + X^-(g)$ and its equivalent. The value of the lattice enthalpy is determined indirectly using a Born–Haber cycle, as shown in Fig. 15.6 (all quantities are given in kJ mol^{-1}). From the cycle it follows that

$$-635\ \text{kJ mol}^{-1} + \Delta H_L = (178 + 1735 + 249 - 141 + 844)\ \text{kJ mol}^{-1}$$

Thus $\Delta H_L = \boxed{3500\ \text{kJ mol}^{-1}}$

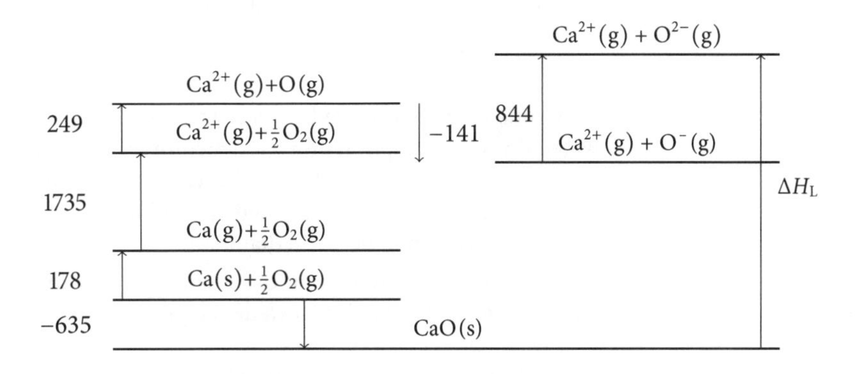

Figure 15.6

Solutions to problems

P15C.1 The packing fraction is $f = NV_a/V_c$, where N is the number of atoms per unit cell, $V_a = 4\pi R^3/3$ is the volume of an atom of radius R, and V_c is the unit cell volume. The structure of diamond is shown in Fig. 15C.15 on page 664: there are 8 atoms at the vertices of the cell (weight $\frac{1}{8}$), 6 atoms at the face-centres (weight $\frac{1}{2}$), and 4 atoms within the unit cell (weight 1), giving a total of 8 atoms per unit cell.

The two nearest-neighbour atoms which touch along the body diagonal are at locations (0,0,0) and $(\frac{1}{4}, \frac{1}{4}, \frac{1}{4})$, where the coordinates are expressed as fractions of the length of the side of the unit cell, a. These two atoms are at the opposite corners of a small cube with edge $a/4$. The body diagonal of a cube is $\sqrt{3}$ times the length of the edge, so it follows that the length of the body diagonal of this small cube is $\sqrt{3}a/4$. As the two atoms touch along this diagonal, this distance is also equal to $2R$, hence $a = 8R/\sqrt{3}$. The packing fraction is therefore given by

$$f = \frac{8 \times 4\pi R^3/3}{a^3} = \frac{8 \times 4\pi R^3/3}{\left(8R/\sqrt{3}\right)^3} = \frac{\sqrt{3}\pi}{16} = \boxed{0.3401}$$

P15C.3 (a) Close-packed spheres form a face-centred cubic structure, which is shown in *Exercise* E15C.2(b) to have a packing density of $f = 0.7405$. A sample of volume V of diamond therefore contains $fV/(4\pi R^3/3)$ carbon atoms, where R is the atomic radius. The mass of these carbon atoms is $fV/(4\pi R^3/3)\times(M/N_A)$, where M is the molar mass of carbon, therefore the mass density is

$$\rho = \frac{\text{mass}}{\text{volume}} = \frac{fV/(4\pi R^3/3) \times (M/N_A)}{V} = \frac{3fM}{4\pi N_A R^3}$$

With the data given

$$\rho = \frac{3fM}{4\pi N_A R^3} = \frac{3 \times (0.7405) \times (12.01\ \text{g mol}^{-1})}{4\pi \times (6.0221 \times 10^{23}\ \text{mol}^{-1}) \times (\frac{1}{2} \times 154.45\ \text{pm})^3}$$
$$= \boxed{7.655\ \text{g cm}^{-3}}$$

(b) The experimentally determined density is significantly lower than that calculated on the assumption of a fcc structure. This implies that atoms which are assumed to be in contact in the fcc structure are in fact further apart, and in turn this can be ascribed to the highly directional (tetrahedal) bonding known to occur in diamond.

In *Problem* P15C.1 it is shown that the packing density for the diamond structure is $f = 0.3401$. With this value the predicted density is reduced to

$$(7.655\ \text{g cm}^{-3}) \times (0.3401/0.7405) = 3.516\ \text{g cm}^{-3}$$

which is in close agreement with the experimental value.

P15C.5 The formation of a band in one dimension results in a set of states which spread, to a finite extent, above and below the energy of the original atomic orbital from which the band is created (Fig. 15C.6(e) on page 658). If the system is extended to two dimensions, each one of these original states itself becomes the starting point for a band arising from overlap in the second dimension; this is illustrated in a highly schematic way in Fig. 15.7.

The original atomic orbital is indicated on the left, and overlap of these results in a one-dimensional band, indicated by the dotted lines. Then, each state in this band itself gives rise to a further band when interactions are allowed in a second dimension. This is illustrated for the states at the very top and bottom of the band (shown by dotted lines), and a selection of levels between. The band clearly increases in overall width, but in addition the density of states increases in the centre of the two-dimensional band as many of one-dimensional bands overlap here. In contrast, at the extremities of the band, fewer one-dimensional bands are overlapping. These are only qualitative arguments, but they are indicative of the origin for the change in the density of states which is indeed observed.

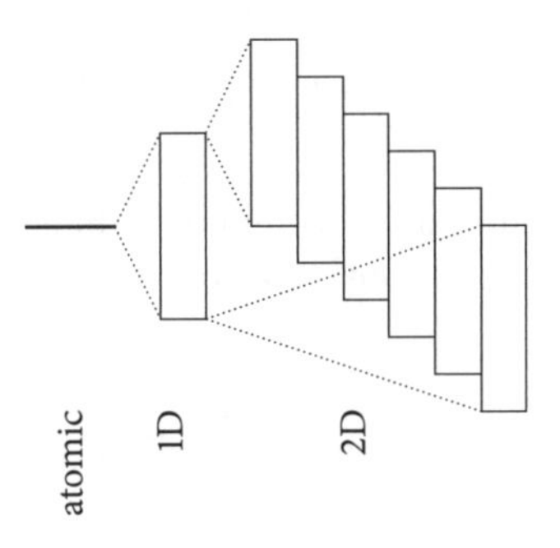

Figure 15.7

P15C.7 (a) The rock salt (NaCl) structure, shown in Fig. 15C.10 on page 660, exhibits sixfold coordination. Let the radius of the chloride ions be r_l and that of the sodium ions be r_s; the radius ratio γ is defined as $\gamma = r_s/r_l$. The lowest value for γ occurs when r_l is as large as possible, and the limit of its value will be when the chloride ions just touch along the face diagonal. In this limit, the length of the face diagonal will be $4r_l$; if the edge of the unit cell is of length a, it follows that $(4r_l)^2 = 2a^2$.

When the chloride ions are just touching, γ is further minimized by making the sodium ions as large as possible, and the limit of this is set by the point at which the sodium ions and chloride ions touch along the edge of a the cube. In this limit $2r_l + 2r_s = a$.

Combining these two results establishes the relationship between the two radii

$$(4r_l)^2 = 2a^2 = 2(r_l + r_s)^2 \quad \text{hence} \quad r_s^2 + 2r_l r_s - r_l^2 = 0$$

The quadratic in r_s is solved to give $r_s = (-1 \pm \sqrt{2})r_l$; of these solutions, only the one with the positive sign is physically reasonable so $r_s = (\sqrt{2} - 1)r_l$ and hence $\gamma = r_s/r_l = \sqrt{2} - 1 = 0.414$.

(b) Eightfold coordination is shown in Fig. 15C.9 on page 660. The limit on the size of the chloride ions is when they touch along the edge, which is when $2r_l = a$. The limit on the size of the caesium ion is when it touches the chloride ions along the body diagonal. Given that the length of the body diagonal is $\sqrt{3}a$, the condition is $2r_l + 2r_s = \sqrt{3}a$. It follows that

$$2r_l + 2r_s = \sqrt{3}a = \sqrt{3}(2r_l)$$

Solving this gives $r_s = (\sqrt{3} - 1)r_l$ and hence $\gamma = r_s/r_l = \sqrt{3} - 1 = 0.732$.

P15C.9 The contribution of the Coulomb interaction to lattice energy, E_p, is given by equation [15C.3–*662*] and the positive contribution due to overlap of atomic orbitals, E_p^*, is given by equation [15C.4–*662*].

$$E_{p,tot} = E_p + E_p^* = -A\frac{|z_A z_B|N_A e^2}{4\pi\varepsilon_0 d} + N_A C' e^{-d/d^*} \qquad (15.1)$$

The minimum in this is found by differentiating $E_{p,tot}$ with respect to d and setting the result equal to zero.

$$\frac{dE_{p,tot}}{dd} = A\frac{|z_A z_B|N_A e^2}{4\pi\varepsilon_0 d^2} - \frac{N_A C'}{d^*}e^{-d/d^*} = 0$$

$$\text{hence} \quad C' e^{-d/d^*} = A\frac{|z_A z_B|e^2 d^*}{4\pi\varepsilon_0 d^2}$$

Note that in this expression the distance d is now that which gives the minimum potential energy. Substituting this expression for $C' e^{-d/d^*}$ into eqn 15.1 gives

$$E_{p,min} = -A\frac{|z_A z_B|N_A e^2}{4\pi\varepsilon_0 d}\left(1 - \frac{d^*}{d}\right)$$

15D The mechanical properties of solids

Answer to discussion question

D15D.1 If, when the stress is removed, an object returns to the same shape it had before the stress was applied, the deformation is said to be elastic. If the result of the stress is a change in shape which remains even after the stress has been removed, the deformation is said to be plastic.

Solutions to exercises

E15D.1(a) The relationship between the applied pressure p, the bulk modulus K, and the fractional change in volume $\Delta V/V$ is given by [15D.1b–666], $K = p/(\Delta V/V)$. For a fractional change of 1%, $\Delta V/V = 0.01$, the pressure is $p = 0.01 \times 3.43 \times 10^9$ Pa = $\boxed{34.3 \text{ MPa}}$.

E15D.2(a) The Young's modulus E is related to the stress σ and the strain ε by [15D.1a–666], $E = \sigma/\varepsilon$. The stress is given by $\sigma = F/A$ where F is the force applied and A is the cross-sectional area. Hence

$$\sigma = \frac{F}{A} = \frac{500 \text{ N}}{\pi(1.0 \times 10^{-3} \text{ m})^2} = 1.59... \times 10^8 \text{ Pa} = \boxed{1.6 \times 10^2 \text{ MPa}}$$

$$\varepsilon = \frac{\Delta L}{L} = \frac{\sigma}{E} = \frac{1.59... \times 10^8 \text{ Pa}}{4.42 \times 10^9 \text{ Pa}} = 0.036...$$

Hence the percentage increase in length L is $\boxed{3.6\%}$

E15D.3(a) Poisson's ratio, ν_P, is defined in [15D.2–667], $\nu_P = \varepsilon_{trans}/\varepsilon_{norm}$, where ε_{trans} is the transverse strain and ε_{norm} is the normal (uniaxial) strain. If the normal strain is 1.0%, it follows that the change in length ΔL_{norm} is

$$\Delta L_{norm} = \varepsilon_{norm} L_{norm} = 0.01 \times (1.0 \times 10^{-2} \text{ m}) = 1.0 \times 10^{-4} \text{ m}$$

The transverse strain is $\varepsilon_{trans} = \nu_P \varepsilon_{norm}$, so the change in dimension in the transverse direction ΔL_{trans} is

$$\Delta L_{trans} = \varepsilon_{trans} L_{trans} = \nu_P \varepsilon_{norm} L_{trans} = 0.45 \times 0.01 \times (1.0 \times 10^{-2} \text{ m}) = 4.5 \times 10^{-5} \text{ m}$$

It is expected that the result of applying the stress will be to decrease the size of the cube in the transverse dimension (that is ΔL_{trans} is negative), and that the decrease will be the same in each transverse direction. The volume after the stress has been applied is therefore

$$(1.0 \times 10^{-2} \text{ m} + 1.0 \times 10^{-4} \text{ m}) \times (1.0 \times 10^{-2} \text{ m} - 4.5 \times 10^{-5} \text{ m})^2 = 1.000930... \times 10^{-6} \text{ m}^3$$

The change in volume is $1.000930... \times 10^{-6} \text{ m}^3 - 1.0 \times 10^{-6} \text{ m}^3 = \boxed{9.3 \times 10^{-4} \text{ cm}^3}$

Solutions to problems

P15D.1

$$E = \frac{\mu(3\lambda + 2\mu)}{\lambda + \mu} \text{ [i]} \quad K = \frac{3\lambda + 2\mu}{3} \text{ [ii]} \quad G = \mu \text{ [iii]} \quad \nu_P = \frac{\lambda}{2(\lambda + \mu)} \text{ [iv]}$$

Rearranging [iv] to make λ the subject gives

$$\lambda = \frac{2\nu_P \mu}{1 - 2\nu_P} \quad \text{[v]}$$

Substituting [v] into [i], and then using [iii] in the final step gives

$$E = \frac{\mu\left(\frac{6\nu_P\mu+2\mu(1-2\nu_P)}{1-2\nu_P}\right)}{\left(\frac{2\nu_P\mu+\mu(1-2\nu_P)}{1-2\nu_P}\right)} = \frac{6\nu_P\mu + 2\mu(1-2\nu_P)}{2\nu_P + (1-2\nu_P)} = 2\mu(1+\nu_P) = 2G(1+\nu_P)$$

It therefore follows that $G = E/[2(1+\nu_P)]$, which is the first relationship to be shown. Substituting [v] into [ii], and recalling that $E = 2\mu(1+\nu_P)$, gives

$$K = \frac{3(2\nu_P\mu) + 2\mu(1-2\nu_P)}{3(1-2\nu_P)} = \frac{2\mu(1+\nu_P)}{3(1-2\nu_P)} = \frac{E}{3(1-2\nu_P)}$$

as required

15E The electrical properties of solids

Answer to discussion question

D15E.1 The Fermi–Dirac distribution takes into account the effect of the Pauli exclusion principle, which is that no more than two electrons may occupy any one state. In contrast, the Boltzmann distribution places no restriction on the number of particles that can occupy a given state; such a distribution cannot, in general, be used to described the behaviour of electrons.

In both the Boltzmann and Fermi–Dirac distributions the probability of a state being occupied depends on its energy and the temperature, and this probability tails off exponentially as the energy is increased. However, in the Fermi–Dirac distribution an additional parameter, the chemical potential μ, appears. At $T = 0$ the probability of states with energy $< \mu$ being occupied is 1, and states at higher energies are not occupied. At a finite temperature, the probability of the state with energy μ being occupied is $\frac{1}{2}$.

Solutions to exercises

E15E.1(a) Assuming that the temperature, T, is not so high that many electrons are excited to states above the Fermi energy, E_F, the Fermi–Dirac distribution can be written as [15E.2b–*671*], $f(E) = 1/[e^{(E-E_F)/kT} + 1]$, where $f(E)$ is the probability of occupation of a state with energy E.

For $E = E_F + kT$, $f(E_F + kT) = 1/[e^{(E_F+kT-E_F)/kT} + 1] = 1/[e^1 + 1] = \boxed{0.269}$

E15E.2(a) The Fermi–Dirac distribution is given by [15E.2b–*671*], $f(E) = 1/[e^{(E-E_F)/kT} + 1]$, where $f(E)$ is the probability of occupation of a state with energy E, and E_F is the Fermi energy. In this case $E_F = 1.00\ \text{eV} = 1.60... \times 10^{-19}$ J, using the conversion factor from inside of the front cover. With some rearrangement of the expression for $f(E)$ it follows that

$$\begin{aligned} E &= kT\ln[1/f(E) - 1] + E_F \\ &= (1.3806\times10^{-23}\ \text{J K}^{-1}) \times (298\ \text{K}) \times \ln(1/0.25 - 1) + (1.60...\times10^{-19}\ \text{J}) \\ &= 1.64...\times10^{-19}\ \text{J} = \boxed{1.03\ \text{eV}} \end{aligned}$$

E15E.3(a) Arsenic is a Group 15 element and germanium is a Group 14 element. Thus, an electron can be transferred from an arsenic atom into the otherwise empty conduction band, thereby increasing the conductivity of the material relative to pure germanium. This type of doping results in an $\boxed{\text{n-type}}$ semi-conductor.

Solutions to problems

P15E.1 The Fermi–Dirac distribution is given by [15E.2b–*671*], $f(E) = 1/[e^{(E-E_F)/kT} + 1]$, where $f(E)$ is the probability of occupation of a state with energy E, and E_F is the Fermi energy. Let $x = (E - E_F)/E_F$ and $y = E_F/kT$ so that $f(E)$ can be written as $f(x, y) = 1/(e^{xy} + 1)$. Note that x can be negative for energies below the Fermi energy, but y must always be positive.

A set of curves for different combinations of x and y are shown in Fig. 15.8. Note that as $T \to \infty$, $y \to 0$ and $f \to \frac{1}{2}$ since all available energy states have the same probability of $\frac{1}{2}$ of being occupied. Also, as $T \to 0$, $y \to \infty$ and f tends towards a step distribution for which $f = 1$ for $x < 0$ and $f = 0$ for $x > 0$.

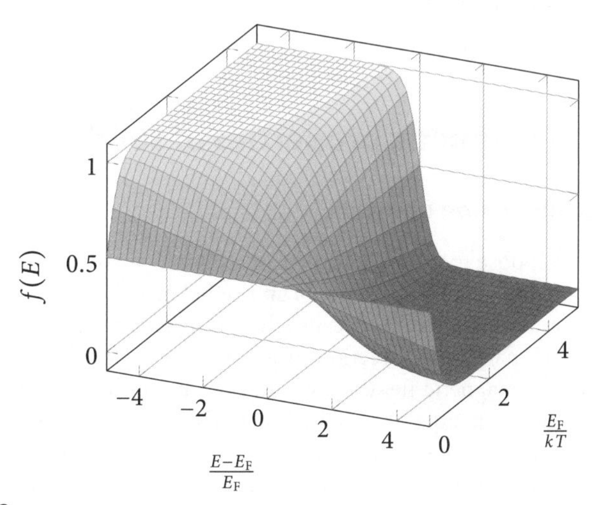

Figure 15.8

P15E.3 Substituting eqn [15E.2a–*670*] into eqn [15E.1–*670*] and integrating over the full energy range gives

$$N = \int_0^\infty dN(E) = \int_0^\infty \rho(E)f(E)\,dE = \int_0^\infty \frac{\rho(E)}{e^{(E-\mu)/kT} + 1}\,dE$$

Assuming that ρ is independent of temperature then in order for N to remain constant as the temperature is increased from $T = 0$ it follows that $e^{(E-\mu)/kT}$ must remain constant. Hence as T is increased, $(E - \mu)$ must increase and therefore the value of μ must decrease.

P15E.5 The arrangement of bands in a semiconductor is shown in Fig. 15E.4 on page 670. An n-type semiconductor consists of a host sample of a Group 14 element doped with Group 15 atoms. The presence of the Group 15 atoms results in occupied donor levels at energies just below the bottom of the conductance band, as shown in Fig. 15E.6 on page 671. The energy gap between the donor levels and the conductance band is significantly smaller than that between the valence band and the conduction band.

At $T = 0$ the the valence band is full and the conductance band empty: the material is therefore an insulator. When the temperature is increased to the point where kT is comparable to the energy separation between the donor states and the conduction band, electrons will be promoted from these states into the conduction band and the material will start to conduct. As the temperature is raised further a point will be reached when just about all the electrons from the donor levels have been promoted. Now the conductivity no longer increases with temperature and a plateau is reached.

If the temperature is raised much higher, electrons will start to be excited from the valence band into the conduction band, and the conductivity will start to rise once more.

15F The magnetic properties of solids

Answer to discussion question

D15F.1 Suppose that the molecules in a sample possess a permanent magnetic dipole moment. In the absence of an applied magnetic field, these dipoles will point in random directions because there is no energetic preference for them to point in any particular direction. The sample therefore has no net magnetic moment. If a magnetic field is applied, then it will be energetically favourable for the dipoles to point in certain directions. When averaged over the whole sample the contributions from the dipoles will not cancel and the result is that the sample will have a net magnetic moment. This is the origin of the magnetization of the sample.

The same idea applies when considering the interaction between a permanent electric dipole and an applied electric field. In the presence of the field the dipoles favour certain directions and so when averaged over the sample they do not cancel. The result is a net electric moment of the sample, called the polarization. In both cases the magnetization or polarization depends on the competition between the randomizing effect of thermal motion and the ordering effect of the applied field.

Solutions to exercises

E15F.1(a) The magnetic moment m is given by [15F.3–675], $m = g_e[S(S+1)]^{1/2}\mu_B$, where $g_e = 2.0023$ and $\mu_B = e\hbar/(2m_e)$. For $CrCl_3$, $3.81\mu_B = g_e[S(S+1)]^{1/2}\mu_B$, the constant μ_B cancels leaving a quadratic which is solved for S

$$S^2+S-(3.81/2.0023)^2 = 0 \qquad S = \tfrac{1}{2}(-1\pm\sqrt{1+4\times1\times3.620...}) = -0.500\pm1.967...$$

Of the two solutions, $S = -2.47$ is non-physical, and the solution $S = 1.47$ is close to $S = \frac{3}{2}$. A reasonable conclusion is therefore that $CrCl_3$ has $\boxed{\text{three}}$ unpaired electrons.

E15F.2(a) The molar susceptibility χ_m of a substance is given by [15F.2–674], $\chi_m = \chi V_m$, where χ is the volume magnetic susceptibility and V_m is the molar volume. The mass density ρ can be written $\rho = M/V_m$, hence $V_m = M/\rho$. With the data given

$$\chi_m = \chi V_m = \frac{\chi M}{\rho} = \frac{(-7.2\times10^{-7})\times(6\times12.01+6\times1.0079)\,\text{g mol}^{-1}}{0.879\,\text{g cm}^{-3}}$$
$$= \boxed{-6.4\times10^{-11}\,\text{m}^3\,\text{mol}^{-1}}$$

E15F.3(a) The molar susceptibility χ_m of a substance is given by [15F.4b–675], the Curie law,

$$\chi_m = \frac{C}{T} \quad \text{where} \quad C = \frac{N_A g_e^2 \mu_0 \mu_B^2 S(S+1)}{3k}$$

This is rearranged to give the spin quantum number as

$$S(S+1) = \frac{3kT\chi_m}{N_A g_e^2 \mu_0 \mu_B^2}$$
$$= \frac{3\times(1.3806\times10^{-23}\,\text{J K}^{-1})\times(294.53\,\text{K})}{(6.0221\times10^{23}\,\text{mol}^{-1})\times(2.0023)^2}$$
$$\times\frac{1.463\times10^{-7}\,\text{m}^3\,\text{mol}^{-1}}{(1.2566\times10^{-6}\,\text{J s}^2\,\text{C}^{-2}\,\text{m}^{-1})\times(9.2740\times10^{-24}\,\text{J T}^{-1})^2}$$
$$= 6.839...$$

Note the conversion of the molar magnetic susceptibility from units of $\text{cm}^3\,\text{mol}^{-1}$ to $\text{m}^3\,\text{mol}^{-1}$. To sort out the units the relations $1\,\text{T} = 1\,\text{kg s}^{-2}\,\text{A}^{-1}$ and $1\,\text{A} = 1\,\text{C s}^{-1}$, hence $1\,\text{C} = 1\,\text{A s}$, are useful. The value of S is found by solving the quadratic

$$S^2+S-6.839... = 0 \qquad S = \tfrac{1}{2}(-1\pm\sqrt{1+4\times1\times6.839...}) = -0.500\pm2.662...$$

The root $S = -3.16$ is non-physical. The other root, $S = 2.16$, implies an effective number of electrons of $2\times2.16 = \boxed{4.3}$ (higher precision is not justified because the expected result is an integer).

The high-spin arrangement of electrons in Mn^{2+} has 5 unpaired electrons. The discrepancy arises because the analysis here considers only the contribution from the electron spins and does not include any possible orbital contribution; in addition, the effect of interactions between the spins is not considered.

E15F.4(a) The spin contribution to the molar magnetic susceptibility is given by equation the Curie law, [15F.4b–675]

$$\chi_m = \frac{C}{T} \quad \text{where} \quad C = \frac{N_A g_e^2 \mu_0 \mu_B^2 S(S+1)}{3k}$$

If octahedral coordination is assumed, the 9 d electrons in Cu^{2+} are arranged as $t_{2g}^6 e_g^3$ so theoretically there is one unpaired electron and $S = \frac{1}{2}$

$$\frac{N_A g_e^2 \mu_0 \mu_B^2}{3k} = \frac{(6.0221 \times 10^{23}\ \text{mol}^{-1}) \times (2.0023)^2}{3 \times (1.3806 \times 10^{-23}\ \text{J K}^{-1})} \times (1.2566 \times 10^{-6}\ \text{J s}^2\ \text{C}^{-2}\ \text{m}^{-1}) \times (9.2740 \times 10^{-24}\ \text{J T}^{-1})^2$$
$$= 6.302... \times 10^{-6}\ \text{m}^3\ \text{K mol}^{-1}$$

To sort out the units the relations 1 T = 1 kg s^{-2} A^{-1} and 1 A = 1 C s^{-1}, hence 1 C = 1 A s, are useful. The molar susceptibility follows as

$$\chi_m = \frac{(6.302... \times 10^{-6}\ \text{m}^3\ \text{K mol}^{-1}) \times (\frac{1}{2})(\frac{1}{2}+1)}{298\ \text{K}} = \boxed{1.59 \times 10^{-8}\ \text{m}^3\ \text{mol}^{-1}}$$

E15F.5(a) Superconductors classed as *Type I* show abrupt loss of superconductivity when an applied magnetic field exceeds a critical value $\mathcal{H}_c$ characteristic of the material. The dependence of $\mathcal{H}_c$ on T is given by [15F.5–676], $\mathcal{H}_c(T) = \mathcal{H}_c(0)\left[1 - T^2/T_c^2\right]$, provided $T \le T_c$. For Nb,

$$\mathcal{H}_c(6.0\ \text{K}) = (158\ \text{kA m}^{-1})\left[1 - \frac{(6\ \text{K})^2}{(9.5\ \text{K})^2}\right] = \boxed{95\ \text{kA m}^{-1}}$$

The material is superconducting at 6.0 K for 95 kA m^{-1} and weaker applied field strengths.

Solutions to problems

P15F.1 The spin contribution to the molar magnetic susceptibility is given by the Curie lawm [15F.4b–675],

$$\chi_m = \frac{C}{T} \quad \text{where} \quad C = \frac{N_A g_e^2 \mu_0 \mu_B^2 S(S+1)}{3k}$$

$$\frac{N_A g_e^2 \mu_0 \mu_B^2}{3k} = \frac{(6.0221 \times 10^{23}\ \text{mol}^{-1}) \times (2.0023)^2}{3 \times (1.3806 \times 10^{-23}\ \text{J K}^{-1})} \times (1.2566 \times 10^{-6}\ \text{J s}^2\ \text{C}^{-2}\ \text{m}^{-1}) \times (9.2740 \times 10^{-24}\ \text{J T}^{-1})^2$$
$$= 6.302... \times 10^{-6}\ \text{m}^3\ \text{K mol}^{-1}$$

To sort out the units the relations 1 T = 1 kg s^{-2} A^{-1} and 1 A = 1 C s^{-1}, hence 1 C = 1 A s, are useful.

For $S = 2$

$$\chi_m = \frac{(6.302... \times 10^{-6}\ \text{m}^3\,\text{K}\,\text{mol}^{-1}) \times (2)(3)}{298\ \text{K}} = \boxed{1.27 \times 10^{-7}\ \text{m}^3\,\text{mol}^{-1}}$$

For $S = 3$

$$\chi_m = \frac{(6.302... \times 10^{-6}\ \text{m}^3\,\text{K}\,\text{mol}^{-1}) \times (3)(4)}{298\ \text{K}} = \boxed{2.54 \times 10^{-7}\ \text{m}^3\,\text{mol}^{-1}}$$

For $S = 4$

$$\chi_m = \frac{(6.302... \times 10^{-6}\ \text{m}^3\,\text{K}\,\text{mol}^{-1}) \times (4)(5)}{298\ \text{K}} = \boxed{4.23 \times 10^{-7}\ \text{m}^3\,\text{mol}^{-1}}$$

The Boltzmann factor, $e^{-E/RT}$, represents the probability of a state of energy E relative to the probability of a state with energy $E = 0$. If the state $S = 3$ has relative energy $E = 0$ with Boltzmann factor $f = 1$ then states $S = 2$ and $S = 4$ have probability

$$f = e^{(-50\times10^3\ \text{J}\,\text{mol}^{-1})/(8.3145\,\text{J}\,\text{K}^{-1}\,\text{mol}^{-1})\times(298\ \text{K})} = 1.72 \times 10^{-9}$$

The populations of the states with $S = 2$ and $S = 4$ is therefore negligible, and hence the molar susceptibility is well-approximated by the molar susceptibility of the state with $S = 3$, $\boxed{2.54 \times 10^{-7}\ \text{m}^3\,\text{mol}^{-1}}$.

15G The optical properties of solids

Answers to discussion questions

D15G.1 See Section 15G.1 on page 678.

Solutions to exercises

E15G.1(a) The energy gap is given by

$$\Delta\varepsilon = h\nu = \frac{hc}{\lambda} = \frac{(6.6261 \times 10^{-34}\,\text{J}\,\text{s}) \times (2.9979 \times 10^8\,\text{m}\,\text{s}^{-1})}{(350 \times 10^{-9}\ \text{m})}$$
$$= 5.67... \times 10^{-19}\ \text{J}$$

Converting to eV the band gap is

$$(5.67... \times 10^{-19}\ \text{J}) \times [1\ \text{eV}/(1.6022 \times 10^{-19}\,\text{J}\,\text{eV}^{-1})] = \boxed{3.54\ \text{eV}}$$

Solutions to problems

P15G.1 (a) To find whether or not a vector is an eigenvector of the hamiltonian matrix, the matrix is allowed to act on the vector

$$\begin{pmatrix} \tilde{\nu}_{\text{mon}} & \tilde{\beta} \\ \tilde{\beta} & \tilde{\nu}_{\text{mon}} \end{pmatrix}\begin{pmatrix} 1 \\ 1 \end{pmatrix} = \begin{pmatrix} \tilde{\nu}_{\text{mon}} + \tilde{\beta} \\ \tilde{\nu}_{\text{mon}} + \tilde{\beta} \end{pmatrix} = (\tilde{\nu}_{\text{mon}} + \tilde{\beta})\begin{pmatrix} 1 \\ 1 \end{pmatrix}$$

Acting on the vector with the hamiltonian matrix regenerates the original vector times a constant, which is the eigenvalue $\tilde{\nu}_+ = \tilde{\nu}_{\text{mon}} + \tilde{\beta}$. Similarly, for the second proposed eigenvector

$$\begin{pmatrix} \tilde{\nu}_{\text{mon}} & \tilde{\beta} \\ \tilde{\beta} & \tilde{\nu}_{\text{mon}} \end{pmatrix}\begin{pmatrix} 1 \\ -1 \end{pmatrix} = \begin{pmatrix} \tilde{\nu}_{\text{mon}} - \tilde{\beta} \\ -\tilde{\nu}_{\text{mon}} + \tilde{\beta} \end{pmatrix} = (\tilde{\nu}_{\text{mon}} - \tilde{\beta})\begin{pmatrix} 1 \\ -1 \end{pmatrix}$$

The vector is indeed an eigenvector with eigenvalue $\tilde{\nu}_- = \tilde{\nu}_{\text{mon}} - \tilde{\beta}$

(b) The normalisation factor N is calculated by evaluating the following integral

$$\begin{aligned} I &= \int \Psi_+^* \Psi_+ \mathrm{d}\tau = \int [\Psi_b^*(1) + \Psi_b^*(2)][\Psi_b(1) + \Psi_b(2)]\,\mathrm{d}\tau \\ &= \int \Psi_b^*(1)\Psi_b(1)\,\mathrm{d}\tau + \int \Psi_b^*(2)\Psi_b(2)\,\mathrm{d}\tau + 2\int \Psi_b^*(1)\Psi_b(2)\,\mathrm{d}\tau \end{aligned}$$

Assuming that $\Psi_b(i)$ is normalised the first and second integrals are = 1, and with the definition $S = \int \Psi_b^*(1)\Psi_b(2)\,\mathrm{d}\tau$ the third term is $2S$; overall $I = 2(1+S)$. Division of the wavefunction by $I^{1/2}$ therefore normalizes the function, so the normalization constant is $N_+ = [2(1+S)]^{-1/2}$. A similar calculation gives the normalization constant for Ψ_- as $N_- = [2(1-S)]^{-1/2}$

(c) The integral $\mu_{\text{dim}} = \int \Psi_\pm^* \hat{\mu} \Psi_0\,\mathrm{d}\tau$ is evaluated by substituting in the given forms of $\Psi_\pm$ and Ψ_0 and using the definition $\mu_{\text{mon}} = \int \Psi_b^*(i)\hat{\mu}\Psi_a(i)\,\mathrm{d}\tau$. Note that $\int \Psi_b^*(2)\hat{\mu}\Psi_a(1)\,\mathrm{d}\tau = \int \Psi_b^*(1)\hat{\mu}\Psi_a(2)\,\mathrm{d}\tau = 0$ because these correspond to transitions from a level of one monomer to a level of the other monomer.

$$\begin{aligned} \mu_{\text{dim}} &= \int \Psi_\pm^* \hat{\mu} \Psi_0 \mathrm{d}\tau \\ &= \int \frac{1}{[2(1\pm S)]^{1/2}}[\Psi_b(1) \pm \Psi_b(2)]^* \hat{\mu} \frac{1}{2^{1/2}}[\Psi_a(1) + \Psi_a(2)]\,\mathrm{d}\tau \\ &= \frac{1}{2(1\pm S)^{1/2}}\Bigg[\overbrace{\int \Psi_b^*(1)\hat{\mu}\Psi_a(1)\,\mathrm{d}\tau}^{\mu_{\text{mon}}} \pm \overbrace{\int \Psi_b^*(2)\hat{\mu}\Psi_a(2)\,\mathrm{d}\tau}^{\mu_{\text{mon}}} \\ &\qquad \pm \overbrace{\int \Psi_b^*(2)\hat{\mu}\Psi_a(1)\,\mathrm{d}\tau}^{=0} + \overbrace{\int \Psi_b^*(1)\hat{\mu}\Psi_a(2)\,\mathrm{d}\tau}^{=0}\Bigg] \\ &= \frac{1}{2(1\pm S)^{1/2}}(\mu_{\text{mon}} \pm \mu_{\text{mon}}) \end{aligned}$$

Hence for the excited state wavefunction Ψ_+, $\mu_{\text{dim}} = \boxed{(1+S)^{-1/2}\mu_{\text{mon}}}$ and for the wavefunction Ψ_-, $\mu_{\text{dim}} = \boxed{0}$.

P15G.3 An incident electric field E induces a dipole moment μ in a material. If the response is non-linear then, according to [15G.1–680], $\mu = \alpha E + \frac{1}{2}\beta E^2$. If there are two electric fields applied at frequencies ω_1 and ω_2, the total electric field is $E = E_1 \cos\omega_1 t + E_2 \cos\omega_2 t$. Expansion of the E^2 factor in the non-linear term $\frac{1}{2}\beta E^2$ gives the non-linear response

$$\begin{aligned} E^2 &= (E_1 \cos\omega_1 t + E_2 \cos\omega_2 t)^2 \\ &= E_1^2 \cos^2(\omega_1 t) + E_2^2 \cos^2(\omega_2 t) + 2E_1 E_2 \cos(\omega_1 t)\cos(\omega_2 t) \\ &= \tfrac{1}{2}E_1^2[1 + \cos(2\omega_1 t)] + \tfrac{1}{2}E_2^2[1 + \cos(2\omega_2 t)] + E_1 E_2 \cos([\omega_1 + \omega_2]t) \\ &\quad + E_1 E_2 \cos([\omega_1 - \omega_2]t) \end{aligned}$$

The trigonometric identities

$$\cos^2 A = \tfrac{1}{2}(1 + \cos 2A) \quad \text{and} \quad \cos A \cos B = \tfrac{1}{2}[\cos(A - B) + \cos(A + B)]$$

are used to generate the final expression.

The induced dipole therefore has components oscillating at $2\omega_1, 2\omega_2, (\omega_1+\omega_2)$ and $(\omega_1 - \omega_2)$, and each of these can result in radiation at that frequency. Thus, a medium with a non-linear response may result in the generation of sum and difference frequencies (as well as harmonics).

Answers to integrated activities

I15.1 The spacing of the $\{hkl\}$ planes in a cubic lattice is given by [15A.1a–645], $d_{hkl} = a/(h^2 + k^2 + l^2)^{-1/2}$. This is used with the Bragg law [15B.1b–648], $\lambda = 2d \sin\theta$, to give

$$\lambda = \frac{2a \sin\theta}{(h^2 + k^2 + l^2)^{1/2}} \quad \text{hence} \quad a = \frac{\lambda(h^2 + k^2 + l^2)^{1/2}}{2 \sin\theta}$$

For the (111) reflection this becomes $a = 3^{1/2}\lambda/2\sin\theta$.

at 100 K $\quad a(100\,\text{K}) = 3^{1/2} \times (154.0562\,\text{pm})/2\sin(22.0403°) = 355.53...\,\text{pm}$

at 300 K $\quad a(300\,\text{K}) = 3^{1/2} \times (154.0562\,\text{pm})/2\sin(21.9664°) = 356.66...\,\text{pm}$

The volume is $V = a^3$, thus the change in the volume is $\delta V = a^3(300\,\text{K}) - a^3(100\,\text{K})$. The thermal expansion coefficient is therefore

$$\begin{aligned} \alpha &= \frac{1}{V}\frac{\delta V}{\delta T} = \frac{1}{a^3(100\,\text{K})}\frac{a^3(300\,\text{K}) - a^3(100\,\text{K})}{\delta T} \\ &= \frac{1}{(355.53...\,\text{pm})^3}\frac{(356.66...\,\text{pm})^3 - (355.53...\,\text{pm})^3}{(300 - 100)\,\text{K}} = \boxed{4.811 \times 10^{-5}\,\text{K}^{-1}} \end{aligned}$$

If the average volume is used in the denominator, $\alpha = 4.788 \times 10^{-5}\,\text{K}^{-1}$.

I15.3 The scattering factor $f(\theta)$ is given by [15B.2–649]

$$f(\theta) = 4\pi \int_0^\infty \rho(r) \frac{\sin kr}{kr} r^2 \,dr \qquad k = \frac{4\pi}{\lambda} \sin\theta$$

The quantity $4\pi r^2 \rho(r)$ is identified as the radial distribution function $P(r)$, which is given in terms of the radial wavefunction of the orbital, $P(r) = R(r)^2 r^2$

$$f(\theta) = \int_0^\infty P(r) \frac{\sin kr}{kr} \,dr$$

The requested plot is of $f(\theta)$ as a function of $\xi = \sin\theta/\lambda$, hence $k = 4\pi\xi$ and

$$f(\theta) = \int_0^\infty P(r) \frac{\sin(4\pi\xi r)}{4\pi\xi r} \,dr = \frac{1}{4\pi\xi} \int_0^\infty R(r)^2 \sin(4\pi\xi r)\, r \,dr \qquad (15.2)$$

From Table 8A.1 on page 306 the 1s hydrogenic orbital the radial wavefunction for $Z = 1$ is $R(r) = 2(1/a_0)^{3/2} e^{-r/a_0}$. A suitable gaussian function which mimics this is $G(r) = Ne^{-\alpha(r/a_0)^2}$, where N is the normalization constant and α is a parameter to be determined. The normalization constant is found by normalizing $G(r)$ in the same way that $R(r)$ is normalized

$$N^2 \int_0^\infty G(r)^2 r^2 \,dr = N^2 \int_0^\infty e^{-\alpha(2r/a_0)^2} r^2 \,dr = 1$$

The integral is evaluated using Integral G.3 to give

$$N = \frac{2\sqrt{2}(\alpha/a_0^2)^{3/4}}{(\pi/2)^{1/4}}$$

The corresponding radial distribution function, $P(r) = r^2 G(r)^2$ is therefore

$$P_g(r) = \frac{8(\alpha/a_0^2)^{3/2}}{(\pi/2)^{1/2}} r^2 e^{-2\alpha(r/a_0)^2} \qquad (15.3)$$

Using a numerical procedure the value of α is adjusted to minimize the difference $(P_g(r) - P(r))^2$, integrated over r; the best fit is obtained with $\alpha = 0.2064$. With this value the scattering factor is evaluated as a function of ξ by numerical integration of eqn 15.2 using the radial distribution function from 15.3.

The parameter ξ is some fraction of $1/\lambda$, where λ is the wavelength of the X-rays used. Typically $\lambda = 100$ pm so ξ is less than 10^{10} m^{-1}. The upper limit of the integration can conveniently be set to a modest multiple of the Bohr radius, say $100a_0$, because beyond this distance the electron density will be negligible. The computed scattering factor for the exact 1s function and the gaussian function are compared in Fig. 15.9. The two plots are similar, but for the gaussian function the scatting drops off more sharply.

The comparison depends very much of the value of α which may be chosen according to other criteria. For example, rather than minimizing the difference between the radial distribution functions this parameter can be chosen to minimize the difference between the radial functions. This gives $\alpha = 0.508$ and a somewhat different scattering factor, as shown in Fig. 15.9.

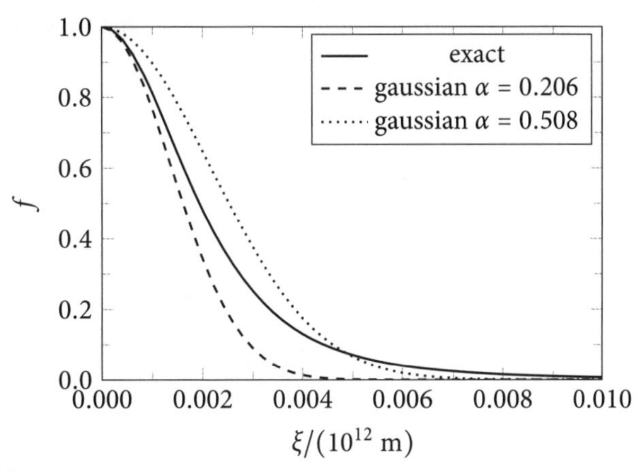
1.0
0.8
0.6
0.4
0.2
0.0
f
0.000
0.002
0.004
0.006
0.008
0.010
ξ/(10^12 m)
exact
gaussian α = 0.206
gaussian α = 0.508

Figure 15.9

16 Molecules in motion

16A Transport properties of a perfect gas

Answers to discussion questions

D16A.1 See Section 16A.2(a) on page 693, and especially *How is that done?* 16A.2 on page 693.

Solutions to exercises

E16A.1(a) For a perfect gas, the collision flux Z_w is [16A.7a–*693*], $Z_w = p/(2\pi mkT)^{1/2}$. The number of argon molecule collisions within area A in time interval t is therefore $N = Z_w At$. The mass m is written in terms of the molar mass M: $m = M/N_A$.

$$N = Z_w At = \frac{p}{(2\pi mkT)^{1/2}}At = \frac{pN_A^{1/2}}{(2\pi MkT)^{1/2}}At$$

$$= \frac{(90\ \text{Pa}) \times (6.0221 \times 10^{23}\ \text{mol}^{-1})^{1/2} \times [(2.5 \times 3.0) \times 10^{-6}\ \text{m}^2] \times (15\ \text{s})}{[2\pi \times (0.03995\ \text{kg mol}^{-1}) \times (1.3806 \times 10^{-23}\ \text{J K}^{-1}) \times (500\ \text{K})]^{1/2}}$$

$$= \boxed{1.9 \times 10^{20}} \quad \text{collisions}$$

E16A.2(a) The diffusion constant is given by [16A.9–*694*], $D = \frac{1}{3}\lambda v_{\text{mean}}$, where λ is the mean free path length $\lambda = kT/\sigma p$ [16A.1a–*690*], and v_{mean} is the mean speed $v_{\text{mean}} = (8RT/\pi M)^{1/2}$ [16A.1b–*690*].

$$D = \frac{1}{3}\frac{kT}{\sigma p}\left[\frac{8RT}{\pi M}\right]^{1/2}$$

$$= \frac{(1.3806 \times 10^{-23}\ \text{J K}^{-1}) \times (293.15\ \text{K})}{3 \times (3.6 \times 10^{-19}\ \text{m}^2) \times (p/\text{Pa})} \times \left[\frac{8 \times (8.3145\ \text{J K}^{-1}\ \text{mol}^{-1}) \times (293.15\ \text{K})}{\pi \times (0.03995\ \text{kg mol}^{-1})}\right]^{1/2}$$

$$= (1.477...\ \text{m}^2\ \text{s}^{-1}) \times \frac{1}{p/\text{Pa}}$$

The flux of argon atoms J_z is related to the diffusion coefficient D and the concentration gradient $\text{d}\mathcal{N}/\text{d}z$ by [16A.4–*691*], $J_z = -D\text{d}\mathcal{N}/\text{d}z$. From the perfect gas equation, $pV = NkT$, the number density is expressed in terms of the pressure as $\mathcal{N} = N/V = p/kT$. With this, the concentration gradient is written in terms of the pressure gradient: $\mathrm{d}\mathcal{N}/\mathrm{d}z = (1/kT)\mathrm{d}p/\mathrm{d}z$, and hence the flow is $J_z = -(D/kT)\mathrm{d}p/\mathrm{d}z$

$$J_z = \frac{-D}{kT}\frac{\mathrm{d}p}{\mathrm{d}z} = \frac{-1}{p/\mathrm{Pa}} \times \frac{(1.47\ldots\ \mathrm{m^2\,s^{-1}}) \times (1.0 \times 10^5\ \mathrm{Pa\,m^{-1}})}{(1.3806 \times 10^{-23}\ \mathrm{J\,K^{-1}}) \times (293.15\ \mathrm{K})}$$
$$= -(3.64\ldots \times 10^{25}\ \mathrm{m^{-2}\,s^{-1}}) \times \frac{1}{p/\mathrm{Pa}}$$

p/Pa	$D/(\mathrm{m^2\,s^{-1}})$	$J_z/(\mathrm{m^{-2}\,s^{-1}})$	$(J_z/N_A)/(\mathrm{mol\,m^{-2}\,s^{-1}})$
1.00	1.48	-3.65×10^{25}	-60.6
1.00×10^5	1.48×10^{-5}	-3.65×10^{20}	-6.06×10^{-4}
1.00×10^7	1.48×10^{-7}	-3.65×10^{18}	-6.06×10^{-6}

E16A.3(a) The thermal conductivity is given by [16A.10c–695], $\kappa = \nu pD/T$, where the diffusion coefficient D is given by [16A.9–694], $D = \lambda v_{\mathrm{mean}}/3$. The mean free path λ is given by [16A.1a–690], $\lambda = kT/\sigma p$, and the mean speed v_{mean} is given by [16A.1b–690], $v_{\mathrm{mean}} = (8RT/\pi M)^{1/2}$. The quantity ν is the number of quadratic contributions to the energy, and this is related to the heat capacity by $C_{V,\mathrm{m}} = \nu kN_A$, hence $\nu = C_{V,\mathrm{m}}/kN_A$. The thermal conductivity is therefore expressed as

$$\kappa = \frac{\nu pD}{T} = \frac{\nu p\lambda v_{\mathrm{mean}}}{3T} = \frac{C_{V,\mathrm{m}}}{kN_A}\frac{p}{3T}\frac{kT}{\sigma p}\left(\frac{8RT}{\pi M}\right)^{1/2} = \frac{C_{V,\mathrm{m}}}{3\sigma N_A}\left(\frac{8RT}{\pi M}\right)^{1/2}$$

$$\text{hence } \kappa = \frac{12.5\ \mathrm{J\,K^{-1}\,mol^{-1}}}{3 \times (3.6 \times 10^{-19}\ \mathrm{m^2}) \times (6.0221 \times 10^{23}\ \mathrm{mol^{-1}})} \times \left(\frac{8 \times (8.3145\ \mathrm{J\,K^{-1}\,mol^{-1}}) \times (298\ \mathrm{K})}{\pi \times (3.995 \times 10^{-2}\ \mathrm{kg\,mol^{-1}})}\right)^{1/2}$$
$$= \boxed{7.6 \times 10^{-3}\ \mathrm{J\,K^{-1}\,m^{-1}\,s^{-1}}}$$

E16A.4(a) The thermal conductivity is given by [16A.10c–695], $\kappa = \nu pD/T$, where the diffusion coefficient D is given by [16A.9–694], $D = \lambda v_{\mathrm{mean}}/3$. The mean free path λ is given by [16A.1a–690], $\lambda = kT/\sigma p$, and the mean speed v_{mean} is given by [16A.1b–690], $v_{\mathrm{mean}} = (8RT/\pi M)^{1/2}$. The quantity ν is the number of quadratic contributions to the energy, and this is related to the heat capacity by $C_{V,\mathrm{m}} = \nu kN_A$, hence $\nu = C_{V,\mathrm{m}}/kN_A$. The thermal conductivity is therefore expressed as

$$\kappa = \frac{\nu pD}{T} = \frac{\nu p\lambda v_{\mathrm{mean}}}{3T} = \frac{C_{V,\mathrm{m}}}{kN_A}\frac{p}{3T}\frac{kT}{\sigma p}\left(\frac{8RT}{\pi M}\right)^{1/2} = \frac{C_{V,\mathrm{m}}}{3\sigma N_A}\left(\frac{8RT}{\pi M}\right)^{1/2}$$

Rearranging gives an expression for σ in terms of the thermal conductivity

$$\sigma = \frac{C_{V,\mathrm{m}}}{3\kappa N_A}\left(\frac{8RT}{\pi M}\right)^{1/2}$$

The value of $C_{p,\mathrm{m}}$ is given in the *Resource section*; $C_{V,\mathrm{m}}$ is found using $C_{p,\mathrm{m}} - C_{V,\mathrm{m}} = R$ for a perfect gas.

$$\sigma = \frac{(20.786\,\mathrm{J\,K^{-1}\,mol^{-1}}) - (8.3145\,\mathrm{J\,K^{-1}\,mol^{-1}})}{3\times(4.65\times10^{-2}\,\mathrm{J\,K^{-1}\,m^{-1}\,s^{-1}})\times(6.0221\times10^{23}\,\mathrm{mol^{-1}})}$$
$$\times\left(\frac{8\times(8.3145\,\mathrm{J\,K^{-1}\,mol^{-1}})\times(273\,\mathrm{K})}{\pi\times(2.018\times10^{-2}\,\mathrm{kg\,mol^{-1}})}\right)^{1/2} = \boxed{0.0795\,\mathrm{nm^2}}$$

The value reported in Table 1B.2 on page 17 is 0.24 nm^2.

E16A.5(a) The flux of energy is given by [16A.3–*691*], $J_z = -\kappa\,\mathrm{d}T/\mathrm{d}z$. The value of the thermal conductivity κ for Ar at 298 K is determined in *Exercise* E16A.3(a) as $7.6\times10^{-3}\,\mathrm{J\,K^{-1}\,m^{-1}\,s^{-1}}$. As is seen in that *Exercise*, $\kappa \propto T^{1/2}$ provided that the heat capacity is constant over the temperature range of interest. It therefore follows that $\kappa_{280\,\mathrm{K}} = (280\,\mathrm{K}/298\,\mathrm{K})^{1/2}\,\kappa_{298\,\mathrm{K}} = 7.40...\times10^{-3}\,\mathrm{J\,K^{-1}\,m^{-1}\,s^{-1}}$. With these data the flux is computed as

$$J_z = -\kappa\,\mathrm{d}T/\mathrm{d}z = -(7.40...\times10^{-3}\,\mathrm{J\,K^{-1}\,m^{-1}\,s^{-1}})\times(10.5\,\mathrm{K\,m^{-1}})$$
$$= \boxed{-0.078\,\mathrm{J\,m^{-2}\,s^{-1}}}$$

E16A.6(a) The flux of energy is given by [16A.3–*691*], $J_z = -\kappa\,\mathrm{d}T/\mathrm{d}z$, where κ is the thermal conductivity and the negative sign indicates flow of heat is towards the lower temperature. The rate of energy transfer is $r = J_zA$, where A is the cross-sectional area. The temperature gradient is approximated as $\mathrm{d}T/\mathrm{d}z = \Delta T/\Delta z$; because 1 W = 1 J s^{-1} it follows that 24 mW K^{-1} m^{-1} is equivalent to $2.4\times10^{-2}\,\mathrm{J\,K^{-1}\,m^{-1}\,s^{-1}}$

$$r = J_zA = -\kappa A\frac{\Delta T}{\Delta z}$$
$$= -(2.4\times10^{-2}\,\mathrm{J\,K^{-1}\,m^{-1}\,s^{-1}})\times(1.0\,\mathrm{m^2})\times\frac{[(-15)-(28)]\,\mathrm{K}}{0.010\,\mathrm{m}}$$
$$= 103\,\mathrm{J\,s^{-1}} = 103\,\mathrm{W}$$

Hence a heater of power $\boxed{103\,\mathrm{W}}$ is required to make good the loss of heat.

E16A.7(a) The viscosity η is given by [16A.11c–*696*], $\eta = pMD/RT$. In turn the diffusion constant is given by [16A.9–*694*], $D = \frac{1}{3}\lambda v_{\mathrm{mean}}$, where λ is the mean free path length $\lambda = kT/\sigma p$ [16A.1a–*690*], and v_{mean} is the mean speed $v_{\mathrm{mean}} = (8RT/\pi M)^{1/2}$ [16A.1b–*690*]. The first step is to find an expression for η as a function of temperature

$$\eta = \frac{pMD}{RT} = \frac{pM}{RT}\frac{kT}{3\sigma p}\left(\frac{8RT}{\pi M}\right)^{1/2} = \frac{M}{3\sigma N_{\mathrm{A}}}\left(\frac{8RT}{\pi M}\right)^{1/2} = \frac{1}{3\sigma N_{\mathrm{A}}}\left(\frac{8RM}{\pi}\right)^{1/2} T^{1/2}$$
$$= \frac{1}{3\times(4.0\times10^{-19}\,\mathrm{m^2})\times(6.0221\times10^{23}\,\mathrm{mol^{-1}})}$$
$$\times\left(\frac{8\times(8.3145\,\mathrm{J\,K^{-1}\,mol^{-1}})\times(0.029\,\mathrm{kg\,mol^{-1}})}{\pi}\right)^{1/2} T^{1/2}$$
$$= (1.08...\times10^{-6}\,\mathrm{kg\,K^{-1/2}\,m^{-1}\,s^{-1}})\times(T/\mathrm{K})^{1/2}$$

where $1\,\mathrm{J} = 1\,\mathrm{kg\,m^2\,s^{-2}}$ has been used to arrive at the units on the final line. Using this expression the following table is drawn up (recall that $10^{-7}\,\mathrm{kg\,m^{-1}\,s^{-1}} = 1\,\mu\mathrm{P}$)

T/K	$\eta/(\mathrm{kg\,m^{-1}\,s^{-1}})$	$\eta/(\mu\mathrm{P})$
273	1.79×10^{-5}	179
298	1.87×10^{-5}	187
1000	3.43×10^{-5}	343

E16A.8(a) In the solution to *Exercise* E16A.7(a) it is shown that

$$\eta = \frac{1}{3\sigma N_A}\left(\frac{8RMT}{\pi}\right)^{1/2} \quad \text{hence} \quad \sigma = \frac{1}{3\eta N_A}\left(\frac{8RMT}{\pi}\right)^{1/2}$$

Recalling that $10^{-7}\,\mathrm{kg\,m^{-1}\,s^{-1}} = 1\,\mu\mathrm{P}$, the cross section is computed as

$$\begin{aligned}\sigma &= \frac{1}{3 \times (2.98 \times 10^{-5}\,\mathrm{kg\,m^{-1}\,s^{-1}}) \times (6.0221 \times 10^{23}\,\mathrm{mol^{-1}})} \\ &\quad \times \left(\frac{8 \times (8.3145\,\mathrm{J\,K^{-1}\,mol^{-1}}) \times (0.02018\,\mathrm{kg\,mol^{-1}}) \times (273\,\mathrm{K})}{\pi}\right)^{1/2} \\ &= \boxed{0.201\,\mathrm{nm^2}}\end{aligned}$$

E16A.9(a) The rate of effusion, r is given by [16A.12–697], $r = pA_0N_A/(2\pi MRT)^{1/2}$; this rate is the number of molecules escaping through the hole in a particular period of time, divided by that time. The mass loss Δm in period Δt is therefore $\Delta m = \Delta t pA_0N_A/(2\pi MRT)^{1/2} \times m$, where m is the mass of a molecule. This mass is written $m = M/N_A$ and so it follows $\Delta m = \Delta t pA_0 M^{1/2}/(2\pi RT)^{1/2}$. Evaluating this with the values given

$$\begin{aligned}\Delta m &= \frac{\Delta t pA_0 M^{1/2}}{(2\pi RT)^{1/2}} \\ &= \frac{(7200\,\mathrm{s}) \times (0.835\,\mathrm{Pa}) \times \pi \times (\tfrac{1}{2} \times 2.50 \times 10^{-3}\,\mathrm{m})^2 \times (0.260\,\mathrm{kg\,mol^{-1}})^{1/2}}{[2 \times \pi \times (8.3145\,\mathrm{J\,K^{-1}\,mol^{-1}}) \times (400\,\mathrm{K})]^{1/2}} \\ &= 1.04 \times 10^{-4}\,\mathrm{kg} = \boxed{104\,\mathrm{mg}}\end{aligned}$$

E16A.10(a) The rate of effusion, r is given by [16A.12–697], $r = pA_0N_A/(2\pi MRT)^{1/2}$; this rate is the number of molecules escaping through the hole in a particular period of time, divided by that time. The mass loss Δm in period Δt is therefore $\Delta m = \Delta t pA_0N_A/(2\pi MRT)^{1/2} \times m$, where m is the mass of a molecule. This mass is written $m = M/N_A$ and so it follows $\Delta m = \Delta t pA_0 M^{1/2}/(2\pi RT)^{1/2}$. This is rearranged to give an expression for p

$$p = \frac{\Delta m(2\pi RT)^{1/2}}{\Delta t A_0 M^{1/2}} = \frac{\Delta m}{\Delta t A_0}\left(\frac{2\pi RT}{M}\right)^{1/2}$$

$$= \frac{2.85\times10^{-4}\,\text{kg}}{(400\,\text{s})\times\pi\times(2.5\times10^{-4}\,\text{m})^2}\left(\frac{2\pi\times(8.3145\,\text{J K}^{-1}\,\text{mol}^{-1})\times(673.15\,\text{K})}{0.100\,\text{kg mol}^{-1}}\right)^{1/2}$$

$$= \boxed{2.15\times10^3\,\text{Pa}}$$

E16A.11(a) The rate of effusion, r is given by [16A.12–697], $r = pA_0N_A/(2\pi MRT)^{1/2}$; this rate is the number of molecules escaping through the hole in a particular period of time, divided by that time. In this experiment the pressure changes so the rate of effusion changes throughout the experiment; nevertheless, the rate is always proportional to $M^{-1/2}$. The two experiments involve comparing the time for the *same* drop in pressure, therefore the only factor that affects this time is the molar mass of the effusing gas. Because the *rate* is proportional to $M^{-1/2}$ the *time* for a given fall in pressure will be proportional to the inverse of this, that is $M^{1/2}$. It follows that

$$\frac{\text{rate for gas A}}{\text{rate for gas B}} = \frac{\text{time for gas B}}{\text{time for gas A}} = \left(\frac{M_B}{M_A}\right)^{1/2}$$

Therefore

$$\frac{42\,\text{s}}{52\,\text{s}} = \left(\frac{M_{N_2}}{M_A}\right)^{1/2} \quad \text{hence} \quad M_A = (28.02\,\text{g mol}^{-1})\left(\frac{52}{42}\right)^2 = \boxed{43.0\,\text{g mol}^{-1}}$$

E16A.12(a) The rate of effusion is given by [16A.12–697], $\text{d}N/\text{d}t = pA_0N_A/(2\pi MRT)^{1/2}$; this is the rate of change of the number of molecules. If it is assumed that the gas is perfect, the equation of state $pV = NkT$ allows the number to be written as $N = pV/kT$, and therefore $\text{d}N/\text{d}t = (V/kT)\text{d}p/\text{d}t$. The rate of change of the pressure is therefore

$$\frac{\text{d}p}{\text{d}t} = -\frac{kT}{V}\frac{pA_0N_A}{(2\pi MRT)^{1/2}} = -\frac{RTA_0}{V(2\pi MRT)^{1/2}}\times p = -\underbrace{\frac{A_0}{V}\left(\frac{RT}{2\pi M}\right)^{1/2}}_{\alpha}\times p$$

The minus sign is needed because the pressure falls with time. This differential equation is separable and can be integrated between $p = p_i$ and $p = p_f$, corresponding to $t = 0$ and $t = t$.

$$\int_{p_i}^{p_f}(1/p)\,\text{d}p = \int_0^t -\alpha\,\text{d}t \quad \text{hence} \quad \ln(p_f/p_i) = -\alpha t$$

The time for the pressure to drop by the specified amount is therefore

$$t = -\ln(p_\mathrm{f}/p_\mathrm{i})/\alpha = \ln(p_\mathrm{i}/p_\mathrm{f})\frac{V}{A_0}\left(\frac{2\pi M}{RT}\right)^{1/2}$$

$$= \ln\left(\frac{8.0\times10^4\ \mathrm{Pa}}{7.0\times10^4\ \mathrm{Pa}}\right)\frac{(3.0\ \mathrm{m^3})}{[\pi(10^{-4}\ \mathrm{m})^2]}\left(\frac{2\pi\times(3.200\times10^{-2}\ \mathrm{kg\,mol^{-1}})}{(8.3145\ \mathrm{J\,K^{-1}\,mol^{-1}})\times(298\ \mathrm{K})}\right)^{1/2}$$

$$= 1.15\times10^5\ \mathrm{s} = \boxed{1.3\ \mathrm{days}}$$

Solutions to problems

P16A.1 In the solution to *Exercise* E16A.7(a) it is shown that

$$\eta = \frac{1}{3\sigma N_\mathrm{A}}\left(\frac{8RMT}{\pi}\right)^{1/2} \quad \text{hence} \quad \sigma = \frac{1}{3\eta N_\mathrm{A}}\left(\frac{8RMT}{\pi}\right)^{1/2}$$

At 270 K and 1.00 bar

$$\sigma = \frac{1}{3\times(9.08\times10^{-6}\ \mathrm{kg\,m^{-1}\,s^{-1}})\times(6.0221\times10^{23}\ \mathrm{mol^{-1}})}$$
$$\times\left(\frac{8\times(8.3145\ \mathrm{J\,K^{-1}\,mol^{-1}})\times(0.0170\ \mathrm{kg\,mol^{-1}})\times(270\ \mathrm{K})}{\pi}\right)^{1/2}$$
$$= 6.00...\times10^{-19}\ \mathrm{m^2}$$

The collision cross-section is $\sigma = \pi(2r)^2$, where r is the molecular radius of NH_3 and $d = 2r$ is the effective molecular diameter. With the value of σ determined above d is found as $\boxed{437\ \mathrm{pm}}$. A similar calculation at 490 K and 10.0 bar gives $\sigma = 4.21...\times10^{-19}\ \mathrm{m^2}$ and $\boxed{d = 366\ \mathrm{pm}}$.

P16A.3 In the solution to *Exercise* E16A.2(a) it is shown that the diffusion constant is given by

$$D = \frac{1}{3}\frac{kT}{\sigma p}\left[\frac{8RT}{\pi M}\right]^{1/2}$$

If the gas is assumed to be perfect then the equation of state $pV = NkT$ can be used to find the number density $\mathcal{N}$ as $\mathcal{N} = N/V = p/kT$. The collision cross section is estimated as $\sigma = \pi(2a_0)^2$ where a_0 is the Bohr radius. A density of 1 atom $\mathrm{cm^{-3}}$ corresponds to $\mathcal{N} = 1\times10^6\ \mathrm{m^{-3}}$.

$$D = \frac{1}{3}\frac{kT}{\sigma p}\left[\frac{8RT}{\pi M}\right]^{1/2} = \frac{1}{3[\pi(2a_0)^2]\mathcal{N}}\left[\frac{8RT}{\pi M}\right]^{1/2}$$
$$= \frac{1}{3\times[\pi(2\times5.2918\times10^{-11}\ \mathrm{m})^2]\times(1\times10^6\ \mathrm{m^{-3}})}$$
$$\times\left[\frac{8\times(8.3145\ \mathrm{J\,K^{-1}\,mol^{-1}})\times(10\times10^3\ \mathrm{K})}{\pi(1.0079\times10^{-3}\ \mathrm{kg\,mol^{-1}})}\right]^{1/2}$$
$$= \boxed{1.37\times10^{17}\ \mathrm{m^2\,s^{-1}}}$$

The thermal conductivity is given in terms of the diffusion constant by [16A.10c–695], $\kappa = \nu pD/T$, which is rewritten using $\mathcal{N} = p/kT$ as $\kappa = \nu\mathcal{N}kD$. For an atom there are just three degrees of translational freedom, $\nu = \frac{3}{2}$.

$$\kappa = \nu\mathcal{N}kD = \tfrac{3}{2}\times(1\times10^{6}\ \mathrm{m^{-3}})\times(1.3806\times10^{-23}\ \mathrm{J\,K^{-1}})\times(1.37...\times10^{17}\ \mathrm{m^{2}\,s^{-1}})$$
$$= \boxed{2.84\ \mathrm{J\,K^{-1}\,m^{-1}\,s^{-1}}}$$

For a gas at ambient temperature and pressure a typical value for the diffusion coefficient is $D = 1.5\times10^{-5}\ \mathrm{m^2\,s^{-1}}$, and a typical value for the thermal conductivity is $\kappa = 0.025\ \mathrm{J\,K^{-1}\,m^{-1}\,s^{-1}}$. The diffusion constant is much higher in interstellar space when compared to ambient conditions because in interstellar space the much higher temperature results in a higher mean speed, and the much lower pressure results in a longer mean free path. Molecules move more quickly and experience fewer collisions, resulting in more rapid diffusion.

Because $\kappa \propto \mathcal{N}D$ and $D \propto 1/\mathcal{N}$, the value of the thermal conductivity is unaffected by the change in number density in going from ambient pressure to interstellar conditions. The higher thermal conductivity in the latter is therefore attributable to the higher mean speed.

The kinetic theory of gases assumes that the rate of atomic collisions is very high such that thermal equilibrium is established quickly. However, at such a dilute concentration, the timescales on which particles exchange energy by collision make this assumption questionable. In fact, atoms are more likely to interact with photons from stellar radiation than with other atoms.

P16A.5 The rate of effusion, r is given by [16A.12–697], $r = pA_0N_A/(2\pi MRT)^{1/2}$. The area of the slit is $A_0 = (10\ \mathrm{mm})\times(1.0\times10^{-2}\ \mathrm{mm}) = 0.1\ \mathrm{mm^2} = 1.0\times10^{-7}\ \mathrm{m^2}$.

$$r = \frac{pA_0N_A}{(2\pi MRT)^{1/2}}$$
$$= \frac{(p/\mathrm{Pa})\times(1.0\times10^{-7}\ \mathrm{m^2})\times(6.0221\times10^{23}\ \mathrm{mol^{-1}})}{[2\pi\times(M/\mathrm{kg\,mol^{-1}})\times(8.3145\ \mathrm{J\,K^{-1}\,mol^{-1}})\times(380\ \mathrm{K})]^{1/2}}$$
$$= (4.27...\times10^{14}\ \mathrm{s^{-1}})\times\frac{(p/\mathrm{Pa})}{(M/\mathrm{kg\,mol^{-1}})^{1/2}}$$

For cadmium, $r = (4.27...\times10^{14}\ \mathrm{s^{-1}})\times0.13/(0.11241)^{1/2} = 1.7\times10^{14}\ \mathrm{s^{-1}}$. Hence there are $\boxed{1.7\times10^{14}}$ atoms per second in the beam.

For mercury, $r = (4.27...\times10^{14}\ \mathrm{s^{-1}})\times12/(0.20059)^{1/2} = 1.1\times10^{16}\ \mathrm{s^{-1}}$. Hence there are $\boxed{1.1\times10^{16}}$ atoms per second in the beam.

16B Motion in liquids

Answers to discussion questions

D16B.1 The ionic radius, as assigned according to the distances between ions in a crystal, is a measure of ion size. The hydrodynamic radius (or Stokes radius) of an ion is its effective radius in solution taking into account all the water molecules it carries in its hydration shell. A hydrodynamic radius of a small ion is typically much larger than the ionic radius. This happens because small ions give rise to stronger electric fields than large ones so the small ions are more extensively solvated than big ones. Thus, an ion of small ionic radius may have a large hydrodynamic radius because it drags many solvent molecules through the solution as it migrates.

Solutions to exercises

E16B.1(a) The temperature dependence of the viscosity η is given by [16B.1–699], $\eta = \eta_0 e^{E_a/RT}$, where η_0 is viscosity in the limit of high temperature and E_a is the associated activation energy. Taking the natural logarithm gives $\ln\eta = \ln\eta_0 + E_a/RT$. Hence

$$\ln\eta_1 - \ln\eta_2 = (\ln\eta_0 + E_a/RT_1) - (\ln\eta_0 + E_a/RT_2) = \frac{E_a}{R}\left(\frac{1}{T_1} - \frac{1}{T_2}\right)$$

Rearranging gives an expression for the activation energy

$$\begin{aligned} E_a &= R\frac{\ln(\eta_1/\eta_2)}{(T_1^{-1} - T_2^{-1})} \\ &= (8.3145\,\mathrm{J\,K^{-1}\,mol^{-1}})\frac{\ln[(1.002\,\mathrm{cP})/(0.7975\,\mathrm{cP})]}{[(293.15\,\mathrm{K})^{-1} - (303.15\,\mathrm{K})^{-1}]} \\ &= \boxed{16.9\,\mathrm{kJ\,mol^{-1}}} \end{aligned}$$

E16B.2(a) According to the law of independent migration of ions, the limiting molar conductivity Λ_m° of an electrolyte is given by the sum of the limiting molar conductivities λ_i of the ions present, [16B.6–701], $\Lambda_m^\circ = \nu_+\lambda_+ + \nu_-\lambda_-$; in this expression ν_+ and ν_- are the numbers of cations and anions provided by each formula unit of electrolyte. For each of the given electrolytes it follows that

$$\Lambda_{\mathrm{AgI}}^\circ = \lambda_{\mathrm{Ag^+}} + \lambda_{\mathrm{I^-}} \quad \Lambda_{\mathrm{NaNO_3}}^\circ = \lambda_{\mathrm{Na^+}} + \lambda_{\mathrm{NO_3^-}} \quad \Lambda_{\mathrm{AgNO_3}}^\circ = \lambda_{\mathrm{Ag^+}} + \lambda_{\mathrm{NO_3^-}}$$

These expressions are manipulated to give $\Lambda_{\mathrm{AgI}}^\circ$

$$\begin{aligned} \Lambda_{\mathrm{AgI}}^\circ &= \lambda_{\mathrm{Ag^+}} + \lambda_{\mathrm{I^-}} \\ &= (\Lambda_{\mathrm{AgNO_3}}^\circ - \lambda_{\mathrm{NO_3^-}}) + (\Lambda_{\mathrm{NaI}}^\circ - \lambda_{\mathrm{Na^+}}) = \Lambda_{\mathrm{AgNO_3}}^\circ + \Lambda_{\mathrm{NaI}}^\circ - \Lambda_{\mathrm{NaNO_3}}^\circ \\ &= (13.34 + 12.69 - 12.16)\,\mathrm{mS\,m^2\,mol^{-1}} = \boxed{13.87\,\mathrm{mS\,m^2\,mol^{-1}}} \end{aligned}$$

E16B.3(a) The ion molar conductivity λ is given in terms of the mobility u by [16B.10–703], $\lambda = zuF$, where z is the charge number of the ion (unsigned) and F is Faraday's constant; it follows that $u = \lambda/zF$. Note that $1\,\mathrm{S} = 1\,\mathrm{C\,V^{-1}\,s^{-1}}$.

$$u_{\mathrm{Li^+}} = \frac{3.87\,\mathrm{mS\,m^2\,mol^{-1}}}{(1)(96485\,\mathrm{C\,mol^{-1}})} = 4.01\times10^{-5}\,\mathrm{mS\,m^2\,C^{-1}} = \boxed{4.01\times10^{-8}\,\mathrm{m^2\,V^{-1}\,s^{-1}}}$$

$$u_{\mathrm{Na^+}} = \frac{5.01\ \mathrm{mS\,m^2\,mol^{-1}}}{(1)(96485\,\mathrm{C\,mol^{-1}})} = 5.19\times10^{-5}\ \mathrm{mS\,m^2\,C^{-1}} = \boxed{5.19\times10^{-8}\ \mathrm{m^2\,V^{-1}\,s^{-1}}}$$

$$u_{\mathrm{K^+}} = \frac{7.35\ \mathrm{mS\,m^2\,mol^{-1}}}{(1)(96485\,\mathrm{C\,mol^{-1}})} = 7.62\times10^{-5}\ \mathrm{mS\,m^2\,C^{-1}} = \boxed{7.62\times10^{-8}\ \mathrm{m^2\,V^{-1}\,s^{-1}}}$$

E16B.4(a) The ion molar conductivity λ is given in terms of the mobility u by [16B.10–*703*], $\lambda = zuF$, where z is the charge number of the ion (unsigned) and F is Faraday's constant. Note that $1\ \mathrm{S} = 1\ \mathrm{C\,V^{-1}\,s^{-1}}$.

$$\lambda = zuF = (1)\times(7.91\times10^{-8}\ \mathrm{m^2\,V^{-1}\,s^{-1}})\times(96485\,\mathrm{C\,mol^{-1}}) = \boxed{7.63\ \mathrm{mS\,m^2\,C^{-1}}}$$

E16B.5(a) The drift speed s of an ion is given by [16B.8b–*702*], $s = u\mathcal{E}$, where $\mathcal{E}$ is the electric field strength. This field strength is given by $\mathcal{E} = \Delta\phi/l$ where $\Delta\phi$ is the potential difference between two electrodes separated by distance l.

$$s = u\mathcal{E} = u\frac{\Delta\phi}{l} = (7.92\times10^{-8}\ \mathrm{m^2\,V^{-1}\,s^{-1}})\times\frac{25.0\ \mathrm{V}}{7.00\times10^{-3}\ \mathrm{m}}$$
$$= 2.83\times10^{-4}\ \mathrm{m\,s^{-1}} = \boxed{283\ \mathrm{\mu m\,s^{-1}}}$$

E16B.6(a) The Einstein relation, [16B.13–*704*], $u = zDF/RT$, gives the relationship between the mobility u, the charge number of the ion z, and the diffusion coefficient D.

$$D = \frac{uRT}{zF} = \frac{(7.40\times10^{-8}\ \mathrm{m^2\,V^{-1}\,s^{-1}})\times(8.3145\,\mathrm{J\,K^{-1}\,mol^{-1}})\times(298\ \mathrm{K})}{(1)\times(96485\,\mathrm{C\,mol^{-1}})}$$
$$= \boxed{1.90\times10^{-9}\ \mathrm{m^2\,s^{-1}}}$$

Solutions to problems

P16B.1 The temperature dependence of the viscosity η is given by [16B.2–*699*], $\eta = \eta_0 e^{E_a/RT}$, where E_a is the activation energy . Taking the natural logarithm gives $\ln\eta = \ln\eta_0 + E_a/RT$. A plot of $\ln\eta$ against $(1/T)$ therefore has slope E_a/R; such a plot is shown in Fig. 16.1.

θ/°C	T/K	η/cP	$1/(T/\mathrm{K})$	$\ln(\eta/\mathrm{cP})$
10	283	0.758	0.003 53	−0.277
20	293	0.652	0.003 41	−0.428
30	303	0.564	0.003 30	−0.573
40	313	0.503	0.003 19	−0.687
50	323	0.442	0.003 10	−0.816
60	333	0.392	0.003 00	−0.936
70	343	0.358	0.002 92	−1.027

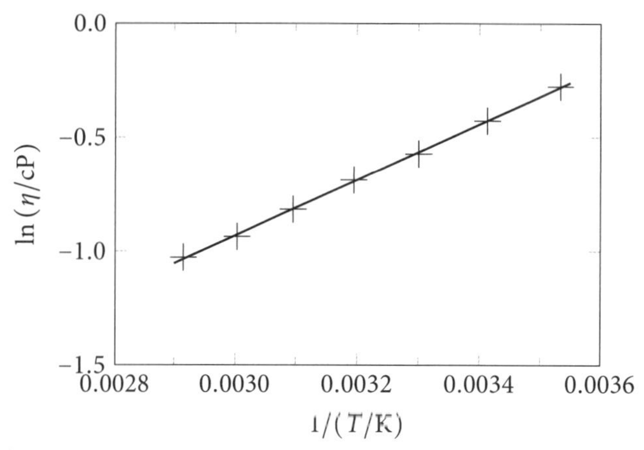

Figure 16.1

The data are a good fit to a straight line with equation

$$\ln(\eta/\,\mathrm{cP}) = (1.2207\times10^{3})\times 1/(T/\mathrm{K}) - 4.5939$$

The activation energy is computed from the slope

$$\begin{aligned} E_\mathrm{a} &= R\times(\text{slope}) \\ &= (8.3145\,\mathrm{J\,K^{-1}\,mol^{-1}})(1.2207\times10^{3}\,\mathrm{K}) = \boxed{10.15\,\mathrm{kJ\,mol^{-1}}} \end{aligned}$$

P16B.3 The molar conductivity Λ_m is defined by [16B.4–*700*], $\Lambda_\mathrm{m} = \kappa/c$, where κ is conductivity and c is concentration. The Kohlrausch law, [16B.5–*700*], gives the variation of the molar conductivity with concentration as $\Lambda_\mathrm{m} = \Lambda^\circ_\mathrm{m} - \mathcal{K}c^{1/2}$. Hence a plot of Λ_m against $c^{1/2}$ has slope equal to $-\mathcal{K}$ and y-intercept equal to the limiting molar conductivity Λ°_m. In computing Λ_m the concentration needs to be converted from mol dm^{-3} to mol m^{-3}. The graph is shown in Fig. 16.2.

$\kappa/\mathrm{S\,m^{-1}}$	$c/\mathrm{mol\,dm^{-3}}$	$\Lambda_\mathrm{m}/\mathrm{mS\,m^2\,mol^{-1}}$	$c^{1/2}/(\mathrm{mol\,dm^{-3}})^{1/2}$
13.1	1.334	9.82	1.155
13.9	1.432	9.71	1.197
14.7	1.529	9.61	1.237
15.6	1.672	9.33	1.293
16.4	1.725	9.51	1.313

The two points corresponding to the highest concentrations seem to be anomolous and are ignored in finding the best-fit line, the equation of which is

$$\Lambda_\mathrm{m}/\mathrm{mS\,m^2\,mol^{-1}} = (-2.5262)\times c^{1/2}/(\mathrm{mol\,dm^{-3}})^{1/2} + 12.737$$

Therefore, $\mathcal{K} = \boxed{2.53\ \mathrm{mS\,m^2\,(mol\,dm^{-1})^{-3/2}}}$ and $\Lambda^\circ_\mathrm{m} = \boxed{12.7\ \mathrm{mS\,m^2\,mol^{-1}}}$

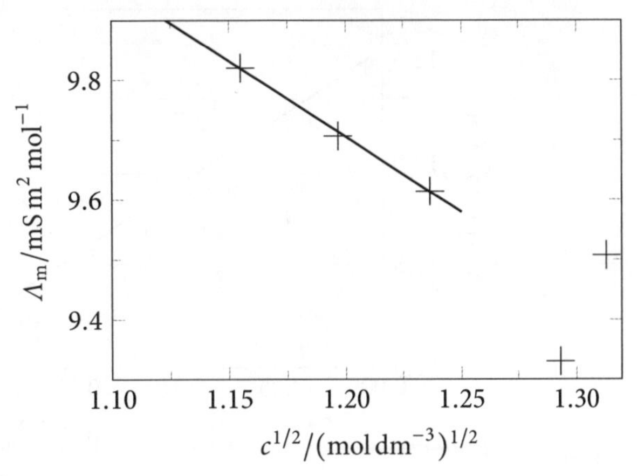

Figure 16.2

P16B.5 (a) The molar conductivity Λ_m is given by [16B.4–*700*], $\Lambda_m = \kappa/c$, where κ is conductivity and c is concentration. The Kohlrausch law, [16B.5–*700*], gives the dependence of the molar conductivity on concentration for strong electrolytes, $\Lambda_m = \Lambda_m^\circ - \mathcal{K}c^{1/2}$. According to this law a plot of Λ_m against $c^{1/2}$ will be a straight line with slope $-\mathcal{K}$ and y-intercept Λ_m°. Given that $\kappa = C/R$ where $C = 0.2063\ \text{cm}^{-1}$, the molar conductivity is computed from $\Lambda_m = C/cR$. The plot is shown in Fig. 16.3.

R/Ω	$c/\text{mol dm}^{-3}$	$\Lambda_m/\text{mS m}^2\,\text{mol}^{-1}$	$c^{1/2}/(\text{mol dm}^{-3})^{1/2}$
3 314	0.000 50	12.45	0.022
1 669	0.001 0	12.36	0.032
342	0.005 0	12.06	0.071
174	0.010	11.85	0.100
89	0.020	11.58	0.141
37	0.050	11.11	0.224

The data fall on a good straight line, as predicted by the Kohlrausch law, and the equation for the best-fit line as

$$\Lambda_m/\text{mS m}^2\,\text{mol}^{-1} = (-6.6551) \times c^{1/2}/(\text{mol dm}^{-3})^{1/2} + 12.5558$$

Thus $\boxed{\mathcal{K} = 6.655\ \text{mS m}^2\,(\text{mol dm}^{-1})^{-3/2}}$ and $\boxed{\Lambda_m^\circ = 12.56\ \text{mS m}^2\,\text{mol}^{-1}}$.

(b) The law of independent migration of ions, [16B.6–*701*], allows the limiting molar conductivity to be calculated from the values for the individual ions, and this is then converted to the molar conductivity using the

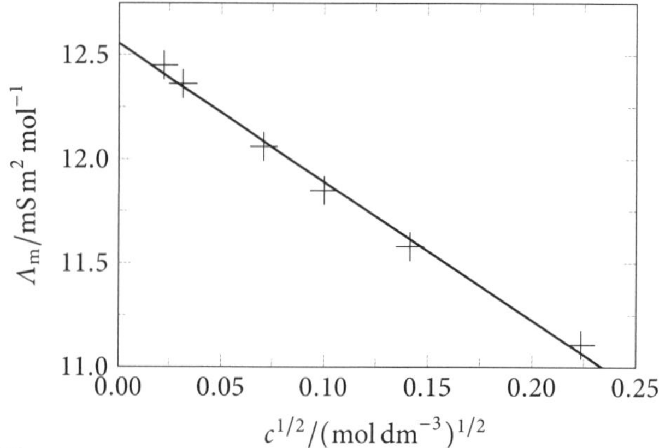

Figure 16.3

Kohlrausch law

$$\begin{aligned}\Lambda_m^\circ &= \nu_+\lambda(\text{Na}^+) + \nu_-\lambda(\text{I}^-)\\ &= (1)\times(5.01\ \text{mS m}^2\,\text{mol}^{-1}) + (1)\times(7.68\ \text{mS m}^2\,\text{mol}^{-1})\\ &= 12.69\ \text{mS m}^2\,\text{mol}^{-1}\\ \Lambda_m &= \Lambda_m^\circ - \mathcal{K}c^{1/2}\\ &= (12.69\ \text{mS m}^2\,\text{mol}^{-1})\\ &\qquad - [6.655\ \text{mS m}^2\,(\text{mol dm}^{-1})^{-3/2}]\times(0.010\ \text{mol dm}^{-3})^{1/2}\\ &= \boxed{12.02\ \text{mS m}^2\,\text{mol}^{-1}}\end{aligned}$$

The conductivity is found using [16B.4–*700*], and the resistance using the given cell constant

$$\begin{aligned}\kappa = c\Lambda_m &= (0.010\ \text{mol dm}^{-3})\times(12.02\ \text{mS m}^2\,\text{mol}^{-1})\\ &= (0.010\times10^3\ \text{mol m}^{-3})\times(12.02\ \text{mS m}^2\,\text{mol}^{-1}) = \boxed{120\ \text{mS m}^{-1}}\end{aligned}$$

$$R = \frac{C}{\kappa} = \frac{0.2063\times10^2\ \text{m}^{-1}}{120\times10^{-3}\ \text{S m}^{-1}} = \boxed{172\ \Omega}$$

where $1\text{S}^{-1} = 1\ \Omega$ is used.

P16B.7 A spherical particle of radius a and charge ze travelling at a constant speed through a solvent of viscosity η has mobility u given by [16B.9–*702*], $u = ze/f$, where f is the frictional coefficient with Stokes' law value $f = 6\pi\eta a$. Hence

$$\begin{aligned}a &= \frac{ze}{6\pi\eta u} = \frac{(1)\times(1.6022\times10^{-19}\ \text{C})}{6\pi\times(0.93\times10^{-3}\ \text{kg m}^{-1}\,\text{s}^{-1})\times(1.1\times10^{-8}\ \text{m}^2\,\text{V}^{-1}\,\text{s}^{-1})}\\ &= 8.30...\times10^{-10}\ \text{m} = \boxed{0.83\ \text{nm}}\end{aligned}$$

This is substantially larger than the 0.5 nm van der Waals radius of a Buckminsterfullerene (C_{60}) molecule because the anion attracts a considerable hydration shell through the London dispersion attraction to the nonpolar solvent molecules and through the ion-induced dipole interaction. The Stokes radius reflects the larger effective radius of the combined anion and its solvation shell.

P16B.9 (a) The initial concentration of AB is c_{AB}. After a fraction α has dissociated the concentration of AB is $c = (1-\alpha)c_{AB}$ and the concentration of A is $c_A = \alpha c_{AB}$, which is equal to the concentration of B, c_B. Therefore, the equilibrium constant K is given by

$$K = \frac{(c_A/c^\ominus)(c_B/c^\ominus)}{c/c^\ominus} = \frac{c_A c_B}{cc^\ominus} = \frac{(\alpha c_{AB})^2}{(1-\alpha)c_{AB}c^\ominus} = \frac{\alpha^2 c_{AB}}{(1-\alpha)c^\ominus}$$

(b) As the solution becomes more dilute, the degree of dissociation increases and, in the limit of infinite dilution, $\alpha = 1$. This is a consequence of the form of K derived in part (a): because K is constant, a decrease in c_{AB} requires an increase in α towards 1.

The concentration of ions in the solution scales directly with α, therefore the conductivity, and hence the molar conductivity, will be proportional to α: $\Lambda_m \propto \alpha$. At infinite dilution the molar conductivity takes the value $\Lambda_{m,1}$ and $\alpha = 1$, therefore $\alpha = \Lambda_m/\Lambda_{m,1}$.

(c) Substitution of this expression for α into the equilibrium expression gives

$$K = \frac{\alpha^2 c_{AB}}{(1-\alpha)c^\ominus} = \frac{\Lambda_m^2 c_{AB}}{\Lambda_{m,1}^2[1-(\Lambda_m/\Lambda_{m,1})]c^\ominus}$$

$$\frac{\alpha^2}{(1-\alpha)} = \frac{\Lambda_m^2}{\Lambda_{m,1}^2(1-\frac{\Lambda_m}{\Lambda_{m,1}})} \quad \text{hence} \quad \frac{\alpha^2\Lambda_{m,1}^2}{(1-\alpha)\Lambda_m} = \frac{\Lambda_m}{(1-\frac{\Lambda_m}{\Lambda_{m,1}})}$$

$$\frac{(1-\alpha)\Lambda_m}{\alpha^2\Lambda_{m,1}^2} = \frac{(1-\frac{\Lambda_m}{\Lambda_{m,1}})}{\Lambda_m} = \frac{1}{\Lambda_m} - \frac{1}{\Lambda_{m,1}} \quad \text{hence} \quad \frac{1}{\Lambda_m} = \frac{1}{\Lambda_{m,1}} + \frac{(1-\alpha)\Lambda_m}{\alpha^2\Lambda_{m,1}^2}$$

16C Diffusion

Answers to discussion questions

D16C.1 See Section 16C.1 on page 706.

Solutions to exercises

E16C.1(a) The root mean square displacement in three dimensions is given by [16C.13b–712], $\langle r^2\rangle^{1/2} = (6Dt)^{1/2}$, where D is the diffusion coefficient and t is the time period.

$$t = \frac{\langle r^2\rangle}{6D} = \frac{(5.0\times10^{-3}\ \text{m})^2}{6\times(6.73\times10^{-10}\ \text{m}^2\,\text{s}^{-1})} = \boxed{6.2\times10^3\ \text{s}}$$

E16C.2(a) The diffusion in one dimension from a layer of solute is described by [16C.10–*710*]

$$c(x,t) = \frac{n_0}{A(\pi Dt)^{1/2}} e^{-x^2/4Dt}$$

where $c(x,t)$ is the concentration at time t and distance x from the layer, and n_0 is the amount in moles in the layer of area A placed at $x = 0$. If the mass of sucrose is m, then $n_0 = m/M$, where M is the molar mass (342.30 g mol^{-1}).

$$c(x,t) = \frac{m}{MA(\pi Dt)^{1/2}} e^{-x^2/4Dt}$$

$$c(10\text{ cm},t) = \frac{(0.020\text{ kg}) \times e^{-(10\times10^{-2}\text{ m})^2/4\times(5.216\times10^{-9}\text{ m}^2\text{ s}^{-1})t}}{(342.30\text{ g mol}^{-1})\times(5.0\times10^{-4}\text{ m}^2)\times[\pi(5.216\times10^{-9}\text{ m}^2\text{ s}^{-1})t]^{1/2}}$$

$$= [(9.12...\times10^2\text{ mol dm}^{-3})\times(t/\text{s})^{-1/2}]e^{-4.79...\times10^5/(t/\text{s})}$$

$$c(10\text{ cm},10\text{ s}) = (9.12...\times10^2\text{ mol dm}^{-3})\times(10)^{-1/2}\times e^{-4.79...\times10^5/(10)}$$

$$= \boxed{0.00\text{ mol dm}^{-3}}$$

$$c(10\text{ cm},24\text{ h}) = (9.12...\times10^2\text{ mol dm}^{-3})[24(3600)]^{-1/2}\times e^{-4.79...\times10^5/[24(3600)]}$$

$$= \boxed{0.0121\text{ mol dm}^{-3}}$$

Diffusion is a very slow process: after 10 s the concentration at a height of 10 cm is zero to within the precision of the calculation. Even after 24 hours only a very small amount of the sucrose has moved up into the liquid.

E16C.3(a) The thermodynamic force $\mathcal{F}$ is given by [16C.3b–*706*]

$$\mathcal{F} = -\frac{RT}{c}\left(\frac{\partial c}{\partial x}\right)_{T,p}$$

Substituting $c(x) = c_0 - \alpha c_0 x$ into the above expression gives

$$\mathcal{F} = -\frac{RT}{c_0 - \alpha c_0 x}(-\alpha c_0) = \frac{\alpha RT}{1-\alpha x}$$

The constant α is found by noting that $c = c_0/2$ at $x = 10\text{ cm} = 0.10\text{ m}$. Hence $c_0/2 = c_0 - \alpha c_0 \times (0.10\text{ m})$ and therefore $\alpha = 5.0\text{ m}^{-1}$. At $T = 298$ K and $x = 10$ cm the force is

$$\mathcal{F} = \frac{(5\text{ m}^{-1})\times(8.3145\text{ J K}^{-1}\text{ mol}^{-1})\times(298\text{ K})}{1-(5\text{ m}^{-1})(10\times10^{-2}\text{ m})} = \boxed{25\text{ kN mol}^{-1}}$$

A similar calculation at $x = 15$ cm gives $\mathcal{F} = \boxed{50\text{ kN mol}^{-1}}$. The force is greater at the larger distance, even though the gradient is the same.

E16C.4(a) The thermodynamic force $\mathcal{F}$ is given by [16C.3b–*706*]

$$\mathcal{F} = -\frac{RT}{c}\left(\frac{\partial c}{\partial x}\right)_{T,p}$$

Substituting $c(x) = c_0\mathrm{e}^{-\alpha x^2}$ into the above expression gives

$$\mathcal{F} = -\frac{RT}{c_0\mathrm{e}^{-\alpha x^2}}(-2\alpha c_0 x\mathrm{e}^{-\alpha x^2}) = 2\alpha xRT$$

The constant α is found by noting that $c = c_0/2$ at $x = 5$ cm $= 0.05$ m. Hence $c_0/2 = c_0\mathrm{e}^{-\alpha(0.05\ \mathrm{m})^2}$ and therefore $\alpha = \ln 2/(0.05\ \mathrm{m})^2 = 277\ \mathrm{m}^{-2}$ The thermodynamic force at $T = 293$ K and $x = 5.0$ cm is

$$\mathcal{F} = 2(277\ \mathrm{m}^{-2})\times(0.050\ \mathrm{m})\times(8.3145\ \mathrm{J\,K^{-1}\,mol^{-1}})\times(293\ \mathrm{K}) = \boxed{67.5\ \mathrm{kN\,mol^{-1}}}$$

E16C.5(a) The root mean square displacement in three dimensions is given by [16C.13b–*712*], $\langle r^2\rangle^{1/2} = (6Dt)^{1/2}$, where D is the diffusion coefficient and t is the time period. Hence,

$$t = \frac{\langle r^2\rangle}{6D} = \frac{(5.0\times10^{-3}\ \mathrm{m})^2}{6\times(3.17\times10^{-9}\ \mathrm{m^2\,s^{-1}})} = \boxed{1.3\times10^3\ \mathrm{s}}$$

E16C.6(a) The Stokes–Einstein equation [16C.4b–*708*], $D = kT/6\pi\eta a$, relates the diffusion coefficient D to the viscosity η and the radius a of the diffusing particle, which is modelled as a sphere. Recall that 1 cP $= 10^{-3}\ \mathrm{kg\,m^{-1}\,s^{-1}}$.

$$a = \frac{kT}{6\pi\eta D} = \frac{(1.3806\times10^{-23}\ \mathrm{J\,K^{-1}})\times(298\ \mathrm{K})}{6\pi\times(1.00\times10^{-3}\ \mathrm{kg\,m^{-1}\,s^{-1}})\times(5.2\times10^{-10}\ \mathrm{m^2\,s^{-1}})} = \boxed{0.42\ \mathrm{nm}}$$

E16C.7(a) The Einstein–Smoluchowski equation [16C.15–*713*], $D = d^2/2\tau$, relates the diffusion coefficient D to the jump distance d and time τ required for a jump. Approximating the jump length as the molecular diameter, then $d \approx 2a$ where a is the effective molecular radius. This is estimated using the Stokes–Einstein equation [16C.4b–*708*], $D = kT/6\pi\eta a$, to give $2a = 2kT/6\pi\eta D$.

Combining these expressions and using the value for viscosity of benzene from the *Resource section* gives

$$\tau = \frac{d^2}{2D} = \frac{1}{2D}\left(\frac{2kT}{6\pi\eta D}\right)^2 = \frac{1}{18D^3}\left(\frac{kT}{\pi\eta}\right)^2$$

$$= \frac{1}{18\times(2.13\times10^{-9}\ \mathrm{m^2\,s^{-1}})^3}\left(\frac{(1.3806\times10^{-23}\ \mathrm{J\,K^{-1}})\times(298\ \mathrm{K})}{\pi\times(0.601\times10^{-3}\ \mathrm{kg\,m^{-1}\,s^{-1}})}\right)^2$$

$$= 2.73\times10^{-11}\ \mathrm{s} = \boxed{27.3\ \mathrm{ps}}$$

E16C.8(a) The root mean square displacement in one dimension is given by [16C.13a–*711*], $\langle x^2\rangle^{1/2} = (2Dt)^{1/2}$, where D is the diffusion coefficient and t is the time period. For an iodine molecule in benzene, $D = 2.13\times10^{-9}\ \mathrm{m^2\,s^{-1}}$

$$\langle x^2\rangle^{1/2} = (2Dt)^{1/2} = [2\times(2.13\times10^{-9}\ \mathrm{m^2\,s^{-1}})\times(1.0\ \mathrm{s})]^{1/2} = 6.5\times10^{-5}\ \mathrm{m} = \boxed{65\ \mu\mathrm{m}}$$

For a sucrose molecule in water, $D = 0.5216\times10^{-9}\ \mathrm{m^2\,s^{-1}}$

$$\langle x^2\rangle^{1/2} = [2\times(0.5216\times10^{-9}\ \mathrm{m^2\,s^{-1}})\times(1.0\ \mathrm{s})]^{1/2} = 3.2\times10^{-5}\ \mathrm{m} = \boxed{32\ \mu\mathrm{m}}$$

Solutions to problems

P16C.1 Thermodynamic force, $\mathcal{F}$, is given by [16C.3b–*706*].

$$\mathcal{F} = -\frac{RT}{c}\left(\frac{\partial c}{\partial x}\right)_{T,p}$$

where c is the concentration.For a linear gradation of intensity, that is concentration, down the tube

$$\mathrm{d}c/\mathrm{d}x = \Delta c/\Delta x = [(0.050-0.100)\times10^3\ \mathrm{mol\,m^{-3}}]/(0.10\ \mathrm{m}) = 500\ \mathrm{mol\,m^{-4}}$$

$$\mathcal{F} = -\frac{RT}{c}\frac{\mathrm{d}c}{\mathrm{d}x} = -\frac{(8.3145\ \mathrm{J\,K^{-1}\,mol^{-1}})\times(298\ \mathrm{K})}{c}\times(-500\ \mathrm{mol\,m^{-4}}) = \frac{1.23...\times10^6\ \mathrm{N\,mol^{-1}}}{(c/\ \mathrm{mol\,m^{-3}})}$$

At the left face, $c = 0.100\ \mathrm{mol\,dm^{-3}}$:

$$\mathcal{F} = (1.23...\times10^3\ \mathrm{kN\,mol^{-1}})/(0.100\times10^3) = \boxed{12.4\ \mathrm{kN\,mol^{-1}}} = \boxed{2.1\times10^{-20}\ \mathrm{N\,(molecule)^{-1}}}$$

In the middle, $c = 0.075\ \mathrm{mol\,dm^{-3}}$:

$$\mathcal{F} = (1.23...\times10^3\ \mathrm{kN\,mol^{-1}})/(0.075\times10^3) = \boxed{16.5\ \mathrm{kN\,mol^{-1}}} = \boxed{2.7\times10^{-20}\ \mathrm{N\,(molecule)^{-1}}}$$

Close to the left face, $c = 0.050\ \mathrm{mol\,dm^{-3}}$:

$$\mathcal{F} = (1.23...\times10^3\ \mathrm{kN\,mol^{-1}})/(0.050\times10^3) = \boxed{24.8\ \mathrm{kN\,mol^{-1}}} = \boxed{4.1\times10^{-20}\ \mathrm{N\,(molecule)^{-1}}}$$

P16C.3 The thermodynamic force $\mathcal{F}$ is given by [16C.3b–706]

$$\mathcal{F} = -\frac{RT}{c}\left(\frac{\partial c}{\partial x}\right)_{T,p}$$

Substituting $c(x) = c_0(1 - e^{-ax^2})$ into the above expression gives

$$\mathcal{F} = -\frac{RT}{c_0(1 - e^{-ax^2})}(2ac_0 x e^{-ax^2}) = -\frac{2axRTe^{-ax^2}}{(1 - e^{-ax^2})} = \frac{2axRT}{(1 - e^{ax^2})}$$

The final step involves multiplying top and bottom of the fraction by e^{ax^2}. For thermodynamic force for $a = 0.10\ \mathrm{cm}^{-2} = 1000\ \mathrm{m}^{-2}$ and $T = 298$ K is

$$\mathcal{F} = \frac{(5.0\ \mathrm{MN\ mol^{-1}}) \times (x/\mathrm{m})}{(1 - e^{1000\times(x/\mathrm{m})^2})}$$

$$= \frac{(8.2 \times 10^{-18}\ \mathrm{N\ molecule^{-1}}) \times (x/\mathrm{m})}{(1 - e^{1000\times(x/\mathrm{m})^2})}$$

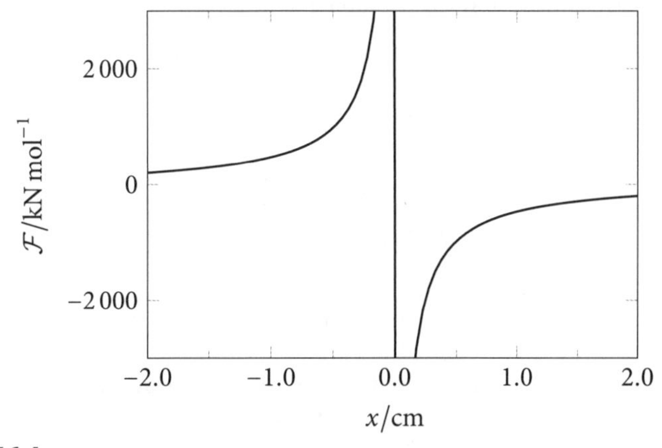

Figure 16.4

A plot of the thermodynamic force per mole against x is shown in Fig. 16.4. It demonstrates that the force is directed such that mass is pushed by the thermodynamic force toward the centre of the tube to where the concentration is lowest. A negative force pushes mass toward the left ($x > 0$) and a positive force pushes mass toward the right ($x < 0$).

At $x = 0$ the gradient of the concentration is zero, so the thermodynamic force is also zero. However, as x approaches zero the modulus of the thermodynamic force increases without limit on account of the concentration becoming smaller and smaller.

P16C.5 The generalised diffusion equation is [16C.6–709], where c is concentration, t is time, D is the diffusion coefficient and x is displacement.

$$\frac{\partial c}{\partial t} = D\frac{\partial^2 c}{\partial x^2}$$

An expression for $c(x,t)$ is a solution of the diffusion equation if substitution of the expression for $c(x,t)$ into each side of the diffusion equation gives the same result. The proposed solution is

$$c(x,t) = \frac{n_0}{A(\pi Dt)^{1/2}} e^{-x^2/4Dt}$$

LHS

$$\frac{\partial c}{\partial t} = \frac{n_0 e^{-x^2/4Dt}}{A(\pi Dt)^{1/2}}\left(\frac{x^2}{4Dt^2} - \frac{1}{2t}\right)$$

RHS

$$D\frac{\partial^2 c}{\partial x^2} = D\frac{\partial}{\partial x}\left[\frac{n_0 e^{-x^2/4Dt}}{A(\pi Dt)^{1/2}}\left(\frac{-x}{2Dt}\right)\right] = D\frac{n_0 e^{-x^2/4Dt}}{A(\pi Dt)^{1/2}}\left[\left(\frac{x}{2Dt}\right)^2 - \frac{1}{2Dt}\right]$$

$$= \frac{n_0 e^{-x^2/4Dt}}{A(\pi Dt)^{1/2}}\left(\frac{x^2}{4Dt^2} - \frac{1}{2t}\right)$$

As required the LHS = the RHS, hence the proposed form of $c(x,t)$ is indeed a solution to the diffusion equation.

As $t \to 0$ the exponential term $e^{-x^2/4Dt}$ falls off more and more rapidly, implying that in the limit $t = 0$ all the material is at $x = 0$. The exponential function dominates the term $t^{1/2}$ in the denominator.

P16C.7 As discussed in Section 16C.2(c) on page 710 the probability of finding a molecule in an interval dx at distance x from the origin at time t is $P(x,t)dx$, where $P(x,t)$ is given by

$$P(x,t) = \frac{1}{(\pi Dt)^{1/2}} e^{-x^2/4Dt}$$

The mean value of x^4 is found by integrating $P(x,t)x^4\,dx$ over the full range of x, which in this case is 0 to ∞

$$\langle x^4 \rangle = \int_0^\infty x^4 P(x)\,dx = \frac{1}{(\pi Dt)^{1/2}} \int_0^\infty x^4 e^{-x^2/4Dt}\,dx$$

$$= \frac{1}{(\pi Dt)^{1/2}} \times \tfrac{3}{8}(4Dt)^2 \times (4\pi Dt)^{1/2} = 12D^2t^2$$

where to go to the final line Integral G.5 is used with $k = 1/4Dt$. Hence, $\langle x^4 \rangle^{1/4} = (12D^2t^2)^{1/4}$.

A similar calculation is used to find $\langle x^2 \rangle$

$$\langle x^2 \rangle = \int_0^\infty x^2 P(x)\,dx = \frac{1}{(\pi Dt)^{1/2}} \int_0^\infty x^2 e^{-x^2/4Dt}\,dx$$

$$= \frac{1}{(\pi Dt)^{1/2}} \times \tfrac{1}{4}\pi^{1/2}(4Dt)^{3/2} = 2Dt$$

where to go to the final line Integral G.3 is used with $k = 1/4Dt$. Hence, $\langle x^2 \rangle^{1/2} = (2Dt)^{1/2}$.

The ratio of the $\langle x^4\rangle^{1/4}$ to $\langle x^2\rangle$ is

$$\frac{\langle x^4\rangle^{1/4}}{\langle x^2\rangle^{1/2}} = \frac{(12D^2t^2)^{1/4}}{(2Dt)^{1/2}} = \left(\frac{12}{4}\right)^{1/4} = \boxed{3^{1/4}}$$

The result is independent of the time.

P16C.9 The probability of being n steps from the origin is $P(nd) = N!/(N-N_R)!N_R!2^N$ where N_R is the number of steps taken to the right and N is the total number of steps. Note that $n = N_R - N_L$ and $N = N_R + N_L$, where N_L is the number of steps taken to the left.

$$N_R = N - N_L = N_L + n \quad \text{hence} \quad N_L = \frac{N-n}{2}$$

$$N_L = N - N_R = N_R - n \quad \text{hence} \quad N_R = \frac{N+n}{2}$$

$$\text{therefore } P(nd) = \frac{N!}{[N - (\frac{N-n}{2})]!\ (\frac{N-n}{2})!\ 2^N} = \frac{N!}{(\frac{N+n}{2})!\ (\frac{N-n}{2})!\ 2^N}$$

The probability of being six paces away from the origin $(x = 6d)$ is

$$P_{\text{exact}}(6d) = \frac{N!}{(\frac{N+6}{2})!\ (\frac{N-6}{2})!\ 2^N}$$

This is the 'exact' value of the probability according to the random walk model.

In the limit of large N the probability density of being at distance x and time t is given by [16C.14–*713*]

$$P(x,t) = \left(\frac{2\tau}{\pi t}\right)^{1/2} e^{-x^2\tau/2td^2}$$

For the present case the value of x is taken as nd, and t/τ is taken as N because the time to take N steps is $N\tau$. With these substitutions

$$P_{\text{lim}}(nd,N) = \left(\frac{2}{\pi N}\right)^{1/2} e^{-n^2/2N}$$

The following table compares the exact values of the probability with those predicted for large N. The discrepancy between the two values falls to less than 0.1% when N is greater than about 53.

N	P_{exact}	P_{lim}	$100(P_{\text{exact}} - P_{\text{lim}})/P_{\text{exact}}$
6	0.0156	0.0162	−3.79
10	0.0439	0.0417	5.09
14	0.0611	0.0590	3.51
18	0.0708	0.0692	2.30
22	0.0762	0.0751	1.55
26	0.0792	0.0783	1.07
30	0.0806	0.0799	0.75
34	0.0810	0.0806	0.54
38	0.0809	0.0806	0.38
42	0.0804	0.0802	0.27
46	0.0797	0.0795	0.19
50	0.0788	0.0787	0.13
54	0.0779	0.0778	0.08
58	0.0768	0.0768	0.05
60	0.0763	0.0763	0.03

P16C.11 The Stokes–Einstein relation [16C.4b–*708*], shows that $D \propto T/\eta$ where D is the diffusion coefficient and η is the viscosity. The temperature dependence of viscosity is given by [16B.2–699], $\eta = \eta_0 e^{E_a/RT}$, it therefore follows that $D \propto Te^{-E_a/RT}$. The activation energy E_a can therefore be determined from the ratio of the diffusion constants at two temperatures

$$\frac{D_{T_1}}{D_{T_2}} = \frac{T_1 e^{-E_a/RT_1}}{T_2 e^{-E_a/RT_2}} = \frac{T_1}{T_2} e^{\frac{E_a}{R}(1/T_2 - 1/T_1)}$$

Solving for E_a gives

$$\begin{aligned} E_a &= \frac{R}{1/T_2 - 1/T_1} \ln \frac{D_{T_1} T_2}{D_{T_2} T_1} \\ &= \frac{(8.3145\,\text{J K}^{-1}\,\text{mol}^{-1})}{1/(298\,\text{K}) - 1/(273\,\text{K})} \times \ln\left(\frac{(298\,\text{K}) \times (2.05 \times 10^{-9}\,\text{m}^2\,\text{s}^{-1})}{(273\,\text{K}) \times (2.89 \times 10^{-9}\,\text{m}^2\,\text{s}^{-1})}\right) \\ &= 6.9\,\text{kJ mol}^{-1} \end{aligned}$$

The activation energy associated with diffusion of therefore $\boxed{6.9\,\text{kJ mol}^{-1}}$.

Answers to integrated activities

I16.1 If it is assumed that viscous flow involves an activated process in which molecules jump from one environment to another, then it follows that the 'rate constant' for this process is inversely proportional to the viscosity because such jumps are more infrequent in a more viscous liquid. From [16B.2–699], $\eta = \eta_0 e^{E_{a,\text{visc}}/RT}$, it follows that the rate constant $k_{r,\text{visc}}$ goes as $k_{r,\text{visc}} = Ae^{-E_{a,\text{visc}}/RT}$, where A is a constant. The definition of the activation energy is then applied

$$E_a = RT^2\left(\frac{d\ln k_{r,visc}}{dT}\right) = RT^2\left(\frac{d(\ln A - E_{a,visc}/RT)}{dT}\right)$$
$$= RT^2\frac{E_{a,visc}}{RT^2} = E_{a,visc}$$

The quantity $E_{a,visc}$ can indeed be identified as an activation energy.

From the expression in *Problem* P16B.2 the viscosity is written $\eta = \eta_{20}\times 10^{f(T)}$, where $f(T) = 1.3272(20 - T/°C) - 0.001053(20 - T/°C)^2]/(T/°C + 105)$. Noting that $k_{r,visc} \propto 1/\eta$, the activation energy is found from its definition using

$$E_a = RT^2\frac{d\ln k_{r,visc}}{dT} = -RT^2\frac{d\ln\eta}{dT} = -\frac{RT^2}{\eta}\frac{d\eta}{dT}$$

Note that $\ln x = \log x\times\ln 10$. Hence $x = 10^{\log x} = e^{\log x\times\ln 10}$. Therefore,

$$E_a = -\frac{RT^2}{\eta}\frac{d[\eta_{20}\times 10^{f(T)}]}{dT} = -\frac{RT^2\eta_{20}}{\eta}\frac{d[e^{f(T)\times\ln 10}]}{dT}$$
$$= -\frac{RT^2\eta_{20}}{\eta}\times e^{f(T)\times\ln 10}\times\ln 10\times\frac{df(T)}{dT}$$
$$= -\frac{RT^2\,\overbrace{\eta_{20}10^{f(T)}}^{\eta}\times\ln 10}{\eta}\times\frac{df(T)}{dT}$$
$$= -RT^2\times\ln 10$$
$$\times\left[\frac{-1.3272 + 0.001053\times(2)\times(20 - T/°C)}{T/°C + 105} - \frac{f(T)}{T/°C + 105}\right]$$

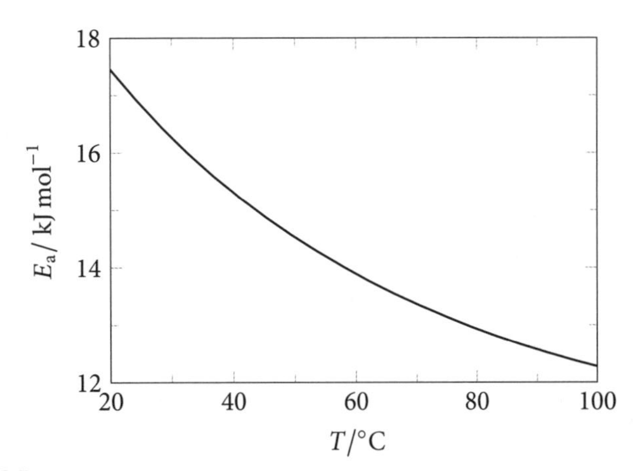

Figure 16.5

A plot of activation energy against temperature is shown in Fig. 16.5; the activation energy decreases from 17.5 kJ mol^{-1} at 20 °C to 12.3 kJ mol^{-1} at 100 °C.

This decrease may be caused by the decrease in the density that occurs as the temperature is increased. The decrease in density implies an increase in the average intermolecular distance which may cause a decrease in the strength of the hydrogen bonds between water molecules. There may also be a decrease in the hydration sphere of a molecule, thereby making movement easier.

Chemical kinetics

17A The rates of chemical reactions

Answers to discussion questions

D17A.1 Many reactions are found to have rate laws of the form $v = k_r[A]^a[B]^b\ldots$. The power to which the concentration of a species is raised in this rate law is the order of the reaction with respect to that species. The power is 0, 1, 2, for zeroth-, first-, and second-order, respectively. The sum of all the powers of the species present in the rate law is the overall order of the reaction.

For a simple rate law of the form $v = k_r[A]^a$ the concentration varies with time in a way which is characteristic of the order. Assuming that A is a reactant, for a zeroth-order reaction [A] falls linearly with time, for a first-order reaction [A] falls exponentially, and for a second-order reaction 1/[A] increases linearly with time.

A reaction is said to be pseudofirst-order when all but one of the reactants are in such large excess that their concentrations remain constant, and the order with respect to the remaining reactant is one.

D17A.3 If an order can be ascribed to a reaction, and hence to the each of the species involved in the rate law, it is then possible to write a relatively simple expression for the rate as a function of the concentrations of the species involved. This rate equation can then be integrated, either analytically or numerically, to give a prediction of how the concentration varies with time.

Solutions to exercises

E17A.1(a) Assuming perfect gas behaviour the total pressure is proportional to the total amount in moles of gas present, provided that the temperature is constant and the volume of the container is fixed. The reaction $2\,ICl(g) + H_2(g) \rightarrow I_2(g) + 2\,HCl(g)$ involves the same number of gas molecules on both sides of the reaction arrow and therefore the total amount in moles of gas present does not change as the reaction proceeds. Consequently there is $\boxed{\text{no change}}$ in the total pressure during the reaction. This means that the composition of the reaction mixture cannot be determined by measuring the total pressure in this case.

E17A.2(a) The stoichiometry of the reaction shows that one mole of Br_2 is formed for every two moles of NO formed. Therefore the rate of formation of Br_2 is half the rate of formation of NO.

$$\frac{d[Br_2]}{dt} = \tfrac{1}{2} \times \frac{d[NO]}{dt} = \tfrac{1}{2} \times (0.24\ \text{mmol dm}^{-3}\,\text{s}^{-1}) = \boxed{0.12\ \text{mmol dm}^{-3}\,\text{s}^{-1}}$$

E17A.3(a) For a homogeneous reaction in a constant volume system the rate of reaction is given by [17A.3b–*726*], $\upsilon = (1/\nu_J)d[J]/dt$, which is rearranged to $d[J]/dt = \nu_J \upsilon$. In these expressions ν_J is the stoichiometric number of species J, which is negative for reactants and positive for products. For this reaction $\nu_A = -1$, $\nu_B = -2$, $\nu_C = +3$ and $\nu_D = +1$.

For A $\quad d[A]/dt = \nu_A \upsilon = (-1) \times (2.7\ \text{mol dm}^{-3}\,\text{s}^{-1}) = -2.7\ \text{mol dm}^{-3}\,\text{s}^{-1}$

For B $\quad d[B]/dt = \nu_B \upsilon = (-2) \times (2.7\ \text{mol dm}^{-3}\,\text{s}^{-1}) = -5.4\ \text{mol dm}^{-3}\,\text{s}^{-1}$

For C $\quad d[C]/dt = \nu_C \upsilon = (+3) \times (2.7\ \text{mol dm}^{-3}\,\text{s}^{-1}) = +8.1\ \text{mol dm}^{-3}\,\text{s}^{-1}$

For D $\quad d[D]/dt = \nu_D \upsilon = (+1) \times (2.7\ \text{mol dm}^{-3}\,\text{s}^{-1}) = +2.7\ \text{mol dm}^{-3}\,\text{s}^{-1}$

The rate of consumption of A is $\boxed{2.7\ \text{mol dm}^{-3}\,\text{s}^{-1}}$, the rate of consumption of B is $\boxed{5.4\ \text{mol dm}^{-3}\,\text{s}^{-1}}$, the rate of formation of C is $\boxed{8.1\ \text{mol dm}^{-3}\,\text{s}^{-1}}$, and the rate of formation of D is $\boxed{2.7\ \text{mol dm}^{-3}\,\text{s}^{-1}}$.

E17A.4(a) For a homogeneous reaction in a constant volume system the rate of reaction is given by [17A.3b–*726*], $\upsilon = (1/\nu_J)d[J]/dt$, where ν_J is the stoichiometric number of species J which is negative for reactants and positive for products. For species C, which has $\nu_C = +2$, this gives

$$\upsilon = \frac{1}{\nu_C}\frac{d[C]}{dt} = \frac{1}{+2} \times (2.7\ \text{mol dm}^{-3}\,\text{s}^{-1}) = 1.35...\ \text{mol dm}^{-3}\,\text{s}^{-1}$$
$$= \boxed{1.4\ \text{mol dm}^{-3}\,\text{s}^{-1}}$$

Rearranging [17A.3b–*726*] then gives

For A $\quad d[A]/dt = \nu_A \upsilon = (-2) \times (1.35...\ \text{mol dm}^{-3}\,\text{s}^{-1}) = -2.70...\ \text{mol dm}^{-3}\,\text{s}^{-1}$

For B $\quad d[B]/dt = \nu_B \upsilon = (-1) \times (1.35...\ \text{mol dm}^{-3}\,\text{s}^{-1}) = -1.35...\ \text{mol dm}^{-3}\,\text{s}^{-1}$

For D $\quad d[D]/dt = \nu_D \upsilon = (+3) \times (1.35...\ \text{mol dm}^{-3}\,\text{s}^{-1}) = +4.05...\ \text{mol dm}^{-3}\,\text{s}^{-1}$

The rate of consumption of A is $\boxed{2.7\ \text{mol dm}^{-3}\,\text{s}^{-1}}$, the rate of consumption of B is $\boxed{1.4\ \text{mol dm}^{-3}\,\text{s}^{-1}}$, and the rate of formation of D is $\boxed{4.1\ \text{mol dm}^{-3}\,\text{s}^{-1}}$.

E17A.5(a) As explained in Section 17A.2(b) on page 726 the units of k_r are always such as to convert the product of concentrations, each raised to the appropriate power, into a rate expressed as a change in concentration divided by time. In this case the rate is given in $\text{mol dm}^{-3}\,\text{s}^{-1}$, so if the concentrations are expressed in mol dm^{-3} the units of k_r will be $\boxed{\text{dm}^3\,\text{mol}^{-1}\,\text{s}^{-1}}$ because

$$\overbrace{(\text{dm}^3\,\text{mol}^{-1}\,\text{s}^{-1})}^{k_r} \times \overbrace{(\text{mol dm}^{-3})}^{[A]} \times \overbrace{(\text{mol dm}^{-3})}^{[B]} = \text{mol dm}^{-3}\,\text{s}^{-1}$$

The rate of reaction is given by [17A.3b–*726*], $v = (1/\nu_J)(d[J]/dt)$, where ν_J is the stoichiometric number of species J. Rearranging gives $d[J]/dt = \nu_J v$. In this case $\nu_C = +3$, $\nu_A = -1$, and $v = k_r[A][B]$ so

$$\frac{d[C]}{dt} = \nu_C v = 3k_r[A][B] \qquad \frac{d[A]}{dt} = \nu_A v = -k_r[A][B]$$

The rate of formation of C is therefore $d[C]/dt = \boxed{3k_r[A][B]}$ and the rate of consumption of A is $-d[A]/dt = \boxed{k_r[A][B]}$.

E17A.6(a) The rate of reaction is given by [17A.3b–*726*], $v = (1/\nu_J)(d[J]/dt)$. In this case $\nu_C = +2$ so

$$v = \frac{1}{\nu_J}\frac{d[C]}{dt} = \frac{1}{+2}k_r[A][B][C] = \boxed{\tfrac{1}{2}k_r[A][B][C]}$$

As explained in Section 17A.2(b) on page 726 the units of k_r are always such as to convert the product of concentrations, each raised to the appropriate power, into a rate expressed as a change in concentration divided by time. In this case the rate is given in $\text{mol dm}^{-3}\,\text{s}^{-1}$, so if the concentrations are expressed in mol dm^{-3} the units of k_r will be $\boxed{\text{dm}^6\,\text{mol}^{-2}\,\text{s}^{-1}}$ because

$$\overbrace{(\text{dm}^6\,\text{mol}^{-2}\,\text{s}^{-1})}^{k_r} \times \overbrace{(\text{mol dm}^{-3})}^{[A]} \times \overbrace{(\text{mol dm}^{-3})}^{[B]} \times \overbrace{(\text{mol dm}^{-3})}^{[C]} = \text{mol dm}^{-3}\,\text{s}^{-1}$$

E17A.7(a) As explained in Section 17A.2(b) on page 726 the units of k_r are always such as to convert the product of concentrations, each raised to the appropriate power, into a rate expressed as a change in concentration divided by time.

(i) A second-order reaction expressed with concentrations in moles per cubic decimetre is one with a rate law such as $v = k_r[A][B]$. If the rate is given in $\text{mol dm}^{-3}\,\text{s}^{-1}$ then the units of k_r will be $\boxed{\text{dm}^3\,\text{mol}^{-1}\,\text{s}^{-1}}$ because

$$\overbrace{(\text{dm}^3\,\text{mol}^{-1}\,\text{s}^{-1})}^{k_r} \times \overbrace{(\text{mol dm}^{-3})}^{[A]} \times \overbrace{(\text{mol dm}^{-3})}^{[B]} = \text{mol dm}^{-3}\,\text{s}^{-1}$$

A third-order reaction expressed with concentrations in moles per cubic decimetre is one with a rate law such as $v = k_r[A][B][C]$. The units of k_r will then be $\boxed{\text{dm}^6\,\text{mol}^{-2}\,\text{s}^{-1}}$ because

$$\overbrace{(\text{dm}^6\,\text{mol}^{-2}\,\text{s}^{-1})}^{k_r} \times \overbrace{(\text{mol dm}^{-3})}^{[A]} \times \overbrace{(\text{mol dm}^{-3})}^{[B]} \times \overbrace{(\text{mol dm}^{-3})}^{[C]} = \text{mol dm}^{-3}\,\text{s}^{-1}$$

(ii) If the rate laws are expressed with pressures in kilopascals then a second-order reaction is one with a rate law such as $v = k_r p_A p_B$ and a third-order reaction is one with a rate law such as $v = k_r p_A p_B p_C$. If the rate is given in $\mathrm{kPa\,s^{-1}}$ then the units of k_r will be $\boxed{\mathrm{kPa^{-1}\,s^{-1}}}$ and $\boxed{\mathrm{kPa^{-2}\,s^{-1}}}$ respectively.

$$\text{For second-order}\quad \overbrace{(\mathrm{kPa^{-1}\,s^{-1}})}^{k_r} \times \overbrace{(\mathrm{kPa})}^{p_A} \times \overbrace{(\mathrm{kPa})}^{p_B} = \mathrm{kPa\,s^{-1}}$$

$$\text{For third-order}\quad \overbrace{(\mathrm{kPa^{-2}\,s^{-1}})}^{k_r} \times \overbrace{(\mathrm{kPa})}^{p_A} \times \overbrace{(\mathrm{kPa})}^{p_B} \times \overbrace{(\mathrm{kPa})}^{p_C} = \mathrm{kPa\,s^{-1}}$$

E17A.8(a) (i) In the rate law $v = k_{r1}[A][B]/(k_{r2} + k_{r3}[B]^{1/2})$ the concentration of A appears raised to the power +1, so the reaction is first order in A, and hence can be assigned an order with respect to A, $\boxed{\text{under all conditions}}$.

(ii) The concentration of B does not appear as a single term raised to a power, so the reaction has an indefinite order with respect to B. However, if $k_{r2} \gg k_{r3}[B]^{1/2}$, which might occur at very low concentrations of B, then the term $k_{r3}[B]^{1/2}$ in the denominator is negligible compared to the k_{r2} term and so the rate law becomes

$$v = k_{r1}[A][B]/k_{r2} = k_{r,\text{eff}}[A][B] \quad \text{where } k_{r,\text{eff}} = k_{r1}/k_{r2}$$

In this effective rate law the concentration of B appears raised to the power +1, so under these conditions the reactions is first order in B. Similarly, if $k_{r2} \ll k_{r3}[B]^{1/2}$, which might occur at very high concentrations of B, the term k_{r2} in the denominator is negligible compared to the $k_{r3}[B]^{1/2}$ term and so the rate law becomes

$$v = k_{r1}[A][B]/k_{r3}[B]^{1/2} = k_{r,\text{eff}}[A][B]^{1/2} \quad \text{where } k_{r,\text{eff}} = k_{r1}/k_{r3}$$

In this effective rate law the order with respect to B is $+\frac{1}{2}$.
To summarize, an order can only be assigned with respect to B if either $\boxed{k_{r2} \gg k_{r3}[B]^{1/2}}$, in which case the order is +1, or $\boxed{k_{r2} \ll k_{r3}[B]^{1/2}}$, in which case the order is $+\frac{1}{2}$.

(iii) An overall order can be assigned only if all of the individual orders can be assigned. Consequently the reaction can only be assigned an overall order if $\boxed{k_{r2} \gg k_{r3}[B]^{1/2}}$ or $\boxed{k_{r2} \ll k_{r3}[B]^{1/2}}$. The overall order in these two cases is +2 and $+\frac{3}{2}$.

E17A.9(a) The gaseous species is denoted A and the order with respect to A as a. The rate law expressed in terms of partial pressure is then $v = k_r p_A^a$. Taking (common) logarithms gives

$$\log v = \log k_r + \log p_A^a = \log k_r + a \log p_A$$

where the properties of logarithms $\log(xy) = \log x + \log y$ and $\log x^a = a \log x$ are used.

This expression implies that a graph of $\log v$ against $\log p_A$ will be a straight line of slope a, from which the order can be determined. However, because there are only two data points a graph is not necessary so an alternative approach is used.

If the initial partial pressure of the compound is $p_{A,0}$ then the partial pressure when a fraction f has reacted, so that a fraction $1-f$ remains, is $(1-f)p_{A,0}$. Data are given for two points, $f_1 = 0.100$ and $f_2 = 0.200$. Denoting the rates at these points by v_1 and v_2 and using the expression $\log v = \log k_r + a \log p_A$ from above gives the equations

$$\log v_1 = \log k_r + a \log[(1-f_1)p_{A,0}] \qquad \log v_2 = \log k_r + a \log[(1-f_2)p_{A,0}]$$

Subtracting the second equation from the first gives

$$\log v_1 - \log v_2 = a \log[(1-f_1)p_{A,0}] - a \log[(1-f_2)p_{A,0}]$$

Hence

$$\log\left(\frac{v_1}{v_2}\right) = a \log\left(\frac{(1-f_1)p_{A,0}}{(1-f_2)p_{A,0}}\right) = a \log\left(\frac{1-f_1}{1-f_2}\right)$$

where the property of logarithms $\log x - \log y = \log(x/y)$ is used and the factor of $p_{A,0}$ is cancelled.

Rearranging for a gives

$$a = \frac{\log(v_1/v_2)}{\log[(1-f_1)/(1-f_2)]} = \frac{\log[(9.71\ \text{Pa s}^{-1})/(7.67\ \text{Pa s}^{-1})]}{\log[(1-0.100)/(1-0.200)]} = \boxed{2.00}$$

Solutions to problems

P17A.1 The rate law is assumed to take the form $v_0 = k_r[C_6H_{12}O_6]^a$ where v_0 is the initial rate and a is the order with respect to glucose. Taking (common) logarithms gives

$$\log v_0 = \log k_r + \log[C_6H_{12}O_6]^a = \log k_r + a \log[C_6H_{12}O_6]$$

where the properties of logarithms $\log(xy) = \log x + \log y$ and $\log x^a = a \log x$ are used.

This expression implies that a graph of $\log v_0$ against $\log[C_6H_{12}O_6]$ will be a straight line of slope a and intercept $\log k_r$. The data are plotted in Fig. 17.1.

$[C_6H_{12}O_6]$ /mol dm^{-3}	v_0 /mol dm^{-3} s^{-1}	$\log([C_6H_{12}O_6]$ /mol dm$^{-3})$	$\log(v_0$ /mol dm^{-3} s$^{-1})$
1.00×10^{-3}	5.0	−3.000	0.699
1.54×10^{-3}	7.6	−2.812	0.881
3.12×10^{-3}	15.5	−2.506	1.190
4.02×10^{-3}	20.0	−2.396	1.301

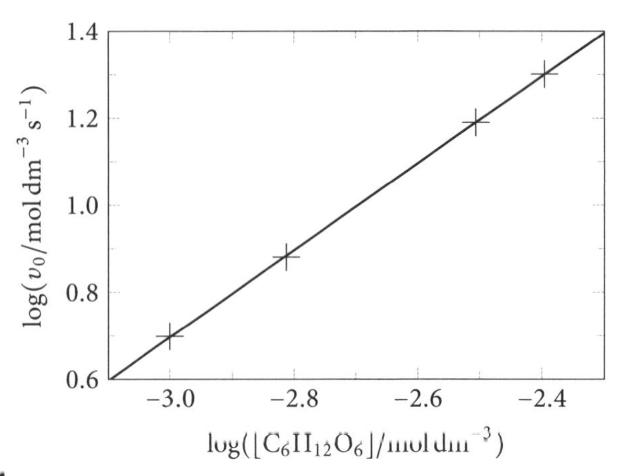

Figure 17.1

The data fall on a good straight line, the equation for which is

$$\log(v_0/\text{mol dm}^{-3}\,\text{s}^{-1}) = 1.00 \times \log([\text{CH}_6\text{H}_{12}\text{O}_6]/\text{mol dm}^{-3}) + 3.692$$

(a) Identifying the order a with the slope gives $a = 1.00$; that is, the reaction is $\boxed{\text{first order}}$ in glucose.

(b) The intercept at $\log([\text{C}_6\text{H}_{12}\text{O}_6]/\text{mol dm}^{-3}) = 0$ is $\log(v_0/\text{mol dm}^{-3}\,\text{s}^{-1}) = 3.692$, which corresponds to $v_0 = 4.92\times10^3\ \text{mol dm}^{-3}\,\text{s}^{-1}$ when $[\text{C}_6\text{H}_{12}\text{O}_6] = 1\ \text{mol dm}^{-3}$. Because $a = 1$, the rate law is $v_0 = k_r[\text{C}_6\text{H}_{12}\text{O}_6]^1$, which is rearranged to give

$$k_r = \frac{v_0}{[\text{C}_6\text{H}_{12}\text{O}_6]} = \frac{4.92\times10^3\ \text{mol dm}^{-3}\,\text{s}^{-1}}{1\ \text{mol dm}^{-3}} = \boxed{4.92\times10^3\ \text{s}^{-1}}$$

P17A.3 (a) Experiments 1 and 2 both have the same initial H_2 concentration, but experiment 2 has an ICl concentration twice that of experiment 1. Because the rate of experiment 2 is also twice that of experiment 1, it follows that the rate is proportional to [ICl] and hence that the reaction is first order in ICl.

Experiments 2 and 3 both have the same initial ICl concentration, but experiment 3 has an H_2 concentration three times that of experiment 2. Because the rate of experiment 3 is approximately three times that of experiment 2, it follows that the rate is proportional to $[H_2]$ and hence that the reaction is first order in H_2.

Therefore the rate law is $\boxed{v = k_r[\text{ICl}][\text{H}_2]}$.

(b) The rate law $v = k_r[\text{ICl}][\text{H}_2]$ implies that a graph of v_0 against $[\text{ICl}][\text{H}_2]$ should be a straight line of slope k_r and intercept zero. The data are plotted in Fig. 17.2.

Expt.	$[\text{ICl}]_0$ /mol dm^{-3}	$[\text{H}_2]_0$ /mol dm^{-3}	$[\text{ICl}]_0[\text{H}_2]_0$ /mol^2 dm^{-6}	v_0 /mol dm^{-3} s^{-1}
1	1.5×10^{-3}	1.5×10^{-3}	2.25×10^{-6}	3.7×10^{-7}
2	3.0×10^{-3}	1.5×10^{-3}	4.50×10^{-6}	7.4×10^{-7}
3	3.0×10^{-3}	4.5×10^{-3}	1.35×10^{-5}	2.2×10^{-6}

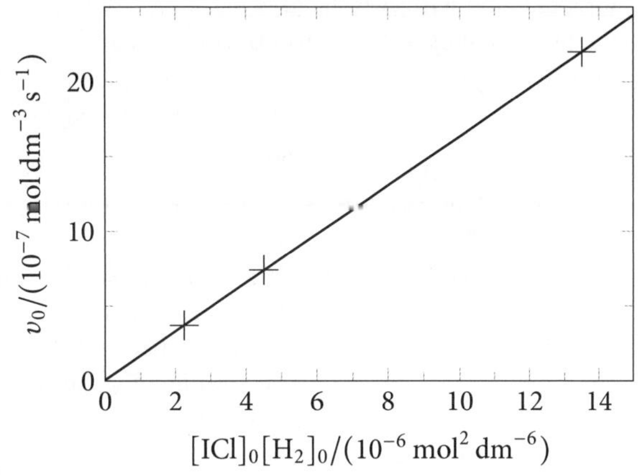

Figure 17.2

The data lie on a good straight line, the equation of which is

$$v_0/\text{mol dm}^{-3}\,\text{s}^{-1} = 0.163 \times \{[\text{ICl}]_0[\text{H}_2]_0/(\text{mol}^2\,\text{dm}^{-6})\} + 6.19 \times 10^{-9}$$

Identifying the slope with k_r gives $\boxed{k_r = 0.16\ \text{dm}^3\,\text{mol}^{-1}\,\text{s}^{-1}}$.

(c) The initial reaction rate for experiment 4 is predicted from the rate law

$$\begin{aligned} v_0 &= k_r[\text{ICl}][\text{H}_2] \\ &= (0.163...\ \text{dm}^3\,\text{mol}^{-1}\,\text{s}^{-1}) \times (4.7 \times 10^{-3}\ \text{mol dm}^{-3}) \\ &\quad \times (2.7 \times 10^{-3}\ \text{mol dm}^{-3}) = \boxed{2.1 \times 10^{-6}\ \text{mol dm}^{-3}\,\text{s}^{-1}} \end{aligned}$$

17B Integrated rate laws

Answers to discussion questions

D17B.1 The determination of a rate law is simplified by the isolation method in which the concentrations of all the reactants except one are in large excess. If B is in large excess, for example, then to a good approximation its concentration is constant throughout the reaction. Although the true rate law might be $v = k_r[\text{A}][\text{B}]$, we can approximate [B] by $[\text{B}]_0$ and write

$$v = k_r'[\text{A}] \quad \text{where} \quad k_r' = k_r[\text{B}]_0$$

which has the form of a first-order rate law. Because the true rate law has been forced into first-order form by assuming that the concentration of B is constant, it is called a pseudofirst-order rate law. The dependence of the rate on the concentration of each of the reactants may be found by isolating them in turn (by having all the other substances present in large excess), and so constructing the overall rate law.

In the method of initial rates, which is often used in conjunction with the isolation method, the rate is measured at the beginning of the reaction for several different initial concentrations of reactants. Suppose that the rate law for a reaction with A isolated is $v = k_r[A]^a$; then its initial rate, v_0, is given by the initial values of the concentration of A and is written $v_0 = k_r[A]_0^a$. Taking logarithms gives

$$\log v_0 = \log k_r + a \log[A]_0$$

For a series of initial concentrations, a plot of the logarithms of the initial rates against the logarithms of the initial concentrations of A should be a straight lime with slope a.

The method of initial rates might not reveal the full rate law, for the products may participate in the reaction and affect the rate. For example, products participate in the synthesis of HBr, where the full rate law depends on the concentration of HBr. To avoid this difficulty, the rate law should be fitted to the data throughout the reaction. The fitting may be done, in simple cases at least, by using a proposed rate law to predict the concentration of any component at any time, and comparing it with the data.

Because rate laws are differential equations, they must be integrated in order to find the concentrations as a function of time. Even the most complex rate laws may be integrated numerically. However, in a number of simple cases analytical solutions are easily obtained and prove to be very useful. These are summarized in Table 17B.3 on page 735. Experimental data can be tested against an assumed rate law by manipulating the integrated rate law into a form which will give a straight line plot. If the data do indeed fall on a good straight line, then the data are consistent with the assumed rate law.

D17B.3 By comparison with Table 17B.3 it is seen that

(a) Zero order: $d[A]/dt = -k_r$

(b) First order: $d[A]/dt = -k_r[A]$

(c) Second order: $d[A]/dt = -k_r[A]^2$

Solutions to exercises

E17B.1(a) (i) The integrated rate law for a zeroth-order reaction is given by [17B.1–*731*], $[A] = [A]_0 - k_r t$ where in this case A is NH_3. If concentrations are expressed in terms of partial pressures, this becomes $p_{NH_3} = p_{NH_3,0} - k_r t$. Rearranging for k_r and using $p_{NH_3} = 10$ kPa when $t = 770s$ with $p_{NH_3,0} =$ 21 kPa gives

$$k_r = \frac{p_{NH_3,0} - p_{NH_3}}{t} = \frac{(21 \times 10^3 \text{ Pa}) - (10 \times 10^3 \text{ Pa})}{770 \text{ s}} = 14.2... \text{ Pa s}^{-1}$$

$$= \boxed{14 \text{ Pa s}^{-1}}$$

(ii) When all the ammonia has been consumed, $p_{NH_3} = 0$. Rearranging the rate law for t gives

$$t = \frac{p_{NH_3,0} - p_{NH_3}}{k_r} = \frac{(21 \times 10^3\ \text{Pa}) - 0}{14.2...\ \text{Pa s}^{-1}} = \boxed{1.5 \times 10^3\ \text{s}}$$

E17B.2(a) The fact that the two half-lives are not the same establishes that the reaction is not first-order because, as explained in Section 17B.2 on page 731, a first-order reaction has a constant half-life. For orders $n \neq 1$ the half-life is given by [17B.6–*734*], $t_{1/2} = (2^{n-1} - 1)/[(n-1)k_r[\text{A}]_0^{n-1}]$. Denoting the two measurements by $t_{1/2,\text{i}}$ and $t_{1/2,\text{ii}}$ and expressing concentration in terms of partial pressure gives the two equations

$$t_{1/2,\text{i}} = \frac{2^{n-1} - 1}{(n-1)k_r p_{\text{A,i}}^{n-1}} \qquad t_{1/2,\text{ii}} = \frac{2^{n-1} - 1}{(n-1)k_r p_{\text{A,ii}}^{n-1}}$$

The second equation is divided by the first to give

$$\frac{t_{1/2,\text{ii}}}{t_{1/2,\text{i}}} = \left(\frac{p_{\text{A,i}}}{p_{\text{A,ii}}}\right)^{n-1} \quad \text{hence} \quad \log\left(\frac{t_{1/2,\text{ii}}}{t_{1/2,\text{i}}}\right) = (n-1)\log\left(\frac{p_{\text{A,i}}}{p_{\text{A,ii}}}\right)$$

where $\log x^a = a \log x$ is used. Rearranging for n gives

$$n = \frac{\log\left(t_{1/2,\text{ii}}/t_{1/2,\text{i}}\right)}{\log\left(p_{\text{A,i}}/p_{\text{A,ii}}\right)} + 1 = \frac{\log[(880\ \text{s})/(410\ \text{s})]}{\log[(363\ \text{Torr})/(169\ \text{Torr})]} + 1 = 2.00$$

Therefore the reaction is $\boxed{\text{second-order}}$.

E17B.3(a) For the reaction $2N_2O_5(g) \rightarrow 4NO_2(g) + O_2(g)$ the rate, as given by [17A.3b–*726*], $v = (1/\nu_J)(d[J]/dt)$, is

$$v = \frac{1}{-2}\frac{dp_{N_2O_5}}{dt}$$

where concentrations are expressed in terms of partial pressures. It is given that the reaction is first-order in N_2O_5, so $v = k_r p_{N_2O_5}$. Combining this with the above expression for v gives

$$\frac{1}{-2}\frac{dp_{N_2O_5}}{dt} = k_r p_{N_2O_5} \quad \text{hence} \quad \frac{dp_{N_2O_5}}{dt} = -2k_r p_{N_2O_5}$$

This has the same form, except with $2k_r$ instead of k_r, as [17B.2a–*731*], $(d[\text{A}]/dt) = -k_r[\text{A}]$, for which it is shown in Section 17B.2 on page 731 that the half-life and the integrated rate law are

$$t_{1/2} = \frac{\ln 2}{k_r} \qquad [\text{A}] = [\text{A}]_0 e^{-k_r t}$$

The expressions for the reaction in question are analogous, but with k_r replaced by $2k_r$.

$$t_{1/2} = \frac{\ln 2}{2k_r} \qquad p_{N_2O_5} = (p_{N_2O_5,0})e^{-2k_r t}$$

The half-life is

$$t_{1/2} = \frac{\ln 2}{2k_r} = \frac{\ln 2}{2 \times 3.38 \times 10^{-5}\ \mathrm{s}^{-1}} = \boxed{1.03 \times 10^4\ \mathrm{s}}$$

The partial pressures at the specified times are calculated from the above integrated form of the rate law. Hence

$$t = 50\ \mathrm{s} \quad p_{N_2O_5} = (500\ \mathrm{Torr}) \times e^{-2\times(3.38\times10^{-5}\ \mathrm{s}^{-1})\times(50\ \mathrm{s})} = \boxed{489\ \mathrm{Torr}}$$

$$t = 20\ \mathrm{min} \quad p_{N_2O_5} = (500\ \mathrm{Torr}) \times e^{-2\times(3.38\times10^{-5}\ \mathrm{s}^{-1})\times([20\times60]\ \mathrm{s})} = \boxed{461\ \mathrm{Torr}}$$

E17B.4(a) The reaction is of the form A + B → products. Assuming that it has rate law $v = k_r[A][B]$, the integrated rate law is given by [17B.7b–734]

$$\ln \frac{[B]/[B]_0}{[A]/[A]_0} = ([B]_0 - [A]_0)k_r t$$

Suppose that after time t the concentration of A has fallen by an amount x so that $[A] = [A]_0 - x$. Because of the stoichiometry of the reaction the concentration of B must fall by the same amount, so $[B] = [B]_0 - x$. Therefore

$$\ln \frac{([B]_0 - x)/[B]_0}{([A]_0 - x)/[A]_0} = ([B]_0 - [A]_0)k_r t$$

Hence

$$\frac{([B]_0 - x)[A]_0}{([A]_0 - x)[B]_0} = e^{([B]_0-[A]_0)k_r t}$$

Rearranging gives

$$[B]_0[A]_0 - x[A]_0 = [B]_0[A]_0 e^{([B]_0-[A]_0)k_r t} - x[B]_0 e^{([B]_0-[A]_0)k_r t}$$

Hence

$$x = \frac{[B]_0[A]_0\left(e^{([B]_0-[A]_0)k_r t} - 1\right)}{[B]_0 e^{([B]_0-[A]_0)k_r t} - [A]_0} = \frac{[B]_0[A]_0(e^{\lambda} - 1)}{[B]_0 e^{\lambda} - [A]_0}$$

where $\lambda = ([B]_0 - [A]_0)k_r t$. Taking A and B as OH^- and $CH_3COOC_2H_5$ respectively, the concentrations at the specified times are

For $t = 20$ s

$$\begin{aligned}\lambda &= ([B]_0 - [A]_0)k_r t \\ &= \left[(0.110\ \mathrm{mol\,dm^{-3}}) - (0.060\ \mathrm{mol\,dm^{-3}})\right] \times (0.11\ \mathrm{dm^3\,mol^{-1}\,s^{-1}}) \times (20\ \mathrm{s}) \\ &= 0.11\end{aligned}$$

$$\begin{aligned}x &= \frac{[B]_0[A]_0(e^{\lambda} - 1)}{[B]_0 e^{\lambda} - [A]_0} = \frac{(0.110\ \mathrm{mol\,dm^{-3}}) \times (0.060\ \mathrm{mol\,dm^{-3}}) \times (e^{0.11} - 1)}{(0.110\ \mathrm{mol\,dm^{-3}}) \times e^{0.11} - (0.060\ \mathrm{mol\,dm^{-3}})} \\ &= 0.0122...\ \mathrm{mol\,dm^{-3}}\end{aligned}$$

Hence the concentration of ester is

$$[\mathrm{B}] = [\mathrm{B}]_0 - x$$
$$= (0.110\ \mathrm{mol\,dm^{-3}}) - (0.0122...\ \mathrm{mol\,dm^{-3}}) = \boxed{0.0978\ \mathrm{mol\,dm^{-3}}}$$

For $t = 15$ min

$$\lambda = ([\mathrm{B}]_0 - [\mathrm{A}]_0)k_r t$$
$$= \left[(0.110\ \mathrm{mol\,dm^{-3}}) - (0.060\ \mathrm{mol\,dm^{-3}})\right] \times (0.11\ \mathrm{dm^3\,mol^{-1}\,s^{-1}})$$
$$\times ([15 \times 60]\ \mathrm{s}) = 4.95...$$

$$x = \frac{[\mathrm{B}]_0[\mathrm{A}]_0(\mathrm{e}^{\lambda} - 1)}{[\mathrm{B}]_0\mathrm{e}^{\lambda} - [\mathrm{A}]_0} = \frac{(0.110\ \mathrm{mol\,dm^{-3}}) \times (0.060\ \mathrm{mol\,dm^{-3}}) \times (\mathrm{e}^{4.95...} - 1)}{(0.110\ \mathrm{mol\,dm^{-3}}) \times \mathrm{e}^{4.95...}) - (0.060\ \mathrm{mol\,dm^{-3}})}$$
$$= 0.0598...\ \mathrm{mol\,dm^{-3}}$$

Hence the concentration of ester is

$$[\mathrm{B}] = [\mathrm{B}]_0 - x$$
$$= (0.110\ \mathrm{mol\,dm^{-3}}) - (0.0598...\ \mathrm{mol\,dm^{-3}}) = \boxed{0.0502\ \mathrm{mol\,dm^{-3}}}$$

E17B.5(a) Using [17A.3b–*726*], $v = (1/\nu_J)(\mathrm{d}[\mathrm{J}]\mathrm{d}t)$, the rate of the reaction $2\mathrm{A} \rightarrow \mathrm{P}$ is $v = -\frac{1}{2}(\mathrm{d}[\mathrm{A}]/\mathrm{d}t)$. Combining this with the rate law $v = k_r[\mathrm{A}]^2$ gives

$$-\frac{1}{2}\frac{\mathrm{d}[\mathrm{A}]}{\mathrm{d}t} = k_r[\mathrm{A}]^2 \quad \text{hence} \quad \frac{\mathrm{d}[\mathrm{A}]}{\mathrm{d}t} = -2k_r[\mathrm{A}]^2$$

This is essentially the same as [17B.4a–*733*], $\mathrm{d}[\mathrm{A}]/\mathrm{d}t = -k_r[\mathrm{A}]^2$ except with k_r replaced by $2k_r$. The integrated rate law is therefore essentially the same as that for [17B.4a–*733*], that is, [17B.4b–*733*] $1/[\mathrm{A}] - 1/[\mathrm{A}]_0 = k_r t$, except with k_r replaced by $2k_r$. Hence for the reaction in question

$$\frac{1}{[\mathrm{A}]} - \frac{1}{[\mathrm{A}]_0} = 2k_r t$$

Rearranging for t gives

$$t = \frac{1}{2k_r}\left(\frac{1}{[\mathrm{A}]} - \frac{1}{[\mathrm{A}]_0}\right)$$
$$= \frac{1}{2 \times (4.30 \times 10^{-4}\ \mathrm{dm^3\,mol^{-1}\,s^{-1}})} \times \left(\frac{1}{0.010\ \mathrm{mol\,dm^{-3}}} - \frac{1}{0.210\ \mathrm{mol\,dm^{-3}}}\right)$$
$$= \boxed{1.1 \times 10^5\ \mathrm{s}} \text{ or } 1.3 \text{ days}$$

E17B.6(a) The integrated rate law for a second-order reaction of the form $\mathrm{A} + \mathrm{B} \rightarrow \mathrm{P}$ is given by [17B.7b–*734*],

$$\ln\frac{[\mathrm{B}]/[\mathrm{B}]_0}{[\mathrm{A}]/[\mathrm{A}]_0} = ([\mathrm{B}]_0 - [\mathrm{A}]_0)\,k_r t$$

(i) In 1 hour the concentration of B falls from 0.060 mol dm^{-3} to 0.030 mol dm^{-3}, so the change in the concentration of B in this time period is –0.030 mol dm^{-3}. It follows from the reaction stoichiometry that the concentration of A must fall by the same amount, so the concentration of A after 1 hour is

$$[\mathrm{A}] = \overbrace{(0.080\ \mathrm{mol\,dm^{-3}})}^{[\mathrm{A}]_0} - (0.030\ \mathrm{mol\,dm^{-3}}) = 0.050\ \mathrm{mol\,dm^{-3}}$$

The rate constant is then found by rearranging the integrated rate equation for k_r and using the values of [A] and [B] at 1 hour, which corresponds to $1\ \mathrm{h} \times (60^2\ \mathrm{s\,h^{-1}}) = 3600\ \mathrm{s}$.

$$k_r = \frac{1}{([\mathrm{B}]_0 - [\mathrm{A}]_0)t} \ln \frac{[\mathrm{B}]/[\mathrm{B}]_0}{[\mathrm{A}]/[\mathrm{A}]_0}$$

$$= \frac{1}{[(0.060\ \mathrm{mol\,dm^{-3}}) - (0.080\ \mathrm{mol\,dm^{-3}})] \times (3600\ \mathrm{s})}$$

$$\times \ln\left(\frac{(0.030\ \mathrm{mol\,dm^{-3}})/(0.060\ \mathrm{mol\,dm^{-3}})}{(0.050\ \mathrm{mol\,dm^{-3}})/(0.080\ \mathrm{mol\,dm^{-3}})}\right)$$

$$= 3.09... \times 10^{-3}\ \mathrm{dm^3\,mol^{-1}\,s^{-1}} = \boxed{3.1 \times 10^{-3}\ \mathrm{dm^3\,mol^{-1}\,s^{-1}}}$$

(ii) The half-life of a particular reactant is the time taken for the concentration of that reactant to fall to half its initial value. The half-life of B is $\boxed{\text{1 hour}}$ because it is given in the question that after 1 hour the concentration of B had fallen from 0.060 mol dm^{-3} to 0.030 mol dm^{-3}, half the original value.

The initial concentration of A is 0.080 mol dm^{-3} so the half-life is the time at which the concentration of A has dropped by 0.040 mol dm^{-3} to 0.040 mol dm^{-3}. It follows from the stoichiometry of the reaction that the concentration of B must also fall by 0.040 mol dm^{-3} during this period, so the concentration of B will be

$$[\mathrm{B}] = 0.060\ \mathrm{mol\,dm^{-3}} - 0.040\ \mathrm{mol\,dm^{-3}} = 0.020\ \mathrm{mol\,dm^{-3}}$$

Rearranging the integrated rate equation then gives

$$t = \frac{1}{k_r}\frac{1}{([\mathrm{B}]_0 - [\mathrm{A}]_0)} \ln\left(\frac{[\mathrm{B}]/[\mathrm{B}]_0}{[\mathrm{A}]/[\mathrm{A}]_0}\right)$$

$$= \frac{1}{3.09... \times 10^{-3}\ \mathrm{dm^3\,mol^{-1}\,s^{-1}}} \times \frac{1}{(0.060\ \mathrm{mol\,dm^{-3}}) - (0.080\ \mathrm{mol\,dm^{-3}})}$$

$$\times \ln\left(\frac{(0.020\ \mathrm{mol\,dm^{-3}})/(0.060\ \mathrm{mol\,dm^{-3}})}{(0.040\ \mathrm{mol\,dm^{-3}})/(0.080\ \mathrm{mol\,dm^{-3}})}\right)$$

$$= \boxed{6.5 \times 10^3\ \mathrm{s}} \text{ or 1.8 hours}$$

Solutions to problems

P17B.1 The concentration of B is given in the question as

$$[B] = n[A]_0(1 - e^{-k_r t}) \quad \text{hence} \quad [B]/[A]_0 = n(1 - e^{-k_r t})$$

The concentration of A for a first-order reaction is given by [17B.2b–*732*],

$$[A] = [A]_0 e^{-k_r t} \quad \text{hence} \quad [A]/[A]_0 = e^{-k_r t}$$

These expressions are plotted in Fig. 17.3

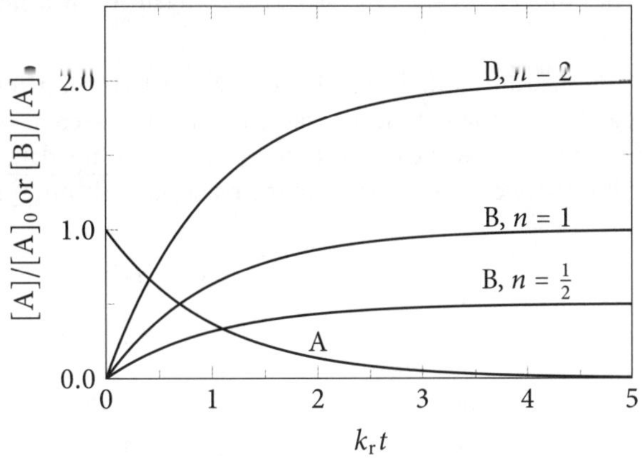

Figure 17.3

P17B.3 The first task is to convert the masses of urea into concentrations of ammonium cyanate A. Because the only fate of the ammonium cyanate is to be converted into urea, the mass of ammonium cyanate m_A remaining at any given time is equal to the original mass of ammonium cyanate minus the mass of urea, $m_A = m_{A,0} - m_{urea}$. In this case $m_{urea} = 22.9$ g. Dividing by the molar mass of the ammonium cyanate, $M_A = 60.0616 \text{ g mol}^{-1}$, gives the amount of A in moles, and division by the volume of the solution then gives the concentration in mol dm^{-3}.

t/min	t/s	m_{urea}/g	m_A/g	[A]/mol dm^{-3}
0	0	0.0	22.9	0.381
20	1 200	7.0	15.9	0.265
50	3 000	12.1	10.8	0.180
65	3 900	13.8	9.1	0.152
150	9 000	17.7	5.2	0.087

The order is determined by testing the fit of the data to integrated rate law expressions. A zeroth-order reaction of the form A → P has an integrated rate law given by [17B.1–*731*], $[A] = [A]_0 - k_r t$, so if the reaction is zeroth-order then a plot of $[A]$ against t will be a straight line of slope $-k_r$. On the other hand, a first-order reaction has an integrated rate law given by [17B.2b–732], $\ln([A]/[A]_0) = -k_r t$, so if the reaction is first-order then a plot of $\ln[A]/[A]_0$ against t will be a straight line of slope $-k_r$. Finally, if the order is $n \geq 2$ the integrated rate law is given in Table 17B.3 on page 735 as

$$k_r t = \frac{1}{n-1}\left(\frac{1}{([A]_0 - [P])^{n-1}} - \frac{1}{[A]_0^{n-1}}\right) \text{ hence } \frac{1}{[A]^{n-1}} = (n-1)k_r t + \frac{1}{[A]_0^{n-1}}$$

where to obtain the second expression the relation $[P] = [A]_0 - [A]$ is substituted and the equation rearranged. This expression implies that if the reaction has order $n \geq 2$ a plot of $1/[A]^{n-1}$ against t will be a straight line of slope $(n-1)k_r$.

The data are plotted assuming zeroth , first , second , and third order in Fig. 17.4 and using the data in the table below. The second-order plot shows a good straight line, while the other three plots show the data lying on distinct curves. It is therefore concluded that the reaction is second-order.

t/s	$[A]$ /mol dm^{-3}	$\ln\frac{[A]}{[A]_0}$	$1/[A]$ /dm^3 mol^{-1}	$1/[A]^2$ /dm^6 mol^{-2}
0	0.381	0.000	2.623	6.879
1 200	0.265	−0.365	3.777	14.269
3 000	0.180	−0.752	5.561	30.928
3 900	0.152	−0.923	6.600	43.562
9 000	0.087	−1.482	11.550	133.410

In an alternative approach visual examination of the concentration data indicates that the half-life is not constant, and comparison of the 0–50 min data and the 50-150 min data suggests that the second half-life is approximately double the first. That is, the half-life starting from half the initial concentration is about twice the initial half-life, suggesting that the half-life is inversely proportional to the initial concentration. According to [17B.6–734] the half-life of a reaction with order $n > 1$ is given by $t_{1/2} \propto 1/[A]_0^{n-1}$, so this result suggests that the reaction may have $n = 2$ as this gives $t_{1/2} \propto 1/[A]_0$. Because it is suspected on this basis that the reaction may be second-order, only the second-order plot from Fig. 17.4 is made, and the fact that it gives a good straight line confirms that the reaction is indeed second-order.

The equation of the line in the second-order plot is

$$[A]^{-1}/\text{dm}^3\,\text{mol}^{-1} = 9.95 \times 10^{-4} \times (t/\text{s}) + 2.62$$

Identifying the slope with $(n-1)k_r$ as discussed above, and noting that $n = 2$ for a second-order reaction, gives $\boxed{k_r = 9.95 \times 10^{-4}\ \text{dm}^3\,\text{mol}^{-1}\,\text{s}^{-1}}$.

The concentration of ammonium cyanate left after 300 min, which is $(300\text{min}) \times (60\text{s}/1\text{ min}) = 18000$ s, is calculated using the integrated rate law for a second-

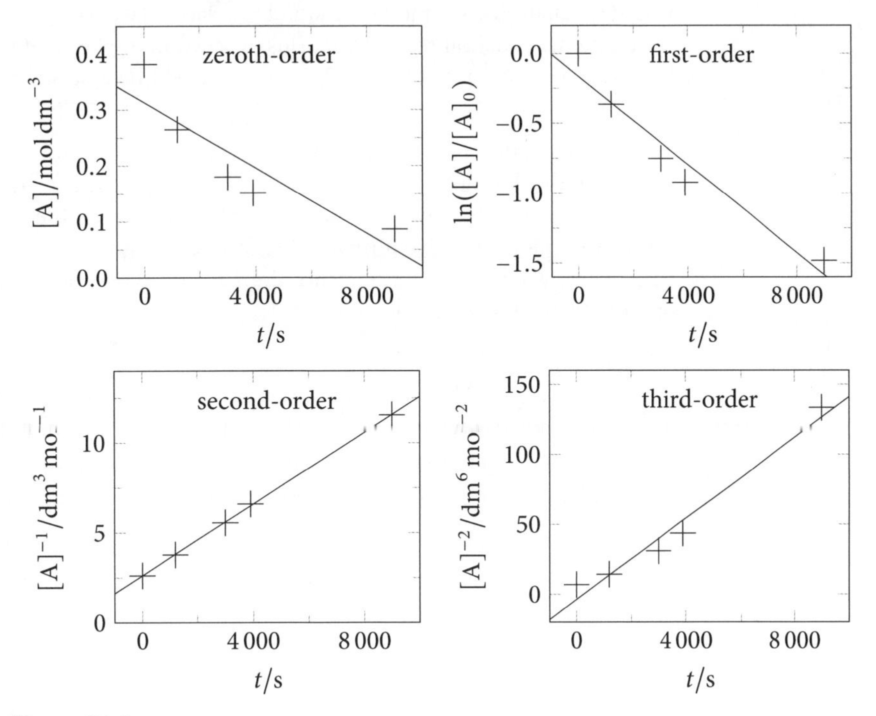

Figure 17.4

order reaction [17B.4b–*733*], $[A] = [A]_0/(1 + k_r t[A]_0)$

$$[A] = \frac{(0.381...\ \mathrm{mol\,dm^{-3}})}{1 + (9.95... \times 10^{-4}\ \mathrm{dm^3\,mol^{-1}\,s^{-1}}) \times (18000\,\mathrm{s}) \times (0.381...\ \mathrm{mol\,dm^{-3}})}$$
$$= 0.0487...\ \mathrm{mol\,dm^{-3}}$$

Multiplication by the volume gives the amount in moles, and multiplication of this by the molar mass gives the mass of A in g.

$$m_A = MV[A] = (60.0616\,\mathrm{g\,mol^{-1}}) \times (1.00\,\mathrm{dm}) \times (0.0487...\ \mathrm{mol\,dm^{-3}}) = \boxed{2.9\,\mathrm{g}}$$

P17B.5 The order is determined by testing the fit of the data to integrated rate law expressions. A zeroth-order reaction of the form A → P has an integrated rate law given by [17B.1–*731*], $[A] = [A]_0 - k_r t$, so if the reaction is zeroth-order then a plot of $[A]$ against t will be a straight line of slope $-k_r$. In this case, A is the organic nitrile. On the other hand, a first-order reaction has an integrated rate law given by [17B.2b–*732*], $\ln([A]/[A]_0) = -k_r t$, so if the reaction is first-order then a plot of $\ln [A]/[A]_0$ against t will be a straight line of slope $-k_r$. Finally, if the order is $n \geq 2$ the integrated rate law is given in Table 17B.3 on page 735 as

$$k_r t = \frac{1}{n-1}\left(\frac{1}{([A]_0 - [P])^{n-1}} - \frac{1}{[A]_0^{n-1}}\right) \text{ hence } \frac{1}{[A]^{n-1}} = (n-1)k_r t + \frac{1}{[A]_0^{n-1}}$$

where to obtain the second expression the relation $[P] = [A]_0 - [A]$ is substituted and the equation rearranged. This expression implies that if the reaction has order $n \geq 2$ a plot of $1/[A]^{n-1}$ against t will be a straight line of slope $(n-1)k_r$.

The data are plotted assuming zeroth-, first-, second-, and third-order in Fig. 17.5. The second-order plot shows the best fit to a straight line, so it is concluded that the reaction is likely to be **second-order**. However, the first-order and third-order plots also give a reasonable fit to a straight line, so experimental data over a wider range of concentrations would be needed to establish the order with greater confidence.

$t/10^3$ s	$[A]$ /mol dm^{-3}	$\ln \frac{[A]}{[A]_0}$	$1/[A]$ /dm^3 mol^{-1}	$1/[A]^2$ /dm^6 mol^{-2}
0	1.500 0	0.000	0.67	4.444×10^{-1}
2	1.260 0	−0.174	0.79	6.299×10^{-1}
4	1.070 0	−0.338	0.93	8.734×10^{-1}
6	0.920 0	−0.489	1.09	1.181×10^{0}
8	0.810 0	−0.616	1.23	1.524×10^{0}
10	0.720 0	−0.734	1.39	1.929×10^{0}
12	0.650 0	−0.836	1.54	2.367×10^{0}

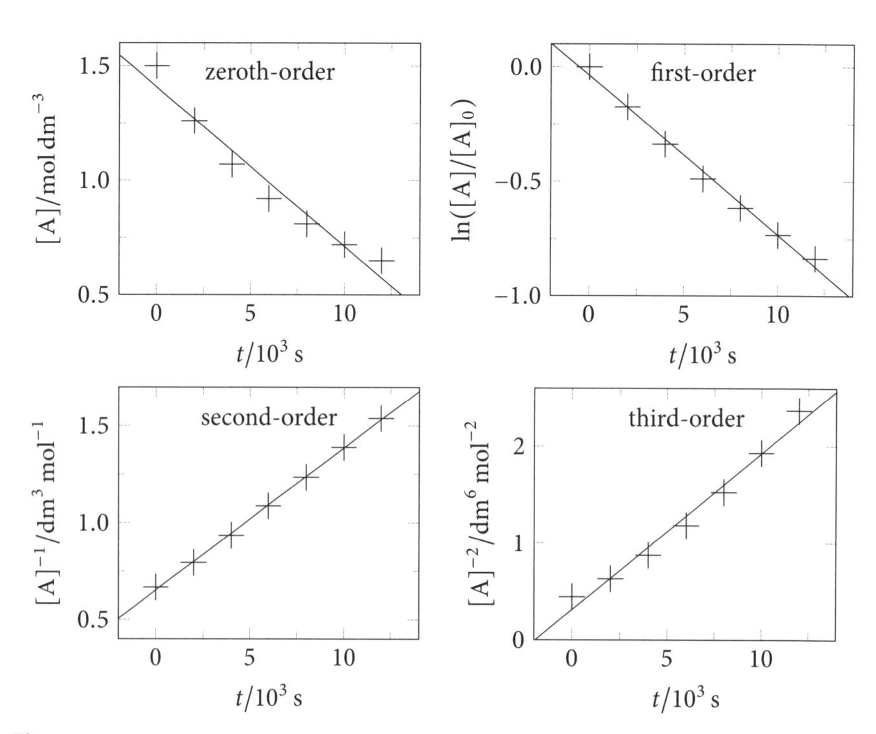

Figure 17.5

The equation of the line in the second-order plot is

$$[\mathrm{A}]^{-1}/\mathrm{dm^3\,mol^{-1}} = 7.33 \times 10^{-5} \times (t/\mathrm{s}) + 0.652$$

Identifying the slope with $(n-1)k_r$ as discussed above and noting that $n = 2$ for a second-order reaction gives $k_r = \boxed{7.33 \times 10^{-5}\ \mathrm{dm^3\,mol^{-1}\,s^{-1}}}$.

P17B.7 The order is determined by testing the fit of the data to integrated rate law expressions. A first-order reaction has an integrated rate law given by [17B.2b–732], $\ln([\mathrm{A}]/[\mathrm{A}]_0) = -k_r t$, or $\ln[\mathrm{A}] - \ln[\mathrm{A}]_0 = -k_r t$, so if the reaction is first-order then a plot of $\ln[\mathrm{A}]$ against t will be a straight line of slope $-k_r$. On the other hand, a second-order reaction has an integrated rate law given by [17B.4b–733], $1/[\mathrm{A}] - 1/[\mathrm{A}]_0 = k_r t$, which implies that if the reaction is second-order then a plot of $1/[\mathrm{A}]$ against t will be a straight line of slope k_r.

The data are plotted in Fig. 17.6. The first-order plot shows a good straight line while in the second-order plot the data lie on a curve. It is therefore concluded that the reaction is $\boxed{\text{first-order}}$.

t/min	$c/\mathrm{ng\,cm^{-3}}$	$\ln(c/\mathrm{ng\,cm^{-3}})$	$1/(c/\mathrm{ng\,cm^{-3}})$
30	699	6.550	1.43×10^{-3}
60	622	6.433	1.61×10^{-3}
120	413	6.023	2.42×10^{-3}
150	292	5.677	3.42×10^{-3}
240	152	5.024	6.58×10^{-3}
360	60	4.094	1.67×10^{-2}
480	24	3.178	4.17×10^{-2}

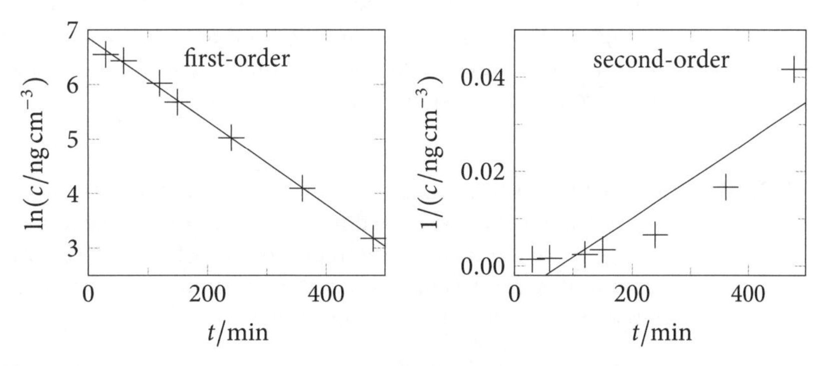

Figure 17.6

The equation of the line in the first-order plot is

$$\ln(c/\mathrm{ng\,cm^{-3}}) = -7.65 \times 10^{-3} \times (t/\mathrm{min}) + 6.86$$

Identifying the slope with $-k_r$ as discussed above gives the first-order rate constant as $k_r = \boxed{7.65 \times 10^{-3}\ \text{min}^{-1}}$.

The half-life of a first-order reaction is given by [17B.3–732], $t_{1/2} = \ln 2/k_r$.

$$t_{1/2} = \frac{\ln 2}{k_r} = \frac{\ln 2}{7.65 \times 10^{-3}\ \text{min}^{-1}} = \boxed{91\ \text{min}}$$

P17B.9 The units of the rate constants show that both reactions are first-order, so their rate equations are assumed to be

$$v_1 = \frac{d[CH_4]}{dt} = k_1[CH_3COOH] \quad \text{and} \quad v_2 = \frac{d[CH_2CO]}{dt} = k_2[CH_3COOH]$$

where [17A.3b–726], $v = (1/\nu_J)(d[J]/dt)$, is used to express the rates in terms of rate of formation of CH_4 and CH_2CO. The ratio of the rate of ketene formation to the total rate of product formation is therefore

$$\frac{k_2[CH_3COOH]}{k_1[CH_3COOH] + k_2[CH_3COOH]} = \frac{k_2}{k_1 + k_2} = \frac{4.65\ \text{s}^{-1}}{(3.74\ \text{s}^{-1}) + (4.65\ \text{s}^{-1})}$$
$$= 0.554...$$

Because this ratio is independent of the CH_3COOH concentration it will be constant throughout the duration of the reaction and will be equal to the ratio of ketene formed to total product formed. The maximum possible yield of ketene is therefore 0.554... or $\boxed{55.4\,\%}$.

Similarly the ratio of the rate of formation of ketene and methane is

$$\frac{k_2[CH_3COOH]}{k_1[CH_3COOH]} = \frac{k_2}{k_1} = \frac{4.65\ \text{s}}{3.74\ \text{s}} = 1.24...$$

This ratio is independent of the CH_3COOH concentration so it will be constant throughout the duration of the reaction and equal to the ratio of the total ketene and methane formed up to any given time. Hence $[CH_2CO]/[CH_4] = 1.24...$ and is $\boxed{\text{constant}}$ over time.

P17B.11 The first task is to calculate the concentrations of the reactant A at each time. The stoichiometry of the reaction $2A \rightarrow B$ means that the initial concentration of A is twice the final concentration of B, $[A]_0 = 2[B]_\infty$. In addition, the amount of A that has reacted at any given time is equal to twice the amount of B that has been formed. It follows that

$$\overbrace{[A]_0 - [A]}^{\text{A that has reacted}} = 2[B] \quad \text{hence} \quad [A] = [A]_0 - 2[B]$$

Substituting $[A]_0 = 2[B]_\infty$ from above gives $[A] = 2([B]_0 - [B])$; this expression is used to calculate the concentration of $[A]$ at each of the times.

The order is determined by testing the fit of the data to integrated rate law expressions. If the rate law is $v = k_r[A]^n$, where n is the order to be determined, expressing v in terms of the rate of change of concentration of $[A]$ using [17A.3b–726], $v = (1/\nu_J)(d[J]/dt)$ gives

$$v = \frac{1}{-2}\frac{d[A]}{dt} = k_r[A]^n \quad \text{hence} \quad \frac{d[A]}{dt} = -2k_r[A]^n$$

Integrated rate laws are given in Table 17B.3 on page 735, but care is needed because these are for reactions of the form A → P but here the reaction is 2A → B.

For $n = 0$, Table 17B.3 on page 735 shows that a reaction A → P with rate law $v = d[P]/dt = k_r$ has integrated rate law $A = A_0 - k_r t$. To adapt this expression for the reaction in question, the rate law for the reaction in the table is first written as $d[A]/dt = -k_r$ using $d[P]/dt = -d[A]/dt$ for a reaction of the form A → P. This rate law matches that found above, $d[A]/dt = -2k_r[A]^n$, for $n = 0$ except that k_r is replaced by $2k_r$. The integrated rate law will therefore be the same except with k_r replaced by $2k_r$, that is, $[A] = [A]_0 - 2k_r t$. This expression implies that if the reaction is zeroth-order a plot of $[A]$ against t will give a straight line of slope $-2k_r$.

Similarly Table 17B.3 on page 735 gives the integrated rate law for a first-order reaction A → P with rate law $v = d[P]/dt = k_r[A]$ as $\ln([A]_0/[A]) = k_r t$, equivalent to [17B.2b–732], $\ln([A]/[A]_0) = -k_r t$. By the same reasoning as above the integrated rate law for the reaction will therefore be $\ln([A]/[A]_0) = -2k_r t$, implying that a plot of $\ln([A]/[A]_0)$ against t will give a straight line of slope $-2k_r$.

Finally, if the order is $n \geq 2$ the integrated rate law for a reaction A → P with rate law $v = d[P]/dt = k_r[A]^n$ is given in Table 17B.3 on page 735 as

$$k_r t = \frac{1}{n-1}\left(\frac{1}{([A]_0 - [P])^{n-1}} - \frac{1}{[A]_0^{n-1}}\right) \text{ hence } \frac{1}{[A]^{n-1}} = (n-1)k_r t + \frac{1}{[A]_0^{n-1}}$$

where to obtain the second expression the relation $[P] = [A]_0 - [A]$ is substituted and the equation rearranged. Adapting this expression for the reaction in question gives $1/[A]^{n-1} = 2(n-1)k_r t + 1/[A]_0^{n-1}$. This expression implies that if the reaction has order $n \geq 2$ a plot of $1/[A]^{n-1}$ against t will be a straight line of slope $2(n-1)k_r$.

The data are plotted assuming zeroth-, first-, second-, and third-order in Fig. 17.7. The first-order plot shows a good fit to a straight line, while the other plots are curved, so it is concluded that the reaction is first-order.

t/min	[B] /mol dm^{-3}	[A] /mol dm^{-3}	$\ln\dfrac{[A]}{[A]_0}$	1/[A] /dm^3 mol^{-1}	1/[A]^2 /dm^6 mol^{-2}
0	0.000	0.624	0.000	1.603	2.57
10	0.089	0.446	−0.336	2.242	5.03
20	0.153	0.318	−0.674	3.145	9.89
30	0.200	0.224	−1.025	4.464	19.93
40	0.230	0.164	−1.336	6.098	37.18
∞	0.312	0.000			

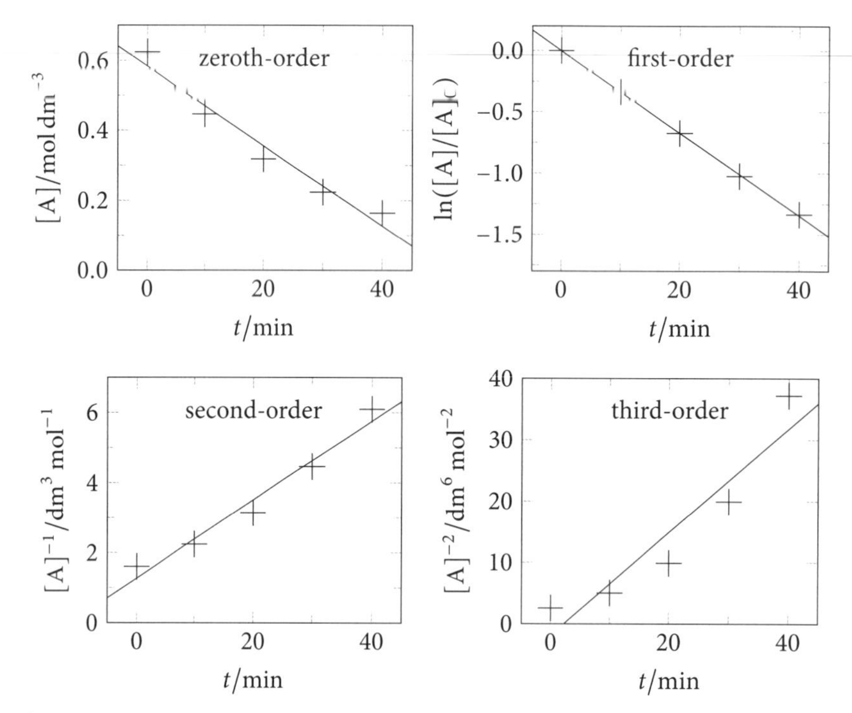

Figure 17.7

The equation of the line in the first-order plot is

$$\ln([A]/[A]_0) = -0.03361 \times (t/\text{min}) - 1.896 \times 10^{-3}$$

Identifying the slope with $-2k_r$ as discussed above gives the first-order rate constant as

$$k_r = -\tfrac{1}{2} \times (-0.03361\ \text{min}^{-1}) = \boxed{0.0168\ \text{min}^{-1}}$$

P17B.13 The order is determined by fitting the data to integrated rate laws. Table 17B.3 on page 735 gives integrated rate laws for the reaction A → P with rate law

$v = \mathrm{d}[\mathrm{P}]/\mathrm{d}t = k_r[\mathrm{A}]^n$ as

$$n = 0 \quad [\mathrm{A}] = [\mathrm{A}]_0 - k_r t \quad \text{hence} \quad k_r = ([\mathrm{A}]_0 - [\mathrm{A}])/t$$

$$n = 1 \quad k_r t = \ln\frac{[\mathrm{A}]_0}{[\mathrm{A}]} \quad \text{hence} \quad k_r = \frac{1}{t}\ln\frac{[\mathrm{A}]_0}{[\mathrm{A}]}$$

$$n \geq 2 \quad k_r t = \frac{1}{n-1}\left(\frac{1}{([\mathrm{A}]_0 - [\mathrm{P}])^{n-1}} - \frac{1}{[\mathrm{A}]_0^{n-1}}\right)$$

$$\text{hence} \quad k_r = \frac{1}{(n-1)t}\left(\frac{1}{[\mathrm{A}]^{n-1}} - \frac{1}{[\mathrm{A}]_0^{n-1}}\right)$$

where in the $n \geq 2$ case, $[\mathrm{A}] = [\mathrm{A}]_0 - [\mathrm{P}]$ is used.

These expressions imply that if the reaction is zeroth order the quantity $([\mathrm{A}]_0 - [\mathrm{A}])/t$ should be a constant, equal to k_r, while if it is first-order $[\ln([\mathrm{A}]_0/[\mathrm{A}])]/t$ should be constant, and if the order is $n \geq 2$ the quantity

$$([\mathrm{A}]^{-(n-1)} - [\mathrm{A}]_0^{n-1})/[(n-1)t]$$

should be constant. In this case A is cyclopropane and the concentrations are expressed in terms of partial pressures. Results assuming $n = 0, 1, 2$ are shown in the following table.

			$n = 0$	$n = 1$	$n = 2$
$p_{\mathrm{A},0}$ /Torr	t/s	p_A /Torr	$(p_{\mathrm{A},0} - p_\mathrm{A})/t$ /Torr s^{-1}	$[\ln(p_{\mathrm{A},0}/p_\mathrm{A})]/t$ /s^{-1}	$(1/p_\mathrm{A} - 1/p_{\mathrm{A},0})/t$ /Torr^{-1} s^{-1}
200	100	186	0.140	7.26×10^{-4}	3.76×10^{-6}
200	200	173	0.135	7.25×10^{-4}	3.90×10^{-6}
400	100	373	0.270	6.99×10^{-4}	1.81×10^{-6}
400	200	347	0.265	7.11×10^{-4}	1.91×10^{-6}
600	100	559	0.410	7.08×10^{-4}	1.22×10^{-6}
600	200	520	0.400	7.16×10^{-4}	1.28×10^{-6}

The values assuming $n = 1$ are approximately constant, while those assuming $n = 0$ or $n = 2$ are not. It is therefore concluded that the reaction is first-order. The average value of $[\ln(p_{\mathrm{A},0}/p_\mathrm{A})]/t$, and hence of k_r, from the table is $7.1 \times 10^{-4}\ \mathrm{s}^{-1}$.

P17B.15 A reaction of the form $\mathrm{A} \rightarrow \mathrm{P}$ that is nth order in A has rate law $v = k_r[\mathrm{A}]^n$. Combining this with [17A.3b–726], $v = (1/\nu_\mathrm{J})(\mathrm{d}[\mathrm{J}]/\mathrm{d}t)$ gives

$$\frac{1}{-1}\frac{\mathrm{d}[\mathrm{A}]}{\mathrm{d}t} = k_r[\mathrm{A}]^n \quad \text{hence} \quad -[\mathrm{A}]^{-n}\,\mathrm{d}[\mathrm{A}] = k_r\,\mathrm{d}t$$

Initially, at $t = 0$, the concentration of A is $[\mathrm{A}]_0$, and at a later time t it is $[\mathrm{A}]$. These are used as the limits of the integration to give

$$\overbrace{\int_{[\mathrm{A}]_0}^{[\mathrm{A}]} -[\mathrm{A}]^{-n}\,\mathrm{d}[\mathrm{A}]}^{\text{Integral A.1}} = \int_0^t k_\mathrm{r}\,\mathrm{d}t \quad \text{hence} \quad \frac{1}{n-1}[\mathrm{A}]^{-(n-1)}\Big|_{[\mathrm{A}]_0}^{[\mathrm{A}]} = k_\mathrm{r}t\Big|_0^t$$

$$\text{hence} \quad \frac{1}{n-1}\left(\frac{1}{[\mathrm{A}]^{n-1}} - \frac{1}{[\mathrm{A}]_0^{n-1}}\right) = k_\mathrm{r}t \qquad (\text{for } n \neq 1)$$

This integrated rate law is equivalent to that given in Table 17B.3 on page 735.

(a) When $t = t_{1/2}$, $[\mathrm{A}] = \frac{1}{2}[\mathrm{A}]_0$. Therefore, on dividing through by k_r, the above integrated rate law gives

$$t_{1/2} = \frac{1}{(n-1)k_\mathrm{r}}\left(\frac{1}{\left(\frac{1}{2}[\mathrm{A}]_0\right)^{n-1}} - \frac{1}{[\mathrm{A}]_0^{n-1}}\right) = \frac{1}{(n-1)k_\mathrm{r}}\left(\frac{2^{n-1}}{[\mathrm{A}]_0^{n-1}} - \frac{1}{[\mathrm{A}]_0^{n-1}}\right)$$

$$= \boxed{\frac{2^{n-1}-1}{(n-1)k_\mathrm{r}[\mathrm{A}]_0^{n-1}}}$$

(b) The time taken for the concentration of a substance to fall to one-third its initial value is denoted $t_{1/3}$. Thus, at $t = t_{1/3}$, $[\mathrm{A}] = \frac{1}{3}[\mathrm{A}]_0$

$$t_{1/3} = \frac{1}{(n-1)k_\mathrm{r}}\left(\frac{1}{\left(\frac{1}{3}[\mathrm{A}]_0\right)^{n-1}} - \frac{1}{[\mathrm{A}]_0^{n-1}}\right) = \frac{1}{(n-1)k_\mathrm{r}}\left(\frac{3^{n-1}}{[\mathrm{A}]_0^{n-1}} - \frac{1}{[\mathrm{A}]_0^{n-1}}\right)$$

$$= \boxed{\frac{3^{n-1}-1}{(n-1)k_\mathrm{r}[\mathrm{A}]_0^{n-1}}}$$

P17B.17 The stoichiometry of the reaction $2\mathrm{A} + \mathrm{B} \rightarrow \mathrm{P}$ implies that when the concentration of P has increased from 0 to x, the concentration of A has fallen to $[\mathrm{A}]_0 - 2x$ and the concentration of B has fallen to $[\mathrm{B}]_0 - x$. This is because each P that forms entails the disappearance of two A and one B. The rate law $v = \mathrm{d}[\mathrm{P}]/\mathrm{d}t = k_\mathrm{r}[\mathrm{A}]^2[\mathrm{B}]$ then becomes

$$\frac{\mathrm{d}[\mathrm{P}]}{\mathrm{d}t} = k_\mathrm{r}([\mathrm{A}]_0 - 2x)^2([\mathrm{B}]_0 - x) \quad \text{hence} \quad \frac{\mathrm{d}x}{\mathrm{d}t} = k_\mathrm{r}([\mathrm{A}]_0 - 2x)^2([\mathrm{B}] - x)$$

where to go to the second expression $[\mathrm{P}] = x$ is used, which implies that $\mathrm{d}[\mathrm{P}]\mathrm{d}t = \mathrm{d}x/\mathrm{d}t$. The expression is rearranged and the initial condition $x = 0$ when $t = 0$ is applied. This gives the integrations required as

$$\int_0^x \frac{1}{([\mathrm{A}]_0 - 2x)^2([\mathrm{B}]_0 - x)}\,\mathrm{d}x = \int_0^t k_\mathrm{r}\,\mathrm{d}t$$

The right-hand side evaluates to $k_\mathrm{r}t$. The left-hand side is evaluated below.

(a) If $[B]_0 = \frac{1}{2}[A]_0$ the left-hand side becomes

$$\int_0^x \frac{1}{([A]_0 - 2x)^2(\frac{1}{2}[A]_0 - x)}\,dx = \int_0^x \frac{1}{([A]_0 - 2x)^2 \times \frac{1}{2}([A]_0 - 2x)}\,dx$$

$$= \int_0^x 2([A]_0 - 2x)^{-3}\,dx = \frac{1}{2}([A]_0 - 2x)^{-2}\Big|_0^x = \frac{1}{2([A]_0 - 2x)^2} - \frac{1}{2[A]_0^2}$$

Combining this with the right-hand side from above gives the integrated rate law as

$$\boxed{\frac{1}{2([A]_0 - 2x)^2} - \frac{1}{2[A]_0^2} = k_r t}$$

(b) If $[B]_0 = [A]_0$ the left-hand side is integrated using the method of partial fractions described in *The chemist's toolkit* 30 in Topic 17B. The integrand is first written as

$$\frac{1}{([A]_0 - 2x)^2([A]_0 - x)} = \frac{A}{([A]_0 - 2x)^2} + \frac{B}{[A]_0 - 2x} + \frac{C}{[A]_0 - x}$$

where A, B, and C are constants to be found. This expression is multiplied through by $([A]_0 - 2x)^2([A]_0 - x)$ to give

$$1 = A([A]_0 - x) + B([A]_0 - 2x)([A]_0 - x) + C([A]_0 - 2x)^2$$

This expression must be true for all x, so the values of A, B and C are most conveniently found by substituting particular values of x.

$$\text{When } x = [A]_0 \quad 1 = C(-[A]_0)^2 \quad \text{hence} \quad C = 1/[A]_0^2$$

$$\text{When } x = \tfrac{1}{2}[A]_0 \quad 1 = A(\tfrac{1}{2}[A]_0) \quad \text{hence} \quad A = 2/[A]_0$$

$$\text{When } x = 0 \quad 1 = A[A]_0 + B[A]_0^2 + C[A]_0^2$$

$$= \frac{2}{[A]_0}[A]_0 + B[A]_0^2 + \frac{1}{[A]_0^2}[A]_0^2$$

$$\text{hence} \quad 1 = 3 + B[A]_0^2 \quad \text{hence} \quad B = -2/[A]_0^2$$

The required integral is therefore

$$\int_0^x \frac{1}{([A]_0 - 2x)^2([A]_0 - x)}\,dx$$

$$= \int_0^x \frac{2}{[A]_0^2([A]_0 - 2x)} - \frac{2}{[A]_0^2([A]_0 - 2x)} + \frac{1}{[A]_0^2([A]_0 - x)}\,dx$$

$$= \frac{1}{[A]_0([A]_0 - 2x)} + \frac{1}{[A]_0^2}\ln([A]_0 - 2x) - \frac{1}{[A]_0^2}\ln([A]_0 - x)\Big|_0^x$$

$$= \left(\frac{1}{[A]_0([A]_0 - 2x)} + \frac{1}{[A]_0^2}\ln([A]_0 - 2x) - \frac{1}{[A]_0^2}\ln([A]_0 - x)\right)$$

$$- \left(\frac{1}{[A]_0^2} + \frac{1}{[A]_0^2}\ln[A]_0 - \frac{1}{[A]_0^2}\ln[A]_0\right)$$

$$= \frac{1}{[A]_0([A]_0 - 2x)} + \frac{1}{[A]_0^2}\ln\frac{[A]_0 - 2x}{[A]_0 - x} - \frac{1}{[A]_0^2}$$

Combining this with the right-hand side integral found above gives the integrated rate law as

$$\boxed{\frac{1}{[\mathrm{A}]_0([\mathrm{A}]_0-2x)}+\frac{1}{[\mathrm{A}]_0^2}\ln\frac{[\mathrm{A}]_0-2x}{[\mathrm{A}]_0-x}-\frac{1}{[\mathrm{A}]_0^2}=k_r t}$$

17C Reactions approaching equilibrium

Answers to discussion questions

D17C.1 This is discussed in Section 17C.2 on page 738. The quantity which can be determined is sum of the forward and reverse rate constants of the equilibrium reaction for a reaction in which both processes are first order.

Solutions to exercises

E17C.1(a) The equilibrium constant in terms of rate constants is given by [17C.8–*738*], $K = k_r/k_r'$. However because the forward and backward reactions are of different order it is necessary to include a factor of $c^{\ominus}$ so that the ratio of k_r, with units $\mathrm{dm^3\,mol^{-1}\,s^{-1}}$, to k_r', with units $\mathrm{s^{-1}}$, is turned into a dimensionless quantity. The equation required is

$$K=\frac{k_r c^{\ominus}}{k_r'}=\frac{(5.0\times10^6\ \mathrm{dm^3\,mol^{-1}\,s^{-1}})\times(1\ \mathrm{mol\,dm^{-3}})}{2.0\times10^4\ \mathrm{s}}=\boxed{2.5\times10^2}$$

E17C.2(a) The relaxation time in a jump experiment is given by [17C.9a–*739*], $\tau = 1/(k_r + k_r')$. This equation is rearranged for k_r'. It is convenient to convert τ to ms.

$$k_r'=\frac{1}{\tau}-k_r=\frac{1}{27.6\times10^{-3}\ \mathrm{ms}}-(12.4\ \mathrm{ms^{-1}})=\boxed{23.8\ \mathrm{ms^{-1}}}$$

Solutions to problems

P17C.1 The expression for $[\mathrm{A}]$ in [17C.4–*737*] is differentiated

$$[\mathrm{A}]=\frac{k_r'+k_r\mathrm{e}^{-(k_r+k_r')t}}{k_r+k_r'}[\mathrm{A}]_0\quad\text{hence}\quad\frac{\mathrm{d}[\mathrm{A}]}{\mathrm{d}t}=-k_r[\mathrm{A}]_0\mathrm{e}^{-(k_r+k_r')t}$$

According to [17C.3–*737*], $\mathrm{d}[\mathrm{A}]/\mathrm{d}t=-(k_r+k_r')[\mathrm{A}]+k_r'[\mathrm{A}]_0$. To verify that the two expressions for $\mathrm{d}[\mathrm{A}]/\mathrm{d}t$ are the same, the expression for $[\mathrm{A}]$ from [17C.4–*737*] is substituted into [17C.3–*737*]

$$\begin{aligned}\frac{\mathrm{d}[\mathrm{A}]}{\mathrm{d}t}&=-(k_r+k_r')\overbrace{\left[\frac{k_r'+k_r\mathrm{e}^{-(k_r+k_r')t}}{k_r+k_r'}[\mathrm{A}]_0\right]}^{[\mathrm{A}]}+k_r'[\mathrm{A}]_0\\&=-k_r'[\mathrm{A}]_0-k_r\mathrm{e}^{-(k_r+k_r')t}[\mathrm{A}]_0+k_r'[\mathrm{A}]_0=-k_r[\mathrm{A}]_0\mathrm{e}^{-(k_r+k_r')t}\end{aligned}$$

Therefore the two expressions for $d[A]/dt$ are the same and so the equation is satisfied.

P17C.3 (a) The forward and backward reactions are

$$A \rightarrow B \quad \frac{d[A]}{dt} = -k_r[A] \qquad B \rightarrow A \quad \frac{d[A]}{dt} = +k_r'[B]$$

The overall rate of change of $[A]$ is therefore

$$\frac{d[A]}{dt} = -k_r[A] + k_r'[B]$$

The stoichiometry of the reaction $A \rightleftharpoons B$ means that the amount of B present at any time is equal to the initial amount plus the amount of A that has reacted. Hence $[B] = [B]_0 + ([A]_0 - [A])$. This is substituted into the above expression to give

$$\frac{d[A]}{dt} = -k_r[A] + k_r'([B]_0 + [A]_0 - [A])$$

Rearranging and integrating with the initial condition that $[A] = [A]_0$ when $t = 0$ gives

$$\int_{[A]_0}^{[A]} \frac{d[A]}{k_r'([A]_0 + [B]_0) - (k_r + k_r')[A]} = \int_0^t dt$$

Performing the integration gives

$$\frac{\ln\left[k_r'([A]_0 + [B]_0) - (k_r + k_r')[A]\right]}{-(k_r + k_r')}\Bigg|_{[A]_0}^{[A]} = t$$

$$\text{Hence} \quad \frac{1}{-(k_r + k_r')} \ln \frac{k_r'([A]_0 + [B]_0) - (k_r + k_r')[A]}{k_r'([A]_0 + [B]_0) - (k_r + k_r')[A]_0} = t$$

Rearranging for $[A]$ yields

$$\boxed{[A] = \frac{k_r'([A]_0 + [B]_0) + (k_r[A]_0 - k_r'[B]_0)e^{-(k_r + k_r')t}}{k_r + k_r'}}$$

(b) As $t \rightarrow \infty$, the exponential term in the expression for $[A]$ decreases to zero and the concentrations reach their equilibrium values. The equilibrium concentration of A is therefore

$$\boxed{[A]_{eq} = \frac{k_r'([A]_0 + [B]_0)}{k_r + k_r'}}$$

Noting from above that the concentration of B is given by $[B] = [B]_0 + [A]_0 - [A]$, the equilibrium concentration of B is therefore

$$[B]_{eq} = [B]_0 + [A]_0 - [A]_{eq} = [B]_0 + [A]_0 - \frac{k_r'([A]_0 + [B]_0)}{k_r + k_r'}$$

$$= \frac{(k_r + k_r')([B]_0 + [A]_0) - k_r'([A]_0 + [B]_0)}{k_r + k_r'} = \boxed{\frac{k_r([A]_0 + [B]_0)}{k_r + k_r'}}$$

This result is alternatively and more simply obtained by noting that at equilibrium the rates of the forward and backward reactions are equal, implying that

$$k_r[A]_{eq} = k_r'[B]_{eq} \quad \text{hence} \quad k_r[A]_{eq} = k_r'([B]_0 + [A]_0 - [A]_{eq})$$

$$\text{hence} \quad [A]_{eq} = \frac{k_r'([B]_0 + [A]_0)}{k_r + k_r'} \quad \text{as before}$$

P17C.5 (a) Application of [17A.3b–*726*], $v = (1/\nu_J)(d[J]/dt)$, to the forward and backward reactions gives

$$\text{Forward } 2A \rightarrow A_2 \quad v = k_a[A]^2 = \frac{1}{-2}\frac{d[A]}{dt} \quad \text{hence} \quad \frac{d[A]}{dt} = -2k_a[A]^2$$

$$\text{Backward } A_2 \rightarrow 2A \quad v = k_a'[A_2] = \frac{1}{2}\frac{d[A]}{dt} \quad \text{hence} \quad \frac{d[A]}{dt} = 2k_a'[A_2]$$

The overall rate of change of A is therefore

$$\frac{d[A]}{dt} = -2k_a[A]^2 + 2k_a'[A_2]$$

If the deviation of $[A]$ from its new equilibrium value is denoted $2x$, so that $[A] = [A]_{eq} + 2x$, the stoichiometry of the reaction implies that $[A_2] = [A_2]_{eq} - x$. These are substituted into the above expression to give

$$\begin{aligned}
\frac{d[A]}{dt} &= -2k_a\left([A]_{eq} + 2x\right)^2 + 2k_a'\left([A_2]_{eq} - x\right) \\
&= -2k_a\left([A]_{eq}^2 + 4x[A]_{eq} + 4x^2\right) + 2k_a'\left([A_2]_{eq} - x\right) \\
&= \overbrace{-2k_a[A]_{eq}^2 + 2k_a'[A_2]_{eq}}^{0} - 8k_a x[A]_{eq} + 2k_a' x + \overbrace{8k_a x^2}^{\text{Neglect}} \\
&= -\left(8k_a[A]_{eq} + 2k_a'\right)x
\end{aligned}$$

In the third line the first two terms cancel because at equilibrium the rates of the forward reaction $k_a[A]_{eq}^2$ and the backward reaction $k_a'[A_2]_{eq}$ are equal. The last term is neglected because x is assumed to be small. Next, because $[A] = [A]_{eq} + 2x$ it follows that $d[A]/dt = 2\,dx/dt$. This is substituted into the above expression to give

$$2\frac{dx}{dt} = -\left(8k_a[A]_{eq} + 2k_a'\right)x \quad \text{hence} \quad \frac{dx}{dt} = -\left(4k_a[A]_{eq} + k_a'\right)x$$

Rearranging, and integrating with the condition that $x = x_0$ when $t = 0$ gives

$$\int_{x_0}^{x} \frac{dx}{x} = \int_0^t -\left(4k_a[A]_{eq} + k_a'\right)dt \quad \text{hence} \quad \ln\frac{x}{x_0} = -\left(4k_a[A]_{eq} + k_a'\right)t$$

$$\text{hence} \quad x = x_0 e^{-(4k_a[A]_{eq} + k_a')t} = x_0 e^{-t/\tau} \quad \text{where} \quad \frac{1}{\tau} = 4k_a[A]_{eq} + k_a'$$

Squaring both sides of the expression for $1/\tau$ gives

$$\frac{1}{\tau^2} = \left(4k_a[\mathrm{A}]_{eq} + k_a'\right)^2 = 16k_a^2[\mathrm{A}]_{eq}^2 + 8k_ak_a'[\mathrm{A}]_{eq} + k_a'^2$$

$$= 16k_a\overbrace{\left(k_a[\mathrm{A}]_{eq}^2\right)}^{k_a'[\mathrm{A}_2]_{eq}} + 8k_ak_a'[\mathrm{A}]_{eq} + k_a'^2 = 16k_ak_a'[\mathrm{A}_2]_{eq} + 8k_ak_a'[\mathrm{A}]_{eq} + k_a'^2$$

$$= 8k_ak_a'\overbrace{\left(2[\mathrm{A}_2]_{eq} + [\mathrm{A}]_{eq}\right)}^{[\mathrm{A}]_{tot}} + k_a'^2 = \boxed{8k_ak_a'[\mathrm{A}]_{tot} + k_a'^2}$$

In the second line, $k_a[\mathrm{A}]_{eq}^2 = k_a'[\mathrm{A}_2]_{eq}$ is used; these quantities are equal because as explained above the rates of the forward and backward reactions are equal at equilibrium. In the third line, the relationship $[\mathrm{A}]_{tot} = [\mathrm{A}] + 2[\mathrm{A}_2]$ is used; this expression is valid at all stages of the reaction including at equilibrium.

(b) The result $1/\tau^2 = 8k_ak_a'[\mathrm{A}]_{tot} + k_a'^2$ implies that a plot of $1/\tau^2$ against $[\mathrm{A}]_{tot}$ should give a straight line of intercept $k_a'^2$ and slope $8k_ak_a'$; from these quantities k_a' and k_a are determined.

(c) The data are plotted in Fig. 17.8.

$[\mathrm{P}]/\mathrm{mol\,dm^{-3}}$	τ/ns	$\tau^{-2}/\mathrm{ns}^{-2}$
0.500	2.3	0.189
0.352	2.7	0.137
0.251	3.3	0.092
0.151	4.0	0.063
0.101	5.3	0.036

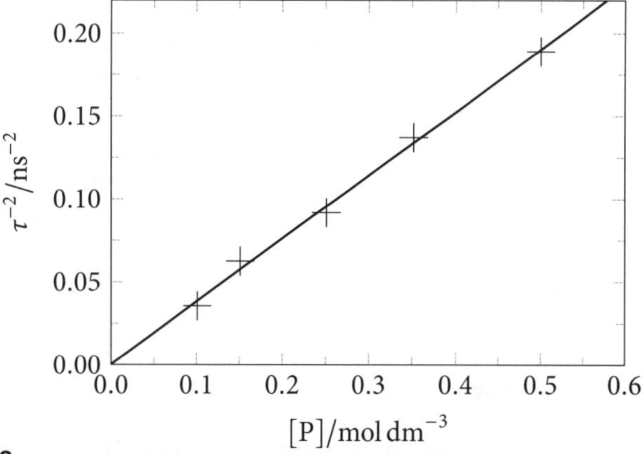

Figure 17.8

The data fall on a reasonable straight line, the equation for which is

$$\tau^{-2}/\mathrm{ns}^{-2} = 0.380 \times ([\mathrm{P}]/\mathrm{mol\,dm^{-3}}) + 2.87 \times 10^{-4}$$

Identifying the intercept with $k_a'^2$ gives $k_a'^2 = 2.87 \times 10^{-4}\ \text{ns}^{-2}$, hence

$$k_a = \sqrt{2.87 \times 10^{-4}\ \text{ns}^{-2}} = 0.0169...\ \text{ns}^{-1} = \boxed{1.7 \times 10^7\ \text{s}^{-1}}$$

Identifying the slope with $8k_a k_a'$ gives

$$8k_a k_a' = 0.380\ \text{dm}^3\ \text{mol}^{-1}\ \text{ns}^{-2}$$

$$\text{hence}\ \ k_a = \frac{0.380\ \text{dm}^3\ \text{mol}^{-1}\ \text{ns}^{-2}}{8k_a'} = \frac{0.380\ \text{dm}^3\ \text{mol}^{-1}\ \text{ns}^{-2}}{8 \times (0.0169...\ \text{ns}^{-1})}$$

$$= 2.80...\ \text{dm}^3\ \text{mol}^{-1}\ \text{ns}^{-1} = \boxed{2.8 \times 10^9\ \text{dm}^3\ \text{mol}^{-1}\ \text{s}^{-1}}$$

The equilibrium constant is given by [17C.8–*738*], $K = (k_a/k_a') \times (k_b/k_b') \times$ ..., but it is necessary to include a factor of $c^{\ominus}$ because the forward reaction is second-order while the backward reaction is first-order.

$$K = \frac{k_a c^{\ominus}}{k_a'} = \frac{(2.80\ ...\ \text{dm}^3\ \text{mol}^{-1}\ \text{ns}^{-1}) \times (1\ \text{mol}\ \text{dm}^{-3})}{(0.0169...\ \text{ns}^{-1})} = \boxed{1.7 \times 10^2}$$

It is noted that the points in Fig. 17.8 do not lie on a perfect straight line, and the intercept is closer to zero than some of the points are to the line. In fact, mathematical software gives the standard error in the intercept as $4 \times 10^{-3}\ \text{ns}^{-2}$, which is an order of magnitude larger than the intercept itself. This indicates that there is considerable uncertainty in the intercept and therefore in the values of the rate constants and equilibrium constant deduced from it.

17D The Arrhenius equation

Answers to discussion questions

D17D.1 In the expression $\ln k_r = \ln A - E_a/RT$ (the Arrhenius equation), k_r is the rate constant, A is the frequency factor, and E_a is the activation energy. The Arrhenius equation is both an empirical expression that often provides a good summary of the temperature dependence of reaction rates and an expression whose parameters have sound physical significance. As an empirical expression, the equation is good, but neither unique nor universal. Over relatively small temperature ranges, other expressions can fit rate data; for example, plots of $\ln k_r$ against $\ln T$ or against T are often linear. Over large temperature ranges, plots of $\ln k_r$ against $1/T$ sometimes exhibit curvature, which may imply that the activation energy is itself temperature dependent.

Solutions to exercises

E17D.1(a) The Arrhenius equation is given by [17D.4–*743*], $k_r = Ae^{-E_a/RT}$. In this case

$$k_r = (8.1 \times 10^{-10}\ \text{dm}^3\ \text{mol}^{-1}\ \text{s}^{-1}) \times \exp\left(-\frac{23 \times 10^3\ \text{J}\ \text{mol}^{-1}}{(8.3145\ \text{J}\ \text{K}^{-1}\ \text{mol}^{-1}) \times (500\ \text{K})}\right)$$

$$= \boxed{3.2 \times 10^{-12}\ \text{dm}^3\ \text{mol}^{-1}\ \text{s}^{-1}}$$

E17D.2(a) The relationship between the values of a rate constant at two different temperatures is given by [17D.2–742], $\ln(k_{r,2}/k_{r,1}) = (E_a/R)(1/T_1 - 1/T_2)$. Rearranging for E_a gives

$$E_a = \frac{R\ln(k_{r,2}/k_{r,1})}{1/T_1 - 1/T_2} = \frac{(8.3145\,\text{J K}^{-1}\,\text{mol}^{-1}) \times \ln\left[(2.67 \times 10^{-2})/(3.80 \times 10^{-3})\right]}{1/([35 + 273.15]\,\text{K}) - 1/([50 + 273.15]\,\text{K})}$$
$$= 1.07... \times 10^5\,\text{J mol}^{-1} = \boxed{108\,\text{kJ mol}^{-1}}$$

The frequency factor is found by rearranging the Arrhenius equation [17D.4–743], $k_r = Ae^{-E_a/RT}$, for A. The data for both temperatures gives the same result.

At T_1 $\quad A = k_r e^{E_a/RT_1}$

$$= (3.80\times10^{-3}\,\text{dm}^3\,\text{mol}^{-1}\,\text{s}^{-1})\times\exp\frac{1.07...\times10^5\,\text{J mol}^{-1}}{(8.3145\,\text{J K}^{-1}\,\text{mol}^{-1})\times([35+273.15]\,\text{K})}$$
$$= \boxed{6.62 \times 10^{15}\,\text{dm}^3\,\text{mol}^{-1}\,\text{s}^{-1}}$$

At T_2 $\quad A = k_r e^{E_a/RT_2}$

$$= (2.67\times10^{-2}\,\text{dm}^3\,\text{mol}^{-1}\,\text{s}^{-1})\times\exp\frac{1.07...\times10^5\,\text{J mol}^{-1}}{(8.3145\,\text{J K}^{-1}\,\text{mol}^{-1})\times([50+273.15]\,\text{K})}$$
$$= 6.62 \times 10^{15}\,\text{dm}^3\,\text{mol}^{-1}\,\text{s}^{-1}$$

E17D.3(a) The relationship between the values of a rate constant at two different temperatures is given by [17D.2–742], $\ln(k_{r,2}/k_{r,1}) = (E_a/R)(1/T_1 - 1/T_2)$. Rearranging for E_a, and using $k_{r,2}/k_{r,1} = 3$ because the rate constant triples between the two temperatures, gives

$$E_a = \frac{R\ln(k_{r,2}/k_{r,1})}{1/T_1 - 1/T_2} = \frac{(8.3145\,\text{J K}^{-1}\,\text{mol}^{-1}) \times \ln 3}{1/([24+273.15]\,\text{K}) - 1/([49+273.15]\,\text{K})} = \boxed{35\,\text{kJ mol}^{-1}}$$

E17D.4(a) The relationship between the values of a rate constant at two different temperatures is given by [17D.2–742], $\ln(k_{r,2}/k_{r,1}) = (E_a/R)(1/T_1 - 1/T_2)$. Hence, taking $T_1 = 37\,°\text{C}$ and $T_2 = 15\,°\text{C}$,

$$\frac{k_{r,2}}{k_{r,1}} = \exp\left[\frac{E_a}{R}\left(\frac{1}{T_1} - \frac{1}{T_2}\right)\right]$$
$$= \exp\left[\frac{87 \times 10^3\,\text{J mol}^{-1}}{8.3145\,\text{J K}^{-1}\,\text{mol}^{-1}} \times \left(\frac{1}{[37+273.15]\,\text{K}} - \frac{1}{[15+273.15]\,\text{K}}\right)\right]$$
$$= \boxed{0.076}$$

The rate constant therefore drops to about $\boxed{7.6\,\%}$ of its original value when the temperature is lowered for 37 °C to 15 °C.

E17D.5(a) As explained in Section 17D.2(a) on page 743 the fraction f of collisions that are sufficiently energetic to be successful is given by the exponential factor $e^{-E_a/RT}$. Rearranging $f = e^{-E_a/RT}$ for T and setting $f = 0.10$ gives

$$T = -\frac{E_a}{R\ln f} = -\frac{50 \times 10^3\ \mathrm{J\,mol^{-1}}}{(8.3145\ \mathrm{J\,K^{-1}\,mol^{-1}}) \times \ln 0.10} = \boxed{2.6 \times 10^3\ \mathrm{K}}$$

Solutions to problems

P17D.1 The definition of E_a in [17D.3–*742*], $E_a = RT^2(\mathrm{d}\ln k_r/\mathrm{d}T)$, is rearranged and integrated.

$$\frac{E_a}{RT^2} = \frac{\mathrm{d}\ln k_r}{\mathrm{d}T} \quad \text{hence} \quad \int \mathrm{d}\ln k_r = \int \frac{E_a}{RT^2}\,\mathrm{d}T$$

The left-hand side integral is simply $\ln k_r$. If E_a does not vary with temperature then the integral on the right is evaluated by taking E_a/R outside the integral to give

$$\ln k_r = \frac{E_a}{R}\int \frac{1}{T^2}\,\mathrm{d}T = -\frac{E_a}{RT} + c$$

This is [17D.1–*741*], $\ln k_r = \ln A - E_a/RT$, once the constant of integration c is identified with $\ln A$.

P17D.3 The Arrhenius equation [17D.1–*741*], $\ln k_r = \ln A - E_a/RT$, implies that a plot of $\ln k_r$ against $1/T$ should give a straight line of slope $-E_a/R$ and intercept $\ln A$. The data are plotted in Fig. 17.9.

T/K	$k_r/\mathrm{dm^3\,mol^{-1}\,s^{-1}}$	$1/(T/\mathrm{K})$	$\ln(k_r/\mathrm{dm^3\,mol^{-1}\,s^{-1}})$
1 000	8.35×10^{-10}	0.001 000	−20.90
1 200	3.08×10^{-8}	0.000 833	−17.30
1 400	4.06×10^{-7}	0.000 714	−14.72
1 600	2.80×10^{-6}	0.000 625	−12.79

The data fall on a good straight line, the equation for which is

$$\ln(k_r/\mathrm{dm^3\,mol^{-1}\,s^{-1}}) = (-2.165 \times 10^4) \times 1/(T/\mathrm{K}) + 0.7457$$

Identifying the slope with $-E_a/R$ gives the activation energy as

$$E_a = -slope \times R = -(-2.165 \times 10^4\ \mathrm{K}) \times (8.3145\ \mathrm{J\,K^{-1}\,mol^{-1}}) = \boxed{180\ \mathrm{kJ\,mol^{-1}}}$$

Identifying the intercept with $\ln A$ gives the frequency factor as

$$A = e^{0.7457}\ \mathrm{dm^3\,mol^{-1}\,s^{-1}} = \boxed{2.11\ \mathrm{dm^3\,mol^{-1}\,s^{-1}}}$$

The units of A are the same as the units of k_r.

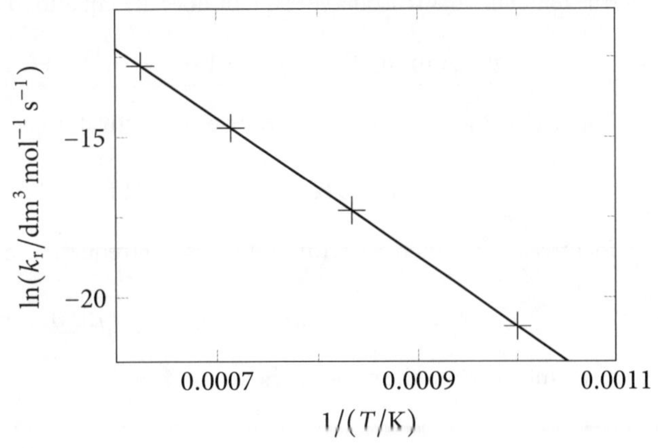

Figure 17.9

P17D.5 The Arrhenius equation [17D.1–*741*], $\ln k_r = \ln A - E_a/RT$, implies that a plot of $\ln k_r$ against $1/T$ should give a straight line of slope $-E_a/R$ and intercept $\ln A$. The data are plotted in Fig. 17.10.

T/K	k_r/dm^3 mol^{-1} s^{-1}	$1/(T/\mathrm{K})$	$\ln(k_r/\mathrm{dm^3\,mol^{-1}\,s^{-1}})$
295	3.55×10^6	0.003 39	15.08
223	4.94×10^5	0.004 48	13.11
218	4.52×10^5	0.004 59	13.02
213	3.79×10^5	0.004 69	12.85
206	2.95×10^5	0.004 85	12.59
200	2.41×10^5	0.005 00	12.39
195	2.17×10^5	0.005 13	12.29

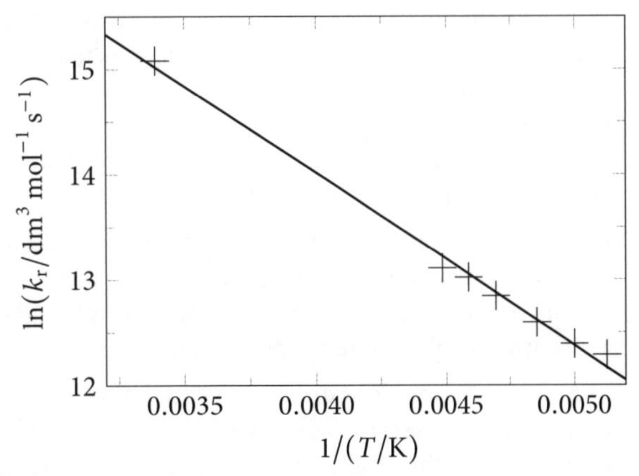

Figure 17.10

The data fall on a reasonable straight line, the equation for which is

$$\ln(k_r/\mathrm{dm^3\,mol^{-1}\,s^{-1}}) = (-1.642\times 10^3)\times 1/(T/\mathrm{K}) + 20.59$$

Identifying the slope with $-E_a/R$ gives the activation energy as

$$E_a = -slope \times R = -(-1.642\times 10^3\ \mathrm{K})\times(8.3145\ \mathrm{J\,K^{-1}\,mol^{-1}}) = \boxed{13.7\ \mathrm{kJ\,mol^{-1}}}$$

Identifying the intercept with $\ln A$ gives the frequency factor as

$$A = \mathrm{e}^{20.59}\ \mathrm{dm^3\,mol^{-1}\,s^{-1}} = \boxed{8.75\times 10^8\ \mathrm{dm^3\,mol^{-1}\,s^{-1}}}$$

The units of A are the same as the units of k_r.

17E Reaction mechanisms

Answers to discussion questions

D17E.1 The overall reaction order is the sum of the powers of the concentrations of all of the substances appearing in the experimental rate law for the reaction; hence, it is the sum of the individual orders (exponents) associated with a each reactant. Reaction order is an experimentally determined quantity.

Molecularity is the number of reactant molecules participating in an elementary reaction. Molecularity has meaning only for an elementary reaction, but reaction order applies to any reaction. In general, reaction order bears no necessary relation to the stoichiometry of the reaction, with the exception of elementary reactions, where the order of the reaction corresponds to the number of molecules participating in the reaction; that is, to its molecularity. Thus for an elementary reaction, overall order and molecularity are the same and are determined by the stoichiometry.

D17E.3 In the pre-equilibrium assumption an intermediate is assumed to be in equilibrium with the reactants. For this assumption to apply it is necessary for the rate at which the intermediate returns to reactants to be fast compared to the rate at which the intermediate goes to products. The rate-determining step is between the intermediate and the products. The rate at which the intermediate is formed from the reactants is not relevant to the establishment of pre-equilibrium provided the other criteria are satisfied.

In the steady-state assumption an intermediate is formed from the reactants and the moment it is formed it goes on to products. The step between reactants and intermediate is therefore the rate-determining step, and necessarily the concentration of the intermediate is low.

The two approximations differ in that in the steady-state approximation the intermediate is necessarily at low concentration, whereas in the pre-equilibrium approximation this condition does not hold – indeed, the intermediate may accumulate. However, if the intermediate reacts immediately by either returning to reactants or going on to products, then the pre-equilibrium assumption also results in a low concentration of the intermediate.

In the pre-equilibrium assumption the apparent rate constant is a product of a rate constant and an equilibrium constant and so the activation energy may be negative (Section 17E.5 on page 750). For the steady-state approximation the activation energy is positive.

D17E.5 Suppose that a reactant R can give alternative products P and Q by different reactions. If the rate constant for the formation of P is greater than that for forming Q, then to start with more P will be formed. However, as time proceeds it may be that the reverse reactions from P and Q back to R start to become significant, and eventually the reactions reach equilibrium. It may be that at equilibrium the amount of Q exceeds that of P, even though initially the amount of P exceeded that of Q.

If the relative proportions of the products are determined by the rate at which they are formed, the reaction is said to be under kinetic control. If the amounts are determined by the relevant equilibrium constants, the reaction is said to be under thermodynamic control. The latter will only occur if the reverse reactions are significant.

Solutions to exercises

E17E.1(a) (i) A pre-equilibrium $A_2 \rightleftharpoons 2A$ between A_2 and A is described by the equilibrium constant K given by

$$K = \frac{([A]/c^\ominus)^2}{([A_2]/c^\ominus)} = \frac{[A]^2}{[A_2]c^\ominus} \quad \text{hence} \quad [A] = (Kc^\ominus[A_2])^{1/2}$$

The equilibrium constant K is written in terms of rate constants using [17C.8–738], $K = (k_a/k'_a) \times (k_b/k'_b) \times ....$ However, in order to make K dimensionless it is necessary in this case to include a factor of $1/c^\ominus$ because k_a is a first-order rate constant with units s^{-1} while k'_a is a second-order rate constant with units $dm^3\,mol^{-1}\,s^{-1}$. Thus $K = k_a/k'_a c^\ominus$, which, on substituting into the above expression for $[A]$ yields

$$[A] = \left(\frac{k_a}{k'_a c^\ominus} c^\ominus [A_2]\right)^{1/2} = \left(\frac{k_a}{k'_a}[A_2]\right)^{1/2}$$

This expression is alternatively obtained by noting that at equilibrium the rates of the forward and reverse reactions are the same (provided that the step to P can be ignored)

$$k_a[A_2] = k'_a[A]^2 \quad \text{hence} \quad [A] = \left(\frac{k_a}{k'_a}[A_2]\right)^{1/2}$$

The rate of formation of P is given by $d[P]/dt = k_b[A][B]$; substituting the above expression for $[A]$ into this gives

$$\frac{d[P]}{dt} = k_b[A][B] = k_b\left(\frac{k_a}{k'_a}\right)^{1/2}[A_2][B] = \boxed{k_b\left(\frac{k_a}{k'_a}\right)^{1/2}[A_2]^{1/2}[B]}$$

(ii) The net rate of change in the concentration of A is

$$\frac{d[A]}{dt} = 2k_a[A_2] - 2k'_a[A]^2 - k_b[A][B]$$

In the steady-state approximation this is assumed to be zero

$$2k_a[A_2] - 2k'_a[A]^2 - k_b[A][B] = 0$$

Hence $2k'_a[A]^2 + k_b[B][A] - 2k_a[A_2] = 0$. This is a quadratic equation in $[A]$, for which the solution is

$$\text{hence} \quad [A] = \frac{-k_b[B] + \left(k_b^2[B]^2 + 16k'_a k_a[A_2]\right)^{1/2}}{4k'_a}$$

where the positive square root is chosen in order to avoid obtaining a negative value for $[A]$. The rate of formation of P is given by $d[P]/dt = k_b[A][B]$; substituting the above expression for $[A]$ into this gives

$$\frac{d[P]}{dt} = k_b[A][B] = k_b[B] \times \frac{-k_b[B] + \left(k_b^2[B]^2 - 16k'_a k_a[A_2]\right)^{1/2}}{4k'_a}$$

$$= \frac{k_b[B]}{4k'_a}\left[-k_b[B] + k_b[B]\left(1 + \frac{16k'_a k_a[A_2]}{k_b^2[B]^2}\right)^{1/2}\right]$$

$$\boxed{= \frac{k_b^2[B]^2}{4k'_a}\left[-1 + \left(1 + \frac{16k'_a k_a[A_2]}{k_b^2[B]^2}\right)^{1/2}\right]}$$

Under certain circumstances this rate law simplifies. If $16k'_a k_a[A_2]/k_b^2[B]^2 \gg 1$ then

$$\frac{d[P]}{dt} \approx \frac{k_b^2[B]^2}{4k'_a}\left[-1 + \left(\frac{16k'_a k_a[A_2]}{k_b^2[B]^2}\right)^{1/2}\right]$$

$$\approx \frac{k_b^2[B]^2}{4k'_a} \times \left(\frac{16k'_a k_a[A_2]}{k_b^2[B]^2}\right)^{1/2} = k_b\left(\frac{k_a}{k'_a}\right)^{1/2}[A_2]^{1/2}[B]$$

which is the same as the rate law derived in part (i) assuming a pre-equilibrium. The condition $16k'_a k_a[A_2]/k_b^2[B]^2 \gg 1$ corresponds to the $A_2 \rightleftarrows A + A$ steps being much faster than the step involving B and k_b; this is precisely the situation corresponding to a pre-equilibrium because the removal of A in the reaction with B is then too slow to affect the maintenance of the equilibrium.

On the other hand, if $16k'_a k_a[A_2]/k_b^2[B]^2 \ll 1$ then the square root is approximated by the expansion $(1+x)^{1/2} \approx 1 + \frac{1}{2}x$ to give

$$\frac{d[P]}{dt} \approx \frac{k_b^2[B]^2}{4k'_a}\left[-1 + \left(1 + \frac{1}{2} \times \frac{16k'_a k_a[A_2]}{k_b^2[B]^2}\right)\right] = \frac{k_b^2[B]^2}{4k'_a} \times \frac{8k'_a k_a[A_2]}{k_b^2[B]^2}$$

$$= 2k_a[A_2]$$

This rate law corresponds to the step $A_2 \rightarrow A + A$ being rate-determining: once A has formed from A_2 in this step it immediately goes on to form product. The factor of 2 arises because each molecule of A_2 that dissociates forms two A molecules which react with two B molecules to form two molecules of product. Hence the rate of product formation is twice the rate of A_2 dissociation. This situation does not correspond to a pre-equilibrium because the immediate removal of A by its reaction with B does not allow A_2 and A to come to equilibrium.

E17E.2(a) The steady-state approximation is applied to the intermediate species O.

$$\frac{d[O]}{dt} = k_a[O_3] - k_a'[O_2][O] - k_b[O][O_3] = 0$$

Rearranging for [O] gives

$$(k_a'[O_2] + k_b[O_3])[O] = k_a[O_3] \quad \text{hence} \quad [O] = \frac{k_a[O_3]}{k_a'[O_2] + k_b[O_3]}$$

The rate of decomposition of O_3 is

$$\frac{d[O_3]}{dt} = -k_a[O_3]+k_a'[O_2][O]-k_b[O][O_3] = -k_a[O_3]+[O]\{k_a'[O_2]-k_b[O_3]\}$$

because O_3 is consumed in steps 1 and 3, but produced in step 2. Inserting the steady-state expression for [O] gives

$$\frac{d[O_3]}{dt} = -k_a[O_3] + \frac{k_a[O_3]\{k_a'[O_2] - k_b[O_3]\}}{k_a'[O_2] + k_b[O_3]}$$

$$= \frac{-k_a k_a'[O_2][O_3] - k_a k_b[O_3]^2 + k_a k_a'[O_3][O_2] - k_a k_b[O_3]^2}{k_a'[O_2] + k_b[O_3]}$$

$$= \frac{-2k_a k_b[O_3]^2}{k_a'[O_2] + k_b[O_3]}$$

If step 3 is rate limiting, such that $k_a'[O_2][O] \gg k_b[O][O_3]$, and hence $k_a'[O_2] \gg k_b[O_3]$, the denominator simplifies to $k_a'[O_2]$ and hence

$$\frac{d[O_3]}{dt} = \frac{-2k_a k_b[O_3]^2}{k_a'[O_2]}$$

As required, the rate of decomposition of O_3 is second order in O_3 and order -1 in O_2.

E17E.3(a) The overall activation energy for a reaction consisting of a pre-equilibrium followed by a rate-limiting elementary step is given by [17E.13–*751*], $E_a = E_{a,a} + E_{a,b} - E_{a,a'}$, where $E_{a,a}$ and $E_{a,a'}$ are the forward and reverse activation energies for the pre-equilibrium and $E_{a,b}$ is the activation energy for the following elementary step. In this case

$$E_a = (25\text{ kJ mol}^{-1}) + (10\text{ kJ mol}^{-1}) - (38\text{ kJ mol}^{-1}) = \boxed{-3\text{ kJ mol}^{-1}}$$

As explained in Section 17E.5 on page 750, negative activation energies such as this are possible for composite reactions.

Solutions to problems

P17E.1 The concentration of I in the reaction mechanism $\mathrm{A} \xrightarrow{k_a} \mathrm{I} \xrightarrow{k_b} \mathrm{P}$ is given by [17E.4b–*747*],

$$[\mathrm{I}] = \frac{k_a}{k_b - k_a}\left(e^{-k_a t} - e^{-k_b t}\right)[\mathrm{A}]_0$$

This expression is plotted in Fig. 17.11 for $[\mathrm{A}]_0 = 1\ \mathrm{mol\,dm^{-3}}$, $k_b = 1\ \mathrm{s^{-1}}$, and various values of k_a. The line for $k_a = 10\ \mathrm{s^{-1}}$ corresponds to part (a) of the question.

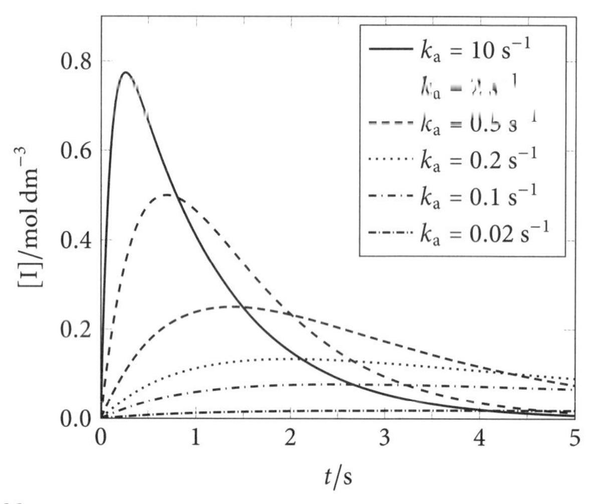

Figure 17.11

If $k_b \gg k_a$, the concentration of I remains low and, apart from the initial induction period, approximately constant during the reaction. Thus the steady-state approximation that $\mathrm{d}[\mathrm{I}]/\mathrm{d}t = 0$ becomes increasingly valid as the ratio k_b/k_a increases.

P17E.3 It is shown in *Example* 17E.1 on page 748 that for the case of two consecutive unimolecular reactions the concentration of the intermediate is greatest at a time given by $t_{max} = (\ln[k_a/k_b])/(k_a - k_b)$. The half-life of a first-order reaction is related to the rate constant according to [17B.3–*732*], $t_{1/2} = \ln 2/k_r$. This is rearranged to $k_r = \ln 2/t_{1/2}$ and used to substitute for the rate constants in the expression for t_{max}.

$$t_{max} = \frac{1}{k_a - k_b}\ln\frac{k_a}{k_b} = \frac{1}{(\ln 2/t_{1/2,a}) - (\ln 2/t_{1/2,b})}\ln\frac{\ln 2/t_{1/2,a}}{\ln 2/t_{1/2,b}}$$

$$= \frac{1}{\ln 2\left[1/t_{1/2,a} - 1/t_{1/2,b}\right]}\ln\frac{t_{1/2,b}}{t_{1/2,a}}$$

Hence $t_{1/2} = \dfrac{1}{\ln 2 \times [1/(22.5\ \mathrm{d}) - 1/(33.0\ \mathrm{d})]} \times \ln\dfrac{33.0\ \mathrm{d}}{22.5\ \mathrm{d}} = \boxed{39.1\ \mathrm{d}}$

P17E.5 For the scheme $\mathrm{A} \underset{k_a'}{\overset{k_a}{\rightleftarrows}} \mathrm{B} \underset{k_b'}{\overset{k_b}{\rightleftarrows}} \mathrm{C} \underset{k_c'}{\overset{k_c}{\rightleftarrows}} \mathrm{D}$ the rates of change of the intermediates B and C are

$$\frac{\mathrm{d[B]}}{\mathrm{d}t} = k_a[\mathrm{A}]-k_a'[\mathrm{B}]-k_b[\mathrm{B}]+k_b'[\mathrm{C}] \qquad \frac{\mathrm{d[C]}}{\mathrm{d}t} = k_b[\mathrm{B}]-k_b'[\mathrm{C}]-k_c[\mathrm{C}]+k_c'[\mathrm{D}]$$

In the steady-state approximation, both of these expressions are equal to zero. Furthermore, because D is removed as soon as it is formed, $[\mathrm{D}] = 0$ and so the expression for $\mathrm{d[C]}/\mathrm{d}t$ becomes

$$k_b[\mathrm{B}] - k_b'[\mathrm{C}] - k_c[\mathrm{C}] = 0 \quad \text{hence} \quad [\mathrm{B}] = \frac{(k_b' + k_c)[\mathrm{C}]}{k_b}$$

The expression for $\mathrm{d[B]}/\mathrm{d}t$ becomes

$$k_a[\mathrm{A}] - (k_a' + k_b)[\mathrm{B}] + k_b'[\mathrm{C}] = 0$$

$$\text{hence} \quad k_a[\mathrm{A}] - (k_a' + k_b)\frac{(k_b' + k_c)[\mathrm{C}]}{k_b} + k_b'[\mathrm{C}] = 0$$

$$\text{hence} \quad \left(\frac{(k_a' + k_b)(k_b' + k_c) - k_b k_b'}{k_b}\right)[\mathrm{C}] = k_a[\mathrm{A}]$$

$$\text{hence} \quad [\mathrm{C}] = \frac{k_a k_b[\mathrm{A}]}{(k_a' + k_b)(k_b' + k_c) - k_b k_b'} = \frac{k_a k_b[\mathrm{A}]}{k_a' k_b' + k_a' k_c + k_b k_c}$$

where on the second line the expression for $[\mathrm{B}]$ derived above is substituted in. Finally, the rate of formation of D is

$$\frac{\mathrm{d[D]}}{\mathrm{d}t} = k_c[\mathrm{C}] - k_c'[\mathrm{D}] = k_c[\mathrm{C}] = \boxed{\frac{k_a k_b k_c[\mathrm{A}]}{k_a' k_b' + k_a' k_c + k_b k_c}}$$

where $[\mathrm{D}] = 0$ is used and the expression for $[\mathrm{C}]$ is substituted in.

P17E.7 The equilibrium constants for the two pre-equilibria are

$$K_1 = \frac{[(\mathrm{HCl})_2]c^{\ominus}}{[\mathrm{HCl}]^2} \quad \text{hence} \quad [(\mathrm{HCl})_2] = \frac{K_1[\mathrm{HCl}]^2}{c^{\ominus}}$$

$$K_2 = \frac{[\text{complex}]c^{\ominus}}{[\mathrm{HCl}][\mathrm{CH_3CH{=}CH_2}]} \quad \text{hence} \quad [\text{complex}] = \frac{K_2[\mathrm{HCl}][\mathrm{CH_3CH{=}CH_2}]}{c^{\ominus}}$$

The factors of $c^{\ominus}$ are needed to make K_1 and K_2 dimensionless. The rate of product formation is

$$v = \frac{\mathrm{d[CH_3CHClCH_3]}}{\mathrm{d}t} = k_r[(\mathrm{HCl})_2][\text{complex}]$$

$$= k_r \times \frac{K_1[\mathrm{HCl}]^2}{c^{\ominus}} \times \frac{K_2[\mathrm{HCl}][\mathrm{CH_3CH{=}CH_2}]}{c^{\ominus}}$$

$$= \boxed{\frac{k_r K_1 K_2}{c^{\ominus 2}}[\mathrm{HCl}]^3[\mathrm{CH_3CH{=}CH_2}]}$$

Thus the reaction is predicted to be first-order in propene and third-order in HCl, as required.

P17E.9 Applying the steady-state approximation to the intermediates OF and F gives

$$\frac{d[OF]}{dt} = k_a[F_2O]^2 + k_b[F][F_2O] - 2k_c[OF]^2 = 0$$

$$\frac{d[F]}{dt} = k_a[F_2O]^2 - k_b[F][F_2O] + 2k_c[OF]^2 - 2k_d[F]^2[F_2O] = 0$$

On adding together these two equations the k_b and k_c terms cancel to give

$$2k_a[F_2O]^2 - 2k_d[F]^2[F_2O] = 0 \quad \text{hence} \quad k_a[F_2O] = k_d[F]^2$$

$$\text{hence} \quad [F] = \left(\frac{k_a}{k_d}[F_2O]\right)^{1/2}$$

Steps a and b lead to the net consumption of one F_2O, while steps d and c lead to no net change. The rate of consumption of F_2O is therefore

$$-\frac{d[F_2O]}{dt} = k_a[F_2O]^2 + k_b[F_2O][F] = k_a[F_2O]^2 + k_b[F_2O]\left(\frac{k_a}{k_d}[F_2O]\right)^{1/2}$$

$$= \underbrace{k_a}_{k_r}[F_2O]^2 + \underbrace{k_b\left(\frac{k_a}{k_d}\right)^{1/2}}_{k'_r}[F_2O]^{3/2}$$

which is the required expression with $k_r = k_a$ and $k'_r = k_b\sqrt{k_a/k_d}$.

17F Examples of reaction mechanisms

Answers to discussion questions

D17F.1 This is discussed in Section 17F.1 on page 753.

D17F.3 The Michaelis-Menten mechanism of enzyme activity models the enzyme with one active site, that weakly and reversibly, binds a substrate in homogeneous solution. It is a three-step mechanism. The first and second steps are the reversible formation of the enzyme-substrate complex (ES). The third step is the decay of the complex into the product. The steady-state approximation is applied to the concentration of the intermediate (ES) and its use simplifies the derivation of the final rate expression. However, the justification for the use of the approximation with this mechanism is suspect, in that both rate constants for the reversible step may not be as large, in comparison to the rate constant for the decay to products, as they need to be for the approximation to be valid. The mechanism clearly indicates that the simplest form of the rate law, $v = v_{max} = k_b[E]_0$, occurs when $[S]_0 \gg K_M$, and the general form of the rate law does seem to match the principal experimental features of enzyme catalysed reactions. It provides a mechanistic understanding of both the turnover number and catalytic efficiency. The model may be expanded to include multisubstrate reactions and inhibition.

Solutions to exercises

E17F.1(a) The effective rate constant for the Lindemann–Hinshelwood mechanism is given by [17F.8–*754*], $1/k_r = k_a'/k_a k_b + 1/k_a[A]$. The difference between the effective rate constant at two pressures is therefore

$$\frac{1}{k_{r,2}} - \frac{1}{k_{r,1}} = \frac{1}{k_a}\left(\frac{1}{[A]_2} - \frac{1}{[A]_1}\right) \quad \text{hence} \quad k_a = \frac{1/[A]_2 - 1/[A]_1}{1/k_{r,2} - 1/k_{r,1}}$$

The rate constant for the activation step, k_a, is therefore

$$k_a = \frac{1/(12 \text{ Pa}) - 1/(1.30 \times 10^3 \text{ Pa})}{1/(2.10 \times 10^{-5} \text{ s}^{-1}) - 1/(2.50 \times 10^{-4} \text{ s}^{-1})} = \boxed{1.9 \times 10^{-6} \text{ Pa}^{-1} \text{ s}^{-1}}$$

or $\boxed{1.9 \text{ MPa}^{-1} \text{ s}^{-1}}$.

E17F.2(a) The fraction of condensed groups at time t of a stepwise polymerisation is given by [17F.11–*755*], $p = k_r t[A]_0/(1 + k_r t[A]_0)$. Hence, after 5.00 h, or 5.00 h $\times$ $(60^2 \text{ s})/(1 \text{ h}) = 1.80 \times 10^4$ s,

$$\begin{aligned} p &= \frac{k_r t[A]_0}{1 + k_r t[A]_0} \\ &= \frac{(1.39 \text{ dm}^3 \text{ mol}^{-1} \text{ s}^{-1}) \times (1.80 \times 10^4 \text{ s}) \times (10.0 \times 10^{-3} \text{ mol dm}^{-3})}{1 + (1.39 \text{ dm}^3 \text{ mol}^{-1} \text{ s}^{-1}) \times (1.80 \times 10^4 \text{ s}) \times (10.0 \times 10^{-3} \text{ mol dm}^{-3})} \\ &= 0.996... = \boxed{0.996} \end{aligned}$$

The degree of polymerisation in a stepwise polymerisation is given by [17F.12a–*755*], $\langle N \rangle = 1/(1 - p)$.

$$\langle N \rangle = \frac{1}{1-p} = \frac{1}{1 - 0.996...} = \boxed{251}$$

E17F.3(a) The kinetic chain length in a chain polymerisation reaction is given by [17F.14c–*757*], $\lambda = k_r[M][In]^{-1/2}$. The ratio of chain length under the two different sets of conditions is therefore

$$\frac{\lambda_2}{\lambda_1} = \frac{k_r[M]_2[In]_2^{-1/2}}{k_r[M]_1[In]_1^{-1/2}} = \left(\frac{[M]_2}{[M]_1}\right) \times \left(\frac{[In]_2}{[In]_1}\right)^{-1/2} = \frac{1}{4.2} \times 3.6^{-1/2} = \boxed{0.13}$$

E17F.4(a) The Michaelis–Menten equation for the rate of an enzyme-catalysed reaction is given by [17F.18a–*759*], $v = v_{max}/(1 + K_M/[S]_0)$. Rearranging for v_{max} gives

$$\begin{aligned} v_{max} &= v\left(1 + \frac{K_M}{[S]_0}\right) = (1.04 \text{ mmol dm}^{-3} \text{ s}^{-1}) \times \left(1 + \frac{0.046 \text{ mol dm}^{-3}}{0.105 \text{ mol dm}^{-3}}\right) \\ &= \boxed{1.50 \text{ mmol dm}^{-3} \text{ s}^{-1}} \end{aligned}$$

E17F.5(a) *Example* 17F.2 on page 760 gives the values $K_M = 10.0\ \text{mmol dm}^{-3}$ and $v_{max} = 0.250\ \text{mmol dm}^{-3}\,\text{s}^{-1}$ for an enzyme concentration of $[\text{E}]_0 = 2.3\ \text{nmol dm}^{-3}$. The catalytic efficiency is defined in the exercise as k_b/K_M, and v_{max} is related to k_b according to [17F.17b–759], $v_{max} = k_b[\text{E}]_0$, hence $k_b = v_{max}/[\text{E}]_0$. Therefore, the catalytic efficiency is

$$\frac{k_b}{K_M} = \frac{v_{max}}{K_M[\text{E}]_0} = \frac{0.250\times10^{-3}\ \text{mol dm}^{-3}\,\text{s}^{-1}}{(10.0\times10^{-3}\ \text{mol dm}^{-3})\times(2.3\times10^{-9}\ \text{mol dm}^{-3})}$$

$$= \boxed{1.1\times10^{7}\ \text{dm}^3\,\text{mol}^{-1}\,\text{s}^{-1}}$$

Solutions to problems

P17F.1 The effective rate constant k_r in the Lindemann–Hinshelwood mechanism is given by [17F.8–754], $1/k_r = k_a'/k_ak_b + 1/k_a[\text{A}]$. This expression implies that a plot of $1/k_r$ against $1/[\text{A}]$ should be a straight line. The data are plotted in Fig. 17.12, using pressure as a measure of concentration.

p/Torr	$10^4\,k_r/\text{s}^{-1}$	$1/(p/\text{Torr})$	$1/(10^4\,k_r/\text{s}^{-1})$
84.1	2.98	0.0119	0.336
11.0	2.23	0.0909	0.448
2.89	1.54	0.346	0.649
0.569	0.857	1.76	1.167
0.120	0.392	8.33	2.551
0.067	0.303	14.9	3.300

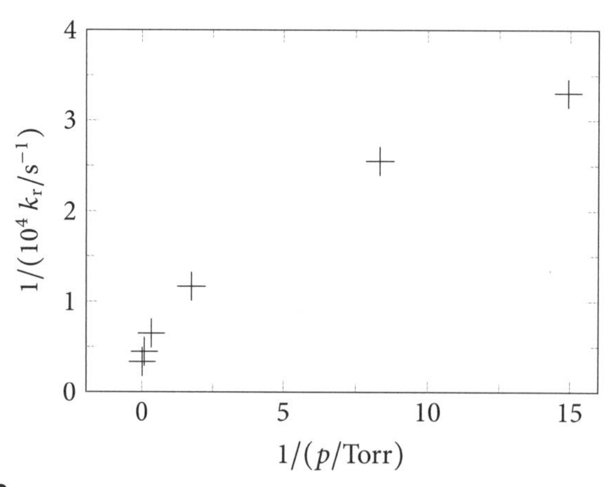

Figure 17.12

The data lie on a curve rather than on a straight line, so it is concluded that the Lindemann–Hinshelwood mechanism does not fit these data.

P17F.3 Each molecule of hydroxyacid has one OH group and one COOH group (A), so $[\mathrm{OH}] = [\mathrm{A}]$. Hence the given rate expression, $\mathrm{d}[\mathrm{A}]/\mathrm{d}t = -k_\mathrm{r}[\mathrm{A}]^2[\mathrm{OH}]$, becomes

$$\frac{\mathrm{d}[\mathrm{A}]}{\mathrm{d}t} = -k_\mathrm{r}[\mathrm{A}]^3 \quad \text{hence} \quad -\frac{1}{[\mathrm{A}]^3}\,\mathrm{d}[\mathrm{A}] = k_\mathrm{r}\,\mathrm{d}t$$

Integration of this expression, with the limits that the concentration is $[\mathrm{A}]_0$ at time $t = 0$ and $[\mathrm{A}]$ at some later time t, gives

$$\int_{[\mathrm{A}]_0}^{[\mathrm{A}]} -\frac{1}{[\mathrm{A}]^3}\,\mathrm{d}[\mathrm{A}] = \int_0^t k_\mathrm{r}\,\mathrm{d}t \quad \text{hence} \quad \left.\frac{1}{2[\mathrm{A}]^2}\right|_{[\mathrm{A}]_0}^{[\mathrm{A}]} = k_\mathrm{r}t\big|_0^t$$

$$\text{hence} \quad \frac{1}{[\mathrm{A}]^2} - \frac{1}{[\mathrm{A}]_0^2} = 2k_\mathrm{r}t$$

Rearranging gives

$$\frac{1}{[\mathrm{A}]^2} = 2k_\mathrm{r}t + \frac{1}{[\mathrm{A}]_0^2} \quad \text{hence} \quad [\mathrm{A}]^2 = \frac{1}{2k_\mathrm{r}t + 1/[\mathrm{A}]_0^2} = \frac{[\mathrm{A}]_0^2}{2k_\mathrm{r}t[\mathrm{A}]_0^2 + 1}$$

To go to the final expression the top and bottom of the fraction are multiplied by $[\mathrm{A}]_0^2$. Taking the square root gives $[\mathrm{A}] = [\mathrm{A}]_0/(2k_\mathrm{r}t[\mathrm{A}]_0^2 + 1)^{1/2}$. As explained in Section 17F.2(a) on page 755, the degree of polymerisation $\langle N\rangle$ is the ratio of the initial concentration of A, $[\mathrm{A}]_0$, to the concentration of end groups, $[\mathrm{A}]$, at the time of interest. Hence

$$\langle N\rangle = \frac{[\mathrm{A}]_0}{[\mathrm{A}]} = \frac{[\mathrm{A}]_0}{[\mathrm{A}]_0/(2k_\mathrm{r}t[\mathrm{A}]_0^2 + 1)^{1/2}} = \boxed{(2k_\mathrm{r}t[\mathrm{A}]_0^2 + 1)^{1/2}}$$

P17F.5 The Michaelis–Menten equation [17F.18a–*759*] is

$$v = \frac{v_\mathrm{max}}{1 + K_\mathrm{M}/[\mathrm{S}]_0}$$

This equation is plotted for fixed v_max with a range of K_M values in Fig. 17.13, and for fixed K_M with a range of v_max values in Fig. 17.14.

P17F.7 The Lineweaver-Burk equation, [17F.18b–*759*], expresses the reciprocal of the velocity as $1/v = 1/v_\mathrm{max} + (K_\mathrm{M}/v_\mathrm{max})(1/[\mathrm{S}]_0)$. This expression implies that a plot of $1/v$ against $1/[\mathrm{S}]_0$ will be a straight line of slope $K_\mathrm{M}/v_\mathrm{max}$ and intercept $1/v_\mathrm{max}$. Such a plot is shown in Fig. 17.15.

$[\mathrm{ATP}]/\ \mu\mathrm{mol\,dm^{-3}}$	$v/\ \mu\mathrm{mol\,dm^{-3}\,s^{-1}}$	$\dfrac{1}{[\mathrm{ATP}]/(\mu\mathrm{mol\,dm^{-3}})}$	$\dfrac{1}{v/(\mu\mathrm{mol\,dm^{-3}\,s^{-1}})}$
0.6	0.81	1.67	1.23
0.8	0.97	1.25	1.03
1.4	1.30	0.71	0.77
2.0	1.47	0.50	0.68
3.0	1.69	0.33	0.59

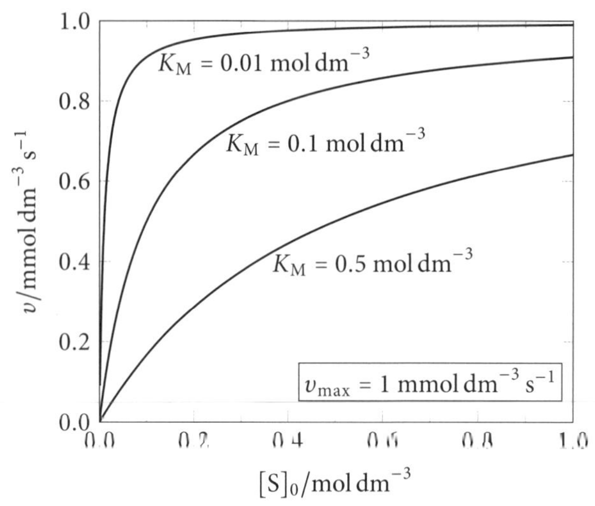

Figure 17.13

The data lie on a good straight line, the equation of which is

$$1/(v/\mu\text{mol dm}^{-3}\,\text{s}^{-1}) = 0.48 \times 1/([\text{ATP}]/\mu\text{mol dm}^{-3}) + 0.43$$

The intercept is identified with $1/v_{max}$ so that

$$v_{max} = \frac{1}{0.43\ \mu\text{mol dm}^{-3}\,\text{s}^{-1}} = 2.32...\ \mu\text{mol dm}^{-3}\,\text{s}^{-1} = \boxed{2.3\ \mu\text{mol dm}^{-3}\,\text{s}^{-1}}$$

The slope is identified with K_M/v_{max} so that

$$K_M = (\overbrace{0.48}^{\text{slope}}\ \text{s}) \times (\overbrace{2.32...\ \mu\text{mol dm}^{-3}\,\text{s}^{-1}}^{v_{max}}) = \boxed{1.1\ \mu\text{mol dm}^{-3}}$$

17G Photochemistry

Answer to discussion question

D17G.1 The time scales of atomic processes are rapid indeed. Note that the times given here are in some way typical values for times that may vary over two or three orders of magnitude. For example, vibrational wavenumbers can range from about 4400 cm^{-1} (for H_2) to 100 cm^{-1}(for I_2) and even lower, with a corresponding range of associated times. Radiative decay rates of electronic states can vary even more widely: times associated with phosphorescence can be in the millisecond and even second range. A large number of time scales for physical, chemical, and biological processes on the atomic and molecular scale are reported in Figure 2 of A. H. Zewail, 'Femtochemistry: Atomic-Scale Dynamics of the Chemical Bond', *Journal of Physical Chemistry A*, **104**, 5660 (2000).

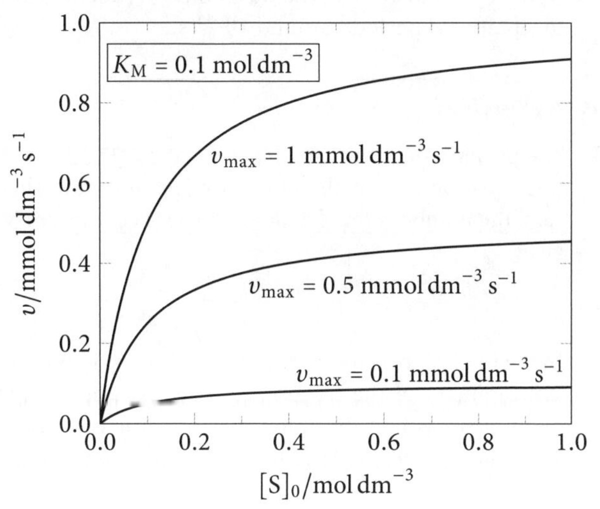

Figure 17.14

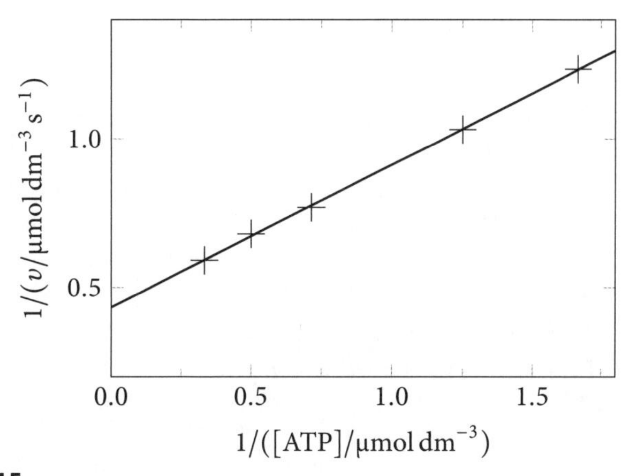

Figure 17.15

Radiative decay of excited electronic states can range from about 10^{-9} s to 10^{-4} s, and even longer for phosphorescence involving 'forbidden' decay paths. Molecular rotational motion takes place on a scale of 10^{-12} s to 10^{-9} s. Molecular vibrations are faster still, about 10^{-14} s to 10^{-12} s. Proton transfer reactions occur on a timescale of about 10^{-10} s to 10^{-9} s, although protons can hop from molecule to molecule in water even more rapidly (1.5×10^{-12} s).

Harvesting of light during plant photosynthesis involves very fast time scales of several energy-transfer and electron-transfer steps in photosynthesis. Initial energy transfer (to a nearby pigment) has a time scale of around 10^{-13} s to 5×10^{-12} s, with longer-range transfer (to the reaction centre) taking about 10^{-10} s. Immediate electron transfer is also very fast (about 3 ps), with ultimate transfer (leading to oxidation of water and reduction of plastoquinone) taking from 10^{-10} s to 10^{-3} s. The mean time between collisions in liquids is similar to vibrational periods, around 10^{-13} s.

Solutions to exercises

E17G.1(a) The primary quantum yield is defined by [17G.1a–*763*], $\phi = N_{\text{events}}/N_{\text{abs}}$. In this equation N_{abs} is the number of photons absorbed and N_{events} is, in this case, the number of molecules of A that decompose, $N_{\text{decomposed}}$. Rearranging gives

$$N_{\text{abs}} = \frac{N_{\text{decomposed}}}{\phi} = \frac{n_{\text{decomposed}}N_{\text{A}}}{\phi} = \frac{(n_{\text{formed}}/2)N_{\text{A}}}{\phi}$$

In the final expression, n_{formed} is the amount in moles of B that is formed. The stoichiometry of the reaction $\text{A} \rightarrow 2\text{B} + \text{C}$ implies that the amount of A that decomposes is half the amount of B that is formed, $n_{\text{decomposed}} = n_{\text{formed}}/2$. The quantum yield is 210 mmol einstein^{-1}, or 0.210 mol mol^{-1} = 0.210, hence

$$N_{\text{abs}} = \frac{(n_{\text{formed}}/2)N_{\text{A}}}{\phi} = \frac{(2.28\times10^{-3}\ \text{mol})/2\times(6.0221\times10^{23}\ \text{mol}^{-1})}{0.210}$$
$$= \boxed{3.27\times10^{21}}$$

E17G.2(a) The fluorescence quantum yield is given by [17G.4–*765*], $\phi_{\text{F},0} = k_{\text{F}}\tau_0$. The observed lifetime τ_0 is given by [17G.3b–*764*], $\tau_0 = 1/(k_{\text{F}} + k_{\text{ISC}} + k_{\text{IC}})$, which is written as $\tau_0 = 1/k_{\text{r}}$ where $k_{\text{r}} = k_{\text{F}} + k_{\text{ISC}} + k_{\text{IC}}$ is the effective first-order rate constant for the decay of the excited state of the fluorescing species. For a first-order process k_{r} is related to the half-life according to [17B.3–*732*], $t_{1/2} = \ln 2/k_{\text{r}}$, and combining this expression with $\tau_0 = 1/k_{\text{r}}$ gives $t_{1/2} = \ln 2/(1/\tau_0) = (\ln 2)\tau_0$. Hence $\tau_0 = t_{1/2}/\ln 2$.

Rearranging [17G.4–*765*] then gives

$$k_{\text{F}} = \frac{\phi_{\text{F},0}}{\tau_0} = \frac{\phi_{\text{F},0}}{t_{1/2}/\ln 2} = \frac{\phi_{\text{F},0}\ln 2}{t_{1/2}} = \frac{0.35\times\ln 2}{5.6\times10^{-9}\ \text{s}} = \boxed{4.3\times10^{7}\ \text{s}^{-1}}$$

E17G.3(a) The Stern–Volmer equation [17G.5–*765*] is $\phi_{\text{F},0}/\phi_{\text{F}} = 1 + \tau_0 k_{\text{Q}}[\text{Q}]$, where ϕ_{F} and $\phi_{\text{F},0}$ are the fluorescence quantum yields with and without the quencher. The rate of fluorescence v, and hence the fluorescence intensity, is directly proportional to the fluorescence quantum yield according to [17G.1b–*763*], $\phi = v/I_{\text{abs}}$. Therefore to reduce the fluorescence intensity to 50% of the unquenched value requires $\phi_{\text{F}} = \frac{1}{2}\phi_{\text{F},0}$ and hence $\phi_{\text{F},0}/\phi_{\text{F}} = 2$. Rearranging the Stern–Volmer equation then gives

$$[\text{Q}] = \frac{\phi_{\text{F},0}/\phi_{\text{F}} - 1}{\tau_0 k_{\text{Q}}} = \frac{2-1}{(6.0\times10^{-9}\ \text{s})\times(3.0\times10^{8}\ \text{dm}^3\ \text{mol}^{-1}\ \text{s}^{-1})}$$
$$= \boxed{0.56\ \text{mol dm}^{-3}}$$

E17G.4(a) The efficiency of resonance energy transfer η_T is defined by [17G.6–767], $\eta_T = 1 - \phi_F/\phi_{F,0}$, and the distance-dependence of the efficiency is given by [17G.7–767], $\eta_T = R_0^6/(R_0^6 + R^6)$, where R is the donor–acceptor distance and R_0 is a constant characteristic of the particular donor–acceptor pair.

In this case a decrease of the fluorescence quantum yield by 10% implies that $\phi_F = 0.9\phi_{F,0}$. Hence the efficiency is $\eta_T = 1 - \phi_F/\phi_{F,0} = 1 - 0.9 = 0.1$. Rearranging [17G.7–767] for R, and taking $R_0 = 4.9$ nm from Table 17G.3 on page 767, gives

$$R = R_0\left(\frac{1-\eta_T}{\eta_T}\right)^{1/6} = (4.9\text{ nm}) \times \left(\frac{1-0.1}{0.1}\right)^{1/6} = \boxed{7.1\text{ nm}}$$

Solutions to problems

P17G.1 The quantum yield is given by [17G.1a–763], $\phi = N_{\text{events}}/N_{\text{abs}}$ where N_{abs} is the number of photons absorbed and N_{events} is, in this case, the number of molecules of the absorbing substance that decomposed. The latter is equal to $n_{\text{decomposed}}N_A$, where $n_{\text{decomposed}}$ is the amount in moles of substance that decomposed.

The number of photons absorbed is found by noting that the energy transferred by each photon is given by [7A.9–241], $\Delta E = h\nu = hc/\lambda$. Therefore the total energy absorbed is $E_{\text{abs}} = N_{\text{abs}}hc/\lambda$. This energy is also given by $E_{\text{abs}} = fPt$, where P is the incident power, t is the time of exposure, and f is the fraction of incident radiation that is absorbed. In this case $f = 1 - 0.257 = 0.743$. Combining these expressions gives

$$fPt = \frac{N_{\text{abs}}hc}{\lambda} \quad \text{hence} \quad N_{\text{abs}} = \frac{fPt\lambda}{hc}$$

This is substituted into $\phi = N_{\text{events}}/N_{\text{abs}}$, together with $N_{\text{events}} = n_{\text{decomposed}}N_A$, to give

$$\phi = \frac{N_{\text{events}}}{N_{\text{abs}}} = \frac{n_{\text{decomposed}}N_A}{fPt\lambda/hc} = \frac{n_{\text{decomposed}}N_A hc}{fPt\lambda}$$

$$= \frac{(0.324\text{ mol})\times(6.0221\times10^{23}\text{ mol}^{-1})}{(0.743)\times(87.5\text{ W})\times(28.0\text{ min})\times(60\text{ s})/(1\text{ min})}$$

$$\times\frac{(6.6261\times10^{-34}\text{ J s})\times(2.9979\times10^{8}\text{ m s}^{-1})}{(320\times10^{-9}\text{ m})} = \boxed{1.11}$$

Note that 1 W = 1 J s^{-1}. The fact that the quantum yield is greater than 1 indicates that each absorbed photon can lead to the decomposition of more than one molecule of absorbing material.

P17G.3 (a) The concentration of the excited dansyl chloride decays with time according to [17G.3a–764], $[S^*] = [S^*]_0e^{-t/\tau_0}$, or $[S^*]/[S^*]_0 = e^{-t/\tau_0}$. The rate of fluorescence is given by $v = k_F[S^*]$ so the rate of fluorescence, and hence the fluorescence intensity I_F, is proportional to $[S^*]$. Therefore $I_F/I_0 = [S^*]/[S^*]_0$, and hence

$$\frac{I_F}{I_0} = \frac{[S^*]}{[S^*]_0} = e^{-t/\tau_0} \quad \text{hence} \quad \ln\left(\frac{I_F}{I_0}\right) = -\frac{t}{\tau_0}$$

This expression implies that a plot of $\ln(I_F/I_0)$ against t should be a straight line of slope $-1/\tau_0$ and intercept zero. The data are plotted in Fig. 17.16.

t/ns	I_F/I_0	$\ln(I_F/I_0)$
5.0	0.45	−0.799
10.0	0.21	−1.561
15.0	0.11	2.207
20.0	0.05	−2.996

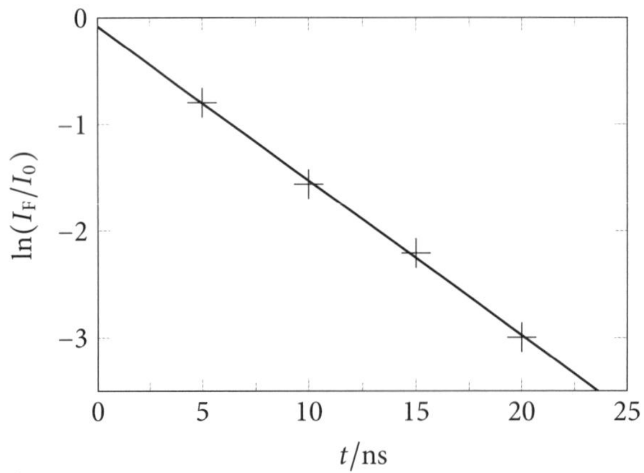

Figure 17.16

The data lie on a good straight line that passes close to the origin. The equation of the line is

$$\ln(I_F/I_0) = -0.145 \times (t/\text{ns}) - 0.081$$

Identifying the slope with $-1/\tau_0$ gives

$$-\frac{1}{\tau_0} = -0.145\ \text{ns}^{-1} \quad \text{hence} \quad \tau_0 = \frac{1}{0.145\ \text{ns}^{-1}} = 6.89...\ \text{ns} = \boxed{6.9\ \text{ns}}$$

(b) The fluorescence quantum yield is given by [17G.4–765], $\phi_{F,0} = k_F\tau_0$. This equation is rearranged for k_F

$$k_F = \frac{\phi_{F,0}}{\tau_0} = \frac{0.70}{6.89... \times 10^{-9}\ \text{s}} = \boxed{1.0 \times 10^8\ \text{s}^{-1}}$$

P17G.5 The Stern–Volmer equation [17G.5–765] is $\phi_{F,0}/\phi_F = 1 + \tau_0 k_Q[Q]$. As explained in Section 17G.4 on page 765, the ratio τ_0/τ, where τ is the lifetime in the presence of the quencher, is equal to $\phi_{F,0}/\phi_F$, so the Stern–Volmer equation becomes

$$\frac{\tau_0}{\tau} = 1 + \tau_0 k_Q[Q] \quad \text{hence} \quad k_Q = \frac{\tau_0/\tau - 1}{\tau_0[Q]}$$

In order to use the equation to calculate k_Q it is necessary to find τ_0 and τ. This is done as follows.

The concentration of an excited species such as Hg* varies with time according to [17G.3a–764], $[Hg^*] = [Hg^*]_0 e^{-t/\tau}$. Rearranging and taking logarithms gives

$$\frac{[Hg^*]}{[Hg^*]_0} = e^{-t/\tau} \quad \text{hence} \quad \ln\left(\frac{[Hg^*]}{[Hg^*]_0}\right) = -\frac{t}{\tau}$$

The rate of fluorescence is $v = k_F[Hg^*]$, so the fluorescence intensity I is proportional to the Hg* concentration. Hence $I/I_0 = [Hg^*]/[Hg^*]_0$, and from the above equation a plot of $\ln(I/I_0)$ against t should therefore be a straight line of slope $-1/\tau$ and intercept zero. The fluorescence intensity data are given relative to the value at $t = 0$ and therefore represent I/I_0.

The data are plotted in Fig. 17.17.

$p_{N_2} = 0$

$t/\mu s$	I/I_0	$\ln(I/I_0)$
0.0	1.000	0.000
5.0	0.606	−0.501
10.0	0.360	−1.022
15.0	0.220	−1.514
20.0	0.135	−2.002

$p_{N_2} = 9.74 \times 10^{-4}$ atm

$t/\mu s$	I/I_0	$\ln(I/I_0)$
0.0	1.000	0.000
3.0	0.585	−0.536
6.0	0.342	−1.073
9.0	0.200	−1.609
12.0	0.117	−2.146

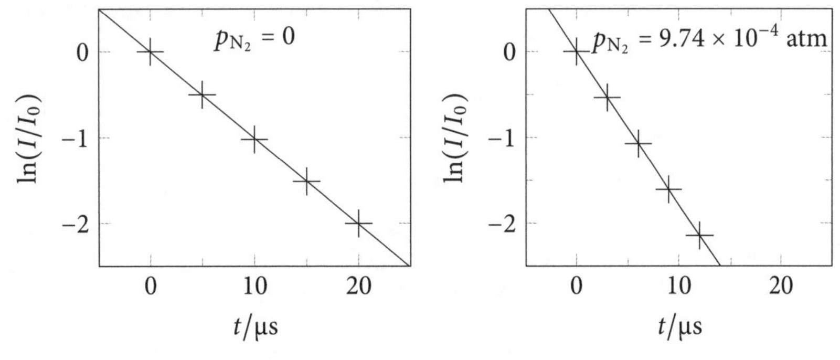

Figure 17.17

The data fall on good straight lines, the equations of which are

$$p_{N_2} = 0 \qquad \ln(I/I_0) = -0.100 \times (t/\mu s) + -4.18 \times 10^{-3}$$

$$p_{N_2} = 9.74 \times 10^{-4}\ \text{atm} \qquad \ln(I/I_0) = -0.179 \times (t/\mu s) + 7.02 \times 10^{-5}$$

The slopes are identified with $-1/\tau_0$ and $-1/\tau$ respectively

$$\tau_0 = -\frac{1}{-0.100\ \mu s^{-1}} = 10.0...\ \mu s \qquad \tau = -\frac{1}{-0.179\ \mu s^{-1}} = 5.59...\ \mu s$$

The rearranged form of the Stern–Volmer equation found earlier, $k_Q = (\tau_0/\tau - 1)/\tau_0[Q]$ is then used to calculate k_Q. The concentration of the N_2 quencher is calculated from the partial pressure using the perfect gas equation [1A.4–8], $pV = nRT$.

$$\begin{aligned}[N_2] &= \frac{n_{N_2}}{V} = \frac{p_{N_2}}{RT} \\ &= \frac{9.74 \times 10^{-4}\ \text{atm}}{(8.3145\ \text{J K}^{-1}\ \text{mol}^{-1}) \times (300\ \text{K})} \times \frac{1.01325 \times 10^5\ \text{Pa}}{1\ \text{atm}} \times \frac{1\ \text{m}^3}{10^3\ \text{dm}^3} \\ &= 3.95... \times 10^{-5}\ \text{mol dm}^{-3}\end{aligned}$$

where 1 Pa = 1 kg m^{-1} s^{-2} and 1 J = 1 kg m^2 s^{-2} are used. Hence

$$\begin{aligned}k_Q &= \frac{\tau_0/\tau - 1}{\tau_0[N_2]} = \frac{(10.0... \times 10^{-6}\ \text{s})/(5.59... \times 10^{-6}\ \text{s}) - 1}{(10.0 \times 10^{-6}\ \text{s}) \times (3.95... \times 10^{-5}\ \text{mol dm}^{-3})} \\ &= \boxed{2.00 \times 10^9\ \text{dm}^3\ \text{mol}^{-1}\ \text{s}^{-1}}\end{aligned}$$

P17G.7 The efficiency of resonance energy transfer is given by [17G.6–*767*],

$$\eta_T = 1 - \frac{\phi_F}{\phi_{F,0}} = 1 - \frac{\tau}{\tau_0}$$

where the second expression comes from the fact that, according to [17G.4–*765*], $\phi_{F,0} = k_F\tau_0$, the lifetime is proportional to quantum yield. The efficiency of resonance energy transfer in terms of donor–acceptor distance is given by [17G.7–*767*], $\eta_T = R_0^6/(R_0^6 + R^6)$. Equating the two expressions for η_T gives

$$1 - \frac{\tau}{\tau_0} = \frac{R_0^6}{R_0^6 + R^6} \quad \text{hence} \quad \frac{1}{1-\tau/\tau_0} = 1 + \frac{R^6}{R_0^6} \quad \text{hence} \quad R = R_0\left(\frac{1}{1-\tau/\tau_0} - 1\right)^{1/6}$$

The distance required to give τ = 10 ps is therefore

$$R = (5.6\ \text{nm}) \times \left(\frac{1}{1 - (10 \times 10^{-12}\ \text{s})/(1 \times 10^{-9}\ \text{s})} - 1\right)^{1/6} = \boxed{2.6\ \text{nm}}$$

Solutions to integrated activities

I17.1 (a) The expressions $[A] = [A]_0 - x$ and $[P] = [P]_0 + x$ are substituted into the rate law to give

$$v = -\frac{d[A]}{dt} = k_r[A][P] = k_r([A]_0 - x)([P]_0 + x)$$

The expression $[A] = [A]_0 - x$ implies that $d[A]/dt = -dx/dt$ so the expression becomes

$$\frac{dx}{dt} = k_r([A]_0 - x)([P]_0 + x) \quad \text{hence} \quad \frac{dx}{([A]_0 - x)([P_0] + x)} = k_r\, dt$$

Integration of this expression, using $x = 0$ at time $t = 0$ gives

$$\int_0^x \frac{dx}{([A]_0 - x)([P_0] + x)} = \int_0^t k_r\, dt$$

The left-hand side is evaluated using Integral A.3

$$\int_0^x \frac{dx}{([A]_0 - x)([P_0] + x)} = -\overbrace{\int_0^x \frac{dx}{([A]_0 - x)(-[P_0] - x)}}^{\text{Integral A.3 with } A = [A]_0 \text{ and } B = -[P]_0}$$
$$= -\frac{1}{(-[P]_0) - [A]_0} \ln\left(\frac{(-[P]_0 - x)[A]_0}{([A]_0 - x)(-[P]_0)}\right)$$
$$= \frac{1}{[A]_0 + [P]_0} \ln\left(\frac{[A]_0([P]_0 + x)}{[P]_0([A]_0 - x)}\right)$$

The right-hand side is $k_r t$, hence the integrated rate law is

$$\frac{1}{[A]_0 + [P]_0} \ln\left(\frac{[A]_0([P]_0 + x)}{[P]_0([A]_0 - x)}\right) = k_r t$$

The expression $[P] = [P]_0 + x$ is rearranged to $x = [P] - [P]_0$. This is used to replace x in the integrated rate law

$$\frac{1}{[A]_0 + [P]_0} \ln\left(\frac{[A]_0[P]}{[P]_0([A]_0 - [P] + [P]_0)}\right) = k_r t$$

$$\text{hence} \quad \ln\left(\frac{[A]_0[P]}{[P]_0([A]_0 - [P] + [P]_0)}\right) = \overbrace{([A]_0 + [P]_0)k_r}^{a}\, t$$

$$\text{hence} \quad [A]_0[P] = [P]_0([A]_0 - [P] + [P]_0)e^{at}$$

$$\text{hence} \quad [P]([A]_0 + [P]_0 e^{at}) = [P]_0([A]_0 + [P]_0)e^{at}$$

$$\text{hence} \quad \frac{[P]}{[P]_0} = \frac{([A]_0 + [P]_0)e^{at}}{[A]_0 + [P]_0 e^{at}} = \frac{(1 + \overbrace{[P]_0/[A]_0}^{b})e^{at}}{1 + \underbrace{([P]_0/[A]_0)}_{b} e^{at}} = \boxed{\frac{(1 + b)e^{at}}{1 + be^{at}}}$$

(b) The quantity $[P]/[P]_0$ is plotted against at in Fig. 17.18 for various values of b.

The plots are sigmoid in shape because the reaction is initially slow because only a small amount of P is present. As more product is formed, the rate of the reaction $v = k_r[A][P]$ increases and the curve becomes steeper,

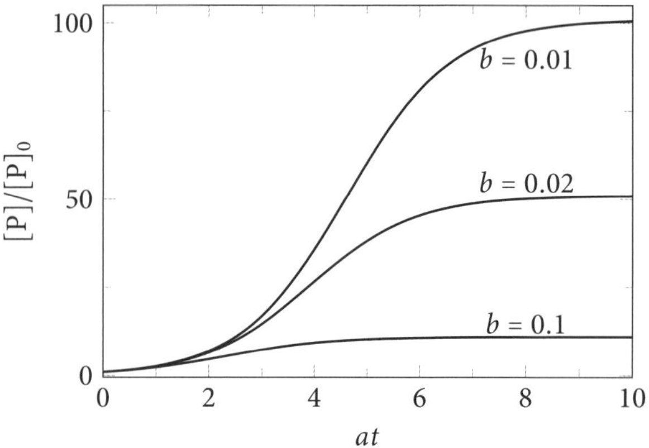

Figure 17.18

until the reaction slows down towards the end due to the reactant A being used up.

The curves level off at different values because $[P]/[P]_0$ is being plotted. In each case the final concentration of P is given by the initial concentration of A, because all the A is eventually converted to P, plus the concentration of P that was present at the start, that is, $[P]_\infty = [A]_0 + [P]_0$. The final value of $[P]/[P]_0$ is therefore

$$\frac{[P]_\infty}{[P]_0} = \frac{[A]_0 + [P]_0}{[P]_0} = \frac{[A]_0}{[P]_0} + 1 = \frac{1}{b} + 1$$

Changing b therefore changes the final value of $[P]/[P]_0$.

The integrated rate equation for a first-order process $A \rightarrow P$ is given in Table 17B.3 on page 735 as $[P]/[A]_0 = 1 - e^{-k_r t}$. In order to facilitate comparison to the autocatalytic reaction it is instructive to re-plot the autocatalytic curves as $[P]/[A]_0$ rather than as $[P]/[P]_0$. Furthermore it is convenient to consider $([P]-[P]_0)/[A]_0$ rather than $[P]/[A]_0$, because in this way the plot reflects the amount of P that is produced in the reaction rather than including any P that was present at the start. The expression for $[P]/[P]_0$ derived above is adapted to give $([P] - [P]_0)/[A]_0$

$$\frac{[P]}{[P]_0} = \frac{(1+b)e^{at}}{1+be^{at}} \quad \text{hence} \quad [P] = \frac{(1+b)e^{at}[P]_0}{1+be^{at}}$$

Hence

$$\begin{aligned}\frac{[P]-[P]_0}{[A]_0} &= \frac{[P]}{[A]_0} - \overbrace{[P]_0/[A]_0}^{b} = \frac{(1+b)e^{at}\overbrace{[P]_0/[A]_0}^{b}}{1+be^{at}} - b \\ &= \frac{b(1+b)e^{at}}{1+be^{at}} - b\end{aligned}$$

This expression is plotted against t in Fig. 17.19 for various values of b, taking $a = 1\ \mathrm{s}^{-1}$ in each case. The quantity $[P]/[A]_0 = 1 - e^{-k_r t}$ is also plotted, taking $k_r = 1\ \mathrm{s}^{-1}$. As already noted, the autocatalytic curves are sigmoid, in contrast to the first-order curve which is not. The autocatalytic curves with larger b, that is a greater initial amount of P relative to the initial amount of A, reach their maximum value faster than those with smaller b. This is because, if less P is present to begin with, the autocatalytic step is initially slower and the amount of P present builds up more slowly.

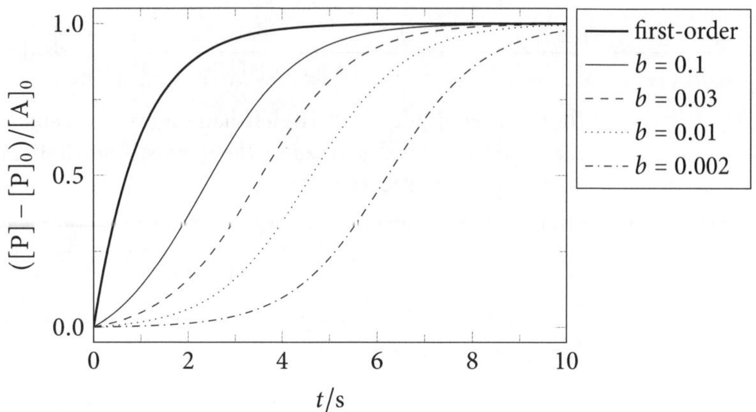

Figure 17.19

(c) The rate law found in part (a), $[P]/[P]_0 = (1+b)e^{at}/(1+be^{at})$, is rearranged to $[P] = (1+b)e^{at}[P]_0/(1+be^{at})$ and differentiated to give an expression for the rate.

$$v = \frac{d[P]}{dt} = \frac{d}{dt}\left(\frac{(1+b)e^{at}[P]_0}{1+be^{at}}\right)$$

$$= \frac{(1+be^{at}) \times a(1+b)e^{at}[P]_0 - (1+b)e^{at}[P]_0 \times abe^{at}}{(1+be^{at})^2}$$

$$= \frac{a(1+b)e^{at}[P]_0}{(1+be^{at})^2}$$

The maximum rate is found by differentiating v with respect to t and setting the derivative equal to zero

$$\frac{dv}{dt} = \frac{(1+be^{at})^2 \times (a^2(1+b)e^{at}[P]_0) - a(1+b)e^{at}[P]_0 \times 2abe^{at}(1+be^{at})}{(1+be^{at})^4}$$

At the maximum, when $dv/dt = 0$, the numerator of this expression is zero

$$a^2(1+b)(1+be^{at})^2 e^{at}[P]_0 - 2a^2 b(1+b)(1+be^{at})e^{2at}[P]_0 = 0$$

Cancelling of terms followed by rearrangement gives

$$1 + be^{at} - 2be^{at} = 0 \quad \text{hence} \quad be^{at} = 1 \quad \text{hence} \quad \boxed{t = -(1/a)\ln b}$$

(d) As in part (a), [A] and [P] are written as $[A]_0 - x$ and $[P]_0 + x$ respectively. The rate law is

$$v = \frac{d[P]}{dt} = k_r[A]^2[P] = k_r([A]_0 - x)^2([P]_0 + x)$$

The expression $[P] = [P]_0 + x$ implies that $d[P]/dt = dx/dt$ so the rate law becomes

$$\frac{dx}{dt} = k_r([A]_0 - x)^2([P]_0 + x) \quad \text{hence} \quad \frac{dx}{([A]_0 - x)^2([P]_0 + x)} = k_r\,dt$$

Integration of this expression, using $x = 0$ at time $t = 0$, gives

$$\int_0^x \frac{dx}{([A]_0 - x)^2([P]_0 + x)} = \int_0^x k_r\,dt$$

The right-hand side is $k_r t$. The left-hand side is evaluated using the method of partial fractions described in *The chemist's toolkit* 30 in Topic 17B. The fraction is expressed as a sum

$$\frac{1}{([A]_0 - x)^2([P] + x)} = \frac{A}{([A]_0 - x)^2} + \frac{B}{[A] - x} + \frac{C}{[P]_0 + x}$$

where A, B, and C are constants to be found. This expression is multiplied through by $([A]_0 - x)^2([P] + x)$

$$1 = A([P]_0 + x) + B([A]_0 - x)([P]_0 + x) + C([A]_0 - x)^2$$

The brackets are expanded and the terms are collected

$$1 = (C{-}B)x^2{+}(A{+}B[A]_0{-}B[P]_0{-}2C[A]_0)x{+}(A[P]_0{+}B[A]_0[P]_0{+}C[A]_0^2)$$

Equating coefficients gives the three equations

$$C - B = 0$$
$$A + B[A]_0 - B[P]_0 - 2C[A]_0 = 0$$
$$A[P]_0 + B[A]_0[P]_0 + C[A]_0^2 = 1$$

The first equation implies that $C = B$. Substituting this into the second two equations gives

$$A - B([A]_0 + [P]_0) = 0$$
$$A[P]_0 + B([A]_0[P]_0 + [A]_0^2) = 1$$

The first equation of these two equations is multiplied by $[P]_0$ and subtracted from the second to give

$$B([A]_0[P]_0 + [A]_0^2) + B([A]_0[P]_0 + [P]_0^2) = 1$$

Rearranging gives

$$B([A]_0^2 + 2[A]_0[P]_0 + [P]_0^2) = 1 \quad \text{hence} \quad B = C = \frac{1}{([A]_0 + [P]_0)^2}$$

This is substituted back into the equation $A - B([A]_0 + [P]_0) = 0$ from above to give

$$A - \frac{1}{([A]_0 + [P]_0)^2} \times ([A]_0 + [P]_0) = 0 \quad \text{hence} \quad A = \frac{1}{[A]_0 + [P]_0}$$

The required integral is therefore

$$kt = \int_0^x \frac{1}{([A]_0 - x)^2([P]_0 + x)}\,dx$$

$$= \int_0^x \left(\frac{1}{([A]_0 + [P]_0)([A]_0 - x)^2} + \frac{1}{([A]_0 + [P]_0)^2([A]_0 - x)} + \frac{1}{([A]_0 + [P]_0)^2([P]_0 + x)} \right) dx$$

$$= \left[\frac{1}{([A]_0 + [P]_0)([A]_0 - x)} - \frac{\ln([A]_0 - x)}{([A]_0 + [P]_0)^2} + \frac{\ln([P]_0 + x)}{([A]_0 + [P]_0)^2} \right]_0^x$$

$$= \left(\frac{1}{([A]_0 + [P]_0)([A]_0 - x)} + \frac{1}{([A]_0 + [P]_0)^2} \ln \frac{[P]_0 + x}{[A]_0 - x} \right) - \left(\frac{1}{([A]_0 + [P]_0)[A]_0} + \frac{1}{([A]_0 + [P]_0)^2} \ln \frac{[P]_0}{[A]_0} \right)$$

$$= \frac{1}{[A]_0 + [P]_0} \left(\frac{1}{[A]_0 - x} - \frac{1}{[A]_0} \right) + \frac{1}{([A]_0 + [P]_0)^2} \ln \frac{[A]_0([P]_0 + x)}{[P]_0([A]_0 - x)}$$

Substituting $x = [P] - [P]_0$ gives the integrated rate law as

$$kt = \frac{1}{[A]_0 + [P]_0} \left(\frac{1}{[A]_0 + [P]_0 - [P]} - \frac{1}{[A]_0} \right) + \frac{1}{([A] + [P]_0)^2} \ln \frac{[A]_0[P]}{[P]_0([A]_0 + [P]_0 - [P])}$$

It is not possible to rearrange this equation to give a simple expression for [P].

To find the time at which the rate reaches a maximum, the expression for the rate, $v = k_r[A]^2[P] = k_r([A]_0 - x)^2([P]_0 + x)$, is differentiated and the derivative is set equal to zero. The chain rule for differentiation implies that $dv/dt = (dv/dx) \times (dx/dt)$, hence

$$\frac{dv}{dt} = \frac{d}{dx}\left[k_r([A]_0 - x)^2([P]_0 + x)\right] \times \frac{dx}{dt}$$

$$= \left[-2k_r([A]_0 - x)([P]_0 + x) + k_r([A]_0 - x)^2\right] \frac{dx}{dt}$$

Setting this equal to zero implies that

$$-2k_r([A]_0 - x)([P]_0 + x) + k_r([A]_0 - x)^2 = 0 \quad \text{or} \quad \frac{dx}{dt} = 0$$

Because $x = [P] - [P]_0$, $dx/dt = d[P]/dt = v$ and so the solution $dx/dt = 0$ corresponds to $v = 0$. This represents a minimum rate rather than a maximum and so is rejected. Examining the other solution gives

$$-2k_r([A]_0 - x)([P]_0 + x) + k_r([A]_0 - x)^2 = 0$$

$$\text{hence} \quad ([A]_0 - x)\left[([A]_0 - x) - 2([P]_0 + x)\right] = 0$$

Hence

$$x = [A]_0 \quad \text{or} \quad [A]_0 - x = 2([P]_0 + x)$$

Because $x = [A]_0 - [A]$, the solution $x = [A]_0$ corresponds to $[A] = 0$. From the rate law $v = k_r[A]^2[P]$ this corresponds to $v = 0$ and therefore to a minimum rate rather than a maximum. The maximum rate is therefore given by the second expression, which is rearranged to yield $x = \frac{1}{3}([A]_0 - 2[P]_0)$.

This is then substituted into the integrated rate law from above

$$k_r t = \frac{1}{[A]_0 + [P]_0}\left(\frac{1}{[A]_0 - x} - \frac{1}{[A]_0}\right) + \frac{1}{([A]_0 + [P]_0)^2}\ln\frac{[A]_0([P]_0 + x)}{[P]_0([A]_0 - x)}$$

$$= \frac{1}{[A]_0 + [P]_0}\left(\frac{1}{\frac{2}{3}[A]_0 + \frac{2}{3}[P]_0} - \frac{1}{[A]_0}\right)$$

$$+ \frac{1}{([A]_0 + [P]_0)^2}\ln\frac{[A]_0(\frac{1}{3}[P]_0 + \frac{1}{3}[A]_0)}{[P]_0(\frac{2}{3}[A]_0 + \frac{2}{3}[P]_0)}$$

$$= \frac{1}{([A]_0 + [P]_0)^2}\left(\frac{3}{2} - \frac{[A]_0 + [P]_0}{[A]_0} + \ln\frac{[A]_0}{2[P]_0}\right)$$

The time at which the rate is at a maximum is therefore

$$\boxed{t = \frac{1}{k_r([A]_0 + [P]_0)^2}\left(\frac{1}{2} + \frac{[P]_0}{[A]_0} + \ln\frac{[A]_0}{2[P]_0}\right)}$$

(e) The rate law is integrated as in part (d). Writing $[A] = [A] - x$ and $[P] = [P]_0 + x$ the rate law is

$$v = \frac{d[P]}{dt} = k_r[A][P]^2 = k_r([A]_0 - x)([P]_0 + x)^2$$

The expression $[P] = [P]_0 + x$ implies that $d[P]/dt = dx/dt$. Therefore

$$\frac{dx}{dt} = k_r([A]_0 - x)([P]_0 + x)^2 \text{ hence } \int_0^x \frac{dx}{([A]_0 - x)([P]_0 + x)^2} = \int_0^t k\,dt$$

The right-hand side is $k_r t$. The left-hand side is evaluated using the method of partial fractions, as in part (d). The fraction is expressed as a sum

$$\frac{1}{([A]_0 - x)([P]_0 + x)^2} = \frac{A}{[A]_0 - x} + \frac{B}{[P]_0 + x} + \frac{C}{([P]_0 + x)^2}$$

where A, B, and C are constants to be found. The expression is multiplied through by $([A]_0 - x)([P]_0 + x)^2$.

$$1 = A([P]_0 + x)^2 + B([A]_0 - x)([P]_0 + x) + C([P]_0 + x)$$

$$= (A - B)x^2 + (2A[P]_0 + B[A]_0 - B[P]_0 - C)x$$

$$+ (A[P]_0^2 + B[A]_0[P]_0 + C[A]_0)$$

Equating coefficients gives

$$A - B = 0$$
$$2A[\mathrm{P}]_0 + B[\mathrm{A}]_0 - B[\mathrm{P}]_0 - C) = 0$$
$$A[\mathrm{P}]_0^2 + B[\mathrm{A}]_0[\mathrm{P}]_0 + C[\mathrm{A}]_0 = 1$$

The first equation implies that $A = B$. Substituting this into the second two equations gives

$$A([\mathrm{A}]_0 + [\mathrm{P}]_0) - C = 0$$
$$A([\mathrm{P}]_0^2 + [\mathrm{A}]_0[\mathrm{P}]_0) + C[\mathrm{A}]_0 = 1$$

The first of these two equations is multiplied by $[\mathrm{A}]_0$ and added to the second equation to give

$$A([\mathrm{A}]_0^2 + 2[\mathrm{A}]_0[\mathrm{P}]_0 + [\mathrm{P}]_0^2) = 1 \quad \text{hence} \quad A = B = \frac{1}{([\mathrm{A}]_0 + [\mathrm{P}]_0)^2}$$

This is substituted back into the equation $A([\mathrm{A}]_0 + [\mathrm{P}]_0) - C = 0$ from above to give

$$\frac{1}{([\mathrm{A}]_0 + [\mathrm{P}]_0)^2} \times ([\mathrm{A}]_0 + [\mathrm{P}]_0) - C = 0 \quad \text{hence} \quad C = \frac{1}{([\mathrm{A}]_0 + [\mathrm{P}]_0)}$$

The required integral is therefore

$$\begin{aligned}
k_r t &= \int_0^x \left(\frac{1}{([\mathrm{A}]_0 + [\mathrm{P}]_0)^2([\mathrm{A}]_0 - x)} + \frac{1}{([\mathrm{A}]_0 + [\mathrm{P}]_0)^2)([\mathrm{P}]_0 + x)} \right. \\
&\qquad \left. + \frac{1}{([\mathrm{A}]_0 + [\mathrm{P}]_0)([\mathrm{P}]_0 + x)^2} \right) \mathrm{d}x \\
&= \left[\frac{-\ln([\mathrm{A}]_0 - x)}{([\mathrm{A}]_0 + [\mathrm{P}]_0)^2} + \frac{\ln([\mathrm{P}]_0 + x)}{([\mathrm{A}]_0 + [\mathrm{P}]_0)^2} - \frac{1}{([\mathrm{A}]_0 + [\mathrm{P}]_0)([\mathrm{P}]_0 + x)} \right]_0^x \\
&= \left(\frac{1}{([\mathrm{A}]_0 + [\mathrm{P}]_0)^2} \ln \frac{[\mathrm{P}]_0 + x}{[\mathrm{A}]_0 - x} - \frac{1}{([\mathrm{A}]_0 + [\mathrm{P}]_0)([\mathrm{P}]_0 + x)} \right) \\
&\qquad - \left(\frac{1}{([\mathrm{A}]_0 + [\mathrm{P}]_0)^2} \ln \frac{[\mathrm{P}]_0}{[\mathrm{A}]_0} - \frac{1}{([\mathrm{A}]_0 + [\mathrm{P}]_0)[\mathrm{P}]_0} \right) \\
&= \frac{1}{([\mathrm{A}]_0 + [\mathrm{P}]_0)^2} \ln \frac{[\mathrm{A}]_0([\mathrm{P}]_0 + x)}{[\mathrm{P}]_0([\mathrm{A}]_0 - x)} + \frac{1}{[\mathrm{A}]_0 + [\mathrm{P}]_0} \left(\frac{1}{[\mathrm{P}]_0} - \frac{1}{[\mathrm{P}]_0 + x} \right)
\end{aligned}$$

Substituting $x = [\mathrm{P}] - [\mathrm{P}]_0$ gives the integrated rate law as

$$k_r t = \frac{1}{([\mathrm{A}]_0 + [\mathrm{P}]_0)^2} \ln \frac{[\mathrm{A}]_0[\mathrm{P}]}{[\mathrm{P}]_0([\mathrm{A}]_0 + [\mathrm{P}]_0 - [\mathrm{P}])} + \frac{1}{[\mathrm{A}]_0 + [\mathrm{P}]_0} \left(\frac{1}{[\mathrm{P}]_0} - \frac{1}{[\mathrm{P}]} \right)$$

As in part (d) it is not possible to rearrange this equation to give a simple expression for $[\mathrm{P}]$.

The time at which the rate reaches a maximum is found by the same method as used in part (d).

$$\frac{\mathrm{d}v}{\mathrm{d}t} = \frac{\mathrm{d}v}{\mathrm{d}x} \times \frac{\mathrm{d}x}{\mathrm{d}t} = \frac{\mathrm{d}}{\mathrm{d}t}\left[k_\mathrm{r}([\mathrm{A}]_0 - x)([\mathrm{P}]_0 + x)^2\right] \times \frac{\mathrm{d}x}{\mathrm{d}t}$$
$$= \left[2k_\mathrm{r}([\mathrm{A}]_0 - x)([\mathrm{P}]_0 + x) - k_\mathrm{r}([\mathrm{P}]_0 + x)^2\right] \times \frac{\mathrm{d}x}{\mathrm{d}t}$$
$$= k_\mathrm{r}([\mathrm{P}]_0 + x)(2[\mathrm{A}]_0 - [\mathrm{P}]_0 - 3x)\frac{\mathrm{d}x}{\mathrm{d}t}$$

At the maximum, this expression is equal to zero. The solution $\mathrm{d}x/\mathrm{d}t = 0$ is discarded for the same reason as in part (d), and the solution $x = -[\mathrm{P}]_0$ is discarded because x must be positive. The remaining solution is

$$2[\mathrm{A}]_0 - [\mathrm{P}]_0 - 3x = 0 \quad \text{hence} \quad x = \tfrac{1}{3}(2[\mathrm{A}]_0 - [\mathrm{P}]_0)$$

This is substituted into the integrated rate law from above

$$k_\mathrm{r}t = \frac{1}{([\mathrm{A}]_0 + [\mathrm{P}]_0)^2}\ln\frac{[\mathrm{A}]_0([\mathrm{P}]_0 + x)}{[\mathrm{P}]_0([\mathrm{A}]_0 - x)} + \frac{1}{[\mathrm{A}]_0 + [\mathrm{P}]_0}\left(\frac{1}{[\mathrm{P}]_0} - \frac{1}{[\mathrm{P}]_0 + x}\right)$$
$$= \frac{1}{([\mathrm{A}]_0 + [\mathrm{P}]_0)^2}\ln\frac{[\mathrm{A}]_0(\frac{2}{3}[\mathrm{A}]_0 + \frac{2}{3}[\mathrm{P}]_0)}{[\mathrm{P}]_0(\frac{1}{3}[\mathrm{A}]_0 + \frac{1}{3}[\mathrm{P}]_0)}$$
$$+ \frac{1}{[\mathrm{A}]_0 + [\mathrm{P}]_0}\left(\frac{1}{[\mathrm{P}]_0} - \frac{1}{\frac{2}{3}[\mathrm{A}]_0 + \frac{2}{3}[\mathrm{P}]_0}\right)$$
$$= \frac{1}{([\mathrm{A}]_0 + [\mathrm{P}]_0)^2}\left(\ln\frac{2[\mathrm{A}]_0}{[\mathrm{P}]_0} + \frac{[\mathrm{A}]_0 + [\mathrm{P}]_0}{[\mathrm{P}]_0} - \frac{3}{2}\right)$$

Hence the maximum rate is reached at

$$\boxed{t = \frac{1}{k_\mathrm{r}([\mathrm{A}]_0 + [\mathrm{P}]_0)^2}\left(\ln\frac{2[\mathrm{A}]_0}{[\mathrm{P}]_0} + \frac{[\mathrm{A}]_0}{[\mathrm{P}]_0} - \frac{1}{2}\right)}$$

I17.3 (a) Because the second step is rate-determining, the first step and its reverse are treated as a pre-equilibrium because the rate of reaction of A^- with AH to form product is assumed to be too slow to affect the maintenance of the pre-equilibrium. As explained in Section 17E.5 on page 750 it follows that

$$K = \frac{k_\mathrm{a}}{k'_\mathrm{a}} = \frac{[\mathrm{BH}^+][\mathrm{A}^-]}{[\mathrm{AH}][\mathrm{B}]} \quad \text{hence} \quad [\mathrm{A}^-] = \frac{k_\mathrm{a}[\mathrm{AH}][\mathrm{B}]}{k'_\mathrm{a}[\mathrm{BH}^+]}$$

The rate formation of product is $v = \mathrm{d}[\mathrm{P}]/\mathrm{d}t = k_\mathrm{b}[\mathrm{A}^-][\mathrm{AH}]$. The expression for $[\mathrm{A}^-]$ is substituted into this to give

$$v = k_\mathrm{b}[\mathrm{A}^-][\mathrm{AH}] = k_\mathrm{b}\frac{k_\mathrm{a}[\mathrm{AH}][\mathrm{B}]}{k'_\mathrm{a}[\mathrm{BH}^+]}[\mathrm{AH}] = \boxed{\frac{k_\mathrm{a}k_\mathrm{b}[\mathrm{AH}]^2[\mathrm{B}]}{k'_\mathrm{a}[\mathrm{BH}^+]}}$$

The same result is alternatively derived using the steady-state approximation. Applying the steady-state approximation to A^- gives

$$\frac{\mathrm{d}[\mathrm{A}^-]}{\mathrm{d}t} = k_\mathrm{a}[\mathrm{AH}][\mathrm{B}] - k'_\mathrm{a}[\mathrm{BH}^+][\mathrm{A}^-] - k_\mathrm{b}[\mathrm{A}^-][\mathrm{AH}] = 0$$

This expression is rearranged

$$[A^-] = \frac{k_a[AH][B]}{k_a'[BH^+] + k_b[AH]}$$

Substituting this into the rate of formation of product gives

$$v = k_b[A^-][AH] = k_b \frac{k_a[AH][B]}{k_a'[BH^+] + k_b[AH]}[AH] = \frac{k_a k_b[AH]^2[B]}{k_a'[BH^+] + k_b[AH]}$$

Finally it is noted that because the second step is rate-determining the rate of conversion of A^- to products, $k_b[A^-][AH]$, is much slower than the rate of its reversion to reactants, $k_a'[A^-][BH^+]$.

$$k_a'[A^-][BH^+] \gg k_b[A^-][AH] \quad \text{hence} \quad k_a'[BH^+] \gg k_b[AH]$$

The term $k_b[AH]$ is therefore neglected in the denominator of the rate law which then becomes $v = k_a k_b[AH]^2[B]/k_a'[BH^+]$ as before.

(b) Because the second step is rate-determining, the formation of HAH^+ from HA and H^+ forms a pre-equilibrium in which the rates of the forward and backward steps are considered to be equal because the gradual removal of HAH^+ to form products is assumed to be too slow to significantly affect the maintenance of the equilibrium. Equating the rates of the forward and backward steps gives

$$k_a[HA][H^+] = k_a'[HAH^+] \quad \text{hence} \quad [HAH^+] = \frac{k_a[HA][H^+]}{k_a'}$$

The rate of product formation is equal to the rate of the second step

$$v = k_b[HAH^+][B] = k_b \frac{k_a[HA][H^+]}{k_a'}[B] = \boxed{\frac{k_a k_b}{k_a'}[HA][H^+][B]}$$

I17.5 A polymer consisting of N monomer units has a molar mass of NM_1, where M_1 is the molar mass of a single monomer unit. The mean molar mass is therefore $\langle NM_1 \rangle = M_1\langle N \rangle$, and likewise the mean square molar mass and mean cube molar mass are

$$\langle M^2 \rangle = \langle (NM_1)^2 \rangle = M_1^2\langle N^2 \rangle \quad \text{and} \quad \langle M^3 \rangle = \langle (NM_1)^3 \rangle = M_1^3\langle N^3 \rangle$$

The task is therefore to find $\langle N^2 \rangle$ and $\langle N^3 \rangle$.

It is supposed that each monomer has one end group A with which it can join to another monomer. In a polymer, only the terminal monomer unit in the chain has a free end group.

The probability P_N that a polymer consists of N monomers is equal to the probability that it has $N-1$ reacted end groups and one unreacted end group. The fraction of end groups that have reacted is p and the fraction of free end groups remaining is $1-p$, so the probability that a polymer contains $N-1$ reacted groups and one unreacted group is $p^{N-1} \times (1-p)$.

It is convenient to begin by evaluating the average value of N.

$$\langle N\rangle = \sum_{N=1}^{\infty} NP_N = \sum_{N=1}^{\infty} Np^{N-1}(1-p) = (1-p)\sum_{N=1}^{\infty} Np^{N-1}$$

To evaluate the sum, it is noted that Np^{N-1} corresponds to the derivative of p^N. Hence

$$\sum_{N=1}^{\infty} Np^{N-1} = \sum_{N=1}^{\infty} \frac{d}{dp}p^N = \frac{d}{dp}\left[\sum_{N=1}^{\infty} p^N\right] = \frac{d}{dp}[p + p^2 + p^3 + ...]$$

The expression in square brackets is a geometric series with first term p and common ratio p; the sum to infinity of this series is therefore $p/(1-p)$. Hence

$$\sum_{N=1}^{\infty} Np^{N-1} = \frac{d}{dp}\left[\frac{p}{1-p}\right] = \frac{(1-p)+p}{(1-p)^2} = \frac{1}{(1-p)^2}$$

The average value of N is therefore

$$\langle N\rangle = (1-p)\sum_{N=1}^{\infty} Np^{N-1} = (1-p) \times \frac{1}{(1-p)^2} = \frac{1}{1-p}$$

This is the same result as [17F.12a–755] which is derived in Section 17F.2(a) on page 755 by a different method. However the approach used here is more easily generalised to find an expression for $\langle N^2\rangle$ and $\langle N^3\rangle$.

$$\langle N^2\rangle = \sum_{N=1}^{\infty} N^2P_N = \sum_{N=1}^{\infty} N^2p^{N-1}(1-p) = (1-p)\sum_{N=1}^{\infty} N^2p^{N-1}$$

The sum $\sum_{N=1}^{\infty} N^2p^{N-1}$ is evaluated by noting that Np^{N-1} is the derivative of p^N

$$\sum_{N=1}^{\infty} N^2p^{N-1} = \sum_{N=1}^{\infty} N \times Np^{N-1} = \sum_{N=1}^{\infty} N \times \frac{d}{dp}p^N = \frac{d}{dp}\sum_{N=1}^{\infty} Np^N$$
$$= \frac{d}{dp}\left[p\sum_{N=1}^{\infty} Np^{N-1}\right]$$

The sum $\sum_{N=1}^{\infty} Np^{N-1}$ was already evaluated above; its value is $1/(1-p)^2$. Hence

$$\sum_{N=1}^{\infty} N^2p^{N-1} = \frac{d}{dp}\left[p \times \frac{1}{(1-p)^2}\right] = \frac{1+p}{(1-p)^3}$$

The mean value of N^2 is therefore

$$\langle N^2\rangle = (1-p)\sum_{N=1}^{\infty} N^2p^{N-1} = (1-p) \times \frac{1+p}{(1-p)^3} = \frac{1+p}{(1-p)^2}$$

The mean value of N^3 is evaluated in a similar way, using the result already deduced that $\sum_{N=1}^{\infty} N^2 p^{N-1} = (1+p)/(1-p)^3$.

$$\langle N^3 \rangle = \sum_{N=1}^{\infty} N^3 P_N = \sum_{N=1}^{\infty} N^3 p^{N-1}(1-p) = (1-p)\sum_{N=1}^{\infty} N^2 \times N p^{N-1}$$

$$= (1-p)\sum_{N=1}^{\infty} N^2 \frac{d}{dp} p^N = (1-p)\frac{d}{dp}\sum_{N=1}^{\infty} N^2 p^N$$

$$= (1-p)\frac{d}{dp}\left[p \overbrace{\sum_{N=1}^{\infty} N^2 p^{N-1}}^{(1+p)/(1-p)^3}\right] = (1-p)\frac{d}{dp}\left[p \times \frac{1+p}{(1-p)^3}\right]$$

$$= (1-p) \times \frac{p^2+4p+1}{(1-p)^4} = \frac{p^2+4p+1}{(1-p)^4}$$

Using the results $\langle M^2 \rangle = M_1^2 \langle N^2 \rangle$ and $\langle M^3 \rangle = M_1^3 \langle N^3 \rangle$ deduced at the start of the question, together with the expressions for $\langle N^2 \rangle$ and $\langle N^3 \rangle$, the mean square and cube molar masses are

$$\langle M^2 \rangle = \frac{M_1^2(1+p)}{(1-p)^2} \qquad \langle M^3 \rangle = \frac{M_1^3(p^2+4p+1)}{(1-p)^3}$$

(a) The required ratio is

$$\frac{\langle M^3 \rangle}{\langle M^2 \rangle} = \frac{M_1^3(p^2+4p+1)/(1-p)^3}{M_1^2(1+p)/(1-p)^2} = \boxed{\frac{M_1(p^2+4p+1)}{(1+p)(1-p)}}$$

(b) The average number of monomers per polymer, that is the chain length, is given by [17F.12a–755], $\langle N \rangle = 1/(1-p)$. This expression is rearranged to $p = 1 - 1/\langle N \rangle$. This is then substituted into the expression derived in (b) to give the ratio in terms of chain length.

$$\frac{\langle M^3 \rangle}{\langle M^2 \rangle} = \frac{M_1(p^2+4p+1)}{(1+p)(1-p)} = M_1\frac{(1-1/\langle N \rangle)^2 + 4(1-1/\langle N \rangle)+1}{(1+[1-1/\langle N \rangle])(1-[1-1/\langle N \rangle])}$$

$$= \boxed{\frac{M_1(6\langle N \rangle^2 - 6\langle N \rangle + 1)}{2\langle N \rangle - 1}}$$

I17.7 A possible mechanism is

$Cl_2 + h\nu \rightarrow 2Cl$	$v = I_a$
$Cl + CHCl_3 \rightarrow CCl_3 + HCl$	$v = k_b[Cl][CHCl_3]$
$CCl_3 + Cl_2 \rightarrow CCl_4 + Cl$	$v = k_c[CCl_3][Cl_2]$
$2CCl_3 + Cl_2 \rightarrow 2CCl_4$	$v = k_d[CCl_3]^2[Cl_2]$

The steady-state approximation is applied to the intermediates Cl and CCl_3.

$$\frac{d[Cl]}{dt} = 2I_a - k_b[Cl][CHCl_3] + k_c[CCl_3][Cl_2] = 0$$

$$\frac{d[CCl_3]}{dt} = k_b[Cl][CCl_3] - k_c[CCl_3][Cl_2] - 2k_d[CCl_3]^2[Cl_2] = 0$$

On adding these two equations together, the terms in k_b and k_c cancel

$$2I_a - 2k_d[CCl_3]^2[Cl_2] = 0 \quad \text{hence} \quad [CCl_3] = \left(\frac{I_a}{k_d[Cl_2]}\right)^{1/2}$$

The rate of formation of CCl_4 product is

$$\frac{d[CCl_4]}{dt} = k_c[CCl_3][Cl_2] + 2k_d[CCl_3]^2[Cl_2]$$

The expression $[CCl_4] = (I_a/k_d[Cl_2])^{1/2}$ is substituted into this

$$\frac{d[CCl_4]}{dt} = k_c\left(\frac{I_a}{k_d[Cl_2]}\right)^{1/2}[Cl_2]+2k_d\left(\frac{I_a}{k_d[Cl_2]}\right)[Cl_2] = \frac{k_c I_a^{1/2}}{k_d^{1/2}}[Cl_2]^{1/2}+2I_a$$

If the chlorine pressure is sufficiently high that the term $(k_c I_a^{1/2}/k_d^{1/2})[Cl_2]^{1/2}$ is large compared with the $2I_a$ term, the latter is neglected and the rate law becomes

$$\boxed{\frac{d[CCl_4]}{dt} = \frac{k_c}{k_d^{1/2}} I_a^{1/2}[Cl_2]^{1/2}}$$

which is the observed rate law $d[CCl_4]/dt = k_r[Cl]^{1/2}I_a^{1/2}$ with $k_r = (k_c/k_d)^{1/2}$.

18 Reaction dynamics

18A Collision theory

Answers to discussion questions

D18A.1 Collision theory expresses a rate of reaction as a fraction of the rate of collision, on the assumption that reaction happens only between colliding molecules, and then only if the collision has enough energy and the proper orientation. Therefore the rate of reaction is directly proportional to the rate of collision, which is computed using kinetic molecular theory. The fraction of collisions energetic enough for reaction also comes from kinetic-molecular theory combined with the Boltzmann distribution.

D18A.3 This is described in Section 18A.1(c) on page 784.

Solutions to exercises

E18A.1(a) The collision frequency is given by [1B.12b–*17*], $z = \sigma v_{\rm rel} p/kT$, where σ is the collision cross-section, given in terms of the collision diameter d as $\sigma = \pi d^2$, and $v_{\rm rel}$ is the mean relative speed of the colliding molecules. This speed is given by [1B.11b–*16*], $v_{\rm rel} = (8kT/\pi\mu)^{1/2}$, with $\mu = m_{\rm A}m_{\rm B}/(m_{\rm A}+m_{\rm B})$. For collisions between like molecules $\mu = m/2$ and $v_{\rm rel} = (16kT/\pi m)^{1/2}$.

$$\begin{aligned}
z &= \frac{\sigma v_{\rm rel} p}{kT} = \frac{\pi d^2 p}{kT}\left(\frac{16kT}{\pi m}\right)^{1/2} = 4d^2 p\left(\frac{\pi}{mkT}\right)^{1/2}\\
&= 4\times(380\times10^{-12}\ {\rm m})^2\times(120\times10^3\ {\rm Pa})\\
&\quad\times\left(\frac{\pi}{(17.03\times1.6605\times10^{-27}\ {\rm kg})\times(1.3806\times10^{-23}\ {\rm J\,K^{-1}})\times(303\ {\rm K})}\right)^{1/2}\\
&= \boxed{1.12\times10^{10}\ {\rm s^{-1}}}
\end{aligned}$$

To confirm the units of z it is useful to recall that $1\ {\rm J} = 1\ {\rm kg\,m^2\,s^{-2}}$ and $1\ {\rm Pa} = 1\ {\rm kg\,m^{-1}\,s^{-2}}$.

The collision density between identical molecules is given by [18A.4b–*781*]

$$Z_{\rm AA} = \sigma\left(\frac{4kT}{\pi m}\right)^{1/2} N_{\rm A}^2[{\rm A}]^2$$

where [A] is the molar concentration of the gas. In turn, this is expressed in terms of the pressure using the perfect gas equation to give $[\mathrm{A}] = n_\mathrm{A}/V = p_\mathrm{A}/RT$.

$$Z_{\mathrm{AA}} = \pi d^2 \left(\frac{4kT}{\pi m}\right)^{1/2} \frac{N_\mathrm{A}^2 p_\mathrm{A}^2}{R^2T^2} = 2d^2\left(\frac{\pi kT}{m}\right)^{1/2} \frac{p_\mathrm{A}^2}{k^2T^2} = 2d^2\left(\frac{\pi}{mk^3T^3}\right)^{1/2} p_\mathrm{A}^2$$

$$= 2 \times (360 \times 10^{-12}\ \mathrm{m})^2 \times (120 \times 10^3\ \mathrm{Pa})^2$$

$$\times \left(\frac{\pi}{(17.03 \times 1.6605 \times 10^{-27}\ \mathrm{kg}) \times (1.3806 \times 10^{-23}\ \mathrm{J\,K^{-1}})^3 \times (303\ \mathrm{K})^3}\right)^{1/2}$$

$$= \boxed{1.62 \times 10^{35}\ \mathrm{m^{-3}\,s^{-1}}}$$

The above expression shows that $z \propto pT^{-1/2}$, but at constant volume $p \propto T$, therefore the overall temperature dependence is $z \propto T^{1/2}$. The percentage increase in z on increasing T by 10 K is therefore

$$\frac{313^{1/2} - 303^{1/2}}{303^{1/2}} = 0.0163... = \boxed{1.6\%}$$

Similarly the final expression for the collision density shows $Z_{\mathrm{AA}} \propto p^2T^{-3/2}$ which, with $p \propto T$, gives $Z_{\mathrm{AA}} \propto T^2T^{-3/2} \propto T^{1/2}$. This is the same dependence as z, so the same percentage increase will result.

E18A.2(a) The collision theory expression for the rate constant is given in [18A.9–*783*]. In this expression, the factor $\mathrm{e}^{-E_\mathrm{a}/RT}$ is identified as the fraction of collisions f having at least kinetic energy E_a along the flight path. For example with $E_\mathrm{a} = 20\ \mathrm{kJ\,mol^{-1}}$ and $T = 350$ K

$$\frac{E_\mathrm{a}}{RT} = \frac{20 \times 10^3\ \mathrm{J\,mol^{-1}}}{(8.3145\ \mathrm{J\,K^{-1}\,mol^{-1}}) \times (350\ \mathrm{K})} = 6.87... \quad f = \mathrm{e}^{-6.87...} = \boxed{1.04 \times 10^{-3}}$$

A similar calculation gives $\boxed{f = 0.069}$ at $T = 900$ K. With $E_\mathrm{a} = 100\ \mathrm{kJ\,mol^{-1}}$ the result is $\boxed{f = 1.19 \times 10^{-15}}$ at $T = 350$ K, and $\boxed{f = 1.57 \times 10^{-6}}$ at $T = 900$ K.

E18A.3(a) The method for calculating the fractions is shown in the solution to *Exercise* E18A.2(a). For $E_\mathrm{a} = 20\ \mathrm{kJ\,mol^{-1}}$ and $T = 350$ K it is found that $f = 1.03... \times 10^{-3}$ and increasing the temperature to 360 K gives $f = 1.25... \times 10^{-3}$. The percentage increase is

$$100 \times \frac{(1.25... \times 10^{-3}) - (1.03... \times 10^{-3})}{1.03... \times 10^{-3}} = \boxed{21\%}$$

A similar calculation gives an increase by $\boxed{3.0\%}$ at 900 K. With $E_\mathrm{a} = 100\ \mathrm{kJ\,mol^{-1}}$ the result is $\boxed{160\%}$ at $T = 350$ K, and $\boxed{16\%}$ at $T = 900$ K.

E18A.4(a) The collision theory expression for the rate constant is given in [18A.9–*783*].

$$k_r = \sigma N_A \left(\frac{8kT}{\pi\mu}\right)^{1/2} e^{-E_a/RT}$$
$$= (0.36 \times 10^{-18}\ m^2) \times (6.0221 \times 10^{23}\ mol^{-1})$$
$$\times \left(\frac{8 \times (1.3806 \times 10^{-23}\ J\,K^{-1}) \times (650\ K)}{\pi(3.32 \times 10^{-27}\ kg)}\right)^{1/2}$$
$$\times\, e^{-(171\times10^3\ J\,mol^{-1})/[(8.3145\ J\,K^{-1}\,mol^{-1})\times(650\ K)]}$$
$$= \boxed{1.0 \times 10^{-5}\ mol^{-1}\ m^3\ s^{-1}}$$

The units are best resolved by realising that $(8kT/\pi\mu)^{1/2}$ is a speed, with units $m\ s^{-1}$. Note that $0.36\ nm^2$ is $0.36 \times 10^{-18}\ m^2$.

E18A.5(a) As described in Section 18A.1(b) on page 781, the reactive cross section may be estimated from the (non-reactive) collision cross sections of A and B: $\sigma_{est} = \frac{1}{4}(\sigma_A^{1/2} + \sigma_B^{1/2})^2$. The steric factor is given by the ratio of the experimental reactive cross section, σ_{exp}, to the estimated cross section

$$P = \frac{\sigma_{exp}}{\sigma_{est}} = \frac{9.2 \times 10^{-22}\ m^2}{[(0.95 \times 10^{-18}\ m^2)^{1/2} + (0.65 \times 10^{-18}\ m^2)^{1/2}]^2/4}$$
$$= \boxed{1.2 \times 10^{-3}}$$

E18A.6(a) In the RRK theory the rate constant for the unimolecular decay of an energized molecule A* is given by [18A.11–*785*],

$$k_b(E) = \left(1 - \frac{E^*}{E}\right)^{s-1} k_b = (1-x)^{s-1} k_b$$

where $x = E^*/E$. For a non-linear molecule with 5 atoms there are $3N - 6 = 3 \times 5 - 6 = 9$ normal modes, so $s = 9$. This expression is rearranged for x to give

$$x = 1 - [k_b(E)/k_b]^{1/(s-1)}$$
$$= 1 - [3.0 \times 10^{-5}]^{1/(9-1)} = \boxed{0.73}$$

E18A.7(a) In the RRK theory the rate constant for the unimolecular decay of an energized molecule A* is given by [18A.11–*785*],

$$\frac{k_b(E)}{k_b} = \left(1 - \frac{E^*}{E}\right)^{s-1}$$

where E^* is the minimum energy needed to break the bond, and E is the energy available from the collision. With the data given

$$\frac{k_b(E)}{k_b} = \left(1 - \frac{200\ kJ\,mol^{-1}}{250\ kJ\,mol^{-1}}\right)^{10-1} = \boxed{5.12 \times 10^{-7}}$$

Solutions to problems

P18A.1 The collision theory expression for the rate constant is given in [18A.9–783]

$$k_r = \sigma^* N_A \left(\frac{8kT}{\pi\mu}\right)^{1/2} e^{-E_a/RT}$$

Here σ^* is interpreted as the reactive cross-section, related to the collision cross-section σ by $\sigma^* = P\sigma$, where P is the steric factor. Comparison of the above expression for k_r with the Arrhenius equation, $k_r = Ae^{-E_a/RT}$, gives the frequency factor as $A = \sigma^* N_A (8kT/\pi\mu)^{1/2}$; this is rearranged to give an expression for σ^*. It is convenient to express the given frequency factor 2.4×10^{10} dm^3 mol^{-1} s^{-1} as 2.4×10^{7} m^3 mol^{-1} s^{-1}. The mass of a CH_3 radical is 15.03 m_u, therefore the reduced mass of the collision is $\mu = \frac{1}{2}\times 15.03\ m_u = 1.24...\times10^{-26}$ kg.

$$\sigma^* = \frac{A}{N_A}\left(\frac{\pi\mu}{8kT}\right)^{1/2}$$

$$= \frac{2.4\times10^{7}\ \mathrm{m^3\,mol^{-1}\,s^{-1}}}{6.0221\times10^{23}\ \mathrm{mol^{-1}}}\left(\frac{\pi(1.24...\times10^{-26}\ \mathrm{kg})}{8\times(1.3806\times10^{-23}\ \mathrm{J\,K^{-1}})\times(298\ \mathrm{K})}\right)^{1/2}$$

$$= 4.34...\times10^{-20}\ \mathrm{m^2} = \boxed{0.043\ \mathrm{nm^2}}$$

The units are best resolved by realising that $(8kT/\pi\mu)^{1/2}$ is a speed, with units m s^{-1}.

To estimate the collision cross-section assume that d is twice the C–H bond length and compute $\sigma = \pi d^2 = \pi(2\times154\times10^{-12}\ \mathrm{m})^2 = 2.98...\times10^{-19}\ \mathrm{m^2}$. The steric factor is $P = \sigma^*/\sigma = (4.34...\times10^{-20}\ \mathrm{m^2})/(2.98...\times10^{-19}) = \boxed{0.15}$.

P18A.3 The collision theory expression for the rate constant is given in [18A.9–783]

$$k_r = \sigma N_A \left(\frac{8kT}{\pi\mu}\right)^{1/2} e^{-E_a/RT}$$

The maximum value for the rate constant is when $E_a = 0$. The collision cross section is taken as $\sigma = \pi d^2 = \pi(308\times10^{-12}\ \mathrm{m})^2 = 2.98...\times10^{-19}\ \mathrm{m^2}$. The mass of a CH_3 radical is 15.03 m_u, therefore the reduced mass of the collision is $\mu = \frac{1}{2}\times15.03\ m_u = 1.24...\times10^{-26}$ kg

$$k_r = \sigma N_A\left(\frac{8kT}{\pi\mu}\right)^{1/2} = (2.98...\times10^{-19}\ \mathrm{m^2})\times(6.0221\times10^{23}\ \mathrm{mol^{-1}})$$

$$\times\left(\frac{8\times(1.3806\times10^{-23}\ \mathrm{J\,K^{-1}})\times(298\ \mathrm{K})}{\pi(1.24...\times10^{-26}\ \mathrm{kg})}\right)^{1/2}$$

$$= \boxed{1.64\times10^{8}\ \mathrm{mol^{-1}\,m^3\,s^{-1}}}$$

The units are best resolved by realising that $(8kT/\pi\mu)^{1/2}$ is a speed, with units m s^{-1}.

For a second-order reaction the integrated rate law is [17B.4b–733], $1/[CH_3] - 1/[CH_3]_0 = k_r t$. Suppose that initially an amount in moles n_0 of C_2H_6 is introduced into the vessel, and that a fraction α dissociates. The amount of C_2H_6 remaining is $n_0(1-\alpha)$ and the amount of CH_3 produced is $2n_0\alpha$. The total amount of gas is $n_0(1+\alpha)$, therefore the mole fraction of CH_3 is $2\alpha/(1+\alpha)$ and hence the partial pressure of CH_3 is $2\alpha p_{tot}/(1+\alpha)$. The molar concentration corresponding to this pressure is found using the perfect gas law as

$$[CH_3] = \frac{n_{CH_3}}{V} = \frac{p_{CH_3}}{RT} = \frac{2\alpha p_{tot}}{RT(1+\alpha)}$$

With the data given this evaluates as

$$[CH_3] = \frac{2 \times 0.1 \times (100 \times 10^3\ \text{Pa})}{(8.3145\ \text{J K}^{-1}\ \text{mol}^{-1}) \times (298\ \text{K}) \times (1+0.1)} = 7.33...\ \text{mol m}^{-3}$$

If recombination proceeds to 90%, the amount of CH_3 remaining is $\frac{1}{10}$ of the initial. The time for this to take place is found by solving

$$\frac{10}{[CH_3]_0} - \frac{1}{[CH_3]_0} = k_r t$$

Hence

$$t = \frac{9}{[CH_3]_0 k_r} = \frac{9}{(7.33...\ \text{mol m}^{-3}) \times (1.64 \times 10^8\ \text{mol}^{-1}\ \text{m}^3\ \text{s}^{-1})}$$
$$= \boxed{7.5\ \text{ns}}$$

P18A.5 The collision theory expression for the rate constant, including the steric factor P, is given in [18A.10–784]

$$k_r = P\sigma N_A \left(\frac{8kT}{\pi\mu}\right)^{1/2} e^{-E_a/RT}$$

As described in Section 18A.1(b) on page 781, the collision cross-section between A and B may be estimated from the collision cross sections of A and B: $\sigma = \frac{1}{4}(\sigma_A^{1/2} + \sigma_B^{1/2})^2$. From the *Resource section* the cross section for O_2 is 0.40 nm^2. No values are given for the ethyl and cyclohexyl radicals, so these will be approximated by the values for ethene (0.64 nm^2) and benzene (0.88 nm^2), respectively. The reactive cross sections are therefore

$$\sigma_{ethyl} = \tfrac{1}{4}[(0.40)^{1/2} + (0.64)^{1/2}]^2 = 0.512...\ \text{nm}^2$$

A similar calculation gives $\sigma_{hexyl} = 0.616...\ \text{nm}^2$

The mass of O_2 is 32.00 m_u, that of the C_2H_5 radical is 29.06 m_u, and that of the C_6H_{11} radical is 83.15 m_u. The reduced mass of the O_2–C_2H_5 collision is

$$\mu = \frac{m_{O_2} m_{C_2H_5}}{m_{O_2} + m_{C_2H_5}} = \frac{32.00 \times 29.06}{32.00 + 29.06} \times (1.6605 \times 10^{-27}\ \text{kg}) = 2.52... \times 10^{-26}\ \text{kg}$$

For the O_2–C_6H_{11} collision the reduced mass is $3.83... \times 10^{-26}$ kg.

Taking the activation energy as $E_a = 0$, the steric factor is given by

$$P = \frac{k_r}{\sigma N_A}\left(\frac{\pi\mu}{8kT}\right)^{1/2}$$

For this calculation it is convenient to express the rate constants in units of $m^3\ mol^{-1}\ s^{-1}$. For the reaction with C_2H_5

$$P = \frac{4.7 \times 10^6\ m^3\ mol^{-1}\ s^{-1}}{(0.512... \times 10^{-18}\ m^2) \times (6.0221 \times 10^{23}\ mol^{-1})} \times \left(\frac{\pi(2.52... \times 10^{-26}\ kg)}{8 \times (1.3806 \times 10^{-23}\ J\,K^{-1}) \times (298\ K)}\right)^{1/2}$$
$$= \boxed{0.024}$$

A similar calculation for the reaction with C_6H_{11} gives $\boxed{P = 0.043}$.

18B Diffusion-controlled reactions

Answers to discussion questions

D18B.1 A reaction in solution can be regarded as the outcome of two stages: the first is the encounter of two reactant species; the second is the actual reaction between the two species. If the rate-determining step is the former, then the reaction is said to be diffusion controlled, if it is the latter which is rate-determining, the reaction is said to be activation controlled.

For a diffusion-controlled reaction the rate constant is approximated by [18B.4–*789*], $k_d = 8RT/3\eta$, where η is the viscosity. The viscosity does vary with temperature according to $\eta \propto e^{E_a/RT}$ with $E_a \approx 15\ kJ\,mol^{-1}$ for water. Thus, diffusion-controlled reactions do show a small activation energy. Reactions which are activation-controlled are expected to show an activation energy.

Solutions to exercises

E18B.1(a) The second-order rate constant for a diffusion-controlled reaction is given by [18B.3–*789*], $k_d = 4\pi R^* D N_A$, where R^* is the critical distance and D is the diffusion constant. As explained in the text, D is the sum of the diffusion constants of the two species, therefore in this case D is twice the value given. With the data given

$$k_d = 4\pi \times (0.5 \times 10^{-9}\ m) \times (2 \times 6 \times 10^{-9}\ m^2\ s^{-1}) \times (6.0221 \times 10^{23}\ mol^{-1})$$
$$= \boxed{4.5 \times 10^7\ m^3\ mol^{-1}\ s^{-1}}$$

E18B.2(a) For a diffusion-controlled reaction the rate constant is approximated by [18B.4–*789*], $k_d = 8RT/3\eta$, where η is the viscosity.

(i) For water

$$k_d = \frac{8 \times (8.3145\,\mathrm{J\,K^{-1}\,mol^{-1}}) \times (298\,\mathrm{K})}{3 \times (1.00 \times 10^{-3}\,\mathrm{kg\,m^{-1}\,s^{-1}})} = \boxed{6.61 \times 10^6\,\mathrm{m^3\,mol^{-1}\,s^{-1}}}$$

In sorting out the units it is useful to recall $1\,\mathrm{J} = 1\,\mathrm{kg\,m^2\,s^{-2}}$.

(ii) For pentane

$$k_d = \frac{8 \times (8.3145\,\mathrm{J\,K^{-1}\,mol^{-1}}) \times (298\,\mathrm{K})}{3 \times (2.2 \times 10^{-4}\,\mathrm{kg\,m^{-1}\,s^{-1}})} = \boxed{3.0 \times 10^7\,\mathrm{m^3\,mol^{-1}\,s^{-1}}}$$

E18B.3(a) For a diffusion-controlled reaction the rate constant is approximated by [18B.4–*789*], $k_d = 8RT/3\eta$, where η is the viscosity. Recall that $1\,\mathrm{P} = 10^{-1}\,\mathrm{kg\,m^{-1}\,s^{-1}}$, so that $1\,\mathrm{cP} = 10^{-3}\,\mathrm{kg\,m^{-1}\,s^{-1}}$. Therefore the rate constant is

$$\begin{aligned} k_d &= \frac{8 \times (8.3145\,\mathrm{J\,K^{-1}\,mol^{-1}}) \times (320\,\mathrm{K})}{3 \times (0.89 \times 10^{-3}\,\mathrm{kg\,m^{-1}\,s^{-1}})} \\ &= 7.97... \times 10^6\,\mathrm{m^3\,mol^{-1}\,s^{-1}} = \boxed{8.0 \times 10^6\,\mathrm{m^3\,mol^{-1}\,s^{-1}}} \end{aligned}$$

The half-life of a second-order reaction is given by [17B.5–*734*], $t_{1/2} = 1/k_r[\mathrm{A}]_0$. The initial concentration is 1.5 mmol dm^{-3} which is 1.5 mol m^{-3}. With the data given

$$t_{1/2} = \frac{1}{(7.97... \times 10^6\,\mathrm{m^3\,mol^{-1}\,s^{-1}}) \times (1.5\,\mathrm{mol\,m^{-3}})} = \boxed{84\,\mathrm{ns}}$$

E18B.4(a) The second-order rate constant for a diffusion-controlled reaction is given by [18B.3–*789*], $k_d = 4\pi R^* D N_A$, where R^* is the critical distance and D is the diffusion constant. As explained in the text D is the sum of the diffusion constants of the two species. The value of D is estimated using the Stokes–Einstein equation, $D = kT/6\pi\eta R$, and with the data given separate values of D are computed for the two species. The critical distance is taken as $R^* = R_A + R_B$.

$$\begin{aligned} k_d &= 4\pi(R_A + R_B)(D_A + D_B)N_A \\ &= 4\pi N_A(R_A + R_B)\frac{kT}{6\pi\eta}\left(\frac{1}{R_A} + \frac{1}{R_B}\right) \\ &= 4\pi \times (6.0221 \times 10^{23}\,\mathrm{mol^{-1}}) \times (655 + 1820) \\ &\quad \times \frac{(1.3806 \times 10^{-23}\,\mathrm{J\,K^{-1}}) \times (313\,\mathrm{K})}{6\pi \times (2.93 \times 10^{-3}\,\mathrm{kg\,m^{-1}\,s^{-1}})}\left(\frac{1}{655} + \frac{1}{1820}\right) \\ &= 3.04... \times 10^6\,\mathrm{m^3\,mol^{-1}\,s^{-1}} \end{aligned}$$

The initial concentrations are $[\mathrm{A}] = 0.170\,\mathrm{mol\,dm^{-3}} = 0.170 \times 10^3\,\mathrm{mol\,m^{-3}}$ and $[\mathrm{B}] = 0.350\,\mathrm{mol\,dm^{-3}} = 0.350 \times 10^3\,\mathrm{mol\,m^{-3}}$. The initial rate is therefore

$$\begin{aligned} \frac{d[\mathrm{P}]}{dt} &= k_d[\mathrm{A}][\mathrm{B}] \\ &= (3.04... \times 10^6\,\mathrm{m^3\,mol^{-1}\,s^{-1}}) \\ &\quad \times (0.170 \times 10^3\,\mathrm{mol\,m^{-3}}) \times (0.350 \times 10^3\,\mathrm{mol\,m^{-3}}) \\ &= \boxed{1.81 \times 10^{11}\,\mathrm{mol\,m^{-3}\,s^{-1}}} \end{aligned}$$

Using [18B.4–*789*], $k_\mathrm{d} = 8RT/3\eta$, the rate constant is

$$k_\mathrm{d} = \frac{8 \times (8.3145\,\mathrm{J\,K^{-1}\,mol^{-1}}) \times (313\,\mathrm{K})}{3 \times (2.93 \times 10^{-3}\,\mathrm{kg\,m^{-1}\,s^{-1}})} = 2.37 \times 10^{6}\,\mathrm{m^3\,mol^{-1}\,s^{-1}}$$

This value would result in a significantly slower initial rate, casting doubt therefore on the validity of the approximations used.

Solutions to problems

P18B.1 To simplify the notation the dependence of [J] and $[\mathrm{J}]^*$ on x and t will not be written explicitly. The proposed solution, [18B.8–*790*], $[\mathrm{J}]^* = [\mathrm{J}]\mathrm{e}^{-k_\mathrm{r}t}$, is substituted into the right-hand side of [18B.7–*790*]

$$D\frac{\partial^2[\mathrm{J}]^*}{\partial x^2} - k_\mathrm{r}[\mathrm{J}]^* = D\frac{\partial^2}{\partial x^2}[\mathrm{J}]\mathrm{e}^{-k_\mathrm{r}t} - k_\mathrm{r}[\mathrm{J}]\mathrm{e}^{-k_\mathrm{r}t}$$
$$= D\frac{\partial^2[\mathrm{J}]}{\partial x^2}\mathrm{e}^{-k_\mathrm{r}t} - k_\mathrm{r}[\mathrm{J}]\mathrm{e}^{-k_\mathrm{r}t}$$

The solution is now substituted into the left-hand side of [18B.7–*790*]

$$\frac{\partial[\mathrm{J}]^*}{\partial t} = \frac{\partial}{\partial t}[\mathrm{J}]\mathrm{e}^{-k_\mathrm{r}t} = \frac{\partial[\mathrm{J}]}{\partial t}\mathrm{e}^{-k_\mathrm{r}t} - k_\mathrm{r}\mathrm{e}^{-k_\mathrm{r}t}[\mathrm{J}]$$

The left-and right-hand sides are now set equal

$$D\frac{\partial^2[\mathrm{J}]}{\partial x^2}\mathrm{e}^{-k_\mathrm{r}t} - k_\mathrm{r}[\mathrm{J}]\mathrm{e}^{-k_\mathrm{r}t} = \frac{\partial[\mathrm{J}]}{\partial t}\mathrm{e}^{-k_\mathrm{r}t} - k_\mathrm{r}\mathrm{e}^{-k_\mathrm{r}t}[\mathrm{J}]$$

The term $k_\mathrm{r}\mathrm{e}^{-k_\mathrm{r}t}[\mathrm{J}]$ cancels to give

$$D\frac{\partial^2[\mathrm{J}]}{\partial x^2}\mathrm{e}^{-k_\mathrm{r}t} = \frac{\partial[\mathrm{J}]}{\partial t}\mathrm{e}^{-k_\mathrm{r}t} \qquad \text{hence} \qquad D\frac{\partial^2[\mathrm{J}]}{\partial x^2} = \frac{\partial[\mathrm{J}]}{\partial t}$$

As specified in the problem, [J] is a solution of [18B.7–*790*] when $k_\mathrm{r} = 0$, and indeed this is precisely the differential equation which has just been generated.

P18B.3 It is first convenient to compute the derivative of $[\mathrm{J}]^*$ with respect to t and its second derivative with respect to x.

$$\frac{\partial[\mathrm{J}]^*}{\partial t} = \frac{\partial}{\partial t}\left[k_\mathrm{r}\int_0^t [\mathrm{J}]\mathrm{e}^{-k_\mathrm{r}t}\,\mathrm{d}t + [\mathrm{J}]\mathrm{e}^{-k_\mathrm{r}t}\right]$$
$$= k_\mathrm{r}[\mathrm{J}]\mathrm{e}^{-k_\mathrm{r}t} + \frac{\partial[\mathrm{J}]}{\partial t}\mathrm{e}^{-k_\mathrm{r}t} - k_\mathrm{r}\mathrm{e}^{-k_\mathrm{r}t}[\mathrm{J}] = \frac{\partial[\mathrm{J}]}{\partial t}\mathrm{e}^{-k_\mathrm{r}t} \tag{18.1}$$

$$\frac{\partial^2[\mathrm{J}]^*}{\partial x^2} = \frac{\partial^2}{\partial x^2}\left[k_\mathrm{r}\int_0^t [\mathrm{J}]\mathrm{e}^{-k_\mathrm{r}t}\,\mathrm{d}t + [\mathrm{J}]\mathrm{e}^{-k_\mathrm{r}t}\right]$$
$$= k_\mathrm{r}\int_0^t \frac{\partial^2[\mathrm{J}]}{\partial x^2}\mathrm{e}^{-k_\mathrm{r}t}\,\mathrm{d}t + \frac{\partial^2[\mathrm{J}]}{\partial x^2}\mathrm{e}^{-k_\mathrm{r}t} \tag{18.2}$$

Recall that [J] is a solution to [18B.7–*790*] with $k_r = 0$

$$\frac{\partial[\mathrm{J}]}{\partial t} = D\frac{\partial^2[\mathrm{J}]}{\partial x^2}$$

This is used to substitute for $\partial^2[\mathrm{J}]/\partial x^2$ in eqn 18.2

$$D\frac{\partial^2[\mathrm{J}]^*}{\partial x^2} = k_r \int_0^t \frac{\partial[\mathrm{J}]}{\partial t}\mathrm{e}^{-k_r t}\,\mathrm{d}t + \frac{\partial[\mathrm{J}]}{\partial t}\mathrm{e}^{-k_r t}$$

where the factor of D has been taken over to the left. Next use of made of the result in eqn 18.1, $\partial[\mathrm{J}]^*/\partial t = (\mathrm{e}^{-k_r t})\partial[\mathrm{J}]/\partial t$ to rewrite the last expression as

$$\begin{aligned}D\frac{\partial^2[\mathrm{J}]^*}{\partial x^2} &= k_r \int_0^t \frac{\partial[\mathrm{J}]^*}{\partial t}\,\mathrm{d}t + \frac{\partial[\mathrm{J}]^*}{\partial t}\\ &= k_r\left\{[\mathrm{J}]^*(t) - [\mathrm{J}]^*(0)\right\} + \frac{\partial[\mathrm{J}]^*}{\partial t}\\ &= k_r[\mathrm{J}]^*(t) + \frac{\partial[\mathrm{J}]^*}{\partial t}\end{aligned}$$

where the initial condition that [J], and hence $[\mathrm{J}]^*$, must be zero at $t = 0$ is used. Rearranging the final equation gives the required differential equation, thus demonstrating that the proposed form of $[\mathrm{J}]^*$ is indeed a solution.

$$\frac{\partial[\mathrm{J}]^*}{\partial t} = D\frac{\partial^2[\mathrm{J}]^*}{\partial x^2} - k_r[\mathrm{J}]^*$$

18C Transition-state theory

Answers to discussion questions

D18C.1 The discarded mode would be the anti-symmetric stretch in which the B–C distance lengthens and the A–B distance decreases.

D18C.3 This is described in Section 18C.2(b) on page 797. If the solvent were altered to one with a lower dielectric constant the interaction between the ions would be greater and this would be manifested in an increased value of $\mathcal{A}$, and hence steeper slopes for the plots of the rate constant against ionic strength.

Solutions to exercises

E18C.1(a) The empirical expression is compared to the Arrhenius equation $k_r = A\mathrm{e}^{-E_a/RT}$, allowing the activation energy to be determined from $E_a/R = 8681$ K; hence $E_a = (8.3145\,\mathrm{J\,K^{-1}\,mol^{-1}}) \times (8681\,\mathrm{K}) = 72.1...\,\mathrm{kJ\,mol^{-1}}$. The frequency factor is $A = 2.05 \times 10^{13}\,\mathrm{dm^3\,mol^{-1}\,s^{-1}} = 2.05 \times 10^{10}\,\mathrm{m^3\,mol^{-1}\,s^{-1}}$.

The relationship between E_a and $\Delta^{\ddagger}H$ for a bimolecular solution-phase reaction is given by [18C.17–796], $\Delta^{\ddagger}H = E_a - RT = (72.1... \times 10^3\ \mathrm{J\,mol^{-1}}) - (8.3145\ \mathrm{J\,K^{-1}\,mol^{-1}}) \times (303\ \mathrm{K}) = \boxed{69.7\ \mathrm{kJ\,mol^{-1}}}$. The relationship between A and $\Delta^{\ddagger}S$ for a bimolecular solution-phase reaction is given by [18C.19b–796]

$$A = \mathrm{e}\frac{kT}{h}\frac{RT}{p^{\ominus}}\mathrm{e}^{\Delta^{\ddagger}S/R}$$

$$\begin{aligned}\text{hence } \Delta^{\ddagger}S &= R\ln\frac{Ap^{\ominus}h}{\mathrm{e}kRT^2} \\ &= (8.3145\ \mathrm{J\,K^{-1}\,mol^{-1}}) \\ &\quad\times \ln\frac{(2.05\times10^{10}\ \mathrm{m^3\,mol^{-1}\,s^{-1}})\times(10^5\ \mathrm{Pa})\times(6.6261\times10^{-34}\ \mathrm{J\,s})}{\mathrm{e}(1.3806\times10^{-23}\ \mathrm{J\,K^{-1}})\times(8.3145\ \mathrm{J\,K^{-1}\,mol^{-1}})\times(303\ \mathrm{K})^2} \\ &= \boxed{25.3\ \mathrm{J\,K^{-1}\,mol^{-1}}}\end{aligned}$$

Note the conversion of the units of A to $\mathrm{m^3\,mol^{-1}\,s^{-1}}$.

E18C.2(a) The empirical expression is compared to the Arrhenius equation $k_r = A\mathrm{e}^{-E_a/RT}$, allowing the activation energy to be determined from $E_a/R = 9134$ K; hence $E_a = (8.3145\ \mathrm{J\,K^{-1}\,mol^{-1}}) \times (9134\ \mathrm{K}) = 75.9...\ \mathrm{kJ\,mol^{-1}}$. The frequency factor is $A = 7.78\times10^{14}\ \mathrm{dm^3\,mol^{-1}\,s^{-1}} = 7.78\times10^{11}\ \mathrm{m^3\,mol^{-1}\,s^{-1}}$.

The relationship between E_a and $\Delta^{\ddagger}H$ for a bimolecular solution-phase reaction is given by [18C.17–796], $\Delta^{\ddagger}H = E_a - RT = (75.9... \times 10^3\ \mathrm{J\,mol^{-1}}) - (8.3145\ \mathrm{J\,K^{-1}\,mol^{-1}}) \times (303\ \mathrm{K}) = +73.4...\ \mathrm{kJ\,mol^{-1}}$. The relationship between A and $\Delta^{\ddagger}S$ for a bimolecular solution-phase reaction is given by [18C.19b–796]

$$A = \mathrm{e}\frac{kT}{h}\frac{RT}{p^{\ominus}}\mathrm{e}^{\Delta^{\ddagger}S/R}$$

$$\begin{aligned}\text{hence } \Delta^{\ddagger}S &= R\ln\frac{Ap^{\ominus}h}{\mathrm{e}kRT^2} \\ &= (8.3145\ \mathrm{J\,K^{-1}\,mol^{-1}}) \\ &\quad\times \ln\frac{(7.78\times10^{11}\ \mathrm{m^3\,mol^{-1}\,s^{-1}})\times(10^5\ \mathrm{Pa})\times(6.6261\times10^{-34}\ \mathrm{J\,s})}{\mathrm{e}(1.3806\times10^{-23}\ \mathrm{J\,K^{-1}})\times(8.3145\ \mathrm{J\,K^{-1}\,mol^{-1}})\times(303\ \mathrm{K})^2} \\ &= +4.88...\ \mathrm{J\,K^{-1}\,mol^{-1}}\end{aligned}$$

Note the conversion of the units of A to $\mathrm{m^3\,mol^{-1}\,s^{-1}}$. $\Delta^{\ddagger}G$ is found by combining the values of $\Delta^{\ddagger}H$ and $\Delta^{\ddagger}S$ in the usual way

$$\begin{aligned}\Delta^{\ddagger}G &= \Delta^{\ddagger}H - T\Delta^{\ddagger}S \\ &= (+73.4...\times10^3\ \mathrm{J\,mol^{-1}}) - (303\ \mathrm{K})\times(+4.88...\ \mathrm{J\,K^{-1}\,mol^{-1}}) \\ &= \boxed{+71.9\ \mathrm{kJ\,mol^{-1}}}\end{aligned}$$

E18C.3(a) The rate constant for a bimolecular gas phase reaction is given by [18C.18a–796]

$$k_r = \mathrm{e}^2\frac{kT}{h}\frac{RT}{p^{\ominus}}\mathrm{e}^{\Delta^{\ddagger}S/R}\mathrm{e}^{-E_a/RT}$$

To use this expression the given rate constant needs to be converted to units of $m^3\ mol^{-1}\ s^{-1}$. From the perfect gas law $[J] = n/V = p/RT$, therefore to convert the rate constant from units of pressure^{-1} to units of concentration^{-1} units requires multiplication by RT

$$\begin{aligned} k_r &= 7.84 \times 10^{-3}\ \text{kPa}^{-1}\ \text{s}^{-1} = 7.84 \times 10^{-6}\ \text{Pa}^{-1}\ \text{s}^{-1} \\ &= (7.84 \times 10^{-6}\ \text{Pa}^{-1}\ \text{s}^{-1}) \times (8.3145\ \text{J K}^{-1}\ \text{mol}^{-1}) \times (338\ \text{K}) \\ &= 0.0220...\ \text{m}^3\ \text{mol}^{-1}\ \text{s}^{-1} \end{aligned}$$

The units can be deduced using 1 Pa = 1 $\text{kg m}^{-1}\ \text{s}^{-2}$ and 1 J = 1 $\text{kg m}^2\ \text{s}^{-2}$.

The above equation rearranges to

$$\begin{aligned} \Delta^{\ddagger}S &= R\ln\left(k_r \frac{hp^{\ominus}}{e^2 kRT^2} e^{E_a/RT}\right) = R\ln\left(\frac{k_r hp^{\ominus}}{e^2 kRT^2}\right) + \frac{E_a}{T} \\ &= (8.3145\ \text{J K}^{-1}\ \text{mol}^{-1}) \\ &\quad \times \ln\left(\frac{(0.0220...\ \text{m}^3\ \text{mol}^{-1}\ \text{s}^{-1}) \times (6.6261 \times 10^{-34}\ \text{J s}) \times (10^5\ \text{Pa})}{e^2(1.3806 \times 10^{-23}\ \text{J K}^{-1}) \times (8.3145\ \text{J K}^{-1}\ \text{mol}^{-1}) \times (338\ \text{K})^2}\right) \\ &\quad + \frac{58.6 \times 10^3\ \text{J mol}^{-1}}{338\ \text{K}} \\ &= \boxed{-91.2\ \text{J K}^{-1}\ \text{mol}^{-1}} \end{aligned}$$

E18C.4(a) In *Example* 18C.1 on page 794 the following expression for the rate constant for a reaction between structureless particles is derived

$$k_r = N_A \left(\frac{8kT}{\pi\mu}\right)^{1/2} \sigma^* e^{-\Delta E_0/RT}$$

The activation energy is obtained from its usual definition, [17D.3–*742*]

$$\begin{aligned} E_a &= RT^2 \frac{d\ln k_r}{dT} \\ &= RT^2 \frac{d}{dT}\left\{\ln\left[N_A\left(\frac{8kT}{\pi\mu}\right)^{1/2}\sigma^*\right] - \frac{\Delta E_0}{RT}\right\} \\ &= RT^2\left(\frac{1}{2T} + \frac{\Delta E_0}{RT^2}\right) = \tfrac{1}{2}RT + \Delta E_0 \end{aligned}$$

Therefore $\Delta E_0 = E_a - \frac{1}{2}RT$ and hence

$$k_r = N_A \left(\frac{8kT}{\pi\mu}\right)^{1/2} \sigma^* e^{1/2} e^{-\Delta E_a/RT}$$

The rate constant for a bimolecular gas phase reaction is given by [18C.18a–*796*]

$$k_r = e^2 \frac{kT}{h}\frac{RT}{p^{\ominus}} e^{\Delta^{\ddagger}S/R} e^{-E_a/RT}$$

Comparing these two expressions gives

$$N_A\left(\frac{8kT}{\pi\mu}\right)^{1/2}\sigma^* e^{1/2} = e^2\frac{kT}{h}\frac{RT}{p^\ominus}e^{\Delta^\ddagger S/R}$$

This is rearranged to give $\Delta^\ddagger S$, noting that for a collision between like molecules $\mu = \frac{1}{2}m$

$$\begin{aligned}\Delta^\ddagger S &= R\ln\left(N_A\left(\frac{8kT}{\pi\mu}\right)^{1/2}\sigma^*\frac{hp^\ominus}{e^{3/2}kRT^2}\right)\\ &= (8.3145\,\mathrm{J\,K^{-1}\,mol^{-1}})\\ &\quad\times\ln\Bigg[(6.0221\times10^{23}\,\mathrm{mol^{-1}})\times\left(\frac{8\times(1.3806\times10^{-23}\,\mathrm{J\,K^{-1}})\times(300\,\mathrm{K})}{\pi\times\frac{1}{2}\times66\times(1.6605\times10^{-27}\,\mathrm{kg})}\right)^{1/2}\\ &\quad\times(0.35\times10^{-18}\,\mathrm{m^2})\\ &\quad\times\frac{(6.6261\times10^{-34}\,\mathrm{J\,s})\times(10^5\,\mathrm{Pa})}{e^{3/2}(1.3806\times10^{-23}\,\mathrm{J\,K^{-1}})\times(8.3145\,\mathrm{J\,K^{-1}\,mol^{-1}})\times(300\,\mathrm{K})^2}\Bigg]\\ &= \boxed{-74\,\mathrm{J\,K^{-1}\,mol^{-1}}}\end{aligned}$$

E18C.5(a) It is convenient to convert the units of the frequency factor and express it as $A = 4.6\times10^9\,\mathrm{m^3\,mol^{-1}\,s^{-1}}$. The relationship between E_a and $\Delta^\ddagger H$ for a bimolecular gas-phase reaction is given by [18C.17–*796*], $\Delta^\ddagger H = E_a - 2RT = (10.0\times10^3\,\mathrm{J\,mol^{-1}}) - 2\times(8.3145\,\mathrm{J\,K^{-1}\,mol^{-1}})\times(298\,\mathrm{K}) = +5.04...\,\mathrm{kJ\,mol^{-1}} = \boxed{+5.0\,\mathrm{kJ\,mol^{-1}}}$. The relationship between A and $\Delta^\ddagger S$ for a bimolecular gas-phase reaction is given by [18C.19a–*796*]

$$A = e^2\frac{kT}{h}\frac{RT}{p^\ominus}e^{\Delta^\ddagger S/R}$$

$$\begin{aligned}\text{hence } \Delta^\ddagger S &= R\ln\frac{Ap^\ominus h}{e^2kRT^2}\\ &= (8.3145\,\mathrm{J\,K^{-1}\,mol^{-1}})\\ &\quad\times\ln\frac{(4.6\times10^9\,\mathrm{m^3\,mol^{-1}\,s^{-1}})\times(10^5\,\mathrm{Pa})\times(6.6261\times10^{-34}\,\mathrm{J\,s})}{e^2(1.3806\times10^{-23}\,\mathrm{J\,K^{-1}})\times(8.3145\,\mathrm{J\,K^{-1}\,mol^{-1}})\times(298\,\mathrm{K})^2}\\ &= -45.8...\,\mathrm{J\,K^{-1}\,mol^{-1}} = \boxed{-46\,\mathrm{J\,K^{-1}\,mol^{-1}}}\end{aligned}$$

$\Delta^\ddagger G$ is found by combining the values of $\Delta^\ddagger H$ and $\Delta^\ddagger S$ in the usual way

$$\begin{aligned}\Delta^\ddagger G &= \Delta^\ddagger H - T\Delta^\ddagger S\\ &= (+5.04...\times10^3\,\mathrm{J\,mol^{-1}}) - (298\,\mathrm{K})\times(-45.8...\,\mathrm{J\,K^{-1}\,mol^{-1}})\\ &= \boxed{+19\,\mathrm{kJ\,mol^{-1}}}\end{aligned}$$

E18C.6(a) The variation of the rate constant with ionic strength is given by [18C.23–797], $\lg k_r = \lg k_r^\circ + 2Az_Az_BI^{1/2}$; at 298 K and for aqueous solutions $A = 0.509$. In the absence of further information assume $z_A = +1$ and $z_A = -1$. Rearranging for $\lg k_r^\circ$ gives

$$\begin{aligned}\lg k_r^\circ &= \lg k_r - 2Az_Az_BI^{1/2}\\ &= \lg(12.2\ \mathrm{dm^6\ mol^{-2}\ min^{-1}}) - 2\times(0.509)\times(+1)\times(-1)\times(0.0525)^{1/2}\\ &= 1.31...\end{aligned}$$

Therefore $\boxed{k_r^\circ = 20.9\ \mathrm{dm^6\ mol^{-2}\ min^{-1}}}$.

E18C.7(a) The effect of deuteration on the rate constant is given by [18C.25–799]

$$\frac{k_r(\text{C–D})}{k_r(\text{C–H})} = e^{-\zeta} \quad \zeta = \frac{\hbar\omega(\text{C–H})}{2kT}\left\{1-\left(\frac{\mu_{\text{CH}}}{\mu_{\text{CD}}}\right)^{1/2}\right\}$$

In this expression $\omega(\text{C–H}) = (k_f/\mu_{\text{CH}})^{1/2}$. It can be adapted for other pairs of isotopes by changing the effective masses and the force constant.

The effective mass for $^{12}\text{C–}^{1}\text{H}$ is

$$\mu_{\text{CH}} = \frac{m_\text{C}m_\text{H}}{m_\text{C}+m_\text{H}} = \frac{12\times1.0078}{12+1.0078}m_\text{u} = 0.929...\ m_\text{u} = 1.54...\times10^{-27}\ \text{kg}$$

Likewise for $^{12}\text{C–}^{3}\text{H}$ (denoted C–T) the effective mass is

$$\mu_{\text{CT}} = \frac{m_\text{C}m_\text{T}}{m_\text{C}+m_\text{T}} = \frac{12\times3.016}{12+3.016}m_\text{u} = 2.41...\ m_\text{u}$$

With the given force constant

$$\omega(\text{C–H}) = \left(\frac{450\ \text{N m}^{-1}}{1.54...\times10^{-27}\ \text{kg}}\right)^{1/2} = 5.39...\times10^{14}\ \text{s}^{-1}$$

At 298 K

$$\begin{aligned}\zeta &= \frac{(1.0546\times10^{-34}\ \text{J s})\times(5.39...\times10^{14}\ \text{s}^{-1})}{2\times(1.3806\times10^{-23}\ \text{J K}^{-1})\times(298\ \text{K})}\left\{1-\left(\frac{0.929...}{2.41...}\right)^{1/2}\right\}\\ &= 2.62...\end{aligned}$$

$$\frac{k_r(\text{C–T})}{k_r(\text{C–H})} = e^{-\zeta} = e^{-2.62...} = \boxed{0.073}$$

Raising the temperature will decrease ζ which will have the effect of increasing the ratio $k_r(\text{C–T})/k_r(\text{C–H})$ and thus moving it closer to 1. That is, the isotope effect will be reduced.

Solutions to problems

P18C.1 It is convenient to convert the units of the frequency factor and express it as $A = 4.07 \times 10^2\ \mathrm{m^3\,mol^{-1}\,s^{-1}}$. The relationship between E_a and $\Delta^\ddagger H$ for a bimolecular gas-phase reaction is given by [18C.17–*796*], $\Delta^\ddagger H = E_\mathrm{a} - 2RT = (65.4 \times 10^3\ \mathrm{J\,mol^{-1}}) - 2 \times (8.3145\ \mathrm{J\,K^{-1}\,mol^{-1}}) \times (300\ \mathrm{K}) = +60.4...\ \mathrm{kJ\,mol^{-1}} = \boxed{+60.4\ \mathrm{kJ\,mol^{-1}}}$. The relationship between A and $\Delta^\ddagger S$ for a bimolecular gas-phase reaction is given by [18C.19a–*796*]

$$A = \mathrm{e}^2 \frac{kT}{h}\frac{RT}{p^\ominus}\mathrm{e}^{\Delta^\ddagger S/R}$$

$$\begin{aligned}\text{hence } \Delta^\ddagger S &= R \ln \frac{Ap^\ominus h}{\mathrm{e}^2 kRT^2} \\ &= (8.3145\ \mathrm{J\,K^{-1}\,mol^{-1}}) \\ &\quad \times \ln \frac{(4.073 \times 10^2\ \mathrm{m^3\,mol^{-1}\,s^{-1}}) \times (10^5\ \mathrm{Pa}) \times (6.6261 \times 10^{-34}\ \mathrm{J\,s})}{\mathrm{e}^2 (1.3806 \times 10^{-23}\ \mathrm{J\,K^{-1}}) \times (8.3145\ \mathrm{J\,K^{-1}\,mol^{-1}}) \times (300\ \mathrm{K})^2} \\ &= -1.80... \times 10^2\ \mathrm{J\,K^{-1}\,mol^{-1}} = \boxed{-181\ \mathrm{J\,K^{-1}\,mol^{-1}}}\end{aligned}$$

$\Delta^\ddagger G$ is found by combining the values of $\Delta^\ddagger H$ and $\Delta^\ddagger S$ in the usual way

$$\begin{aligned}\Delta^\ddagger G &= \Delta^\ddagger H - T\Delta^\ddagger S \\ &= (+60.4... \times 10^3\ \mathrm{J\,mol^{-1}}) - (300\ \mathrm{K}) \times (-1.80... \times 10^2\ \mathrm{J\,K^{-1}\,mol^{-1}}) \\ &= \boxed{+115\ \mathrm{kJ\,mol^{-1}}}\end{aligned}$$

The relationship between ΔU and ΔH is given by [2B.4–*48*], $\Delta H = \Delta U + \Delta\nu_\mathrm{g} RT$, where $\Delta\nu_\mathrm{g}$ is the change in stoichiometric coefficients for gaseous species. In this case $\Delta\nu_\mathrm{g} = 1 - 2 = -1$, hence

$$\begin{aligned}\Delta^\ddagger U &= \Delta^\ddagger H - \Delta\nu_\mathrm{g} RT = \Delta^\ddagger H + RT \\ &= (+60.4... \times 10^3\ \mathrm{J\,mol^{-1}}) + (8.3145\ \mathrm{J\,K^{-1}\,mol^{-1}}) \times (300\ \mathrm{K}) \\ &= \boxed{+62.9\ \mathrm{kJ\,mol^{-1}}}\end{aligned}$$

P18C.3 The starting point is the expression

$$k_\mathrm{r} = \kappa \frac{kT}{h}\frac{RT}{p^\ominus}\left(\frac{N_\mathrm{A}\Lambda_\mathrm{A}^3\Lambda_\mathrm{B}^3}{\Lambda_{\mathrm{C}^\ddagger}^3 V_\mathrm{m}^\ominus}\right)\frac{2IkT}{\hbar^2}\mathrm{e}^{-\Delta E_0/RT}$$

The first step is to realise that because $pV = nRT$, $RT/p^\ominus V_\mathrm{m}^\ominus = 1$; the N_A is also taken out of the bracket to give

$$k_\mathrm{r} = \kappa \frac{kT}{h} N_\mathrm{A}\left(\frac{\Lambda_\mathrm{A}\Lambda_\mathrm{B}}{\Lambda_{\mathrm{C}^\ddagger}}\right)^3 \frac{2IkT}{\hbar^2}\mathrm{e}^{-\Delta E_0/RT}$$

The thermal wavelength is $\Lambda = h/(2\pi mkT)^{1/2}$; substituting this and cancelling over the fraction gives

$$= \kappa \frac{kT}{h} N_\mathrm{A}\left(\frac{h^2 m_{\mathrm{C}^\ddagger}}{2\pi kT m_\mathrm{A} m_\mathrm{B}}\right)^{3/2} \frac{2IkT}{\hbar^2}\mathrm{e}^{-\Delta E_0/RT}$$

Now note that $m_{C^‡} = m_A + m_B$, so that $m_{C^‡}/m_A m_B = 1/\mu$. In addition, the moment of inertia is written as $I = \mu r^2$

$$= \kappa \frac{kT}{h} N_A \left(\frac{h^2}{2\pi kT\mu}\right)^{3/2} \frac{2\mu r^2 kT}{\hbar^2} e^{-\Delta E_0/RT}$$

The reactive cross section is identified as $\sigma^* = \kappa\pi r^2$; using this, and tidying up all the constants gives the required expression

$$k_r = N_A \left(\frac{8kT}{\pi\mu}\right)^{1/2} \sigma^* e^{-\Delta E_0/RT}$$

P18C.5 The rate constant is given by the Eyring equation, [18C.10–*794*]

$$k_r = \kappa \frac{kT}{h} \frac{RT}{p^⦵} \frac{N_A \bar{q}^⦵_{HD_2^‡}}{q^⦵_H q^⦵_{D_2}} e^{-\Delta E_0/RT}$$

To simplify the notation, the overline, double dagger and standard symbols will be omitted. With such a complex calculation it is best to break it down into parts and then assemble them to give the final result. First, the pre-multiplying constants (assuming $\kappa = 1$)

$$\frac{kT}{h}\frac{RT}{p^⦵}N_A = \frac{(1.3806\times10^{-23}\,\text{J K}^{-1})\times(400\,\text{K})}{6.6261\times10^{-34}\,\text{J s}} \times \frac{(8.3145\,\text{J K}^{-1}\,\text{mol}^{-1})\times(400\,\text{K})}{10^5\,\text{Pa}} \times (6.0221\times10^{23}\,\text{mol}^{-1})$$
$$= 1.66...\times10^{35}\,\text{m}^3\,\text{mol}^{-2}\,\text{s}^{-1}$$

Next consider the ratio of the translational partition functions: each is given by $q^⦵ = V^⦵_m/\Lambda^3 = RT/p^⦵\Lambda^3$, with $\Lambda = h/(2\pi mkT)^{1/2}$. For the purpose of this approximate calculation it is sufficient to use integer masses

$$\frac{q_{HD_2}}{q_H q_{D_2}}_{\text{trans}} = \frac{p^⦵}{RT}\left(\frac{h^2}{2\pi kT}\right)^{3/2}\left(\frac{m_{HD_2}}{m_H m_{D_2}}\right)^{3/2}$$
$$= \frac{10^5\,\text{Pa}}{(8.3145\,\text{J K}^{-1}\,\text{mol}^{-1})\times(400\,\text{K})} \times \left(\frac{(6.6261\times10^{-34}\,\text{J s})^2}{2\pi(1.3806\times10^{-23}\,\text{J K}^{-1})\times(400\,\text{K})}\right)^{3/2} \times \left(\frac{5}{1\times4\times1.6605\times10^{-27}\,\text{kg}}\right)^{3/2}$$
$$= 2.79...\times10^{-29}\,\text{mol}$$

Next consider the ratio of the vibrational partition functions: each is given by $q = (1 - e^{-hc\tilde{\nu}/kT})^{-1}$. The vibrational frequency of D_2 is so high that $q = 1$.

Each normal mode of the activated complex has ν = 1000 cm^{-1}n thus the contribution to the partition function is computed as

$$\frac{hc\tilde{\nu}}{kT} = \frac{(6.6261\times10^{-34}\,\mathrm{J\,s})\times(2.9979\times10^{10}\,\mathrm{cm\,s^{-1}})\times(1000\,\mathrm{cm^{-1}})}{(1.3806\times10^{-23}\,\mathrm{J\,K^{-1}})\times(400\,\mathrm{K})}$$
$$= 3.59...$$
$$q = (1-\mathrm{e}^{-3.59...})^{-1} = 1.02...$$

The ratio of vibrational partition functions is therefore

$$\left.\frac{q_{\mathrm{HD_2}}}{q_{\mathrm{H}}q_{\mathrm{D_2}}}\right|_{\mathrm{vib}} = (1.02...)^{n_{\mathrm{vib}}}$$

where n_{vib} is the number of vibrational normal modes of the activated complex.

To find the rotational partition function for D_2 requires a knowledge of the rotational constant, which is computed from the given bond length of H_2. The effective mass for D_2 is 1 m_u thus

$$\tilde{B} = \frac{h}{8\pi^2 c\mu R^2}$$
$$= \frac{6.6261\times10^{-34}\,\mathrm{J\,s}}{8\pi^2(2.9979\times10^{10}\,\mathrm{cm\,s^{-1}})\times(1\times1.6605\times10^{-27}\,\mathrm{kg})\times(74\times10^{-12}\,\mathrm{m})^2}$$
$$= 30.7...\,\mathrm{cm^{-1}}$$

The rotational partition function is therefore

$$q = \frac{kT}{\sigma hc\tilde{B}} = \frac{(1.3806\times10^{-23}\,\mathrm{J\,K^{-1}})\times(400\,\mathrm{K})}{2\times(6.6261\times10^{-34}\,\mathrm{J\,s})\times(2.9979\times10^{10}\,\mathrm{cm\,s^{-1}})\times(30.7...\,\mathrm{cm^{-1}})}$$
$$= 4.51...$$

The rotational partition function for the activated complex depends on the model chosen, so for the moment it is simply written $q^{\mathrm{R}}_{\mathrm{D_2H}}$.

The exponential term $\mathrm{e}^{-\Delta E_0/RT}$ evaluates to $2.68...\times10^{-5}$. The H atom has a doublet ground state (one unpaired electron) as does the activated complex (three electrons in total, one unpaired); the ground state of D_2 is not generate. The electronic partition functions therefore cancel. Putting this all together gives the following expression for the rate constant

$$k_r = (1.66...\times10^{35}\,\mathrm{m^3\,mol^{-2}\,s^{-1}})\times(2.79...\times10^{-29}\,\mathrm{mol})\times(1.02...)^{n_{\mathrm{vib}}}$$
$$\times\left[q^{\mathrm{R}}_{\mathrm{D_2H}}/(4.51...)\right]\times(2.68...\times10^{-5})$$
$$= (27.7...\,\mathrm{m^3\,mol^{-1}\,s^{-1}})\times(1.02...)^{n_{\mathrm{vib}}}\times q^{\mathrm{R}}_{\mathrm{D_2H}}$$

(a) The next step is to compute the moments of intertia, and hence the rotational constants, for the isoceles geometry of the activated complex. From the data given the D–D distance is 88.8 pm and the H–D distance is 96.2 pm. Figure 18.1 shows the geometry: the filled circles are D and the open circle is H. For this approximate calculation integer masses are used.

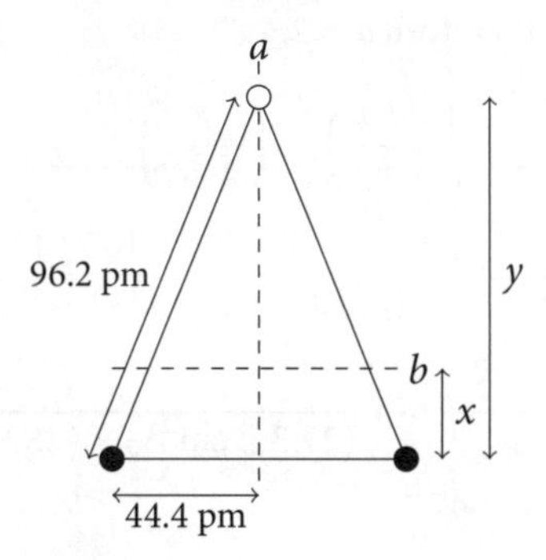

Figure 18.1

The centre of mass lies along the vertical bisector a, and the moment of inertia about this axis is

$$I_a = 2 \times 2 \times m_u \times (44.4 \times 10^{-12}\ \text{m})^2 = 1.30... \times 10^{-47}\ \text{kg m}^2$$

The corresponding rotational constant is computed as

$$\tilde{A} = \frac{h}{8\pi^2 cI} = \frac{(6.6261 \times 10^{-34}\ \text{J s})}{8\pi^2(2.9979 \times 10^{10}\ \text{cm s}^{-1}) \times (1.30... \times 10^{-47}\ \text{kg m}^2)}$$
$$= 21.3...\ \text{cm}^{-1}$$

The centre of mass lies at the intersection of the axes a and b. Taking moments about axis b gives $4x = 1 \times (y - x)$. By Pythagoras $y = (96.2^2 - 44.4^2)^{1/2} = 85.3...$ pm, using which gives $x = 17.0...$ pm. The moment of inertia about the b axis is therefore

$$I_b = 2 \times 2 \times m_u \times x^2 + 1 \times m_u(y - x)^2$$
$$= 4 \times (1.6605 \times 10^{-27}\ \text{kg}) \times (44.4 \times 10^{-12}\ \text{m})^2$$
$$+ 1 \times (1.6605 \times 10^{-27}\ \text{kg}) \times ([85.3... - 17.0...] \times 10^{-12}\ \text{m})^2$$
$$= 9.67... \times 10^{-47}\ \text{kg m}^2$$

The corresponding rotational constant is $\tilde{B} = 28.9...\ \text{cm}^{-1}$. The third moment of inertia is most easily found using the property of planar bodies that the moment of inertia about the axis perpendicular to the plane of the body is equal to the sum of the other two moments of inertia. Using this $I_c = 2.27... \times 10^{-47}\ \text{kg m}^2$ and $\tilde{C} = 12.2...\ \text{cm}^{-1}$. The rotational partition function is given by [13B.14–*545*]; the molecule has a two-fold axis therefore $\sigma = 2$.

$$q = \frac{1}{\sigma}\left(\frac{kT}{hc}\right)^{3/2}\left(\frac{\pi}{\tilde{A}\tilde{B}\tilde{C}}\right)^{1/2}$$

$$= \frac{1}{2}\left(\frac{(1.3806\times10^{-23}\,\mathrm{J\,K^{-1}})\times(400\,\mathrm{K})}{(6.6261\times10^{-34}\,\mathrm{J\,s})\times(2.9979\times10^{10}\,\mathrm{cm\,s^{-1}})}\right)^{3/2}$$

$$\times\left(\frac{\pi}{(21.3...\,\mathrm{cm^{-1}})\times(28.9...\,\mathrm{cm^{-1}})\times(12.2...\,\mathrm{cm^{-1}})}\right)^{1/2}$$

$$= 47.1...$$

For this triangular activated complex there are three normal modes, one of which corresponds to the reaction coordinate. Thus, the rate constant is

$$k_r = (27.7...\,\mathrm{m^3\,mol^{-1}\,s^{-1}})\times(1.02...)^{n_{vib}}\times q^{R}_{D_2H}$$

$$= (27.7...\,\mathrm{m^3\,mol^{-1}\,s^{-1}})\times(1.02...)^{2}\times(47.1...)$$

$$= 1.4\times10^{3}\,\mathrm{m^3\,mol^{-1}\,s^{-1}} = \boxed{1.4\times10^{6}\,\mathrm{dm^3\,mol^{-1}\,s^{-1}}}$$

(b) Now consider a linear geometry for the activated complex, H–D–D, with the H–D distance as 96.2 pm and the D–D distance as 88.8 pm. The first task is to locate the centre of mass. Assuming that this is a distance x from the right-hand D it follows that $2x = 2(88.8 - x) + 1(96.2 + 88.8 - x)$. which gives $x = 72.52$ pm. The moment of inertia is therefore

$$I = \left\{2\times(72.52\,\mathrm{pm})^2 + 2\times([88.8-72.52]\,\mathrm{pm})^2 + 1\times([96.2+88.8-72.52]\,\mathrm{pm})^2\right\}$$

$$\times(1.6605\times10^{-27}\,\mathrm{kg})\times[(10^{-12}\,\mathrm{m})/(1\,\mathrm{pm})]^2$$

$$= 3.93...\times10^{-47}\,\mathrm{kg\,m^2}$$

The corresponding rotational constant is $\tilde{B} = 7.11...\,\mathrm{cm^{-1}}$. The rotational partition function is

$$q = \frac{kT}{hc\tilde{B}}$$

$$= \frac{(1.3806\times10^{-23}\,\mathrm{J\,K^{-1}})\times(400\,\mathrm{K})}{(6.6261\times10^{-34}\,\mathrm{J\,s})\times(2.9979\times10^{10}\,\mathrm{cm\,s^{-1}})\times(7.11...\,\mathrm{cm^{-1}})} = 39.0...$$

For this linear activated complex there are four normal modes, one of which corresponds to the reaction coordinate. Thus, the rate constant is

$$k_r = (27.7...\,\mathrm{m^3\,mol^{-1}\,s^{-1}})\times(1.02...)^{n_{vib}}\times q^{R}_{D_2H}$$

$$= (27.7...\,\mathrm{m^3\,mol^{-1}\,s^{-1}})\times(1.02...)^{3}\times(39.0...)$$

$$= 1.2\times10^{3}\,\mathrm{m^3\,mol^{-1}\,s^{-1}} = \boxed{1.2\times10^{6}\,\mathrm{dm^3\,mol^{-1}\,s^{-1}}}$$

The effect of changing the geometry is rather small.

(c) The calculations performed so far both give rate constants in excess of the target (by a factor of about 3). The only contribution which can feasibly change by this much is the rotational partition function of the activated complex. To make this smaller, the rotational constants need to be increased, which is achieved by generally shrinking the size of the complex. Some trial calculations suggest that shrinking the H–D distance to 80% of the H_2 distance achieves the desired result, but this seems a rather implausible structure of the activated complex.

P18C.7 The variation of the rate constant with ionic strength is given by [18C.23–797], $\lg k_r = \lg k_r^\circ + 2Az_Az_BI^{1/2}$; at 298 K and for aqueous solutions $A = 0.509$. A plot of $\lg(k_r/k_r^\circ)$ against $I^{1/2}$ is used to explore whether or not this relationship applies.

I	$I^{1/2}$	k_r/k_r°	$\lg(k_r/k_r^\circ)$
0.0100	0.1000	8.10	0.908
0.0150	0.1225	13.30	1.124
0.0200	0.1414	20.50	1.312
0.0250	0.1581	27.80	1.444
0.0300	0.1732	38.10	1.581
0.0350	0.1871	52.00	1.716

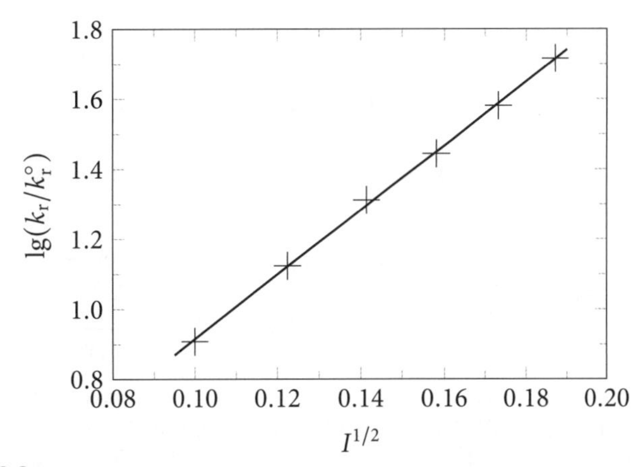

Figure 18.2

The plot is shown in Fig 18.2: the data fall on a good straight line with slope +9.18. Such a value implies

$$2 \times (0.509) \times (z_Az_B) = +9.18 \quad \text{hence} \quad (z_Az_B) = +9.02$$

A plausible interpretation is $z_A = +3$ and $z_B = +3$, which is consistent with the protein being cationic.

P18C.9 Some experimentation with various graphs shows that a plot of $\lg k_r$ against I (not $I^{1/2}$) gives a good straight line, as shown in Fig 18.3.

I	$k_r/(\mathrm{dm^3\,mol^{-1}\,s^{-1}})$	$\lg[k_r/(\mathrm{dm^3\,mol^{-1}\,s^{-1}})]$
0.0207	0.663	−0.1785
0.0525	0.670	−0.1739
0.0925	0.679	−0.1681
0.1575	0.694	−0.1586

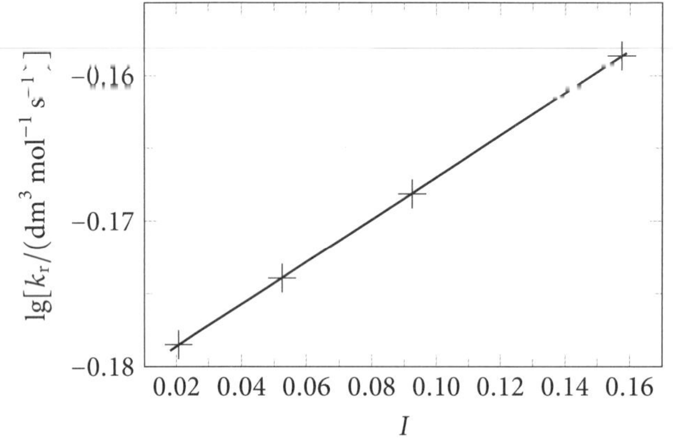

Figure 18.3

The equation of the best-fit line is

$$\lg[k_r/(\mathrm{dm^3\,mol^{-1}\,s^{-1}})] = 0.1451 \times I - 0.1815$$

In the limit of zero ionic strength $\lg[k_r^\circ/(\mathrm{dm^3\,mol^{-1}\,s^{-1}})] = -0.1815$, hence $\boxed{k_r^\circ = 0.658\ \mathrm{dm^3\,mol^{-1}\,s^{-1}}}$.

From the text [18C.21b–797] gives the dependence of $\lg k_r$ on the activity coefficients

$$\lg k_r = \lg k_r^\circ - \lg \frac{\gamma_{C^\ddagger}}{\gamma_A \gamma_B}$$

In this case one reactant (say A) is known to have a charge of −1, and the other reactant (say B) is neutral. Therefore, the charge on the activated complex is also −1 and hence, if the Debye–Hückel limiting law applies to these species, $\gamma_{C^\ddagger} = \gamma_A$. It therefore follows that

$$\lg k_r = \lg k_r^\circ + \lg \gamma_B$$

The data given show a linear dependence of $\lg k_r$ on I, therefore it is concluded that $\lg \gamma_B \propto I$. The constant of proportion is the slope of the graph: $\boxed{\lg \gamma_B = 0.145\, I}$.

P18C.11 The effect of deuteration on the rate constant is given by [18C.25–799]

$$\frac{k_r(\text{C–D})}{k_r(\text{C–H})} = e^{-\zeta} \quad \zeta = \frac{\hbar\omega(\text{C–H})}{2kT}\left\{1-\left(\frac{\mu_{\text{CH}}}{\mu_{\text{CD}}}\right)^{1/2}\right\}$$

In this expression $\omega(\text{C–H}) = (k_f/\mu_{\text{CH}})^{1/2}$.

In the case described in the problem the H (or D) is attached to a much heavier fragment (the rest of the molecule, X), thus the effective mass for the vibration is well approximated by the mass of the lighter atom (H or D), giving

$$\frac{k_r(\text{D})}{k_r(\text{H})} = e^{-\zeta} \quad \zeta = \frac{\hbar\omega(\text{X–H})}{2kT}\left\{1-\left(\frac{m_{\text{H}}}{m_{\text{D}}}\right)^{1/2}\right\}$$

Given that $k_r(\text{D})/k_r(\text{H}) = 1/6.4$ it follows that $\zeta = 1.86...$. With this value, and using integer masses for H and D, the expression for ζ becomes

$$1.86... = \frac{(1.0546\times10^{-34}\,\text{J s})\omega}{2(1.3806\times10^{-23}\,\text{J K}^{-1})\times(298\,\text{K})}\{1-(1/2)^{1/2}\}$$

hence $\omega = 4.95...\times10^{14}\,\text{s}^{-1}$

The force constant is therefore

$$k_f = \omega^2 m_{\text{H}} = (4.95...\times10^{14}\,\text{s}^{-1})\times(1\times1.6605\times10^{-27}\,\text{kg}) = \boxed{408\,\text{N m}^{-1}}$$

18D The dynamics of molecular collisions

Answers to discussion questions

D18D.1 These are discussed in Section 18D.2(a) on page 804.

D18D.3 This is discussed in Section 18D.4(b) on page 808.

Solutions to exercises

E18D.1(a) Refer to Fig. 18D.18 on page 808, which shows an attractive potential energy surface as well as trajectories of both a successful reaction and an unsuccessful one. The trajectories begin in the lower right, representing reactants. The successful trajectory passes through the transition state (marked as ‡ ○). This trajectory is fairly straight from the lower right through the transition state, indicating little or no vibrational excitation in the reactant. Therefore most of its energy is in translation. Since it has enough total energy to reach the transition state, the reactant can be described as being high in translational energy and low in vibrational energy.

This successful trajectory moves from side to side along the valley representing products, so the product is high in vibrational energy and relatively lower in translational energy. The unsuccessful trajectory, by contrast, has a reactant high in vibrational energy; it moves from side to side in the reactant valley without reaching the transition state.

E18D.2(a) The numerator of [18D.6–809] is integrated as follows with the assumption that $\overline{P}(E)$ is independent of E and can therefore be written $\overline{P}$

$$\int_0^\infty \overline{P}(E)\,\mathrm{e}^{-E/kT}\,\mathrm{d}E = \overline{P}\int_0^\infty \mathrm{e}^{-E/kT}\,\mathrm{d}E$$
$$= -\overline{P}kT\,\mathrm{e}^{-E/kT}\Big|_0^\infty = \boxed{\overline{P}kT}$$

The effect of this term is to make the rate constant increase with temperature.

Solutions to problems

P18D.1 The change in intensity of the beam, $\mathrm{d}I$, is proportional to the number of scatterers per unit volume, $\mathcal{N}$, the intensity of the beam, I, and the path length $\mathrm{d}L$. The constant of proportionality is the collision cross section σ, the 'target area' of each scatterer.

$$\mathrm{d}I = -\sigma\mathcal{N}I\,\mathrm{d}L \quad \text{hence} \quad \frac{1}{I}\,\mathrm{d}I = -\sigma\mathcal{N}\,\mathrm{d}L \quad \text{hence} \quad \mathrm{d}\ln I = -\sigma\mathcal{N}\,\mathrm{d}L$$

If the incident intensity at $L = 0$ is I_0, and the intensity after scattering through length L is I, integration gives

$$\int_{I_0}^{I} \mathrm{d}\ln I = \int_0^L -\sigma\mathcal{N}\,\mathrm{d}L$$
$$\text{hence } \ln I/I_0 = -\sigma\mathcal{N}L$$

The result may be expressed as $\boxed{I = I_0\,\mathrm{e}^{-\sigma\mathcal{N}L}}$.

P18D.3 Following *Brief illustration* 18D.1 on page 805

$$k_\mathrm{r} = \frac{k_\mathrm{r}^\circ}{q}\sum_{v,v'}\delta_{vv'}\,\mathrm{e}^{-\lambda v}\,\mathrm{e}^{-vh\nu/kT} = \frac{k_\mathrm{r}^\circ}{q}\sum_{v'}\mathrm{e}^{-\lambda v'}\,\mathrm{e}^{-v'h\nu/kT} = \frac{k_\mathrm{r}^\circ}{q}\sum_{v'}\mathrm{e}^{-v'(\lambda+h\nu/kT)}$$

Writing out the first few terms of the sum shows that it is a geometric progression with common ratio $\mathrm{e}^{-(\lambda+h\nu/kT)}$

$$= \frac{k_\mathrm{r}^\circ}{q}\left[1 + \mathrm{e}^{-(\lambda+h\nu/kT)} + \mathrm{e}^{-2(\lambda+h\nu/kT)} + \dots\right]$$
$$= \frac{k_\mathrm{r}^\circ}{q}\,\frac{1}{1-\mathrm{e}^{-(\lambda+h\nu/kT)}}$$

where to go to the final line the fact that the sum to infinity of a geometric series is $a/(1-r)$, where r is the common ratio and a is the first term, is used. This relationship requires $r < 1$, which is the case provided $\lambda > 0$.

18E Electron transfer in homogeneous systems

Answers to discussion questions

D18E.1 This is discussed in detail in Section 18E.2 on page 811 and Section 18E.3 on page 812.

D18E.3 The inverted region is discussed in Section 18E.4 on page 813. In this region the rate constant for the electron-transfer reaction decreases even though the reaction itself has a increasingly negative $\Delta_r G^\ominus$, a behaviour which is predicted by [18E.8–*814*]. One way of understanding this behaviour is to refer to Fig. 18E.3 on page 813. Here it is seen that the Gibbs energy of activation, $\Delta_r G^\ddagger$, is determined by the crossing point between the Gibbs energy curves for the product P and reactant R. As $\Delta_r G^\ominus$ becomes more negative the parabola representing P is lowered, and therefore the Gibbs energy of the crossing point decreases until the minimum in curve R is reached; after this, the energy of the crossing point increases as it travels up the left-hand side of curve R.

Solutions to exercises

E18E.1(a) The distance dependence of $H_{et}(d)^2$ given by [18E.4–*812*], $H_{et}(d)^2 = H_{et}^{\circ 2} e^{-\beta d}$.

$$\frac{H_{et}(d_2)^2}{H_{et}(d_1)^2} = e^{-\beta(d_2 - d_1)}$$
$$= e^{-(9\ \mathrm{nm}^{-1})[(2.0\ \mathrm{nm})-(1.0\ \mathrm{nm})]} = 1.23... \times 10^{-4}$$

Increasing the distance from 1.0 nm to 2.0 nm reduces $H_{et}(d)^2$ to about $\boxed{0.01\%}$ of its initial value.

E18E.2(a) The rate constant for electron-transfer is given by [18E.5–*812*] together with [18E.6–*813*]

$$k_{et} = \frac{1}{h}\left(\frac{\pi^3}{RT\Delta E_R}\right)^{1/2} H_{et}(d)^2\, e^{-\Delta^\ddagger G/RT} \qquad \Delta^\ddagger G = \frac{(\Delta_r G^\ominus + \Delta E_R)^2}{4\Delta E_R}$$

With the given data there is only one unknown quantity, ΔE_R, but it is not possible to find an analytical expression for this in terms of the other parameters. However, mathematical software is able to find a solution numerically. Before embarking on such a calculation it is important to make sure that the units of the various quantities are consistent.

The choice is made to express the energies as molar quantities ($\mathrm{J\,mol^{-1}}$). It therefore follows that the term in parentheses has units ($\mathrm{J^{-1}\,mol}$), and given that the units of $1/h$ are ($\mathrm{J^{-1}\,s^{-1}}$), it follows that for the rate constant to have the expected units of $\mathrm{s^{-1}}$, $H_{et}(d)^2$ must be in $\mathrm{J^2\,mol^{-1}}$. Using the conversion factors from inside the front cover

$$H_{et}(d)^2 = \left((0.04\ \text{cm}^{-1}) \times \frac{1.9864 \times 10^{-23}\ \text{J}}{1\ \text{cm}^{-1}}\right)^2 \times (6.0221 \times 10^{23}\ \text{mol}^{-1})$$
$$= 3.80... \times 10^{-25}\ \text{J}^2\ \text{mol}^{-1}$$

$$\Delta_r G^{\ominus} = (-0.185\ \text{eV}) \times \frac{96.485 \times 10^3\ \text{J mol}^{-1}}{1\ \text{eV}} = -1.78... \times 10^4\ \text{J mol}^{-1}$$

The constant factor evaluates to

$$\frac{1}{h}\left(\frac{\pi^3}{RT}\right)^{1/2} H_{et}(d)^2 = \frac{1}{6.6261 \times 10^{-34}\ \text{J s}}\left(\frac{\pi^3}{(8.3145\ \text{J K}^{-1}\ \text{mol}^{-1}) \times (298\ \text{K})}\right)^{1/2}$$
$$\times (3.80... \times 10^{-25}\ \text{J}^2\ \text{mol}^{-1})$$
$$= 6.41... \times 10^7\ \text{J}^{1/2}\ \text{mol}^{-1/2}\ \text{s}^{-1}$$

The equation to be solved is therefore

$$(37.5\ \text{s}^{-1}) = (6.41... \times 10^7\ \text{J}^{1/2}\ \text{mol}^{-1/2}\ \text{s}^{-1}) \times \Delta E_R^{-1/2} \times e^{-\Delta^{\ddagger}G/(2.47...\times 10^3\ \text{J mol}^{-1})}$$
$$\text{with } \Delta^{\ddagger}G = \frac{[(-1.78... \times 10^4\ \text{J mol}^{-1}) + \Delta E_R]^2}{4\Delta E_R}$$

In solving this equation it is helpful to know that the result is likely to be of the order of tens of kJ mol^{-1} so as to guide the numerical solution. An alternative is to plot the right-hand side of the above expression for over a range of values of ΔE_R and look for the value which gives the required k_r. The final result is $\boxed{\Delta E_R = 2\ \text{kJ mol}^{-1}}$.

E18E.3(a) The rate constant for electron-transfer is given by [18E.5–812] together with [18E.6–813]

$$k_{et} = \frac{1}{h}\left(\frac{\pi^3}{RT\Delta E_R}\right)^{1/2} H_{et}(d)^2 e^{-\Delta^{\ddagger}G/RT} \qquad \Delta^{\ddagger}G = \frac{(\Delta_r G^{\ominus} + \Delta E_R)^2}{4\Delta E_R}$$

For the two reactions given, ΔE_R and $\Delta^{\ddagger}G$ are assumed to be the same. The distance dependence of $H_{et}(d)^2$ is given by [18E.4–812], $H_{et}(d)^2 = H_{et}^{\circ 2} e^{-\beta d}$, therefore

$$\frac{k_{et,2}}{k_{et,1}} = \frac{(H_{et}(d)^2)_2}{(H_{et}(d)^2)_1} = e^{-\beta(d_2 - d_1)}$$
$$\text{hence } \ln(k_{et,2}/k_{et,1}) = -\beta(d_2 - d_1)$$
$$\text{therefore } \beta = -\frac{\ln(k_{et,2}/k_{et,1})}{(d_2 - d_1)}$$
$$\beta = -\frac{\ln[(4.51 \times 10^4\ \text{s}^{-1})/(2.02 \times 10^5\ \text{s}^{-1})]}{(1.23\ \text{nm}) - (1.11\ \text{nm})}$$
$$= \boxed{12.5\ \text{nm}^{-1}}$$

Solutions to problems

P18E.1 This *Problem* is somewhat ill-posed and some of the references to equations in the text are incorrect. The key missing item is that it should be assumed that the electron-transfer process is rate limiting so that $k_r = Kk_{et}$, where K is the equilibrium constant for the diffusive encounter; this point is discussed in Section 18E.1 on page 810. In addition, the first-order rate constant for the electron-transfer process is written, using transition-state theory, as

$$k_{1st} = \kappa\nu^{\ddagger}K^{\ddagger} = \kappa\nu^{\ddagger}e^{-\Delta^{\ddagger}G/RT} \quad (18.3)$$

This relationship is analogous to those developed in Section 18C.1 on page 792 for second-order reactions. In parts (c) and (d) of the problem eqn 18.3 should be used. The description of the dependence of f is somewhat misleading.

(a) The Gibbs energy of activation is given by [18E.6–*813*], $\Delta^{\ddagger}G = (\Delta_r G^{\ominus} + \Delta E_R)^2/4\Delta E_R$. For the DA electron transfer

$$\Delta^{\ddagger}G_{DA} = (\Delta_r G^{\ominus} + \Delta E_{R,DA})^2/4\Delta E_{R,DA} \quad (18.4)$$

For the AA self exchange $\Delta_r G^{\ominus} = 0$ therefore

$$\Delta^{\ddagger}G_{AA} = (\Delta E_{R,AA})^2/4\Delta E_{R,AA} = \tfrac{1}{4}\Delta E_{R,AA}$$

and likewise $\Delta^{\ddagger}G_{DD} = \tfrac{1}{4}\Delta E_{R,DD}$

(b) The square on the right-hand side of eqn 18.4 is expanded and the terms separated to give

$$\Delta^{\ddagger}G_{DA} = \frac{\Delta_r G^{\ominus 2}}{4\Delta E_{R,DA}} + \frac{2\Delta_r G^{\ominus}\Delta E_{R,DA}}{4\Delta E_{R,DA}} + \frac{\Delta E_{R,DA}{}^2}{4\Delta E_{R,DA}}$$
$$= \frac{\Delta_r G^{\ominus 2}}{4\Delta E_{R,DA}} + \tfrac{1}{2}\Delta_r G^{\ominus} + \tfrac{1}{4}\Delta E_{R,DA} \approx \tfrac{1}{2}\Delta_r G^{\ominus} + \tfrac{1}{4}\Delta E_{R,DA}$$

where the approximation $|\Delta_r G^{\ominus}| \ll \Delta E_{R,DA}$ is used in the last step. As indicated in the *Problem*, $\Delta E_{R,DA}$ may be written

$$\Delta E_{R,DA} = \tfrac{1}{2}(\Delta E_{R,AA} + \Delta E_{R,DD})$$

Using the above relationships $\Delta^{\ddagger}G_{DD} = \tfrac{1}{4}\Delta E_{R,DD}$ and $\Delta^{\ddagger}G_{AA} = \tfrac{1}{4}\Delta E_{R,AA}$, the expression for $\Delta E_{R,DA}$ becomes

$$\Delta E_{R,DA} = \tfrac{1}{2}(4\Delta^{\ddagger}G_{AA} + 4\Delta^{\ddagger}G_{DD}) = 2(\Delta^{\ddagger}G_{AA} + \Delta^{\ddagger}G_{DD})$$

Using this in the above expression for $\Delta^{\ddagger}G_{DA}$ gives the required expression

$$\Delta^{\ddagger}G_{DA} = \tfrac{1}{2}\Delta_r G^{\ominus} + \tfrac{1}{4}\Delta E_{R,DA} = \tfrac{1}{2}\Delta_r G^{\ominus} + \tfrac{1}{2}(\Delta^{\ddagger}G_{AA} + \Delta^{\ddagger}G_{DD}) \quad (18.5)$$

(c) The overall rate constant for the self-exchange process are written $k_{AA} = K_{AA}k_{et,AA}$ and likewise for DD. The first-order rate constant for the electron transfer process is written using eqn 18.3 as $k_{et,AA} = \kappa\nu^{\ddagger}e^{-\Delta^{\ddagger}G_{AA}/RT}$

and likewise for $k_{\text{et,DD}}$. As indicated in the *Problem*, it is assumed that $\kappa\nu^{\ddagger}$ is the same for all reactions. Therefore, the expressions for the overall rate constants are

$$k_{\text{AA}} = K_{\text{AA}}\kappa\nu^{\ddagger}\text{e}^{-\Delta^{\ddagger}G_{\text{AA}}/RT} \qquad k_{\text{DD}} = K_{\text{DD}}\kappa\nu^{\ddagger}\text{e}^{-\Delta^{\ddagger}G_{\text{DD}}/RT}$$

(d) By analogy with the expression for k_{AA}

$$k_{\text{r}} = k_{\text{AD}} = K_{\text{AD}}\kappa\nu^{\ddagger}\text{e}^{-\Delta^{\ddagger}G_{\text{AD}}/RT}$$

(e) Next, eqn 18.5 is used to substitute for $\Delta^{\ddagger}G_{\text{AD}}$ in the expression for k_{r}

$$\begin{aligned} k_{\text{r}} &= K_{\text{AD}}\kappa\nu^{\ddagger}\text{e}^{-\Delta^{\ddagger}G_{\text{AD}}/RT} \\ &= K_{\text{AD}}\kappa\nu^{\ddagger}\text{e}^{-(\Delta_{\text{r}}G^{\ominus}+\Delta^{\ddagger}G_{\text{AA}}+\Delta^{\ddagger}G_{\text{DD}})/2RT} \\ &= K_{\text{AD}}\text{e}^{-\Delta_{\text{r}}G^{\ominus}/2RT}\left[\kappa\nu^{\ddagger}\text{e}^{-\Delta^{\ddagger}G_{\text{AA}}/RT}\right]^{1/2}\left[\kappa\nu^{\ddagger}\text{e}^{-\Delta^{\ddagger}G_{\text{DD}}/RT}\right]^{1/2} \end{aligned}$$

The term is the first bracket is recognised as $k_{\text{AA}}/K_{\text{AA}}$, and that in the second as $k_{\text{DD}}/K_{\text{DD}}$ to give

$$k_{\text{r}} = K_{\text{AD}}\text{e}^{-\Delta_{\text{r}}G^{\ominus}/2RT}\left[k_{\text{AA}}/K_{\text{AA}}\right]^{1/2}\left[k_{\text{DD}}/K_{\text{DD}}\right]^{1/2}$$

The equilibrium constant for the overall reaction is written $K = \text{e}^{-\Delta_{\text{r}}G^{\ominus}/RT}$, therefore $\text{e}^{-\Delta_{\text{r}}G^{\ominus}/2RT} = K^{1/2}$ and hence

$$k_{\text{r}} = \frac{K_{\text{AD}}}{(K_{\text{AA}}K_{\text{DD}})^{1/2}}(k_{\text{AA}}k_{\text{DD}}K)^{1/2}$$

which is of the required form.

P18E.3 The variation of the electron-transfer rate constant with $\Delta_{\text{r}}G^{\ominus}$ is given by [18E.8–814]

$$\ln k_{\text{et}} = -\frac{RT}{4\Delta E_{\text{R}}}\left(\frac{\Delta_{\text{r}}G^{\ominus}}{RT}\right)^2 - \tfrac{1}{2}\left(\frac{\Delta_{\text{r}}G^{\ominus}}{RT}\right) + \text{const.}$$

A plot of $\ln k_{\text{et}}$ against $-\Delta_{\text{r}}G^{\ominus}$ is expected to be an inverted parabola and, as described in the text, the maximum occurs at $-\Delta_{\text{r}}G^{\ominus} = \Delta E_{\text{R}}$. The plot is shown in Fig 18.4.

The data are a good fit to the second-order polynomial

$$\lg k_{\text{et}} = -2.828 \times (-\Delta_{\text{r}}G^{\ominus}/\text{eV})^2 + 5.942 \times (-\Delta_{\text{r}}G^{\ominus}/\text{eV}) + 7.129$$

which is shown on the plot. The maximum of this function occurs when the derivative is zero, that is when $2\times-2.828\times(-\Delta_{\text{r}}G^{\ominus}/\text{eV})+5.942 = 0$; this occurs at $(-\Delta_{\text{r}}G^{\ominus}/\text{eV}) = 1.05....$ Therefore $\boxed{\Delta E_{\text{R}} = 1.05\ \text{eV}}$.

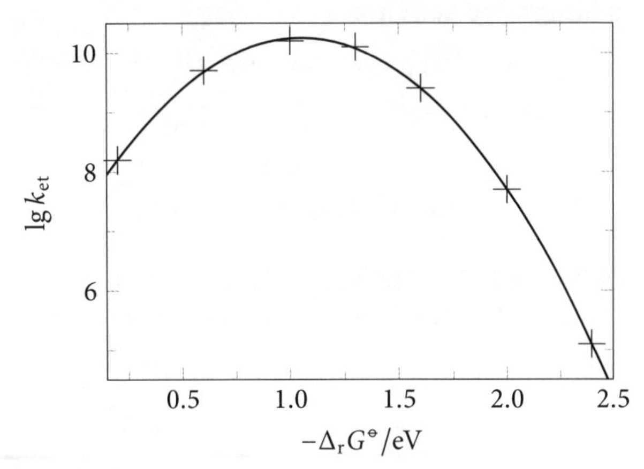

Figure 18.4

P18E.5 The variation of the electron-transfer rate constant with distance is given by [18E.7–*813*], $\ln k_{et} = -\beta d + \text{const}$. This relationship is tested by plotting $\ln k_{et}$ against d; the plot is shown in Fig 18.5.

d/nm	k_{et}/s^{-1}	$\ln(k_{et}/\text{s}^{-1})$
0.48	1.58×10^{12}	28.1
0.95	3.98×10^{9}	22.1
0.96	1.00×10^{9}	20.7
1.23	1.58×10^{8}	18.9
1.35	3.98×10^{7}	17.5
2.24	63.1	4.14

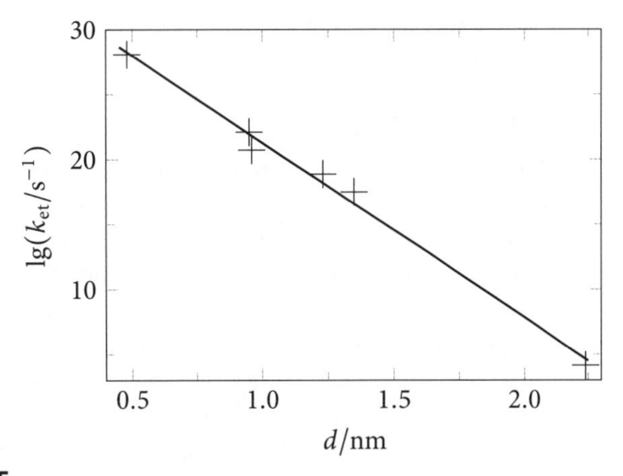

Figure 18.5

The data are a reasonable fit to a straight line with slope −13.43. The slope is identified as $-\beta$, therefore $\boxed{\beta = 13\ \text{nm}^{-1}}$.

Answers to integrated activities

I18.1 Using Stirling's approximation, $\ln P$ is developed as

$$\begin{aligned}\ln P = n\ln n - n + (n - n^* + s - 1)\ln(n - n^* + s - 1) - (n - n^* + s - 1)\\ - (n - n^*)\ln(n - n^*) + (n - n^*)\\ - (n + s - 1)\ln(n + s - 1) + (n + s - 1)\end{aligned}$$

following the hint by writing $\ln(n - n^* + s - 1) \approx \ln(n - n^*)$ and also $\ln(n + s - 1) \approx \ln(n)$, and also gathering together the linear terms gives

$$\begin{aligned}\ln P = -n - (n - n^* + s - 1) + (n - n^*) + (n + s - 1)\\ n\ln n + (n - n^* + s - 1)\ln(n - n^*)\\ - (n - n^*)\ln(n - n^*) - (n + s - 1)\ln(n)\end{aligned}$$

The linear terms cancel, and then gathering together terms gives

$$\begin{aligned}\ln P &= [n - (n + s - 1)]\ln(n) + [(n - n^* + s - 1) - (n - n^*)]\ln(n - n^*)\\ &= -(s - 1)\ln(n) + (s - 1)\ln(n - n^*)\end{aligned}$$

hence

$$P = \left(\frac{n - n^*}{n}\right)^{s-1}$$

I18.3 Both the Marcus theory of photo-induced electron transfer and the Förster theory of resonance energy transfer examine interactions between a molecule excited by absorption of electromagnetic energy (the chromophore S) and another molecule Q. They explain different mechanisms of quenching, that is, different ways that the chromophore gets rid of extra energy after absorbing a photon through intermolecular interactions. Another common feature of the two is that they depend on physical proximity of S and Q: they must be close for action to be efficient.

In the Marcus theory, the rate of electron transfer depends on the reaction Gibbs energy of electron transfer, $\Delta_r G$, and on the energy cost to S, Q, and the reaction medium of any concomitant molecular rearrangement. The rate is enhanced when the driving force ($\Delta_r G$) and the reorganization energy are well matched.

Resonant energy transfer in the Förster mechanism is most efficient when Q can directly absorb electromagnetic radiation from S. The oscillating dipole moment of S is induced by the electromagnetic radiation it absorbed. It transfers the excitation energy of the radiation to Q via a mechanism in which its oscillating dipole moment induces an oscillating dipole moment in Q. This energy transfer can be efficient when the absorption spectrum of the acceptor (Q) overlaps with the emission spectrum of the donor (S).

19 Processes at solid surfaces

19A An introduction to solid surfaces

Answers to discussion questions

D19A.1 (a) These are described in Section 19A.1 on page 824.

(b) Dislocations, or discontinuities in the regularity of a crystal lattice, result in surface steps and terraces. The edge dislocation can be envisioned by imagining small clumps of crystalline matter sticking together from either a melt or a solution. The lowest energy pattern sticks them together with valence requirements satisfied or atoms in a close-packed arrangement. This process is expected to form surface terraces because terraces yield the maximum possible number of nearest neighbours at a surface and the lowest possible surface energy. However, the very process of small clumps of matter rapidly sticking is very unlikely to always produce a perfect spacefilled, crystalline structure. Crystal defects such as the halfplane of atoms shown in Fig. 19A.2 on page 824 may form near the surface of the growing crystal. This edge dislocation distorts adjacent planes into a high energy configuration that is inherently unstable, but thermal agitations of the growth process cause the dislocation to propagate to the surface, thereby, forming a step.

Solutions to exercises

E19A.1(a) The collision flux, Z_W, is given by [19A.1–*825*], $Z_W = P/(2\pi MkT/N_A)^{1/2}$ where p is the pressure of gas, M is the molar mass of the molecule, k is Boltzmann's constant, T is the temperature and N_A is Avogadro's constant. From inside the front cover, 760 Torr = 1 atm = 1.01325×10^5 Pa, therefore 1 Torr is 133.32 Pa.

(i) For a hydrogen molecule, the molar mass $M = 2 \times (1.0079\ \mathrm{g\,mol^{-1}}) = 2.0158\ \mathrm{g\,mol^{-1}}$.

$$Z_W = \frac{p}{(2\pi MkT/N_A)^{1/2}}$$

$$= \frac{(0.10 \times 10^{-6}\ \mathrm{Torr}) \times (133.32\ \mathrm{Pa\,Torr^{-1}}) \times (6.0221 \times 10^{23}\ \mathrm{mol^{-1}})^{1/2}}{\left[2\pi \times (2.0158 \times 10^{-3}\ \mathrm{kg\,mol^{-1}}) \times (1.3806 \times 10^{-23}\ \mathrm{J\,K^{-1}}) \times (298.15\ \mathrm{K})\right]^{1/2}}$$

$$= 1.43... \times 10^{18}\ \mathrm{m^{-2}\,s^{-1}} = \boxed{1.4 \times 10^{14}\ \mathrm{cm^{-2}\,s^{-1}}}$$

(ii) For propane, the molar mass $M = 3\times(12.011\,\mathrm{g\,mol^{-1}})+8\times(1.0079\,\mathrm{g\,mol^{-1}}) = 44.096\,\mathrm{g\,mol^{-1}}$.

$$\begin{aligned} Z_\mathrm{W} &= \frac{p}{(2\pi MkT/N_\mathrm{A})^{1/2}} \\ &= \frac{(0.10\times10^{-6}\,\mathrm{Torr})\times(133.32\,\mathrm{Pa\,Torr^{-1}})\times(6.0221\times10^{23}\,\mathrm{mol^{-1}})^{1/2}}{\left[2\pi\times(44.096\times10^{-3}\,\mathrm{kg\,mol^{-1}})\times(1.3806\times10^{-23}\,\mathrm{J\,K^{-1}})\times(298.15\,\mathrm{K})\right]^{1/2}} \\ &= 3.06...\times10^{17}\,\mathrm{m^{-2}\,s^{-1}} = \boxed{3.1\times10^{13}\,\mathrm{cm^{-2}\,s^{-1}}} \end{aligned}$$

E19A.2(a) The collision flux, Z_W, is given by [19A.1–825], $Z_\mathrm{W} = p/(2\pi MkT/N_\mathrm{A})^{1/2}$ where p is the pressure of gas, M is the molar mass of the molecule, k is Boltzmann's constant, T is the temperature and N_A is Avogadro's constant.

The collision rate, z, is given by $z = AZ_\mathrm{W}$ where A is the surface area. Hence,

$$z = AZ_\mathrm{W} = \frac{Ap}{(2\pi MkT/N_\mathrm{A})^{1/2}}$$

For an argon atom, the molar mass $M = 39.95\,\mathrm{g\,mol^{-1}}$. Thus, for $A = \pi(d/2)^2$, where d is the diameter of the circular surface, rearranging the above expression gives

$$\begin{aligned} p &= \frac{r(2\pi MkT/N_\mathrm{A})^{1/2}}{A} \\ &= \frac{4.5\times10^{20}\,\mathrm{s^{-1}}}{\pi\times(0.5\times1.5\times10^{-3}\,\mathrm{m})^2} \\ &\quad\times\left(\frac{2\pi\times(39.95\times10^{-3}\,\mathrm{kg\,mol^{-1}})\times(1.3806\times10^{-23}\,\mathrm{J\,K^{-1}})\times(425\,\mathrm{K})}{6.0221\times10^{23}\,\mathrm{mol^{-1}}}\right)^{1/2} \\ &= 1.25...\times10^{4}\,\mathrm{Pa} = \boxed{0.13\,\mathrm{bar}} \end{aligned}$$

E19A.3(a) For a perfect gas, and at constant temperature, $p \propto 1/V$, where V is the volume occupied by the gas at pressure p. Therefore

$$\frac{p_2}{p_1} = \frac{V_1}{V_2} \quad \text{hence} \quad V_2 = \frac{V_1 p_1}{p_2}$$

The surface coverage θ is given by $\theta = V/V_\infty$ where V is the volume of gas adsorbed at a particular pressure p and V_∞ is the volume of gas which gives a complete monolayer, but where the volume has been corrected to the same pressure p.

At 5.0 bar, the volume adsorbed is a complete monolayer and thus $V_\infty = 22\,\mathrm{cm^3}$ at 5.0 bar. At 0.1 bar, this same volume is

$$V_{\infty,0.1\,\mathrm{bar}} = \frac{(22\,\mathrm{cm^3})\times(5.0\,\mathrm{bar})}{0.1\,\mathrm{bar}} = 1100\,\mathrm{cm^3}$$

Hence the surface coverage is

$$\theta = \frac{10\,\mathrm{cm^3}}{1100\,\mathrm{cm^3}} = \boxed{9.1\times10^{-3}}$$

E19A.4(a) For a process to be spontaneous it must be accompanied by a reduction in the Gibbs energy, that is $\Delta G < 0$. The adsorption of a gas on a surface is likely to be accompanied by a significant reduction in entropy on account of the loss of translational degrees of freedom, therefore $\Delta S < 0$. Given that $\Delta G = \Delta H - T\Delta S$, a process with $\Delta S < 0$ can only have $\Delta G < 0$ if ΔH is sufficiently negative, that is the process must be exothermic.

Solutions to problems

P19A.1 The Coulombic energy of interaction of the test ion with a section of lattice is determined by summing the interaction energy of this ion with each of the ions in the section of the lattice. Interactions between ions of opposite charge make a negative contribution to the energy of $-C/r$, where C is a positive constant and r is the distance between the test ion and an ion in the lattice. Similarly, interactions between ions of the same charge make a positive contribution to the energy of $+C/r$. Define a_0 as the distance between nearest neighbours in the lattice, that is the lattice spacing.

(a) For a Type 2 section, and considering the nearest 10 ions only, the interaction energy with the test ion is

$$E_2 = \frac{C}{a_0} \times \left(-1 + \frac{1}{2} - \frac{1}{3} + \ldots + \frac{1}{10}\right) = \left(\frac{C}{a_0}\right) \sum_{n=1}^{10} \frac{(-1)^n}{n} = \boxed{-0.646\left(\frac{C}{a_0}\right)}$$

(b) For a Type 1 section of lattice, with 10 atoms in each direction, the interaction energy with the test ion is

$$E_1 = \left(\frac{C}{a_0}\right) \sum_{n=1}^{10} \sum_{m=1}^{10} \frac{(-1)^{n+m}}{(n^2+m^2)^{1/2}} = \boxed{+0.259\left(\frac{C}{a_0}\right)}$$

(c) To calculate the energy of interaction between the test ion and the lattice in arrangement (a), observe that there is one Type 2 interaction and two Type 1 interactions. Hence the interaction energy is given by

$$E_{(a)} = E_2 + 2E_1 = \left(\frac{C}{a_0}\right) \times (-0.646 + 2 \times 0.259) = \boxed{-0.128\left(\frac{C}{a_0}\right)}$$

To calculate the energy of interaction between the test ion and the lattice in arrangement (b), observe that there are two Type 2 interactions and three Type 1 interactions. Hence the interaction energy is given by

$$E_{(b)} = 2E_2 + 3E_1 = \left(\frac{C}{a_0}\right) \times [2(-0.646) + 3(0.259)] = \boxed{-0.516\left(\frac{C}{a_0}\right)}$$

The energy of interaction of the probe cation is much lower for (b) than for (a), therefore $\boxed{\text{(b) is the more favourable arrangement}}$.

P19A.3 From inside the front cover, 760 Torr = 1 atm = 1.01325×10^5 Pa, therefore 1 Torr is 133.32 Pa. The unit cell for a face-centred cubic lattice is shown in Fig. 15A.8 on page 643 (cubic F) and how the planes are identified using Miller indices is described in Section 15A.2(a) on page 643.

(a) The (100) plane is the face of the cube and the arrangement of the atoms in the plane is shown in Fig. 19.1. There two atoms in this face, being the total of one in the centre and a quarter of each of the four atoms at the corners. Each atom has surface area πr^2, where r is the atomic radius, and the area of the face is $a^2 = (352 \times 10^{-12}\ \text{pm})^2 = 1.24 \times 10^{-15}\ \text{cm}^2$.

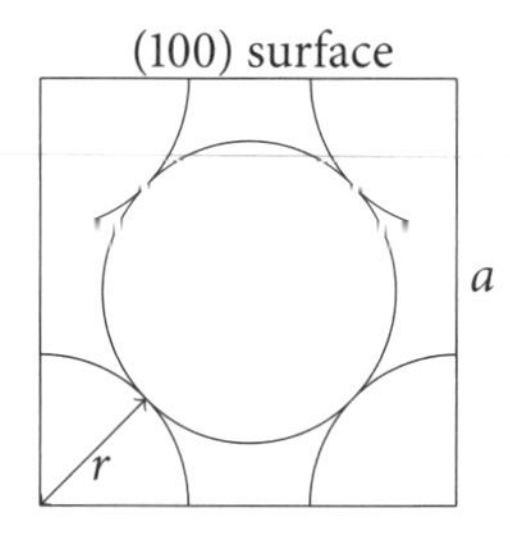

Figure 19.1

The surface number density n is the number of atoms divided by the area of the face

$$n = \frac{2}{1.24 \times 10^{-15}\ \text{cm}^2} = \boxed{1.61 \times 10^{15}\ \text{cm}^{-2}} = 1.61 \times 10^{19}\ \text{m}^{-2}$$

$A = 1/n$ is therefore the area occupied by atoms within this face.
The collision flux, Z_W, is given by [19A.1–825], $Z_\text{W} = p/(2\pi MkT/N_\text{A})^{1/2}$. For a hydrogen molecule, the molar mass $M = 2 \times (1.0079\ \text{g mol}^{-1}) = 2.0158\ \text{g mol}^{-1}$, so at $T = 298.15$ K and $p = 100$ Pa the frequency, f, of molecular collisions with the atoms exposed on this face is

$$f = AZ_\text{W} = \frac{Ap}{(2\pi MkT/N_\text{A})^{1/2}}$$

$$= \frac{(1.61 \times 10^{19}\ \text{m}^{-2})^{-1} \times (100\ \text{Pa}) \times (6.0221 \times 10^{23}\ \text{mol}^{-1})^{1/2}}{\left[2\pi \times (2.0158 \times 10^{-3}\ \text{kg mol}^{-1}) \times (1.3806 \times 10^{-23}\ \text{J K}^{-1}) \times (298.15\ \text{K})\right]^{1/2}}$$

$$= \boxed{6.7 \times 10^5\ \text{s}^{-1}}$$

At $p = 0.1 \times 10^{-6}$ Torr, which is $(0.1 \times 10^{-6}\ \text{Torr}) \times (133.32\ \text{Pa Torr}^{-1}) = 1.33... \times 10^{-5}$ Pa

$$f = AZ_\text{W} = \frac{Ap}{(2\pi MkT/N_\text{A})^{1/2}}$$

$$= \frac{(1.61 \times 10^{19}\ \text{m}^{-2})^{-1} \times (1.33... \times 10^{-5}\ \text{Pa}) \times (6.0221 \times 10^{23}\ \text{mol}^{-1})^{1/2}}{\left[2\pi \times (2.0158 \times 10^{-3}\ \text{kg mol}^{-1}) \times (1.3806 \times 10^{-23}\ \text{J K}^{-1}) \times (298.15\ \text{K})\right]^{1/2}}$$

$$= \boxed{8.9 \times 10^{-2}\ \text{s}^{-1}}$$

For propane, the molar mass $M = (3 \times 12.01 + 8 \times 1.0079)\ \mathrm{g\,mol^{-1}} = 44.09\ \mathrm{g\,mol^{-1}}$. Similar calculations give the corresponding collision rates as $\boxed{f = 1.42 \times 10^5\ \mathrm{s^{-1}}}$ at 100 Pa, and $\boxed{f = 1.9 \times 10^{-2}\ \mathrm{s^{-1}}}$ at 0.10 μTorr.

(b) The (110) plane is a diagonal plane taken from corner to corner along one face, and perpendicular to the face. It has the surface structure shown in Fig. 19.2. There are again two atoms in this face.

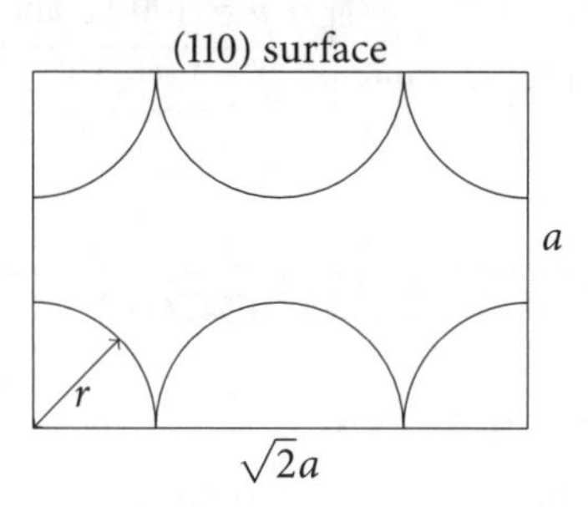

Figure 19.2

The area of the face is $\sqrt{2}a^2$, therefore the surface number density n is

$$n = \frac{2}{\sqrt{2}a^2} = \frac{2}{\sqrt{2} \times (352 \times 10^{-12}\ \mathrm{pm})^2}$$
$$= 1.14 \times 10^{19}\ \mathrm{m^{-2}} = \boxed{1.14 \times 10^{15}\ \mathrm{cm^{-2}}}$$

Similar calculations to those above give, for H_2, $\boxed{f = 9.4 \times 10^5\ \mathrm{s^{-1}}}$ and $\boxed{f = 0.13\ \mathrm{s^{-1}}}$ at $p = 100$ Pa and $p = 0.10$ μTorr, respectively. For propane the rates are $\boxed{f = 2.0 \times 10^5\ \mathrm{s^{-1}}}$ and $\boxed{f = 2.7 \times 10^{-2}\ \mathrm{s^{-1}}}$.

(c) The (111) plane has the surface structure shown in Fig. 19.3; there is one atom on this face.

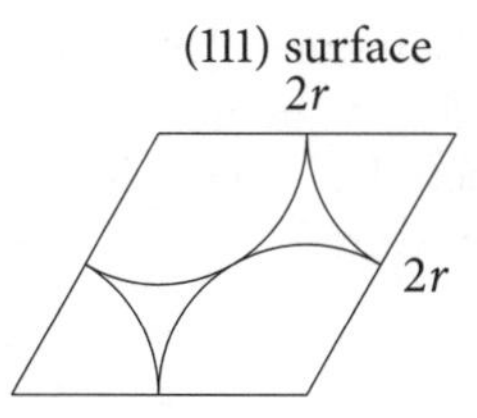

Figure 19.3

The area of this rhombus is $2r \times \sqrt{3}r = 2\sqrt{3}r^2$. Because the atoms touch along the face diagonals it follows that $4r = \sqrt{2}a$, hence $r^2 = a^2/8$. Using this the area of the face is $2\sqrt{3}r^2 = \sqrt{3}a^2/4$, and thus the surface number density is

$$n = \frac{1}{\sqrt{3}a^2/4} = \frac{1}{\sqrt{3} \times (352 \times 10^{-12}\ \text{pm})^2/4}$$
$$= 1.86 \times 10^{19}\ \text{m}^{-2} = \boxed{1.86 \times 10^{15}\ \text{cm}^{-2}}$$

Similar calculations to those above give, for H_2, $\boxed{f = 5.8 \times 10^5\ \text{s}^{-1}}$ and $\boxed{f = 7.7 \times 10^{-2}\ \text{s}^{-1}}$ at $p = 100$ Pa and $p = 0.10$ μTorr, respectively. For propane the rates are $\boxed{f = 1.2 \times 10^5\ \text{s}^{-1}}$ and $\boxed{f = 1.6 \times 10^{-2}\ \text{s}^{-1}}$.

19B Adsorption and desorption

Answers to discussion questions

D19B.1 The assumptions made in deriving the Langmuir isotherm are:

(1) Adsorption cannot proceed beyond monolayer coverage.

(2) All sites are equivalent and the surface is uniform.

(3) The ability of a molecule to adsorb at a given site is independent of the occupation of neighbouring sites.

For the BET isotherm assumption (1) is removed so that multi-layer coverage is possible. For the Temkin isotherm assumption (2) is removed and it is assumed that the energetically most favourable sites are occupied first. The Temkin isotherm corresponds to supposing that the adsorption enthalpy changes linearly with pressure.

The Freundlich isotherm removes assumption (2) but this isotherm corresponds to a logarithmic change in the adsorption enthalpy with pressure.

Solutions to exercises

E19B.1(a) The Langmuir isotherm is [19B.2–833], $\theta = \alpha p/(1 + \alpha p)$, with $\alpha = k_a/k_d$. The surface coverage may be written in terms of the volume of gas adsorbed V, $\theta = V/V_\infty$, where V_∞ is the volume corresponding to complete coverage. For two different pressures

$$\frac{V_1}{V_\infty} = \frac{\alpha p_1}{1 + \alpha p_1} \qquad \frac{V_2}{V_\infty} = \frac{\alpha p_2}{1 + \alpha p_2}$$

Inverting both sides

$$\frac{V_\infty}{V_1} = \frac{1}{\alpha p_1} + 1 \qquad \frac{V_\infty}{V_2} = \frac{1}{\alpha p_2} + 1$$

To eliminate α first multiply the left-hand equation by $1/p_2$ and the right-hand equation by $1/p_1$

$$\frac{V_\infty}{p_2 V_1} = \frac{1}{\alpha p_1 p_2} + \frac{1}{p_2} \qquad \frac{V_\infty}{p_1 V_2} = \frac{1}{\alpha p_1 p_2} + \frac{1}{p_1}$$

Subtracting the two equations gives then eliminates α

$$\frac{V_\infty}{p_2 V_1} - \frac{V_\infty}{p_1 V_2} = \frac{1}{p_2} - \frac{1}{p_1}$$

$$\text{hence} \quad V_\infty = \frac{1/p_2 - 1/p_1}{1/p_2 V_1 - 1/p_1 V_2} = \frac{p_1 - p_2}{p_1/V_1 - p_2/V_2}$$

where for the last step top and bottom are multiplied by $p_1 p_2$.

With the data given

$$\begin{aligned} V_\infty &= \frac{p_1 - p_2}{p_1/V_1 - p_2/V_2} \\ &= \frac{(145.4\ \text{Torr}) - (760\ \text{Torr})}{(145.4\ \text{Torr})/(0.286\ \text{cm}^3) - (760\ \text{Torr})/(1.443\ \text{cm}^3)} = \boxed{33.6\ \text{cm}^3} \end{aligned}$$

E19B.2(a) The residence half-life is given by [19B.14–839], $t_{1/2} = \tau_0 e^{E_{a,des}/RT}$. The activation energy for desorption, $E_{a,des}$, is approximated as minus the enthalpy of adsorption.

$$\begin{aligned} t_{1/2} &= (1.0 \times 10^{-14}\ \text{s})\, e^{(120\times10^3\ \text{J mol}^{-1})/[(8.3145\ \text{J K}^{-1}\ \text{mol}^{-1})\times(400\ \text{K})]} \\ &= \boxed{47\ \text{s}} \end{aligned}$$

E19B.3(a) The Langmuir isotherm is [19B.2–833], $\theta = \alpha p/(1 + \alpha p)$. The surface coverage may be written in terms of the volume of gas adsorbed V, $\theta = V/V_\infty$, where V_∞ is the volume corresponding to complete coverage. Equivalently, θ may be expressed in terms of the mass adsorbed, $\theta = m/m_\infty$, where m_∞ is the mass corresponding to complete coverage. For two different pressures

$$\frac{m_1}{m_\infty} = \frac{\alpha p_1}{1 + \alpha p_1} \qquad \frac{m_2}{m_\infty} = \frac{\alpha p_2}{1 + \alpha p_2}$$

The aim is to find m_∞, and the algebra to do this is just the same as the method for finding V_∞ in *Exercise* E19B.1(a) with volumes replaced by masses. The result is

$$\begin{aligned} m_\infty &= \frac{p_1 - p_2}{p_1/m_1 - p_2/m_2} \\ &= \frac{(26.0\ \text{kPa}) - (3.0\ \text{kPa})}{(26.0\ \text{kPa})/(0.44\ \text{mg}) - (3.0\ \text{kPa})/(0.19\ \text{mg})} = 0.531...\ \text{mg} \end{aligned}$$

The surface coverage at the first pressure is therefore

$$\theta_1 = \frac{m_1}{m_\infty} = \frac{0.44\ \text{mg}}{0.531...\ \text{mg}} = \boxed{0.83}$$

At the second pressure $\theta_2 = (0.19\ \text{mg})/(0.531...\ \text{mg}) = \boxed{0.36}$.

E19B.4(a) The Langmuir isotherm is [19B.2–833], $\theta = \alpha p/(1+\alpha p)$, inverting both sides gives

$$\frac{1}{\theta} = \frac{1}{\alpha p} + 1 \quad \text{hence} \quad \frac{1}{\alpha p} = \frac{1}{\theta} - 1 = \frac{1-\theta}{\theta}$$

Inverting again gives

$$\alpha p = \frac{\theta}{1-\theta} \quad \text{hence} \quad p = \frac{\theta}{\alpha(1-\theta)}$$

With the data given

$$p = \frac{0.15}{(0.75\ \text{kPa}^{-1}) \times (1-0.15)} = \boxed{0.24\ \text{kPa}}$$

A similar calculation for $\theta = 0.95$ gives $\boxed{25\ \text{kPa}}$.

E19B.5(a) The isosteric enthalpy of adsorption is define as [19B.5b–*834*]

$$\left(\frac{\partial \ln(\alpha p^{\ominus})}{\partial(1/T)}\right)_{\theta} = -\frac{\Delta_{\text{ad}} H^{\ominus}}{R}$$

From the Langmuir isotherm is follows that $\alpha = \theta/p(1-\theta)$ but, because an isosteric process is being considered (θ is constant), this reduces to $\alpha = C/p$, where C is a constant. With just two sets of data the derivative is approximated as the finite interval to give

$$\frac{\ln(Cp^{\ominus}/p_2) - \ln(Cp^{\ominus}/p_1)}{(1/T_2) - (1/T_1)} = -\frac{\Delta_{\text{ad}} H^{\ominus}}{R}$$

$$\text{hence } \ln p_1/p_2 = -\frac{\Delta_{\text{ad}} H^{\ominus}}{R}\left(\frac{1}{T_2} - \frac{1}{T_1}\right)$$

$$\text{hence } \ln p_2 = \ln p_1 + \frac{\Delta_{\text{ad}} H^{\ominus}}{R}\left(\frac{1}{T_2} - \frac{1}{T_1}\right)$$

The data gives the enthalpy of desorption as +10.2 J for 1.00 mmol of gas, therefore the molar enthalpy of adsorption is $-10.2\ \text{kJ mol}^{-1}$.

$$\ln(p_2/\text{kPa}) = \ln(12\ \text{kPa}) + \frac{-10.2\ \text{kJ mol}^{-1}}{8.3145\ \text{J K}^{-1}\ \text{mol}^{-1}}\left(\frac{1}{313\ \text{K}} - \frac{1}{298\ \text{K}}\right) = 2.68...$$

Therefore $(p_2/\text{kPa}) = e^{2.68...}$, giving $\boxed{p_2 = 15\ \text{kPa}}$.

E19B.6(a) The isosteric enthalpy of adsorption is define as [19B.5b–*834*]

$$\left(\frac{\partial \ln(\alpha p^{\ominus})}{\partial(1/T)}\right)_{\theta} = -\frac{\Delta_{\text{ad}} H^{\ominus}}{R}$$

From the Langmuir isotherm is follows that $\alpha = \theta/p(1-\theta)$ but, because an isosteric process is being considered (θ is constant), this reduces to $\alpha = C/p$, where C is a constant. With just two sets of data the derivative is approximated as the finite interval to give

$$\frac{\ln(Cp^{\ominus}/p_2) - \ln(Cp^{\ominus}/p_1)}{(1/T_2) - (1/T_1)} = -\frac{\Delta_{\text{ad}}H^{\ominus}}{R}$$

$$\text{hence } \ln(p_1/p_2) = -\frac{\Delta_{\text{ad}}H^{\ominus}}{R}\left(\frac{1}{T_2} - \frac{1}{T_1}\right)$$

$$\text{hence } \Delta_{\text{ad}}H^{\ominus} = \frac{R\ln(p_2/p_1)}{1/T_2 - 1/T_1}$$

With the data given

$$\Delta_{\text{ad}}H^{\ominus} = \frac{(8.3145\,\text{J K}^{-1}\,\text{mol}^{-1}) \times \ln[(3.2\times 10^6\,\text{Pa})/(490\times 10^3\,\text{Pa})]}{1/(250\,\text{K}) - 1/(190\,\text{K})}$$

$$= \boxed{-12.4\,\text{kJ mol}^{-1}}$$

E19B.7(a) The rate constant for desorption is assumed to follow an Arrhenius law, $k_{\text{des}} = Ae^{-E_{\text{a,des}}/RT}$. Recall that for a first order process the half life is simply proportional to the inverse of the rate constant, therefore the time needed for a certain amount to desorb is also inversely proportional to the rate constant. Thus

$$\tau_1/\tau_2 = \text{e}^{-(E_{\text{a,des}}/R)(1/T_2 - 1/T_1)}$$

$$\text{hence } \ln(\tau_1/\tau_2) = -\frac{E_{\text{a,des}}}{R}\left(\frac{1}{T_2} - \frac{1}{T_1}\right)$$

$$\text{hence } E_{\text{a,des}} = \frac{-R\ln(\tau_1/\tau_2)}{(1/T_2 - 1/T_1)}$$

With the data given

$$E_{\text{a,des}} = \frac{-(8.3145\,\text{J K}^{-1}\,\text{mol}^{-1})\ln[(2.0\,\text{min})/(27\,\text{min})]}{1/(1978\,\text{K}) - 1/(1856\,\text{K})}$$

$$= 6.51...\times 10^5\,\text{J mol}^{-1} = \boxed{651\,\text{kJ mol}^{-1}}$$

The times for desorption at different temperatures are computed using

$$\tau_1/\tau_2 = \text{e}^{-(E_{\text{a,des}}/R)(1/T_2 - 1/T_1)} \quad \text{hence} \quad \tau_2 = \tau_1 \text{e}^{(E_{\text{a,des}}/R)(1/T_2 - 1/T_1)}$$

The time needed at 298 K is related to that at 1856 K

$$\tau_2 = (27\,\text{min})\,\text{e}^{[(6.51...\times 10^5\,\text{J mol}^{-1})/(8.3145\,\text{J K}^{-1}\,\text{mol}^{-1})][1/(298\,\text{K}) - 1/(1856\,\text{K})]}$$

$$= 1.7\times 10^{97}\,\text{min}$$

Effectively, the gas does not desorb at this temperature. Repeating the calculation at 3000 K

$$\tau_2 = (27\,\text{min})\,\text{e}^{[(6.51...\times 10^5\,\text{J mol}^{-1})/(8.3145\,\text{J K}^{-1}\,\text{mol}^{-1})][1/(3000\,\text{K}) - 1/(1856\,\text{K})]}$$

$$= 2.8...\times 10^{-6}\,\text{min} = \boxed{0.17\,\mu\text{s}}$$

At the higher temperature the gas leaves very rapidly indeed.

E19B.8(a) The average time that a species remains adsorbed is proportional to its half-life, given by [19B.14–839], $t_{1/2} = \tau_0 e^{E_{a,des}/RT}$. Therefore, if the two times are τ_1 and τ_2 at temperatures T_1 and T_2

$$\tau_2/\tau_1 = e^{(E_{a,des}/R)(1/T_2 - 1/T_1)}$$

$$\text{hence } \ln(\tau_2/\tau_1) = \frac{E_{a,des}}{R}\left(\frac{1}{T_2} - \frac{1}{T_1}\right)$$

$$\text{hence } E_{a,des} = \frac{R\ln(\tau_2/\tau_1)}{1/T_2 - 1/T_1}$$

With the data given

$$E_{a,des} = \frac{(8.3145\,\text{J K}^{-1}\,\text{mol}^{-1})\ln[(3.49\,\text{s})/(0.36\,\text{s})]}{1/(2362\,\text{K}) - 1/(2548\,\text{K})}$$

$$\boxed{611\,\text{kJ mol}^{-1}}$$

E19B.9(a) The half-life for a species on the surface is given by [19B.14–839], $t_{1/2} = \tau_0 e^{E_{a,des}/RT}$.

(i) With $E_{a,des} = 15\,\text{kJ mol}^{-1}$

$$\text{at 400 K} \quad t_{1/2} = (0.1\,\text{ps})\,e^{(15\times10^3\,\text{J mol}^{-1})/[(8.3145\,\text{J K}^{-1}\,\text{mol}^{-1})\times(400\,\text{K})]} = \boxed{9.1\,\text{ps}}$$

$$\text{at 1000 K} \quad t_{1/2} = (0.1\,\text{ps})\,e^{(15\times10^3\,\text{J mol}^{-1})/[(8.3145\,\text{J K}^{-1}\,\text{mol}^{-1})\times(1000\,\text{K})]} = \boxed{0.61\,\text{ps}}$$

(ii) With $E_{a,des} = 150\,\text{kJ mol}^{-1}$

$$\text{at 400 K} \quad t_{1/2} = (0.1\,\text{ps})\,e^{(150\times10^3\,\text{J mol}^{-1})/[(8.3145\,\text{J K}^{-1}\,\text{mol}^{-1})\times(400\,\text{K})]} = 3.86...\times10^{18}\,\text{ps} = \boxed{3.9\times10^{6}\,\text{s}}$$

$$\text{at 1000 K} \quad t_{1/2} = (0.1\,\text{ps})\,e^{(150\times10^3\,\text{J mol}^{-1})/[(8.3145\,\text{J K}^{-1}\,\text{mol}^{-1})\times(1000\,\text{K})]} = 6.83...\times10^{6}\,\text{ps} = \boxed{6.8\,\mu\text{s}}$$

Solutions to problems

P19B.1 (a) The Langmuir isotherm is [19B.2–833], $\theta = \alpha p/(1+\alpha p)$, inverting both sides gives $1/\theta = 1/\alpha p + 1$. Figure 19.4 shows a plot of $1/\theta$ against $1/p$ for three different values of α.

(b) The Langmuir isotherm for adsorption with dissociation is [19B.4–833], $\theta = (\alpha p)^{1/2}/[1+(\alpha p)^{1/2}]$. Figure 19.5 shows a plot of $1/\theta$ against $1/p$ for the same values of α used in Fig. 19.4. In contrast to the straight lines seen in Fig. 19.4, for the case of adsorption with dissociation the $1/\theta$ against $1/p$ plot shows pronounced curvature. Such a plot may therefore in principle make it possible to distinguish between dissociative and non-dissociative adsorption.

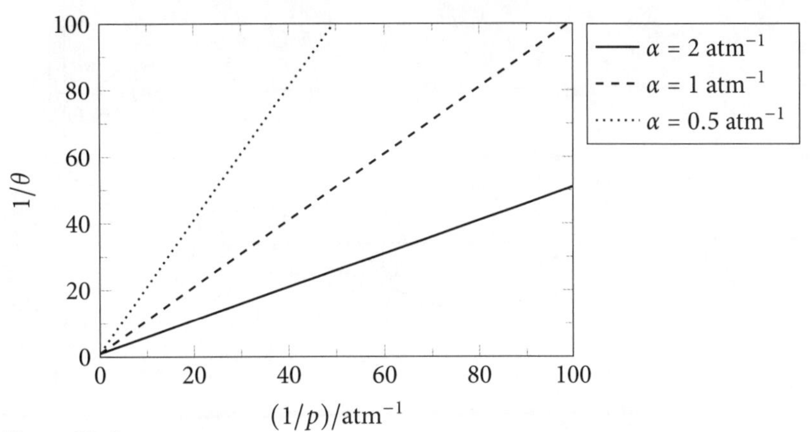

Figure 19.4

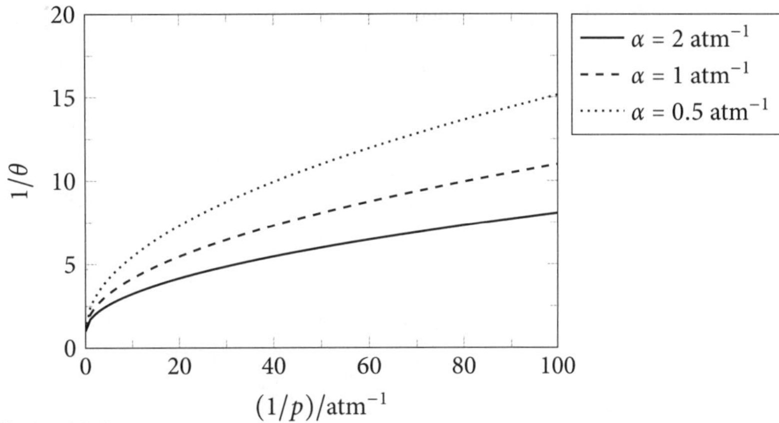

Figure 19.5

(c) In [19B.7–*836*] the BET isotherm is manipulated into a straight-line plot

$$\frac{z}{(1-z)V} = \frac{1}{cV_{\text{mon}}} + \frac{(c-1)}{cV_{\text{mon}}}z \quad \text{or} \quad \frac{zV_{\text{mon}}}{(1-z)V} = \frac{1}{c} + \frac{(c-1)}{c}z$$

Thus a plot of $zV_{\text{mon}}/(1-z)V$ against z is expected to be a straight line. Figure 19.6 shows such a plot for three different values of c. Note that when $c \gg 1$, the slope becomes independent of c, and tends to 1.

P19B.3 In [19B.7–*836*] the BET isotherm is manipulated into a straight-line plot

$$\frac{z}{(1-z)V} = \frac{1}{cV_{\text{mon}}} + \frac{(c-1)}{cV_{\text{mon}}}z \qquad z = p/p^*$$

Thus a plot of $z/(1-z)V$ against z is expected to be a straight line with slope $(c-1)/cV_{\text{mon}}$ and intercept $1/cV_{\text{mon}}$; note that (slope)/(intercept) = $c-1$. For brevity the term $z/(1-z)V$ is denoted y.

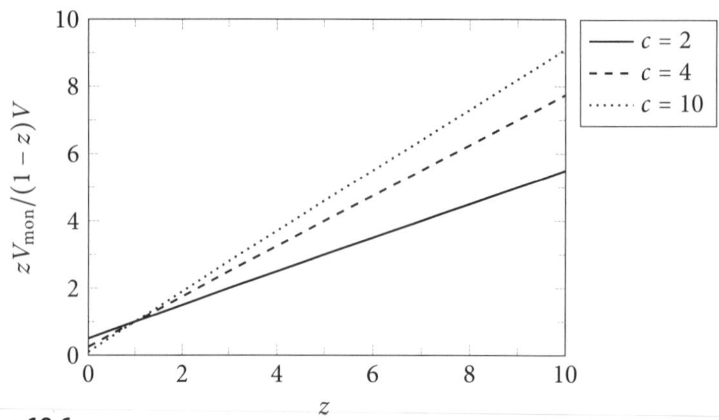

Figure 19.6

(a) The data are shown in the table below and the plot is shown in Fig. 19.7.

p/kPa	V/cm^3	z	$y/(\text{cm}^{-3})$
14.0	11.1	0.0326	0.00303
37.6	13.5	0.0875	0.00711
65.6	14.9	0.153	0.0121
79.2	16.0	0.184	0.0141
82.7	15.5	0.193	0.0154
100.7	17.3	0.234	0.0177
106.4	16.5	0.248	0.0200

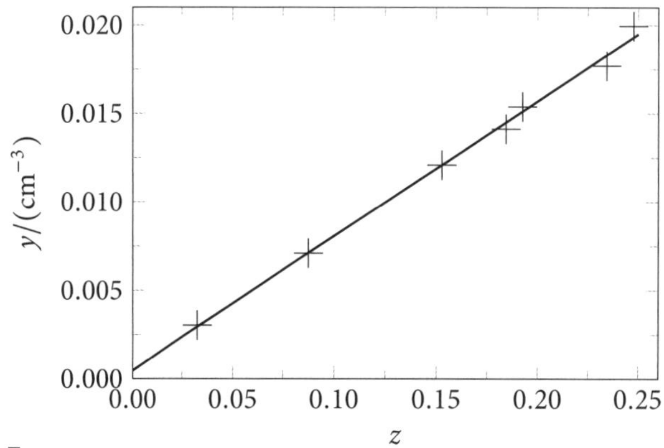

Figure 19.7

The data are a reasonable fit to a straight line with equation

$$y/(\text{cm}^{-3}) = 0.07612 \times (z) + 4.638 \times 10^{-4}$$

The parameter c is found using

$$\frac{\text{slope}}{\text{intercept}} = c - 1 \quad \text{hence} \quad c = 1 + \frac{0.07612}{4.638 \times 10^{-4}} = \boxed{165}$$

The intercept is $1/cV_{\text{mon}}$, therefore

$$V_{\text{mon}} = \frac{1}{c \times (\text{intercept})} = \frac{1}{165 \times (4.638 \times 10^{-4}\ \text{cm}^{-3})} = \boxed{13.1\ \text{cm}^3}$$

(b) The data are shown in the table below and the plot is shown in Fig. 19.8.

p/kPa	V/cm^3	z	y/(cm^{-3})
5.3	9.2	0.0065	0.00071
8.4	9.8	0.0102	0.00106
14.4	10.3	0.0176	0.00174
29.2	11.3	0.0356	0.00327
62.1	12.9	0.0758	0.00635
74.0	13.1	0.0903	0.00758
80.1	13.4	0.0977	0.00808
102.0	14.1	0.1244	0.01008

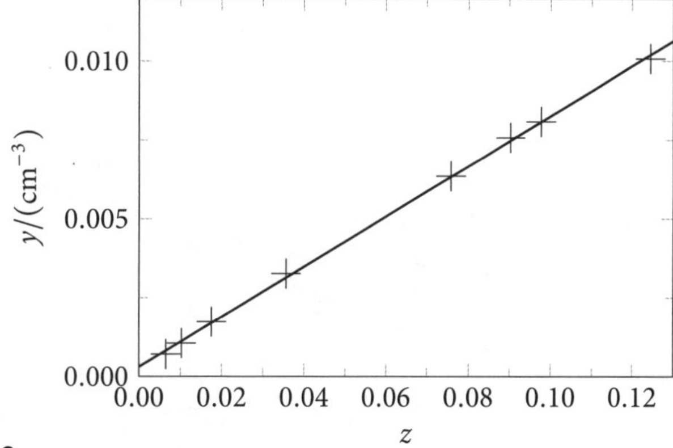

Figure 19.8

The data are a reasonable fit to a straight line with equation

$$y/(\text{cm}^{-3}) = 0.07953 \times (z) + 3.036 \times 10^{-4}$$

The parameter c is found using

$$\frac{\text{slope}}{\text{intercept}} = c - 1 \quad \text{hence} \quad c = 1 + \frac{0.07953}{3.036 \times 10^{-4}} = \boxed{263}$$

The intercept is $1/cV_{\text{mon}}$, therefore

$$V_{\text{mon}} = \frac{1}{c \times (\text{intercept})} = \frac{1}{263 \times (3.036 \times 10^{-4}\ \text{cm}^{-3})} = \boxed{12.5\ \text{cm}^3}$$

P19B.5 The Langmuir isotherm is [19B.2–833], $\theta = \alpha p/(1+\alpha p)$; the fractional coverage can be expressed as n/n_∞, where n is the amount in moles covering the surface, and n_∞ is the amount corresponding to a monolayer. The same argument as developed in *Example* 19B.1 on page 833 then applies, but with n instead of V. A suitable plot to fit data to the Langmuir isotherm is therefore of p/n against p; such a plot has intercept $1/\alpha n_\infty$ and slope $1/n_\infty$. The table of data is given below and the plot is shown in Fig. 19.9.

p/kPa	$n/(\text{mol kg}^{-1})$	$(p/n)/(\text{kPa mol}^{-1}\text{ kg})$
31.00	1.00	31.00
38.22	1.17	32.67
53.03	1.54	34.44
76.38	2.04	37.44
101.97	2.49	40.95
130.47	2.90	44.99
165.06	3.22	51.26
182.41	3.30	55.28
205.75	3.35	61.42
219.91	3.36	65.45

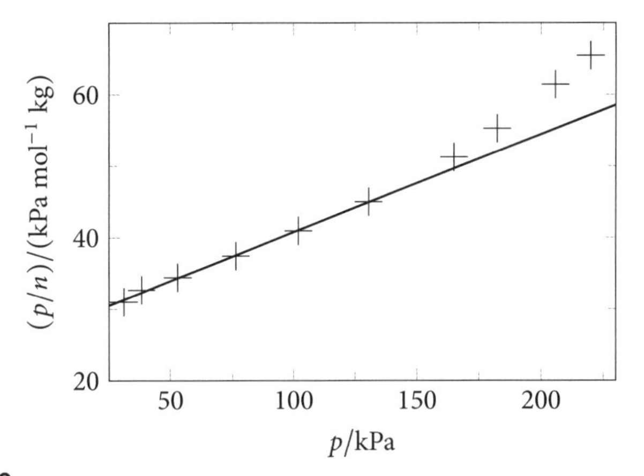

Figure 19.9

The data plainly fall on a curve, rather than the expected straight line. On the assumption that the isotherm is more likely to hold at low pressure, the first six data points are used to construct the line, the equation of which is

$$(p/n)/(\text{kPa mol}^{-1}\text{ kg}) = 0.1368 \times (p/\text{kPa}) + 27.08$$

The limiting coverage is $n_\infty = 1/\text{slope} = 1/(0.1368)\text{ mol kg}^{-1} = \boxed{7.3\text{ mol kg}^{-1}}$.

The value of α is found from (slope)/(intercept), $\alpha = (0.1368)/(27.08)\text{ kPa}^{-1} = \boxed{5.1 \times 10^{-3}\text{ kPa}^{-1}}$.

P19B.7 The isosteric enthalpy of adsorption is define as [19B.5b–*834*]

$$\left(\frac{\partial \ln(\alpha p^{\ominus})}{\partial(1/T)}\right)_{\theta} = -\frac{\Delta_{\text{ad}}H^{\ominus}}{R}$$

For the data in this *Problem* $p^{\ominus}$ is replaced by $c^{\ominus}$. This equation implies that a plot of $\ln(\alpha c^{\ominus})$ against $1/T$ will have slope $-\Delta_{\text{ad}}H^{\ominus}/R$, for data at constant θ.

T/K	$(10^{-11}\,\alpha)/(\text{mol}^{-1}\,\text{dm}^3)$	$(1/T)/(10^{-3}\,\text{K}^{-1})$	$\ln(\alpha c^{\ominus})$
283	2.642	3.53	26.30
298	2.078	3.36	26.06
308	1.286	3.25	25.58
318	1.085	3.14	25.41

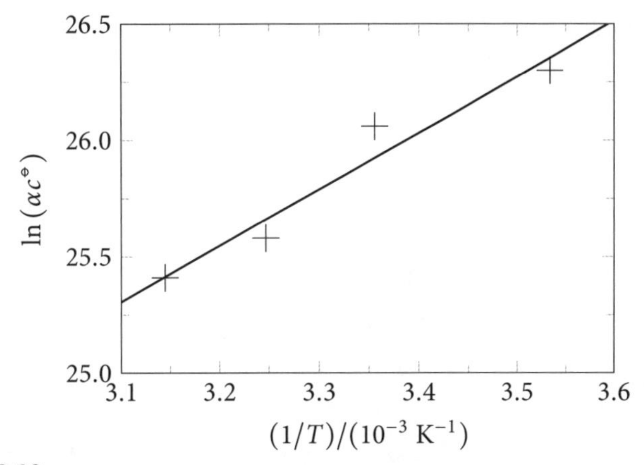

Figure 19.10

The data are a modest fit to a straight line, the equation of which is

$$\ln(\alpha c^{\ominus}) = 2.42 \times (1/T)/(10^{-3}\,\text{K}^{-1}) + 17.81$$

The value of $\Delta_{\text{ad}}H^{\ominus}$ is obtained from the slope (the scatter means that high precision on the result is not warranted)

$$\begin{aligned}\Delta_{\text{ad}}H^{\ominus} &= -R \times (\text{slope}) = -(8.3145\,\text{J}\,\text{K}^{-1}\,\text{mol}^{-1}) \times 10^3 \times (2.42\,\text{K}^{-1}) \\ &= \boxed{-20\,\text{kJ}\,\text{mol}^{-1}}\end{aligned}$$

With the given data

$$\begin{aligned}\Delta_{\text{ad}}G^{\ominus} &= \Delta_{\text{ad}}H^{\ominus} - T^{\ominus}\Delta_{\text{ad}}S^{\ominus} \\ &= (-20\,\text{kJ}\,\text{mol}^{-1}) - (300\,\text{K}) \times (+0.146\,\text{kJ}\,\text{K}^{-1}\,\text{mol}^{-1}) = \boxed{-64\,\text{kJ}\,\text{mol}^{-1}}\end{aligned}$$

P19B.9 The Freundlich isotherm is given in [19B.11–837], $\theta = c_1 p^{1/c_2}$. For the purposes of analysing this data set the isotherm is rewritten by assuming that the surface coverage is proportional to the mass adsorbed, and by replacing the pressure by the concentration divided by the standard concentration, to give $w_a = c_1([A]/c^\ominus)^{1/c_2}$. The units of c_1 are adjusted accordingly. Taking logarithms gives $\ln w_a = \ln c_1 + (1/c_2)\ln([A]/c^\ominus)$, implying that a plot of $\ln w_a$ against $\ln([A]/c^\ominus)$ should be a straight line of slope $1/c_2$ and intercept $\ln c_1$. The data are given below and the plot is shown in Fig. 19.11.

$[A]/c^\ominus$	w_a/g	$\ln([A]/c^\ominus)$	$\ln(w_a/g)$
0.05	0.04	−3.00	−3.22
0.10	0.06	−2.30	−2.81
0.50	0.12	−0.69	−2.12
1.00	0.16	0.00	−1.83
1.50	0.19	0.41	−1.66

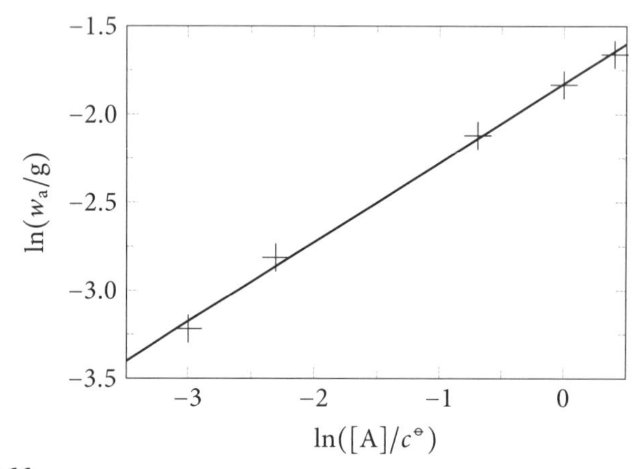

Figure 19.11

The data fall on a reasonable straight line, the equation of which is

$$\ln(w_a/g) = 0.450 \times \ln([A]/c^\ominus) - 1.83$$

The slope is $1/c_2$, therefore $c_2 = 1/\text{slope} = 1/(0.450) = \boxed{2.22}$. The intercept gives $\ln c_1$ and hence $\boxed{c_1 = 0.16\ \text{g}}$.

P19B.11 The Langmuir isotherm is [19B.2–833], $\theta = \alpha p/(1+\alpha p)$; the fractional coverage can be expressed as s/s_∞, where s is the amount in moles covering the surface (per g of charcoal), and s_∞ is the amount corresponding to a monolayer. For adsorption from solution the pressure is replaced by the concentration c. The same argument as developed in *Example* 19B.1 on page 833 then applies, but with s instead of V, and c instead of p. A suitable plot to fit data to the Langmuir isotherm is therefore of c/s against c; such a plot has intercept $1/\alpha s_\infty$ and slope $1/s_\infty$.

The table of data is shown below and the plot is shown in Fig. 19.12; for brevity the units of c and s are omitted throughout. It is clear that the data fall on a curve and so do not conform to the Langmuir isotherm.

c	s	c/s
15.0	0.60	25.0
23.0	0.75	30.7
42.0	1.05	40.0
84.0	1.50	56.0
165	2.15	76.7
390	3.50	111.4
800	5.10	156.9

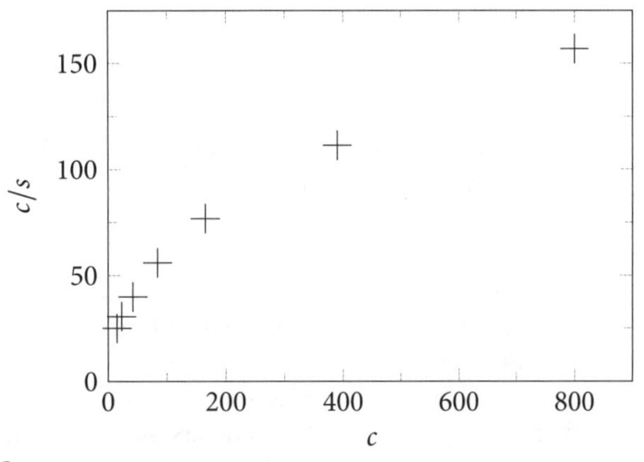

Figure 19.12

The Freundlich isotherm [19B.11–837] is written $s = Kc^{1/n}$; taking logarithms gives $\ln s = \ln K + (1/n)\ln c$. This implies that a plot of $\ln s$ against $\ln c$ should be a straight line of slope $1/n$ and intercept $\ln K$. The data are given below and the plot is shown in Fig. 19.13.

c	s	$\ln c$	$\ln s$
15.0	0.60	2.71	−0.511
23.0	0.75	3.14	−0.288
42.0	1.05	3.74	0.049
84.0	1.50	4.43	0.405
165	2.15	5.11	0.765
390	3.50	5.97	1.253
800	5.10	6.68	1.629

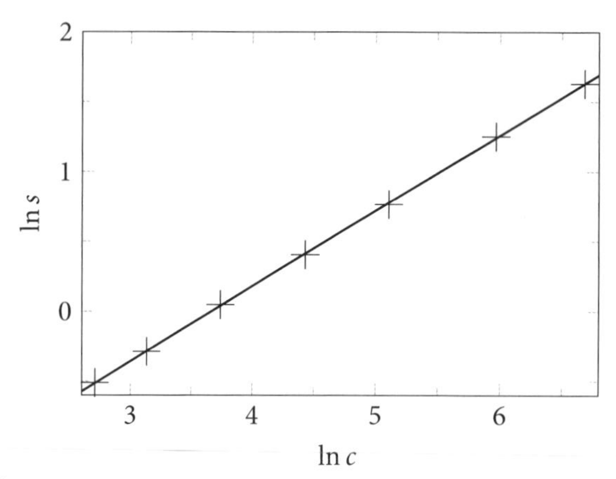

Figure 19.13

The data fall on a good straight line, the equation of which is

$$\ln s = 0.539 \times \ln c - 1.975$$

The slope is $1/n$, therefore $n = 1/\text{slope} = 1/(0.539) = \boxed{1.9}$. The intercept gives $\ln K$ and hence $\boxed{K = 0.14 \text{ mmol acetone/g charcoal}}$.

The Temkin isotherm [19B.10–*837*] is written in this case as $s = K\ln(nc)$. There is no straight-line plot for testing the data against this isotherm, but mathematical software can be used to find the best-fit parameters K and n. The result of such a fit is $s = 1.083 \ln(0.0738\,c)$. The table below compares the values of s predicted by this relationship and those predicted by the Freundlich isotherm (using the best-fit parameters found above). It is evident that the Freundlich isotherm reproduces the data far more precisely than does that Temkin isotherm.

c	s	s_{Temkin}	$s_{\text{Freundlich}}$
15.0	0.60	0.11	0.60
23.0	0.75	0.57	0.75
42.0	1.05	1.23	1.04
84.0	1.50	1.98	1.51
165.0	2.15	2.71	2.18
390.0	3.50	3.64	3.46
800.0	5.10	4.42	5.09

19C Heterogeneous catalysis

Answers to discussion questions

D19C.1 These are discussed in Section 19C.1(b) on page 842 and Section 19C.1(c) on page 843.

Solutions to exercises

E19C.1(a) The amount in moles of N_2 gas is found using the perfect gas law; close attention to the units is needed.

$$n = \frac{pV}{RT} = \frac{(760\ \text{Torr}) \times (1.01325 \times 10^5\ \text{Pa})}{760\ \text{Torr}} \times \frac{3.86 \times 10^{-6}\ \text{m}^3}{(8.3145\,\text{J}\,\text{K}^{-1}\,\text{mol}^{-1}) \times (273.15\ \text{K})}$$
$$= 1.72... \times 10^{-4}\ \text{mol}$$

which corresponds to $N_A n = (6.0221 \times 10^{23}\ \text{mol}^{-1}) \times (1.72 \times 10^{-4}\ \text{mol}) = 1.03... \times 10^{20}$ molecules.

A rough calculation of the surface area notes that the collision cross section is $\sigma = \pi d^2$, where d is the diameter of the colliding spheres. Therefore $d = (\sigma/\pi)^{1/2}$, and hence $r = \frac{1}{2}(\sigma/\pi)^{1/2}$. The area of one molecule is $\pi r^2 = \pi \frac{1}{4}\sigma/\pi = \frac{1}{4}\sigma$. The surface area is therefore $(1.03... \times 10^{20}) \times \frac{1}{4} \times (0.43 \times 10^{-18}\ \text{m}^2) = \boxed{11\ \text{m}^2}$.

In fact circles do not cover a plane completely, and it can be shown that the highest coverage which can be achieved is one in which the circles cover 0.91 of the area of the plane. The estimate of the area therefore needs to be scaled up by a factor of $1/0.91 \approx 1.1$ to give $12\ \text{m}^2$.

Solutions to problems

P19C.1 (a) The Langmuir–Hinshelwood rate law is given in [19C.2b–*842*]

$$v = \frac{k_r \alpha_A \alpha_B p_A p_B}{(1 + \alpha_A p_A + \alpha_B p_B)^2}$$

(b) When the partial pressures of the reactants are low, $\alpha_A p_A \ll 1$ and $\alpha_B p_B \ll 1$, and therefore these terms may be ignored in the denominator to give

$$v_{\text{low}} = k_r \alpha_A \alpha_B p_A p_B$$

The order with respect to A and B is now 1, and the overall order is 2.

(c) If, compared to B, A is strongly adsorbed or is present at high pressure, then $\alpha_A p_A \gg 1$ and $\alpha_A p_A \gg \alpha_B p_B$. The denominator simplifies to $(\alpha_A p_A)^2$ and

$$v_{\text{high A}} = \frac{k_r \alpha_A \alpha_B p_A p_B}{(\alpha_A p_A)^2} = \frac{k_r \alpha_B p_B}{\alpha_A p_A}$$

In this limit the rate law is −1 order in A, first order in B, and therefore overall zeroth order. It does not appear to be possible to achieve zeroth order for either A or B alone.

P19C.3 (a) The Langmuir isotherm is [19B.2–833], $\theta = \alpha p/(1+\alpha p)$. The fraction of uncovered sites, θ_u, is $\theta_u = 1 - \theta$

$$\theta_u = 1 - \frac{\alpha p}{1+\alpha p} = \frac{1+\alpha p - \alpha p}{1+\alpha p} = \frac{1}{1+\alpha p} \approx \frac{1}{\alpha p}$$

where the approximation holds in the limit $\alpha p \gg 1$, that is strong adsorption of the gas.

(b) If a gas is weakly adsorbed, meaning $\alpha p \ll 1$, the fractional coverage is $\theta \approx \alpha p$.

If hydrogen is strongly adsorbed, the fraction of uncovered surface sites goes as $p_{H_2}^{-1}$: these are the sites available for ammonia to bind to and to react, therefore the rate is expected to go as $p_{H_2}^{-1}$. The ammonia only binds weakly, so its surface coverage, and hence the rate of reaction, goes as p_{NH_3}. Overall, the rate law is expected to go as p_{NH_3}/p_{H_2}.

(c) The stoichiometric equation is $NH_3 \longrightarrow \frac{1}{2}N_2 + \frac{3}{2}H_2$. Assume that the initial pressure is p_0 and due solely to NH_3. After some time, suppose that the partial pressure of NH_3 has fallen to $p_{NH_3} = p_0 - \delta$; the partial pressure of H_2 is then $\frac{3}{2}\delta$. It follows that $\delta = p_0 - p_{NH_3}$ and hence $p_{H_2} = \frac{3}{2}\delta = \frac{3}{2}(p_0 - p_{NH_3})$. The rate law can therefore be written in terms of p_0 and p_{NH_3}

$$\frac{dp_{NH_3}}{dt} = -k_c \frac{p_{NH_3}}{\frac{3}{2}(p_0 - p_{NH_3})} = -\tfrac{2}{3}k_c \frac{p_{NH_3}}{p_0 - p_{NH_3}}$$

The differential equation is separable and can be integrated in a straightforward way; to simplify the notation p_{NH_3} is written as p

$$\int_{p_0}^{p} \frac{p_0 - p}{p}\, dp = -\tfrac{2}{3}k_c \int_0^t dt$$

$$\int_{p_0}^{p} \left[\frac{p_0}{p} - 1\right] dp = -\tfrac{2}{3}k_c \int_0^t dt$$

$$[p_0 \ln p - p]\big|_{p_0}^{p} = -\tfrac{2}{3}k_c t$$

$$p_0 \ln(p/p_0) - (p - p_0) = -\tfrac{2}{3}k_c t$$

(d) A plot of $y = p_0 \ln(p/p_0) - (p - p_0)$ against t is expected to be a straight line. The table of data is shown below and the plot is shown in Fig. 19.14.

t/s	p/kPa	y/kPa
0	13.3	0.000
30	11.7	−0.105
60	11.2	−0.186
100	10.7	−0.293
160	10.3	−0.400
200	9.9	−0.527
250	9.6	−0.636

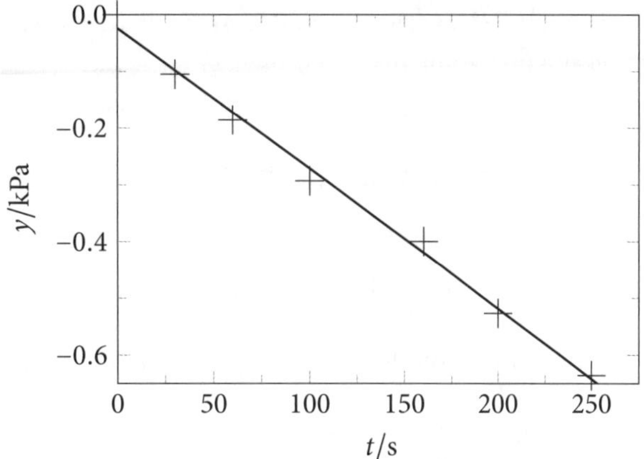

Figure 19.14

The data fall on a reasonable straight line with equation

$$y/\text{kPa} = -2.474 \times 10^{-3} \times (t/\text{s}) - 0.0237$$

A simple statistical analysis suggests an error of about 8×10^{-5} on the slope and 0.01 on the intercept. The intercept should be zero, and given the estimated errors the data are pretty much consistent with this.
From the slope it follows that $k_c = (-\frac{3}{2}) \times (-2.474 \times 10^{-3}\ \text{kPa s}^{-1})$ hence $\boxed{k_c = 3.7 \times 10^{-3}\ \text{kPa s}^{-1}}$.

19D Processes at electrodes

Answers to discussion questions

D19D.1 These are described in Section 19D.1 on page 845.

Solutions to exercises

E19D.1(a) If the anodic process is dominant, the current density is given by [19D.5a–*850*], $\ln j = \ln j_0 + (1-\alpha)f\eta$, where $f = F/RT$. At 298.15 K

$$f = (96485\ \text{C mol}^{-1})/[(8.3145\ \text{J K}^{-1}\ \text{mol}^{-1}) \times (298.15\ \text{K})] = 38.921\ \text{V}^{-1}$$

where the units are resolved by recalling 1 V = 1 J C^{-1}. Taking the difference of two expressions for ln j for different overpotentials gives

$$\ln(j_2/j_1) = (1-\alpha)f(\eta_2-\eta_1)$$

$$\text{hence } \eta_2 = \frac{\ln(j_2/j_1)}{(1-\alpha)f} + \eta_1$$

$$= \frac{\ln(75/55.0)}{(1-0.39)\times(38.921\,\text{V}^{-1})} + 0.125\,\text{V} = \boxed{0.14\,\text{V}}$$

E19D.2(a) If the anodic process is dominant, the current density is given by [19D.5a–*850*], $j = j_0\text{e}^{(1-\alpha)f\eta}$, where $f = F/RT$. At 298.15 K, f = 38.921 V^{-1}. Rearranging for j_0 and then using the data given

$$j_0 = j\text{e}^{-(1-\alpha)f\eta}$$

$$= (55.0\,\text{mA cm}^{-2})\,\text{e}^{-(1-0.39)\times(38.921\,\text{V}^{-1})\times(0.125\,\text{V})} = \boxed{2.8\,\text{mA cm}^{-2}}$$

E19D.3(a) If the anodic process is dominant, the current density is given by [19D.5a–*850*], $j = j_0\text{e}^{(1-\alpha)f\eta}$, where $f = F/RT$. At 298.15 K, f = 38.921 V^{-1}. Taking the ratio of two expressions for j for different overpotentials gives

$$j_2/j_1 = j_0\text{e}^{(1-\alpha)f\eta_2}/j_0\text{e}^{(1-\alpha)f\eta_1}$$

$$\text{hence } j_2 = j_1\text{e}^{(1-\alpha)f(\eta_2-\eta_1)}$$

$$= (1.0\,\text{mA cm}^{-2})\,\text{e}^{(1-0.5)\times(38.921\,\text{V}^{-1})\times[(0.60-0.40)\,\text{V})]} = \boxed{49\,\text{mA cm}^{-2}}$$

The current density increases dramatically with this increase in overpotential.

E19D.4(a) (i) The Butler–Volmer equation is [19D.2–*848*], $j = j_0(\text{e}^{(1-\alpha)f\eta} - \text{e}^{-\alpha f\eta})$. For H^+ on Ni $j_0 = 6.3\times10^{-6}$ A cm^{-2} and α = 0.58; at 298.15 K, f = 38.921 V^{-1}. For an overpotential of +0.20 V the current density is

$$j = (6.3\times10^{-6}\,\text{A cm}^{-2})$$
$$\times\,(\text{e}^{(1-0.58)\times(38.921\,\text{V}^{-1})\times(0.20\,\text{V})} - \text{e}^{-0.58\times(38.921\,\text{V}^{-1})\times(0.20\,\text{V})})$$
$$= \boxed{1.7\times10^{-4}\,\text{A cm}^{-2}}$$

(ii) If the current is entirely anodic, only the first term is needed

$$j = (6.3\times10^{-6}\,\text{A cm}^{-2})\times\text{e}^{(1-0.58)\times(38.921\,\text{V}^{-1})\times(0.20\,\text{V})}$$
$$= \boxed{1.7\times10^{-4}\,\text{A cm}^{-2}}$$

The result confirms that the current is indeed dominated by the anodic term, which is the term for which the power of the exponential is positive.

E19D.5(a) The Butler–Volmer equation is [19D.2–*848*], $j = j_0(e^{(1-\alpha)f\eta} - e^{-\alpha f\eta})$; at 298.15 K, $f = 38.921\ \text{V}^{-1}$.

(i) With the given data and for an $\eta = +0.010$ V the current density is

$$j = (0.79\ \text{mA cm}^{-2}) \times (e^{(1-0.5)\times(38.921\ \text{V}^{-1})\times(0.010\ \text{V})} - e^{-0.5\times(38.921\ \text{V}^{-1})\times(0.010\ \text{V})}) = \boxed{0.31\ \text{mA cm}^{-2}}$$

(ii) For $\eta = +0.100$ V the current density is

$$j = (0.79\ \text{mA cm}^{-2}) \times (e^{(1-0.5)\times(38.921\ \text{V}^{-1})\times(0.100\ \text{V})} - e^{-0.5\times(38.921\ \text{V}^{-1})\times(0.100\ \text{V})}) = \boxed{5.4\ \text{mA cm}^{-2}}$$

(iii) For $\eta = -5.0$ V the current density is

$$j = (0.79\ \text{mA cm}^{-2}) \times (e^{(1-0.5)\times(38.921\ \text{V}^{-1})\times(-5.0\ \text{V})} - e^{-0.5\times(38.921\ \text{V}^{-1})\times(-5.0\ \text{V})}) = \boxed{-1.4\times10^{42}\ \text{mA cm}^{-2}}$$

Such a current density would be quite impossible to achieve in practice.

E19D.6(a) At equilibrium, only the exchange current flows, therefore for an electrode with area A the current is $j_0 A$, and thus the charge passing in time t is (current × time): $q = j_0 At$. If each species passing through the double layer carries one fundamental change, the number of charges is $N = q/e = j_0 At/e$. Thus the number per second through an area of 1.0 cm^2 is, for H^+/Pt,

$$N/t = j_0 A/e = (7.9\times10^{-4}\ \text{A cm}^{-2})\times(1\ \text{cm}^2)/(1.6022\times10^{-19}\ \text{C}) = 4.93...\times10^{15}\ \text{s}^{-1} = \boxed{4.9\times10^{15}\ \text{s}^{-1}}$$

A similar calculation for Fe^{3+}/Pt gives $\boxed{1.6\times10^{16}\ \text{s}^{-1}}$, and for H^+/Pb the result is $\boxed{3.1\times10^{7}\ \text{s}^{-1}}$.

The number of atoms covering 1 cm^2 of electrode is $(10^{-4}\ \text{m}^2)/(280\times10^{-12}\ \text{m})^2 = 1.27...\times10^{15}$. Therefore for H^+/Pt the number of times per second that each atom is involved in a electron transfer event is (number of such events)/(number of atoms) $= (4.93...\times10^{15}\ \text{s}^{-1})/(1.27...\times10^{15}) = \boxed{3.9\ \text{s}^{-1}}$. Similar calculations for Fe^{3+}/Pt and H^+/Pb give $\boxed{12\ \text{s}^{-1}}$ and $\boxed{2.4\times10^{-8}\ \text{s}^{-1}}$, respectively. For H^+/Pb the time between events is more than 1 year.

E19D.7(a) In the linear region the current density and overpotential are related by [19D.4–*849*], $\eta = RTj/Fj_0$, therefore the current density is $j = \eta F j_0/RT$. For an electrode of area A the current is $I = jA$, and therefore the resistance is

$$r = \frac{\eta}{I} = \frac{\eta}{\eta F j_0 A/RT} = \frac{RT}{F j_0 A}$$

For H^+/Pt

$$r = \frac{(8.3145\,J\,K^{-1}\,mol^{-1}) \times (298\,K)}{(96485\,C\,mol^{-1}) \times (7.9 \times 10^{-4}\,A\,cm^{-2}) \times (1.0\,cm^2)} = \boxed{33\,\Omega}$$

The units are resolved by using (from inside the front cover) $1\,V = 1\,J\,C^{-1}$ and $1\,\Omega = 1\,V\,A^{-1}$. A similar calculation for H^+/Hg gives $\boxed{3.3 \times 10^{10}\,\Omega}$.

E19D.8(a) Because the standard potential of Zn^{2+}/Zn is -0.76 V, under standard conditions Zn metal will only be deposited when the applied potential is more negative than -0.76 V. The current density is given by [19D.2–*848*], $j = j_0(e^{(1-\alpha)f\eta} - e^{-\alpha f\eta})$, but under these conditions only the second term (the cathodic current) is significant. Using the data given for H^+, assuming $\alpha = 0.5$, and recalling that, at 298.15 K, $f = 38.921\,V^{-1}$

$$\begin{aligned} j_{H^+} &= -j_0 e^{-\alpha f\eta} \\ &= -(50 \times 10^{-12}\,A\,cm^{-2})\,e^{-0.5 \times (38.921\,V^{-1}) \times (-0.76\,V)} = -1.3 \times 10^{-4}\,A\,cm^{-2} \end{aligned}$$

It is usually considered that the metal can be deposited if the current density for discharge of H^+ is less than about $1\,mA\,cm^{-2}$, which is satisfied in this case, but not by a large margin. The expectation is that zinc metal will be deposited, but accompanied by significant evolution of H_2 due to discharge of H^+.

Solutions to problems

P19D.1 (a) The current density is given by [19D.2–*848*], $j = j_0(e^{(1-\alpha)f\eta} - e^{-\alpha f\eta})$, but for positive η the second term (the anodic current) dominates and therefore $\ln j = \ln j_0 + (1-\alpha)f\eta$. A plot of $\ln j$ against η will have slope $(1-\alpha)f$ and intercept $\ln j_0$. Such a plot is shown in Fig. 19.15.

η/V	$j/(mA\,cm^{-2})$	$\ln[j/(mA\,cm^{-2})]$
0.050	2.66	0.978
0.100	8.91	2.19
0.150	29.9	3.40
0.200	100	4.61
0.250	335	5.81

The data fall on a good straight line with equation

$$\ln[j/(mA\,cm^{-2})] = 24.18 \times (\eta/V) - 0.230$$

From the slope it follows that $(1-\alpha) \times (38.921\,V^{-1}) = 24.18\,V^{-1}$ hence $\boxed{\alpha = 0.38}$. The exchange current density is computed from the intercept as $\boxed{j_0 = 0.79\,mA\,cm^{-2}}$.

(b) For negative overpotentials the cathodic current dominates and $j = -j_0 e^{-\alpha f\eta}$. The following table is drawn up using the results from (a).

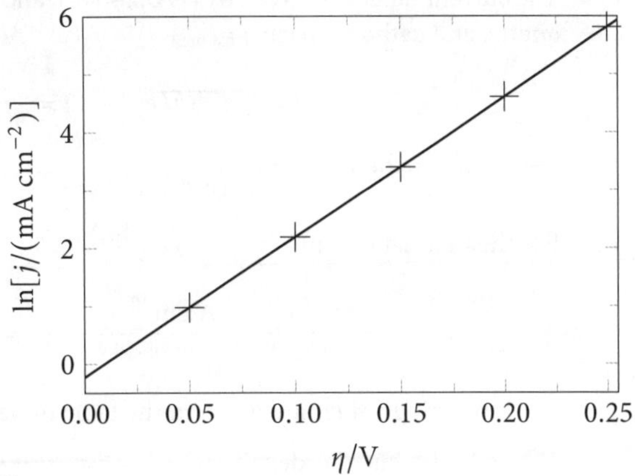

Figure 19.15

η/V	$j/(\text{mA cm}^{-2})$
−0.050	−1.65
−0.100	−3.47
−0.150	−7.26
−0.200	−15.2
−0.250	−31.9

P19D.3 (a) The Nernst equation [6C.4–*221*] for the half cell is

$$E(\text{Fe}^{2+}/\text{Fe}) = E(\text{Fe}^{2+}/\text{Fe})^{\ominus} + \frac{RT}{2F}\ln a_{\text{Fe}^{2+}}$$

Therefore with the given concentration, the potential is

$$E(\text{Fe}^{2+}/\text{Fe}) = (-0.44\text{ V}) + \frac{(8.3145\,\text{J K}^{-1}\,\text{mol}^{-1})\times(298\text{ K})}{2\times(96485\,\text{C mol}^{-1})} \times \ln\frac{1.70\times10^{-6}\text{ mol dm}^{-3}}{1\text{ mol dm}^{-3}} = \boxed{-0.611\text{ V}}$$

The overpotential is thus computed as $\eta = E' - E = E' - (-0.611\text{ V}) = E' + (0.611\text{ V})$.

(b) The current density is the rate of deposition in moles, multiplied by the Faraday constant (to give the charge) and divided by the area of the electrode: $j = 2\upsilon F/A$; the factor of two is needed as a divalent ion is being discharged. For the first data point

$$j = 2\times(1.47\times10^{-12}\text{ mol s}^{-1})\times(96485\,\text{C mol}^{-1})/(9.1\text{ cm}^2)$$
$$= 3.11...\times10^{-8}\text{ A cm}^{-2} = 31.1...\text{ nA cm}^{-2}$$

The current density is given by [19D.2–848], and can be separated into an anodic and cathodic part: $j = j_a + j_c = j_0(e^{(1-\alpha)f\eta} - e^{-\alpha f\eta})$. Thus

$$j = j_0\, e^{-\alpha f\eta}(e^{f\eta} - 1) = -j_c(e^{f\eta} - 1)$$

$$\text{hence } j_c = \frac{j}{1 - e^{f\eta}}$$

For the first data point

$$j_c = \frac{31.1\ldots \text{ nA cm}^{-2}}{1 - e^{(38.921\text{ V}^{-1})\times[(-0.702+0.611)\text{ V}]}} = 32.1\ldots \text{ nA cm}^{-2}$$

The remaining values are given in the table in part (c).

(c) The cathodic current density is $|j_c| = j_0 e^{-\alpha f\eta}$, therefore a plot of $\ln|j_c|$ against η should have slope $-\alpha f$ and intercept $\ln j_0$. The data are tabulated below and such a plot is shown in Fig. 19.16.

E'/V	η/V	$v/(\text{pmol s}^{-1})$	$j/(\text{nA cm}^{-2})$	$\lvert j_c\rvert/(\text{nA cm}^{-2})$	$\ln[\lvert j_c\rvert/(\text{nA cm}^{-2})]$
−0.702	−0.091	1.47	31.2	32.1	3.47
−0.727	−0.116	2.18	46.2	46.7	3.84
−0.752	−0.141	3.11	65.9	66.2	4.19
−0.812	−0.201	7.26	154	154	5.04

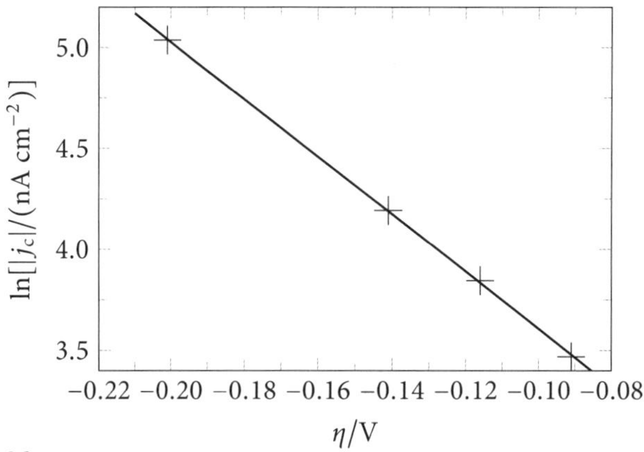

Figure 19.16

The data fall on a good straight line with equation

$$\ln[|j_c|/(\text{nA cm}^{-2})] = -14.20 \times (\eta/\text{V}) + 2.187$$

From the slope it follows that $-\alpha \times (38.921\text{ V}^{-1}) = -14.20\text{ V}^{-1}$ hence $\boxed{\alpha = 0.365}$. The exchange current density is computed from the intercept as $\boxed{j_0 = 8.91\text{ nA cm}^{-2}}$.

P19D.5 The data given correspond to positive overpotentials, so the anodic current will dominate and hence $\ln j = \ln j_0 + (1-\alpha)f\eta$. A plot of $\ln j$ against η will have slope $(1-\alpha)f$ and intercept $\ln j_0$. Such a plot is shown in Fig. 19.17.

η/V	j/(mA m^{-2})	ln[j/(mA m^{-2})]
0.60	2.9	1.06
0.65	6.3	1.84
0.73	28	3.33
0.79	100	4.61
0.84	250	5.52
0.89	630	6.45
0.93	1 650	7.41
0.96	3 300	8.10

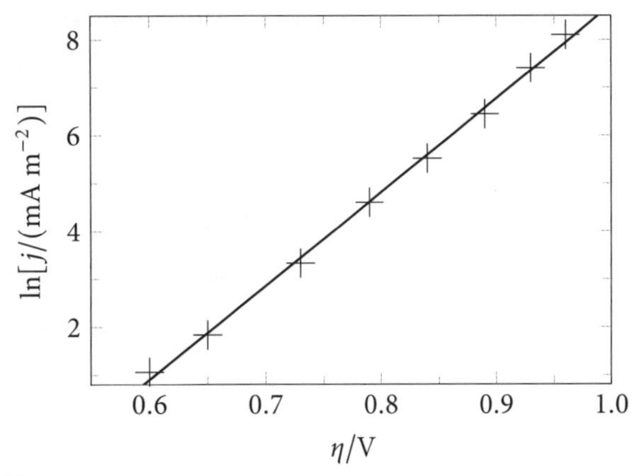

Figure 19.17

The data fall on a good straight line with equation

$$\ln[j/(\text{mA m}^{-2})] = 19.55 \times (\eta/\text{V}) - 10.83$$

From the slope it follows that $(1-\alpha)\times(38.921\ \text{V}^{-1}) = 19.55\ \text{V}^{-1}$ hence $\boxed{\alpha = 0.50}$. The exchange current density is $\boxed{j_0 = 1.99 \times 10^{-5}\ \text{mA m}^{-2}}$ (computed from the intercept).

P19D.7 Because the overpotential is always positive and 'high', the current is entirely anodic and given by $j = j_0 e^{(1-\alpha)f\eta}$. Imagine that the potential ramps linearly from η_- to η_+ in a time τ: over this period the time-dependent overpotential is $\eta(t) = \eta_- + (\eta_+ - \eta_-)t/\tau$. For this rising part of the ramp the current is

$$j_{t=0\to\tau} = j_0 e^{(1-\alpha)f[\eta_- + (\eta_+ - \eta_-)t/\tau]} = j_0 e^{(1-\alpha)f\eta_-} e^{(1-\alpha)f(\eta_+ - \eta_-)t/\tau}$$

This is a current which rises from $j_0 e^{(1-\alpha)f\eta_-}$ at $t = 0$ to $j_0 e^{(1-\alpha)f\eta_+}$ at $t = \tau$ with an exponential dependence on time.

For the falling part of the ramp the time-dependent overpotential is $\eta(t) = (2\eta_+ - \eta_-) + (\eta_- - \eta_+)t/\tau$, giving a current

$$j_{t=\tau\to 2\tau} = j_0 e^{(1-\alpha)f[(2\eta_+-\eta_-)+(\eta_--\eta_+)t/\tau]} = j_0 e^{(1-\alpha)f(2\eta_+-\eta_-)} e^{(1-\alpha)f(\eta_--\eta_+)t/\tau}$$

This is a current which falls from $j_0 e^{(1-\alpha)f\eta_+}$ at $t = \tau$ to $j_0 e^{(1-\alpha)f\eta_-}$ at $t = 2\tau$ with an exponential dependence on time. After 2τ the system is back at its starting position.

Answers to integrated activities

I19.1 The model is to treat the solid argon (mass density $\rho = 1.784 \text{ g cm}^{-3}$) as a continuum of matter that has the number density

$$\mathcal{N} = \rho N_A/M = 2.689 \times 10^{28} \text{ m}^{-3}$$

The arrangement used for the calculation is shown in Fig. 19.18. The atom is at a distance R from the surface of the solid, shown by the shaded area. An annulus of material, of radius r and radial thickness dr, and height dz is located at depth z within the solid.

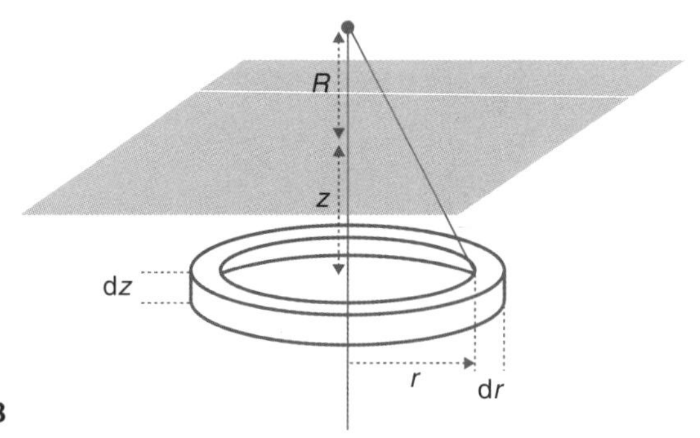

Figure 19.18

The Lennard-Jones (6,12)-potential for the interaction between an Ar atom and a point within the solid at the distance ξ from the atom is

$$V(R,z,r) = 4\varepsilon\left[(r_0/\xi)^{12} - (r_0/\xi)^6\right] \quad \text{where} \quad \xi = [(R+z)^2 + r^2]^{1/2}$$

The number of atoms in the annulus is $2\pi\mathcal{N} r\,dr\,dz$ so the interaction energy between the adsorbate atom and the atoms in the annulus is

$$dU = 8\pi\varepsilon\mathcal{N} \times \left[(r_0/\xi)^{12} - (r_0/\xi)^6\right] \times r\,dr\,dz$$

To find the total interaction energy it is necessary to integrate dU over the ranges $0 \leq z \leq \infty$ and $0 \leq r \leq \infty$.

$$U = 8\pi\varepsilon\mathcal{N}\int_0^\infty\int_0^\infty \left[\left(\frac{r_0}{[(R+z)^2+r^2]^{1/2}}\right)^{12} - \left(\frac{r_0}{[(R+z)^2+r^2]^{1/2}}\right)^6\right] r\,dr\,dz$$

The integral over r is found using the standard integral

$$\int_0^\infty \frac{r}{(a^2+r^2)^{n/2}}\,\mathrm{d}r = \frac{1}{(n-2)a^{n-2}}$$

to give

$$U = 8\pi\varepsilon\mathcal{N}\int_0^\infty \left[\frac{r_0^{12}}{10(R+z)^{10}} - \frac{r_0^6}{4(R+z)^4}\right]\mathrm{d}z$$

The integral over z is found using the standard integral

$$\int_0^\infty \frac{1}{(R+z)^{n-2}}\,\mathrm{d}z = \frac{1}{(n-3)R^{n-3}}$$

to give

$$\boxed{U = \tfrac{4}{3}\pi\varepsilon r_0^3\mathcal{N}\left[\frac{1}{15}\left(\frac{r_0}{R}\right)^9 - \frac{1}{2}\left(\frac{r_0}{R}\right)^3\right]}$$

The negative term of the above expression is the attractive interaction and this does indeed go as R^{-3}, as required.

The position of equilibrium corresponds to the minimum in U which is found by setting the derivative to zero

$$\frac{\mathrm{d}U}{\mathrm{d}R} = \tfrac{4}{3}\pi\varepsilon r_0^3\mathcal{N}\left[-\frac{9r_0^9}{15R^{10}} + \frac{3r_0^3}{2R^4}\right] = 0$$

which is solved for $R_{\mathrm{eq}} = (2/5)^{1/6}r_0$. For argon the Lennard-Jones parameters are $\varepsilon = 128\ \mathrm{kJ\,mol^{-1}}$ and $r_0 = 342\ \mathrm{pm}$, therefore $\boxed{R_{\mathrm{eq}} = 294\ \mathrm{pm}}$. The energy at this separation is

$$U = \tfrac{4}{3}\pi(128\ \mathrm{kJ\,mol^{-1}}) \times (342\times10^{-12}\ \mathrm{m})^3 \times (2.689\times10^{28}\ \mathrm{m^{-3}})$$
$$\times\left[\frac{1}{15}\left(\frac{1}{(2/5)^{1/6}}\right)^9 - \frac{1}{2}\left(\frac{1}{(2/5)^{1/6}}\right)^3\right]$$
$$= \boxed{-304\ \mathrm{kJ\,mol^{-1}}}$$

I19.3 The Coulombic potential between two charges Q_1 and Q_2 at a distance r is

$$V = \frac{Q_1Q_2}{4\pi\varepsilon_0 r}$$

The force is given by $F = -\mathrm{d}V/\mathrm{d}r$

$$F = -\frac{\mathrm{d}}{\mathrm{d}r}\frac{Q_1Q_2}{4\pi\varepsilon_0 r} = \frac{Q_1Q_2}{4\pi\varepsilon_0 r^2}$$
$$= \frac{(1.6022\times10^{-19}\ \mathrm{C})^2}{4\pi\times(8.8542\times10^{-12}\ \mathrm{J^{-1}\,C^2\,m^{-1}})\times(2.00\times10^{-9}\ \mathrm{m})^2}$$
$$= \boxed{57.7\ \mathrm{pN}}$$

I19.5 The approach is to compute the standard reaction Gibbs energy of the combustion reaction using tabulated standard Gibbs energies of formation. The standard cell potential is then computed using $\Delta_r G^\ominus = -\nu F E^\ominus$; the number of electrons involved in the reaction is identified by considering the oxidation numbers of the products and reactants. For brevity the phases of the species are omitted from the chemical equations.

(a) The reaction is $H_2 + \frac{1}{2}O_2 \rightarrow H_2O$. The oxygen goes from oxidation number 0 in O_2, to -2 in H_2O, that is a change of 2. Because there is just one oxygen atom involved, $\nu = 1 \times 2 = 2$. For this reaction $\Delta_r G^\ominus$ is equal to $\Delta_f G^\ominus(H_2O)$, which is -237.13 kJ mol^{-1}. Hence $E^\ominus = -\Delta_r G^\ominus/\nu F = -(-237.13 \times 10^3\,\text{J mol}^{-1})/[(2) \times (96485\,\text{C mol}^{-1})] = \boxed{+1.23\,\text{V}}$.

(b) The reaction is $CH_4 + 2\,O_2 \longrightarrow CO_2 + 2\,H_2O$. The oxygen goes from oxidation number 0 in O_2, to -2 in both CO_2 and H_2O, that is a change of 2. Because there are 4 oxygen atoms in total, $\nu = 4 \times 2 = 8$.

$$\begin{aligned}\Delta_r G^\ominus &= \Delta_f G^\ominus(CO_2) + 2\Delta_f G^\ominus(H_2O) - \Delta_f G^\ominus(CH_4)\\ (\Delta_r G^\ominus/\text{kJ mol}^{-1}) &= (-394.36) + 2 \times (-237.13) - (-50.72) = -817.9\\ E^\ominus &= -\Delta_r G^\ominus/\nu F\\ &= -(-817.9 \times 10^3\,\text{J mol}^{-1})/[8 \times (96485\,\text{C mol}^{-1})]\\ &= \boxed{+1.06\,\text{V}}\end{aligned}$$

(c) The reaction is $C_3H_8 + 5\,O_2 \longrightarrow 3\,CO_2 + 4\,H_2O$. The oxygen goes from oxidation number 0 in O_2, to -2 in both CO_2 and H_2O, that is a change of 2. Because there are 10 oxygen atoms in total, $\nu = 10 \times 2 = 20$.

$$\begin{aligned}\Delta_r G^\ominus &= 3\Delta_f G^\ominus(CO_2) + 4\Delta_f G^\ominus(H_2O) - \Delta_f G^\ominus(C_3H_8)\\ (\Delta_r G^\ominus/\text{kJ mol}^{-1}) &= 3 \times (-394.36) + 4 \times (-237.13) - (-23.49)\\ &= -2108.11\\ E^\ominus &= -\Delta_r G^\ominus/\nu F\\ &= -(-2108.11 \times 10^3\,\text{J mol}^{-1})/[20 \times (96485\,\text{C mol}^{-1})]\\ &= \boxed{+1.09\,\text{V}}\end{aligned}$$